Handbook of Health Hazard Control in the Chemical Process Industry

Handbook of Health Hazard Control in the Chemical Process Industry

Sydney Lipton
Consultant
Port Townsend, Washington

Jeremiah Lynch
Exxon Chemical Company
East Millstone, New Jersey

A WILEY-INTERSCIENCE PUBLICATION
John Wiley & Sons, Inc.
NEW YORK / CHICHESTER / BRISBANE / TORONTO / SINGAPORE

This text is printed on acid-free paper.

Library of Congress Cataloging in Publication Data:

Lipton, Sydney.
Handbook of health hazard control in the chemical process industry / Sydney Lipton, Jeremiah Lynch
p. cm.
Rev. ed. of: Health hazard control in the chemical process industry. c1987.
"A Wiley-Interscience publication."
Includes index.
ISBN 0-471-55464-2 (acid-free)
1. Chemical industry—Safety measures. 2. Chemical industry—Health aspects. I. Lynch, Jeremiah. II. Lipton, Sydney. Health hazard control in the chemical process industry. III. Title.
TP149.L57 1994
660′.2804--dc20 94-2360

Printed in the United States of America

10 9 8 7 6 5 4 3 2 1

Contents

Preface

In recent years a substantial body of technology has been developed relating to the control of toxic hazards to human health in typical closed-unit chemical process plants. Most of this technology resides in company files and in government reports that are not readily available. As new evidence of health effects is discovered and exposure standards become progressively lower, it is increasingly important that this body of technology be screened, organized, and presented to provide a resource in the control of occupational health hazards. This book summarizes the technology in a way that will be useful to both industrial hygienists and chemical engineers.

The need to control chemical emissions that can cause health problems in the plant and in the neighboring community was intensified by the passage of the Clean Air Act Amendments (CAAA) of 1990. This law and the regulations promulgated under its authority are discussed in detail in the text. In summary, they require a step change downward in allowable release rates for a large listed category of "air toxic" chemicals. These lower levels of allowable release are based on an assumed "maximum achievable control technology" (MACT). This book describes the specific engineering practice available to meet the requirements of the CAAA.

Traditionally, the education of chemical engineers has not dwelt on health issues, but today most engineers and executives in the chemical industry are spending an increasing amount of time anticipating and preventing the adverse consequences of chemical hazards. This book fills the need of those in industry who have responsibility for the design and modification of chemical manufacturing plants with a known degree of risk control at an acceptable cost. In addition, the book will serve the needs of the following:

- Industrial hygienists in industry who review engineering plans for health hazard controls and must anticipate and monitor the results.

- Government agencies that are conducting research in health hazard control or that need to understand the limitations of control technology.
- Manufacturers of the chemical process industry equipment whose design and application is critical to emissions control.

To serve the needs of this varied audience, the book begins by covering the broad industrial hygiene subject of the anticipation, recognition, evaluation, and control of health hazards with particular reference to the chemical industry. Next, fugitive emission control regulations designed to minimize the release of criteria pollutants and air toxics are presented along with their relation to exposures and control options in chemical plants. Specific control options for most of the equipment used in chemical plants are discussed in detail, and the criteria for the most cost-effective option are presented. An exposure prediction technique aids in the selection of control options. Finally, certain special topics such as sampling, drainage, sewer emissions, and major process hazards complete presentation of the major aspects of health hazard control in the chemical process industry.

SYDNEY LIPTON
JEREMIAH LYNCH

Port Townsend, Washington
East Millstone, New Jersey
March 1994

Glossary

ACGIH	American Conference Governmental Industrial Hygienists
AIChE	American Institute of Chemical Engineers
AIHA	American Industrial Hygiene Association
ANSI	American National Standards Institute
API	American Petroleum Institute
ASME	American Society of Mechanical Engineers
ASTM	American Society for Testing and Materials
BAAQMD	Bay Area Air Quality Management District
BHRA	British Hydromechanics Research Association
BID	Background Information Document (EPA)
CAAA	Clean Air Act Amendments of 1990
Censored Source	Screening value > 100,000 ppmv; actual value unknown
CFR	Code of Federal Regulations (US)
CMA	Chemical Manufacturers Association
CSHIB	Chemical Safety and Hazard Investigation Board
CTG	Control Techniques Guideline (EPA)
CWA	Clean Water Act
DMS	Dual Mechanical Seal
EPA	Environmental Protection Agency (US)
FR	Federal Register (US)
HAP	Hazardous Air Pollutant
HON	Hazardous Organic National Emission Standards for Hazardous Air Pollutants
LDAR	Leak Detection and Repair
MACT	Maximum Achievable Control Technology
MTI	Materials Technology Institute of the Chemical Process Industries
NAAQS	National Ambient Air Quality Standards

NESHAP	National Emission Standards for Hazardous Air Pollutants
NSPS	New Source Performance Standards
OSHA	Occupational Safety and Health Administration (US)
OVA®	Organic Vapor Analyzer
PVRC	Pressure Vessels Research Committee of the Welding Research Council
QAP	Quality Assurance Program
QIP	Quality Improvement Program
RAMC	Risk Assessment and Management Commission
SARA	Superfund Act Reauthorization Amendments
SCAQMD	Southern California Air Quality Management District
SIP	State Implementation Plan
SMS	Single Mechanical Seal
SOCMI	Synthetic Organic Chemical Manufacturing Industry
TACB	Texas Air Control Board
TOC	Total Organic Compounds
TSDF	Treatment, Storage, and Disposal Facility
VHAP	Volatile Hazardous Air Pollutant
VMA	Valve Manufacturers Association
VOHAP	Volatile Organic Hazardous Air Pollutant
VOC	Volatile Organic Compound

Handbook of Health Hazard Control in the Chemical Process Industry

1

Occupational Health Hazards

Occupational health hazards are those factors arising in or from the occupational environment that adversely impact health. They may cause sudden acute effects like carbon monoxide poisoning or long-term chronic effects like lung cancer. They may be gases, vapors, dusts, or physical agents like ionizing radiation. They may enter the body by inhalation or ingestion or by passing through the skin. They may be found in traditional heavy industries, modern high-tech industries, or even office environments. While much is known about occupational health hazards, new relationships between exposure and disease are continually being discovered.

This chapter discusses the ways in which chemicals can enter and affect the body, the concept of dose response, the relationship between dose and exposure, and how exposure limits are set and interpreted.

1.1 HEALTH EFFECTS

There are many ways in which health hazards can affect the body. Yet, given the enormous number of potentially hazardous agents and the finite number of ways in which the body can respond, the same response is often seen for many agents. Agents causing the same effect are often chemically related (most hydocarbons affect the nervous system), but some quite different agents may also cause similar effects (cadmium and carbon tetrachloride are both kidney poisons). Also, a single agent can cause different effects at the same dose or one effect at a low dose and another at a higher dose.

What is known about the effects of chemicals and physical agents is recorded in the extensive literature on the subject which is accessed through computer databases such as National Library of Medicine MEDLARS and TOXLINE. Other sources are in published works listed in the references at the end of this chapter. Some of the health effects referred to in these sources can be classified as follows:

Acute Poisoning. Some chemicals act very quickly to shut down some critical body function and thus cause death. The classic poison cyanide inhibits certain enzymes and causes chemical asphyxia at the cellular level. Arsenic causes severe gastrointestinal corrosion leading to shock, severe water loss, and death. Industrial acute poisons include carbon monoxide, which prevents the transfer of oxygen at the blood level, and hydrogen sulfide, which can cause respiratory paralysis.

Irritation. Irritation may occur as a local sensation in either the eyes or upper respiratory tract or may occur in the deep lung where gas transfer takes place. This latter form of irritation can be very serious. The severe deep lung irritation caused by nitrogen dioxide can cause acute pulmonary edema, which is a sudden flooding of the lungs, and death. Longer term, lower level or "chronic" exposure can cause shortness of breath (dyspnea) and other decrements in lung function. Thus, while simple sensory irritation may only be minor, troublesome discomfort, other more serious effects may occur. The American Conference of Governmental Industrial Hygienists holds that "limits based on physical irritation should be considered no less binding than those based on physical impairment. There is increasing evidence that physical irritation may initiate, promote or accelerate physical impairment."

Dermatitis. The skin is the largest organ of the body and serves several critical functions, the most important of which is keeping body fluids inside and harmful substances outside. This barrier can be destroyed by trauma (physical injury that cuts the skin), corrosion by strong acids or bases, or skin diseases or dermatoses. Occupational dermatoses range from a mild reddening of the skin (erythema) to effects so serious as to destroy the barrier and threaten life. Most dermatoses begin with repeated or prolonged exposure to an agent, such as a defatting solvent, which diminishes the skin's natural protection. This may lead to other effects such as pigmentation, acne, fissuring, and ulceration. Some chemical agents are allergic sensitizers (amine epoxy hardeners), photosensitizers (chlorobenzols), or skin carcinogens (coal tar).

Neuropathy. Many chemicals affect some part of the nervous system. Volatile organic solvents can cause central nervous system (CNS) effects ranging from a "high" feeling like drunkenness to narcosis (stupor). These effects are usually reversable, but "glue sniffers" and others who voluntarily take massive overdoses for the narcotic effect suffer irreversable brain damage. Other solvents such as methyl *n*-butyl ketone are capable of causing peripheral neuropathy, a loss of nervous function of the arms and legs characterized by a loss of sensation and partial paralysis. Mild cases of peripheral neuropathy may be reversible, but permanent loss of function may occur in more severe cases. Other chemicals like mercury and magnesium can cause serious nervous system and brain disorders resulting in personality changes and psychoses.

Cancer. Classic occupational diseases like lead poisoning and black lung still occur but they are much less common and at present more concern is directed

TABLE 1.1 Some Known Human Occupational Carcinogens

Substance	Site
Asbestos	Lung, elsewhere
β-Naphthylamine	Bladder
Benzene	Leukemia
Bis(chloromethyl) ether	Lung
Chromium	Lung
Coal Tar	Skin, Lung
Vinyl chloride	Liver

toward cancer. While cigarette smoking is the major cause of cancer in the general population, certain workplace substances are also known to cause cancer in humans (Table 1.1). In addition to about 50 substances that are known to cause cancer in humans, there are many more that cause cancer in laboratory animals and are therefore suspect human carcinogens. A difficulty in determining if a chemical that causes cancer in animals can also cause cancer in humans is that many types of cancer can be caused by either occupational or nonoccupational factors. Lung cancer caused by a workplace chemical would probably be masked by smoking. Another difficulty is that there is a long delay or latency period between the onset of exposure and the first appearance of disease. This latency of often 20 years or more makes it difficult to associate exposure with disease.

Reproductive Effects. A less certain area of industrial toxicology is damage to reproductive functions caused by chemicals. Reduction or elimination of male sperm by dibromochloropropane is known to cause sterility. Other chemicals can act on the female to prevent conception or cause spontaneous abortion. Another serious possibility is that a chemical, like thalidomide, could act on the fetus, particularly during the germinal stage of pregnancy, to produce birth defects.

1.2 ROUTES OF ENTRY

For a chemical to affect health it must come in contact with or enter the human body. There are several ways in which this can happen. Some routes, such as inhalation, are more important occupationally, while others, such as ingestion or injection, are of importance in chemical therapy. All, however, can occur in the industrial environment.

1.2.1 Skin Contact

Chemicals that cause dermatitis usually do so through direct contact with the skin. Some chemicals, like corrosive acids, can damage the skin by a single contact while others, like organic solvents, may cause damage by repeated defatting. However, in addition to the direct action of chemicals on the skin, the skin may also serve as a route of entry into the body. Most chemicals can enter the body where the skin has been damaged by a wound. Some chemicals, however, are able to alter or dissolve in skin constituents and then pass through intact skin. The dose received in this manner will vary according to the contact area, time, and rate of absorption. Dimethylformamide is rapidly absorbed through the skin while benzene is absorbed slowly. Bodily immersion will obviously result in much greater intake than slight hand contact. Volatile substances will probably not be in contact with the skin long but less volatile materials will remain until washed off. Wet clothing will hold a substance in contact with the skin. Overall, skin absorption can range from an insignificant contributor to total systemic dose to the major dose route and, in some cases, such as phenol, can cause death.

Physical state is a major factor in determining skin absorption. Hygroscopic solids may cause skin damage by skin dehydration and, when in solution, may be absorbed to some extent. Liquids are the most readily absorbed whether they come in contact with the skin in bulk from a splash, or thinly from wetting by a liquid aerosol. Gases that condense on the skin behave as liquids, but absorption of noncondensing gases is sharply limited by their dilution in air and because they do not readily partition into the skin constituents.

1.2.2 Ingestion

Unlike drugs, chemicals in the workplace are not eaten intentionally but some amount may enter the mouth and stomach unintentionally and contribute to the body burden. As will be discussed below, some solid materials when inhaled are cleared from the respiratory tract into the mouth and end up in the stomach. A more common source of exposure by ingestion is contamination of food or cigarettes consumed at work. For this reason modern factories separate eating and smoking areas from contaminated work-places. Small amounts are also taken in from the hands, particularly when personal hygiene is poor. Overall, these sources are small and almost always rank behind inhalation and skin contact as contributors to total dose.

1.2.3 Injection

While injection is often used as a convenient means of dosing experimental animals, it is a rare dose route in industry. About the only unintentional introduction of chemicals into the body by injection occurs when materials are ejected from high pressure systems at high enough velocity to break through the skin and enter the body. Some examples are high-pressure paint systems and leaks from flanges on high-pressure piping. These exposures occur as acute, accidental episodes rather than as a routine source of chronic exposure.

1.2.4 Inhalation

By far the most common and most significant source of workplace exposure to chemicals and the most difficult to control is inhalation. An adult male will breath about 10 m^3 of air during a normal work day. Much of it comes in contact with the lung, which has the largest surface area exposed to the air of any part of the body, much larger than the skin. Whatever is in the air may enter the lung along with the air and may pass into the body by this route.

Airborne chemicals are classified as follows:

Gases—chemicals that are in the gaseous state at room temperature and ambient pressure (e.g., chlorine, radon, nitrogen dioxide).

Vapors—airborne gases from chemicals that are liquids (or even solids) at room temperature (e.g., solvents, hydrazine, iodine).

Aerosols—particles of solids or liquids suspended in air. These are usually divided into:

- Dusts—solid particles,
- Fogs—liquid droplets, and
- Fumes—very fine particles from combustion or condensation processes.

Note that while vapors evaporating from liquids are commonly called "fumes" the definition given above corresponds to accepted technical usage.

The function of the lung is to transfer gases between air and blood. As a consequence the cell barrier in the alveolar air sacs of the lung where the transfer occurs is very thin, only one cell thick. Thin cell walls can be easily damaged by corrosive gases. Further, some irritating chemicals can cause "leakage" and flooding of the lung. Certain solid particles like quartz can cause lung scar tissue to form and this reduces the flexibility of the lung and its ability to transfer oxygen and carbon dioxide. Other aerosols such as cigarette smoke and asbestos fibers can cause lung cancer.

Apart from these toxic effects of inhaled chemicals on the lung itself, the ability of the lung to rapidly transfer materials into the blood provides a path for systemic poisons. Indeed poisoning by inhalation is much faster than by ingestion or skin absorption. Further, while only about a third of inhaled air reaches the deep lung, quantitative equilibrium of partial pressures of gases in air and in blood are reached very quickly. The rate of transfer of a substance from air to blood (dose rate) will depend partly on breathing rate (work rate) but mostly on solubility in blood (approximately water). The concentration in the blood will be diminished by metabolism or deposition in other parts of the body. For some chemicals such as alcohols, uptake is rapid and complete, whereas for other less soluble substances equilibrium is quickly reached and transfer into the blood and back into the air are equal.

1.3 DOSE RESPONSE

A fundamental principle of industrial toxicology is that the intensity of toxic action is a function of concentration of the toxic agent that reaches the site of action. The relationship between this concentration and the intensity of effect is the dose–response relationship for the substance (Figure 1.1). Even the most nontoxic substances, like water, have a dose at which death will occur while even the most poisonous substances have a dose below which nothing of importance happens. To understand the potential for harm of a workplace exposure, knowledge of the mere presence of, or contact with, the substance is not enough. It is necessary to estimate the dose to appreciate the effect.

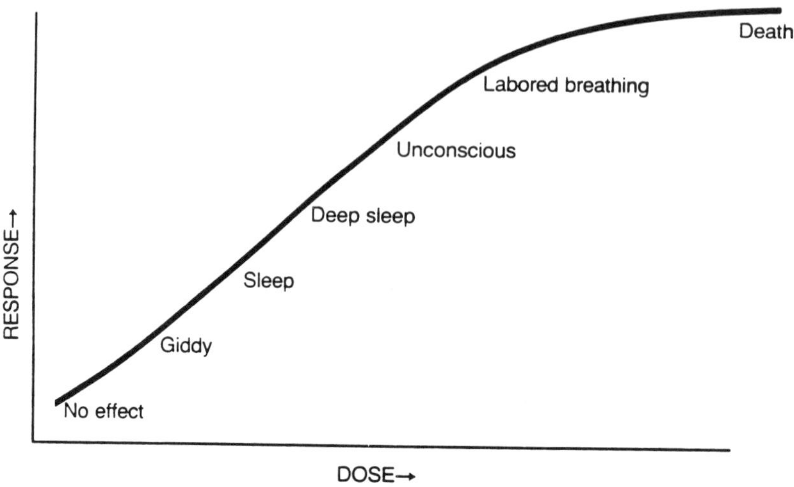

Figure 1.1 *Dose – response curve for ethyl alcohol (from Marczewski and Kamrin).*

In a single individual, increasing dose will produce increasing effect. At some very low dose no effect will occur. As dose is increased effects may occur that are measurable by very sensitive chemical or behavioral methods but that have no health significance. As dose is further increased there are early subclinical indicators of disease, then frank disease, then serious degenerative processes leading to permanent impairment and death. The low-level effects may be on quite different organs than the effects that occur at higher levels.

When the effect of a dose on a population of people or test animals is studied it is seen that there is some variability among individuals. Thus whatever the end point from the least measurable effect to death, a given dose may produce that end point in some individuals but not in others. Starting at low doses no individuals will show the effect and as the dose is increased more and more will be affected until at some high dose all will have reached the end point. A common way of describing the acute toxicity of chemicals is to determine the dose at which 50% reach a particular end point. For death as the end point this dose is called the lethal dose for 50% or the LD_{50} (see Table 1.2). It is about the highest of the doses used to characterize the toxicity of a substance and is obviously very much higher than a "safe" dose.

The rate at which a substance is taken into the body, or dose rate, is the product of the concentration to which the person is exposed times some uptake factor that is dependent on work rate, solubility, metabolism, and so on, as discussed under inhalation. The chemical taken into the body is partitioned between various compartments (blood, fat, bone marrow, etc.), metabolized, and excreted. The portion of the substance or its active metabolite reaching the critical site of toxic action and acting on it over time determines the toxic consequence. Unfortunately it is not possible to measure dose at the critical site directly so surrogate measures must be used. Biological measurements of a substance or its metabolites in blood or urine are coming into increasing use as dose (and response) estimators, but in most instances it is necessary to rely on exposure measurement as the bridge

TABLE 1.2 Acute Toxicity[a]

Chemical	Rat LD_{50}
Sugar (sucrose)	29,700
Alcohol	14,000
Vinegar (acetic acid)	3,310
Salt (sodium chloride)	3,000
Malathion	1,375
Aspirin	1,000
2,4-D	375
Ammonia	350
Carbaryl	250
DDT	113
Arsenic (arsenic acid)	48
Strychnine sulfate	2
Nicotine	1
Dioxin (TCDD)	0.001
Botulinum toxin	0.00001

[a]From Marczewski and Kamrin.

between conditions at work and toxic effect. Exposure means contact by the individual with the substance as either a gas, liquid, or solid, over time, so that the intake of a dose may occur through one of the routes of entry. To use exposure as a surrogate for dose some factors or assumptions are needed to relate the duration of an exposure to the expected effect. The simplest assumption, called Haber's rule, is that the effect is proportional to cumulative dose as estimated by exposure concentration times duration of exposure. This relationship results in the use of the 8-hr averaging time to deal with intraday variations in exposure in most exposure dose–response predictions. This simple relationship is adequate, given other sources of error, for most substances at usual occupational exposure levels. For some very slow-acting substances, like mineral dusts, even longer averaging times could be used. However, for some fast-acting substances peak exposures that do not exceed an acceptable daily average could cause damage. For these substances short-term limits or limits on dose rate, in addition to longer averaging time limits, are necessary.

1.4 HEALTH STANDARDS

A toxic response results from a dose of a chemical acting at a critical site over time. We estimate this dose by measurement of exposure. In the industrial environment the preponderant route of exposure is inhalation. Standards are set to limit exposure by inhalation and thereby prevent or limit significant toxic effect. These standards are usually based on an 8-hr averaging time although other, shorter averaging times are used for fast-acting toxic substances.

The term "standard" is used here to include both voluntary limits set by professional societies or other nongovernmental bodies and mandatory limits enforced by government regulatory agencies. Some of the major standard setting groups are as follows:

- **Occupational Safety and Health Administration (OSHA), United States Department of Labor.** Standards set by OSHA are called Permissible Exposure Limits (PEL) and are legally enforceable in the United States under the Occupational Safety and Health Act of 1970. There are about 400 substances on the OSHA list of limits and they are taken from the 1968 TLV list (see below) or from American National Standards Institute limits of about the same time period. The exceptions are the few substances, like asbestos, cotton dust, and vinyl chloride, that have made it all the way through the very long rule-making process.
- **American Conference of Governmental Industrial Hygienists (ACGIH).** This private, nongovernmental professional society is made up for the most part of industrial hygienists who work for one or another State or Federal government agency. A committee of this group, called the Chemical Substances Threshold Limit Value (TLV) Committee, publishes an annual list of about 600 substances. While this list has no regulatory authority in the United States, it is used voluntarily by many employers since it is the most up-to-date list. It is incorporated into the laws and regulations of many foreign countries.
- **American Industrial Hygiene Association (AIHA).** This organization is the largest industrial hygiene society and has a Workplace Environmental Exposure Limit (WEEL) Committee which sets limits for workplace exposure. The WEEL Committee has set limits for about 100 substances and their list does not, in general, overlap the TLV list.
- **National Institute for Occupational Safety and Health (NIOSH).** In the course of developing criteria for standards NIOSH sometimes makes recommendations for occupational exposure limits. Despite the fact that NIOSH is an agency of the U.S. Government, these recommendations, called Recommended Exposure Limits or REL's do not carry the force of law unless they are adopted by OSHA.
- **United Kingdom Health and Safety Executive (HSE).** The HSE publishes a series of Guidance Notes, one of which, EH 40 Occupational Exposure Limits, gives a set of limits that have legal status in the United Kingdom. This list includes "control limits," which are known to be achievable and are enforced in all industries, and "recommended limits," which represent good practice and realistic criteria for the control of exposure.

In addition to these lists there are other limits set by various countries and international organizations such as the World Health Organization and the International Labor Organization.

1.4.1 Bases for Standards

The intention of most standards-setting bodies is to select a level of exposure below which no significant effect is expected to occur over a working life-time. Traditionally, these groups have used the concept of a threshold below which it is thought no adverse effect occurs. Based on alternate metabolic pathways at different dose rates and on the ability of the body to repair minor damage, it is clear that as dose rate increases qualitative differences in effect may occur. Below certain dose rates for certain substances no adverse process may occur and the equilibrium of the body, or homeostasis, is maintained. The efforts of the standards committees are directed at finding these dose rates and converting them into threshold limits. While this threshold concept doubtless holds true for certain substances for normally healthy individuals, its applicability to populations of people is less clear. Indeed the preface to the TLV list acknowledges that some workers may experience discomfort or more serious effects below the TLV because of individual susceptibility or a preexisting condition. It is not always practical or possible to identify these individuals who are at higher risks and for this reason, as well as because of the uncertainty discussed below, it is good practice to maintain exposures as far below the specified limit as is practicable.

For carcinogens the concept of a threshold is less clear. Where different metabolic pathways are taken at different dose rates, and where the carcinogenic activity is the result of the action of a metabolite, it can be seen that the same reasoning for a threshold would apply as for other toxic effects. However, where the substance itself is a direct acting carcinogen there may be no threshold but instead there may be some risk down to zero dose. Even for these nonthreshold, genotoxic carcinogens the concept of a dose response still applies. In the case of an event such as cancer, dose response applies in the sense that the higher the dose the larger the fraction of the exposed population that will get cancer. Looked at another way, the higher the dose to an individual the higher the probability of that individual acquiring a malignant disease. As dose is reduced, the probability is reduced but, if there is no threshold, the probability does not go to zero until the dose goes to zero. However, at very low dose the risk will be very small and can seem to vanish when compared to other risks. Also, for some carcinogens the latency, or time from exposure to onset of disease, increases as dose decreases. For very low doses the latency may be so long that disease would not be expected until beyond the end of a normal life span. Thus, setting exposure limits for carcinogens, while very difficult and fraught with uncertainties, can be done by choosing levels that result in vanishingly small risks or very long latency. The acceptance of some small risk, in a world that can never be risk free, is not philosophically different from accepting the risks to the unidentified hyposusceptible from exposure to substances that have a threshold.

Another consideration in setting exposure limits is the issue of whether all effects on the exposed should be prevented or only significantly adverse effects. In the case of carcinogens it is not a problem since cancer is clearly a significantly adverse effect. For some substances, however, present limits have been chosen such that exposure under the limit may result in temporary or even permanent damage to the body which is judged not to be adverse or not to accumulate to a disabling level. For example, it is possible to measure changes in response time resulting from exposure to a neurotoxic substance at below the accepted exposure limits. However, these effects, which are transient, are not thought to be a part of any disease process that would result in significant harm to the exposed individual. Similarly, iron fume will discolor the lung and be seen on an X ray, but this change apparently has no health importance. In the case of noise it is accepted that exposure even at commonly recommended limits does result in some hearing loss, but less hearing loss, cumulatively, than would create hearing disability. The decision as to what are acceptable and unacceptable effects is critical in standard setting and is the subject of a continuing debate.

The problem that all standard setters face is to identify the threshold or minimal risk level for the effect of concern using the best information available. The data is almost always indirect because human experimentation is severely limited. Human experiments are not performed using any substance suspected of being a carcinogen and the same restraint applies to other substances where permanent harm might result. While some experiments are possible where the effects are minor and transient (behavioral changes), researchers are reluctant because of uncertainty about more significant effects that might be discovered later. Thus essentially all the data on which limits are based are not the result of direct experimentation, but come from unintentional human experience or animal experiments.

Another problem faced by standard setters is the quality of the information available. Given the prohibition of human experimentation, the paucity of incidental human experience, particularly for new chemicals, and the difficulty of interpreting data from animals, the standard setter may not be able to arrive at a limit that has a sound scientific justification. Yet when people are exposed to a substance an effective limit is set by default at whatever level exists even when limit setters decline to choose a number because of a lack of complete scientific evidence. For this reason most limit-setting groups will estimate levels based on the "best available information" to quote the TLV introduction, even when that information is less complete than they would like.

The kinds of data used by standard setters fall into several categories.

Human Experience. For many common chemicals there have been unfortunate, unintentional episodes where workers have been exposed at levels that have produced acute illness or death. Records of these events form a major part of the body of knowledge on the toxic effects of such chemicals as chlorine, hydrogen sulfide, and nitrogen dioxide. A weakness of this data is that the exposure is almost always unknown and can only be estimated. Another weakness is the uncertainty of extrapolation from episodes involving serious overexposure to levels of exposure that are believed to be safe for most people over long periods of time. While human experience has been a key data point for some few common chemicals, there is no recorded experience for most chemicals and obviously would be none for any new chemical.

Epidemiology. Epidemiology is a special case of the use of human experience. By means of the science of epidemiology populations of exposed workers are observed, usually over long periods, and health effects are related to exposure. Two common kinds of studies are (1) cohort studies where disease occurring in a defined group of workers is related to some unexposed base population and (2) case control studies where individuals who have a disease are compared with others who do not. Cohort studies are usually done on very large populations to generate hypotheses about possible cause and effect relationships, while case control studies are done on smaller groups of workers as a means of testing causal hypotheses. Both types of studies usually involve looking backward over the years since the effects of interest, typically cancer, may take many years to occur. Because of the "retrospective" nature of these studies good, quantitative data on exposure is not usually available. Another weakness is that for common diseases like lung cancer, which may be of occupational origin but is also very likely to be related to smoking, a larger number of excess cases would have to occur to demonstrate an occupational component.

Animal Tests. Although man and laboratory animals seem quite different, biologically they are very much alike in terms of organs and their physiology. Thus it is possible to use tests on animals to estimate human effects. The validity of these comparisons, about which books have been written, cannot be discussed here. It can be said that these data, while useful and often essential, are only an approximation of human response and their use contributes to the uncertainty of limits.

Structute – Activity Relationships. Human experience often does not exist and animal tests are expensive and not always helpful. Another way to estimate the toxicity of a chemical is to compare it with other chemicals with similar structure. This process of structure–activity relationships (SAR) analysis is increasingly used and shows great promise.

Occupational exposure limits occupy a position intermediately between community air pollution limits and emergency limits. Air pollution standards are intended to limit the "involuntary" exposure of the general population to certain specific air contaminants (SO_2) and some generic classes of compound (volatile organic compounds—VOC) based on either health effects or nonhealth effects such as obscuration or damage to the ozone layer. The health-based air pollution standards are intended to protect the whole population, including children and people in ill health, from continuous 24 hr per day, 7 day per week exposure. Therefore, these standards are invariably lower than occupational standards. Some jurisdictions have arbitrarily divided the occupational limit by 100 to obtain an air pollution limit.

For emergency planning purposes the National Research Council (NRC) has estimated the exposure levels that cause serious, usually acute, health effects. These numbers, which the National Academy of Science (NAS) calls Emergency Exposure Guidance Levels (EEGLs), are much higher than occupational exposure limits. The American Industrial Hygiene Association sets similar limits called Emergency Response Planning Guides (ERPGs) which are discussed in Chapter 15. Both EEGLs and ERPGs are used to help identify facilities for which emergency plans are needed and to define the area that could potentially be impacted by a catastrophic release. They are not in any sense "recommended" exposure limits.

1.4.2 Interpretation of Standards

Given the uncertainties involved in the selection of occupational exposure standards and the variability of the occupational environment it would be unreasonable to interpret occupational limits as rigidly as one might interpret an engineering standard or specification. In fact, the introduction to the TLV list states, "These limits are not fine lines between safe and dangerous concentrations" and goes on to recommend the use of professional industrial hygiene judgment. What this tells us is not to make one measurement and then declare the workplace unsafe if the result is just over the limit, and not to declare if safe if it is just under the limit. While how not to interpret TLVs is clear, it is not clear, aside from the injunction to use professional judgment, how TLVs should be interpreted. In practice the use of judgment means to take several measurements and, more importantly, observe the conditions and situations that give rise to the results found. Using all this information and knowledge of the origin of the limit and what it is intended to prevent, the industrial hygienist can then decide what action, if any, is needed to prevent overexposure. In this manner of practice it might be decided that a single measurement above the limit was atypical and not a cause for concern, or that a series of measurements below the limit were too close to the limit and did not provide sufficient assurance that unsampled periods were safe.

A recent development in the practice of industrial hygiene is that these somewhat subjective judgments have been made more scientific and uniform by the application of statistics. Analysis of environmental variability predicts that some few measurements in excess of the limit will occur with a certain probability in a series of measurement even when the mean exposure is well below the limit. If the substance in question has a long biological half time and no acute effects occur within several times the TLV then some fraction of measurements over the TLV or "exceedances" can be tolerated, provided we are confident that the mean is under the limit. The degree of confidence that the mean is under the limit can vary according to the seriousness of the consequences of overexposure. It is not necessary to be anywhere near as certain that irritation will not occur as that cancer will not occur.

The preceding statistical reasoning for substances with long biological half times can be developed for short half-time substances and for acute exposures, Based on this logic it is possible to develop decision-making strategies that will prescribe how many samples to take, where and when to take them, and how to treat the data so as to be able to make a decision about safety at an acceptable risk of being wrong. In general, it is usually necessary to get expert industrial hygiene advice in the development and application of these strategies.

REFERENCES

ACGIH. *The Documentation of Threshold Limit Values and Biological Exposure Indices*, 6th ed. Cincinnati: American Conference of Governmental Industrial Hygienists, 1991 (with annual supplements).

ACGIH. *Threshold Limit Values for Chemical Substances and Physical Agents and Biological Exposure Indices for 1992–1993*. Cincinnati: American Conference of Governmental Industrial Hygienists, 1992.

Adams, R. M. *Occupational Skin Diseases*. New York: Grune & Stratton, 1983.

Gosselin, R. E., R. P. Smith, and H. C. Hodge, Eds. *Clinical Toxicology of Commercial Products*, 5th ed. Baltimore: Williams & Wilkins, 1984.

Klaassen, C. D., M. O. Amdur, and J. Doull, Eds. *Cassarett and Doull's Toxicology*. Macmillan. New York, 1986.

Kusnetz, S. and M. K. Hutchison, Eds. *A Guide to the Work Relatedness of Disease*, Rev. Ed. Cincinnati: NIOSH, 1979 (GPO S/N 017-033-00177-4).

Lewis, R. J. *Carcinogenically Active Chemicals: A Reference Guide*. New York: Van Nostrand Reinhold, 1990.

Lewis, R. J., Ed. *Sax's Dangerous Properties of Industrial Materials*, 8th ed. New York: Van Nostrand Reinhold, 1992.

La Dow, J., Ed. *Occupational Health Law: A Guide for Industry*. New York: Dekker, 1981.

Mackison, E. W., R. S. Tricoff, and L. J. Partridge, Jr., Eds. *NIOSH/OSHA Ioccupational Health Guidelines for Chemical Hazards*. Washington DC: NIOSH, 1981 (GPO S/N 017-033-00337-8).

Maibach, H. I., Ed. *Occupational and Industrial Dermatology*, 2nd ed. Chicago: Year Book Medical Publishers. 1986.

Marczewski, A. E. and M. Kamrin. *Toxicology for the Citizen*, 2nd ed. Michigan State University: Center for Environmental Toxicology. 1987.

Monson, R. R. *Occupational Epidemiology*, 2nd ed. Boca Raton, FL: CRC Press, 1989.

NIOSH. *Occupational Diseases: A Guide to Their Recognition*, 2nd ed. Cincinnati: NIOSH, 1977 (GPO S/N 017-03-00266-5).

NIOSH. *Pocket Guide to Chemical Hazards*, Pub. No. 90-117, June 1990.

NIOSH. *Registry of Toxic Effects of Chemical Substances*. Cincinnati: NIOSH, Annual.

NIOSH. *The Industrial Environment: Its Evaluation and Control*. Cincinnati: NIOSH, 1973 (GPO S/N 017-001-00396-4).

Olishifski, J., Ed. *Fundamentals of Industrial Hygiene*, 2nd ed. Chicago: National Safety Council, 1979.

Permeggiani, L., Ed. *Encyclopedia of Occupational Health and Safety*, 3rd rev. ed. Geneva: International Labour Office, 1983.

Patty's Industrial Hygiene and Toxicology, 3rd rev. ed. New York: Wiley, 1978–1988.

Proctor, N. H., J. P. Hughes, and M. L. Fischman, Eds. *Chemical Hazards of the Workplace*, 2nd ed. Philadelphia: Lippincott, 1988.

Roach, S. *Health Risks from Hazardous Substances at Work*. New York: Pergamon, 1992.

Weeks, J. L., B. S. Ley, and G. R. Wagner, Eds. *Preventing Occupational Disease and Injury*. Washington DC: American Public Health Association, 1991.

2

Sources of Exposure

The relationship between dose and disease was discussed in Chapter 1. For a chemical to cause a disease or other adverse response it is necessary for the individual to receive a dose of the chemical. That dose may be external, as in the case of skin contact, or may enter the body via on of the routes of entry (inhalation, injection, etc). The term exposure is used in industrial hygiene to describe contact with a substance such that a dose may be delivered. An individual is exposed when a chemical comes in contact with the skin or is in the air inhaled by the individual. The object of health hazard control is to limit dose by preventing exposure. It is not practical to measure or control dose directly so instead exposure is measured and limits are set for the control of exposure. Figure 2.1 illustrates the sources of exposure that must be combined to completely assess exposure.

In industry the most important route of entry is inhalation. Workers are exposed to a substance by inhalation when the substance contaminates the air they breathe. Our understanding of the sources of contaminants to which workers are exposed is important to the recognition, evaluation, and control of occupational health hazards.

2.1 MODES OF AIR CONTAMINATION

The ways in which substances become dispersed so as to contaminate workplace air depend on how they are generated or handled and on their physical and chemical properties.

2.1.1 Solids

Mechanical abrasions of solid material by cutting, grinding, or drilling can produce small particles, which can form an airborne dust cloud or solid aerosol. This is the main mode of dust generation in mining and quarrying and is also important in foundry casting cleaning, woodworking, and metalworking. The quantity and size

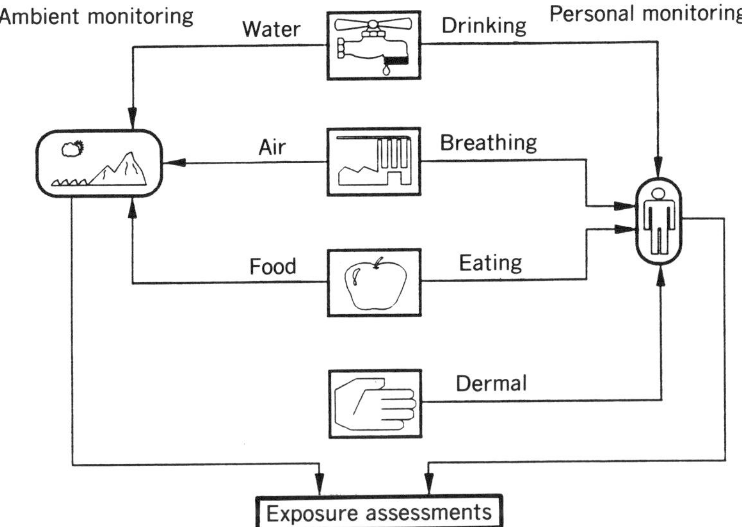

Figure 2.1 Sources of exposure in a complete exposure assessment.

distribution of the dust produced by this primary mode of generation is dependent on such factors as the energy applied and the strength and hardness of the solid is being abraded.

Secondary dispersion of solid particles occurs when settled dust is stirred up or when fine powders are blown into the air. This mode of air contamination by solid aerosols is common to all material handling operations that transport dusty materials and also results from poor housekeeping. The degree to which a material can produce a dust cloud that will stay suspended in the air and be transported by air currents depends on the type of powder being dispersed. It must be finely divided and the individual particles must be free to separate rather than stick to each other or adhere due to electrostatic charge or liquid films. Some material, although finely divided, will not create a dust cloud unless considerable energy is expended, while other materials fly into the air with any stimulus.

Some solids, such as iodine, have sufficient vapor pressure that they will sublime and become airborne by going directly from a solid to a gas.

2.1.2 Liquids

Liquid aerosols may be produced by any process that supplies enough energy to overcome the surface tension of the liquid. Obviously a liquid aerosol has enormously more surface area than the same quantity of bulk liquid so work has to be done to separate and disperse the droplets. This process occurs intentionally in spray coating and nebulizing and unintentionally when oil mist is generated from the lubricants or coolants used on high-speed machinery. Given their large surface area to volume ratio and the increase in vapor pressure caused by surface curvature, liquid droplets tend to evaporate quickly unless the vapor pressure of the material is quite high. For example, aliphatic hydrocarbons with less than 20 carbon atoms will not usually exist as aerosols for long.

Liquid aerosols are also produced by condensation. When a high boiling liquid at high temperature is released into ambient air as from a leak, it may flash to a vapor. The vapor when mixed with the cooler air will then condense into very small droplets that will aglomerate into larger liquid droplets. This source of exposure can occur whenever hot liquids leak from pipes or liquids are heated in open vats.

Liquids in contact with the air will evaporate to some degree depending on their volatility. Some liquids, like heavy oils, have such low vapor pressure that even at saturation at room temperatures, the concentration will be very low, well below their permissible limit. Liquids rarely reach even one-tenth of saturation when evaporating into workplace air. Appreciable amounts of such liquids could become airborne as liquid aerosols, but it is important to note that aerosols exposure limits are much lower in general than limits for less toxic substances as vapors. Thus 15 mg/m^3 of an aerosol is considered an upper acceptable level even for a nontoxic substance because of housekeeping and obscuration considerations. On the other hand, nonane, at its TLV of 200 ppm, would amount to over 1000 mg/m^3 on a mass basis.

Not all heavy liquids are unable to exceed their exposure limits by evaporation. Elemental mercury vapor can exceed its TLV of 0.05 mg/m^3 by slow, room temperature evaporation. A significant source of mercury poisoning is exposure in workrooms where mercury has been spilled and been caught in cracks where it slowly evaporates.

Obviously, highly volatile liquids are usually manufactured, transported, and used in closed systems. Flourocarbons are kept contained when used as refrigerants, but are allowed to evaporate when used as cleaning agents. All solvents used in coatings are intended to evaporate. In some instances exhaust ventilation is applied to remove solvents from coatings but when solvent-containing paints are applied to walls the evaporating solvent will contaminate the workplace before it is removed by ventilation. Degreasing solvents may be used in either open or semiclosed systems. Parts are often cleaned with a solvent wet brush on a bench or over a tank. Well-designed vapor degreasing tanks condense the vapor on the surface to be cleaned and retain most of the solvent in the tank. Solvents used for dry cleaning are recovered to the maximum extent possible for economic reasons and to avoid air pollution. All solvent use is conditioned by its toxicity. Thus benzene and carbon tetrachloride are no longer used as solvents in open systems where they would be released into the workplace. Instead, substitutes such as toluene and 1, 1, 1-trichloroethane are used and release is kept as small as possible.

2.1.3 Gases

Substances that are gases at room temperature are usually kept in closed systems. For these gases to contaminate workplace air they must either be intentionally released or there must be a loss of containment such as a leak or accidental venting. As discussed elsewhere, virtually all closed systems leak to some degree. The tightness of a system is determined by the engineering and leak monitoring effort applied, which in turn depends on the consequences of these fugitive emissions. High-value and very toxic materials are usually very tightly controlled. Increasingly, substances that create a nuisance in neighboring communities are controlled even when they are not valuable or particularly toxic. Flammable materials are generally controlled because a leak may become a fire and lead to total disaster. For some gases, like carbon dioxide or nitrogen, which are neither toxic, valuable, odorous, nor flammable, substantial leaks may be tolerated.

Toxic gases often evolve from processes that are not closed systems. Most processes that involve combustion can produce carbon monoxide. This is a major concern in steel mills and in the metal-smelting industries. Toxic gases may also evolve when organic materials are thermally degraded in processing. Hot metal poured into molds in foundries will cause organic binders to break down into carbon monoxide, aldehydes, acrolein, polynuclear aromatics, and other toxic chemicals. Without adequate ventilation these substances will contaminate the air of the foundry and result in a significant exposure of workers.

Some wet processes release gases as a result of chemical reactions. Fermentation gives off large amounts of relatively nontoxic carbon dioxide. Hydrogen will evolve from electroplating and may carry with it liquid droplets containing toxic plating-solution components. Plating and chemical cleaning can also give off acid gases such as hydrogen chloride.

2.2 SOURCES OF EXPOSURE

Chemicals are usually manufactured in closed systems because of the heat and pressure involved, to contain gaseous and volatile materials, and to avoid product contamination and loss. Unlike industries such as mining or trades like painting where the worker is closely involved with the material, the worker in the chemical process industry is usually separated from the process materials. While routine, direct worker-exposure operations such as bagging and drumming do occur, they are not typical of the exposure of the chemical process industry worker.

Table 2.1 summarizes the kind of exposure sources typically found in the chemical process industry and rates each source in terms of the following characteristics which determine the degree of exposure.

Intermittent or continuous. Does the exposure source release a contaminant continuously when the process is in operation or do the releases occur over discrete intervals of time?

Episodes. If releases are intermittent do they occur randomly or are they the result of certain intentional or unintentional events?

Worker activity. Is the release and consequent exposure directly related to some activity on the part of the exposed worker?

Importance. What is the relative importance of a particular source as a contributor to overall exposure? This will vary from plant to plant so the ratings given represent an industry average.

Control. Typical measures of source release reduction applied are: M, Maintenance—primarily the monitoring and repair of leaks by replacing pump seals, repacking valve glands, tightening flanges, sealing holes in ductwork, and so on.

P, Personal protection—the use of an air purifying or supplied air respirator, usually for a short period of time, for a particular hazardous operation.

TABLE 2.1 Exposure Sources in the Chemical Process Industry

	Intermittent or Continuous	Episodic	Worker Activity	Importance	Control
Fugitive emissions					
Pump seal leaks	Either	No	No	High	M
Flange leaks	Cont.	No	No	Low	M
Agitator seal leaks	Either	No	No	Med	M
Valve stem leaks	Cont	No	No	High	M
Process operations					
Sampling	Int.	Yes	Yes	Med	W,E
Filter change	Int.	Yes	Yes	Low	W,P
Gauging	Int.	Yes	Maybe	Low	E.W
Venting and flaring	Either	No	No	Med	E
Extruding	Either	No	Yes	Med	E
Material handling					
Solid addition	Int.	Yes	Yes	Med	
Liquid transfer	Either	No	No	High	E
Bagging	Cont.	No	Yes	High	E
Drumming	Cont.	No	Yes	High	E,W
Bag dumping	Int.	No	Yes	High	E,W
Screening	Cont.	No	No	Med	E
Open mixing	Int.	No	No	Med	E,P
Baumbury mixing	Either	No	Yes	High	E,P
Milling	Either	No	No	Med	E,P
Maintenance					
Equipment opening	Int.	Yes	Yes	High	W,P
Instrument line draining	Int.	Yes	Yes	Med	W,P
Welding	Int.	Yes	Yes	High	E,W,P
Painting	Int.	Yes	Yes	Med	W,P
Sandblasting	Int.	Yes	Yes	High	E,P
Insulating	Int.	Yes	Yes	High	W,P[a]
Insulation removal	In.	Yes	Yes	High	W,P
Chemical cleaning	Int.	Yes	Yes	Med	W,P
Degreasing	Int.	Yes	Yes	Low	E
Cutting and burning	Int.	Yes	Yes	Med	W,P
Catalyst handling	Int.	Yes	Yes	High	W,P
Waste Handling					
Bag house cleaning	Int.	Yes	Yes	High	P
Drain and sewer venting	Either	No	No	High	E
Spill clean up	Int.	Yes	Yes	Med	P
Sweeping	Int.	Yes	Yes	Low	W
Incineration	Either	No	Maybe	Med	E
Waste water treatment	Cont.	No	No	Med	E
Sludge handling	Int.	Yes	Yes	Med	W,P

[a] Substitution of less toxic materials for asbestos is the most common control.

W, Work practices—staying up wind of a release source; not spilling volatile liquids on the ground; keeping the work area clean to avoid redispersion of dusty materials.

E, Engineering—equipment or process modifications to prevent or contain releases such as welded pipe joints, hermetic pumps, vent scrubbers, sealed drains, or local exhaust ventilation.

2.2.1 Fugitive Emissions

Fugitive emissions are "leaks" that occur wherever there are discontinuities in the barrier that maintains containment. While flange and seal leaks are small by themselves, they can cumulatively amount to the main source of loss from a unit. Even when they are very small and cannot be detected as losses in a material balance they can still result in high local concentrations of contaminants which can lead to overexposure. Further, some leak sources, such as valve stem leaks, tend to gradually increase over time and can become major if not corrected. Other leaks, such as pump seal leaks, while usually small, can become very large in the event of total seal failure. Overall, in typical well-maintained plants pumps and valves are more important sources than flanges. For that reason, leak detection and repair efforts usually focus on pumps and valves unless there is reason to suspect flanges. Taken as a whole, fugitive emissions, even without catastrophic seal failure, are the origin of the continuous background exposure of workers. This source of exposure may not, by itself, result in overexposure, but its presence reduces the margin within which other exposures may vary while still remaining under the accepted limit.

2.2.2 Process Operations

The operation of a modern chemical plant is typically computer controlled and does not involve any routine operator contact with the feedstock, intermediates, or product. There are, however, a few actions the operators may need to take which could involve contact with process materials. Sampling of process streams is one such task. While on-line analyzers have substantially reduced the need for operator-collected samples, they are still necessary to check on the on-line analyzer or where on-line analyzers are not used. Exposure during sampling can be very high if the sampling line is flushed by running a quantity of a volatile liquid out on the pad. On the other hand, exposure can be very low where the sample is collected in a bomb from a closed loop. Worker care in following prescribed practices is important.

Many filters in chemical process units are either changed very rarely or are back flushed automatically so there is hardly any exposure. Some filters, however, still require frequent manual changing or cleaning. Where this is done, significant exposure may occur unless operators follow the proper procedure. The filter container should be drained of any toxic material and then flushed and purged as needed so that when it is opened there will only be minimal exposure. Zero exposure is difficult to achieve in situations where a disposable paper filter cartridge may retain and slowly release a toxic material that cannot be removed by multiple flushes and purges.

Gauging is often done automatically but there are still occasions where it needs to be done manually with a tape dropped through a hatch on the top of a tank. Even where automatic systems are installed, manual gauging may be used as a check. Depending on the nature of the liquid in the tank, vapors can be released more or less actively while the ullage hatch is open. Short of using respiratory protection, the only exposure control applicable to open-hatch gauging is the work practice of standing upwind if the platform at the hatch permits.

Vents and flares are intended to take contaminants released from safety valves away from work areas. However, it sometimes happens that an elevated vent is at the level of an occupiable platform on the same or an adjacent unit. Thus a worker on a platform may, under certain wind conditions, be subject to the nearly undiluted effluent of a vent. While such elevated platforms may rarely be occupied, a heavy exposure from a vent could incapacitate a worker or cause a fall. Tanks that only vent when being filled are common causes of this concern. The usual solution is to raise the vent above any occupiable platform or, at greater cost, to scrub the vent effluent.

Extrusion is a common way for solid products such as plastics to emerge from closed manufacturing systems. Normally the polymer is hot when extruded and may contain additives and oligomers that are volatile at these elevated temperatures. The result is fuming at the extruder head. These fumes can result in employee annoyance, housekeeping problems, and, at worst, depending on composition, health hazards. Aerosol control is accomplished by careful temperature control backed up by local exhaust ventilation if necessary.

In all of these process operations except venting and flaring the exposure is related to worker activity and to some extent is dependent on worker behavior and the work practices applied. The same will be true of those sources discussed below for which "yes" is shown under worker activity in Table 2.1. The distinction between those exposures that are impacted by worker behavior, and those that, barring the use of respirators, are not, is important. The types of control methods to be applied and the methods of exposure measurement to be used are influenced by this difference.

2.2.3 Material Handling

The continuous movement of materials through a process unit does not in itself result in any opportunities for release and consequent exposure. However, some material-handling steps are difficult to accomplish with total containment. Wherever quantities of material are allowed to build up and be drawn from tankage or at temporary or permanent storage points, the possibility of release needs to be considered. Liquids entering fixed tanks displace air, which must be vented to avoid overpressuring the vessel. Control of liquid-transfer operations can be achieved by variable volume vessels, such as those with floating roofs, which do not require venting, or by scrubbing, flaring, or recovering the vented gas stream.

In drumming and the filling of tank cars and trucks, where the vessel is initially empty, the amount of material being transferred that could be released by displacement will depend on how much evaporates during the filling. Rarely would a material evaporate so quickly that the entire volume of displaced gas would be saturated. More likely the initial release at the start of filling will contain only a small amount and the concentration will increase toward saturation as the filling proceeds. How quickly the concentration in the vented gas increases will depend on the temperature and volatility of the material and on the loading mode. Splash loading, where the material leaves the filling spout at the top of the tank and free-falls to the bottom, will result in much more evaporation of a volatile material than bottom loading or submerged spout loading where the liquid level rises gently without splashing.

Some very volatile liquids and gases that are liquids under pressure are allowed to autorefrigerate by controlled evaporation during transit.

Vapors vented from liquid transfer can be collected and sent to an elevated vent or flare, returned to the discharging vessel head space, reliquified by refrigeration and returned to storage.

One other opportunity for vapor release in material handling is air open mixing. In batch blending or production operations it may be the practice to add liquids or solids to an open vat that already contains some liquid. As the liquid evaporates and as the gas in the head space of the vat is displaced by material addition, some vapor may escape out of the hatch or vat top. The amount of release will depend on the temperature, volatility of the liquid, degree of mixer agitation, rate of addition of material, and openness of the vat top or hatch. Control is usually accomplished by local exhaust ventilation systems that maintain a negative pressure in the vat or collect vapors as they escape from the vat, or by closed addition systems such as rotary valves.

Solids handling is often done by open means both because the hazard is perceived to be less and because it is more difficult to design totally closed solids handling systems. Where solids are handled in closed systems, often as fluids in pneumatic conveying systems, the potential release problems are similar to those for liquid transfer. Air displaced from hoppers and silos as they are filled may contain dust. Also, the conveying air must be released from the vessel where the conveyed materials is deposited. These streams are sometimes sent to an air cleaning system such as a baghouse or other filter or an electrostatic precipitator. While such devices are effective in removing particulates from the air, maintenance of the air cleaner itself sometimes results in a secondary exposure.

Semiclosed systems for handling solids often involve the use of "big bags" or tote bins. In these systems the solid is shipped in a large container (as much as a ton or more), which is then lifted into place over a closed hopper or feed mechanism and a sealed connection is made. It is sometimes necessary to rap or agitate the big container to keep the solid moving and this action can result in deterioration or damage to the container or seals with consequent leaks. Also, there is some release of solids when containers are connected or disconnected with opportunities for dust generation.

Solids handling in small containers such as bags and drums can be either automated or manual. Automated bag- and drum-filling machines can position the container, fill it with a weighed amount of product, seal it, and label it. Some bag-filling machines exhaust dust generated inside the bag during filling. Dust generation and dispersion during bag and drum filling depend on the dustiness of the solid, the rate and manner of container filling, the degree of care taken in the filling process, the maintenance of machinery, and the cleanup of spills.

Bag dumping can be done in open or closed systems. In open bag dumping there are potential releases when bags rupture during handling, when they are dumped, and when the empty bag is collapsed for disposal. In manual bag-handling operations spills from ruptured bags damaged by fork lifts and other traffic can be a significant source of exposure. Blowing dust off machinery and clothing with compressed air adds to the problem. A means of control is frequent cleaning with wet sweeping or vacuuming with heavy-duty installed systems.

Automatic bag dumping can be done by machinery that conveys the bag into a closed chamber where it is split open and dumped. The empty bag is then transported to a compactor. The whole system is enclosed and exhausted so that all leaks will be inward.

Handling of small quantities of solids may be altogether manual. Materials may be scooped out of drums, weighed on scales, and dumped into mixers entirely by hand. Exposure may be insignificant or a major concern depending on the toxicity and dustiness of the material. Some users of small quantities of solid additives may arrange for the supplier to package the material in preweighed batches packed in a bag so that the whole prepackaged container can be added to the mix.

Mixing of solids by means of a banbury, muller, calender, or mill is either an open or incompletely enclosed process. Except for the loading of a mixer there is usually little opportunity for dust created by the escape of powders since the mixing solids are generally wet or tacky. However, mixing of solids often requires considerable energy and generates heat which results in "fumes" of evaporated and condensed particulate. While local exhaust ventilation is effective in removing these fumes, close manual handling may still result in exposure.

2.2.4 Maintenance

Closed systems will contain process materials except when they leak and release fugitive emissions or when they are opened for maintenance. Open-system maintenance can add to exposure by disturbing and dispersing deposits of materials in equipment. Most maintenance is done while the plant is in operation so the maintenance workers need to be in close proximity to operating equipment for long periods of time. Local contaminant releases and physical hazards such as noise or thermal radiation need to be considered. In addition, the valves or other barriers blocking off the operating parts of the plant may leak into the maintenance work area and there is the possibility of failure of the barrier.

The piece of equipment being maintained should be cleaned as necessary to reduce exposure before it is opened and repaired. Where highly toxic process materials are present it may be necessary to flush with a low-toxicity stream, strip with steam, and purge with nitrogen. Where this is necessary the equipment design should include the special fittings needed for the flush and purge line connections. Even when cleaning prior to opening is done as completely as possible, it may be necessary to use respirators at least for the initial opening to guard against overexposure resulting from trapped toxic substances. Proper cleaning and opening of equipment lines and vessels where toxic material may be present is a complex requiring careful planning and attention to detail in execution.

Turnarounds, or major periodic overhauls of chemical plant units, are a special case of plant maintenance. Since the units are shut down during turnarounds some risks are avoided, but because the unit is out of production there is time pressure to complete the work. Contractors and other workers who may not be familiar with the unit may be brought in and many maintenance activities proceed simultaneously. In this environment there is the potential for disorganization and mishap with resulting unanticipated releases of chemicals. To conduct a safe turnaround it is necessary to plan the event carefully in advance. Contingencies should be anticipated to the extent possible and plans made to deal with them. All of the workers involved should be specially instructed in their duties and closely supervised during the entire turnaround from shutdown through start-up and back to normal operation.The materials and operations used in maintenance may present a set of hazards quite separate from the hazards of the feedstocks, intermediates, and products of the chemical unit.

Welding. Any of the metals in the rod or the alloy being welded can become airborne in the welding fume. Zinc and some others can cause metal fume fever which is a frequent problem for welders. Other metals such as cadmium can produce systemic effects. Chromium, under certain conditions, can be released in the potentially carcinogenic hexavalent form. In addition to these metal fumes the welding process produces oxides of nitrogen, ozone, and ultraviolet radiation. All of the emissions can be controlled by general or local ventilation or by respirators if necessary. The welder's face mask provides only slight respiratory protection. Welding in confined spaces is particularly hazardous owing to the difficulty in delivering clean air to the site of the welding and in exhausting welding fumes.

Painting. While lead paints are no longer used for domestic painting they are still occasionally used in industry. Frequently surfaces being prepared for painting may have remnants of old lead or chrome coatings which could become airborne during scraping or grinding. The solvents used in paints, while not highly toxic, can reach excessive concentrations in poorly ventilated spaces. Low rates of paint application as from brushing will produce lower solvent release rates than intermediate rate application by roller or high rate spraying. Certain kinds of modern coatings such as polyurethanes and epoxies present special toxic hazards.

Sandblasting. While some modern corrosion-resistant treatments do not require the removal of all rust, sandblasting to clean metal surfaces prior to coating is very common. In addition to the metal dust, the very fine fragments broken off from the abrasive particles may be respirable, that is, capable of reaching the deep lung where they may cause damage. The degree of risk depends greatly on the type of abrasive used. Steel balls and walnut shells produce relatively nontoxic dust, as does aluminum oxide. On the other hand, fine dust from sand, which is typically composed of silicon dioxide, is very toxic and can produce a serious lung disease. The degree of dust exposure from sandblasting will depend on the degree of enclosure and the use of personal protective equipment. Small pieces can be cleaned in fully enclosed blast cabinets with local exhaust ventilation to maintain negative pressure. Large objects, like truck bodies, which are too large to be done in cabinets, are often cleaned in large booths with down draft local exhaust ventilation. For structures and fixed piping sandblasting is done out in the open. When blasting in either the booths or in the open the operator should be protected by a special sandblaster's supplied-air hood. A common problem is that the operator will use the hood for physical protection but will not connect the hood to a supply of clean air. When a hood is used in this manner fine dust will enter the worker's breathing zone under the hood and the hood will not provide respiratory protection.

Insulation. While asbestos-containing insulation is not widely used at present, because of the health hazard, it was widely used in the past. It is common for insulation workers replacing insulation at older plants to encounter asbestos. For that reason it is necessary to be certain of the composition of old (and sometimes inadequately labeled new) insulation to be sure that it does not contain asbestos or to be sure that proper procedures are followed if it does. The removal of asbestos-containing insulation is a complex and difficult process beyond the scope of this book. To do it safely requires personal protective equipment, monitoring, containment, special disposal procedures, stringent work practices, and record keeping. Many companies elect to have asbestos removal done by specialized contractors.

Modern nonasbestos-containing insulation frequently incorporates man-made mineral fibers (MMMF) for strength at high temperatures. The glass or rock wool fibers are usually silicates and are usually large compared to asbestos, too large to be inhaled into the deep lung. Also they do not split the long way and thus do not produce as many fine fibers. For these reasons, even if they are as potent as asbestos fiber for fiber, they are less hazardous because there will be fewer respirable fibers. Ceramic fibers, however, are made in respirable size ranges and are therefore more hazardous. Also, when they are used inside high-temperature furnaces, they may be converted to christobalite, which is more toxic than quartz.

Chemical Cleaning. Removal of deposits from inside vessels and pipes is often done with acids, caustics, or strong solvents. The handling of these chemicals can cause a number of hazards. Transfers and mixing of small quantities is usually done manually from drums or tank trucks. Application involves pumping materials through hoses or temporary piping. Sometimes strong cleaners are applied as a pressure spray or jet with consequent spattering. The reaction of the chemical with the deposit materials and the metal of the pipe or vessel can produce dangerous gaseous air contaminants. Even when cleaning is done *in situ*, removal and disposal of the cleaner and flushing fluid can cause exposure. Since these operations are infrequent, installed control equipment is rarely used. In most cases workers rely on personal protection equipment plus detailed handling precautions.

2.2.6 Catalyst Handling

A great many chemical manufacturing reactions are made possible by the use of catalysts (Table 2.2). Catalysts may be divided into two categories: homogenous catalysts, which are dispersed in the reactant mix so that the entire reaction takes place in a single phase, and heterogeneous catalysts, where the catalysis occurs at phase interfaces. Homogeneous catalysts are added to reaction streams the same as any other process chemical and they, or their reaction products , are removed by the usual finishing separations process. From the point of view of worker exposure these chemicals are no different than any other chemical additive. Heterogeneous catalysts, on the other hand, are often used in the form of fixed beds, which must periodically be regenerated or remove and replaced. These tasks present significant exposure opportunities.

Catalyst charging and topping is an occasional task typically done at the top of a reactor using temporary handling facilities. For this reason local exhaust ventilation is rarely used even though the operation may be dusty and the catalyst toxic.

TABLE 2.2 Catalyst Health Effects

Catalyst	Health Effect
Aluminum oxide (Al_2O_3)	Nuisance
Aluminum chloride	Decomposes to HCl; irritation
Aluminum alkyls	Acute thermal burns from contact, lung damage
Chromic oxide (CrO_3)	Cr^{3+}, low toxicity; may convert to Cr^{6+}, toxic and carcinogenic
Cobalt	Lung irritation
Cobalt hydrocarbonyl ($CoH(CO)_4$)	Acute respiratory failure
Ferric oxide (Fe_2O_3)	Siderosis; low toxicity
Molybdenum compounds	Pneumoconiosis
Nickel compounds	Carcinogenic; nickel-subsulfide (Ni_3S_2)
Nickel carbonyl	Acute respiratory failure; carcinogenic
Platinum and compounds	Low toxicity; dermatitus
Thorium oxide	Low toxicity; radioactive
Uranium	Kidney damage
Vanadium	Respiratory irritation

Scrupulous use of personal protective equipment and adherence to work practices is essential. Respiratory protection can be so critical as to require air-line respirators. Skin protection may include full protective suits. When catalysts are dumped from a reactor they may be very dusty because of particle size reduction occurring in the reactor and because of handling. Dumping is often done via chutes, which do a poor job of containing the dust. In addition to protecting the workers it may be necessary to erect a temporary enclosure to prevent contamination of adjacent work areas. Some catalysts being dumped are pyrophoric. Water sprays used to prevent fires also help control the dust. When catalyst beds are inerted there is the danger of the release of large amounts of the inerting gas into the work area, which may cause asphyxiation.

Most catalyst removal operations are carried out by experienced contractors using special equipment and techniques. It is important for plant personnel to recognize their limitations in this regard and not undertake any job for which they are not properly trained or equipped. It is also important to keep in mind that the hazard of what comes out of a reactor may be quite different and much more severe than the catalyst that went into the reactor.

2.2.7 Waste Handling

Housekeeping procedures in general can have a significant impact on employee exposure, and certain waste handling procedures, although intermittent, can result in very serious exposure without proper precautions.

The best way to keep a plant clean is to not spill in the first place. Management reviews of the origins of spills and accumulated debris not only keep the plant cleaner but prevent loss of valuable material, save cleanup manpower, and reduce fire and safety hazards. Spilled materials in aisles and on walkways can become airborne by redispersion and can be spread onto surfaces and cause skin contact.

Dry powders are best cleaned up with either installed or portable industial vacuum cleaners. Liquid spills can be soaked up with a number of available solvents and scraped or shoveled into containers. Careful consideration should be give to the methods used to clean floors. Serious overexposures have resulted from the use of volatile solvents on large floor areas inside buildings.

Air cleaning systems are often used to remove dust or vapors from plant or process exhaust streams. Dust collecting systems such as filters or electrostatic precipitators that handle heavy loads of dust are usually designed to be self-cleaning but it is still necessary to enter the air cleaner periodically for inspection or repair. Dust deposits inside the equipment are likely to be stirred up and inhaled by unprotected workers. Bag houses are particularly likely to cause exposure since large amounts of dust may be retained in the cloth and released when the bags are handled.

Waste water treatment facilities may receive chemical process wastes and spills. These wastes may volatilize on emerging from a closed sewer system into open waste treatment tanks particularly if hot streams have heated the tank. These releases can occur without warning and result in unexpected employee exposure. Covering reduces the hazard and can also reduce air emissions but does require careful design to avoid creating an explosion hazard.

Toxic substances trapped in separator or biological oxidation sludge may be released when sludge is filtered and skin contact can result from sludge handling.

2.3 MATERIAL BALANCE

The substances to which workers are exposed are released from the production process. Usually they are raw materials, intermediates, or products or they are wastes that originated from raw materials, intermediates, or products. As such, these losses should show up in a loss-control material balance. In fact the emissions that result in worker exposure are among the losses accounted for in a material balance and actions taken to reduce loss, such as fugitive emission control, also tend to reduce exposure. Unfortunately, a low loss rate as measured by a material does not automatically equate to a low level of exposure. The reason is that the amount of material required to produce significant exposure of a few workers in a local area is very small, too small to be seen in the large overall mass of material dealt with in a balance. For example the amount of benzene required to produce 1 ppm in a 1000-m^3 space with an air change rate of six changes per hour is only 3g/min. While a leak of this size would often be quite obvious, this small a loss would be lost in the round-off in a material balance.

2.4 OFF THE JOB EXPOSURE

In accounting for chemically induced illness that could be of occupational origin it is important to consider that significant chemical exposure may occur off the job.

The following are some of the home maintenance and hobby activities that can cause chemical exposure.

- *Painting*. Although most home painting is done with water-base paints, exposure to a full range of solvents is possible with specialty coatings. Also present are isocyanates in polyurethane and amines in epoxies. Paint removers may contain methylene chloride. Sanding will produce wood and paint dust, which may contain lead.
- *Guns*. Noise exposure indoors and outdoors and lead exposure in indoor firing ranges is likely.
- *Auto Repair*. Gasoline, paint, adhesive solvents, and carburetor cleaners (glycolethers) are common hazards.
- *Arts and Crafts*. Paint solvents, adhesive solvents, epoxies, lead, silica, isocyanates, and welding fumes are among the almost limitless variety of materials and processes used by artists.
- *Plumbing*. Lead solder and acid fluxes are used in making pipe connnections and caustics are used for cleaning drains. Solvents are used for connecting PVC pipe.
- *Gardening*. A wide variety of pesticides are used in gardening and fertilizer may also be hazardous.

While there are many opportunities for chemical exposure from off the job activities, the likelihood of significant harm from these exposures is probably low. Toxic chemicals like benzene and carbon tetrachloride are no longer available as consumer products. Lead-based paints are not generally available, and the range of substances occuring in domestic paints is much smaller than in industrial coatings. Also, while hobby activities are often undertaken with great enthusiasm and little care for safety and health, the period of time spent is comparatively short and intermittent.

REFERENCES

Burgess, W. A. *Recognition of Health Hazards in Industry*, New York: Wiley-Interscience, 1981.

Chemical Manufacturers Association. *U.S. Chemical Industry Statistical Handbook*, Washington: CMA, 1992.

Cralley, L. V. and L. J. Cralley, *In-Plant Practices for Job Related Health Hazards Control*, Vol. 2, *Engineering Aspects*. New York: Wiley, 1988.

Cralley, L. V. and L. J. Cralley, *In-Plant Practices for Job Related Health Hazards Control*, Vol. 1, *Production Processes*. New York: Wiley, 1988.

Faucett, H. H. and W. S. Wood. *Safety and Accident Prevention in Chemical Operations*, 2nd ed. New York: Wiley-Interscience, 1982.

Kent, J. A. *Handbook of Industrial Chemistry*, 9th ed. New York: Van Nostand Reinhold, 1992.

Kirk, H. and R. Othmer, Eds. *Encyclopedia of Chemical Technology*. New York: Wiley-Interscience, 1978–1983.

March, J. *Advanced Organic Chemistry*, 3rd ed. New York: Wiley, 1985.

McKelta, J. *Chemical Technology: An Encyclopedic Treatment*. Totowa, NJ: B + N Imports, 1968–1979.

National Institute for Occupational Safety and Health. *Information Profiles on Potential Occupational Hazards*. Available from the National Technical Information Service.

National Safety Council. *Personnel Safety in the Chemical and Allied Industries*. Chicago: National Safety Council, 1979.

National Safety Council. *Fundamentals of Industrial Hygiene*, Chicago: NSC, 1988.

Peck, T., Ed. *Chemical Industries Information Sources*. Detroit: Gale Research, 1979.

3

Exposure Evaluation

For health hazards to be controlled, they must first be recognized and evaluated. Recognition is an orderly process combining detailed knowledge of process materials with knowledge of actual and potential sources of exposure. This mass of information is used to estimate likely levels of exposure and, based on these estimates, to set priorities for evaluation. By evaluation we mean the actual measurement of the quantity of an air contaminant in the air employees breathe at such times and periods as to represent exposure in such a way that risk may be estimated. Based on our understanding of the risk, decisions are made regarding control, up to and including very costly plant modifications. The entire process is complex because of the long list of scenarios that must be considered, difficult because of the many opportunities for error that must be avoided, and continuous because plant physical conditions, work practices, and the health standards themselves all change.

3.1 INVENTORY

A logical place to start the process of health hazard recognition is with a total inventory of all substances present. While it is true that at the end of the process of recognition and evaluation only a few substances will in fact require control action, it is not possible to leap ahead to this conclusion without error. All substances present are capable of producing toxic effects depending on the dose. Even when substances are present in only trace amounts, there is more than enough present to produce a hazardous situation in a localized work area. While the feedstocks and products of a process are well known, intermediates, by-products, and wastes may be less obvious and may not have been identified. Other, not so obvious, materials present are catalysts, additives, cleaning agents, and maintenance materials. The inventory sources and content will vary for the several different categories of materials.

3.1.1 Process Materials

Generally all feeds, isolatable intermediates, products, and wastes are known and are listed on process material balance tables. The description of these materials, however, may not be complete and may be misleading. A feed listed under the name of a single isomer may in fact be a wide mix of isomers and may also contain compounds boiling in the same range and impurities. Later separation processes may concentrate these minor constituents. Some process intermediates, which come and go as the stream moves through the process, cannot be considered isolatable but could still result in worker exposure if leaks occur at particular locations. Wastes are often poorly described, although evolving waste handling legislation is requiring ever increasing detail. Thus, while the process material balance makes a good starting point for the process materials inventory, it is necessary to look beyond it in many cases for detailed analyses that may have been performed in either the plant or the suppliers laboratory. In some cases special analyses may be necessary when the subsequent prioritization process reveals that a hazardous situation may exist.

The information needed to assemble the process material inventory should already exist in various files. These files, however, will often be widely dispersed and difficult to find, and the information may be in a variety of different forms and units. For this data to be usable as an inventory it is necessary to bring it together and tabulate it in a common format. While any kind of "file" could be used, a computer database under the control of a database management program is most efficient for all but the smallest inventories. Each inventory database will be constructed differently depending on the size and nature of the plant (batch or continuous), the means used to update the data, and how the data will interact with other files. A typical file might contain the following common elements:

Plant name and location
Name of production unit within the plant
Specific chemical stream and location
Designation of stream type (feedstock, product, etc.)
Where the stream comes from and goes to
Composition of the stream (quantitative or qualitative)
Chemical Abstract Service (CAS) number for each component of the stream
Some indication of the degree of concern for each substance or stream
Date of most recent update

3.1.2 Maintenance Materials

An inventory of maintenance materials should also be assembled, but the information collected will obviously be different from the process inventory. Most of these materials will be purchased, brand-name products of unknown or only partly known composition. The inventory could contain such data as the name, source, use, location, volume used, and quantity on hand. Hazard and composition information would most likely be taken from the Material Safety Data Sheet provided by the supplier. These data need to be collected and retained for other purposes such as employee information as part of a Hazard Communication Program. It may therefore be useful to integrate the inventory system with the Hazard Communication Program and have them both fed by transaction records from purchasing. However the inventory is built, if the substance product names and CAS numbers are in the same format as in the process materials inventory it may be possible to merge these elements of inventory into a single comprehensive listing for some uses.

3.1.3 Laboratory Chemicals

A complete listing of all chemicals present in laboratories is very difficult to create and maintain. Laboratory scientists are often permitted to buy chemicals without central control and, even when purchases are funneled through a single checkpoint, the number and variety of materials can be overwhelming. Also, in research laboratories new chemicals may be synthesized and these new chemicals may not be fully characterized in all cases. While assembling a proper inventory in a laboratory is often a lot of work, it yields benefits beyond the present health hazard recognition issue:

- The inventory process stimulates housekeeping measures to reduce the number of chemicals on hand.
- Awareness of chemicals present and procured can anticipate potentially difficult and costly hazardous waste disposal problems.
- An inventory facilitates security measures taken to prevent the misappropriation of chemicals for use in the manufacture of designer drugs.

A laboratory inventory would of course be much simpler than a process material inventory and might consist only of basic information such as name, CAS number, location, quantity, class, and name of person responsible.

3.2 HAZARD DATA

Some estimate of the toxicity or intrinsic hazard is needed for each item in the inventory. For the majority of chemicals, which are either purchases or products, a Material Safety Data Sheet (MSDS), as required by the Occupational Safety and Health Administration (OSHA) Hazard Communication Standard (or similar requirements outside the United States) will be available. For specific chemical species present in wastes or as contaminants, standard hazard data sources may supply the needed information even when MSDSs are not available.

Mixtures often present difficulties in obtaining adequate hazard data. If a mixture, such as a by-product, is unique to a plant, it may be necessary to characterize it analytically and then estimate its hazard based on its components. Industrial chemical mixtures usually come with MSDSs. Suppliers of consumer product mixtures are not required to provide MSDSs, but these products usually do not contain chemicals of significant toxicity.

3.3 EXPOSURE ESTIMATION

To do a risk assessment to prioritize exposure measurement effort, it is necessary to do an approximate initial exposure potential assessment. Sources of exposure were described in Chapter 2. For each chemical present and for each source of exposure for that chemical an estimate of exposure can be made. These exposure estimates combined with a toxicity estimate from the hazard data can be combined to yield a risk estimate which can serve as a basis for prioritization of concern.

It is not necessary to make an exposure estimate for every chemical/exposure source combination. Most will obviously be of low significance and can be ignored. For those chemical/exposure source combinations that could be near the top of the priority list, the exposure estimate is probably needed only to an order of magnitude. Methods of making this type of estimate have been extensively developed by the Environmental Protection Agency (EPA) for the purpose of evaluating Premanufacturing Notifications (PMNs). They are made by a combination of first principles (heat and mass transfer, etc.) and analogy to similar situations that have been measured. Considerable error is likely, but it is not important in this screening process as long as potential error is taken into account when using the priority list. While formal estimation and prioritization procedures are useful, experienced industrial hygienists tend to do the whole process judgmentally with entirely acceptable results. Either way, given the uncertainty in the interpretation of the hazard data and the experience needed to synthesize the data needed to make exposure estimates, it is generally preferable to have this risk assessment/prioritization step done by a professional industrial hygienist. Even in small plants, where the other tasks in a complete hazard control program would normally be done by engineers or technicians, it is worthwhile to use a qualified health professional to do the initial hazard recognition and program start-up tasks.

3.4 EXPOSURE EVALUATION

Before deciding what, when, and how to measure exposure, it is first necessary to consider the reasons for making measurements and some of the constraints.

3.4.1 Objectives of Occupational Exposure Evaluation

Measurements of the concentration of an air contaminant in a workplace or of the exposure of an employee to a contaminant might be made for several different reasons. Measurement might be a means of gathering data on the identity of substances present in air as an extension of the recognition/inventory process. A routine program of health surveillance might require periodic measurements of exposure to certain key contaminants so as to document the fact that exposure levels are insignificant from the health point of view. This documentation of the negative is common for potential exposure to ionizing radiation where film badges or other "dosimeters" are used routinely to demonstrate that exposure did not occur. For some substances a routine surveillance program may be required by regulations.

Epidemiological studies that look for relationships between health effects and workplace exposure to chemicals also require some means of exposure assessment. To be most useful for health effects studies, these assessments would ideally include exposure of all employees to essentially all chemicals present over all of the time period of the study. Since it is not usually practical to completely fill this data matrix by direct measurement, exposure estimating systems that combine measurements with estimates are used to provide exposure assessments for epidemiology.

When engineering means are applied to reduce or remove emissions to control exposure, the effectiveness of these engineering controls can be monitored by means of workplace contaminant measurements. These measurements may range from simple, instantaneous, nonspecific leak detection as part of a monitoring and maintenance program to elaborate multifactor analyses of variance to sort out the individual contributions of combinations of source controls and local exhaust ventilation. These control evaluation measurements can be used as "before and after" tests of a newly installed control or periodic checks of the continuing satisfactory performance of an installed system.

While all of the goals of evaluation are useful and important, the present intention is to focus on measurements made to evaluate the safety of an exposure or potential exposure for the purpose of determining whether additional health hazard control is necessary. These measurements are made to permit an inference to be made about the degree of exposure that is occurring and based on that inference to draw a conclusion about the safety of the exposure. A conclusion that conditions are unsafe would lead to a decision to make changes that would rectify the situation and bring the health hazard under control. This decision-making process can take place over a short period of time, as in the case of a decision to enter or not enter a confined space, or over a long time to develop plans for the elimination of overexposure to a chronic hazard resulting from many sources in a large plant. The decision-making process may be a rapid judgment based on a few data points or a complex statistical exercise requiring thousands of measurements. In all cases the important point is that from the beginning the whole data collection and analysis process be conducted so as to permit a decision to be made at the end with an acceptable level of confidence. The demands of the decision-making process in terms of what is considered safe or unsafe and how much certainty is needed drive the choices of numbers, time, location, and analytical method.

Decision making does not always require measurement. The observation of a visible cloud of an acutely toxic gas escaping from a vessel may be all the information required to arrive at a decision that corrective action is required. Similarly, in other situations consideration of quantities and release rates may make it perfectly clear without making any measurements that overexposure could not occur. Measurements, therefore, are needed for the close calls and, in particular, statistical analysis of measurement data is necessary where the correct interpretation of the result may be obscured by error and environmental variability.

3.4.2 Measurement Constraints

The way in which exposure occurs, the limitations of the sampling and analytical methods available, and the rules for interpretation of the standards all narrow the options available and determine the optimum combination of method and sampling conditions.

Typical workplace environment contaminant concentrations are highly variable. The background level of exposure in a chemical plant is probably the result of a large number of small fugitive emissions. Each of these emissions may vary with time. The contaminants released from each of these sources will move through the plant environment and be mixed with the contaminants from other sources or be dispersed. In addition certain events, such as equipment opening or vessel venting, will occur at regular or irregular intervals and contribute to the level of a contaminant present in the workplace. Workers move through this variable contaminant field in various patterns. All of these "degrees of freedom" result in a wide-ranging amount of exposure of an individual over time or of different individuals doing nominally the same job. These sets of exposure values, either between individuals or within an individual over time, are found to be best described by a distribution that is Gaussian or normal (bell-shaped) when the data are converted to their logarithms. These "log-normal" distributions of exposure data are characterized by the antilogs of the mean of the logs (geometric mean) and the antilog of the standard deviation of the logs (geometric standard deviation). Geometric standard deviations (GSD) are typically in the range of 2 to 4, meaning that exposure values of, for example, several times the geometric mean (GM) are not at all unlikely.

Some other statistical characteristics of industrial hygiene exposure data are also important. Obviously, since exposure is a continuous event it is likely that each instantaneous value of exposure will be similar to the value of exposure immediately preceding it. Thus, values in a series of measurements are not independent in a statistical sense but are described as "auto-correlated." This fact leads to a number of mathematical consequences beyond the scope of this book, but one important consequence is that the maximum value found with a given likelihood will vary with the averaging time. The shorter the averaging time, the higher the maximum; for longer averaging times the maximum tends to converge on the true long-term mean.

These aspects of the variability of the work environment and of the exposure of workers combine to require that when evaluating exposure, a sufficient number of samples be taken to permit characterization of the statistical distribution and to permit estimation of exposure over the appropriate averaging time. The appropriate averaging time is the period defined by the standard that matches the rapidity of toxic effects. In mathematical terms the averaging time should not be longer than the biological half time of a substance acting in the body. Longer half times would "average out" and thus lose biologically significant events. Some substances, like hydrogen cyanide, act very quickly, while other substances, such as coal dust, usually take months or years to produce significant damage. Although the range of

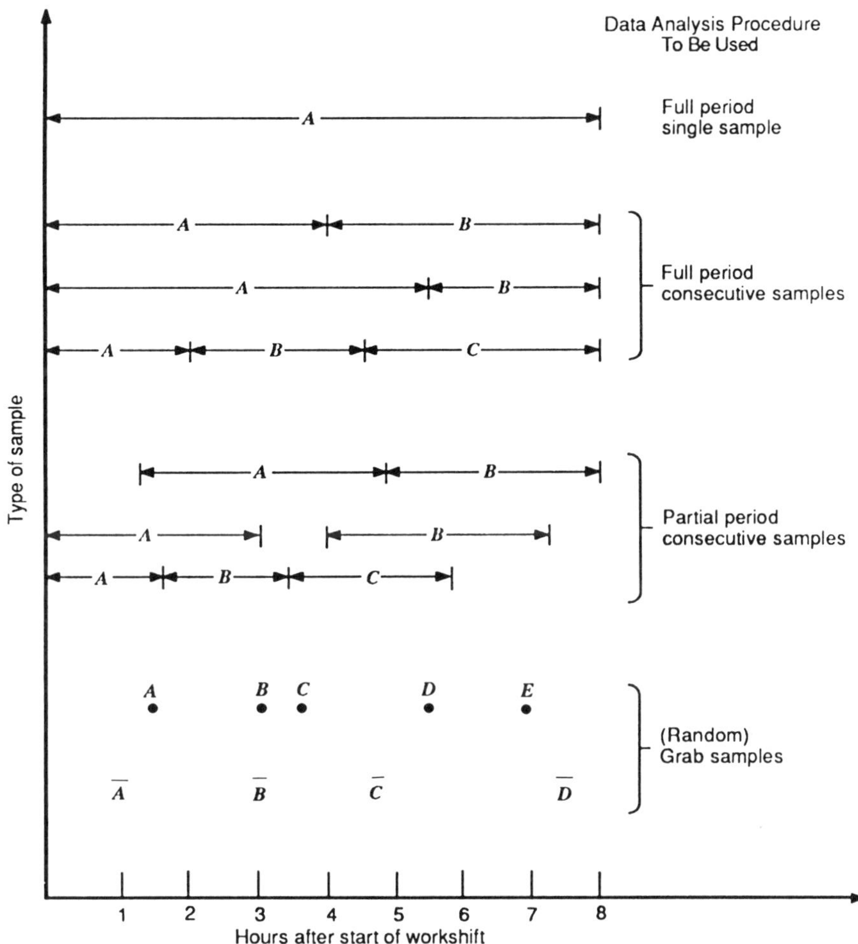

Figure 3.1 *Exposure measurements for an 8-hr average exposure limit.*

biological half times is continuous, for simplicity only a few discrete averaging times are commonly used. For fast-acting substances 15 and 30 min are used, while 8 hr is most often used for substances with biological half times longer than 8 hr.

The 8-hr averaging time is usually called the 8-hr time weighted average or TWA, a name carried over from a traditional way of calculating the mean. Modern TWA values are usually calculated from samples collected as illustrated in Fig. 3.1. The average exposure over an 8-hr period could be measured by a single sample of 8-hr duration, called a "full period single sample." If this is not practical, various combinations of shorter samples taken consecutively can make up a "full period consecutive sample." Even shorter period samples or nearly instantaneous (grab) samples can be used to estimate the 8-hr TWA provided some assumptions are made about the unsampled periods. Multiple sample estimates of the 8-hr average are more precise because the standard deviation of the average of such samples is the standard deviation of the samples divided by the square root of n, the number of samples.

Another constraint on the selection of sampling and analytical methods and strategies for the evaluation of worker exposure is that occupational exposure limits, since they are health limits, are limits on the actual exposure of the worker. That means that the limits apply to the air being inhaled by the worker, as nearly as that can be measured, rather than the general air at fixed locations in the workplace. While it might be thought that the general ambient concentration of an air contaminant would be an unbiased estimator of the air breathed by a worker, we know from empirical evidence that such is not the case. The worker's exposure can be a factor of 2 or more different from the ambient workroom air. In some cases, where a worker's own actions are generating the contaminant, as when cleaning, grinding, or taking a sample, the local concentration breathed by the worker is higher than the nearby ambient concentration. In other, badly contaminated atmospheres, the worker may take action to avoid inhaling the contaminant by finding clean locations (near a window, upwind) or by leaving the workplace occasionally. These defensive actions, which may not be obvious, can result in lower personal exposure.

Since most workers in chemical process plants are mobile, that is, not confined to a fixed work station as, for example, workers on an assembly line, the sampling

Figure 3.2 *High-volume sampler (courtesy of BGI Inc.).*

system needs to move with them. This necessity imposes some requirements on industrial hygiene sampling methods that are difficult to achieve. Early sampling devices often required motor-operated pumps which had to be plugged in for power and which were moved with the worker with great difficulty. Later, hand-operated pumps were developed which enabled the sample collector to move with the worker, but the sample period was limited by the endurance of the sample collector to usually 20 or 30 min. Also, the presence of the sample collector was likely to influence the behavior of the worker and render the sample unrepresentative. Fortunately, in recent years a number of personal sampling devices, which can be worn or carried by the worker, and which can sample over long periods of time, have been developed. These will be discussed below.

Even with the availability of convenient personal samplers, there are incentives to use general air sampling. General air samples may be less disruptive and more convenient. They may permit the use of high-volume samplers (Fig. 3.2) which provide great sensitivity, complex simultaneous samples or continuous monitors with remote, computer-controlled data acquisition, and management systems. Attempts to relate general air measurements to personal exposure have been made. In general, these relationships are not precise enough to permit predictions of safety or the lack of it when exposures are close to the limit but may be useful as backup or routine monitoring systems where exposures are well below the limit. Of course, fixed-station measurements may serve other purposes such as control system performance analysis. However, an adequate number of direct measurements of worker exposure by personal samples yielding an estimate over an appropriate averaging time are required to give credible evidence for health risk evaluation or compliance purposes.

3.5 SAMPLING METHODS

The following sections will discuss the various categories of sampling devices available for making measurements to evaluate exposure and give some comments on their advantages and disadvantages (see Table 3.1). Each of the sampling methods meets some need, in terms of the ability to sample a particular substance, usefulness as a personal sampler, instantaneously available result, or cost, but none can be described as universal samplers, suitable for all purposes. It is the purpose here to survey the field rather than to provide enough information to permit a definitive selection of a sampler for a particular purpose. The intent is to facilitate

TABLE 3.1 Sampling and Analytical Method Selection Considerations

Specificity the ability to measure the contaminant of interest in the presence of interferences.
Sensitivity the ability to measure low enough levels to permit evaluation with respect to the exposure limit.
Accuracy precision and freedom from bias sufficient for decision-making purposes.
Averaging time the ability to sample for periods relevant to the exposure limit.
Convenience ease of handling in the field and in transport to the laboratory.
Intrusiveness the degree to which the sampler influences worker behavior.
Cost low enough to permit taking an adequate number of samples for statistical purposes.

communication with the industrial hygienist responsible for monitoring program design.

3.5.1 Particulates

The "dusty trades" were among the several industries producing occupational disease in almost epidemic proportions during the early years of the industrial revolution. These trades included hard rock mining and quarrying, ceramic and glass making, foundries, and grinding and polishing. The most toxic dust was crystalline-free silica, which produces silicosis, a progressive and often fatal lung disease. In the later years of this period, during which the most control was achieved, the dust was measured by count methods on samples collected with a thermal precipitator in the United Kingdom and by a Greenberg Smith or "midget" impinger (Fig. 3.3) in the United States. These methods are rarely used now, having been almost completely replaced by the methods discussed below, but the count limits were not dropped from the TLVs until the 1985–1986 list.

Filters have always been potential collection devices for particulates and various types of paper filters have been used for many years. They were not completely satisfactory, however, because they had high resistance to air flow, inadequate efficiency for small particles, and high and variable weight and ash. All of this changed with the development of membrane filters. These porous films of cellulose esters or polyvinyl chloride have very small pores, low resistance to airflow, and are light enough to be weighed on very sensitive balances. Weight change with humidity is small or can be controlled. Some filters can be "cleared" or rendered transparent to permit microscopic examination of the collected dust. Some filters have essentially no residue when ashed and some can be dissolved to allow transfer

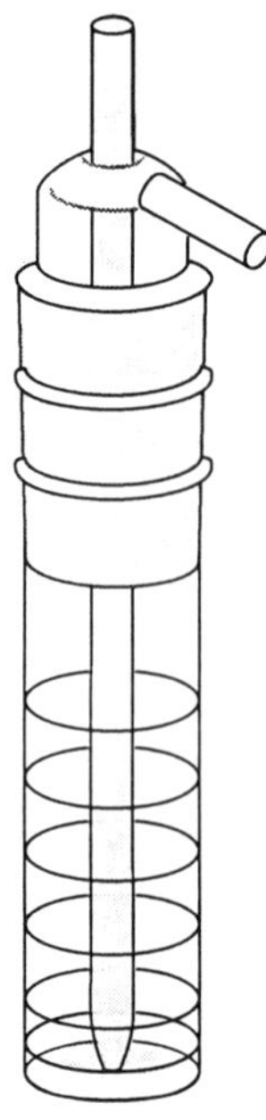

Figure 3.3 *Midget impinger (courtesy of SKC Inc.).*

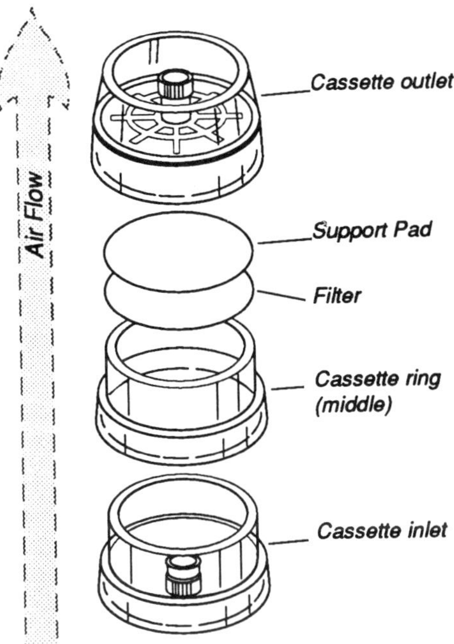

Figure 3.4 Membrane filter cassette (courtesy of SKC Inc.).

of the collected dust to electron microscope grids. Overall, the adoption of membrane filters as dust samplers in the 1960s revolutionized the evaluation of exposure to particulates.

Gross dust samples are collected for materials like lead, where the concern is for all of the material taken into the body, not merely that taken into the deep lung. However, these samples should not include very large particles, which may be airborne for an instant but have little or no chance of being inhaled, since they will bias the measurement. For the collection of gross dust samples the filters are usually contained in plastic holders or "cassettes" (Fig. 3.4), which are used as protectors to transport the clean and used filter. The filters may be tared before use or not, depending on the analytical technique. For most purposes 25- or 37-mm filters are used and air is sucked through the filter by personal sampling pumps (Fig. 3.5) operated at several liters per minute. In the course of an 8-hr work day it is easy to sample about 1 m^3 of air and thus achieve microgram per cubic meter sensitivity when using a sensitive balance or analytical technique. A wide variety of filter types is available and selection is made based on pore size, weight stability, clearance properties, and ash. While pore sizes as low as 0.2 micrometer (μm) are available, flow resistance is higher for small pore sizes. The small sizes are rarely necessary since most membrane filters do not act like sieves but tend to be nearly 100% efficient for airborne particles much smaller than the pore size. Thus one 5-μm membrane filter is specified to be better than 99.9% efficient for 0.3-μm dioctyl phthalate aerosol droplets.

Respirable dust, that is, dust capable of reaching the deep lung, should be measured to evaluate exposure to materials, like quartz, whose primary site of biological activity is in the deep lung. Various size-selective presamplers have been

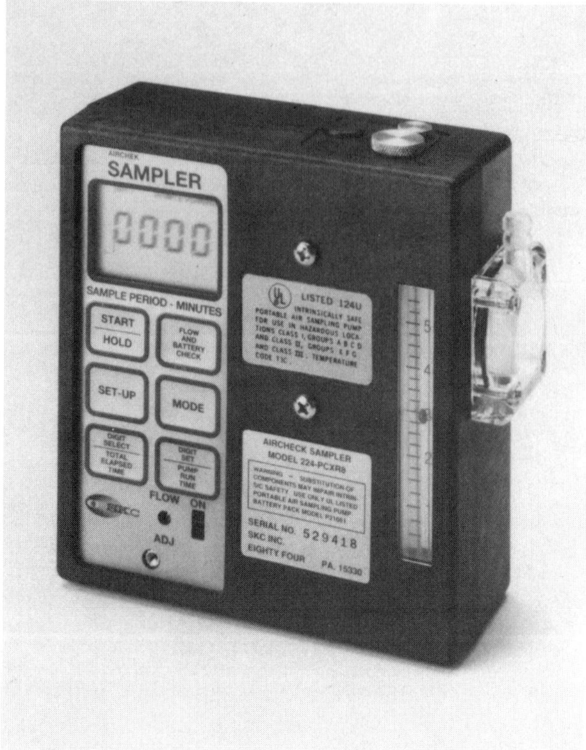

Figure 3.5 *Personal sampling pump (courtesy of SKC Inc.).*

developed to sort out the larger particles in the air so that only the small "respirable" particles, typically under 10 μm aerodynamic size are collected on the filter. These size-selective presamplers are either cyclones (United States) (Fig. 3.6) or horizontal elutriators (United Kingdom) and are used for both free silica and for coal dust. Unlike with gross dust samplers, where only the total volume of air sampled need be accurately known, with size-selective presamplers it is also necessary for the flow rate to be known and constant since the performance of the size selector depends on flow rate. This places much more stringent requirements on battery-powered personal sampling pumps which are usually equipped with electronic flow rate regulation for this purpose (Fig. 3.5). Flow rate pulsation caused by the action of the pump's diaphragm or piston also affects the size selector and needs to be damped out.

Asbestos is a special case among particulates. Since there is no analytical means of telling the difference between asbestos fibers, which are toxic, and small bits of rock of the same composition, which are not, it is necessary to look at each sample and count the fibers. For that reason asbestos and some other fibers are the only materials whose occupational health limits are still based on count. For asbestos the technique calls for collecting dust on an open-face filter. No size selector is needed since the small fibers dominate the count. Since membrane filters are surface filters rather than depth filters (like fiberglass filters), all of the collected dust lies on the surface in the same plane and thus is close to in focus even for a

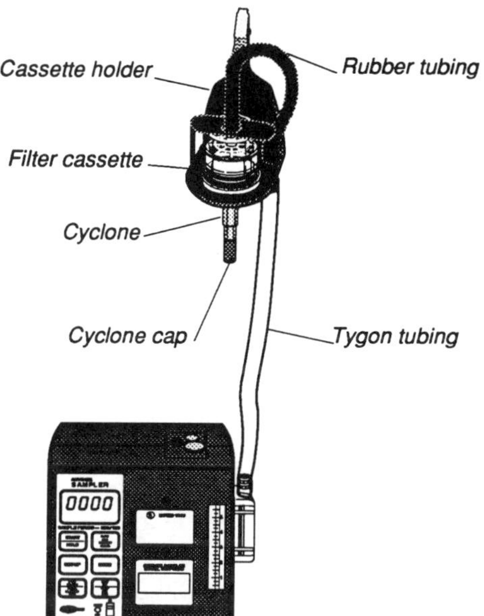

Figure 3.6 *Cyclone size-selective sampler (courtesy of SKC Inc.).*

shallow depth of field 430 × magnification lens system. Thus when the filter is rendered transparent by various solvents and mounted on a microscope slide, it is possible to distinguish fibers from other shaped particles and to quantitatively count them. Where it is necessary to determine the mineralogical identity of the fibers, other light and electron microscope techniques are available for samples collected on filters.

Direct reading instruments for the near instantaneous measurement of the mass of airborne particulate are available. One type of device (Fig. 3.7) electronically

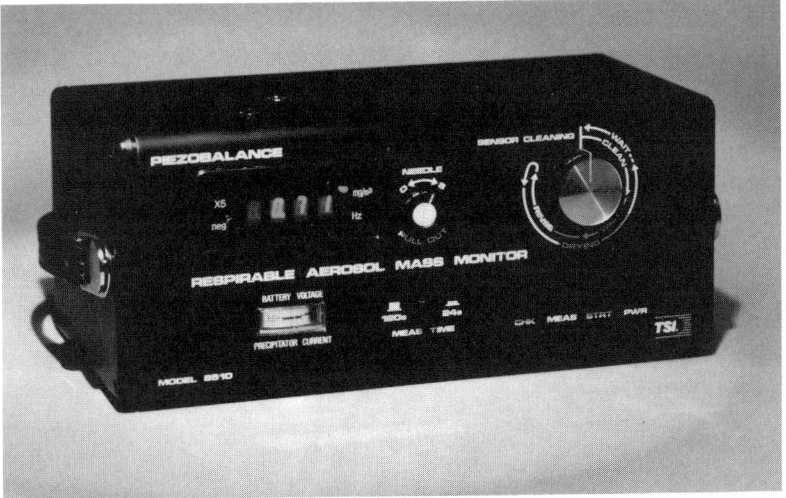

Figure 3.7 *Respirable aerosol mass monitor (courtesy of TSI Inc.).*

precipitates particles on an oscillating crystal and measures the change in resonant frequency caused by the mass. These instruments are useful for tasks such as the evaluation of control systems, but are not so valuable for the evaluation of exposure against a standard since most dust limits are 8-hr TWAs.

3.5.2 Gases and Vapors

As impingers were used for dust, so bubblers were commonly used for gases and vapors years ago. These wet collectors produced a liquid sample which was then analyzed by the wet chemistry techniques that were the stock-in-trade of analytical chemistry before the days of instrumental methods. These wet collectors, which were insensitive and difficult to handle, have all but vanished from the industrial hygiene sampling scene.

Solid sorbents come as near as anything to being the universal samplers. Some of the commonly used sorbents are activated carbon, silica gel, and alumina. For sampling, the sorbent is usually contained in a tube through which the air is drawn. The contaminant is adsorbed onto the sorbent and then desorbed in the laboratory either by a desorbing chemical, typically carbon disulfide for charcoal, or by heat. The contaminant thus collected and concentrated is then analyzed, usually by gas–liquid chromatography. For a particular sorbent to be successful as a sampling

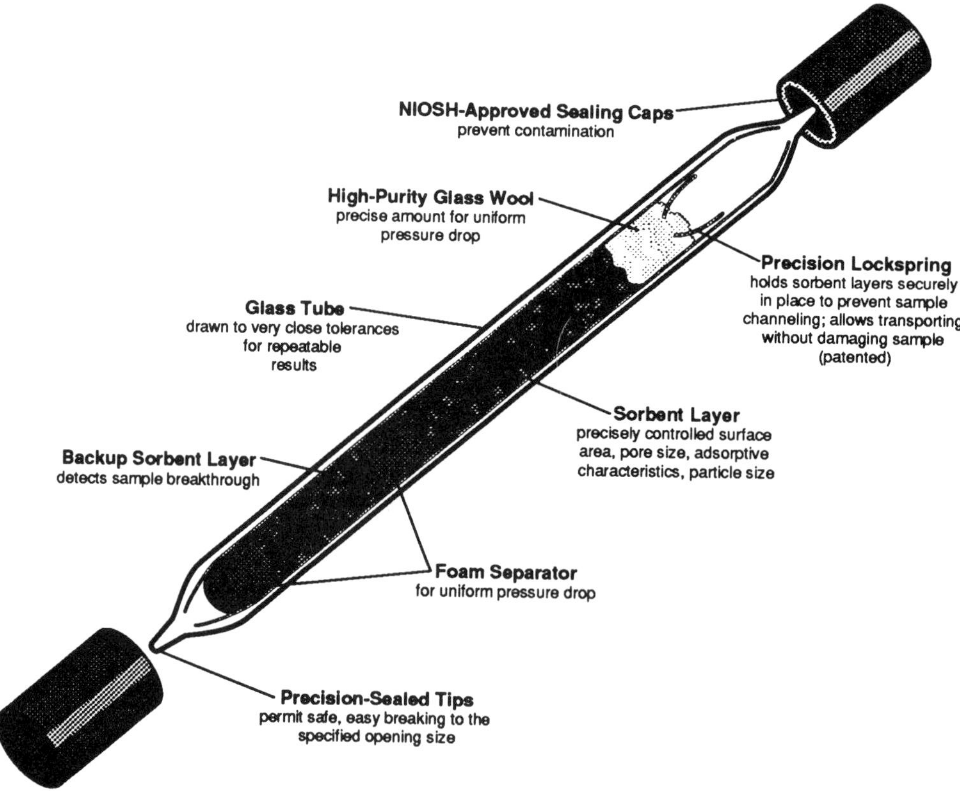

Figure 3.8 *Solid sorbent tube (courtesy of SKC Inc.).*

medium for a gas or vapor it is necessary that the volume of air sampled be small enough that breakthrough does not occur, that is, that the adsorbed material does not migrate down the sorbent column and be lost out the back end. Most sorbent tubes (Fig. 3.8) have a backup section to determine if breakthrough has occurred, although, if it has, there is no way to know how much may have been lost. In developing and testing sorbent methods for particular chemicals, breakthrough quantities, along with sorption and desorption efficiencies, are among the important parameters determined. Sorbent tubes come in a wide variety. The most common solid sorbent sampling device is a glass tube with a 50-mg charcoal front section and a 25-mg backup section. Other larger tubes are available for difficult-to-collect low-molecular-weight gases. When thermal desorption is to be used the charcoal is contained in a specially designed metal tube which fits snugly into a heater on the desorption apparatus. Thermal desorption is a simple but usually "one-shot" process, whereas a sample eluted into a liquid desorbent can be analyzed repeatedly.

While pumps (Fig. 3.9) are used to bring the air through the sorbent tube and into contact with the sorbent, passive samplers have been developed which do not

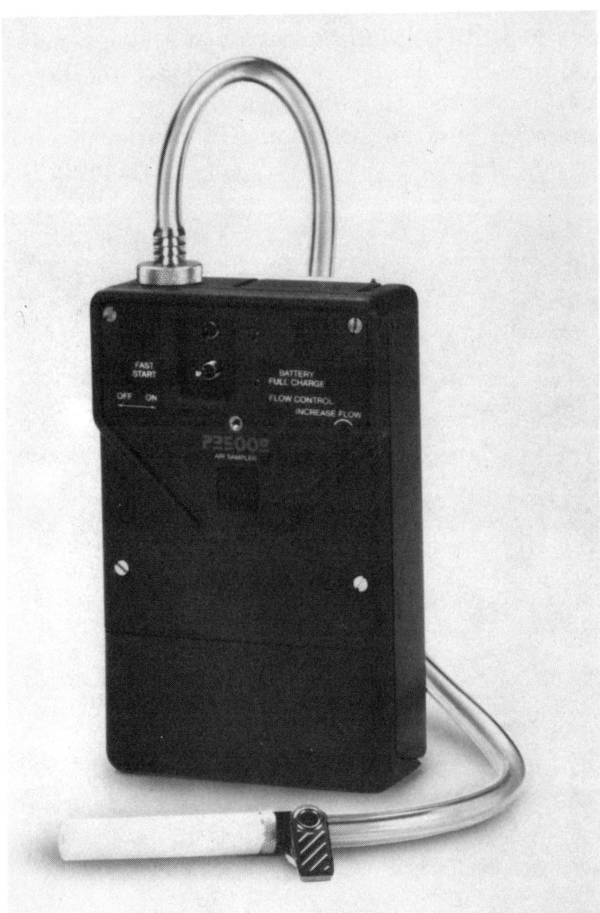

Figure 3.9 *Personal sampler pumps with solid sorbent tube holder (courtesy of AMETEK).*

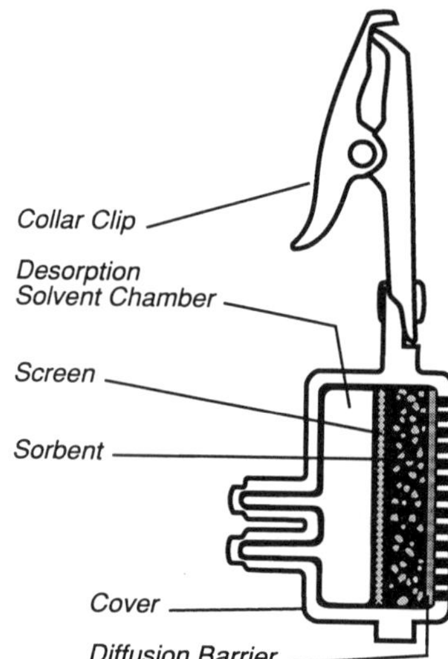

Figure 3.10 *Passive sampler (courtesy of SKC Inc.).*

use pumps. These devices consist of a bed of sorbent separated from the outside environment to be sampled by a diffusion barrier through which the contaminant must diffuse in order to be collected (Fig. 3.10). Since the rate of diffusion is a quite constant function of the dimensions of the device, the properties of the substance being sampled, and temperature, it is possible to relate the amount collected over a period of time to the average concentration in the air during that

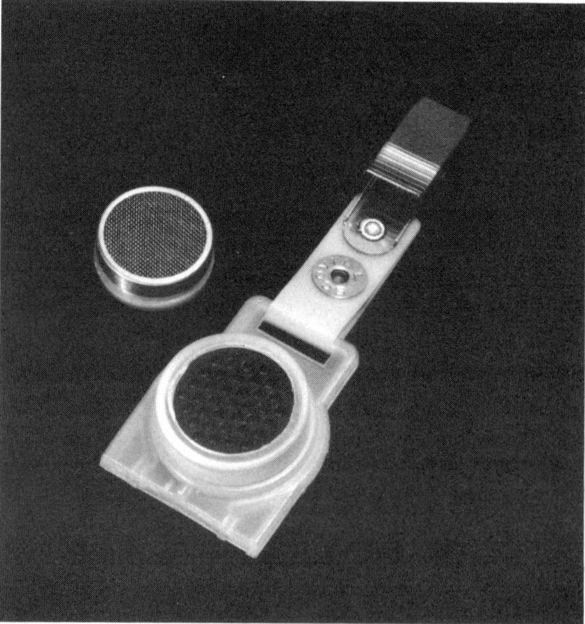

Figure 3.11 *Passive sampler badge (courtesy of SKC Inc.).*

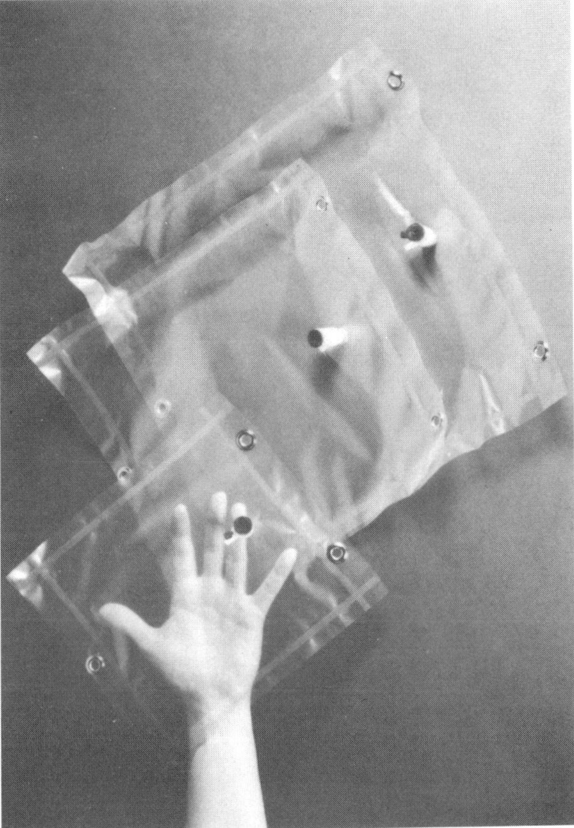

Figure 3.12 *Tedlar® sample bags (courtesy of SKC Inc.).*

interval. While there are certain sources of inaccuracy and these passive "dosimeters," as they are called, are not suitable for all contaminants, they are enormously convenient and are widely accepted by workers and by industrial hygienists. They are quite small, so that they can be pinned to the clothing and worn like a badge (Fig. 3.11). Some passive badges include both a front section and a backup section to detect breakthrough and some are impregnated with reagents that help trap and retain specific contaminants such as formaldehyde. Because of questions of accuracy in complex, dynamic environments, their use is often coupled with other, better validated, sampling methods.

Not all gases and vapors can be concentrated in a sample by adsorption and desorption from a solid. For some small molecules, such as ethylene, the collection efficiency is too low and too variable to be useful. One alternative for these substances is the collection of bulk air samples in gas bags (Fig. 3.12). Special bags made of various impermeable plastics are available for this purpose and have largely replaced the glass gas-sampling flasks that had been used. While air samples collected in a bag do not concentrate the contaminant, analytical instrument techniques for concentrating may be used. In general, the enormous improvements in instrument sensitivity have made both gas bags and even sorbent samplers much more useful than they would have been.

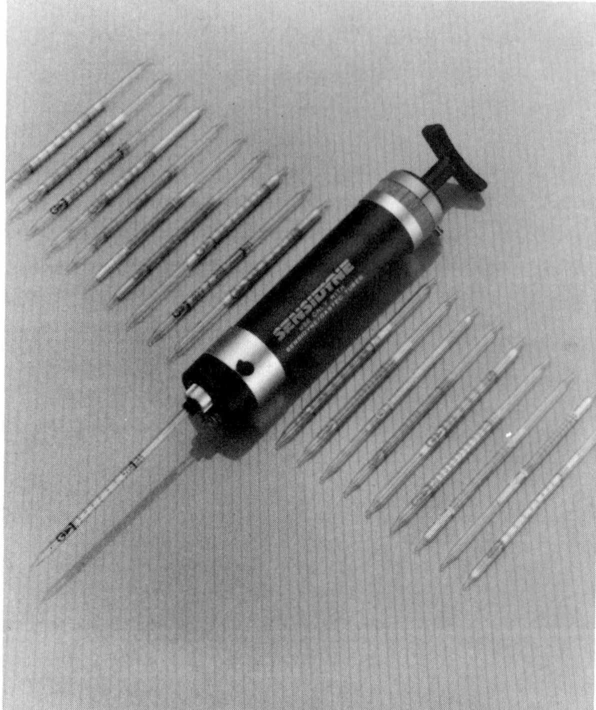

Figure 3.13 Detector tubes and pumps (courtesy of Sensidyne®).

Direct-reading measurement devices of several different types are available for the measurement of gases and vapors. Detector tubes (Fig. 3.13) are the most common of the specific methods of measurement and tubes are manufactured for a long list of air contaminants. The tubes consist of a sorbent impregnated with reagents that react, with more or less specificity, with the contaminant to be measured and produce a visible color change that is related to the amount of contaminant collected. The early National Bureau of Standards carbon monoxide tube produced a change of color that was compared to a set of color standards. Most modern tubes produce a color stain whose length is read off a scale and converted to the original concentration based on the number of strokes of the sampling pump (Fig. 3.14).

The accuracy of detector tubes is limited and their essential simplicity does not allow for protection against many interferences. Another, perhaps more important, limitation is that they are inherently grab samples although some long period tubes are being made. Detector tubes are most useful when used as screening, presurvey measurement devices. If measurements made by detector tubes in the worst location at times of maximum exposure show that levels are sure to be way below the limit, taking into account the accuracy of the detector tube and possible interferences, then further measurements are unnecessary. On the other hand, if results are equivocal, then more exact evaluation tools should be used.

Direct-reading instruments for measuring combustible gases have been available for many years. These devices, often called "explosion meters," were designed to

Figure 3.14 *Length-of-stain detector tube and pump (courtesy of Dräger).*

read in percentages of the lower explosive limit (LEL) and were intended to be used to estimate the risk of explosions, particularly in confined spaces. Since the lower explosive limit for typical hydrocarbons is about 1%, these meters were reading about 10,000 ppm full scale; that is, 1% on the meter scale was equal to about 100 ppm. A hydrocarbon with a relatively high TLV of, say, 500 ppm would read 5% at the TLV, which would be barely readable on the low end of the scale. For *n*-hexane at 50 ppm there would be essentially no reading. Thus, while these meters were useful for the purpose originally intended, attempts to use them in the toxic rather than explosive range were unsuccessful. To meet the need in the toxic

Figure 3.15 *Confined space monitors (courtesy of GASTECH).*

range, instruments with extended range scales were developed. While these direct-reading meters are useful for detecting the presence of hydrocarbons as low as 10 ppm, accurate measurement is difficult because of a tendency to drift and because of unequal response to different hydrocarbons.

Where a variety of toxic gases may be present and an instantaneous measurement is needed, as in confined space entry, several toxic gas monitors (Fig. 3.15) are available.

Accurate and sensitive meters based on photo ionization have been developed and are widely used. They are commonly used for fugitive emission surveys as part of monitoring and maintenance programs for screening purposes as described above, and, when coupled with data acquisition systems, as sensors for charting the time history of a contaminant concentration.

3.5.3 Combined Vapors and Aerosols

Vapor and aerosol of the same substance in air at the same time present a difficult measurement problem. If it is not necessary to know how much of the substance was in the air as aerosol and how much as vapor, then a filter followed by a solid sorbent can be used to give total concentration. It might seem that the filter would measure the aerosol and the sorbent the vapor, but in reality if the material has any appreciable vapor pressure, as it must to exist as a vapor, some of the aerosol collected on the filter will evaporate and be collected as vapor on the sorbent. Accurate and reliable means of separately measuring coexisting aerosol and vapor do not presently exist.

3.6 ANALYTICAL METHODS

The most commonly used methods for the analysis of samples of airborne contaminants are listed in Table 3.2. Any method used for a particular contaminant must be appropriate for the sampling media, have sufficient sensitivity, and be reasonably free from interference. The ultimate confidence that can be placed in an analytical result depends in part on the accuracy of the method, but to a greater extent on how well the method has been validated for the particular purpose and on the reliability of the laboratory.

The National Institute for Occupational Safety and Health (NIOSH) publishes sampling and analytical methods for the evaluation of exposure. These methods

TABLE 3.2 Air Analysis Methods

Method	Substance
Atomic absorption spectroscopy	Metals
Gas chromatography	Volatile organics
Gravimetric	Nuisance dust, coal
Particle count	Asbestos
Ion-specific electrode	Halogens, HCN, NH_3
X-ray diffraction	Silica
Colorimetry	Miscellaneous

TABLE 3.3 Method Validation

Least validated
Single literature reference in with no testing
Tested on air samples with apparent success
Commonly used but not validated
Lab validated for accuracy and reliability
Field validated in the presence of interferences
Round-robin tested and ruggedized
Most validated

TABLE 3.4 Criteria for Accreditation of Industrial Hygiene Laboratories[a]

1. Laboratory Director
   The laboratory shall be under the overall direction of a qualified person. For purposes of this document, this person is designated as the laboratory director. The director must be on site and actively involved in laboratory administration.
   a. Qualifications — Either an earned degree in a basic science and a minimum of five (5) years industrial hygiene experience beyond the bachelor's level, or full certification by the American Board of Industrial Hygiene.
2. Laboratory Supervision
   The day-to-day operation of the laboratory shall be supervised by a qualified person. For purposes of this document, this person is designated as the supervisor.
   a. Qualifications — Should either have an earned degree in basic science and a minimum of five (5) years experience in industrial hygiene chemistry or related procedures; or full certification by the American Board of Industrial Hygiene in the chemical aspects of industrial hygiene.
3. Special Circumstances
   a. It is possible for a single person to serve as both laboratory director and supervisor as long as the qualifications and responsibilities of both roles are satisfied.
   b. Experience beyond the bachelor's level may be a combination of academic and work experience.
4. Quality Control Coordinator
   a. Qualifications — The quality control coordinator must have a minimum of Baccalaureate degree in the basic sciences and be knowledgeable about statistics and quality control procedures. He or she should be independent of the analyses and be capable of effectively running a quality control program.
5. Laboratory Personnel
   Laboratory Analysts shall be qualified by education and/or experience to produce reliable analytical testing.
6. Proficiency Testing
   a. Participation in the AIHA designated proficiency testing program is required. Currently AIHA is using the NIOSH PAT (Proficiency Analytical Testing) Program.
   b. The data resulting from this program must be made available to the Laboratory Accreditation Committee and the evaluation of these results, insofar as it affects the accreditation of a laboratory, will be the responsibility of that committee.
7. Quality Control and Equipment
   Routine quality control procedures must be an integral part of laboratory procedures and functions. These must include:
   a. Routine introduction of samples of known content included along with samples of analysis.
   b. Routine checking and calibrating of equipment and instruments to insure adequate performance.
   c. Routine checking of procedures and reagents.
   d. Good housekeeping, cleanliness, and general orderliness.
   e. The applicant laboratory must have a written quality control plan as well as hard copy of any quality control data in an easily accessible place.

TABLE 3.4 (*Continued*)

f. The applicant laboratory must have a designated person who has the quality control responsibility in the laboratory.
g. Examples of completed, actual industrial hygiene quality control records must be included in the application.

8. Facilities. The laboratory must have space, facilities, and equipment adequate for the services provided. These must include:
   a. Adequate workbench and instrumentation space in a well-lighted facility with proper temperature control.
   b. Proper ventilation of laboratory hoods and instruments.
   c. Adequate services, such as electricity, water, compressed air, and vacuum, in suitable locations.
   d. Safe facilities for chemical storage and for the disposal of containers, chemicals, and refuse.
   e. Appropriate safety equipment, such as goggles, shields, protective clothing, deluge showers, and fire extinguishers.
   f. Suitable locations of laboratory in relation to potable drinking water, eating facilities, toilets, and showers.
9. Records. The laboratory must maintain proper and adequate records and files. These must include:
   a. The identification and numbering of incoming samples.
   b. An adequate and systematic numbering system relating laboratory samples to incoming samples.
   c. An adequate record system of internal logistics for each sample, including incoming sample date, analysis and procedures, and reporting of data.
   d. A record of the checking system for the calibration and standardization of equipment and internal control samples.
   e. Completed, actual industrial hygiene sample records.
10. Methods
   a. The laboratory must use accepted industrial hygiene methods.
   b. Methods must be approved by the laboratory director and available to all analysts.
   c. A methods manual must be maintained by the laboratory. There must be a record that changes are made when necessary and that these changes are approved by the laboratory director.

[a]Courtesy of the American Industrial Hygiene Association.

and methods developed by other laboratories can be rated according to how well their performance has been tested. Table 3.3 presents a typical ranking scale. While a better method should usually be selected in preference to a less well-validated method, even the less well-validated methods can be useful. Method selection, and the amount of money that should be spent on method improvements, depend on the quality of method needed to make a decision. For an exposure to a substance substantially below the occupational exposure limit where the toxicity is such that the consequences of overexposure are not serious, it may not be necessary to have a well-validated method.

The laboratory in which the analysis is performed has as much to do with confidence in the result as the method of sampling and analyses used. It has been found from quality control testing that industrial laboratories that usually do quite good work may do poorly on air analyses. This may be because these laboratories do not routinely analyze air samples and the sensitivity required is below that usually practiced. To help provide some assurance that a laboratory is proficient at the analysis of air samples, the American Industrial Hygiene Association (AIHA) operates a program for the accreditation of industrial hygiene laboratories. The aim of the AIHA accreditation program is to recognize those laboratories that achieve and maintain a high level of professional performance in providing the data upon which decisions of great health and economic import are made. The key criteria for the accreditation of industrial hygiene laboratories are given in Table 3.4.

3.7 MONITORING PROGRAM DESIGN

Evaluation of exposure on a continuing basis is done by making measurements at times and in places and numbers that permit making decisions about safety as they are needed to ensure worker protection. The design of the monitoring program to accomplish this are beyond the scope of this book, but some general considerations are appropriate.

Monitoring programs are not unlike quality control programs and many of the same concepts apply. For efficiency, monitoring programs usually include some element of sequential sampling. The use of screening samples was discussed above. When it looks likely that a full-scale monitoring program may be needed, a presurvey is usually done to collect enough data to estimate the various means and variances to provide essential data for the design of the full-scale program.

As in most data-collecting activities, the reason for collecting the data tends to get lost with time and immersion in the details of the sampling and analytical problem. It is necessary to keep in the forefront of our thinking that the continuing purpose of exposure evaluation is to draw conclusions about safety so as to make decisions about control. If this is done it may often be evident that there are simpler ways of accomplishing the end objective. For example, if gross dust measurements are such that the exposure would be safe even if all the dust were the substance of concern, then there is no reason to analyze the dust for that substance. Similarly, a less accurate method may be entirely adequate if it permits a conclusion regarding safety at an adequate level of confidence.

3.7.1 Informing Employees

Finally, it is generally a requirement of law and usually a matter of policy to communicate the results of exposure measurement to the employees whose exposure was evaluated. This serves the double purpose of reassuring them when results are low and enlisting their cooperation when exposure reduction action is needed.

REFERENCES

American Conference of Governmental Industry Hygienists. *Air Sampling Instruments*, 7th ed. Cincinnati: ACGIH, 1989.

Attio, V., V. Riihimaki, and H. Vainio, Eds. *Biological Monitoring and Surveillance of Workers Exposed to Chemicals*. Washington, DC: Hemisphere, 1984.

Beaulien, H. J. and R. M. Buchan. *Quantitative Industrial Hygiene*. New York: Garland, 1981.

Crist, S. Solid Sorbent Gas Samplers. *Ann. Occup. Hyg.* **23**, 47 (1980).

Hawkins, N. C., S. K. Norwood, and J. C. Rock, Eds. *A Strategy for Occupational Exposure Assessment*. American Industrial Hygiene Association, 1991.

Johnsen, M. A. *Aerosol Handbook*, 2nd ed. Mendham, NJ: Dorland, 1982.

Lauwerys, R. T. and P. Hoet, Eds. *Industrial Chemical Exposure: Guidelines for Biological Monitoring*, 2nd ed. Boca Raton, FL: Lewis, 1993.

Lawrence, J. F., Ed. *Trace Analysis*, Vol. 1. New York: Academic, 1981.

Leidel, N. A., K. A. Busch, and J. R. Lynch. *Occupational Exposure Sampling Strategy Manual*. NIOSH Pub. No. 77-173. US Government Printing Office, 1977.

Lodge, J. P., Ed. *Methods of Air Sampling and Analysis*, 3rd ed. Chelsea, MI: Lewis, 1988.

National Academy of Science. *Prudent Practices for Handling Hazardous Chemicals in Laboratories*. Washington, DC: National Academy Press, 1981.

National Institute for Occupational Safety and Health. *NIOSH Manual of Analytical Methods*. Cincinnati: NIOSH, 1990.

Natrella, M. G. *Experimental Statistics*. National Bureau of Standards Handbook 91, US Government Printing Office, 1966.

Nothstein, G. S. *The Law of Occupational Safety and Health*. New York: Free Press, 1981.

Phalen, R. F. et al. Rationale and Recommendations for Particle-Size Selective Sampling in the Workplace. *Appl. Ind. Hyg.* **1**, 3 (1986).

Posner, J. C. and W. J. Woodfin. Sampling with Gas Bags I. Losses of Analyte with Time. *Appl. Ind. Hyg.* **1**, 163 (1986).

Rahjans, G. S. and J. Sullevan. *Asbestos Sampling and Analysis*. Ann Arbor, MI: Ann Arbor Science, 1981.

Shaw, D. T. *Fundamentals of Aerosol Science*. New York: Wiley-Interscience, 1978.

4

Emission Regulations

Through the Clean Air Act in the United States, the U.S. Environmental Protection Agency (EPA) has been regulating the discharge of emissions to the atmosphere. The overall objective has been the improvement of public health and general welfare in large areas where overall air quality has deteriorated. Regulations have been expanded and amended through the years to meet the air quality standards in the basic Air Act promulgated by Congress (1970). These standards, termed the National Ambient Air Quality Standards (NAAQS) are listed in Table 4.1 for six basic pollutants. The concentrations shown are the maximum allowable levels, with the primary standard established to protect the general public health and secondary standards for protection of the general welfare (wildlife, vegetation, etc.).

One of the six pollutants in Table 4.1 is ozone, whose formation is dependent upon the presence of photochemically reactive volatile organic compounds (VOCs). Most compounds in the petroleum and chemical industries are highly reactive photochemically. Some of the poorly reactive chemical compounds are shown in Section 4.1. The VOC emissions are released into the atmosphere from point sources such as stacks or vents and diffuse or nonpoint sources. Leakage from valves, pumps, flanges, and other dispersed equipment are the source for diffuse VOC emissions, which are normally termed "fugitive emissions." These fugitive emissions contribute significantly to the ozone concentration and are also a major source of exposure in the workplace. Regulations controlling these fugitive emissions have been promulgated by the EPA to reduce photochemical reactivity. The VOC regulations cover petroleum refineries and the Synthetic Organic Chemical Manufacturing Industry (SOCMI).

The Occupational Safety and Health Agency (OSHA) also issues regulations to control exposures in the workplace. Some OSHA regulations recommend the specific installation of equipment controls, but most of the regulations grant the plant operator flexibility in achieving a maximum allowable exposure level.

In 1990 the Clean Air Act was amended to further reduce emissions. The amendments included directives to further reduce VOC emissions and an itemized list of hazardous chemicals that must be controlled. In addition, hazard risk assessments and a safety management system are required. These regulations are principally concerned with process equipment leaks and accidental release, but emissions control regulations also cover tankage, sewers, wastewater treatment, waste handling, tanktruck loading, and barge and ship loading, along with other equipment installations.

This chapter briefly reviews the major regulations issued by the EPA and OSHA along with their potential impact on workplace exposures. An understanding of the regulations is important in determining the type and effectiveness of mandated workplace controls and any additional controls necessary to meet regulated, industry, or company exposure concentration levels. Effective control technology often requires a combination of regulatory requirements along with modifications to achieve desired workplace concentration levels and reduced exposures.

4.1 U.S. ENVIRONMENTAL PROTECTION AGENCY

The basic Clean Air Act was promulgated in the United States in 1970 to control the concentration of the six contaminants shown in Table 4.1 to concentrations less than the maximum designated concentration levels. The primary standard was defined to protect health and the secondary standard covers the general welfare (wildlife, vegetation, etc.) with both standards essentially applicable to nonprocess plant areas. Where the standards of compliance are achieved, the area is designated an attainment area, with a nonattainment designation for those areas not attaining compliance. All standards must be met for an attainment designation. In addition to the six base contaminants, Section 112 of the Clean Air Act requires control of extremely hazardous pollutants. Pollutants are classified hazardous if they are carcinogens or if exposure may cause serious problems to the public. After studying individual chemicals, the EPA may designate a compound extremely hazardous and then promulgate regulations to control emissions of the particular chemical. These hazardous compounds are controlled under regulations termed "National Emission Standards for Hazardous Air Pollutants" (NESHAP). The

TABLE 4.1 National Ambient Air Quality Standards[a]

		NAAQS Standards ($\mu g/m^3$)	
Pollutant	Averaging Time	Primary	Secondary
Carbon monoxide	8 hr	10,000 (9 ppm)	—
	1 hr	40,000 (35 ppm)	—
Lead	Calendar Quarter	1.5	Same
Nitrogen oxide	Annual	100 (0.053 ppm)	Same
Ozone	1 hr	235 (0.12 ppm)	Same
Sulfur dioxide	Annual	80 (0.030 ppm)	—
	24 hr	365 (0.140 ppm)	—
	3 hr	–	1300 (0.5 ppm)
PM-10[b]	Annual	50	Same
	24 hr	150	Same

[a]Adapted from Ref. 1.
[b]Particle diameter $\leq \mu m$.

NESHAP regulations are equipment regulations designed to control emissions and are similar to the New Source Performance Standards (NSPS) regulations for control of VOC emissions described in Section 4.1.2. Prior to 1990, only seven materials were designated extremely hazardous, with specific regulations promulgated for control of these emissions in process plants. These seven materials are arsenic, benzene, vinyl chloride, mercury, asbestos, beryllium, and radionuclides.

4.1.1 State Implementation Plans

In order to meet NAAQS standards, the states developed State Implementation Plans (SIP) which were reviewed for adequacy and enforcement by the EPA. Through the latter part of the 1970s these plans were implemented by the states, and concentrations of the various contaminants were reduced. However, the reductions did not always achieve complete compliance with the NAAQS standards. In highly populated areas, a particularly difficult contaminant to reduce has been ozone. Volatile organic compounds are a major contributor to high ozone levels as a result of photochemical reactivity. Major sources of VOCs are petroleum refineries, chemical plants, and mobile sources. Various EPA test programs indicated that fugitive emissions are a major factor in the gross VOC emission estimates from refineries and chemical plants. Mobile source emissions from automobiles and trucks were improved in accordance with other portions of the Clean Air Act and energy efficiency laws.

Reductions of VOCs in chemical plants and petroleum refineries were undertaken in various states through SIP laws by requiring monitoring and maintenance programs (M&M). These programs are also termed LDAR programs for leak detection and repair which is currently used in the EPA literature. The LDAR programs reduced fugitive VOC emissions, and these programs were then incorporated into New Source Performance Standards (NSPS) for chemical plants that were promulgated in October 1983 and in May 1984 for petroleum refineries.

Although most compounds in the petroleum and chemical industries are photochemically reactive, a number of compounds have a very low reactivity level, and the concentrations of these compounds are normally omitted in VOC determinations. However, states may vary in their interpretation of low reactivity and guidance. The actual status of a compound must be obtained from a state environmental department. A tentative list of low reactivity compounds includes:

Dichlorotrifluorethane
Chlorodifluoromethane
Tetrafluoroethane
Dichlorofluorethane
Chlorodifluoroethane
Perchloroethylene

Other compounds that have low reactivity are 1, 1, 1 trichloroethane, methylene chloride, trichlorofluoromethane, methane, and ethane. However, ethane is generally included in VOC compound groups, with methane generally excluded from VOC controls. The list of nonreactive chemicals has been expanded as companies have successfully demonstrated very low reactivity for other compounds and obtained agreement on a low reactivity status from the EPA. However, the overall list of low reactivity compounds is relatively small but should be carefully investigated when developing emission control programs.

4.1.2 NSPS Standards

New Source Performance Standards (NSPS) limit equipment emission rates through regulation in petroleum refineries and chemical facilities. In turn, the controlled emissions reduce workplace exposures through lower contaminant concentrations in the workplace. For certain contaminants or pollutants, a lower average concentration in the workplace may not achieve the desired exposure standard. These exposure standards may be issued by OSHA, the American Conference of Governmental Industrial Hygienists (ACGIH), individual companies, and governmental (state) or other authorized entities within the Unites States. Exposure standards elsewhere in the world may be issued by federal or regional (province) authorities. Where NSPS regulations do not achieve the desired concentrations or exposure levels, the additional controls or expenditures necessary to meet the exposure criteria may be minor. Work practice changes may be sufficient in some cases along with NSPS requirements.

While the term New Source Performance Standards indicates these regulations are applicable to new plants, the NSPS standards may also be required in reconstructed or modified facilities. The definition of new is quite clear, but the definition of modified and reconstructed must be carefully analyzed.[2,3] Many older facilities require upgrading to NSPS control levels as a result of these definitions in addition to plants that have been changed or modified in more recent years. The general objective of these regulations was the eventual conversion of all petroleum and chemical facilities to NSPS equipment control levels.

4.1.3 Clean Air Act Amendments 1990

Since the NAAQS objectives were not met in a large number of areas within the United States and OSHA has been particularly slow in adding chemicals or materials to the NESHAP list, a group of amendments to the Clean Air Act were added to ensure satisfactory reductions in contaminants and hazardous pollutants within a 5- to 10-year period. These amendments were signed into law November 15, 1990 and have been termed the Clean Air Act Amendments (CAAA).

The CAAA has a number of subtitles or sections, but the portions of major concern are related to VOC and hazardous pollutants. Implementation of VOC controls to meet ozone standards requires a rapid response by industry and the states. However, responsibility for actual equipment control requirements has not been clarified. Since equipment control requirements for hazardous pollutants is a major EPA responsibility and proposed equipment control regulations for the chemical industry were issued late in 1992, the overlap with various VOC compounds also classified as hazardous may result in similar control requirements for general VOC operations. Application of hazardous air pollutant (HAP) equipment controls for VOC control would definitely reduce VOC emissions of non-HAP chemicals. Since regulations will be issued in 1994 for emission control in petroleum refineries, more restrictive fugitive emission controls covering most pollutants can be anticipated. Moreover, the CAAA specifically requires reviewing the number of HAP compounds periodically, and the original list of 189 HAP compounds will probably be expanded to cover a large number of HAP compounds that are photochemically reactive. Moreover, these HAP emission-type control regulations may also be necessary to achieve attainment status for ozone in conformance with the NAAQS standards.

The following brief review of the CAAA provides some background on the scope of the amendments and areas applicable to exposure control. A more detailed analysis of equipment control requirements is presented later in this chapter. A brief outline of the CAAA 1990 follows:

Title I. Attainment and Maintenance of National Ambient Air Quality Standards. In general, areas must be categorized for nonattainment, including the out of compliance contaminants. All reasonable available control measures should be implemented as soon as possible and an inventory of emissions from all sources must be developed in each state. Nonattainment areas for ozone must be classified from marginal to extreme, as shown in Table 4.2. Some other major points relating to these classifications are:

1. *Marginal.* Inventory of emissions from *all* sources within 2 years after the CAAA enactment date. A revised inventory shall be submitted every 3 years after the original inventory until the area is in attainment. The operator of each stationary source will submit an inventory of actual nitrogen oxide and VOC emissions from classes or categories of sources defined by the state. The first inventory for these emissions is required 3 years after the CAAA enactment date and yearly thereafter. If less than 25 tons per year of VOC or nitrogen oxides is emitted, the inventory emission requirement may be waived.

2. *Moderate.* Submissions under marginal to the SIP are included in this category. No later than 3 years after the CAAA enactment date, the state will revise the SIP to provide a procedure for VOC emissions reductions within 6 years after the CAAA enactment date of $\geq 15\%$ from baseline emissions (1990 is the baseline year). This accounts for any emission growth after 1990, and the SIP will provide for specific annual reductions to achieve the air quality standard by the attainment date.

TABLE 4.2 Classification and Attainment Dates for 1989 Nonattainment Areas[4]

Area Class	Range Classification[a] (ppm)	Primary Standard Attainment Date[b] (yr)
Marginal	0.121 – 0.138	3
Moderate	0.138 – 0.160	6
Serious	0.160 – 0.180	9
Severe 1	0.180 – 0.190	15
Severe 2	0.190 – 0.280	17
Extreme	$\geq$ 0.280	20

[a]Actual concentration range.
[b]Years after enactment date of amendments.

A percentage $< 15\%$ is permitted if all sources within an area that emit or have the potentials to emit at least 5 tons/yr of VOCs are included, reasonably available control technology (RACT) is required for all major existing sources, and the plan contains all control measures that feasibly can be included. The RACT revision to the state SIP must include implementation plans encompassing all Control Techniques Guidelines (CTG) for VOC sources issued before the CAAA enactment date and CTG requirements issued between the CAAA enactment and attainment dates. Also, RACT must be installed on all other major VOC stationary sources.

In addition, vapor recovery emissions during motor vehicle refueling is required.

3. *Serious.* Submissions under moderate are included in this category. Major and major stationary sources are those with the potential to emit at least 50 tons/yr of VOC. Enhanced monitoring rules will be promulgated to obtain more comprehensive and representative data.

Within 4 years after the CAAA enactment date, the state shall demonstrate analytically that the air quality standard will be achieved by the applicable attainment date. Also, the state shall demonstrate beginning 6 years after the CAAA enactment date, VOC emissions averaged over a 3-year period until the attainment date will be $\geq 3\%$ of baseline emissions each year. If 3% cannot be achieved, the state must demonstrate that all measures that can be feasibly implemented technologically are included. Alternatively, NO_x control may be substituted for VOC control or combined with NO_x control for maximum ozone reduction.

Where a major stationary source emits < 100 tons/year of VOC, any modification of the source that results in an emissions increase (other than de minimus) shall not be considered a modification if the increase is offset by a VOC emissions reduction from another operation of at least 1.3 : 1. If an offset is not selected, then best available control technology (BACT) as defined in Section 169 of the Act shall be substituted. When a major stationary source that emits or has the potential to emit ≥ 100 tons/year of VOC, any change in the source that results in an emissions increase (other than de minimus) of VOC shall not be considered a modification if the increase is offset by a VOC emissions reduction from other operations or within the source $\geq 1.3 : 1$, the lowest achievable emissions rate (LAER) as defined in Section 173 (a) (2) of the Act shall not apply.

Improved vehicle inspection and maintenance, clean fuels, and transportation control are included.

4. *Severe.* Submissions under serious are included in this category. Major source and major stationary source include any source or group of sources that have the potential to emit at least 25 tons/year of VOC. Any offset requirements for VOC shall be at least 1.3 : 1. If all major sources use best available control technology for VOC, the offset required is at least 1.2 : 1.

Measures to reduce vehicle miles are included in this nonattainment designation.

5. *Extreme.* Submissions under severe are included in this category. Clause provisions relating to VOC reductions less than 3% per year and a reduction of at least 15% by the attainment date are not applicable. Also, source modifications with the de minimus rule are not applicable. Major source and major stationary

source in an extreme area cover any stationary source or group of sources that emit or have the potential to emit ≥ 10 tons/year of VOC.

For VOC offsets, the ratio of total VOC emissions reductions to total increased emissions is at least 1.5 : 1. If all major existing sources use best available control technology for VOC control, the offset ratio is 1.2 : 1. Any change at a major stationary source that results in a VOC emissions increase shall be considered a modification. It will not be considered a modification with an offset ratio of at least 1.3 : 1. Offsets are not applicable if the modification is required to comply with an implementation plan.

An implementation plan may be acceptable to the EPA which anticipates the developments of new technology or improvements in existing technology with an attainment demonstration. However, the state must demonstrate the technology is not necessary to achieve the incremental emissions reductions required during the first 10 years after the CAAA enactment date, and the state has enforceable contingency measures for implementation if the new technology does not achieve the required emissions reductions.

Clean fuel technology for utilities and boilers are included.

6. *Other.* Six years after the CAAA enactment date and every 3 years thereafter, the state shall determine whether nonattainment areas excluding marginal and moderate have achieved the required VOC emissions reductions. These reductions are termed applicable milestones. The state not later than 90 days after each applicable milestone submits to the EPA sufficient information to demonstrate that milestone conditions have been achieved.

If the milestone information is unsatisfactory or milestone conditions have not been attained, the state may reclassify the area to a more stringent classification, implement measures recommended by the EPA, or adopt an economic incentive program. The EPA is required to publish rules regarding an economic incentive program.

Title II. Provisions Relating to Mobile Sources. As described by the title , this section of the CAAA is only concerned with vehicles and does not affect the process industry workplace.

Table III. Hazardous Air Pollutants. Section 112 of the original Clean Air Act established the National Emission Standard for Hazardous Air Pollutants (NESHAP). However, the EPA identified only seven NESHAP pollutants through the years. To expand the number of hazardous pollutants and associated emission controls, Section 112 of the Act was amended with 189 chemicals listed as Hazardous Air Pollutants (HAP). In addition, various other rules and requirements were included in the amended section. A brief review of this section is presented in the following paragraphs with more detailed equipment requirements outlined late in this chapter.

An important consideration in this title is the definition of sources. A "major source" is any stationary or group of stationary sources under common control that emits or has the potential to emit (considering controls) 10 tons/year or more of any hazardous air pollutant or 25 tons/year or more of any combination of hazardous air pollutants. An "area source" is any stationary source of HAP that is not a major source.

The initial list of hazardous air pollutants containing 189 chemicals and compound classes shown in Appendix A was specifically included in Title III of the CAAA. The general compound classes near the bottom of the table are primarily metal compounds, but other general compounds are included that should be carefully reviewed to ensure compliance with the HAP list.

The CAAA also provides for periodic review and revision of the list by rule. The EPA may add pollutants that present adverse human health through inhalation or other exposure routes. Adverse human health effects include but are not limited to carcinogenic, mutagenic, teratogenic, and neurotoxic substances which cause reproductive dysfunction or may be acutely or chronically toxic. Also substances that may cause adverse environmental health effects can be included in the list covering ambient concentrations, bioaccumulation, or deposition.

The EPA is required to publish 12 months after the CAAA enactment a list of all categories and subcategories of major sources and area sources for the pollutants in the HAP list. In addition, the EPA must establish emissions standards for these categories and subcategory sources. For area sources, within 5 years after the CAAA enactment date, a list of categories and subcategories will be issued representing 90% of the area source emissions for 30 HAP compounds in Section 112 (C) (6) that present the greatest threat to human health. Regulations for this group will be promulgated within 10 years after the CAAA enactment date. Also, a similar list will be established for specific compounds that cover 90% of the aggregate emissions of each pollutant and are regulated under Maximum Allowable Control Technology (MACT) requirements. Moreover, additional categories and subcategories may be added to these lists at any time.

Emission Standards. Under this section of Title III, the emissions standards and requirements for control of HAP are outlined. A summary of these requirements follows:

1. *General.* The EPA shall promulgate emission standards for each category or subcategory of major and area sources of HAP as described above. No delay in compliance dates are permitted.

2. Emission standards in this section applicable to new or existing sources require maximum HAP emissions reduction considering the cost of achieving such reductions and nonair quality health or environmental impacts. The reductions include, but are not limited to the following measures:

   Process changes, material substitution, or other modifications to eliminate or significantly reduce the HAP emissions

   Enclose various systems or processes to eliminate emissions

   Collect, capture, or treat HAP emissions when released from process, stack, storage, or fugitive emissions sources

   Design or equipment requirements

   Work practice or operational standards

   A combination of the above

This general definition or requirement is termed Maximum Achievable Control Technology or MACT when regulatory details have been promulgated.

3. Maximum emissions reduction for a new source shall not be less stringent than emissions controls actually achieved in practice by the best controlled operations in a similar source.

Emission standards for existing sources may be less stringent than new source standards but shall not be less stringent and can be more stringent than: The average emission limitation achieved by the best performing 12% of existing sources, with categories and subcategories having 30 or more sources. The average emission limitation achieved by 5 of the best performing sources with categories and subcategories having less than 30 sources.

4. For area sources, the EPA may require specific control technologies or management practices to reduce HAP emissions.

5. Emission standards ≥ 40 categories and subcategories (excluding coke oven batteries) shall be promulgated 2 years after the CAAA enactment date. Emission standards for *all* categories and subcategories will be promulgated no later than 10 years after the CAAA enactment date.

6. No later than 6 years after the CAAA enactment date, the EPA is required to investigate the effect on public health after the HAP emissions reductions achieved as a result of the emission control standards. If a risk remains, the EPA will recommend legislative changes to the congress. Should the legislature fail to act, the EPA can promulgate standards to provide an ample margin of safety in protecting the public health within 8 years after standards were originally issued for each category or subcategory of sources. Revisions may also be promulgated to prevent an adverse environmental effect, considering costs, energy, safety, and relevant factors.

7. Following the implementation date of a permit program (Title V), a major source cannot be *modified* unless MACT requirements for an existing source are observed. Also, a major source may not be *constructed or reconstructed* unless MACT requirements for new sources are observed.

Work Practice Standards. Where compliance with an emission standard is not feasible and the HAP standards cannot be enforced, the EPA may specify or promulgate design, equipment, work practice, or operational standards. A combination of these emission control techniques may also be developed and promulgated as standards. When a design or equipment standard is issued, the EPA will include requirements to ensure proper operation and maintenance of the required equipment.

Any control standard in this subsection shall be promulgated in terms of an emission standard whenever feasible.

Other Standards. Standards in effect before the CAAA enactment date remain in force unless specifically modified. Any of these standards must be reviewed and, if necessary, revised to ensure HAP emissions compliance within 10 years after the CAAA enactment date.

Prevention of Accidental Releases. The objective of this subsection is to prevent the accidental release of any extremely hazardous substances. Accidental release is an unanticipated emission of a regulated or other extremely hazardous substance into the atmosphere from a stationary source. The latter refers to structures, equipment, installations, or substance-emitting sources. Owners or operators must identify hazards from these releases using hazard assessment techniques that also aid in designing and maintaining equipment that minimize releases and the consequences of releases that do occur.

A list of 100 substances must be promulgated 24 months after the CAAA enactment data containing substances which, when accidentally released, may cause death, injury, or serious adverse human health or environmental affects. This list is not limited to the extremely hazardous substance list issued as part of the Emergency Planning and Community Right-to-Know Act of 1986. The CAAA specifically requires the inclusion of the following substances:

Chlorine	Hydrogen sulfide
Anhydrous ammonia	Toluene diisocyanate
Ammonia	Phosgene
Methyl chloride	Bromine
Ethylene oxide	Anhydrous hydrogen chloride
Vinyl chloride	Hydrogen fluoride
Methyl isocyanate	Anhydrous sulfur dioxide
Hydrogen cyanide	Sulfur trioxide

This list differs from the 189 HAP substances included in the CAAA which is shown in Appendix A. The accidental release list may be revised as required by the EPA and must be reviewed every 5 years. This accidental release list of substances will also include threshold quantities of the substance present in the equipment. The threshold quantity (TQ) considers toxicity, reactivity, volatility, dispersiblity, combustibility, or flammability and the quantity of material that may be accidentally released causing death, injury, or serious adverse health or environmental effects. This list will not include any substances that have a National Primary Air Quality Standard (NAAQS). Criteria established for substances considered for this list are:

1. Severity of acute adverse health effects associated with accidental releases.
2. Potential for accidental releases of the specific substance.
3. Magnitude of potential human exposure to accidental releases.

The CAAA in this subsection establishes a Chemical Safety and Hazard Investigation Board (CSHIB). This board investigates *any accidental release* that results in a fatality, serious injury, or substantial property damage and provides a public report on the facts, conditions, circumstances, and probable cause of any accidental release. The CSHIB must periodically issue reports recommending measures and corrective steps to reduce the accidental release potential. Proposed rules would be issued by the EPA or OSHA. The EPA is specifically authorized to issue regulations that control the accidental releases of regulated substances covering release prevention, detection, and correction requirements that may include:

Monitoring	Secondary containment
Record keeping	Design requirements
Reporting	Equipment
Training	Work practices
Vapor recovery	Operational requirements

The regulations are scheduled for promulgation 3 years after the CAAA enactment date and will cover storage as well as operations. In addition, the regulations will require a risk management plan to detect and prevent or minimize accidental releases where a regulated substance containment quantity is greater than the threshold quantity for the substance. This risk management plan will be based upon a risk management report issued by the CSHIB 1 year earlier than the regulatory promulgation date. The risk management plan must be registered with the EPA and submitted to the CSHIB, state, and responsible local entities. An *overall risk management plan must include* the following:

1. A hazard assessment that evaluates the potential effects of an accidental release. The assessment should include an estimate of potential release quantities, downwind effects, and potential human exposures.
2. A program for the prevention of accidental releases that includes safety precautions, maintenance, monitoring, and employee training.
3. A response program defining action procedures following an accidental release to protect human health and the environment. Procedures must be included that inform public and local agencies.

Risk Assessment and Management Commission. The Risk Assessment and Management Commission (RAMC) is required to commence proceedings 18 months after the CAAA enactment date. Commission responsibilities cover the appropriate use of risk assessment and risk management of various federal regulatory programs concerned with the prevention of cancer and other chronic human health effects related to hazardous substance exposures. Policy implications related to risk are a further extension of the RAMC mandate. Specifically, the RAMC is charged with the consideration of:

1. The use and limitations of risk assessment in establishing standards for emissions, effluents, ambient concentrations, exposure criteria, concentration levels, tolerances, or other environmental criteria for HAP that present *potential* carcinogenic or other chronic health effect risks.

2. The most appropriate methods for measuring and describing cancer or chronic health effects.

3. Methods of reflecting uncertainties in measurement and estimating techniques, interactive effects among HAP substances, and the accuracy of extrapolating human health risks from animal exposure data.

4. Risk management policy issues including the application of lifetime cancer risks to the most exposed individuals, incidence of cancer, cost and technical

feasibility of exposure reduction measures and use of site-specific actual exposure information as a basis for determining emission standards.

5. The possibility of developing a consistent risk assessment methodology or consistent acceptable risk standard for various federal programs.

Chemical Process Safety Management. In this subsection, OSHA has the responsibility for promulgating a chemical process safety standard that is designed to *protect employees* from the hazards associated with accidental releases of highly hazardous chemicals in the workplace. The accidental release prevention program previously described is based upon recommendations for release prevention by the Chemical Safety and Hazard Investigation Board to minimize acute adverse health effects in the general population. The EPA is responsible for promulgation of the standards in this latter program. While the EPA program has not been proposed or promulgated, OSHA has promulgated a final rule for Process Safety Management of Highly Hazardous Chemicals, Explosives, and Blasting Agents.[5]

The major requirements for this standard are briefly described in the following list with additional comments in another subsection.

1. A list of highly hazardous chemicals that are toxic, flammable, highly reactive, and explosive will be included in the standard. The list may include chemicals under Section 302 of the Emergency Planning and Community Right to Know Act of 1986.

2. A safety standard at a minimum requires employers to:

a. Develop and maintain written safety information, identifying workplace chemical and process hazards, process equipment, and process technology.

b. Conduct a workplace hazard assessment.

c. Consult with employees on hazard assessment and chemical accident prevention plans.

d. Establish a system for workplace hazards addressing prevention, mitigation, and emergency response.

e. Periodically review the workplace hazard assessment and response program.

f. Implement written operating procedures for operational phases, operating limitations, and safety and health considerations.

g. Train employees in operating procedures emphasizing hazards and safe practices.

h. Contractors and contract employees must be provided appropriate information and training.

i. Train and educate employees and contractors in emergency response.

j. Establish a quality assurance program to ensure that new equipment, maintenance materials, and spare parts are fabricated and installed in accordance with design specifications.

k. Establish maintenance systems for critical process-related equipment to ensure continuing mechanical integrity.

l. Conduct prestartup safety reviews of all new or modified equipment.

m. Establish procedures to manage changes in process chemicals, technology, equipment, and facilities.

n. Investigate every incident that results in or could have resulted in a major workplace accident. Install or modify the equipment if appropriate.

Title IV. Acid Deposition Control. Acidic compounds and their precursors of concern in this section are emissions of sulfur and nitrogen oxides from the combustion of fossil fuels. The emission control requirements in this section are primarily concerned with utilities and large furnace or boiler emissions. These combustion sources emit their pollutants from tall stacks that normally disperse over a number of miles. Chemical plant stacks may be somewhat shorter, but dispersion also occurs with the smaller stacks. Workplace exposure is not a major concern in this section.

Title V. Permits. A variety of permits have been required in chemical plants and refineries covering single process units, multiple process units, and various emission sources. In this section permit program requirements are outlined with a major change in the need for multiple permits. A single permit can be issued to cover a great number of sources, process units, and an entire plant. Permits are based upon emissions quantification, which is described in Chapter 5. Workplace exposure is not a concern in this section.

Title VI. Stratospheric Ozone Protection. This section is concerned with chlorofluorocarbon release to the upper atmosphere. Workplace exposure is not a concern.

Title VII. Provisions Relating to Enforcement. This section is only concerned with regulatory enforcement.

Title VIII. Miscellaneous Provisions. Of concern in workplace exposure are the fugitive or leak emissions from equipment that result in workplace exposures. Extensive efforts by the EPA and the Chemical Manufacturers Association (CMA) have been undertaken through the years to develop equipment emission factors. Variations between the organizations regarding emission factors have been observed and several methods of calculating fugitive emissions totals have been developed. In subsection 804 of the CAAA, emission factors have been addressed:

1. Six months after the CAAA enactment date and at least every 3 years thereafter, the EPA will review and revise as necessary the equipment emission factors, specifically carbon monoxide, VOC, and nitrogen oxides from stationary and mobile sources. Also, the EPA will establish emission factors for sources that do not have emission factors.

2. Improved emission estimating techniques developed by non-EPA personnel can be presented for review by the EPA. If satisfactory, the EPA will authorize use of the improved technique. However, approval is only granted after appropriate public participation.

3. Until the EPA formally reviews and revises the emission factors, emission factors used prior to the CAAA enactment date will remain valid.

Title IX. Clean Air Research. The research effort in this title is not specifically directed toward workplace exposure. However, there are some areas that may provide technical information of assistance in reducing workplace exposure.

1. A liquified gaseous fuels spill test will commence operation. A minimum of two chemicals will be field tested each year. Highest priority is given to those chemicals that present the greatest potential risk to human health as a result of an accidental release. This research will improve dispersion modeling and will also evaluate the effectiveness of hazard mitigation. Emergency response technology will also be evaluated as part of the program. Accidental release models will also be evaluated and improved periodically.
2. The EPA will conduct a basic engineering research and technology program to develop, evaluate, and demonstrate nonregulatory strategies and technologies for air pollution prevention. Emphasis is placed on pollutants that pose a significant risk to human health and the environment. While much of the effort is directed toward fossil fuel power plants, some effort will be directed toward reducing air emissions from area sources. In addition, the program also includes improvements in nonregulatory strategies and technologies for preventing, detecting, and correcting accidental release of hazardous air pollutants.
3. The EPA is required to conduct a study of air pollution control technologies in selected countries to determine whether there are other air pollution control technologies that may have application in the United States. Of particular interest is the control of toxic air emissions. A report on this investigation is required 2 years after the CAAA enactment date.

Titles X and XI. These sections are non concerned with workplace exposures of air pollution.

4.2 OSHA

The Occupational Safety and Health Act (OSHA) is designed to control exposures in the workplace to toxic and hazardous substances. The toxic substances of concern in the OSHA regulations may also be covered by the hazardous air pollutant (CAAA) and/or the volatile organic compound regulations. While these are apparently overlapping regulatory requirements, the EPA does not consider the regulations overlapping or contradictory since EPA and OSHA objectives differ. In general, EPA is concerned with public health and the overall environment, which differs from OSHA concentration on the control of exposures in the workplace.

4.2.1 OSHA Standards

The EPA regulations are generally equipment oriented to reduce emissions that contribute to the basic pollutants regulated by NAAQS standards. Also, the HAP emissions are specifically controlled through regulation. The NESHAP regulations are essentially superseded by the new HAP control regulations in most areas. Obviously, the equipment controls reduce emissions, thereby reducing workplace pollutant concentrations and worker exposure. Consequently, knowledge of the regulations indicates where controls are required and have been installed, their effectiveness, and the need for additional exposure controls.

The OSHA requirements have not been particularly equipment oriented to control workplace exposures. Accomplishment of OSHA objectives may be achieved through various exposure control techniques or combinations of the following general control schemes:

Substitution of less hazardous material in the process
Process modifications
Equipment controls
Worker rotation
Ventilation controls
Process or worker isolation
Modifications of work practices

The list indicates that worker exposure can be controlled without significant reductions in emissions to the atmosphere. As an example, ventilation controls do not minimize emissions from a source, but generally transmit the emissions for discharge to the atmosphere from one location to another. Moreover, combining some equipment controls with changes in work practices may only partially reduce emissions with a greater dependence on work practice changes. The latter may require personal protective equipment as the primary method of controlling workplace exposure.

As described in section 4.1.3, OSHA is responsible for regulations covering process safety management of highly hazardous chemicals. These regulations were promulgated in February 1992.[5] The purpose of these regulations is the establishment of procedures for process safety management that protect employees by preventing or minimizing the consequences of chemical accidents involving highly hazardous chemicals. These requirements are briefly outlined in section 4.1.3 and additional details are presented elsewhere in this chapter.

4.3 OTHER REGULATIONS

The previous comments referred to fugitive emissions from process equipment and operations in the processing area. A number of other VOC and hazardous air pollutant emission sources mentioned in the chapter introduction have regulations controlling emissions and are briefly reviewed in this section. Regulations covering emissions in Sections 4.1 and 4.2 were U.S. EPA or federal regulations. While these regulations are standard requirements, there are some differences in other areas of the United States.

4.3.1 Regional Regulations

The EPA regulations are the basic environmental control standards, but they are not the only regulatory standards in the United States. In states that have developed satisfactory State Implementation Plans (SIP) of EPA regulations, the

Clean Air Act permits the states to directly enforce EPA regulatory requirements. Also, the Clean Air Act permits the establishment of regional air management districts that have the authority, when properly constituted, to issue control regulations. These regional bodies may have full regulatory control and enforcement responsibility in attainment or nonattainment areas. The most notable air control districts are in California, namely, the Bay Area Air Quality Management District (BAAQMD) and the South Coast Air Quality Management District (SCAQMD). The BAAQMD is located in the San Francisco area and the SCAQMD is located in the Los Angles Area.

Since these are nonattainment areas, the BAAQMD and SCAQMD have issued regulations for the control of fugitive emissions that are more stringent for various items than the EPA fugitive emission control regulations. For duly authorized groups, more stringent regulations can be issued than the promulgated EPA regulations. However, the regulatory requirements cannot be relaxed or reduced below the regulatory controls promulgated by the EPA. States with nonattainment areas may also issue more stringent regulations than the EPA to reduce pollutant quantities.

In addition to EPA and OSHA regulatory requirements, state and regional requirements must be investigated to determine whether more stringent regulatory controls are required. The more stringent controls tend to reduce emissions and lower potential exposure levels.

4.3.2 Other Regulations

In addition to process equipment regulations, other regulations have been promulgated by the EPA and various states to control VOC and hazardous pollutant emissions from various in-plant sources. These additional restrictions also reduce concentration levels near the equipment and potential workplace exposures. Some of the regulations promulgated to control overall emissions are:

1. *Storage Tank Emission Control Standards for Volatile Organic Liquids.* These standards cover VOC emissions from both petroleum refinery and chemical storage tanks.[6–8]. In addition, regional tankage emission control standards can be applicable and may possibly be more stringent than EPA regulations.[9] The more stringent regulatory requirements must then be observed requiring an understanding of the applicable regulations. In addition, more stringent tankage emission control standards may be required in other regulations such as the NESHAP benzene regulations, the HON regulations[31] described in the following paragraph, or the NESHAP for benzene storage.[10]

2. *Wastewater Emissions Control Standards.* A refinery wastewater VOC emissions NSPS standard was promulgated in 1988.[11] As implementation of these regulations began, promulgation of a NESHAP standard for Benzene Waste Operations in 1990 resulted in overlap and preemption of the NSPS regulations.[12,13] The reason for consideration of preemption is that wastewater streams with a benzene concentration ≥ 10 ppmw are subject to the NESHAP regulation. For petroleum refineries, the NESHAP regulations may be the governing set of restrictions.[14]

In chemical plants, the benzene wastewater NESHAP may not be the controlling set of regulations since benzene is considered an HAP and the more restrictive HON regulations[31] may supersede the NESHAP. A careful review of wastewater streams from process units is necessary to determine regulatory status. Where HON regulations are not applicable, the most stringent of the applicable regulations require compliance.

In addition to federal regulations, state and regional regulations are also applicable in wastewater emissions control with rules covering oil/water separators, impoundment basins, and holdup tanks. Emissions and restrictions through these various regulations significantly reduce air pollutant concentrations.

Regulations cover hazardous waste treatment and waste storage ≥ 500 ppmw of volatile organic compound material.[16] Emissions reduction are the regulatory objectives, and these rules cover a wide variety of nonbenzene-containing waste streams. In addition, benzene-containing wastes have separate regulatory requirements.[12,13] The VOC regulations specifically address emissions from tanks, surface impoundments, drums, tanktrucks, railcars, barges, and ships. While these regulations are required under the Resource Conservation and Recovery Act (RCRA), they are effective in expanding emission controls and reduced exposure levels in waste facilities.

3. *Tanktruck Loading.* The EPA regulations cover VOC emissions during tanktruck loading at bulk gasoline terminals,[17] and a NESHAP Benzene Standard covers tanktruck, railcar, and marine vessel[18] benzene emissions. In addition, HON regulations also include emission controls for tanktruck and railcar loading.[31] Although these standards apparently control tanktruck loading emissions in petroleum refineries and chemical plants, various state and regional authority regulations also have emission control regulations applicable to loading operations. Typical control regulations covering organic liquid loading for the BAAQMD and SCAQMD are shown in references[19–21]. For emission control requirements in tanktruck loading, the latest regulatory requirements by various authorities should be established. These control system requirements significantly reduce emissions and potential exposure concentration levels in loading operations, which frequently have personnel in the loading areas.

4. *Other.* A number of regulations cover other additional sources of emission leakage. Barge and ship loading regulations are included in some of the emission control regulations with various state and regional authorities issuing stringent emission control loading regulations. The loading regulations may also include unloading regulations and should, therefore, be reviewed for unloading emission control requirements.

The large number of regulations require a general understanding of equipment control regulations and the implications for potential exposure control requirements. Detailed emission control quantities and specific control requirements are prepared for permit reviews with EPA, state, or regional authorities. From the emission estimates, detailed exposure control needs can be determined. As a result of these various VOC and hazardous pollutant regulations, fugitive emission controls are extended to essentially all operating areas, significantly reducing potential workplace exposures.

4.4 EQUIPMENT EMISSION CONTROL REGULATIONS

A major requirement in Section III or the original Clean Air Act was the control of emissions that contribute to air pollution and potentially endanger public health. Volatile organic compounds affect the NAAQS standard for ozone and, in addition, potentially endanger the public health requiring emission control under Section III. During an extensive investigation of VOC emissions, the EPA found that fugitive or disperse emissions were a major component of overall VOC emissions losses to the atmosphere. Control of these emissions then generated a series of regulations to control equipment leakage or fugitive emissions.

The regulations resulting from this investigation for control of VOC emissions were the NSPS regulations described in Section 4.1.2. These regulations require that new,reconstructed, and modified process units comply with the regulatory standards to achieve a low VOC emissions level and compliance with the NAAQS ozone standard. In addition, control of hazardous air pollutants required in Section 112 of the Clean Air Act resulted in the promulgation of NESHAP regulations for emissions control of vinyl chloride and benzene. The NESHAP and NSPS regulations resulted in a decrease in emissions over a period of time in both petroleum refineries and chemical plants.

These regulations were concerned with controlling emissions from specific equipment items such as valves, pumps, flanges, compressors, relief valves, and sampling points. Although considerable attention has been concentrated on these items, other sources have also been regulated to control fugitive emissions from various systems such as tankage, sewers, wastwater treating, and tanktruck loading regulations. These latter regulations also include emission control requirements for equipment such as valves, pumps, and so on associated with off-site operations that also contribute to fugitive VOC and hazardous pollutant emissions.

The effort to control fugitive emissions requires monitoring and maintenance programs (M & M or LDAR), which have assisted significantly in fugitive emission control. All regulations now include LDAR programs to control fugitive emissions. As described in Section 4.1.3, the Clean Air Act Amendments of 1990 hasten the date for compliance with the NAAQS standards. Also, the CAAA increased the number of hazardous air pollutants to 189 from 7 in the older NESHAP regulations to expand emission control of HAP compounds. The new CAAA amendments increase emission control severity—proposed regulations issued in late 1992 that will be promulgated in the early part of 1994. Since the HON equipment emission control regulations were a result of joint industry–EPA consultations, the final regulations for equipment will not change from the proposed regulations. The general reduction in VOC and HAP required by the CAAA will significantly reduce fugitive emissions in the future as the restrictions gradually cover all process facilities. In addition, a variety of other fugitive emission sources will be reviewed, and revised regulations are scheduled for promulgation from 1993 through 1996. As fugitive emissions are significantly reduced, potential workplace exposure concentrations will decrease.

4.4.1 Emissions Control Basis

From 1975 to 1983, the EPA conducted a number of tests in chemical plants and petroleum refineries to quantify actual equipment emission loss rates. The data after analysis provided the basis for the NSPS, NESHAP, and HON or HAP regulations to control ozone and HAP concentrations. The overall test program resulted in the EPA establishing emission factors for valves, pumps, flanges, compressors, relief valves, sampling points, and open-ended lines which provided a basis for estimating industry losses and economic evaluations of alternative control recommendations. The factors were published in AP-42, an EPA document that provides a basis for estimating average fugitive emission losses from a wide variety of sources.[23] Originally, the EPA assumed petroleum refinery and chemical plant emission factors were identical. Further testing indicated there were significant differences in average emission factors for the two industries and average emission factors were published for the Synthetic Organic Chemical Manufacturing Industry (SOCMI). Initially, average equipment emission factors were used in estimating overall losses from existing process facilities or in developing anticipated losses for new facilities. However, other methods of estimating losses have been developed with specific rather than average emission factors and are described in Chapter 5, along with other procedures for estimating emissions.

During the early EPA test program, valve emission leakage was approximately 70% of the estimated fugitive emission loss totals. The remaining losses were distributed among the other equipment leak sources. Test data indicated the average valve stem mass emission leak rate was significantly lower than the average leak rates for mechanical pump seals and compressors. However, the large number of valves in a process unit result in very large losses compared with other sources.

Since valve stems were the major fugitive emissions source in chemical or refining processing units, the EPA contractor analysis concentrated on the control of valve stem emissions. Through an extensive data analysis effort, the EPA concluded that valves with valve stem vapor concentrations of 10,000 ppm or greater should be classified as leaking valves. The 10,000-ppm concentration, termed an action level by the EPA, was selected based upon the mass emission fraction affected by the selected action level, the proportion of valves that can be successfully repaired, and the ineffectiveness of repairing leaking valves in the 1,000- to 10,000-ppm range. The EPA estimated 98% of mass emissions from valves in vapor service would be controlled with a 10,000-ppm action level, and 84% of mass emissions from valves in light liquid service would be controlled at the 10,000-ppm action level.

In addition to defining an action level, the EPA had investigated monitoring and maintenance (LDAR) programs that aided in controlling emissions. As a result, the regulations have both an action level and LDAR programs to ensure that lower mass emissions rates are achieved. A comparison of average SOCMI emission factors before fugitive emission NSPS regulation with emission factors after implementation of the regulations is presented in Table 4.3. In addition to NSPS regulations, a few states and authorities increased restrictions over typical EPA fugitive emissions rules in State Implementation Programs (SIP), further reducing overall mass emission loss rates. The lower fugitive VOC emissions rates result in reduced concentration levels in the workplace, thereby reducing potential workplace exposures.

The average emission factors in Table 4.3 indicate there is a reduction in fugitive emissions particularly in valve and flange leakage. Although the reduction is evident in the table, emissions have been falling since the mid-1980s, as shown in the emissions inventories required by the EPA in various regulations. Overall toxic

TABLE 4.3 Fugitive Emission Factor Changes Before and After Regulation in the Synthetic Organic Chemical Manufacturing Industry

Fugitive Emission Source	Noncontrolled[a] Emission Factor (kg / hr)	Controlled[b] Emission Factor (kg / hr)	Controlled[b] Factor / Excluding Ethylene Units (kg / hr)
Valves			
Gas	0.021	0.0056	0.0027
Light liquid	0.010	0.0071	0.00295
Heavy liquid	0.0003	0.00023	0.0003
Flanges	0.0003	0.00082	0.00023
Pumps			
Light liquid	0.12	0.0495	0.0397
Heavy liquid	0.02	0.021	0.023
Compressors	0.44	0.227	0.227
Relief valves	0.16	0.104	0.104

[a]Adapted from Ref. 22
[b]Adapted from Ref. 24.

releases to the atmosphere fell 26% in the 4-year period from 1987 through 1990,[25] and equipment emissions were significantly reduced as described in a broad study covering 25 chemical companies.[26]

As a result of the emissions reductions, valve emissions are currently roughly equal to flange emissions in processing units.

4.4.2 Regulatory Considerations and Definitions

Fugitive emission control is regulated under NSPS, NESHAP, and state implementation plans or authorities. Additional control requirements will be promulgated in early 1994 by the EPA for hazardous air pollutants in accordance with the Clean Air Act Amendments previously described in regulations termed the Hazardous Organic NESHAP (HON).[31] Initially, the new control regulations are directed toward SOCMI, with petroleum refinery regulations scheduled for release in late 1994. While many of the HON regulations are concerned with equipment such as pumps and valves, the initial regulations termed MACT (Maximum Achievable Control Technology) will also cover control technology for wastewater and sewer systems, storage, transfer (tanktruck and railcar), and process vents.

Prior to describing the equipment control and system regulations, a number of definitions are presented which are common or standard for the various regulations:

Closed Loop System. An enclosed circulating system that returns process fluid to the process and is not vented to the atmosphere except through a closed-vent system.

Closed Vent System. A system composed of pipe, connections, or other equipment that is not open to the atmosphere. The system transports gas or vapor from equipment to a control device or returns the stream to the process.

Connector. Flanged, threaded, or joined fittings that connect piping sections or pipe to process equipment. A common connector is a flange. Joined fittings completely welded circumferentially are not considered in the regulations.

Control Device. An enclosed combustion device, vapor recovery system, or flare. Vapor recovery includes devices such as carbon adsorption, absorption, and condensers.

Distance Piece. Open section through which a piston rod moves in a reciprocating compressor separating the compressor cylinder from the driver.

Double Block and Bleed System. Two block valves in series with a bleed valve connected to the line between the blocks that can vent the line.

In Gas / Vapor Service. An equipment item contains process fluid in the gaseous state at operating conditions. For MACT, this refers to gas/vapor in volatile hazardous air pollutant (VHAP) service for chemical compounds. Compounds are listed in Appendix 4B.

In Heavy Liquid Service. The equipment is not in gas/vapor or light liquid service for VOC or VHAP materials.

In Light Liquid Service (Chemical). For VOC and VHAP compounds, the vapor pressure of one or more of the components is greater than 0.3 kPa at 20°C. Also, the total concentration of the pure components having a vapor pressure greater than 0.3 kPa is $\geq 20\%$ by weight. In addition, the fluid is a liquid at operating conditions.

In Light Liquid Service (Petroleum). The percentage evaporated is greater than 10% at 150°C.

In-Situ Sampling Systems. Nonextractive or in-line samplers.

In VHAP Service. The equipment contains or contacts liquid or gas that is at least 5% VHAP by weight.

In VOC Service. The equipment contains or contacts a process fluid that is at least 10% VOC by weight.

Open-Ended Valve or Line. Any valve (excluding relief valves) with one side of the valve seat in contact with process fluid and one side open to the atmosphere, either directly or through open piping.

Pressure Release. The emission of materials resulting from a system pressure greater than the set pressure of a pressure-relief device.

Product Accumulator Vessel. Any distillate or bottoms receiver, surge control vessel, or product separator in VHAP service vented directly to the atmosphere without venting through a pressure-relief device or vacuum system.

Set Pressure. The pressure at which a properly operated pressure relief device begins to open for relief during typical operating conditions.

Waste Stream. The waste generated by a specific process unit, product tank, or waste management unit. The stream characteristics are determined at the waste generation point. Examples include process wastewater, tank drawdown, sludge, and slop oil from waste management units.

4.4.3 Monitoring

Leak detection and repair (LDAR) procedures were initiated to reduce fugitive emissions or leakage and have been a major factor in reducing VOC emissions and controlling NESHAP emissions. Monitoring is normally conducted periodically on equipment in accordance with regulations such as NSPS, NESHAP, and regional authority requirements. The new HON regulations will also require periodic monitoring for both specific equipment items and other areas. Leakage action levels differ in some of the regulations and must be understood along with the specific regulation applicable.

While monitoring insures identification of leaking items such as valves or pumps, maintenance is necessary to repair the equipment to meet satisfactory emission levels. In addition, some recommendations on maintenance repairs are included for certain types of equipment. The regulations generally specify a maximum time period following leak identification for commencing repairs. After the repairs begin, a definitive time is allotted for completing repairs and placing the equipment on stream. The maintenance repair schedule varies somewhat with the various regulations, but a repair delay is granted if maintenance requires a process shutdown.

Although, specific monitoring action levels that define leaking equipment vary, the monitoring procedures, including monitor types, general calibration, and methods of measuring emissions concentrations are standard. This monitoring standard is referred to in all regulations as Method 21—Determination of Volatile Organic Compound Leaks and is in Appendix A—Test Methods Code Federal Regulations Title 40, Part 60. Monitoring details are presented in Chapter 5. The general requirements for measuring equipment leaks are outlined as follows:

1. *Principle.* A portable instrument is used to detect VOC or VHAP from sources identified in the regulation. The measured leak concentration accuracy is based upon calibration of the instrument with a specific reference compound.

2. *Leak Definition.* The concentration of VOC/VHAP specified by regulation at a surface or opening indicates that a leak exists. The leak definition of NSPS equipment regulations is nominally $\geq$ 10,000 ppmv with NESHAP and HON leak definitions of either $\geq$ 500 ppmv or $\geq$ 1000 ppmv, based upon the equipment type. Leak concentrations are measured with portable instruments acceptable to the EPA that are referenced to leak concentration levels measured with a methane calibrated OVA® instrument (see Chapter 5). Since the leak concentration definition level for various leak sources differs in the various regulations, the leak definitions must be understood by operating personnel.

3. *Response Factor.* This adjustment factor is the ratio of the known concentration of a VOC/VHAP compound to the observed meter reading based upon the instrument calibration reference gas. Response factors are found in EPA documents or can be determined in the laboratory. More detailed comments on response factors are presented in Chapter 5.

4. *Monitoring Instrument.* The EPA suggests selecting instruments that respond to the VOC/VHAP vapors or compounds processed in the plant, but recommends the use of instruments with catalytic oxidation, flame ionization, infrared absorption, or photoionization detectors. Only instruments approved by the EPA can be used in monitoring and leak detection service. Most of the EPA data were obtained with an OVA® instrument (see Chapter 5).

5. *Calibration and Precision.* Method 21 defines EPA requirements for the instrument including response time. The method also presents the calibration before the instrument is operated each day. In practice, the instrument is frequently calibrated after 4 hours and at the end of the working day to monitor drift.

6. *Valve Leak Measurement.* The typical portable instruments used in leak detection have internal pumps that continuously draw gas through a probe mounted on a flexible hose. For valves, the probe inlet is placed against the interface of the stem and the packing gland follower. The probe tip is moved around the stem periphery at the interface, the leak concentration readings are observed with the maximum concentration located, and the concentration measurement is recorded. If the leak concentration less the background concentration is greater than the leak definition specified in the controlling regulation for valves, the valve is recorded as leaking. The probe is also placed at the interface of the packing gland follower and the follower seat or bonnet, with the circumference concentration measured in a procedure identical to the upper measurement procedure described above. The highest concentration measurement of the upper or lower interfaces is used to determine whether a valve is leaking in accordance with the leak definition.

7. *Pumps and Compressors.* A circumferential detection traverse is required at the interface of the compressor or pump shaft and seal. For rotating shafts, the probe is positioned within 1 cm of the shaft–seal interface. The leak definition for pumps differs considerably from the leak definition for either centrifugal or reciprocating compressors and should be established based on the regulations.

8. *Other Leak Sources.* Method 21 also outlines the detection measurement procedure for measuring leakage in other equipment items listed in the regulations. Applicable regulations must be reviewed to determine the leak definition basis for these various items. Some of these items are flanges, connections, relief valves, process drains, open-ended lines or valves, access door seals, and seal system degassing vents.

The objective of the maintenance requirement is rapid repair of the leaking device or equipment unit following identification through the monitoring procedure. The time requirements for repair are detailed in the regulations and must be observed by chemical plants and refineries to ensure compliance. Detailed records of monitoring and maintenance data are required to permit EPA or state review of plant compliance and the effectiveness of fugitive emission control regulations.

General maintenance schedule and procedure requirements are shown in an example of valve maintenance procedures outlined in the regulations:

1. Following leak detection, the valve must be repaired within 15 days.
2. Initial repair shall commence within 5 calendar days after leak detection.
3. Types of repairs initially recommended are described. Repair procedures are presented in Chapter 7.
4. The valve is monitored after repair completion to ensure emission standards are observed and repairs continue until the valve is in compliance
5. Valve maintenance can be deferred until a unit turnaround if maintenance requires a process shutdown and other conditions are met.
6. Repair and monitoring results must be permanently recorded.

Since there are a large number of valves in a process plant, record keeping of valve and equipment LDAR status becomes a significant undertaking. Moreover, all of the valves and equipment must be tagged *or* identified for tracking and regulatory inspection, ensuring a low average contaminant concentration in the workplace.

4.5 NSPS EQUIPMENT CONTROL REGULATIONS

Section 4.4 presented an outline of monitoring procedures along with *general* maintenance requirements that control fugitive VOC and some associated HAP emissions *from* process plants. A more complete review of the NSPS regulations demonstrates the scope of regulatory authority that controls VOC fugitive emissions and significantly reduces exposures in the workplace. These regulations are applicable to both petroleum refineries and chemical plants. Although the apparent thrust of the regulations is targeted at new processes in the NSPS, older processing units are also affected through the modification and reconstruction requirements of the law. Many of the older facilities must now comply with the NSPS regulations, and eventually all process facilities will be controlled in accordance with the provisions of the NSPS regulations.

When hazardous air pollutants are present in the chemical plants, the NSPS regulations are superseded by the HAP regulations (termed Hazardous Organic National Emission Standards for Hazardous Air Pollutants) for those HAP compounds covered in the regulations. However, the NSPS regulations are still applicable in processing units where regulated HAP materials are not present or are present in very low concentrations. A review of the NSPS regulations provides a basis for understanding the changes in controls that are evolving with the HON regulations and the adjustments necessary to upgrade from NSPS to HON standards. As the HAP list expands from 189 hazardous materials and other facilities such as petroleum refineries must comply with HON, along with an increased need to further reduce VOCs to meet NAAQS targets, most process units will be upgraded to the more restrictive HON standards.

Since workplace exposures are effectively reduced through compliance with these regulations, an examination of the regulations identifies potential leak sources and associated methods of controlling exposures. However, the VOC emission NSPS does not completely control all of the fugitive emission sources, and workplace exposures may still exceed desirable levels owing to high leak definition levels. Observance of regulatory requirements does not insure elimination of overexposure, since the regulations do not completely eliminate leakage. However, these control regulations significantly reduce concentration levels and potential exposures. The need for additional controls must be evaluated when actual workplace concentrations are determined.

4.5.1 Valves in Gas / Vapor and Light Liquid Service

Valves must be monitored monthly in the base NSPS regulations and an instrument reading of 10,000 ppmv or greater on the portable meter identifies a leak. When the leak is detected, an initial repair attempt must be conducted within 5 days with repair completion within 15 days after leak detection. Repair is not required if a complete repair requires a process shutdown, and in this situation the repair is deferred until a unit shutdown or turnaround. Monitoring results and repair progress for a leaking valve must be recorded in accordance with the regulatory requirements in the NSPS.

Monthly monitoring of valves that do not exhibit leakage for two successive months can be replaced with a quarterly monitoring program until leaks are detected. Alternative LDAR programs have also been provided in the regulations for plant owners operating very low valve leakage units. Where the number of leaking valves in gas/vapor and light liquid service is less than 2.0% of the total number of valves in this service, quarterly leak monitoring requirements may be deferred for several quarters. Compliance with the 2.0% leakage deferral alternatives in the law requires that an owner notify the EPA prior to implementing this procedure.

Since the basic regulation requires a monthly monitoring program, significant savings can be realized in the overall LDAR effort when the fraction of leaking valves is reduced to less than 2.0%. Savings are observed in a reduction in losses and lower maintenance costs as valve leakage is brought under control.

4.5.2 Pumps in Light Liquid Service

Although the EPA estimates that valves contribute significantly to total VOC fugitive emission losses to the atmosphere, leakage from a single pump seal is considerably greater than emissions from a single valve. Emissions from a large number of valves are dispersed over a considerable area, but pump emissions are concentrated in few locations. Thus higher leak rates from pump seals are a potential source of workplace exposure. However, pump leakage is also controlled under NSPS regulation through an LDAR program.

Pump seals must be monitored monthly, and an instrument reading of 10,000 ppmv or greater on the portable meter identifies a leak. An initial repair attempt must begin within 5 days and repairs must be completed within 15 days after identification of a leak. In addition to instrument monitoring, a weekly visual inspection of the pump seal must be conducted for indication of liquid leakage or dripping.

Pumps equipped with dual mechanical seals that include a barrier fluid and meet specific requirements are exempt from the monthly monitoring program, although the weekly visual inspection program is required. The NSPS regulations detail the specific mechanical requirements for the dual seal systems and are reviewed in Chapter 10. One of the major requirements is a sensor system that indicates a failure in the seal system. Specific sensors are not specified in the regulation, but the plant owner must provide a reliable sensor. Leaks are identified by the sensors, visual observation, or monitoring.

Sealless pumps which do not have shafts extending through the pump housing are exempt from the monitoring program. In addition, a pump with a monitoring measurement less than 500 ppmv may be exempt from monitoring by identifying the pump and notifying the EPA, but visual inspection is still required.

4.5.3 Compressors

The NSPS regulation specifically requires a seal system with a barrier fluid that prevents leakage of the gas or vapor to the atmosphere. The barrier fluid cannot be a light liquid or a gaseous VOC material, and the barrier fluid system must be designed in accordance with one of three recommended procedures. In any barrier seal system design, a sensor is required that indicates a failure in the seal system, barrier fluid system, or both.

When a leak is detected, the initial repair must begin within 5 days after the leak is identified and the repairs completed within 15 days after leak detection. Repairs may be delayed if they require a unit shutdown, but the compressor must be repaired during the next unit shutdown. The regulation does not require monitoring of the seal, since the sealing system has sensors. The specific regulatory equipment design requirements are not mandatory if seal monitoring indicates a concentration less than 500 ppmv above the background level. This particular situation results in a special exemption, but the EPA must be notified of the compressor emissions status. Monitoring is required on a special basis by the EPA for this situation.

The same requirements previously outlined are also applicable to reciprocating compressors. If the compressor has monitoring concentrations above 500 ppmv, a special exemption cannot be obtained, and a control system is required. For many reciprocating compressors, a seal barrier system is not a completely effective emission control solution for reciprocating piston connecting rods. The EPA in various background documentation suggests enclosing or sealing the distance piece and purging the sealed section to a VOC destruction device. The distance piece separates the gas compression end of the compressor from the driver main frame section. By installing a closed vent system, the compressor is exempt from the barrier fluid requirements. Where existing compressors do not have distance pieces and the owner must recast a section of the machine to effectively meet the VOC control requirements, the EPA permits an exception to the regulations.

Compressors in hydrogen service are exempt from regulatory requirements.

4.5.4 Pressure Relief Devices in Gas / Vapor Service

The regulations require that each pressure relief device be operated with no detectable emissions, which is defined as an instrument reading of less than

500 ppmv above the background level. The measurement is obtained at the center of the exhaust end of the relief valve vapor discharge line. Following a pressure release, the exhaust end or pressure relief devise is monitored within 5 calendar days to insure the VOC concentration does not exceed 500 ppm above the background level. If it does, replacement of the valve to meet the regulatory requirement may be necessary.

Valves are exempt from the regulations if they are connected through a closed venting system to a control device. The EPA indicates a control device can be a flare, incinerator, absorbers, and so on designed to reduce VOC emissions to the atmosphere by a 95% or greater efficiency. Trace leakage from a pressure relief device before or following a release does not present any emission problems with a closed vent system and control device.

4.5.5 Sampling Systems

Manual sampling of process streams was a noticeable source of controlled emissions during EPA studies of VOC emissions. However, emissions were periodic rather than continuous, reducing the average or hourly emissions rate to a relatively low level. The manual sampling connection was generally a short pipe length with an open end, with the other end directly connected to a vessel or process line, which permitted drainage of fluid or vapor into a container through a valve in the sampling line. The operator or technician normally placed an open container below the valve and then flushed material through the valve into the container, which effectively purged the line with fresh process fluid and VOC vapor to the atmosphere. Opening a line directly to the atmosphere for flushing or purging was also observed. Following the purge, a container was placed beneath the line opening to obtain a sample. In both situations, a large quantity of VOC process fluid was discharged directly to the atmosphere. Another sampling procedure consists of a bomb or closed container that is connected to the sample line. A quantity of process fluid is then purged to the atmosphere from a valve at the other end of the container. The bomb discharge valve is then closed, and the container is filled with fluid. An inlet valve on the bomb is closed, and the bomb is removed for analysis. Purging this container to the atmosphere is also a source of VOC emissions.

The NSPS requires the installation of a closed-purge or closed-vent system to eliminate purging the system directly to the atmosphere. The type of system preferred by the EPA is described in EPA Background Documents and generally consists of a circulating loop with a section that permits insertion of a closed-sample container. The inlet side of the loop is connected to the desired sampling location with the downstream side attached to a lower-pressure location, permitting fluid flow through the sampler to a receiver. Valves permit flow shutoff and removal of the sampling container.

4.5.6 Open-Ended Valves or Lines

The NSPS regulations require that all open-ended valves or lines be equipped with a cap, blind flange, plug, or second valve. These alternatives effectively seal the valve located on a line in contact with the process fluid. When two valves are installed to seal the system and process fluid flows through the valves, the regulations require closing the valve on the process side before closing the second valve.

The only exception to these requirements is a double block and bleed installation, which is normally installed in the process industries as a safety device to prevent leakage of fluid in either direction. The bleed valve is normally a small valve connected through a tee to the large line between the two block valves. During normal operation, the bleed valve is opened as a vent to prevent fluid buildup between the block valves and leakage of fluid from the high-pressure valve to the low-pressure valve fluid. When the system is placed in operation, the bleed valve is closed and the two block valves are opened.

4.5.7 Light and Heavy Liquid Service

Regulatory standards covering certain equipment in light and heavy liquid service were also included in the NSPS regulations. The definitions of light and heavy liquids were previously given. The specific equipment and the services covered in this section of the regulations are:

1. Pumps and valves in heavy liquid service
2. Pressure relief devices in light and heavy liquid service
3. Flanges and connectors in light and heavy liquid service

When a leak is observed visually, audibly, or through odor from equipment, the specific equipment item must be monitored with a portable monitor. If an instrument reading of 10,000 ppmv or greater is shown on the meter, a leak is identified. The first repair attempt is required within 5 calendar days after detection and must be completed within 15 calendar days after detection as described for other equipment.

4.5.8 Closed-Vent Systems and Control Devices

The closed transport systems have been separated into the closed-vent system that is a closed line between the emissions point and the control device. The latter destroys the transported VOC or efficiently reduces atmospheric emissions through a recovery system.

The closed-vent system consists of piping and other connections that transport the VOC emissions to the control device. As an example, the closed vent system installed on a pressure-relief device would extend from the relief valve exhaust nozzle to a common header and thence to an incinerator inlet. The vent system is monitored annually and must not have an emissions concentration greater than 500 ppm above the background level. In addition, the system must be in operation whenever there is a possibility that emissions may be vented.

The control devices required by the EPA either destroy the VOC emissions in a combustion device or remove the VOC from the carrier gas at an efficiency greater than 95%. Combustion devices must have a minimum destruction efficiency of 95% with a minimum residence time of 0.75 sec and a minimum temperature of 816°C for enclosed combustion systems. The general control device systems that follow are acceptable to the EPA providing they meet the operating standards just described:

1. Vapor recovery adsorption systems such as activated carbon
2. Chilled vapor recovery
3. Solvent absorption systems
4. Flare destruction
5. Enclosed combustion devices such as incineration

Flare destruction is acceptable to the EPA providing specific operating criteria are observed by the plant operator to ensure meeting the 95% or greater destruction efficiency. The operating criteria cover stream heat content (Btu/scf or MJ/scm), flare stack exit velocity, and the adjustments necessary to these criteria based on the type of flare. The latter refers to steam-, air-, or nonassisted flares. Moreover, flares must have a pilot flame at all times with a sensor indicating when the pilot is out, and the flare must be operated without visable emissions.

4.6 NESHAP

Under Section 112 of the Clean Air Act, the U.S. Environmental Protection Agency is authorized to issue emission standards for specific pollutants that are considered extremely hazardous to human health. Following a listing or publication of a substance that is termed a "hazardous air pollutant," the EPA reviews the scientific basis, establishing the pollutant as a substance that may increase mortality or serious illnesses through air pollution. The pollutants listed under Section 112 of the Clean Air Act are not controlled through ambient air quality standards in the community. Pollutants under Section 112 are normally controlled through equipment restrictions and monitoring controls on new and existing plant sources specified in regulations termed National Standards for Hazardous Air Pollutants (NESHAP). These emission control regulations presently cover asbestos, benzene, beryllium, mercury, vinyl chloride, inorganic arsenic, and radionuclides.

With passage of the CAAA, the list of seven pollutants was expanded to 189, as shown in Appendix 4A. New control regulations will be issued covering most of the organic pollutants, as shown in Appendix 4B. Benzene and vinyl chloride, which were regulated under the NESHAP control regulations, are included in the VHAP list, and if the new regulations required are more stringent than the NESHAP regulations, they will determine the required emission controls. Currently the new CAAA regulations are termed HON and are considered MACT. The HON regulations were proposed in late 1992[31] and will be issued as final regulations after comment and review in early 1994. Since the equipment control section of the regulations was a joint effort by the chemical industry and the EPA, the equipment control requirements will not change.

The differences between the NSPS and benzene NESHAP are presented in this section, which is followed by a review of the MACT regulations that are applicable to not only benzene and vinyl chloride but numerous other chemicals. The fugitive emission controls required in these regulations significantly reduce emissions and potential workplace exposures.

4.6.1 Benzene NESHAP

The benzene NESHAP and NSPS regulations are compared to demonstrate differences in the regulations and emissions control requirements. In general, the emission control regulations are similar, indicating that the NESHAP regulations covering benzene are essentially applied in the petroleum refining and chemical industries through NSPS regulations. As a result, workplace contaminant concentrations and exposures are controlled through both regulatory standards. The benzene and VHAP emission control requirements are considerably more restrictive in the new HON regulations. Where HON regulations do not apply, the NSPS and benzene NESHAP provide emission controls.

The benzene NESHAP is applicable where equipment contains or contacts a liquid or gas having a benzene concentration of 10% or more by weight. Benzene is considered a VHAP in regulatory terminology and is included in the HON regulations. For concentrations below 10%, the NSPS regulations must be reviewed along with state regulatory requirements. The basis for NESHAP regulatory emissions control of benzene vapors is a monitoring and maintenance program similar to the program required by the NSPS and described in Section 4.4.3.

4.6.2 Valves

In the benzene NESHAP, valves are monitored monthly, which is identical to the NSPS requirement, and leaks are defined as $\geq$ 10,000 ppm in both sets of regulations. The monitoring frequency for valves in either regulation can be reduced to quarterly inspections if a valve does not leak for two successive months. The NESHAP monitoring frequency can be further reduced if valve leak frequency is less that 2%, which is also similar to the NSPS regulation.

The major difference in regulations is that the valve program in the NESHAP is applicable to any stream containing more than 10% benzene, in contrast to the NSPS, which applies only to gas and light liquid service.

4.6.3 Pumps

For both the NESHAP and NSPS regulations, pumps are monitored monthly, and maintenance or repair requirements are identical. Dual-seal system requirements and exemptions are also similar. The only difference is that the NESHAP monitoring requirement is applicable to all pumps in liquid service containing 10% or more benzene; the NSPS regulations are applicable only to pumps in light liquid service as defined in the regulations.

4.6.4 Compressors

The NESHAP and NSPS regulations are essentially identical and do not require monitoring at specific intervals. Leaks are normally indicated in the sealing system by sensors installed in accordance with the regulations. Reciprocationg requirements have the same regulatory emission restrictions as centrifugal compressors in both sets of regulations. However, the NESHAP does not exempt close-coupled reciprocating compressors from the regulatory restrictions.

4.6.5 Pressure-Relief Devices in Gas / Vapor Service

Requirements in NESHAP and NSPS regulations are identical for pressure-relief valves, with each regulation specifying a 500-ppm vent concentration for valves that release to the atmosphere.

4.6.6 Sampling Systems

Both NESHAP and NSPS regulations are identical for the manual sampling systems covered by this regulatory restriction. Details of the regulation are presented in Section 4.5.5.

4.6.7 Open-Ended Valves or Lines

The NSPS and NESHAP regulations are identical for controlling leakage from open-ended valves or lines. Comments on the regulations are presented in Section 4.5.6.

4.6.8 Liquid Service

The NESHAP regulation refers only to pressure-relief devices, flanges, and other connectors. The NSPS regulations cover pumps and valves in heavy liquid service, pressure-relief devices, including flanges, and other connectors in light and heavy liquid service. The monitoring requirements for those equipment items in NSPS service are reviewed in Section 4.5.7 and are identical for the NESHAP regulation.

Although a distinction between light and heavy service is specified in the NSPS regulations, this difference is omitted in the NESHAP since the only service criteria is a benzene concentration greater than 10% for NESHAP compliance.

4.6.9 Closed-Vent Systems and Control Devices

The NESHAP and NSPS regulations are identical for closed-vent systems and control devices. Details of the regulatory requirements are presented in section 4.5.8.

4.6.10 Product Accumulator Vessels

This requirement in the NESHAP regulation is not included in the NSPS regulations. The NESHAP requires venting of the vessel through a closed system to a control device. Atmospheric discharge of vapors from this vessel is not permitted.

4.6.11 Vinyl Chloride NESHAP

This set of regulations differ to some extent from the control procedures in the benzene NESHAP. The benzene NESHAP is only concerned with specific equipment emissions. Additional equipment sources and some operational procedures are included in the vinyl chloride NESHAP. Also certain features of the equipment control systems differ. A more detailed review is necessary to adequately address the differences. However, the major criteria for leak definition of valves and flanges ($\geq$ 10,000 ppm) is identical to the leak definition for the benzene NESHAP.

Since vinyl chloride is on the volatile hazardous air pollutant list,[27] emission control requirements in the HON regulations will supersede most of the vinyl chloride regulations. Some specific regulatory rules only pertaining to certain vinyl chloride functions in the NESHAP will probably be required along with the HON regulations. Consequently, an understanding of both sets of regulations is necessary to define the complete set of emission controls for vinyl chloride or polyvinyl chloride plants.

4.6.12 Other NESHAP Regulations

The benzene NESHAP regulations described in this section are equipment emissions control regulations for valves, pumps, flanges, and so on that are applicable where this equipment is installed. Also, this equipment regulation is not specific to benzene but may be applied to any chemical specified as a NESHAP material unless a chemical NESHAP has special requirements.

Other benzene NESHAP regulations cover wastewater and drain/sewer systems, transfer operations (tanktruck, railcar, barge loading), and tankage. While this subsection has focused on equipment emissions, other chapters review emission control regulations and requirements on transfer operations, wastewater/drain systems, and tankage.

4.7 HAZARDOUS AIR POLLUTANT REGULATIONS

The CAAA amended Section 112 of the original Clean Air Act which covered hazardous air pollutants. This new section (Section 301) expanded the number of HAP materials significantly and listed all of the chemicals and compounds considered hazardous air pollutants (see Appendix 4A). The CAAA also required promulgation of emission standards that require the "maximum degree of reduction in emissions of the hazardous air pollutants in this section."[4] The hazardous air pollutant list includes benzene and vinyl chloride, which become subject to the regulations issued in accordance with the CAAA.

In developing the regulations to comply with the revised requirements, the EPA negotiated with chemical industry representatives to establish MACT standards for equipment leaks. A joint EPA–chemical industry committee developed a set of standards to significantly reduce leakage and leak frequency which are available in reference.[27] These standards were issued as proposed regulations in December 1992[31] and will become final in early 1994. Changes in equipment control technology are not anticipated owing to the collaboration of industry and the EPA. The regulations are termed HON, as described previously, and basically include MACT.

The proposed standards have been published and include other regulated areas in addition to the equipment control regulations in the HON document. These other regulations are basically systems-oriented covering drains and wastewater streams, wastewater treating steps, transfer operations, tankage, and venting. Requirements are presented in the chapters covering these topics.

4.7.1 General

The overall approach to emission control of VHAPs in the HON regulations is different than control requirements in the NSPS and NESHAP regulations. The latter regulations had a leak definition $\geq$ 10,000 ppmv for valves and pumps with prescribed monitoring and maintenance requirements. Monitoring and repair are termed leak detection and repair (LDAR) which was investigated by the EPA in the early 1980s. Based upon the 10,000-ppmv leak definition and controlled repair, the EPA estimated emissions would be reduced 60–70%, and leak frequencies would average about 5%. Leak frequency is the number of identified leakers in an overall population of valves, pumps, and so on. Emission data accumulated in chemical plants during the late 1980s indicated that significantly lower leak frequencies and lower emission rates can be achieved.[26] The HON regulations are designed to achieve very low leak frequencies and low emission rates.

The regulations covering pumps and valves are similar in concept, requiring standard implementation in three phases to reduce emissions and leak frequencies. In addition, quality improvement programs (QIP) have been included to upgrade equipment and improve compliance with the regulations. These regulations have established a standard performance level in the third phase that is expected from all chemical plants affected by the HON regulations. Incentives for further improvements are included that lengthen monitoring periods and reduce LDAR costs.

In those situations where standard performance levels are not achieved, plants must enter a QIP program for pumps and either a QIP or monthly LDAR program for valves. The purpose of the QIP is to enable plants to achieve the standard performance level and improve emissions control. As described in this section, the QIP consists of information gathering, the identification of superior performing technologies, and the replacement of poorly performing equipment with superior technology until the standard performance level is achieved.

With the exception of connectors (flanges, etc.), most of the other equipment standards are similar to the other regulations. The HON standards and regulations are outlined in the following sections.

4.7.2 Pumps in Light Liquid Service

There are three phases in the standard for *existing process units* with Phase I beginning on the specified applicability date. Phase II begins 1 year after the applicability data and Phase III begins 2 1/2 years after the applicability date. For *new process units*, Phase II begins after startup and Phase III begins 1 year after startup.

1. *Monitoring.* Each pump is monitored monthly to detect leaks in accordance with Method 21. Since lower concentrations are required in each successive phase, the concentration levels must be measured accurately, requiring instrument calibration gas concentrations carefully defined for each phase. In addition, instru-

ment response factors are a major concern at the lower concentrations, requiring careful investigation and documentation. For characterization as non-VHAP service, the stream of concern must have a VHAP content that is less than 5% by weight. Leak definitions for the various phases are:

Phase I	≥ 10,000 ppmv
Phase II	≥ 5,000 ppmv
Phase III	≥ 1,000 ppmv
	≥ 5,000 ppmv for pumps in polymerizing monomer service
	≥ 2,000 ppmv for pumps in food or medical service

Each pump is inspected visually each week for indications of liquids dripping from the pump seal. If any indications of leaking liquid are present, the pump is classified as a leaking pump.

2. *Repair.* When a leak is detected, the first attempt at repair is begun no later than 5 calendar days after leak detection. The leak shall be repaired as soon as practicable, but no later than 15 calendar days after leak detection. For pumps where a 1000-ppmv leak definition is applicable, repairs are not required until a monitoring reading of 2000 ppmv or greater is detected.

3. *Leaking Pump Percentage.* During the first monitoring period, the plant owner or operator will determine whether the leaking pump percentage is determined on a process unit or plantwide basis. All subsequent calculations are completed on the same basis.

In Phase III, the leaking pump percentage is calculated on a 6-month rolling average. If the leaking percentage is greater than either 10% of the pumps (on the basis previously selected) or more than three pumps on the same basis, a QIP for pumps is instituted. Details of the QIP program are presented in Chapter 10.

4. *Dual Mechanical Seals.* Pumps equipped with dual mechanical seals and a barrier fluid system are exempt from the monthly monitoring program, although weekly visual inspection for indications of liquids dripping from pump seals is required. While dual pump seals are not monitored, HON regulations have specific mechanical requirements for dual seal systems. One of the major requirements is a sensor that detects failure of the seal system, the barrier fluid system, or both.

When there are indications of liquid dripping from the seal, the pump is monitored to determine the presence of VHAP in the barrier fluid. If an instrument reading of 1000 ppm or greater is measured, the pump is classified as a leaking pump. For leaks, the repair requirements duplicate the repair schedule outlined in item 2.

5. *Sealless Pumps.* Any pump designed without a shaft penetrating the pump housing is exempt from monitoring.

6. *Closed-Vent System.* Any pump with a closed-vent system capable of capturing and transporting any leakage from the sealing system to a control device is exempt from monitoring.

7. *Remote Pumps.* Any pump located within the boundary of an unmanned plant site is exempt from weekly visual inspections and daily seal sensor alarm inspections provided the pump is visually inspected as often as practicable and at least once per month.

4.7.3 Valves in Gas / Vapor and Light Liquid Service

There are three phases in the standard for *existing process units*, with Phase I commencing on the specified applicability date. Phase II begins 1 year after the applicability date and Phase III begins 2 1/2 years after the applicability date. For *new process units*, Phase II begins after startup and Phase III begins 1 year after startup.

1. *Monitoring.* All valves are monitored on a schedule with the exception of unsafe- and difficult-to-monitor valves. These valve definitions are explained in item 4. In contrast to pump seal monitoring requirements, valve monitoring differs considerably, and monitoring requirements are outlined herein.

Leaks are detected in accordance with Method 21. Since lower concentrations are required in successive phases, the concentration levels must be measured accurately, requiring instrument calibration gas concentrations carefully defined for each phase. In addition, instrument response factors are a major concern at the lower concentrations, requiring careful investigation and documentation. For characterization as non-VHAP service, the stream of concern must have a VHAP content that is less than 5% by weight. Leak definitions for the various phases are:

Phase I	$\geq$ 10,000 ppmv
Phase II	$\geq$ 500 ppmv
Phase III	$\geq$ 500 ppmv

While Phases II and III have identical leak definitions, the phases have differing monitoring requirements, as shown below:

Phase I	Monitored quarterly
Phase II	Monitored quarterly
Phase III	Leaking valves $\geq$ 2% of total valves
	Monitor each valve once a month or 1 year after applicability implement QIP program and monitor quarterly
	Leaking valves $<$ 2% of total valves
	Monitor each valve quarterly (if $>$ 1%)
	Leaking valves $<$ 1%
	Monitor each valve once every 2 quarters
	Leaking valves $<$ 0.5%
	Monitor each valve once every 4 quarters

2. *Repair.* When a leak is detected, the first attempt at repair is begun no later than 5 calendar days after leak detection. The leak shall be repaired as soon as practicable, but no later than 15 calendar days after leak detection. When a leak is repaired, the valve shall be monitored at least once within the first 3 months after repair.

3. *Leaking Valve Percentage.* The percent leaking is calculated as a rolling average of two consecutive periods for monthly, quarterly, or semiannual monitoring programs. Nonrepairable valves are included in leaking valve percentage

calculations the first time a valve is identified as leaking and nonrepairable. Those valves that are nonrepairable from a previous period to a maximum of 1% of the total VHAP valves may be excluded from the calculations for subsequent monitoring periods. If the number of nonrepairable valves exceeds 1% of the total VHAP valves, those nonrepairable valves in excess of 1% are included in the leaking valve calculation.

4. *Nonscheduled Monitored Valves.* Unsafe-to-monitor valves are exempt from scheduled monitoring requirements. These are valves considered unsafe when monitoring personnel may be exposed to an immediate danger, such as an extremely high temperature in a very large valve. Monitoring is required when practicable during safe monitoring periods.

Difficult-to-monitor valves are also exempt from scheduled monitoring requirements but must be monitored at least once per year. This terminology is applied to valves where monitoring personnel are elevated more than 2 m above a support surface. However, this classification is limited to existing units with valves designed for accessibility in new units.

4.7.4 Connectors in Gas / Vapor Service and in Light Liquid Service

Connectors cover flanged, threaded, or other joined fittings as described under definitions. One of the most common fittings in process facilities is a flange, and the large number of flanges in a plant result in a significant source of fugitive emissions. These connectors join piping or join a pipe with a piece of equipment. The flange faces in all of these joints are separated by gaskets which are reviewed in Chapter 9.

1. *Monitoring.* All connectors are monitored as described in Method 21. During the initial 12 months after the applicability date, all connectors in existing process units must be monitored with the exception of those connectors that are unsafe to monitor, unsafe to repair, inaccessible, glass, or glass lined. New process unit connectors are monitored within the first 12 months after startup with the exception of those connectors previously described. When an instrument monitor reading $\geq$ 500 ppm above background is measured, a leak is detected. After conducting the initial monitoring program, subsequent monitoring frequencies are determined as follows:

Percent leaking connectors $\geq$ 0.5% during the last annual biennial monitoring period	Monitoring required once per calendar year
Percent leaking connectors $<$ 0.5% in the last monitoring period	Monitoring required once every 2 calendar years (40% may be monitored in the first year and 60% in the second year)
After a 2-year monitoring period and leaking connectors $<$ 0.5%	Monitoring can be reduced to once every 4 years (20% minimum each year until all are monitored in 4 years)

When a process or plant is on a 4-year monitoring interval and has a leaking connector percentage less than 1% but $\geq 0.5\%$, the monitoring frequency is increased to once every 2 years. Annually when $> 1\%$.

Inaccessible glass or glass-lined connectors are exempt from the general monitoring requirements outlined above. An inaccessible connector may be characterized by one or more of the following descriptions:

- Buried
- Covered with insulation preventing monitoring
- Obstructed by equipment or piping preventing monitor access
- Unable to be monitored from a 25-ft portable scaffold on the ground and the connector is > 2 m above a support surface.

2. *Leak Repair.* When a leak is detected, the first attempt at repair shall be conducted within 5 calendar days after the leak is detected. The leak shall be repaired as soon as practicable, but no later than 15 calendar days after the leak is detected. Unsafe-to-repair connectors are exempt from the repair requirement if personnel are potentially exposed to danger during repair or the connector will be repaired during the next scheduled process unit shutdown. Following repair of the connector, the connector is monitored for leaks during the first 3 months after the repaired connector is placed onstream.

When inaccessible, glass, or glass-lined connectors are observed leaking through visual, audible, olfactory, or other methods, the first attempt at repair shall commence within 5 calendar days after the leak is detected. The leak should be repaired as soon as practicable, but no later than 15 days after the leak is detected.

Connector repair delays are permitted in addition to those described under the following circumstances:

- When a repair is not feasible without a process unit shutdown. The equipment must be repaired during the next process unit shutdown.
- The equipment can be isolated from the process operation and can be removed from VHAP service.
- The emissions of purged material resulting from immediate repair are greater than the fugitive emissions that will accumulate during a repair delay and during repair the purged material is collected and destroyed or recovered.

4.7.5 Compressors

1. *Seal Requirements.* A seal system is required with a barrier fluid system that prevents leakage of process gas or vapor to the atmosphere. The barrier fluid cannot be in light liquid service, and the barrier fluid system must be designed in accordance with one of three recommended procedures. In addition, a reliable sensor is required to indicate failure of the seal system, barrier fluid system, or both.

2. *Monitoring.* The regulation does not specifically require monitoring since the sealing system has a failure alarm system. The specific seal requirements in this subsection are not mandatory if seal monitoring indicates a concentration less than

500 ppmv above background levels. Where the compressor monitoring levels are less than 500 ppmv, a special exemption is permitted, but the EPA must be notified of the compressor emissions status. For this status condition, monitoring is required at the specific request of the EPA.

The same requirements outlined in the previous paragraph are also applicable to reciprocating compressors. For most reciprocating compressors, the type of barrier fluid seal system described above and normally installed in centrifugal compressors is not a practical solution for reciprocating compressor piston connecting rods. Where the distance piece surrounding the connecting rod can be sealed and leakage purged to a control device, the compressor is exempt from barrier fluid sealing system requirements. These systems have monitoring concentrations less than 500 ppmv above background and with the exemption may only have periodic EPA requests for monitoring.

3. *Repair.* If the sensor in a barrier fluid sealing system indicates a failure in the seal system, barrier fluid system, or both, the seal is considered leaking. A first repair attempt must be conducted within 5 calendar days after detection and the repair completed no later than 15 days after leak detection. However, a repair delay is permitted if the repair is not feasible without a process unit shutdown. For processes with single compressors as an example, repair of the equipment can be deferred until the next shutdown.

4.7.6 Pressure-Relief Devices in Gas / Vapor Service

The regulations require that each pressure-relief device in gas/vapor service is operated with an instrument reading of less than 500 ppm above background. For relief devices equipped with an enclosed extension or horn, the monitoring probe inlet is placed at the center of the exhaust end. Where these extensions are not installed, the monitoring probe samples at the sealing seat interfere.

After each pressure release, the exhaust outlet of the extension section is monitored to insure that any emissions concentration is less than 500 ppmv above background level within 5 calendar days. If a 500-ppmv reading is not obtained within 5 calendar days after the pressure release, the relief valve may require replacement or repair to meet the 500-ppmv monitoring constraint.

Relief valves are exempt from the regulations if they are connected through a closed-vent system to a control device. The EPA considers flares, incinerators, absorbers, and other systems control devices that must operate with an efficiency of 95% or better. Thus VHAP emissions to the atmosphere must be reduced by 95% or more. Trace leakage through a pressure-relief device does not present any emission control problems before or after a pressure release with a closed-vent system connected to a control device.

The closed-vent system, which in most instances is a piping system, must be monitored annually or at EPA request. The emissions monitoring reading should be less than 500 ppm for the closed-vent system.

4.7.7 Sampling Systems

The VHAP regulations are similar to those described in Section 4.5.5 covering NSPS regulations for sampling systems. The HON regulations for a manual sampling system require a closed-purge, closed-loop, or closed-vent system. These systems are based upon one of the following requirements:

1. The purged process fluid in the sampling loop is returned directly to the process line.
2. The purged process fluid for cleaning a line is collected and recycled to the process.
3. All of the purged fluid is collected and sent to a control device.

The MACT regulations do not specify actual sample container requirements, but the sample container should be closed. This controls emissions of VHAP and also minimizes exposure during the manual sampling operation. Sampling is reviewed more extensively in Chapter 11.

In situ sampling systems are exempt at this time from these requirements. This terminology applies to automated sampling systems that withdraw samples for analysis such as chromatographs, spectrometers, and infrared systems. Sample or vent disposal for these general systems is reviewed in Chapter 11.

4.7.8 Open-Ended Valves or Lines

The HON regulations for VHAP are similar to those in the NSPS regulations described in Section 4.5.6. The only difference is in the double block and bleed system. The basic regulations require that each open-ended valve or line shall be equipped with a cap, blind flange, plug, or a second valve. With a double block and bleed system, the bleed valve is opened when the main blocks are closed. For flow-through operations, the bleed valve is closed and the two main block valves are opened. However, there is now process fluid on the process side of the bleed valve with one side of the valve opened to the atmosphere. During operations when the bleed valve is closed, the open end side of the valve must be capped, plugged, or fitted with a second valve.

4.7.9 Light and Heavy Liquid Service

Regulatory standards defining light and heavy liquid service were previously described along with criteria for VHAP service. These regulations cover liquid in VHAP service with VOC service regulations presented in Section 4.5.7. The specific equipment and liquid service covered in this section of the regulations are:

1. Pumps, valves, connectors, and agitators in heavy liquid service
2. Instrumentation systems in light and heavy liquid service
3. Pressure relief devices in light and heavy liquid service

When evidence of a potential leak is found by visual, audible, olfactory, or any other method, the leak shall be monitored within 5 calendar days after leak detection. If an instrument reading above background $\geq$ 10,000 ppmv for agitators, $\geq$ 1,000 ppmv for pumps, or $\geq$ 500 ppmv for valves, connectors instrumentation systems, or pressure relief devices is shown, a leak is detected. The first attempt at leak repair should commence no later than 5 calendar days after the leak is detected, and repair should be completed within 15 calendar days. Some repair delay is permitted under certain conditions.

Instrument systems with potential leaks are repaired to ensure satisfactory and accurate instrument performance. Should a liquid leak be identified, a repair is acceptable where visual, audible, olfactory. or other evidence of a leak have been eliminated. In addition, leak tests are acceptable where application of a soap solution does not produce bubbles at numerous potential leak sites or the instrument system will hold a test pressure.

4.7.10 Agitators in Gas / Vapor Service and in Light Liquid Service

The regulation does not indicate whether agitators are in a particular location or configuration in the application. Requirements are applicable to all agitators.

1. *Monitoring.* Each agitator is monitored monthly to detect leaks. When monitoring instrument readings are $\geq$ 10,000 ppmv, the agitator is leaking. In addition, each agitator is visually observed each week for indications of liquid leakage. The presence of liquid characterizes the agitator as leaking.

When an agitator is equipped with a closed-vent system that captures any leakage from the seals and transports the leakage to a control device, the agitator is exempt from monitoring requirements.

2. *Leak Repair.* Following leak detection, the first attempt at a repair must begin within 5 calendar days after leak detection and shall be completed within 15 calendar days after leak detection. Repair delays will be permitted in accordance with the delays outlined in Section 4.7.4 (item 2).

4.7.11 Closed-Vent Systems and Control Devices

These regulations are essentially identical to the NSPS regulatory requirements for closed-vent systems and control devices outlined in Section 4.5.8. Some additional requirements are placed on control devices in various applications in the HON regulations.

4.7.12 Product Accumulator Vessels

All product accumulator vessels defined in this chapter shall be equipped with a closed-vent system capable of capturing and transporting any leakage from the vessel to a control device.

4.7.13 Valve Quality Improvement Program

The valve leak definition decreases from 10,000 ppmv in Phase I to 500 ppmv in Phase III as described in Section 4.7.3. In addition, monitoring requirements are a function of the valve leaking percentage in Phase III which is implemented $1\text{–}2\frac{1}{2}$ years after the applicability date, depending on whether the process unit is existing or new. A valve leaking percentage $> 2\%$ requires monitoring once per month or a QIP, which is instituted within the first year after the Phase III monitoring operation begins. Obviously, monthly monitoring is burdensome, and the objective of the QIP program is to reduce the number of leaking valves and gross emissions. Emission reductions of VHAP also reduces exposure in the workplace.

A plant may elect one of two available QIPs to improve valve emission control performance. Requirements of the selected QIP are observed until the process unit has $< 2\%$ leaking valves, calculated as a rolling average of two consecutive quarters. When the process unit has less than 2% leaking valves, the plant may elect to comply with the requirements outlined in Section 4.7.3 or continue to comply with the QIP. Another alternative is compliance with both the QIP and the Phase III requirements. Further progress or performance trials under the specified QIP is no longer required if the valve leaking percentage is $< 2\%$. When the QIP is discontinued by choice and the leaking valve percentage exceeds 2%, the QIP is no longer an option for the process unit. At this point, the process unit must begin monthly monitoring. The two QIP programs are defined as a QIP to demonstrate further progress and a QIP of technology review and improvement.

QIP Demonstrating Further Progress. The basic requirements of Section 4.7.3 are observed with the exception that each valve is monitored quarterly. In addition, data collection and record keeping for each valve are required subject to the QIP and continue while the unit remains in the program. The following general data and information on each valve are collected:

1. Maximum monitoring reading in each monitoring inspection before repair:
   a. Stream response factor
   b. Instrument model number
   c. Observation date
2. Valve installation in gas or light liquid service.
3. For leaks, the repair method and monitoring readings after repair.

The owner or operator *must demonstrate progress* in reducing the leaking valve percentage each quarter as follows:

- Each quarterly demonstration of progress is a reduction of at least 10% in the leaking valve percentage from the preceding monitoring period.
- The percent leaking valves is a rolling average of two consecutive quarters of monitoring data.
- The percent reduction is calculated using the rolling average leaking valve percentage. If the process unit fails to demonstrate a 10% (or greater) reduction per quarter in the leaking valve percentage based on two consecutive rolling averages, or the overall average reduction is less than 10% per quarter, the unit will comply with monitoring (see Section 4.7.3) or the process unit must comply with the QIP of technology review and improvement.

QIP of Technology Review and Improvement. The basic requirements of Section 4.7.3 are observed with the exception that monthly monitoring does not apply. In addition, data collection and record keeping for valves subject to the QIP are required and continue while the unit remains in the program. The data required for each valve include:

1. *Valve Data.* Type (ball, gate, etc.), manufacturer, design basis, materials of construction, packing material, and installation year
2. *Service Characteristics.* Operating pressure, temperature, line diameter, corrosivity, and gas or light liquid service conditions
3. *Monitoring Readings.* Maximum reading each monitoring inspection before repair including:
   a. Stream response factor
   b. Instrument model number
   c. Observation date
4. *Leaks.* The repair method and monitoring readings after repair completion.
5. *Other.* Description of maintenance or QIPs for use in data analysis. When leaking valves are removed from service, the valves will be inspected. Any parts of the valve that failed in service are identified and recorded. Inspections will also recommend design or specifications to reduce potential emission leakage.

The process unit operator or owner will analyze the data outlined above to determine the services, operating or maintenance practices, valve designs, or technologies that result in poor valve emission control performance and identify the criteria that result in better than average emission control performance. Also, the analysis will investigate and identify other factors that may contribute to poor performance. The analysis will also include:

- Identification of superior performing valve technologies applicable to poorly performing valves based on the data. The superior valve technology demonstrates a leak frequency $< 2\%$ for specific applications. A *candidate* with superior performing valve technology is demonstrated or reported in the literature that has low emission performance and is capable of achieving $< 2\%$ leaking valves.
- Information from the literature and other plant sites identifying valve designs or technologies associated with low emission performance.
- Service limitations on valve designs and information on maintenance procedures to ensure continued low emission performance.
- Data from intercompany or intracompany programs.

The first data analysis shall be completed within 18 months after the initiation of Phase III. The initial analysis will have a minimum of two quarters of data. A data analysis is conducted yearly while the process unit is in the QIP.

A trial evaluation program will be conducted when the data analysis does not identify superior performing valve designs or technologies that can be applied to poorly performing valves. The trial program will evaluate the feasibility and performance of valve designs and technologies identified by others that demonstrate low emissions performance. This trial program includes online trials of valves or operating and maintenance practices identified in the literature or in analysis by others that have the ability to control leak rates to levels less than 2% in similar services. The trial program will specify and document the following:

- Candidate superior performing valve designs or technologies for evaluation
- Stages for evaluating candidate performance
- Estimated test period for evaluation
- Monitoring frequency
- Operating conditions
- Assessment of valve design and technology performance

Within 24 months after the start of Phase III, the owner or operator shall have identified valve designs or technologies operating and/or maintenance practices that demonstrate low emission performance. If a superior technology has not been identified, performance trials will continue until a superior performing design or technology has been identified. These trials can also be discontinued if there are no technically feasible candidate superior technologies. In the latter situation, the owner or operator shall prepare an evaluation report that provides the basis for concluding that superior emission control performance is not feasible.

Valve Quality Assurance Program. When a QIP for technology review and improvement is selected, a valve quality assurance program (QA) must be instituted. This quality assurance program covers purchasing specifications and maintenance procedures for all valves in the process unit. This program is reviewed and, where appropriate, updated for all valves in the process unit each year until the unit has less than 2% leaking valves. This program consists of:

- Establishment of minimum design standards for each valve category
- Requirement that all equipment orders must specify the design standard
- A written procedure for bench testing valves
- Monitoring of all valves repaired on line
- Audit procedure of valve purchasing quality
- Detailed valve maintenance and repair procedures

The valve quality assurance program will begin at the start of the third year after Phase III implementation for plant sites with 400 or more valves or owned by a corporation with 100 or more employees. Alternatively, for plant sites with less than 400 valves and owned by a corporation with less than 100 employees, the QA program may begin at the start of the fourth year. Also, in association with this implementation schedule, any valve that is removed will be replaced with a new or modified valve that complies with the QA standards and is characterized as a superior emission performance technology valve. If a superior emission performing technology valve cannot be identified, the valve replacement shall be one of the lowest emission performance technology valves identified for the application.

4.7.14 Pump Quality Improvement

The pump leak definition decreases from 10,000 ppmv in Phase I to 1000 ppmv in Phase III for most pumps. A 5000-ppmv leak definition in Phase III is permitted for pumps in polymerizing monomer service. These criteria are described in Section 4.7.2 covering pumps in light liquid service. In contrast to valve monitoring, pump monitoring is required monthly. Phase III for existing process units begins 2 1/2 years after the applicability date, and for new process units Phase III is instituted 1 year after unit startup.

When the pump leaking percentage is calculated in Phase III, the percentage is determined on a 6-month rolling average basis. If the greater of either 10% of the pumps in a process unit or 3 pumps in the unit leak, the owner or operator must implement a QIP. The leaking pump criteria is the same for all pumps on a plant site or a process unit basis. This general requirement for a QIP with a high percentage of leakers differs from the valve QIP. In the valve regulations, when a large leaking percentage occurs, the owner or operator can either select monthly monitoring or a QIP.

Once the required leaking percentage performance level is achieved, the pump QIP is rescinded and pumps are monitored on a monthly basis. The actual pump QIP is quite similar to the valve QIP and will not be outlined in this subsection. Complete details are found in reference 31. A major concern in pump technology is seal performance, which is specifically identified in the regulations and is a major component in the pump QA program that is part of the overall pump QIP.

4.7.15 Batch Process Emission Limitation Alternative

Since batch processing differs considerably from continuous processing, the HON regulations permit an alternative method of controlling fugitive emissions in batch processing operations. The regulations also recognize that batch plants are frequently reconfigured and may only operate for relatively short time frames. Consequently, the previous regulatory requirements outlined in Sections 4.7.2–4.7.4 and 4.7.9, including the QIPs, are replaced by specific requirements. There are two alternatives for batch processing emission control standards–one is based on pressure testing and the second is a leak monitoring program.

Pressure Testing. In this procedure the batch process equipment is pressure tested with a gas or liquid to determine the presence of leaks, prior to permitting VHAP flow into the equipment. Details of pressurizing are included in the regulations[31] and are briefly outlined in the following paragraphs:

1. *Gas Pressure Testing.* The process equipment train is pressurized with air or nitrogen to the equipment operating pressure, but the pressure does not exceed the lowest relief valve setting. When the required pressure is attained, gas flow is blocked, and the pressure drop over a period of time is observed. The minimum time period is 15 min, and the pressure drop per hour is calculated. If the pressure

loss is greater than 1 psig/hr, or there is evidence of leakage through visual, audible, or olfactory means, a leak is established.

2. *Liquid Pressure Testing.* The process equipment train or section is pressurized with water or alcohol by completely filling the equipment with liquid, and the feed stream is then blocked. Although a pressure level is not specifically mentioned in the requirement, completely filling the equipment and venting any vapor will result in a pressure equivalent to the inlet pressure in the equipment. This pressure should be less than the lowest relief valve setting. The test pressure will be maintained for 60 min (minimum). Any indications of a fluid drip or fluid loss indicates that a leak is established.

3. *Pressure Test Repair.* When a leak is detected, the equipment is repaired and the batch process equipment is retested before VHAP flow to the equipment is initiated. Should the batch process equipment fail two consecutive tests, the repairs should be completed as soon as practicable, but within 30 calendar days after the equipment is in VHAP service.

4. *Reconfiguring Equipment.* After each reconfiguration of the equipment, the batch process equipment train is pressure tested for leaks before placing the unit in VHAP service. When a seal is broken between two equipment items or when equipment is changed in a section, pressure testing is only required in the new or disturbed equipment.

5. *Normal Operation.* Any batch process in VHAP service during a calendar year must be pressure tested at least once during the calendar year.

Leak Monitoring. The method of monitoring equipment is outlined in the HON regulations and Method 21, which are continuous monitoring requirements. Leak monitoring can be conducted when the batch equipment is in VHAP service, during operation with an acceptable surrogate VOC compound, or during operation with a detectable gas or vapor. Compliance with actual requirements of the various equipment subsections under the HON regulations is required. However, the continuous monitoring frequency is modified for batch processing as follows:

1. Following each equipment reconfiguration, the reconfigured equipment is monitored for leaks within 30 days after return to VHAP service.

2. Connectors (flanges, etc.) are monitored in accordance with continuous requirements outlined in Section 4.7.4.

3. Nonconnector equipment is monitored at frequencies specified in the regulation but is a modification of continuous process monitoring frequencies. The batch monitoring frequency is based upon the *proportion of the year* the batch process train operates with VHAP present. As an example, if the continuous monitoring requirement is quarterly, but the batch process operates less than 25% of the annual time period, batch equipment monitoring is changed to annually.

4. Where equipment is in VHAP service less than 2190 hours per year, valves are monitored once per year, and the pumps and agitators are monitored once per quarter. Where leaks are detected, the leak is repaired as soon as practicable, but within 15 calendar days after leak detection.

4.7.16 Chemical Process Safety Management

Under Title III of the CAAA, OSHA must develop and promulgate a chemical process safety standard designed to protect employees from accidental releases of highly hazardous chemicals in the workplace. Since OSHA was working on a chemical process safety standard, a full regulation was rapidly completed and promulgated in February 1992.[5] The CAAA requirements for this process safety standard are carefully defined in the act[4] and have been incorporated into the regulations. The OSHA regulation is included in general OSHA Standard 29CFR 1910. This brief review outlines important regulatory requirements.

1. *Purpose.* The regulation contains requirements for the prevention or mitigation of the consequences of unwanted (accidental) releases of toxic, reactive, flammable, or explosive chemicals into locations that could potentially expose employees and others to serious hazards. This regulation applies to chemicals and quantities in processes at or above the specified quantities in Appendix 4C.

2. *Employee Participation.* Employers must develop written plans on implementation of employee participation in the development and conduct of hazard analysis along with the development of the other elements in this standard. Also, employee access will be provided to process hazard analyses and all information developed for compliance with the regulations.

3. *Process Safety Information.* A complete written compilation of process safety information is developed by the employer. This information is to aid in the identification of potential hazards posed by processes with highly hazardous chemicals. The information included in this compilation follows:

   a. Highly hazardous chemical characteristics:

      Toxicity information

      Permissible exposure limits

      Physical data

      Reactivity data

      Corrosivity data

      Thermal and chemical stability data

      Hazardous effects of inadvertently mixing different chemicals

   b. Process technology:

      Block flow diagram of process

      Process chemistry

      Maximum planned inventory

      Safe upper and lower limits for temperatures, pressures, flows, and compositions

      Evaluation of deviations and consequences of variables

   c. Process equipment:

      Materials of construction

      Piping and instrument diagrams (P & I)

      Electrical classification

Relief system design and basis

Ventilation system design

Design codes and standards

Material and energy balances after May 26, 1992

Safety system interlocks, detection or suppression systems

For codes or standards no longer in use, employer must document it is designed, inspected, operated, maintained, and tested in a safe manner

4. *Process Hazard Analysis.* This is an organized and systematic effort to identify and analyze the significance of potential hazards. The PHA analyzes potential causes and consequences of fires, explosions, releases of toxic and flammable chemicals, and major spills of hazardous chemicals. The PHA focuses on equipment, instrumentation, utilities, human actions (routine and nonroutine) and external factors that may impact the process. Employers shall determine and document the priority order for conducting process hazard analyses which include such considerations as extent of the process hazards, number of potentially affected employees, age of the process unit, and operating history of the unit. The general requirements are as follows:

   a. The initial process hazard analyses shall be conducted as rapidly as possible but not later than the following schedule:

   - A minimum of 50% of the analyses shall be completed by May 26, 1994
   - No less than 50% of the analyses shall be completed by May 26, 1995
   - No less than 75% of the analyses shall be completed by May 26, 1996
   - All analyses shall be completed by May 26, 1997
   - Analyses completed after May 26, 1997 which meet the requirements of this subsection are acceptable. However, they must be updated and revalidated every 5 years based on their completion date

   b. One or more of the following methodologies shall be used:

   What if

   Check list

   What if/checklist

   Hazard and operability study (HAZOP)

   Failure mode and effects analysis (FMEA)

   Fault tree analysis

   Appropriate equivalent methodology

   c. The PHA shall address the following:

   Process hazards

   Identification of previous incident that has potential for catastrophic consequences

   Engineering and administrative controls, interrelationships including detection methods of early release warnings

   Consequences of engineering and administrative control failures

   Facility siting

   Human factors

Qualitative assessment on safety and health of employees resulting from controls failures

d. Employer will establish system to address PHA team findings and recommendations including documentation and communications with employees.

e. Every 5 years after completion of the initial PHA, the analysis shall be updated and revalidated.

f. All documentation is retained through the life of the process.

5. *Operating Procedures*. Written operating procedures will be developed and implemented for the following items:

a. Operations:

Initial startup

Normal operations

Temporary operations

Emergency shutdown

Emergency operations

Normal shutdown

Startup after turnaround or emergency shutdown

b. Operating limits:

Consequences of deviation

Steps to correct or avoid deviation

c. Safety and health considerations:

Properties and hazards of process chemicals

Precautions for preventing exposure

Control measures for physical contact or airborne exposure occurrences

Quality control of raw materials and inventory control

Any special and unique safety systems and their functions including safe work practices

Operating procedures will be readily available to employees. These procedures will be reviewed frequently to insure they reflect current operating practices, any changes in process chemicals, technology, and equipment changes. The employer must certify annually that the operating procedures are current and accurate.

6. *Incident Investigation*. An employer will investigate each incident which resulted in or could have resulted in a catastrophic release of a highly hazardous chemical in the workplace. The incident investigation will be initiated promptly but no later than 48 hr after the incident. An incident team will investigate the problem and prepare a report that includes the factors contributing to the incident and any recommendations. The employer will devise a system to promptly address and resolve the findings and recommendations. Resolutions and corrective actions will be documented. All investigation reports shall be retained for 5 years.

7. *Mechanical Integrity*. Equipment that processes or stores highly hazardous chemicals must be designed, constructed, installed, and maintained to minimize chemical release risks. These objectives require a mechanical integrity program

to assure continued process equipment integrity. Employers must review maintenance programs to determine whether "breakdown" maintenance is the usual practice rather than an ongoing mechanical integrity program.

The primary line of defense requires operating and maintaining the process as designed along with chemical containment. This line is supported by the controlled release of chemicals through vents to flares, scrubbers, surge vessels, or overflow tanks. The secondary line of defense includes fire monitors, sprinklers, deluge systems, dikes, drainage systems, and other systems that control or mitigate hazardous chemicals should an accidental release occur. Emergency preparedness is the third line of defense. A brief summary of mechanical integrity requirements follows:

a. Regulatory requirements apply to:

Pressure vessels and storage tanks

Piping systems including valves and other components

Relief and vent systems along with control devices

Pumps

Controls (including monitoring devices, sensors, alarms, and interlocks)

Emergency shutdown systems

b. Establishment of written procedures to maintain process equipment integrity

c. Process maintenance training including awareness of potential hazards

d. Equipment inspection and testing procedures

e. Deficiency correction of equipment outside acceptable limits

f. Quality assurance of equipment and new plant construction

8. *Hot Work Permits*. For safety and health reasons, a permit procedure is now required that incorporates the fire and protection requirements in other OSHA regulations (28).

9. *Training*. Process operators in existing units and those prior to transfer to a newly assigned unit will be trained in operating procedures and an overview of the process as described in item 5. Emphasis is placed on safety and health hazards, emergency operations including shutdown, and safe work practices. In lieu of this training, an employer can certify in writing that an experienced process operator has the knowledge, skills, and abilities to safely carry out the duties and responsibilities specified in the operating procedures.

Refresher training is provided at least once every 3 years. The employer must ascertain that each employee has received and understood the training program. A record covering each employee includes the methods that verify the employee understands the content of the training program.

10. *Contractors*. Contractor employees performing maintenance, repair, turnaround, major renovation, or specialty work are covered by this subsection. The plant owner will inform contract employees of known potential fire, explosion, or toxic release hazards. In addition, the owner will explain the provisions of the emergency action plan in item 11 and will maintain a log of contract employees injuries and illnesses. The plant owner will periodically evaluate the performance of contract employers in fulfilling the obligations outlined below:

a. Assurance each contract employee is trained in work practices to safely perform their assigned job.

b. Assurance each contract employee is instructed in potential fire, explosion, or toxic release hazards related to their job and the overall process including applicable provisions of the emergency action plan.

c. The contract employer will document that each employee has received and understood the training program.

d. Assurance that each contract employee observes the safety rules and safe work practices.

e. The contract employer shall inform the plant owner of any hazards present during the contractor's work including unique hazards.

11. *Emergency Planning and Response*. The plant owner or operator shall establish and implement an emergency action plan for the entire plant in accordance with previous OSHA regulations.[29] The plan will also include procedures for handling small releases. Also, plant owners may be subject to hazardous waste and associated emergency response requirements in OSHA regulations.[30]

12. *Compliance Audits*. Plant owners must certify they have evaluated compliance with the regulations in this subsection every 3 years through a compliance audit. A report documenting the audit findings will be prepared. The plant owner will document a response to each audit finding and also document that the deficiencies reported have been corrected.

4.8 SUMMARY

This chapter presents regulations that reduce VOC and hazardous air pollutant (HAP) emissions to the atmosphere. These regulations were designed to improve public health and the environment. The VOC emission control reduces ozone concentrations, and the HAP regulations further reduce fugitive emissions consisting of both VOC and VHAP pollutants as the new CAAA regulations are implemented. Although the regulations are designed to protect the public health, they also reduce potential exposures in the workplace through lower pollutant concentration levels. Monitoring and maintenance required in state and federal regulations have been a major factor in VOC emissions reductions prior to the HON regulations.

While VOC reductions have been reduced in nonattainment areas, ozone remains at an unsatisfactory level in various industrial areas. The Clean Air Act Amendments of 1990 require attainment of the National Ambient Air Quality Standards at a more rapid pace, which requires more stringent fugitive emission controls. State plans to meet these objectives require changes, as the new HON regulations near promulgation. These new regulations are more stringent than the VOC regulations, as shown in Table 4.4, but only apply to the hazardous air pollutants listed in Appendix 4A. However, the HAP list will expand as additional compounds are added periodically and will probably encompass a large number of VOC compounds. Consequently, state and regional authority regulations will probably approach HON control requirements *or* the HON requirements may be

TABLE 4.4 Equipment Leak Definition Comparison of NSPS (VOC) and HON (Hazardous Air Pollutants) Regulations

Equipment	Service Condition	Emission Leak Definition (ppmv)	
		NSPS	HON[a]
Pumps	Light liquid	10,000	1,000[b]
	Heavy liquid	10,000[c]	1,000[c]
Valves	Gas / vapor	10,000	500
	Light liquid	10,000	500
	Heavy liquid	10,000[c]	500[c]
Pressure relief	Gas / vapor	500	500
Valves	Light liquid	10,000[c]	500
	Heavy liquid	10,000[c]	500
Connectors[d]	Gas / vapor	10,000[c]	500
	Light liquid	10,000[c]	500
	Heavy liquid	10,000[c]	500[c]
Compressors	Gas / vapor	500	500[e]
Agitators	Gas / vapor	—	10,000
	Light liquid	—	10,000
	Heavy liquid	—	10,000[c]

[a]All HON definitions are Phase III operations.
[b]5000 ppmv for monomer polymerization pumps.
[c]Emissions monitored after leakage observed by visual, audible, olfactory, or other methods.
[d]Flanges, threaded connections, and so on.
[e]Below 500 ppm – exempt from regulations. Above 500 ppmv — barrier fluid systems required.

adopted to more rapidly achieve ozone attainment status. Another advantage of adopting HON rules for VOC control is that a single set of controls are more efficiently administered since the states and regional authorities administer the regulations.

In Table 4.4 the regulatory comparisons present NSPS and HON regulations but do not include NESHAP regulations. The compounds or materials in the NESHAP standards are included in the HAP list, which supersedes the NESHAP regulations. Moreover, the HON standards are considerably more stringent or severe than the NSPS regulations, which were essentially identical to the NESHAP regulations as described in Section 4.4.

The HON regulations will significantly reduce workplace concentrations and workplace exposures. However, the exposure levels may not be satisfactory for certain toxic materials or may not be acceptable in particular locations. Additional controls may be required and methods of further reducing emissions are necessary to ensure workplace exposures are within permissible limits. Some of the control techniques are presented in other chapters, along with further explanations of regulatory requirements for various equipment items.

The emission regulations discussed in this chapter are primarily concerned with vapor or gas emissions, which generally present inhalation problems to the general public. Skin or dermal protection, although required for many compounds, is not specifically required in the VOC or HON regulations since the general public is not exposed to liquids. However, the stringent HON regulations can present potential exposure problems as a result of an increase in maintenance activities to ensure continued VHAP control compliance. Consequently, engineering modifications to existing plants or new plant designs should include control equipment to minimize dermal and vapor exposure when repairing or removing equipment for repair.

The OSHA regulations for process safety management in Section 4.7.16 are concerned with employee exposure for normal operations and emergency situations. A major requirement is the development of operating procedures that are concerned not only about process unit operations but also employee safety and health considerations. The regulations suggest engineering and administrative controls along with personal protective equipment to prevent exposure. Detailed requirements for engineering controls were not addressed, although they are recognized, with any control recommendations included in the written procedures. However, various daily functions and emergencies require careful consideration to develop adequate exposure control. Some additional background information on OSHA considerations is available in reference 5.

The regulatory review in this chapter is based upon U.S. regulations, but other countries also have fugitive emission regulations which vary in severity. More stringent regulations may be promulgated in the future in other countries and regions as environmental concerns expand.

REFERENCES

1. *National Primary and Secondary Ambient Air Quality Standards*, Title 40, Part 50, pp. 693–774, July 1, 1991.
2. CFR, *General Provisions*, Title 40, Parts 60.14 and 60.15, pp. 231–233, July 1, 1992.
3. CFR, *Standards of Performance for Equipment Leaks of VOC in the Synthetic Chemical Manufacturing Industry*, Title 40, Part 60.481, p. 484, July 1, 1992.
4. Clean Air Act Amendments of 1990, Classifications and Attainment Dates, Public Law 101-549, Title I, Sec. 181.a. (Nov. 15, 1990).
5. FR, *Process Safety Management of Highly Hazardous Chemicals*; *Explosives and Blasting Agents*, Final Rule, Vol. 57, No. 36, pp. 6356–6417, February 24, 1992.
6. CFR, *Standards of Performance for Storage Vessels for Petroleum Liquids for Which Construction, Reconstruction, or Modification Commenced After June 11, 1973 and Prior to May 19, 1978*, Title 40, Subpart K, Parts 60.110–60.113, pp. 346–348, July 1, 1992.
7. CFR, *Standards of Performance for Storage Vessels for Petroleum Liquids for Which Construction, Reconstruction, or Modification Commenced After May 18, 1978 and Prior to July 23, 1984*, Title 40, Subpart Ka, Parts 60.110a–60.115a, pp. 348–352, July 1, 1992.
8. CFR, *Standards of Performance for Volatile Organic Liquid Storage Vessels (Including Petroleum, Liquid Storage Vessels) for Which Construction, Reconstruction, or Modification Commenced After July 23, 1984*, Title 40, Subpart Kb, Parts 60.110b–60.117b, pp. 352–361, July 1, 1992.
9. BAAQMD, Regulation 8, Rule 5, pp. 8-5-1 through 8-5-10, November 16, 1990.
10. CFR, *National Emission Standard for Benzene Emissions from Benzene Storage Vessels*, Title 40, Subpart Y, Parts 61.270–61.277, pp. 144–153, July 1, 1992.

11. FR, *Standards of Performance for New Stationary Sources: VOC Emissions from Petroleum Refinery Wastewater Systems*, Final Rule, Vol. 53, No. 226, pp. 47616–47629, November 23, 1988.

12. FR, *National Emission Standards for Hazardous Air Pollutants: Benzene Emissions from Chemical Manufacturing Vents, Industrial Solvents Use, Benzene Waste Operations, Benzene Transfer Operations and Gasoline Marketing System*, Final Rule, Vol. 55, No. 45, pp. 8292–8361, March 7, 1990.

13. CFR, *National Emission Standard for Benzene Waste Operations*, Title 40, Subpart FF, Parts 61.340–61.358, pp. 162–189, July 1, 1992.

14. Daniel, B. M., S. Kapoor, G. A. Smalley, Jr., and M. E. Norman, *Oil & Gas J.* **90**(21), 54–64 (1992).

15. FR, *National Emission Standards for Hazardous Air Pollutants, Benzene Waste Operations, Final Rule and Proposed Rule*, Vol. 57, No. 44, pp. 8011–8016, March 5, 1992.

16. CFR, *Solid Wastes*, Title 40, Parts 260, 264, 265, 270, 271, pp. 4–967, July 1, 1992.

17. CFR, *Standards of Performance for Bulk Gasoline Terminals*, Title 40, Subpart XX, Parts 60.500–60.506, pp. 500–509, July 1, 1992.

18. CFR, *National Emission Standard for Benzene Emissions from Benzene Transfer Operations*, Title 40, Subpart BB, Parts 61.300–61.305, pp. 142–151, July 1992.

19. BAAQMD, Regulation 8, Rule 6, pp. 8-6-3 through 8-6-5, November 30, 1983.

20. BAAQMD, Regulation 8, Rule 33, pp. 8-33-1 through 8-33-5, July 20, 1988.

21. SCAQMD, Amended Rule 462—Organic Loading, pp. 1–3, December 7, 1990.

22. MSA Research Corp./EPA, *VOC Fugitive Emissions in Synthetic Organic Chemicals Manufacturing Industry—Background Information for Promulgated Standards*, EPA 450/3-80-033, pp. 5-1 through 5-5, November 1980.

23. EPA, *Compilation of Air Pollutant Emission Factors*, AP-42, Volume 1: Stationary Point and Area Sources, 4th ed., September 1985; Supplement A, October 1986; Supplement B, September 1988; Supplement C, September 1990; Supplement D, September 1991.

24. Lipton, S., *Chem. Eng. Prog.* **85**, 42–47 (1989).

25. Hanson, D., *C&EN* **70**, 13 (1992).

26. Burglund, R. L., *Chem. Proc.* **55**, 40–44 (1992).

27. FR, *National Emission Standards for Hazardous Air Pollutants, Announcement of Negotiated Regulation for Equipment Leaks*, Vol. 56, No. 44, pp. 9315–9339, March 6, 1991.

28. CFR, *Welding, Cutting and Brazing*, Title 29, Subpart Q, Part 1910, 252, Section (a), pp. 603–611, July 1, 1991.

29. CFR, *Employee Emergency Plans and Fire Protection Plans*, Title 29, Subpart E, Part 1910.38, Section (a), pp. 134–135, July 1, 1991.

30. CFR, *Hazardous Waste Operations and Emergency Response*, Title 29, Subpart H, Part 1910.120, Sections (a) (p) (q), pp. 366–401, July 1, 1991.

31. FR, *National Emission Standards for Hazardous Air Pollutants for Source Categories: Organic Hazardous Air Pollutants from the Synthetic Organic Chemical Manufacturing Industry and Seven Other Processes*, Vol. 5, No. 252, pp. 62608–62808, December 31, 1992.

APPENDIX 4A Hazardous Air Pollutant Identification List, Clean Air Act Amendments of 1990

CAS No.	Chemical Name	CAS No.	Chemical Name
75070	Acetaldehyde	132649	Dibenzofurans
60355	Acetamide	96128	1,2-Dibromo-3-chloropropane
75058	Acetonitrile	84742	Dibutylphthalate
98862	Acetophenone	106467	1,4-Dichlorobenzene(p)
53963	2-Acetylaminofluorene	91941	3,3-Dichlorobenzidene
107028	Acrolein	111444	Dichloroethyl ether [Bis(2-chloroethyl) ether]
79061	Acrylamide	542756	1,3-Dichloropropene
79107	Acrylic acid	62737	Dichlorvos
107131	Acrylonitrile	111422	Diethanolamine
107051	Allyl chloride	121697	*N,N*-Diethyl aniline (*N,N*-Dimethylaniline)
92671	4-Aminobiphenyl	64675	Diethyl sulfate
62533	Aniline	119904	3,3-Dimethoxybenzidine
90040	*o*-Anisidine	60117	Dimethyl aminoazobenzene
1332214	Asbestos	119937	3,3-Dimethyl benzidine
71432	Benzene (including benzene from gasoline)	79447	Dimethyl carbamoyl chloride
92875	Benzidine	68122	Dimethyl formamide
98077	Benzotrichloride	57147	1,1-Dimethyl hydrazine
100447	Benzyl chloride	131113	Dimethyl phthalate
92524	Biphenyl	77781	Dimethyl sulfate
117817	Bis(2-ethylhexyl) phthalate (DEHP)	534521	4,6-Dinitro-*o*-cresol and salts
542881	Bis(chloromethyl) ether	51285	2,4-Dinitrophenol
75252	Bromoform	121142	2,4-Dinitro-toluene
106990	1, 3-Butadiene	123911	1,4-Dioxane (1,4-Diethyleneoxide)
156627	Calcium cyanamide	122667	1,2-Diphenylhydrazine
105602	Caproloctam	106898	Epichlorohydrin (1-Chloro-2,3-epoxypropane)
133062	Captan	106887	1,2-Epoxybutane
63252	Carbaryl	140885	Ethyl acrylate
75150	Carbon disulfide	100414	Ethyl benzene
56235	Carbon tetrachloride	51796	Ethyl carbamate (Urethane)
463581	Carbonyl sulfide	75003	Ethyl chloride (Chloroethane)
120809	Catechol	106934	Ethylene dibromide (Dibromoethane)
133904	Chloramben	107062	Ethylene dichloride (1,2-Dichloroethane)
57749	Chlordane	107211	Ethylene glycol
7782505	Chlorine	151564	Ethylene imine (Aziridine)
79118	Chloroacetic acid	75218	Ethylene oxide
532274	2-Chloroacetophenone	96457	Ethylene thiourea
108907	Chlorobenzene	75343	Ethylidene dichloride (1,1-Dichloroethane)
510156	Chlorobenzilate	50000	Formaldehyde
67663	Chloroform	76448	Heptachlor
107302	Chloromethyl methyl ether	118741	Hexachlorobenzene
126998	Chloroprene	87683	Hexachlorobutadiene
1319773	Cresols / cresylic acid (isomers and mixture)	77474	Hexachlorocyclopentadiene
95487	*o*-Cresol	67721	Hexachloroethane
108394	*m*-Cresol	822060	Hexamethylene-1,6-diisocyanate
106445	*p*-Cresol		
98828	Cumene		
94757	2,4-D, salts and esters		
3547044	DDE		
334883	Diazomethane		

APPENDIX 4A ***(Continued.)***

CAS No.	Chemical Name	CAS No.	Chemical Name
680319	Hexamethylphosphoramide	1336363	Polychlorinated biphenyls (Aroclors)
110543	Hexane	1120714	1,3-Propane sultone
302012	Hydrazine	57578	β-Propiolactone
7647010	Hydrochloric acid	12386	Propionaldehyde
7664393	Hydrogen fluoride (Hydrofluoric acid)	114261	Propoxur (Baygon)
123319	Hydroquinone	78875	Propylene dichloride (1,2-dichloropropane)
78591	Isophorone	75569	Propylene oxide
58899	Lindane (all isomers)	75558	1,2-Propylenimine (2-Methyl aziridine)
108316	Maleic anhydride	91225	Quinoline
67561	Methanol	106514	Quinone
72435	Methoxychlor	100425	Styrene
74839	Methyl bromide (Bromomethane)	96093	Styrene oxide
74873	Methyl chloride (Chloromethane)	1746016	2,3,7,8-Tetrachlorodibenzo-*p*-dioxin
71556	Methyl chloroform (1,1,1-Trichloroethane)	79345	1,1,2,2-Tetrachloroethane
78933	Methyl ethyl ketone (2-Butanone)	127184	Tetrachloroethylene (Perchloroethylene)
60344	Methyl hydrazine	7550450	Titanium tetrachloride
74884	Methyl iodide (Iodomethane)	108883	Toluene
108101	Methyl isobutyl ketone (Hexone)	95807	2,4-Toluene diamine
624839	Methyl isocyanate	584849	2,4-Toluene diisocyanate
80626	Methyl methacrylate	95534	*o*-Toluidine
1634044	Methyltertbutylether	3001352	Toxaphene (chlorinated camphene)
1001144	4,4-Methylene bis(2-chloroaniline)	120821	1,2,4-Trichlorobenzene
75092	Methylene chloride (Dichloromethane)	79005	1,1,2-Trichloroethane
101688	Methylene diphenyl diisocyanate (MDI)	79016	Trichloroethylene
101779	4,4′-Methylenedianiline	95954	2,4,5-Trichlorophenol
91203	Naphthalene	88062	2,4,6-Trichlorophenol
98953	Nitrobenzene	121448	Triethylamine
92933	4-Nitrobiphenyl	1582098	Trifluralin
100027	4-Nitrophenol	540841	2,2,4-Trimethylpentane
79469	2-Nitropropane	108054	Vinyl acetate
684935	*N*-Nitroso-*N*-methylurea	593602	Vinyl bromide
62759	*N*-Nitrosodimethylamine	75014	Vinyl chloride
59892	*N*-Nitrosomorpholine	75354	Vinylidene chloride (1,1-Dichloroethylene)
56382	Parathion	1330207	Xylene (isomers and mixture)
82688	Pentachloronitrobenzene (Quintobenzene)	95476	*o*-Xylene
87865	Pentachlorophenol	108383	*m*-Xylenes
108952	Phenol	106423	*p*-Xylenes
106503	*p*-Phenylenediamine	0	Antimony compounds
75445	Phosgene	0	Arsenic compounds (inorganic including arsine)
7803512	Phosphine	0	Beryllium compounds
7723140	Phosphorus	0	Cadmium compounds
85449	Phthalic anhydride	0	Chromium compounds
		0	Cobalt compounds

APPENDIX 4A *(Continued.)*

CAS No.	Chemical Name	CAS No.	Chemical Name
0	Coke Oven emissions	0	Fine mineral fibers[c]
0	Cyanide compounds[a]	0	Nickel compounds
0	Glycol ethers[b]	0	Polycyclic organic matter[d]
0	Lead compounds	0	Radionuclides (including radon)[e]
0	Manganese compounds	0	Selenium compounds
0	Mercury compounds		

NOTE: For all listings above that contain the word "compounds" and for glycol ethers, the following applies: Unless otherwise specified, these listings are defined as including any unique chemical substance that contains the named chemical (i.e., antimony, arsenic, etc.) as part of that chemical's infrastructure.

[a]XCN where X = H or any other group where a formal dissociation may occur. For example, KCN or $Ca(CN)_2$.

[b]Includes mono- and diethers of ethylene glycol, diethylene glycol, and triethylene glycol R-$(OCH2CH2)_n$-OR where $n = 1$, 2, or 3; R = alkyl or aryl groups R^1 = R, H, or groups which, when removed, yield glycol ethers with the structure: R-$(OCH2CH)_n$-OH. Polymers are excluded from the glycol category.

[c]Includes mineral fiber emissions from facilities manufacturing or processing glass, rock, or slag fibers (or other mineral derived fibers) of average diameter 1 μm or less.

[d]Includes organic compounds with more than one benzene ring, and which have a boiling point greater than or equal to 100°C.

[e]A type of atom which spontaneously undergoes radioactive decay.

APPENDIX 4B Volatile Hazardous Air Pollutant List for National Emission Standards for Hazardous Air Pollutants (HON)

CAS No.	Chemical Name	CAS No.	Chemical Name
75070	Acetaldehyde	105602	Caprolactam
60355	Acetamide	75150	Carbon disulfide
75058	Acetonitrile	56235	Carbon tetrachloride
96862	Acetophenone	463581	Carbonyl sulfide
53963	2-Acetylaminofluorine	120809	Catechol
107028	Acrolein	79118	Chloroacetic acid
70961	Acrylamide	532274	2-Chloroacetophenone
79107	Acrylic acid	108907	Chlorobenzene
107131	Acrylonitrile	67663	Chloroform
107051	Allyl chloride	107302	Chloromethyl methyl ether
92671	4-Aminobiphenyl	126998	Chloroprene
62533	Aniline	1319773	Cresols / cresylic acid (Isomers and mixture)
90040	*o*-Anisidine		
71432	Benzene	95487	*o*-Cresol
92875	Benzidine	108394	*m*-Cresol
98077	Benzotrichloride	106445	*p*-Cresol
100447	Benzyl chloride	98828	Cumene
92524	Biphenyl	94757	2,4-D, salts and esters
117817	Bis(ethylhexyl)phthalate (DEHP)	3547044	DDE
		334883	Diazomethane
542881	Bis(chloromethyl) ether	132649	Dibenzofurane
75252	Bromoform	96128	1,2-Dibromo-3-chloropropane
106990	1,3-Butadiene	84742	Dibutylphthalate

APPENDIX 4B ***(Continued.)***

CAS No.	Chemical Name	CAS No.	Chemical Name
106467	1,4-Dichlorobenzene(p)	74873	Methyl chloride (Chloromethane)
91941	3,3-Dichlorobenzidene	71556	Methyl chloroform (1,1,1-Trichloroethane
111444	Dichloroethyl ether [Bis(2-chloroethyl)ether]	78933	Methyl ethyl ketone (2-Butanone)
542756	1,3-Dichloropropene	60344	Methyl hydrazine
111422	Diethanolamine	74884	Methyl iodide (Iodomethane)
121697	*N*,*N*-Diethyl aniline (*N*,*N*-Dimethylaniline)	108101	Methyl isobutyl ketone (Hexane)
64675	Diethyl sulfate	624839	Methyl isocyanate
119904	3,3′-Dimethoxybenzidine	80628	Methyl methacrylate
60117	Dimethyl aminoazobenzene	1634044	Methyl tert butyl ether
119937	3,3′-Dimethyl benzidine	101144	4,4-Methylene (bis(2-chloroaniline)
79447	Dimethyl carbamoyl chloride	75092	Methylene chloride (Dichloromethane)
68122	Dimethyl formamide	101688	Methylene diphenyl diisocyanate (MDI)
57147	1,1-Dimethyl hydrazine	101779	4,4′-Methylenedianiline
131113	Dimethyl phthalate	91203	Naphthalene
77781	Dimethyl sulfate	98953	Nitrobenzene
534521	4,6-Dinitro-*o*-cresol and salts	92933	4-Nitrobiphenyl
51285	2,4-Dinitrophenol	100027	4-Nitrophenol
121142	2,4-Dinitrotoluene	79469	4-Nitropropane
123911	1,4-Dioxane (1,4-Diethyleneoxide)	684935	*N*-Nitroso-*N*-methylurea
122667	1,2-Diphenylhydrazine	62759	*N*-Nitrosodimethylamine
106898	Epichlorophydrin (1-Chloro-2,3-epoxypropane	59892	*N*-Nitrosomorpholine
106887	1,2-Epoxybutane	108952	Phenol
140885	Ethyl acrylate	106503	*p*-Phenylenediamine
100414	Ethyl benzene	75445	Phosgene
51796	Ethyl carbamate (Urethane)	85449	Phthalic anhydride
75003	Ethyl chloride (Chloroethane)	1336363	Polychlorinated biphenyls (Aroclors)
106934	Ethylene dibromide (Dibromoethane)	1120714	1,3-Propane sultone
107062	Ethylene dichloride (1,2-Dichloroethane)	57578	β-Propiolactone
107211	Ethylene glycol	123386	Propionaldehyde
75218	Ethylene oxide	114261	Propoxur (Baygon)
96457	Ethylene thiourea	78875	Propylene dichloride (1,2-Dichloropropane
75343	Ethylidene dichloride (1,1-Dichloroethane)	75569	Propylene oxide
50000	Formaldehyde	75558	1,2-Propylenimine (2-Methyl aziridine)
0	Glycol ethers	106514	Quinone
118741	Hexachlorobenzene	100425	Styrene
87683	Hexachlorobutadiene	96093	Styrene oxide
67721	Hexachloroethane	1746016	2,3,7,8-Tetrachlorodibenzo-*p*-dioxin
822060	Hexamethylene-1,6-diisocyanate	79345	1,1,2,2-Tetrachloroethane
680319	Hexamethylphosphoramide	127184	Tetrachloroethylene (Perchloroethylene)
110543	Hexane	108883	Toluene
302012	Hydrazine	95807	2,4-Toluene diamine
123319	Hydroquinone		
78591	Isophorone		
108316	Maleic anhydride		
67561	Methanol		
74839	Methyl bromide (Bromoethane)		

APPENDIX 4B ***(Continued.)***

CAS No.	Chemical Name
584849	2,4-Toluene diisocyanate
95534	*o*-Toluidine
120821	1,2,4-Trichlorobenzene
79005	1,1,2-Trichloroethane
79016	Trichloroethylene
95954	2,4,5-Trichlorophenol
88062	2,4,6-Trichlorophenol
121448	Triethylamine
1582098	Trifluralin
540841	2,2,4-Trimethylpentane
108054	Vinyl acetate
593602	Vinyl bromide
75014	Vinyl chloride
75354	Vinylidene chloride (1,1-Dichloroethylene)
1330207	Xylenes (isomers and mixture)
95476	*o*-Xylenes
108383	*m*-Xylenes
106423	*p*-Xylenes

APPENDIX 4C Highly Hazardous Chemicals List — OSHA Process Safety Management Rule 29 CFR Part 1910

Chemical Name	CAS No.	Threshold[a] Quantity (lb)
Acetaldehyde	75-07-0	2,500
Acrolein (2-Propenal)	107-02-8	150
Acrylyl chloride	814-68-6	250
Allyl chloride	107-05-1	1,000
Allylamine	107-11-9	1,000
Alkylaluminums	Varies	5,000
Ammonia, anhydrous	7664-41-7	10,000
Ammonia solutions (> 44% ammonia by weight)	7664-41-7	15,000
Ammonium perchlorate	7790-98-9	7,500
Ammonium permanganate	7787-36-2	7,500
Arsine (also called arsenic hydride)	7784-42-1	100
Bis(chloromethyl) ether	542-88-1	100
Boron trichloride	10294-34-5	2,500
Boron trifluoride	7637-07-2	250
Bromine	7726-95-6	1,500
Bromine chloride	13863-41-7	1,500
Bromine pentafluoride	7789-30-2	2,500
Bromine trifluoride	7787-71-5	15,000
3-Bromopropyne (also called propargyl bromide)	106-96-7	100
Butyl hydroperoxide (tertiary)	75-91-2	5,000
Butyl perbenzoate (tertiary)	614-45-9	7,500
Carbonyl chloride (see Phosgene)	75-44-5	100
Carbonyl fluoride cellulose nitrate (concentration > 12.6% nitrogen)	9004-70-0	2,500
Chlorine	7782-50-5	1,500
Chlorine dioxide	10049-04-4	1,000
Chlorine pentrafluoride	13637-63-3	1,000
Chlorine trifluoride	7790-91-2	1,000
Chlorodiethylaluminum (also called diethylaluminum chloride)	96-10-6	5,000
1-Chloro-2,4-dinitrobenzene	97-00-7	5,000
Chloromethyl methyl ether	107-30-2	500
Chloropicrin	76-06-2	500
Chloropicrin and methyl bromide mixture	None	1,500
Chloropicrin and methyl chloride mixture	None	1,500
Cumene hydroperoxide	80-15-9	5,000
Cyanogen	460-19-5	2,500
Cyanogen chloride	506-77-4	500
Cyanuric fluoride	675-14-9	100
Diacetyl peroxide (concentration > 70%)	110-22-5	5,000
Diazomethane	334-88-3	500
Dibenzoyl peroxide	94-36-0	7,500
Diborane	19287-45-7	100
Dibutyl peroxide (tertiary)	110-05-4	5,000
Dichloro acetylene	7572-29-4	250
Dichlorosilane	4109-96-0	2,500
Diethylzinc	557-20-0	10,000
Diisopropyl peroxydicarbonate	105-64-6	7,500
Dilauroyl peroxide	105-74-8	7,500
Dimethyldichlorosilane	75-78-5	1,000
Dimethylhydrazine 1,1-	57-14-7	1,000

APPENDIX 4C (*Continued.*)

Chemical Name	CAS No.	Threshold[a] Quantity (lb)
Dimethylamine, anhydrous	124-40-3	2,500
2,4-Dinitroaniline	97-02-9	5,000
Ethyl methyl ketone peroxide (also methyl ethyl ketone peroxide; concentration > 60%)	1338-23-4	5,000
Ethyl nitrite	109-95-5	5,000
Ethylamine	75-04-7	7,500
Ethylene fluorohydrin	371-62-0	100
Ethylene oxide	75-21-8	5,000
Ethyleneimine	151-56-4	1,000
Fluorine	7782-41-4	1,000
Formaldehyde (formalin)	50-00-0	1,000
Furan	110-00-9	500
Hexafluoroacetone	684-16-2	5,000
Hydrochloric acid, anhydrous	7647-01-0	5,000
Hydrofluoric acid, anhydrous	7664-39-3	1,000
Hydrogen bromide	10035-10-6	5,000
Hydrogen chloride	7647-01-0	5,000
Hydrogen cyanide, anhydrous	74-90-8	1,000
Hydrogen fluoride	7664-39-3	1,000
Hydrogen peroxide (52% by weight or greater)	7722-84-1	7,500
Hydrogen selenide	7783-07-5	150
Hydrogen sulfide	7783-06-4	1,500
Hydroxylamine	7803-49-8	2,500
Iron, pentacarbonyl	13463-40-6	250
Isopropylamine	75-31-0	5,000
Ketene	463-51-4	100
Methacrylaldehyde	78-85-3	1,000
Methacryloyl chloride	920-46-7	150
Methacryloyloxyethyl Isocyanate	30674-80-7	100
Methyl acrylonitrile	126-98-7	250
Methylamine, anhydrous	74-89-5	1,000
Methyl bromide	74-83-9	2,500
Methyl chloride	74-87-3	15,000
Methyl chloroformate	79-22-1	500
Methyl ethyl ketone peroxide (concentration 60%)	1338-23-4	5,000
Methyl fluoroacetate	453-18-9	100
Methyl fluorosulfate	421-20-5	100
Methyl hydrazine	60-34-4	100
Methyl iodide	74-88-4	7,500
Methyl isocyanate	624-83-9	250
Methyl mercaptan	74-93-1	5,000
Methyl vinyl ketone	79-84-4	100
Methyltrichlorosilane	75-79-6	500
Nickel carbonyl (nickel tetracarbonyl)	13463-39-3	150
Nitric acid (94.5% by weight or greater)	7697-37-2	500
Nitric oxide	10102-43-9	250
Nitroaniline (*p*-nitroaniline)	100-01-6	50,000
Nitromethane	75-52-5	2,500
Nitrogen dioxide	10102-44-0	250
Nitrogen oxides (NO; NO_2; N_2O_4; N_2O_3)	10102-44-0	250
Nitrogen tetroxide (also called nitrogen peroxide)	10544-72-6	250
Nitrogen trifluoride	7783-54-2	5,000
Nitrogen trioxide	10544-73-7	250
Oleum (65 – 80% by weight; also called fuming sulfuric acid)	8014-94-7	1,000
Osmium tetroxide	20816-12-0	100
Oxygen difluoride (fluorine monoxide)	7783-41-7	100
Ozone	10028-15-6	100
Pentaborane	19624-22-7	100
Peracetic acid (concentration > 60% acetic acid; also called peroxyacetic acid)	79-21-0	1,000
Perchloric acid (concentration > 60% by weight)	7601-90-3	5,000
Perchloromethyl mercaptan	594-42-3	150
Perchloryl fluoride	7616-94-6	5,000
Peroxyacetic acid (concentration > 60% acetic acid; also called peracetic acid	79-21-0	1,000
Phosgene (also called carbonyl chloride)	75-44-5	100
Phosphine (hydrogen phosphide)	7803-51-2	100
Phosphorus oxychloride (also called phosphoryl chloride)	10025-87-3	1,000
Phosphorus trichloride	7719-12-2	1,000

APPENDIX 4C (***Continued.***)

Chemical Name	CAS No.	Threshold[a] Quantity (lb)	Chemical Name	CAS No.	Threshold[a] Quantity (lb)
Phosphoryl chloride (also called phosphorus oxychloride)	10025-87-3	1,000	Sulfuric anhydride (also called sulfur trioxide)	7446-11-9	1,000
			Tellurium hexafluoride	7783-80-4	250
Propargyl bromide	106-96-7	100	Tetrafluoroethylene	116-14-3	5,000
Propyl nitrate	627-3-4	2,500	Tetrafluorohydrazine	10036-47-2	5,000
Sarin	107-44-8	100	Tetramethyl lead	75-74-1	1,000
Selenium hexafluoride	7783-79-1	1,000	Thionyl chloride	7719-09-7	250
Stibine (antimony hydride)	7803-52-3	500	Trichloro (chloromethyl) silane	1558-25-4	100
Sulfur dioxide (liquid)	7446-09-5	1,000	Trichloro (dichlorophenyl) silane	27137-85-5	2,500
Sulfur pentafluoride	5714-22-7	250			
Sulfur tetrafluoride	7783-60-0	250			
Sulfur trioxide (also called sulfuric anhydride)	7446-11-9	1,000	Trichlorosilane	10025-78-2	5,000
			Trifluorochloroethylene	79-38-9	10,000
			Trimethyoxysilane	2487-90-3	1,500

[a]Amount present requiring coverage under this rule.

EMISSION REGULATIONS

5

Emissions Measurement and Estimation

Emission and exposure control are based upon leak measurement techniques that have been standardized by the EPA. These measurements are generally obtained at the emission source with an instrument that provides the concentration in air of the contaminant.[2,3] Through various procedures, this measurement can then be converted to mass flow rate. Alternatively, the mass flow leak rate can be measured directly. While actual measurement of the mass flow is preferable, this procedure is cumbersome, and the concentration or screening measurements are normally obtained. Since screening or monitoring is required to insure compliance with the various environmental regulations, rapid screening with estimating techniques of mass flow are generally employed. From these mass flow rates, dispersion calculation procedures (Chapter 18) provide a method of determining potential exposure concentrations for industrial hygienists.

In this chapter emission measurement procedures are presented for obtaining screening concentrations and mass flow rates with actual measurement schedules and leak level definitions provided in the chapters associated with equipment items and general systems. Also, several methods for estimating emissions are presented, although some of the estimating procedures are not described in detail because of the extensive statistical analytic background required. However, sufficient information is provided for an assessment based upon EPA procedures that are outlined in accordance with an EPA protocol.[1]

Prior to passage of the Clean Air Act of 1990, the leak concentration definition for petroleum refinery vapors and most chemicals was 10,000 ppmv or greater for VOC compounds. A few NESHAPS regulations (benzene, vinyl chloride, etc.) along with certain states or authorities had lower leak definitions. However, the proposed regulations for the chemical industry that have been issued in accordance with the CAAA termed HON (hazardous organic NESHAPS) reduces the EPA leak concentration definition to 1000 ppmv or greater for pump seals and 500 ppmv for most of the other equipment. These lower leak definitions under the new regulations, applicable to hazardous air pollutants (HAP), generally cover equipment and systems in gas/vapor and light liquid services. While the HON regulation is a proposed rule, few changes are expected in the leak definition levels, which are expected to become final in early 1994 when the revised regulations will be issued by the EPA. In accordance with the 1990 Clean Air Act, regulations for petroleum refineries will be proposed in 1994. Further restrictions on VOCs will probably result in a reduction of leakage definitions to these new levels or possibly lower ones, as seen in certain state and authority regulations.

Screening or monitoring of leakage sources is required in these various regulations to determine whether a leak exists. Consequently, mass flow measurements are not required to conform with the regulations. However, mass emissions estimates are required to determine annual emission losses in accordance with SARA (Superfund Act Reauthorization Amendments) regulations, losses anticipated from new facilities, and reductions in emissions losses with time. A reduction in losses to the atmosphere of about 90% is an EPA target in accordance with the new Clean Air Act. Proposed regulations in the European Community (EC) are also designed to reduce emissions by 90%. As a result, a basis for determining mass emission rates is necessary for each plant and a procedure must be selected to ensure consistency in the estimating procedure. Different procedures are presented in this chapter along with comments and restrictions by the EPA on estimating techniques.

5.1 EMISSION MEASUREMENTS

Screening or monitoring of equipment for conformance with regulatory requirements is reviewed in this section along with mass flow measurement procedures. While selection of equipment and measurement is presented on the basis of a process unit, the overall plant must be reviewed and monitored in accordance with federal, state, and regional requirements. Vapor and light liquid services are monitored on a scheduled basis in accordance with regulations, with heavy liquid service monitoring generally limited to those sources where leakage is physically evident. Screening is required for both VOC and HON services by federal, state, and regional authorities, but regulations can differ significantly among these groups. Federal governments provide a minimum control basis (although restrictive) that can be increased in stringency by the state and regional authorities to meet emission requirements.

5.1.1 Screening and Monitoring Basis

Most plants have identified the equipment that must be monitored through the years as a result of existing regulations. In general the equipment included in monitoring and screening programs required by regulation are shown in Table 5.1. For determining actual monitoring or screening requirements, equipment types, numbers of each type, measurement locations, and fluid characterization are necessary. The equipment items do not require tagging in the HON regulations, but tagging the equipment readily identifies specific equipment items and simplifies record keeping. Nontagged equipment requires descriptions for record keeping

TABLE 5.1 Equipment Monitoring Leak Sources[a]

Pump seals	Flanges, threaded connections, and so on
Compressor seals	Open-ended lines
Valves	Drains, vents, doors
Control valves	Agitator seals
Pressure relief valves	Closed vent systems
Sampling connections	Sampling loops

[a]Adapted from reference 1.

which can become quite extensive and confusing in process units with large numbers of valves, flanges, and so on.

Stream identification requires a rather detailed analysis of various streams to determine equipment service status covering gas/vapor, light liquid, or heavy liquid and whether the stream is in VOC or HON service. For this general investigation, up-to-date flow plans with clear stream identification markings are desirable, including identification of equipment. This information may also be required to comply with certain requirements in the regulations, but does provide additional information for analyzing equipment performance in specific services. Where units or plants enter QIPs for specific types of equipment (valves, pumps), all of this information is necessary to adequately assess equipment operating performance. Moreover, equipment identified as unsafe to monitor or difficult to monitor is readily identifiable on detailed flow plans for regulatory review. These terms are defined in Chapters 7–9.

5.1.2 Monitoring Instruments

The criteria for portable detection instruments required for screening or monitoring are defined in EPA Reference Methods 21.[2] In addition, Method 21 also includes the measuring procedure or techniques applicable to the various types of process equipment that must be monitored in accordance with the regulations. Although Method 21 was developed for VOC compounds, it is also the regulatory basis for the measurement of HAP emissions.

In these requirements, the instrument detector type is not specified but must meet the criteria defined below. Basically, many portable instruments are permitted provided the instruments can measure the organic compounds in the specific streams requiring emission monitoring and have been approved by the EPA. In addition, instrument monitoring results must be related to the database developed by the EPA which was generated with an instrument having a flame ionization detector and was calibrated to a methane standard. Any instruments on a nonmethane reference compound must have response factors that correlate screening measurements to concentration.

Operating Principles and Limitations of Portable Detection Devices. Screening instruments used in these services determine emission leakage through various detection methods such as ionization, infrared absorption, and catalytic combustion, which measures thermal conductivity or heat of combustion.

Ionization detectors measure the charge or number of ions produced following ionization of the sample. Two general ionization methods used in screening instruments are flame ionization and photoionization. In flame ionization the sample is burned and the flame ionization detector (FID) in the instrument measures the carbon content of the organic vapor sampled. Generally, a carrier gas of hydrogen supplies the basic flame in the instrument and does not interfere with the detector. However, the FID units will react to water vapor at low sensitivities although this is not usually significant unless there is water condensation in the sampling system which may result in erratic readings. In addition, a particulate filter is normally placed in the system to prevent disturbances in the flame burner or the detector. The FID unit may respond to halogens, although the response may be reduced considerably. Nitrogen or oxygen also result in reduced responses. Moreover, various organics will produce a very small response or the detection unit may not show any response. Consequently, response factors must be developed for each compound that is measured and preferably with any major inert gas that is present, such as nitrogen. Stream analysis is necessary to identify the compounds present and develop response factors (Section 5.2.5) since these analyzers are nonspecific.

Photoionization detectors (PID) use ultraviolet light to ionize organic vapors. Detector response will vary with the functional groups in the chemical compounds. These instruments work well with certain compounds such as formaldehyde that are not detectable with FID or combustion detectors. The PID units will measure chlorinated hydrocarbons, aromatics, aldehydes, ketones, hydrocarbons, and other substances.

Nondispersive infrared (NDIR) instrument detection is based upon wavelength absorption characteristics of certain gases or vapors. These instruments are affected by other vapors or gases such as water vapor and carbon dioxide that may absorb the light at the same wavelength as the compound of interest. Consequently, these instruments are generally used for the detection and measurement of single components. This requires setting the wavelength for the compound of interest, which absorbs infrared at a specific wavelength. Very small changes of wavelength will affect absorption and the concentration measurement. A 3.4-μm infrared wavelength will detect and measure petroleum fractions including gasoline and naphtha.

Combustion analyzers are based on measurement of thermal conductivity or the heat produced through gas combustion with the heat of combustion the more common detection method. These analyzers are generally termed hot wire detectors or catalytic oxidizers. However, these analyzers are also nonspecific and are similar to most of the other detectors in this regard. Also, combustion analyzers have a very low response to gases that are not readily combustible such as formaldehyde and carbon tetrachloride.

Performance Criteria. Any portable detector may be used as a screening or monitoring device provided the detector meets the performance criteria specified in Method 21.[2] However, Method 21, which is applicable to many organic compounds, cannot be used with all compounds and performance tests would require development to qualify an instrument for service as a detector.

TABLE 5.2 Performance Criteria for Portable Detectors[a]

Criteria	Requirement	Test Performance
Instrument response factor	≤ 10 unless correction curve is applied	Once, prior to placement in service
Instrument response time	≤ 30 sec	Once, prior to placement in service. Any modifications to sample pump or flow configuration requires a new test
Calibration precision	≤ 10% of calibration gas value	Prior to placement in service and at 3-month intervals or next use period, whichever is later

[a]Adapted from reference 1.

Performance criteria are presented in Table 5.2 but must be combined with instrument specifications to insure a satisfactory detector is qualified for field application. A major criteria for performance is the response factor, which must be < 10. Thus the ratio of the actual concentration to the observed detector concentration, which is the response factor, should be quite low. Preferably, the response factor is 1.0, but will vary with different compounds. However, the response factor requirement may be relaxed where a response curve is available, preferably curves that have been developed in the plant laboratory. This factor must be developed before the instrument is placed in monitoring service. In addition, the response time should be ≤ 30 sec for the complete system, including any dilution probe, instrument sample probe, pump, and probe filter. Again this criteria must be satisfied before the instrument can be placed in field service. The final criteria in Table 5.1 requires a calibration precision that is ≤ 10% of the calibration gas value.

Instrument specifications in addition to the performance criteria are:

a. The detector will respond to the organic compounds measured as determined by the response factor.
b. The analyzer shall be capable of measuring the leak definition specified in the regulations, covering 500, 1000, or 10,000 ppmv.
c. All analyzer scales shall be readable to ±2.5% of the specified leak definition.
d. Instruments shall be equipped with a pump to insure samples are provided to the detector at a constant flow rate. This flow rate should be 0.10–3.0 L/min under all conditions.
e. The analyzer shall be intrinsically safe for explosive atmospheres and shall meet Class 1, Division 1, and Class 2, Division 1 criteria.

Performance Evaluation Requirements. In addition to performance criteria, additional evaluation requirements placed upon the analyzers are described.

Response Factor. A response factor must be determined for each compound that is measured with the analyzer. These factors vary with the compound and composition of the sample vapor. The standard response factor is defined by the following relationship:

$$\text{Response factor} = \frac{\text{Actual compound concentration}}{\text{Observed detector concentration}} \tag{5.1}$$

The response factors can also be used as a guide in determining whether a particular type of analyzer is recommended for screening or monitoring. Where the response factor is ≥ 0.7, an analyzer will generally be satisfactory. In situations where the response factor is close to zero, the analyzer should not be used. Basically, the detector in the analyzer determines the response factor with the EPA preferring the flame ionization detector as a method of screening where this detector is applicable.

Response factors are normally determined for pure compounds. Initially the analyzer is calibrated with a reference gas such as methane. Then a mixture of the pure compound (test gas) in air is measured with the analyzer and the reading is recorded. This prepared mixture must be quite accurate and standardized gas–air mixtures can be obtained from various suppliers of industrial gases. Following the measurement of the gas–air mixture, the instrument is purged and zeroed with air. The test mixture is again measured, and a final total of three measurements is obtained. A response factor for each measurement is determined, and the final response factor is an average of the three measurements. A large number of response factors were determined by the EPA for various compounds and are listed in reference 3.

While these data indicate there is a single response factor value for a pure compound, test data from other sources indicate response factors are not necessarily a constant value.[4] Above compound concentration levels of about 1000 ppmv in the region investigated by the EPA, the response factors are generally constant for most compounds, which has been satisfactory for screening and monitoring requirements. However, below 500 ppmv, the response factor frequently changes, although the CMA considers the response factor measured at 150 ppmv a satisfactory average value for concentrations less than 150 ppmv.[4] While a reasonable approximation, accuracy for concentrations < 150 ppmv should be investigated since the response factors can either rise or fall rapidly below this level. The EPA response factors at concentration levels < 500 ppmv are not considered reliable, which requires more accurate factors for the new stringent leak definition values of 500 and 1000 ppmv. In locations where the leak definition is less than 500 ppmv, accurate response factors should be established to assist in defining the actual emissions concentration level, since detector accuracy at very low levels may also change. Low-level accuracy is important since much of the equipment that must comply with the 500-ppmv leak definition will have leak concentrations less than 25 ppmv. Instrument accuracy aids in determining losses and mass leak rates that are realistic.

An important consideration in screening or monitoring is the stream composition which affects the response factor. Since most streams in chemical plants and petroleum refineries are mixtures, a response factor based on a single compound results in an erroneous measurement. Consequently, the response factor for a mixture must be determined. One consideration in determining the response factor for a mixture is that the stream composition is generally constant in continuous processes. Obviously there may be minor composition changes, but the overall composition is essentially constant and response factors will not change. Consequently, an important step in determining a response factor is to define the stream composition, whether liquid or vapor. When the vapor composition is known, the leak composition is identical to the stream composition. Where the stream consists of light liquids and the composition of the stream is known, the vapor composition of a leak can be determined from flash calculations. Stream conditions such as pressure, temperature, and composition should be known, which readily permits a vapor or flash calculation.

With known vapor mixtures or compositions, standard mixtures of the vapor and air can be used to determine the response factor with the technique previously described. For this analysis, the air–vapor mixture must be carefully standardized, and these precise mixtures can be obtained from various industrial gas suppliers. However, several air–vapor concentrations should be used to obtain a response factor curve, which will indicate whether the response factor is constant or varies with the concentration. As an alternative, an EPA study evaluating response factors for mixtures indicated that mixture response factors fall between the response factors for pure components, permitting the development of a weighted response factor.[5] The EPA suggested a response factor based on the following equation where the subscripts indicate the pure gas component:

$$\mathrm{RF} = (\text{mol fract}_1)(\mathrm{RF}_1) + (\text{mol fract}_2)(\mathrm{RF}_2) + \cdots \tag{5.2}$$

The CMA in reference 4 has additional information on mixture response factors along with representation response factor curves for several pure compounds. For mixtures, experimental data with vapor mixtures should be obtained, particularly where the overall vapor concentration in air is less than 500 ppmv.

Response Time. Analyzer response time is defined by the EPA as the time interval from a step change in the VOC or HAP concentration at the inlet of the sampling system to the time where the concentration on the analyzer meter reaches 90% of the final concentration value. This response time as shown in Table 5.2 must be ≤ 30 sec for the analyzer used in screening or monitoring service. All analyzers must have a response time test prior to placement of the analyzer in service.

Calibration Precision. This characteristic describes the degree of agreement between measurements of the same known concentration or repeatability. The analyzer is tested first with a zero vapor concentration, and the analyzer reading is then adjusted to zero. A specified calibration (or reference) gas, which is a mixture of a vapor and air, is introduced into the analyzer and the meter reading is recorded. This procedure is repeated three times. Following the tests, the average algebraic difference between the meter readings and the calibration gas are determined and the average difference is divided by the known calibration gas concentration. The resulting figure, which is expressed as a percentage, is the calibration precision.

The calibration precision must be $\leq 10\%$ of the calibration gas value. This test is also required before the analyzer is placed in service and at 3-month intervals when used continually. If used intermittently, the analyzer must be tested at the 3-month interval or before the next usage period, whichever occurs later. Calibration testing should be performed more frequently where the analyzer is in continual service to meet the new stringent leak definition regulations of 500 and 1000 ppmv.

In addition to the general calibration precision test, the EPA also has specifications on the calibration gases. The zero gas should consist of air and a VOC or HAP concentration < 10 ppmv. The calibration gas concentration should be approximately equal to the leak definition specified in the various regulations. Where cylinders of calibration gases are obtained from industrial gas suppliers, the gas composition must be analyzed and certified by the manufacturer for a $\pm 2\%$ composition accuracy. In addition, a shelf life must also be specified by the gas supplier. Alternatively, the calibration gases can be prepared by the user in accordance with accepted procedures that will yield a mixture accuracy of $\pm 2\%$.

Other Monitoring Situations. There are some situations where an analyzer will not meet the requirements previously outlined. In these cases, the EPA has alternate considerations which are detailed in reference 1. Also, one screening instrument may not be satisfactory for screening, particularly where the analyzer does not respond to one or more constituents in the vapor. For these cases, a combination of instruments or detectors can be used and examples are presented in reference 4.

5.1.3 Screening and Monitoring Procedures

Calibration Procedures. When an analyzer is warm and has been internally calibrated for a zero concentration, the analyzer should be calibrated with a specific calibration gas. The instrument meter is then adjusted to the actual calibration gas concentration. When the meter cannot be adjusted, the instrument is considered as malfunctioning and corrective measures are required before the analyzer can be used. The calibration gas should be certified as described in the previous section.

An important consideration with analyzers is frequent calibration to insure the instrument does not drift. When an analyzer is used continuously, the instrument should be calibrated before use in the morning and after the morning's monitoring program is completed. Following completion of the afternoon monitoring program, the instrument should again be calibrated. This procedure not only indicates drift, but provides a basis for analyzing and adjusting the meter readings. In addition, the instrument always starts a monitoring round properly calibrated. The CMA suggests calibration early in the morning and after the completion of the monitoring program in the afternoon.[4] However, midday drift can affect the accuracy of the readings.

Quality assurance is provided by more frequent calibration, and the EPA suggests instrument adjustment and calibration where the reading is off more than $\pm 5\%$ with a concentration standard and more than $\pm 20\%$ with a low concentration standard.[1] Since the low concentration standard becomes important with the new regulations, calibration with a low concentration standard should also have maximum limits of $\pm 5\%$.

Screening Equipment. Details of the screening procedure are presented in Method 21.[2] A general overview of screening or monitoring is presented in the next few paragraphs, based on the analyzers previously described. Each analyzer has a probe or extension tube connected to a flexible hose, which in turn is connected to the analyzer. A pump in the analyzer draws a sample through the probe and into the detector. The gas effluent from the detector is discharged to the atmosphere through the pump. A small vacuum is produced in the system to obtain the sample.

In general, the probe tip is placed at the interface where a leak can occur. Location of the probe tip a small distance from the interface significantly reduces the concentration and produces erroneous readings. Also, the probe should be perpendicular to the interface and not tangential to the interface. After placing the probe at the interface, the probe tip is moved along the interface until the probe reaches the initial probe start location. If an increased leak reading is observed during the probe transit of the interface, the probe should be left at the maximum reading location a length of time that is approximately twice the instrument response time. The maximum reading is then recorded as the screening value. When the concentration exceeds the instrument scale, a dilution probe can be used to allow measurement of the concentration. However, the instrument must also be calibrated with a dilution probe.

The probe tip should not be fouled with grease, dust, or liquids. Frequently, a short flexible piece of PTFE tubing is used at the end of the probe which can also be cut when the tip becomes fouled. Where the probe may encounter particulates, the tubing can be packed with fiberglass as a filter or a filter can be placed in the probe. These filters must be cleaned frequently to prevent fouling, but washing with water or any fluid requires drying the filter before reuse to ensure the detector is not affected.

General screening procedures are outlined for various equipment types:[2]

1. *Valves.* The interface between the valve stem and the gland follower is a leak source. The probe tip is placed at the interface and then moved around the stem circumference. A maximum reading is recorded as the screening value. In addition, the opening between the follower gland and the bonnet should be screened along the entire periphery of the bonnet.

2. *Flanges and Other Connections.* For flanges, the probe tip is placed at the outer edge of the flange–gasket interface and the circumference of the flange is monitored. Threaded connections are also screened with a circumferential traverse at the threaded connection interface.

3. *Pumps and Compressors.* When the leak source is a rotating shaft, the probe tip is positioned within 1 cm of the shaft–seal interface. Should the housing configuration prevent a complete traverse of shaft periphery, all accessible portions are sampled. All other joints on the pump or compressor housing where leakage may occur are monitored.

4. *Pressure Relief Devices.* Screening of these devices is required where the relief device discharges to the atmosphere. The relief valve should be sampled

around the sealing seat interface. Where the discharge opening is extended with pipe or horn, the probe tip is placed approximately in the center of the exhaust opening to the atmosphere.

Since the relief valve or device can discharge to the atmosphere, caution should be exercised in monitoring the valve discharge. During process upsets, screening should be delayed until the unit is under control and operating satisfactorily.

5. *Process Drains.* For open drains, the probe tip is placed in approximately the center of the atmospheric opening. Where the drains are covered, the probe tip is placed at the intersection of the cover or where an interface exists and the periphery is screened.

6. *Degassing and Accumulator Vents.* The probe tip is placed approximately in the center of the opening to the atmosphere.

7. *Access Doors.* These doors may be located in enclosures and are usually gasketed. However, the probe tip is placed at the door seal interface and a peripheral traverse is conducted.

Recordkeeping. A considerable amount of data are accumulated during monitoring surveys and must be recorded. Since the new regulations have low leak definition values, the actual screening values should be recorded for leak frequency analysis, trend changes, satisfactory equipment performance, and background in the event a QIP is necessary. A satisfactory database will minimize the effort necessary to achieve the QIP goals. Field data collected should include the following:

- Data
- Hydrocarbon detector type and instrument number
- Process unit or area
- Source identification—while tags are not specifically required in the HON regulations, a permanent identification system is recommended to simplify maintenance, retested, and performance comparisons.
- Screening value in ppmv
- Source equipment (valve, pump, etc.)
- Equipment type, such as valve type (control valve, check valve, etc.)
- Stream identification—vapor, light, or heavy liquid with reference to stream composition on detailed flow sheets or accompanying analysis tables
- Response factors—these are developed separately and should be included from a master response factor program base
- Final screening value based on the response factor in ppmv
- Instrument calibration periods during the day

5.1.3 Mass Emission Measurements

An important consideration in emissions quantification is determining actual mass leakage rates from valves, flanges, pumps, and so on. The methods described in this section for mass emission measurement are time consuming, but through measurement of selective equipment, components, and screening or monitoring data, various correlation procedures provide methods of estimating the emissions from various equipment components. Although the regulations covering equipment leak controls are based upon leak concentration definitions and leak frequency, actual or estimated mass emission rates are necessary to meet other regulatory VOC and HAP requirements. In addition, mass emission estimates also provide the regulatory authorities with a basis for observing the reduction in emission rates associated with regulatory controls.

Mass emission measurement from an equipment component is generally accomplished by enclosing the valve or flange, and measuring the emissions rate. This encapsulation technique is generally termed "bagging" to describe the procedure, which differs from screening or monitoring. The two mass emission measurement systems acceptable to the EPA are described in references 1 and 4. The measurement methods are described as the vacuum technique or the blow-through procedure. The vacuum procedure was developed by the EPA and has provided mass emission data for many years in various EPA research programs. A second technique was developed (blow-through) that is somewhat faster than the older method and is a safer procedure where potentially explosive or flammable mixtures may occur. Both procedures are described in the following paragraphs and are based upon a bag or enclosure.

Vacuum Technique or Method. The basic vacuum system is shown in Figure 5.1 where valve mass emissions are measured. A bag or tent is placed around the valve packing box and is sealed above and below the packing section. Although ostensibly sealed, the bag is not completely sealed because of the difficulty of completely sealing the bag around a valve bonnet. The vacuum pump draws air through these openings in the sealed portions along with emissions from the valve packing into the line or tubing leading to the dry gas meter. The stream then flows through the vacuum pump to the atmosphere. From this vacuum pump feed line, a small stream is diverted into a sample bag that is taken to the laboratory for analysis of the vapor. A cold trap is shown in the line adjacent to the bag or tent to condense and remove any material that may otherwise condense in the line or meter without this feature. This trap is only required for vapors emitted from the equipment component that are close to the vapor dew point and on the verge of condensation. The magnahelic mater indicates the vacuum pressure in the tent and the manometer indicates the pressure at the dry gas meter. The latter pressure provides a basis for determining the actual flow rate through the system.

This overall system or train can be mounted on a portable cart for movement through the process facilities. However, the train should be a compact installation on the cart, permitting access in various locations, and be relatively light, allowing two men to carry the cart and equipment. While the cart and train are not overly cumbersome, the use of electrical vacuum and diaphragm pumps increases the amount of equipment since the motors require starter/controllers which must also be mounted on the cart. All electrical equipment must meet the applicable electrical safety codes for the process units or plant. Many of the chemicals and vapors are explosive and pumps are frequently air driven, only requiring a hose for connection to a utility station. The design of the train and cart are a function of the facilities in the plant, safety requirements, and flexibility necessary to allow temporary use of this system in all areas of the plant.

The tent must be carefully selected to ensure the tent material is impermeable to vapors and is unaffected by the vapors that will be encapsulated. In many

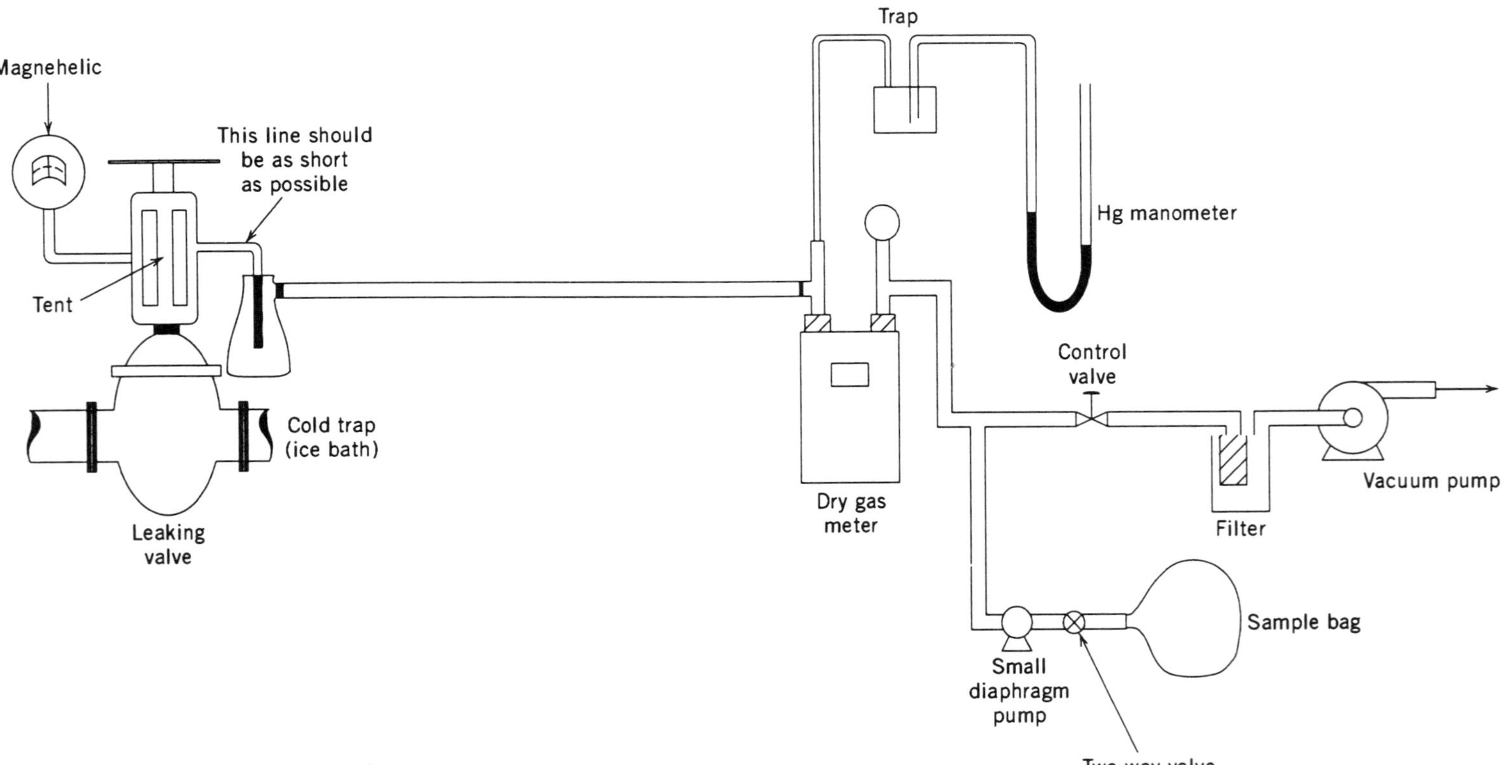

Figure 5.1 *Vacuum emission rate measurement train.*[1]

applications, polymeric bags or film are selected including PTFE and Mylar® that have very low permeability ratings. However, the allowable bag operating temperatures may not be sufficient in all applications.[4] Also, the bag or tents should be sealed with a tape that has a high tensile strength, is flexible, and can be tightened to an essentially fixed position; duct tape is frequently used. For applications at higher temperatures, aluminum bags or tents can be installed which are satisfactory for most of the high temperature applications. Although high-temperature measurements are not always required, the proper bags or tents must be selected and the remaining components of the train should be satisfactory for operating conditions. This covers the tape necessary for sealing the tent, tubing, pumps, and dry gas meter. Where the train cannot be installed for higher-temperature emission measurement without excessive cost or other problems, the blow-through procedure may be preferable.

The size of the vacuum pump, which should be a dry pump according to the EPA, must be sufficient to maintain a maximum flow rate of about 2.5 cfm (0.06 m^3/min) through the sampling train.[1] However, this capacity seems excessive for most leakage measurements and vacuum pumps of 0.5 cfm (0.012 m^3/min) have been satisfactory unless the leak is excessive.[4] With tight sealing of the bag around an equipment component, a vacuum pump will reduce the pressure in the tent significantly. In these situations, a hole is punched in the tent or bag on the side opposite to the outlet tubing line. The EPA indicates the usual vacuum within the tent is about 0.1 in. H_2O, which is slightly less than atmospheric pressure. This pressure is a result of leakage into the tent since sealing is not completely effective and sufficient air leaks into the tent to minimize the vacuum without a hole. A suggested leak rate from the atmosphere by the CMA is 0.01 cfm.[4] The vacuum system should be tested before use to insure the vacuum at the dry gas meter does not exceed 3 in Hg (7.62 cm Hg) and preferably is less.[4] In addition to the vacuum pump standard flow rate, the flow rate can be throttled by the control valve to the 0.5 cfm level. Throttling provides some flexibility in air–vapor mixture control, but the existence of a flammable or explosive mixture is potentially unsafe, and the blow-through system is preferable in these situations.

When sampling through the train, any condensate collecting in the ice trap must be included in calculating the total leak rate. This requires careful timing to insure excessive condensate is not recovered, thus biasing the leakage rate. When the vacuum pump is placed in operation, the concentration is measured at the vacuum pump discharge with a screening or monitoring analyzer. The vapor concentration increases with time and when the concentration stabilizes, the overall system is at equilibrium. At this point, a gas sample can be obtained in a lined sample bag. In addition to the process sample, another sample bag should be filled with ambient air to determine the background concentration. When the vapor sample from the train has been obtained, the vacuum pump is stopped, and the condensate or cold trap is sealed. The cold trap and the bag samples are then sent to the laboratory for analysis.

Following the test, the tent is removed, and the equipment component is screened. The screening value combined with the mass leak rate is important in the emissions analysis.

Blow-Through Method. In this system, nitrogen is blown through a tent or bag to eliminate explosive or flammable concentrations, providing a safe method of measuring emissions flow. In addition, considerably less train equipment is necessary in this system compared with the size of the cart system required for the vacuum pump method. The stand is portable and can be moved by one individual. Also, it can be rapidly placed in most plant locations, resulting in a more flexible system. The EPA indicates the blow-through method is preferable for high emission rates, but the system also works with low emission rates and should be used wherever the leak constituents are explosive or flammable.

The basic below-through system sends nitrogen through a tent which leaks to the atmosphere through the tent seal. While the tent is sealed in this procedure, a very tight seal is not required. In this procedure, nitrogen fed into the bag leaks to the atmosphere and a pump or other equipment is unnecessary. Nitrogen flow dilutes the oxygen in the tent and reduces the oxygen concentration to levels that are less than 5%. Since the tent seal is not leakproof, a low oxygen equilibrium level is attained over a short time period. However, leakage from the equipment into the bag does not develop flammable mixtures.

In this system, nitrogen obtained from a plant or process unit header is passed through a desiccant tube followed by an activated carbon tube to dry and remove any contaminants from the nitrogen stream. The treated stream flows through a rotameter and exits through a "Y" fitting into two flexible tubing lines that connect with the tent. Nitrogen flow can be controlled by a pressure regulator at the dryer inlet or a manual control valve. The pressure in the tent does not exceed 1 psig (6.9 kPag) and is only slightly higher than the atmospheric pressure in most installations. Nitrogen flow enters the tent through both tubing lines, with entrance openings on opposite sides of the tent. A third hole is placed in the tent for insertion of probes. This hole is normally covered with tape when not in use.

The drying and carbon tubes along with a rotameter are bolted to the stand which is located adjacent to the equipment component under test. Tubing connecting the stand and nitrogen source is installed and a short tubing length is placed on the rotameter outlet. A "Y" connection is inserted into the tubing and the "Y" or "T" is connected to two tubes for the injection of nitrogen into the bag. The selected bag material is roughly cut to size, and the tubes from the rotameter are connected to the bag material. Nitrogen flow through the tubes is started at roughly 40 L/min to purge the bag and the bag or tent is then is placed around the equipment component.[1] The initial flow of nitrogen aids in rapidly reducing the oxygen concentration within the bag enclosure. After the bag or tent enclosure is complete, tape or wire seals the bag opening. At this point, the third opening for the probes is placed in the tent, normally between the tubing inlet openings, and is covered with a piece of tape. After a short time, an oxygen probe is inserted into the opening or third hole. The oxygen concentration should be falling at this point, and if the concentration does not fall below 5%, the seal tightness should be adjusted (lessened) until the oxygen concentration is $< 5\%$. When the oxygen concentration is satisfactory, the temperature within the tent is measured through the third hole with a platinum–RTD thermocouple.[1] This thermocouple can be changed in accordance with recommendations from the instrument group.

Following the temperature measurement, the analyzer probe is inserted through the hole and the VOC or HAP concentration is measured at several points to ensure the concentration is similar throughout the tent. The VOC or HAP concentration in the EPA and CMA procedures[1,4] is generally measured with an

OVA® analyzer (Foxboro organic vapor analyzer). Concentrations are then measured within the tent at three different nitrogen flow rates, roughly 40, 30, and 20 L/min. Lower flow rates aid in measuring the concentration at low emission rates. Between the flow rate changes the oxygen concentration is measured to insure the oxygen concentration is less than 5%. Any condensation on the inside of the tent is collected in a graduated cylinder. Liquid dropping from the equipment component is collected and measured but is not included in the vapor leak rate. The equipment component should be screened before the tent system is installed and can be monitored after the bag or tent is removed.

All of the instruments used in this procedure have a Class 1, Division 1 rating to ensure an ignition source is not present. With bags or tents frequently installed in congested areas, safety is very important and all equipment should be reviewed with safety personnel prior to use. In addition, the probe should be calibrated for methane or hexane at a known concentration in nitrogen. The analyzer has a different response to nitrogen as the dilution gas compared with air. In addition, a response factor based on the compound with nitrogen is also necessary, particularly at low VOC or HAP concentrations. The EPA and CMA references[1,4] frequently refer to calibration with a dilution probe on the OVA. The dilution probe reduces the concentration of the sample entering the detector to a measurement zone that is acceptable for the analyzer. A dilution probe is normally placed on the OVA for high emission concentration applications and is not required where the leak concentration is < 9000 ppmv. For the more stringent regulatory leak definitions (500 and 1000 ppmv) in the new regulations, the number of very large leakers will be significantly reduced, and the necessity for a dilution probe will probably be quite limited. Where the dilution probe is used, calibration with the dilution probe is necessary, and the instrument should again be calibrated when the dilution probe is removed.

Since direct volumetric readings are not applicable with this mass emission measurement procedure, a procedure for determining the emissions rate has been developed. The basic flow rate through the tent is adjusted for the oxygen concentration followed by the emission rate calculation.

$$Q = \frac{0.06\mathrm{N}_2}{1 - (\mathrm{O}_c/21)} \tag{5.3}$$

where

Q = flow rate into tent (m^3/hr)
N_2 = nitrogen flow rate (L/min)
O_c = oxygen concentration (%)
0.06 = conversion factor (1/m to m^3 hr)

The actual mass emissions rate is determined from the following relationship:

$$\mathrm{ER} = \frac{4.836 \times 10^{-5}(Q)(\mathrm{MW})(\mathrm{OVA})(\mathrm{RF})}{T + 460} \tag{5.4}$$

where

ER	= emissions rate (lb/hr)
MW	= gas molecular weight (lb/lb-mol)
T	= tent temperature (°F)
OVA	= instrument reading minus background reading (ppmv)
RF	= response factor
4.836×10^{-5}	= conversion factor including gas constant and tent pressure of 1 atm

$$\text{Gas mixture MW} = (\text{mol fract}_1)(\text{MW}_1) + (\text{mol fract}_2)(\text{MW}_2) + \cdots \quad (5.5)$$

where the subscript refers to vapor component $1, 2, \ldots$.

The gas response factor for the mixture calculated at the OVA concentration is determined from Eq. 5.2:

$$\text{RF} = (\text{mol fract}_1)(\text{RF}_1) + (\text{mol fract}_2) + \cdots$$

Industrial Hygiene. While tented procedures are necessary to determine the mass emission rate from equipment components after completion of the measurements, the tent concentration of VOC or HAP remains elevated compared to the basic screening concentration. Opening and removing the tent can result in an elevated short-term exposure particularly with highly toxic compounds. In addition, in the vacuum system the pump may have a somewhat elevated concentration if immediately shut down without purging the system. Consequently, in the vacuum system, after the tent is opened, the pumps should be run on air to purge them. For the blow-through system, continuous nitrogen injection during tent removal may dissipate the vapors. However, both methods should be reviewed by the industrial hygienist to insure potential exposures are minimized.

In addition, these measurements may be obtained in small crowded areas surrounded by equipment at high-temperature levels. Emergencies under these conditions can be a potential concern and operating conditions and procedures should be reviewed with measurement personnel.

5.2 ESTIMATING EMISSIONS

A very important consideration in any petroleum refinery or chemical plant is obtaining actual emissions rates to meet various EPA, state, and regulatory authority regulations for existing and new plants, such as the annual emissions estimate that must be compiled for each operating facility in accordance with SARA regulations, which is then published by the EPA. These estimates initially covered VOC compounds, but an annual estimate of total emissions that includes HAP compounds is now required. In addition, estimates of emissions from new process units or facilities are necessary for review by the various authorities to ensure the emissions rates are within regulatory requirements. Regional or state emission control requirements may be considerably more restrictive than federal regulations for new facilities. While the term emission rates can cover a large number of sources, in this chapter it refers to equipment such as valves, pumps, and so on. Emissions from other sources such as tanks or wastewater points are calculated on a different basis and the emission estimating procedures are described in their respective chapters.

There are various methods of estimating VOC and HAP emissions, and these procedures are described in this section. However, some of these techniques are based upon statistical evaluations and can become quite complex. Increasing complexity also increases the amount of data and analytic evaluation required. As a result, the CMA published a document that describes some of the statistical evaluation procedures.[4]

The basic estimation methods have been developed by the EPA and are considered protocols.[1] These are standard procedures with the objective of providing procedures from generating unit-specific emission estimates. While the EPA has developed the methodologies for specific process units or groups of sources, it has provided methods of developing estimates that are not limited by the group of base sources. The EPA has essentially devised five methods for estimating emissions which are included in the protocol[1] and are listed below:

- Average emission factor method
- Leak/no leak emission factor method
- Three strata emission factor method
- Application of EPA correlations
- Development of new correlations

These methods start with a straightforward equipment count and average emission factor and become increasingly complex. An overall approach or strategy developed by the EPA to quantify emissions is shown in Figure 5.2.

5.2.1 Equipment Information

A very important parameter in any of the emissions estimating procedures is an equipment count. Although seemingly simple, there can be numerous valves and flanges on a process unit and several sources of estimating the amount of equipment in a category are available in various references.[4,6,7] However, actual counts of the equipment should be available from detailed flow plans or field information. Many plants have had monitoring programs to control VOC emissions or NESHAP chemicals that require tagging and valve identification. Consequently, the number of valves should be included in various documents. Flanges may not have been counted, but a relatively close count can be accumulated in each unit based on detailed flow diagrams and an understanding of process unit piping systems. Other equipment counts can readily be determined and do not present any equipment count concerns.

While equipment counting may be somewhat difficult, HON emissions control is based upon leakage frequency and total equipment quantities in a category are required to comply with EPA regulations. However, additional criteria are also

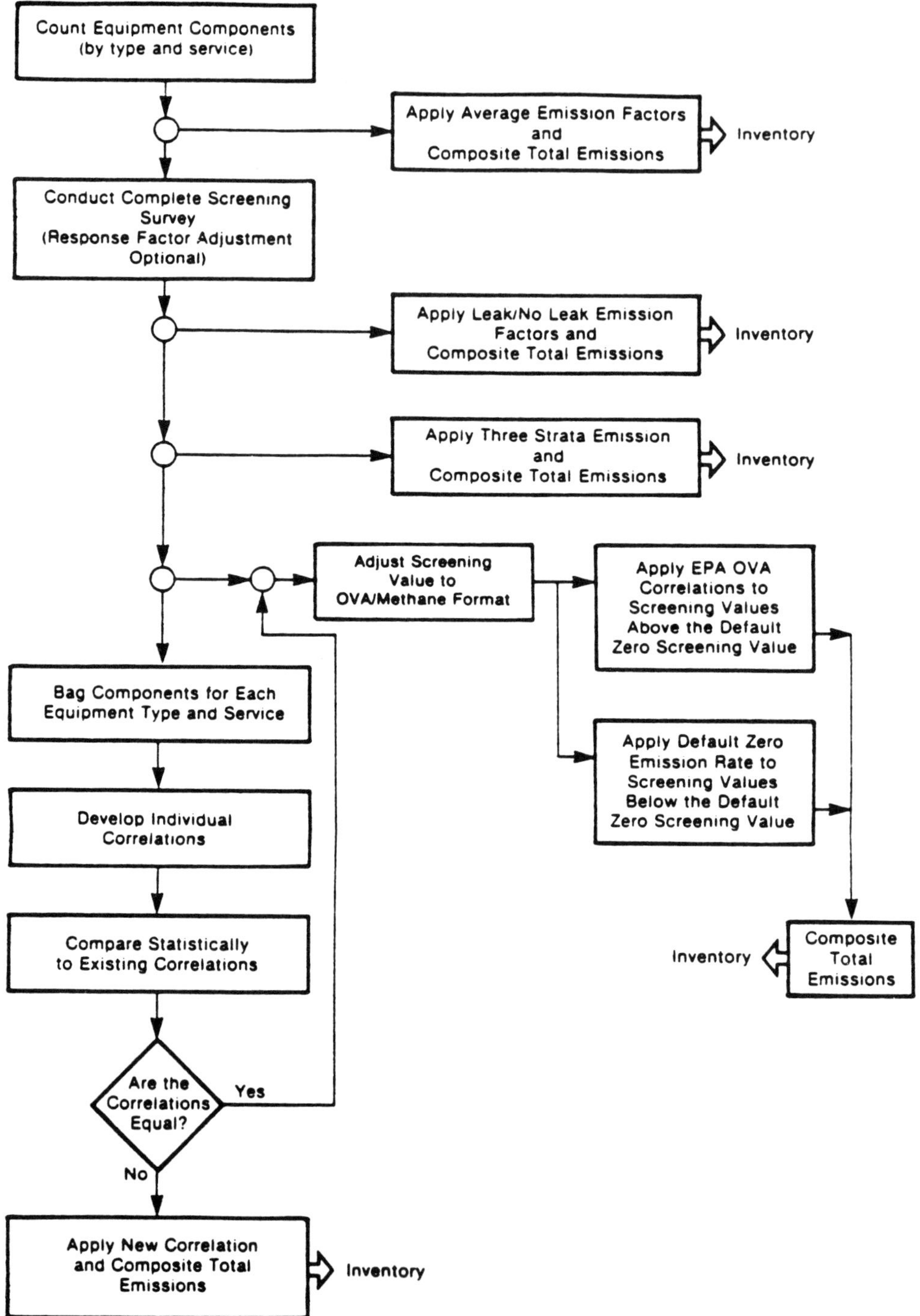

Figure 5.2 *EPA strategy for estimating equipment emissions.[1] (Reference 1)*

TABLE 5.3 Average Emission Factors for Equipment

Equipment Category	Service	Emission Factors (kg / hr) Refineries[a]	SOCMI[b]
Valves	Gas	0.27	0.0056
	Light liquid	0.011	0.0071
	Heavy liquid	0.00022	0.00023
Pump seals	Light liquid	0.11	0.0494
	Heavy liquid	0.021	0.0214
Compressors	Gas/Vapor	0.64	0.228
Pressure relief valves	Gas/Vapor	0.16	0.104
Flanges	All	0.00025	0.00083
Open-ended lines	All	0.0023	0.0017
Sampling connections	All	0.015	0.015

[a]Adapted from reference 10.
[b]Adapted from reference 1.

necessary, such as light and heavy liquid designations. Equipment in heavy liquid service, when identified, is not included in leak frequency determinations. While these HON requirements are limited to the initial HAP chemicals list, the overall list will be expanded through the years and complete equipment counts will probably be required for most process units. Petroleum refineries have been monitoring various equipment categories for VOC emissions control and equipment counts should be available. Additionally, new Clean Air Act regulations for petroleum refineries are expected in 1994, and those units without an equipment count for leak frequency monitoring will need a reasonably accurate equipment count estimate to provide an analytical base.

Based upon equipment counts, the amount of monitoring information and bagging tests can be defined to develop emission correlations. In those situations where the category equipment count is quite small, the unit may consider bagging all of the equipment. This procedure completely defines the total emissions rate for an equipment category.

5.2.2 Average Emission Factor Method

From an accurate equipment category count, the emission rate for a category is determined by multiplying the average emission factor shown in Table 5.3 by the equipment count. This is a relatively simple method of estimating emissions that was developed in an EPA petroleum refinery study.[8,9] The factors were developed from screening and mass leak rate correlations that are explained in a subsequent section. In addition, the EPA coordinated a study of 24 units in the chemical industry[10] along with a six-unit maintenance study[11] that provided an emission factor for valves in gas service. Leak rate and screening correlations were developed for light liquid valves and pumps and were also converted into average emission factors, as shown in Table 5.3. Other emission factors are used by the states and regional authorities.[12] Allowable values can be obtained from the various regulatory groups.

While this procedure provides a rapid method of estimating emissions, the total emission rate will be considerably larger than emissions totals calculated with other methods. However, the procedure does provide a quick estimate that can be used as a general indication of emissions, although it is not particularly accurate. One reason for its disparity with other procedures is the lack of emission control during the EPA test periods. Several years after the EPA tests were concluded, petroleum refineries and chemical plants were required to monitor equipment categories to control VOC emissions. These regulations also defined the leak as $\geq$ 10,000 ppmv for valves, flanges, and pumps. Lower leak levels were required for benzene and vinyl chloride in the NESHAP regulations. These regulations were promulgated after the study, including state and regional authority requirements, and resulted in significant reductions in the number of leaking items in each equipment category. As a result, the average emission factor in a process unit has been reduced considerably, but the EPA has not changed these average emission factors. However, the EPA has recognized the reduction in leakage associated with leakage monitoring by providing adjustments for these average factors when estimating emissions for new plants.[11] For valves in gas service monitored on a monthly basis, the emissions are reduced 74%, which changes when they are monitored on a quarterly basis. Moreover, emission factor adjustments are also available for pumps, flanges, and safety valves.

In addition to monitoring requirements, equipment changes have also occurred which are not reflected in these various factors. Pump seals have improved along with increasing numbers of dual seal installations. Improved valve packings have reduced emissions for block and control valves. Consequently, the uncontrolled emission factors should be used where quick, rough estimates are satisfactory. Modification of the uncontrolled emission factors with monitoring adjustment factors will provide a more accurate emissions estimate.

5.2.3 Leak/No-Leak Emission Factor Method

This method is based upon the separation of equipment components into two populations that leak or do not leak, where the leak is defined as $\geq$ 10,000 ppmv. Consequently, the number of valves, as an example, can be separated into leaking sources and nonleaking sources with an emission factor for each source classification. The EPA developed leaking and nonleaking emission factors on a basis similar to that provided for the average emission factor, which is described in reference 11. This development was based upon emissions factors generated from empirical screening distribution data and leak rate screening correlations, leak frequencies associated with the emission factors, and the mass emissions percentage associated with the leaking sources. The leakage factors are shown in Table 5.4 where the gas/vapor SOCMI values were obtained from early EPA chemical plant studies.[1]

As an example of the use of these factors, a hypothetical process unit is evaluated in Table 5.5, which indicates the number of sources screened and leaking ($\geq$ 10,000 ppmv). The emission estimate was calculated for each source category based on the following relationship and the emission factors from Table 5.4:

$$\text{Emission estimate} = \frac{\text{LEF} \times \text{PCL} + \text{NLEF} \times (100 - \text{PCL})}{100} \tag{5.6}$$

TABLE 5.4 Leaking and Nonleaking Emission Factors for Equipment[a]

Equipment Category	Service	Emission Factors (kg / hr) Leaking (≥ 10,000 ppmv)	Nonleaking (< 10,000 ppmv)
Valves	Gas	0.0451	0.00048
	Light liquid	0.0852	0.00171
	Heavy liquid	0.00023[b]	0.00023
Pump seals	Light liquid	0.437	0.020
	Heavy liquid	0.3885	0.0135
Compressors[c]	Gas	1.608	0.0894
Pressure relief valves	Gas	1.691	0.0447
Flanges	All	0.0375	0.00006
Open-ended lines	All	0.01195	0.00150

[a]Adapted from reference 1.
[b]Assumed equal since computed leaking factors is less than nonleaking factor.
[c]Reflects 60% control level in industry through barrier fluid / degassing system and control device.

where

Emission estimate = leakage (kg/hr)
LEF = leakage emission factor
NLEF = nonleaking emission factor
PCL = percent leaking sources

While this approach is also straightforward, based upon developed factors, the emissions calculated with this procedure are less than the emissions calculations based upon the average emission factors (Section 5.2.2). Although based on a 10,000-ppmv leak definition, the EPA does not plan on changing these values for

TABLE 5.5 Leak/No Leak Emissions Estimate Example[a]

Source	Number Screened	Number Leaking	Percent Leaking	Estimated Emissions (kg / hr) (Source Basis)[b]	Total
Valves					
Gas/vapor	625	19	3.0	0.0018	0.3765
Light liquid	1180	13	1.1	0.0026	3.1030
Heavy liquid	64	0	0	0.00023	0.0147
Pumps					
Light liquid	47	3	6.4	0.0391	1.8390
Heavy liquid	3	1	33.3	0.1385	0.4155
Flanges	2880	20	0.7	0.00032	0.9216
Pressure relief valves					
Gas/vapor	31	1	3.2	0.0978	3.032
Open-ended lines	278	9	3.2	0.0018	0.5096
Compressor seals	4	0	0	0.0894	0.3576
Sampling connections	70	—	—	0.0150	1.05[c]

[a]Adapted from reference 1.
[b]Average based on total number of sources in category.
[c]Intermittent.

TABLE 5.6 Stratified Emission Factors[a]

Source	Service	Screening Value Range (ppmv)		
		0 – 1000	1001 – 10,000	> 10,000
Valves	Gas/vapor	0.00014[b]	0.00165[b]	0.0451[b]
	Light liquid	0.00028	0.00963	0.0852
	Heavy liquid	0.00023	0.00023	0.00023
Pumps	Light liquid	0.00198	0.0335	0.437
	Heavy liquid	0.00380	0.0926	0.3885
Flanges	All	0.00002	0.00875	0.0375
Compressor seals	Gas/vapor	0.01132	0.264	1.608
Pressure relief valves	Gas/vapor	0.0114	0.279	1.691
Open-ended lines	All	0.00013	0.00876	0.01195

[a]Adapted from reference 1.
[b]Emission factors are given in kg/hr.

the new leak definitions of 500 and 1000 ppmv in the HON regulations. These emissions estimates will not be accurate, but they would have application in quickly reviewing improvements in performance and estimating emissions on existing plant operations for new facilities on a general basis. The EPA prefers plant emissions estimates based on plant data (including mass leak rates) and correlations for a statistical analysis.

5.2.4 Stratified Emission Factor Method

This method is a refinement of the leak/no leak approach where the emission factors are developed for three screening (monitoring) ranges. These ranges and factors are determined in identical fashion to the previous emission factors and are based upon extensive screening and mass leak rate data accumulated by the EPA. As a result, various population groups and emission factors can be determined. The EPA screening ranges are shown in Table 5.6 and cover screening ranges of 0–1000, 1001–10,000, and > 10,000 ppmv. With the changes in leak definitions for the new regulations, the stratified emission factors and screening ranges can be modified to reflect the lower emissions levels that will be achieved through compliance with the 500- and 1000-ppmv leak definitions. However, considerable screening and leak rate data are required in the 0–1000 ppm range to separate this region into two or more populations. At present, the EPA is considering discarding this estimating method in favor of statistical emissions correlations, which are a more accurate estimating procedure.

While the EPA may not recognize this estimating procedure in the future, it provides a better emissions estimate than the previous procedures. In addition, for rough estimation the ranges provide a better basis for analyzing leakage performance and general emissions rate. However, the screening data must be properly recorded to adequately define equipment counts within a range. When these counts have been prepared, the emissions estimate is determined by multiplying the equipment counts in each range with the associated emission factor. An example prepared by the EPA is presented in Table 5.7 for a hypothetical situation.

TABLE 5.7 Stratified Emissions Factor Estimate Example[a]

Source	Number Screened	Screening Value Ranges (ppmv) 0–1000	1001–10,000	> 10,000	Estimated Emissions (kg / hr)
Valves					
Light liquid	1180	1020[b]	147[b]	13[b]	2.81
Heavy liquids	64	63	1	0	0.147
Pumps					
Light liquid	47	32	12	3	1.791
Heavy liquid	3	1	1	1	0.485
Flanges	2880	2600	160	20	2.202
Compressor seals	4	3	1	0	0.298
Pressure relief valves Gas/vapor	31	25	5	1	3.656
Open-ended lines	278	236	33	9	0.399

[a]Adapted from reference 1.
[b]Source number in each range.

5.2.5 Leak Rate/Screening Value Correlations

The EPA amassed a considerable amount of data from various test programs and developed correlations or equations that relate screening values to mass emission rates. Equations that describe the correlations for valves, pump seals, and flanges are shown in Table 5.8 and can be applied to all equipment types and services. This basis for estimating is the most accurate of the previous methods described for estimating emissions. However, the database for these equations includes a considerable amount of emission and screening data obtained prior to the promulgation of restrictive emission control regulation. Consequently, these equations may not be a relatively accurate emissions estimating basis where emission control becomes even more restrictive and leakage frequency becomes an important control parameter. However, use of other correlation techniques requires adequate data and analysis to convince the EPA their correlations are not applicable to certain process units andfacilities. The EPA is not adverse to alternative correla-

TABLE 5.8 EPA Emission Prediction Leak Rate Equations for SOCMI Valves, Flanges, and Pump Seals[a, b]

Source	Instrument	Least Squares Equation[c]	Number Data Pairs	Correlation Coefficient (r)	Standard Deviation of Estimate
Valves[d]					
Gas/vapor	OVA®	$\text{nmLk} = 1.68(10^{-5})(\text{max OVA})^{0.693}$	99	0.66	0.716
Light liquid	OVA®	$\text{nmLk} = 3.74(10^{-4})(\text{max OVA})^{0.47}$	129	0.47	0.902
Flanges[e]	OVA®	$\text{nmLk} = 3.731(10^{-5})(\text{max OVA})^{0.82}$	52	0.77	0.520
Pump seals[d]	OVA®	$\text{nmLk} = 1.335(10^{-5})(\text{max OVA})^{0.898}$	52	0.81	0.650

[a]Adapted from reference 1.
[b]Equations applicable for nonmethane leak rates.
[c]nmLK = nonmethane leak rate (lb / hr); max OVA = maximum screening value (ppmv).
[d]See reference 13.
[e]See reference 9.

tions, but a satisfactory basis for change must be established based on actual plant emission data, which is apparently an EPA objective. This alternate program is described in Section 5.2.6.

Correlation Basis. The EPA correlations are based upon OVA measurements obtained in accordance with Method 21[2] and an instrument calibrated with methane. All screening measurements for use in these correlations must have the same instrument and calibration basis. Where other screening instruments have been used and not calibrated with methane, the data must be converted to an OVA–methane basis. There are published transformation relationships for this conversion situation, but where transformation information is not available, an instrument response curve must be developed that relates screening values to the concentration values in an OVA–methane system. Preferably, the proper screening transformation relationship is developed prior to field data collection.

Response Factors. These factors were described in Section 5.1.2 and response factors are applicable to all instruments that comply with Method 21. Where instruments are used that do not comply with Method 21, specific response factors are required. As described, response factors are not completely linear and where the correlation equations are used, estimates of screening concentration must be relatively accurate. Consequently, screening adjustments for nonlinear response factors are necessary, which requires developing response factor correlations in the laboratory.

Zero Screening Rates. Although seemingly unimportant, zero screening values and the associated mass rate are quite important. Instruments frequently show zero screening values (pegged), but some leakage has been found during bagging tests. The screening instrument sensitivity does not provide any basis for correlating the screening value with a mass emissions rate. Moreover, the correlation equations predict zero emissions at a reading of zero on the portable screening instrument. The EPA has noted this anomaly in various tests and as a result has defined a "default zero" screening value of 8 ppmv and established mass emission rates for this default condition. In effect, all components screening at or below the zero default level of 8 ppmv are assumed to have an emission rate equal to the default zero emission rate.

The emission rates for zero default were obtained for valves in SOCMI gas/vapor service although the data appear questionable owing to the limited number of samples.[4] For all equipment types and services, default zero emission rates were established with the correlative equations in Table 5.8 for an 8-ppmv screening value. The default zero mass emission rates are shown in Table 5.9.

Although this does not appear significant for large numbers of items in an equipment service category, these values can become important. In Table 5.8, assuming the 1180 valves in light liquid service were in the zero default category, the estimated leak rate would be 0.532 kg/hr, which is about 20% of the total emissions estimated in the stratified emissions sample. As a result, the CMA recommends bagging a number of these valves or equipment components to establish a process-specific zero emissions rate. Bagging the equipment requires a

TABLE 5.9 Zero Default Mass Emission Rates[a, b]

Source	Zero Default Screening Value (ppmv)	Zero Default Emission Rate (kg / hr)
Valves		
Gas	8	0.000033
Light liquid	8	0.000451
Flanges	8	0.000093
Pumps	8	0.000039
All other components	8	0.000039

[a]Adapted from reference 1.
[b]Applies between zero instrument reading and the zero default screening value of 8 ppmv.

statistically valid sampling approach to establish acceptable emission rates. Both the EPA and CMA have procedures for developing statistically valid zero emission mass rates.[1,4]

The zero default emission factors are applied to equipment sources that screen between 0 and 8 ppmv with the correlations applicable to all sources with screening connections > 8 ppmv. A total emissions rate is developed by combining the emissions rates generated through screening correlations (> 8 ppmv) and the sum of the zero default emissions rates.

5.2.6 Unit Specific Correlations

For the new HON regulations (SOCMI) and others that will be issued during 1994 and subsequent years, the EPA apparently prefers the development by plant facilities of their own emission correlations from plant data. This general approach improves estimating accuracy and includes the reduction in emissions that accompany more stringent regulations. Lower monitoring leak definitions, controlled leakage frequency, and effective LDAR programs will have an impact upon emissions estimating procedures. The various factors developed through the years by the EPA from data accumulated under entirely different conditions are not applicable to operations under these more restrictive conditions. As a result, plants may have to develop the capability of statistically analyzing the plant emission data or obtain the assistance of statistical specialists. While the EPA and CMA have provided information on statistical emission evaluation procedures[1,4] a considerable statistical understanding is necessary to insure the emission rates are calculated properly and the procedures are acceptable to the EPA. The EPA requires rigorous evaluations and defense of a selected statistical emission calculation procedure. A computer program under development by the CMA (Possee®) will be of assistance in the evaluation of various emission calculation procedures.[14]

The EPA permits a plant to develop emission correlations for a process unit where the leak rates are statistically different from the EPA correlations. This requires a rigorous evaluation of plant data and analysis to demonstrate the difference between EPA and plant emission results and provide the basis for acceptance of the analytical procedures by the EPA. For the analysis, various information and steps are necessary which are outlined in the following summation:

- Mass emission data accumulation and leak rate calculations
- Development of leak rate and screening value calculations
- Statistical evaluation of leak rate/screening value correlations
- Application of the correlation to the empirical screening data
- Emissions predictions

These various steps are reviewed in more detail in the following paragraphs.

Mass Emissions Data Generation. When a process unit has been completely screened, the EPA indicates selected sources can be bagged to measure the mass emission rate. However, the source should also be rescreened prior to the bagging emission test. The mass emission rates and rescreening values can then be used in validating the EPA correlations or demonstrating the process unit does not conform with the correlations. With a statistically designed program, correlations can be developed for the specific process unit.

The screening instrument and measurement procedures should meet the procedures previously described and referenced. All screening instrument results must be converted to those that would be obtained with an OVA instrument calibrated with methane. The EPA correlations are based upon methane-calibrated OVA instruments. In addition, any screening source that has readings greater than the saturation point of the detector should be bagged to quantify the emission rate.

While bagging is necessary, the number of bagging tests depends upon the objective of the data collection program. Where the purpose of the procedure is investigating the application of EPA correlation equations to a specific process unit, the EPA indicates that four leak rate measurements for a particular source category may be satisfactory.[1] However, the CMA suggests that at least six bagging tests should be conducted,[4] including data from screening sources less than the lower detection unit, within the range of detection, and measurements where the maximum screening value is exceeded.[4,14] Further information on the bagging test procedure requirements is outlined in reference 4.

In cases where new correlation equations will be developed, the EPA recommends that 30 mass leak rate measurements should be obtained. The EPA indicates that a lesser number of measurements results in an inability to detect differences between correlations with less than 30 data pairs (bagging and screening). However, the statistical goal is the generation of estimates that are within 50% of the mean value with a 95% confidence level and when demonstrated that the objective can be achieved with less than 30 data pairs, the EPA will accept this basis. Statistically significant data pairs (bagging and screening) must be developed to control the amount of bagging, and various procedures are described for this condition.[4]

The EPA has developed some potential cases for demonstrating statistical requirements and one example assumes a large population of sources having screening values that range from 0 to 100,000 + ppmv.[1] For the development of statistically valid leak rate/screening correlations, bagging data should be collected over the entire screening range from individual sources. For each source type (valves, etc.) and service (gas, etc.), a random sample of six sources should be selected for bagging in the following screening range (ppmv):

1–100
101–1000
1001–10,000
10,001–100,000
> 100,000

When the maximum instrument detection limit is 100,000 ppmv, then 20 or all sources, whichever is less, should be bagged. When six sources are not available in a screening range, additional sources from the nearest range should be bagged. A considerable amount of bagging is required in this procedure and a detailed statistical evaluation may change the bagging requirements. Moreover, bagging equipment sources that have screening values > 100,000 ppmv may be a potential source of exposure, and bagging requirements at high screening values should be minimized.

The EPA also indicated that a situation may occur in a process unit where the maximum screening values do not exceed 10,000 ppmv.[1] This level may be a more practical example of current emission performance in well-operated plants. Moreover, the new regulations will probably result in the majority of process units being in this category. Consequently, the previous groupings can be changed to the following screening ranges in ppmv:

1–100
101–300
301–1000
1001–3000
3001–10,000

In addition, where the maximum screening levels are < 1000 ppmv, the EPA indicates that similar groups can be developed. Regulatory emissions reductions will probably result in upper screening levels between 1000 and 3000 ppmv in well-operated plants. However, where statistical evaluations demonstrate that estimates within 50% of the mean value and a 95% confidence level can be achieved with less than 30 data pairs, the EPA recommends the bagging measurements shown in Table 5.10.

The previous EPA measurement groups and testing range numbers are only considered guidelines, but are based upon EPA experience measuring leak rates and developing equation correlations.[1] However, other strategies can be employed, but the plant must provide a strategy and statistical basis that are acceptable to the EPA.

TABLE 5.10 Total Leak Rate Measurements with Less than 30 Data Pairs[a]

	Total Number of Leak Rate Measurements			
Screening Range (ppmv)	4	8	12	20
1 – 100	2	2	3	4
101 – 1000		1	2	4
1001 – 10,000		1	2	4
10,001 – 100,000	2	2	3	4
> 100,00		2	2	4

[a]Adapted from reference 1.

Leak Rate/Screening Value Correlations. Based upon screening data and mass emission rates, leak rate/screening value correlations are developed by converting both screening and mass emission rates to logarithms. Plotting these variables and applying regression techniques results in a linear relationship described by the following equation:

$$\log(\text{leak rate}) = B_0 + B_1 \log(\text{screening concentration}) \qquad (5.7)$$

where B_0 and B_1 are parameters obtained from the curve.

The EPA indicates least-squares regression analyses are conducted to obtain the log relationship. This may be accomplished through other regression techniques such as linear regression, which is based upon the normal distribution of residuals (actual–predicted). Various regression techniques can be applied, but detailed evaluations of these methods are required before they can be used. The basic formulation of the curves provides some background for these correlations.

When an arithmetic plot or histogram is developed for a population, the distribution forms a curve that is skewed from the traditional bell-shaped curve. This is a result of emission rate scatter with many values near zero, some with rates a few orders of magnitude greater, and a few that are a significantly higher than the second group. As a result, the logarithm of each point on a curve results in a normal distribution. Consequently, the data are log normalized before being used in various statistical analyses and comparisons.

The log–log data points or equation can be converted to the original units for determining emission factors but cannot be converted directly. This requires converting the equation through a scale bias correction factor.[16] In all of the EPA emission factor and confidence limit developments, scale bias correction factors are included. These factors are basically a correction factor that is applied to the geometric mean producing the unbiased maximum likelihood mean. The scale bias correction factor is a method of obtaining the true mean of the measured population or the emission factor. Where the analysis is unfamiliar, the EPA suggests consulting a statistician.

Bagged sources with screening values > 100,000 ppmv, but where actual screening values cannot be determined, should not be included in the regression line previously described. Sources censored at 100,000 ppmv (actual screening valves cannot be determined) and not bagged should have leak rates estimated from the average leak rate for censored sources that were bagged.

Leak Rate/Screening Value Correlations and Statistical Considerations. In the second step a least-squares regression line was developed as a correlation that must be statistically evaluated. Where new correlations are developed, they are compared with the refinery and SOCMI correlations previously developed by the EPA. If new correlations differ statistically from the existing EPA correlations, the new correlations may be used in the development of unit specific emission estimates for the specific class of sources, providing the EPA accepts the statistical analysis. These correlations can be compared graphically or mathematically. If the new correlation does not differ significantly from the existing EPA correlations, the EPA correlations must be applied.

Graphical Comparisons of Correlations. When sufficient data provide a basis for the development of a correlation equation, this equation should be compared with the EPA predictive equation. This comparison can be conducted graphically, comparing the new equation with a graph of the EPA process specific equation shown in Figure 5.3 for valves in gas service. This EPA graph was originally obtained from reference 13 and represents early SOCMI data, which is also shown as the SOCMI curve in Figure 5.7 for gas valves.

In addition to the graph of the predictive equation (Fig. 5.3), a comparison can be shown with a log–log plot of the equation where the equation is linear. Standard EPA linear relationships[13] are shown in Figures 5.4–5.6 for pumps and

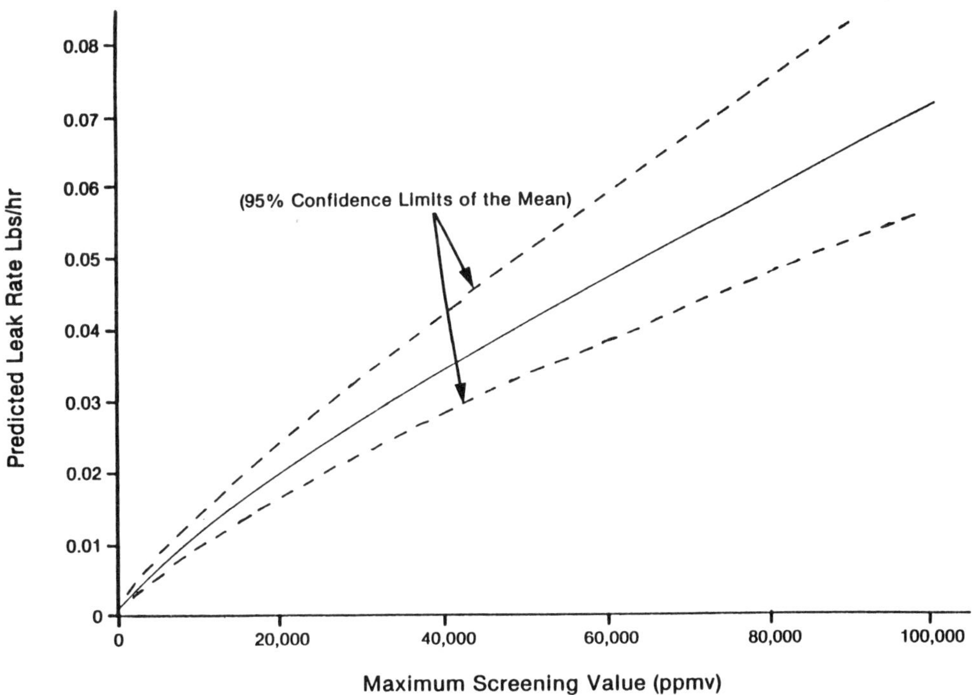

Figure 5.3 *EPA predictive equation for leak rate and OVA® screening values for gas service valves.*[1]

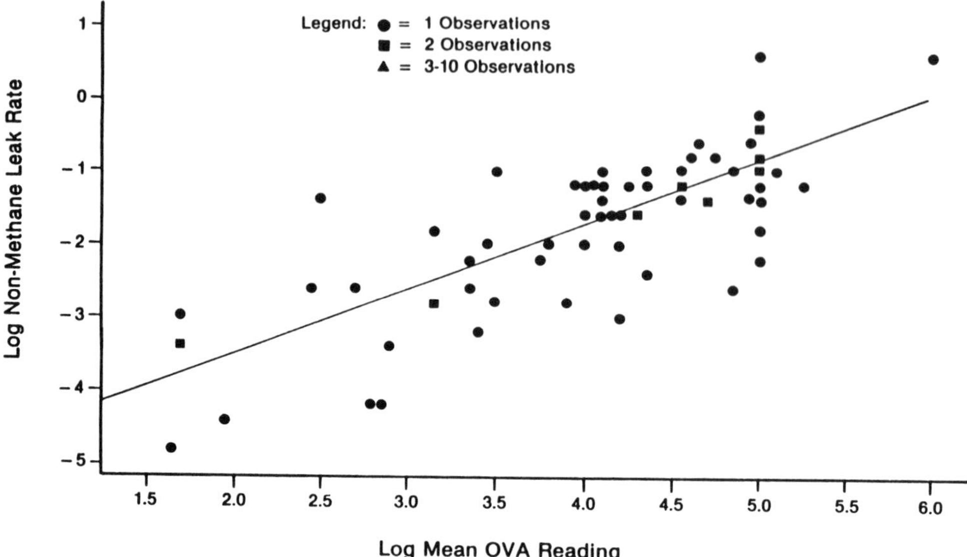

Figure 5.4 *EPA $log_{10} - log_{10}$ relationship for pump seals in light liquid service.*[1]

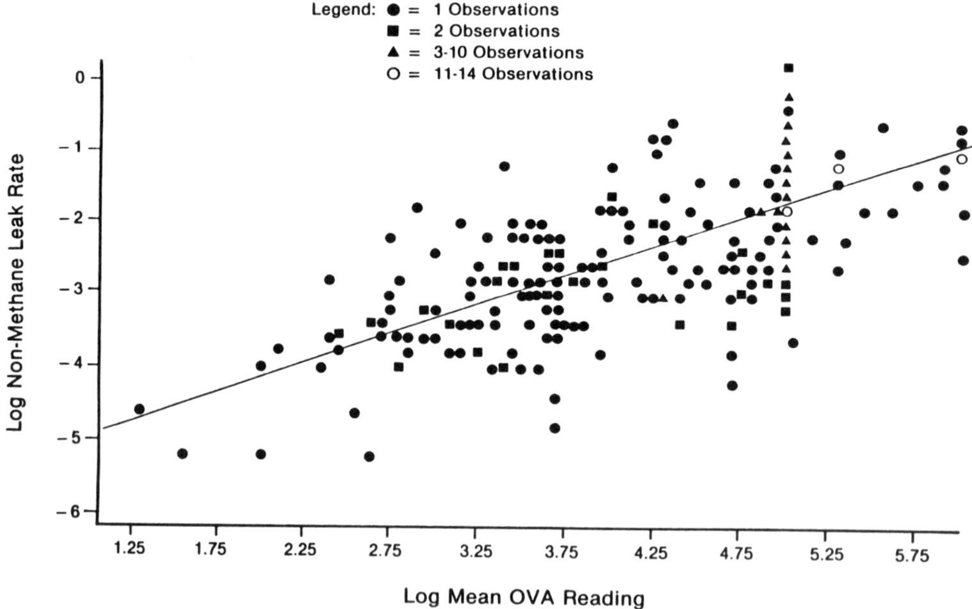

Figure 5.5 *EPA $log_{10} - log_{10}$ relationship for valves in gas service.*[1]

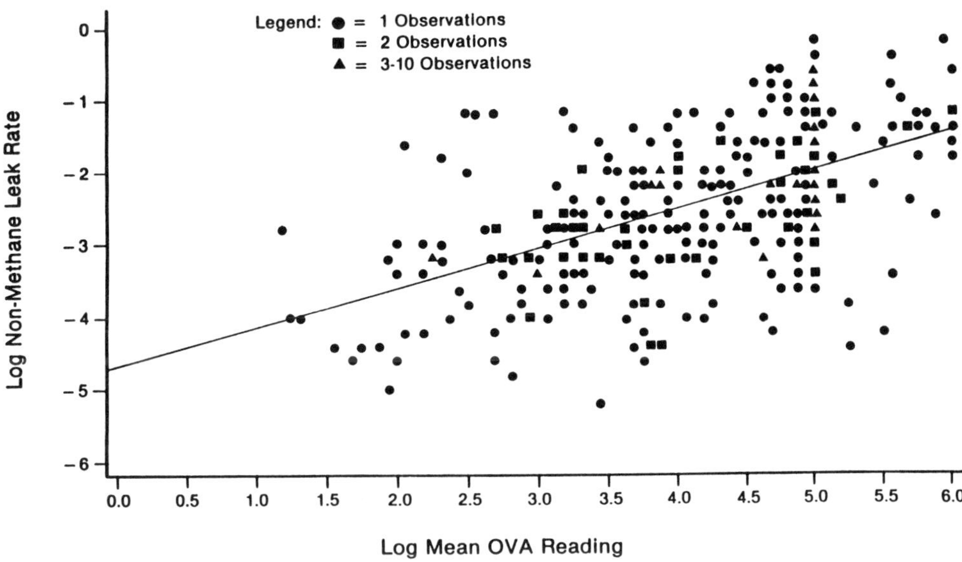

Figure 5.6 *EPA log_{10} – log_{10} relationship for valves in light liquid service.*[1] *(Reference 1)*

valves originally developed in 1980. The greater portion of data for these relationships was originally based on SOCMI data that was modified to include censored sources that apparently did not significantly affect the equations. While satisfactory relationships have been developed, the emission monitoring relationship does not reflect more than a decade of emission control improvements. Consequently, equations developed from current data may demonstrate a considerable difference from the EPA equations. These differences may increase further as the impact of more stringent regulatory requirements further reduces emissions.

When the actual process data are placed on the various curves representing the correlation equations, the EPA considers the distribution of the points in determining application of the predictive equations. Should the data points distribute randomly on both sides of the equation line and fall within the general range of data points used to develop the equation, the EPA considers use of the existing equation a reasonable assumption. In addition, statistical tests aid in determining whether application of the EPA equation is reasonable for the plant data accumulated from the test program. The EPA suggests the Wilcoxon Paired Replicate Rank Test for this investigation of appropriateness,[16] but other tests can be performed, and a statistical evaluation should be undertaken that will be satisfactory to the EPA.

In providing examples of differences in the data and associated equations, the EPA compares two equations developed from EPA data for petroleum refineries and chemical plants (SOCMI) which are shown in Figure 5.7. In the comparison of valves in gas service, the refinery correlation curve estimates or predicts larger leak rates than the SOCMI curve. Also, the confidence limits for the equations do not overlap, indicating the equations are different and represent two distinctly dissimilar types of process units. Although refinery valves in light liquid service also have

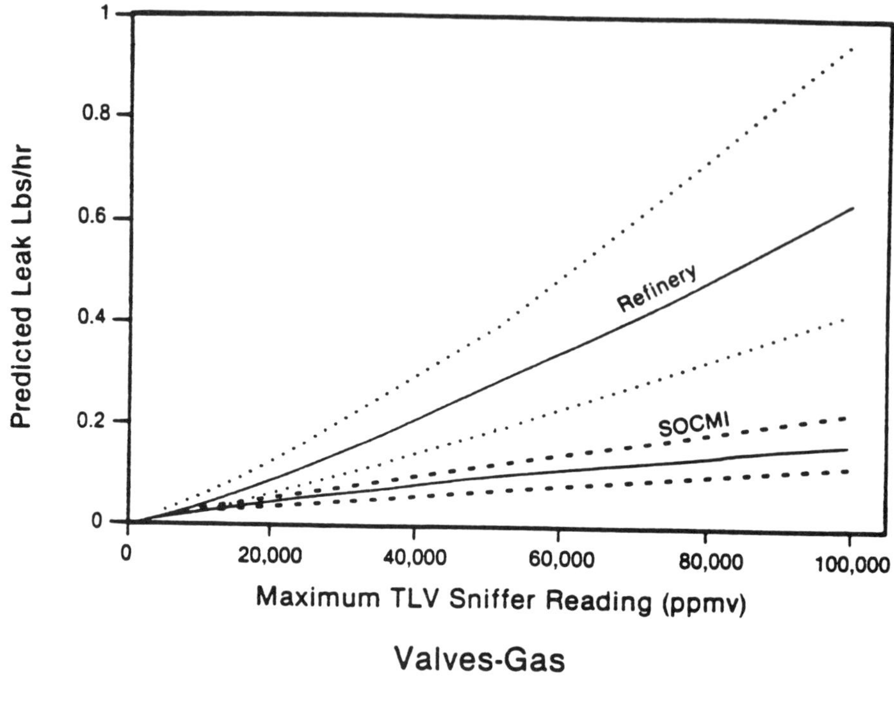

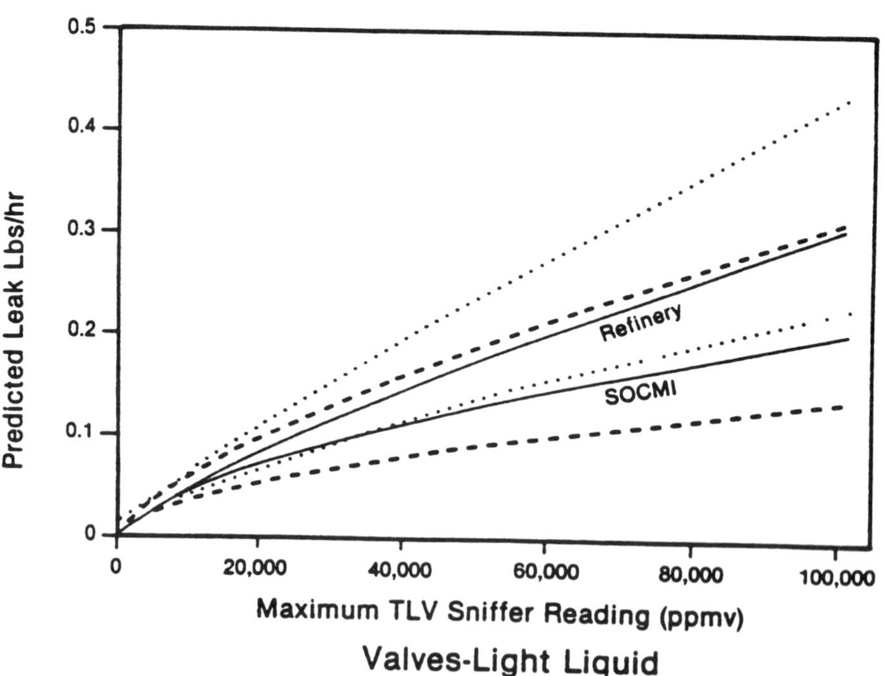

Dotted lines are 90% confidence bounds for the "Refinery" model
Dashed lines are 95% confidence bounds for the SOCMI model

Figure 5.7 EPA comparison of SOCMI and petroleum refinery leak rate correlations.[1]

a prediction equation that predicts larger leak rates than the SOCMI valve equation, the confidence bands of the two prediction equations overlap. Below 30,000 ppmv, the SOCMI equation is within the lower refinery confidence band. The EPA agrees the comparison for valves in light liquid service does not provide a decisive conclusion on whether separate equations should be used. Consequently, the EPA concluded the SOCMI data were sufficient to develop SOCMI emissions estimates[4] based upon 300 data pairs. Where a small number of data pairs are analyzed, the confidence limits for the equation change.

The EPA suggests there may be situations with most of the screening data less than 10,000 ppmv. Correlation equations can be expanded on a 0- to 10,000-ppmv scale and then converted to a log–log scale for linearity. However, the scale bias correction factors do not apply in this situation and comparisons of equations are not recommended. For low emission process units, correlation curves must be developed in accordance with statistical requirements.

Statistical Comparison of Correlations. In the development of the correlation (linear) curve, specific parameters were identified that were termed model parameters in Eq. 5.7. From the correlations in Figures 5.4–5.6, which are log–log curves, the linear correlation curve is described by the intercept on the leak rate or Y axis (B_0), the slope of the line (B_1), and the standard error (S_e). Statistical tests for the slope and intercept can be conducted and are described in reference 17.

For the slope and intercept, 95% confidence intervals were developed for the SOCMI leak range/screening value equations and are described in reference 18. The slope for valves in gas service with an OVA (Fig. 5.5) was estimated at 0.693 and a 95% confidence interval calculated for the slope was 0.53–0.85. Where a slope developed from data generated in a particular process unit is within this confidence limit range, the EPA equation provides a reasonable slope for a predictive equation. When the slope estimate is outside the confidence interval for the EPA correlation equation, the confidence interval for the correlation equation should be evaluated based on procedures in reference 17. This confidence interval evaluation will determine which estimating correlation is used.

These correlation equations have standard errors, which can be determined from references 15 and 17. Standard errors can be compared based on an F-test[15] or other statistical procedure that defines acceptability. The standard error is an important constituent of the scale bias correlation factor and provides a comparison of the scale bias correction factors for the different correlation equations. However, the scale bias correction factor is also affected by population distribution and comparison of the factors must be carefully evaluated.

Other Statistical Analyses. Various analyses are applicable in the development, evaluation, and comparison of leak rate/screening value equations. The EPA recommends a general linear model analysis[15] for equation development and the testing of various relationships for groups of data. While this analytical procedure is satisfactory, other approaches are also applicable, but knowledgeable statistical experts are necessary to evaluate and analyze various statistical techniques.

Generation of Emission Estimates. The final product of the statistical analyses are the generation of emission rates for various equipment or sources. Based upon the leak rate/screening value correlation combined with the screening data, the emission rates can be predicted. For all sources, the emission rate included bagged sources is predicted by the following relationship based upon Eq. 5.7.

$$\text{Predicted emission rate} = \exp[B_0 + B_1 \log(\text{screening concentration})] \times \text{SCBF} \tag{5.8}$$

where SCBF is the scale bias correction factor and B_0 and B_1 are coefficients available from the EPA or developed from linear regression analysis for specific sites.

The *sum* of the individual source predicted leak rates is the total mass emission estimate for a particular source type and service. An emission estimate per source is determined by averaging the total emission estimate over the number of sources in the category. A 95% confidence interval can be determined for the predicted emissions estimate, which is presented in references 1 and 4.

5.3 OTHER COMMENTS

The review in this chapter describing emissions estimating indicates the need for statistical analysis of the emissions analytical evaluation programs. In many instances, experienced statistical experts are needed to aid in the development of sampling programs and analyze the data obtained. This becomes quite important where the emission correlation equations differ from the EPA programs. Emissions estimating of equipment leakage is an important component of overall VOC and HAP emissions losses and requires accurate analysis to describe the leakage. With the EPA and various countries expecting large decreases in emissions losses, estimating accuracy is an important consideration in meeting regulatory objectives and analyzing actual equipment performance.

Restrictive regulatory requirements for HAP will be promulgated in 1993–1995 along with further requirements in VOC controls. As a result of improved emission controls, particularly in the chemical industry since the early 1980s, and more stringent regulations (HON), EPA correlation equations should be compared with current operating results. This requires screening and bagging data to provide a satisfactory basis. The CMA information[4] and the POSSEE® system provide methods of determining sampling requirements and analyzing the data. In addition to the emission changes through the years as a result of LDAR and improved equipment controls, additional changes in equipment will further reduce emission rates and tend to drive the emission rates close to zero in a greater number of sources. Consequently, the default zero analysis requires further investigation to provide a more accurate assessment of leakage at screening values < 8 ppmv.

The log–log curve of leak rate versus screening values has been converted to the arithmetic curve shown in Figure 5.3 for gas valves. The scale bias factor multiplied by the geometric mean of the log normal distribution results in an *unbiased maximum likelihood mean* that is essentially the true mean and is an emission factor. In addition, the scale bias factor converts the linear regression line or correlation equation to the arithmetic curve of Figure 5.3. Where there are screening values for a number of gas valves, the screening concentrations can be averaged and the average leak rate per valve estimated from Figure 5.3. This average leak rate is then multiplied by the number of valves in the unit to estimate the total leak rate for the valve sources in the unit. This curve should not be used to determine the leak rate for each valve which are then totalled to give the leak rate for all of the valves. While the former procedure is a statistically acceptable procedure, the EPA may not accept this method of estimating, preferring the procedure described in Section 5.2.6.

Emission factors can be determined with confidence limits through statistical techniques to insure they are the most accurate assessment of the mean leak rate for a source population. However, the requirements for developing and defending non-EPA emission factors varies with the EPA, the states, and regional authorities. Consequently, any statistically developed emission estimating program must be reviewed with the EPA to insure acceptance.

REFERENCES

1. EPA. *Protocols for Generating Unit-Specific Emission Estimates for Equipment Leaks of VOC and VHAP*, Publ. No. EPA-450/3-88-070, October 1988.
2. CFR. *Method 21—Determination of Volatile Organic Compound Leaks*, Title 40, Part 60, App. A, pp. 1020–1023, July 1, 1992.
3. EPA. *Response Factors of VOC Analyzers Calibrated with Methane for Selected Organic Compounds*, Publ. No. EPA-600/2-81-022, September 1980.
4. Chemical Manufacturers Association. *Improving Air Quality: Guidance for Estimating Fugitive Emissions from Equipment*, Washington, DC: CMA, 1989.
5. EPA. *Response of Portable VOC Analyzers to Chemical Mixtures*. Publ.. No. EPA-600/2-81-051, 1981.
6. EPA. *VOC Fugitive Emissions in Synthetic Organic Chemicals Manufacturing Industry—Background Information for Promulgated Standards*, Publ. No. EPA-450/3-80-036, June 1982.
7. EPA. *VOC Fugitive Emissions in Petroleum Refining Industry—Background Information for Proposed Standards*, Publ. No. EPA-450/3-80-015a, November 1982.
8. EPA (Radian Corp.). *Assessment of Atmospheric Emissions from Petroleum Refining:* Vol. 4, Appendix C, D, E, EPA Publ. No. EPA-600/2-80-075d, July 1980.
9. EPA (Radian Corp.). *Assessment of Atmospheric Emissions from Petroleum Refining*: Vol. 3, Appendix B, EPA Publ. No. EPA-600/2-80-075c, April 1980.
10. EPA. *Compilation of Air Pollutant Emission Factors*, AP-42, 4th ed., September 1985.
11. EPA. *Emission Factors for Equipment Leaks of VOC and HAP*, Publ. No. EPA-450/3-86-002, January 1986.
12. Chemical Manufacturers Association. *Guidance for Estimating Fugitive Emissions*, Washington, DC: CMA, July 1987.
13. EPA. *Evaluation of Maintenance for Fugitive VOC Emissions Control*, Publ. No. EPA-600/52-81-080, pp. 54–56, May 1981.
14. Schaich, J. R. *How to Estimate Fugitive Emissions from Equipment*, American Institute of Chemical Engineers Meeting, Boston, MA, August 21, 1990.
15. Finney, D. J. *J. R. Stat. Soc.* Series B, **7**, 155–161 (1941).

16. Rickmers, A. D. and H. N. Todd, *Statistics, An Introduction*, New York: McGraw-Hill, 1967.

17. Draper, N. R. and H. Smith, *Applied Regression Analysis*, 2nd ed., New York: Wiley, 1981.

18. EPA (Radian Corp.). *Revisions of Emission Factors for Nonmethane Hydrocarbons from Valves and Pump Seals in SOCMI Processes*, Research Triangle Park, NC: EPA, November 1981.

6

Hazard Control

Preceding chapters have dealt with the nature of occupational health hazards, their recognition, and their evaluation. When it is concluded that an exposure problem exists, decisions are made regarding the implementation of hazard control measures for the purpose of reducing exposure and reducing risk.

A given set of exposure conditions does not lead to a fixed set of control remedies. There are many options. Since zero risk is not attainable, a decision must be made regarding the degree of risk reduction to be achieved. Then a series of choices must be made from the wide range of options available to achieve the desired risk reduction. This choice of options is a complex judgmental decision since the precise degree of risk reduction achievable by a particular option is not usually known in advance. Furthermore, many other factors such as impact on production and operability, safety, worker acceptability, and regulatory requirements bear on the decision. This chapter will discuss the options available, the factors to consider in selecting options, and the planning process in which the risk reduction decisions are made. In addition, these decisions on reducing risks to worker health will be put in the broader context of the overall occupational and environmental health and safety issues involved in the planning of industrial facilities.

6.1 CONTROL PRINCIPLES

As stated above, risk is reduced by reducing exposure. Exposure occurs because a substance is allowed to get from where it is intended to be to a location where it comes in contact with the worker. All control strategies are based on some means of intervening along this pathway to the worker. The points of intervention can be divided into three broad categories:

1. *Source Controls*. Measures taken to prevent the release of the toxic substance into the atmosphere.

2. *Transmission Barriers*. Means of capturing or blocking the contaminant along the path from the source of the release to the worker.
3. *Personal Protection*. Devices worn by the worker that prevent contact with the toxic substances.

Each of these general approaches is at least theoretically possible in each situation and should be considered in the early, broad, "brain-storming" search for possible solutions. Where source controls can be applied, they can be almost completely effective, but they often impose unacceptable burdens on the process. Transmission barriers are probably the most commonly used, although they are rarely completely effective. Personal protection for the prevention of skin contact is common and effective but inhalation protection by respirators is less effective and therefore used less often. In many cases successful controls involve a mix of all three elements and different elements may be applied at different times in a stepwise solution to a control problem.

6.2 CONTROL OPTIONS

Detailed consideration of the physical and procedural means of accomplishing the interventions described above leads to a list of categories of specific measures. While source controls are most commonly used in the chemical process industry, knowledge of the complete range of controls is necessary in selecting a cost-effective mix of options, since source controls alone will not always be applicable or completely effective.

6.2.1 Process Change

Basic or detailed changes in the way the process itself operates can eliminate or reduce exposure. Rather than handle a material as a dry powder, it might be transported as a molten liquid or slurry or in solution. Coatings could be applied by dipping instead of spraying. Electric forklifts can take the place of internal combustion engine forklifts to eliminate carbon monoxide. Hydroblast cleaning may be used instead of sandblasting. Centrifuges may be used in place of plate and frame filters. Process changes are usually fundamental changes that can have many desirable and undesirable side effects. A process change may present an easy, low-cost opportunity to prevent a problem, but it needs to be considered early in the design of a project and all of the ramifications must be thought out.

6.2.2 Substitution

A special case of process change is the substitution of a less hazardous material for a more hazardous one. The risks from the use of asbestos are so great that substitutes have been found for it in all but a few applications. Carbon tetrachloride has been replaced by less toxic halogenated and aliphatic hydrocarbons in open solvent use. Steel shot, ground walnut hulls, and carborumdum are less toxic than sand for abrasive cleaning operations. Substitution may greatly reduce the need for control by replacing a high-toxicity material with a low-toxicity material. It may also be useful to reduce the hazard by substituting a low-volatility material for a volatile material of similar toxicity and thus reduce the release rate and exposure. A danger in material substitution is that the substituted, presumably low-toxicity material may later be found to be more toxic than was originally thought, perhaps as toxic as the material for which it was substituted. This unfortunate occurrence can come about because health data, particularly regarding human health effects, is inadequate or nonexistent on many substances.

6.2.3 Containment

Theoretically, if a process could be designed to keep all of the process materials in the process vessels and piping, no exposure would occur. As discussed elsewhere in this book, this is not entirely possible because of fugitive emissions and open process operations. However, containment eliminates most opportunities for exposure and is the preferred method of control in chemical manufacturing. In many chemical plants containment is dictated by pressure, temperature, fire, or product loss control requirements and is not really a health hazard control option. It is an option for certain batch-type processes conducted in open vessels which could be designed as closed continuous systems. Containment may also be an option for material handling segments of processes that could be either open or closed. Containment, when it can be used, seems to be the complete solution and engineers are sometimes misled to believe that absolute control has been achieved. It is necessary to understand that containment is never perfect, releases and exposure opportunities will still occur, and additional control will probably be necessary.

6.2.4 Isolation

By isolation is meant the physical separation of the source of exposure and the worker. In a sense, containment is a form of isolation, but what is usually meant are control arrangements where releases into the atmosphere may occur, but in places not occupied by workers. Isolated systems have been used for years for the manufacture of such high-toxicity materials as prussic acid (HCN) or tetraethyl lead; the entire manufacturing operation is isolated in a separate unoccupied building or sealed-off area. The development of computer process control has reduced the need for workers to be in attendance near the production equipment and robotics may reduce the need even further. When mishaps occur and it is necessary to enter the isolation area, the process can be shut down and aired out, or workers can enter using respiratory protection, typically self-contained breathing apparatus. Although still regarded as only applicable to very high-toxicity processes, as process control technology advances, isolation may be the control of choice in medium-toxicity situations.

6.2.5 Enclosure

Enclosure is intermediate between containment and isolation in that the substance is released into the air but the air is kept inside the process machinery by a hood or enclosure. Most commonly, enclosure is used to control dry powdery material handling operations where conveyors, transfer points, bucket elevators, screens, and so forth are inside sheet metal boxes. Air is usually exhausted from these boxes so as to maintain a negative pressure inside so that any leakage will be inward. In this sense, sheet metal boxes are a special case of local exhaust ventilation. Enclosure is a common but awkward control since access to machinery is restricted. Fire and explosions are also a concern since air is present where flammable materials are confined. Some commonly used enclosures are modern drum and bag filling equipment and abrasive blast cabinets. In laboratories, glove boxes are becoming a common means of total enclosure.

6.2.6 Local Exhaust Ventilation

Local exhaust ventilation controls exposure by establishing a control surface or barrier between the emission source and the worker such that the contaminant is captured and is not allowed to reach the worker's breathing zone. This is accomplished by some arrangement of fans and ducts that suck air into a hood across a control surface at such a rate as to arrest the gas or vapor and cause it to be pulled into the hood and be captured. Local exhaust ventilation is applied as a means of control for releases that have already occurred to prevent them from causing exposure. Hoods are usually added on or attached to pieces of equipment to cover release points. Local exhaust ventilation is rarely 100% effective since capture is not complete and not all release points are covered. Some of the most common uses of local exhaust ventilation are for processes that cannot be completely enclosed, such as open surface plating tanks, grinding and polishing machines, and foundry shakeouts. In chemical plants local exhaust hoods are sometimes used for sampling points or for unavoidable seal leaks. Local exhaust ventilation is cumbersome and inconvenient and requires considerable maintenance, but it is an effective form of control which can be retrofitted to an existing plant and thus solve a problem that was not anticipated in design. For this reason it is often used where simpler and cheaper controls would have worked if the need had been recognized.

6.2.7 Dilution Ventilation

While local exhaust ventilation attempts to capture an air contaminant before it reaches the breathing zone, dilution ventilation removes air containing a contaminant from a workplace after it has become mixed and been breathed by workers. The object of dilution ventilation is not to prevent any exposure but to keep exposure to acceptable levels by dilution. Obviously, dilution ventilation should not be used for highly toxic materials. It is most appropriate for workplaces with many workers and many sources of low-toxicity materials, such as aliphatic hydrocarbons. Some examples are hand painting of small objects, machine shops, and synthetic textile mills. The volume of air to be exhausted is based on the release rate of the contaminant, the volume of workplace, and some assumed mixing factor. Since the release rate and the mixing factor are not usually known, design parameters tend to be empirical estimates and are sometimes too low. Also, while it may be possible to achieve an acceptable average level of exposure, there may be individuals in locations where mixing is inadequate who are being overexposed. Dilution ventilation should be used only in low release rate, low-toxicity (low-hazard) situations.

6.2.8 Work Practices

Work practices are the procedures and precautions taken by the worker that minimize exposure while doing the job. While the controls described above are the main means of reducing exposure and accomplish most of the reduction, there are some things workers can do to lessen exposure:

- Vacuum cleaning instead of sweeping or blowing
- Wetting a potentially dusty task
- Moving portable local exhaust hoods to the proper position
- Carefully handling empty bags with residues of dusty materials
- Cleaning up all spilled material promptly
- Replacing covers and enclosure panels as quickly as possible

There are probably hundreds of examples of things workers can do to reduce exposure. Work practices do not generally eliminate a hazard by themselves but are necessary to prevent overexposure by sources not controlled by engineering means of exposure reductions.

6.2.9 Personal Protection

Personal protection against exposure by inhalation is accomplished by respirators. Respirators, which come in a wide variety of models, are selected based on the nature of the contaminant (particulate, organic vapor, acid gas, etc.), the level of the contaminant in the workplace air, and the situation under which the respirator will be used (routine, emergency rescue, escape, etc.). Respirators are capable of providing considerable protection when selected and used properly. The protection provided by a respirator is measured by its "protection factor," which is the ratio of the concentration outside the respirator to that inside. A simple, single-use, half-face air purifying respirator might provide a protection factor of 5 or 10, a full-face device could provide a factor of 50, and an air-supplied respirator with positive pressure inside the mask could provide a protection factor of over 1000. An example of how OSHA uses these factors to construct respirator selection tables is shown in Tbl 6.1

All respirators have advantages and disadvantages. The air purifying devices are simple and inexpensive and do not require an air supply, but the purifying element becomes exhausted with time and must be replaced and they do not provide a high level of protection. A supplied-air respirator such as the positive pressure self-contained breathing apparatus (SCBA) does provide a high level of protection but is heavy and expensive and the air supply usually only lasts about 30 min. Supplied-air respirators that work off a compressed air supply have long duration, but the worker is tethered by the hose.

Emergency respirators that are worn for only short periods of time in life-threatening circumstances are usually successful. Also, air purifying respirators are

TABLE 6.1 OSHA Respirator Selection

Airborne Concentration or Condition of use	Respirator Type
Less than or equal to 10 × PEL	Half-mask respirator
Less than or equal to 50 × PEL	Full facepiece respirator
Less than or equal to 1000 × PEL	Full facepiece, powered air-purifying respirator
Greater than 1000 × PEL or unknown concentrations	Self-contained breathing apparatus with full facepiece in positive pressure mode
	Full facepiece, positive pressure demand, supplied-air respiratory with auxiliary self-contained air supply
Escape	Any full facepiece air-purifying respiratory
	Any positive pressure or continuous flow self-contained breathing separators with full facepiece or hood
Firefighting	Full facepiece, self-contained breathing apparatus in positive pressure demand mode

appropriate for brief maintenance jobs. Unfortunately the theoretical protection respirators are capable of providing is rarely achieved when they are used routinely. The difficulty with routine use of respirators is that, unlike other items of personal protective equipment, such as hard hats and gloves, respirators are so uncomfortable to wear that it is difficult to get worker cooperation in an effective respiratory protection program. Worker cooperation is obviously needed to insure that the respirator is worn and is properly fitted since an ill-fitting respirator can easily leak so badly that it will provide almost no protection. Respirators seem to be a very inexpensive way to provide protection but if they fail, as they are likely to do when used routinely, the retrofit of other means of control can prove more costly than the control that could have been included when the unit was built.

6.3 CONTROL SELECTION

Sets of feasible control options or mixes of options need to be selected to accomplish the purpose of reducing exposure from the level determined by the exposure evaluation to an acceptable level. The best option or mix of options is then selected from these sets based on cost and some other factors described below.

Cost–benefit analysis, in the sense of balancing the degree of protection achieved against the cost of control, is not usually a practical or desirable technique. Since the benefits are in terms of health effects they are difficult to compare with costs. Also the benefits are not known with any precision. As discussed earlier, the health data available for the delineation of dose response are meager and difficult to extrapolate to humans exposed at low doses. Further, the amount of exposure reduction achieved by a given set of controls is not usually known. When controls are applied to specific emission sources they are usually close to 100% efficient for those sources, so exposure after control will come from the remaining smaller, possibly unknown, and probably unquantified sources. Because of these uncertainties, cost–benefit analysis is not a realistic means of factoring cost into the control option selection decision. The more common and successful cost analysis is to compare sets of options that have approximately equal probability of reducing exposure below an appropriate occupational exposure limit. In this way costs will be compared on an equal protection basis even though the exact degree of protection, that is, the magnitude of the residual risk, is not known, except that it is acceptably small. Costs, including capital and expense, of the several option sets may then be compared using such standard parameters as present net worth ar annualized cost.

Some additional factors should be considered in making the choice between option sets of roughly comparable cost:

- *Process Alterations*. Although any direct change in production caused by a control would be factored into the cost analysis, some less direct, less obvious changes may occur. For example, local exhaust hoods or enclosures may limit access and visibility. On the other hand, better ventilation may yield a cleaner workplace and better product quality.
- *Unscheduled Adjustment*. For a control system to continue to be effective it must not be subject to alteration by "tinkering." For that reason, adjustable valves, dampers, and blast gates should be avoided.
- *Ease of Maintenance*.The installed controls should not add any more than necessary to the burden of maintenance and should themselves be easy to maintain.
- *Air Handling*. Exhaust systems remove heated or cooled air which must be replaced. If makeup air is not provided, the building will become "air starved" and the effectiveness of the exhaust system will be reduced. Untempered makeup air will cause drafts and worker discomfort.
- *Air Cleaning*. Dirty air captured by an exhaust system may need to be cleaned because of environmental concerns or regulations. The operation of these air cleaners can create secondary exposure risks.
- *Noise*. Exhaust system noise is not usually intense enough to cause hearing loss but it can be a speech interference problem.
- *Reliability*. If the system fails, will it "fail safe?" Will there be any warning of failure?
- *Safety*. Does the control of a health hazard create a safety hazard? For example, capturing a flammable solvent in an exhaust system raises the possibility of explosive concentrations occurring.
- *Human Behavior*. Engineering systems may be easier to keep at a high level of effectiveness over long periods of time than systems that rely on the periodic reinforcement of human behavior.

6.4 CONTROL SYSTEM MAINTENANCE AND EVALUATION

Initial evaluation of a newly installed control system and regularly scheduled maintenance are required to insure continuing effectiveness. Unfortunately, since controls are not directly involved in production, they are sometimes neglected and become inoperable.

6.4.1 Initial Evaluation

An engineering system or work practice that is installed for the purpose of preventing a health effect could be evaluated by the degree to which it reduces the occurrence of the health effect. It is in fact common to evaluate hearing conservation programs by measuring hearing loss in the population protected. A system installed to reduce exposure to an acute hazard that has been causing instances of worker irritation should be judged by its effectiveness in eliminating complaints. However, systems that are intended to prevent low-level risk to major consequence health events which, like cancer, take a long time to happen cannot be realistically evaluated by a change in health status. For these systems a more immediate, less ultimate, measure of effectiveness is needed.

As discussed above, the system was selected to achieve a high probability of staying under the appropriate occupational health standard. Therefore, measurements of exposure after control could be compared with measures of exposure before control to see if the desired degree of exposure reduction was achieved. Some number of measurements will be needed to obtain an accurate estimate because of the large variability in the industrial environment described earlier. In fact one measure of control effectiveness is the reduction of the standard deviation of a set of "after' exposure measurements because controlled environments are inherently less variable than uncontrolled environments.

Measurements of exposure, for use in comparison with occupational exposure limits, need to be made over the averaging time appropriate to the standard. The most common averaging time, the 8-hr TWA, would lead to a sample of a much longer duration than is necessary to evaluate an emission control or capture system whose variations are of a much shorter period. Further, samples for comparison with occupational exposure limits should be personal samples, while system performance could be more easily evaluated by fixed-station samples. For these reasons it is often preferred to use short-period, fixed-station before and after samples to evaluate control performance even though these measures are even less direct in that they cannot be compared with occupational exposure limits.

The least direct but very useful method of evaluating control system performance is to measure some system operating parameter and compare it to the intended design value. When enclosure is used as a control, the negative pressure inside the "box" could be measured. For dynamic local exhaust systems either the static pressure or the velocity in a duct or hood could be used as a reference point.

6.4.2 Maintenance

Once initial performance has been established by the most direct means practical, continued performance can be verified by some less direct measure. Periodic instantaneous measurement of overall volatile organic compound concentration can be used to be sure that seals are not leaking. Measurements of static pressure or velocity used for initial evaluation can be used for periodic monitoring to be sure that "like new" performance is maintained. Exhaust systems are particularly vulnerable to performance degradation from duct work damage, duct plugging, fan blade deposits, and fan belt slippage. Instruments to make performance measurements are often permanently installed and routinely read locally or the data are transmitted to a central location. For critical controls, monitors that sound an alarm are used. These alarms should be set to sound when the monitored parameter leaves its normal range, indicating that control is being lost, rather than when unsafe conditions have already occured.

6.5 HEALTH HAZARD CONTROL PLANNING

Insuring continuing and consistent compliance with a variety of environmental, health, safety, and community concerns, regulations, and policies is a major consideration in the design and operation of chemical process units today. A checklist (Appendix A) of the broad range of planning considerations emphasizes environmental, health, and occupational health issues. These considerations need to be translated into design and operation by engineering and plant staff who have not been trained in the biological sciences underlying most of these issues and could not possibly be expert in all the issues, regulations, and technology involved. These decision makers must rely on timely and appropriate advice from specialized experts. For the purpose of controlling occupational health hazards, this means close interaction with the responsible industrial hygienist.

6.5.1 Industrial Hygiene Interaction

Industrial hygienists most often have a chemical or biological undergraduate education. Only a small fraction have an engineering background. On this base is built the professional skills of recognition, evaluation, and control of occupational health hazards. The knowledge acquired in industrial hygiene professional training includes an understanding of the effects of physical and chemical agents on the body which enables the industrial hygienist to explain why health hazard controls are needed. Beyond this is an understanding of the standards, their origin, and how to evaluate exposure with respect to them. In addition the industrial hygienist will have a basic understanding of control strategies and options and will, with experience, gain judgement as to their practical effectiveness. With the exception of local exhaust ventilation, most industrial hygienists do not have engineering design skills. Consequently, it is necessary for the design engineer to ask the right questions and for the industrial hygienist to be aware of the kind of information the engineer needs for interaction to be successful. The engineer needs to recognize that the biological sciences are not precise, that health effects and exposure data contain much uncertainty, and that health hazard control is a matter of risk reduction, not elimination. The industrial hygienist needs to recognize that engineers need to fit health hazard control in with hundreds of other design considerations. Jointly they need to convert uncertainty into hard design parame-

TABLE 6.2 Sequences of Steps in Developing a Project[a]

Time Sequence Project Steps
Chemical route
Laboratory process
Reaction flow diagram
Draft process flow scheme
Economic evaluation
Site selection
Process development
Heat / mass balance
Process flow scheme
First investment estimate + profitability
Design frozen
Construction plan
Definitive calculation of investment and profitability
Engineering flow scheme
Basic design
Detail design
Procurement
Construction
Start-up

[a]Reproduced from *Ann. Occup. Hyg.* **30**, 232–240 (1986) by permission of the British Occupational Hygiene Society.

ters. All of this must be done by frequent contact during the many stages of the design process. The design development proceeds according to a rigid schedule that severely penalizes delays.

6.5.2 Design Stages

An example of the sequence of steps in the development of a project is shown in Table 6.2. The enormity of the task confronting the engineer is illustrated by the fact that each of the items in Appendix A needs to be considered at each of the stages in Table 6.2. For the optimization of health hazard control the industrial hygienist should interact with the project team at almost all of these stages. As will become apparent, interaction at only one point, say at the end of the design, will not work because decisions have already been irreversibly made that foreclose what might have been the optimum control option. Experience has shown that continuous, generally informal, contact leads to the most cost-effective design. Some examples of the kinds of interactions and hazard control choices at various stages in Table 6.2 are as follows:

- *Chemical Route*. If feedstock substitution is a possible control option it needs to be considered when decisions about basic chemistry are made. In rare instances some chemical involved in the process, or possibly even the product itself, may be of such high toxicity or regulatory concern that the whole project needs to be reconsidered.

- *Laboratory Process*. Laboratory and pilot plant operators need protection from health hazards the same as plant workers, so initial evaluation of the toxicity of the materials involved will need to be made and put to use at this point. Also, by-products and wastes may now be produced and identified and their impact evaluated.
- *Reaction Flow Diagram*. Options involving intermediates, solvents, and catalysts should be evaluated by this point.
- *Draft Process Flow Scheme*. The plant design basis should begin to be apparent by this point and selections of types of process units made.
- *Economic Evaluation*. Most developed countries now require some kind of a premanufacturing or premarketing notification under the provisions of a toxic substance control act. Some testing may be done in connection with this notification and, based on the results of these tests, more testing may be needed. These tests cost money (millions of dollars for a chronic inhalation study) and adverse results may limit the use of the product.
- *Site Selection*. Space, climate, and adjacent activities influence the difficulty and success of hazard control measures.
- *Process Development*. Selection of specific equipment types fixes fugitive emission sources and open-system operations. From this point on hazard control becomes transmission control rather than source elimination.
- *First Investment Estimate and Profitability*. For some high-hazard processes involving extensive source controls, flush and purge facilities, installed monitoring equipment, and other expensive features, the cost of hazard control can be a major fraction of the total project cost. For this reason, the major hazard control options must be selected by this point and costs defined so that an accurate economic estimate may be made.
- *Design Frozen*. A signoff by safety, industrial hygiene, and so on is often required at this point so any unresolved hazard control issues must now be settled.
- *Basic Design*. Certain critical aspects of the hazard control scheme which were flagged earlier in the review should now be rechecked.
- *Detail Design*. Coatings and insulation types are specified now as are locations of installed monitors and safety facilities.
- *Procurement*. The burden of achieving certain controls may be shifted to an equipment supplier by specifying some attribute such as an emission rate. Some tests by the supplier may be required as a check.
- *Construction*. Health hazard control during construction is a complex subject beyond the scope of this book. Even when construction is done entirely by contractors, the owner may have some liability and needs to monitor, but not to control, compliance with regulations and practices.
- *Start-up*. Start-up is usually done by the owner, although contractor personnel may still be on site. Unanticipated events that can create health hazards are common so it is prudent to have continuous industrial hygiene coverage until the plant is lined out. Also, start-up is the time to test the precautions in the operator's manual and to establish a safe operating procedures.

- *Post-start-up*. When the plant has been settled down to a steady state a baseline industrial hygiene survey should be conducted to insure that hazards are under control and to establish expected levels for comparison with future surveys.

APPENDIX 6A

Plant Planning Considerations, Emphasizing Environmental Health and Occupational Health Issues*

I. Outside-plant considerations: location, site qualities and use, availability of services

  A. Site

    1. Location variables

      a. Worker access from homes to plant (and labor availability)

        i. Most efficient travel time solution (if choices exist)

        ii. Mass transit available (or planned)

        iii. Private transportation environmental impacts on surroundings (e.g., air quality)

      b. Emergency services

        i. Fire services availability and reliability in normal access and high-traffic conditions, and in possible high-demand situations (i.e., other fires at the same time)

        ii. Availability of emergency medical treatment and ambulance access; brief emergency room staffs of nearby hospitals on processes, exposures, hazards

        iii. At least two approaches to plant and to each process area

      c. Raw materials access

        i. Location w.r.t. supplies, materials, economical transportation

        ii. Rail or truck traffic—impact on surroundings (air, noise)

        iii. Hazards—spills, fires, etc in materials transport

      d. Product shipping

        i. Location w.r.t. users and w.r.t. means of transport

        ii. Loading facilities—docks, rail spurs, etc—w.r.t. efficiency; also spills and releases during loading, and car, truck, or train damage to facility during accidents

        iii. Transport impacts on surroundings and within plant

        iv. Use of transport as storage (e.g., significant part of inventory is in rail cars at a point in time)

        v. Hazards—spills, accidents—of product transport

      e. Location w.r.t. catastrophic explosion or release, vs. probability of population exposure

i. Probability of unacceptable emergency exposure levels staying within site, or with minimum off-site exposure

ii. Conversely, probability that fire/explosion/toxic releases on adjacent industrial sites will be contained within those sites without threat to plant

iii. Location of access and evacuation routes upwind of plant, at least for prevailing winds

f. Rights-of-way to site and space used by on-site pipelines and/or power lines

g. Local climate

i. Location in floodplain and/or electrical storm, hurricane, tornado risks to site

ii. Any future changes likely in hydrology of watershed, floodplain in which site is located (e.g., heavy urbanization upstream, subsidence)

iii. Does climate require processes to be indoors, with associated workplace air quality problems?

h. Security—can it be obtained and maintained?—site factors involved; adequate fencing and gates

i. Site size and layout sufficient for expansion of plant, parking, on-site services

j. Any height restraints (e.g., airport approaches)

2. Isolation, separation of on-site and adjacent hazard areas; consideration of quantities of materials, separation, and barriers

a. Ample land is needed for separation and spacing of processes and future expansion

b. Explosives storage, solvent bunkers—containment, blast deflectors, slopes

c. Chemical bulk storage

i. Raw materials

ii. Wastes

d. Fuel dumps, storage tanks, for long-range stockpile capacity if needed

e. Drainage to contain or direct spills

i. Evaporation suppression for such spilled or released liquid materials, if volatile

ii. Capability to sample and monitor soil and groundwater for contamination

iii. Flammables spills—directed away from ignition sources

f. Access roads and space for fire-fighting access and other emergency equipment

g. Separation of processes required for safety and insurance purposes
    i. Separation of hazards from offices, utilities, control rooms, computers, labs
    ii. Separation of hazardous units from other such units

3. Protection of materials, storage, plant access, and other critical areas against earthquakes, hurricanes, tornadoes, vandalism, air crashes, sabotage, etc., to practical limits
4. Ice, snow, and water removal and control or storage, grounds management, site drainage
5. Air quality aspects of site
    a. Location of site w.r.t. receptors or other sources such as residences, schools, major highways, power plants, or adjacent industrial plants
    b. Intake air quality—adequate? Can possibly contaminated exhaust air enter intakes under any conditions?
    c. Types of emmissions, control attainable, emission monitoring in place
    d. Damage to plant and site from emissions, also surroundings and region
    e. Are local meteorological patterns known w.r.t. dispersal of air contaminants and fire spread, under usual and worst-case patterns? Are weather stations needed?
    f. Regualtory conflicts between ventilation of workspace contaminants to environment vs. emission controls for environmental protection
    g. Contingency plans for catastrophic emissions; dispersion patterns

B. People, materials, and energy flow into/out of site and within site—optimize patterns and reduce interferences
1. Raw materials—inflow, handling, storage, distribution
2. Products—outflow, handling, storage, packaging
3. Workers
    a. Access—speed and ease
    b. Parking and vehicle security
    c. Parking-to-plant personnel transit, on large sites
    d. Worker transit between buildings—buses,cars, trains, bicycles, carts, foot traffic, etc.
    e. Provisions for emergency evacuation without hindering access of emergency crews
4. Power service required
    a. Electric
        i. Voltages, current, phase, transients, above or below surface, grounding
        ii. Two feeder circuits if possible for critical uses
        iii. Location, protection of transformers and substations from fire and other risks

b. Gas—availability, line safety
c. Power or heat generated on site
   i. Burning of combustible wastes: steam, heat, hot water
   ii. Cogeneration of power with waste process heat
      Sales of heat or power
      Community or district heating
   iii. Potential added air emissions and related control costs
d. Provision for emergency backup power and generation, esp. for critical processes and operations, such as boiler feed, cooling water, and fire pumps

5. Water
   a. Supply and quality adequate for each use
      i. Cooling
      ii. Process
      iii. Conveyance
      iv. Potable uses
      v. Solvent
      vi. Washing
      vii. Steam feed
      viii. Air scrubbing
      ix. Fire fighting
   b. Are on-site sources of water needed, possible?
   c. Equipment needed—treatment, distillation, deionization
   d. Steam generation
6. Liquid wastes—wastewater, chemicals
   a. Treatment facilities, methods, space if appropriate on-site, (e.g., space for settling, waste ponds, lagoons)
   b. Sludge disposal and/or recovery
   c. Toxic waste materials handling, treatment, disposal, including off-specification products and effluents from emergency releases
   d. Radioactive wastes—high- and low-level
   e. Routine effluent disposal, including laboratory wastes
   f. Provision for waste analysis, record keeping, plans for treatment failure or overload, protection of waste workers
   g. Consideration of groundwater effects in case of spills or liquid releases
7. Solid wastes—collection, management, and disposition
   a. Types and analysis; include solid wastes from air cleaners
   b. Reuse and recovery
   c. Ultimate disposal of residuals
   d. Plans for collection and disposal systems, including laboratory wastes

8. To complete mass and heat flow balances—materials and heat lost to air, water

II. Plant shell, structure, and services considerations: provide maximum flexibility in a fully serviced envelope of controlled space

A. Flexible layout of space

1. Power available throughout, without circuit overloads; local power takeoffs equipped with lockout provisions
2. Water available throughout, and drainage, for cleaning
3. Roof capable of supporting necessary suspended loads, plus extremes of snow/water
4. Foundation load capacity
   a. Adequate flexibility in locating equipment, later relocating
   b. Can withstand and damp vibration
   c. Adequate for dead loads of filled vessels, including at hydrostatic testing
5. Vertical space use and access
   a. Not a limiting factor (e.g., can mezzanines, etc. be added, and/or overhead systems; access to upper floors available)
   b. Advantages of vertical layout
      i. Gravity feed of materials
      ii. Develops usable thermal head
      iii. Building itself contributes to stack height
      iv. Minimizes dust-gathering horizontal surfaces
   c. Disadvantages of vertical layout
      i. Heavy equipment access and replacement more difficult
      ii. Inefficient for some processes
      iii. Personnel movement more difficult
      iv. Roof usage, access more difficult
6. Horizontal spatial layout
   a. Advantages
      i. Free unobstructed flow of materials
      ii. Better accomodation to movement of a large labor force
      iii. Excellent roof access (e.g., for many smaller air handlers)
   b. Disadvantages
      i. More difficult to control migration of contaminated air
      ii. Does not develop usuable thermal head
      iii. Extensive horizontal ductwork may be required—inefficient
      iv. Packaged building structures usually are not capable of additional loads

7. Steam available as needed
8. Compressed air and/or vacuum available as needed; including different breathing air system if needed (breathing air intake point not contaminated); also inerting gas systems as needed
9. Central coolant systems provided, as needed
   a. With adequate expansion capacity, including coolant processing equipment and space
   b. Waste coolant holding capacity sufficient for whole-system dump
10. Drainage in plant—adequate for cleaning, doesn't leave standing water, etc., or slick floors—size to include sprinkler flow
11. Adequate preplanning for expansion

B. Materials maintenance and selection
   1. Fewest feasible dust-collecting surfaces and spaces; access for cleaning of surfaces, around and under machines
   2. Durable, noncorroding, cleanable structure and surfaces, resistant to chemicals used
   3. Fewest possible materials or fixtures requiring routine hand maintenance or cleaning
   4. Provision for cleaning outside and inside of structure
      a. Occupational health aspects of cleaning materials and tasks
      b. Window and wall washing; equipment designed with building
   5. Central and dispersed space for maintenance equipment and supplies, and repair of maintenance equipment

C. General environmental control system
   1. Thermal environment control
      a. No added contamination (e.g., CO from heaters)
      b. Acceptable supply air contamination levels
      c. No reentry of exhausted air
   2. General lighting levels—intensity, color, glare, contrast, etc.
   3. General ventilation for odor, humidity, temperature control
   4. Adequate makeup air to prevent indoor air quality problems
   5. Where feasible, general ventilation design for low-level control of toxic air contaminants
   6. Buildings, processes, utility lines, process connections, and controls all have appropriate degree of environmental protection (e.g., against freezing or corrosion)

D. Minimum construction requirements (usually building codes)
   1. Construction types
      a. Soils—load bearing; heave tendencies, seismic issues
      b. Fire resistance
      c. Strength, height and roof loads, wind, and climate (including tornado and hurricane loads if appropriate)

d. Sufficient construction quality control as is warranted by hazard (e.g., QC of structural welding)

2. Allowable areas—those in codes are minimums, usually not optimum
3. Allowable heights—are minimums, usually not optimum
4. Fire separations, and susceptibility to ignition from adjacent fires
5. Exterior wall protection—meets adequate over- or underpressure requirements, explosion-relieving sash, walls, roof, as appropriate
6. Fire-fighting requirements—water
7. Fire-resistant materials or insulation, absence of openings to transmit flames
   a. Insulation and construction materials not hazardous (e.g., asbestos)
   b. Consider decomposition products of burned building metals
8. Interior finishes—cleanable with reasonable methods
9. Means of egress
10. Fire protection systems—sprinklers, adequate water, extinguishers (halocarbons, CO_2, or special firefighting procedures for processes used), fire trucks, carts, etc.
11. Vertical openings—controlled or minimized to reduce fire spread
12. Hazardous areas
13. Light and ventilation—minimum and optimum
14. Sanitation, including drainage, water, sewerage
15. Electrical wiring and grounding; lightning arrestors
16. Provisions for handicapped
17. Energy conservation
18. High-hazard provisions

E. Vector and vermin control

III. In-plant physical and organizational considerations

A. Design of work—implications for workplace design

1. Teams vs. assembly lines
2. Individual team facilities
3. Underlying work organization vs. space organization
4. Materials handling system to coincide with designed work patterns
5. Worker participation in work and workplace design; worker participation in planning phase and ongoing hazard anticipation and recognition

B. Process health and safety review: preconstructed written occupational health assessment of each worksite or process or situation, prepared prior to final design

1. Consider material or process substitutions possible to reduce hazard and risk
2. Review of toxicity of feedstocks and products and their typical impurities, by-products, intermediates, effluents, catalysts, and solvents,

additives of all types, and unexpected products generated under abnormal process conditions

3. Normal exposure potential
4. Start-up, shutdown, turnaround emergency, exposure potential
5. Must peak or average exposures be most strongly controlled, i.e., can an occasional excursion be tolerated under extreme conditions, or not?
6. Occupational health of maintenance staff considered, as well as operators
7. Health and safety issues reviewed in major modification or automation planning
8. Avoid process overcrowding, esp. w.r.t. changes in plant; maintain adequate headroom esp. under equipment
9. Failure modes and repairs considered in industrial hygiene evaluations
   a. Switch, breaker, and valve designs suitable for blanks, lockouts
   b. Permit systems for hot work, confined space entry, other hazardous jobs
   c. Plan confined spaces to facilitate entry procedures
10. Positive or negative pressure areas planned to control flow of contaminated air
11. Does toxicity or flammability, stability, etc., justify extreme engineering controls, e.g., much higher standards, special facilities for dump, blowdown, quench, or deluge?
12. Microscale dispersion processes vs. concentration
    a. Pasquill class D (neutral) is a conservative design condition for all but occasional night conditions
    b. Determine downwind distance in normal and emergency conditions to concentrate isopleth of interest (TLV–TWA, TLV-STEL, ceiling or emergency limits), and the population exposed inside site and outside site
    c. In areas where critical releases can occur, are clear and graphic wind direction indicators present?
13. Needed automatic monitoring/sampling systems included?
14. Hierarchical emmission control strategy
    a. Process control and sampling connections; also, insure these do not introduce hazards (e.g., are explosion-proof)
    b. Centrifugal pump seals—packing, single, or double seals
    c. Diaphragm pumps
    d. Safety valve/vent release and flare, quench, or scrubbers
    e. Analytical instrumentation waste disposal
    f. Valve and flange connections—gaskets, vented covers
    g. Catalyst loading/unloading

h. Solids unloading
i. Liquid loading/unloading
j. Seals for centrifugal and reciprocating compressors
k. Equipment cleaning connections, ports, for drain, flush, purge, vent; full means for blanking, blocking, isolating during maintenance
l. Toe curbs
m. Drains
n. Strainers
o. Storage tanks
p. Oil—water separators
q. Grounding and static control
r. Catch pans
s. Shielding/barricading
t. Malfunction alarms
u. Bleed valves, expansion chambers, shock chambers
v. Piping system review for
   i. Supports and vibration effects
   ii. Weld quality
   iii. Inadequate inspection due to inaccessibility
   iv. Temperature and pressure stresses
   v. Localized reactions

15. Select design and maintenance standards adequate for risk category; equipment designed to appropriate recognized standards
16. Detailed considerations of specific process conditions and safety (do not treat as all-inclusive)
    a. Have proper scale-up tests been conducted?
    b. Have all reaction and flow sheets been prepared and reviewed, including piping flow sheets if applicable?
    c. Raw materials, charge weights, rates, and order confirmed
    d. Temperatures and pressures; shock sensitivity
    e. Reaction times and rates
    f. Solvents, alternative solvents
    g. By-products, amounts, composition, etc.
    h. Gas, vapor, or mist evolution
    i. Use and/or production of potentially combustible dusts
    j. Exotherms, spontaneous polymerization
    k. Foaming, sticking
    l. Analysis during reaction—procedures, ports, etc.

m. Precipitation
n. Viscosity changes; color changes
o. Phase separation or change
p. Use of inerting atmospheres
q. Stability
r. Isolation difficulties
s. Physical properties
t. Volume increases
u. Erosion and corrosion; cathodic protection as needed
v. Moving equipment hazard zones; vibration
w. Process alarms, as needed and functioning
x. Extra care at start-up
y. Liquid expansion relief
z. Physical characteristics of process w.r.t. exposure generation, esp. airborne materials; e.g. spraying, welding, grinding, and agitating

17. Insure written operating procedures include safety checks, actions, maintenance
18. Insure operator training, formal and informal, includes safety and health training, including emergency procedures, protective equipment use, and hazard communication
19. Are normal staff numbers and distribution adequate for fire fighting and emergencies, esp. in small plants

C. Isolation and separation of risks to minimize exposure
1. Identification of toxic or radiation areas, minimize number of people exposed, degree of exposure
2. Noise
    a. Absorption by structural mass or other means planned, including massive foundations and/or vibration isolation where needed
    b. Noise specifications on all new equipment e.g., vents, valves, motors, compressors, furnaces, flares
    c. Site emergency relief valves/vents vertically, away from operator areas (II)
3. Radiation
    a. Reflection of nonionizing types where possible
    b. Absorption or shielding
    c. Layout for separation for distance control of radiation
4. Fire-explosion risks separated from ignition hazards, other such risk areas—fire walls, curbs, dikes, barriers, etc.
5. Boilers, other major pressure vessels
6. Storage areas for hazardous materials, also compressed gasses storage and restraint systems

7. Carcinogen or biohazard areas
8. Emergency chemical or other exposure refuge points
9. Stored amounts of materials less than acceptable hazard amounts and areas affected if explosion or fire occur

D. Ergonomics considerations
1. Job analysis needed for repetitive motion, biomechanical stresses, etc.; can machines take over repetitive tasks?
2. Acceptable information flow and control design at human-machine interfaces
3. Workstations and materials handling evaluated for above, redesigned as needed

E. Material flow and handling systems and organization
1. Horizontal transport systems—hilo's and other vehicles mechanical and pneumatic conveyors, carts, robot tugs, piping, augers
2. Vertical transport—elevators, piping, etc.; gravity flow where possible
3. Review transfer, measurement, and packaging points for exposure potential
4. Storage—long term, and temporary ready-to-use
5. Proper location of controls, valves, etc., including access during emergencies
6. Review materials packaging w.r.t. high-risk quantities and risk reduction
7. Temperature and moisture control systems for dry bulk materials

F. People flow
1. Worker access to worksites and facilities (carts, bicycles, scooters, foot)
2. Visitors
3. Conflicts between worker flows/locations and material flows/storage

G. Occupational health input into automation and mechanization
1. Programming of machines to include safe movements, software interlocks, etc., as well as hardware interlocks
2. Robots require $2\times$ to $3\times$ more space

H. Industrial sanitation and services
1. Water—quantity and quality as needed
   a. No cross-connections
   b. Adequate drains, sewerage
2. Food handling, eating, lounge/rest space away from work site; vending machine sanitation
3. Solid waste collection and handling in-plant
   a. Materials collection sites, bins, carts, and sufficient space for these
   b. Waste movement systems and storage of collected wastes in plants; reclamation if possible
   c. Incineration or other means of destruction/disposal

4. Vector control—rats, insects
5. Air cleaning where required
   a. Air recirculation if feasible
   b. Special needs—clean rooms, air supply for critical processes
6. Sanitary facilities: toilets, washrooms—adequate number, size and distribution; internal circular traffic patterns, not in busy areas
7. Personnel services, where required or useful
   a. Lockers for street clothes and change rooms
   b. Showers, washing facilities
   c. Lockers for work clothes, change rooms; facilities for work clothes handling, cleaning, decontamination, or disposal
   d. Receipt, cleaning, maintenance, and redistribution of personal protective equipment
   e. Child-care facilities, if desired
8. First aid and medical services space, and access
9. Facilities for industrial hygiene staff, laboratories, information handling
10. Space, facilities, fixtures planned throughout plant for health and safety equipment

I. Hazard communication within plant—consistent system of signage, placarding, content labeling, etc.; planned facilities for MSDS access

REFERENCES

Alden, J. L. and J. M. Kane. *Design of Industrial Ventilation Systems*. New York: Industrial Press, 1982.

American Conference of Governmental Industrial Hygienists. *Industrial Ventilation, A Manual of Recommended Practice*, Lansing , MI: American Conference of Governmental Industrial Hygienists, 1991

ASHRAE. 1992 *ASHRAE Handbook—HVAC Systems and Equipment*. American Society of Heating, Refrigerating and Ventilating Engineers, 1992.

ASHRAE. *ASHRAE Handbook—Fundamentals*. American Society of Heating, Refrigerating and Ventilating Engineers, 1989.

Buffalo Force C.., *Fan Engineering*, 8th ed. Buffalo, NY: Buffalo Forge Co., 1983.

Burgess, W. A., M. J. Ellenbecker, and R. T. Treitman. *Ventilation for Control of the Work Environment*. New York: Wiley, 1989.

Clapp, D. E., D. J. Groh, and C. M. Nenadic. Ventilation Design by Microcomputer. *Am. Ind. Hyg. Assoc. J.* **43**, 212 (1982).

Goldfield, J. Contaminant Concentration Reduction: General Ventilation versus Local Exhaust Ventilation. *Am. Ind. Hyg. Assoc. J.* **41**, 812 (1980).

Goodfellow, H. D. and J. W. Smith. Industrial Ventilation—A Review and Update. *Am. Ind. Hyg. Assoc. J.* **43**, 175 (1982).

Heinsohm, R. J. *Industrial Ventilation*, *Engineering Principles*. New York: Wiley 1991.

McDermott, H. J. *Handbook of Ventilation for Contaminant Control*. Ann Arbor, MI: Ann Arbor Science Publications, 1985.

NIOSH. *Guide to Industrial Respiratory Protection*. Cincinnati, OH: ACGIH, 1987.

Ruch, W. E. and B. J. Held. *Respiratory Protection: OSHA and the Small Businessman*. Ann Arbor MI: Ann Arbor Science Publishers, 1975.

Schwope, A. D., P. P. Costas, J. O. Jackson, and D. J. Weitzman, *Guidelines for the Selection of Chemical Protection Clothing*, 3rd ed. Cincinnati, OH: ACGIH 1987.

Western, N. J. Hygiene Assessment of New Products—A Company View. *Ann. Occ. Hyg.* **30**, 237 (1986).

Whitehead, L. W. Planning Considerations for Industrial Plants Emphasizing Occupational and Environmental Health and Safety Issues. *Appl. Ind. Hyg.* **2**, 1987.

7

Valves

The Environmental Protection Agency estimates that valves contribute about 40% of total fugitive emissions (VOC) from chemical plants and petroleum refinery on-site processing units. The valve and flange fraction of total fugitive emissions is significantly higher than that of other emission sources and in practical terms a figure of about 80% for both categories is a reasonable assumption. The valve category estimate includes emissions from all types of block valves, control valves, and pressure relief valves. Control valves are not included in this chapter but are included in overall valve emission estimates.

The uncontrollable emission rates for different plants vary widely but are an important consideration in workplace concentration levels. Since the workplace concentration level is an industrial hygiene concern, reduction and control of fugitive emissions from valves is a major objective in reducing potential exposures. The emission constraints imposed on these sources through VOC and hazardous air pollutant EPA regulations reduces emissions, but industrial hygienists must understand regulatory requirements and the emission reductions achieved through these techniques to determine whether additional emission reductions are necessary for specific workplace concentrations. Although stringent regulatory constraints are not limited to the United States, the various regulations provide a basis for controlling workplace emissions in other areas of the world.

Since valves are major contributors to fugitive emission losses, EPA efforts have been concentrated on measuring and controlling valve emissions in both chemical plants and petroleum refineries. This overall effort resulted in VOC regulations in the mid-1980s requiring extensive monitoring and maintenance (M & M) emission control programs to reduce fugitive emission losses through state implementation programs (SIP) and federal new source performance standards (NSPS). The M & M term has been replaced by LDAR, an acronym for leak detection and repair. The NSPS regulations are not only applicable to new plants, but also control emissions from older plants through a plant modification investment level.

As a result of these emission control regulations, valve emissions have been reduced, accompanied by a reduction in workplace exposure concentration levels. The newly promulgated MACT (maximum achievable control technology) in the HON (hazardous organic NESHAP) regulations issued by the EPA to reduce hazardous air pollutant (HAP) emissions in chemical plants will further reduce valve emissions and potential exposure concentrations. Although limited in regulatory constraint to 189 hazardous chemicals initially, the number of chemicals will be expanded periodically. In addition, emission control effectiveness must be reviewed every 8 years with the possibility that further restrictions in chemical plants may be required if an overall emissions reduction of 90% is not achieved. Also, increased restrictions are anticipated for refineries with promulgation of new regulations expected in 1994 to control HAP emissions. Modifications in VOC regulations to reduce VOC emissions in accordance with the Clean Air Act amendments of 1990 are also anticipated to meet compliance targets for nonattainment areas. Regional air control authorities in the next several years may require further emissions reductions through regulations that are more stringent than the federal regulations.

While the HON and other regulations require further reductions or limitations on valve emissions, workplace overexposures may still be a potential problem. Although the HON equipment requirements reduce allowable maximum valve leakage concentrations from 10,000 to 500 ppm, valves with concentration levels close to 500 ppm are not classified as leakers, but may pose potential emission concerns with potentially hazardous or toxic materials. In addition, valves are often located in clusters, which can increase the potential for overexposure when a number of valves have leak concentrations close to the leak definition level.

This chapter reviews the effect of regulations on emissions and control improvements necessary to meet the regulations and reduce workplace exposures. The new HON regulations do not require specific emission control techniques, but provide emission targets and methods of achieving the required emissions concentration operating levels based upon maximum achievable control technologies investigated by the EPA. Revised VOC regulations will probably follow these general requirements along with the HON petroleum refinery regulations. The HON regulations require varying monitoring periods to achieve the desired valve leak frequency along with specified repair periods and maintenance procedures. Monitoring and maintenance (LDAR) effectively reduces overall emissions by significantly reducing the number of leaking valves.

7.1 AIR EMISSION REGULATIONS

States issue regulations (SIPs) to meet the National Ambient Air Quality Standards (NAAQS) for six contaminants, including ozone. Fugitive emissions or VOCs contribute to ozone formation through photochemical reactions and are regulated by the states through the SIP program, including enforcement. The NSPS regulations, which are the minimum required for state SIP regulations, along with the NESHAP regulations described in Chapter 4, are promulgated directly by the federal government. The new HON regulations supersede the NESHAP regulations for chemical plants and will also be administered by the states. In plants where HON regulations are currently not applicable, such as petroleum refineries, the NESHAP regulations are applicable. In addition, regional authorities have promulgated emission regulations considerably more stringent than the federal regulations which are enforced by these authorities. Industrial hygienists should have some familiarity with these regulations since the regulations mandate emission control through various alternatives. Moreover, the regulations provide information along with EPA background documents on the type of field effort necessary to achieve the emission control targets.

Great emphasis is placed upon LDAR programs which are accompanied by a potential for overexposure in monitoring and repairing the equipment. The industrial hygienist should have an understanding of repair procedures in this general area because of the large exposure potential associated with numerous repairs. In addition, the removal and replacement of valves during operations is a potential source of exposure, particularly where the valve is in VHAP service (volatile hazardous air pollutant). The regulations provide a basis for determining when a valve can be removed that minimizes atmospheric emissions. This requirement is particularly important in controlling potential exposures.

At present VOC regulations in the form of NSPS requirements and the new chemical plant HON regulations are the principal regulatory restrictions on valve emissions. The NSPS regulations for VOC are considerably less stringent than the HON regulations, but the Clean Air Act Amendments (CAAA) of 1990 clearly require changes in VOC regulations to rapidly comply with the ozone NAAQS standard. The VOC regulations will gradually tighten and approach the HON regulatory restrictions to achieve the NAAQS standard in nonattainment areas. In addition, HON standards for valve emission control in petroleum refineries will be similar to the HON chemical plant regulations. Consequently, the valve NSPS regulations for VOC operations are presented in Section 7.2, followed by the HON regulations in Section 7.3, showing the change to more restrictive regulations. Also, valve emission restrictions for regional authorities are included where these regulations are more stringent than the federal regulations.

7.1.1 Valves (NSPS)

The basic VOC regulations are termed new source performance standards (NSPS)[1,2]. New, modified, and reconstructed facilities are included in these regulations with investment criteria the basis for determining whether the modified or reconstructed facilities are specifically covered by the NSPS regulations. Over a long-term period, all modified or reconstructed units will be subject to the NSPS regulations which are the basis for the SIPs. Some of the states with nonattainment areas have issued more stringent regulations than those included in the NSPS regulations.

The NSPS regulations may exclude a very small group of compounds that are not photochemically reactive as described in Chapter 4 and are generally defined in state SIP requirements. Consequently, most streams near petroleum refineries and chemical plants contain VOCs. Where valve emissions are not controlled under HON regulations, they must comply with VOC regulations (NSPS). A valve is in VOC service when the valve contacts or contains a process fluid with 10% or greater VOC content on a weight basis. The basic NSPS regulations for valves in gas/vapor or light liquid service follow:

1. Each valve is monitored monthly with the technique described in Chapter 4. An instrument reading of 10,000 ppm or larger *indicates a valve is leaking*.

2. When a valve is not leaking according to the definition given above for two successive months, monitoring may be changed to one measurement each quarter until a leak is detected. Following leak identification, the valve is again monitored monthly until two successive no-leak months are achieved; monitoring can then be changed to quarterly measurements.

3. When a leak is detected, initial valve repair must commence within 5 calendar days after detection and repairs must be completed within 15 calendar days after detection. Repair delay is permitted if:

   a. The repair cannot be undertaken without a unit shutdown, but the repair must be completed during the next unit shutdown.

   b. The owner demonstrates to the EPA that purged emissions resulting from immediate repair are greater than fugitive emissions resulting from a repair delay, and when the valve is repaired, purged material is collected and destroyed or recovered in a control device.

   c. Valve assembly parts are not available, providing sufficient inventory had been available before depletion of the supplies. This delay pertains to valves that cannot be repaired during a shutdown.

4. Initial repair attempts should include any or all of the following when practicable:

   a. Tightening of bonnet bolts
   b. Replacement of bonnet nuts
   c. Tightening of packing gland nuts
   d. Injection of lubricant into lubricated packing

5. If a valve reading is less than 500 ppm above the background level, monthly or quarterly monitoring is not required if:

   a. The valve has no external actuating mechanism in contact with the process fluid.

   b. The valve is monitored annually and as required by the EPA.

6. Any valve that is "unsafe to monitor" is exempt from normal monitoring requirements. This designation refers to monitoring a valve that is classified as potentially dangerous because of excessive temperature or pressure conditions. However, the valve must be monitored during safe periods.

7. Any valve that is "difficult to monitor" is exempt from the normal monitoring requirements if:

   a. The valve cannot be monitored without elevating monitoring personnel more than 2 m above a support surface.

   b. The number of valves designated "difficult to monitor" are less than 3% of the total valve count.

   c. The valve is monitored at least once a year.

8. If less than 2% of all the valves leak, the plant can apply for a classification that exempts valves from the normal LDAR requirements. The valves must then

be monitored annually to ensure leakage is less than 2%. A higher leak rate automatically reinstitutes a monthly LDAR program.

9. As an alternative to item 8, when leaking valves are equal to or less than 2% of the total number of valves for two consecutive quarterly periods, the plant may skip one of the quarterly leak detection periods. After five consecutive quarterly leak detection periods, with a maximum valve leak frequency equal to or less than 2%, the plant may skip three quarterly detection periods. A higher leak rate automatically reinstitutes a monthly LDAR program.

The following is applicable to valves in heavy liquid service:

10. If evidence of a potential leak is observed through visual, audible, or olfactory means, the valve must be monitored in accordance with the portable leak detector method presented in Chapter 4. The monitoring test must be completed within 5 days after the leak was identified.

11. A leak is detected if a portable instrument reading of 10,000 ppm or greater is measured.

12. Where a leak is detected, the initial repair must begin within 5 calendar days after detection, with repair completion 15 calendar days after detection. Repair delay is permitted in accordance with item 3.

7.1.2 Pressure Relief Devices (NSPS)

Pressure relief devices or safety valves are installed for overpressure protection in gas/vapor and liquid services. The regulatory requirements for these two services are separate, and the regulatory summary is also separated in this section for clarity.

The following regulatory summary is applicable to gas/vapor service;

1. This regulation applies to a relief valve that emits process vapors to the atmosphere under emergency conditions. During normal operation when the pressure relief device is not discharging vapors in an emergency situation, the fugitive (VOC) emission concentration measured on the portable monitoring instrument must be less than 500 ppm above background. The emission concentration is measured in the center of the atmospheric opening and in the plane of the opening. Relief valves without a discharge pipe are measured at the flange face. When a relief valve has a length of pipe connected to the outlet flange the measurement is obtained at the outlet end of the extension pipe.

2. After an emergency or overpressure release, a monitoring reading must demonstrate less than 500 ppm above the background level within 5 calendar days after the emergency release.

3. A pressure relief device connected through a closed system to a control device is exempt from the previous requirements.

Regulatory rules for pressure relief devices in *light or heavy liquid* service require monitoring within 5 calendar days after leakage is observed through visual, audible, or olfactory methods. A leak is identified when the monitor reading is 10,000 ppm or greater. The regulations for valves in heavy liquid service presented in Section 7.1.1 are identical to the relief valve regulations for liquid service. A closed discharge line to an enclosed system for relief valves in liquid service eliminates or exempts the valves from the regulations.

7.1.3 Open-Ended Valves or Lines (NSPS)

The regulations are concerned with leakage through a valve that terminates an open-ended line or a length of open-ended line that extends beyond the final block to the atmosphere. The general regulations that follow are applicable in all liquid or vapor services.

1. All single valves or open-ended lines should have a blind flange, cap, plug, or second valve. The open end shall be sealed at all times, except during operations requiring process flow through the valve or line..
2. In double block and bleed systems, the bleed valve or line may remain open during operations requiring venting the line between the block valves. When not in operation, the bleed or vent line must be closed as described in reference 1.
3. When there are two valves at the open end of a line, the valves shall be operated by closing the valve on the process fluid side before the valve on the open end is closed.

7.1.4 Valves (HAP)

Valves in HAP service must conform with the HON regulations issued in accord with the CAAA. Valves in existing process units or reconstructed prior to the final regulatory date in late 1993 must conform with the HON regulations, including valves in process units constructed or reconstructed after this date. HAP service refers to valves that contain or contact fluids (gas or liquid) that contain 5% or more HAP on a weight basis. The HAP list of chemicals shown in Appendix 9A (Chapter 9) covers any compound used as a reactant or chemical produced as an intermediate or product in the SOCMI processes shown in Appendix 9B. However, these lists will be expanded through the years and most chemicals will eventually be included in this list.

All valves larger than a nominal 1/2-in. size must conform with the HON regulations, including those valves in instrumentation service. When valves are installed in vacuum service and operate at an internal pressure 5 kPa below ambient pressure, they are excluded from HON regulatory restrictions. These regulations also do not apply to valves in VHAP service that operate less than 300 hr/yr.

Although process units and plants contain a large number of valves, physical tagging of the equipment is not required. However, an alternate method of identification is necessary through site plans, log entries, weatherproof identification, or detailed flow plans. The basic HON regulations for valves in gas/vapor and light liquid service (see Chapter 4) follow:

1. Chemical processes are divided into groups[3] and implementation of the regulations is based upon the group applicability date. Promulgation refers to the issuance date of the final regulations, and examples of various groups and

associated applicability date follow:

Group	Years After Promulgation	Chemical
I	1/2	Benzene, styrene, xylenes, etc.
II	3/4	Butadiene, glycerol, vinyl acetate, etc.
III	1	Bisphenol A, cresols, phenol, etc.
IV	1 1/4	Methanol, methyl bromide, phosgene, etc.
V	1 1/2	Chlorophenols, polyethylene glycol, isobutylene, etc.

2. For each group of processes, the *compliance monitoring program* is separated into phases for *existing* process units based upon the applicability dates determined in item 1. The phases and monitoring criteria leak levels are shown in Figure 7.1 and the following table for all of the HAP chemicals with the exception of benzene and vinyl chloride. The basic NESHAP regulations are more stringent until Phase II is initiated.

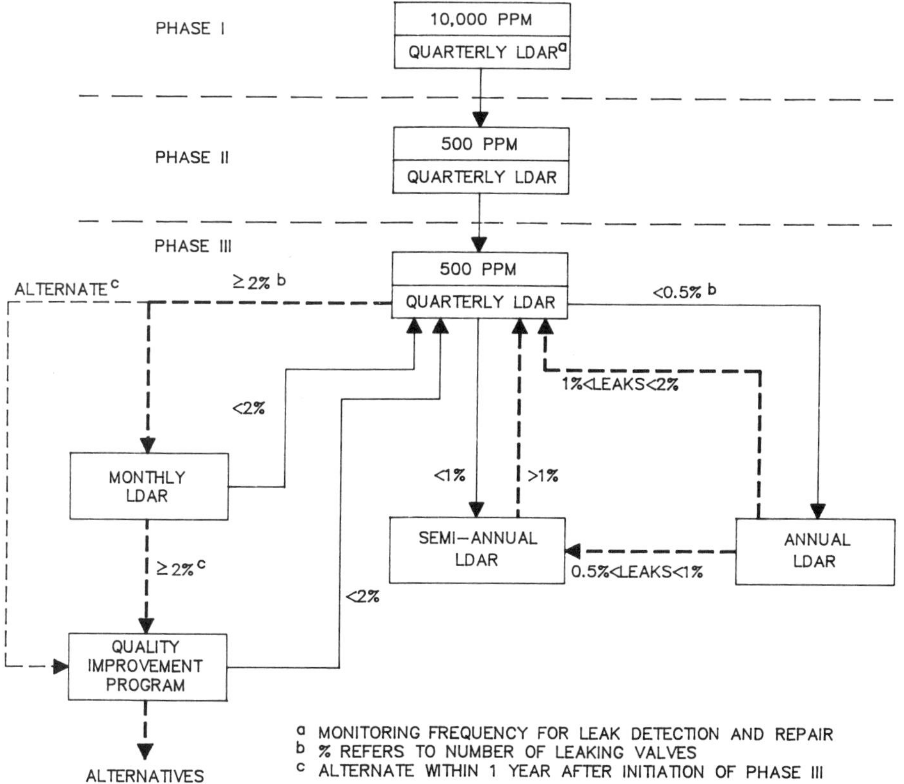

Figure 7.1 *MACT valve compliance monitoring program requirements showing leak frequency criteria.*

Phase I	Begins on applicability date	$\geq$ 10,000 ppm
Phase II	Begins 1 year after applicability date	$\geq$ 500 ppm
Phase III	Begins 2 1/2 years after applicability date	$\geq$ 500 ppm

For all *new* process units, irrespective of groups, following startup the applicable phases are:

Phase II	Begins immediately after initial startup
Phase III	Begins 1 year after startup

The phase leak criteria levels are identical for existing and new processes.

3. Monitoring is conducted quarterly for Phases I and II. In Phase III, monitoring requirements and leakage frequency are associated and apply to new and existing units.

$<$ 2% leaking valves	Quarterly monitoring
$<$ 1% leaking valves	Monitor once in 2 quarters (semiannual)
$<$ 0.5% leaking valves	Monitor once in 4 quarters (annually)

Where the number of leaking valves are $\geq$ 2%, one of the following is required until the leak frequency is $<$ 2%:

a. Monitoring each valve monthly

b. Within 1 year after Phase III initiation, implement a QIP and monitor monthly. Further details on QIPs are presented in items 9–12.

4. Percent leaking valves for determining monitoring frequency are calculated as a rolling average of two consecutive monitoring periods for monthly, quarterly, or semiannual monitoring programs and as an average of any three out of four consecutive monitoring periods for annual monitoring programs. The number of nonrepairable valves excluded from this calculation are detailed in reference 3.

5. When a leak is detected, initial valve repair must commence within 5 calendar days after detection and repairs must be completed 15 days after detection. Following leak repair, the valve shall be monitored once within the first 3 months after the repair. Repair delay is permitted if:

a. The repair is not feasible without a process unit shutdown, but the repair must be completed during the next unit shutdown.

b. The valve is isolated from this process and does not remain in VHAP service.

c. The owner determines that emissions from purged material during an immediate repair are greater than the fugitive emissions resulting from a repair delay, and when the valve is repaired, purged material is collected and destroyed or recovered in a control device.

d. During a process unit shutdown, valve assembly parts are not available, providing sufficient inventory had been available before depletion of supplies. However, the delay will not be permitted beyond the next process unit shutdown unless a process unit shutdown occurs within 6 months after the initial shutdown.

6. Initial valve repair attempts should include any or all of the following when applicable:

   a. Tightening of bonnet bolts

   b. Replacement of bonnet bolts

   c. Tightening of packing gland nuts

   d. Injection of lubricant into lubricated packings

7. Any valve that is unsafe to monitor is exempt from the requirements of items 2 and 3 if:

   a. Monitoring personnel would be exposed to an immediate danger.

   b. A written plan has been developed that requires valve monitoring as frequently as possible during safe monitoring periods.

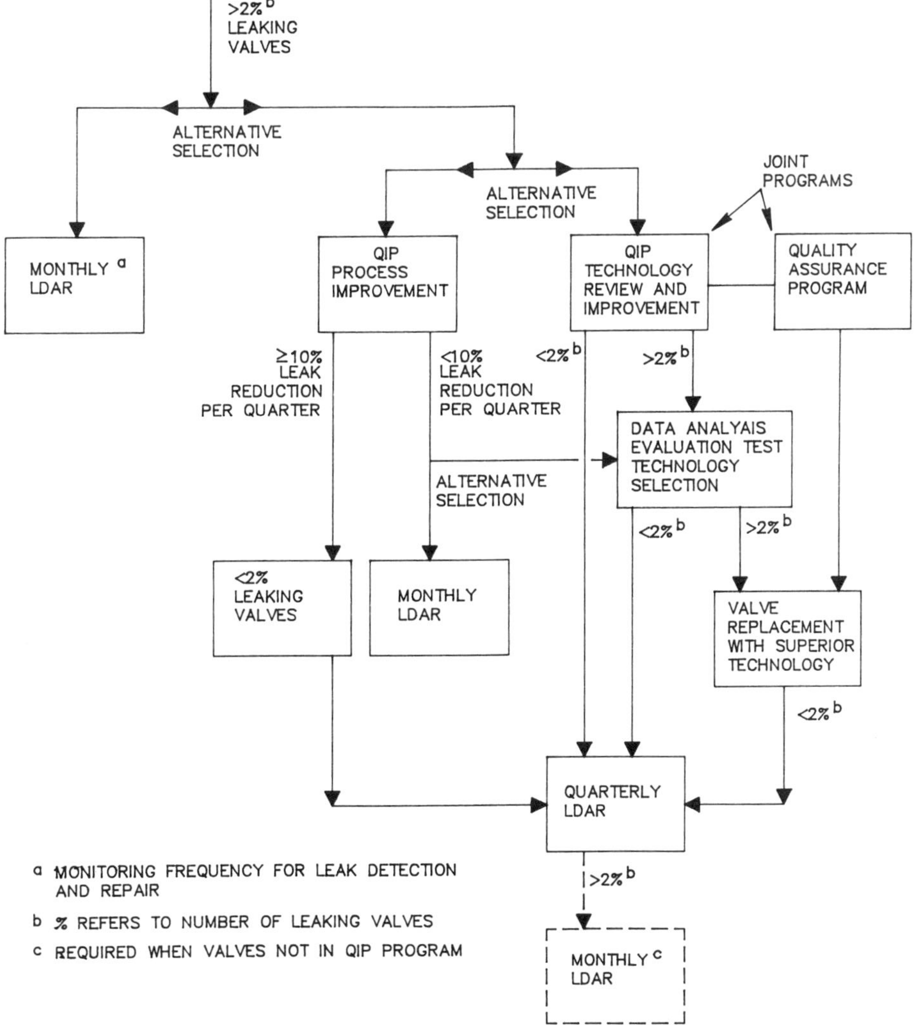

Figure 7.2 *MACT valve alternative programs where the number of leaking valves exceeds 2%.*

8. Any valve that is difficult to monitor is exempt from the requirements of items 2 and 3 if:

a. The valve cannot be monitored without elevating monitoring personnel more than 2 m above a support surface.

b. The valve is located within an *existing* process unit.

c. The valve is monitored at least once a year.

As shown in item 3 and Figure 7.1, a QIP is necessary where the number of leaking valves is $\geq 2\%$ or monthly monitoring is required until the number of leaking valves is $< 2\%$. The QIP must be instituted within 1 year after Phase III is initiated. In general, the overall programs are predicated on reducing monitoring requirements through significant reductions in the number of leaking valves. Conversely, when the number of leaking valves increases, the monitoring frequency increases. Since monitoring is costly, there is an incentive to reduce leak frequency, particularly in those process units where monthly monitoring is required.

The basic valve QIP is outlined in the following items and in Figure 7.2:

9. There are two QIPs, and the decision to comply with either program must be established for new and existing programs within 1 year after Phase III is initiated.

10. Owners selecting either QIP must comply with program requirements until there are $< 2\%$ leaking valves based on a rolling average of two consecutive quarters. At this point, the process unit can comply with item 3, continue to comply with the selected QIP, or both. Certain requirements are omitted if the process unit remains in a QIP.

Where the owner does not continue the QIP and the process unit again exceeds 2% leaking valves, the QIP can no longer be selected as an option and monthly monitoring is required.

11. A QIP that demonstrates further progress in reducing the percent leaking valves is one alternate and requires the following:

a. Valves are monitored quarterly and specific information is required covering monitoring and repair methods. Data are continually collected during the QIP.

b. The owner must demonstrate in each quarter a 10% reduction in the percentage of leaking valves from the preceding monitoring period. The leaking valve percentage is calculated as a rolling average of two consecutive quarters.

c. If a 10% reduction per quarter in the percent leaking valves cannot be demonstrated, then the owner must revert to monthly monitoring or comply with a technology review QIP described in item 12.

12. The *technology review QIP* also considers valve improvements and requires the following:

a. Monthly monitoring is not required, but quarterly monitoring is required.

b. Extensive data and record keeping information is required for *each* valve in the process unit subject to this program as follows:

Valve type (ball, gate, etc.)

Valve design (activating mechanism, flanged body)

Materials of construction

Packing material and year installed

Stream characteristics (pressure, temperature, line diameter, corrosivity)

Gas or light liquid service

Monitoring readings before repair, response factor, instrument model, date

Repair methods and monitor readings after repair

Any quality assurance or maintenance programs for emission performance improvements

Information collection continues during the entire period the process unit remains in the QIP.

c. All leaking valves removed for leakage are inspected and any improvements or correction recommendations are prepared for incorporation in design changes or specifications.

d. The data on valve operations, valve designs, technologies, and maintenance practices are analyzed in accordance with the following objectives:

- Identification of systems that have lower or higher than average emissions performance. Specific problem areas should be identified where possible covering service, operating conditions, maintenance practices, equipment design or other factors.

- The analysis will identify any superior performing valve technologies. The criteria for superior valve technology is a leak frequency $< 2\%$. The superior performing valve technology can be demonstrated or identified in the literature or through a study.

- This analysis will include:

Data from inspections of valves removed due to leakage

Information from available literature and experiences at other sites identifying low emission valve designs or technologies for specific services

Limitations on service and operating conditions for valve designs

Maintenance procedures that ensure continued low emission performance

- Data analysis can be an intercompany or intracompany program for a single or group of process units.

- The first analysis must be completed within 18 months after the initiation of Phase III operations. This analysis is required each year the process unit is in the QIP.

e. Where superior emission control technologies have not been identified, an evaluation trial program will be conducted with valve designs or technologies identified by others as low-emission performance installations with leak rates $< 2\%$. The trial will be conducted on valves identified as poor emission control operations. If certain superior technologies that had been identified are not included in the trial, reasons for rejecting these technologies must be documented.

The number of valves in the trial evaluation for a single process unit shall be the lowest number based on 1% of the total or 20 valves. For multiple units, the number of trial valves are the lowest number based on 1% of the total or 50 valves. Documentation for the trial programs are required.

Performance trials are conducted for a minimum period of 6 months and must begin within 18 months after the start of Phase III. Superior technologies or techniques must be identified within 24 months after the start of Phase III. Performance trials will continue until a superior performing technology is identified. If a superior performing technology cannot be identified, an engineering evaluation is conducted to document the reasons the various technologies were ineffective.

f. Selection of the technology review and improvement QIP outlined in items a–e and shown in Figure 7.2 requires implementation of a *valve quality assurance program*. The QA program covers detailed valve purchasing specifications and maintenance procedures. This program incorporates results of the analytical program outlined in item d and any applicable information from item e.

- This QA program is reviewed and updated each year while process unit leaking valves are $\geq 2\%$.
- This program requires the following:

  Establish minimum design standards for each valve category covering manufacturer, tolerances, construction materials, or other critical parameters.

  Include design standards for valves in all equipment purchase orders.

  Develop written procedures for bench testing valves and specifying acceptance performance criteria including test apparatus criteria for precision and accuracy. When the QA plan is prepared, all valves removed and repaired will be bench tested for leaks.

  Valves will only be installed after certification on the bench test apparatus that they are nonleakers.

  Valves repaired online must be monitored for leaks two successive months after repair.

- An audit program must be established to ensure quality control of purchased equipment.
- Off-line valve maintenance and repair procedures must be detailed to ensure rebuilt or refurbished valves meet design specifications for the valve type.
- A QA program shall be established no later than the start of the third year of Phase III operations.
- Each valve replaced with a new or modified valve must comply with the QA standards and be classified as superior emission performance technology. The latter requirement covers valves that will have a frequency leak rate $< 2\%$.

  Valves shall be maintained as specified in the QA program. If a superior emission technology cannot be identified, leaking valves will be replaced with the lowest emission performance valves identified for the service.

The previous items covered valves in gas/vapor and light liquid service. The following HON requirements are applicable to valves in *heavy liquid service*:

13. If evidence of a potential leak is observed through visual, audible, olfactory, or any other detection method, the valve must be monitored in accordance with

the requirements shown in reference 4 or Method 21. The monitoring test must be conducted within 5 calendar days after the leak is identified.

14. A leak is detected if a portable instrument reading of 500 ppm or greater is measured.

15. Where a leak is detected, the initial repair must begin within 5 calendars after detection. Repair delay is permitted in accordance with item 5.

7.1.5 Pressure Relief Devices (HON)

Pressure relief devices or safety valves were described in Section 7.1.2. These devices are normally designed for gas/vapor and liquid service. However, the HON regulatory requirements for these two services are separate and in this section the services are also separated for clarity.

The following regulatory summary is applicable for gas/vapor service:

1. During normal operation when the pressure relief device is not discharging vapors to the atmosphere, the emission concentration measured with a portable instrument must be less than 500 ppm above the background concentration level. This requirement applies to a pressure relief valve that emits HAP to the atmosphere when the device is activated through an overpressure condition. The emission concentration, when the valve is not activated, is measured in the center of the atmospheric opening and in the plane of the opening as described in reference 4. Pressure relief valves without a discharge pipe are measured at the flange face. When a relief valve has a length of pipe connected to the outlet flange, the measurement is obtained at the outlet end of the extension pipe.

2. After an emergency or pressure release, a monitoring reading must demonstrate less than 500 ppm above the background level within 5 calendar days after the emergency release.

3. A pressure relief device connected through a closed system to a control device is exempt from the requirements in items 2 and 3.

For pressure relief devices in liquid service, the following regulations are applicable:

1. If evidence of a potential leak is observed through visual, audible, olfactory, or any other detection method, the valve must be monitored in accordance with the requirements shown in reference 4 or Method 21. The monitoring test must be conducted within 5 calendar days after the leak is identified.

2. A leak is detected if a portable instrument reading of 500 ppm or greater is measured.

3. When a leak is detected, the initial repair must begin within 5 calendar days after detection, with repair completion 15 calendar days after detection. Repair delay is permitted in accordance with Section 7.1.4.[5]

7.1.6 Open-Ended Valves or Lines (HON)

Requirements for open-ended lines and double block/bleed systems for HON regulations are identical to the regulations described in Section 7.1.3 for VOC operations under NSPS standards.

7.2 EMISSIONS

Valve emissions are a significant contributor to the overall fugitive emission rates from chemical plants and refineries. The exact proportion of emissions attributed to valves is somewhat difficult to quantify at the present time because of regulation variations throughout the U.S. increased LDAR programs and health concerns associated with some chemicals. In the HON chemical emissions regulations, the EPA believes over 50% of total SOCMI fugitive emission quantities emanate from process vents and wastewater emissions, as described in the preface to reference 3. In contrast, process vents are not substantial emission contributors in petroleum refineries. Consequently, the valve emission proportion will differ due to emission source changes and a generally higher level of average valve emission factors for refineries compared with chemical plants.

A comparison of average uncontrolled valve emission factors shown in Table 7.1 indicates higher emission factors for petroleum refineries compared with SOCMI uncontrolled emission factors. The petroleum factors are average values over a wide range of processes and services while the SOCMI factors were developed by the EPA from a combination of petroleum data, a SOCMI study of 24 units, and a special six-unit maintenance study.[5–8] The SOCMI factors were adjusted from the average petroleum factors following information indicating lower equipment emission rates in chemical plants. However, the SOCMI factors were investigated through the Chemical Manufacturers Association (CMA) in three specific types of chemical process units that result in lower emission factors than the SOCMI factors shown in Table 7.1.[9] A comparison of emission factors is presented in Table 7.2 for ethylene oxide, butadiene, and phosgene plants with petroleum refinery and standard SOCMI factors. Significant average emission factor differences exist between the three chemical process units and the average SOCMI factors owing to

TABLE 7.1 Comparison of Uncontrolled Emission Factors for Valves in Petroleum Refineries and Chemical Plants (SOCMI)

	Average Emission Factors, kg / hr (1b / hr)[a]		
Valve Service	Refineries	SOCMI	SOCMI[b]
Gas	0.027 (0.059)	0.0056 (0.0123)	0.0027 (0.0060)
Light liquid	0.011 (0.024)	0.0071 (0.0157)	0.0030 (0.0065)
Heavy liquid	0.000023 (0.00051)	0.00023 (0.00051)	0.0003 (0.0007)

[a]Adapted from references 5 – 8.
[b]Excludes ethylene plants; from reference 10.

TABLE 7.2 Comparison of Valve Emission Factors in Chemical Plants

	Average Emission Factors, kg / hr (1b / hr)[a]			
Valve Service	SOCMI	Ethylene Oxide	1, 3 Butadiene	Phosgene
Gas	0.0056 (0.0123)	0.0002 (0.00044)	0.0005 (0.0011)	0.000001 (0.0000022)
Light liquid	0.0071 (0.0157)	0.00025 (0.00055)	0.0014 (0.0031)	0.0000009 (0.000002)
Heavy liquid	0.00023 (0.00051)	—	—	—

[a]Adapted from reference 9.

TABLE 7.3 Estimated Valve Emission Leakage Reduction 5 Years After HON Becomes Effective[a]

	Baseline Emissions (mg / yr)		Emissions After 5 Years (mg / yr)	
Source[b]	HAP	VOC	HAP	VOC
Total equipment leakage	66,000	84,000	13,000	16,000
Total leakage				
Existing units			10,920	13,440
New units			2,080	2,560
Estimated valve leakage	33,000	42,000	5,850	7,200
Valve leak reduction (%)			82.3	82.3

[a]Adapted from reference 3.
[b]Equipment does not include process vents, storage tanks, wastewater collection, wastewater treatment, and transfer loading operations.

extensive LDAR programs and potential exposure concerns with these chemicals. These emission data along with other information indicated to the EPA that more stringent emission controls can be widely applied in the chemical process industry.

In addition to these data, the EPA obtained information on valve leak performance from Texas, where air control regulations for certain chemical units were more stringent than existing EPA regulations, requiring a 500-ppm leak definition standard. Leak frequencies less than 2% were achieved in some process units, providing a basis for EPA regulations.[3]

A reduction in HAP also reduces VOC emissions and the combined estimated reduction effect of the HON program was determined by the EPA for a 5-year period beginning with baseline emissions when the program becomes effective.[3] In addition the EPA also assumed new units would be added to the industry during the 5 years and the emission quantity at the end of 5 years includes new unit emissions. An overall SOCMI growth rate of 3%/yr was assumed for these calculations. Since fugitive emissions include various elements, the emissions shown in Table 7.3 are limited to equipment leakage and do not include process vents, storage tanks, wastewater collection, and treatment or transfer loading operations. The primary sources of equipment leakage in this table are valves, flanges, pumps, compressors, pressure relief devices, and open-ended lines.

The SOCMI sources generate more VOC than HAP emissions according to EPA calculations.[3] When all of the fugitive emission sources are included, VOC emissions are more than twice the annual quantity of HAP emissions. Consequently, improved valve emission controls significantly affect workplace concentration levels, reducing potential exposure levels.

7.3 SAFETY CONSIDERATIONS

An important consideration in the specification of valves and packings is a requirement for fire-safe valves. Although valves are normally specified for process service at particular conditions, they must operate during and after a fire. Where flammable and/or toxic materials are processed, fire-safe equipment, including packings, is a serious consideration in process plant valve repair, replacement, and new facility design. Existing equipment should be examined for fire safety and replacement or new equipment should meet fire test specifications. Guidelines on fire safety should be established in plants to control potentially toxic emissions from valves during and after a fire emergency in accordance with OSHA requirements and Title III of the CAAA.[11] The objective of these regulations is the prevention of accidental releases of toxic materials, fires, and explosions that may result from releases of hazardous chemicals in the workplace and surrounding community.

These new regulations, which are considered Chemical Process Safety Management, require process hazards evaluations or risk analysis. The OSHA regulations define the process hazards analyses or methodologies acceptable to the agency that companies must conduct. In addition, the regulation also requires immediate accident investigations and revisions to the hazards analysis where accident causes have not been included in the risk analysis. Consequently, valves are a consideration in risk or hazard analysis for OSHA and the EPA since the valves must prevent or minimize releases in severe operating environments such as fire, maloperation, high temperature, high pressure, and corrosive conditions. As a base consideration, valves should not leak to the atmosphere at high rates under fire conditions. In addition, closure and tightness characteristics, which were the reasons for the development of valve fire testing, are very important under fire conditions to block lines that may potentially feed a fire or seal off large sections and equipment.

Industry organizations concerned about fire safety and test standards provide a laboratory test basis for selecting fire-safe valves.[29] Currently there are five standards covering fire testing: API-607, Factory Mutual 6033, Exxon EXES 3-14-1-2A, British Standard BSI 5146 (OCMA FSV-1), and an alternate Exxon test BP3-14-1.[12,13] A typical fire-testing facility, shown in Figure 7.3, is based upon a hydrocarbon test fluid in the valve during the test. The test valve in all of the fire safety tests is directly exposed to fire from the burners during the test procedure. The API-607 and FM6033 valve test facilities use water for the test medium with Exxon and OCMA test standards based on hydrocarbon. Other differences in the tests exist, with the OCMA and Exxon tests more restrictive in leakage requirements and operational conditions. The Exxon test (EXES 3-14-1-2A) also requires water quenching to simulate a fire-fighting water stream and partial exposure to flames as a more realistic simulation of actual fire conditions.

The Exxon and OCMA standards are similar, with a hydrocarbon pressure during the test at 25–30 psig (172–207 kPag). The Exxon metal test temperature is a minimum of 1200°F (649°C) during the test to insure the destruction of soft seats or 0-rings in the valve. The OCMA requirement for destruction of the soft seat and ring seals is similar. During the fire test, the valve is open and leakage around the stem steal (packing) should not produce a flame more than 4 in. (10.2 cm) high and maximum leakage through the closed valve seat of 10 mL/min/in. of valve port diameter. In contrast to the Exxon and OCMA procedure, the valve is shut during the API-607 test. Within 5 min after quenching in the Exxon test, the valve is proven operable by opening and closing the valve for three cycles. The internal pressure of 25–30 psig can be reduced after the first cycle. Following completion of the opening and closing cycle test, the valve is subjected to a hydrostatic pressure of 10 psig (69 kPag) and the leakage is measured across the valve seat. The valve seat leakage rate cannot exceed 10 mL/min/in. of diameter and stem leakage

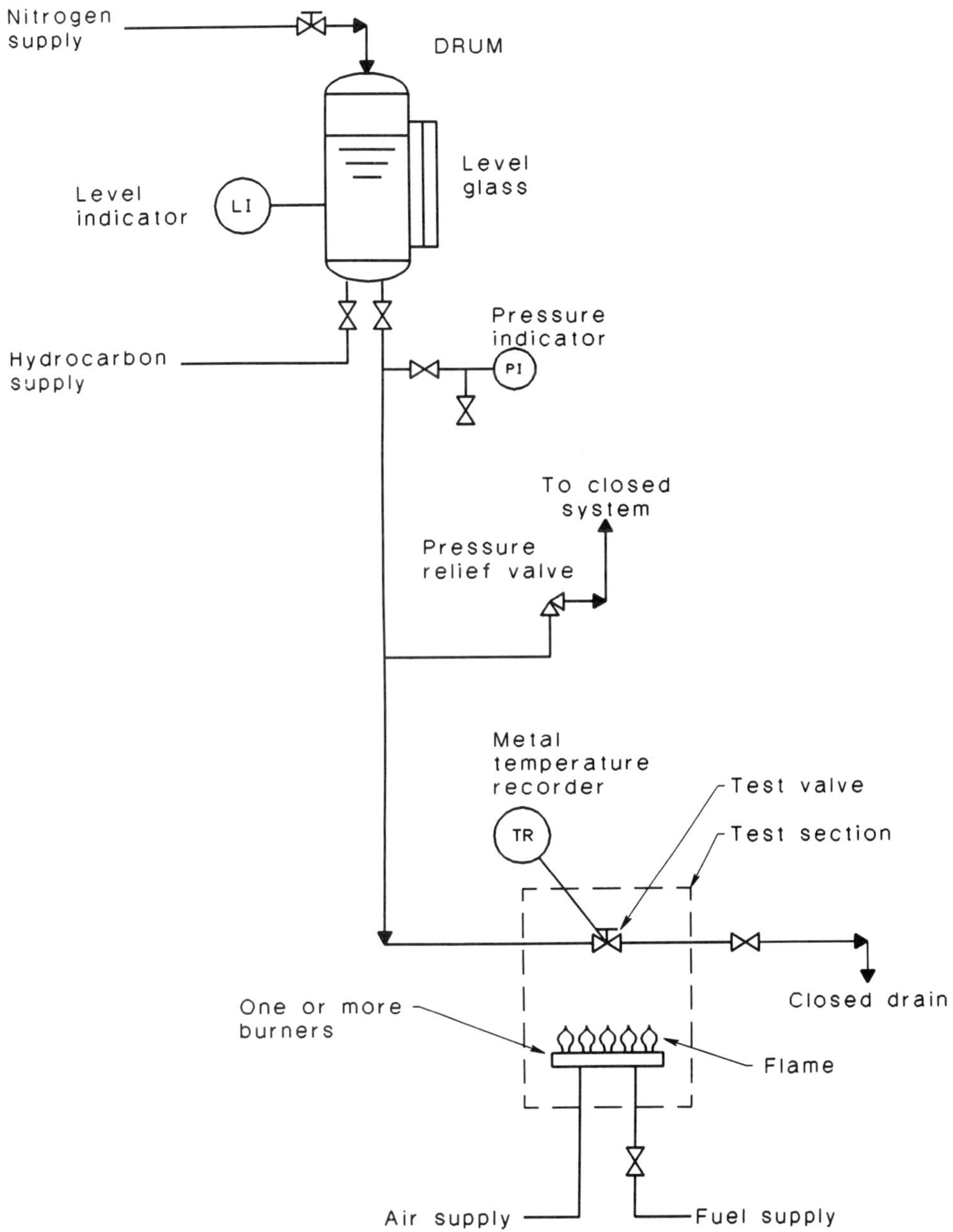

Figure 7.3 *Typical fire-testing facility.*

must be negligible. A second hydrostatic test is performed at about 60% of the valve pressure rating with a third test pressure set at the original seal pressure rating, which is considerably higher than the 10 psig initial test pressure. Seat leakage should not exceed the 10 mL/min/in. of diameter flow rate. The stem seal leakage should also be negligible during the pressure tests. During all of the tests, the packing gland on the valve should not be adjusted. In addition to the full burn test where the flames completely envelope the valve, a partial burn is also conducted to determine whether the valve can operate properly after large changes in stress conditions across the valve due to partial flame exposure.

The fire tests place severe operating requirements on valve stem packings and other valve components. Although the tests are normally conducted with metal valves, nonmetal valves or valves recommended in flammable services should be tested in an overall fire safety evaluation program. In addition, valves that are not in flammable service may be affected by nearby fires and fire resistance should be considered in these applications or installations. All new packings and valves should be reviewed for fire safety with process design and maintenance groups since workplace exposure of personnel during and after the fire are a major concern. Also, guidelines on equipment fire safety standards should be established by the plant for repair of existing equipment and specification of new equipment.

Some basic valve types will not pass the fire tests, but modifications have been developed by the manufacturers to permit these valves to meet test requirements. This overall assessment of plant operations requires specifying valves in accordance with plant requirements. Valve packings or combinations of packings vary considerably in fire test performance, requiring selection of valve and packing combinations that meet fire test conditions. These considerations are included in the following sections covering valves and packings.

7.4 VALVE TYPES

A variety of valves are commercially available for use as manually or remotely operated block valves. In addition, some of these block valves may be used to control flow. Control valves, reviewed in Chapter 8, are normally designed for this operation and generally differ in various respects from block valve designs. While representative block valve configurations or designs are reviewed in the following sections, these valve types do not cover all of the variations in valve designs commercially available, but provide a basis for comparing valve designs. A brief review of hermetically sealed valves presents other valve types that do not leak or emit vapors. However, hermetically sealed valves require packings as a backup, and the comments in Section 7.5 on valve packings are applicable to both hermetically sealed and packed valves.

In the petroleum and petrochemical industry, gate valves are the standard block valve and are widely used through large temperature and pressure ranges. A simplified sketch of a gate valve is shown in Figure 7.4. Globe valves used in a much smaller number, particularly in bypass piping configurations to permit manual control of bypass flow streams, are reviewed in Section 7.4.2. Other valve types are also installed in small quantities in petroleum refineries. In contrast, a wide variety of valves have been used in the chemical industry through the years, such as gate, plug, ball, butterfly, and globe valves. However, the chemical industry has significantly reduced the use of gate valves in process operations and replaced the gates in many applications with plug, ball, and butterfly valves. Globe valve usage in the chemical plants has apparently not changed through the years. A major process valve in widespread use in the chemical industry is the lined plug valve, with considerably less lined ball valve usage. These linings are in the

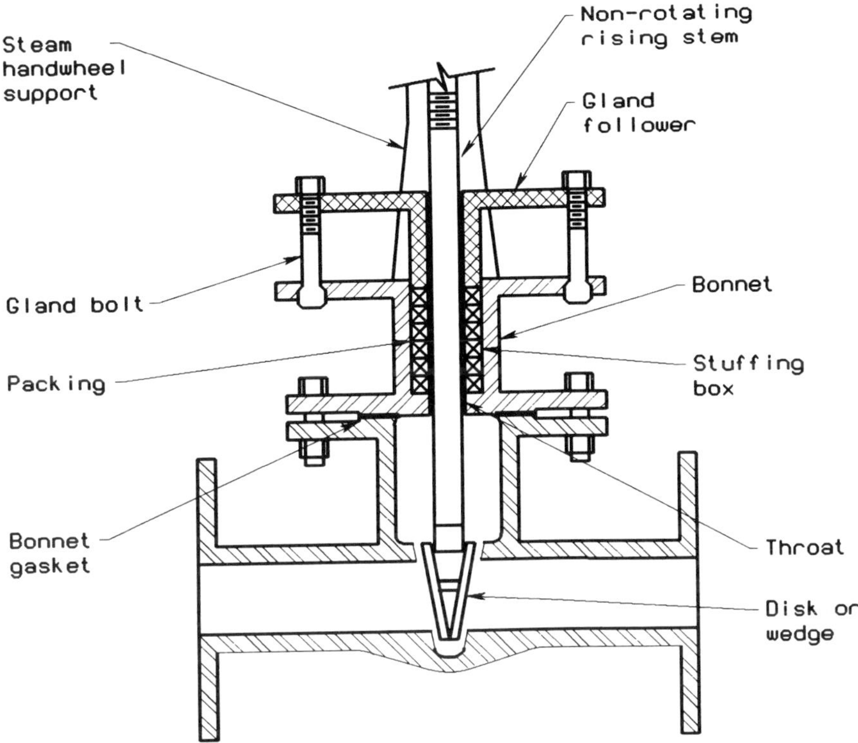

Figure 7.4 *Simplified sketch of a typical gate valve.*

polymer family and generally consist of polypropylene, PTFE®, or polyethylene. All of the internal surfaces may be lined in the plug valve or the plug may be the only surface that is lined. Some lined butterfly valves have also been installed, but apparently in relatively low numbers. There are also a large number of stainless steel valves that are installed where linings are unnecessary or undesirable. The variation in valve types and construction materials in the chemical industry is required to meet a more corrosive environment and higher stream purity requirements than is necessary in the petroleum or petrochemical industries. In addition, a large number of valve applications in the chemical industry occur at lower temperatures and pressures than those generally found in the petroleum industry, which also affects valve selection.

The chemical industry has apparently moved toward nongate valve installations to reduce valve emissions. Valve emission data associated with particular valve types were not investigated by the EPA, but an API study indicated globe, plug, and ball valves had lower emissions than gate valves.[14] Recent information, although not extensive, indicates plug, ball, and butterfly valves have lower emission rates than gate valves.[15–18] Descriptions of the nongate valves do not describe the valves in any detail, and comparisons of effectiveness for specific valve types such as plug valves are difficult owing to differences in valve designs by the manufacturers. Also, valve packings are not defined and emission comparisons with different packings result in emission variations. Member companies of the

CMA have conducted extensive valve emission tests in which associated valve process and mechanical details are included. Following data analysis, the CMA will publish the results of this extensive investigation.

While lower valve emissions are associated with plug and ball valves, the reasons for these lower emissions are unclear. A major consideration that has not been examined in detail is the type of processing operation where the very low emission rates have been observed. A number of these low emission rates are associated with valves in batch processing services containing highly toxic materials such as phsogene and ethylene oxide.[9] Moreover, turnarounds in the phosgene units occur after 1–2 months of operation and in ethylene oxide units generally after 12–14 months of operation. The combination of short time frames between turnarounds and batch processing permits repacking, repairs, and replacement of leaking valves, where a leak has been defined at levels very much lower than 10,000 ppm.

In addition to the advantage of frequent turnarounds, very extensive monitoring programs are characteristic of these plants, which have been significantly greater in scope than the monitoring programs required by the EPA and state programs prior to implementation of the CAAA. Fixed monitors have also been installed to supplement mobile monitors that frequently provide the initial indication of increasing emissions rates. This emphasis on monitoring is accompanied by valve adjustments, repacking, and replacement until valve emission monitoring concentrations are considerably lower than 1 ppm to minimize potential exposures to toxic materials in the plants. Manpower and supervision are allotted to this program to insure that low emissions targets are maintained. In addition, stringent quality control procedures on valves and packings are maintained by the individual companies to provide a standard level of quality.

Batch operations with short run lengths between turnarounds differ from the petrochemical and other chemical industry practices where run lengths are generally 4–5 years and the processes are continuous. These factors hinder valve removal or repacking of valves with high leakage concentrations during the long process runs. In this situation, packing and valve quality control becomes very important in achieving and maintaining low emissions levels.

As described, globe, plug, and ball valves were tested[14–18] for emissions and apparently had lower emissions rates than gate valves. However, in the EPA studies separate emission factors for the various valves were not developed. The database necessary to investigate differences was insufficient for a satisfactory emission factor analysis. All of the valve data were combined and averaged as emission factors for valves in either gas or liquid service. The EPA claims the only variable that affects valve emissions is the phase or state of the stream materials, meaning liquid or gas. Industry data indicates valve size may also affect emissions rates.[18] However, variables such as packing type, configuration, or compression, which have a significant impact on emissions control, were not investigated.

Valve descriptions are presented in the following sections and valve packings are reviewed in Section 7.5.

7.4.1 Gate Valves

The original EPA valve emission test data that served as the basis for the analysis and control of valve emissions were obtained from petroleum refineries during

1977–1978 by an EPA contractor. The typical petroleum refinery gate valve utilized during this period and currently in use, shown in the simplified drawing in Figure 7.4, was probably packed with the standard industry packing, as were most chemical and petroleum industry valves. The standard packing, which may still be installed in various valves, consisted of an asbestos base enclosed within a braided asbestos jacket that contained inconel wire for reinforcement and was impregnated with graphite particles. A number of manufacturers produced this industry standard packing material, which was installed in new valves and also used as a replacement packing. With increasing asbestos health concerns, valve packing manufacturers began reducing the number of asbestos packings in their product lines and introduced a variety of nonasbestos packings as replacements. Valve packing manufacturers have eliminated asbestos packings from their product lines as a result of restrictions[19–21] and packing manufacture by these producers is now concentrated on nonasbestos products.

Although nonasbestos packings were available during the EPA refinery valve emission test program, the EPA test data obtained from existing processing units were probably gathered from asbestos-packed gate valves. The EPA emission data gathered from valves in petrochemical units were probably identical to the asbestos-packed refinery valves and valves in chemical processing units were probably packed with a large proportion of asbestos packings. Since the EPA did not investigate the effect upon emissions of alternative valve packing materials, all of the uncontrolled and controlled valve emission data obtained in the early 1980s are probably based upon emission rates from asbestos-packed valves. Little comparative data are available on emissions from asbestos- and nonasbestos-packed valves.

Packings in a gate valve are compressed by tightening the nuts on the gland bolts shown in Figure 7.4. Under compression, the packing expands outward against the packing box wall and against the valve stem, sealing the valve stem and wall against leakage. When overcompressed, packing friction against the valve stem increases and movement of the stem becomes difficult. Historically, severely tightening the gland bolts generally reduced leakage, particularly with asbestos packings. However, *excessive* packing pressures are not desirable in a number of nonasbestos packings. Packing requirements are described in Section 7.5.

7.4.2 Globe Valves

A simplified sketch of a globe valve cross section is shown in Figure 7.5. The basic differences between the globe and gate valves is the seat and wedge plug in the globe valve that results in a more restricted globe valve opening. In contrast, the gate valve is designed to minimize any flow restriction since the gate valve wedge is completely withdrawn into the upper housing to minimize pressure drop. The gate valve is essentially an open or closed valve and is not used to manually control flow. In contrast, the design of the globe valve permits manual throttling of the flow through the wedge and seat. Frequently, globe valves are installed as block valves, which permits manual use of the valve to control the flow rate if required. In situations where equipment is blocked and bypassed, the bypass valve is usually a globe valve to provide the pressure drop and flow control that accompanies equipment operation. Tight shutoff is also a characteristic of globe valves.

Stuffing box design and packing requirements for the globe valve are identical to the design and packing requirements for gate valves. The EPA assumptions that

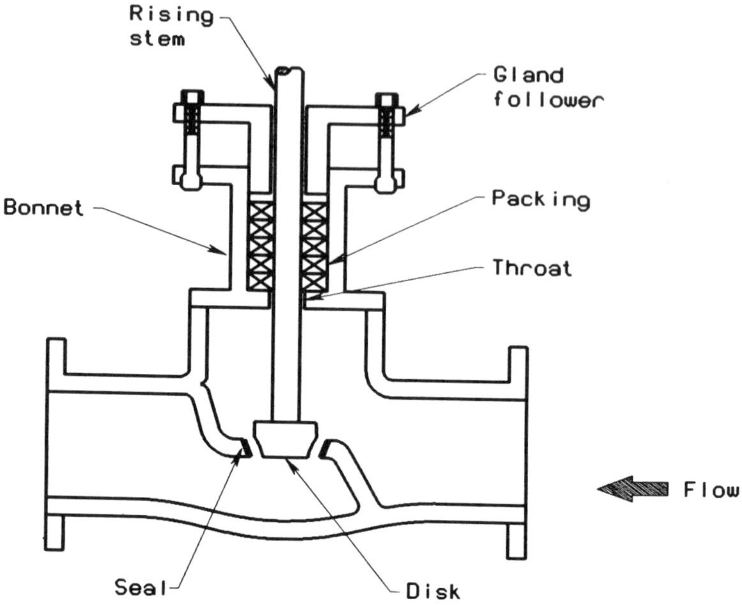

Figure 7.5 Simplified globe valve cross section.

gate and globe valve emissions should be combined for emission factor development is a reasonable assumption based upon valve design. The comments on packings presented in Section 7.5 are also applicable to globe valves since they are a source of emissions.

Globe valves of metal construction are fire-safe valves and are similar to gate valves in this characteristic.

7.4.3 Ball Valves

The ball valve has widespread application in the process industries. A simplified sketch of a ball valve is shown in Figure 7.6. Since there are significant differences in ball valve designs, the sketch presents only the major ball valve elements, and the industrial hygienist must become knowledgeable about the type of valve and sealing mechanisms in existing and new valves.

As shown in Figure 7.6, the ball rides between two vertical ring seats that are constructed of PTFE, filled PTFE or other material, polyetheretherketone or other ketone (PEEK), high molecular weight polyethylene, Viton®, and so on. The balls are normally of stainless steel or a metal compatible with the fluid process application. While most centered ball elements are generally stainless steel with polymeric seat rings, in some valves, the seats for the ball may be constructed of metal, resulting in a metal to metal fit. However, the metal contact surfaces are carefully machined to ensure that metal scarring or galling does not occur, with leakage around the ball seats minimized and a fire-safe valve design. The large variety of ball valve designs precludes a detailed review of the different designs in

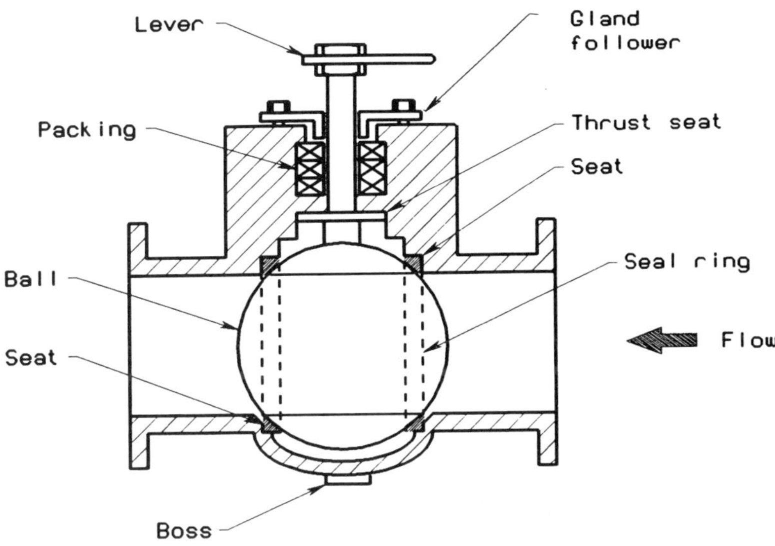

Figure 7.6 *Simplified ball valve cross section.*

this book, but one other aspect of general valve construction must be considered in specifying a valve—fire safety.

As described, the ball seat rings are generally constructed of nonfire-safe materials and standard ball valves with these seats will not pass stringent fire-safety tests. Manufacturers have developed various backup seating systems that permit the valves to meet fire-safety test requirements. However, these fire-safety sealing mechanisms vary widely and a detailed review is necessary to analyze and compare seat seal designs. The fire-safety test ensures that, when closed, the ball and backup rings will block fluid flow in accordance with fire-safety test requirements. Stem emissions leakage is a function of the type of packing system installed. In addition to the ball seat design, ball valves may be lined with a polymeric coating on all liquid exposed faces for process fluid resistance or the prevention of contamination. In these valves, the ball may not be coated, but the valve should be fire tested to ensure the linings will not affect fire-safety performance.

The ball has a passageway that is essentially equivalent to the piping diameter. This minimizes pressure drop and permits cleaning of the line or piping system with a pig. The pig is a device that fills the line and is pressured through the piping system to clean the piping walls or remove any material buildup such as sludge. Some ball valves have venturi-like designs where the ball opening diameter is less than the inner pipe diameter at the outer flanges. Pressure drop exists across this valve, but this valve design is installed to meet specific processing requirements. Since the ball has a through liquid passage, the ball turns 90° from a full open position to the fully closed position. A lever placed on the valve stem manually indicates the position of the ball whether closed or open.

Ball Support Design. In general, the ball is located and supported with two different bases. The overall description presented in the previous paragraphs indicates that the ball fits somewhat snugly between the two seats shown in Figure 7.6 and does not move. This is somewhat misleading in one design where the ball extends into the valve body, similar to a ball on the end of a cantilever beam. When the ball in this situation between the two seats is in the closed position, the upstream pressure moves the ball against the downstream seat, sealing the interface. The seats are fixed in position in this design and do not move. This type of valve is termed a floating ball valve.[25]

The other basic design is a trunnion-mounted ball valve, where the trunnion system bearings do not permit sideways motion of the ball and the ball remains in a fixed position. In the trunnion valve, the seats are adjusted to seal against the ball through the use of springs and line pressure. These adjustments are automatic, with the higher upstream line pressure aiding in forcing the upstream seat against the ball.

A metal to metal seated ball valve has been developed that may have application in the chemical process industry.[35] In this valve, the metal ball sits firmly on the metal seat, as described previously, with the contacting surfaces prepared for sealing under pressure. The *contact surface* on the ball is a special seat that is termed an integral part of the ball since the seat is not a separate piece. Recent advances in metallurgy permit a tight interface that does not gall as the ball is turned. In addition, this tight seal results in little leakage through the seat. This valve may not have been fire tested, which should be done to ensure the valve will operate in accordance with fire-test requirements. While the seats seal tightly, a packing is still required in the valve.

Overpressuring. Basic differences in ball seat design are important when considering ball valve operation. In ball valves, the ball is placed in a cavity and separated from the housing by the two seat rings. During normal flow-through operation in liquid service, the inner ball passageway or ball cavity is filled with liquid and as the ball is turned to close the valve, the liquid continues flowing through the ball into the outer annular volume surrounding the ball. When the ball is in the closed position, the outer annular cavity and the ball passageway are filled with liquid.

With the valve closed, liquid within the valve can increase in pressure under various conditions, posing potential safety and health concerns with toxic fluids. A line can be blocked for either operating or maintenance purposes, resulting in containment of the fluid in the ball valve for a period of time. If the liquid operating temperature is quite low, natural heating will increase the internal valve pressure. This can also occur where a relatively high vapor pressure material entrapped in the valve slowly increases in temperature as a result of solar heating. The outer cavity pressure, which was at the upstream pressure level when the valve was closed, increases well above the upstream line pressure and pressure relief can become a serious problem

When a valve cavity is overpressured, pressure relief is necessary. In floating ball valves that have self-relieving seats, the overpressure is relieved to the upstream side where the seat pressure is lower. Design of self-relieving seats is difficult and relieving set pressures are apparently unreliable.[25] There is a possibility the pressure will be released on the low pressure side creating potential exposure and safety problems in maintenance situations.

Trunnion-mounted ball valves tend to relieve toward the low-pressure side which is a potential problem. Both floating ball and trunnion-mounted valves can be supplied with upstream vent passageways. Many valve manufacturers include internal relieving vents as a standard design item or as an option. These vents relieve the overpressure to the upstream line and are a reliable method of controlling the outer cavity pressure. Since the upstream line pressure will normally have a pressure relief system, the valve overpressure can be controlled.

Bleeding down the outer cavity pressure externally is costly and complicates operations. The internal relief vent is a satisfactory solution but must be reviewed with the valve manufacturer where solids are present in the stream. In addition, depending on a block valve to prevent leakage during maintenance operations can still present potential risks. Where lines are flanged, a blank or blind should be placed in a flange between the valve and the downstream operation. This is a positive method of preventing any leakage through the line. However, the internal vent should be installed irrespective of other operating procedures to ensure the valve is not overpressured.

The boss on the bottom of the ball valve can be drilled and threaded for a valve connection that provides a method of draining the outer annular cavity around the ball. In certain situations the drain valve or bleeder can effectively turn the ball valve into a double block and bleed installation. However, a bleed at the bottom of the ball valve can present a potential source of leakage unless the bleeder is connected to a closed system.

Emissions. In Figure 7.6 the thrust seat is wider than the stem, which prevents blowout of the stem. This is a safety requirement that various companies and organizations require. Valve packing around the stem above the thrust seat controls emissions to the atmosphere. Three rings of packing are shown with a gland follower for packing compression and adjustment. At the present time there are a great number of existing valves that have very short packing boxes and may only have sufficient depth for two or three full-size rings. Since five rings are apparently an optimum quantity for sealing,[17] the short packing boxes may not provide sufficient packing or sealing depth. In existing valves, an extended packing box may be available for certain valves, permitting the installation of standard packing systems. Where the packing box cannot be extended, there are modified packings for short boxes that can be installed as described in Section 7.5.

Where emissions are a major concern, a number of extended packing box valves are now available. Many of these valves have dual sets of packing rings as shown in Figure 7.7. The lantern ring, constructed of metal with center ribs and slots, separates the two sets of packing. Lantern rings have much less annular volume than the packing rings and are provided to permit withdrawal of process material that may have leaked through the lower packing set. The upper packing set prevents any emissions to the atmosphere from the lantern ring. In the ball valve arrangement the bleed or telltale line from the lantern ring area can be used to withdraw any leakage or the line can be used to indicate whether the lower packing set is leaking without any system for continually withdrawing leakage. This function is the source of the name "telltale," with specific telltale connections provided by some manufacturers. However, in toxic or hazardous applications, some leakage can occur during the telltale check which is a potential source of

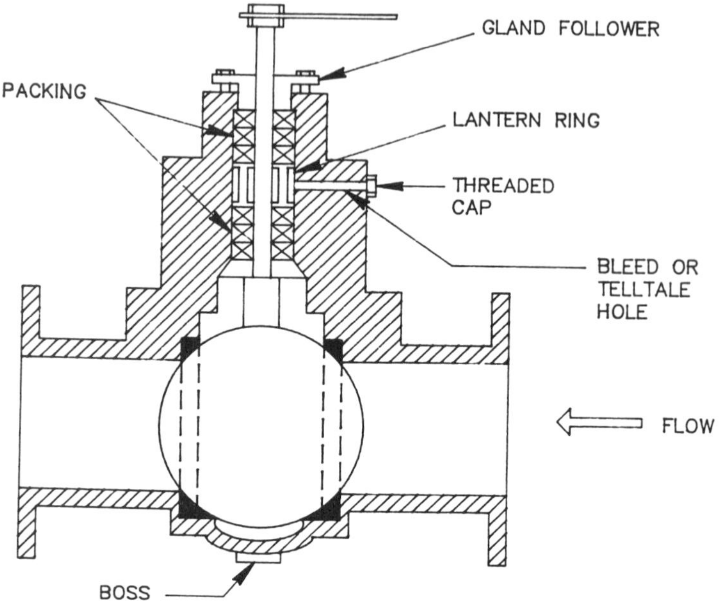

Figure 7.7 Ball valve with extended packing box.

exposure. Where stem leakage is undesirable, a permanent enclosed bleed installation is required to ensure any leakage is controlled and potential exposures are minimized.

Emission data on ball valves compared with other valves are limited. General observations indicate ball valve leakage is less than gate valve leakage,[15,16] although specific emissions comparisons are not available. Data should be available in the future from the CMA data collection study, which will provide a basis for estimating and comparing ball valve emissions with other valves.

Bellows. To control stem emissions, several ball valve manufacturers have developed valves with bellows that prevent stem emissions to the atmosphere. One bellows system that can be used with ball valves is shown in Figure 7.13. Another system is based on installing an external bellows handle on either new or existing valves, as shown in Figure 7.8.

Leakage from the valve system is contained within the external bellows, resulting in an internally pressurized bellows. The handle casing is designed to withstand process pressure and thereby acts as a secondary containment device. Any leakage through the bellows can be determined with the telltale in the handle. Although the bellows and handle form an arc, movement of the arm does not mechanically stress the bellows. The arm rotates 90° for fully opened and closed positions. Before installation, the external bellows should be reviewed for fire safety, which may require a fire test. Application of the external bellows should be reviewed with safety and operating personnel to ensure safe operation during a fire. In addition, removal of the external bellows for repair should be carefully reviewed to ensure that liquid remaining in the bellows will not present a potential health problem

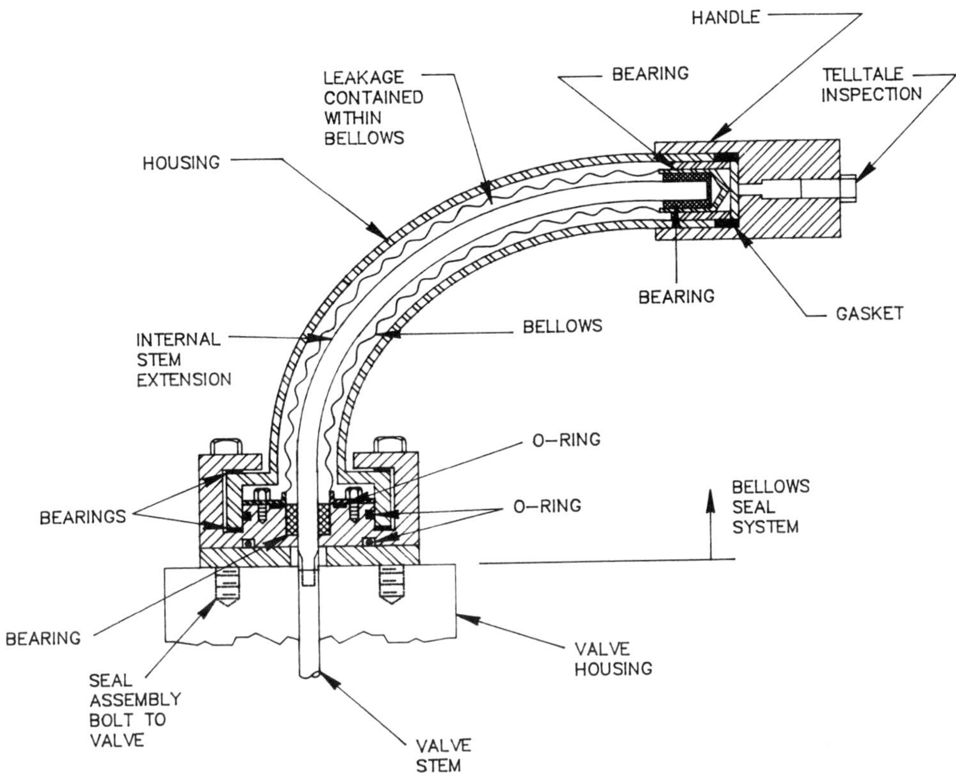

Figure 7.8 External bellows seal for quarter turn valves (courtesy of Kerotest).

during the maintenance program. In addition to these bellows designs, other bellows designs are also available from various valve manufacturers.

7.4.4 Plug Valves

The plug valve has been used in the process industries for many years, and a simplified sketch of the basic valve is shown in Figure 7.9. There are significant variations in plug valve design and the simplified sketch presents the major plug valve elements. The industrial hygienist must become knowledgeable about the type of valve and sealing mechanism in existing valves and proposed replacements or new valves.

Although the details are not shown in Figure 7.9, the plug is confined by close tolerance at the upper and lower ends of the plug. Two rings of packing are shown in this sketch, with a gland follower for packing compression and adjustment. The plug turns only 90°, and either a removable wrench-type or fixed handle moves the stem. On some valves, a boss at the bottom of the valve can be drilled and threaded for a valve connection that provides a method of draining the cavity below the plug. In these plug valve designs, the bleeder can effectively turn the plug valve into a double block and bleed installation.

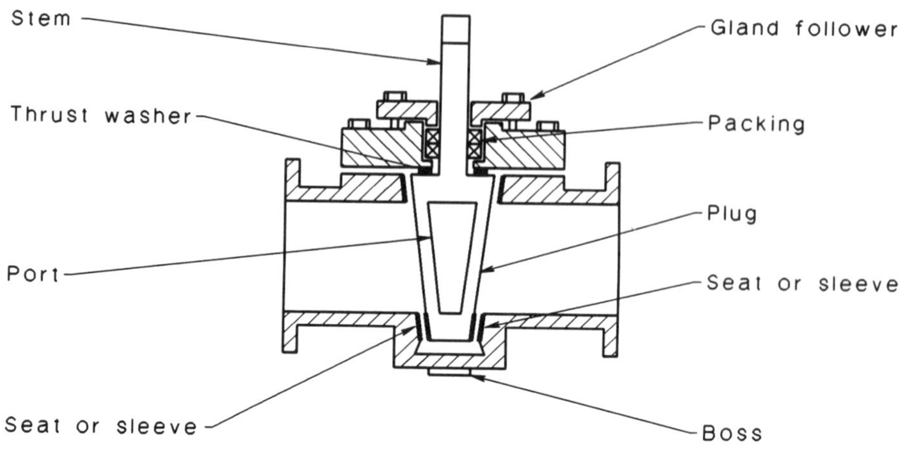

Figure 7.9 *Simplified all-metal plug valve cross section.*

The valve shown is an all-metal plug valve that has been used for may years in the petroleum and petrochemical industries. Since the metal plug generally fits very tightly against the seats or sleeves in normal operation, the plug is lubricated with a grease that is injected through special fittings attached to the valve. The grease lubricates and eases movement of the plug, aids in sealing the plug against emission losses, and assists in sealing the plug against fluid leakage when the plug is closed. Lubricated plug valves have tight shutoff and low torque requirements.[22] The plug valve has been installed in many services to provide tight shutoff with a short 90° turn in contrast to a rising stem valve, which must be turned a number of revolutions to completely open or close the valve. The plug valve can also provide some throttling control since the plug has a flow adjustment coefficient. In this operation the plug position is set in accordance with flow conditions and is adjusted infrequently.

While the plug valve just described has been used for many years in the petroleum and petrochemical industries, the chemical industry has successfully installed a large number of plug valves that differ in some respects from the all-metal valves. One valve type has a polymeric insert or sleeve, usually PTFE, that surrounds the plug, as shown in Figure 7.10, easing movement of the plug in the casing, thereby eliminating lubrication. Operating these valves has become much easier and the insert design apparently reduces emissions leakage. In the all-metal plug valve lubrication grease aids in controlling emissions, but symmetrical distribution of a constant thickness film is difficult to achieve, which frequently reduces emission control effectiveness. Other plug valves are completely lined with a polymer to prevent corrosion or contamination, as shown in Figure 7.11, which also improves valve operation.

Application. Plug valves are widely used throughout the chemical industry with indications that the plug valve population is the largest of the various valve types found in the chemical industry.[10,23] These plug valves are either the sleeve or fully lined valves shown in Figures 7.10 and 7.11. Gate valves are apparently a small

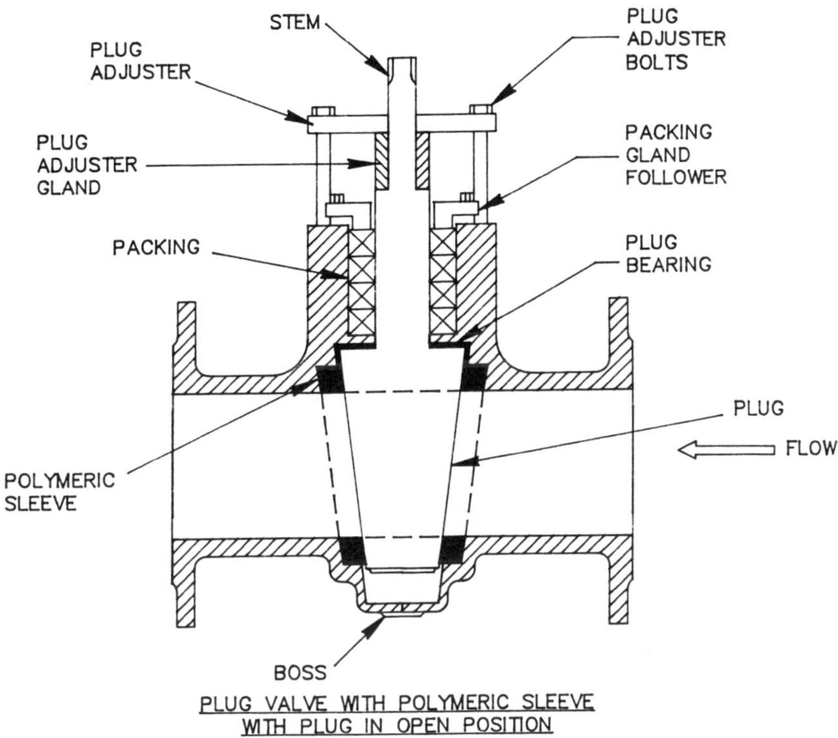

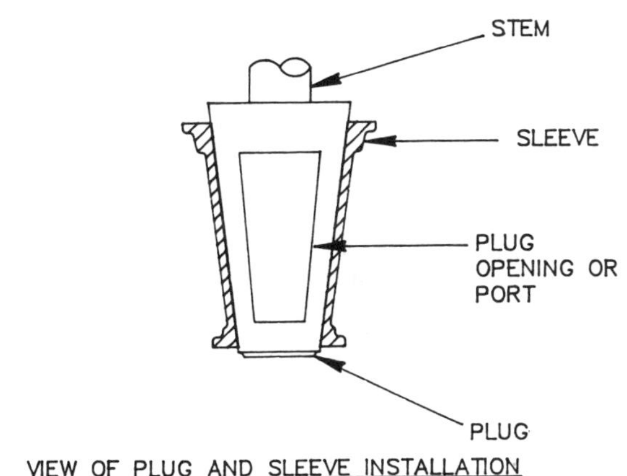

Figure 7.10 *Simplified sleeved plug valve (courtesy of Durco).*

percentage of the total valve population and are being replaced with plug and ball valves at a rapid rate.

Specific emissions data on plug valves compared with emissions from other valves are not available, but some information will be available in the future from the CMA. Installation of lined plug valves in chemical processes is apparently based on improved emissions control performance compared with other valves.

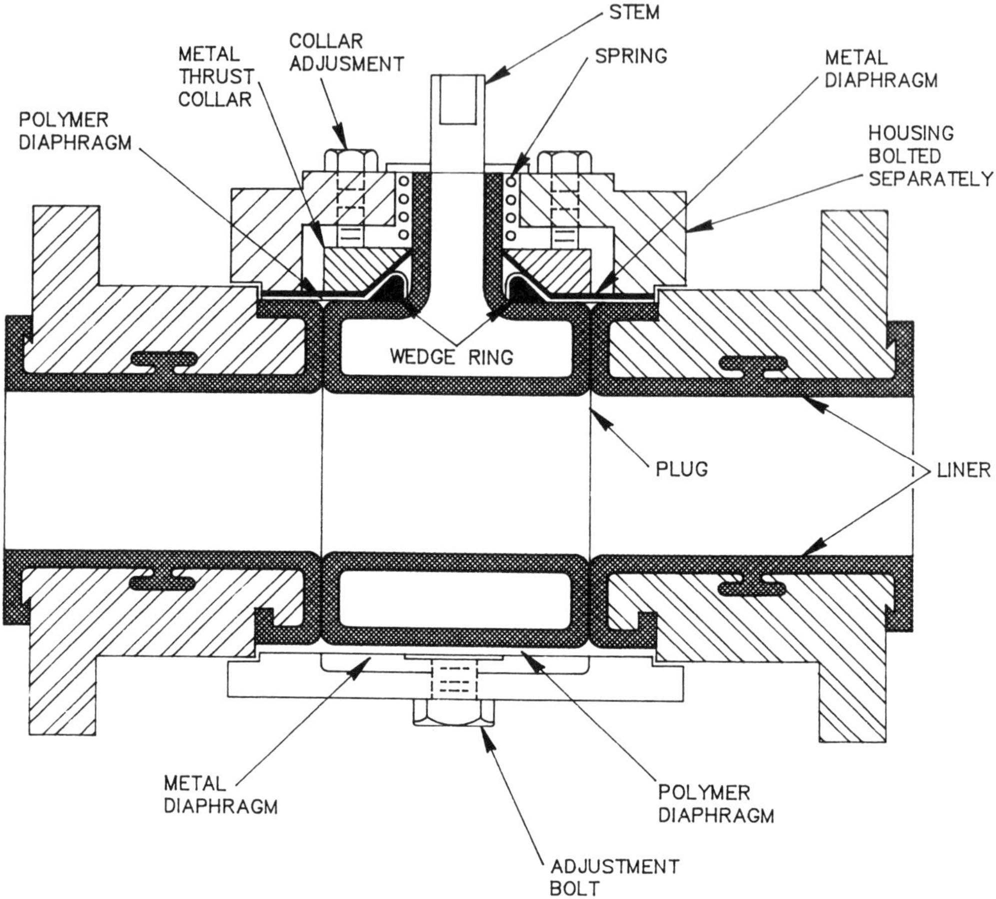

Figure 7.11 *Lined plug valve (courtesy of XOMOX).*

This assumption is partly verified in plant tests on valve emissions where plug valves are a large proportion of the valve population.[9, 15] While emissions leakage and valve leakage frequency in ethylene oxide and phosgene plants are quite low, higher emission rates and leakage frequencies are reported in butadiene plants.[9, 24] Butadiene plants are concentrated in the petrochemical industry and are characterized by very long continuous runs between planned turnarounds (3–4 years) and normally use gate valves as the standard block valve. In contrast, phosgene plants operate with planned turnarounds every 1 or 2 months and ethylene oxide plant turnarounds are generally every 12–14 months. In addition, valves are inspected daily and can be repaired or replaced as required in plants with short operating periods.

While frequent inspections and repairs maintain a low leak frequency in the described chemical plants, the lined plug valves apparently had lower leak frequencies than gate valves in the same services. The data on valve emissions, although qualitative, tends to indicate lower emissions with plug valves compared with gate valves. Further information is also necessary to clarify the specific types of packings or emission control devices on the plug valves because of the various control configurations on the valves.

Other Valve Modifications. In Figure 7.10 a shallow packing box is shown that only permits the installation of a few packing rings. Deep packing boxes are available for sleeved plug valves from various manufacturers that permit the installation of two sets of packings separated by a lantern ring. The lantern ring provides a method of determining whether the lower packing set is leaking and acts as a telltale. Alternatively, the lantern ring tube can be connected to a collection system which is more cumbersome and expensive than a packing system but essentially eliminates atmospheric emissions. Frequently, plug valves with extended packing boxes are installed in acutely hazardous process services such as acids, chlorine, or other similar toxic materials. However, the extended packing system also provides good emission control for other highly toxic compounds. As described in Section 7.4.3 for ball valves, extensions to existing packing boxes are possible along with packing modifications to control emissions from shallow packing boxes.

Some plug valve designs have eliminated the packing boxes and replaced the packings with diaphragms. A fire-resistant diaphragm system is shown in Figure 7.12 which is similar in principle to nonfire-resistant diaphragm installations. Graphite rings and seals are normally not included in the nonfire-seal designs but can be supplied by the valve manufacturers whose standard seal components are polymeric. The diaphragm seal design apparently results in low emissions valves. However, comparisons with other plug valve seal emissions rates are not available.

An internal bellows is also available in a plug valve design that combines the bellows with an internal gearing system for turning the plug, as shown in Figure 7.13. The plug valve has a standard sealing system that is termed a primary seal. Depending upon the seal design, multiple seals can be installed for a primary and secondary basic seal system. Normally, the bellows chamber will not be exposed to any significant quantity of leakage from the seals. If the basic seal system leaks, the leakage would enter the gear housing where the process fluid is contained by the sealed bellows. In the rack and pinion gear system, the rack moves linearly and the pinion converts the linear motion to rotational motion which is transmitted to the plug. Consequently, bellows motion is linear and the bellows ends are welded to provide the sealed chamber shown in Figure 7.13. Linear motion is transferred through the jacking rod to the rack with the jacking rod moved by the handle.

The bellows is in contact with the process fluid on the outside surface of the bellows. Consequently, the external pressure on the bellows reduces the tendency of the bellows to fail. Actual leakage conditions in the basic seal and in the gear housing can be determined through a threaded monitoring plug which acts as a telltale. However, in toxic or hazardous service, inspections for leakage requires careful testing to minimize potential exposures. Overall, this design prevents emissions or leakage to the atmosphere under normal operating conditions.

Fire Safety. While a fire-safe diaphragm seal is shown in Figure 7.12, the entire valve is not considered fire safe. The top seal will prevent excess leakage to the atmosphere in the event of fire, minimizing potential exposures to the valve

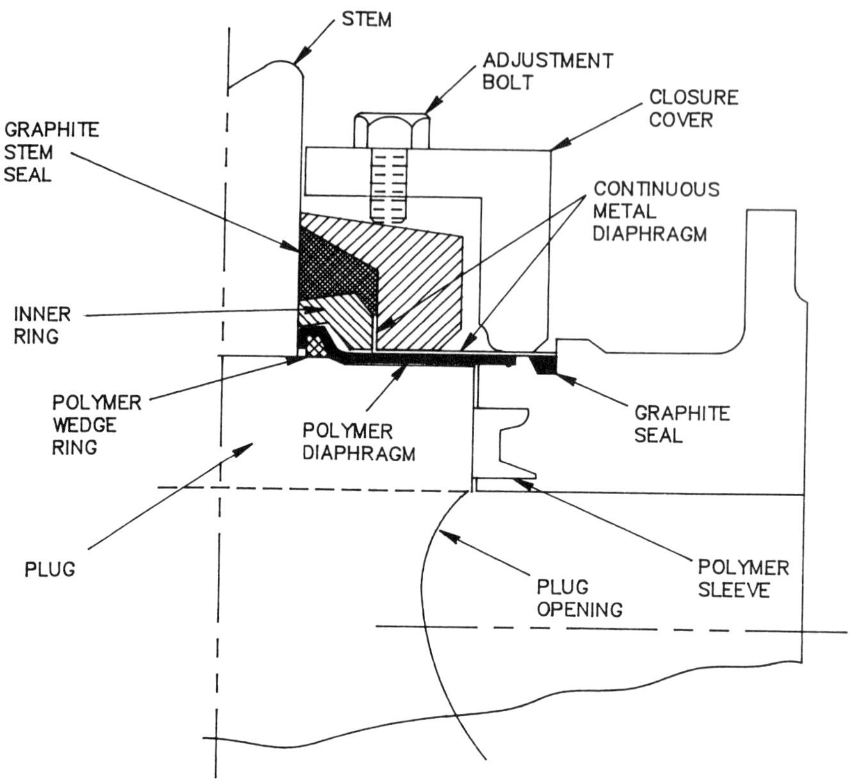

Figure 7.12 *Cross section of plug valve fire-protection diaphragm (courtesy of XOMOX).*

contents. However, the internal sleeve and valve lining may not survive a complete fire test, and the valve is not able to control leakage through the plug when the plug is in a closed position. At this time, manufacturers apparently have not developed a method of sealing or blocking the plug under fire conditions. Metal backup seal rings in ball and butterfly valves permit these valves to meet stringent fire-test requirements. Where fire safety is a consideration, plug valves that meet the fire-test standards are generally all-metal plug valves.

A fire-resistant coating has been developed that enables external valve actuators to withstand fire-test conditions. In addition, the coating will apparently protect a valve that is not fire safe. During installation the coating is provided with small holes that permit valve leak detection. Should a fire occur, the holes disappear as the coating expands and closes the holes. Apparently valve fire tests on this coating have been successful. However, a detailed fire safety assessment of a valve installation with this coating is necessary to assure operating personnel of valve and plant safety.

Installation of plug valves apparently controls valve emissions to low levels accompanied by low valve leakage frequencies. However, these valves may not pass the stringent fire test with kerosene or hydrocarbon fluids described in Section 7.3, although effective fire safe top-sealing devices have been developed by various valve manufacturers that prevent extensive leakage through the stem under fire

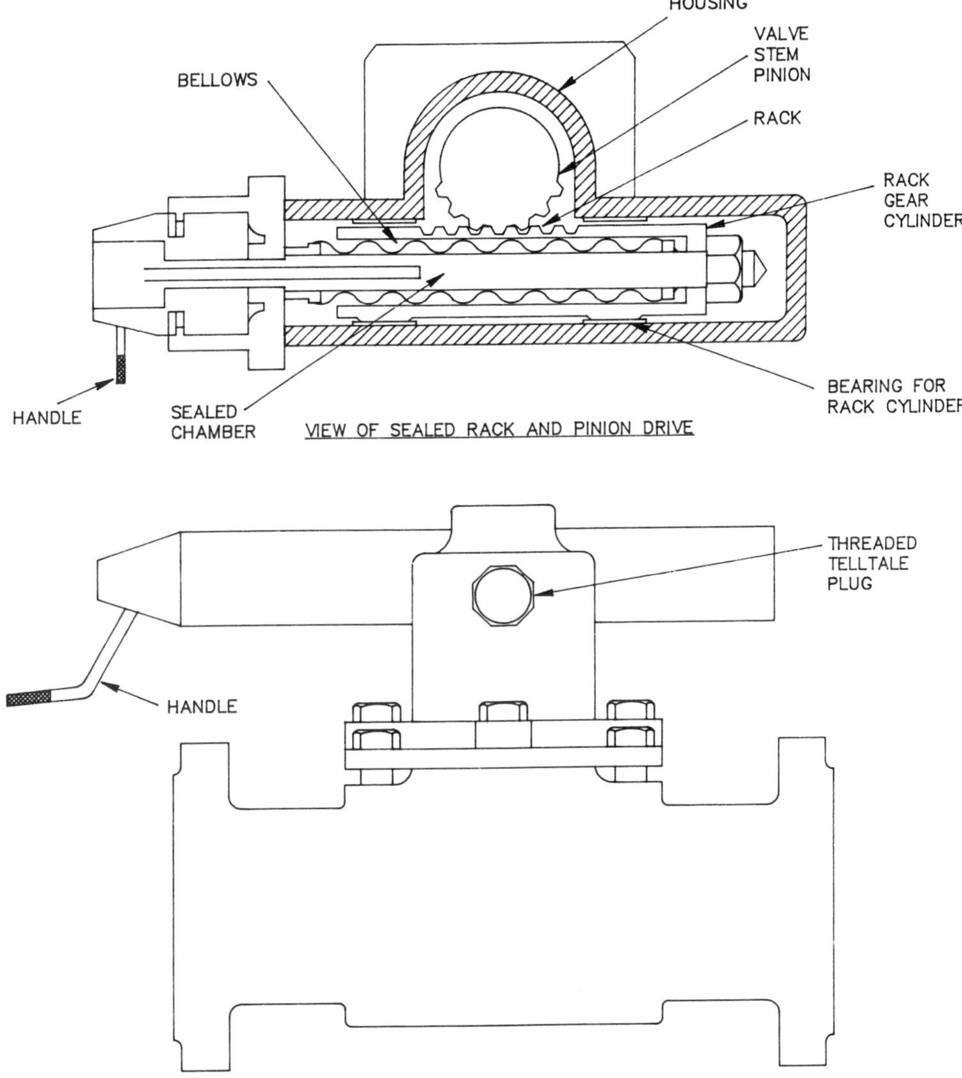

Figure 7.13 *Bellowseal® plug valve (courtesy of XOMOX).*

conditions. Application of these valves in various services should be carefully reviewed with safety and operating personnel.

7.4.5 Butterfly Valves

These valves have been installed in increasing numbers through the years, particularly in the chemical industry. Compared with gate valves, butterfly valves are smaller, lighter, and less expensive, have a tight shutoff and can be used for throttling. The type of butterfly valve currently preferred for process service is a

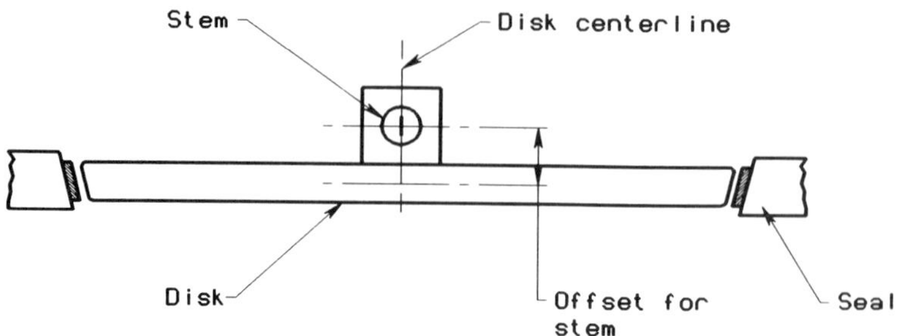

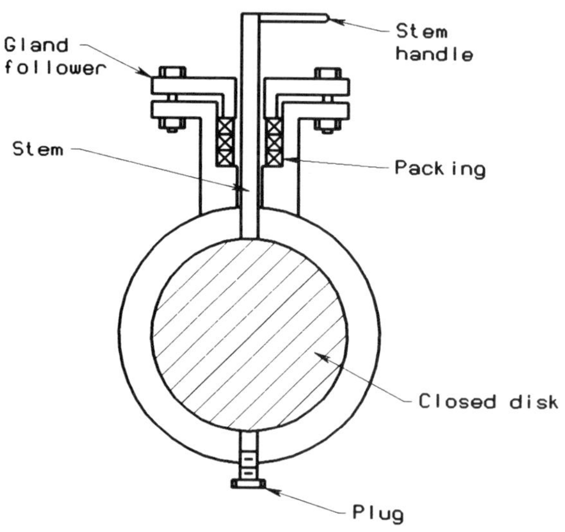

Figure 7.14 *Simplified butterfly valve cross section.*

high-performance valve. These butterfly valves are termed "offset disk valves"[26] to distinguish them from the older butterfly valves, which have been likened to "stove pipe dampers." A cross section of the offset disk valve in Figure 7.14 shows a stem that is offset from the centerline.

As seen in Figure 7.14, the disk swivels around the stem eccentrically and contact of the disk edge against the seal creates a tight shutoff condition. Valve packings around the stem then control valve emissions to the atmosphere. The seal or seats that provide tight disk shutoff differ in design for each manufacturer and are generally proprietary. A valve requirement of concern in a number of plants is that the butterfly valves meet the fire-safe valve tests. Many valves now meet these specifications, which increases potential application of the butterfly valve in a variety of services.

Application. Butterfly valves are used in chemical plants, petrochemical units, and some refinery applications. Apparently butterfly valve installations are less numerous then plug valve installations in the chemical industry.[15,23] However, the improved valve designs along with improved packing and emissions control are increasing application of this valve in various chemical services.

Butterfly valves have tighter shutoff than most of the other valves, and the disk provides a manual method of controlling flow through the valve. In addition, butterfly valve costs are generally considerably lower than costs for gate valves. Design simplicity also tends to reduce maintenance costs.

Originally, butterfly valves were limited to all-metal components. Valves are now available that have all welded parts coated or covered with other materials such as polymers. This covering also includes the disk. As a result, butterfly valves are used in a wide variety of services.

Emissions Data. Little published data are available on butterfly valve emissions and particularly in comparison with other valves. Quarter-turn valves apparently have lower emissions than full-turn valves[18] and may be considered similar to plug valves. Emission data on various valves have been accumulated by the CMA and should be available following a detailed analysis. However, a variety of factors must be incorporated into the analysis to ensure a satisfactory comparison with other valves such as packing type, amount of packing, and lantern rings.

Fire Safety. One of the major concerns in fire-safety tests is sealing the disk when closed to minimize leakage through the line. Many butterfly valve manufacturers have developed seal systems that are effective under fire conditions although these systems vary considerably in design. Moreover, some valves have metal seats that are fire safe and seal well under either normal flow or fire test conditions. In general, the fire-safe valves have internal surfaces of metal with special disk seals.

Packing. Butterfly valve stems are packed as shown in Figure 7.14 or may have somewhat extended packing boxes for full packing sets. A dual packing set separated by a lantern ring to indicate leaking or drawoff capability may also be installed in the packing box. In addition, external bellows devices may be added to the butterfly valve to control emissions. In this situation fire safety is a consideration, as described in Section 7.4.3.

7.4.6 Hermetically Sealed Valves

In certain applications valves that do not leak as described in previous sections have been installed. These have obvious advantages in eliminating leaks, reducing potential exposures, and minimizing LDAR regulatory requirements. However, some of these valves are limited in service applications owing to pressure, temperature, flow, or size restrictions. New valve designs are beginning to appear and these restrictions may be reduced for various applications. A few of the hermetically sealed valves are described in the following sections.

Bellows Valves. Bellows valves were developed to contain toxic fluids and in some instances valuable materials. As described, bellows devices successfully

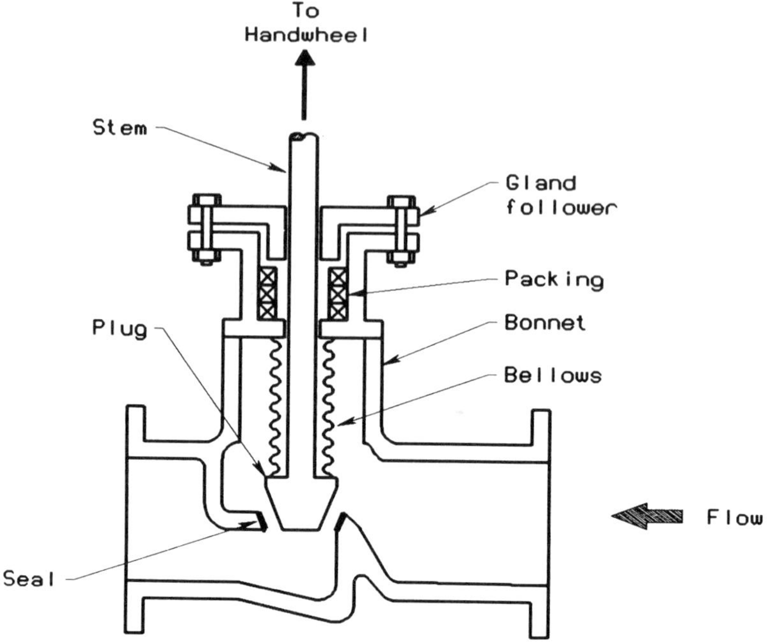

Figure 7.15 *Simplified bellows globe valve cross section.*

contain valve stem emissions. A simplified sketch of a typical bellows globe valve is shown in Figure 7.15. Bellows globe valves are supplied by a number of valve manufacturers, and the manufacturers normally include a stuffing box with packing as a backup should the bellows leak. The bellows may leak at a low rate if a crack develops in the bellows as a result of metal fatigue or failure under stress in the presence of a particular compound. Since any failure in the bellows would produce an excessive amount of emissions, backup packings should be installed in these valves. Bellows of the same type are also installed in gate valves and a number of valve manufacturers supply rising stem bellows valves.

Bellows Operation. The bellows is an accordion-type tube that contracts under pressure and is elongated by a pressure release or extended under force. As shown in Figure 7.15, the bellows follows the movement of the stem. The bellows prevents the passage of fluid from the process into the stem region by providing a continuous impervious metal shield to the process fluid. As shown, the bellows is sealed to the plug and against the bonnet. In this design, process fluid pressure exists in the annular volume between the valve bonnet and bellows. The pressure exerted on the bellows is external, which is the preferred installation for the bellows design thickness.

An alternate bellows configuration attaches the bellows to the lower housing, and the upper end of the bellows is fastened to the stem. This results in fluid within the bellows, requiring a strengthening of the bellows to contain the internal pressure. Larger thicknesses are required for this design which increase the compression or expansion forces required to move the stem.

Bellows life is affected by a number of variables including metal fatigue resulting from flexing of the bellows or cycling, bellows construction or fabrication, stroke length, corrosion, and temperature. Cycling has been a cause of bellows failure, but in block valve applications the valve is not cycled frequently. For valves 3 in. and larger, gate and globe valve allowable cyclic frequencies are 3000–5000 cycles. The allowable cyclic frequency increases with smaller valves. Since bellows are susceptible to failure, stem packing backup is necessary on all valves to prevent large releases to the atmosphere. A bellows failure detector should be installed to determine when the bellows is leaking. Since many of the fluids are toxic, a sensing detector is preferred to minimize potential exposure rather than a manual investigation. The sensor can be a pressure switch or a much more elaborate system.

Bellows Design. The bellows tube is either formed in a press or constructed by welding a group of thin metal rings or plates to construct a bellows. Welding a bellows is somewhat difficult, particularly internally, and fatigue failure in the welded bellows generally occurs in the weld or adjacent to the weld. The welding procedure results in narrow spaces and fissure-type separations of the plates compared with a smooth sinusoidal formed bellows. Thus the welded bellows are more likely to trap solids and create stagnant zones that increase potential corrosion locations. Solids entrapment or stagnant zones are not a problem with the formed bellows.

Since the bellows material is thin to reduce resistance to compression or expansion, the bellows is more susceptible to corrosion than the valve components. Consequently, a more highly resistant metal is required than the body material necessary in the standard valve parts. In chemical service, bellows are frequently constructed of Inconel® or Hastelloy.®[27] These requirements significantly increase valve cost.

The number of folds for a welded bellows is much larger per unit length than the fold number per unit length for the formed bellows. Although the stroke length per fold is about the same for each bellows type, a welded bellows must be about three times the length of a press-formed bellows for the same overall stroke length. This requirement increases the valve length or height for the same stroke length. Reducing the stroke length decreases the valve length but may present process problems.

A cost comparison of bellows valves with packed valves indicates that bellows valve costs begin to diverge from packed valve costs at the 1 1/2-in. size for 150 lb ANSI valves.[28] Bellows valve prices rapidly increase beyond this size, and for a 4-in. valve the bellows valve cost is about 3 times the cost of a packed valve. However, a bellows valve may be preferable to straight-packed valves for particular applications based upon process requirements or eliminating potential exposures and leakage. In all bellows, valve applications packing is also required to contain any leakage that may occur in the bellows and minimize potential exposures.

Bellows valves had previously been restricted in size, but gate valve sizes up to 12 in. are available as standard valves from some manufacturers. Larger valves, although not considered standard, are also available upon request.

Diaphragm Valves. Diaphragm valves have been in service for a number of years. They are widely used in throttling, particularly when a tight shutoff is required. A

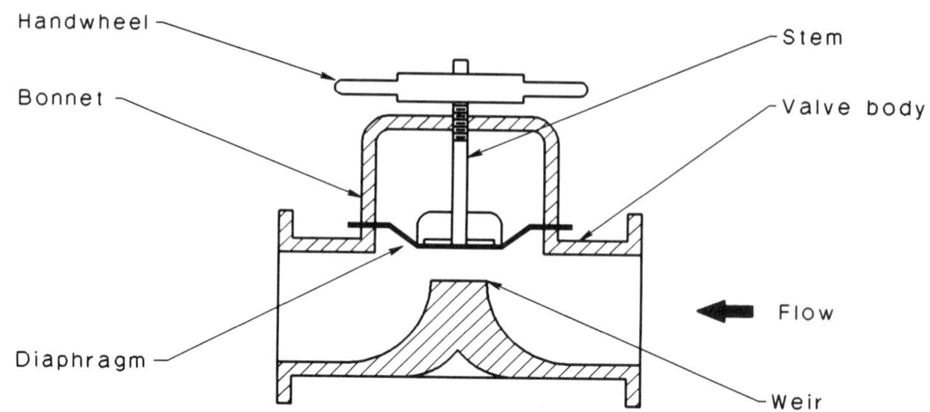

Figure 7.16 *Simplified diaphragm valve cross section.*

wide range of diaphragm materials is used to meet corrosive fluid requirements. In addition, the valve can be constructed of various materials or lined with polymers, elastomers, glass, and so on. The diaphragm in the valve shown in Figure 7.16 prevents any leakage into the stem/bonnet section during normal operation. Cracks or openings permits leakage through the diaphragm and emissions through the stems. Leaking corrosive fluids usually attack the bonnet and stem, which are normally constructed of noncorrosion resistant materials. Consequently, any leakage through the diaphragm should be determined as early as possible and the diaphragm replaced. If possible, diaphragm valves with packings around the stem provide some control of leakage and should be installed in toxic and corrosive stream service.

Diaphragm valve applications are limited to services at low pressure levels and within the diaphragm temperature ratings. These valves are generally not fire safe, but are used successfully in various applications where their characteristics are advantageous. Careful operation and maintenance aid in minimizing potential leakage and exposures.

Pinch Valves. A sketch of a typical pinch valve is shown in Figure 7.17. This valve has a flexible sleeve with a clamp action that throttles the fluid flow rate. The valve is widely used in liquid service to throttle sludges, slurries, and pulps, or in those services where clogging can be a problem. The flexible sleeve is impervious to the fluid and prevents leakage into the bonnet. These valves do not have any leakage or emissions.

Although the valve does not leak, there are limitations on allowable operating pressure and temperature ratings. The valve is used principally in liquid service at relatively low pressures. This particular valve is not a fire-safe valve, but is used in a number of applications.

Other Valve Types. With the emphasis on controlling toxic emissions and losses from valves, a number of newly developed valves are becoming available commercially. These new valves are hermetically sealed to prevent leakage to the atmosphere. The new valves tend to cost more than gate valves and their application

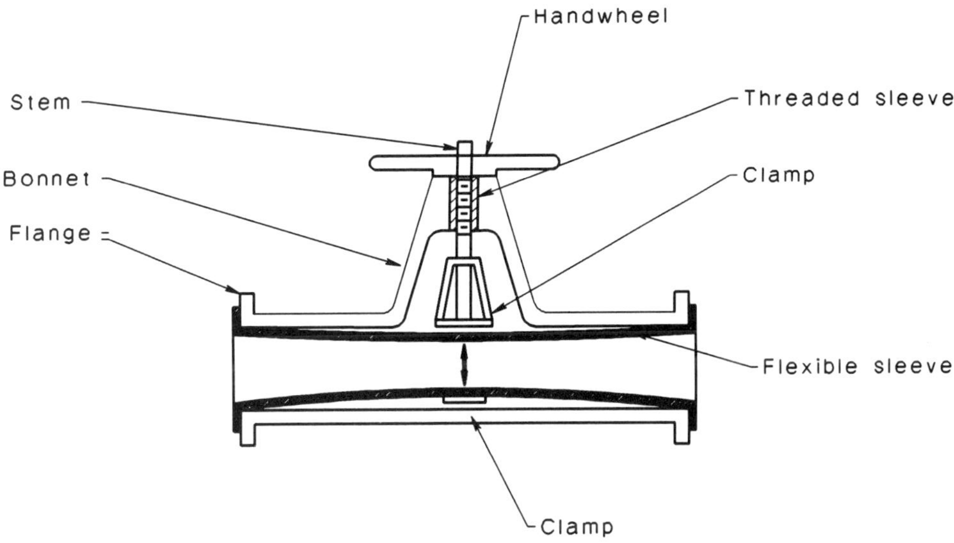

Figure 7.17 Simplified pinch valve cross section.

initially is limited to controlling the emissions of severely toxic materials or containing corrosive fluids. The new designs are generally covered by the following three class designations.[26]

1. Magnetically coupled
2. Ferrofluidic (magnetic fluids)
3. Hybrid designs

The industrial hygienist should examine the design information on hermetically sealed valves as they are purchased and installed to become familiar with typical applications, costs, and potential hazardous situations that may arise upon a failure in the valve.

7.5 VALVE PACKINGS

Although asbestos packings had been used for many years in the process industries, stringent restrictions on asbestos manufacturing and distribution have essentially eliminated the installation of asbestos packings in new valves or as replacement packings in the U.S.[19–21] Since the gradual replacement of asbestos began in the late 1970s, a wide variety of nonasbestos packings became available commercially. Current nonasbestos packings consist in general of graphite, polymeric, and mixed graphite–polymer packings. Within these general classes there are various packing configurations and a large number of different polymers.

The widely diverse group of materials, configurations, and complete packing sets have different emission control characteristics. However, comparative emission data among the various packings are not readily available in the published literature. Moreover, standard packing emission tests have not been developed that permit an evaluation of packing emission control effectiveness or comparison. The CMA valve emission data may provide information on emission control for some of the different packings. Most of the emission data available on packing effectiveness have been provided by the packing and valve manufacturers under conditions that vary with each testing facility. Plant facilities generally test various packings in actual process facilities to select a packing that meets their emission control objectives.

One of the major criteria in assessing packing performance is fire safety. Refineries and petrochemical plants typically require fire-safe packings. Chemical plants have not concentrated on this specific characteristic in packing installation. However, the chemical industry is becoming interested in fire-safe properties as a result of hazard risk assessment requirements issued by OSHA and required in the CAAA along with the various fires and explosions that have occurred in chemical industry plants during the last decade.[11,27] As a result of this interest in fire-safe properties, packing manufacturers have begun including a graphite packing ring in various polymeric packing sets to meet fire-test requirements.

Consequently, this section on packings will cover graphite packings and some of the various polymeric packings commercially available.

7.5.1 Valve Packing Configuration

Most valve packings have a square cross section or in some instances a vee or modified vee packing was installed. However, other packing shapes have been developed as packing manufacturers work to improve emissions control with nonasbestos packings. In addition to these basic packing configurations, cup and cone designs, wedge-shaped, and inverted u-shaped packings are found in packing sets. The EPA emission data, particularly in the late 1970s, was generally based upon square cross-section packings. Since the basic refinery and petrochemical plant emission data were obtained on square-shaped asbestos packings, many of the emissions reductions noted by the EPA recently were measured with nonasbestos packings having various configurations.

7.5.2 Asbestos Packings

The typical industry packing applicable in a wide range of services has been asbestos packing impregnated with graphite, which is the packing constructed with a braided jacket and inner core shown in Figure 7.18. These packings have been replaced or eliminated by packing suppliers in accordance with U.S. regulations and are being replaced in other countries concerned about asbestos exposure. These packings may not be eliminated completely in all areas of the world because of their lower initial material cost compared with the purchased price of nonasbestos packings.

Although asbestos packings in the U.S. are not being installed in new or existing valves, a considerable number of existing valves have asbestos packings remaining in packing boxes. Removal of these packings has not been considered a health hazard. However, a recent paper on the installation and removal of asbestos packings (reference 30) indicates both procedures result in air fiber concentrations

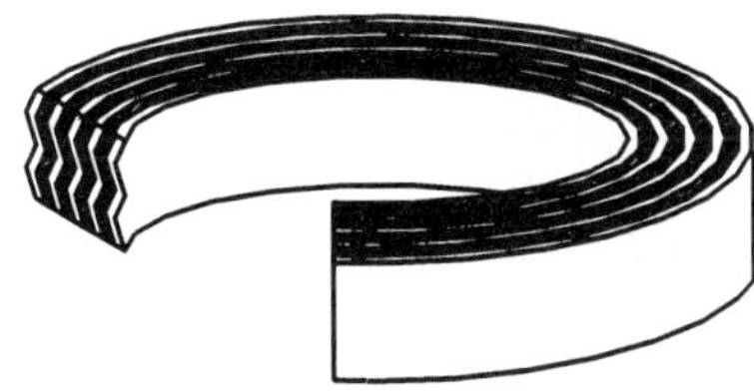

Figure 7.18 *Simplified sections of various packing types.*

much greater than the allowable exposure concentration in the U.S. (0.2 fibers/cm^3) or United Kingdom (0.2 fibers/cm^3 containing any crocidolite or amosite). Further information on fiber exposure is presented in Chapter 17.

In the packing removal procedure, the asbestos packings were removed with picks and threaded rods followed by wet wiping and vacuuming in preparation for the replacement packings. Another method of removal that reduces potential exposures and significantly reduces removal time is with a high-pressure water jet blaster. The jet stream under high pressure both wets the fibers and flushes the packings from the packing box without marring the stem or packing box wall. This water jet procedure is rapidly becoming a standard packing removal technique for either old asbestos or graphite packings.

Disposal of waste packings is regulated in various countries, particularly asbestos packings. In the United states, all asbestos-containing materials must be placed in an impervious sack or bag which is then permanently stored in a secure landfill. All bags or containers have identification labels to insure recognition of the contents of the container. During the collection and encapsulation procedure, personal protection may be required to minimize potential health exposures.

Where asbestos packings may still be acceptable, installation of nonasbestos packings generally requires an economic evaluation by operations to assure management that the normally higher-cost nonasbestos packings are justified. This economic comparison is generally limited to the packing material costs as received from the supplier and nonasbestos packings suffer in the comparison. A more realistic evaluation of alternative packings by operations and engineering would consider the following items:

1. *Material Cost*. The actual material cost is the obvious first step in evaluating materials.

2. *Packing Removal*. The cost of removing used packings from valves can be difficult, particularly with asbestos packings, when performed manually.

3. *Loss Control*. Where a large number of valves will be repacked, the change in emission losses should be evaluated. The losses for nonasbestos packings can be estimated, and the value of the loss difference can then be determined.

4. *Disposal Costs*. A distinct cost is associated with the disposal of asbestos materials and the asbestos packing must be debited for current waste disposal costs along with any future or continuing costs associated with final disposal in landfills and so forth.

5. *Packing Life*. An important consideration is packing life. Information on the life of various nonasbestos packings indicates that nonasbestos packing life may be considerably longer than asbestos packing life in the same service. Asbestos packings may be replaced several times during the life of a comparable nonasbestos packing, which increases the cost of the asbestos packings. Asbestos loses fibrous water with time, and the packing becomes hard, losing resiliency and emission control effectiveness. Data on asbestos life for various services should be available from the plant maintenance group.

6. *LDAR Costs*. Nonasbestos packings with lower emissions can potentially reduce LDAR costs by reducing the monitoring frequency, which is accompanied by a reduction in maintenance.

7.5.3 Nonasbestos Packings

A variety of nonasbestos packings are now available up to a temperature of about 550°F. Above this temperature level the types of nonasbestos packings acceptable for high-temperature, high-pressure operations are limited. Base materials commercially available in packings for this service are flexible graphite, carbon/graphite fibers, a few specific polymer fibers, and metals. Other specialized materials such as ceramic fibers are being developed for high-temperature operations. This diverse group of packing materials successfully control valve emissions, but comparative data on their effectiveness in controlling emissions and loss reduction are not available. The reasons for their overall success in valve packing applications are the inherent disadvantages of asbestos packings:

1. When exposed to high temperatures over extended time periods asbestos loses its water of crystallinity, resulting in a loss of packing volume. High pressure further compresses the fibers, which fractures the material and produces dust. As a result, there is a considerable shrinkage of asbestos in the stuffing box and loss of sealing effectiveness.

Hardening of asbestos packings has also been observed over a long period of time at temperatures of about 350°F (117°C) accompanied by shrinkage.

2. Asbestos is abrasive, but the impregnation of graphite on the outer packing surface reduces this tendency.

3. In high-temperature service, dissipation of heat away from the stem results in the packing operating at a lower temperature and increases packing life and performance. Asbestos is an insulator with poor thermal properties and tends to trap heat in the stem packing interface. The heat flow in the stem gradually increases the interface temperature to a relatively constant level through the stuffing box.

4. Under the combined effect of temperature and pressure, asbestos densifies to a hard mass with virtually no recovery or resiliency. When the packing begins leaking, the valve must be repacked. However, removal of hard asbestos packing manually from a valve stuffing box is difficult and often requires a considerable effort. The removal of jammed and hardened asbestos packing often scores the stem and packing box wall which affects the emission control efficiency of the replacement packings.

7.5.4 Graphite Packings

One of the most important replacement packings for asbestos is graphite packing with installations spreading throughout the world. Although a variety of packing materials have been developed, graphite packing has two major advantages compared with alternative packings. The first characteristic is that graphite packings are fire safe, which is an important requirement for refineries and petrochemical plants. Fire resistance is also becoming a major consideration in the chemical industry, where hazard risk assessments are regulatory requirements and safe operations are a major concern.

In addition, graphite temperature resistance is much higher than the allowable temperatures for alternative packings, which are primarily polymeric. In nonoxidizable atmospheres, the maximum graphite operating temperature is about 1250°F (677°C) and in inert atmospheres such as nitrogen and argon, the maximum operating temperature is about 1650°F (900°C). Higher operating temperatures have been indicated by some packing manufacturers, but the temperatures shown provide a basis for design.

While graphite packings perform well in a wide variety of services, graphite is affected by oxidizing atmospheres, as described above, and service in an oxidizing stream should be reviewed with the packing manufactuerer. A "threshold of oxidation" is the maximum temperature a carbon or graphitic product can withstand without subliming to carbon monoxide or carbon dioxide. A more defintive threshold description is the temperature where 1 cu. in. of material loses 1% of its weight in 24 hr. In hot air the threshold has been established at 650–800°F (343–427°C) with a maximum safe limit of about 650°F (343°C). In steam service the threshold is 1292°F (700°C) and a limit of 1200°F (649°C) has been established. A threshold of 1652°F (900°C) was determined for carbon dioxide that is oxygen free.

The temperature limitation for graphite in oxidizing media is normally determined by the manufacturer for continuous service. Depending upon the oxidizing compound and concentration, maximum operating temperatures of 900°F (482°C) or more can be established. For reducing or inert media, temperatures well in excess of 1200°F (649°C) may be recommended by the packing manufacturer. However, combining different types of graphite packings results in a maximum temperature established by that packing with the lowest threshold temperature. Graphite also performs well in low-temperature operations with minimum temperatures of −200°F (129°C) or lower, as quoted by one packing manufacturer.

Graphite packing performs well in high-pressure service for long periods of time. Installations in steam service at a pressure well above 2000 psig (139 bar) have reported satisfactory performance over a number of years. Packing manufacturers also indicate graphite packings perform well at high pressures in nonsteam service.

Standard Packing. A standard packing was defined some years ago for service in refineries and petrochemical plants by the American Petroleum Institute (API) that would meet operating conditions of 1200°F (649°C) and 2500 psig (173 bar). This standard was satisfied by a graphite-impregnated asbestos packing with a wire insert. Graphite-type replacements have been developed by packing suppliers as substitutes for this standard packing.

Substitutes developed by the manufacturers in the late 1970s are quite similar to the standard asbestos packing, but do not contain any asbestos. The new packing consists of an outer braided carbon yarn jacket containing inconel wire reinforcement over an asbestos-free core which is similar in design to the standard asbestos packing and comparable to the braided jacket and core shown in Figure 7.18. The standard asbestos packing consists of braided asbestos yarn in the outer jacket with inconel wire intertwined with the asbestos yarn. This jacket surrounds a core of asbestos that may be impregnated with graphite.

The outer jacket of the graphite asbestos substitute is composed of a braided yarn and the yarn in turn consists of a carbon fiber jacket spun around an inconel wire and a yarn core. This outer braided jacket surrounds an inner core that is asbestos free, but whose composition is proprietary. The core body does not become rock hard and generally retains some resiliency at higher temperatures. Since the outer jacket is a carbon fiber, the packing is nonabrasive and self-lubricating to valve stems. Cores supplied by various manufacturers may contain carbon, inert materials, or other fibers and an inhibitor.

Resiliency remaining in the packing is important in permitting packing adjustments to meet EPA leak criteria and in turn maintain a lower emission level. Shrinkage is considerably less than the reduction in size exhibited by asbestos packings and packing life is longer for the carbon packing. The maximum combined packing operating conditions for this type of packing is about 1200°F (649°C) and 1200 psig (84 bar). However, the packing may not control leakage if operated at maximum temperature and increasingly higher pressures. Actual operating conditions should be reviewed with the manufacturer.

Carbon Fiber. A carbon fiber packing was developed and is considered superior to the fiber-jacketed packing previously described. This packing consists of a high-purity carbon yarn treated with a proprietary blocking and lubricant system which is then lattice braided. The high-purity carbon yarn has a carbon concentration of about 95%

Carbon fibers are produced from either pitch or rayon where the pitch is apparently obtained from a refinery bottoms fraction. The pitch base essentially consists of carbon with a small quantity of impure material removed during the heating phase. This results in an amorphous carbon structure that has randomly oriented particles and a yarn with a high tensile strength. In addition, pitch-based carbon yarns have a degree of flexibility.

Rayon-based carbon fiber has a large amount of impurities that are removed, followed by extrusion through spinnerettes and drawing or stretching. The drawing produces linear chains compared to randomly distributed molecules in the amorphous fiber. The rayon-based carbon fiber is lower in tensile strength than the pitch-based carbon fiber and has a lower degree of flexibility. Under high compressive and dynamic loads, pitch-based yarns generally have less fracturing and breakage than rayon-based yarns.

The carbon fibers are formed into a lattice braiding system where each strand passes through the cross section resulting in a more uniform cross section than other packing types. A simplified sketch of lattice braid is shown in Figure 7.18. Although the braided system is completely interwoven, a blocking system is necessary to prevent media flow through or weepage. The carbon yarn acts as the base for a lubrication-blocking system. Yarns are coated initially and dried to ensure complete and uniform impregnation of each yarn. After the yarns are braided and interwoven, the packing is again impregnated with the original lubrication-blocking materials, filling the interstices and eliminating any voids. The final addition provides a high degree of resistance to media flow or weepage.

Lubrication and blocking systems are proprietary, with each packing manufacturer developing their own systems. The lubrication/blocking systems must be compatible with the carbon yarn, inert to the fluid media, and satisfactory for high temperature/high pressure service applications. The compression requirement for braided packing is less than that required for the jacketed packing. The latter requires a higher compression pressure to densify the core material in order to resist weepage of the media through the packing. Packings must be impermeable to the process fluid and passages through fibers and yarns are generally eliminated through compression.

While a combination of carbon fiber rings have been installed in valves, these packing sets have not been as effective in controlling emissions as those described in the following sections. However, they perform very useful functions in combination with other types of packing sets based on the following characteristics:

1. Braided carbon is nonhardening and nonabrasive in service through its operating temperature/pressure range. Valve stems are not scored.

2. Braided carbon is thermally stable through its operating envelope. Moreover, the braided packing performs extremely well in thermal cycle operations.

3. The expansion contraction coefficient is almost identical to steel, which minimizes any effects from heating/cooling.

4. Carbon braided rings are heat conductors and heat is transmitted to the outer valve walls away from the stems.

5. This material is applicable across a pH range of 0–14 for chemicals that are not strong oxidizing agents. Braided carbon should not be used with nitric acid, aqua regia, and sulfuric acid above a 95% concentration. For oxidizing service, applications should be reviewed with the manufacturer.

6. Leachable chlorides are < 200 ppm for standard braided end rings and this can be reduced to < 50 ppm when specifically requested.

7. Braided carbon normally operates to a maximum temperature of 1200° F (650° C) and several thousand pounds pressure. Since braided carbon rings are also used in packing sets, allowable operating temperatures may well be higher. The carbon rings also operate to $-200°$ F ($-130°$ C).

Graphite braided rings are also available. The graphite yarns are laminar, crystalline molecular structures and are more susceptible to fracture and breakage under pressure with less flexibility than carbon yarns. Graphite braided rings were developed to meet purity standards, particularly in the nuclear industry where 99+% carbon purity is required along with a leachable chloride concentration < 50 ppm. For process applications, carbon-based end rings are preferable.

End Rings. The previous section indicated that carbon braided rings are combined with graphite packing sets for optimum performance. These braided rings are used as end rings or are placed at each end of the main packing set. The reasons for placing the carbon braided rings in these positions follow:

1. If flexible graphite rings are subject to compressive loading, they will "cold flow" into any opening. In a valve stuffing box, openings or clearances are located at the top of the box around the gland follower and at the bottom of the box shown as the throat in Figures 7.4 and 7.5.

The braided end rings are installed at the bottom and top of the flexible graphite preformed rings to prevent extrusion of the graphite material into these openings. Thus the graphite material is confined within the graphite ring zone and does not extrude.

2. Graphite has a high affinity for metal surfaces. Particles of graphite will adhere to metal surfaces such as valve stems. If the stems were moved (valve actuation) without braided end rings, minute particles attached to the stem would be exposed to the atmosphere, or conversely in gate or globe valves, closing the valve would expose the particles on the stem to the fluid. Once exposed to the environment or the fluid, these particles are lost, resulting in a volume loss of graphite ring material. This loss eventually leads to the creation of leakage paths through the set and excessive emissions as additional graphite particles migrate to the stem.

The braided carbon end rings act as "wiper" rings at the top and bottom of the set, preventing particle transport out of the packing zone. These particles remain within the zone, and a general equilibrium develops between the particles and rings. As a result, ring dimensions do not change, and packing emission control does not vary.

3. Braided rings are quite resilient and, when placed at the ends of a packing stack, cushion and maintain a consistent load in the presence of pressure and temperature changes. In effect, the braided rings act as integral spring or live-loading devices.

Graphite. A major development was the use of an essentially pure graphite to make a packing ring. Graphite was initially marketed as a tape, which was then wound concentrically around a valve stem until the distance between the stem and the wall of the packing box was filled, and this annular spiral was compressed to a disk form, as shown in Figure 7.18. This disk forms a consistent homogeneous mass of material. However, forming the disks at different levels in the packing box requires careful control. The lowest disk along with a few others above this bottom disk can normally not be reached by the gland follower. As a result, a bushing or collar is placed between the gland follower and the uncompressed tape. The tape is then compressed as the gland follower in contact with the bushing transmits compression pressure by tightening the follower bolts. When the ring has been compressed, the process is repeated until the required number of rings have been formed in the packing box.

The original tape and most competitive tapes are corrugated and when compressed are tightly meshed at the corrugations, as shown in the figure. One manufacturer (Garlock) produces a diamond texturized tape, which apparently increases tape interlocking and is considered an improvement over the corrugations. Interlocked and compressed graphite material has a high thermal conductivity, resulting in heat conductance radially from the stem to the stuffing box wall. This characteristic improves packing life and operating characteristics.

Graphite packing can be received from the manufacturer preformed into a packing ring and is termed flexible graphite packing. The manufacturer forms these rings in a die to meet the various valve dimensions specified by the purchaser. Formation of these die-formed rings under pressure permits control of the finished packing ring density. In contrast to the tape installation in the valve, where the tape is converted to a ring within the stuffing box and ring density varies, the preformed ring density can be carefully controlled in the manufacturer's shop and is now preferred for consistency and improved emission control.

A typical flexible graphite ring packing set or stack is shown in Figure 7.19. The end rings for the packing set are braided carbon and three graphite rings are placed between the braided end rings. For optimum performance of a packing set, the flexible graphite rings should be preformed with an overall total of five rings. This number contrasts with older practices where a large number of packing rings were considered preferable in controlling emissions. Recent information in various references recommend a total of five rings[31,32] and packing manufacturers also recommend five ring-packing sets[33] for optimum performance. However, differences exist in the preformed density of the packing ring set available from the packing manufacturers.

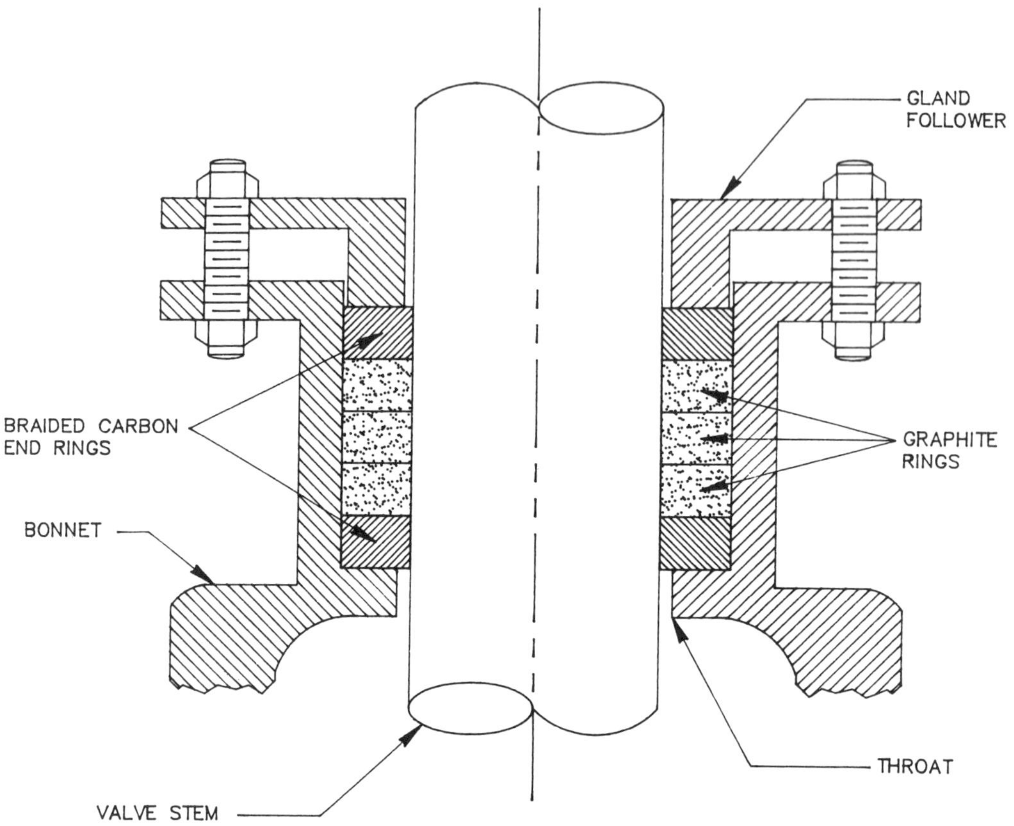

Figure 7.19 *Typical graphite packing installation*

The preformed flexible graphite rings have good cold flow properties and expand under compression to fill the voids in the stuffing box area or annular openings around the stem and any gap between the outer ring diameter and the stuffing box wall. In an ideal installation, the packing would be elastomeric and compression is translated into outer and inner radial expansion that would completely fill and seal all openings. However, compressibility and radial expansion is a function of the ring density with higher-density materials less compressible than the low-density rings. Initial preformed flexible packing ring densities available from packing manufacturers vary from 70 to a little over 100 lb/cf (1.12–1.60 gm/cm^3) with most packing densities in the range of 70 to 87.5 lbs/of (1.12–1.40 gm/cm^3). The difference in packing densities is a function of the initial tape density and general packing compression philosophy. Comments on preformed flexible packing ring sets follow:

Initial tape density. Before formation into rings, basic graphite corrugated or textured tapes are normally obtained from graphite manufacturers with densities of about 46 and 68 lb/cf (0.74–1.09 gm/cm^3). The 46-lb/cf tape is preformed in a die to a density of 70 lb/cf (1.12) and the 68-lb/cf tape is preformed to the desired

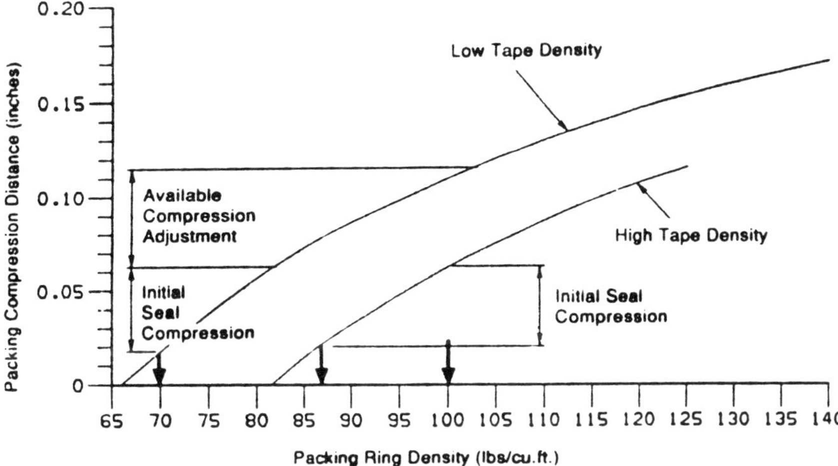

Figure 7.20 *Graphite density and compression distance as a function of high and low initial tape densities (courtesy of Chemical Engineering Progress).*

tape density[34]. A number of preformed rings are provided by manufacturers in the 87.5–90.0 lb/cf (1.40–1.44 gm/cm^3) range.

The preformed flexible graphite rings are placed in the valve packing boxes and compressed in accordance with manufacturers instructions.

Compression. The amount of compaction or compression and radial expansion is a function of the compression pressure applied with the gland follower, as shown in Figure 7.20[34]. In the figure, the low initial tape density curve (46 lb/cf) differs from the compression curve developed for the high initial tape density (68 lb/cf) material. The tapes are placed in dies and compressed to 70 lb/cf (1.12 gm/cm^3) and 87.5 lb/cf (1.40 gm/cm^3), respectively, which are the initial seal ring densities as shown by the arrows.

From the preformed densities indicated by the arrows, the rings were compressed in a test fixture to develop the curves shown. Initial seal compression refers to the change that occurs as the packing is compressed from the initial seal ring density with the gland follower to a desired compaction level. Following installation, the 70-lb/cf packing was compressed 25% in height and the 87.5-lb/cf set was also compressed 25%. This level of compaction resulted in a ring density of about 83 lb/cf for the low-density ring after compression, but the density of the 87.5-lb/cf ring increased to 100 lb/cf for the same compaction distance. While compression movement is an indicator of resilience, the upper curve for the low initial ring density allows more compression movement to achieve equivalent ring densities for compacted high initial ring density materials.

Figure 7.20 indicates more resilience for flexible graphite packings with lower initial preformed ring densities, resulting in greater radial movement and presumably better sealing than can be obtained with high initial preformed ring densities. This difference is also accompanied by significant differences in the amount of compression pressure required to achieve compaction changes for the different

TABLE 7.4 Graphite Packing Ring Compression Pressure and Ring Densities[a]

Base Preformed Ring Density lb/cf (gm/cm^3)	Initial Ring[b]/Compression Ring Density lb/cf (gm/cm^3)	Compression Pressure psig (bar)
70.0 (1.12)	82–83 (1.31–1.33)	1200 (84)
87.5 (1.40)	100–102 (1.60–1.63)	3800 (263)
100.0 (1.60)	115–117 (1.84–1.88)	6500 (449)
106.0 (1.70)	120–123 (1.92–1.97)	8000 (552)

[a]Adapted from reference 34.
[b]Typical density after recommended compression procedure.

initial seal ring densities, as shown in Table 7.4. The initial compression change is based upon the packing manufacturers' recommendations and indicates the required compression pressures are much higher as the density of the base or initial preformed ring increases.

In addition, Table 7.5 indicates the final densities that nominally result when the packings have been tightened to about their maximum field tightness. The ring compression pressures associated with final compaction increase significantly as the base ring density increases. This table also indicates that compression movement is less as the base ring density increases, which shows less resiliency with higher-density material and lower radial expansion changes, a measure of sealability.

The very high compression pressures necessary at high densities requires large gland follower loadings. Gland followers have been deformed as the bolt loadings were increased on seal packing sets where the ring densities had been compressed to levels much greater than 120 lb/cf (1.92 gm/cm^3). Although high ring density materials do not compress well, this characteristic forms the basis for specialized packing designs that significantly improve sealing capability and are described in the following sections. While improved flexible graphite packings are being installed in many applications, the basic five-ring set still performs well in many services. With satisfactory LDAR programs, these sets will meet regulatory requirements in various applications.

TABLE 7.5 Typical Graphite Packing Ring Loadings and Ring Compaction[a] Ring Density

	Ring Density, lb / cf (gm / cm^3)			
Base Preformed Ring	Initial Ring Compression[b]	Typical Final Compaction Density[c]	Final Ring Pressure psig (bar)	Overall Compaction Movement inch (mm)
70.0 (1.12)	82 (1.31)	103 (1.65)	3,600 (249)	0.054 (1.37)
70.0 (1.12)	82 (1.31)	120 (1.92)	6,900 (477)	0.082 (2.08)
87.5 (1.40)	100 (1.60)	120 (1.92)	7,600 (525)	0.044 (1.12)
100.0 (1.60)	116 (1.86)	125 (2.00)	9,400 (649)	0.016 (0.41)
106.0 (1.70)	121 (1.94)	130 (2.08)	12,000 (828)	0.014 (0.36)

[a]Adapted from reference 34.
[b]Typical density after recommended compression procedure.
[c]Maximum densities.

Graphite Packing Inhibitors. Corrosion of the valve stem may result when there is a conducting fluid such as water in contact with carbon or graphite packing and the valve stem. Frequently, the corrosion is a result of water traces remaining in the valve after hydrostatic testing of the valve and a delay in installation. Should corrosion occur, a roughened stem can provide minute pathways for leakage between the valve stem and packing, resulting in an undesirable increase in the emissions concentration.

A method of protecting the metal is necessary since the metal is anodic and the packing is cathodic. Active sacrificial corrosion inhibitor systems are normally recommended which have an inhibitor that is more anodic than the metal stem. In this situation, the inhibitor is preferentially attacked, protecting the metal stem. An alternative protection solution is a passive system or an inhibitor that coats the interface area on the metal stem facing the packing. While this system may appear satisfactory, the packing surface must be completely coated to form an unbroken barrier. Moreover, the coating system must remain intact, which is difficult to maintain when the valve stem is moved. As a result, an active inhibitor system is preferred which requires corrosion inhibitors for both carbon braided end rings and dieformed graphite rings. To meet the requirements for sacrificial material, zinc and aluminum have been specified through the years for this service. These metals are normally applied by the valve manufacturers as powders.

Normally, zinc is specified for most applications. However, this material must be compatible with processing conditions and the fluid stream. In applications where processing conditions are potential problems, aluminum powder has been selected as the replacement for zinc. The aluminum powder performs satisfactorily, but is apparently not completely equivalent to zinc in protection performance according to some packing manufacturers. Where zinc as a low-melting metal is considered detrimental to the process, aluminum powder has apparently been a successful replacement. Other materials have also been used as inhibitors and packing manufacturers have considerable experience with inhibitors and should be consulted for various applications. Methods of coating graphite and carbon braided rings have been developed by the individual packing manufacturers.

As an alternative to corrosion-inhibited packings, washers between rings have also been used as corrosion inhibitors. Both zinc and aluminum washers can be used in this service. However, testing shows they only protect the small stem area facing the washer and essentially are not particularly effective. Metal packing inhibitor powders such as zinc and aluminum are generally considered superior corrosion inhibitors compared to other corrosion protection techniques.

Standard Packing Installations. As described, the majority of packing installations consist of three flexible graphite rings and a carbon braided ring at the bottom and top of the ring set for a total of five packing rings. Two flexible graphite rings with braided carbon fiber end rings have also operated satisfactorily in high-temperature (1250° F) and high-pressure tests.[35] While a five-ring installation set provides optimum service performance, in long stuffing boxes a bushing is installed in the stuffing box between the bottom of the box and the packing set to fill the excess volume. As an alternative, the excess volume can be filled with carbon braided rings at the bottom of the stuffing box and then covered with the packing set. However, the bushing at the bottom of the box is preferred.

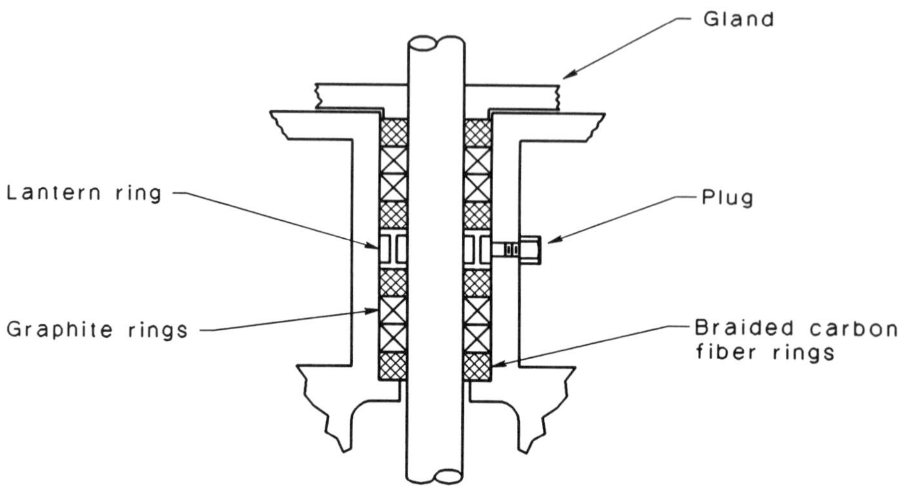

Figure 7.21 *Graphite packed lantern ring installation (courtesy of Garlock).*

In valves with lantern rings, the recommended installation is two five-ring combination packing sets with one set above and one set below the lantern ring. In Figure 7.21 the stuffing box sketch shown with a lantern ring cannot accommodate two five-ring packing sets and two shorter sets are installed. For this installation, a flexible graphite ring in each set is deleted to insure that braided carbon fiber end rings are included. In situations where six or more graphite rings are installed without a lantern ring, a braided carbon ring should be included in the center of the rings as an internal wiper ring.

The packing set previously recommended contains carbon end rings and 70-lb/cf (1.12 gm/cm^3) density rings and is compressed 25% for fugitive emission control and containment. This packing set is satisfactory to an operating pressure level of 2000 psig (139 bar) according to Garlock, the packing manufacturer of 70-lb/cf packing rings. For operating pressures above 2000 psig, the manufacturer recommends a flexible graphite ring packing density of 87.5 lb/cf (1.40 gm/cm^3), but the manufacturer should be consulted about compression set. Typical asbestos packings were often tightened to produce a very high compression load and maintenance personnel became accustomed to expending considerable effort in adjusting the gland follower bolts. The extended adjustment effort is not required when compressing the combined carbon and flexible graphite packing set, but the movement of the gland must be carefully measured to insure compaction is within the recommended 20–25% compression range or a compaction pressure is placed on the packing in accordance with recommendations. This requires an understanding of the physical changes in graphite packing characteristics and adjustment procedures by maintenance personnel.

The initial packing installation is important for insuring proper ring seating and the ability to adjust packing compression, which controls valve emissions rates, thereby minimizing exposures that can occur as a result of high leakage rates. The proper packing installation permits adjustment if the valves are leaking above the

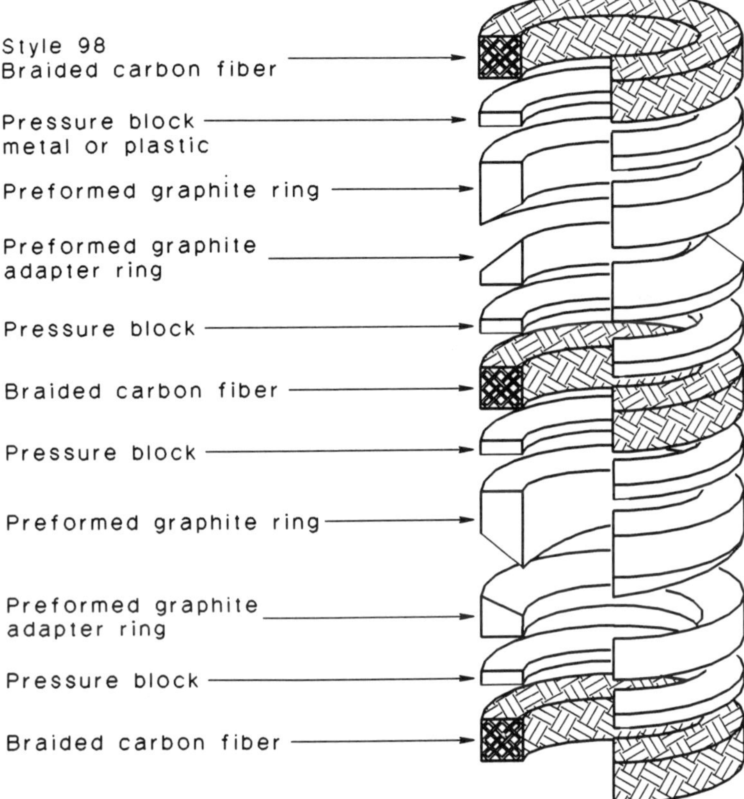

Figure 7.22 *A 9000 EVSP® valve packing set (courtesy of Garlock).*

permitted EPA monitoring level. Following startup, some flexible graphite packed valves may have leak concentrations higher than regulatory limits; the packing is then adjusted, but packing compression changes require careful adjustment. Gland follower *bolt nuts* are only moved one face or at the most one quarter turn at a time, and the concentration around the valve stem is measured after each adjustment until the monitor reading is well below the desired level. This is also in accordance with EPA regulations, which require a monitoring reading after packing adjustments are completed. However, the allowable concentration levels for hazardous or toxic materials should be understood prior to adjusting the packing to prevent potential exposures.

9000 EVSP® Packing Set. A different flexible graphite-based packing set was developed by Garlock in the late 1970s to specifically control emissions in highly regulated areas where emissions levels are limited by air pollution regulations. This complete packing set is identified as Garlock 9000 EVSP®, which is an acronym for expandable valve stem packing and is shown in Figure 7.22.

The basic sealing mechanism in this packing set is based upon different graphite densities where high-density graphite is less compressible than a low-density

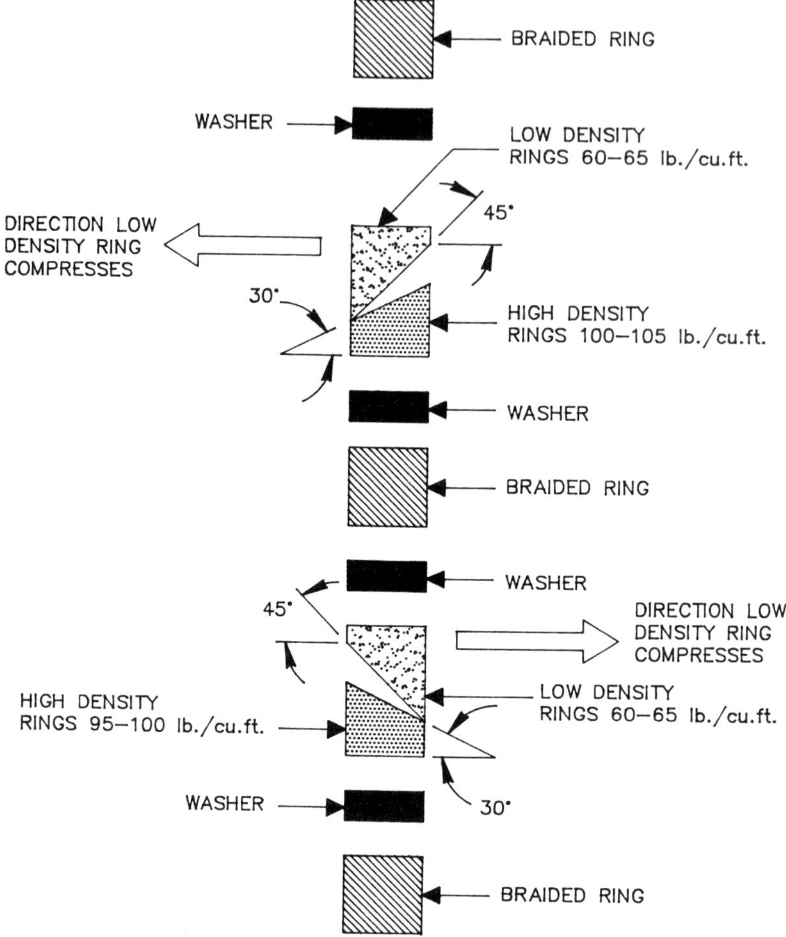

Figure 7.23 *Simplified cross section of 9000 EVSP® packing set (courtesy of Garlock).*

graphite set described previously. This set is a combination of braided carbon fiber rings (Garlock 98)®, preformed flexible graphite rings specifically molded for this set and metal washers. While the general packing set is shown in Figure 7.22, a simplified cross section of the 9000 EVSP® set is shown in Figure 7.23 to describe the basic sealing functions.

Two beveled pairs of opposed molded graphite rings are shown in Figure 7.23 and in each pair the opposed rings have different densities. In addition, the bevels in the opposed graphite rings have different angles. The purpose of these differences is to shape and direct the natural cold-flow properties of the molded graphite rings to the outside edge of the packing interface with the stuffing box and the interface of the packing set with the valve stem. Compression of the EVSP® set is translated into outward and inner radial growth to the stuffing box and stem interfaces rather than densification of the set. The braided carbon fiber rings act as wiper rings in this packing set.

In each molded graphite pair, one flexible ring has a density of 60–65 lb/cf (0.96–1.04 gm/cm^3) and the other about 100–105 lb/cf (1.60–1.68 gm/cm^3). In addition, the bevel on the low-density ring is at an angle of 45°, whereas the bevel angle on the high-density graphite ring is 30°. The compressive load on the packing set preferentially directs expansion of the low-density rings in different directions, as shown in Figure 7.23. One low-density ring seals against the valve stem and the other against the stuffing box wall.

When dieformed rings, such as those in a five-ring stack, are compressed, most of the compression energy is absorbed within the rings in further densification or direct compression of the ring. There is an estimate that $< 40\%$ of the compressive energy is translated into radial expansion to form a seal with square cross-sectional packing. In contrast, the difference in densities, accompanied by the angular design that translates compression into radial expansion in the 9000 EVSP® set is a more efficient design. Approximately 70% of the compression energy is converted into radial expansion compared with $< 40\%$ for the flat ring design.

The assigned directional expansion in the 9000 EVSP® set *insures* an inner and outer seal compared with the flat graphite ring in a standard stack that must radially expand in two directions for good sealing. Radial expansion capability of 0.125 in. (3.8 mm) for 9000 EVSP® is considerably higher than the 0.02-in. (0.51 mm) radial expansion normally achieved with standard performed graphite rings.[17] This difference translates into a tighter seal and lower emission rates.

Braided Carbon. The Garlock 98 braided carbon rings perform the functions in the 9000 EVSP® set that were previously described for braided end rings. These rings prevent extrusion of graphite into any annular openings and act as wiper rings to prevent particle transport to the atmosphere or process fluid and as resilient springs within the set that adjust to thermal and pressure changes, maintaining a consistent seal. The braided carbon ring at the center of the set separates the two sealing zones.

Washers. The washers adjacent to the braided carbon rings distribute the compression loading evenly throughout the set. In addition, the solid washers also act as block or dam preventing media weepage through the body of the set. The washers are supplied as aluminum or alternative metal based upon the application.

Installation. There are 11 basic pieces in a complete 9000 EVSP® set. For deep stuffing boxes, additional braided carbon (Garlock 98) rings may be installed. While these rings can be installed, metal bushings are preferred. Since the rings and all components are circular, installation of all the packing set components can be accomplished by two different methods. In the first procedure, the valve upper part is dismantled to place each ring on the stem in the proper sequence and press the rings into position. This can be accomplished during valve manufacture or if the valve is completely disassembled for repair. While this general procedure is satisfactory for a valve that is disassembled, a very high proportion of valves must be packed in position. A valve that remains in the piping is packed during a turnaround or the valve must be completely isolated during operation. In the latter situation, flange connections are blinded to prevent any leakage and the valve assembly is purged for safety and exposure prevention. In repacking, the gland follower is unbolted, the existing packing is removed, and the packing or stuffing box is cleaned. Packing removal is a difficult operation and should not score or mar the valve stem or packing well. The stem prior to repacking should be inspected to ensure it is in satisfactory condition for packing installation. Badly scored stems or stems that have a slight curve may require a replacement stem.

In field repacking, the rings must be split or cut to allow placement of the rings around the stem and in the stuffing box opening. The carbon braided ring (Garlock 98) is quite flexible and only requires one split, which is a vertical split, to permit the ends to butt against one another. In contrast to the flexible carbon braided rings, the metal spacers are not flexible and require double splits. The flexible graphite rings have one angled split at 30° that permits installation of the rings. Splits should be separated from a superimposed ring by 180°. Garlock normally supplies the 9000 EVSP® set with one split in each carbon and graphite ring.

While preformed flexible graphite rings have a split, the angled split under compression pressure results in rebonding the split. However, all ring components should be installed to prevent splits being aligned in successive rings. The splits should be offset 180° from the ring below the newly placed ring. With this procedure, the overall set would be properly aligned. For double-split spacer rings, a 90° offset is necessary.

Bushings. Although packings ideally fit properly into a stuffing box after compression, many of these boxes are much longer than the compressed packing set. As a result, bushings are frequently required as an extension or filler piece in the lower part of the box. These bushings have been constructed of various materials including carbon and various metals.

Frequently, pure carbon rings are installed as bushings that are solid, hard, and generally brittle. The brittleness makes handling difficult, and where the bottom of a stuffing box is angled or may contain clumps of particles remaining from previous packings, there is a distinct possibility the bottom of the carbon bushing will fracture. Other types of mixed carbon bushings are available and brittleness characteristics may change somewhat. However, metals are normally recommended for bushing service, and the packing manufacturer will provide the bushing or measurements for a locally produced bushing.

The bushings have two vertical splits 180° apart. While insertion of the bushing into the packing box is not difficult, removal of the bushing requires preparation for this step. Since the bushing is about the width of the packing box, two threaded holes are placed in the top flat surface of each bushing section. Threaded rods extended into the stuffing box permit easy removal of the bushing sections.

Ports or Stuffing Box Openings. In deep boxes, ports or openings for potential lantern rings are often found in valves. The 9000 EVSP® set should be located above or below the opening in the wall. Otherwise, extrusion of the flexible graphite rings into the port significantly reduces the effectiveness of the packing in controlling leakage and emissions.

If bushing rings cover the port and the 9000 EVSP® set is located above the port, the integrity of the packing is maintained and the installation will satisfactorily control emissions. Alternatively, locating the packing set below the port is satisfactory, with a bushing filling the volume between the packing and the gland follower. Where an opening cannot be averted by relocating the packing, the hole should be filled and the stuffing box wall ground smooth.

Surface Finish. The smoothness of the wall and stem are a factor in packing performance. An overly smooth or rough stem finish reduces graphite leakage control effectiveness. For 9000 EVSP® graphite service and most valves, a 32 RMS finish is recommended. This RMS finish is a standard smoothness designation used in various industries and basically indicates roughness through microscopic measurement of ridge height. In this system, very low numbers are extremely smooth and high numbers indicate a rough surface.

While the stem finish is 32 RMS, the stuffing box wall roughness recommendation is 125–150 RMS. These finishes may vary with different manufacturers, and for nongraphite packings other finishes may be suggested.

Oxidation. Since 9000 EVSP® is a graphite packing, service conditions must be investigated to ensure the process does not contain an oxidant that affects the packing set. Packing manufacturers have assembled a considerable quantity of information covering the effects of various chemicals on graphite. In addition, 9000 EVSP® has washers that must also be compatible with the process fluid. Inhibitors as described should be compatible with the process application.

In considering this aspect of packing application, the effect of the various chemicals must be determined for the process temperature, which can be a major concern.

Emissions and Application. Emission tests reported by the manufacturer indicate that actual emissions are carefully controlled by 9000 EVSP® packing in petroleum refinery hydrocracker service. The hydrocracker is a high-temperature, high-pressure petroleum refinery operation, and 9000 EVSP® packings are apparently superior to other packings in controlling emissions and packing life. Operation in other high-temperature refinery services has been successful along with nuclear and very high pressure/temperature steam systems. A similar result was obtained in a 650° F kerosene service at 200 psig (14.8 bar). In a laboratory, the packing set in an actuated valve in propane service after 50,000 cycles had a monitoring concentration < 200 ppmv.

The 9000 EVSP® packing can maintain a very low emission rate (< 500 ppm), but initially the valve stem should be monitored frequently and the gland follower adjusted when required. Following packing installation, the packing must be carefully compressed in accordance with the manufacturer's instructions. Frequent valve stem movement will affect emission levels, but minor gland adjustments are very effective in controlling emissions. Gland adjustments should be carefully controlled through the procedure outlined in the previous subsection on graphite packing.

Although this packing exhibits excellent emission control characteristics, the industrial hygienist should be aware of the relatively high cost of this packing compared with the standard five-ring braided carbon preformed graphite combination. However, in difficult services where emissions can present potential hazards in the workplace, the 9000 EVSP® packing should be considered as a method of minimizing workplace exposures. An alternative is the simplified EVSP® described in the following section.

9000 EVSP Simplified Set®. While the 9000 EVSP® set has been a highly successful emissions control packing, this packing set has some drawbacks, particularly from a cost standpoint. Also, the 11-ring set is a complicated packing to install and compress. As a result of this assessment, Garlock, the packing manufacturer, developed a modified packing set termed "Simplified EVSP" to overcome the apparent 9000 EVSP® problems. The 9000 EVSP® set has been extremely successful controlling emissions and the Simplified EVSP® set, based on the same principles, has also been successful in controlling emissions to the same levels.

The basis for sealing with the Simplified EVSP® set shown in Figure 7.24 is essentially identical to the sealing mechanism in the standard 9000 EVSP® set.

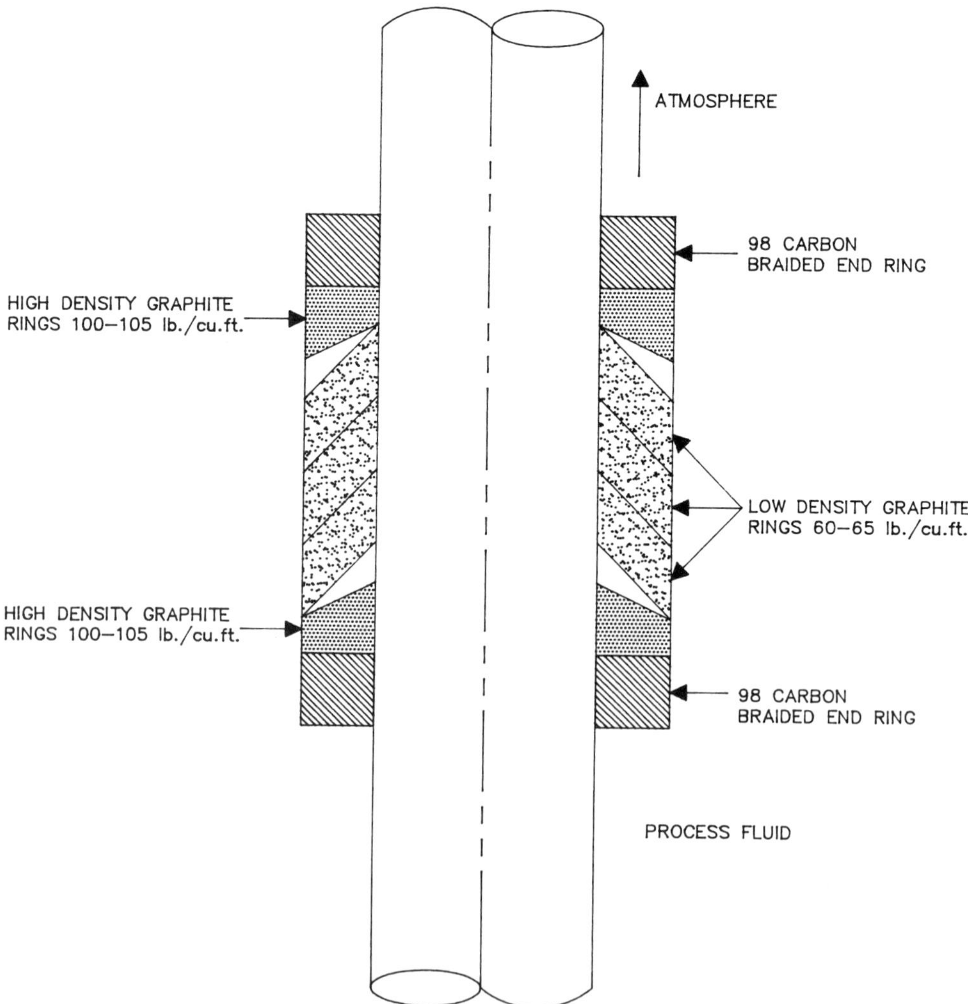

Figure 7.24 *Cross section of simplified EVSP® packing set (courtesy of Garlock).*

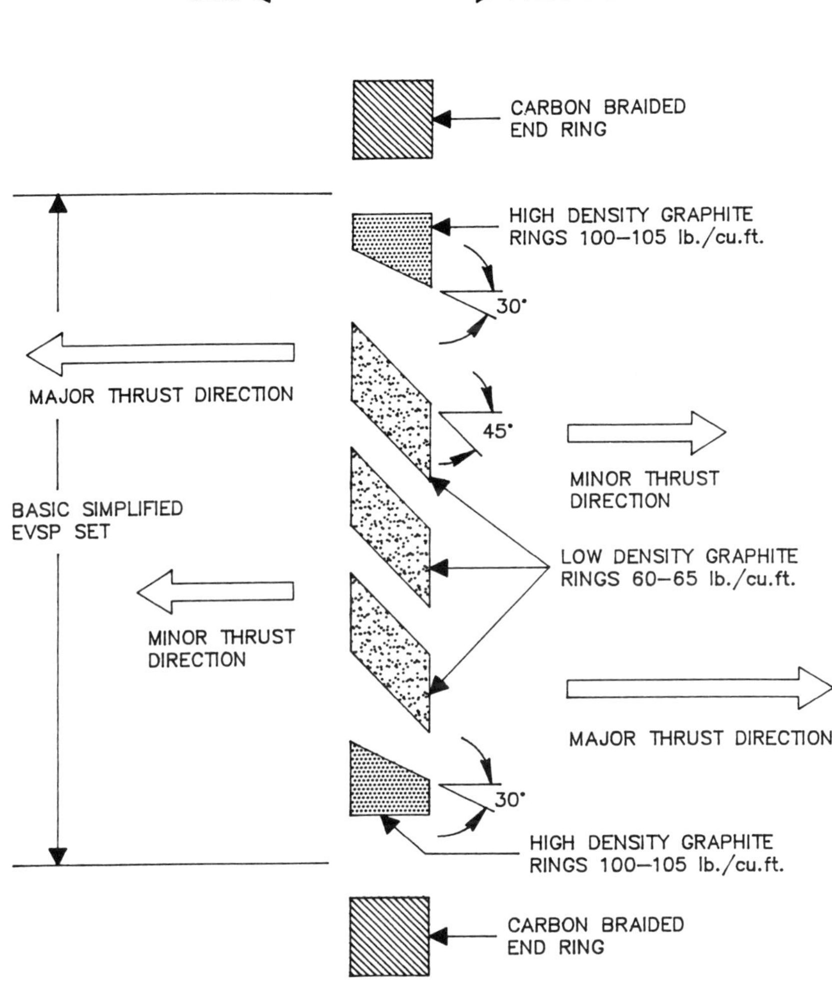

Figure 7.25 *Construction and sealing with Simplified EVSP® packing set (courtesy of Garlock).*

The general concept is to compress and laterally expand a lower density graphite ring with a higher density graphite ring, while controlling the direction of expansion to obtain a higher compaction/expansion efficiency. A large difference in density between the two flexible graphite rings of about 40 lb/cf (0.64 m/cm^3) is important in preferentially compressing and expanding the lower density flexible graphite ring which has a density of 60–65 lb/cf (0.96–1.04 gm/cm^3). The high-density flexible graphite ring of 100–105 lb/cf (1.60–1.68 gm/cm^3) does not compress significantly compared with the low-density ring at the compression pressure set by the gland follower. Expansion is directed by the graphite ring angles shown in Figure 7.25, where the high-density ring has a 30° angle to the horizontal plane and the low-density ring a 45° angle. The basic thrust of the low-density flexible graphite rings are shown along with minor thrusts or expansions. The middle low-density flexible graphite ring apparently moves equally in both directions.

An important consideration with the simplified EVSP set is that the set must be installed with braided carbon end rings. These end rings (Garlock 98) perform the same end ring functions described previously, namely, prevent ring extrusion, wipe the shaft, and provide some resiliency. With the end rings, the overall Simplified EVSP® set contains 7 rings compared with 11, including the end rings for the 9000 EVSP® packing set.

In general, the advantages for the Simplified EVSP® compared with the base 9000 EVSP® packing set are:

Cost. The Simplified EVSP® set is about half the cost of 9000 EVSP® and is considered an equivalent packing for controlling emissions across a broad spectrum of services. In a general cost comparison of a five-ring standard graphite packing stack, Simplified EVSP® and 9000 EVSP®, the costs are roughly in a ratio of 1 : 2 : 4, respectively. Simplified EVSP® packing costs, although somewhat higher than the standard graphite packing set, are more effective in controlling emissions, which tends to offset the differential cost.

Packing Box Depth. In each packing box, regardless of cross section, optimum performance from a 9000 EVSP® set can only be achieved with a full packing set. In long or deep boxes, bushings or bushing rings successfully fill the nonpacked void. However, in shallow boxes, when the compressed 9000 EVSP® set does not fit in the box, eliminating any ring affects performance. This minimum stack height is a particular characteristic of 9000 EVSP® packing that prevents broad application in shallow boxes.

The Simplified EVSP® with the three center graphite rings, shown in Figure 7.25, can be reduced in height by removing a central ring. If necessary, this set can be reduced to one center ring. This flexibility, not available in 9000 EVSP®, is attractive in the simplified set. The set ordered on an emergency basis can be adjusted onsite to fit the packing box. In addition, rings can also be added to the set if the box is longer than the compressed set. This attribute is of importance when the packing box depth is unknown and bushing ring sizes cannot be premeasured and ordered. Extra center rings can easily be added to the Simplified® set, increasing set height.

Complexity. The 9000 EVSP® set contains 11 components which must be installed in the *correct sequence* for optimum set performance. This frequently requires constant referencing by a mechanic to the instructions for proper installation. The 9000 EVSP® design contrasts with the Simplified® set, which only has seven components that include braided carbon end rings. The basic five-ring graphite components of the Simplified® set are "much" simpler to install, since the three center rings can only be installed in the correct sequence, which eliminates the possibility of a packing failure.

Compression Procedure. The 9000 EVSP® set must be installed and carefully compressed in accordance with compression values in special tables or compression pressures converted to gland follower bolt torques. In addition, the 9000 EVSP® set is installed and compressed in two separate stages, which is a complicated procedure compared with the Simplified® set compression operation.

The Simplified® set is designed for installation and compression in one step, eliminating the two-stage compression procedure. In addition, overall compression is based upon a 25% reduction in the uncompressed packing set height or a gland bolt torque requirement. For most installations, compression pressure settings are preferable.

1. *Installation in Deep Packing Boxes.* Where deep boxes are encountered, current information for the standard five-ring flexible graphite packing set indicates additional packing rings are not very effective. As a result, bushings and bushing rings are installed to fill the extra packing box volume. In addition, packing box ports requires movement of the packing stack above or below the port to prevent graphite extrusion into the hole. Another alternate is to fill the port. In general, bushing rings are installed to place the packing set above the port.

In deep boxes, the packing manufacturer suggests adding an additional low-density center ring to the Simplified EVSP® sets. While this contradicts the comments regarding the addition of packing rings to the standard five-ring set, there is a basis for this recommendation. In the standard packing sets, the rings are square in cross section, and most of the compression energy is absorbed in compacting the rings, rather than expanding the rings and providing additional sealing on the ID and OD surfaces. The Simplified® and 9000 EVSP® sets have a much higher ratio of compression pressure translated into expanding or sealing movement. Consequently, an additional low-density center ring may be added to the basic seven-ring stack, which contains three center rings.

While the packing manufacturer indicates one additional low-density center ring can be added, another alternate is to install two simplified sets with one set installed above the first set. This alternate is based on information the manufacturer received on packing performance in various field installations, but the information is qualitative and actual emission data are not available. However, this approach must be viewed with caution until firm data are obtained on the performance of dual Simplified EVSP® packings. The following installations are suggested for deep packing boxes until further information is available:

- A maximum of four low-density center rings.
- Since deep boxes generally have ports or openings, bushings or bushing rings should cover the port or hole to prevent graphite extrusion. An alternative is a lantern ring to cover the port with a packing set above and below the lantern ring.
- With a lantern ring in a packing box, Simplified EVSP® sets can be installed above and below the lantern ring and in the event a void volume still exists, the void should be filled with bushing rings.
- For a Simplified EVSP® set in the lower portion of a deep packing box, the entire set is installed in the box at its proper location and then compressed. Extenders are required between the gland follower and the packing set to compress the set. Following compression, the extenders are removed and a lantern ring is installed with another packing set installed above the lantern ring. The upper set is then compacted as required.

2. *Short Packing Boxes*. While rising stem block valves have average to very deep packing boxes, quarter-turn valves (plug, ball, butterfly) frequently have short or shallow depth boxes. These short boxes require modifications to the packings and two center preformed rings can be removed from the Simplified EVSP set. However, this results in two braided carbon end rings and the shortened center section, which may still be excessive for the packing box. As a result, a thinner braided end ring was developed, as shown in Figure 7.26, and has been termed the simplified short set.

The braided end ring can be a typical carbon braided ring (Garlock 98) that is about $\frac{1}{8}$ in. in thickness. In addition, Garlock also has a braided end ring constructed of PBI® or polybenzimidazole. This material covers an inconel wire to form a strand, which in turn is woven into a braided covering with a center core of carbon braided 98. The PBI ring apparently has better weepage resistance and a $\frac{1}{8}$-in. thick end ring of this material is also available. The PBI ring is fire safe and

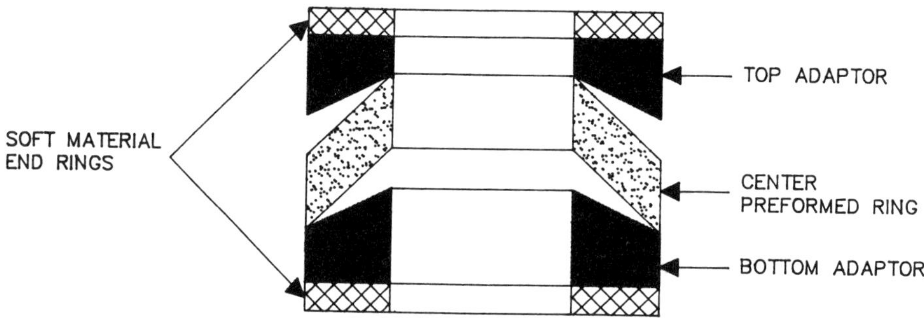

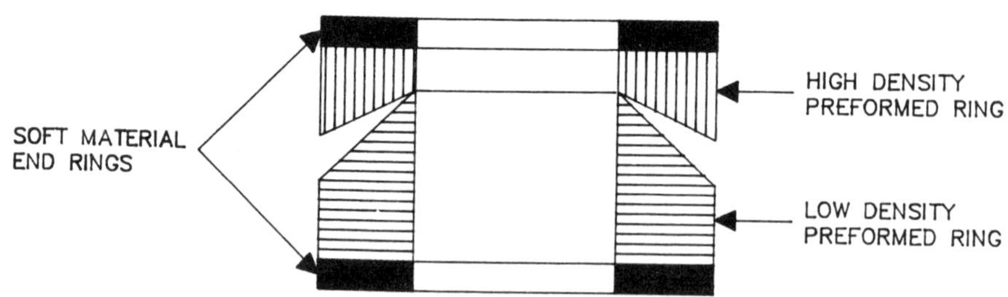

Figure 7.26 *Short or shallow depth EVSP® packing sets (courtesy of Garlock).*

can be used in various services. The packing manufacturer can provide recommendations on end ring requirements. Inhibitors used with PBI are generally moly or tungsten disulfide and compatibility with the process fluid is necessary.

A further reduction in packing height can be achieved with the Simplified® packing design to further reduce packing height for short packing boxes. This newer development, termed Modified EVSP®, is shown in Figure 7.26. There are only two preformed graphite rings with the 45° angle ring density of 100–105 lb/cf (1.6–1.68 gm/cm^3) and the 30° angle ring density of 60–65 lb/cf (0.96–1.04 gm/cm^3). This large difference in densities of about 40 lb/cf results in significant radial expansion and good sealing. In addition, the seal ring is also about $\frac{1}{8}$ in. in thickness and is constructed of copper, minimizing seepage. The very short Modified EVSP® set is a recent development and emissions data for comparison with the other packing sets in similar installations are not available. However, this particular packing set has been installed in various applications.

7.5.5 Nongraphite Packings

A wide variety of graphite packings is produced for high-temperature, high-pressure service, and few packings are available for these service conditions that do not contain graphite.

Below the 550° F (288° C) level, a large number of packing materials and combinations of packing materials are available commercially. The major alternatives to graphite packings in high-temperature service are glass fibers, polybenzimidazole (PBI), and metallics. However, little published information is available on emissions from these packings. In addition, some ceramic packings have apparently reached the commercialization stage. The packing suppliers may have emission data that would be helpful in defining control efficiencies and estimates of potential concentration level reductions in the workplace.

Glass Fiber. This packing is of braided construction with the braids consisting of fiberglass with an inconel insert wound around a proprietary nonasbestos core impregnated with graphite. This packing has a temperature limitation of 1200° F (649° C) and a pressure limitation of 2500 psig (173 bar). However, the allowable pH application range is limited (4–8) and chemical resistance must be reviewed with the manufacturer. In general, this packing should have a reasonably wide range of applications.

Polybenzimidazole (PBI)®. This is high-temperature-resistant fiber that is wound around an inconel wire and other materials to form braids and was described in Section 7.5.4. These braids then form a jacket around a core containing graphite, inhibitors, and other fibers. The maximum temperature for this material is 1200° F (steam), but the packing can also be installed in services with pressures up to 2500 psig (173 bar). Chemical resistance is considered superior, and the packing has various applications in the chemical industry. This packing is also fire safe.

Metal. Metal packings consisting of braids and foil have been available from packing suppliers through the years. These packings may contain a metal insert and are applicable in high-temperature/high-pressure service with allowable temperature levels to 1800° F (982° C). Metal packings can withstand high pressures at elevated temperatures, which is not always possible with nonmetal packings. Very little emission data are available on packings of this type.

In applications with process stream temperatures below 550° F (288° C), a wide range of packings are available in addition to graphite packings, although many packing materials are combined with graphite. However, little published emission or comparative emission data are available on the effectiveness of these lower temperature nonasbestos packings. Various valve packing suppliers have emission test data, but the purchaser must compare the test emission data with EPA uncontrolled emission factors to determine control efficiencies for estimating emissions rates. Some of the major constituents of these packings are TFE, PTFE, Kevlar®, Neoprene®, Viton®, Aramid® fibers, and Kalrez®.

Although a large number of packings are available, a potential problem with several packings is their inability to remain unaffected in retaining sealability during a fire. Any packing recommended for emission control should be reviewed with operations and design groups to insure that it meets plant safety standards. In several of these packing sets, a graphite ring has been added to meet fire test requirements.

Some of these packings are reviewed in Chapter 8, which covers control valves.

7.5.6 Live Loading

In valves, compression loadings are placed on packings with the gland follower that is tightened with the nuts on the gland follower bolts shown in the various figures. The pressure on the packing may be several thousand pounds per square inch, when initially compressed. With time the packing gradually consolidates and the actual pressure on the packings may decrease, thereby permitting an increase in leakage and emissions. Maintenance of compression pressure is accomplished by placing disk springs between the gland follower and the gland follower bolt nuts. These springs maintain a pressure on the gland follower and in turn the packing. These springs have resulted in the term "live loading" since the spring continually applies pressure.

Live loading began to appear some years ago when asbestos was the standard packing. Asbestos loses water of hydration and shrinks in volume, requiring gland follower adjustments to maintain packing pressure and control leakage. The disk springs or Belleville washers were helpful for maintaining compression pressure on the asbestos packings, particularly in deep boxes with a large number of asbestos packing rings.[36] However, their effectiveness on graphite packings is questionable, since the maximum consolidation observed with graphite packing is 2.5%, which may occur after extensive service and cycling. A very large number of cycles will only occur in control valves or on–off valves that are fully cycled at least several times per hour. In block valve service, the block valves may only be opened and closed in a full cycle a few times per year in a continuous process operation. Where batch operations are part of the process, blocks may be cycled much more frequently, or in plants that have shutdowns or turnarounds every month. However, block valve cycling is considerably lower than control valve cycling, with consolidation considerably less than the 2.5% maximum mentioned for control valves.

Current operations require monitoring valve leakage on a periodic schedule and block valves are adjusted as required to control excess emissions. Live loaded valves must also be monitored and adjustments may still be required, depending upon the compression pressure placed on the packing initially. If a graphite combination packing set is compressed about 25% or to a predetermined pressure, the compression pressure placed on the packing cannot exceed the pressure necessary to achieve this degree of compaction. Obviously, the springs will not increase the pressure level should the packing require further compression. In this situation, the gland bolt nuts must be tightened. A very high initial compression pressure reduces the amount of adjustment possible in graphite packings and does not aid in controlling leakage. Disk springs may maintain in approximate fashion the original loading pressure, but will not exceed this pressure since spring expansion basically results in a lower pressure when a packing set consolidates. Additional pressure is necessary to increase radial expansion and seal the valve. For graphite packings, the live loading springs or washers apparently do not provide any advantage when the graphite packing sets have been properly designed, constructed, installed, and compressed.[36, 37]

Live loading is of assistance in certain applications and should be considered under certain circumstances. Thermal cycling with some packings requires springs to maintain a more or less consistent loading. Other nongraphite materials and packing configurations can be more effective with live loaded gland followers such as vee-type rings. When valve stems cannot be moved or actuated during an outage, the valve location is inaccessible for adjustment, standard installation procedures cannot be observed, and springs aid in improving the packing fit and adjustment. Where the valve stem is in a state of constant modulation or a large number of complete stem actuations (> 5000 cycles) may occur before repacking, live loading can be helpful.

The addition of live loading to a packing installation significantly increases costs and should be carefully investigated before being implemented.

7.5.7 Sealant Injection

Sealant injection is a recognized method of controlling valve packing emissions in onstream valves and is included in EPA regulations as a recommended emission control technique (see Section 7.1.1). All valves that are on line and leaking by EPA definition, but cannot be removed from process service without a unit shutdown, are candidates for sealant injection. These valves can be in liquid or vapor service and cover a wide variation in process operating pressures and temperatures. Although relatively new (early 1980s), this control technique has been used extensively in the petroleum refining, petrochemical, and chemical industries as a method of controlling on-line leakage from valves.

While sealant injection has reduced the monitoring concentration well below the 10,000-ppm leakage level, its efficacy for valves categorized as leakers at 500 ppm is less well understood. Basically, the injection should reduce emissions to a level below 100 ppm, but published information at this level is not available. In addition, the time duration the injection is effective is an important consideration. Where valve emissions concentrations are reduced from 10,000 to 1000 ppm or less, sealant effectiveness can last from 6 months to several years in continuous operations until the monitoring concentration reaches the 10,000-ppm level. At this point, reinjection of sealant is required until the valve is repacked during turnaround. The effectiveness time period is a function of the type of sealant and injection procedures.

When sealant is injected into 500-ppm leakers, the duration effectiveness of sealing is unknown. However, reinjections will probably be required somewhat more frequently with the same sealant. This type of maintenance operation becomes costly and frequent sealant injections increase the probability of an accident or safety incident since sealant injection occurs with the valve on line.

The leaking valves are frequently gate and globe valves that may not have lantern ring holes or ports. Injection ports are necessary for the sealant and where ports are not available in the leaking valves, the valve bonnet is drilled and sealant is injected. Companies experienced in these operations generally supply the drilling and sealing services necessary for leaking valves. However, sealant compatibility with the process fluid is required and must be established before injecting any material into the valve. This is of particular importance in chemical process units.

7.5.8 Lantern Rings

Lantern rings have been shown with various packing combinations and are frequently recommended in control valve service, which is reviewed in Chapter 8. Lantern rings in packings have frequently been installed to split the packing rings and allow drain or collection points for process fluids leaking through the packings on the process side. The second set of packing rings between the gland follower and lantern ring is normally provided as a safety ring set to prevent leakage to the atmosphere. Although lantern rings are frequently installed, they are only connected infrequently on block valves through the external opening to a collection system. Lantern ring collection systems are often removed during maintenance, down time, or repairs, and the valve is not reconnected to the collection system. In certain situations, a large number of lantern ring collection lines may be required, and the system becomes rather extensive and cumbersome, resulting in decisions to eliminate the system. Moreover, collection systems are a significant investment, increasing overall control costs. As a result of the lack of collection systems, data are generally not available on the efficacy of this device in block valves to control atmospheric leakage although control valve manufacturers have actively investigated lantern ring performance and frequently recommend lantern ring installations.

With lantern rings, two sets of packings can be installed in the valve, and the ring acts as a spacer. Although the additional packing set presumably increases packing effectiveness, a properly designed and installed single packing combination set is highly effective as described and can readily meet a maximum emission concentration of 500 ppm.[15] In reference 15, the only valves with an excessive leak frequency are gate valves, indicating that the large majority of valves are satisfactory with single packing sets. Where valves have excessive leakage, the problem may not be completely solved with lantern rings, particularly in valves having bowed stems or misaligned packing boxes. Valve mechanical condition is very important in obtaining optimum packing performance. Consequently, valves require a thorough inspection, which may indicate the need for a complete overhaul or replacement before emissions can be properly controlled.

In special situations where the process fluid is highly toxic, a lantern ring with a bleedoff system may be desirable. With a very low TLV fluid, bleedoff can be effective in maintaining an essentially zero emissions concentration. However, the bleedoff system must be properly designed to provide a receiving location with a much lower absolute pressure level than the valve fluid pressure. In addition, this collection system must be in operation continually when the valve is in operation.

Additionally, lantern ring ports have provided a method of determining when the lower packing set is leaking excessively. The lantern ring operates as a telltale for this function. Although this may be useful in some situations, inspecting or monitoring the external port operation for leakage can be a potential source of exposure and is somewhat hazardous. For telltale operation, the port outlet must be equipped with a reliable bleed valve that has a satisfactory support system. In determining leakage with a telltale, a pressure indicator will indicate leakage changes but may not be sufficiently sensitive at low emission rates. Sensitive pressure devices may be required for satisfactory indication of changes. A sampling bomb connected to the bleed valve over a defined time period will provide an indication of liquid leakage accumulation against a base level. Vapor leakage comparisons are also possible by determining the rise in pressure in the sampling bomb over a fixed timeperiod. The sampling bomb is normally evacuated and in a fixed volume the final pressure level permits a determination of the vapor accumulated.

As described in Section 7.5.7, the lantern ring and associated inlet also provide a sealant injection port. In addition, lantern rings are frequently used as spacer bushings in deep packing boxes. These spacer rings are often provided with control valves which usually have deep packing boxes. The alternative is the installation of bushings that are generally machined or obtained by the local plant.

7.6 PRESSURE RELIEF DEVICES

Pressure relief devices or pressure relief valves are also termed "safety valves" in the chemical and petroleum industries. The valves are installed to provide overpressure relief in liquid or vapor service. Relief in mixed vapor liquid service is unusual, but liquids can vaporize in the relief valve as the pressure falls through the valve. Relief valve designs vary, but the valves operate on the same general principle. A disk or surface is held against process pressure by a calibrated spring and when the spring set pressure is exceeded by a high-process pressure, the disk moves, releasing process fluid. This action relieves the process and the disk is reseated when the process pressure falls below the set pressure. However, disks do not always reseat properly, and small openings result in emission leakage. The EPA regulations reviewed in Section 7.1 are designed to control leakage before and after an emergency release. These regulations are similar to the regulations issued in the NSPS for VOC emissions control.

Liquid relief valves are generally discharged through piping into a closed system. Liquid discharge to the atmosphere is a potential source of exposure in the workplace and can also be a fire hazard. Moreover, a large volume of liquid released to the atmosphere requires containment and cleanup which can be time consuming. Industrial hygiene supervision is necessary to minimize and control exposure in this situation. Following liquid discharges, the valve must be monitored after an emergency release to the atmosphere. In contrast to liquid discharges, which are normally contained in a closed system, gas or vapor discharges from pressure relief valves are either sent to a closed system or discharged to the atmosphere.

Gas or vapor discharge to a closed system is not a problem from a regulatory standpoint, but a pressure relief valve discharge to the atmosphere is regulated because of fugitive emission losses to the atmosphere. The fugitive emission concentration in a vapor outlet to the atmosphere cannot exceed 500 ppm above the background level in the NSPS and HAP regulations. Following an emergency release, the relief valve vent is tested and must be in compliance with the 500-ppm restriction within 5 calendar days after the emergency release. This is a difficult level to attain in typical pressure relief outlets. Methods of achieving compliance are presented in the following sections.

7.6.1 Emergency Atmospheric Releases

Although regulatory requirements restrict fugitive emission flow to the atmosphere, the emission control regulations do not restrict emergency releases to the atmosphere. The emergency release is a very infrequent occurrence, but requires consideration of potential workplace exposures during the short release period. Dispersion of toxic materials from the relief vent can impinge on the ground and the concentration depends upon various factors including temperature, vent velocity, mass rate, and vent exit location. The emergency relief valve release period may extend for a relatively short time (5–15 min) but a dispersion analysis for each atmospheric emergency relief vent on the unit is necessary. The ground level concentration (glc) in the plant is estimated, and the concentration in the adjacent community should be determined to ensure excessive concentrations will not occur. Since community and worker risk and hazard assessments are required, glc's are necessary along with the assessment results. If the glc and assessments indicate potential exposure concerns, then the vent stack must be redesigned to reduce the glc to appropriate levels. Redesign or modification may require a change in vent diameter, vent exit location, or both. Frequently, a change in vent exit location requires raising the vent exit elevation.

In addition to determining the emergency ground dispersion pattern from a relief vent, the general vapor dispersion pattern throughout the unit should be investigated. The relief valve vent exit may be located below the elevation of nearby work platforms and workers on an elevated platform can be exposed to an emergency vent gas release. The emergency relief vent dispersion pattern across these platforms can be calculated, and an estimate of the platform concentration determined. If the concentration level is higher than recommended short-term exposure limits, various alternatives are investigated to insure a safe, cost-effective solution:

1. *Elevating the Vent Exit Location*. Although relatively simple, adding height to a vent can be expensive. The vent exit size on the relief valve is much larger than

the inlet and installing sections of vent to increase vent height may require a large structural support system.

2. *Workplace Personnel Exposure Protection*. This approach requires consideration of vent gas toxicity, type of personal protective equipment, and implementation of control procedures to insure personnel observance of recommended industrial hygiene practices. As an example, personnel ascending the structure above a certain level could be required to carry a portable respiratory mask suitable for protection against the vent gas. In an emergency vent release, the worker would don the mask and proceed to a lower elevation. The noise from the safety release should be sufficient to warn the worker. As a supplemental warning, a local vent gas alarm can be located on an exposed platform to provide a specific alarm.

3. *Enclosing the Relief Valve Vent*. A full discussion of this solution is presented in Section 7.6.4.

7.6.2 Rupture Disks

One of the alternatives considered by the EPA in meeting the 500-ppm relief valve vent concentration level is the installation of rupture disks upstream of the pressure relief valve as shown in Figure 7.27. The rupture disk is a circular metal disk, slightly dish shaped, that does not leak and is designed to burst at a specified pressure. The disk separates the process fluid from the relief valve and thereby

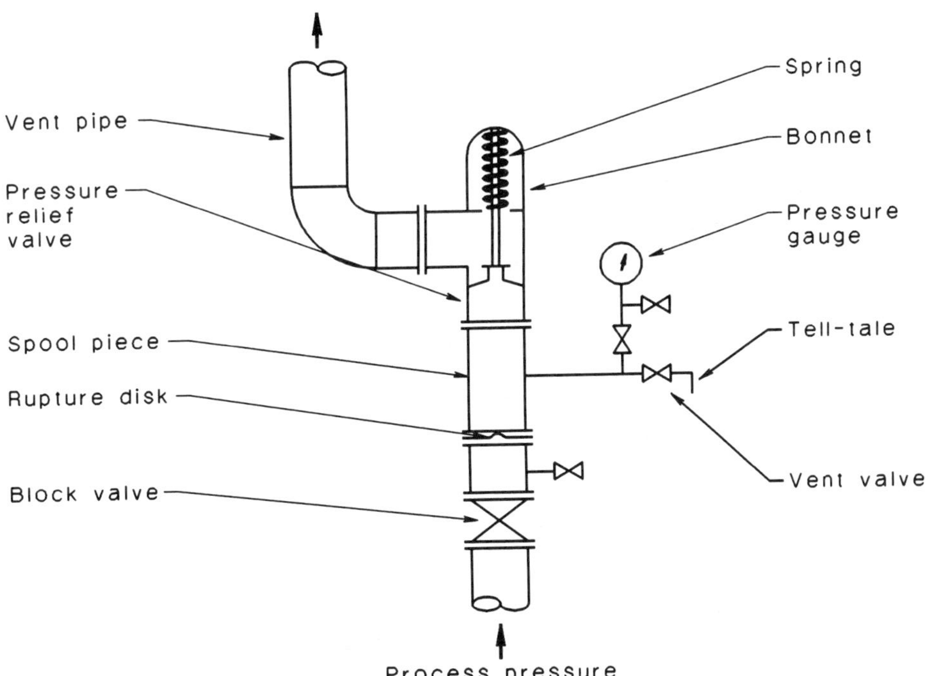

Figure 7.27 *Pressure relief valve with upstream rupture disk.*

prevents leakage through the safety valve. This leak prevention capability is a method of achieving the EPA target level of a 500-ppm maximum emission concentration of VOC or HAP in the vent pipe.

The rupture disk is placed in a special flange that is generally installed between two regular flanges. An opening between the piping is specified by the rupture disk supplier to accept insertion of the rupture disk flange. The rupture disk is in turn separated from the relief valve inlet by a short spool piece, as shown in the figure. Upon rupture, the disk splits into a number of wedges that remain attached to the disk. These pieces could potentially interfere with the relief valve inlet section, but the spool section inserted between the relief valve and rupture disk provides extra space or volume. Since rupture disk technology has improved, the spool piece may be eliminated with certain disk types, and the telltale shown in Figure 7.27 can be connected through a flange to the zone above the disk. Telltale operation is described in the next paragraph.

The rupture disk is normally a thin metal disk that does not leak as described previously. However, the disk may develop a minute leak and the spool piece may fill with vapor. Liquid in the form of condensation can form owing to a relatively cool spool piece or vapor expansion. A buildup of pressure occurs in the spool piece until the pressure is equivalent to the upstream process pressure. As process material fills the spool piece, the loss of disk integrity must be identified to insure replacement of the rupture disk. A telltale, shown in Figure 7.27, has been recommended by the EPA to indicate leakage through the disk. The vent valve can be opened directly to determine whether any leakage has occurred, but this method increases the probability of exposure since the senses may fail to identify leaking material. One technique to indicate rupture disk leakage is a measurement of the pressure in the spool piece with the pressure gauge shown in the figure. Leakage results in a pressure buildup that is readily identifiable on the gauge. The gauge is then depressurized after a reading through the bleeder valve. Venting through the bleeder and opening the telltale valve must be carefully managed to prevent overexposure when the process fluid is highly toxic. Discharge of any vapors during bleedoff into a sealed, portable container minimizes potential exposures. The sealed container can also be used as a method of obtaining a sample that would also indicate leakage.

If the rupture disk must be replaced as a result of leakage or an emergency release, the upper and lower spool pieces can be filled with vapor, condensed liquid, or a combination of both. A small bleeder valve just above the block valve should be installed to drain any liquid. The bleeder valve can be connected to a special container to prevent the release of toxic fumes or toxic liquid to the atmosphere. This may require several container fillings to effectively reduce the liquids and fumes from the spool pieces. Alternatively, a connection to process equipment if acceptable would minimize process material in the spoon sections and is a desirable solution. Evacuation of the spool piece where possible is preferable if the fluid is highly toxic. In some block valves, a boss on the valve permits drilling and insertion of the bleeder directly into the block valve at a lower level than shown, resulting in removal of additional fluid from the valve. When the pressure falls to the atmospheric pressure level, the small quantity of vapor remaining in the spool piece can be purged to the atmosphere, although evacuation is preferred, and both the leaking or burst rupture disk and safety valve can be removed for replacement and maintenance.

Another method of determining rupture disk failure has been developed which does not require opening the telltale valve. A flat sensor similar to a thin sheet is placed immediately above the rupture. This sensor consists of a conductive foil laminated between thin polymeric sheets. The edge of this material is fitted with flat gaskets to minimize thickness. A small electric current flows through the foil and any size leak in a rupture disk will also break the conductive foil and interrupt current flow, indicating a rupture has occurred. When a rupture is indicated, both the rupture disk and the sensor are replaced. Although this device minimizes potential exposures that may occur with a telltale, the system becomes complicated in a unit with a large number of disks requiring maintenance and attention. However, the sensor unit can be installed where disks have been a problem. Temperatures for the sensor are limited to a maximum of 500° F (260° C).

In the event of leakage or pressure relief, the rupture disk installation must be replaced, which is a potential source of exposure. However, the rupture disk can be replaced with another device that does not require removal. The nonspring relief valve described in Section 7.6.5 can be used as a rupture disk rather than a relief valve. Should the nonspring valve lift in an overpressure situation, the extension pin bends and the valve remains open until the external pin is replaced. This procedure eliminates the need to replace the mechanism. The major concern in this system is the holdup of liquid or vapor in the intervening volume between the nonspring valve and the relief valve after the nonspring valve has been reset. Safety may require removal of this material, but the simpler installation compared with the rupture disk should have a smaller holdup volume, reducing potential disposal and exposure problems.

7.6.3 Seat Sealing

A simplified sketch of a relief valve is shown in Figure 7.28, which contains the essential elements of a conventional relief valve. The disk opens when the spring force is overcome by the differential pressure across the disk. The spring pressure on the disk normally results in a relatively close fit of the disk and seat. However, some leakage generally occurs since the disk and seat are not mated perfectly under normal conditions. Polishing the faces will reduce leakage, and the concentration standard can be achieved. Relief valve manufacturers have improved disk seating to reduce leakage. As a result, a concentration less than 500 ppm above the background level in the relief valve outlet is possible, but requires considerable effort.

Machining and surface grinding the disk and seat faces particularly on existing relief valves can successfully reduce outlet emission concentrations below the maximum EPA specification. Polishing and grinding is a tedious process since the relief valve must be tested to demonstrate that it does not leak prior to installation. Following an emergency release, the relief valve disk rarely reseats completely, and the emission leak generally exceeds the 500-ppm monitoring limit. The relief valve must then be removed and resurfaced to meet the emission standards.

The EPA earlier suggested on O-ring in the relief valve disk or seat to control fugitive emissions. However, the O-ring has temperature limitations that reduce the general applicability of this type of relief valve unless the valve is carefully classified and installed to insure the temperature is not exceeded. Also, any O-ring must be compatible with the process fluid. Information on relief valve emissions is

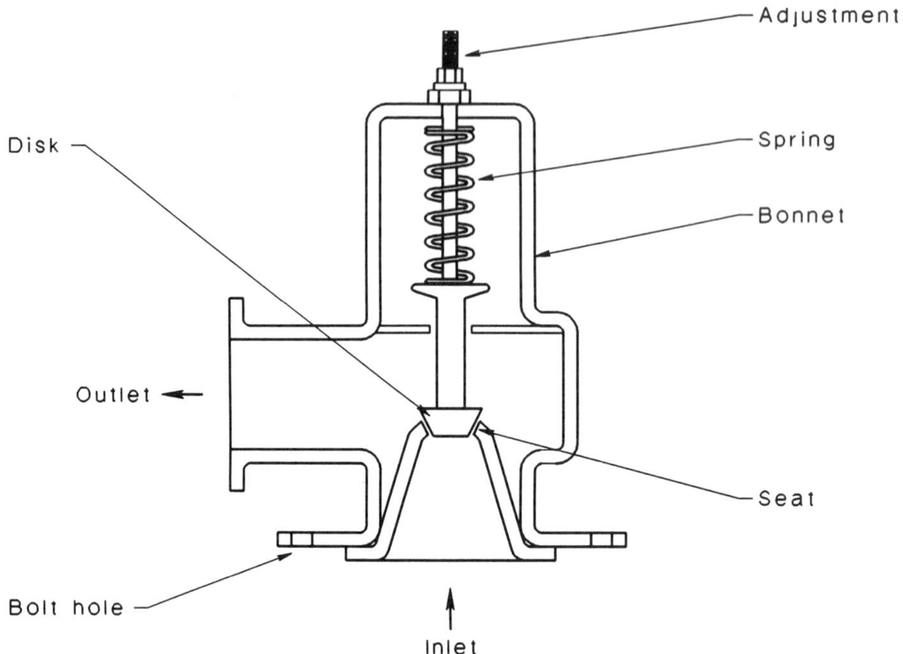

Figure 7.28 Simplified cross section of a conventional pressure relief valve.

limited, and the manufacturers of relief valves with disk and seat seals may have additional data on leakage emission.

7.6.4 Closed Discharge

Relief valve discharge into a closed system and a control device exempt relief valves from regulatory EPA requirements (NSPS and HON) and eliminate potential exposures associated with open atmospheric vents. Open vent exposures may occur as a result of monitoring, emergency release dispersions, or relief valve purging and testing in the rupture disk configuration. Although recommendations for closed discharge connections are obvious for highly toxic materials (acute, carcinogenic) or HAP, closed discharge connections and emission control devices require a considerable investment compared with atmospheric relief valves. Industrial hygiene considerations and investment costs should be carefully analyzed in formulating recommendations for relief valve emission controls since emergency releases are involved in the analysis. Decisions on relief valve discharge controls are more difficult for existing units than discharge recommendations for new units. Installation of closed discharge systems in existing units can be difficult mechanically and a control device is necessary, which generally results in a large investment covering the expansion of the existing control device (flare, incinerator, etc.) or construction of a new control device. Historically, existing control devices have operated near capacity levels and increased loadings have required expansion or purchase of new equipment. Closed relief systems in new units are normally included in the design since emergency releases must be controlled in many locations to meet local, state, or federal regulations. The installation of additional closed relief systems to eliminate atmospheric discharge during the design of new units is generally evaluated on an incremental basis. Normally, the overall investment increase is not very large.

Wherever possible, closed discharge systems are recommended to minimize potential workplace and community exposures. Moreover, recovery of a considerable portion of the leakage and discharge vapors are possible through various systems in the relief valve vent lines.

Closed Emergency Releases. Although closed relief systems are attractive as a means of encapsulating fugitive emissions, a major consideration is the emergency release. As an example, if fugitive emissions were sent to a separate flare, the design basis for the actual fugitive emission flow rate would result in a small diameter flare. The flare diameter in this situation, although quite small, would probably be increased to insure mechanical rigidity. A 2-in. flare may be satisfactory for destruction of the emissions, but a 4- or 6-in. diameter pipe would probably be installed to meet the requirements for a self-supporting rigid structure necessary for an elevated flare. The emergency release rate contrasts dramatically with the fugitive emission leakage rate from relief valves. The estimated pressure relief valve fugitive emission leakage rate in chemical plants is 0.228 lb/hr/source (0.104 kg/hr/source) compared with emergency release rates that may be as high as 100,000 lb/hr (45455 kg/hr) or greater. This huge disparity in rates is the major concern in designing closed pressure relief valve discharge systems. The emergency relief condition is the obvious basis for system design. Separating the emergency relief flow from the small constant fugitive emission flow rate is a possibility with the emergency stream discharged directly to the atmosphere. However, the fugitive emission flow must be returned to the small flare or control device after the large emergency release is dissipated and the emergency is under control. The emergency flow release in this situation must discharge at a location that meets dispersion and risk assessment requirements. However, closed emergency relief has an advantage in minimizing potential exposure conditions or losses to the atmosphere and should be considered wherever possible.

Although flares were described as a control device in the example, various types of incinerators or other control systems are also alternatives. However, any control device or system selected for fugitive emission control must be capable of processing the enormous change in emergency flow rates previously described. Flares and various control devices are designed specifically for wide flow variations but must be specified properly to insure they operate under all conditions. The control devices or recycling equipment installed in accordance with the regulations should be investigated by the industrial hygienist to determine whether potential exposures may occur in the operation and maintenance of these systems. Review and recommendations for industrial hygiene control must encompass the complete system and should not be limited to the pressure relief valve.

Pressure Relief Valves. In closed systems, relief valves are generally not of concern since both fugitive and emergency emissions from the valves are enclosed. However, an understanding of valve operation is important since potential exposure problems can arise from the type of relief valve selected for the service. In the typical relief valve shown in Figure 7.28, the emergency release inlet pressure desired for relief is normally set on the valve in the maintenance shop. The backpressure on the valve is atmospheric pressure and the valve disk differential pressure is essentially the set pressure. With an inlet set pressure of 100 psia (6.9 bar) and an atmospheric backpressure of 14.7 psia, the differential valve pressure would be 85.3 psi (5.8 bar). The major criteria in valve operation is the differential pressure that provides the necessary force to compress the spring and open the disk. When the backpressure rises above the atmospheric pressure level, the addition of the differential disk relief pressure to the increased backpressure results in an increase in the pressure at which the valve disk will open and relieve the process system. An example based on the differential pressure of 85.3 psi (5.8 bar) just described assumes a backpressure increase to 24.7 psia (1.7 bar), which then effectively raises the set pressure to 110 psia (7.6 bar). The increased set pressure endangers process system safety and is a major design concern.

Relatively high backpressures occur in closed systems, particularly in existing plants where a number of units may have relief valves connected to a common control device such as a flare or an incinerator. Higher processing rates and unit expansions generally increase control system loadings and during emergency conditions the backpressure on the relief valve disk from the closed control system may rise to unacceptable levels. Although the backpressure is relatively low and may possibly reach a level of 25–30 psia (1.72–2.1 bar), an original set pressure of 25 psig (2.76 bar) would be increased to 35 or 40 psig (3.4–3.8 bar), which may pose a potential safety problem. Backpressure fluctuations in the downstream circuits can be eliminated by the use of a bellows relief valve which has an atmospheric backpressure and is indifferent to downstream or exit pressure variations. A sketch of a bellows relief valve is shown in Figure 7.29.

The major difference between the relief valves shown in Figures 7.28 and 7.29 is the bellows attached to the disk in Figure 7.29. The bellows prevents contact of the upper disk face with the backpressure vapors from the closed system. The face of the disk enclosed by the bellows is open to the atmosphere through a vent in the bonnet and in this design the disk always has atmospheric pressure on one face and process pressure on the other face. Thus a set pressure for the relief valve will not change, and the setting can be specified precisely for the valve. However, a potential exposure problem exists with bellows valves that are installed in both liquid and vapor service. The bellows may develop a small leak or fail and various materials can be emitted to the atmosphere through the vent. The material leaking through the bellows are contaminants from the downstream or closed-system side of the relief valve, which may contain a wide variety of toxic materials. Although a bellows may leak or fail, the valve will still relieve the process system since the spring will function and the valve will continue to operate as a relief valve until removed for maintenance. Should a bellows valve be actuated when there is a small leak in a bellows, process material from the high pressure side will also tend to leak. Identification of the leak in the early stages initiates maintenance and replacement procedures.

Since the bellows may leak or fail, location of the bonnet vent is important as a leak identification indicator for operations and a potential exposure source for industrial hygienists. A metal tube or very small diameter piping is run from the

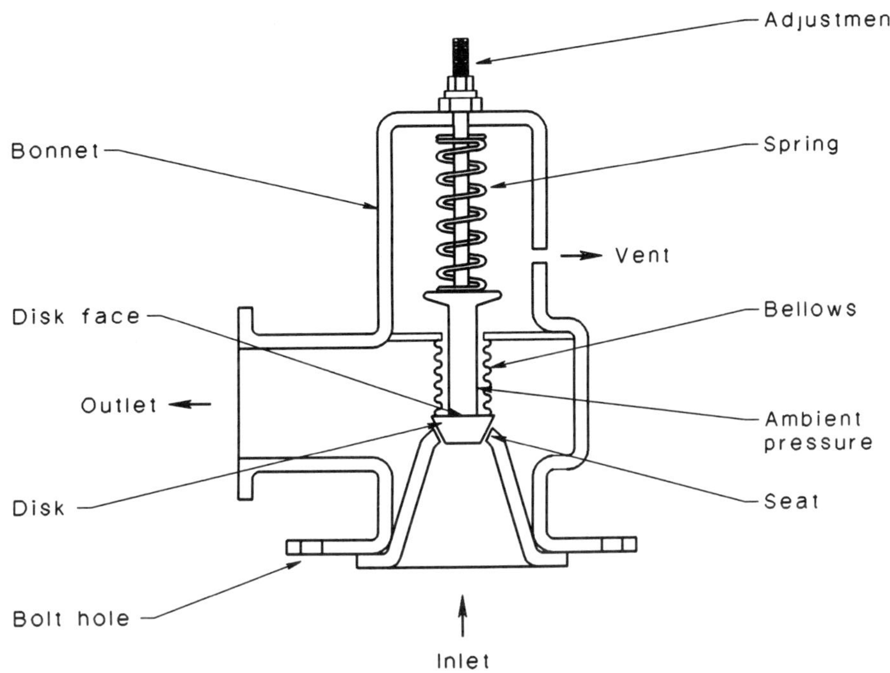

Figure 7.29 *Simplified cross section of a bellows-type pressure relief valve.*

relief valve bonnet to a location where the open tube end can be observed periodically. Frequently, the pipe end is located near a structure and elevated well above ground level. Since bellows valves are installed in both liquid and gas or vapor service the following guide provides recommendations on controls for the bellows vent:

1. *Nontoxic vapor service (air, steam, etc.).* Discharge should be located to prevent direct impingement on personnel.

2. *Toxic vapor service (no immediate danger to life and health, and low chronic toxicity).* The vent discharge point should be located where dispersion of vapors over work platforms and work areas is minimized for very low leak rates or a bellows failure. These rates can be estimated in consultation with the process designers. However, vent testing may still require protective equipment, depending upon the vent installation. The final decision on protective equipment can only be formulated based upon the proposed design. Instrumentation to detect leaks is a possibility, and an instrument with a high degree of reliability is necessary or the worker is then dependent upon an instrument that can easily malfunction.

3. *Toxic liquids (vapor IDLH is not a concern and the liquid is not an acute dermal toxicity material).* Where the liquid does not vaporize or form aerosols, the vent can discharge into a large funnel that is connected to a process drain or sewer via a sealed connection. If the liquid forms an aerosol or vapor/liquid mixture, the vent line should be discharged into a small enclosed pot. Aerosol carryover from

the pot can be minimized by installing a crinkled wire mesh screen in the pot above the bonnet vent inlet. The vapor is then discharged at a location that minimizes exposures.

The discharge side of the relief valve is generally filled with liquid from the low-pressure receiver to which the relief valve is designed to discharge high-pressure fluid. Consequently, when the bellows is cracked or leaking, the fluid from the discharge line leaks into the bellows and is discharged from the vent port. The vent port line should be designed for a bellows failure flow rate with the downstream liquid and this liquid should be discharged into a closed pot that is connected to the closed drain system through a seal. A small level glass can be installed on the pot to indicate leakage, which can be shown by placing a seal weir above the bottom of the pot.

4. *Acutely toxic vapors and liquids, high chronic toxicity* (immediate danger to life and health, carcinogens). For acutely toxic vapors, the vent should be located well above the highest work platform with the final height and location defined by dispersion calculations. In vapor vents, a reliable instrument should be installed to indicate vapor leakage.

For acutely toxic liquids, the line should discharge into a small pot and the liquid from the pot is then connected through a seal to a closed drain. Instrumentation and a small level glass are installed to determine whether the bonnet vent is leaking. The pot and drain should be designed for a bellows failure. The vapor vent from the pot is discharged to the atmosphere at a location that minimizes exposures.

Bonnet vents normally do not leak, and EPA emissions regulations (HON and NSPS) for vents are not applicable to this type of installation. However, a review of the installation with regulatory authorities will clarify any requirements. When a leak develops, the valve should be removed and the bellows replaced. As a result, the amount of overall leakage will be very small. Industrial hygienists should examine vent discharge locations, disposal, and designs.

7.6.5 Nonspring Relief Valve

Normally, relief valves depend upon a spring force as the sealing mechanism which provides the resistance that must be exceeded for pressure relief. A relief valve is now available that replaces the spring with a long metal rod or pin. This pin holds a piston against the valve seat, as shown in Figure 7.30, until the set pressure is reached and the pin bends or buckles, raising the piston to the full open position and thereby relieving the pressure.

This design is based upon pin operation within the metal elasticity range, which has been termed Euler's Law. There is a constant modulus of elasticity within the range of pin operation. The pin and replacement pins are machined within close tolerances to insure that the valve relieves at the proper pressure. Following the relief operation, the bent pin is replaced and the valve is resealed. As shown in Figure 7.30, the valve seat has a piston seal to prevent leakage and a stem seal that controls any leakage around the stem. The holding nut releases the bent pin and is adjusted when a new pin is installed.

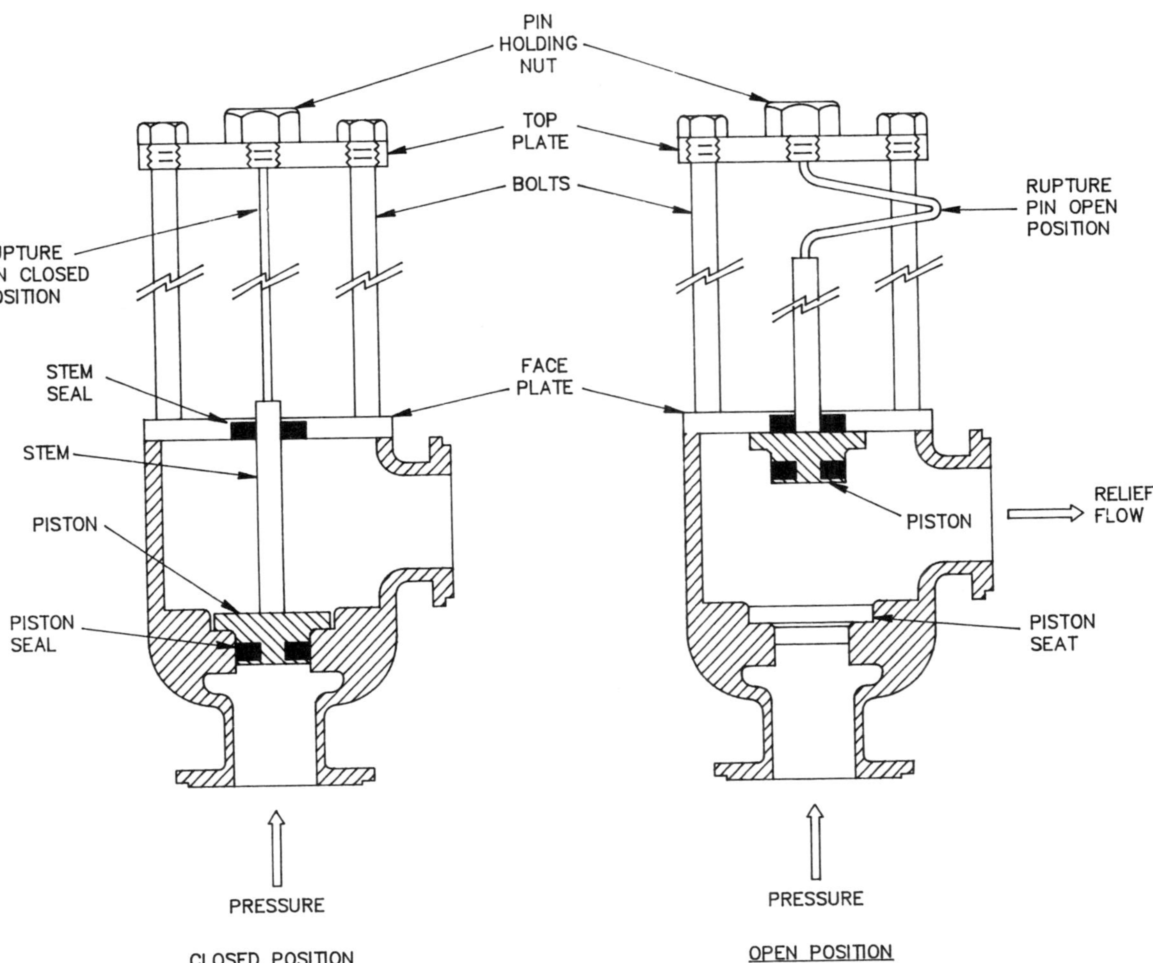

Figure 7.30 *Simplified cross section rupture pin relief valve (courtesy of Rupture Pin Technology).*

Currently, the relief valve operates to temperatures of about 600° F (315° C) and pressures well in excess of 10,000 psig (190 bar). At the 600° F level, the seals are Kalrez®, but metal to metal sealing has been under development and is apparently available. For various applications, the relief valves must be identified to ensure they are installed in the proper location.

The advantages of using this type of valve are:

1. Accurate set point pressure. Normal set point pressure is ±5%, but accuracies as low as ±1% are available from the manufacturer. In addition, the set point pressures can readily be changed by installing a different pin. These valve set pressures also have high set point accuracies for pressures less than 50 psig (4.5 bar).
2. Very low fugitive emissions leakage. Current installations are apparently well below the 500-ppmv leakage regulation.
3. Pin replacement does not require valve removal and retesting. This is a problem in locations where regulations restrict opening lines.
4. Pin removal and replacement is completed within about 30 min, significantly reducing maintenance time. Spare rupture pins are stored in the valve housing for rapid repair.
5. Rupture disks are not required with this valve, eliminating disk replacement should the disk leak or the valve open for relief. Also, the definition of set point pressures based upon a combination of rupture disk and relief valve setting accuracy is not a concern. Removal of the rupture disk also eliminates the need for telltale systems. Potential disk exposures associated with rupture disk operations are eliminated. This device can also function as a rupture disk (Section 7.6.2).

In addition to the standard valve shown in Figure 7.30, a bellows system is also available with the nonspring valve that permits relief valve operation where the valve must relieve against a backpressure. The enlarged piston in this design only has system pressure in the vertical direction against a constant atmospheric pressure at the other end of the piston. This balanced valve design, shown in Figure 7.31, is frequently installed in liquid service. A distinct advantage with a nonremoval relief valve is that liquid leakage associated with valve removal is eliminated, and it can be reset in position. The balanced valve in gas or vapor service eliminates emergency releases to the atmosphere when connected to a closed system.

7.7 OPEN-ENDED VALVES

The regulatory requirements for open-ended valves and lines are presented in Section 7.1.4. Installation of a blind or cap on the end of a line should be recommended as standard practice where NSPS or HON regulations are not applicable. Leakage through the last valve on the line can present potential localized exposure concerns. The requirement for caps or blinds to reduce fugitive emissions will not have a serious impact on overall fugitive emission losses since

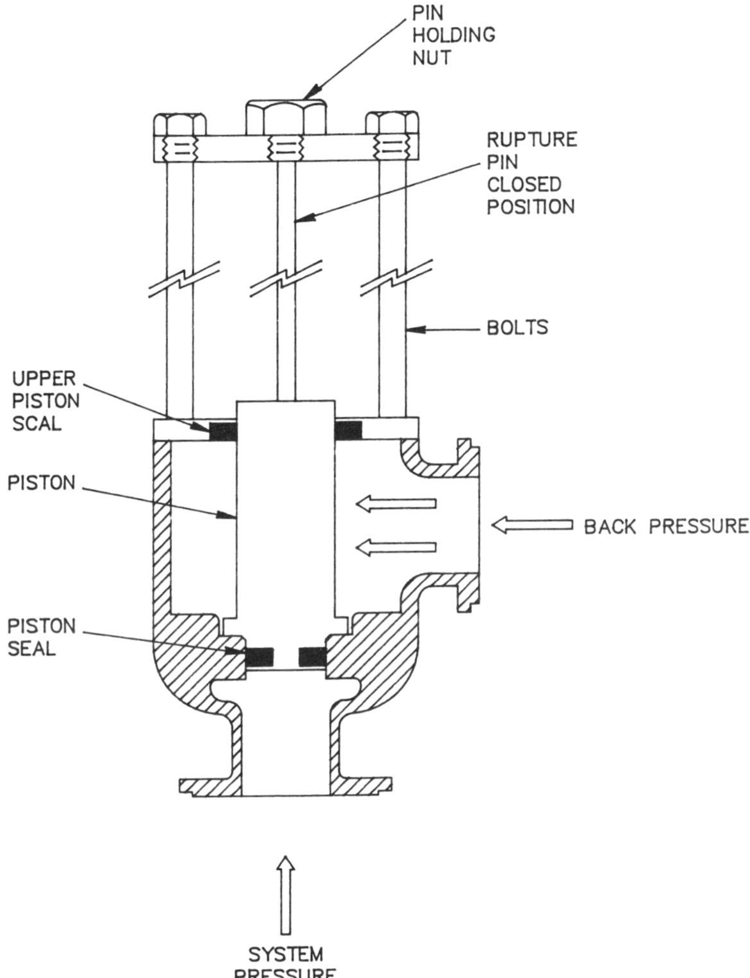

Figure 7.31 *Simplified cross section balanced rupture pin relief valve (courtesy of Rupture Pin Technology).*

these losses are quite small. The emission rate per source is estimated as 0.0034 lb/hr (0.0017 kg/hr) and there are only a few sources in process units.

In double block or bleed operations, the bleed valve is permitted to remain open when the blocks are closed to vent and indicate any leakage. This valve block system is a telltale arrangement that prevents leakage of high-pressure material into a low-pressure system by discharging any leakage to the atmosphere or a receiver, as shown in Figure 7.32. In typical installations, leakage through the bleeder to the atmosphere is a sign or telltale indication that one or both of the block valves leak. The bleed operation is an industry standard that has been used for many years as a safety measure. The regulations permit the use of double block and bleed valves reflecting this safety concern. However, an open bleed valve may permit emission leakage to the atmosphere from a high-pressure source and potential exposures may occur at the bleed valve opening. All bleed valves in

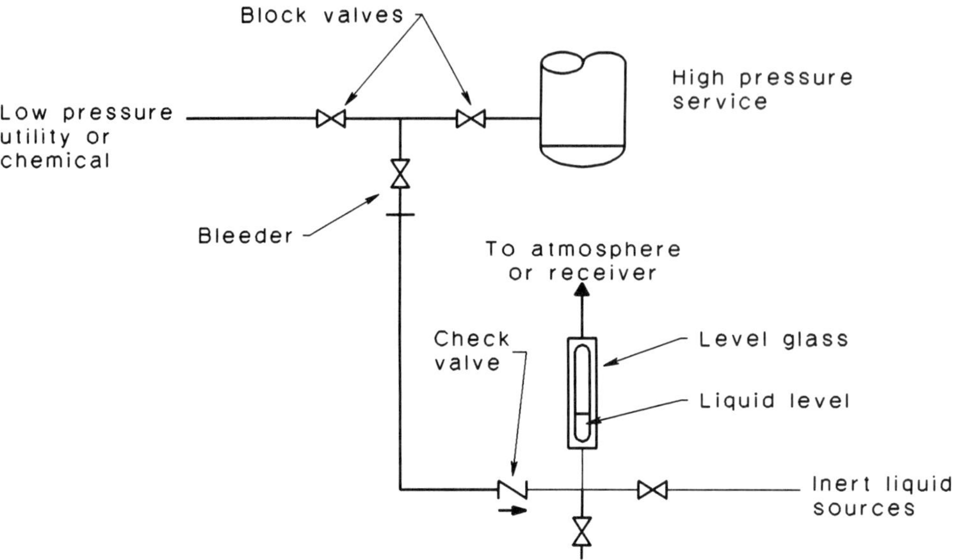

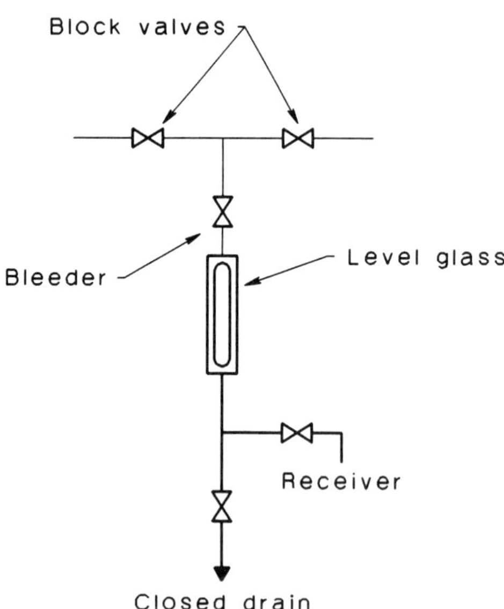

Figure 7.32 *Leak detection in double block and bleed installations.*

double block service should be examined by the industrial hygienist to determine the contaminant concentration level in the atmosphere. Double block and bleed units in liquid service may discharge liquids through the bleed valve, presenting a potential dermal exposure problem for acutely toxic liquids, but leakage is apparent and some provision to prevent splatter is required.

Where toxic vapors or liquids may potentially leak from a bleed valve, the industrial hygienist should determine whether the leakage presents a potential exposure problem. These systems are important safety controls and are inspected often to insure that leakage is not excessive and the bleed is in operation. If the leaking fluid is a health hazard, the bleeder outlet should have a closed drain. However, the bleeder outlet is normally examined visually and a closed drain prevents visual observation of leakage. One solution is the installation of a level glass or rotameter in the bleeder outlet of a vapor line where bubbling through an inert liquid indicates valve leakage. The level glass or rotameter (without a float) can be obtained with a shield protection device and should be specified for a satisfactory pressure level as a safety measure although the pressure probably will be close to the atmospheric pressure level. This general system is shown in Figure 7.32. The gas can be discharged to the atmosphere from an elevated vent, assuming satisfactory dispersion of any vent gas, or can be sent to a low-pressure level enclosed vent system with a control device. The liquid in the level glass should be inert chemically and not react with the gas or vapor.

Liquid leakage from the bleed valve may also be a potential health problem, and one solution for a closed-drain system with leakage indication is also shown in Figure 7.32. The bleed valve drain line discharges into a level glass or small flow sight glass and flows down the glass to the closed drain outlet. Liquid flow can be observed through the sight glass and, as a safety precaution, the level or sight glass should have a protective shield and be specified for pressure operation. To indicate liquid flow, the outlet valve can be cracked open and level changes observed in the sight glass as an alternate method of indicating and determining leakage. The liquid draining from the level glass block valve outlet can be sent to a closed central drain system or collected in a small receiver.

REFERENCES

1. CFR. Standards of Performance for Equipment Leaks of VOC in the Synthetic Organic Chemicals Manufacturing Industry, *Title 40*, Subpart VV, Parts 60.480–60.489, pp. 484–499, July 1, 1992.
2. CFR. Standards of Performance for Equipment Leaks of VOC in Petroleum Refineries, *Title 40*, Subpart GGG, Parts 60.590–60.593, pp. 580–582, July 1, 1992.
3. FR. National Emission Standards for Organic Hazardous Air Pollutants from Synthetic Organic Chemical Manufacturing Equipment Leaks, Proposed Rule, *Title 40*, Subpart H, Parts 63.160–63.184, pp. 62765–62767, December 31, 1992.
4. CFR. Determination of Volatile Organic Compounds Leaks, *Title 40*, Part 60, Appendix A, Method 21, pp. 1020–1023, July 1, 1992.
5. EPA. *Fugitive Emission Sources of Organic Compounds—Additional Information on Emissions. Emission Reductions, and Costs*, Publ. No. EPA 450/3-82-010. April 1982.

6. EPA. *Evaluation of Maintenance for Fugitive VOC Emissions Control*, Publ. No. EPA-600/52-81-1080, May 1981.
7. EPA. *Problem-Oriented Report: Frequency of Leak Occurrence for Fittings in Synthetic Organic Chemical Plant Process Units*, Publ. No. EPA-600/2-81-003, September 1980.
8. EPA. *Emission Factors for Equipment Leaks of VOC and HAP*, Publ. No. EPA-450/3-86-002, January 1986.
9. Burglund, R. L., J. A. Mullins, L. S. Parrett, and R. R. Romano. *Fugitive Emissions from Equipment Leaks: Results of Industry Wide Studies in the Butadiene. Ethylene Oxide, and Phosgene Production Industries*, National AIChE Mtg., Philadelphia, PA, August 20–23, 1989.
10. Lipton, S. *Chem. Eng. Prog.* **85**(6), 42–47 (1989).
11. FR. Process Safety Management of Highly Hazardous Chemicals; Explosives and Blasting Agents, OSHA, *Title 29*, Part 1910, Final Rule, pp. 6356–6417, February 24, 1991.
12. Burns, R. *Chem. Proc.* **55**(9), 51–53 (1992).
13. Blicky, G. J. *Control Eng.* **32**(9), 118–120 (1985).
14. Rockwell International for American Petroleum Institute. Fugitive Hydrocarbon Emissions from Production Operations, API *4322*, March 1990.
15. Frikken, D. and C. Schreckenberg. *Finalizing and Fixing Valve Stem Leaks—Monsanto's Experience*. Valve Manufacturers Association Mtg., Philadelphia, PA, November 18–19, 1992.
16. Davis, R. *A Report on Chemical Plant Valve Usage in Light of the Clean Air Act*, Fugitive Emissions Seminar, VMA, Houston, TX, September 25–26, 1991.
17. Offutt, W. *A Valve Stem Packing Overview of New Packing Techniques to Meet Low Leakage Requirements*, Fugitive Emissions Seminar, VMA, Philadelphia, PA, November 18–19, 1992.
18. Kittleman, T. *Approaches to Equipment Monitoring for Regulatory Compliance*, Fugitive Emissions Seminar, VMA, Philadelphia, PA, November 18–19, 1992.
19. CFR. Occupational Safety and Health Standards, Subpart Z—Toxic and Hazardous Substances, *Title 29*, Part 1910.1001, pp. 34–77, July 1, 1991.
20. CFR. National Emission Standard for Asbestos, *Title 40*, Subpart M, Parts 61.140–61.157, pp. 76–104, July 1, 1992.
21. FR. Asbestos; Manufacture, Importation, Processing and Distribution; Effect of Court Decision, EPA, 40CFR Part 763, pp. 11364–11365, April 2, 1992.
22. Weismantel, G. *Chem. Eng.* **91**(17), 71–83 (1984).
23. Lipton, S. *Variations in Petroleum and Chemical Industry Practices Result in Fugitive Emission Factor Differences*, National AIChE Mtg., New Orleans, LA, March 6–10, 1988.
24. Davis, B. C. *Comparison of the Fugitive Emissions from Two Butadiene Plants*, Air and Waste Management Association, National Meeting, Anaheim, CA, June 25–30, 1989.
25. Ridey, M. *Chem. Eng.* **98**, 98–101 (1991).
26. Weismantel, G. *Chem. Eng.*, **92**(4), 71–124 (1985).
27. Grumstrup, B. F. *Chem. Eng.*, **99**(4), 124–129 (1992).
28. Snyder, P. *Chevron's Experience with Fugitive Emissions Regulations and the Cost of Compliance*, Fugitive Emissions Seminar, VMA, Houston, TX, September 25–26, 1991.
29. Barns, D. *Occup. Health Safety*, **61**(10), 36–37 (1992).
30. McKinnery, W. N., Jr. and R. W. Moore. *J. Am. Ind. Hyg. Assoc.*, **53**, 531–532 (1992).

31. Rockwell Corp. Rockwell Flow Control Division, Report: V-Rep-82-1, 1982.

32. Electric Power Research Institute. Report: EPRI NP5697, May 1988.

33. Drago, J. *Plant Eng.* **46**(8), 71–73 (1992).

34. Lipton, S. *Chem. Eng. Prog.* **86**(8), 70–76 (1990).

35. Beasley, M. E. *Metal Seated Ball Valves for High Pressure Vent and Drain Service*, EPRI 4th Valve Technology Symposium, Dallas, TX, August 11–13, 1992.

36. Harrelson, A. L. *Live-Loading and Packing Compression Uniformity*, Fugitive Emissions Seminar, VMA, Philadelphia, PA, November 29–30, 1990.

37. Harrelson, A. L. *Live-Loaded Packing*, Fugitive Emissions Seminar, VMA, Houston, TX, September 25–26, 1991.

8

Control Valves

Control valves are installed in a variety of applications in petroleum refineries and chemical plants to perform stream control functions that can be automatic, remotely operated, locally operated, and so on. Although a number of control valves are observed in process units, they are a small fraction of the total number of valves in typical chemical or petroleum units. The function of the great mass of valves in the process industries is primarily a blocking operation, requiring valves in a completely open or closed position. In contrast, the majority of control valves generally operate continuously in an automatic mode with a sensor controlling the valve position through a controller, transmission system, and valve operator. As a result, control valve stems are in almost continuous movement to adjust flow through the valve for maintenance of the conditions required by the sensor such as flow rate, temperature, or pressure. There are a number of valve sensor and operator configurations, but the emphasis in this chapter is on a control valve with an actuator or operator that receives signals from a controller in response to sensor conditions.

In a typical situation, the sensor signal is transmitted to a control instrument located in a remote control room that analyzes and compares the signal with a desired operating point. Instructions are then relayed as a signal to the valve operator. The signal for valve adjustment can be electronic, air, or optical, with air transmission systems essentially displaced by electronic instrumentation in most facilities. New and modified facilities are normally constructed with electronic instrumentation to facilitate computer control and operation analysis. At the valve, electronic signals are frequently converted to an air signal in an electronic/air interface transducer, which is needed to drive the valve. The change in air pressure to the control valve actuator then adjusts the valve stem position. This seemingly cumbersome system transmits instructions rapidly and the control valve stems have a rapid response. Electrical drives or activators directly controlled by electronic signals that function have become widely available and are beginning to replace air-actuated valves, particularly in new installations. However, air as a source of motive power is reliable, relatively inexpensive, and still used widely. A positioner may be included on the air-actuated valve which decreases response time through improved air action but does not basically change the valve action.

8.1 REGULATIONS

While control valves have completely different functions than the manually operated block valves reviewed in Chapter 7, from a regulatory standpoint all valves, with the exception of pressure relief valves (devices), are considered a single emission category or classification. When the original field emissions tests were considered for the EPA, all valve emission data were combined and analyzed as a single category termed "valves." As a result, valve regulations developed through the years, including the new HON regulations, cover both block and control valves as one single classification. The NSPS and HON regulations described in Section 7.1, therefore, are also applicable to control valves.

Control valve leak criteria, as defined in the new HON regulations (Section 7.1), is 500 ppm and all of the other regulations require compliance. Since leak detection and repair (LDAR)l requirements are identical, additional emission control efforts may be required to meet the new valve regulations. Normal control valve operation results in considerable packing wear due to stem cycling and a wide variety of packings and control systems have been developed to meet the more stringent emission regulations. Specific control valve regulatory requirements are presented in Section 7.1.

8.2 TYPES OF CONTROL VALVES

There are two major stem movement systems in control valve operations with numerous variations in actual control valve designs and applications. The general motions of *control valve stems* are linear (reciprocating) or rotary, with most rotary motion valves limited to a 90° turning movement. A simplified drawing of a linear (reciprocating) motion control valve is shown in Figure 8.1. The reciprocating motion is also termed "sliding-stem" action. A linear action air-activated valve of the type shown in Figure 8.1 was the general refinery service control valve included in the EPA fugitive emission test program.[1] Leakage from this type of valve is relatively large, although control valve emissions were not separated from gate (block) valve emissions, but were included in the overall EPA valve emissions summary. The database for analyzing control valve emissions in the test programs has not been large enough to define a separate control valve emission factor.

As shown in Figure 8.1, the control valve is separated into the control valve and the actuator. Although the actuator in Figure 8.1 is an air actuator, an electronic/electrical actuator mounted on the valve does not change the valve mechanism or design of the valve. The valve body design varies somewhat with different manufacturers, but the major elements in the sketch are present in essentially all sliding stem control valves. The control valve stuffing or packing box in a sliding stem valve contains packing rings and spacers, which are essentially wide lantern rings. The spacers provide a method of detecting liquid or vapor

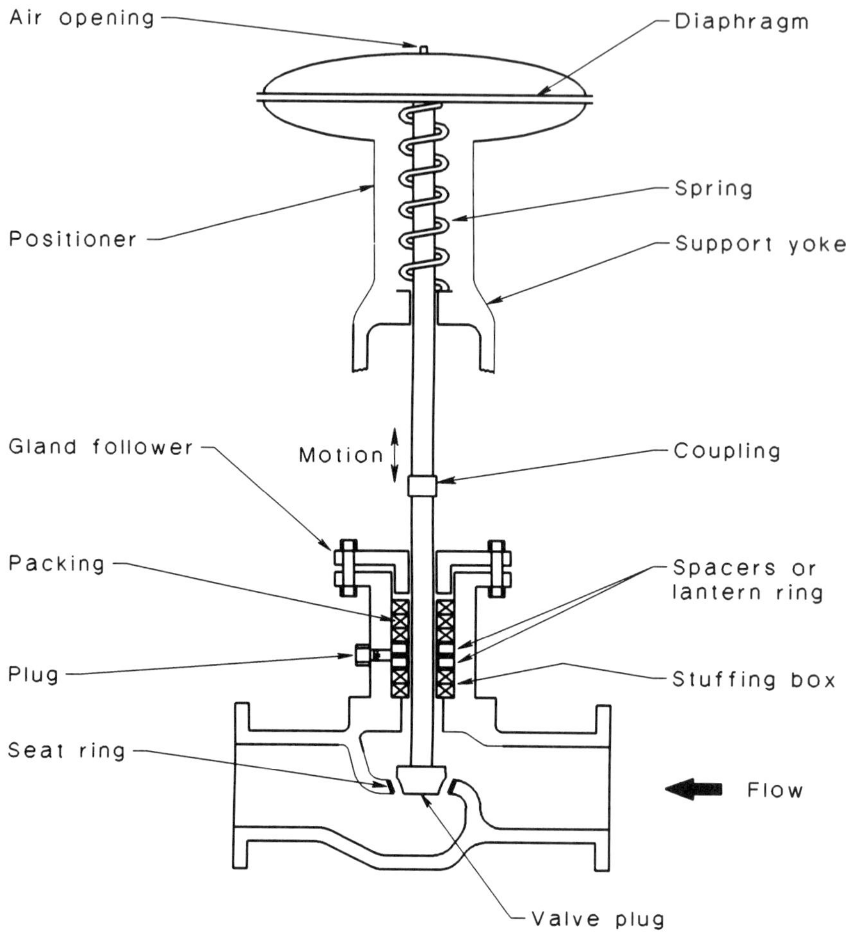

Figure 8.1 *Simplified cross section of a linear or sliding stem control valve.*

leakage through a plug connection of the lower packing set and fill the packing box, which is considerably longer than block valve packing boxes. In the event leakage is a problem, the plug is removed and a closed connection is provided to a receiver at low pressure. The plug connection has also been used as an inlet opening for injecting a lubricant into the packing area to stop leakage. Both methods will work for a period of time, but in the event of a packing failure neither method will completely control emissions.

There are a variety of rotary motion control valves installed in various chemical and petrochemical plants. These rotary control valves are similar to the quarter-turn

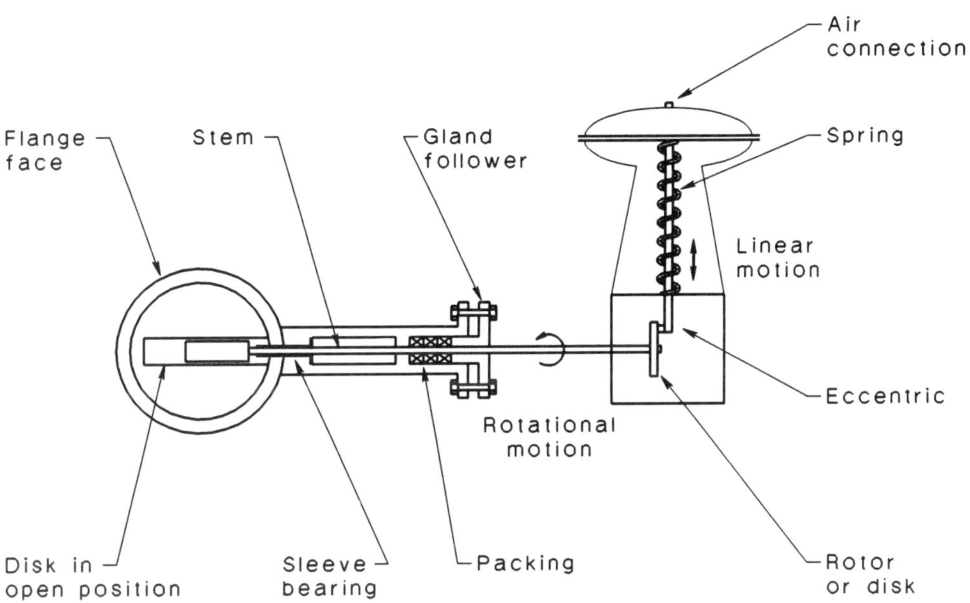

Figure 8.2 Simplified drawing of a rotary action control valve and actuator.

valves reviewed in Chapter 7, with an actuator or operating device to turn the valve stem. In general, rotary motion control valves have a maximum turning angle of 90° and typical rotary valves in control valve service include ball, butterfly, modified ball, and eccentric plug.[2] A simplified sketch of a rotary control butterfly valve is shown in Figure 8.2, which presents the basic elements in a rotary control valve design. The actuators are either pneumatic or electric and the motion of the actuator is converted into rotary motion through various mechanical devices available from a number of actuator and valve manufacturers.

A large number of manufacturers produce rotary control valves and the packing box size and configuration vary considerably. The packing system in Figure 8.2 is a distance from the process fluid, but in other valve designs the packing box may be considerably closer to the process material. In addition, the length of the packing box varies, some have considerable depth and are provided with spacers. Other packing boxes may only accommodate a single packing set. Older rotary control valves frequently have short boxes and many of these control valves are still operating. Packing configurations for the short boxes generally contain packings of the type shown in Chapter 7 for quarter-turn manual valves.

In addition to the standard rotary and sliding stem valves, other types are available to specifically control fugitive emission leakage. Bellows valves are a type of emissions control valve that has been marketed for a number of years in sliding stem configurations. Some bellows control systems are also installed in rotary valves. The bellows devices in control valves are similar to those found in the bellows valves described in Chapter 7 and are briefly reviewed in another part of this chapter. Although bellows valves have been available for many years, other valve designs for specifically controlling emissions are available commercially. As

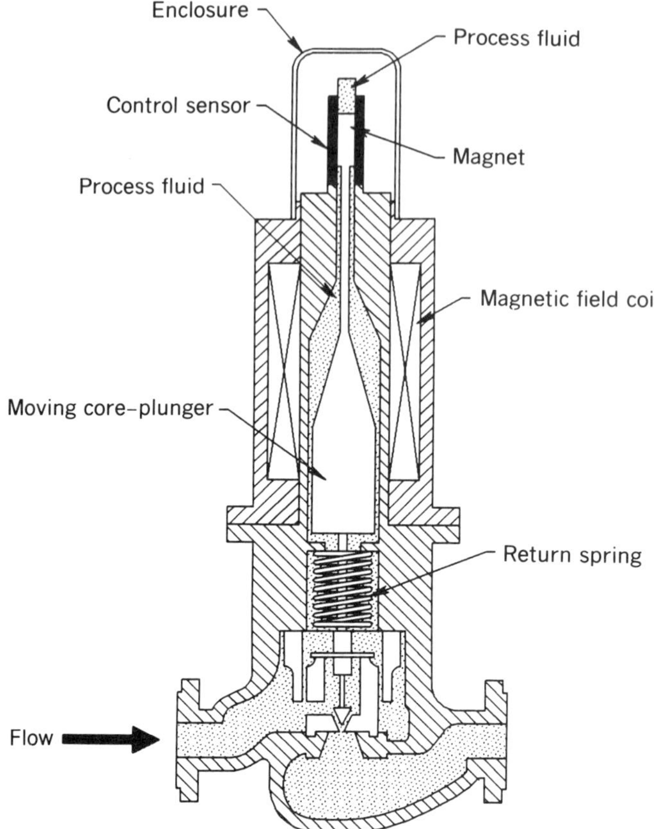

Figure 8.3 Simplified drawing of a hermetically sealed control valve (courtesy of Target Rock Corp.).

an example, a newer valve that is completely enclosed or hermetically sealed is shown in Figure 8.3 and is termed a "zero emission valve." In this design, a spring maintains closure pressure on the stem and an electric field force moves the stem core, overcoming the spring force. The stem is completely encapsulated, resulting in a hermetically sealed valve. A stem sensor controls the plunger position in accordance with the electronic control signals.

8.3 SAFETY

A very important consideration with control valves is fire safety. Although control valves are a small percentage of the total valve population in processing facilities, these valves are also a potential hazard source in a fire situation. Should a significant hazard develop during a fire from a control valve, the valve must be considered in a process hazard analysis required by OSHA[3] and the CAAA. Consequently, the process facility must determine whether fire-safe control valves are necessary for each installation. Where they are required, control valves and packings should be tested in accordance with the fire test recommended in Chapter 7.

8.3.1 Control Valve Installations

Control valves must be installed in accordance with requirements in the process facility. This may require control valves to have a tight shut-off capability during an emergency shutdown or fire. In effect, the control valve automatically performs the function of a manually operated valve by either closing or opening completely. Although valve closure is generally considered the fail-safe position, valves also may be designed for a fully open condition during an emergency. Valve manufacturers supply control valves to fail either closed or open in accordance with plant specifications. Where a tight shutoff capability is not required, valves will have some leakage through the process line when fully closed. However, the valves should meet all leakage specifications under fire conditions in accordance with installation specifications.

Many valve packings are composed of nonfire safe packing materials to control emissions. To meet fire-safety standards, a graphite ring is added on the outboard end of the packing set. During the fire test, the polymeric and nonfire safe rings are consumed or destroyed in the test, but the graphite ring will remain intact. Although the graphite ring remains, packing compression is lost without the other rings and control is questionable. Moreover, this type of packing system will pass the fire test when the packing is new, but has not been tested when the packing has aged and the number of valve cycles has increased significantly.

Control valves may be installed with blocks and bypass valves, as shown in Figure 8.4, or as a single valve. This difference frequently reflects the type of processing, such as continuous or batch, where a consideration in continuous processing is the length of time between unit turnarounds. When the time between

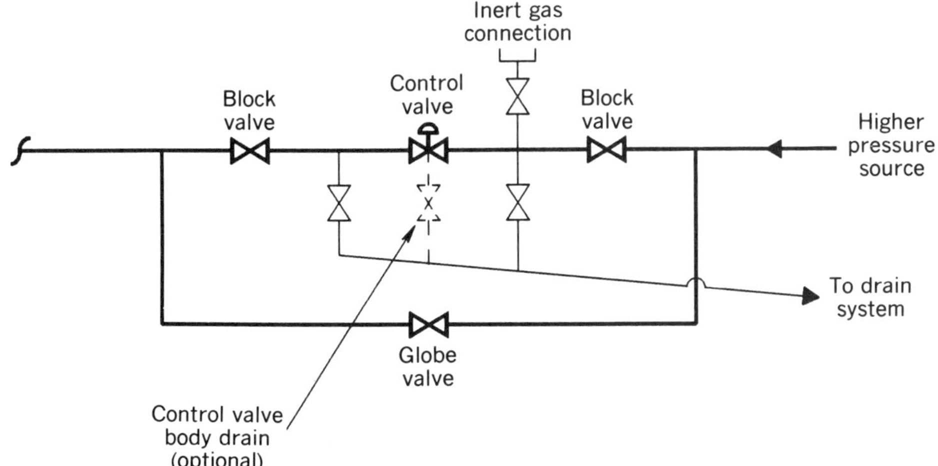

Figure 8.4 *Typical control valve installation with block and bypass valves showing drain connections.*

turnarounds is several years, block and bypass installations permit continuous operation while the control valve undergoes maintenance or repairs. A manually operated globe valve in the bypass line provides flow control during the repair operation. The control valves may also be removed and replaced in a block and bypass installation. However, block and bypass installations are considerably more expensive than a single control valve installation. Although blocks in the bypass configuration can be closed in a fire situation, a fire that affects the control valve may prevent manual operation of the block valves. Consequently, a control valve must be fire safe and have a tight shutoff capability in this type of situation.

8.3.2 Actuators and Accessories

Control valves have actuators to provide power to the control valve stem and may also have other accessories, such as a positioner, that provides additional capability in rapid stem movement. These mechanisms may also be exposed in a fire situation and should have fire protection to prevent destruction and permit valve operation for emergency control. Fire resistance may be a specific characteristic of the equipment or the equipment can be fireproofed separately. In the latter situation a fireproof coating can be sprayed on the equipment. These coatings are about $\frac{1}{2}$ in. (12.7 mm) thick and are resistant to most chemicals. Moreover, the coatings are generally self-extinguishing. The fireproof coatings have generally passed various fire tests, including those in several other countries.

8.3.3 Rotary Valves

A number of rotary valves for control valve service also have seat ring backup seals to provide shutoff capability during a fire. These backup seals or rings are similar to those described in Chapter 7 for rotary or quarter-turn block valves. Since tight shutoff is available with rotary valves, this characteristic should be considered in all rotary valves. In applications where a tight shutoff is not required, the allowable leakage flow on a closed valve is quite small. However, loss of a seat seal through fire will increase the flow rate to very high levels, which does not occur in rotary valves with backup seal rings. In general, these valves should have a tight shutoff capability and any application where tight shutoff is not required during an emergency should be reviewed with safety groups.

In addition, all rotary valves should have a blowout prevention system to eliminate the possibility of the stem escaping from its normal position. This is an important criteria, particularly under fire conditions.

8.3.4 Valve Drainage

A potential source of exposure is control valve repair or replacement in an operating process unit. This procedure is only possible where the control valve is installed in a block and bypass arrangement, as shown in Figure 8.4. However, where the fluid is toxic, the control valve must be drained and purged or flushed to remove all traces of fluid contamination. When a control valve is blocked and bypassed, the piping and control valve between the blocks are filled with the process fluid. In Figure 8.4 permanent drains on each side of the control valve are shown that permit drainage of the process fluid to a sealed plant drainage system. Since complete liquid drainage requires a vapor or gas to prevent a vacuum during the drainage operation, a valve connection is provided that permits the injection of an inert gas. Alternatively, where the fluid does not present a potentially flammability concern, the vacuum breaker valve can be opened to the atmosphere. Following process fluid drainage, the valve should be flushed to remove any remaining traces of process fluid.

If the control valve is in gas or vapor service, a single drain connection to a control vent or collection system from the downstream side of the control valve is generally sufficient for decreasing the contained pressure. Opening a purge valve to inert gas purges the control valve to the closed collection drain system. A valve body drain can also be installed for either liquid or vapor valves, as shown in Figure 8.4. Where a control valve has dead space, volume, or a depressed region that collects liquid or vapor, a valve connection to the bottom of the control valve permits drainage or venting of the zone. A ball control valve has a spherical annular chamber which gradually fills with process fluid and can be drained with a valve connected to a boss at the bottom of the valve. This method of draining a manual ball block valve, described in Chapter 7, is also applicable to control valves.

For some control valves, particularly rising stem valves, process fluid does not become trapped within certain enclosed volumes. However, in liquid service where a control valve can have the lower body configuration of the control valve shown in Figure 8.3, a drain valve may be required at the low point of the valve to completely drain the valve. This separate drain valve is particularly important in toxic service where the fluid does not readily vaporize at ambient temperature and pressure. Vapor valves do not present a problem in this regard unless there is some vapor condensation.

The drainage system described is considered "hard piped" or permanent and minimizes potential exposures to hazardous streams or HAP. Where control of VOC emissions from control valve drainage is not a major problem and the stream material is not toxic, an alternate method of drainage is possible. Hose connections to a closed drain system can be used. However, this requires careful control of hose movements to prevent liquid leakage from the hose when disconnected. Moreover, hoses cannot be indiscriminently used in various services without thoroughly cleaning the hose or limiting use to applications where contamination or reaction with the remaining traces in the hose is not a potential problem. Drained liquid and any purged material can be discharged into a closed container, providing the container holds all of the drained and purged materials, can be tightly sealed, and is an acceptable safety container for materials that are flammable or explosive. The container contents must then be discharged safely and with minimal exposure into an acceptable receiver. This procedure may not be acceptable over the long term, when additional materials will be classified as toxic and tighter control of VOC emissions can be expected. Industrial hygienists should follow disposal procedures for all drained and purged material to insure that potential exposures are minimized through final disposal.

In general, the previous recommendation are based on liquids that do not readily volatilize. However, volatile liquids in the valve will vaporize upon depressuring and present potential problems where the drainage system is not hard piped. Additionally, liquids that are potential skin hazards should be hard piped to reduce potential exposure situations. Any installations require careful review by industrial hygienists.

8.4 GRAPHITE PACKINGS

Control valve packings have been classified as fire-safe or nonfire-safe packings, which is similar to the manual block valve classification. Flexible graphite packings are the standard fire-safe packings that have been selected to replace asbestos packings in both block and control valves. The term flexible graphite also covers carbon and various carbon–graphite combinations that have been developed to control emissions. Both packing and control valve manufacturers have developed a number of flexible graphite and nongraphite packings, and a variety are available commercially. Although many of the control valve manufacturers have developed proprietary packings for their control valve products, existing control valves can also be retrofitted with some of these packings sets. Apparently some of these proprietary manufacturers packing products can be installed in control valves produced by other manufacturers. Since there are various control valve manufacturers of graphite packings, a few graphite packings are described in this section that provide a basis for comparison with alternative graphite packing sets.

In all control valves, the valve stem moves almost constantly to maintain the required process control condition in accordance with signals from the process sensor. This essentially continuous stem movement or cycling increases wear on packings and other components such as guide bushings. There have been indications that emissions increased as the number of cycles increased. Cyclic testing of packings by control valve and packing manufacturers generally extends through several hundred thousand cycles for both rising stem and rotary control valves.

With the advent of flexible graphite and carbon packings, corrosion or pitting at valve stems has occurred, which may be related to the additives in the packings, although definitive reasons for this problem have not been identified.[4] This general stem corrosion problem is usually prevented by the installation of a corrosion-resistant stem. In addition, the finish on the stem must conform with the type of packing that is installed. Typical flexible graphite packings normally require a stem finish of 28–32 RMS, with low numbers representing much smoother finishes. Some packing systems may require a stem finish of 4–8 RMS, which is a difficult machining operation and is more costly.

Friction due to the frequent movement of the valve stem can be a potential problem in control valves. When friction increases with certain packings, the size or power of the actuator may be insufficient in an existing valve and a new actuator may be required. In new control valves, a higher power requirement does not pose a serious problem. However, the amount of friction or change in friction associated with a packing change must be understood to adequately match the actuator power with the packing set requirements. Information on friction differences in packings is presented in the various packing sections.

There are two basic types of packed control valves currently installed in processing facilities—rotary and rising stem valves. Also, within these two classifications, various types of rotary valves are installed in process plants. However, both rotary and rising stem sealed bellows control valves are also available commercially and have been installed in some services. Comments on bellows control valves are presented in Section 8.6.

8.4.1 Valve Construction

In general, of the two basic types of control valves (rising stem and rotary), the rising stem or reciprocating control valve was the standard control valve for many years, but has been replaced in many services, particularly in the chemical industry by rotary valves. Consequently, a mixture of these valves exists in various process plants and emission control or leakage requirements must consider the differences that exist in these valves.

Rising Stem Valves. A typical simplified cross section of a rising stem valve is shown in Figure 8.1. These valves are produced by a number of manufacturers throughout the world and various differences exist among these valves. However, the figure indicates the basic packing, system which is similar for most valves. Mechanical construction details and requirements of this particular valve type follow.

Packing Boxes. Most of the packing boxes in rising stem valves are quite long and have a port or leak opening from the packing box through the bonnet wall. This opening is normally plugged when the valve is received from the manufacturer. In addition, the packing box normally contains a packing set with the rest of the packing box filled with metal spacer rings. These spacer rings can also be used as lantern rings in some valves. When flexible graphite packing is installed in the packing box, the graphite rings should not be in contact with the port opening. If this cannot be avoided, the hole should be plugged and then smoothed until all of the burrs are removed.

Short packing boxes are similar to those described for rotary valves.

Stem Design. Since most of the stems in rising stem valves are quite long, the stem must be properly designed to prevent excessive wear of the packing or stem. As an example, a stem may flex during movement, resulting in wear on the packing rings. As a result of flexing the packing rings may gradually assume an oval shape or be worn, albeit somewhat unevenly. This type of operation results in leakage and is very difficult to control.

This general problem can be overcome by ensuring an adequate stem thickness to prevent flexing and a straight stem that will not affect the packing during movement. The stem is then ground to the required smoothness. In addition, the packing box wall must be free of burrs and ground to a smoothness of 125–150 RMS. Certain packings may require a smoother packing box wall finish.

As an alternate, guide bushings can be installed in the packing box that act as sleeve bearings and minimize stem flexing. The guide bushing in this service must not wear and is compatible with the stem material. However, the packing box must be long enough to install the extra bushings that can replace some of the spacers. This is a less costly alternative and can be implemented easily. While guide bushings can contain shaft flexing, shafts or stems must be straight, which requires quality control during manufacture and plant inspection.

Rotary Valves. Various types of manually operated rotary valves were described in Chapter 7. Most of the quarter-turn or rotary valves in control valve service are ball and butterfly or disk-type valves. Plug-type control valves are normally not found in process operations, but some control valve models are now available. However, a wide variety of ball-type control valves are commercially available that are widely used in process applications. All of these control valves have a 90° turning arc and are also termed quarter-turn valves. In general, many of the valves are of a construction similar to the quarter-turn control valve shown in Figure 8.2.

Packing Boxes. Packing box locations vary in control valves from the type shown in Figure 8.2, to very long boxes that almost extend the length of the bonnet. In addition to these packing box configurations, some valve designs have packing boxes similar to those found in rotary valves designs shown in Chapter 7. Thus the rotary control valve boxes vary in depth from relatively short to long boxes with spacers. The long boxes may also have a port opening which permits use of a spacer as a lantern ring or telltale. Where a port or hole exists, flexible graphite rings should not be in contact with the opening. If contact cannot be avoided, the hole should be plugged and smoothed until all of the burrs are removed.

Stem Design. Many quarter-turn valves are installed in the horizontal position, which frequently affects the packing. For large valves, the stem weight results in a slight bending of the stem, which causes uneven wear of the packing, and the packing eventually forms an oval. This phenomenon can also occur from certain types of actuators and may cause a slight sideways movement of the stem, resulting in a similar oval formation in the packing. The out of round packing formation increases emission leakage.

Increasing the stem diameter to prevent shaft bending would eliminate excess leakage from the valve. Although this design change would improve emissions control, major changes in valve design are necessary. As an alternative, close tolerance guide bushings can be installed in the packing box, replacing some of the spacers. These guide bushings act as sleeve bearings, supporting the shaft and preventing sideways movement. However, a satisfactory box length is necessary for the installation of guide bushings. This alternative can be implemented and is less costly, but the guide bushings must not wear or gall the valve stem, requiring a compatible bushing. In addition, the valve stem must be straight, which requires quality control during manufacture and plant inspection.

8.4.2 Standard Graphite Packing Set

A standard flexible graphite packing set, described in Section 7.5.4, consists of braided end rings and preformed flexible graphite rings for a total of five rings. A braided ring is placed at each end of the packing set. Although termed graphite braided end rings, these are generally carbon braided end rings, which are more flexible than graphite braided rings. The preformed graphite rings vary in density from 70 to 100 lb/cf (1.12–1.60 gm/cm^3), which has been described in Chapter 7.

Long Box Packing Sets. Standard packing sets may be installed in long boxes as a single or a double set. Where a single set is installed, the remaining volume in the packing box is filled with spacers or lantern rings. Frequently, dual packing sets

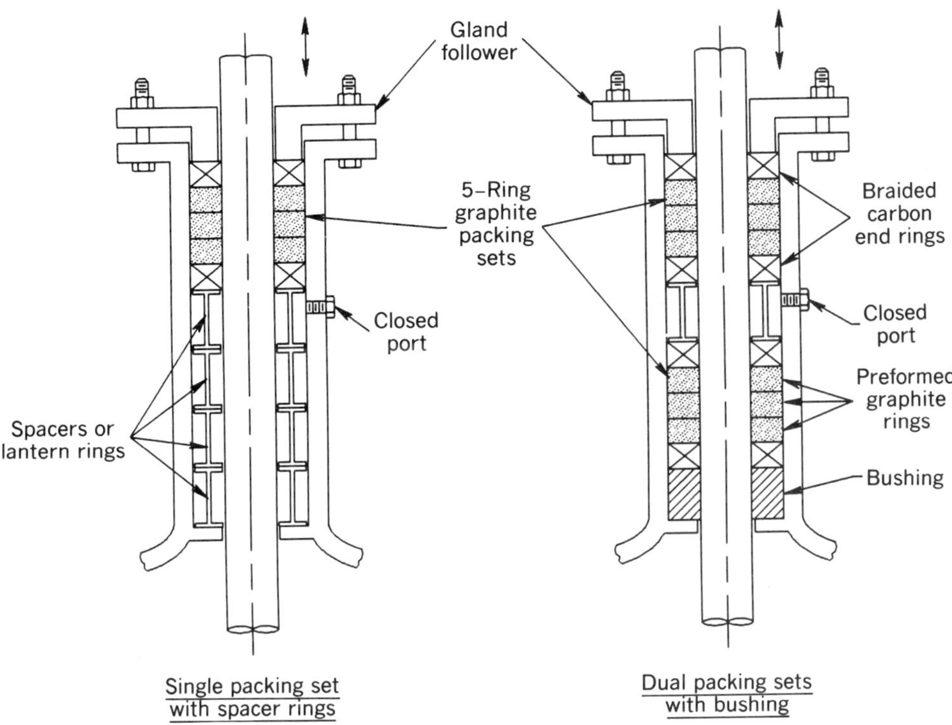

Figure 8.5 *Standard graphite packings in deep-control valve packing box.*

are installed that are separated by a lantern ring or spacer, as shown in Figure 8.5. However, the entire packing box volume may not be filled and a bushing may be installed to fill the void. All spacers, lantern rings, and bushings are constructed of material compatible with the stem or are designed not to contact the stem. Wide clearances are filled by the compressed expansion of the carbon braided end rings, but excessive clearance may interfere with its functions. Consequently, the allowable clearance for rings and spacers should be reviewed with the packing manufacturer.

Leakage. The ports have previously been mentioned as a method of controlling leakage or emissions from the valve gland. This is accomplished with two packing sets in a packing box, where leakage through the lower set is drawn off or vented through the port to another location. The process pressure on the upper packing set is reduced significantly with little material left for leakage through the upper packing set. While this procedure is possible in manual or block valves, the large number of block valves precludes installation of this method for leakage or emission control. However, venting or bleeding off leakage from a much smaller number of control valves through a lantern ring is encountered in various chemical plants. This procedure varies from leakage bleedoff control on all control valves to bleedoff controls on selected valves. This is a successful method of controlling leakage and should be considered in plants where valve leak frequency is a potential problem. Both manual and control valves are included in controlling valve emissions in a single category by the EPA. Where control valve leakage and the frequency of control valve leakage is high, bleedoff system should be considered.

Valve bleedoff through a port requires a receiver that should be at a much lower pressure level, providing a differential pressure driving force. However, different materials from control valves may not be compatible and would require disposal in various receivers, which becomes complex and costly. Consequently, any consideration of bleedoff control requires careful investigation. Leakage material, whether vapor or liquid, can be recovered through bleedoff systems and recycled, which should be included in an economic evaluation. In addition, an inert purge through a lantern ring can also be used to sweep any leakage into the discharge port. The inert gas is separated in the receiving system and may or may not be recycled, depending on the design of the overall purge–leak system.

Packing Installation. As shown in Figure 8.5, one or two packing sets can be installed in a deep or long packing box. However, installation of the packing set or sets should be carefully reviewed with the manufacturer. When one packing set is installed, frequently the entire set is completely installed and then compressed as required. This may result in an uneven load distribution in the packing set, with the upper preformed ring compressed more than the ring beneath and so on throughout the set. As a result, there is maximum radial expansion in the upper ring which is reduced in each successive ring below the preceding ring. This gradient, which may be considerable,[5] can be modified significantly be installing the five-ring packing set in two separate groups and compressing each stage. As an example, the bottom two or three rings are installed and compressed to the required depth. The remaining rings are then installed and compressed, which results in a more even distribution of compression among the packing rings.

An alternative is compressing the entire set to about 3800 psi (262 bar) based upon gland follower torque requirements that can be provided by the packing manufacturer or developed by the mechanical group. When this compression pressure is achieved, the pressure can be reduced to 1.5 times the process, but never less than 1500 psi (103 bar). This general procedure expands the graphite under pressure and then permits packing adjustment due to flexibility in the braided end rings and slight relaxation in the preformed graphite rings.

When the packing consists of two packing sets, the packing in each set should be compressed separately. The lower set is compressed in one stage, followed by installation of the spacer and/or lantern rings. The upper packing set is then installed and compressed in one stage. Installation of the dual packings and rings in one step with a single overall compression effort results in a wide disparity in compression across both sets of packing rings. While compression in two stages is more effective in emission control, added effort and cost is involved in this procedure. In addition, metal extension rings similar to long bushings are necessary when compressing the lower set since the gland follower is not sufficiently deep to extend into the box to the lower ring set. The extension rings or bushings are split in half, placed around the stem, and slid into the packing box. Threaded holes in the top surface of these extension rings permit removing rings with threaded rods. These sets can be compressed to the pressures previously described.

Since this is one procedure, the manufacturer supplying the packings should be consulted for a recommended packing compression procedure.

Friction. The standard five-ring packing set requires a greater stem actuation torque than a Teflon® or PTFE packing set, which has generally been considered the base in comparing friction effects. Compared with PTFE, the standard five-ring flexible graphite set friction resistance is more than the PTFE frictional resistance. The graphite resistance can be much greater depending upon the preformed ring density and compression pressure.

Where control valve actuators are designed for PTFE® packings, a larger actuator may be required when the packing is changed to a standard five-ring flexible graphite packing set. Any packing changes in a control valve require a careful analysis of actuator capability. In addition, friction also affects control valve hysteresis, which is a lag in stem movement that occurs when the valve stem is moved in either direction. This lag results in delay in adjustment of the valve position and affects process control. This hysteresis effect is more pronounced with flexible graphite than with PTFE.[4] Hysteresis is also affected by graphite ring density and compression pressure.

Short Box Packing Sets. In a number of valves the packing boxes are short and cannot accommodate a standard five-ring graphite packing set. Packing box extensions may be available for existing valves and longer boxes can be specified in new valves. Where the box length cannot be adjusted, shorter versions of the five-ring set can be obtained. In situations where the packing box, as an example, is only two full rings in depth, a single preformed flexible graphite ring can be installed with two modified braided rings. The standard carbon braided end ring can be reduced in height about 50% when purchased specifically for this application. With two modified braided end rings and a standard flexible graphite preformed ring, this packing combination can be installed and compressed to provide emission control. The effectiveness of this sort packing set has not been described in any publications, but these sets are installed in control valves.

For somewhat longer boxes, although still relatively short, the number of preformed rings can be increased by including modified braided end rings. Two preformed flexible graphite rings with the modified graphite braided rings described above can be installed in somewhat longer boxes. Another alternative is a combination of preformed graphite rings with PBI braided end rings, as described in Chapter 7. These rings can be reduced in thickness to about $\frac{1}{8}$ in. (3.18 mm), permitting the installation of two full-size flexible graphite preformed rings with very thin end rings. Other combinations are available for packing boxes shorter than the box depths necessary for standard five-ring sets.

Packing Dimensions. In all of these packing combinations, box depth, uncompressed packing ring depth, and compressed packing set dimensions must be understood to properly specify and install a flexible graphite packing set. The packing manufacturers can provide information on compression and individual ring dimensions. For short packing boxes, uncompressed and compressed dimensions are necessary to define packing specifications and requirements.

Rotary Valves. The standard graphite packings can also be installed in rotary valves. In this service, friction is not a severe problem compared with rising stem valves and does not normally require an increase in actuator torque on existing values. Actuators in new valves are sized in accordance with specific friction requirements.

Packing box depths in rotary valves vary considerably based on the type of valve and manufacturers' design. As a result, there can be long boxes or boxes of sufficient depth for a full packing set, but the actual box may be located some distance from the process fluid. Also, short boxes can be found in new and existing valves. Short graphite packings previously described are also applicable in rotary valves.

Other Packing Considerations. In long boxes or rotary valves with rather long stems, an important consideration is stem flexing or bending slightly. This stem movement away from the center line of the stem wears packing rings, leading to an emissions increase. In addition, the stem must be straight and without any curvature, which should be ascertained before accepting the control valve for service. Since less flexible stems are difficult to obtain, bushings or sleeve-bearing bushings should be considered to minimize stem flexing. This is not difficult in long packing boxes, but short packing boxes do not have a sufficient space to install these close-fitting bushings. Some control valve manufacturers supply guide bushings on the inside of the gland follower to aid in positioning the stem and minimize stem flexing. All of the guide bushings in sleeve-bearing service should have a minimum clearance of about 0.005 in. (0.127 mm) and a maximum clearance of 0.015 in. (0.331 mm).

In addition to the guide bushings, the stem finish is important, along with the bushing material. For the standard graphite packing sets, a stem finish of 28–32 RMS is satisfactory. However, the guide bushing material must not gall or score the stem, requiring a material compatible with the service. Some metal and nonmetal materials are available commercially and are frequently installed in this type of application. In addition to the stem finish required with these guide bushings, a packing box wall finish of 125 RMS is usually found on the box walls.

Live Loading. This is a controversial subject; some packing manufacturers favor spring loading and others recommend graphite packing installations without spring loading the gland follower.[4,6] The subject of live loading was reviewed in Chapter 7 and the conclusion was that live loading is not required with a preformed graphite packing density of 70 lb/cf (1.12 g/cm^3) in a standard valve. With high-density graphite rings greater than 87.5 lb/cf (1.4 gm/cm^3), a much higher gland pressure is required to compress the graphite rings and expand the rings against the stem and packing box wall. In these situations, live loading may be recommended, particularly in inaccessible locations.

In situations where valve modulations are > 5000 cycles, live loading is attractive to maintain packing consolidation, particularly in rising stem valves. This cycle estimate is based upon full cycle movement of the stems, but control valve stems rarely have a full cycle movement. Stem movement for process adjustment is quite small, but an estimate of cycle movement has been developed by some control valve manufacturers. In processes with long periods between turnarounds

(3–5 years), the number of full cycles should exceed 5000 by a considerable margin. For short turnaround periods, full cycle life may be shorter.

Live loading is expensive and in flexible graphite packing applications should be carefully evaluated to ensure the system will provide an additional measure of emissions control compared as described in Chapter 7.

8.4.3 Expandable Valve Stem Packing®

The Expandable Valve Stem Packing (EVSP)® sets described in Chapter 7 for manually operated block valves are also installed in both rising stem and rotary control valves. The configuration of these packings results in a large amount of radial expansion, as shown in Figure 7.23 and 7.25. Radial expansion capability is about $\frac{1}{8}$ in. -(3.2 mm) for EVSP® packing compared with 0.02 in. (0.51 mm) for standard flexible graphite rings,[4] although the latter can be improved somewhat by lower density preformed rings. As a result, emissions control is considerably superior for EVSP® packings compared to standard preformed flexible graphite ring packings.

The EVSP® packings have been installed in rising stem and rotary control valves along with installations in large automatic remotely controlled block valves operated at high temperatures and pressures. However, most of the packing applications are Simplified EVSP® packings which have 7 rings in their basic configuration compared with 11 rings in the standard EVSP® packing set. The standard EVSP® set has largely been supplanted by the Simplified® set, and installations for the former are generally limited to specialized applications.

Long Box Packing Sets. Simplified EVSP® packing sets are installed in long packing boxes as a single or double set in rising stem control valves. There is excess volume in the long stuffing boxes, particularly with a single packing set installation, and the excess for one packing set is filled in with spacers and lantern rings. These spacer rings are generally installed by the control valve manufacturer when the type of packing has been defined by the purchaser. When two packing sets are installed, they are separated by a spacer or lantern ring, as shown in Figure 8.5. When extra volume still exists in the packing box, the volume is filled with a bushing. The control valve packing manufacturer will provide assistance in determining the layout of the packing rings, spacers, and bushings.

As noted in Chapter 7, carbon braided end rings are necessary to prevent extrusion of the graphite rings and contain or wipe graphite materials adhering to the valve stems. This is important at guide bushings, the gland follower, and the spacer or lantern rings. The spacer and lantern rings must have some clearance at the stem and the ring materials must be compatible with the stem. Galling or interfering with stem travel must be prevented; selecting the proper metals and following the recommended clearances eliminate potential problems.

Leakage. In control valves, a single Simplified EVSP® set has been sufficient to control emissions or leakage in a long packing box set to levels below 5000 ppmv. However, with the new HON emission leak standard of 500 ppmv and maximum valve leakage frequencies of 2%, dual packing set installations may be preferable in long boxes to ensure compliance. Dual packing sets with a lantern ring provide an alternative in the event leakage through the packing sets reaches unacceptable levels. Although this situation is somewhat improbable, leakage may occur as a result of other factors, such as stem flexing. Connections through a port would be similar to those described in Section 8.4.1 for standard flexible graphite packing systems.

Packing Installation. As shown in Figure 8.5, one or two Simplified EVSP® packing sets can be installed in a deep or long packing box. However, the packing installation should be carefully reviewed with the packing manufacturer, particularly the dimensions of the packing box and compression procedures. Actual placement of the rings in the packing box for the Simplified EVSP® packing is straightforward since the rings can only be installed in one direction. This contrasts with the standard 11-ring EVSP® set where the rings can be installed in an improper sequence.

The entire Simplified EVSP® set is installed in the packing box and then compressed the compaction distance recommended by the manufacturer. Stroking of the valve should occur after the packing is installed and the packing is recompressed if necessary. However, recompressing does not require a large compression change. Frequently air entrapment or small misalignments occur and stroking improves valve operation. When adjustments are required after valve operation commences, the gland follower nuts are turned one flat at a time in conjunction with the monitoring concentrations.

In dual packing systems, the lower set is completely installed and compressed before the upper set is installed. Compression of the lower set is conducted with the extension rings previously described. The spacer or lantern rings are then installed and the upper packing set is placed on the topmost spacer ring and the upper packing set is then compressed to the proper compaction level. Compression is conducted in accordance with the manufacturer's recommendations. This will also include stroking the valve and adjusting the compression pressure if necessary.

Friction. Although constructed of flexible graphite, the Simplified EVSP® packing does not require as much stem actuation torque as the standard flexible graphite packing. An increase of about 30% above PTFE® packing torque requirements is estimated by the manufacturer as necessary to compensate for increased friction. Although an increase in actuator requirements, the increase is significantly less than the added torque required for standard graphite packing rings described in Section 8.4.2. This lower torque requirement for a graphite set is accomplished with 28–32 RMS stem and 125 RMS packing box wall finishes.

The reason for this difference in the performance of Simplified EVSP® compared with the standard flexible graphite packing is not completely apparent. However, a difference in the graphite packing surface area that is in contact with the stem may account for some of the difference in friction. In the standard flexible graphite packing, all of the rings are compressed at relatively high pressure where the initial packing densities are greater than 87.5 lb/cf (1.40 m/cm^3) to ensure adequate emission control. The three flexible graphite packing ring inner annular surfaces are pressed against the stem, which contrasts with Simplified EVSP®, where one ring is fully pressed against the stem and the other two rings have a minor thrust against the stem. In addition, the low-density graphite rings of

65 lb/cf (1.04 m/cm^3), although compressed, may be more flexible than the preformed graphite rings under compression.

Since friction is lower with the Simplified EVSP® set, the hysteresis effect is much less pronounced than hysteresis effects with the standard flexible graphite packing set. However, comparisons with PTFE® hysteresis have apparently not been published and a direct comparison of PTFE® and Simplified EVSP® hysteresis is unavailable. Hysteresis effects somewhat close to those observed with PTFE® packings are indicated by the packing manufacturer.

Short Box Packing. In some rising stem valves, the packing boxes are short and cannot accommodate the full Simplified EVSP® sets. Where the box length cannot be adjusted, shorter versions of the Simplified EVSP® set available as described in Chapter 7. The short set arrangement shown in Figure 7.26 is installed in most of the short boxes where the braided end rings are PBI® braided thin rings approximately $\frac{1}{8}$ in. (3.2 mm) thick. If somewhat larger end ring thicknesses are permitted, a modified braided carbon end ring can be installed.

For applications with very short packing box lengths the modified EVSP® set shown in Figure 7.26 can be installed. This packing set has soft metal end rings that will not gall the valve stem. Both the short and modified sets have been installed in various valves with apparent success, but a comparison of emission capability is not available.

Rotary Valves. The various EVSP® packings can be installed in rotary valves. These packings do not increase friction sufficiently to warrant an increase in actuator torque on existing valves. New valve actuators can be sized in accordance with actual friction requirements.

As described in Section 8.4.2, packing box depth varies considerably in rotary control valves based upon manufacturers' designs. Full-size packing sets or short sets may be required for the valve. In addition, long packing boxes permit dual packing sets with ports from leakage control through lantern rings. The packing box may also be located a distance from the process fluid in some of the horizontal valve designs. Packing box designs and stem movements or wear effects should be investigated to ensure satisfactory, long-term packing operation and emission control.

Other Packaging Considerations. In rising stem and rotary valves, stem flexing or bending has an effect upon packing and emission control, as described in Section 8.4.2. Long box rising stem valves have very long stems and horizontal rotary valves frequently have long stems that are extended to connect with the turning mechanism. A significant number of rotary valve designs are based on a horizontal design. Since stem flexing can affect the packings by gradually wearing the packings in specific areas and/or significantly reducing packing life, emissions control may be difficult. Where heavier nonflexing shafts are unavailable, close tolerance guide bushings or sleeve bearings should be installed to minimize shaft stem flexing. Moreover, the stems should be straight before commencing control valve operation, requiring quality control by the manufacturers. In short packing boxes, insufficient space may prevent the installation of sleeve bushings, but some manufacturers provide guide bushings on the inside of the gland follower to aid in positioning the stem and controlling stem flexing. Again, guide bushings in sleeve bearing service should have a minimum clearance of about 0.005 in. (0.127 mm) to 0.015 in. (0.381 mm).

Since many of the rotary valves are horizontal, total stem weight and gravity tend to wear the packing into an oval slope, which increases emissions around the stem. Close fitting guide or sleeve bushings reduce the amount of flexing. However, a stiffer shaft stem will reduce the flexing problem. In the horizontal rotary valves, some of the *driver mechanisms* may also produce an oval-type packing opening. Some data or information on horizontal control valve packing life and emissions should be obtained from the control valve manufacturer.

The guide or sleeve bushings should not gall or score the stem, requiring a bushing material compatible with the stem. Both metal and nonmetal guide bushing materials are available for this service. However, the bushings and packings should perform satisfactorily with stem finishes of 28–32 RMS and packing box wall surface finishes of 125 RMS.

Live Loading. With EVSP® packings live loading or springs on the gland followers may not be necessary. The correct loading is placed on the packings and very small adjustments over long operating periods will maintain emissions monitoring concentrations below the 500-ppm leakage criteria level.

8.4.4 ENVIRO-SEAL®

This graphite-based packing system, a proprietary product of Fisher Controls, was originally developed for high temperature and high pressure steam service and was termed HIGH-SEAL® packing. However, the manufacturer improved the packing system to meet VOC and HON requirements on a consistent basis. The improved packing system has been designated ENVIRO-SEAL® to signify effective emission control in HAP service, specifically where a 500-ppmv leak concentration definition is applicable.

Since the ENVIRO-SEAL® packing is similar to the HIGH-SEAL® packing, the HIGH-SEAL® packing is described initially, followed by the modification that differentiates the two packings. The information on the packings indicate this is an extended packing set which will not fit into a short packing box. Fisher manufactures control valves and their proprietary packings are designed for the packing boxes in Fisher values. The graphite-based packings are normally installed in rising stem valves, with nongraphite packings installed in either rising stem or rotary valves.

HIGH-SEAL® Packing System. Fisher graphite systems were originally a typical five-ring flexible graphite packing set with four preformed rings. In high-temperature and particularly high-pressure service (4000 psig, 276 bar), this packing set was not considered satisfactory, apparently because of deficiencies in the braided graphite end rings. The braided graphite appeared to break down at the high-pressure level and deformation of the braided graphite rings increased stem friction. Details on the preformed flexible graphite and braided graphite rings were not provided.[5]

Fisher then developed a graphite packing set for high-temperature, high-pressure service based upon composite rings replacing the braided graphite end rings. The composite rings are an extruded mixture of graphite and amorphous carbon that are considerably less resilient than the braided graphite end rings, but are still somewhat pliable. Composite rings confine the preformed flexible graphite rings from expanding through the annular openings against the stem and the packing box wall. However, the composite rings in turn are controlled or confined by carbon bushings, which are very hard and do not have any flexibility. As a result, the packing arrangement shown in Figure 8.6 has a gradient of deformability, with the preformed graphite rings the most pliable material.

In this packing system, the carbon bushings control the extrusion of composite material, which in turn controls the extrusion of the preformed flexible graphite rings. In addition, the tight or close clearance of the carbon bushing against the stem contains the graphite materials in the stem area between the carbon bushings. A third function of the carbon bushings is in acting as a stem guide or sleeve bearing. The packing system in Figure 8.6 also indicates a long guide bushing is placed at the bottom of the packing that acts as a spacer. In the graphite packing

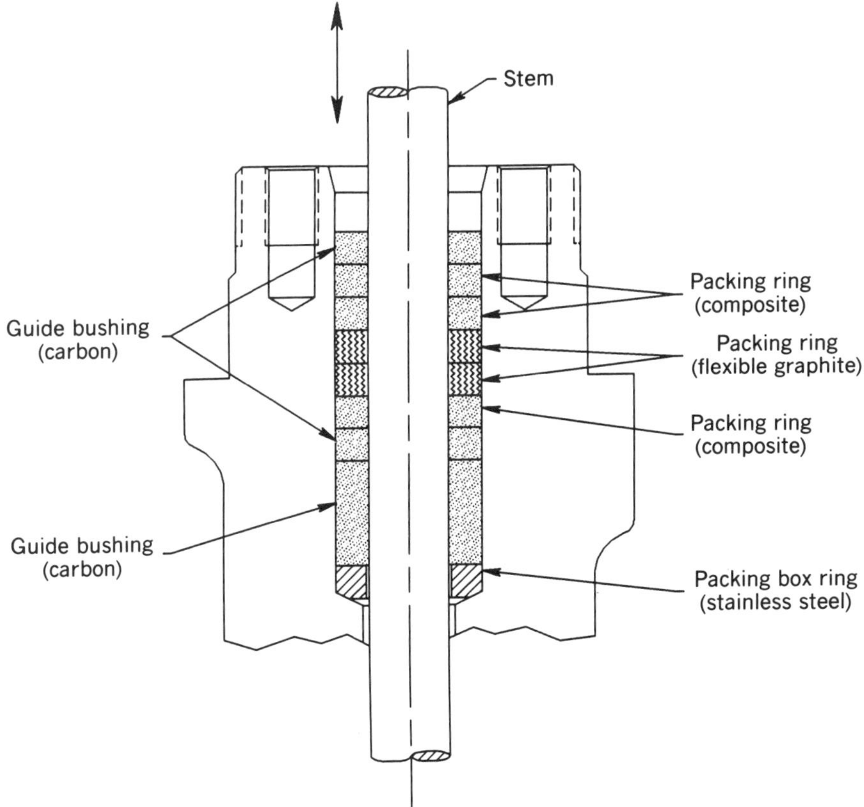

Figure 8.6 *HIGH-SEAL® graphite packing system rising stem control valve (courtesy of Fisher Controls).*

systems, metal spacers are generally not installed and carbon bushings are the standard spacers that also act as a guide bushing.

Stem Guides. Fisher indicates very high compression pressures are necessary with preformed flexible graphite packing rings to institute sealing which tends to move or bend the stem. The guide bushings minimize or eliminate stem flexing and improve emission control. A lined gland follower can be used as a stem guide, but these linings in the follower are limited to PTFE inserts and cannot be used where the temperatures are greater than 500°F (260°C). Consequently, stem guides are limited to bushings in the packing box.

The coefficient of thermal expansion for the carbon bushings are similar to the expansion coefficients of the stem and metal housing. This results in maintaining the clearance tolerances under various temperature conditions. However, the carbon bushings are also brittle and may crack under high compression loadings. Fisher indicates the bushings in this situation maintain their integrity although cracks exist and continue to perform their function.

Friction. Stem friction as described previously for the standard 5-ring preformed flexible graphite packing is considerably higher than the friction with PTFE packing. This higher friction load may require a larger actuator in an existing valve or specification of a large actuator on a new valve. Installation of the HIGH-SEAL® graphite packing system reduces the friction load below the friction levels demonstrated with the standard five-ring packings. Data published by Fisher indicates[5] that the friction load for the standard 5-ring set is about twice the friction load exerted by the HIGH-SEAL® packing at the same packing stress level.

One of the requirements for the HIGH-SEAL® and ENVIRO-SEAL® graphite packings is a stem finish of 4 RMS, which minimizes friction from the carbon bushings and the graphite packing rings.

Live Loading. Spring systems for live loading HIGH-SEAL® and ENVIRO-SEAL® packings are shown in Figure 8.7 Live loading in this packing system is considered effective in maintaining compression stress on the packing. According to the manufacturer this loading is required to compensate for the loss of packing volume due to packing extrusion, loss of entrapped air, and friction. In addition, the live load compensates for thermal cycling and volume variations under these conditions. As the losses in volume are experienced, the actual spring force on the packing falls somewhat as the springs expands. However, the spring is normally designed for much higher loadings and the small spring extension will not significantly lower the compression pressure.

The springs are generally compressed about 85% of full travel. An indicator on spring installations shows when the required compression level is attained. An example of an indicator is shown with the high temperature springs in Figure 8.7. Other spring systems are compressed to the desired level with a torque wrench based on torque recommendations from the control valve manufacturer. The sprints in these live loading systems are Belleville springs formed from washers that are flattened truncated cones. The washers are assembled into the right number for a particular size with loadings determined by the valve manufacturer. Although only two springs are shown in Figure 8.7, a larger number may be used, depending upon control valve design and compression requirements.

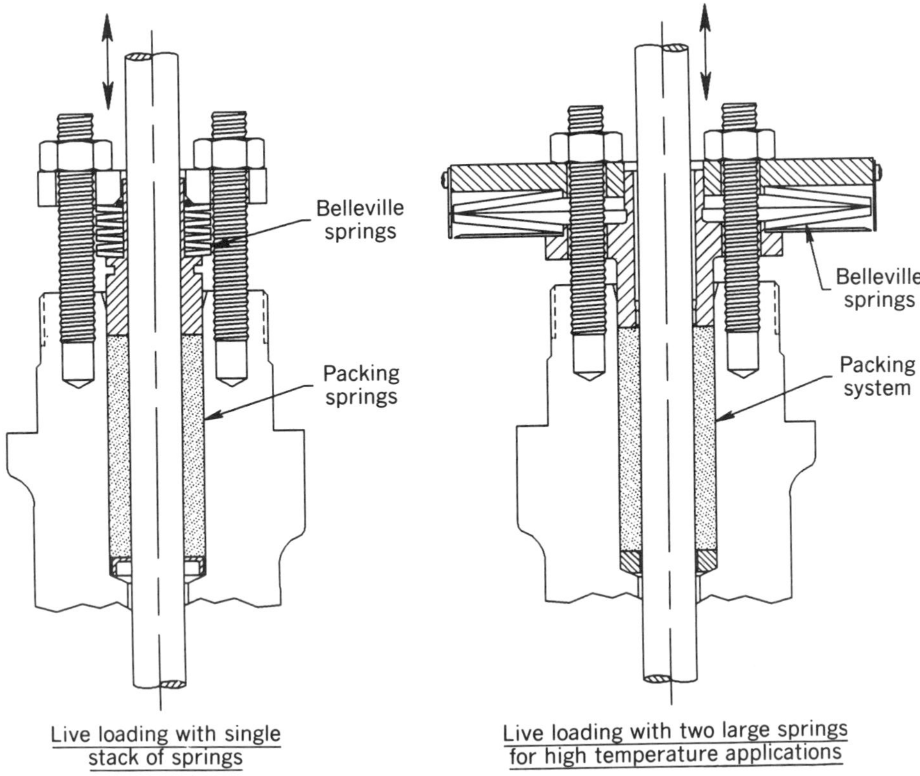

Figure 8.7 Live loading spring systems rising stem control valve (courtesy of Fisher Controls).

Since many packings were developed with live loading, optimum performance for the packings may require the application of springs.

ENVIRO-SEAL® Packing Systems. The control valve manufacturer indicated that the leak rate for the HIGH-SEAL® graphite packing set would not consistently conform with the new EPA definition of a 500-pmv leak. The HIGH-SEAL® graphite packing system was then modified to provide lower leakage values on a more consistent basis for packings required in fire-safe or high-temperature applications.

Packing Modifications. The basic modifications in the HIGH-SEAL® packing system is the insertion of thin PTFE wafers between the various rings in the packing set. This change resulted in packing emission concentrations consistently lower than the 500-ppm leak standard. However, the PTFE thermal decomposition temperature varies between 650 and 700°F (343–371°C), and the valve operating temperature should be reviewed with the manufacturer where the operating temperature is greater than 650°F. Moreover, PTFE becomes elastic and mechanical properties change well below the decomposition temperature, suggesting a review of operating conditions with the control valve manufacturer. Thus the maximum valve temperature operating conditions must be clarified to define temperature limitations.

Fire Safety. Although PTFE wafers are incorporated in the packing, the PTFE only constitutes 2.3% of the total packing set volume. Consequently, under fire test conditions decomposition of the PTFE will not affect the graphite and a seal will remain that prevents excessive leakage.

Sealing Mechanism. The actual sealing mechanism that controls emissions has not been determined.[5] However, the manufacturer indicates that under high compression stress and elevated temperatures, the PTFE will flow into microscopic channels or imperfections in the stem that aids in sealing these leak paths. Since the PTFE is on the stem surface, PTFE lubricity maintains a smooth interface between the packing and stem that minimizes stem and packing wear. Few channels or changes occur that create leak paths. By acting as a lubricant, the amount of graphite adhering to the stem is reduced, lowering graphite material loss.

Stem Finish. The stem finish recommended for the high temperature ENVIRO-SEAL® packing is 4 RMS, which results in an extremely smooth stem.

Emission Data. Control valve manufacturers frequently develop emission loss data in the laboratory and data were obtained that compared the ENVIRO-SEAL® graphite packing with emissions from the standard five-ring flexible graphite preformed packing set. In these tests, an integral part of the testing procedure is cycling the valve stem, which tends to increase leakage as the number of cycles increases.

Fisher tested a control valve with a 1-in. (25 mm) travel range at a pressure of 300 psig (22 bar). The media was air at 70°F (21°C) and air leakage was 0.001 mL/min for 10,000 cycles, which gradually increased to 0.002 mL/min at 100,000 cycles. For the standard five-ring packing set at 3000 cycles and 100 psig (7.9 bar), the leakage rate was 7.5 mL/min. At 300 psig (22 bar) the leakage was 21 mL/min at 1500 cycles. These data show a significant difference in emission leakage for the two different packings with the ENVIRO-SEAL® considerably superior under these conditions.

8.4.5 Argo Packing

This graphite-based packing system was developed by the Argo Packing Company independently of other packings and resembles the ENVIRO-SEAL® system. The original basic valve seal packing was the standard five-ring flexible graphite packing set consisting of preformed graphite rings and graphite braided end rings. However, this packing set did not meet emission control requirements and performance debits were apparently a result of problems associated with the braided graphite end rings. The braided graphite yarns in the end rings fractured and the compression pressure was converted to radial load within the braided ring, thereby reducing the axial load on the preformed rings. In addition, the expansion of the braided end ring against the stem increased friction. Other operating debits were also found with the graphite braided rings.

Stem friction requirements are important in determining or sizing actuator power loads. A method of calculating stem friction was developed based on the ratio of axial to radial stresses for packings. In addition, the conversion of axial to radial stresses provides a measure of packing ring effectiveness in meeting individual ring performance objectives. An equation for estimating stem friction follows:

$$F = 3.1415 S_1 f Y D_5 L \tag{8.1}$$

where

F = stem friction (lb)

S_1 = packing set compression stress (psi)

f = coefficient of friction between the stem and packing

D_s = valve stem diameter (in.)

Y = axial to radial stress ratio in the packing

L = packing set uncompressed length (in.)

The ratio of axial and radial stresses can be determined from the curves shown in Figures 8.8 and 8.9 developed by Argo in defining the effectiveness of individual rings and packing sets. In Figure 8.8 the braided graphite end ring essentially converts all of the compressive stress to radial loadings, as shown by the transfer coefficient. The composite ring in the normal compressive stress range converts a little more than 50% of the axial load to radial stress. A complete packing set in

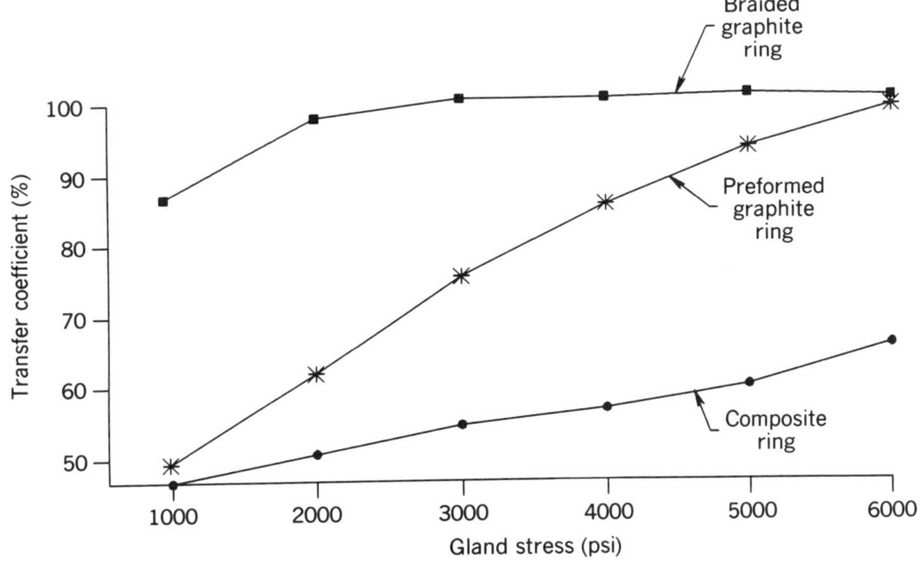

Figure 8.8 *Graphite packing transfer ratio (courtesy of Argo Packing Co.).*

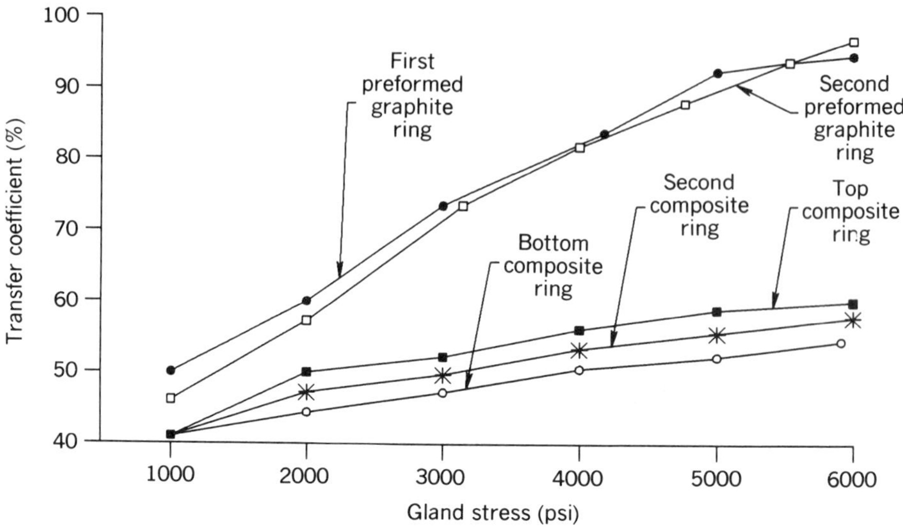

Figure 8.9 *Transfer ratio of five-ring graphite packing sets (courtesy of Argo Packing Co.).*

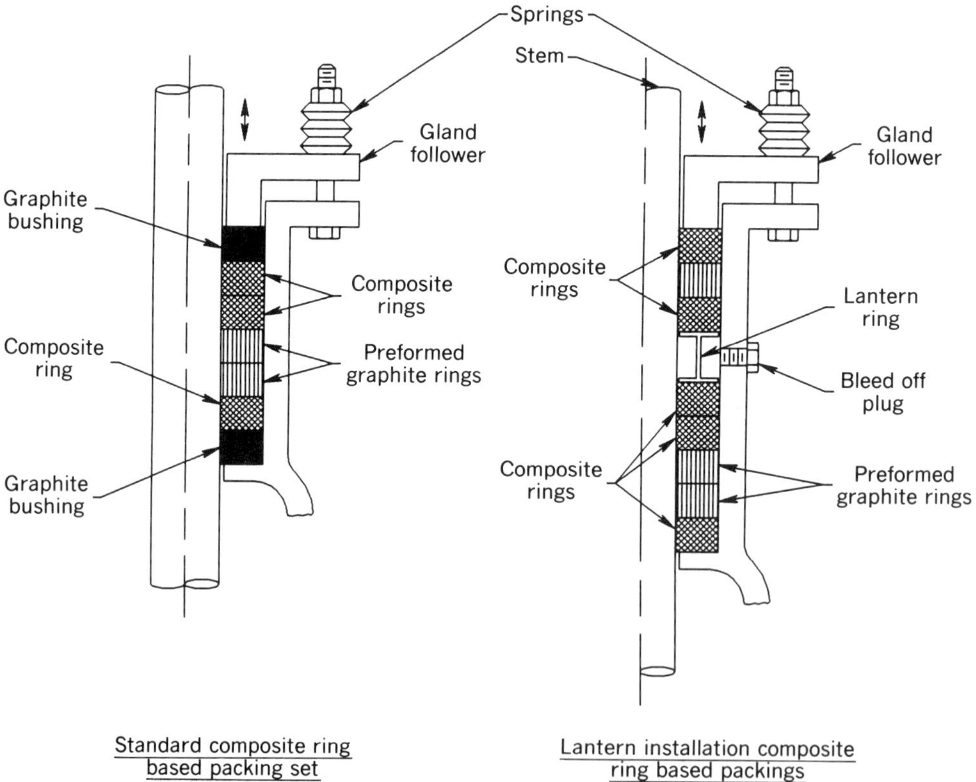

Figure 8.10 *Live-loaded composite ring packing systems rising stem control valve (courtesy of Argo Packing Co.).*

Figure 8.9, similar to the packing set in Figure 8.10, converts axial loads to radial stress in the rings designed for expansion and sealing. As the compressive or gland stress increases, the composite rings transmit axial loads at a relatively constant rate. The preformed flexible graphite rings that are designed for expansion show an increasing conversion from axial to radial stress as the compression stress increases. The preformed flexible graphite density curve is based upon an 89 lb/cf (1.43 gm/cm^3) ring density. This curve will shift with changes in the preformed density.

Based upon this effort and additional development work, a basic packing set was designed, which is shown in Figure 8.10. This new packing set has been an improvement over the original five-ring graphite set originally marketed with preformed graphite ring densities of about 89 lb/cf (1.43 gm/cm^3). The components of the new packing system are described in the following paragraphs.

Composite Rings. These rings form the basis for replacement of the braided end rings. The composite rings are extruded graphite rings that may have some fibrous material in the ring. This composite ring, similar to the composite ring described in Section 8.4.4, is hard, nonpliable, and somewhat brittle. However, the composite ring has certain advantages:[7]

1. The composite rings do not fracture in cyclic operation, which was a problem with braided graphite.
2. Axial load is transmitted directly to the preformed flexible graphite rings, increasing the axial load in these rings and improving sealing.
3. The composite ring has a low axial loading effect on the stem.
4. Packing life has been extended.

Preformed Graphite Rings. The preformed flexible graphite ring densities in the packing set are 89 lb/cf (1.44 gm/cm^3), which require a minimum pressure of 3000 psi (207 bar).[7] Since the preformed flexible graphite rings expand, the composite ring prevents extrusion of the graphite rings and the composite ring performs a necessary function.

Graphite Bushings. These bushings are guide bushings that prevent flexing or movement of the stem and are essentially sleeve bearings. Since clearances are small, the stem of the valve must have a smooth surface and the finish on the stem is generally in the 4- to 8-RMS range, where 8 RMS is a slightly rougher surface.

For optimum control of stem movement, a bushing is placed at each end of the set. This configuration is installed for both vertical and horizontal valve stems.[7] When the packing box is long, the guide bushings are lengthened and provide a large containment area to control stem movement. Side or radial load produced by the stem is apparently exacerbated in pneumatically operated valves where the actuator springs induce a side loading. Since the flexible graphite packing rings are not bearings, both sealing effectiveness and packing life were reduced. This has been overcome with the bushings.

Purge or Leakage System. Where a lantern ring leakage control system is required or specified, dual packing sets are installed. However, the dual sets increase friction significantly in this system and the packing installation has been modified to ensure the lower set in Figure 8.8 seals properly.[7] As a result, the packing sets have been modified, with an upper or second packing set that minimizes the number of packing rings and consists of one preformed flexible graphite ring and two composite rings. The rings in the upper set should have a density that is about 10% higher than the ring densities in the lower or primary set.[7]

The primary or lower set can be the standard five-ring set previously described without the bushings, as shown in Figure 8.10. Alternatively, a four-ring set can be installed where two preformed flexible graphite rings are placed between composite end rings. This basic configuration does not have any bushings to control stem movement. When the lantern ring combination is installed in a long box, guide bushings added to the packing set will fill the added void volume and minimize the stem movement. These guide bushings are installed between the gland follower and top composite ring in the upper set and at the bottom of the packing box in the lower set.

Live Loading. The rationale for all of live loading with this packing system is similar to the comments presented in Section 8.4 and all of these packing systems are live loaded.

Rotary Valves. While the description of this packing set concentrated on rising stem valves, this packing is also installed in rotary valves with satisfactory results.

8.4.6 Other Graphite Packings

A few of the graphite packings commercially available were described in this section. However, alternative packing designs have been commercialized and can be evaluated preferably through plant testing or other test procedures when plant testing cannot be conducted. Since graphite packings are fire safe, any evaluation must consider whether fire-safe packings are required. Where graphite packings are considered necessary, comparative emission data are required for an objective analysis. However, selection of a packing from a carefully conducted test program will not ensure satisfactory operation unless the packing is installed properly, maintained, and placed in a packing service where the stem and packing box meet the mechanical requirements for satisfactory packing performance.

8.5 NONGRAPHITE PACKINGS

Teflon® or PTFE has been used for many years in control valve packings where operating temperatures are less than 425°F (218°C). Excellent resistance to a broad list of chemicals, very low friction characteristics, and reasonable operating life at temperatures relatively close to ambient levels with satisfactory emissions control were attractive characteristics for various applications. These PTFE packings were generally braided packings that were also reinforced with asbestos in some configurations. In addition, V-type PTFE rings also became available that served as an alternative packing. In rotary valves, V-type PTFE rings have lower emission concentrations compared with braided PTFE packing as the number of cycles increases.[4] Moreover, stem friction is lower with V-type rings.[4] However, in rising stem valves V and braided PTFE packings demonstrate identical emission concentrations at ambient temperature conditions.

A major problem with PTFE packings is thermal expansion at elevated temperatures since thermal expansion of PTFE is considerably greater than the thermal expansion characteristics of steel. In this situation, temperatures less than the maximum operating range of 425–450°F (218–232°C) may be recommended for PTFE. At these elevated temperatures, PTFE expands, increasing the load on the packing and resulting in extrusion. For constant operating temperatures, the gland load can be adjusted, maintaining a reasonable amount of packing load and emissions control. If the valve has cyclic temperature changes, lower operating temperatures result in packing load decreases as the PTFE packing contracts and leakage can occur. With increasing temperature the packing load will increase but will be lower than the previous packing load level, demonstrating a loss of packing material.[5]

Since the control of valve emissions has become increasingly important, braided and standard V-type PTFE packings were considered inadequate for operations with the more stringent emissions requirements of the CAAA. Consequently, packing and control valve manufacturers have developed a number of different packing designs to meet the 500-ppmv valve emission leak definition. Several valve packing designs are presented in this section that provide a basis for comparing various packing sets. Identification of these packings is difficult because a single material is used in different configurations, manufacturer's designations are used to identify the packings. However, the nongraphite packings have operating temperature restrictions which should be reviewed carefully prior to specifying packings.

8.5.1 Fire Safety

In general, nongraphite packings are based on polymers that are not fire safe and do not pass the more stringent hydrocarbon fire tests. Fire safety is an important consideration in a variety of processes and applications, as described in Section 8.3. Since hazard risk analysis is required by OSHA and the HON regulations, any packing choice must be carefully selected.

Packing and control valve manufacturers aware of fire safety concerns have added a preformed flexible graphite ring between the top packing ring and the gland follower. During fire tests, the graphite ring is unaffected, but nongraphite rings disintegrate and may be consumed in the test fire. As a result, only one ring is left which is unsupported and without any compression pressure. This ring may hold temporarily in the packing box and meet fire test requirements, but the extent of emission control under full process pressure has not been defined. Moreover, these fire tests are generally conducted on a fresh packing set that is specifically installed for the test, including a new preformed flexible graphite ring. Since the preformed flexible graphite ring is not aged, packing extrusion is not a problem. However, after several months of operation the preformed flexible graphite ring will extrude around the gland follower owing to the lack of an antiextrusion ring between the gland follower and the preformed graphite ring.

Extrusion of the graphite ring has not been considered in fire tests because of new packing set installed for the test. In addition, aging a packing set for fire-test conditions is an expensive procedure and requires some months to prepare the set. Consequently, long-term fire resistance of a nongraphite packing set with a graphite protective ring has not been established. These observations should be considered when specifying packing sets for fire-safety operations.

8.5.2 ENVIRO-SEAL® Packings

The ENVIRO-SEAL® packing systems are a development of Fisher Controls, which is a control valve manufacturer. These packings are designed for Fisher valves, but have also been installed in non-Fisher valves. Somewhat different packing designs were developed for rising stem and rotary control valves, as shown in the following sections.

Rising Stem Control Valves. In conventional V-shaped PTFE packing sets, voids exist between the packing rings when they are installed and compressed. As a result, the compressive load falls in about 1 hr after the load is applied and the packing requires adjustment.

In the ENVIRO-SEAL® design, the V rings are shaped to mate closely, minimizing void consolidation. Although termed "V rings," the actual rings have an arched shape which is slightly different than the rings shown in Figure 8.11. While the main ring in the ENVIRO-SEAL® system is PTFE, an antiextrusion ring is installed on both sides of the PTFE ring. The antiextrusion rings are carbon-filled PTFE rings that do not have the expansion characteristics of pure PTFE materials. These antiextrusion rings confine the PTFE ring and wipe the stem of material collected externally or internally. The antiextrusion rings do not scratch the stem which is highly polished. Normally, the stems have a 4-RMS or better finish.

Although the antiextrusion rings are carbon-filled PTFE rings, there is a very small expansion which results in some extrusion that must be controlled. Hence a washer is placed on each end of the packing set. These outside washers or flat antiextrusion rings are polymeric and are designed for a close fit with the sliding stem. The polymer or possibly composite ring does not mar or scratch the stem.

Packing System. The single packing set shown in Figure 8.11 will work well, but the manufacturer has essentially standardized on double packing set installations. Packing box lengths are generally quite long in control valves, permitting dual packing sets. Although the packing box in Figure 8.11 is not particularly long, additional space or lantern rings can be installed in longer boxes with a packing set at each end of the packing box.

A lantern ring permits bleedoff of any leakage from the lowest packing set. For optimum performance the manufacturer minimizes the number of pure PTFE rings installed in the sets as shown.

Stem Alignment. A follower lined with PTFE acting as a guide bushing is normally installed which minimizes lateral stem motion in the upper part of the packing box.

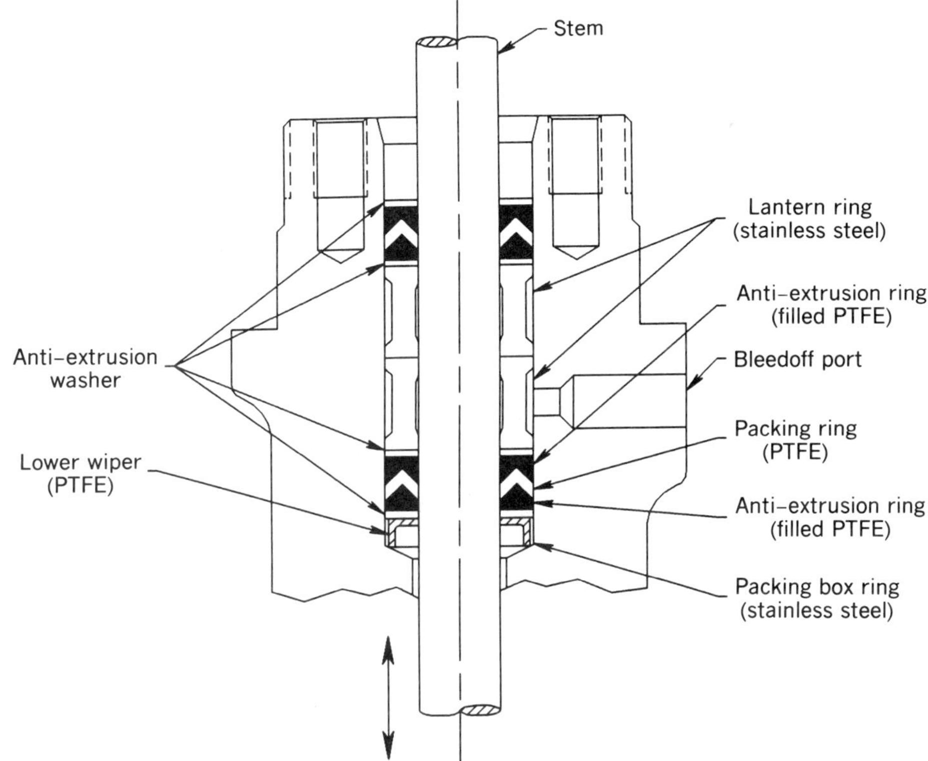

Figure 8.11 ENVIRO-SEAL® PTFE packing system for rising stem control valves (courtesy of Fisher Controls).

Live Loading. A live-loading spring system is installed to ensure and maintain low leakage rates. The springs automatically compensate for increased packing stress resulting from thermal expansion that occurs with higher temperatures.

Operation. The manufacturer indicates this packing system controls emissions (without lantern ring bleedoff) to levels well below the HON mandated level of 500 ppmv. In addition, maintenance is low and service life has been extended considerably, with 750,000 cycles cited as an example.

Rotary Stem Control Valves. Generally rotary valves have a lower ratio of packing to stem distance traveled for each full stroke, compared with a rising stem valve, which results in a much lower material loss as a result of the friction. In addition, dirt and other materials do not enter the packing area from the stem, which can occur in rising stem valves. However, shaft alignment must be maintained with bushings to minimize packing wear. Shaft deflection is a significant problem with many rotary valves, resulting in packing extrusion and a very low packing service life. Attempts to overcome this problem include highly stressing the packing, which permits the packing to deform and follow the stem flexing or deformation. The high packing stress, along with stem deflection, increases the tendency for packing extrusion and antiextrusion control is considered much more difficult in rotary valves than in rising stem valves. Additional information on the ENVIRO-SEAL® Systems for rotary valves is given in the following sections.

Packing Systems. Test results indicated that a two-component antiextrusion ring system similar to the antiextrusion system utilized in the rising stem packing system is an effective method of controlling extrusion. As a result, the packing system shown in Figure 8.12 has demonstrated low emission concentrations and a long service life. This basic packing set for a rotary control valve consists of two V-shaped pure PTFE rings contained between carbon filled antiextrusion rings.

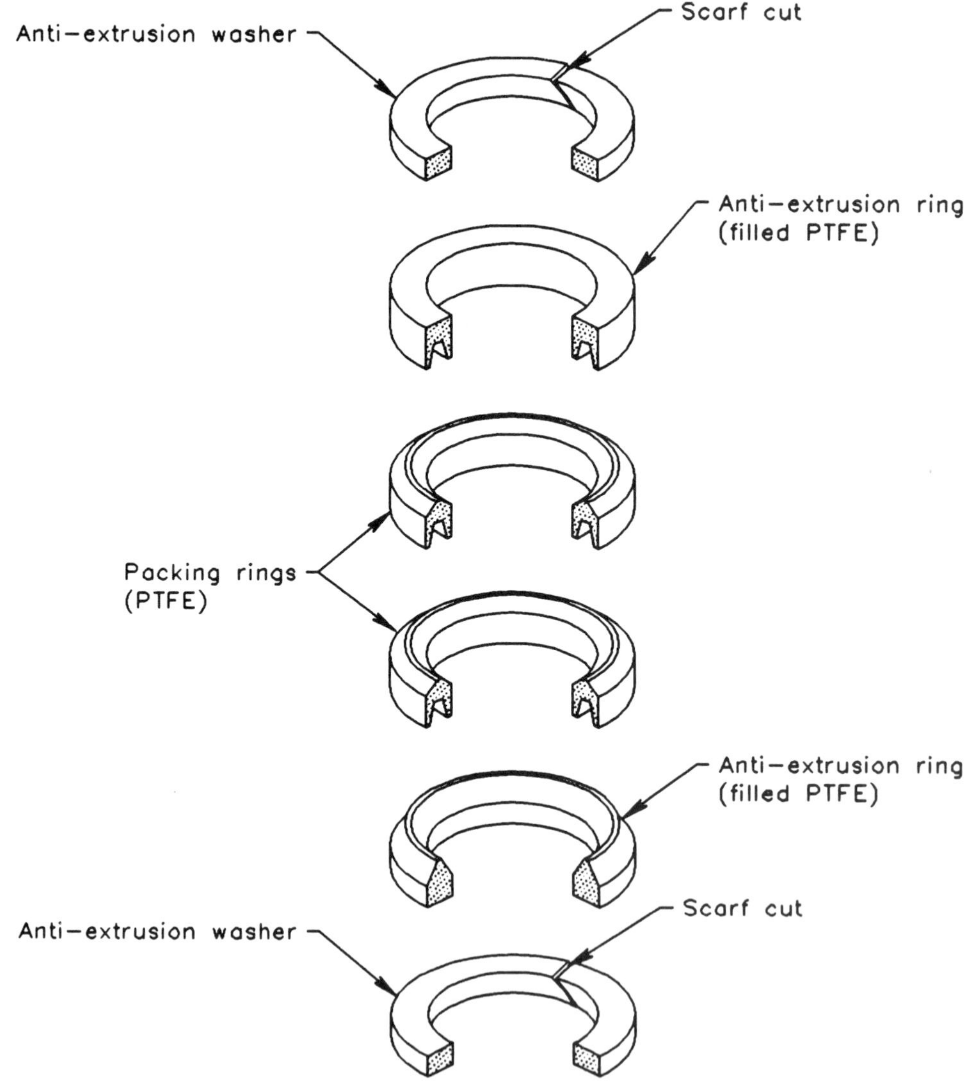

Figure 8.12 *ENVIRO-SEAL® PTFE packing system for rotary stem control valves (courtesy of Fisher Controls).*

The end rings or antiextrusion washers are flat washers constructed of a high-strength polymer.

The antiextrusion washer rings are sufficiently sturdy to prevent extrusion, but the washers will not score or mar the shaft. The clearance tolerance between the washer and shaft is considered zero. However, the washer ring does not increase friction. Moreover, the scarf cut in the washer does not permit extrusion in the opening since the scarf gap closes when the packing is compressed.

Stem Alignment. This problem has been overcome as described by highly stressed packings.

Live Loading. The general problems of the PTFE-based materials and stem alignment require a very high loading pressure, which is provided by springs on the gland follower. The basic requirements for live loading are similar to the requirements for live loading in rising stem valves.

Double Packing Sets. In many rotary valves, the packing box length is limited and a single set is installed. However, lantern ring installations are frequently requested, but the packing boxes may not be much longer and the packing system was modified to reduce the overall length required by two full packing sets. The lantern ring set developed by the manufacturer is shown in Figure 8.13.

In the lantern ring set, a single carbon-filled PTFE ring on each side of the lantern ring is the packing ring, which is controlled by a scarf cut, high-strength antiextrusion washer, and scarf cut adaptor antiextrusion ring of a high-strength polymeric material. For sealing, the packing must be highly stressed to deform the carbon-filled ring.

Operation. The manufacturer indicates the lantern ring system performed well and meets the new EPA requirements without a bleedoff operation.

8.5.3 Valtek® Packings

Valtek is a manufacturer of control valves and has developed nongraphite packings to control emissions from their valves. Although graphite packings can be installed in the control valves, nongraphite packings are preferred, and with the addition of a graphite ring are considered fire safe. This latter consideration eliminates one of the major advantages for graphite. Along with the development of the nongraphite packings, a considerable developmental program confirmed the efficacy of these emissions control packings.

The Valtek packings have been separated into two different classifications on the basis of the specific type of polymer packings. These packings, termed SafeGuard® and SureGuard®, are also differentiated by the need for live loading.

SafeGuard® Packing. The packings for rotary and rising stem control valves are shown in Figure 8.14. The packing set consists of virgin PTFE V rings separated by a carbon-filled PTFE ring and set between transition rings. This basic set consists of five rings in either rotary or rising stem valves. In the rising stem valve, an additional set of rings is placed near the fluid end of the shaft and these

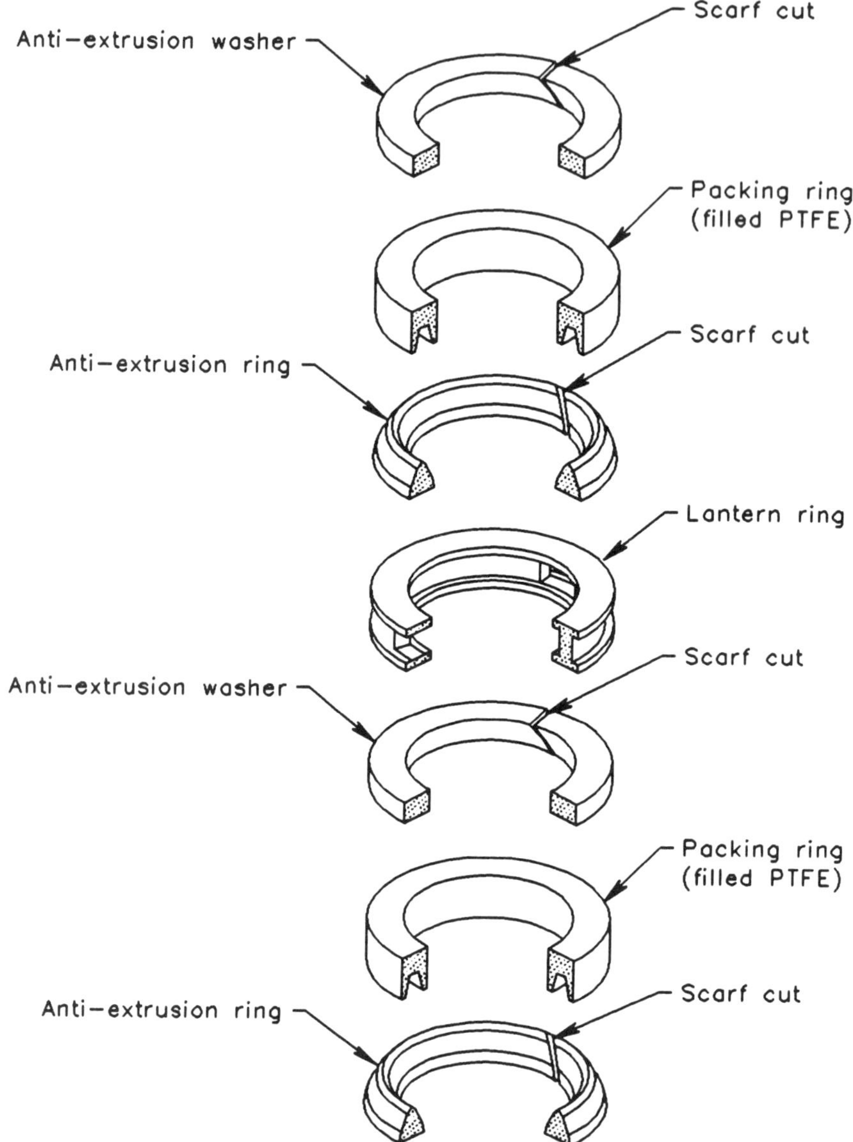

Figure 8.13 *ENVIRO-SEAL® PTFE double packing lantern ring systems for rotary stem control valves (courtesy of Fisher Controls).*

carbon-filled PTFE rings continually wipe the stem, minimizing carryover of process fluid along the stem into the packing box.

The SafeGuard® packing set is also installed in the rotary control valve, as shown in Figure 8.14. A double packing set with a lantern ring and purge port, although shown in the drawing, is not always installed in the rotary valve. The drawing indicates a typical double packing set installation for a rotary control valve. In addition, where fire safety is required, a high-density preformed flexible

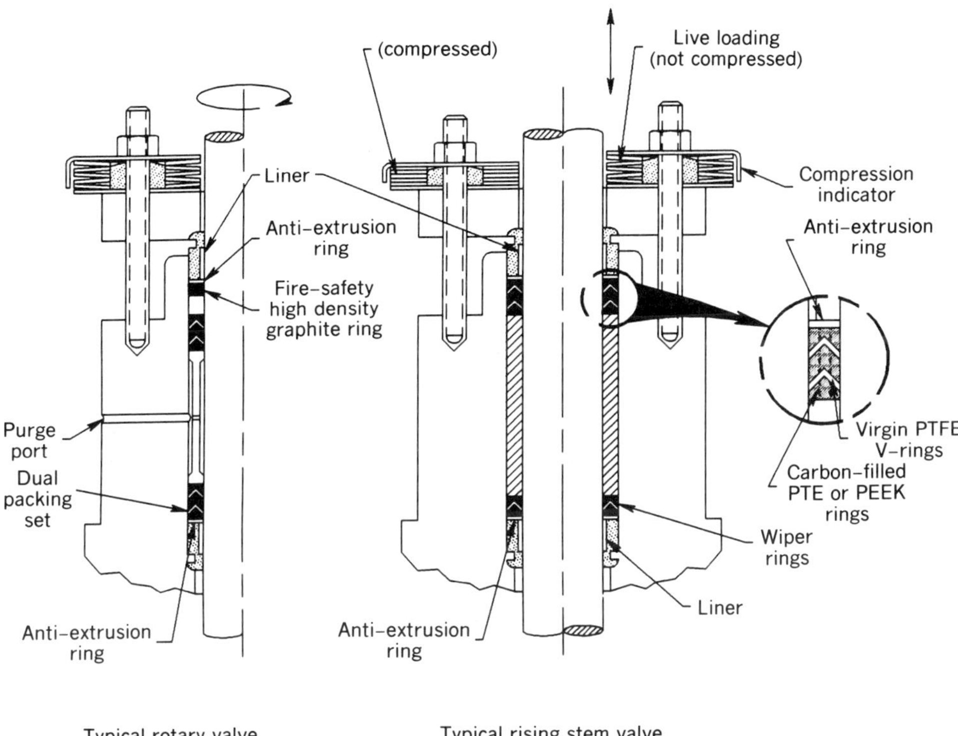

Figure 8.14 *SafeGuard® live-loaded control valve packing system (courtesy of Valtek).*

graphite ring is installed below the gland follower. This high-density graphite ring is separated from the packing set by a solid spacer.

Live Loading. Springs are installed with SafeGuard® packings to obtain optimum performance from the packing set. This packing set can gradually lose compression load over a period of time, but the spring system maintains a reasonably high loading as volume in the packing box contracts with time. Friction and wear along with cold flow of the PTFE ring results in volume changes. Spring compression is set by the manufacturer and the amount of compression required is determined by the indicator or pointer on the spring set.

Stem Alignment. As described in previous sections, stem flexing in both rising stem and rotary control valves affects the packing, resulting in emission losses. Guide bushings or sleeve bearings are frequently installed to control stem movement, particularly in rotary valves that have horizontal stems. While stem movement is a potential problem in most control valves, this control valve manufacturer minimizes the problem by installing larger diameter stems that have a much higher stiffness coefficient or index than stems in other types of control valves.

In addition to less flexible stems, Valtek also has liners that contain the stem and prevent any eccentric motion. As shown in Figure 8.14, there are liners in the upper gland follower and in the lower bushing, which is between the bottom throat opening and the wiper rings or lower packing set. These liners may be PTFE, graphite or a metal, depending upon the installation and processing conditions.

Packing System. As described, the SafeGuard® set consists of virgin PTFE V rings, a carbon-filled PTFE V ring, and V transition rings. Although carbon-filled PTFE cold flow characteristics are much smaller than those for pure PTFE, antiextrusion rings are placed between the packing and the follower or guide bushing. This prevents any extrusion into the liner at the top and bottom of the packings. However, antiextrusion rings are not placed at the other end of the packing set. The low cold flow characteristics and the close clearing of the spacer of filler bushing in these valves prevents extrusion at the other end of the packing set.

The packing system is common for both rotary and rising stem valves. In addition, the installation of a preformed flexible graphite ring for fire safety in a rising stem valve is exactly the same as the safety ring installation in the rotary valve (Fig. 8.14). Also, the wiper rings are not installed in rotary valves as a replacement for the lower packing set shown in Figure 8.14. For single packing sets in a rotary valve, a set of wiper rings is not required.

One unusual aspect of these packing installations is that the manufacturer will install extrusion bonnets on the valve. For high-pressure or high-temperature service, the added bonnet length increases the distance of the packing set from the fluid. In addition, an extended bonnet from a valve in high-temperature service results in a packing temperature that is considerably lower than the fluid temperature. The bonnet extrusion is not insulated and heat is dissipated to the atmosphere directly from the bonnet. Consequently, operation and emission control in difficult service is not affected significantly. However, each application should be reviewed to ensure optimum performance is obtained from the packing. The maximum operating temperature for the packing is about 400°F (204°C).

Dual Packing Sets. Although a dual packing set is shown in Figure 8.14, this manufacturer does not particularly recommend dual seals. With the addition of a second packing set, there is a greater potential for problems due to an increase in the number of rings and misalignments that may occur. Consequently, a single set is normally recommended for most applications, including installations with difficult sealing problems. If specifically requested, dual packing sets will be installed.

Friction. Since PTFE has a very low friction coefficient, additional actuator power is not required for emission control, including installations with an extra preformed flexible graphite ring for fire safety. In contrast, tests with graphite packings have shown that an increase in actuator power is required.

Fire safety. The preformed flexible graphite ring provides fire safety according to the manufacturer, meeting the API-607 fire-test standard and other fire-test requirements. The flexible graphite ring is a high-density preformed ring that has a relatively close tolerance to the stem. However, graphite ring extrusion over a period of time and the subsequent effect on fire safety has not been investigated. Observations indicate that emission monitoring concentrations are stable as cycle age increased. The fire-safety ring installed for rotary valves shown in Figure 8.14

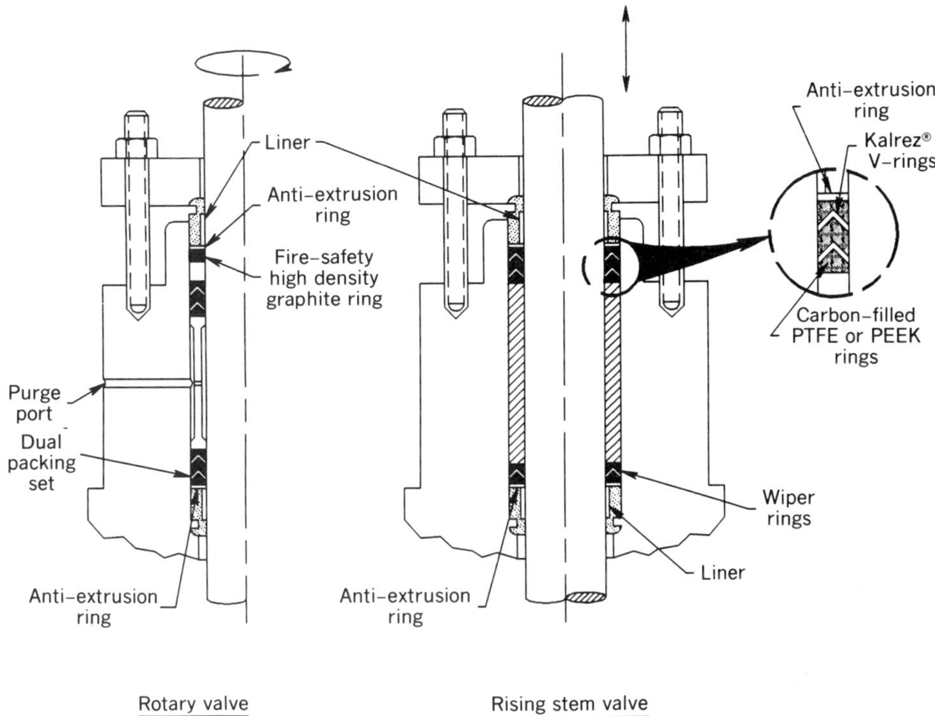

Figure 8.15 *SureGuard® live-loaded control valve packing system (courtesy of Valtek).*

is also installed in the rising stem valve, where a preformed flexible graphite ring adjacent to the gland follower is separated from the packing set by a metal bushing.

Emissions. The SafeGuard® packing system has been successful in meeting the new EPA–HON emissions monitoring requirement of 500 ppmv. The manufacturer apparently guarantees performance during the first year of operation to ensure meeting the specifications. This general operating standard continues over a number of years.

SureGuard® Packing. This packing for both rotary and rising stem valves is shown in Figure 8.15, and is based on Kalrez® packing material. The SafeGuard® set previously described is based upon pure or virgin PTFE material. While one SureGuard® packing set is shown in the rising stem valve, a double packing set is shown in the rotary valve (Fig. 8.15). This difference and the other characteristics of this particular packing configuration, which is similar to the PTFE based polymer, are presented in the following paragraphs.

Packing System. This packing set consists of Kalrez V rings separated by a carbon-filled PTFE ring and set between carbon-filled transition rings. The basic set consists of five rings in either the rotary or rising stem valves. The Kalrez material is perfluorelastomer that is resistant to most chemicals and has a higher operating temperature than PTFE. Elastomeric characteristics are important in a highly efficient emission control system.

The SureGuard® system cold flow characteristics are essentially negligible. Moreover, wear is considerably less for Kalrez compared with PTFE. Cold flow in PTFE packings requires a high packing pressure to insure sealing, which tends to increase wear, since stem sealing is critical in emission controls. In addition, the elastomeric characteristics of Kalrez result in this material acting as an internal spring. Since cold flow is essentially nil and there is an internal spring effect, external live-loaded spring systems are not required.

With Kalrez and carbon-filled PTFE, the operating temperature can be increased to 500°F (260°C). By replacing the carbon-filled PTFE with PEEK, the maximum operating temperature can be increased to 600°F (316°C) or slightly higher. These temperature ratings can be increased to a 700°F (371°C) fluid temperature with the use of an extended bonnet. The PEEK is a polyether-etherketone polymer that can be obtained either as a somewhat flexible material or a very hard, nonresilient material. In the packing set, PEEK is in a somewhat flexible state that permits movement under pressure. Moreover, it does not score or mar the stem, which can occur when the PEEK is hard and inflexible.

Wiper rings of carbon-filled PTFE or PEEK are installed in the rising stem valve, but are not installed in the rotary valve. The rotary valve stem does not carry fluid along the stem and does not require a wiper ring. In addition, antiextrusion rings are also located in the positions previously described for the SafeGuard® packing. These prevent any material cold flowing into the small opening between the liner and the stem.

For the SureGuard® packing sets, the manufacturer will also install extrusion bonnets on the valve. In high-pressure or high-temperature service, the added bonnet length increases the distance of the packing set from the fluid. Also, an extended bonnet from a valve in high-temperature service results in a packing temperature that is considerably lower than the fluid temperature. The bonnet extrusion is uninsulated and heat is dissipated directly to the atmosphere from the bonnets. Consequently, this packing can efficiently perform in a difficult service. However, each application should be reviewed to ensure optimum performance is obtained from the packing.

Stem Alignment. Stem flexing or bending is not a problem in these valves, which have larger diameter stems than most control valves, increasing overall stem stiffness. In addition, any stem movement is contained by liners in the follower or bushing as shown in Figure 8.15. Where SureGuard® is used in high-temperature service, the lower liner against the fluid would be graphite or metal.

Live Loading. As described above, live loading is not required with the Kalrez packing. Moreover, frequent adjustments are apparently not required to maintain emission control.[8] The manufacturer indicated emission concentrations are well below the 500-ppmv regulatory level, with 10 ppmv considered typical in most applications.

Dual Packing Sets. The comments on the SafeGuard® packings and dual packing set installations are also applicable to the Kalrez-based SureGuard® packing. Dual sets for the SureGuard® packing are not normally recommended and are generally installed at the customer's request.

Friction. The SafeGuard® packing based on PTFE V packing has a very low coefficient of friction and is generally considered a base value in friction comparisons. In control valve friction analysis, hysteresis is a major concern. When a packing has a high coefficient of friction, stem movement is affected and this is reflected in a lag in stem movement compared with signal commands. This lag in reaction is termed hysteresis, and data on hysteresis effects in control valves have been determined for various packings. With a PTFE V ring as a basis for stem hysteresis in a rising stem valve, a Kalrez ring packing set has a much lower frictional effect than PTFE and hysteresis is considerably less than the hysteresis effect with a PTFE V-ring packing set.[4] The leak concentration during these studies was less than 100 ppm. As a result, actuator requirements are not changed for the SureGuard® packings.

Fire Safety. The SureGuard® packing set is also offered with a high-density preformed flexible graphite ring for fire-safe applications. Comments on the SafeGuard® fire-safe packing system also apply to the SureGuard® packing set (see Section 8.5.3).

Emissions. According to the manufacturer, the SureGuard® packing set has very low emissions concentration levels and a 10-ppmv monitoring value can be expected.[8] In addition, this packing has a long cyclic life and does not require frequent adjustment. During long runs between turnarounds, little packing adjustment is necessary. Moreover, adjustments are infrequent with changes in pressure and/or temperature.

Emissions data on Kalrez packings are also supported by field information on this packing.[9] Data indicate benzene concentrations during a test on a control valve in a chemical plant over an 8-week period was 5 ppmv. In butadiene service, one control valve during the same period had an emission concentration of 7 ppmv and a second valve had a concentration of 19 ppm.[9] All of these concentrations are well below the 500-ppmv HON regulatory requirements. A more difficult test on cyclic valves in a pressure swing absorption unit in hydrogen service did not indicate any leakage after 6 months. In this service, the valves cycle three times per minute, and in a 6-month period fully cycle more than 500,000 times.[9]

Other. While Kalrez cold flow is negligible, cold flow in the supporting rings must be similar to achieve long life and reliability. When constructed of PTFE, the supporting V rings are reinforced with carbon fibers, approximately 1.4 in. (6.35 mm) in length, forming a composite used in packing sets such as SureGuard®. This composite has negligible creep characteristics up to 500°F (260°C) and 2000 psi (138 bar), resulting in a packing set whose overall component creep rate is quite similar.

8.5.4 Masoneilan® Low Emissions Packing

Masoneilan is a control valve manufacturer that has developed packing systems to provide emissions packings with their control valves[10, 11]. These packing systems were developed for rotary and rising stem valves. There are two basic systems termed low-emission packing (LE®) and emissions-free packing (EF Seal®). The basis for the seal developments are described in the following sections.

***LE Packing*®.** A variety of packings and packing sets were investigated to determine the optimum packing configuration for emission control. The major packing sets investigated were PTFE V rings, braided PTFE or Teflon,® and a combination of Kalrez® and carbon-filled PTFE. This latter material is termed Zymaxx® and is used in packing sets produced by other manufacturers, which were described in Sections 8.5.2 and 8.5.3. The various packing sets were tested at ambient temperatures and a 100-psi (7.9 bar) methane pressure. Emission results in a reciprocating valve installation for these various packings is shown in Figure 8.16. The emissions concentrations with braided PTFE and PTFE V rings climb rapidly within a few thousand cycles compared with the Kalrez®/Zymaxx® combination packing set that does not change emissions concentration over the full cycle test program.

In rotary valves the same packings were tested with somewhat different results, as shown in Figure 8.17[11,13]. The braided PTFE gradually increases in emission concentration as cycle life increases, although the concentration does not exceed 1800 ppmv. However, the PTFE V-ring emissions concentration does not exceed 500 ppmv until the number of cycles reach approximately 175,000. The final PTFE V-ring emissions concentration reaches a maximum of about 550 ppmv, where the concentration remains constant until the end of the test. Although 550 ppmv is not excessive and can be reduced to 500 ppmv, for practical operation under the HON regulations the emissions concentration for operating valves should not exceed 100 ppmv. Preferably, the control valve monitoring concentrations during plant operation should be less than 50 ppmv. This objective was achieved in the rotary valve with the Kalrez®/Zymaxx® packing set, as shown in Figure 8.17[10,11].

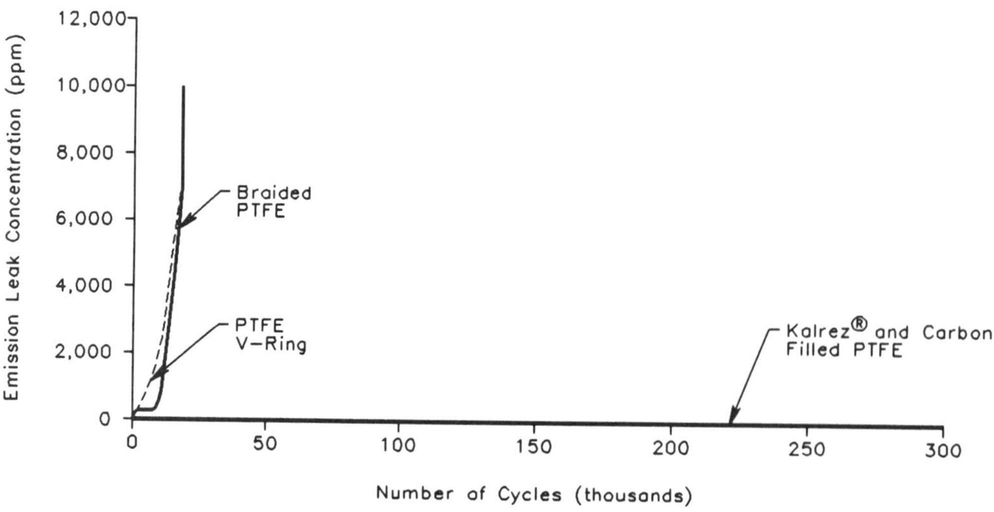

Figure 8.16 *Comparison emission leak concentration in rising stem control valves for various packings (courtesy of Masoneilan).*

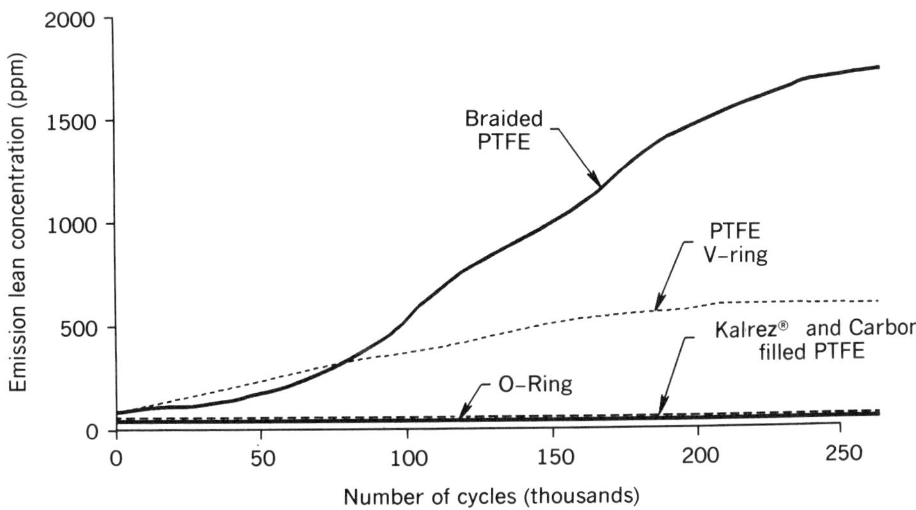

Figure 8.17 Comparison emission leak concentration in rotary control valves for various packings (courtesy of Masoneilan).

Packing Configurations. Based upon these results the Kalrez®/Zymaxx® packing set was selected for the LE® system. However, leakage changes with cyclic temperatures since the expansion and contraction coefficient for the metal differs considerably from the coefficients for the polymeric rings, resulting in a loss in emission control. This indicated the need for live loading and Belleville springs were added to maintain a high pressure on the packing set. The manufacturer tested both internal and external springs to determine the optimum installation and concluded the internal spring system was superior to the external spring system. In addition, internal springs would not be exposed to the elements or subject to damage. The final packing configuration is shown in Figure 8.18.

Live Loading. Although the difference that may exist between internal and external live loading is unclear, testing indicates there is more fluctuation as a result of thermal cycling conditions with an external spring system. These observations were the basis for selecting internal live loading as a standard for the LE® packing system[10, 11]. In addition, gland retightening is normally required periodically to maintain a low emissions concentration with the internal live-loading system maintaining sufficient pressure to control emissions.

Antiextrusion. The long carbon fiber filled PTFE end transition rings butt against the bottom of the box and the spacer rung. Antiextrusion rings are not included in this set. Spacer ring clearance is quite small and provides some antiextrusion capability.

Stem Alignment. Guide bushings are not provided with this packing set, but the close tolerance spacer ring provides guide bushing capability. However, the metal in either rising stem or rotary service must be selected to prevent metal scoring

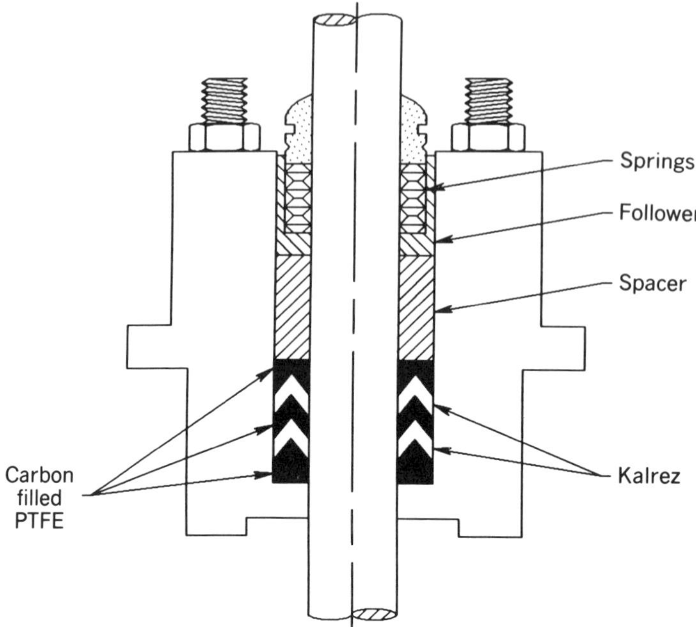

Figure 8.18 *LE® packing (courtesy of Masoneilan).*

and galling. The spacer apparently does not mar the stem and various metals are available for different applications. In this service, stem smoothness is about 4–7 RMS which is comparable in smoothness to other control valves.

While specific guide bushings are not installed in the packing box to control stem alignment or flexing, the valves contain guide bushings in other areas of the valves to control stem alignment. The various rising stem and rotary valves must be examined in some detail to determine where stem guide bushings or sleeve bearings are located. In addition, the actuators may also have a guide bushing. As a result, the Masoneilan packing systems differ somewhat in this respect from the packing systems described in Sections 8.5.2 and 8.5.3. Control valve designs should be analyzed with all proprietary packing systems to ensure that stem alignment and flexing are controlled.

Friction. Data obtained in a comparison of graphite braided PTFE and the Kalrez®/Zymaxx® packing sets indicated the lowest friction occurred with the Kalrez® packing set. In addition, emissions are lower with the Kalrez® packing set. The deadband test which is similar to a hysteresis test, demonstrated a much lower range compared with braided PTFE.

Fire Safety. For a fire-safe packing system, the packing installation in Figure 8.18 is modified by installing a preformed flexible graphite ring between the spacer and the follower. The spacer separates the graphite ring from the packing set and is installed without any antiextrusion rings. LE® packing with the preformed flexible graphite ring passes the API-607 fire test. Apparently, the long carbon fibers in the

PTFE V ring are released as the PTFE decomposes and flow around the graphite ring interfaces, further improving the seal. Although the spacer clearance tolerance is considered quite small, the annular opening is apparently sufficiently large to permit carbon fiber movement.

Another reason for the small spacer tolerance is to prevent graphite particles from flowing between the spacer and stem. Graphite particles leaking through this annulus enter the packing box and interfere with the Kalrez® rings, thereby reducing emission control.

The fire-safety test for this graphite modified system was also performed with a fresh or new packing installation. Extrusion of the preformed flexible graphite ring after a considerable number of cylces generally occurs without any antiextrusion rings. As a result, fire-safe resistance after the packing set has been in operation for a relatively long number of cycles has not been demonstrated. Also, fire testing with the more stringent hydrocarbon fluids was not undertaken.

Dual Packing Sets. These packing sets can be installed with a lantern ring, similar to packing set installations previously described. Normally, dual packing sets are not recommended since the single packing set is considered sufficient for emission control. The dual set with a packing ring also permits testing leakage through the primary set into the lantern ring volume. However, a number of dual sets have been installed with purge streams through the lantern rings. A nontoxic purge stream pressure in the lantern ring has been set at relatively high pressures to minimize leakage into the lantern ring and any leakage from the stem is nontoxic. This requires a backpressure controller, which increases system complexity. Alternatively, a reasonably large flow of low pressure nontoxic purge gas through the lantern ring minimizes potential leakage of toxic material into the secondary packing set. Where necessary for dual packing set installations, an extension bonnet can be installed.

A major concern with dual packing sets and a lantern ring purge or leak is the disposition of the leakage stream. Disposal recommendations include venting to a flare, pressure sewer, or the atmosphere at a safe location.[11] The latter presents potential environmental problems and probably requires careful investigation, but is not a long-term solution. Another alternative is discharge to a control device such as an incinerator or a thermal oxidizer. Since a number of control valves generally have similar packing systems, the various leak streams are normally combined prior to final disposition. However, this requires a review of the various potential leak streams to ensure the streams will not react or present other potential problems.

In addition to the disposal alternative presented, there may be opportunities to recycle the leak stream to the process. Again, this requires careful analysis and installation. Since purge or leak stream piping is small, maintenance requires careful attention to ensure procedures are safe and toxic potential is minimized. Potential exposure to leakage during maintenance may occur more readily in continuous processing units where valve operation generally continues during maintenance or repair in the leak system. Valves can be blocked and bypassed during maintenance in some installations, but the pressure in the control valve will initially be the process pressure, unless the control valve section is drained or bled down to a closed system.

While the dual packing set system is more expensive than a single set, there is a high reliability factor in emission control due to the dual packings, and a purge or bleedoff system further improves emission control performance.

Application. The LE® packing can be installed in either rotary or rising stem valves with temperatures to about 450°F (232°C). For higher temperatures, an extended bonnet can be installed that permits operation at much higher fluid temperatures. While this packing set can be installed in rotary valves, the EF Seal packing is generally used.

According to the manufacturer, the LE® packing system has a life expectancy that exceeds 200,000 full stroke cycles with emission leak concentrations considerably less than 500 ppmv to pressures of 750 psig (53 bar).

EF Seal® Packing. In developing improved packings for rotary valves, Masoneilan investigated the gland follower and installed O rings in the gland follower as a method of controlling leakage[10, 11]. This general O-ring system is shown in Figure 8.19 and the effectiveness of the O-ring follower in emission control appears in Figure 8.17 for rotary valves[10, 11]. In the development of the EF Seal system, braided PTFE (Fig. 8.17) leakage increased as the number of cycles increased. The installation of the O-ring follower reduced the emission leak concentration with braided PTFE to an emission control level identical to that obtained with the Kalrez® and carbon-filled PTFE combination packing set (Fig. 8.17). Emission leakage remained constant for the entire test.

Packing Configurations. The basic backing configuration for rotary valves is presented in Figure 8.19. A support ring is provided at the bottom of the box that

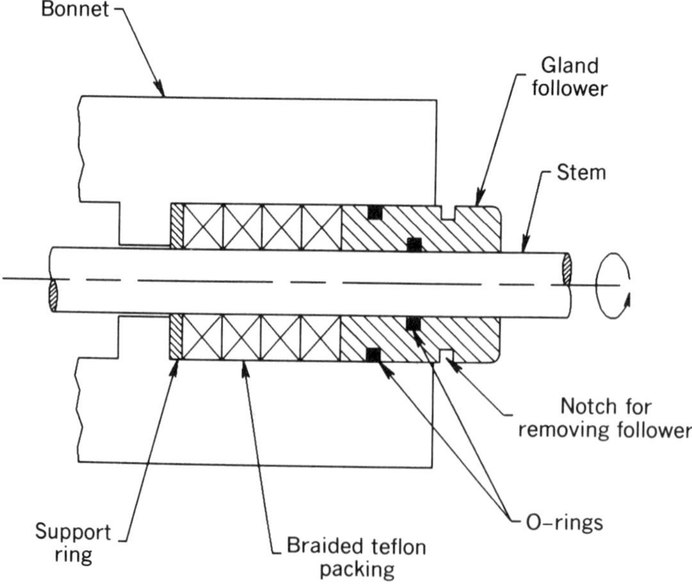

Figure 8.19 *EF Seal® for rotary control valves (courtesy of Masoneilan).*

is constructed of metal. The major difference in any packing configuration is the type of O ring installed in the follower. The O-ring materials can be selected from Viton®, nitrile rubber, Kalrez®, or other standard material that is resistant to the stream material and whose follower temperature is within the temperature operating envelope of the selected O-ring material. While braided PTFE is the packing of choice for the EF Seal® set, the packing can probably be changed if another packing is desired.

The EF Seal® packing system with the braided PTFE is considerably less expensive than the LE® packing and consequently is preferred for rotary valves.

Live Loading. The EF Seal® packing system is not live loaded and apparently needs little adjustment for long operating periods in rotary valves. Where the LE® packing system is installed in rotary valves, internal springs are included in the packing system.

Antiextrusion. The braided PTFE packing rings do not present an extrusion problem against the gland follower. At the process end of the packing box a supporting ring is provided for the packing (Fig. 8.19) that is constructed of metal. The tolerance between the stem and supporting ring is quite small, which prevents any extrusion. Although of metal construction, the metal does not score the shaft.

Stem Alignment. Stem movement is controlled by bushing and sleeve bearings located in other parts of the control valve. Consequently, guide bushings are not required in the packing box.

Friction. Frictional data show somewhat lower frictional effects for braided PTFE compared with the Kalrez® carbon-filled PTFE packing set.[4] Both packing sets have a considerably lower frictional impact than the frictional effects generated by standard PTFE V-ring packings. There are apparently some differences in emission leakage related to the deadband hysteresis effect between the EF Seal® and LE® packings. However, this difference is about 10 ppmv, which is negligible in this comparison.

Fire Safety. For fire safety a preformed flexible graphite ring is placed between the gland follower and the braided PTFE rings. During fire testing the graphite ring remained intact while the braided PTFE rings disintegrated. However, the O rings in the gland follower will decompose and essentially disappear, leaving the graphite ring as the only intact material in the packing system[11].

Testing of these packings with the graphite ring has been limited to fresh noncycled packing. Fire-test data and emission control on packings after cycling through a large number of cycles are unavailable.

Dual Packing Sets. The EF Seal® packing system can also be installed as a dual packing system. However, the control valve manufacturer does not believe that dual packing sets are necessary[10].

Applications. The EF Seal® packing system is capable of operating at 750 psig (53 bar). However, the operating temperature is a function of the packing location in the valve. Where the valve packing set is close to the process fluid, the maximum process temperature is limited to about 450°F (232°C). In some rotary valves where the packing system is located some distance from the process fluid, a much higher fluid temperature can be sealed. For this type of installation, the control valve manufacturer can define the actual packing box temperature, which should not exceed the 450°F (232°C) PTFE operating limit. The O rings must be compatible with the temperature in the packing box.

8.6 BELLOWS SEALS

Bellows seals have been available for many years and have been installed to prevent leakage of toxic materials in chemical processing and nuclear facilities. These seals have been successful in many applications, but failures in some of the bellows seals require backup packings to ensure control of leakage and potential exposure.[12] However, backup packings must be carefully selected to prevent packing failure or reduced capability over the life of the bellows seal.[12]

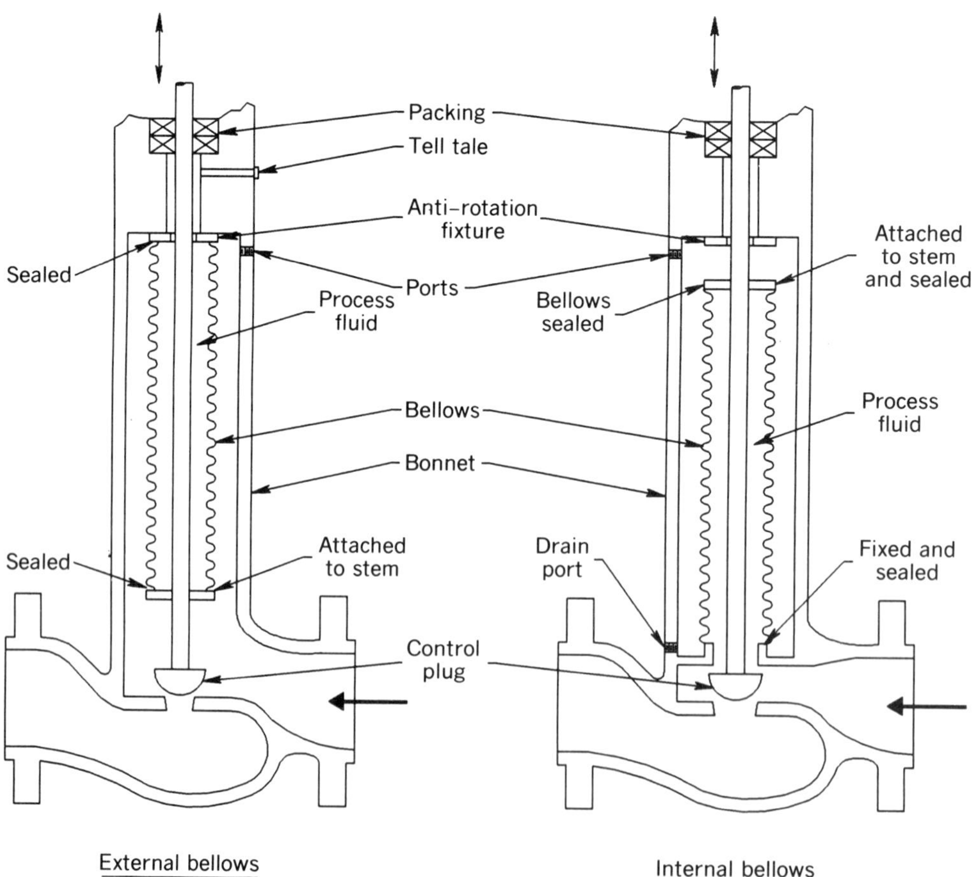

Figure 8.20 Simplified cross sections of typical control valve bellows seals.

In general, there are two types of bellow seals, as shown in Figure 8.20 for rising stem valves—internal and external bellows seals. Process fluid is encapsulated within the bellows as shown and any leakage through the bellows is contained within the bonnet or casing. The external bellows seal installation contains the process fluid in the annular volume between the bonnet and the bellows. In both installations the bellows prevents process fluid from entering the passage leading into the packing.

Rotary control valves may also have bellows seals and several were described in Chapter 7. These valves are automated for remote control operation with actuators that have been developed for this function. The external bellows seal designed specifically for rotary valves (Kerotest®) and shown in Figure 7.8 is also adapted for control valve operation. This external bellows is offered as original control valve equipment by some manufacturers. Since the external bellows has a relatively long arm and can also be installed on existing rotary valves, the actuator support mounts are modified. The actuator spindle or stem is generally welded to the bellows arm housing, providing the necessary turning motion.

8.6.1 Rising Stem Valve Bellows

The bellows in many rising stem valves of the type shown in Figure 8.20 are about 20 in. (500 mm) in length and require stem travel that is at least equivalent to the stem travel required for a nonbellows sealed control valve. The bellows length and convolutions affect the amount of compression along with other bellows design characteristics including bellows materials and whether the bellows is preformed or built of welded construction. Additional information on bellows characteristics is presented in Chapter 7.

One of the problems with bellows seals in control valves is the stem rotation that frequently occurs as a result of actuator operation. This rotation twists the bellows, which may result in fracturing or creating stress cracks. Antirotation systems installed in the upper end of the stem to prevent rotation can be of simple construction. In addition to twisting, cycling may result in metal fatigue and cracks in the bellows. However, improved bellows designs have reduced the incidence of this problem, but the overall bellows seal design should be carefully reviewed with the control valve manufacturer. General stem movement is controlled with guide bushings located between the control plug and the bottom of the bellows. A guide bushing may also be installed between the packing box and the top of the bellows.

The bellows must be constructed of material that is unaffected by corrosive process fluids. For an external pressure bellows, the bonnet enclosing the bellows should be of anticorrosive material. With internal pressure bellows, the bonnet normally does not require anticorrosive materials. Should a leak occur, the bonnet will be attacked and a rapid response is required to remove the valve from service. If the valve is not removed from service under these bellows leaking conditions, a hazardous situation may develop. Consequently, inspecting or testing the valve is required. For an externally pressured bellows, leakage into the bellows may then result in leakage into the packing system through the stem passageway. However, this situation is a function of stem design, which varies with the different control valve manufacturers.

The rising stem valves shown in Figure 8.20 have a relatively long stroke. Moreover, the actuator must be sized to permit rapid fluctuation of the stem against compression resistance of the bellows seals, in addition to the differential pressure across the control valve plug. To reduce the overall size of the bellows system and lower actuator requirements, short-stroke bellows seal control valves have become available. The height of the valves have been reduced significantly along with the length of the bellows. The bellows are of welded construction with a different configuration than the preformed bellows in Figure 8.20. For an installation with a bellows in a rising stem valve, detailed comparisons of the valves should be undertaken to select the optimum control valve for the service. Potential leakage problems and maintenance are considered in an overall assessment of control valve capacity.

8.6.2 Operation and Potential Exposures

An important aspect of operating a bellows seal control valve is determining when a bellows is leaking. Some of the potential problems have been described. However, potential exposure situations may occur when a bellows leaks in either an internal or external pressure bellows.

External Bellows. With this bellows the process fluid surrounds the bellows and the internal volume of the bellows is free of contamination. Should a very small or pinhole leak occur in the bellows, process fluid enters the bellows. Purging the valve will empty or remove process fluid from the annular volume surrounding the bellows. However, process fluid within the bellows may not be removed during this step. When the bellows is then removed for repair, the maintenance technician can be exposed to process fluid in the shop.[13] Consequently, a detection system is necessary to ensure the bellows can be tested for leakage to prevent a buildup of process fluid during normal operations. This potential exposure problem should be reviewed with the control valve manufacturer to determine whether modifications to the design are necessary or improved maintenance repair procedures will essentially eliminate this concern.

When bellows seal leakage occurs on a relatively large scale, the process fluid will enter the packing box where it is prevented from direct release to the atmosphere. A major bellows crack results in both the bellows and the surrounding volume filled with fluid. When drained, most of the process material will be removed from the bellows and the outer volume or bonnet surrounding the bellows. However, complete removal of process fluid between the bellows compartment and the packing box may not be possible, particularly with fluids that are not volatile at ambient temperatures. Special drain passageways between compartments aid in draining all fluids. In addition, these drain openings are of assistance when purging with a liquid for final removal of the remaining process fluid in the bonnet volume. Purging when a large bellows crack exists will remove a considerable amount of process fluid from the bonnet. Removal of a failed bellows from a valve should be reviewed with an industrial hygienist to minimize potential exposures.

Ports should also be placed as shown in Figure 8.20 to completely flush and purge the volume surrounding the bellows. In the external bellows, all of the process material from the bonnet volume surrounding the bellows drains into the valve body. To determine whether the bellows is leaking, a telltale connection can be used periodically. However, the telltale must be opened with procedures that minimize potential exposure and do not permit the discharge of material directly to the atmosphere. Any material leaking through the bellows can have full inlet process pressure at the telltale port. Alternatively, a lantern ring at the packing box can perform the telltale function, particularly where the lantern ring is at the bottom of the packing box. A lantern ring in this position may need a specific port for the telltale function.

Internal Bellows. Process fluid is contained within the bellows and in the event of leakage is contained within the surrounding volume. Vapor or liquid leakage from leaking bellows can be controlled with the packing system. However, the material buildup in the outer annular volume must be removed before the bellows can be removed and repaired. This requires draining the outer volume and a drain port should be provided in the bonnet chamber surrounding the bellows. In addition, a port in the upper part of the bellows chamber (Fig. 8.20) permits purging the outer chamber. The process fluid in the bellows will drain into the valve body and repairs can be performed on the bellows in a shop without overexposing the maintenance mechanic.

For large-scale leaks, the leakage will rapidly enter the packing box. Consequently leak detection is necessary to indicate when either a large or small bellows leak occurs. The design of the internal bellows readily permits leak detection through an upper port (Fig. 8.20), although a separate telltale connection similar to the one shown for the external bellows can be installed. A port at the top of the bellows chamber can serve two functions—as a telltale and a connection for purge or flushing material. For telltale operations, a connection should be provided that minimizes potential exposure during inspection.

A major consideration is draining, purging, or flushing the outer annular chamber of process fluid. The drain port or ports for this annular chamber must be placed to completely drain all liquid from the chamber. Moreover, any variations in the bottom of the chamber must be designed for complete drainage. The drain port should be permanently piped to a receiver, which ensures that exposure to the fluid is minimized.

Telltale Inspection. Since bellows integrity must be maintained to control leakage, periodic inspection should be scheduled to observe whether there are any indications of leakage through the telltale. The inspection procedures should be reviewed with the industrial hygienist to minimize potential exposures. Leakage fluid, which can be toxic, may be at elevated pressures, requiring careful analysis of inspection procedures.

8.6.3 Cost

Bellows valves are somewhat more expensive than standard packed control valves up to about a 2-in. valve size. Above this size the difference in cost escalates rapidly and at the 6-in. valve control size, the cost of the bellows valve is approximately 20 times greater than the cost of the packed control valve.[12] These costs are based upon steel control valves, and may differ considerably when the material is a stainless steel.

Although bellows valves may be more expensive than packed control valves, the bellows valve may be the solution for difficult applications when emission control and toxic exposure are major considerations.

8.6.4 Other

All bellows valves should be installed with packings as a backup control system in the event there is a failure in the bellows. The packing permits emission control until the valve can be removed from service for repair.

8.7 OVERVIEW

Control valve packing systems have changed considerably in the last decade and further changes can be expected as control valve and packing manufacturers continue to improve their packings. These improvements can be expected in both graphite and nongraphite systems. The two basic packing systems presented in this chapter provide a basis for understanding some of the packing technology currently available. However, a number of packing systems and modifications are commercially available that should be investigated when improvements in emission control are necessary.

An important aspect of an emission control system is the installation of improved packing systems in existing valves. While such systems should be obtained to meet emission targets, packing system performance is also a function of control valve condition. Valve stems should be straight, aligned properly, and polished to the required finish level. These stems must not be scored or marred, which interferes with packing emission control performance. In addition, the packing boxes must be in good condition, with dimensions close to the original valve dimensions. These general mechanical requirements ensure optimum packing performance and repairs or modifications should be considered to meet satisfactory valve mechanical standards.

REFERENCES

1. Dimmick, W. F. and K. C. Hustvedt. *Equipment Leaks of VOC: Emissions and Their Control*. Air Pollution Control Association Meeting, Pittsburgh, PA, June 24–29, 1984.
2. Adams, M. *Valve Mag*. **2**(1), 20–24 (1990).
3. FR. Process Safety Management of Highly Hazardous Chemicals. Explosives and Blasting Agents. OSHA. Title *20*, Part 1910, Final Rule, pp. 6356–6417, February 24, 1991.
4. Offutt, W. A. *A Valve Stem Packing Overview of New Packing Techniques to Meet Low Leakage Requirements*. Fugitive Emissions Seminar, VMA, Philadelphia, PA November 18–19, 1992.
5. Bresel, R., W. Hutchens, and C. Wood. *Control Valve Packing Systems*. Fisher Controls, Technical Monograph No. 38, Marshalltown, IA, 1992.

6. Harrelson, A. L. *Live-Loaded Packing*. Fugitive Emissions Seminar, VMA, Houston, TX, September 25–26, 1991.
7. VanTassel, D. M. *Advances in Graphite Valve Packing Systems*. EPRI 4th Valve Technology Symposium, Dallas, TX, August 11–13, 1992.
8. Valtek®. *Fugitive Emissions Control*. Company publication, p. 5, July 1991.
9. Wett, T. *Chem. Proc.* **55**(2), 28–32 (1992).
10. Rozenas, B. *Fugitive Emissions*. Masoneilan publication, August 1991.
11. Ekelmans, A. J. *Fugitive Emission and Control Valves*. Chemical Processing Seminar, Houston, TX, November 6, 1991.
12. Snyder, P. *Chevron's Experience with Fugitive Emissions Regulations and the Cost of Compliance*. Fugitive Emissions Seminar, VMA, Houston, TX, September 25–26, 1991.
13. Ekelmans, A. J. *Fugitive Emission and Control Valves*. Educational Seminar, VMA, Sarnia, Canada, October 25, 1990.

9

Flanges and Connections

Although valves have received considerable attention as a source of fugitive emissions in chemical plants and petroleum refineries, piping flanges and connections are also emission source. Prior to environmental regulation, uncontrolled valve emissions ranged from 40 to 70% of total estimated emissions, with flange emissions approximately 19–27% of total emissions. These figures indicate that uncontrolled valve emissions were significantly larger than flange emissions. With the promulgation of environmental regulations in the mid-1980s and emphasis on valve emission control, actual valve emissions have decreased significantly along with leak frequency. As a result of these regulatory restrictions, the overall emissions loss quantities in plants have decreased, with valve emissions becoming a smaller percentage of total emissions losses.[1] A reduction in flange emissions has also occurred, but total fugitive emission flange losses are now similar to the current valve emission loss quantity and the percentage of overall total losses for each category is also similar. Since flange or connector emissions have become an important contributor to emissions losses, the new MACT or HON regulations also require further emission reductions in flange or connector services.

Flange emission factors—although quite low for an individual flange compared with an average valve emission factor, as shown in Chapter 5—turn flanges and connectors into a relatively large emission source as a result of the large number of flanges in process units. A comparison of estimated valve and flange counts for various chemical units and a typical small complex refinery is shown in Table 9.1 The model chemical process unit was developed by the EPA as a typical example of a large chemical processing unit.[3] The EPA evaluations of equipment populations indicated that there is a flange–valve ratio of 1.6 : 1 in chemical units, which is an average value since the ratio changes considerably, as shown in Table 9.1. During the EPA evaluation of flange emissions, threaded connections were not included in the emission total, which probably influenced the overall valve–flange ratio compared with the petroleum refinery ratio. Piping in chemical units is generally smaller than process piping in petroleum units, where flanges are

TABLE 9.1 Estimated Equipment Counts in Typical Process Operations

Source	Petroleum Refinery[a]	Ethylene Unit[b]	Methlyl Ethyl Ketone Unit[b]	Vinyl Chloride Monomer Unit[b]	Model Unit C[b]
Valves	11,500	4,915	682	2,811	4,468
Flanges	46,500	15,980	1,140	2,840	7,200

[a]Adapted from reference 2.
[b]Adapted from reference 3.

normally required in process piping connections and higher valve–flange ratios exist.

In processing units the wide profusion of flanges and connectors can affect the general contaminant concentration level in the process areas. Connector or flange emission control reduces the general emission level, fugitive emission losses, and potential exposure levels. In this chapter methods of controlling connector emissions are presented in chemical and petroleum processing facilities. The EPA documentation does not provide any general guidance in emission control but does mention welding flange openings and threaded connections. Welding flanges presents potential mechanical problems and is not the preferred solution. In addition, flange gaskets in many applications must be fire resistant, which affects the type of gaskets installed and in some instances the types of flanges. Also, flange gaskets must be designed and installed to prevent gasket blowout. Since any flaw in a gasket installation can result in leakage and potential exposures to toxic chemicals, changes in gasketing materials or flanges must be carefully investigated.

9.1 EMISSION REGULATIONS

There are two major groups of environmental regulations applicable to emission control from flanges and connectors. The NSPS regulations[4] cover VOC emissions, and the HON or MACT regulations issued under the Clean Air Act Amendments of 1990[5] are applicable in VHAP chemical service. A list of the VHAP chemicals on the toxic list is shown in Appendix 9A and the regulations are also applicable to chemicals manufacturing processes shown in Appendix 9B that manufacture, consume, or produce as a byproduct any of the chemicals. While this initial list of chemicals covers 147 compounds, the list will be expanded periodically and probably include many of the chemicals not currently listed. Although the MACT or HON regulations[5] are only applicable to the hazardous chemicals in the toxics list, similar regulations will be promulgated for petroleum refineries in 1994, and the toxics list will be expanded periodically.

The MACT or HON regulations are applicable to hazardous air pollutants and essentially supersede the NESHAPS regulations that covered benzene and vinyl chloride processing facilities since the MACT or HON requirements are more stringent. In flange and connector emission control, the HON regulations are more restrictive.

In addition to the federal regulations, states and regional control boards have also issued flange and connector emissions regulation. As a result, regulatory requirements will probably be similar to the HON chemical regulations in most areas, with a few regions requiring somewhat lower emissions through lower leak definitions.

9.1.1 NSPS Regulations

The NSPS regulations for VOC control in reference 4 refer to the synthetic organic chemical manufacturing industry. However, Subpart GGG of the same general regulation covering equipment leaks of VOC in petroleum refineries requires conformance with the chemical regulations.[4] The chemical and petroleum refinery regulation only refers to flanges and connectors in light liquid and heavy liquid service. Gas or vapor service is not included in this regulation.

Flanges or connectors are not monitored on a specific basis. When evidence of a leak is observed, the leak is monitored with an instrument within 5 days. A leak is defined as $\geq$ 10,000 ppm VOC when measured with a vapor monitor. The leak should then be repaired as soon as practicable within 15 days after detection. If the leak cannot be repaired without a process shutdown, then the repair can be delayed until the next process unit shutdown.

In VOC service refers to equipment that contains or contacts a process fluid that has a VOC concentration equal to or greater than 10% VOC on a weight basis. The VOC compounds are all materials that are photochemically active. In various regions a list of negligible photochemically reactive compounds is available. Additionally, *light liquid service* is defined by the following three conditions:

1. The vapor pressure or one or more components of the stream is greater than 0.3 kPa at 20°C.
2. The total concentration of pure components with a vapor pressure greater than 0.3 kPa at 20°C is $\geq$ 20%.
3. The fluid is a liquid at operating conditions.

Heavy liquid service indicates the equipment is not in light liquid or gas/vapor service. This flange connection service essentially covers all liquid streams in VOC service. The VOC regulations are applicable to petroleum refineries and chemical plants. However, the new HON regulations supersede the VOC regulations in chemical plants where the equipment is in HAP service, as defined by the list of chemicals in Appendix 9A of Chapter 9. A similar situation will exist in petroleum refineries when the refinery HON regulations are promulgated in 1994.

9.1.2 HON Regulations

As described in other chapters, the HON regulations are applicable to chemical equipment in toxic or VHAP service. A list of the VHAP chemicals shown in Appendix 9A contains 147 chemicals and will be expanded periodically by the EPA. In addition, those process units that use any of the chemicals on the VHAP list as a reactant or produce them in the form of intermediates or products are shown in Appendix 9B and must confirm to the HON regulations. When a facility manufacturers chemical on the Appendix 9B chemical processing list, but does not use or produce any of the VHAP chemicals (Appendix 9A), the process unit is exempt from the HON regulations.

The HON regulations also do not apply to petroleum refinery processes or ethylene process units at this time. Also, processes that must conform with the HON regulations but have equipment that does not contain VHAP chemicals can exclude the equipment from the HON requirements. The specific equipment in this chapter is normally termed connectors in EPA regulations, but this term actually covers flanged, threaded, or other types of joined fittings that connect piping or a line and a piece of equipment. Where connectors are completely welded around the circumference of an interface, the fittings are not considered connectors and are excluded from regulatory requirements. In addition, glass or glass-lined connectors are excluded from the regulations.

As described previously, there are a large number of flanges in process units (Table 9.1), but threaded connections were not included in this summary, which can significantly increase the overall number of connectors. However, these connectors do not have to be tagged but must be adequately described in accordance with record-keeping requirements. Methods of identification can be site plans, log entries, weatherproof identification, or detailed flow plans.

Regulations for Gas / Vapor and Light Liquid Service

1. The chemical process units are divided into groups as indicated in Appendix 9B and implementation of the regulations is based upon the applicability date where promulgation refers to the issuance date of the final regulations:

Process Group	Applicability Date (Years after Promulgation)
I	1/2
II	3/4
III	1
IV	1 1/4
V	1 1/2

In addition to the processes in the various groups, the following process units and VHAP compounds are included in Group I for implementation 1/2 year after promulgation:

Processes	Chemicals
Styrene–butadiene rubber (SBR)	Butadiene, styrene
Polybutadiene rubber (PBR)	Butadiene
Chlorine production	Carbon tetrachloride
Pesticide production	Carbon tetrachloride, methylene chloride, ethylene dichloride
Chlorinated hydrocarbon use	Carbon tetrachloride, methylene chloride, tetrachloroethylene, chloroform, and ethylene dichloride
Pharmaceutical production	Carbon tetrachloride, methylene chloride
Miscellaneous butadiene use	Butadiene

2. The definition of light liquid service is identical to the light liquid service VOC definition (Section 9.1.1) with the exception that the constituents refer to volatile hazardous air pollutants.

3. Monitoring is conducted in accordance with EPA Method 21.[6] A monitoring probe is placed at the interface where leakage can occur and traverses the interface. For flanges, the probe is placed at the outer edge of the gasket–flange interface and the circumference of the flange is traversed. Threaded connections are monitored circumferentially at the thread and coupling interface.

4. When a monitoring instrument reading is equal to or greater than 500 ppmv above the background level, a leak is detected.

5. Monitoring is conducted in accordance with the applicability or promulgation dates described in item 1 as a basis:

a. For existing units, all connectors shall be monitored within 12 months after the specified applicability date for each group.

b. For new units, within 12 months after startup begins or within 12 months after the applicability date (whichever is later) all connectors shall be monitored.

6. When the initial monitoring survey of all connectors is completed (item 5), monitoring is conducted thereafter on the following basis:

a. When the connector leakage frequency (f) is 0.5% or greater, monitoring is conducted once per calendar year during the last annual or biennial monitoring period.

b. When the connector leakage frequency (f) is less than 0.5% during the previous monitoring period, the connectors can be monitored once every 2 yr (see Fig. 9.1). Monitoring is not required in one year, but 40% (minimum) can be monitored in the first year, and the remainder in the second year. Leakage frequency is the sum of both monitoring periods.

c. For process units in a biennial leak monitoring program, a leak frequency (f) less than 0.5% permits a change in monitoring requirements from biennial to once every 4 yr. The unit is permitted to monitor at least 20% of the connectors each year until all of the connectors have been monitored within 4 yr.

d. Process units in a 4 yr monitoring program interval remain in the program when the leak frequency is less than 0.5%. When the leak frequency is equal to or exceeds 0.5%, but is less than 1%, the monitoring program is changed to monitoring once every 2 yr (see Fig. 9.1). The unit is again permitted to monitor at least 40% of the connectors in the first year and the remainder in the second year. Should the leak frequency fall below 0.5%, the monitoring frequency can be changed to once every 4 yr. If the leak frequency in the biennial monitor program is equal to or greater than 0.5%, the monitoring frequency is increased to an annual basis.

e. For process units in a 4-yr monitoring program interval where the leak frequency (f) increases and is equal to or greater than 1%, the monitoring frequency is changed to an annual monitoring program (Fig. 9.1).

7. Connectors can be eliminated by welding the circumference of the interface or by removing the connector and welding the piping. When either procedure is

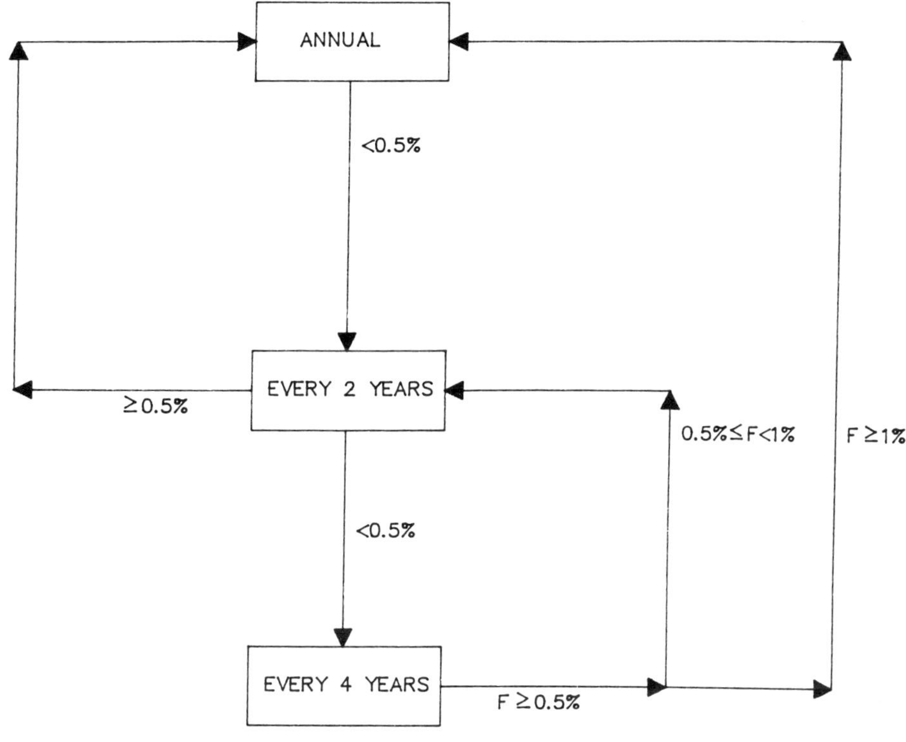

LEAK DEFINITION ≥ 500PPM

Figure 9.1 *Monitoring frequency standards for connectors and flange(s).*

completed, the weld integrity must be investigated and verified. This is accomplished through monitoring or weld inspection by X-ray, acoustic monitoring, hydrotest, or other acceptable procedures. A schedule for welding inspection and verification has been established as follows:

Welded after Dec. 31, 1992 and before promulgation date	Monitored or tested 3 months after promulgation date
Welded after promulgation date	Monitored or tested 3 months after welding completion

Where welded connectors are not properly welded or not completely welded around the circumference, the connector is unsatisfactory and not exempt from monitoring requirements. The inadequately welded connector is then monitored on the schedule previously described. If welding is satisfactory, the welded connector can be removed from the overall connector list and receive a connector credit, which is described in item 11.

8. Where a connector has been opened or the seal is broken, the connector is monitored for leaks 3 months after the connector has been returned to VHAP service. If the connector is considered a leaking connection, it should be repaired.

If the repair is unsatisfactory, the connector is considered nonrepairable, which is included in the overall calculation for the leaking connector percentage. Alternatively, the leaking connector term C_{AN} in Eq. 9.2 can be fixed as zero for all monitoring periods.

For threaded connectors with a nominal size of 2 in. (5.08 cm) or less installed prior to December 31, 1992, the installation can be considered a connector in heavy liquid service. However, the connector must be monitored for leaks within 3 months after returning to VHAP service.

9. When a leak is detected, the connector shall be repaired within 15 calendar days. The initial repair attempt must be inaugurated within 5 calendar days after the leak is detected. Following repair, the connector must be monitored for leaks within 3 months after repair completion.

For connectors described in item 10, repair scheduling is modified in accordance with an assessment of conditions.

10. In addition to the previous general connector description, certain connectors have additional classifications which are termed "unsafe to monitor" and "unsafe to repair." These are described as follows:

a. Connectors designated "unsafe to monitor" are exempt from the monitoring requirements of item 6 when the process unit owner or operator establishes that personnel would be exposed to an immediate danger conducting monitoring procedures. The owner or operator must have a plan that requires monitoring as frequently as practicable during safe-to-monitor periods.

b. Connectors designated "unsafe to repair" are exempt from the monitoring and repair requirements of items 6 and 9 when the process unit owner or operator establishes that personnel would be exposed to an immediate danger attempting a connector repair. However, the connector must be repaired during the next scheduled shutdown.

In addition to these two categories, connectors that are inaccessible, glass, or glass-lined are exempt from the monitoring requirements of item 6 along with record-keeping and reporting requirements. The inaccessible connectors can be defined by the following:

a. Buried

b. Insulated in such fashion that the connector cannot be monitored with a probe

c. Obstructed by equipment or piping preventing access for a monitoring probe

d. Cannot be reached from a 25-ft (7.6 m) portable scaffold on the ground and is higher than 2 m above a support surface

Where an inaccessible, glass, or glass-lined connector is observed, leaking through visual, audible, olfactory, or other methods, the leak must be repaired as soon as practicable but no later than 15 calendar days after the leak is detected. The first repair attempt should be instituted within 5 calendar days after the leak is detected. This repair equipment can be delayed by the unsafe-to-repair guidelines or a determination that emissions of purged material resulting from an immediate repair would be greater than the fugitive emissions leakage that accompany repair delay. Also in the latter situation, during repair procedures purged material is collected and destroyed or recovered in a control device.

11. The leak frequency for the initial monitoring period is determined from Eq. 9.1. From the initial leak frequency, the first monitoring period and associated requirements are established.

$$\%C_L = C_L/(C_T + C_C) \times 100 \tag{9.1}$$

where

$\%C_L$ = leaking connectors (%)
C_L = number of leaking connectors ($\geq$ 500 ppmv)
C_T = Total number of connectors monitored
C_C = optional credit for removed connectors; 0.67(total removed − total added) in VHAP service after applicability date (existing units) and after startup date for new units.
C_C = 0 where credits are not recorded

The following equation for determining leak frequency is used for all monitoring periods after the initial monitoring period:

$$\%C_L = (C_L - C_{AN})/(C_T + C_C) \times 100 \tag{9.2}$$

where

C_L = leaking connectors ($\geq$ 500 ppmv) including nonrepairables
C_{AN} = allowable nonrepairable connectors: maximum permitted is 2% of C_T
C_T = Total number connectors monitored including nonrepairables
C_C = Optional credit for removed connectors: 0.67(total removed − total added) in VHAP service after applicability date (existing units) and after startup dates for new units; C_C = 0 where credits are not recorded

Regulations for Connectors in Heavy Liquid Service. Heavy liquid service is defined as not in light liquid or gas/vapor service, with the general EPA definition of light liquid presented in Section 9.1.1. The heavy VHAP liquids have a low vapor pressure and connectors are not monitored on a standard schedule basis, as described for light liquid and gas/vapor connectors. Potential leaks are first indicated or identified in heavy liquid connectors through visual, audible, olfactory, or other detection methods. When a potential leak is identified, the leak must be monitored with a probe within 5 calendar days after leak identification.

Should the monitor indicate a concentration of 500 ppmv or greater, the connector is defined as leaking. The connector should be repaired as soon as practicable, but not later than 15 calendar days after leak detection. The initial repair attempt should be conducted within 5 calendar days after the leak is detected with the monitor. Some delay in repair is permitted as described in item 10 of the previous section.

9.2 EMISSIONS AND LEAKAGE

Although considerable effort has been concentrated on valve leakage and leak frequency, information on flange leakage is not extensive. In addition, the data currently available are difficult to assess because of the many changes that have occured in both gasket materials and designs in an effort to replace asbestos gaskets. As a result, the emission and leak frequency data currently available do not indicate the type of gasket tested in the flanges and generally do not indicate either the size of the flange or the process fluid phase. There is little comparative data on gasket effectiveness in gas/vapor service or in light liquid service available in the literature.

As a result, during the development of new packing materials and gasket designs various packing manufacturers have developed tests to determine leakage. Test procedures developed by industry to improve gasket performance are also used to determine gasket leakage under various conditions.[7,8] From industry test procedures developed through the Welding Research Council (WRC) and the Materials Technology Institute (MTI), some estimates of potential leakage quantities can be calculated.[9] However, this estimating procedure, while quite promising, has not been correlated with actual field data to determine overall accuracy. Moreover, estimates of leakage frequency that can be anticipated with certain gaskets, bolting techniques, and so on have not been developed. Wide variations exist in leakage frequencies owing to a number of variables in the general flange system that contribute to the overall problem of estimating these frequencies.

While there is a lack of accurate leakage and leak frequency estimating techniques, the various test procedures have been successful in comparing the effectiveness of individual gaskets. These comparisons cover a wide variety of tests and also include fire safety testing, which is an important criteria for many applications. As a result of industry test procedures developed through the WRC and MTI programs, the ASME piping code will be modified in the next few years to incorporate the methods developed for the establishment of flange and gasket performance criteria.

Some available flange emission data is presented in the following sections, but the leakage definition for most of the published data is 10,000 ppmv, which is considerably higher than the 500-ppmv leak definition established in the new HON chemical industry regulations. Consequently, a large number of variables in the data have not been identified. In addition, the flange leak definition for the petroleum industry will probably be reduced to 500 ppmv and estimating leakage for refineries will be difficult. Also, the leak data cover flanges, couplings, and other connectors, which further complicates data analysis. The data in the following sections have various criteria outlined where possible and may differ significantly from current field estimates. Moreover, cautious use of the data for estimating is suggested.

In most petroleum refineries and chemical plants, flanges are normally installed in the process piping. Couplings or nonflange connectors are generally a small proportion of the overall piping connections in the process industries.

9.2.1 Leakage Frequency

An important aspect of connector emission control in the HON regulations is leakage frequency. As described in Section 9.1.2, the base or target is a leakage frequency of less than 0.5% for VHAP service. Since the number of chemicals will expand in the future, a larger number of connectors will be included in the group with a 500-ppmv leak definition. Also, monitoring and emission control requirements for petroleum refinery connectors will probably be similar to the current chemical HON requirements. Some data available in this area, along with the variables that affect leakage, are presented to provide information on the status of leak frequency. However, certain data are from the early 1980s, before stringent environmental controls were promulgated. Consequently, early emission data will probably not be representative of the improved emission control performance in plants through the intervening years to the advent of the new HON regulations.

A 1981 study of emissions in various chemical process units by several contractors was funded by the EPA and determined the connector leak frequency.[10] The connectors in this study were generally flanged but did contain a few other connector types. Based on a leak definition of 10,000 ppmv, the leakage frequency for flanges in gas service was 4.6% and 1.2% for flanges in light liquid service. The flanges did not leak in heavy liquid service. In addition to the 10,000-ppmv leak basis, most of the gaskets in the piping were probably constructed of asbestos.

The CMA conducted a survey of emission leakage during 1988 from butadiene, ethylene oxide, and phosgene processing units.[11] A 10,000-ppmv leakage definition was the basis for the study, and the connector leak frequency for each class of units was determined. The data compared the average leak frequencies with an average frequency established by the EPA during an early study of 24 chemical units termed the SOCMI study by the EPA. An average connector leak frequency of 2.1% for the SOCMI study was considerably higher than a 0.1% leak frequency for butadiene units and 0.6% for ethylene oxide units. Phosgene plant connectors had a zero leak frequency at the 10,000-ppm leak level. Leakage in the phosgene process units is considerably lower than the leak frequencies in other units as a result of phosgene toxicity. While this study indicates a much lower average leak frequency in 1988 than the SOCMI and other early studies, performance at lower leakage levels was not reported. The improvement in connector leak frequency was a result of increased monitoring and changes in flange gasket materials. These changes were explained in general qualitative terms and the extent of gasket modifications cannot be identified.[11]

A more recent study investigating fugitive emissions provides information on leak frequencies that can be attained in ethylene oxide (EO) connectors.[1] This study monitored flanges in two EO units to determine whether a difference in performance could be expected due to age differences or other operating factors. Following initial monitoring results, the data were reviewed and changes were instituted to improve flange performance. As a result, leaking frequency data

TABLE 9.2 Flange Leak Frequencies in Ethylene Oxide Units for 500-PPM Leak Level[a]

Stage	Ethylene Oxide Unit I (%)	Ethylene Oxide Unit II (%)
Reaction, initial	3.7	10.0
Recovery, initial	1.6	2.7
Reaction, modified	0	3.3
Recovery, modified	0.3	0.15

[a]Adapted from reference 1.

before and after the flange improvements indicate the type of leakage frequency performance changes that can be expected. The EO plants were separated into reaction and recovery sections with the EO stream concentration in the former section about 1% and close to 100% in the latter section. The reduction achieved with the flange modifications are shown in Table 9.2 for a 500-ppmv leak rate. For the ethylene oxide units at 500 ppmv, the flange leak frequencies were reduced significantly. However, achieving less than 0.5% leakage frequency may require additional improvements in the reaction system of EO Unit II, although the overall flange leakage frequency at 500 ppmv was apparently 0.8 and 1.0%.

These recent studies indicate that the leak frequency can be reduced significantly, and the target of less than 0.5% leaking flanges at the 500-ppmv leak definition level can be achieved as shown by the operational improvements of the recovery section in Table 9.2. Since the flange gaskets were changed, detailed information on the older gaskets is not complete, although Teflon® and Teflon®-containing gaskets apparently demonstrated cold flow and signs of failure.[1] A single spiral-wound gasket type replaced the older gaskets, but the gasket filler material was not described. In addition, the actual flange installations were inspected and modified where required in a detailed quality control program.

9.2.2 Flange and Connector Emissions

Flange and threaded or other connector emissions are generally considered as a single group of connectors. However, most of the emission studies concentrate on flanges and overall connector emission data are not defined. Since threaded pipe connections in process service throughout various process facilities are normally welded and in considerably fewer numbers than flanges, emission figures for connectors are generally considered flange emission values. Also, threaded connections in process service are welded to prevent leakage, particularly in piping of stainless steel construction where galling frequently occurs.

The various methods for determining emissions are described in Chapter 5. Most of the methods require monitoring and/or bagging data to determine total emissions from a category such as flanges. While average emission factors provide an overall estimate of emissions leakage, they are much less accurate than the other procedures in estimating total emissions. In addition, the average emission factors published in the literature are generally based on a 10,000-ppmv leak definition with a SOCMI value for flanges of 0.00183 1b/hr (0.00083 kg/hr). Obviously, a 500-ppmv leak definition results in a reduction in the overall average emission factor. Where a screening or monitoring versus emissions rate correlation is available, the leakage rate can be determined as described in Chapter 5. However, these correlations can change based upon the gasket material in the flange[12] and other factors.

At this time, average flange emission data for 500 ppmv monitoring leak definitions are not available. Some data can be found in the literature for specific chemicals such as ethylene oxide.[1] However, these data are also based upon specific types of flange gaskets which provide a screening (monitoring) versus mass rate correlation. The emission data will change for different gaskets and will also vary with internal pressure and gasket stress.[12,13,14,34] Consequently, estimating emissions for new plants can be difficult and may require a complete analysis of emissions and leak frequencies based upon existing unit operations. These data can then serve as a basis for new plant performance. Preferably, the existing data are based upon flange operations with the proposed flange gasket for the new plant. Since the EPA does not have any adjustment for improved controls on flange emissions, existing plant data is the most reliable source of flange emission performance.

9.2.3 Flanges

Various flange types are used in the chemical and petroleum industries, as shown in Figure 9.2 which displays three major types of metal flanges. Polymeric lined, glass lined, and glass piping flanges may differ somewhat from the flanges in the figure. Process piping is designed in accordance with the ASME Code for Pressure Piping B31 which are national standards that are recognized in the Code of Federal Regulations. These flanges have standard dimensions that are listed in the ASME/ANSI B16.5 code and are available from the gasket manufacturers.

Raised face flanges are generally installed in most applications or services. However, the flanges must meet pressure and temperature limitations which are defined in the B16.5 code. As a result, raised face flanges are classified as 150, 300, 400, 600, 900, 1500, and 2500, with 150 designated for low pressure and temperatures. The pressure–temperature ratings increase as the class numbers increase. Since the metal piping can be constructed of various materials, the ratings are adjusted in accordance with the piping material. The wall and flange thicknesses increase with higher ratings and the flanges become quite large. Flange thickness has an effect upon the bolting stress that can be placed upon the flange, which in turn affects the gasket stress and leakage. In addition, temperatures must be carefully observed since thermal gradients may change the configuration of the flange slightly and possibly reduce the bolt loading, which can lead to an increase in leakage. A 400°F (204°C) temperature has been cited in the B16.5 code for 150 class flanges where severe external loads and thermal gradients should be avoided.

The 150 class flanges also present potential problems due to their lighter design and may bend when an excessive bolt stress is placed on the flange ends. These excessive stresses are generally a result of attempts to reduce leakage by increasing the gasket stress, particularly on nonasbestos gaskets.[10] The bolt stress permitted with 150 class flanges has been reduced in the latest code issue, which may have an effect upon leakage.[20] To control joint leakage, 300 class flanges are recommended in reference 19. However, these restrictions on flange rating were recommended with spiral wound gaskets, which require a high bolt stress to control leakage. A

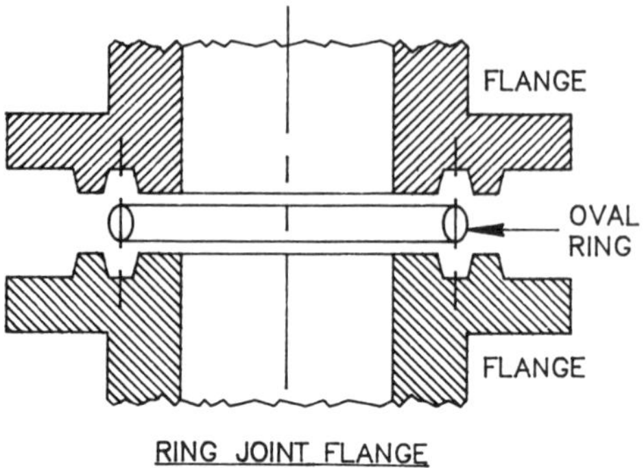

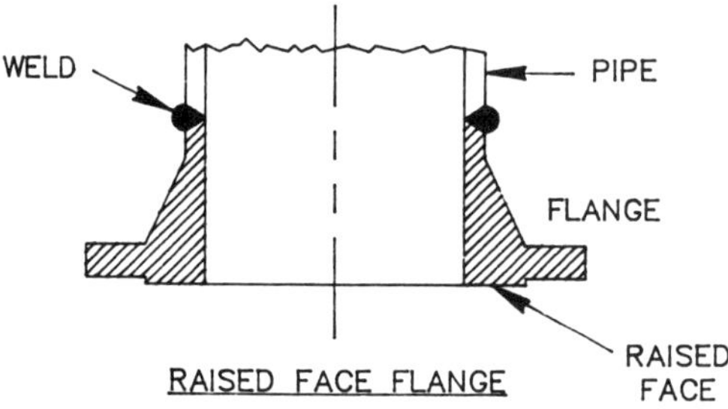

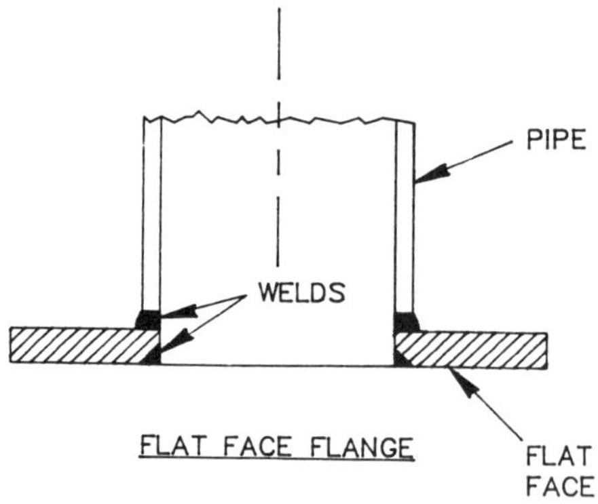

Figure 9.2 *Typical process piping flanges.*

new spiral wound gasket is available that significantly reduces bolt stress requirements.

A ring joint flange is shown in Figure 9.2 that is rated on a similar basis as that described for the raised face flanges in the B16.5 code. The ring is metal, and these flanges are normally installed in high-temperature service of $> 1000°F$ (538°C). This design reduces the possibility of gasket blowout and is frequently installed in petroleum process and petrochemical facilities. However, leakage or emissions data from ring-type joints are only available from laboratory tests[23] and emissions control effectiveness has not been identified by the EPA. The PVRC laboratory tests indicate that metal O-ring or ring joint leakage changes considerably with very small changes in the gasket stress.[23] This suggests that careful attention to gasket or ring sealing, along with the correct gasket stress, is necessary to ensure a low leak rate. These data also indicate lower leakage rates with properly seated ring joints compared with spiral wound gaskets at identical operating conditions.[23] Field operating data are necessary to substantiate these conclusions.

Flat face flanges are found in a few chemical services in contrast to petroleum and petrochemical plants where raised face flanges are the standard connection. The large flat surface can be somewhat uneven, which affects gasket stress and in turn may affect leakage. In addition, raised face flange construction has been considered sturdier, resisting deformation under load.

Although the general flange ratings are termed classes, older terminology still prevails in most facilities, and the flanges are identified as 150 lb, 300 lb, and so on. References also use the older terminology, but these terms are not a change from the code classification system.

9.3 ASBESTOS GASKETS

The standard gasket for many years in chemical and petroleum processing facilities has been the asbestos gasket. This was an attractive gasket for a number of reasons, particularly its resistance to a very broad range of fluids, high-temperature application, fire resistance, and low cost. Moreover, maintenance was simplified since the one-gasket type essentially covered all flange installations in a process facility. The health implications of asbestos usage resulted in the U.S. EPA banning the manufacture or importation of asbestos products in August 1994 with the possible exceptions of a few applications in special circumstances. However, implementation of this regulation has been delayed by a court ruling and final resolution of this impasse is unclear. As a result, asbestos gaskets can still be installed in the United States, but asbestos exposure and handling requirements must conform with the appropriate environmental and health regulations.

Asbestos exposure regulations issued by OSHA limit exposure levels to 0.2 fibers/cm^3 for an 8-hr time-weighted average.[15] This regulation is applicable to asbestos, actinolite, tremolite, anthophyllite, or a combination of these materials. Personnel cannot be exposed to more than 1 fiber/cm^3 for a 30-min sampling period. In general, these regulations cover exposure to any type of asbestos minerals. Other countries and regions differ in exposure regulations; for example, the United Kingdom and a number of other countries have asbestos exposure regulations based upon the specific minerals found in asbestos. A simplified version of the U.K. asbestos regulations follows:

1. Where the asbestos contains crocidolite or amosite, the maximum exposure level is 0.2 fibers/cm^3 averaged over a 4-hr time period. The maximum concentration is 0.6 fibers/cm^3 averaged over a 10-min period.

2. For asbestos materials that do not contain crocidolite or amosite, the maximum exposure level increases to 0.5 fibers/cm^3 averaged over a four-hr time period. The maximum concentration is 1.5 fibers/cm^3 averaged over a 10-min period.

3. Cumulative exposures are also determined over a 12-wk period for long-term exposure assessment. This is calculated by multiplying the fiber concentration by the time period and adding the individual totals for each period. The overall total should not exceed 48 fiber-hr/cm^3 for asbestos containing crocidolite or amosite. This number is termed the action level and when exceeded initiates response. Where crocidolite or amosite are not present, the action level is increased to 120 fiber-hr/cm^3.

In addition to the general EPA exposure regulations, disposal of waste asbestos in the form of used gaskets or scrap from asbestos sheet gasket operations must be collected and placed in sealed impermeable bags or a sealed impermeable container.[15] This sealed bag or container must then be labeled for identification and placed in a secure landfill. This method of disposal must be used with all asbestos gaskets removed from flanges. Replacement of asbestos gaskets with nonasbestos materials eventually results in the removal of all asbestos gaskets. The nonasbestos gasket disposal procedures are less stringent, which results in lower disposal costs.

The asbestos gasket consists of about 80% asbestos fibers, a 10–15% elastomer compound, and inorganic fillers as the remaining materials.[13] The finished asbestos material, termed "compressed asbestos," forms a matrix with the cross-linked elastomer and the finished sheet is composed of encapsulated asbestos fibers and fillers in the elastomeric material. Since the rubber has been compounded and cross-linked, the sheet is dry and not tacky. Gaskets can be purchased from a supplier in the required sizes or large sheets can be obtained with gaskets cut to size in local plant facilities. The latter has been economically attractive where a large number of gaskets are required. In addition, specialty gaskets can be more readily obtained.

When gaskets are initially installed before operation in high-temperature service (> 600°F or 316°C), the gaskets are somewhat flexible. After a period of service that may last several years at high temperature, removal of the gasket can be difficult. The gasket may bond to the flange surface and must then be scraped from the flange mating surfaces, resulting in pieces and fragments of the old gasket. In addition, those pieces removed from the flanges are brittle and break easily. At stream operating temperatures < 600°F (316°C) the gaskets can be readily removed, although the same bonding effect has been observed at somewhat lower temperatures. The bonding effect is a result of elastomer breakdown at elevated temperatures and asbestos brittleness results from the loss of the water of hydration in the asbestos fibers.

9.3.1 Asbestos Exposure

Since asbestos is still installed in some facilities, exposure during installation is a concern along with the removal of old asbestos gaskets. While many facilities have been replacing asbestos gaskets, this type of gasket still remains in various facilities

TABLE 9.3 Asbestos Sheet Gasket Fabrication Exposures[a]

Gasket Cutting Task	Sampling Duration (min)	Exposure Concentration (fibers/cm^3)
Power shear and hammer punch	330–359	0.005–0.015
Power shear and wheel cutter	408	0.017
Knife on lead table	490	0.012
Saber saw	55	0.33–0.39
Power shear and wheel cutter (short term)	30	0.34–0.49

[a]Adapted from reference 16.

awaiting replacement. Removal of the used or old gaskets to install nonasbestos replacements still presents potential exposure problems. Recent investigations of asbestos gasket exposure during installation and removal indicate where any potential exposure problems exist.[16,17]

In most installations, asbestos sheet gasket material is used, which is obtained on a bulk or large sheet basis. Gaskets are then cut or fabricated from these sheets in a plant shop. Consequently, gasket fabrication is a potential source of exposure and data obtained in a fabrication shop is shown in Table 9.3. The 8-hr average exposure level is well below the 0.2 fiber/cm^3 regulatory limit. Where a short-term exposure sample was obtained during a specific operation, the exposure concentration increases. The exposures during the 30- to 55-min operations did not exceed 0.49 fibers/cm^3 which is less than the excursion exposure limit of 1.00 fibers/cm^3 for 30 min mandated by OSHA.[15] These exposures reflect the matrix composition characteristics of asbestos sheet gaskets and the exposures may possibly have been lowered even further with additional ventilation.

Following operations that may last several years, the gaskets are removed during turnaround. Gasket removal may be straightfoward, but frequently the gaskets are bound to the flanges and must be physically removed. After removal of the major process gasket, pieces or lumps remain on the flange surface and must be removed through scraping and/or grinding. When the gasket material is dry, fiber emissions are relatively high compared with emissions of wetted gasket material. These emissions are compared in Table 9.4 which demonstrates the suppression effect of wetting agents on fiber emissions. The wetting agents were either a solvent or penetrating oil.

In a laboratory installation simulating gasket removal and field installation, exposure data are presented in Table 9.5. The short-term OSHA exposure limits of 1.0 fibers/cm^3 (30 min) are not exceeded in this operation. However, these data were obtained with phase contrast microscopy (PCM), but transmission electron microscopy (TEM) results indicate exposure levels 5–10 times greater.[17] The large difference in exposure concentrations also indicate the variation in results that occur in asbestos fiber concentration analyses. However, averaging the two results produces an average exposure level that exceeds the OSHA excursion limit. These data indicate that a potential exposure problem may occur and personnel should have sufficient protection to prevent an overexposure. Since a wetting agent was not used in the simulation program, a considerable reduction in the airborne fiber

TABLE 9.4 Exposure During Asbestos Sheet Gasket Removal and Replacement[a]

Gasket Task	Sampling Duration (min)	Exposure Concentration (fibers/cm^3)
Dry scraping/brushing flanges for pump and valve gasket replacement	19–46	0.11–0.19
Dry-brushed flanges for gasket replacement	55	0.33
Power-sanded flanges for gasket replacement	25	1.4
Wet removal and cleaning flanges for pump gasket replacement	30	< 0.06
Wet removal and surface cleaning for field flange gasket replacement	15	< 0.03

[a]Adapted from reference 16.

concentration can be expected, which would probably meet the OSHA standard based upon the results in Table 9.4 for either analytical technique.

The previous exposure information was based upon the installation of sheet gaskets in flanges. Asbestos is also a filler material in a spiral-wound gasket in which the asbestos fill separates the metal spiral winding. This gasket is described in more detail in another section of this chapter. The asbestos is firmly held between the metal spiral ring winding where the distance between the metal windings is about 1/16 in. (1.00 mm). Between the flanges, the metal receives the compression load from the flanges, and the asbestos does not bond to the flanges. This type of gasket is easily removed and does not require extensive brushing and grinding of the flange faces. As a result, the fiber concentration level is < 0.06 fibers/cm^3, which does not pose an exposure problem.

9.3.2 Asbestos Gasket Overview

While asbestos gaskets may be used in the U.S., a complete EPA ban on asbestos will probably become effective at some point. However, the use of low-cost asbestos gaskets may be somewhat misleading from the perspective of an overall economic evaluation. An important consideration is waste disposal, which requires placing the gaskets in sealed, identified containers that are deposited in secure,

TABLE 9.5 Simulated Exposures During Asbestos Gasket Removal and Installation[a]

Gasket type	Sampling Duration (min)	Exposure Concentration[b] (fibers/cm^3)
Removal with some flange cleaning	32	0.05–0.44
Installation	31	0.13–0.29

[a]Adapted from reference 17.
[b]Phase contrast microscopy.

toxic landfills. Charges for these types of landfill operations are increasing, raising overall disposal costs.

The asbestos gaskets have been replaced in many installations to reduce emissions and leakage frequency. Improvements in connector performance in the ethylene oxide facilities was attributed to gasket material changes[1] and lower emission rates are desirable to meet the new MACT or HON regulations, which will gradually be expanded to most chemical facilities and petroleum refineries. High leak frequencies result in increased monitoring costs since all of the flanges must be monitored more often, as shown in Figure 9.1. Consequently, nonasbestos gaskets that have lower emission rates and low leak frequencies may be required although not specifically mandated.

When estimating overall costs of asbestos versus nonasbestos gaskets, waste disposal and leakage are important considerations. In addition, exposure information on asbestos gasket use should be determined along with the modifications or operational changes necessary to control fiber emissions. These costs must be included with all other costs.

9.4 NONASBESTOS GASKETS

A wide variety of nonasbestos gaskets are available commercially as replacements for asbestos gaskets. Although these gaskets have been developed in reaction to the potential health problems associated with asbestos, many appear superior in emission control to the asbestos gaskets. These data have been obtained in various process facilities and in numerous laboratory test programs.[1,11,12] Although the test programs have not specifically targeted emissions in all cases, much of the mechanical data developed to measure gasket performance aid in assessing gasket tightness and subsequently emission control effectiveness.[12] Industry has sponsored a research effort that has been underway for a number of years through the Pressure Vessel Research Committee (PVRC) of the Welding Research Council and the Materials Technology Institute (MTI). The methods of determining gasket and bolting requirements through the industry program will be incorporated into the ASME piping code.

9.4.1 Nonasbestos Gasket Performance Objectives

Through extensive laboratory efforts by gasket manufacturers, industry, and individual companies a number of gasket parameters and requirements were investigated. With the information available from this work, various criteria for gasket installations have been developed to provide a basis for assessing the potential operating performance of a gasket in a process application. The major gasket performance objectives are described in the following sections.

Fire safety. A critical characteristic for a gasket is the ability to maintain seal integrity during a fire. Few fire tests have been conducted on bolted flange gaskets, but one gasket manufacturer has developed a fire test which is used to evaluate gaskets with a kerosene fluid.[13] Other tests are simulations based upon placing a gasket within a fixture and heating the overall fixture to a specific temperature.

One test was developed during the PVRC–MTI study which places gasket strips or sections in a fixture that simulates a flange, with the entire fixture placed in an oven for an actual fixture temperature of 1200°F (650°C). This procedure is described in reference 18.

The general procedure described above, termed the Fire Simulation Screen Test (FIRS), revealed that asbestos gaskets remained intact, but aramid, carbon, and glass fiber gaskets with NBR and SBR binders essentially disintegrated. Additional tests were conducted on four PTFE sheet gaskets and six graphite materials. The graphite material was not oxidized and survived the test, but graphite bonded to a metal sheet was delaminated.[7, 18, 21] The PTFE gaskets were destroyed, not unexpectedly.

A more sophisticated test simulation system (FITT) was developed in the industry study which evaluated tightness and helium leakage at the 1200°F (650°C) level.[18] The results of the gasket test revealed that asbestos is satisfactory. However, a reinforced aramid fiber sheet (NBRD) was unstable at both the 1200°F level and a lower temperature (1000°F or 538°C). These general results were supported in a study based upon ASTM D-2863 flammability tests.[21] In additional FITT tests, flexible graphite sheet and a spiral-wound gasket with graphite filler were satisfactory. A spiral wound mica–graphite filled gasket passed the test but was not considered the equivalent of the other gaskets. A PTFE-filled spiral-wound gasket lost the filler and fire resistance was considered unacceptable.

Nonpolymeric gaskets are superior in fire resistance according to these tests. The need for fire-safe gaskets should be carefully considered in determining gasket requirements.

Service Life and Reliability. Asbestos gaskets have demonstrated satisfactory performance in a variety of services, with gasket life measured in years. This contrasts with service life data for nonasbestos gaskets, where little long-term operating data are available, particularly on sheet-type gaskets, metal reinforced gaskets, and various new gasket designs. Although aging tests have been performed in the industry study, these tests are simulations and do not indicate the performance consistancy that can be expected for a 4- to 5- yr period between turnarounds.

Some gaskets have been in service for many years, such as spiral-wound gaskets, Teflon®, and other polymeric gaskets, but gasket life for other designs is not as extensive. Gasket data on actual service life for specific gaskets can be obtained from the manufacturers.

Leak performance. The data on emissions and leakage previously described in Section 9.2 were broadly obtained from asbestos and other gaskets, but the types of gaskets were not specifically identified. However, nonasbestos gaskets have frequently been installed to reduce emissions levels and leak frequency.[1] Actual leakage from a specific gasket is not available in the literature and none of the large scale emission studies conducted by the EPA have described the type of gasket operating conditions, flange ratings, or bolt loadings. As a result, operating data on specific gaskets may require plant tests or obtaining information from process units where nonasbestos gaskets have been installed and performance data are available.

Gasket manufacturers have generally developed leakage data for their specific gasket products, but testing procedures differ and comparisons with actual field installations are difficult. Some data are also available from the industry test program.[12] However, data from laboratory tests are generally obtained with helium, methane, or other inert gases at room temperature. While the tests at room temperature provide some basis for a qualitative assessment, simulation of design pressure and temperature with a fluid similar to a processing fluid is desirable. Laboratory tests at elevated temperatures are normally conducted with an inert gas such as helium. The relationship of an inert gas to actual fluids is tenuous at these temperatures and overall assessments are difficult. Recent data on gasket leakage with various light hydrocarbon gases and helium demonstrates different leakage rates for each gas.[35] In addition, little information is available on gasket performance in liquid service, requiring actual plant operating information.

A review of actual plant data is necessary to establish leak performance criteria in the absence of published operation data. Moreover, in-plant data also provide information on maintenance and potential problems that accompany use of a specific gasket, such as bonding to a flange. This may require grinding or solvent application, which pose potential health exposure problems. In addition, bolt tightening during operation requires precautions to prevent overtightening or overstressing the bolts, which can result in increased emissions.

Elevated Temperature Service. There is a lack of emission and performance data at temperatures greater than room temperature and particularly in applications where the temperature is > 500°F (260°C). Gasket aging above the 500°F temperature level while simulated has not been reported. For an accurate assessment of gasket performance, field emission data on the nonasbestos gaskets are necessary.

Broad Applications. One of the advantages of asbestos gaskets is their general application in a wide variety of services, since asbestos is essentially unaffected by chemicals, steam and so on. Unfortunately, nonasbestos gaskets are not completely resistant to the same broad spectrum of process materials and process conditions, particularly high-temperature service (> 500°F or 260°C). As a result, process facilities probably cannot standardize to one gasket design. A number of nonasbestos substitutes may be required in a process plant to cover the various services, which presents opportunities for installing the wrong gasket in a service.

A minimum number of gasket types should be selected to minimize misapplication and improve gasket identification. This consideration will improve quality control and reduce potential safety or health problems that can arise with the installation of an incorrect gasket.

Cost. Since nonasbestos gaskets are more expensive than asbestos gaskets, selecting the least expensive gasket is an attractive method of controlling costs. However, other factors should be examined before gaskets are selected to ensure an optimum gasket is installed. In addition, gaskets must be selected that will not blow out during operation.

An important element in considering gasket costs is leakage performance. Under the new HON regulations, poor gasket leak performance results in increased monitoring requirements. For example, monitoring on an annual basis for poorly performing gaskets versus monitoring on a 4-yr basis is considerably more expensive and should be included in an overall assessment of life cycle costs.

9.4.2 Types of Nonasbestos Gaskets

A wide variety of nonasbestos gaskets are available commercially from a number of gasket manufacturers. In addition, there are differences in the gaskets from various manufacturers even when the base material is identical. Differences in reinforcement systems, binders, and fillers and further variations in the composition produce gaskets with varying physical characteristics. A general summary of these gaskets is now presented.

Sheet Gaskets. The sheets are similar to the asbestos sheet operation where gaskets are cut from the sheets as required. Although sheets are supplied to process facilities for manual preparation, cut gaskets composed of the same material can be supplied directly by the manufacturer. Where the cut gasket is supplied, it is still termed a sheet gasket. While some typical examples of sheet gaskets follow, the list is not complete, and a variety of other gaskets are available.

1. *PTFE*. Numerous types of PTFE or Teflon® sheet gaskets are available commercially. The PTFE gaskets also have a variety of fill materials such as silica, barium, sulfate, glass fibers, carbon, or graphite. In addition, the differing methods for producing these sheeting materials affect tensile stress and leakage control characteristics. Another form of sheeting has a perforated metal insert.

2. *Aramid*. This material is used in fiber form and is bound with a rubber material similar to asbestos sheeting. The nitrile butadiene rubber (NBR) sheet is resistant to hydrocarbons, some chemicals, and steam. Styrene butadiene rubber (SBR) resistance is limited and this material is generally installed in steam service. Neoprene is also used, which has a somewhat broader resistance to various materials than NBR.[22]

3. *Fiberglass*. This is woven into a cloth and normally has a rubber binder. Other materials may be present in the cloth along with metal wire. Temperatures and pressure ratings on this material should be investigated to ensure the application is within recommended gasket operating limits.

4. *Rubber*. A variety of elastomers are available commercially, but there are temperature and pressure limitations on these materials. Many of these sheets are cloth-rubber or reinforced materials.

5. *Laminated Flexible Graphite*. Since graphite is fire safe, a variety of gaskets have been developed with graphite bonded to a metal surface, forming laminations. In addition to fire safety, these gaskets have been generally satisfactory in controlling leakage. Some of the general graphite gasket types described in the following paragraphs are supplied as sheet or cut gaskets.

   a. *Sheet Reinforced*. In this gasket sheet, a graphite coating is placed on both sides of a thin metal strip and bonded to the metal on both sides with an adhesive. These sheets are available with various metals, although a stainless steel sheet is generally preferred. Leakage control appears satisfactory, but some reports of delamination at high temperatures (> 750°F, 399°C) have been reported.[18, 21] In addition, long-term aging effects indicate load reduction behavior at 750°F (see item 6).

b. *Wire Reinforced*. Wire replaces the thin metal sheet with graphite bonded to the wire. Little data have been published on this gasket.

c. *Tang Bonded*. In this gasket, graphite is bonded mechanically to a thin metal sheet. The bonding is accomplished by punching holes through the sheet prior to pressing graphite against the sheet. Sheet material moved aside by the punch projects above the sheet, forming tabs. These tabs provide the mechanical bonding since a large number of holes are punched into the sheet.

6. *Carbon and Graphite Fiber*. Sheet gasketing is available with either a carbon or graphite fiber base. Although this is a fibrous-based sheet, the quantity of graphite or carbon fibers differs from the fiber quantity in asbestos sheeting. The quality of material is based upon weight, with a typical asbestos sheet containing approximately 80% asbestos, 10–15% rubber compound, and a filler. For carbon–graphite fiber sheets, fiber content is generally 20–50%, with total fiber and filler concentration about 90 wt %.[13] The general elastomer content on this basis is approximately 10 wt %.

The allowable operating temperatures for these carbon or graphite gaskets are considerably higher than the allowable temperatures for the sheet gaskets previously described. Some packing manufacturers indicate these fiber gaskets are satisfactory at temperatures greater than 700°F (371°C). However, 7000-hr stress relaxation tests at 750°F (399°C) show that flexible graphite sheets demonstrated load reductions between 5000 and 7000 hr.[21] While these tests were apparently run on laminated type sheets (item 5), the data are probably also applicable to fiber sheets, indicating a potential problem above 700°F (371°C). Where the flange is not insulated, the process fluid temperature may be somewhat greater than 700°F (371°C) since the outer edge of the gasket will be 700°F or less and air oxidation is not a problem. When the flange is insulated, the maximum process flow temperature should probably not exceed a 700°F (371°C) temperature.

Concerns regarding maximum operating temperature can be reviewed with the gasket manufacturers where actual operating data, particularly in longer-term operations, can be established.

Preformed Gaskets. Some preformed gaskets are described in this section which generally require manufacturing in a factory by the supplier since they cannot be cut from a sheet. In addition, the gasket manufacturers will also provide precut gaskets from the various sheets described in the previous section. Although several gaskets are mentioned, this is not a complete category of available gaskets.

1. *Spiral Wound*. These gaskets have been available commercially for many years throughout the world. In general, the gaskets consist of a metal strip that is wound into a flat spiral with filler material separating the coiled winding. The coil is formed by placing several thin layers of filler material against one side of a metal strip and coiling or winding the metal strip. A cross section of a spiral wound gasket is shown in Figure 9.3. While these gaskets are produced by various companies, they are also described as Flexitallic® gaskets as a result of the

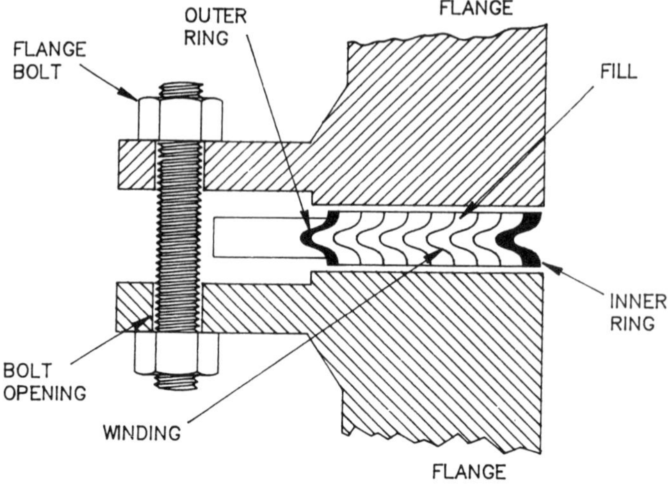

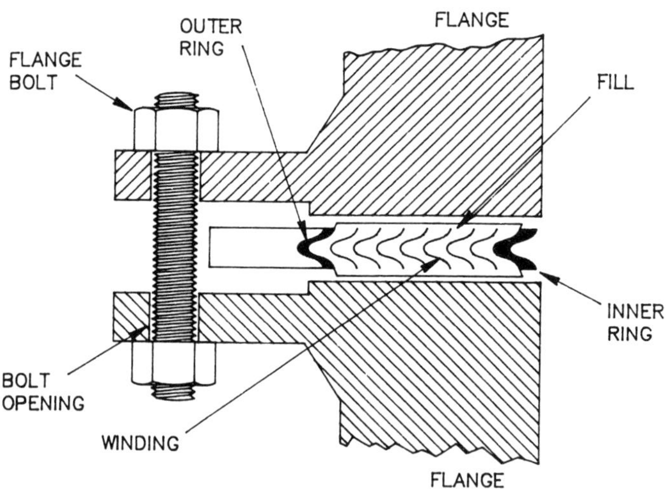

Figure 9.3 *Flexitallic® spiral-wound gaskets (courtesy of Flexitallic® Company).*

introduction of these materials to industry by the Flexitallic® Company. API Bulletin 601 covers the specification of spiral-wound gaskets.

A variety of fill materials is used in these gaskets. The initial fill material was asbestos, which has been replaced with materials such as mica, mica-graphite, graphite, and PTFE. For many current applications, graphite-filled spiral-wound gaskets are installed. With graphite the gasket is considered fire safe, which has been an important characteristic. In addition, these gaskets have performed well in various tests investigating leakage,[7,12,23] although detailed plant operating data comparisons are unavailable.

In Figure 9.3 two types of spiral-wound gaskets are shown. In the standard spiral gasket, the top and bottom edges of the metal winding strip are flush with the outer edges of the gasket. When compressed, the metal flanges press against the vertical metal winding strips, which have a slight bend permitting compression of the gasket and sealing against the flange faces. The metal spiral-wound gasket requires a high bolt loading to obtain the necessary gasket stress for loading. Since the maximum bolt stress permitted for 150-lb flanges is 25,000 psi (1723 bar), the spiral-wound gasket was modified to reduce the required sealing bolt stress for 150-lb flanges. The modified gasket termed LS® in Figure 9.3 has been changed by narrowing the metal winding strip and extending the fill material to cover the windings between the end rings. When the LS® gasket is compressed, the flanges press directly against fill material, sealing the gasket against the flange faces. The lower sealing stress requirements permit installation of this gasket in 150-lb flanges from extremely small to 24-in.flanges. Leakage emission control data on these new gaskets are quite limited and comparisons with other flanges are unavailable.

2. *Graphite Corrugated Metal*. A thin corrugated metal ring coated on both sides with graphite, as shown in Figure 9.4, has been termed a Graphonic® gasket by the manufacturer. The design is relatively new in pipe flange applications. However, the gasket has been installed in exchanger flanges for a number of years with very credible results. This application led to the installation of the gasket in pipe flanges, particularly in various chemical processing facilities along with some utility applications. To determine gasket sealing effectiveness, gasket tests were conducted by the study group that developed the PVRC–MTI gasket test procedures. Room temperature tightness tests (termed ROTT by the PVRC) on these gaskets indicated very low leakage rates at low gasket stress levels. In a comparison with laminated graphite gaskets, the initial leak rates were 100 times less at an initial compressive stress of 8000 psi (551 bar). With stress levels raised to 11,600 psi (800 bar) and an 800-psi (55 bar) internal pressure, leakage was difficult to measure with an extremely sensitive detector on the corrugated graphite gasket.[30]

Since this type of gasket is a combination of metal and graphite, the PVRC confirmed this gasket had excellent fire resistance in the fire simulation screening (FIRS) and simulated fire tightness tests (FITT).[30] In the second test, the tightness of the corrugated graphite gasket measured after the fire exposure period exceeded flexible graphite laminate and was equivalent to graphite-filled spiral-wound gaskets. Overall, this type of gasket seals flanges at lower bolt loads than alternative gaskets, which helps prevent bolt overstress and provides further compression if some leakage develops. Additionally, this gasket has been used successfully to control leakage in flanges that had grooves cut by vapor leaking through the flange. While the actual sealing mechanism is unclear, the ability to seal under these conditions provides the plant with a gasket that will seal under poor flange face conditions.

In addition to the Graphonic® gasket, similar gaskets are available in Europe, but the peaks are sharp instead of the smooth sinusoidal wave form shown in Figure 9.4. A corrugated gasket from a competitive manufacturer has a smooth metal wave form, but graphite is concentrated in one ring root as a small block on

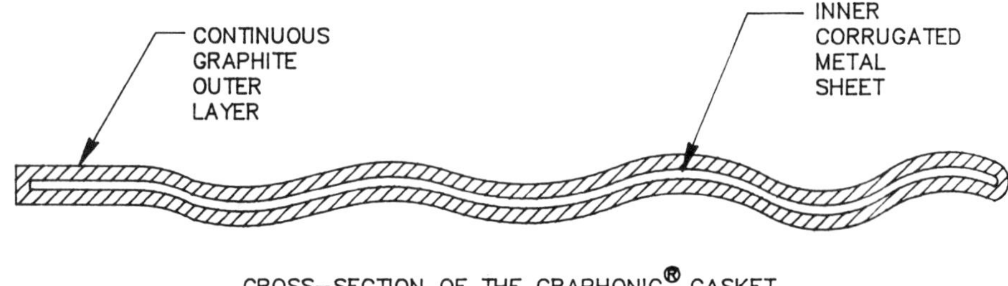

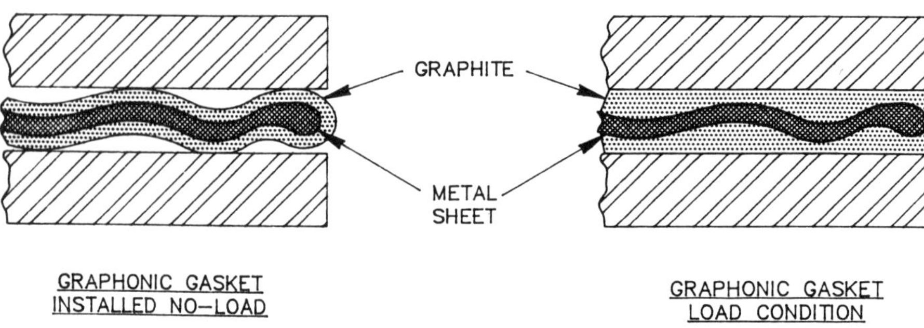

Figure 9.4 *Graphonic® corrugated gasket (courtesy of Marine & Petroleum Mfg. Co.).*

either side of the corrugations. Under compression, the graphite spreads but does not completely cover the metal surfaces on both sides of the gasket. Emission data are apparently available from the manufacturer.

3. *Double Jacketed*. The typical double-jacketed API gasket shown in Figure 9.5 consists of corrugated upper and lower flat surfaces enclosing a fill material. One metal side is extended and crimped around the other metal surface to enclose the ends of the gasket. Various fill materials can be used in this gasket, with typical fill consisting of asbestos millboard, mica, or graphite. The asbestos millboard is being replaced with other materials in the new gaskets. This gasket with any of these fill materials is fire safe.

Overall leak and tightness tests indicate that double-jacketed gasket performance is similar to spiral wound and perhaps superior in some respects to other gaskets.[7,23] Actual operating emission data are not available for comparison. These gaskets are more expensive than other preformed gaskets and are not widely used. However, continuous operating temperatures generally recommended for this type of gasket are higher than those cited for other gaskets. Manufacturer's data on this aspect of operating performance is available along with API recommendations.

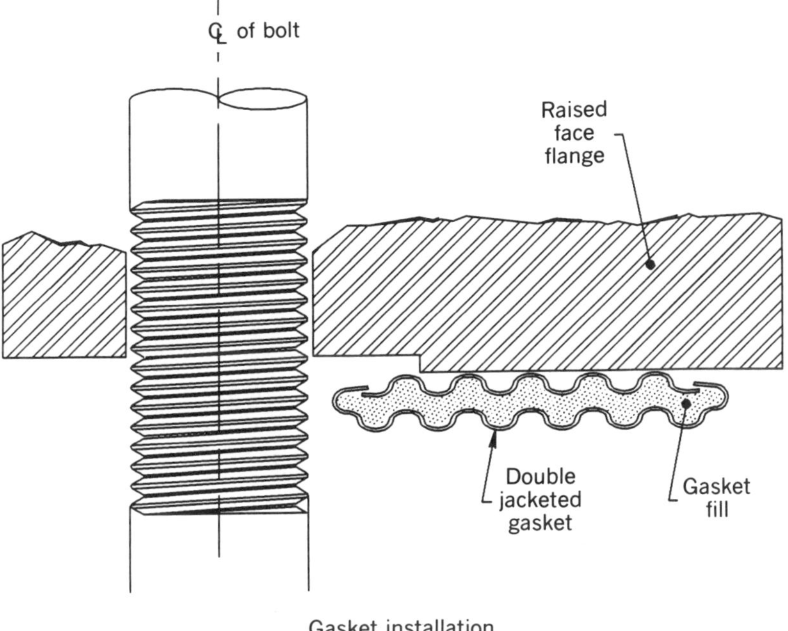

Gasket installation

Gasket detail

Figure 9.5 *Double-jacketed API-type gaskets.*

For heat exchanger gasket service, flat surfaces replace the corrugated surface of this double-jacketed gasket design. Although generally not mentioned, heat exchangers have flanges that are generally much larger than piping flanges. Sealing heat exchanger flanges can be troublesome and leakage at the flanges should be monitored by industrial hygienists to ensure leakage is a minimum or the flanges can become potential exposure sources.

4. *Envelope*. While the double-jacketed API gasket is not completely sealed, envelope-type gaskets are sealed. Some of these gaskets have PTFE envelopes with PTFE or other types of fill material. Other outer envelope materials including metal are available. However, temperature applications for these gaskets are limited, and leakage data are not readily available. These gaskets are not fire safe,

as is an entirely polymeric gasket, and usually find application in specialty services. The metal envelope gasket may be fire safe, but a fill material is necessary that does not disintegrate in fire-test conditions.

5. *Solid Metal*. Solid flat metal gaskets have been installed in various applications for a number of years and normally require a high bolt stress for sealing. However, actual or laboratory emission data on these gaskets has not been published, and comparisons of effectiveness cannot be defined. Modifications including combinations with other gaskets have been installed, reducing the required bolt force. These gaskets have been used in valve bonnets fittings, tongue and groove joints, and so on. Corrugated or profile-type metal gaskets are also available. These gaskets are used when bolting or clamping forces are insufficient to seat a flat-type gasket.

The O rings are another type of metal gasket that are installed in ring-type joints or flanges and were reviewed in Section 9.2.3.

9.5 FLANGE GASKET EVALUATION AND EMISSION ESTIMATING

The previous sections reviewed various gaskets and their effectiveness, particularly the availability of actual emissions data. As described previously, an extensive research and development effort has been conducted under the sponsorship of the PVRC and MTI that was initiated during the early part of the 1970s. This work successfully developed procedures to determine gasket resistance to fire, to estimate gasket leakage, and to establish gasket and flange bolting stress requirements. These procedures are expected to be incorporated into the ASME piping code, which establishes the standards for petroleum refineries and the chemical industry, during the mid-1990s.

A brief review of the procedures developed are included in this section, with emphasis on the method of estimating emissions.

9.5.1 Background

Raised flanges are the standard flange in the petroleum and chemical industries. Although other flanges are used. the majority are raised face and the development effort has been concentrated on this type of flange. While one flange type is the basis for the overall program, a large number of different gasket types has been tested and investigated. Asbestos gasketing was initially investigated in some detail and the asbestos gasket performance data served as a basis for comparing the performance of other gaskets.

Flange Description. A typical raised-face flange and gasket is shown in Figure 9.6 to illustrate the various forces that affect a flange installation. The flanges are bolted together as shown to compress the gasket and prevent leakage. However, the internal pressure of the fluid at the flange interface places a force on the flange that tends to drive the flanges apart. This force has been termed the hydrostatic end force and is one of the forces that must be overcome in determining bolt requirements.

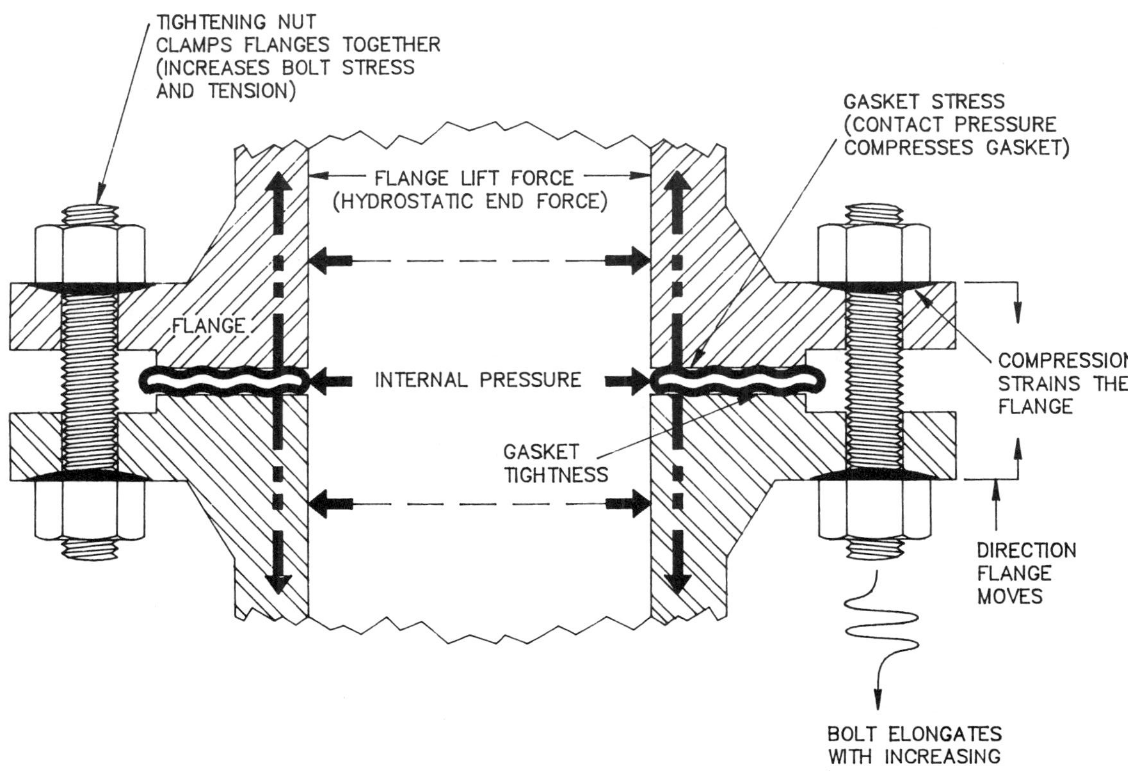

Figure 9.6 *Gasketed flange forces (courtesy of Marine and Petoleum Mfg. Co.).*

Gasket compression is of course a major requirement in obtaining a tight interface seal. The gasket is compressed by tightening the nuts on the flange bolts and the overall pressure developed by the flange force is the gasket stress, which is a major variable in gasket analysis. While this appears straightforward, the force on the bolt required to compress the gasket places considerable stress on the bolt. This bolt stress or bolt load tends to elongate the bolts, which are under a tensile loading. An important aspect of bolt elongation is that the bolt extension provides a basis for determining the stress in the bolt and whether it is in agreement with the design bolt stress. Careful measurement of bolt length can be converted into bolt stress through Hooke's law while the bolt is in the elastic range. Bolt elongation and stress should be measured periodically to determine whether connector conditions have changed which may result in increased emissions leakage.

An important consideration is internal pressure in a gasket system. In addition to providing some hydrostatic end lift force, the internal pressure also places some force on the gasket. When the gasket stress and internal pressure difference decreases to rather small levels or is roughly equal, the total bolt load is almost equal to the hydrostatic end lift force. At this point, the joint is forced to spread apart and a large fluid release occurs as the internal pressure can blow the gasket out of the joint or permit a large flow of fluid past the gasket. This has been demonstrated in tests by the PVRC and has been termed "blowout."

Flange Design. In installations where a high bolt stress exists, the outer edges of the flanges under compression may begin to bend toward the interface. When this occurs, the *flange face* may also begin to bend and leakage can become a problem as the gasket stress across the gasket face forms a gradient due to seating changes. This problem occurs more frequently with 150 class flanges, which have the lightest design of the flange range.[21] As a result, the allowable bolt stress for these flanges has been decreased to 25,000 psi (1723 bar). Flanges should be carefully selected in conjunction with the selected gasket and required bolt stress.

Flange Finish. The flange faces are serrated to prevent gasket movement. Although various PVRC test work does not indicate that this is a major variable, a serious scratch across the control area of the flange face can significantly increase leakage from the flange.[7] The serrations in raise-face flanges are a series of grooves that have a standard height based upon the ASME/ANSI piping Code[20] and the number of grooves per inch are specified. The code specifies 30–55 grooves per inch, which provides a range for the serrations. Data on leakage compared with flange finish indicate the leak at 40 grooves per inch is about 10-fold less than the leakage at 32 grooves per inch.[13] Both data points are considerably lower than the leakage determined where the face finish is 24 grooves per inch. Although this effect has not been investigated by the PVRC, these data suggest that the flange finish should be carefully specified to ensure satisfactory leak performance.

Bolt Load Changes. When the bolt loading decreases, leakage through the flanges increase. In addition to the factors mentioned, bolt loadings can change for other reasons. Apparently the majority of leaks occur after thermal changes which include startup, shutdown, thermal upsets, or rain on exposed joints.[7] In the last situation, major temperature changes can be observed in uninsulated equipment during a rainstorm. Also, time may have had an effect upon the gasket with changes in gasket creep, bolt creep, or gasket degradation.[7, 21]

Various factors can also affect bolt loading, which in turn increases flange leakage. Bolt contraction during a thermal change may result in nut depression into the flange. This situation affects the gasket stress and tightness during subsequent thermal changes. In addition, other effects that may impact upon the bolt loading are:

1. *External Piping Stress*. When piping is not properly supported, the effective bolt loading on the gasket can be decreased, increasing emission losses. The bolts in certain situations must also carry the weight of the pipe, which changes the basic mechanical design.

2. *Bolt Changes*. When bolt stresses exceed the elastic range, their elasticity changes and effectiveness decreases. Increases in internal pressure may contribute to this situation.

3. *Temperatures*. A temperature gradient across the flange can affect the flange and the bolts. Rain has been previously identified as a cause of transient temperature change. Short-term heating or cooling can result in temperature lags in various components. Where piping is thermally cycled on a continual basis, the gasket or bolt can be affected. Deformation in the gasket can occur and the bolt stress may exceed the elastic range. Where a differential expansion occurs across a flange connection, various axial and radial effects may occur.

Overall Design Improvements. Gasket selection based upon testing as described in the following sections minimizes potential gasket problems.[7, 9] Preventing transient temperature effects can be reduced by insulating the flanges and piping or installing thermal shields. The insulation improves process control by eliminating upsets and reduces energy consumption variations, thereby providing an economic incentive to control transient effects. A reduction in flange leakage with insulation improvements also reduces maintenance costs and potential increases in monitoring requirements as a result of increased leakage.

Increasing joint flexibility will also aid in controlling transient temperature effects. Belleville washers in this type of installation provide the flexibility necessary for the various thermal changes. Alternate bolting can aid in providing flexibility. In addition, the overall mechanical design of the piping system must prevent undue stress on the flanges during normal operation, startup, and shutdown.

9.5.2 General PVRC and MTI Objectives

For many years, the gasket design factors in the ASME Boiler and Pressure Vessel Code were the basis for the design of bolted gasketed flanged joints.[24] Although these gasket design factors, defined as m and y are listed in reference 24, use of these factors is not compulsory. The factors are considered suggested values for determining required bolt loads for gasket seating and operating conditions. Since the actual bolt stresses are much greater than the design values developed from the ASME code design bases, the m and y criteria have been considered successful.[25] In this general scheme, the y or seating factor is based on the gasket clamping pressure necessary to seal at an internal pressure of 2 psi (0.01 MPa). The m or maintenance factor is a multiplier used to calculate the bolt loading necessary at room temperature to maintain the sealing pressure under a specified internal pressure. Calculated methods for these two factors are found in Appendix 2 of the ASME code.[24]

The usefulness of the m and y factors has been questioned through the years because of problems in test verification and their limited ability to make adjustments for different products and materials. A lack of standard workable test procedures has been a severe hindrance. In addition, the m and y factors do not consider the leak rate of gasketed joints, which is an extremely important characteristic of gasket performance. As a result, the Pressure Vessel Research Committee (PVRC) instituted a program at room temperature conditions to develop:

1. Gasket leakage data with applicable gasket constants
2. A reliable measure of gasket tightness performance

These objectives have been realized and the procedures developed by the PVRC and MTI will be included as Appendix 2A in the ASME code,[24] which is nonmandatory, eventually replacing Appendix 2. Also, the tests will be included in the ASTM standards. Additional efforts have focused on fire testing, gasket leak performance at elevated temperatures, aged bolt tension and relaxation tests, and a general quality criteria comparison. An important consideration in these various test and evaluation procedures is the ability to screen new gaskets for the characteristics necessary to achieve satisfactory performance in plant applications.

9.5.3 Tightness Characterization

The development of tightness parameters and calculation techniques through the PVRC program has been formulated over a period of time and is still relatively new. All of these procedures at room temperature are based upon leakage data. Test procedures are presented in the following summary, followed by the analytical procedures.

Gasket Tests. The room temperature tightness test (ROTT) procedure to obtain gasket constants consists of two parts. In Part A the data developed are associated with gasket seating and the seating load requirement is determined, which is similar to the y factor. During the Part A test the gasket stress is increased through five stress levels from 1025 psi (71 bar) to 15,160 psi (1045 bar). At each stress level the leak rate is measured at internal pressures of 400 psig (29 bar) and 800 psig (56 bar). For these standard tests, leak rate is measured with helium. Other inert gases have been used for comparison, but the standard test is based upon helium. A typical curve obtained for Part A is shown in Figure 9.7 (the long unbroken line), which is representative of various tests.

While conducting the Part A test at three of the five stress levels (8090, 11,640, and 15,160 psi), the gasket is unloaded to a pressure level of 1025 psi (71 bar). The

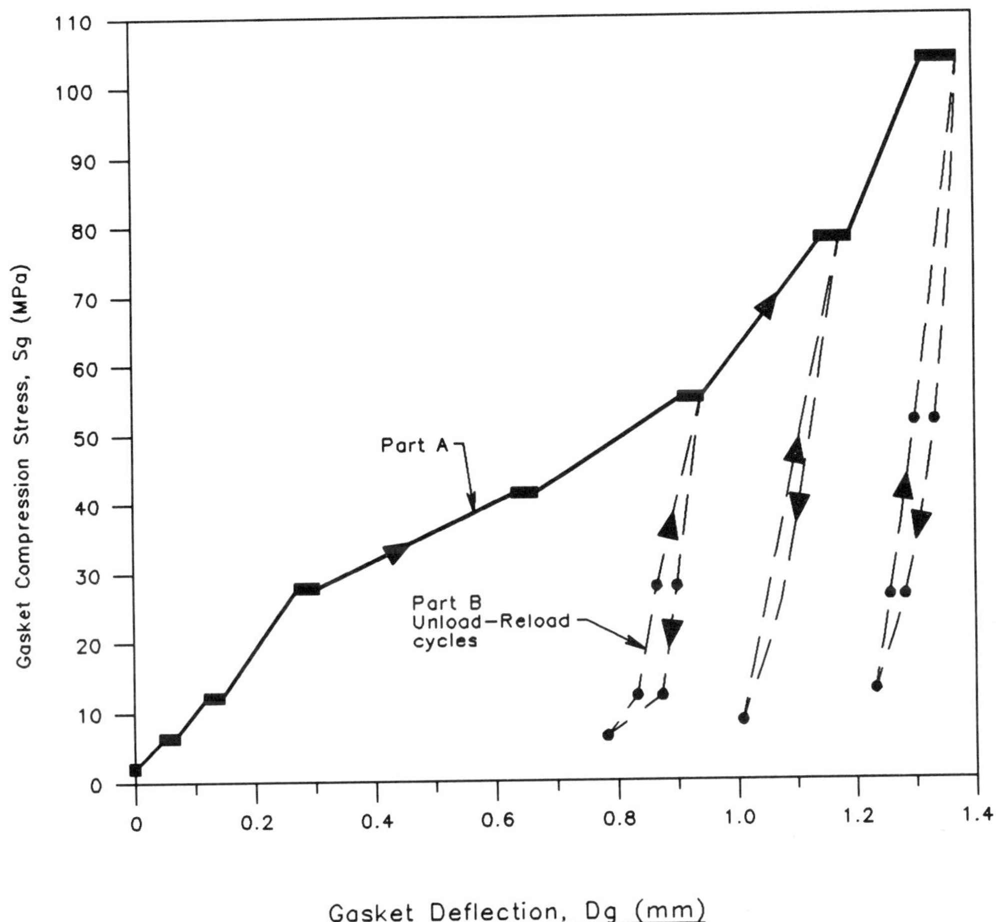

Figure 9.7 Typical room temperature tightness test (ROTT) measurements for Parts A and B (courtesy of J. Payne Associates. JPAC Inc.).

gasket is then reloaded to the original gasket stress level. This unload–reload cycle is conducted at an 800-psig (56 bar) internal pressure and has been termed Part B. Following this cycle, the Part A procedure continues to the next stress level where the unload–reload cycle is repeated. The Part B unload–reload cycles consist of three separate cycles, as shown in Figure 9.7.

Mass leak rate is measured during these tests, and a log–log plot of leak rate versus internal pressure is developed based upon gasket stress. For a constant gasket stress, parallel lines essentially occur on the plot, which permits combining leak rate and pressure in a tightness parameter. In addition, mass leak rate versus gasket stress can also be transformed into curves.

Tightness. Based upon these tests a tightness concept has evolved where tightness is considered a measure of a gasket's ability to control leak rate at a given stress level. In general, a tighter gasket requires a much higher internal pressure to produce a leakage rate equivalent to that observed in a less tight gasket. With tight gaskets, lower leak rates result.

From their analysis, the PVRC group developed a tightness parameter T_p, which is the internal pressure required to produce a helium leak of 1 mg/sec from a 150-mm outside diameter (OD) gasket in a flange. This size gasket is essentially equivalent to a 4-in. pipe gasket which is termed an NPS4 joint. The tightness parameter is nondimensional and is defined as[7]

$$T_p = \left(\frac{P}{p^*}\right)\left(\frac{1}{L_r}\right)^{0.5} \qquad (9.3)$$

where

P = design pressure (same units as p^*)
p^* = atmospheric pressure
L_r = mass leak rate per unit diameter (mg/sec-mm (1b/hr-in.))

This equation is based upon a ratio of the leak rate of 1 mg/sec to the leak rate in a test for a specific gasket. While the 1 mg/sec-mm rate aided in defining the tightness parameter, a base level of leakage acceptable in an operating system must be defined. This tightness criteria, termed the minimum tightness ($T_{p\,\min}$), has been defined as a standard tightness class ($T2$) which is a leak of 0.002 mg/sec-mm (0.0004 1b/hr-in). Classifications are shown in Table 9.6. The total leakage rate is a function of the gasket outside circumference. From Table 9.6, the minimum tightness equation is derived:

$$T_{p\,\min} = 1.82574cP_r \qquad (9.4)$$

or

$$T_{p\,\min} = 0.124cP \qquad (9.5)$$

where

P_r = pressure ratio, $P/14.7$ (psi)
c = constant
P = internal pressure, psi (MPa)

TABLE 9.6 Gasket Tightness Criteria for Flange Design

Tightness Classification	Mass Leak Rate Per Unit Circumference		Constant	Tightness Parameter Pressure Ratio
	mg/sec-mm	lb/hr-in.	(C)	($T_{p\,\min}/P_r$)
T1 (Economy)	0.2	0.04	0.1	0.18257
T1–T2	0.02	0.004	0.316	0.57734
T2 (Standard)	0.002	0.0004	1	1.8257
T2–T3	0.0002	0.00004	3.16	5.7734
T3 (Tight)	0.00002	0.000004	10	18.257

[a]Adapted from reference 9.

TABLE 9.7 General Gasket Tightness Classification Considerations for Flange Design[a]

General Consideration	Tightness Classification[b]				
	T1	T1–T2	T2	T2–T3	T3
Carcinogenicity	Noncarcinogenic	Noncarcinogenic	Suspected human carcinogens	Mutants Embryotoxins Teratogens NESHAP Mat'l.	Human carcinogens
Health hazard	Skin irritation	Skin irritation	Skin or eye contact harmful	Skin or eye contact dangerous	Skin or eye contact dangerous; potential burns on exposure
Fire and stability	Low fire hazard	Combustible	Combustible	Decomposes violently Water reactive Flammable	Decomposes violently Pyrophoric Flammable
Constant[c]	0.1	0.316	1.0	3.16	10
Process conditions[b]	$-32 < T < 150°F$ $P < 100$ psi	$150 < T < 250$ $100 < P < 200$	$250 < T < 600$ $200 < P < 600$	$600 < T < 900$ $600 < P < 1000$	$T > 900$ $P > 1000$

[a]Adapted from reference 9.
[b]All conditions in psi and °F.
[c]For Eqs. 9.4 and 9.5.

In general, Eq. 9.5 is used to rapidly define the $T_{p\,min}$ value for different pressures. The value of c is 1 for the standard tightness $T2$, with the $T_{p\,min}$ a function of c at a constant internal pressure level. For a specific gasket material, the minimum tightness parameters can be determined over a range of pressures and diameters. In general, a parallel series of curves will result, separated by the tightness classification. While this provides an insight into various applications, the maximum allowable pressure levels for a gasket should be determined.

The criteria for the tightness classifications developed by Winter[9] have been used as a reference in various papers.[7,25] These classifications are based upon hazard and safety considerations and a simplified relationship is shown in Table 9.7. This is not a definitive classification table, but can be used as a basis for determining classification by industrial hygienists and safety personnel. The temperature and pressure ranges can change considerably in accordance with plant conditions and an assessment of the hazards.

Constants and Bolt Loading. Constants are necessary for determining bolt loading. These constants are determined from the leakage data transferred to curves of gasket stress (s_g) versus the tightness parameter (T_p). The curves are placed on log–log paper and identified as Part A and Part B in establishing the constants. An ideal curve of these relationships is shown in Figure 9.8.

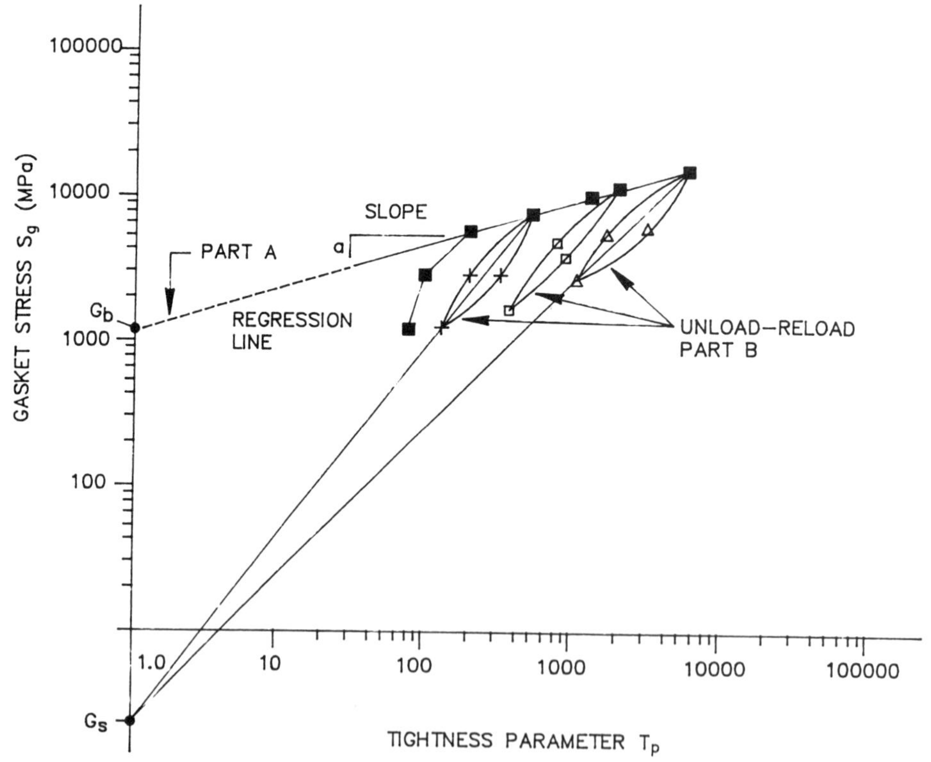

Figure 9.8 Ideal gasket tightness and gasket stress relationship for determining constants (courtesy of J. Payne Associates, JPAC Inc.).

TABLE 9.8 Typical Loading Constants for Various Gaskets[a]

	Gasket Types			
Constant	LGSB[b]	SSG[c]	CAP[d]	SSA[e]
G_b (psi)	816	2300	2500	3400
a	0.377	0.237	0.15	0.30
G_s (psi)	0.066	13	117	7

[a]Adapted from reference 7.
[b]Flexible graphite laminate bonded to stainless steel strip.
[c]Spiral-wound gasket, graphite filled.
[d]Compressed asbestos sheet.
[e]Spiral-wound gasket, asbestos filled.

The G_b and a constants provide information on gasket seating load since these constants are developed from the Part A portion of the gasket test. These constants can be related to the older y factor in the ASME code, which is also related to the seating stress. For the constants developed with various gaskets, typical constants are shown in Table 9.8. The G_s is a function of gasket operating performance since the unload and reload cycles describe leakage effects. A low G_s indicates a small decrease in tightness for a large bolt load reduction, which is an important characteristic where pressure or thermal variations occurs.

In general, low values of G_b, a, and G_s are desirable. In addition, a factor developed in the program is used to compare seating properties of gaskets for constant T_p values. This seating factor is

$$G_b(T_p)^a \tag{9.6}$$

A comparison of the seating values is shown in Table 9.9 for the gaskets in Table 9.8.

TABLE 9.9 Typical Values of $G_b(T_p)^a$ for Gaskets[a]

	$G_b(T_p)^a$ (psi)			
T_p	LGSB[b]	SSG[b]	CAP[b]	SSA[b]
100	4631	6851	4988	13536
1000	11033	11823	7046	27007

[a]Adapted from reference 7.
[b]See table 9.8 for description.

The *bolt loading procedure* provides an overview of the general calculation technique based upon the constants obtained from the test data. These calculations should be conducted by mechanical engineers, but gasket modifications which may affect emissions should be reviewed with an industrial hygienist. The bolt loading calculation is more extensive than the previous tightness calculations, and the following procedure is based upon $T_{p\,\min}$, G_b, a, and G_s:

1. Determine the minimum tightness from Eq. 9.5 based on the tightness class.
2. The required seating stress (S_{ya}) is determined from $T_{p\,\min}$

$$S_{ya} = G_{b/e}\left[1.5(T_{p\,\min})\right]^a \tag{9.7}$$

where e is the boltup factor and is 0.75 for manual bolting. Other e factors are assigned based upon the bolting procedure.[9,26] Factors for boltup can also be estimated from 0.85 for a calibrated torque wrench, 0.95 for multipoint hydraulic tensioning, and 1.0 where bolt stretch is measured.

For equal code allowable operating and seating bolt design stresses ($S_b = S_a$) and S_{ya} is determined from Eq. 9.7

In elevated temperature joints the 1.5 factor is increased by the S_a/S_b ratio where S_a is greater than S_b. A more exact method for this factor is determined through iteration where $S_{m1} = S_{m2}$ (see item 3 following).

3. Determine the minimum design gasket stress (S_m) required to maintain $T_{p\,\min}$ during operation. Here S_m is the greater of S_{m1}, S_{m2}, or $2P$, the operating and seating requirements.

Seating stress:

$$S_{m2} = S_{ya}/1.5 - P(A_i/A_g) \tag{9.8}$$

where

P = total bolt lead

A_i = $0.785G^2$ area against which pressure acts

A_g = $2b(3.14G)$ gasket area over which minimum stress is maintained

G = gasket OD less twice the effective width of the gasket

b = gasket width

factor = 1.5 is increased by ratio S_a/S_b when $S_a > S_b$ for elevated temperature joints

Operating stress:

$$S_{mi} = G_s\left[(e/G_s)(S_{ya})\right]\left(\frac{1}{T_r}\right) \tag{9.9}$$

$$T_r = \log(1.5T_{p\,\min})/\log(T_{p\,\min}) \tag{9.10}$$

and S_m is then determined from the final evaluation as described above.

4. Determination of minimum bolt load:

$$W_m = P(A_i) + S_m(A_g) \tag{9.11}$$

This represents the sum of the pressure load and the gasket loading necessary to maintain a seal and adequate seating.

While this procedure provides a bolt loading, there are other practical considerations that affect the bolt loading values determined through these calculations:

1. A practical maximum seating stress is necessary since S_{ya} can be extrapolated from test data. The value of $e(S_{ya})$ should not exceed 10–20% of the maximum test stress.

2. For compression stress, the gasket load should have a compression stop to limit the displacement of the sealing element.

3. A minimum assembly load estimate is necessary since there is a difference between the design bolt load and the assembly load. The design bolt load based upon the ASME code sets flange requirements in accordance with the code-allowable stress philosophy. In existing flange joints, the design loading is less than minimum tightness requirements. The assembly load is an estimate of the actual bolt loading necessary to ensure the loading is greater than minimum tightness requirements. The minimum assembly loadings can be determined from reference 7.

4. The maximum assembly loading is the highest bolt loading that can be safely applied consistent with bolt, flange, and gasket limitations.

A difference between the minimum and maximum assembly loadings should exist to ensure a satisfactory installation. Where a spread does not exist, the flange system and gasket are inadequate. Alternatives that can be considered for this condition are tighter gaskets, a lower tightness class, or flange replacement.

An emissions estimate can be developed from the gasket assembly stress S_{ya} and the G_s constant or from the tightness test data. While these estimates are based upon room temperature conditions, emissions estimates at actual operating conditions are desirable for a more accurate assessment of leakage. Hot tightness tests have been developed and are described in the following sections.

Elevated Temperature Tightness Tests (HOTT and AHOT)

A hot operational Tightness Test (HOTT) was developed that is similar to the room temperature tightness test previously described. This test was designed to reproduce operating conditions to 800°F (427°C) over a 1-wk period, although tests for 3 wk have been reported.[7] While the test conditions were originally limited to 800°F, the testing temperatures have been increased to 1200°F. (649°C). In addition, the test system was modified and designed for gasket tests of 1 month or more. This modified test is the aged hot operational test (AHOT), and a considerable number of gasket tests have been conducted in this test equipment.

In the HOTT test, the gasket is placed in the test rig at room temperature, stressed, and then heated at constant stress to a desired temperature where it is maintained for a period of time, permitting stress relaxation. The relaxation is accompanied by gasket stress changes. When the stress and temperature are restored to 8090 psi (558 bar) after the relaxation period, unloading begins, which reduces gasket stress to about the internal pressure to achieve a near blowout condition. Alternatively, a leak rate of 8 mg/sec of helium is considered a blowout condition. The gasket stress is then increased to about 3000 psi (208 bar), where a thermal disturbance is placed on the gasket, which results in different temperatures on the platens pressed against the gasket. Gasket temperature and stress are increased to initial preunloading conditions. This cycle procedure is then repeated twice with leakage testing for all cycles at 400 psi (28 bar) and 800 psi (55 bar) conducted at various points. The thermal disturbance places considerable shear across the gasket, which indicates gasket mechanical integrity following test completion.

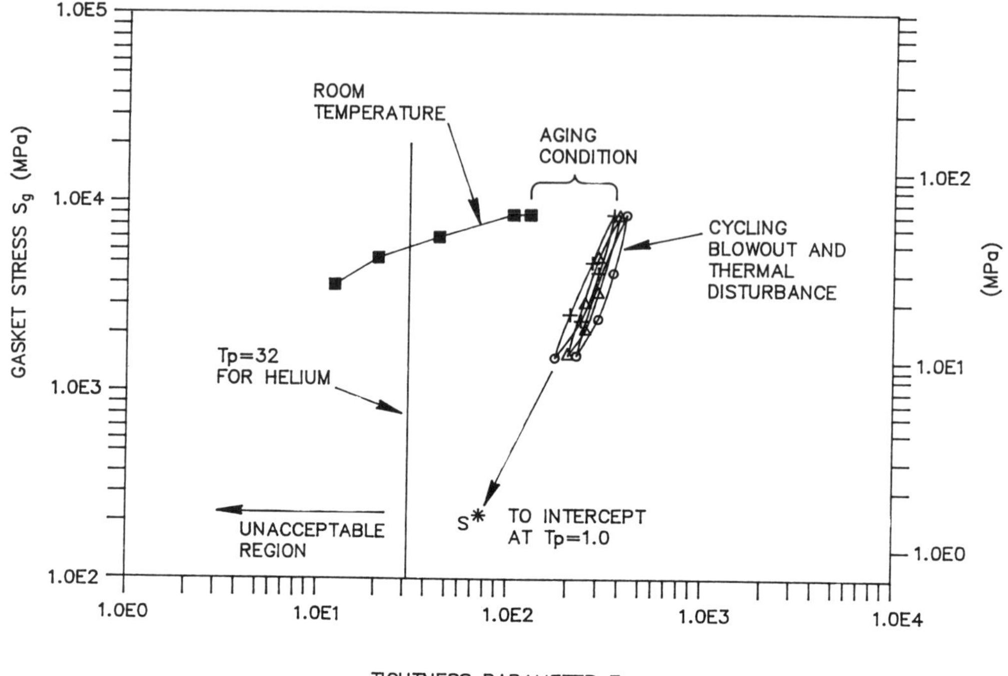

Figure 9.9 *Ideal AHOT test results (courtesy of J. Payne Associates, JPAC Inc.).*

The AHOT procedure duplicates the initial part of the previous test in the initial conditions to the defined temperature level. At this point, the test gasket (held between platens) is removed from the test rig and placed in an oven where it is held at temperature for a number of days or weeks to provide the necessary aging time. When the aging period is completed, the platens are removed from the oven and placed in the heated test rig. The general tests then follow the HOTT test procedures, including blowout cycles and thermal disturbance effects. Mass leakage of helium is measured at comparable points in the test procedure

A general figure covering the type of test results obtained in an AHOT test is presented in Figure 9.9. For helium, a minimum tightness (T_p) of 32 is shown, which has been selected as an acceptable performance level from numerous test data.[7] The emissions can be estimated from this figure based on the cycles generated through the blowout, thermal disturbance effects as follows:

1. Place a slope on the figure based on the minimum and maximum tightness associated with the minimum and maximum gasket stress as shown in Figure 9.9.

2. The intercept of the slope with the T_p ordinate at 1.0 defines the intercept gasket stress S^*. This stress is determined in similar fashion to G_s in the room temperature tests.

3. An operating tightness parameter defined as (T_p) is determined from the slope, intercept, and operating gasket stress:

$$T_p = (S_g/S^*)(1/\text{slope}) \tag{9.12}$$

where

S_g = operating stress
S^* = intercept
Slope = slope of the line extending from the cycles to the intercept; $\log(S_{g\,max}/S_{g\,min}/\log(T_{p\,max}/T_{p\,min})$

4. Based on Eq. 9.3, the leak rate calculated for a 150-mm OD gasket is adjusted for the selected gasket size:

$$\text{Estimated leak rate} = (\text{gasket OD}/150)(P/T_p p^*)^2 \tag{9.13}$$

where

leak rate = mg/sec
P = internal pressure (psi or MPa)
p^* = atmospheric pressure (psi or MPa)
T_p = from Eq. 9.12

The estimated leak rate from the aged elevated temperature tests provides a basis for determining leak rates at operating conditions. Room-temperature leakage rates also aid in estimating low-temperature emissions from new gaskets. For new gaskets at elevated temperatures, the HOTT tests provide a basis for determining emission rates. Where gaskets operate at temperatures in the range of 150°F (66°C) to 550°F (288°C), there is a lack of leakage data and tightness parameter relationship figures. Consequently, test data may be required for operating conditions in this range, although testing of various gaskets by the PVRC is still underway and the accumulated test data should be reviewed prior to specifying gaskets.

9.5.4 Fire Tests

Although fire testing was reviewed in Section 9.4.1, these tests must be considered as part of the overall evaluation procedure for gaskets and are briefly described in this section. In addition, these tests are also related to tensile/relaxation tests described in the next section. The basic method of fire testing developed in the overall PVRC/MTI program is to simulate fire conditions by heating gaskets under compression stress to a temperature considered equivalent to a fire temperature within a flange. These tests are designed to evaluate the gasket rather than the flange joint which contains the gaskets since the flange may be affected by bolt elongation, warpage, or other effects. In addition, the assessment indicates that the actual temperature of the gasket is affected by the mass of the joint, the pressure of liquid in the piping, and the fire intensity.[7] Although these are valid reasons for not conducting an actual fire test, gasket manufacturers have developed fire tests that are considered quite reliable.[13] Should a flange open slightly during an actual fire test, the performance of the gasket under these conditions may be an important consideration in the evaluation of gasket performance.

Fire Simulation Screening Test (FIRS). The fire simulation screening test consists of an initial fire testing screening apparatus where a few strips from a sheet gasket were placed on a flat surface and then compressed under a single bolt loading. The fixture is heated rapidly and simulates fire conditions by holding the fixture for 30 min at 1200°F (649°C). Following the test, the retained stud load, thickness change, and residual tensile strength of the gasket strips in the weak direction are measured. In these tests compressed asbestos and graphite flexible sheets are satisfactory, but PTFE and PTFE reinforced with various fiber materials failed.

The FIRS test is the initial screening test for a material and is very important in gasket evaluation. Success in this test requires the material to remain intact to permit handling and tensile testing. Suggested tensile stress for the sample should be approximately 100 psi (6.9 bar) with little evidence of extrusion, lateral flow, or excessive spread. If the gasket material does not pass the FIRS test, the gasket is normally considered a failure and is discarded.[18]

Fire Tightness Test (FITT). This test is conducted on equipment similar to the test equipment in the HOTT test described in Section 9.5.3. This test is an extension of the FIRS test, which was unsatisfactory for nonsheet-type gaskets such as spiral-wound, double-jacketed, and corrugated gaskets. In addition, a leakage rate is measured during the test which cannot be performed in the FIRS test rig. The FITT test exposes a 150-mm OD or NPS4 gasket to rapid heatup followed by a heat soaking at 1200°F (649°C) for 15 min.

The measurement of helium leakage during an initial preload stage and at the soak temperature permits an evaluation of tightness performance at both room temperature and soak conditions. Tightness is defined for the beginning of the 1200°F (649°C) soak period and at the end of the soak period. A comparison of tightness indicated several gasket types were below the T_p 32 minimum tightness criteria at the beginning of the soak temperature period.[7] Others had a decrease in tightness over the soak period but were acceptable. Since a large number of gaskets have been tested, specified data should be obtained on the gasket of concern. In addition, FITT results are shown as a gasket stress versus tightness curve[7,18] which provides an indication of performance under test conditions.

9.5.5 Other Screening Tests

In addition to fire screening tests, other tests have been developed to further extend screening assessments and aid in qualifying acceptable gaskets.

Aged Tensile/Relaxation Screen Test (ATRS). This test investigates stud (bolt) load retention and the creep or relaxation characteristics of gasket material. The fixture is similar to the FIRS test equipment when a single bolt compresses test samples between compression platens. Since the fixture is small, it is placed in an oven for both aging and temperature conditions as required. In addition, bolt loading can be applied as required with bolt loading and temperature simulating actual process conditions. Following aging over a long period at process temperature, the retained stud loading, gasket material thickness change, and residual gasket material tensile strength are measured.

Initially, a compressive stress of 5000 psi (345 bar) is placed on the material. Dumbell-type sample strips are placed within the platens prior to compression. Three specimens or test strips are installed as layers with two separate layers in the test fixture.[7] Over a period of time, which can be measured in days or weeks, the compressive stress gradually falls or relaxes as the gasket materials creep and in some instances deteriorate. Generally, bolt stress relaxation of 30–60% has been observed.[7] For temperatures higher than 750°F (399°C), a high-temperature fixture or test rig is used which has been termed HATR for hot aged tensile relaxation test.

This test investigates mechanical performance at conditions resembling operation and aids in determining suitability for AHOT testing (Section 9.5.3). Although similar to plant operation, air in and around the gasket material does enhance oxidation effects, unless the fixture is placed in an inert atmosphere. The ATRS test is limited to sheet gasketing, which reduces its use in evaluating formed ring gaskets such as spiral-wound gaskets.

Aged Relaxation Leakage Adhesion Test (ARLA). This test fixture, which is a recent development, is a lighter-weight device than the ATRS test fixture and is more versatile.[7] In contrast to the rectangular ATRS fixture, the ARLA fixture is circular and permits investigation of both sheet and formed gaskets. In addition, spring loading aids in evaluation of relaxation and post test tightness properties along with information on gasket adhesion and weight loss.[7,27] For example, adhesion characteristics of graphite were demonstrated in some ATRS tests.[7] Weight loss measurement appears to be a promising screening parameter.[27] While this test is basically a screening procedure, post test tightness evaluations are conducted which can be combined with relaxation and weight loss as further descriptions of gasket acceptability and performance. This test system is limited to a 750°F (399°C) temperature, but aging is conducted as required.

9.5.6 Screening Test Evaluations

Hundreds of tests have been conducted on gasket sheet materials with the FIRS, ATRS, and ARLA procedures. These data indicate that PTFE (within its temperature limits) is not affected by aging, and laminated graphite sheet at temperatures less than 700°F (370°C) shows little change during long-term tests. In contrast, elastomeric materials demonstrate aging and temperature effects where changes are observed with increased time or temperature. As a result, time-temperature parameters have been developed to describe these interactive variables.[7,27] These parameters in general are an assessment of quality and are described as follows.

Equivalent Aged Exposure (A_e). This parameter relates time-exposure effects and was devised for compressed asbestos sheets with elastomeric binders, aramid, and glass-reinforced materials. The parameter is described in the following equation:[7]

$$A_e = 0.0502\,(T_e - 300)\,H^{0.2} \tag{9.14}$$

where

T_e = exposure temperature (°F)
H = exposure time (hr)

Here A_e provides a basis for comparing test conditions in an air medium and is used to extrapolate ATRS test data for the prediction of trends in service life.[7,28] From the current relationship, an A_e of 100 is equivalent to an exposure of 1000 hr at 800°F (427°C). This parameter is generally not used as an individual assessment but is combined with a quality parameter.

Quality Parameters. Since asbestos has been the standard gasket for many years and a large body of data on asbestos sheet gaskets in the PVRC/MTI tests were accumulated, quality parameters were developed. These parameters use asbestos as the quality basis, with a value of 1 for comparison.

Load Retention Quality (Q_r). This evaluation indicates bolt load retention by the gasket after exposure conditions.[7]

$$Q_r = \frac{(\%\ \text{stud load retained test gasket})^2}{(\%\ \text{stud load retained aged sheet asbestos})} \tag{9.15}$$

Tensile Quality (Q_x). This quality parameter is based upon the sheet gasket tensile strength after exposure. Tensile testing is in the weak fiber direction.[7]

$$Q_x = T_{sx}/1000 \tag{9.16}$$

The denominator is a 1000-psi (69 bar) reference tensile strength for compressed asbestos sheet with an elastomeric binder.

Overall Quality Parameter (Q_p). This parameter combines the properties of gasket materials in a flanged joint[28]

$$Q_p = Q_x Q_r \tag{9.17}$$

The A_e factor previously described is correlated with the quality parameter Q_p and provides a basis for comparison with asbestos sheet. Above an equivalent aging parameter (A_e) of about 175, the long-term value of Q_p is about 0.5. Correlation for some other sheet materials indicate that the quality parameter Q_p falls to zero at an aging exposure factor (A_e) of about 65.[7] Satisfactory materials should be equivalent or superior to asbestos (0.5) for long-term operation.

Tightness Quality (Q_t). This parameter is based on tightness data.[7]

$$Q_t = T_{p\,\min}/32 \text{ for helium} \tag{9.18}$$

$$= T_{p\,\min}/12 \text{ for air and nitrogen} \tag{9.19}$$

For acceptance, a Q_t equivalent to or greater than 1 is recommended. In these criteria, the values of 32 and 12 are minimum T_p values for aged compressed asbestos. The different gases affect the tightness level, and comparisons with some process materials would aid in assessing gasket performance over a broader spectrum.

9.5.7 Overall Evaluation Procedure

Based on these tests, evaluation procedures can be established for new gasket materials or existing materials that have not been extensively tested. The screening tests, including fire testing, can be combined into the following overall evaluation scheme.

$$\text{Screening tests} \longleftarrow \text{(FIRS, FITT, and ATRS)}$$

From these data, gaskets that are successfully fire tested move to the FITT test and then proceed to the ATRS test. When $Q_p \geq 0.5$ and $Q_t \geq 1.0$, the gasket material is evaluated with the equivalent exposure parameter A_e. For A_e values equivalent to the asbestos value of 0.5 or greater, the gasket material is satisfactory for long-term operation. With all parameters acceptable, the testing then proceeds to the next stage:

$$\text{Tightness tests} \longleftarrow \text{(HOTT and AHOT)}$$

In these tests, the tightness parameters are developed for the flange system and leakage calculations. The HOTT test provides the basis for AHOT tests which demonstrate long-term aging effects.

Other Tests. For additional assessments, two other tests are under development by the PVRC study group:[7]

An emissions hot tightness test (EHOT) is a modified hot operational tightness test (HOTT) designed to evaluate gasket emission performance in the 200–400°F (93–204°C) range.[7] This range covers a great number of applications in the chemical, petrochemical and petroleum refining industries. Before and after the EHOT test, room operational tightness tests (ROTT) are conducted providing G_b, a, and G_s constants. The EHOT is currently under development and should be considered for gaskets in this operational range.

A hot blowout test (HOBT) is also under development that is primarily designed to assess PTFE gasket tightness under extensive relaxation conditions.[7] Following heating to the test temperature, creep relaxation occurs and the internal pressure is increased until a very large leak occurs (blowout). This test is in development.

Although the test follow a general path, considerable assessment of test results is conducted during each screening test to define operating conditions for succeeding tests. In addition, the data are evaluated to ensure the acceptance criteria are met and the aging test provides long-term information. Typical general testing procedures for gasket categories or classes follow.

Elastomer Bound Sheets. Tests are conducted on these sheets to determine an acceptable A_e performance level. These tests begin with a FIRS test to ensure heat resistance. Following this test, a set of three ATRS tests are conducted, covering different time periods and different temperatures.[7,28] The purpose of the tests is to screen for an indicated service temperature of approximately 400°F (204°C) for 10 years. Additional tests may be necessary to define the limiting A_e value, that is, the basis for a HOTT test with air. Frequently an AHOT test with nitrogen follows the HOTT test and is based on HOTT tests results.

PTFE and Laminated Graphite Sheets. When PTFE and graphite sheets operate at temperatures less than 700°F (371°C), the testing program is simplified compared with the program outlined for the elastomer bound sheets.[7] Also, aging is not considered a problem and testing requirements in this area are omitted. A hot tightness test is not conducted since PTFE seals very well. Moreover, a fire test is not conducted unless there are specific fire resistance claims. The new hot blowout test is designed for PTFE testing and is recommended when development has been completed.

While flexible graphite sheets under 700°F (371°C) are not a particular problem, at operating temperatures above this level degradation with aging has occured. Consequently, the ATRS and accompanying high-temperature test HATR are expanded to define long-term aging effects. The aging tests reported in reference 21 indicated fall-off in gasket stress around 7000 hr, suggesting additional information is necessary to fully characterize long-term flexible graphite performance.

Formed Gaskets. These gaskets consist of spiral-wound, jacketed, graphite laminated, corrugated, and other preformed types. Since the gaskets are generally composed of fire-resistant materials, basic fire tests are not required. Where the gaskets contain materials that have not been adequately defined such as mica or PTFE fillers, a FITT test is generally considered.[7] In addition, aging tests for graphite and mica fillers are necessary to adequately determine long-term performance.

For high-temperature testing of mica or graphite-filled gaskets, the HATR test covering varying temperatures and time exposures provides a basis for tightness tests (HOTT and AHOT).

9.5.8 Temperature Limitations

Many gasket tests have been conducted at elevated temperature levels for relatively long testing periods to determine optimum long-term performance characteristics. While gaskets may operate at relatively high temperatures for short time periods, short-term operations are normally associated with thermal cycling conditions that typically occur where fixed beds are regenerated frequently or with short cycle batch operations. Long-term continuous operations at high temperatures may require a lower maximum operating temperature level. Test data for laminated graphite sheet gasket at 750°F (399°C) indicates a fall-off in stress relaxation load after 7000 hr.[21] Should this characteristic be confirmed in other tests, a long-term maximum continuous operating temperature lower than 750°F (399°C) is required. The maximum operating temperature may require further investigation to establish the reason for the change in gasket performance and test data to define the maximum operating temperature.

Graphite oxidation can affect maximum temperature performance. Operation of a graphite material in the presence of air is normally restricted to a maximum temperature of about 800°F (427°C). However, laminated graphite gaskets only have the outer edge of the gasket exposed to air during high-temperature operation and the amount of oxidation should not be significant. Moreover, where the flange is insulated, outer edge contact with air is reduced considerably and the allowable gasket operating temperature may be somewhat higher. In addition to the sheet gaskets, graphite filler is a standard material in spiral-wound gaskets. Since many spiral-wound gaskets have outer and inner rings, oxidation effects do not appear to be a problem and higher operating temperatures are justified.[7, 29]

For continuous operations, Table 9.10 is a tentative list of gasket types and suggested maximum temperatures. These temperatures should be modified or adjusted based on operating experience, gasket manufacturer data, or data from study sources such as PVRC/MTI and EPRI. Field data, where available, provides a reliable basis for defining maximum continuous temperatures. Spiral-wound gaskets have been in service for many years; however, asbestos may have been the standard filler material and the data may not adequately describe spiral-wound graphite-filled gasket performance.

TABLE 9.10 Suggested Maximum Gasket Continuous Operating Temperatures[a]

Type	Maximum Temperature, °F (°C)
Laminated flexible graphite gasket	700 (371)[b]
Carbon / graphite with elastomer binder	700 (371)[b]
Aramid, glass fiber with elastomer binder	400 (204)[c]
Spiral-wound, graphite filler	950 (570)[d]
Double-jacketed, graphite filled	800 (427)
Corrugated metal with laminated graphite	700 (371)[b,c]

[a]Adopted from references 7, 21, 27, 28, and 29.
[b]Conservative; based on reference 7, can change with further information.
[c]Conservative; can change with reformulation and additional data.
[d]Based on flexible graphite data; further information may change level.

9.6 EXISTING GASKET TEST PROCEDURES

In contrast to the newer test procedures described in Section 9.5, a number of older or existing test procedures are still used to describe gasket performance. Various countries in the world have developed test systems through the years, but some of the more widely used standard tests are from the American Society for Testing and Materials (ASTM) and the Deutsche Institut Sur Normung (DIN). These gasket testing methods are frequently used by gasket manufacturers to describe their products. Results are simpler to analyze and report. Brief summaries of several tests are presented in the following subsections.

While the data from these various tests can be of assistance, placing the PVRC/MTI analytical procedures in the ASME code indicate the newer procedures have received considerable support from manufacturers and industrial consumers. This widespread support suggests a gradual change from older procedures to the new analytical procedures and tests. The newer tests also provide methods of estimating leakage that are not available from the older gasket test procedures. In addition, aging tests and gasket testing with plant process conditions provide an important base for predicting operating performance. The various test equipment and test procedures developed by the PVRC/MTI study will also be placed in the ASTM list of test procedures for gaskets.

9.6.1 Sealability of Gasket Materials

In this ASTM test (designation F37), a load is placed on a gasket installed between two flanges. The flanges are bored to permit fluid injection into the fixture for testing leakage. Both gas and liquids are used in the test, with the gas generally nitrogen and isooctane a popular liquid test medium. This is a room-temperature test with the internal pressure about 30 psig (3 bar) and gasket stress pressures about 3000 psi (207 bar).

Sealability is measured by the leakage rate through the gasket, flange, or both. The leakage is a function of time, temperature, and internal and gasket compression pressures. The internal fluid pressure in these tests is limited and the temperature is generally room temperature since a more elaborate test device is necessary for temperature testing.

9.6.2 Gas Permeability

This DIN test (designation 3535), measures gas leakage through the flange joint in a more elaborate test fixture than the test ring in ASTM F37. The DIN test measures leakage of nitrogen gas at an internal pressure of 580 psig (41 bar) with a gasket loading of 4640 psi (320 bar). This test is normally conducted at room temperature, although it can be modified for operating at elevated temperatures. This apparatus permits testing gaskets of various size, thicknesses, internal pressures, and loadings and, when modified at higher temperatures.

Leak rates are direct relationships in this system and estimated leakage rates are based upon operating conditions similar to test conditions. However, aging is not normally considered in this test, although the system can probably be operated for a 60- to 100-hr period. The long-term tests conducted by the PVRC[7] or EPRI[21] for 7000 hr cannot be duplicated without significant changes.

9.6.3 Creep Relaxation

In this ASTM test (designation F38 Method B) sheet gasket strips are placed between two circular disks. The disks are bolted together with a single calibrated bolt that measures the creep relaxation of the gasket strips. Measurements of the bolt are taken at a specified time after the compressive stress has been applied.

These data are obtained without a fluid present. Loss of bolt torque in this test results from creep relaxation of the gasket material.

Since torque loss is also a function of bolt elongation and flange distortion, further testing is also recommended, particularly in the hot compression test.

9.6.4 Hot Creep Relaxation Test

This is a DIN test (designation 52913) where the gasket stress can be adjusted to a maximum of 7250 psi (500 bar) at different temperatures. In addition, the test period for a specific condition can be extended to 100 hr. The maximum temperature for this testing device is 572°F (300°C). Results provide a basis for determining relaxation and sealability. However, the test is conducted in air without an internal pressure and does not investigate leakage. Both the ambient and elevated creep/relaxation tests provide information on physical characteristics and gasket changes.

9.6.5 Other Tests

Various other tests can be conducted on gaskets, but estimated leakage calculations of gaskets based on the tests are questionable unless specific gasket conditions similar to process requirements have been tested. Moreover, the aging factor, which appears to be an important variable, particularly at operating temperatures, is not included in these tests. However, some of these general ASTM and DIN test results can aid in screening new gasket materials. In addition, some gasket manufacturers have modified various tests and developed new tests to further characterize gasket performance. All of this information, combined with the PVRC/MTI procedures, provide improved assessments of gasket performance in process operations.

9.7 BOLTING

All of the gaskets tested in PVRC/MTI and other programs are placed in standard test jigs and carefully bolted. Flange deflection measurements, hydraulic bolt tension systems, and carefully instrumented mechanical bolting procedures in the laboratory ensure accurate bolt loadings and gasket stresses. Overall these control/procedures result in essentially a constant circumferential stress, minimizing leakage. Data obtained from laboratory tests under these conditions provide leakage information on each major gasket type, which was reviewed in Section 9.5. Since all of the gasket data were obtained under well-controlled conditions, laboratory leakage and gasket data are a basis for comparison with gasket performance in the field. An extremely important consideration, therefore, is the actual field installation of a gasket and the bolting procedures required to achieve a relatively constant circumferential gasket stress at the desired bolt loading. Bolting procedures and methods of controlling leakage from operating flanges are presented in the following sections.

9.7.1 Background

Emission leakage control from gasketed flanged joints requires satisfactory flange tightness in field installations, which is not a simple problem. A bolted joint is technically complex and often several assembly and maintenance problems arise. Moreover, attempting to control emissions from these joints may result in exacerbating the emission problem.

In general, a bolt is a clamp, and joint behavior depends upon the clamping force exerted by bolts on the joint. Excessive force presents problems by crushing gaskets, damaging flanges, and causing bolt failure through stress corrosion cracking or bolt splitting in assembly. A less than desirable clamping force can result in leakage, slippage, blowouts, or other problems. The main objective in any assembly operation is the attainment of the desired pretension or bolt preloading force. The preload bolt force is the joint bolting assembly load placed on the flange at ambient conditions prior to process unit startup. This normally requires the use of a torque wrench or other system to control bolt tension and loading.

Bolts and joint members are similar to springs, with small springs (bolts) compressing a large spring (joint). When an individual bolt is tightened perfectly, the bolt will relax somewhat as an adjacent bolt is tightened or loaded. This is an example of spring-like effects and has been termed "elastic interaction."[33] Since the bolt has relaxed in this situation, the amount of relaxation is difficult to measure. Altering bolt loadings to the required stress level requires careful adjustments.

As a result of elastic interactions among bolts, the actual bolt preload can vary widely for a flanged joint, although cross-bolting multipass procedures are used.[30] The bolt loadings can vary by more than 3 to 1, but not as a result of a specific tightening procedure. Patterns similar to those in bolt loading are observed with a wide variety of tightening procedures. The scatter in bolt loading is a function of bolt interaction rather than the actual tightening method or system.[33] In addition, other factors contribute to preload bolt variation such as postassembly creep and relaxation of gaskets and metal surfaces, cramped or misaligned joint members, differential thermal expansion, and the nonuniform effect of external loads on the joint.

A large number of variables affect the flanged joint, bolting and gasket tightness. Consequently, bolting at present is still somewhat of an experimental science based principally on experience rather than theory or prediction. This lack of knowledge is demonstrated in technical specifications where bolted joints receive significantly less attention than welded joints. These comments indicate the difficulties associated with assembling and bolting flanges. Methods of bolting flanges are presented in other sections.

9.7.2 Flange Leakage Considerations

There are various reasons for flange leakage, with proper bolt tightening one of the major considerations. A large number of potential causes has been identified in various publications, but only major items are considered in the following paragraphs. Although not included in the following list, inspection is a very important consideration since various boltup problems can be reduced by careful inspection and installation procedures. All of the various flange boltup procedures become very important in meeting the compliance requirements associated with the new EPA flange leakage definition of 500 ppm. Since gasket flange leakage has increased in importance, the following observations should be carefully reviewed to aid in improving flange installations.

Uneven Bolt Stress. As described in Section 9.7.1, unequal bolt stresses lead to variations in gasket compression and can result in flange leak rates greater than desired. Moreover, uneven bolt loads can crush the gasket as a result of concentrated localized loads that further increase leakage. These tendencies can be exacerbated at high temperatures where the bolts relax and the flanged joints loosen.

Flange Alignment. Improper alignment, particularly where flange faces are not parallel, can result in uneven gasket compression with symptoms similar to those described for uneven bolt stress occurring, specifically increased leakage and partial crushing of the gasket. Proper alignment or centering of the flanges is also important to achieve constant compression distribution on the gasket. Various alignment criteria for flanges are published in codes and general publications. Where alignment can be a problem, flanges specifically designed for this condition are available commercially and a typical example is the TAPER-LOK® connection described in Section 9.9.

Gasket Alignment. A gasket that is not installed with a gasket centerline that coincides with the flange centerline will be unevenly compressed and probably result in leakage. Certain gaskets such as the spiral-wound gasket have an extension to the inner diameter of the bolt circle that assists in centering the gasket. Similar extensions are available for a wide variety of gaskets. Laminated sheet gaskets do not have centering rings but can be specified with a gasket outer diameter that corresponds to the inner diameter of the bolt circle. This specification will automatically center the gasket.

Flange Faces. Rust, dirt, or foreign materials on a flange face can create a leak path along the interface between the gasket and flange face. Some gasket material may remain on the flange face after the previous gasket has been removed following a lengthy run between turnarounds. This frequently occurred with asbestos gaskets and was described in Section 9.3. Other gasket material may adhere to the flange facing, depending upon the operating temperature. In addition, another material that can accumulate on a flange is sealing compound, which may be injected around the gasket to control leakage after a leak is discovered. The leaking flange set is enclosed and sealing compound can be injected into the enclosure, as described in Section 9.8. Sealing compound is not placed on the gasket or flanges prior to boltup for leakage control and is generally not recommended. All material on flange faces should be removed prior to the installation of a new gasket. Although foreign matter must be removed, scratches and surface imperfections also affect gasket sealing and can affect leakage control with many gaskets.

Raised-face flanges are the standard flange for chemical plants and petroleum refineries for most of the medium temperature and pressure applications with ring-type joints installed at higher temperatures and pressures. The raised-face flanges have serrations that provide a satisfactory seating surface for sheet- and nonsheet-type gaskets in accordance with ANSI piping standard B16.5. This roughness is necessary to develop sufficient friction to aid in preventing gasket blowout and excessive creep. The piping standard requires either serrated concentric or a serrated spiral finish having from 24 grooves/in. to 40 grooves/in. and also requires cutting tools with a radius of 0.06 in. (1.5 mm) or larger. The resultant surface finish should be in the range of 125 to 500 AARH. This is a standard method of measuring roughness that is an average value of the profile variation from a center line area reference length and is also named the center line average (CLA) or arithmetic average. These units are in micro-inches with raised-face flanges in the 125- to 500-micro-inch range. Another roughness term is RMS or root mean square, which is 1.11 times the AARH value in micro-inches.

In Figure 9.2 a ring tight joint is shown where an oval metal ring fits into a groove cut in the flange face. ANSI B16.5 limits the surface roughness on the side wall of the groove to 63 RMS, indicating the roughness should not be greater than this value. The oval ring is a relatively smooth material and is generally constructed of soft iron or a low carbon content steel. Ring hardness is lower than the flange hardness, permitting a tight seal.

Gasket Selection. The selection of the gasket for the service is important in obtaining sealing and leakage control in addition to the other considerations in this section. These selections are conducted with the procedures described in Section 9.5.

Piping System Mechanical Forces. When piping system design is faulty or excessive force is necessary to align flange faces, bending moments on the flanges build to high levels and result in flange distortion which affects sealing and leakage. In addition, improper location, design, or lack of piping support can result in high bending moments and forces on flanges that change bolt loadings and reduce gasket stresses. As an example, a flange in a vertical pipe run that is not properly supported will have the weight of the pipe held by the flange bolts reducing gasket compression and increasing potential leakage. Moreover, tension on the bolts increase, which may adversely affect the bolts if the metal yield point is reached.

Vibration in the piping system can also affect flanges since the vibrations affect sealing, which in turn can influence leakage. Methods of isolating vibratory effects and dampening these conditions are necessary to ensure a satisfactory gasket seal.

Thermal Shock. Rapid temperature fluctuations may cause temporary flange warping until steady-state conditions are again achieved. In regenerative fixed-bed units, wide temperature changes generally occur during the regeneration phase which may affect gasket seals. This general approach is partially included in the AHOT tests, where some changes do occur under ideal control conditions. Also, actual cycling is much more frequent than the three or four cycles in the AHOT tests, which can contribute significantly to flange leakage problems.

While regeneration is a typical source of thermal shock, another thermal shock may occur during rainstorms. Sudden rains chill flanges when they are uninsulated. If a line is at 400–450°F and is suddenly drenched with rain, the uninsulated flange temperature drops significantly. This change in temperature can affect the gasket seal, particularly where the temperature drop is not equal around the entire flange. As a result, bolt loading becomes uneven and the gasket seal is affected. Where the rain effect is a potential problem, a simple sheet metal shield over the flanges or insulation will eliminate flange temperature effects and metal distortion.

Training. While all of the aforementioned items require an understanding of the proper gasket flange requirements and the methods of proper gasket boltup, a proper installation cannot be achieved unless mechanical personnel are thoroughly trained in flange and gasket technology. Consequently, training becomes a very important facet of emission control from flanges. Personnel involved in flange/gasket installations should have sufficient training to ensure consistent and satisfactory quality control of flange installations.

9.7.3 Boltup Procedures

The boltup procedures described in this section are a general approach to bolting. Various facilities and plants have developed specific procedures that satisfactorily meet leakage and control requirements. These boltup procedures are followed by mechanical personnel where detailed requirements have been issued. The different steps in bolting flanges are described in the following sections.

Inspection. As described in Section 9.7.2, inspection is necessary to ensure that a satisfactory flange joint installation can be achieved. Inspection generally covers the following:

1. Flange face conditions should be inspected for solid matter, dirt, scale, scratches, or deep indentations. If there are deep scratches or dents, the flange surface may require resurfacing with a facing machine that meets ANSI B16.5 requirements. Dirt or other materials on the flange face should be wire brushed or removed to provide a satisfactory surface.
2. All of the flange dimensions should be investigated, such as flange parallelism, pipe alignment, flange rotation, and bolt holes. If the flange is warped, which can be checked with a straight edge or other device, the flange should be rejected.
3. Gaskets must be inspected to determine whether the gasket type is the one specified. Also, the gasket size should be carefully measured. The gasket OD should be equivalent to the diameter of the bolt hold circle less one bolt hole diameter. The gasket ID should be slightly larger than the pipe ID. If the incorrect gasket or dimensionally incorrect gaskets are supplied, the gaskets should be rejected.
4. Bolts should be inspected to ensure they are the proper material, diameter, and length.

5. When a joint had been opened, gaskets should be replaced with a new one. Rings for RTJ flanges do not have to be replaced, but should be inspected prior to closing the flanges.

Bolting Procedures. A very important and critical aspect of flange boltup is the bolt stress recommended for a tight gasket fit. This figure is generally developed by mechanical engineering based upon the desired gasket stress. The bolt material and the bolt elongation which defines actual bolt stress are specified by mechanical engineering, since bolt elongation, bolt stress, and the modulus of elasticity are related through Hooke's Law. In addition, a torque can also be defined for obtaining the proper bolt stress. For applications with 300 class and higher rating flanges, the bolt stress is in the 40,000–50,000 psi (2800–3500 bar) range. The 150 class flange bolt stress ratings are 25,000 psi (1723 bar).

Direct methods of determining elongation can be conducted in various ways as described in the section on uniformity or bolt load. A standard method is the direct measurement of the length change with a caliper-like device. This direct measuring system has been used for many years, but newer devices are replacing the direct measurement procedure. As an example, ultrasonic devices measure bolt length and will automatically calculate the bolt stress. For large flanges in critical service, the actual measurement of bolt length change is important. However, measurements on each flange bolt significantly increase boltup time and costs since this technique is also cumbersome. Where a few flange joints have been bolted, bolt measurements on a number of tightened bolts act as a check on general bolt tightening procedures.

Standard procedures for obtaining the correct bolt stress have generally been considered satisfactory for leak rates defined as 10,000 ppm. These procedures are based upon boltup use of a wrench and hammer, air impact wrench, or torque wrench (a hydraulic jack is an alternate). More recently, hydraulic bolt tensioners have been used successfully and are described in the following section. The bolt tensioner is similar to the hydraulic bolt adjuster on the AHOT test rig.[26, 29]

While the hydraulic bolt tensioner is quite accurate and reliable, the overall tensioner system was rather cumbersome, but was advantageous in tightening bolts on very large flanges. Newer hydraulic tensioners are considerably smaller and are an improved method of tightening bolts in various flange sizes.

Preliminary Boltup. The initial bolting procedure requires the insertion of the bolts and gasket. This general phase is conducted as follows:

1. Where the nuts contact the flange, the surfaces of the nut and flange should be clean and smooth. The bolt threads and nut are lubricated with a recommended lubricant for the service. This aids in bolting and standardizing bolt tensioning.
2. About half of the flange bolts are inserted, and the flange faces are aligned with sufficient open distance between the flanges for the gasket.
3. The gasket is inserted between the flange faces, without any sealing compound on the gasket or the flange face.
4. The gasket must be properly centered.
5. All of the remaining bolts are inserted, and the nuts are threaded on the bolts. If flange alignment is satisfactory, the nuts are finger tightened.

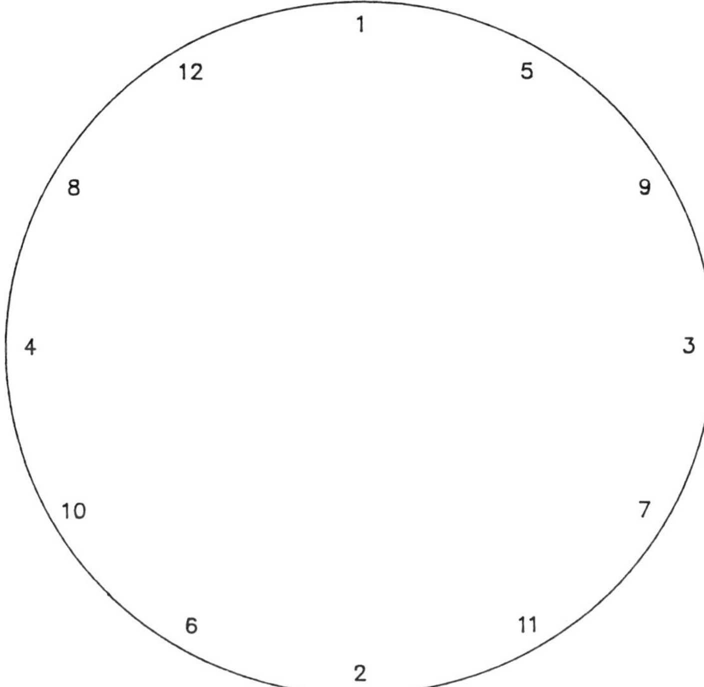

Figure 9.10 *Suggested tightening sequence for a 12-bolt flange.*

Boltup. Bolt tightening is based upon a crisscross tightening sequence described in ANSI B16.5 and other publications where the bolts are tightened in accordance with a number of sequences shown for a typical flange in Figure 9.10. In addition, the mechanic *should not* tighten one or two bolts completely before tightening the other bolts. Completely tightening one or two bolts will cause local gasket crushing and/or pinching that will result in leaks. The bolts should be tightened gradually in the crisscross pattern to simulate an even distribution of flange pressure against the gasket. After each round (or pass) of tightening, the bolt alignment may be checked by measuring the distance between flange faces.

One of the major problems in tightening flange bolts has been the large variation in bolt stress observed after the bolts have been considered tight.[31] Frequently, bolt tightening is considered satisfactory after three full passes have been completed. However, investigations of bolt tightening indicate that the variations in bolt stress and elongation can only be improved by increasing the number of passes.[31] Where elongation or stress can be measured or monitored, bolt stress scatter can potentially be reduced about $\pm 10\%$. However, these procedures may be overly expensive and cumbersome, although improvements in monitoring and indication have been commercialized. A few of these modifications are described in Section 9.7.5. Alternatively, bolt tightening is dependent upon mechanical skill for quality control.

Although a considerable effort has focused on bolt tightening procedures, little information is available on bolt quality. Variations in bolt quality can also have a significant effect upon bolt stresses. Improvements in bolt quality control procedures can reduce the potential effects of this variable.

In addition to standard bolt tightening methods, some other alternatives are presented that improve evaluation and control of bolt loadings.

Wrench and Hammer. The bolts are generally tightened with a short wrench in the suggested crisscross pattern until all bolts are tightened equally. A second pass is conducted with a longer wrench or spanner and a light hammer in a crisscross sequence for equal load distribution. The bolts are again tightened in a third pass with the wrench and a heavier hammer until all of the bolts are completely tight. Additional passes may be necessary to achieve complete tightness with all passes in a crisscross pattern.

Torque Wrench. A variety of manual torque wrenches are available to tighten the bolts, and hydraulic types are also available. With a defined torque, loads theoretically can be applied evenly across all of the bolts. The torque loadings can be obtained from gasket manufacturers, literature references, or plant engineering departments. Torque requirements can be separated into loads for the first tightening pass, with subsequent increases for each pass until the final torque load is achieved. The torque wrench is normally not used until the bolts have been tightened with the short wrench.

Torque loadings may not be accurate for the final bolt stress which can be 40,000 psi (2757 bar). Bolt elongation defines bolt stress and the stress should be compared with the torque recommended to produce the desired bolt stress. Elongation measurements are not required for every flange joint or whenever a flange is bolted. However, changes in torque recommendations should be verified with elongation measurements. When elongation measurements are obtained periodically, bolt tightening consistency can be checked along with potential mechanical changes in bolting materials that may affect the bolt stress–torque relationship.

Bolt Tensioner. This device is hydraulically operated and is similar to the bolt tensioner in the AHOT test rig. The tensioner head is clamped on a bolt nut and the nut is tightened through a mechanical linkage that is driven by high-pressure fluid from a pump in the attached hydraulic system. Bolt stress and the desired tension is readily achieved. Two separate tensioner heads may be obtained with the tensioner device rather than one permitting tightening of two bolt nuts at the same time. Thus, two adjacent nuts can be tightened to the same target bolt loading. In addition, a number of nut tightening heads can be combined in one overall head to tighten in a much more rapid fashion single bolts in large flanges. In some devices, the hydraulic bolt head is much larger than the enclosed nut and may not fit on smaller flanges. Moreover, the overall hydraulic system including the bolt nut tensioning heads is expensive. However, improved bolt tightening procedures may result in reducing the number of leaking flanges ($\geq$ 500 ppm), which would justify the use of this equipment.

The bolt tensioners are applied after the bolt nuts have been tightened with a short wrench in a first pass. Since various numbers of bolts can be tightened with this system, the actual tightening procedure will be a function of the bolt number in the tensioner head. Tightening is also conducted on a pass basis until the desired bolt load is attained.

Tension Control Bolts. Bolts, have been developed and are commercially available that include self-contained tension indicators. With all flange bolts of a self-indicating type, a more even distribution of bolt load is possible. However, these bolts are expensive and applications have been limited because of cost, but they are apparently effective in evenly distributing the bolt load. One type of indicating bolt essentially consists of a narrow diameter rod in a small diameter hole through the center of bolt. When the bolt is elongated through stress, the internal rod displacement provides a measurement of bolt elongation. Another type of tension control bolt indicates bolt tension by a change in liquid color observed at the bolt end. A bladder contains fluid which normally is pressed into the outer part of the bladder by an internal rod. This rod compresses the center of the bladder and as the bolt is elongated, the rod gradually reduces bladder compression and fluid slowly enters the outer portion of the bladder. This results in a color change which can be seen through a small port at the bolt end. The color change with bolt expansion indicates the amount of bolt elongation. Application of the color bolt is limited by the maximum allowable temperatures for the bladder and fluid. Temperature and cost are considerations with the color bolt, but improved bolt load distribution may justify the use of these bolts.

Hydraulic Nuts. These nuts provide a means of gradually stressing all bolts on a flange in unison. This system consists of a large nut containing small pistons that press against the flange. The pistons are moved against the flange by a high-pressure fluid connection. In operation, the nut is finger tightened against the flange before hydraulic pressure is raised. When the nut is satisfactorily placed, the hydraulic pressure is raised and the piston in the nut presses against the flange. In reaction to the pressure, the hydraulic nut stresses the bolt, which gradually elongates. The pressure is increased until the desired pressure level is attained. At this point, the final pressure is maintained on the pistons and hydraulic nut by a check valve or other device.

A number of pistons are installed in each nut, requiring a larger nut than the standard bolt nut. Sufficient space must be available on the flange for these nuts. The hydraulic system attached to all of the nuts on one flange permits equivalent pressure changes in all of the nuts. As a result, bolt stress is equivalent on all of the bolts. However, this system is more expensive than the standard bolt system and may have temperature limitations based on hydraulic fluid composition. Where critical flanges exist and other flanges are continuous leakers ($>$ 500 ppm), this type of bolting system may be attractive.

Torquenut®. The Torquenut shown in Figure 9.11 is also termed a Superbolt® Torquenut. While appearing bulky, these nuts fit in the space normally occupied by a standard unit. The standard Torquenuts, as shown in Figure 9.11, have external jackbolt heads with the recessed head as an alternate to minimize clearance. Each standard nut on one end of a stud bolt is replaced with these specialized nuts in a flange joint. In flange joints, the Torquenuts are located on the same flange with standard nuts on the mated flange.

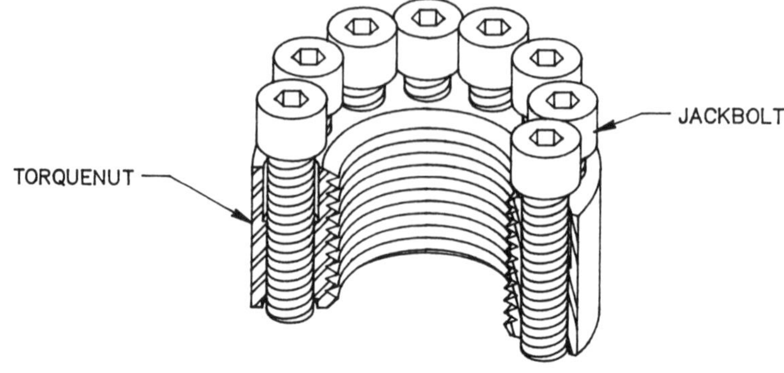

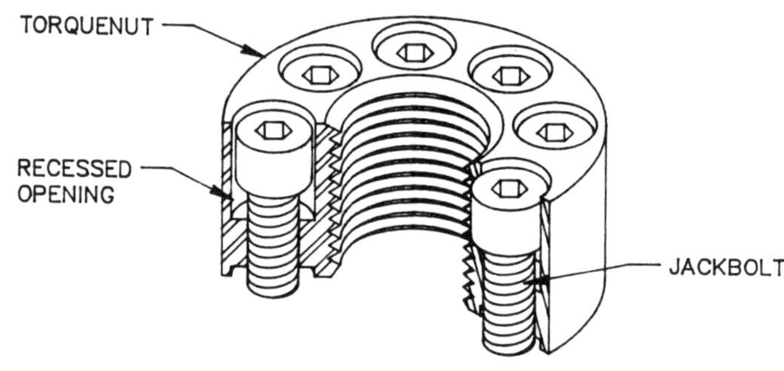

Figure 9.11 *Standard and recessed Torquenuts® (courtesy of Superbolt® Inc.).*

A typical single bolt installation is shown in Figure 9.12 with a torque wrench used for tightening the jackbolts. The Torquenut is threaded on the bolt or stud, similar to threading a standard nut. Once in position, actual tension on the stud bolt is achieved by tightening the jackbolts that press against the washer, which is described as a heavy flat ring. As the jackbolts are tightened, the stud bolt is stressed and is gradually elongated. This is a simple, straightfoward procedure where the tightening of the jackbolts is accomplished with hand tools. An ordinary air impact wrench or torque wrench is sufficient to preload studbolts considerably above 100,000 psi (6892 bar). Normally, bolts are preloaded to 40,000–50,000 psi (2757–3446 bar) for flanges with class ratings of 300 lb or greater. A bolt stress of 25,000 psi (1723 bar) is applied to 150 class flanges. An important consideration is the flat ring or hardened washer against the flange, which takes the full thrust of the jackbolts. Without the flat ring, the jackbolts would mar and score the flange surface. An incentive for the use of Torquenuts is the *lower* torque necessary to stress stud bolts to a desired level. Moreover, the load on each jackbolt is consistent, resulting in the desired stress on each bolt and an evenly distributed stress on the gasket. The equal stress on each bolt aids in controlling leakage and

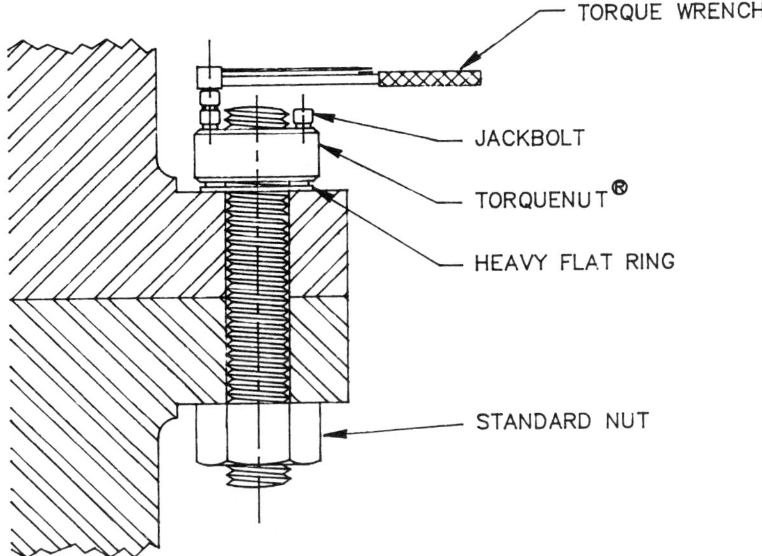

Figure 9.12 *Assembled studbolt with Torquenut® and manual torque wrench (courtesy of Superbolt® Inc.).*

meeting the new leak definition level of 500 ppm. These special nuts are also used to control leakage in exchanger flanges and very large diameter piping flanges. Accompanying improved flange performance is manpower savings resulting from improved bolt tightening procedures, particularly in large flanges.

Actual bolt torque requirements for the jackbolts are considerably smaller than the torque requirements for the standard bolt nuts. As an example, a bolt stress target of 60,000 psi (4135 bar) normally requires a 16,000 lb-ft torque for a 3 in. (about 76 mm) stud bolt. A Torquenut only requires 160 lb-ft of torque for each jackbolt. For small jackbolt torques as shown, very high preloadings can be achieved. As a result, a manual torque wrench is used for torque accuracy and aids in rapidly tightening bolts. A considerable amound of information on jackbolt torque requirements has been assembled by the manufacturer for various size bolts and loadings. These values can vary if the system is constantly reused or washed in solvent. The latter removes lubricant and galling will result, which affects the preload jackbolt relationship.

The Torquenuts are available in a wide range of sizes and normally have a temperature limit of 600°F (316°C). However, the temperature limit is extended to 1200° (649°C) with changes in the materials of construction. These considerations also apply to the washers, where the materials are changed to match the Torquenut temperature ratings. These nuts have been installed in applications where contained pressures are 1500 psi (103 bar) at elevated temperatures.

9.7.4 Hot Bolting

This technique is particularly important at operating temperatures greater than 500°F (260°C). When flanges after mechanical assembly are heated during startup, bolts, flanges, and gaskets work or deform to some extent. These changes affect the gasket, which generally results in relaxation of the bolt load. Consequently, tightening of the bolts should be considered during startup, at a process temperature of about 500°F, and after the full process temperature level is attained. Since the bolt stress must be maintained at a reasonable level to control emissions, hot bolting is necessary to meet the repair requirements in the MACT regulations. This can potentially expose personnel to health and safety hazards; an industrial hygiene overview of this type of work function is necessary.

If bolt tightening is necessary during the process run, the procedure should be performed after a process temperature swing is completed or after a regeneration. The tightening may be necessary as a result of gradual relaxation and/or changes in emissions. Obviously, the nuts should not be in a seized condition as a result of oxidation or improper lubrication. For effective tightening, the mechanic should ensure the nut can move relative to the bolt.

Tightness Procedure. When leaks occur, bolt tightening procedures described for the wrench and hammer or torque wrench should be applied. For the wrench procedure, a heavy hammer may be necessary to stop the leak. Tightening of the bolt nuts should begin with the nuts located at the leak. Following this step, a crisscross pattern is again followed to place an even load on the gasket.

While laminated gaskets and laminated corrugated metal gaskets can be tightened considerably, joints with spiral-wound gaskets can probably be tightened to the centering metal ring thickness, but further tightening of the spiral wound may not be effective. Double-jacketed API gaskets can be tightened, but the corrugations and metal thickness can make tightening difficult.

Continuous Leakage. If the leakage cannot be stopped by tightening, there is a possibility of sealing the leak through various methods which are described in Section 9.8. However, prior to any sealing procedure, the potential control technique should be reviewed for safety. High-temperature process fluids pose significant safety problems compared with low-temperature fluids and if leakage containment through sealing is not permitted, the line is isolated where possible and the joint broken to determine the cause of joint failure. The following procedure is a general guide for investigating and modifying the joint for improved hot temperature operations:

1. Examine flange facings for damage, warping, foreign matter.
2. Check flange alignment and cut or realign piping where necessary.
3. Inspect the gasket to ensure the proper material, dimensions, and type is installed.
4. Analyze gasket deformation and determine whether the gasket was centered.
5. Obtain a new gasket for reassembling the joint.
6. Reassemble the joint.
7. Place the joint onstream and inspect for leakage.
8. Should leakage persist, piping supports and flexibility must be examined. A revamped support system or new hangers may be required to lower bending moments.

9. If leakage occurs during rainstorms, warpage has occured as a result of uneven temperature distribution. For uninsulated flanges, the only simple solution is a rain shield. These can easily and effectively be installed on horizontal piping with vertical flanges. Where the piping run is vertical and the flanges are horizontal, rain shields will not be effective. Another shield is necessary and, as an example, one similar to a large truncated cone placed around the pipe and above the flange can be effective.

9.7.5 Bolt Load Uniformity

Bolt load uniformity is necessary to properly seal flanges and minimize leakage. The loading can be determined by measuring bolt elongation following tightening. This frequently results in the use of various mechanical extensometers that are relatively inexpensive. While each bolt does not have to be measured, random measurements on an adequate basis are important for achieving the correct bolt loading consistently for the various bolts in a particular flange. In one study, average elongation measurements of installed bolts were lower than the target elongation distance and were accompanied by wide variations in bolt load uniformity.[32] While mechanical measurements are frequently used and have been described in various references, this procedure requires careful detailed attention to the actual measurement technique for reasonably accurate results. Consequently, other methods of measurement are frequently used, such as ultrasonic techniques and the tension control bolts described in Section 9.7.3.

Ultrasonic Measurement. Ultrasonic systems can be used to measure either the change in stud length or the change in bolt stress. Moreover, these data can be obtained when the bolt has been stressed. However, training and experience are required to ensure the equipment is operated properly. While the early equipment was cumbersome, somewhat bulky, and very expensive, these characteristics are changing drastically with advances in electronics. Consequently, interpretation of results have changed from individual analysis to miniaturized computer analysis with essentially instantaneous results. Application of these devices to improve bolting and minimize leakage is becoming widespread in process plants.

Typical Ultrasonic Application. The general procedure to improve equal bolt loading distributions, resulting in a consistent circumferential gasket stress, is outlined in the following approach:

1. Initally measure bolt lengths with the ultrasonic extensometer at ambient temperature.
2. Bolts are tightened with a wrench and hammer or torque wrench and, after two passes, bolt elongation in two bolts can be measured to determine whether reasonable bolt loads develop with somewhat similar elongation characteristics.
3. Measure all bolt lengths with the ultrasonic extensometer after the third pass. Generally, considerable variation in the bolt stretch occurs with the

average elongation value for all of the bolts less than the desired target range.

4. The bolt loads are then adjusted based upon third pass measurements to achieve a more consistent loading and ensure the bolts are in the target elongation range. Normally, these objectives can be reached with a fourth pass and the employment of the ultrasonic extensometer.

Bolt loading consistency may require a more concerted effort for flexible flanges compared with stiff or inflexible flanges. Lower class rated flanges are more flexible than those with a higher class rating. The mean bolt loading for the flanges must provide sufficient gasket compression as described in Section 9.5 for variations in gasket stress that can occur as a result of any changes in internal pressure or temperature.

9.8 SEALING FLANGE LEAKS

The previous sections concentrated on selecting the proper gaskets and bolting; another area of concern is the control of flange leaks from operating installations. This is quite important for meeting the emission control and repair requirements in the new HON regulations and changes in various VOC regulations. In this section, several techniques are reviewed that control leakage emissions. However, this is not a complete review of all the commercially available techniques and equipment. With mounting interest in emission control, new and different procedures are frequently available for flange leak control. As a result, a broad spectrum of products are available worldwide on a commercial basis that cannot be reviewed adequately in this section. The control techniques discussed in the following sections have been available for some time and industry experience indicates these systems perform well in emission control.

Prior to describing some of the sealing techniques, a general approach to sealing is presented. An understanding of the many factors involved aids in determining whether a leaking flange joint should be sealed and paramount in these considerations are safety concerns. Also, in determining whether a particular flange should be sealed, an understanding of emission regulations regarding monitoring, record keeping, and turnaround requirements is desirable. An overall review of the need to seal a flange should, therefore, consider a variety of factors.

9.8.1 Flange Leakage Regulations

Under the new HON regulations, all flanges in vapor and light liquid service are monitored initially, depending upon the number of leaking flanges as described in Section 9.1, and the periodic inspection monitoring frequency is determined. Flange leak frequencies greater than 1% for leaks ($\geq$ 500 ppm) require annual monitoring programs. When a flange leak is found, the flange must be identified and initial repairs undertaken within a specific time frame ($<$ 5 days) and completed not later than 15 days after leak determination. For flanges in heavy liquid service where leaks are initally observed either visually, audibly, or through an olfactory occurrence, the leak must be monitored within 5 days after the observation. If the monitor indicates a concentration $\geq$ 500 ppm, the flange is classified as leaking. Where the leaks cannot be repaired without a process unit shutdown, flange leak repair can be deferred until the next scheduled turnaround. However, the number of leaking flanges contribute to the leak frequency, which may require annual monitoring. The leak frequency is determined by the equations in Section 9.1.

While VOC regulations for the NSPS regulations may be less stringent, in some regions the VOC regulations are similar to the new HON regulations. Moreover, HON regulations will be promulgated for petroleum refineries which will probably be similar to those covering chemical processes, and further reductions in VOC emissions may be required to meet NAAQS standards. Consequently, emission control of leaking flanges is a major consideration in process plants.

9.8.2 Leak Evaluation

Following identification of a flange leak, an immediate inspection of the flange and any mechanical records pertaining to the particular flange should be undertaken. Records should indicate whether the flange is a frequent leaker, bolt studs have posed problems in the past, corrosion is a concern, and so on. At this point, potential health or safety problems should be reviewed prior to any action on the flange joint. A decision to control or not control the leakage until turnaround is then developed based upon these considerations and potential environmental requirements. If a decision to proceed is recommended, than the initial step is tightening of the flange studs, which is also recommended by the EPA to quickly control the leakage, although the tightening step is not always sucessful. However, prior to stud tightening, flange conditions should be investigated and a general approach is described in the following paragraphs:

Process Conditions. Flange process conditions should be checked against flange design conditions. If flange process conditions have changed, repair attempts may be deferred or a more extensive flange repair program may be necessary. Alternatively, where process conditions have not changed, the repair program may proceed.

Flange Inspection. Flange inspection will indicate whether the flanges are corroded. Also, flange distortion or warpage should be investigated, which can be determined with alignment techniques. Flanges in poor mechanical condition present potential adjustment and repair problems along with potential safety problems.

Bolt Studs. The condition of the studbolts is very important, and a visual examination may indicate whether any of the studs are corroded. In addition, the studbolts should be tested for cracks when information may indicate a studbolt cracking potential exists or processing conditions may be sufficiently severe that studbolt cracks can develop. The studbolts can be tested ultrasonically for cracking, and this test will also indicate whether severe corrosion exists. While this procedure is somewhat expensive, a more cautious approach suggests most studs should be tested for mechanical failure with nonintrusive devices prior to tightening since onstream bolt tightening efforts will probably increase to meet the low leaking frequency targets in the regulations.

In addition to these considerations, a longer-term view should be considered with regard to a flange leak. If a leak should cause brittle fracture of the studbolts, steel corrosion, or stress corrosion cracking, these factors must be considered in a general appraisal of flange repair possibilities. Where studbolts have cracked, a potential exists for stress corrosion cracking, or if a high stress level exists in the stud, it may be necessary to change the studbolt prior to tightening the flange. Bolt removal and replacement procedures developed in the process plant should then be observed. For services where the process temperature is elevated, bolt changing and tightening should follow hot-bolting practices. Where bolt removal and replacement along with tightening procedures are not available, these procedures should be developed with safety and industrial hygiene participation.

Gasket Assessment. Along with flange and bolt condition assessments, the gasket condition is also important. However, qualitative inspection of the gasket is difficult and an observation can only indicate whether the outside edge of the gasket is intact. Obviously, the disappearance of the outer edge or significant changes in appearance would suggest the gasket is in poor condition and tightening the flanges may not be effective. Moreover, the gasket may suffer further mechanical degradation with the possibility the emissions may increase. If the installed gasket is apparently intact and the gasket is the proper material normally performing well in this service, then tightening can proceed. If the installed gasket is the wrong gasket for the service or has other potential problems, this will have an effect upon a decision regarding tightening or sealing the flange.

9.8.3 Leak Emission Control

Following the general flange leak evaluation investigation, a method of controlling the leak is selected for rapid implementation. This decision will determine whether the flange can be tightened, sealed, or rebolted with new bolts. Alternatively, repairs may not be possible and a recommendation not to attempt repairing or sealing the flange could be proposed, which would defer flange emission controi to the turnaround period when the flange joint is opened and examined. The flange joint is repaired or replaced, and a new gasket is installed before the flange joint is tightened for startup operations.

There are two basic leak control methods. One technique is bolt tightening to reduce leakage, which may also include studbolt replacement. This may be acceptable after detailed reviews are completed. The other method is leak sealing; a variety of sealing methods are available and a few are described in the following sections.

Safety and Health Considerations. Where the flange will be tightened or repaired on stream, the following general procedure should be considered, along with a complete review of all procedures with safety personnel:

Stream Fluid. Flammability and explosion hazards should be identified and safety requirements determined. In addition, potential toxicity considerations are reviewed with industrial hygiene personnel along with protective equipment requirements and other protective measures.

General Safety and Health Requirements. These are summarized in the following list:

- Ensure adequate working space around the leaking flange.
- Escape routes must be identified and cleared of all impediments.
- Protective equipment where necessary should be provided along with goggles, gloves, and ear protection. Other industrial hygiene recommendations may be proposed which should be incorporated into the overall protective equipment list.
- Consideration should be given to steam lance accessibility in the event of fire requirements.
- Procedures for conducting these operations should follow plant requirements for work on operating equipment, including notification of all personnel.

Flange and Bolt Tightening. Where a bolt does not have to be replaced, the studs are tightened in the crisscross pattern recommended for flange tightening procedures. Where a bolt must be replaced, a general procedure is outlined in the following paragraphs, but plant procedural requirements may differ and local procedures that have been developed during the years should be observed:

Safety and Health Requirements. The safety and health requirements previously outlined should be observed for this procedure. The only addition is determining whether sufficient protective equipment is available if the leak worsens during bolt removal. Obviously, protective equipment should be sufficient to cover a worst case situation.

Clamp. A clamp should be installed on the flange near the bolt scheduled for removal. This clamp must be capable of carrying the bolt load and is tightened to the required loading.

Bolt Removal. Following placement and tightening of the clamp, the bolt stud nuts can be loosened and the bolt removed. The replacement stud bolt is installed and then the studbolt nuts can be tightened to a desired bolt stress level that is determined by mechanical engineering, which considers the temperature of the process system. An exessive bolt stress can result in flange and/or bolt problems when the piping is cooled during shutdown operations.

When the desired bolt stress is attained, the studbolts are generally tightened in a crisscross pattern to ensure an equal gasket stress distribution. Since bolt stress targets are defined prior to stud replacement, methods or procedures for determining the bolt stress must be defined prior to boltup and tightening. Any equipment necessary for bolt stress determination is available at the repair location prior to initiation of the work effort.

Leak Sealing. An important aspect of leak sealing is the selection of the sealing method, since a variety of sealing techniques are available. Generally, a ring is

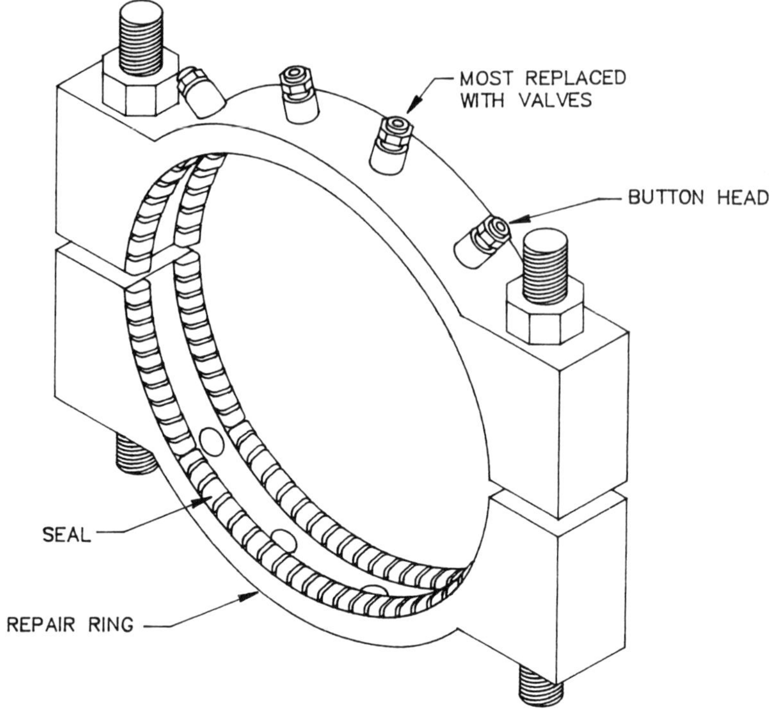

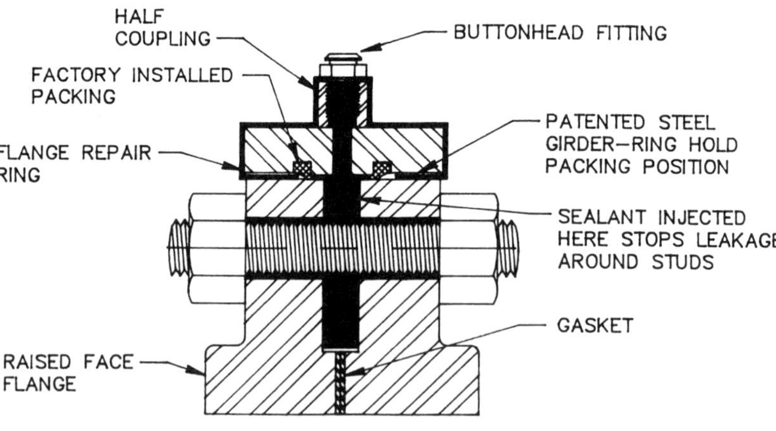

Figure 9.13 *Flange repair or sealing ring (courtesy of PLIDCO® Int'l Co.).*

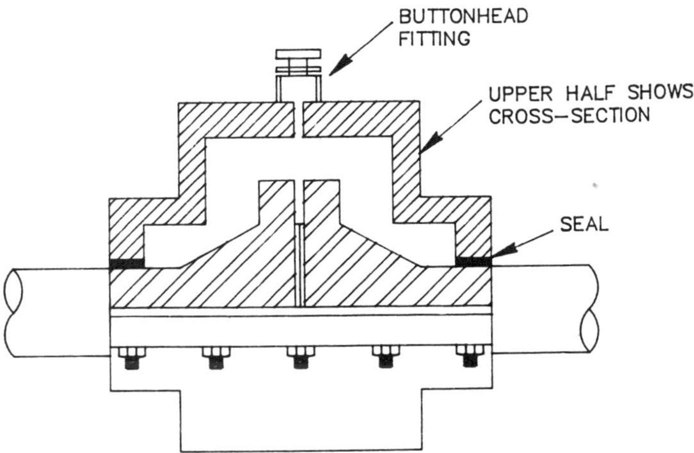

Figure 9.14 *Flange repair enclosure with split sleeve (courtesy of PLIDCO® Int'l. Co.).*

placed around the perimeter of the flange and a sealing compound is injected into the space between the flanges and the enclosure ring. Large metallic housings are also placed around the entire flange joint and sealed to the piping on either side of the flanges. These oversized containers or housings can also be filled with sealing compound. While several sealing systems are described in the following paragraphs, a number of other systems are available commercially and should be investigated prior to selecting a sealing system.

Sealant Retention Rings. Some retention rings are shown in Figures 9.13 and 9.14. In general, numerous sealing systems are based on a ring clamped against the outer circumference of the mated flanges with sealant injected through the ring into the open volume between the flanges and around the bolts. For satisfactory sealing, the outer rings should be constructed of the same material as the flanges to minimize relative movement between the flanges and ring resulting from differential thermal expansion. Although buttonhead fittings are shown in Figure 9.13, the fittings should be removed and replaced with valves to permit sealant injection. The valves can also be used as vents when the sealant fills the void space created by the ring.

Another type of ring is shown in Figure 9.14. This is basically a flange enclosure since the entire flange is enclosed within the housing. The housing can be either completely welded or bolted in place, but a seal is required around the piping, as shown in the figure. The large enclosure is split in half and bolted together, permitting easy removal of the enclosure. In this design, the enclosure is designed as though it were a pressure vessel. If the flange studs corrode or crack, the enclosures must carry the piping and flange loads. Again, this enclosure is filled with sealant or can remain a pressure vessel without a sealant.

Nonpressure retaining enclosures may be used, where a large pressure retaining closure cannot be installed. For nonpressure housings, the enclosure must be vented to prevent pressure accumulation. However, the vent should be connected to a closed system for disposal or recovery of any vapors. Similarly, if any liquid leaks from the flange, the leakage must be enclosed and disposal requires complete containment. Where there are leaks and a vapor disposal point can be used, a flange retaining ring can be used to collect the vapor and vent the vapors to the receiver. With venting systems, sealants are not used.

An important consideration in enclosing or sealing a leak is the allowable temperature rating of the sealing system. All components of the seal system must be satisfactory for the operating and design temperatures.

Sealants. Sealants are obtained from various manufacturers and sealant composition is proprietary. This requires an investigation of sealant suitability for the application. Since sealant is injected to thoroughly fill all voids and prevent leakage to the atmosphere around the stud bolts, the sealant must be compatible with the internal process fluid. Consequently, sealants should not react with the process fluid and should not contain any materials that result in bolt stress corrosion cracking. If any reaction occurs with the process fluid, reaction products can be corrosive and/or explosive. Nuclear quality sealants may be preferable since the contaminants in the sealant materials applied in nuclear plants are a minimum. This eliminates any possibility of reaction of process fluid with sealing contaminants, which may happen.

The preferred sealant will gradually become a flexible solid, which minimizes any potential increase in stud loading and permits thermal movement without leakage. Manufacturers generally have a number of sealants available that differ in characteristics and composition. The sealant is normally selected by a manufacturer based upon the process fluid and operating conditions. Pressure is not a variable in sealant selection. In addition to other considerations, sealants from different vendors should probably not be used in the same flange since they may not be compatible.

Sealant Injection. Sealant is injected into an encapsulated void space with a hydraulic hand-pump sealant gun. The hand-pump guns are preferred because they provide the operator with "feel" for flow resistance and pressure buildup. Sealant injection may increase bolt loadings, since the sealant flows between the flanges. Consequently, the maximum sealant pressure should not exceed about 6500 psig (449 bar) and should be specified prior to injection, preventing overpressuring.

During injection, all unused ports are generally open to permit venting any enclosed vapor. When all ports have been injected except the one remaining port, the final port vent is permitted to remain open and the sealant within the ring then sets for a period of time defined by the manufacturer, allowing the sealant to reach equilibrium and the bolt load to stabilize. Finally, the sealant is injected into the last port to complete the sealing process.

The amount of sealant required should be calculated before injection. Excess sealant can be forced into the process stream where the sealant may interfere with the equipment and instrumentation downstream of the sealing point. Alternatively, the pressured sealant could move the gasket, which could result in increased leakage.

Sealant Reinjection. Although a joint has been sealed, there is a possibility the joint will leak after some period of time. While the joint could be reinjected with sealant, this potential solution should be carefully investigated to determine whether sealant reinjection is a problem. A careful inspection of the flange joint and the bolt studs should be instituted prior to reinjection. Where potential deterioration of the studs is possible as a result of corrosion, erosion, or stress corrosion cracking, the studs should be examined ultrasonically. Any damaged studs are replaced prior to sealant reinjection. This procedure is straightfoward with a flange retaining ring, but cannot be performed with a flange enclosure. The flange retaining ring may be preferred if a flange leak must be controlled for a long period of time until turnaround. If the sealant has a long life and does not affect the bolts, and if the process fluid will not affect the bolts, an enclosure is satisfactory. Since sealant technology continually improves, performance or operating data may indicate that newer sealants have a long life and do not affect the bolts.

For leak control and sealing, industrial hygiene personnel should carefully review sealant materials and potential exposures during the sealing operation. Following shutdown and removal of enclosures, sealant compound generally covers the flange. In addition, the flange joint must be opened and the flanges completely cleaned prior to installation of a new gasket. These operations should be reviewed along with disposal of the sealant.

9.9 ALTERNATIVES

While some of the newer developments in gasket design were reviewed new flange designs were not mentioned. A variety of new flanges is available in the commercial market, with relatively little plant operating information about their performance, which is of concern with regard to emission control. New flanges that meet the various mechanical design code requirements should be tested for emissions leakage. One new flange that may have potential emission control benefits is reviewed in this section.

9.9.1 TAPER-LOK® Flanges

A standard TAPER-LOK® flange set is shown in Figure 9.15. These flange joints have male and female ends with a metal, cone-shaped gasket. The standard flange has a self-aligning capability that allows for a maximum of 2° in axial pipe misalignment. Special flange designs provide for handling 10° and 20° pipe misalignments with a flange set capable of handling a 10° misalignment shown in Figure 9.15. This design has curved faces to allow for misaligned piping centerlines. The 20° flange set has a somewhat different design.

These flanges are smaller and lighter than conventional flanges and due to their design, the self-aligning flange sets apparently maintain seal integrity to 1500°F temperatures and 20,000-psig pressures. The flanges are designed for piping sizes that range from 1 in. (25.4 mm) to greater than 72 in. (1830 mm). Flange TAPER-LOK applications in process operations are generally concentrated in high-temperature and high-pressure service, with temperatures to 1400°F (760°C) and pressures between 8000 and 10,000 psi (551–690 bar). Some of the advantages

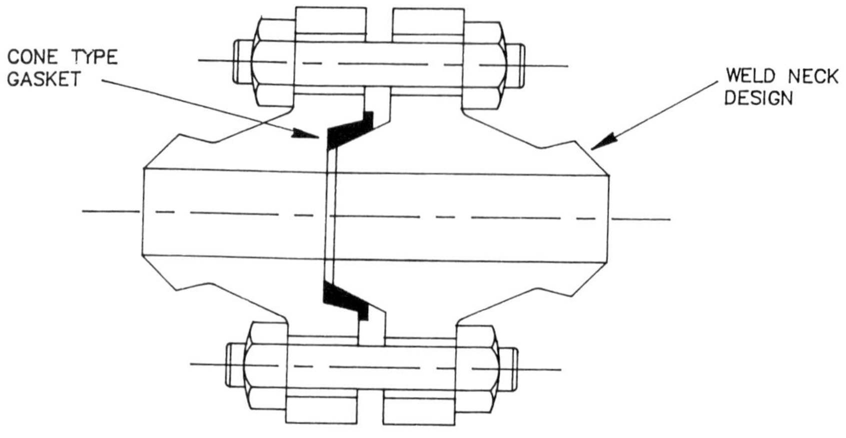

STANDARD TAPER–LOK® FLANGE

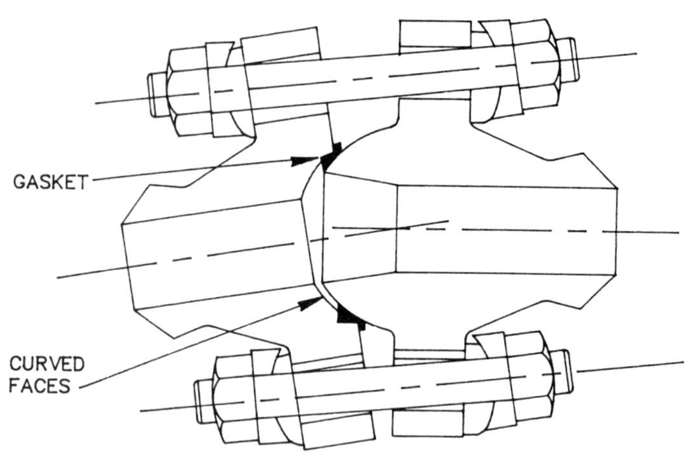

TAPER–LOK® FLANGES FOR A 10" PIPING MISALIGNMENT

Figure 9.15 *TAPER – LOK® flanges (Courtesy of TAPER – LOK® Corp.).*

claimed for this flange in addition to the previous comments are:

1. Reduced space and weight compared with standard flanges.
2. Smaller bolt sizes than the bolts required in standard flanges.
3. Lower torque requirements.
4. Blowout-proof gasket.
5. Very low emissions leakage. Some early laboratory data are available indicating emissions leakage was essentially nil at room temperature. Published data are lacking on leakage performance.

6. The flanges are self centering.
7. The gaskets are constructed of metal and are reusable.
8. Lower installation costs owing to smaller flanges and lower torque requirements.

Although this flange design differs from standard flanges, the potential low emissions performance in process service should be evaluated. The flange meets the various code requirements according to the manufacturer and is apparently acceptable for installation since the flanges have been installed in services with mechanical code specifications.

9.9.2 Welding

Various methods of controlling leakage have been mentioned in the chapter; welding the flanges is another such technique which is described in reference 32 and has been used successfully. However, there are limitations on seal welding since flange beveling is required and the technique is expensive. This procedure is generally limited to flange installations where leakage is a consistent problem. Moreover, beveled flanges must be analyzed to determine whether they are mechanically acceptable for process temperature and pressure conditions after the beveling action has been completed.

Since process units have large numbers of flanges, a reduction in the number of flanges is desirable as a method of reducing overall loss and the number of leaking flanges. However, elimination of flanges is difficult and requires changes in piping designs and configurations. The reduction in flanges should begin with process units in the design stage as a method of developing criteria for the elimination of flanges.

For threaded fittings, seal welding eliminates leakage and seal-welded connections are not included in the number of connectors that must be reported to the EPA under the HON regulations. In existing plants where the regulations will become applicable, seal welding will reduce monitoring requirements.

9.10 SPECIFICATIONS

For improved emission control, plants should consider gasket specifications based on performance criteria. These criteria provide a method of comparing various gaskets under standard conditions. With the new PVRC/MTI gasket tests and procedures, standard test procedures are available for gasket evaluation.

Gasket specifications issued by process facilities would require gasket manufacturers to supply the test information requested in the specification. Typical specifications might include the following:

1. *Fire Resistance*. A gasket manufacturer would be requested to supply test information from FITT tests conducted in accordance with PVRC procedures described in Section 9.5. The tests must be conducted on NPS 4 gaskets. In addition, a $T_{p\,\min}$ value must be determined on a postexposure basis.

2. *Room Temperature Tightness Test*. This test (ROTT) was described in Section 9.5 and is conducted with standard size gaskets. From this test, the various operating constants and stresses are determined. This provides a basis for evaluating the gasket along with seating factors.

In addition, the ROTT test can also determine the maximum stress level that indicates or defines resistance to excessive bolt loads. For each flange class rating, a maximum stress can be specified that must be applied and the gasket must have adequate resistance.

3. *High Temperature Tightness Tests*. For determining gasket capability at elevated temperatures, HOTT and AHOT tests can also be required. Plant requirements can determine the conditions necessary for their particular facilities. The results of these tests combined with ROTT tests provide a basis for evaluating and comparing gasket performance.

REFERENCES

1. Berglund, R. L., W. J. Mollere, C. R. Ritter, and L. W. Salathe. *A Quality Program to Reduce Episodic and Fugitive Emissions from an Ethylene Oxide Unit*. Fugitive Emissions Seminar, VMA, Houston TX, September 25–26, 1991.
2. EPA. *Compilation of Air Pollution Emission Factors, Vol.1, Stationary Point and Area Sources*. AP-42, 4th ed., September 1985.
3. EPA. *Fugitive Emission Sources of Organic Compounds—Additional Information on Emissions, Emission Reductions, and Costs*, Publ. No. EPA-450/3-82-010, April 1982.
4. CFR. Standards of Performance for Equipment Leaks of VOC in the Synthetic Organic Chemicals Manufacturing Industry, *Title 40*, Subpart VV, Part 60.482-8, p. 490, July 1, 1992.
5. FR. National Emission Standards for Organic Hazardous Air Pollutants from Synthetic Organic Chemical Manufacturing Industry Equipment Leaks. Proposed Rule. *Title 40*, Subpart H, Parts 63.169 and 63.174, pp. 62770–62773, December 31, 1992.
6. CFR. Method 21—Determination of Volatile Organic Compound Leaks. *Title 40*, Part 60, App. A. pp. 1020–1023, July 1, 1992.
7. Payne, J. R. *Bolted-Joint Improvements Through Gasket Performance Tests*. National Petroleum Refiners Association Maintenance Conference, San Antonio, TX, May 19–22, 1992.
8. Bazergui, A., L. Marchand, and H. D. Raut. *Further Gasket Leakage Behavior Trends*. Bull. 325, Welding Research Council, New York, July 1987.
9. Winter, J. R. *Gasket Selection Flowchart*. ASME—Pressure Vessel and Piping Conference, Nashville, TN, June 18–21, 1990.
10. Harris, G. E. and B. A. Tichenor. *Frequency of Fugitive Emissions in Synthetic Organic Chemical Plants*. 74th Mtg., APCA, Philadelphia, June 21–26, 1981.
11. Berglund, R. L., J. A. Mullins, L. S. Perrett, and R. R. Romano. *Fugitive Emissions from Equipment Leaks: Results of Industry-Wide Studies in the Butadiene, Ethylene Oxide and Phosgene Production Industries*. AIChE National Meeting, Philadelphia, PA, August 20–23, 1989.
12. Payne, J. R. and J. R. Winter. *Fugitive Emission Estimates from Gasket Tightness Test Data*. ASME Pressure Vessel and Piping Conference, New Orleans, LA, June 21–25, 1992.

13. Ore, J. *Qualifying Gasketing Materials to Reduce Fugitive Emissions*. Fugitive Emissions Seminar, VMA, Philadelphia, PA, November 18–19, 1992.
14. McKillop, G. R. and S. F. Thomas. *Oil Gas J.* **84**(21), 78–86, May 26, 1986.
15. CFR. Asbestos, Tremolite, Anthophyllite, and Actionolite. *Title 29*, Subpart Z, Sec. 1910.1001, pp. 34–77, July 1, 1991.
16. Cheng, R. T., H. J. McDermott, E. H . Nicolls, and T. V. Smith. *Asbestos Gaskets: Exposures and Substitutes*. American Industrial Hygiene Conference, Orlando, FL, May 1989.
17. Mckinnery, W. N. and R. W. Moore. *Am. Ind. Hyg. Assoc.* **53**(8), 531–532, August 1992.
18. Bazergui, A., M. Derenne, L. Marchand and J. R. Payne. *Development of Test Procedures for Fire Resistance Qualification of Gaskets*. 13th International Conference on Fluid Sealing, Brugge, Belgium, April 7–9, 1992.
19. Short, W. E. *A Parametric Study of Class 150 Lb. Flanges with Spiral Wound Gaskets*. ASME Pressure Vessel and Piping Conference, New Orleans, LA, June 21–25, 1992.
20. *ASME/ANSI B31.3 Section Code for Chemical Plant and Refinery Piping*. ASME, 1990 ed. (A91B), New York, 1990.
21. Jones, W. F. and B. B. Seth. *Evaluation of Asbestos Free Gasket Materials*. ASME/IEEE Power Generation Conference, Boston, MA, October 25–28, 1990.
22. Singh, B. *Hydrocarbon Proc.* **71**(4), 79–81, April 1991.
23. Bazergui, A. and L. Marchand *PVRC Milestone Gasket Tests, First Results*. *Weld. Res. Counc. Bull.* **292**, 1–36, February 1984.
24. ASME Boiler and Pressure Vessel Code. ASME, Sec. VIII, Div. 1, New York, 1990.
25. Crowley, E. D. and D. G. Hart. *Minimizing Fugitive Emissions Through Implementation of New Gasket Constants*. Fugitive Emissions Seminar, VMA, Houston, TX, September 25–26, 1991.
26. Bazergui, A., G. F. Leon, and J. R. Payne. *Getting New Gasket Design Constants from Gasket Tightness Data*. Special Supplement Experimental Techniques, pp, 22–27, November 1988.
27. Bazergui, A., M. Derenne, and L. Marchand. *The Influence of Thermal Degradation on Sealing Performance of Compressed Sheet Gasket Materials with Elastomer Binder, Part I: Experiment, Part II: Analysis*. 2nd International Sympsium on Fluid Sealing of Gasketed Joints, LaBaule, France, September 18–20, 1990.
28. Bazergui, A. and J. R. Payne. *Evaluation of Test Methods for Asbestos Replacement Gasket Materials*. MTI Publ. 36, Materials Technology Institute, New York, 1990.
29. Bazergui, A. and L. Marchand. *Development of Tightness Test Procedures for Gaskets in Elevated Temperature Service*. Weld. Res. Counc. Bull. **339**, 1–20 (1988).
30. Payne, J. R. and M. Derenne. *Room Temperature Tightness Tests (ROTT) and Determination of Gasket Constraints for Marine & Petroleum Graphonic® (TM) Gaskets*. Marine and Petroleum Mfg. Co. Inc., Orange, TX, 1991.
31. Bickford, J. *Introduction to the Design and Behavior of Bolted Joints*, 2nd ed. New York: Marcel Dekker, 1990, pp. 184–190.
32. Lipton, S. and J. R. Lynch. *Health Hazard Control in the Chemical Process Industry*. 1st ed. New York: Wiley, 1987, pp. 192–193.
33. Bibel, G. *Uniform Bolt Loads*, Marine and Petroleum Mfg. Co., Houston, TX, 1992.
34. Corbett, R. A. *Oil Gas J.* **84**(46), 39–44, November 18, 1986.
35. Derenne, M. and L. Marchand. *Fugitive Emission Characteristics of Gaskets*. PVRC Progress Report, PVRC Project No. 92-95, October 18, 1993.

APPENDIX 9A List of volatile hazardous air pollutants[a]

Chemical Name	CAS No.	Chemical Name	CAS No.
Acetaldehyde	75070	Dimethylformamide	68122
Acetamide	60355	1,1-Dimethylhydrazine	57147
Acetonitrile	75058	Dimethyl phthalate	131113
Acetophenone	98862	Dimethyl sulfate	77781
2-Acetylaminofluorine	53963	4,6-Dinitro-*o*-cresol, and salts	534521
Acrolein	107028	2,4-Dinitrophenol	51285
Acrylamide	79061	2,4-Dinitrotoluene	121142
Acrylic acid	79107	1,4-Dioxane (1,4-Diethyleneoxide)	123911
Acrylonitrile	107131	1,2-Diphenylhydrazine	122667
Allyl chloride	107051	Epichlorohydrin (1-chloro-2,3-epoxypropane)	106898
4-Aminobiphenyl	92671		
Aniline	62533	1,2-Epoxybutane	106887
o-Anisidine	90040	Ethyl acrylate	104885
Benzene	71432	Ethylbenzene	100414
Benzidine	92875	Ethyl carbamate (urethane)	51796
Benzotrichloride	98077	Ethyl chloride (chloroethane)	75003
Benzyl chloride	100447	Ethylene dibromide (dibromoethane)	106934
Biphenyl	92524		
Bis(2-ethylhexyl)phthalate DEHP	117817	Ethylene dichloride (1,2-dichloroethane)	107062
Bis(chloromethyl)ether	542881		
Bromoform	75252	Ethylene glycol	107211
1,3-Butadiene	106990	Ethylene oxide	75218
Caprolactam	105602	Ethylene thiourea	96457
Carbon disulfide	75150	Ethyldene dichloride (1,1-dichloroethane)	75343
Carbon tetrachloride	56235		
Carbonyl sulfide	463581	Formaldehyde	50000
Catechol®	120809	Glycol ethers	0
Chloroacetic acid	79118	Hexachlorobenzene	118741
2-Chloroacetopheonone	532274	Hexachlorobutadiene	87683
Chlorobenzene	108907	Hexachloroethane	67721
Chloroform	67663	Hexamethylene-1,6-diisocyanate	822060
Chloromethyl methyl ether	107302	Hexamethylphosphoramide	680319
Chloroprene	126998	Hexane	110543
Cresols and cresylic acids (mixed)	1319773	Hydrazine	302012
Cresol and cresylic acid (*o*-isomer)	95487	Hydroquinone	123319
Cresol and cresylic acid (*m*-isomer)	108394	Isophorone	78591
Cresol and cresylic acid (*p*-isomer)	106445	Maleic anhydride	108316
Cumene	98828	Methanol	67561
2,4-D, salts and esters	94757	Methyl bromide (bromomethane)	74839
DDE	3547044	Methyl chloride (chloromethane)	74873
Diazomethane	334883	Methyl chloroform (1,1,1-trichloroethane)	71556
Dibenzofurans	132649		
1,2-Dibromo-3-chloropropane	96128	Methyl ethyl ketone (2-butanone)	78933
Dibutylphthalate	84742	Methyl hydrazine	60344
1,4-Dichlorobenzene(*p*-)	106467	Methyl iodide (iodomethane)	74884
3,3-Dichlorobenzidine	91941	Methyl isobutyl ketone (hexone)	108101
Dichloroethyl ether bis(2-chloroethyl)ether)	111444	Methyl isocyanate	624839
		Methyl methacrylate	80626
1,3-Dichloropropene	542758	Methyl tert butyl ether	1634044
Diethanolamine	111422	4,4-Methylene bis(2-chloroaniline)	101144
N,*N*-Dimethylaniline	121697	Methylene chloride (dichloromethane)	75092
Diethyl sulfate	64675		
3,3′-Dimethoxybenzidine	119904	Methylene diphenyl diisocyanate (MDI)	101688
Dimethyl aminoazobenzene	60117		
3,3′-Dimethylbenzidine	119937	4,4^1-Methylenedianiline	101779
Dimethyl carbamoyl chloride	79447	Naphthalene	91203

APPENDIX 9A ***(Continued.)***

Chemical Name	CAS No.	Chemical Name	CAS No.
Nitrobenzene	98953	1,1,2,2-Trichloroethane	79345
4-Nitrobiphenyl	92933	Tetrachloroethylene (perchloroethylene)	127184
4-Nitrophenol	100027	Toluene	108883
2-Nitropropane	79469	2,4-Toluene diamine	95807
N-Nitroso-*N*-methylurea	684935	2,4-Toluene diisocyanate	584849
N-Nitrosodimethylamine	62759	*o*-Toluidine	95534
N-Nitrosomorpholine	59892	1,2,4-Trichlorobenzene	120821
Phenol	108952	1,1,2-Trichloroethane	79005
p-Phenylenediamine	106503	Trichloroethylene	79016
Phosgene	75445	2,4,5-Trichlorophenol	95954
Phthalic anhydride	85449	2,4,6-Trichlorophenol	88062
Polychlorinated biphenyls (aroclors)	1336363	Triethylamine	121448
1,3-Propane sultone	1120714	Trifluralin	1582098
β-Propiolactone	57578	2,2,4-Trimethylpentane	540841
Propionaldehyde	123386	Vinyl acetate	108054
Propoxur (Baygon)	114261	Vinyl bromide	593602
Propylene dichloride (1,2-dichloropropane)	78875	Vinyl chloride	75014
Propylene oxide	75569	Vinylidene chloride (1,1-Dichloroethylene)	75354
1,2-Propylenimine (2-methyl aziridine)	75558	Xylenes (not otherwise specified)	1330207
Quinone	106514	Xylene (*o*-isomer)	95476
Styrene	100425	Xylene (*m*-isomer)	108383
Styrene oxide	96093	Xylene (*p*-isomer)	106423
2,3,7,8-Tetrachlorodibenzo-*p*-dioxin	1746016		

[a]From Chemical Abstract Service.

APPENDIX 9B List of Chemicals Produced in Chemical Manufacturing Processes[a]

Chemical Name	CAS No.	Chemical Name	CAS No.
Group I		Chloroacetophenone (2-isomer)	532274
Acetone	67641	Chloroaniline (*o*-isomer)	95512
Acetonitrile	75058	Chlorobenzene	108907
Acetophenone	98862	Chlorodifluoromethane	75456
Acrylamide	79061	Chloroform	67663
Acrylonitrile	107131	Chloronitrobenzene (*o*-isomer)	88733
Adiponitrile	111693	Chloronitrobenzene (*p*-isomer)	100005
Allyl alcohol	107186	1-Chloro-3-nitrobenzene	121733
Aminophenol (*p*-isomer)	123308	Cumene hydroperoxide	80159
Aniline	62533	Cumene (isopropyl benzene)	98828
Azobenzene	103333	Cyclohexane	110827
Benzene	71432	Cyclohexanol	108930
Benzenedisulfonic acid	98486	Cyclohexanone	108941
Benzenesulfonic acid	98113	Cyclohexene	110838
Benzidine	92875	Dichloroaniline (mixed isomers)	27134276
Benzophenone	119619	Dichlorobenzene (*p*-isomer) (PDB)	106467
Biphenyl	92524	Dichlorobenzene (*m*-isomer)	541731
Bis(chloromethyl)ether	542881	Dichlorobenzene (*o*-isomer)	95501
Bromobenzene	108861	Dichlorobenzidine (3,3′-isomer)	91941
Butanediol (1,4-isomer)	110634	Dichloroethane (1,2-isomer) (EDC)	107062
Butyrolacetone	96480	Dichloroethylether (bis(2-chloroethyl)ether)	111441
Carbon tetrachloride	56235		

APPENDIX 9B ***(Continued.)***

Chemical Name	CAS No.
Dichlorodifluoromethane	75718
Diethanolamine	111422
Diethylene glycol	111466
Diethylene glycol dibutyl ether	112732
Diethylene glycol diethyl ether	112367
Diethylene glycol dimethyl ether	111966
Diethylene glycol monobutyl ether acetate	124174
Diethylene glycol monobutyl ether	112345
Diethylene glycol monoethyl ether	111900
Diethylene glycol monomethyl ether	111773
Dimethyl sulfate	77781
Dimethylaminoethanol (2-isomer)	108010
Dinitrobenzenes	25154545
Dioxane (1,4-diethyleneoxide)	123911
Dioxolane (1,3-isomer)	646060
Diphenyl methane	101815
Diphenyl oxide	101848
Dipropylene glycol	25265718
Dodecylbenzene (*n*-isomer)	121013
Epichlorohydrin (1-chloro-2,3-epoxypropane)	106898
Ethanolamine	141435
Ethylbenzene	100414
Ethylene carbonate	96491
Ethylene dibromide (dibromoethane) (EDB)	106934
Ethylene glycol	107211
Ethylene glycol diacetate	111557
Ethylene glycol diethyl ether (1,2-diethoxyethane)	629141
Ethylene glycol dimethyl ether	110714
Ethylene glycol monobutyl ether acetate	112072
Ethylene glycol monobutyl ether	111762
Ethylene glycol monoethyl ether acetate	11159
Ethylene glycol monoethyl ether	110805
Ethylene glycol monomethyl ether acetate	110496
Ethylene glycol monomethyl ether	109864
Ethylene glycol monophenyl ether	122996
Ethylene glycol monopropyl ether	2807309
Ethylene oxide	75218
Formaldehyde	50000
Fumaric acid	110178
Hexamethylenetetramine	100970
Hydroquinone	123319
Linear alkylbenzene (no CAS number assigned)	
Maleic anhydride	108316
Maleic hydrazide	123331
Malic acid	6915157
Metanilic acid	121471
Methionine	63683
Methylene chloride (dichloromethane)	75092
Methylene dianiline (4,4′-isomer)(MDA)	101779
Methylstyrene (*a*-isomer)	98839
Nitroaniline (*o*-isomer)	88744
Nitroaniline (*p*-isomer)	100016
Nitrobenzene	98953
Octene-1	111660
Paraformaldehyde	9002817
Pentaerythritol	115775
Perchloroethylene (tetrachloroethylene)	127184
Phenylenediamine (*p*-isomer)	106503
Piperazine	110850
Propiolactone (β-isomer)	57578
Propionic acid	79094
Propylene glycol	57556
Propylene glycol monomethyl ether	107982
Propylene oxide	75569
Resorcinol	108463
Styrene (vinyl benzene)	100425
Succinic acid	110156
Succinonitrile	110612
Tartaric acid	526830
Tetrachlorobenzene (1,2,4,5-isomer)	95943
Tetraethylene glycol	112607
Tetrahydrofuran	109999
Toluene	108883
Trichlorobenzene (1,2,4-isomer)	102821
Trichloroethylene	79016
Trichlorofluoromethane	75694
Trichlorotrifluoroethane	76131
Trichlorophenol (2,4,5-isomer)	95954
Triethanolamine	102716
Triethylene glycol	112276
Triethylene glycol dimethyl ether (glycol ether)	112492
Triethylene glycol monomethyl ether	112356
Trimethylolpropane	77996
Vinyl chloride (chloro ethylene)	75014
Xylenes (not otherwise specified)	1330207
Xylene (*o*-isomer)	95476
Xylene (*p*-isomer)	106423
Xylene (*m*-isomer)	108383
Group II	
Acetaldehyde	75070
Acetaldol	107891
Acetamide	60355
Acetanilide	103844
Acetic acid	64197
Acetic anhydride	108247
Anisidine (*o*-isomer)	90040
Butadiene (1,3-isomer)	106990

APPENDIX 9B ***(Continued.)***

Chemical Name	CAS No.	Chemical Name	CAS No.
Butylene glycol (1,3-isomer)	107880	Toluene diisocyanates (mixture)	26471625
Caprolactam	105602	Toluidine (*o*-isomer)	95534
Carbon tetrabromide	558134	Trichloroethane (1,1,1-isomer)	71556
Carbon tetrafluoride	75730	Trichloroethane (1,1,2-isomer)	79005
Chloral	75876	(vinyl trichloride)	
Chloroacetic acid	79118	Vinyl acetate	108054
Chloroaniline (*p*-isomer)	106478	Vinylcyclohexene (4-isomer)	100403
Chlorophenol (*m*-isomer)	108430	Vinylidine chloride	75354
Chlorophenol (*p*-isomer)	106489	(1,1-dichloroethylene)	
Chlorophenol (*o*-isomer)	95578	*Group III*	
Chloroprene (2-chloro-1,3-butadiene)	126998	Acetoacetanilide	102012
Chlorotrifluoromethane	75729	Aniline hydrochloride	142041
Crotonaldehyde	4170300	Anthraquinone	84651
Cyanoacetic acid	372098	Benzaldehyde	100527
Cycloctadienes	29965977	Benzil	134816
Dichloro-1-butene (3,4-isomer)	760236	Benzilic acid	76937
Dichloroethylene (1,2-isomer)	540590	Benzoic acid	65850
Dichloropropene (1,3-isomer)	542756	Benzoin	119539
Diethyl sulfate	64675	Benzonitrile	100470
Dimethylbenzidine (3,3′-isomer)	119937	Benzotrichloride	98077
Dimethylformamide (*N*,*N*-isomer)	68122	Benzoyl chloride	98884
(DMF)		Benzyl acetate	140114
Dimethylhydrazine (1,1-isomer)	57147	Benzyl alcohol	100516
Dimethyl terephthalae	120616	Benzyl benzoate	120514
Ethyl acrylate	140885	Benzyl chloride	100447
Ethyl chloroacetate	105395	Benzyl dichloride	98873
Ethylenimine (aziridine)	151564	Bisphenol A	80057
Ethbylenediamine	107153	Butylbenzyl phthalate	85687
Ethylhexyl acrylate (2-isomer)	130117	Chlorotoluene (*m*-isomer)	108418
Formamide	75127	Chlorotoluene (*o*-isomer)	95498
Formic acid	64186	Chlorotoluene (*p*-isomer)	106434
Glycerol	56815	Cresols and cresylic acids (mixed)	1319773
Glycerol tri(polyoxypropylene)ether	25791962	Cresol and cresylic acid (*o*-isomer)	95487
Glycine	56406	Cresol and cresylic acid (*m*-isomer)	108394
Glyoxal	107222	Cresol and cresylic acid (*p*-isomer)	106445
Hexachlorobenzene	118741	Cyclohexylamine	108918
Hexachlorobutadiene	87683	Dichlorophenol (2,4-isomer)	120832
Hexachloroethane	67721	Diethylaniline (*N*,*N*-isomer)	91667
Hexadine (1,4-isomer)	592450	Diethyl phthalate	84662
Methyl formate	107313	Diisodecyl phthalate	26761400
Methyl phenol carbinol	98851	Dimethyl phthalate	131113
m-Nitroaniline	99092	Dimethylaniline-*N*,*N*	121697
Nitropropane (2-isomer)	79469	Dinitrophenol (2,4-isomer)	51285
Paraldehyde	123637	Dinitrotoluene (2,4-isomer) (DNT)	121142
Peracetic acid	79210	Di-*o*-tolyguanidine	97392
Picoline (β-isomer)	108996	Diphenyl thiourea	102089
Pyridine	110861	Diphenylamine	122394
Sulfolane	126330	Dodecylphenol	27193868
Terephthalic acid	100210	Ethylaniline (*n*-isomer)	103695
Tetrachloroethane (1,1,2,2-isomer)	79345	Ethylaniline (*o*-isomer)	578541
Tetrahydrophthalic anhydride	85438	Isophthalic acid	121915
Tetramethylenediamine	110601	Isopropylphenol	25168063
Toluene 2,4 diamine	95807	Methylaniline (*n*-isomer)	100618
Toluene 2,4 diisocyanate	584849	Methylcyclohexane	108872

APPENDIX 9B ***(Continued.)***

Chemical Name	CAS No.
Methylcyclohexanone	1331222
Methylene diphenyl diisocyanate (4,4-isomer)(MDI)	101688
Nitroanisole (*o*-isomer)	91236
Nitroanisole (*p*-isomer)	100174
Nitrophenol (4-isomer)	100027
Nitrophenol (*o*-isomer) (2-nitrophenol)	88755
Nitrotoluene (all isomers)	1321126
Nitrotoluene (2-isomer)	88722
Nitrotoluene (3-isomer)	99081
Nitrotoluene (4-isomer)	99990
Octylphenol	27193288
Pentachlorophenol	87865
Phenetidine (*p*-isomer)	156434
Phenol	108952
Phenolphthalein	77098
Phenolsulfonic acids (all isomers)	1333397
Phenyl anthranilic acid	91407
Phloroglucinol	108736
Phthalic acid	88993
Phthalic anhydride	85449
Phthalimide	85416
Phthalonitrile	91156
p-tert-Butyl toluene	98511
Quinone	106514
Salicylic acid	69727
Sodium phenate	139026
Stilbene	588590
Sulfanilic acid	121573
Tetrabromophthalic anhydride	632791
Tetrachlorophthalic anhydride	117088
Toluenesulfonic acids (all isomers)	104154
Toluenesulfonyl chloride	98599
Trichloroaniline (2,4,6-isomer)	634935
Vinyl toluene	25013154
Xylene sulfonic acid	25321419
Xylidine	1300738
Group IV	
Acrolein	107028
Acrylic acid	79107
Allyl chloride	107051
Allyl cyanide	109751
Bromonaphthalene	27497514
Carbon disulfide	75150
Chloronaphthalene	25586430
Decahydronaphthalene	91178
Diallyl phthalate	131179
Diethylamine	109897
Dimethyl ether	115106
Dimethylamine	124403
Ethyl chloride (chloroethane)	75003
Glutaraldehyde	111308
Hexanetriol (1,2,6-isomer)	106694
Isophorone	78591
Methanol	67561

Chemical Name	CAS No.
Methyl acetate	79209
Methyl bromide (bromomethane)	74839
Methyl chloride (chloromethane)	74873
Methyl hydrazine	60344
Methyl isobutyl carbinol	108112
Methyl isobutyl ketone (hexone)	108101
Methyl isocyanate	624839
Methyl mercaptan	74931
Methyl methacrylate	80626
Methylamine	74895
Naphthalene	91203
Naphthalene sulfonic acid (α-isomer)	85472
Naphthalene sulfonic acid (β-isomer)	120183
Napthol (α-isomer)	90153
Napthol (β-isomer)	135193
Nitronaphthalene (1-isomer)	86577
Perchloromethyl mercaptan	594423
Phosgene	75445
Propionaldehyde	123386
Propylene dichloride (1,2-dichloropropane)	78875
Sodium methoxide	124414
Tetraethyl lead	78002
Tetrahydronaphthalene (tetralin)	119642
Triethylamine	121448
Trimethylamine	75503
Trimethylcyclohexanol	933482
Trimethylcyclohexanone	2408379
Group V	
Acenaphthene	83329
Acetal	105577
Acetone cyanohydrin	75865
Alizarin	72480
Alkyl anthraquinones	008
Alkyl naphthalenes (no CAS number assigned)	
Aminophenol sulfonic acid	0010
Anthracene	120127
Bromoform	75252
Butyl acrylate (*n*-isomer)	141322
Carbaryl	63252
Carbazole	86478
Chlorodifluoroethane	25497294
Chrysene	218019
Diacetoxy-2-butene (1,4-isomer)	0012
Diaminophenol hydrochloride	137097
Dibromomethane	74953
Dibutoxyethyl phthalate	117839
Diethylaniline (2,6-isomer)	579668
Diisoocytyl phthalate	27554263
Dodecylaniline	28675174
Dodecyl benzene (branched)	123013
Dodecyl phenol (branched)	0013
Ethylamine	75047
Ethylcellulose	9004573

APPENDIX 9B ***(Continued.)***

Chemical Name	CAS No.	Chemical Name	CAS No.
Ethylcyanoacetate	105566	Phenanthrene	85018
Ethylenediamene tetracetic acid	60004	Polyethylene glycol	25322683
Glyceraldehyde	367475	Polypropylene glycol	25322694
Hydroxyadipaldehyde	0016	Propylene carbonate	108327
Iminodiethanol (2,2-)	111422	Pyrene	129000
Isobutyl acrylate	106638	*n*-Vinyl-2-pyrrolidine	88120
Isobutylene	115117	Resorcylic acid (dihydroxybenzoic acid)	27138574
Isophorone nitrile	0017		
Lead phthalate	0018	Tetraethylenepentamine	112572
Mesityl oxide	141797	Tetramethyl lead	75741
Methacryllic acid	79414	Tetramethylethylenediamine	110189
Methyl acrylate	96333	Thiocarbanilide	102089
Methyl ethyl ketone (2-butanone)	78933	Triethylene glycol monethyl ether	112505
Methyl tert-butyl ether	1634044	Triethylcyclohexylamine	007
Methylcyclohexanol	25639423	Trimethylpentane (2,2,4-isomer)	540841
Methylionones (α-)	79696	Tripropylene glycol	24800440
Methylpentynol	77758	Xanthates	140896
Naphthosulfonic acid (1-)	567180	Xylene	1300716
Naphthylamine sulfonic acid (1,4-)	84866	Xylenol (2,3-isomer)	526750
Naphthylamine sulfonic acid (2,1-)	81163	Xylenol (2,4-isomer)	105679
Naphthylamine (1-)	134327	Xylenol (2,5-isomer)	95874
Naphthylamine (2-)	91598	Xylenol (2,6-isomer)	576261
Nitroxylene	25168041	Xylenol (3,4-isomer)	95658
Nonylbenzene (branched)	1081772	Xylenol (3,5-isomer)	108689
Nonylphenol	25154523		

[a]From Chemical Abstract Service.

10

Rotating Equipment

Pump and compressor seals are sources of leakage which contribute to fugitive and toxic emissions, thereby increasing potential workplace exposures. Although other rotating equipment sources are present in chemical plants and refineries, pump and compressor seals exist in much greater number and this chapter focuses on control of emissions from these sources. Some of the engineering control systems presented are applicable to other types of rotating equipment such as mixers. Since a wide variety of pump types and seals are currently available, only *centrifugal pumps* and controls are presented, but the control principles are broadly applicable.

Although seal emissions or leakage must be controlled to minimize potential exposures, the impact of regulatory requirements on equipment leakage should be considered prior to recommending engineering controls. The need or type of emission control necessary to establish an acceptable workplace concentration level may be affected by regulatory considerations. Moreover, an understanding of the fugitive emission (VOC) and toxic (HON) regulations provide the industrial hygienist with background information on EPA-mandated controls, practical methods for controlling emissions, and a basis for recommending additional modifications necessary to further reduce exposures. The various regulatory controls and their potential effects upon exposures are presented in this chapter.

10.1 AIR EMISSION REGULATIONS

Both NSPS (VOC) and NESHAP or toxic (HON) regulations are issued directly by the EPA, with changes added as required by states and regions to meet the National Ambient Air Quality Standards (NAAQS). Various regulations are described in Chapter 4, with state and federal regulations normally administered by state authorities. Since state VOC and toxic emission regulations may be more restrictive than the basic federal regulations, industrial hygienists should have some familiarity with the various regulations that mandate emission control through various alternatives, providing economic options for local plant operators.

In locations where workplace exposure controls are the major consideration, a knowledge of applicable emission laws and potential changes in these laws is helpful in recommending specific exposure controls. As an example, fugitive emissions may be controlled in a particular process by state regulation requiring pump seal monitoring every 3 months along with repair procedures for excessive emission sources. A review of potential exposures from a leaking pump seal in this location by an industrial hygienist may indicate that a revised single mechanical seal with a backup gland plate and throttle bushing or equivalent is satisfactory to reduce emissions and potential exposures. However, process modifications scheduled later in the year can place the unit within the NSPS, or the pump may be covered by the HON regulations, indicating a review with process designers for the problem seal suggests a dual seal installation is preferable for unit pumps under the new requirements. Installation of the revised seal system would then be reviewed and a temporary repair proposed as a solution until the plant turnaround or until a dual seal can be installed to meet near- and long-term exposure and emission.

Summaries of the NSPS and HON regulations for pumps and compressors presented in Chapter 4 outline the various types of emission controls and control alternatives available. The type of control scheme selected for a pump or compressor to meet these regulations is an economic decision, but the regulatory requirements can be used as a basis for workplace exposure control by industrial hygienists. The HON regulations are generally based on maximum achievable control technologies (MACT) selected by the EPA prior to issuance of the proposed regulations. However, continual improvements in control technology results in the application of superior control technologies with time requiring an awareness of technology advances by industrial hygienists that aid in further reducing emissions and potential workplace exposures.

10.1.1 Pumps (NSPS)

The current EPA regulatory fugitive (VOC) emission control requirements for centrifugal pumps are as follows:[1,2]

1. Each pump in light liquid service (Chapter 4) shall be monitored monthly to detect leaks as described earlier. Also, each pump shall be visually checked each week for indications of liquid dripping from the seal.

2. A monitoring instrument reading equal to or greater than 10,000 ppm leakage at 1 cm from the shaft and packing interface indicates a leak.

3. A leaking pump shall be repaired within 15 calendar days after the leak is detected, with initial repairs commencing no later than 5 days after leak identification. Repair can be deferred if the repair requires a process unit shutdown. In this situation, repairs must be completed before the end of the next shutdown.

4. Pumps equipped with dual mechanical seal systems that include barrier fluid systems are exempt from monitoring requirements providing the sealing system

installations meet the following guidelines:

a. The barrier fluid pressure is greater than the stuffing box pressure, *or* the barrier fluid has a degassing reservoir that is connected by a closed vent to a control device, *or* the barrier fluid is purged through the dual seals directly into the process with zero VOC emissions to the atmosphere.

b. The barrier fluid is a heavy liquid or is not a VOC fluid.

c. The dual-seal system must be equipped with sensors that detect a failure in the seals, barrier fluid, or both. Each sensor is checked daily or must have an audible alarm.

d. Each pump must be checked weekly for indications of liquid leakage.

e. Liquid leaking or a sensor alarm indicates seal failure and the pump seal must be repaired within 15 calendar days. Some delay may be permitted based upon various criteria in the regulations.

5. A pump is exempt from the monitoring requirements if the pump has no external shaft penetrating the housing (sealless) and does not have a monitor reading of 500 ppm or greater above the background level.

6. A pump equipped with a closed-vent system is exempt from monitoring requirements provided the emissions or leakage are transported to a control device (Chapter 4).

10.1.2 Compressors (NSPS)

The regulations are designed to cover both centrifugal and reciprocating machines with a specific reference to reciprocating compressor design.

1. A compressor should be equipped with a barrier fluid seal system that prevents leakage of VOC or fugitive emissions to the atmosphere. The compressor seal system shall meet the following guidelines:

a. The barrier fluid operating pressure shall be higher than the compressor stuffing box pressure, *or* the barrier fluid system is connected through a closed-vent system to a control device, *or* the barrier fluid is purged into a process stream with zero VOC emissions to the atmosphere.

b. The barrier fluid is a heavy liquid or is not a VOC fluid.

c. The barrier fluid system should have sensors that detect failures in the seals, barrier fluid system, or both.

d. Each sensor shall be checked daily or is equipped with an audible alarm.

e. A failure or sensor alarm in the seals or barrier systems indicates a leak and the sealing system must be repaired with 15 calendar days. However, repairs may be delayed to the next process unit shutdown if the repair entails an immediate process shutdown.

2. A compressor is exempt from the requirements outlined in (1) if the machine is equipped with a closed-vent system that captures all of the emissions and transfers the emissions to a control device.

3. A compressor may be exempt from all requirements if emission monitoring readings are 500 ppmv or less above background level.

4. An existing reciprocating compressor is exempt from these requirements (1–4) if the distance piece or spacer must be recast to bring the machine into compliance with the regulations.

5. Compressors in hydrogen service are exempt from these requirements if the concentration of hydrogen in the process stream always exceeds 50% by volume in petroleum refinery units.[2]

10.1.3 Pumps and Compressors (BAAQMD)

The following rules and regulations are applicable to pumps and compressors in the Bay Area Air Quality Management District in California after January 1, 1993:[3]

1. Pumps and compressors subject to this rule are monitored at least once per quarter. Following repairs for leakage, the seal points are monitored for leaks within 7 days after the equipment has been returned to service.

2. All pumps and compressors processing organic compounds are visually inspected for leaks daily. For a liquid leak, the seal must be monitored to determine the concentration. Total organic compounds are measured, including methane (after January 1995).

3. Organic compounds cover all compounds of carbon, excluding carbon monoxide, carbon acid, metallic carbides or carbonates, ammonium carbonate, and methane (until January 1, 1995).

4. Any pump or compressor that has a leak in excess of 1000 ppmv is considered in violation of the rule, unless the leak was minimized within 24 hr and repaired within 7 days. Alternatively, if the leak was discovered by the BAAQMD and was repaired within 24 hr, the equipment is not in violation of the rule.

5. On January 1, 1997 the leakage standard for pumps and compressors is 500 ppmv, but conforms with the other requirements in item 4.

6. For pumps and compressors that are nonrepairable, the pump or compressor must be repaired or replaced within 5 yr or at the next scheduled turnaround, whichever occurs first. In addition, the number of pumps and compressors awaiting repair should not exceed 10% of the total quantity installed. After January 1, 1997 this figure is reduced to 1%. Spare pumps and compressors must be repaired and are not included in the percent awaiting repair.

7. After January 1, 1995 any *repaired*, replaced, new pump or compressor shall not leak in excess of 500 ppmv for four consecutive quarters. This rule is not applicable to reciprocating compressors or gear pumps.

8. Exemptions in the regulations:

a. The rule does not apply to pump seals vented to vapor recovery or disposal systems that reduce organic emissions to less than 0.3 lb/day (0.136 kg/day).

b. Pumps and compressors that process organic liquids with an initial boiling point greater than 302°F (150°C).

c. Compressors with seals vented to vapor recovery or disposal systems that reduce organic emissions to less than 1.7 lb/day (0.77 kg/day).

10.1.4 Pumps and Compressors (SCAQMD)

The regulations of the Southern California Air Quality Management District for liquid and gas leakage are based on minor leakage at 1000 ppmv and major gas leaks at 10,000 ppm. A maximum number of major leaks is established, with time repair limits for major and minor gas leaks. Major leaks require repair or replacement within 5 calendar days with the best available control technology (BACT) or retrofit technology (BART). An alternate is venting to an emission control device.

10.1.5 Pumps (HON)

The hazardous organic NESHAP (HON) regulations are applicable to pumps in the Synthetic Organic Chemical Manufacturing Industry (SOCMI), with light and heavy liquid pump services included in the regulations. The light liquid pump service regulations are quite extensive and are presented in this section. For completeness, the few heavy liquid pump service requirements have also been included.

Initial regulatory requirements specify the volatile hazardous air pollutants (VHAP) covered by the light liquid service pump regulations. The list of VHAP compounds presented in Appendix 9A (Chapter 9) indicates that any of these compounds used as a reactant, intermediate, or final product processed through a pump must conform with the HON regulations. In addition, these regulations are applicable to those process units shown in Appendix 9B (Chapter 9) that process the VHAP chemicals or produce the compounds as products. The list of process units has been separated into five distinct groups that must conform with regulatory requirements on a schedule based upon the final HON regulatory promulgation date as follows:

Group I	1/2 yr after promulgation
Group II	3/4 yr after promulgation
Group III	1 yr after promulgation
Group IV	1 1/4 yr after promulgation
Group V	1 1/2 yr after promulgation

The pump regulations also apply to the processes and compounds shown in Table 10.1 six months after promulgation. The process groups are combined with the phases in the standard to determine the applicability date for the new allowable leakage requirements. There are three phases with increasingly stringent emissions requirements which are shown in Figure 10.1 to indicate the change in leakage requirements with time. Applicability dates for the phases of each group of

TABLE 10.1 Additional HAP Chemicals and Processes Covered by the HON Pump Regulations[a]

Process	Chemicals
Styrene – butadiene rubber production units (SBR)	Butadiene, styrene
Polybutadiene production units	Butadiene
Chlorine production	Carbon tetrachloride
Chlorinated hydrocarbon use	Carbon tetrachloride, methylene chloride, tetrachloroethylene, chloroform, ethylene dichloride
Pesticide production	Carbon tetrachloride, methylene chloride, tetrachloroethylene, ethylene dichloride
Pharmaceutical production	Carbon tetrachloride, methylene chloride
Miscellaneous butadiene use	Butadiene

[a]Adapted from reference 5.

existing processes are as follows:

Phase I	Initiated on the applicability date
Phase II	Initiated 1 yr after applicability date
Phase III	Initiated 2 1/2 yrs after the applicability date

Thus Group I applicability date for Phase I begins 1/2 yr after the HON promulgation date, but Phase I for Group III does not begin until 1 yr after promulgation.

For *new* process units, compliance is initiated for the varying phases as follows:

Phase I	This phase is omitted
Phase II	After initial startup, comply with regulations
Phase III	Initiate 1 yr after startup

Definitions. Prior to describing the HON regulations, certain definitions are presented as a basis for understanding regulatory applications and limitations.

Light Liquid Service. The vapor pressure of one or more of the VHAP compounds is > 0.3 kPa at 20°C (0.45 psi at 68°F) and the total concentration of the pure VHAP compounds having a vapor pressure > 0.3 kPa at 20°C is equal to or greater than 20% by weight of the entire stream. In addition the fluid is a liquid at operating conditions.

In VHAP Service. For this category, a piece of equipment contains or contacts a fluid (liquid or gas) that contains at least 5 wt % of VHAP material.

Volatile Hazardous Air Pollutant. A substance or compound listed in Appendix 9A (Chapter 9).

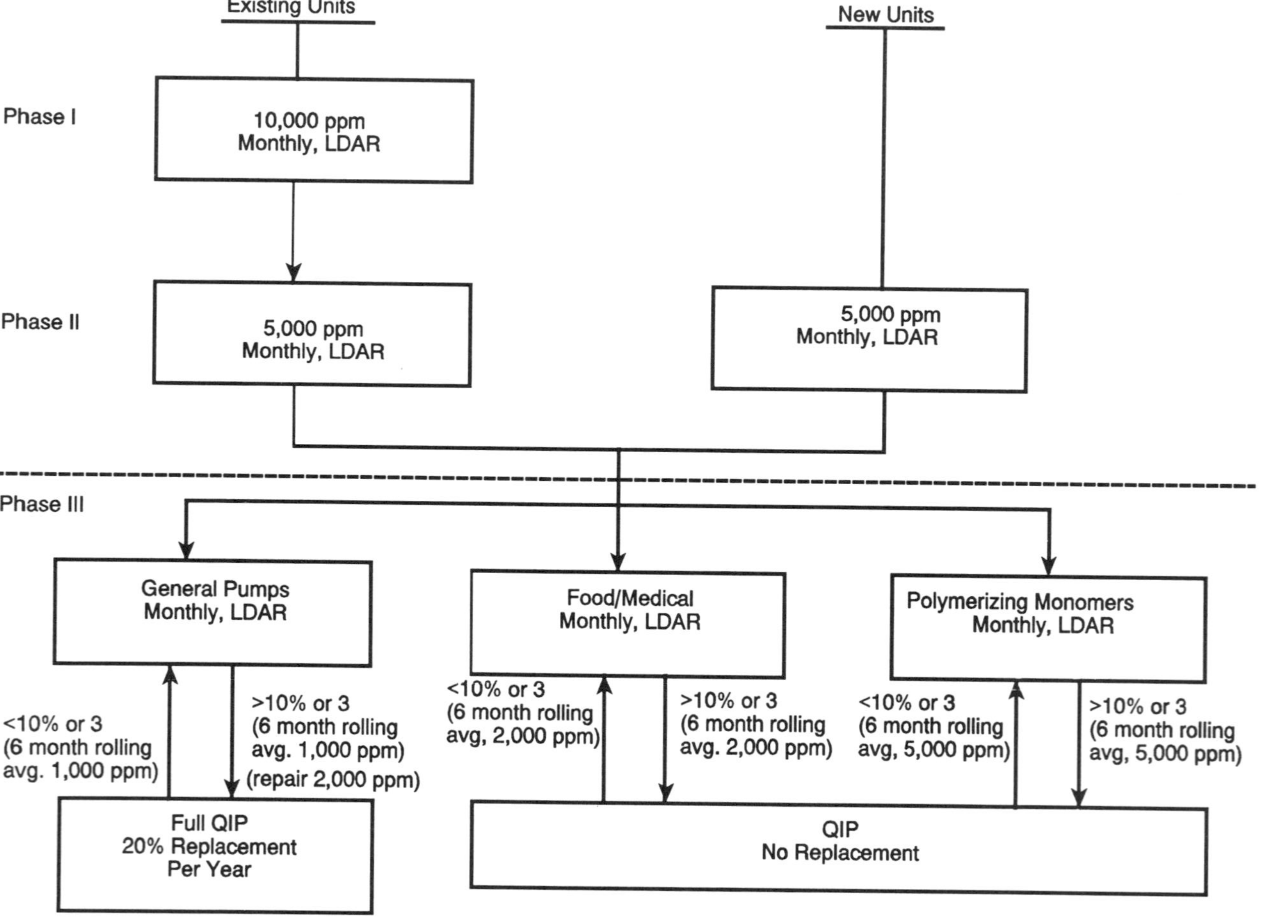

Figure 10.1 *HON regulatory standards for pumps.*[5]

Closed-Loop System. An enclosed system that returns process fluid to the process and is not vented to the atmosphere except through a closed-vent system.

Closed-Vent System. A system that is not open to the atmosphere and is composed of piping, connections, or flow-inducing devices that transport gas or vapor from equipment to a control device or recycle the vapor to the process.

Regulations. The following regulations are applicable to pumps in light liquid VHAP service in accordance with the various requirements in the proceeding paragraphs. Item 8 includes requirements for pumps in heavy liquid service.

1. The pump seals are monitored monthly with the phase requirements and leak definitions shown below:

Phase I	≥ 10,000 ppm
Phase II	≥ 5,000 ppm
Phase III	≥ 1,000 ppm
	≥ 5,000 ppm (polymerizing monomer pumps)
	≥ 2,000 ppm (food/medical service pumps)

An overall view of pump phases, leak criteria, and requirements are shown in Figure 10.1, where LDAR refers to leak detection and repair.

2. All pumps are visually inspected each calendar week. Any indications of liquid dripping classifies the pump as leaking.

3. When a leak is detected, the seal is repaired within 15 calendar days after leak detection. An initial repair attempt must be undertaken within 5 calendar days after leak detection, where initial repair attempts may include but are not limited to the following practices:

   a. Tighten gland nuts.

   b. Ensure seal flush operation is at proper pressure and temperature level.

Repairs may be delayed if the repair requires a unit shutdown, but must be completed during the next turnaround. In addition, deferments are also granted for other mechanical changes.

In Phase III which has a leak definition level of 1000 ppmv, *repairs are not required* unless the monitoring concentration ≥ 2000 ppm.

4. The percentage of pumps leaking is determined on either a process unit or plantwide basis. When the basis has been selected, this calculation procedure cannot be changed. In Phase III the pump leaking percentage can be determined on a rolling average basis. When the number of leaking pumps is greater than either 10% of the total number of pumps or three pumps, a quality improvement program for pumps must be instituted, as shown in Figure 10.1. The number of pumps for either a process unit or plantwide basis, as determined earlier, provides the number of base pumps used in calculating the leaking pump percentage.

5. Pumps equipped with a dual mechanical seal system that includes a barrier fluid system are exempt from the monitoring requirements described in items 1 and 2, providing *one* of the following requirements is observed:

a. Each dual mechanical seal system is operated with a barrier fluid pressure that is higher at all times than the pump stuffing box.

b. The barrier fluid system is equipped with a degassing drum that is connected at all times through a closed-vent system to a control device. The latter must meet the control device restrictions described in other chapters.

c. The pump is equipped with a closed-loop system that purges barrier fluid into the process stream.

In addition, the dual mechanical seal system must also comply with the following:

a. The barrier fluid is not in light liquid VHAP service.

b. Each barrier fluid system is equipped with a sensor that detects failure of the seal system, the barrier fluid system, or both.

c. Each sensor is observed daily or is equipped with an alarm.

d. Each pump is visually inspected each calendar week for indications of liquid dripping from the pump seal. Where there is evidence of liquid leakage, the seal is monitored and concentrations $\geq$ 1000 ppm indicate the seal is leaking.

e. When a leak is detected it shall be repaired within 15 calendar days unless a repair requires a process shutdown and the repair is then deferred until the next planned shutdown.

6. Any pump designed with a shaft penetrating the pump housing is exempt from the monitoring requirements in items 1 and 2.

7. Any pump equipped with a closed-vent system capable of capturing and transporting leakage from the seals to a control device is exempt from the requirements of items 1–5e.

8. Pumps in *heavy liquid service* are monitored within 5 calendar days after evidence of a potential leak is indicated through visual, audible, olfactory, or other detection methods. A monitor concentration $\geq$ 1000 ppm identifies a leaking pump, which must be repaired with 15 calendar days after detection. The initial repair must be conducted with 5 calendar days after detection. Repairs are deferred where a repair requires a process unit shutdown and the repair is deferred until the next scheduled process unit shutdown.

9. A quality improvement program (QIP) is required in Phase III where the greater of either 10% of the pumps or three pumps leak on either a process unit or plantwide preselected basis. Where pumps are in food/medical or polymerizing service, the requirements outlined in the following list must also be observed with the exception of item 11. The general QIP requirements follow:

a. All of the QIP requirements must be observed until the number of leaking pumps is less than the greater of 10% of the pumps or three pumps calculated on a six-month rolling average basis. When a satisfactory performance level is achieved, the owner can then comply with the monthly monitoring requirements in item 1 and the other requirements through item 5.

b. In subsequent monitoring periods after returning the pumps to standard monthly monitoring tests, should the leaking pumps become greater than 10% of the pumps or three pumps, calculated on a six-month rolling average basis, the plant resumes a QIP, commencing with performance trials.

c. The basic QIP requires the following:

- Various data must be collected and recorded for each pump in the program covering:

Pump type	Piston, horizontal or vertical, centrifugal, gear Seal type and manufacturer Pump design (flanged body, external shaft) Materials of construction Barrier fluid or packing material Year each installed
Service	Discharge pressure, temperature, flow rate, Corrosivity, annual operating hours
Monitoring	Records of readings (prior to repair), stream response factor
Repair	Repair methods used and monitored readings after repair

- Data will continue to be collected on the pumps as long as the process unit or plant site is in the quality improvement program.
- All frequent seal failures are investigated to determine probable cause of failure. Recommendations are developed to improve performance.
- The data collected on pump type, service, and so on are reviewed and analyzed to determine the services, operating or maintenance practices, pump or pump seal designs, or technologies that exhibit poor average emission performance. Also, those pump services with superior average emission control performance are identified. The analysis will determine whether specific detrimental factors can be identified.

An analysis shall also determine whether superior pump or seal technologies are applicable in services exhibiting poor emission control performance. Superior technologies are those that have a leak frequency $< 10\%$ of the total pump base. Also, candidates for superior performing pump or pump seal technologies are investigated. The improved technologies are characterized as having low emission performance and the capability of achieving a leak frequency of less than 10% of the base pump group.

The data analysis shall include pump and seal inspection data from pumps removed from service due to leakage, along with identification of pump designs and technologies associated with low emission performance for specific services. All of the information developed shall include any limitations or special requirements on service conditions, operating conditions, and maintenance practices to ensure continued low-emission performance.

- The first data analysis investigation will be completed no later than 18 months after initiation of the QIP. A minimum of 6 months of data are required for the analysis. Data analysis is required each year the process unit is in the QIP.

- Where the data analysis *has not identified* superior pump or pump seal technology, a *trial evaluation program* shall be conducted with pump or pump seal technology identified by others that apparently exhibit low-emission performance characteristics.

The *trial program* is an online performance test of superior technologies identified by others as having pump leak frequency rates $< 10\%$ for similar pump services. These technologies have a low probability of failure and do not have an external activating mechanism in contact with the process fluid.

For those superior technologies that have been identified, but were not included in the performance trials, reasons for rejecting specific technologies must be documented.

- The trial evaluation program shall specify and include documentation of candidate pump and pump seal designs for evaluation. Program stages for evaluation include (1) the test period necessary for defining applicability, (2) monitoring frequency and equipment inspection, (3) range of operating conditions for component evaluation, and (4) performance evaluation conclusions including the appropriate operating conditions.
- Performance trials shall initially be conducted for a 6-month period and begin not later than 18 months after the QIP is initiated.

No later than 24 months after the start of the QIP, superior pump or pump seal technologies, including operating and maintenance practices, will have been identified that meet low-emission performance criteria.

Where superior technology has not been demonstrated, the trials will continue until all of the potentially superior technology has been tested.

When all alternative superior technologies have been investigated, performance trial tests can be discontinued, whether superior performance has or has not been demonstrated. In the situation where technically feasible technologies are not satisfactory, the plant shall prepare an engineering report providing the analysis that demonstrates the technology is unsatisfactory and will not meet emissions control requirements.

10. A quality assurance program (QAP) details purchasing specifications and maintenance procedures for all pumps and seals in a process unit. The QAP can be established for classes or categories that distinguish operating condition and services associated with poor as well as superior emission performance. This program is developed based on the data analysis described in item 9 along with the evaluation trial test results. The QAP is updated every year until the pump leak frequency is less than 10% of the total pump number or three, whichever is greater. The QAP shall:

a. Establish minimum design standards for each category of pumps or pump seals, including tolerances, materials of construction, and manufacturer.

b. Require all purchasing orders to include the design standard (item a) for the pump or pump seal.

c. Provide an audit procedure for quality control of the purchased equipment.

d. Define offline pump maintenance and repair procedures. These procedures include meeting pump and seal design specifications.

The QAP shall be established no later than the start of the third year of the QIP, where a plant has 100 employees or 400 valves or a greater number.

11. For plants with 400 more valves or has at least 100 employees, at the start of the third year of the QIP those pumps or pump seals with poor performance shall be replaced with superior emission performance pumps or pump seals as described in the following paragraphs. Superior emission performance covers pump or pump seal technology combined with appropriate process, operating, and maintenance practices resulting in less than 10% pump leakage. This may include installation of multiple seals, pump replacement material, or design changes.

a. Pumps or pump seals shall be replaced at the rate of 20% per year based on the total number of pumps in light liquid service. Pump replacement will continue until all pumps are determined to have superior performance technology.

b. The replacement of pumps or pump seals with superior emission technology may be delayed until the next process unit shutdown, but the replacement number must be equivalent to the 20% or annual replacement rate.

c. The pumps shall be maintained as described in the quality assurance program.

10.1.6 Compressors (HON)

These regulations are applicable to compressors in HON service or VHAP service. A list of VHAP compounds presented in Appendix 9A (Chapter 9) indicates that any of these compounds used as a reactant, intermediate, or final product processed through a compressor requires conformance of the compressor with the following HON regulations:

1. Each compressor must be equipped with a seal system that includes a barrier fluid system preventing leakage of process fluid to the atmosphere.

2. In accordance with the seal requirements in item 1, the seal system will also meet *one* of the following requirements:

a. The seal is operated with the barrier fluid at a pressure greater than the stuffing box pressure.

b. The seal is equipped with a barrier fluid system that is connected through a closed vent system to a control device.

c. A closed-loop system purges barrier fluid directly into the process stream.

3. The barrier fluid must not be in light liquid service.

4. Barrier fluid systems will be equipped with a sensor that detects failure of the seal system, barrier fluid system, or both.

5. The sensor required in item 4 must be observed daily or must be equipped with an alarm. Based on design considerations and operating experience, a criteria is established that will indicate failure of the seal system, the barrier fluid system, or both.

6. Should the sensor indicate a failure of the seal system, barrier fluid system, or both, based on item 5, a leak is defined or detected.

7. When a leak is detected it shall be repaired within 15 calendar days after detection. A first attempt at repair must be undertaken within five calendar days after detection. Repairs can be deferred until the next planned shutdown, if a compressor repair requires a process shutdown.

8. A compressor is exempt from the requirements of items 1 and 2 if the compressor is equipped with a closed-vent system capable of capturing and transporting any leakage from the seal to a control device.

9. When a compressor can operate with seal leakage at less than 500 ppmv above the background level, the machine does not have to conform with the requirements shown in 1–8. However, the machine leakage tests are designated by the EPA and are generally conducted a short time after startup and annually. Other tests can also be required.

10.2 EMISSIONS

In the last decade actual emission factors for pump seals have been difficult to ascertain. Through this decade environmental regulations have become more stringent and federal, state, and regional requirements have reduced overall emissions. As a result, pump emissions have been reduced and are characterized by significant changes in the average pump emission factors. However, regulatory changes have also been accompanied by changes in sealing equipment, reduced monitoring time periods, and more severe maintenance requirements. The maintenance and monitoring programs have been termed LDAR for leak detection and repair. Consequently, pump emission factors have not been adequately evaluated for characteristics such as process conditions, dual seals, high and low barrier fluid pressures, pump revolutions per minute, pumping seals, auxiliary seals, cartridge seals, seal face materials, and sealing face widths. This section will provide some background information on pump emissions and recommendations for estimating emission factors.

10.2.1 Emission Factors

Emission factors for various methods of calculating emission losses are described in Chapter 5. Within each specific emission calculation procedure, the actual emissions data are determined on a statistical basis along with the monitoring data and the overall emissions basis is then calculated through one of several approved calculation procedures.[6,7] The existing plant emission measurement programs indicate current emission losses and establish average overall emission factors if desired. However, these data rarely define pump seal losses for specific types of seals or other variables. As a result, the emission factors are difficult to apply in predicting emissions factors and losses for new or modified pumps.

One approach in predicting emissions losses and factors is the application of the emission factors determined through a field test program for new or modified pumps. However, this technique requires identical LDAR program requirements to ensure the correct application of the emissions factors. Frequently, field data are based upon LDAR programs that differ significantly from the programs required in new or modified installations. Moreover, predicting emission factors resulting from regulatory changes can be difficult. The EPA in the HON and NSPS regulations essentially eliminates monitoring requirements for dual mechanical seal pumps that meet the EPA installation requirements, by assuming emissions factors for these pump seals are essentially nil. Visual monitoring of the mechanical seals is the only requirement that can be of assistance where the fluid does not vaporize. In a light liquid service pump, fluid volatility may preclude an accurate visual assessment of leakage. Some approaches to defining pump emission factors are presented in the following sections.

10.2.2 Sealing Effectiveness

Since various emissions data sources have not determined the emissions rates for specific sealing systems, a comparison of different sealing configurations is helpful for qualitatively selecting a sealing system for installation. Limited information from reference 8, presented in Table 10.2, indicates that the double mechanical seal has a very low emissions rate, although a bellows seal with auxiliary packing was presumably a superior seal. There have been significant seal improvements since the test that formed the basis for the qualitative comparisons in Table 10.2, and the index for the mechanical seals may change significantly.

In addition to changes and improvements in the basic seals described in the table, new seal designs have been commercialized and improved auxiliary seals are available that are designed to control emissions. An important development in seals has been the introduction of pumping seals, which are combined with an outboard seal to provide an essentially tandem seal configuration. The inner or pumping seal has a face plate with a shallow configuration designed to provide some pumping action of the barrier fluid from the seal system into the seal box process fluid, thereby preventing any process fluid flow into the mechanical seal chamber. Fluid leaking from the outboard seal will not contain any process fluid. Barrier fluid flow through the pumping seal is extremely small, although in pump services where the process fluid cannot be contaminated pumping seals may not be desirable. This seal, however, can be installed in a wide variety of pump services.

TABLE 10.2 Centrifugal Pump Emission Rate Comparison[a]

Seal Type	Emissions Rate Index
Packing without sealant	100
Single mechanical seal, flushed[b]	1.2
Single mechanical with auxiliary packing	1.2
Tandem mechanical seals[c]	0.15
Double mechanical seals[d]	0.004
Bellows seals with auxiliary packing	nil

[a]Adapted from reference 8.
[b]Self- or external flushing.
[c]Low barrier fluid pressure.
[d]High barrier fluid pressure.

An alternate to the tandem pumping seal installation has been developed by the manufacturer of the pumping seal. Two back to back pumping seals are installed with an inert gas buffer fluid. The gas, acting as a lubrication film, is pumped at a very low rate into the process fluid and prevents the process fluid from leaking into the inner volume between the seals. The buffer gas is frequently nitrogen, which does not react or contaminate the process fluid. In most applications, the inert vapor is separated from the process fluid in the processing system. A very small leakage flow of inert vapor is also discharged to the atmosphere from the outboard seal. This seal installation does not have any emissions and has a long operating life, minimizing maintenance and leakage that accompanies seal failures.

During the 1980s significant improvements in canned and magnetic pumps along with more restrictive regulations resulted in the installation of these pumps in various chemical and petroleum services. Their major advantage is the pumps do not leak and monitoring or LDAR programs are not required under existing NSPS regulations or the proposed HON regulations. The proposed HON equipment regulations including pumps will become law since they were a joint development of industry and the EPA. The pumping seal, the dual gas lubricated seals, and the canned or magnetic pumps will probably find increased acceptance as the emissions regulations become more restrictive.

10.2.3 Leak Frequency

The leak frequency or the number of pumps that exceed the specified regulatory leak definition varies in accordance with the leak definition. Emission data from the EPA for ethylene oxide and butadiene process facilities investigating leak frequency indicate that the number of leaking pumps increases as the leak definition level decreases, which is shown in Table 10.3. However, in certain process units none of the pumps leak, but the type of pumps and seals have not been described. In other units the leaking percentage does not change as the leak definition decreases, which indicates the other pumps in the units leak at levels less than 1000 ppmv. These data cover process conditions that vary from 10 to 288°C and 136 to 3550 kPa or 50 to 550°F and 5 to 500 psig, with line sizes that vary from 0.5 to 6.0 in. nominal pipe size.

TABLE 10.3 Pump Seal Leak Frequency at Different Leak Definitions[a]

	Leak Definition Level (ppmv)				
Process Units	500	1000	2000	5000	10,000[b]
Ethylene oxide	23	18	14	9	
	36	27	9	9	
	0	0	0	0	
Butadiene	24	21	15	13	
	10	10	10	10	
	2.9	0	0	0	
Acrolein	4	4	4	4	
Average SOCMI pumps					9

[a]Adapted from reference 5. Percentage of leaking pumps is given for each level.
[b]Column for SOCMI pumps only. Process units not investigated at this level.

The data on these units were obtained from the CMA and included information on plant work practices, operations, and equipment design specifications. However, the leak frequencies reported were not correlated with the operational and design details available from the CMA. The EPA concluded that the recent pump leak frequency data obtained from the CMA demonstrates that pump leak frequencies are considerably lower than the leak frequencies gathered during EPA studies conducted in the 1970s and 1980s. According to the EPA, CMA data confirms that effective leak control involves a combination of factors since specific process parameters or equipment design do not correlate with leak frequencies.

Pump leak frequency data from the Texas Air Control Board[5] for cumene, phenol/acetone, and butyraldehyde/butanol processes in an LDAR program with a 500-ppmv leak definition level covered a period extending from 1 to 6 quarters. During this program, the average pump leak frequency varied from 0 to 20% in the first period and from 3.6 to 5.0% in the second period.[5] Obviously, more restrictive LDAR programs reduce the leak frequency at low leak definition levels.

During the development of the pump HON regulations, the committee reviewing the data investigated leakage data from single mechanical seals (SMS) and dual mechanical seals (DMS). While the results have not been published, DMS monitoring data indicated emission concentrations of less than 500–1000 ppmv. The SMS pumps had monitoring concentrations that varied from < 100 to about 9100 ppmv, with the majority of pumps in the 100- to 9100-ppmv range for those pumps $< 10{,}000$ ppmv in performance. Consequently, actual leak performance of SMS pumps is somewhat unpredictable, requiring higher maintenance costs to ensure low leak frequencies.[5] Moreover, improvements in auxiliary sealing equipment and cartridge seals may significantly reduce SMS leak frequencies. These various test results cannot be adequately evaluated until the data become widely available.

Leak frequency prediction is rather difficult, but the reported results by the EPA indicate that the dual mechanical seal leak frequency will be considerably less than the 10% target value in the regulations for the 1000-ppmv Phase III leak definition level. However, satisfactory performance of any seal system requires careful operation and maintenance.

10.2.4 Emission Factors

In Chapter 5 emissions were determined based upon data obtained from an operating process unit with various factors for certain conditions such as leak/no-leak or stratified emissions. When an operating unit has pump monitoring concentrations that meet the HON emission requirements, the emission data from the existing unit can be used as a basis for determining emission rates and leak frequency. The leak frequency is of particular importance since the regulations specifically address this variable based upon a 1000-ppm leak detection level. For a new unit with pump designs and seals equivalent to or an improvement upon the equipment in the existing unit, the emission and frequency data from the existing unit that meets EPA regulations can be used. When a new unit is constructed and emission data from comparable existing units are not available, a new emissions estimate is required. One approach for estimating emissions is presented in the following paragraphs.

TABLE 10.4 Uncontrolled Pump Emission Factors[a]

	Emission Factors (kg/hr/source)		
Service	Refineries	Chemical Plants	Chemical Plants, No Ethylene Units
Light liquid	0.114	0.0495	0.0397
Heavy liquid	0.021	0.021	0.0225

[a]Adapted from reference 9, 10.

The EPA has established a series of uncontrolled emission factors, and in various states and regions emission factors have also been defined. In addition, adjustments are provided for various LDAR programs based on the uncontrolled pump emissions. Uncontrolled pump emissions that were developed by the EPA and then adjusted for certain processes are shown in Table 10.4. These uncontrolled factors, when used as average pump emission factors, result in very large overall emissions values. A method of establishing emission factors for a more realistic evaluation of emission losses is based upon adjustments to the uncontrolled emission factors. As an example, for an LDAR program based on monthly monitoring of light liquid pumps, the uncontrolled emission factor can be reduced by 62%. Where the LDAR basis is quarterly monitoring, the uncontrolled emission factor is reduced by 33%. While these adjustment factors may not be applicable in all locations, states and regional authorities have developed adjustments factors.

Since the uncontrolled emission factors were developed for various pump designs and seal systems, it is difficult to assign specific values to a seal category. In addition, leak frequency is rather general, as shown in Section 10.2.3. A combination of leak frequency data based on Table 10.3 and general comments in reference 5 can provide a basis for establishing an average leak frequency for a 1000-ppmv leak definition level. Although average, a stratified emission factor in the 0- to 1000-ppm range can then be used as a general factor. These factors are shown in Chapter 5 and in references 10 and 11.

Where all of the pumps are dual mechanical seal pumps, the average emissions monitoring concentrations are less than 500 ppm.[5] However, a leak frequency has not been established but will be less than 10% of the pumps in a unit or less than 3 pumps, which are the criteria for the HON regulations. For a leak frequency of 1 in 200 pumps, which is not uncommon for DMS seals that are properly maintained, an estimated emission factor can be developed based on an emission concentration level for the leaking pump and either a stratification factor or the EPA correlation for pump seals.[7] A leaking pump in the HON regulation must be repaired at 2000 ppmv. For the nonleaking DMS pumps, concentrations of 10–50 ppmv at the seal can be converted to leak rate with the EPA correlation equation. Emissions for other pump seal types should be based on field data for leak frequency under the required LDAR standards.

General recommendations for seal installations have been developed based upon process liquid specific gravity and emissions requirements.[13] For services where the liquid specific gravity is less than 0.40 gm/cm^3, double mechanical seals are recommended to control emissions to very low levels. Where the liquid gravity is greater than 0.40, tandem or dual seals are recommended for emissions control concentration between 0 and 500 ppm. Single seals are recommended for emission control concentrations between 500 and 1000 ppm. However, single mechanical seals have also been effective to emission concentration levels that are less than 500 ppm. Tandem seal emission concentrations can reach the nil level where the buffer fluid has vapor recovery or control,[13] which is required in the regulations. Double seals have a higher buffer liquid pressure level than the process fluid in the stuffing box and the VOC or HON emissions are also nil.

10.2.5 Control System Considerations

Regulatory requirements, particularly for the HON regulations, reduce emissions to the atmosphere and potential exposures. In contrast, the NSPS regulations based on a 10,000-ppmv leak monitoring concentration level permit larger emission rates than the 1000-ppmv leak concentration level in the HON regulations. Further regulatory restraints in the HON regulations limit the number of pumps that may exceed the allowable leak level (1000 ppmv). Excessive numbers of leaking pumps require a QIP to control the number of leaking pumps, which also controls leakage well below the levels observed with 10,000-ppmv regulations.

These variations in regulatory constraints affect emissions and maintenance control efforts and affect potential personnel exposure situations. In general, the basic emission control programs revolve about the leak detection and repair programs, which are the starting point for emission controls. Such LDAR programs not only require monitoring and maintenance, but also include extensive recordkeeping requirements, resulting in expensive control programs. The installation of dual mechanical seals (DMS) to control emissions rather than LDAR programs appears preferable based on the emissions control superiority of the DMS systems. The low emissions rate indices for dual mechanical seals presented in Table 10.2 are considerably lower than the indices for compared single mechanical seals. However, new auxiliary seals and seal improvements have changed this relationship, requiring detailed knowledge of seal performance to ensure a satisfactory seal emission control system is installed. Cost comparisons also enter seal selections and other factors that must be considered in pump seal comparisons are seal life and repair or maintenance costs.

While sealed pumps require careful investigation, seallesss pumps such as canned or magnetic pumps that do not have shafts extending through the pump housing are exempt from regulatory requirements. Installation of these sealless pumps in new facilities can be considered where applicable and as replacements for existing pumps that are difficult to seal. However, some limitations exist with canned or magnetic pumps and these are reviewed in another section of this chapter.

Some economic considerations that are related to the various control systems described in Section 10.3 are presented in the following paragraphs. An overall economic basis for consideration of pump control systems is presented in Section 10.4.

LDAR Program. As described in Section 10.1, pump seals are monitored every month to meet NSPS and HON regulations. Repairs must be initiated within 5 calendar days after leak detection in both regulations. The monitoring effort may be less frequent under some state VOC programs (SIP), possibly every 3 or 6 months.

Although the monitoring and repair requirements are not in themselves particularly troublesome, all programs require manpower to monitor, maintain, and document emission data and maintenance records. Documentation is a major consideration since all pump leakers must be recorded along with information on repair initiation, completion, and acceptance after repair. Normally, a team is required for mandated valve LDAR programs and a pump LDAR effort is an addition to the basic valve LDAR program. Consequently, the economics or cost of a pump seal LDAR program versus installation of dual seals is not a straightforward comparison. When an LDAR program is not required by law, a valve LDAR effort will probably be a consideration since valve emissions are a much greater percentage of fugitive emission losses than pump seal emissions. However, nonattainment areas require emission controls and with the new HON regulations, which will be broadened with additional regulations for petroleum refineries, few plant sites will be exempt from regulatory oversight.

Dual Seals. The NSPS or HON regulations do not require monthly monitoring of pumps with dual seal installations. Although this change essentially eliminates the LDAR program, leaks and repair records must be maintained in accordance with regulatory requirements. As described in the previous paragraph, record cost is real, but is an incremental effort for an established LDAR team and program.

Installation of dual seals is costly since an auxiliary barrier fluid system is required and sensors mandated by law in the NSPS and HON regulations must be installed to indicate failure of the inner or outer seal. Dual seals and barrier fluid systems are described in Section 10.3. In addition to the large auxiliary dual seal installations, dual seals are more expensive than single seals and are considered more difficult to install in pump stuffing boxes. Cartridge seals eliminate many of the mechanical seal installation problems by inserting a completed assembled single or double seal system. This development has been a considerable improvement in installation compared to building a seal system since it places properly mounted and aligned seals in a stuffing box. However, dual seal repair requires a considerable maintenance effort and the auxiliary barrier fluid facilities require maintenance of instruments and equipment. The costs associated with these considerations can be determined in conjunction with design and maintenance groups to provide an economic basis for comparison with other alternatives.

In addition to emission control characteristics, dual seals provide added protection from a safety standpoint. Should a single seal "blow out" with a flammable liquid, fire may occur and exposures are a major concern. The added seal provides some safety factor, including the situation where an outboard seal is lost, but the inner seal will contain the process fluid until the pump is shut down. An auxiliary seal has a certain amount of leakage and does not completely control emissions. Consequently, dual seals provide an added safeguard.

Dual noncontacting gas lubricated seals have a long life and minimize the possibility of blown seals. Moreover, eliminating most of the auxiliary equipment and extending seal life significantly reduces maintenance costs. These seals are described in Section 10.8.3.

Sealless Pumps. The installation of sealless pumps exempts a facility from LDAR programs and the documentation normally required. However, installation of a new pump can be costly in an existing facility due to piping changes and must be compared in detail with other alternatives. While detrimental comments were issued by the EPA in 1982 about sealless pumps,[12] improvements to sealless pump designs by the pump manufacturers have increased their attractiveness for installation in many services. However, pump repair and maintenance costs for these pumps should be included in any evaluation. If the economics are not attractive for replacing all of the standard single-seal centrifugal pumps with sealless pumps, then the installation of a sealless pump should be considered for a particularly toxic service or as a replacement for a pump with a particularly troublesome seal. Sealless pump economics are reviewed in Section 10.3.

Closed Vents. The NSPS and HON regulations exempt pumps from LDAR requirements if a closed-vent system captures the seal leakage and transports the emissions to a control device. This system requires enclosing the opening around a seal and either purging the emissions with a gas into a flare, incinerator, or other device, or installing a blower to move the vapors into the control device. This general approach for control of pump emissions is unusual and an acceptable design must be developed prior to any economic evaluation. The design must be reviewed with safety personnel to insure the design is safe and operable. In addition, maintenance personnel should review the design to determine whether specific design modifications are required, and the entire system should be included in any economic evaluation of alternatives. A further complication is that closed-vent vapors must be sent to a control device under all conditions, and shutdown of a control device (flare, incinerator, etc.) requires an alternative disposition point for the emissions. Installation costs for equipment necessary to meet these requirements will increase investment.

In general, ventilation is not a solution to an emission control problem since the ventilation system only moves the vapors from one location to another. Emissions from a leak in a building are gathered into the ventilation inlet opening and are normally discharged to the atmosphere at the ventilation system exhaust port in a different location. Emission releases to the atmosphere are not acceptable to the EPA and this stream must be controlled to remove the VOC and toxic components from the vapor before the stream can be discharged to the atmosphere.

Vapor Recovery. The various systems described either reduce emission losses or destroy the fugitive emissions. Any economic comparison should consider the debits and credits associated with the destruction or recovery of the emissions. Moreover, utilities consumption changes, such as a change in pump efficiency or the addition of a blower, are a consideration for the various control systems and must be included in any economic evaluation.

10.3 PUMP SEALS

The previous sections described emission losses, emission factors, and regulatory requirements to minimize VOC and HON emission losses from pump seals. Various types of seals and barrier fluid auxiliary systems are presented in this section to indicate potential sources of workplace exposure and methods of controlling and minimizing exposures. This review will provide the industrial

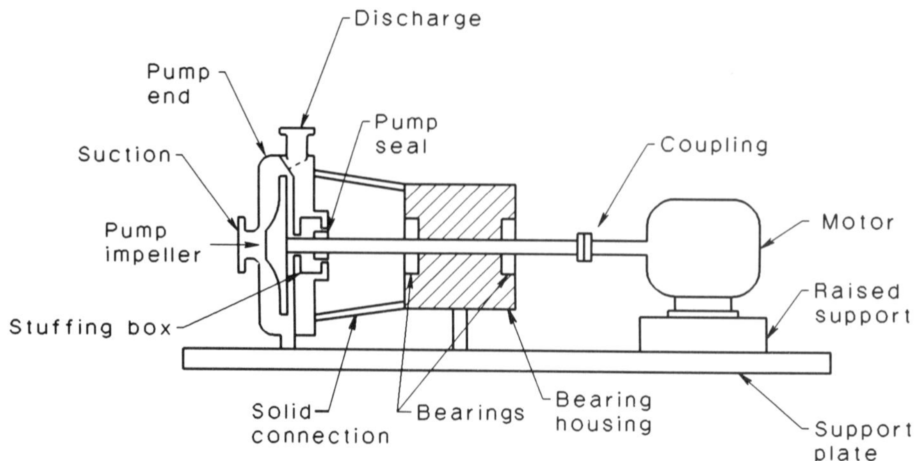

Figure 10.2 *Simplified sketch of a horizontal centrifugal pump.*

hygienist with a basis for selecting the type of control system necessary to meet workplace exposure criteria.

The majority of pumps in the chemical and petroleum industries are horizontal centrifugal pumps and the information in this section concerns these pumps. Much of the seal information is also applicable to vertical centrifugal pumps or the vertical centrifugal in-line process pumps. A simplified sketch of a process pump typical of horizontal centrifugal pumps is shown in Figure 10.2 and consists of the pump end, bearing housing, motor, and support plate. The term pump generally refers to both the pump and bearing housing, which are purchased as a complete item from the pump supplier and may or may not include the motor. The pump and bearing housing are mounted on the support plate, which is normally supplied with the pump. A pump seal is generally specified by the purchaser to meet anticipated pump service conditions.

1.3.1 Packed Seals

The packed stuffing box is the oldest and simplest type of pump shaft seal and a typical seal installation is shown in Figure 10.3. The stuffing box cavity normally consists of a number of packing rings, a restrictive throat bushing on the impeller side, and a follower or gland, but does not always have a lantern ring. The packed stuffing box normally leaks at a small flow rate between the packing and shaft sleeve to provide lubrication for the packing. A shaft sleeve covers or encloses the main shaft to protect the main shaft from scoring.

The actual leak rate from packing glands varies considerably and is generally a function of the pumped fluid and pump service conditions, the condition and type of packing, the mechanical condition of the pump, and the gland adjustment frequency. In addition, a water or stream quench may be sprayed on thc gland face to flush leakage away from the seal face or absorb any leakage material.

A hollow spacer or lantern ring is often provided in the stuffing box to permit the injection of a lubricant or sealant. As shown in Figure 10.3, the sealant or

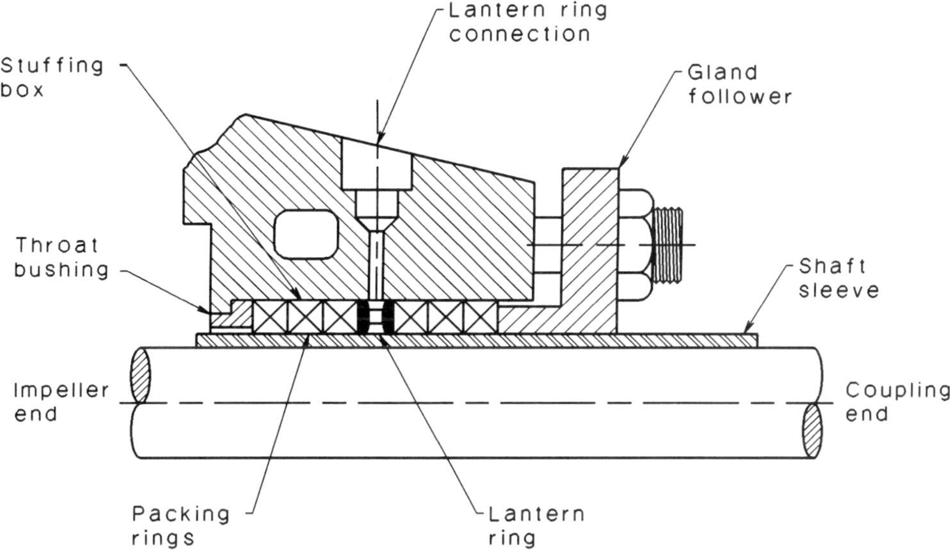

Figure 10.3 *Typical packed-pump stuffing box.*

lubricant is added through the lantern ring connection and dispersed against the shaft where it is distributed. The lubricant or sealant is added to a pump where the packing has poor lubricity or the pumped material is potentially corrosive, toxic, or a foulant. Lubricant or sealant is also added in services where the pump fluid contains solid particles or a positive seal pressure is needed to operate the pump under vacuum.

The injected lubricant flows toward both the throat bushing and packing gland to seal and lubricate the packing through adjustment of the gland follower. However, there will be liquid leakage at the gland and selection of a lubricant with a relatively low toxicity is important. Dripping and maintenance from the gland can result in skin or dermal exposure to maintenance personnel. In addition to liquid leakage, packed seal vapor emissions are considerably larger than the SMS emission seal index shown in Table 10.2. An approximated lubricated packing index is about 10.0

Packed seals still exist in a number of older pump installations, although mechanical seals have largely replaced many packings in seal applications. Packing emissions potentially present leakage problems in monitoring and maintenance situations. Replacement of packings with mechanical seals to reduce the number of leaking pumps is also an opportunity to minimize potential workplace exposures.

Variations of the conventional packed stuffing box are now available in modified pumps from the pump vendors and leakage or emissions may be considerably lower than the packed emission rates measured by the EPA. Moreover, new packings such as graphite and polymers are commercially available and may be superior to the older style packings in controlling emissions. However, little data are available on emission rates from the new packings and their general emission control efficiency compared with older packing emission factors.

10.3.2 Single Mechanical Seals

Single mechanical seals are widely used in the chemical and petroleum industries across a broad range of applications. Two seals may be combined and the installation is termed a dual mechanical seal (DMS). The combination is described in Section 10.3.3. Mechanical seals are considered superior to a packed stuffing box, as shown in Table 10.2, although seal emission data vary considerably. However, mechanical seals normally have relatively low emissions rates when properly installed. Much of the contradictory information on seal leakage may be a result of poor seal installations or errors made by maintenance personnel during maintenance or repairs.[18] New seal modifications and innovations such as cartridge seals have significantly improved installation procedures, resulting in improved performance and emission control.

A simplified drawing of a single mechanical seal is shown in Figure 10.4. An element is mounted on a shaft and rotates against a stationary member. The faces of the rotating element and the stationary member are pressed together by spring force and hydraulic pressure in the stuffing box, resulting in a low leak seal. A thin film of the liquid from the stuffing box forms between the faces, providing lubrication of the seal faces. As shown in the figure, a number of springs create a force on the rotating element. The rotating element and the face of the stationary member can be composed of a variety of materials.[13]

The stuffing box has a flush connection which permits flushing the box to improve seal operations under various conditions. Flushing systems are shown in API 610[14] and are selected in accordance with seal box operating objectives. Generally seals are flushed for the following reasons:

1. To prevent vaporization in the stuffing box by increasing stuffing box pressure, which also requires a close clearance throat bushing.
2. To minimize entrance into the stuffing box of solids in the pumped stream.
3. To lubricate seal faces when the pumped fluid is a poor lubricant.
4. To cool the seal parts.

The type of flushing material is determined based upon stream characteristics and the purpose of the flushing liquid. In applications with clear liquids, a small stream from the pump discharge is recycled as a flush, and in other services a separate flush liquid is required. A flush liquid is generally not required where the pumped stream is a light, clear liquid and the stream temperature will not create seal problems. The flushing liquid system determines the design of the throat bushing opening.

An important aspect of seal design is the gland, which normally has a throttle bushing in the petroleum refining industry. The throttle bushing is provided to minimize leakage if there is a complete failure of the seal. The throttle bushing is required in API 610[14] for refinery service and is optional in the chemical industry, but a throttle and gland plate are recommended where a safety or potential hazard may exist. The throttle bushing will not prevent leakage during normal operation and emissions will not be reduced. However, the throttle bushing will minimize leakage during a complete seal failure, preventing excessive exposure.

The throttle bushing is normally bronze, but can be replaced by two rings of packing, a floating bushing, or a polymeric bushing, depending on plant service and

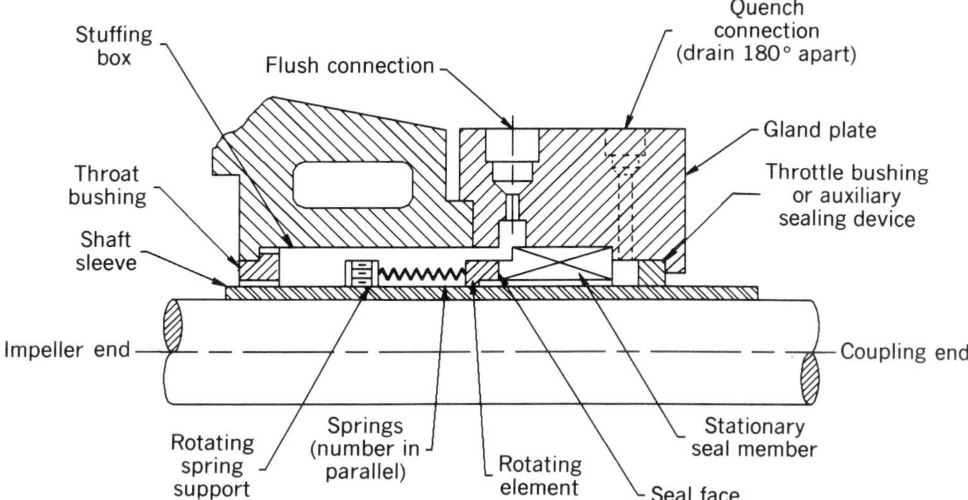

Figure 10.4 *Simplified sketch of a single mechanical seal.*

maintenance experience. The throttle bushing can also be replaced by a polymeric lip seal which maintains contact with the shaft and minimizes leakage along the shaft during normal operation. The lip seal generally has a temperature limitation of about 400°F (204°C) but can be used in a large number of applications. Another development is an auxiliary seal, similar to a lip seal, that will expand in the event of a mechanical seal failure and completely contain the pump contents. This emergency is generally placed outboard in most installations, but has also been located between the mechanical seal and throttle bushing to provide maximum containment and minimum exposure during high leakage situations, with particularly toxic materials. A more extensive review of lip and auxiliary seals is presented in Section 10.3.9.

Flushing the seal box was previously described, but the gland plate is also flushed for various reasons. Flush material is connected to the quench opening and a drain connection at the bottom of the gland permits closed drainage of the gland cavity. Steam, water, oil, or a chemical may be used, but drainage containment meets HON regulatory requirements and is preferred to minimize emissions, although some drain systems may be open, depending upon the quench (flush) media and the toxic characteristics of the leakage material.

Drainage from the gland plate, shaft, or seal may drip directly upon the base plate if the drainage is not piped to a collection point. For containment of liquid leakage, the base plate should also be considered in general drainage considerations. API 610[14] specifies base plates with a slight slope and a rim or lip to contain any fluid, thereby permitting flow of the retained fluid to the low end of the base plate where it is drained through an opening. While leakage should be contained, various situations arise where liquid material drains from the pump or bearing housing to the base plate and the plates should be designed for drainage. Chemical pump base plates are also constructed in one piece, but do not necessarily have rims to contain leakage, which may also consist of oil from the bearing housing. Without a gland plate against the seal, pump leakage will discharge directly onto the base plate. A collection pan can be placed against the seal, but emissions will not be contained. The gland plate permits drainage through a closed pipe system to a container or closed toxic liquid sewer connection, which minimizes exposures to vapors and liquids. A rimmed base plate with drainage will contain leakage of used bearing lubricating oil to the pad or other leaking liquids, which will reduce potential skin exposures.

10.3.3 Bellows Mechanical Seals

A major improvement in seal technology was the development of the bellows seal. A simplified drawing of a single bellows seal is shown in Figure 10.5. The bellows is an unbroken, expandable, or flexible metal joint that does not have an opening in the bellows and is sealed to the rotating element and the retaining ring. The bellows acts as a single large spring and provides the force necessary to press the rotating element against the stationary member. This contrasts with the spring seal, which has a large number of springs that provide the force and pressure for the rotating element. The bellows seal eliminates the need for movement of the rotating element along the shaft, which is required for the spring type seal and may result in misalignment.

The spring seal also requires an O ring at the interface of the shaft sleeve and the rotating element to prevent leakage along the shaft. The bellows provides an unbroken face to the fluid in the stuffing box and the only leak path is through the seal faces. This is an advantage when pumping particulate laden fluids, where centrifugal action prevents accumulation of the particles between the sawtooth recesses of the bellows and migration to the shaft and rotating element.

10.3.4 Dual Mechanical Seals

Dual mechanical seals refer to either two opposed mechanical seals on a shaft (double) or two mechanical seals in series, indicating the seals are aligned in one direction (tandem). Simplified drawings of these two seal configurations are shown in Figures 10.6 and 10.7. A liquid barrier fluid fills the cavity between the seals,

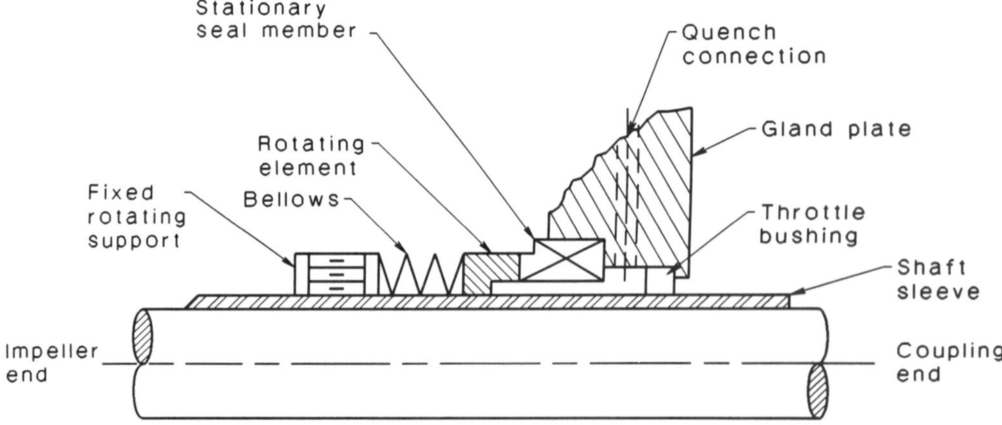

Figure 10.5 Simplified sketch of a single bellows mechanical seal.

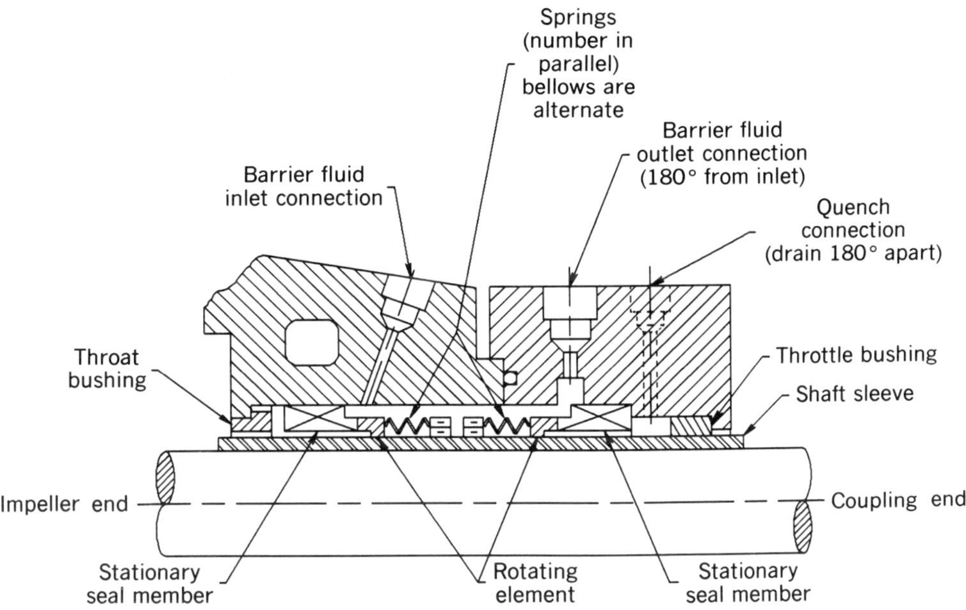

Figure 10.6 *Simplified sketch of a double mechanical seal installation.*

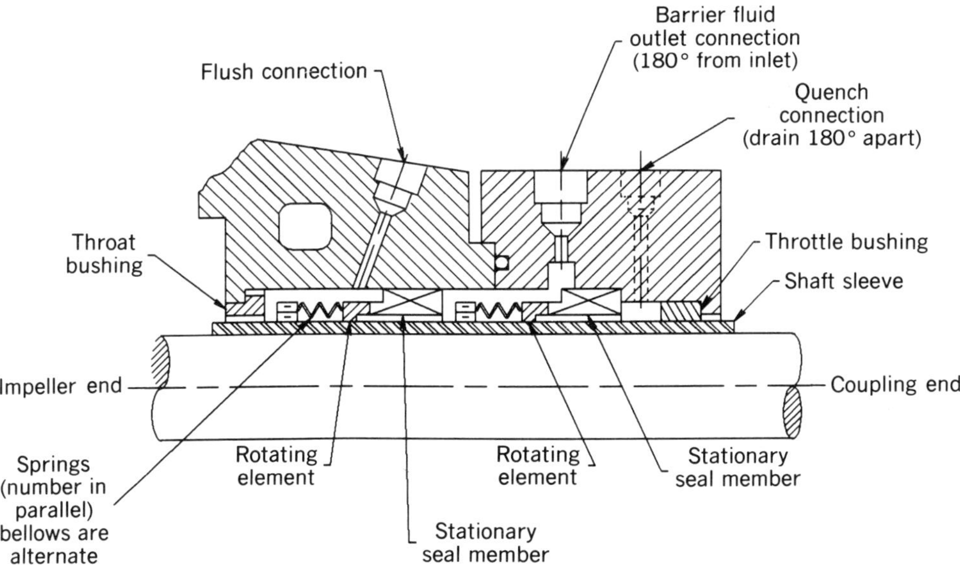

Figure 10.7 *Simplified sketch of a tandem mechanical seal installation.*

with the barrier fluid pressure for the double seal installation higher than the pumped fluid pressure in the stuffing box. In tandem seals, the barrier fluid pressure between seals is lower than the pumped fluid pressure. The API 610[14] also specifies an end plate with a throttle bushing to minimize leakage in the event there is an outer seal failure.

The fugitive emission data for dual seals published by the EPA is somewhat contradictory on actual emissions. Uncontrolled fugitive emission factors for single and dual mechanical seals are identical in the EPA AP-42 emission data book,[15] which has been revised, but never changed. However, published emission factors as shown in Table 10.2 and reference 13 indicate a lower emission factor for double seals compared with the single-seal emission factor. The EPA indicated in an older document that the control techniques for dual seal pumps result in 100% control efficiency in chemical plants and petroleum refineries. Fugitive emission leakage for dual seal pumps is essentially eliminated according to these older data, but for practical operation, dual seal emissions are essentially nil where the seals are installed properly and have vapor recovery systems.

The difference or improvement in dual performance is apparently credited to the requirements for dual seal controls in the final NSPS fugitive emission (VOC) regulations described in Section 10.1.1. These are quite similar to the HON dual seal requirements in Section 10.1.5. The three major changes that apparently resulted in the essential elimination of emissions, which are also applicable in HON service, are:

1. The *barrier* fluids are either heavy liquids or are not in VOC or HON service. If the barrier fluid is a VOC or HON fluid, only one seal separates the fluid from the atmosphere and the emission factor for a single seal is applied. A non-VOC or non-HON barrier fluid results in two seals between the pump fluid and the atmosphere.

2. The barrier fluid is operated at a higher pressure than the pump stuffing box pressure in double mechanical seals. This results in some leakage into the process fluid if the seal on the inboard or impeller side leaks, and leakage of non-VOC or non-HON fluid to the atmosphere if the outer seal leaks.

3. If the barrier fluid is not at a higher pressure (tandem seals) than the pump stuffing box pressure, a barrier fluid degassing reservoir is required that is vented to a control device. Should the seal on the impeller side leak, pump fluid flows into the lower pressure barrier liquid and the barrier fluid is vented through the reservoir, which is essentially at atmospheric pressure to a control device. The outer seal remaining intact prevents emissions to the atmosphere. Leakage of non-VOC or non-HON barrier fluid to the atmosphere through the outer seal will not affect the VOC or HON emissions factor.

The industrial hygienist can use a 100% control efficiency factor or essentially zero emissions with heavy liquid or non-VOC and non-HON barrier fluids when estimating fugitive emission leakage from dual seal pumps. Tandem and double seals are more expensive installations than single seals because of multiple seal costs and the auxiliary equipment associated with barrier fluid systems. Also, maintenance costs for the dual seal system can be greater owing to the added complexity of the entire dual seal system.

Typical double and tandem seal systems are discussed in the following sections, but general guidelines for the selection of a double or tandem seal installation by an industrial hygienist to minimize exposures are based on the following considerations:

1. If the pumped fluid should leak through the seal on the impeller side *into* the barrier fluid, the barrier fluid becomes contaminated and leakage through the outer seal can pose a potential health problem when the process fluid is toxic. In this situation only one seal remains as a barrier between the toxic fluid and the atmosphere. While light material can be vaporized in the recovery drum, the barrier fluid is not pumped in a tandem seal and the concentration of toxic material can be increased significantly during an inboard seal leak.
2. If the pumped fluid is a serious health concern, then the barrier fluid pressure should be greater than the pump stuffing box pressure. Inner seal leakage would then result in barrier fluid flow into the process fluid, preventing contamination of the barrier fluid. Leakage of the outer seal is not a problem in this situation.
3. When the process fluid is only mildly toxic, although included in the HAP list, and the major concern is environmental, requiring a general reduction in VOC emissions, a barrier fluid at a pressure less than the stuffing box pressure is satisfactory. A tandem seal is generally applicable for this service, where the process fluid is volatile. For process fluids that have a medium volatility, observing and controlling the increase in volume of the barrier fluid is necessary when the inboard seal leaks.
4. If the pumped fluid is toxic, but contamination of the pumped stream by a barrier fluid cannot be tolerated, then a tandem seal is required with the barrier fluid at a lower pressure than the stuffing box pressure. The emergency outboard seal described in Section 10.3.9 should be installed to contain any contaminated barrier fluid in the event the outer seal fails completely.
5. A dual seal system with an inboard pumping seal or a dual gas lubricated seal system should also be considered in these services. Further information on these seal systems is presented in Sections 10.3.7 and 10.3.8.

10.3.5 Double Mechanical Seals

Double mechanical seals are opposed seals with a barrier fluid pressure that is higher than the process fluid in the stuffing box. Leakage of the process side seal results in the barrier fluid entering the process and contaminating the process fluid. Frequently contaimination can be accepted in various services, but stringent limits on contamination may not permit a double seal installation. In this situation the potential barrier fluid contamination problem must be resolved with operations and design teams prior to specifying a double mechanical seal.

A type of circulating barrier fluid system frequently installed for double mechanical seals is shown as a thermosyphon unit in Figure 10.8. Heat and impulse from the rotating shaft create a circulation of barrier fluid from the storage drum to the pump. The barrier fluid pressure must be greater than the process pressure in the stuffing box and the storage drum is pressured to the desired barrier system

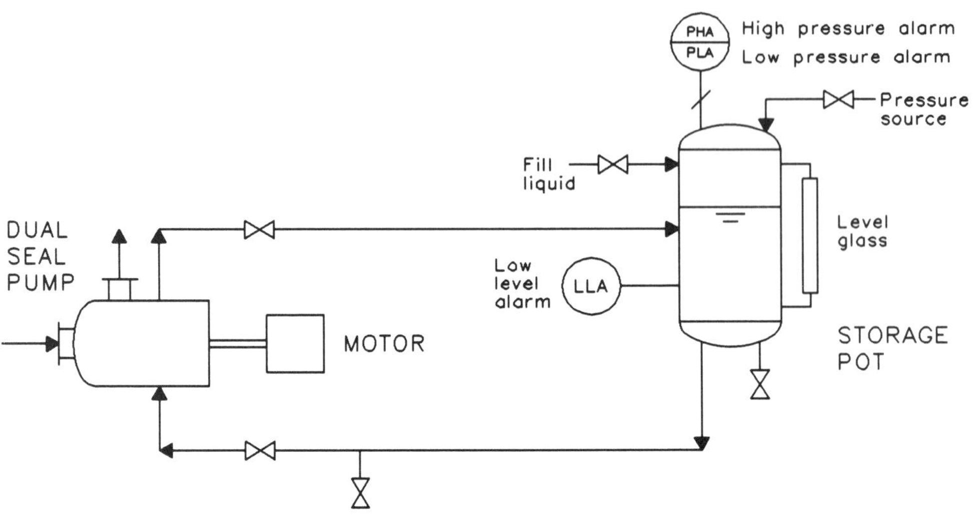

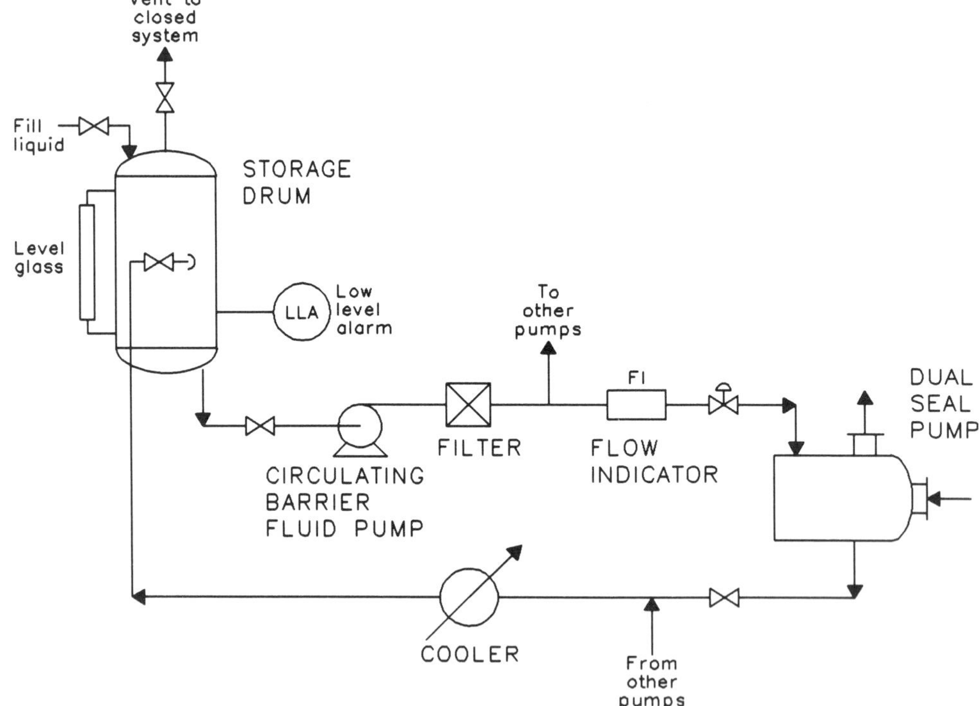

Figure 10.8 *Circulating barrier fluid systems for double mechanical seals.*

pressure with an inert gas such as nitrogen or CO_2. In accordance with NSPS and HON sensor requirements, instruments are also shown on the drum. The low-level alarm (LLA) indicates when a seal is leaking and barrier fluid is lost from the circulation system. A low-pressure alarm (PLA) indicates a fall in pressure and supplements the low-level alarm. These two alarms indicate that one or two seals have failed and are leaking heavily. An external or gland side leak requires another indicator to show that the leak is not the inner or pump side seal. A pressure indicator in the pump gland pocket will show a higher pressure if the outer seal is leaking and an alarm can be activated by the pressure indicator. An examination of the gland drain will also reveal whether the outer seal is leaking, but the NSPS and HON requires an instrument. An instrument can be installed in the drain from the gland but may not function properly where the outboard seal face is flushed. However, an instrument can be selected to determine the presence of process fluid traces in the face flush stream.

Another barrier fluid circulation system installed where there are a number of pumps with double seals is also shown in Figure 10.8. A separate barrier fluid pump circulates barrier fluid through the double seals in the various pumps, a cooler, and a surge or storage drum. A cooler is often installed to control the temperature in this multiple service system. The low-level alarm on the storage drum indicates a seal failure in one of the pumps, but there is a lag in this system due to storage inventory and the alarm is only a general warning. The regulations require instrumenting each pump with an alarm to show when a seal has failed and additional instrumentation is necessary on the pump. A pressure indicator with an alarm on the seal side of the hand control valve will indicate that an inner seal is leaking if a secondary pressure indicator in the pump gland plate does not alarm. When an outer seal fails, both instruments signal an alarm. Some of these instruments can be locally or manually read, but the regulations require alarms. Although several alarm systems are described, others may be preferred and a flow indicator can also serve as an alarm.

In the forced circulation system the surge drum is vented through a block valve. It is unlikely that volatile, toxic material will enter the barrier fluid circulation system and contaminate the storage drum causing the emission of potentially toxic vapors. However, the possibility exists that an unusual circumstance may cause contaimination of the barrier fluid, and the vent should be connected through a closed a line to a control device such as a flare. The EPA does not require a vent connection where the barrier fluid pressure is greater than the stuffing box pressure. Another consideration is that a seal failure can result in a loss of liquid level in the storage drum, causing a reduction in drum pressure. The connection to a control device results in vapor entering the drum from the control device system. Since the storage drum is not large, vapor replacement will be small but the vapors may contain contaminants that require consideration and review with the operational group. The thermosyphon drum is pressured with a gas and if a complete seal failure occurs, an automatic valve on the gas line will maintain sufficient pressure to prevent any major backflow of contaminant into the storage pot.

The storage drums in these systems are under pressure and temperature, requiring ASME coding for the drums. Where the pressure is low, the possibility of an increase in pressure due to changes in the control device will also increase the pressure. In addition, many of the materials in the drum are flammable. The

ASME coded drums significantly increase the cost of the auxiliary barrier fluid system.

Following shutdown, the barrier fluid system must be drained and flushed. An understanding of the toxicity of the barrier fluid and the potential toxicity of containinated barrier fluid is important to insure working personnel are not exposed during this procedure. Permanently installed drain and flush connections are recommended where barrier fluids are toxic, contain toxic additives, or may become containinated.

10.3.6 Tandem Seals

Tandem seals are dual seals installed in series or in the same direction with a barrier fluid pressure that is lower than the process fluid pressure in the stuffing box. Leakage through the process side seal or inboard seal results in contamination of the barrier fluid by the process fluid. A small leak occurring in the outer seal would then potentially transmit contaminated barrier fluid to the workplace atmosphere. In addition to inner seal failure and process material entering the barrier fluid, the barrier fluid system volume increases along with the liquid level in the barrier fluid storage drum. Light components in the barrier fluid storage drum can vaporize from the barrier liquid mixture and, as the liquid volume increases, can overflow the storage drum.

The circulating barrier fluid system normally installed for tandem seals is the thermosyphon unit shown in Figure 10.9, which is similar to the thermosyphon system shown in Figure 10.8 for double seals. The major difference in the two systems is the pressure in the tandem barrier fluid system is much lower than the process liquid pressure in the stuffing box which differs from the pressure balance in the double mechanical seal installation described in Section 10.3.5. The lower pressure in the barrier fluid system requires a vent on the storage drum, but compliance with the NSPS and HON regulations requires a closed vent on the storage pot that is connected to a control device. In addition, the basis for instrumenting the barrier fluid storage pot also changes in the tandem seal system.

Based on the NSPS and HON regulations that require connection of the degassing reservoir or storage pot to a closed-vent system and control device, the storage pot in Figure 10.8 is connected to a closed vent that transfers vapors to a flare, vessel, or other control device. The vent line has a restriction orifice that limits vapor flow and raises the barrier fluid system pressure when the inner seal begins to leak a relatively volatile process fluid. The high-pressure alarm (PHA) is activated in this situation to indicate the inner seal is leaking. A higher system pressure, resulting from the vent flow restriction, tends to reduce the pressure drop across the inner seal and the flow rate of process fluid into the barrier fluid system.

If the pumped fluid has a relatively low volatility, the liquid level in the storage pot will rise and the high-level alarm (LHA) will be activated, indicating that the pump side or inner seal has failed. The high-level alarm becomes very important in this situation since a rapid liquid buildup will overflow the drum and enter the vent system. Liquid in the vent system can present operating problems in other portions of the vent system and should be prevented from entering the overflow line. A permanent drain connection on the storage pot that is normally closed permits

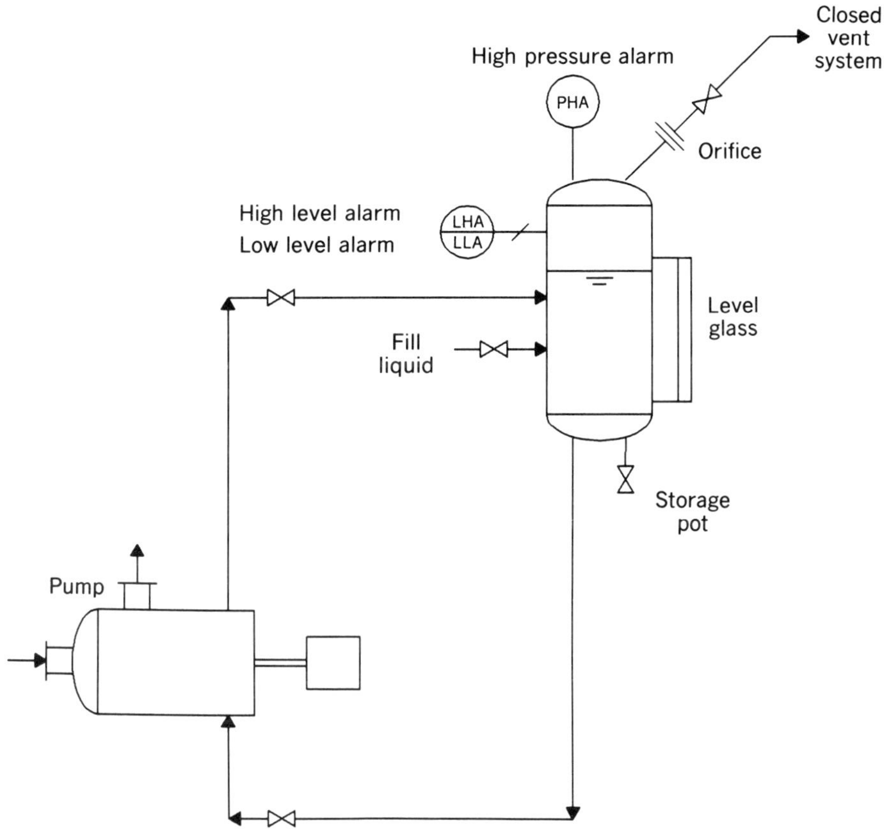

Figure 10.9 *Circulating barrier fluid system for tandem mechanical seals.*

drainage and prevents the overflow condition previously described. A falling liquid level in the storage pot indicates that the outer seal to the atmosphere is leaking and a low liquid level alarm is also necessary. The instrumentation shown on the storage pot and in the circulating barrier fluid system is minimum, but alarms are also required by the regulations.

In the tandem seal system, the barrier fluid will be contaminated with the pumped fluid when an inner seal fails. Leakage through the outer pump seal of contaminated barrier fluid then presents potential skin exposure problems if the pumped fluid is not particularly volatile. A highly volatile process fluid will vaporize in the barrier fluid storage pot with little contaimination remaining. Pumped liquid volatility characteristics in the barrier fluid permit calculation of the vapor/liquid relationship and the concentration of the contaminant in the barrier fluid. Glands with throttle bushings are not required for tandem seals in API 610,[14] but should be purchased with tandem seals to control potential leakage of contaminated barrier fluid through the outer seal. Liquid leakage can then be drained to a collection system. When the inner seal fails, only the outer seal remains as a buffer between contaminated barrier fluid and the workplace.

Contamination of tandem seal barrier fluid is a distinct possibility in the tandem seal system and procedures for draining and flushing the barrier fluid system at shutdown are required to minimize potential workplace exposures. Permanent connections for draining and flushing the system are recommended to prevent HON or VOC emissions and potential exposures to the emissions or liquid.

10.3.7 Pumping Seal

The normal double seal systems previously described have a high pressure barrier fluid which may leak at very low rates through the inboard seal into the process fluid, which is at a lower pressure. In addition, a large differential pressure exists across the outboard seal and an extremely low leakage rate may exist across this seal. Properly installed double seal systems have emissions that are very small when monitored as described in Section 10.2.4.

A new double seal system is now available commercially that is based on a different concept. An inner seal with pumping capability is combined with a standard mechanical seal in the outboard position as a replacement for a typical double mechanical seal installation.[17] In this system, a liquid barrier fluid at low pressure is added to the seal chamber and the inner seal pumps the barrier fluid through the pumping seal faces into the process fluid. Barrier fluid in this system is lost at a very low, finite rate to the process. Through pumping action, the barrier fluid seals the faces and prevents process fluid leakage into the seal chamber and thence to the atmosphere.[16,17]

Description. The inner seal resembles a standard mechanical seal. However, there is a difference in the seal faces, which is characteristic of the pumping seal. In the pumping seal, the rotating face has an outline of vanes etched into the face, which resemble the curved backward vanes normally found in centrifugal pump impellers. The vane outline in the seal face has been termed etched since the vane outline is only a few thousandths of an inch deep.[17]

When the shaft begins to rotate, the etched face gradually begins to lift at 90 rpm, creating a pumping action between the faces that pumps barrier fluid into the process fluid in the stuffing box. Although the rotating face lifts, the lift is limited and a very small opening is created. Actual pumping capacity is very small, but the pumping action is distributed evenly around the seal and the equal pressure distribution effectively seals the chamber from leakage or process fluid entering the chamber. In addition, the narrow opening between the seal faces is lubricated with barrier fluid, which minimizes face contact and prolongs seal life.

Since barrier fluid is pumped into the process stream, the barrier fluid must be compatible with the process fluid and the process system. Although pumping rates are small, some chemical streams can be contaminated if extremely high purity streams are required with minute contaminant traces. In other services, the small pumped capacities can readily be tolerated, such as hydrocarbon services in petroleum refineries. For estimating purposes, pumping rates of about 1–2 gal/day or 0.00069–0.00139 gpm (0.0026–0.0052 Lpm) are apparently average values although figures of < 1 gal/day have apparently been observed. The higher pumping figures provide a more realistic basis for determining the effects of barrier fluid contamination on the process stream and downstream processing. However, the very low flow rates should contaminate few streams. Assuming a 100-gpm pump flow rate for a seal pump rate of 0.000069 gpm, the potential contamination level is 6.19 ppmv at the 1-gal/day leak level.

Regulatory Requirements. The pumping action of the inner seal maintains a low pressure in the seal chamber where the barrier fluid is a heavy liquid or low boiling non-VOC material. Depending upon barrier fluid composition, the pressure in the seal chamber may be a function of the sealpot head, which would be slightly above atmospheric pressure. Since this is a pumping action, there is a possibility that the seal chamber pressure in some situations could be slightly less than atmospheric pressure because of the pumping action.

Since this seal system is basically a dual seal system and has a barrier fluid, it is exempt from monthly monitoring requirements. However, the regulations must be observed to obtain the monthly monitoring exemption and regulatory requirements are outlined as a basis for defining the overall seal system:

- If the barrier fluid pressure is less than the suction box pressure, a degassing reservoir connected to a closed vent and control device is required or the barrier fluid is purged into the process system with zero HAP or VOC emissions.
- The barrier system must be equipped with a sensor that will detect failure of the seal system, the barrier fluid system, or both.
- The barrier fluid system is in heavy liquid service or is not in VOC or HAP service.

Installation. Considering the seal system, a sealpot or barrier fluid drum installation is required in accordance with the regulations and is also a means of controlling and monitoring the barrier fluid/seal system. In this installation, the barrier fluid would be dead-ended into the seal chamber from a small sealpot. With a normal seal pumping rate of 1 gal/day or 0.00069 gpm (0.026 Lpm), the sealpot size is small and is defined by other considerations such as instrumentation and level changes for metering flow.[17] In addition, the sealpot must be connected to the flare line or a control device through a closed line.

Should the primary pumping seal fail, process fluid will enter the seal chamber and the sealpot. Therefore a high-level alarm on the sealpot will indicate a primary seal failure. Fluid backup in the sealpot and seal chamber will also increase the pressure in the sealpot with an orifice in the vent line. An outboard seal failure can be shown with a low-level alarm in the sealpot. Although the liquid level falls normally in the sealpot, a seal failure will cause a rapid decrease in the sealpot liquid level to a predetermined action level activating the alarm. In general, the sealpot would resemble the storage pot shown in Figure 10.9.

The seal system is a double seal system and, in accordance with API 610,[14] an auxiliary seal is required in the gland plate, but a drain and closed vent are required to control leakage. While only a single leg from the sealpot to the seal chamber is necessary, a closed drain in the seal chamber is required along with a closed drain from the sealpot.

One sealpot per pump should be installed with this sealing system. Although this requirement increases instrumentation and associated costs, the regulations requiring alarms and emission control indicate individual pump sealpots are necessary. In addition, a barrier fluid *supply* system should be designed as a header with takeoffs for individual sealpots. Since the barrier fluid supply is also critical, alarms should be installed on the barrier fluid system to ensure makeup fluid is available when needed. In addition, the need for an ASME coded sealpot must be determined, although the sealpot is quite small. Moreover, potential pressure elevation in the sealpot during a primary seal failure may require a pressure relief valve. If a valve is required, it can discharge into the vent line.

Process Stream Solids. With solids in the process stream, the pumping seal effectively prevents solids from entering the gap between the seal faces. However, the process stream solids concentration should have some limitations to minimize other problems in the process system that may develop, although this consideration is not critical compared with the potential solids concentration in the barrier fluid. Since the barrier fluid flows through the pumping seal, solids in this stream will affect seal operation. Consequently, the barrier fluid makeup to the sealpot should be filtered to reduce solids concentration to particles $< 20\mu m$ in size. Moreover, the filters should be designed to prevent a breakthrough of solids in the event a filter cloth tears. In addition, the metal parts of the barrier fluid system should be cleaned prior to filling the system with fluid.

10.3.8 Dual Noncontacting Gas-Lubricated Pump Seals

The noncontacting seal is similar to the pumping seal described in Sections 10.3.7 and the dry running compressor seal described in Section 10.5.6. In each seal one face has spiral grooves that are designed for operation with either a gas or liquid layer between spiral groove face and a flat-face mating seal. The grooves are extremely shallow with a depth of several microns or nanometers and the groove patterns are basically logarithmic spirals, as shown in Figure 10.19. The overall spiral groove design was developed by one manufacturer (John Crane) and has been designed for installation in both pumps and compressors. For pumps, the manufacturer has installed a number of pumping seals, providing dual seal capability with a gas barrier fluid. This latest design for pumps contains two back to back gas seals, as shown in Figure 10.10, with an inert gas barrier fluid.

The advantage of this seal system compared with other dual seal systems are:

1. Emissions to the atmosphere of HAP or VOC compounds are zero owing to the inner barrier system of gas and the pressure developed between the seal faces.

2. With wear eliminated between the seal faces, very long seal lives are attainable. Seal operation in centrifugal compressors of 8–9 yr has been demonstrated, with estimated seal life considerably longer than this time period. Since seal conditions are similar in pumps, very little seal face wear is anticipated. This consideration significantly reduces life cycle costs.

3. Since the seal faces are not in contact and a liquid film is not used for lubrication, frictional resistance through the seals is eliminated and power loss through the seals is essentially nil. This improves pump efficiency and also reduces life cycle costs.

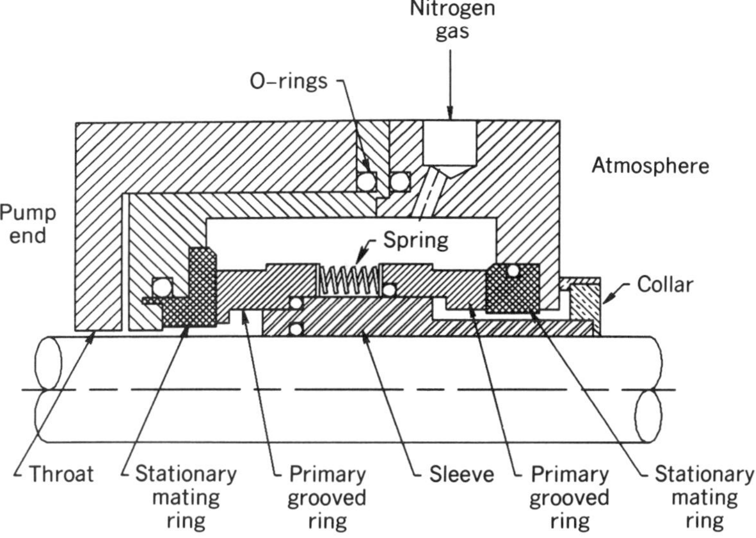

Figure 10.10 *Simplified sketch of dual gas lubricated seals for pumps (courtesy of John Crane, Inc.).*

4. Nitrogen as the barrier fluid is inert and is generally available at various plant locations.

5. A gas barrier fluid eliminates the liquid storage pots and connecting systems for the pumps, reducing investment compared with liquid barrier installations.

6. These seals can be retrofitted into existing pump stuffing boxes.

7. The dual seals are also supplied as cartridge units, simplifying installation and dimensional stability.

Basic System. In this dual system, the normal inert buffer gas is nitrogen, which is at a pressure 10–20 psi (0.7–1.3 bar) higher than the process section pressure. While nitrogen is generally used as the buffer fluid, other inert gases can be used as a buffer fluid seal and may be preferable in some applications.

As shown in Figure 10.10, the primary grooved rings seated on the shaft sleeve rotate with the shaft against the mating rings. The grooved and mating rings are constructed of dissimilar materials and are selected in accordance with the manufacturer's recommendations. Contact between the seal faces is maintained by a spring at rest conditions, and with increasing shaft rotation a layer of gas builds between the seal faces, maintaining a small but distinct opening between the faces. As shown in the figure, nitrogen fills the seal chamber, and at the inboard seal the face grooves in the rotating ring pump the gas toward the dam (Fig. 10.19) or inner diameter, creating a gas cushion and gas seal between the seal faces. This gas seal acts as a barrier to process liquid leakage and, based on the pressure balance, results in an extremely small flow of nitrogen into the fluid.

The outboard seal functions somewhat differently than the inboard seal, having a nitrogen pressure in the seal box considerably higher than the atmospheric pressure. This large pressure differential falls across the outboard seal face in contrast to a differential of about 10–20 psi across the inboard seal. As a result, the spiral grooves are designed to build a small amount of air pressure between the seal faces with air from the atmospheric side of the seal. The nitrogen pressure is retained in this system with a very small amount of nitrogen leakage to the atmosphere.

With this basic dual system, leakage to the atmosphere of VOC or HAP is zero, but there is some external and internal nitrogen leakage. The total amount of nitrogen leakage is quite small and can be estimated from some general curves published by the manufacturer.[33] For example, with a pump pressure of 125 psig (9 bar) and a peripheral wheel velocity of 1700 ft/min (534 m/min), total nitrogen leakage is 1.2 SCFH (34 L/hr). However, these figures are for estimation purposes and accurate figures are provided by the manufacturer. Field data indicate that actual leakage is considerably less than the loss numbers determined with the emission curves.

In addition to the nitrogen seal system, the manufacturer has provided a hydraulically balanced inner ring that is designed for upset conditions when the barrier pressure is lost. As a result, the ring withstands pressure reversal and *prevents* process liquid leakage into the seal box.

Regulations. The nitrogen or inert gas system meet the HON and VOC equipment regulations and exempts the pump from monthly monitoring. For overall HON requirements, the exemption requirements for dual mechanical seal systems applicable to the noncontacting seals are:

1. The pump is operated with a barrier fluid pressure that is at all times greater than the pump stuffing box pressure,
2. The barrier fluid is not in VOC or VHAP service.
3. Each barrier fluid system (set of seals) is equipped with a sensor that detects failure of the seal system, the barrier fluid system, or both.
4. Each pump is inspected visually every calendar week for indications of liquid dripping from the pump seal.
5. Each sensor is observed daily or is equipped with an alarm.
6. Criteria based on design considerations and operating experience are developed for the sensor to insure failures are indicated.

Application. Noncontacting dry running seal performance in field applications has duplicated test results with zero emissions at the outboard seal. However, sustained field performance requires a pump in satisfactory condition. Moreover, the pump should be aligned properly with a minimum of vibration to achieve optimum performance and lengthy seal life. Prior to installation of these seals in an existing pump, the pump should be overhauled to insure the pump will operate for very long periods without a mechanical failure in nonseal pump components such as bearings. A combination of noncontacting seals and a well-maintained pump will result in a long seal life.

Life Cycle Costing. When comparing pump costs as described in Section 10.4.4, the important components are seal life and maintenance. A long seal life reduces maintenance and also eliminates the necessity of supplying and installing new dual seal sets. By removing these major cost components, this type of sealing system is attractive in a comparative analysis. Moreover, installing the seals in existing pumps minimizes the costs associated with converting existing pumps to dual seal liquid barrier fluid installations. Although not considered in life cycle costs, changes in existing pumps necessary to comply with the new emission regulations should be included in any economic evaluation.

10.3.9 Other Seal Information

Cartridge Seals. Typical mechanical seals and their functions were described in previous sections. These seals require careful installation to insure that seal face contact is correct and that a satisfactory seal life will be realized.[18] Proper installation of mechanical seals is a time-consuming effort and the seal assembly may not perform properly immediately after startup, necessitating seal removal and replacement. This mechanical work load is reduced considerably by the installation of fully assembled seals in a cartridge-like form. Single or dual seal cartridge assemblies are available for installation. Economic evaluations and comparisons of cartridge seals should also include installation and removal costs for both noncartridge and cartridge seal systems.

Since cartridge seals are completely assembled in the factory, clearances and tolerances critical for proper operation and performance are achieved under quality control conditions resulting in well-sealed pumps. Moreover, service life appears longer with the cartridge seal systems. For economic evaluations, service life is an important parameter, as shown in the economic evaluation of sealless pumps.

Gland or Auxiliary Seals. Since the barrier fluid pressure for double mechanical or tandem seals is greater than the atmospheric pressure, an outflow of fluid can be expected should an outboard seal fully or partially fail. Both the NSPS and HON regulations require alarms if the inner or outboard seal fails. While an alarm can be installed on the barrier fluid sealpot, an alarm sensor in the outer gland may not be completely effective, delaying pump shutdown and accompanied by a large outflow of fluid. Moreover, any toxic fluid leakage from an inboard seal in a tandem seal system will be present in the barrier fluid because of changes in pressure across the inboard seal. For double mechanical seals, the sudden release of barrier fluid pressure as a result of an outboard seal failure may also result in leakage of process fluid into the barrier fluid. Consequently, a large barrier fluid flow may present toxic (HON), VOC, and safety concerns. As a result, auxiliary seals should be considered for both tandem and double mechanical seal installations. Since mechanical seals (SMS) are also subject to potential seal failures that result in large quantities of process fluid released into the atmosphere, auxiliary and abeyant seals should be considered for all SMS installations.

Outboard of the mechanical seal and mounted in the gland plate, as shown in Figures 10.5–10.7, are auxiliary sealing devices shown as throttling bushings. The auxiliary seal is not a primary sealing device but is a secondary device with the following general functions:

1. Provides a sealed chamber on the outside of the major seal. Leakage from the major seal can increase the pressure level in the chamber between the major and auxiliary seals, triggering a sensor that indicates failure of the major or primary seal.
2. The open channel between the outboard and auxiliary seal permits drainoff of process leakage to a drain or venting of vapor to a flare or control device. This reduces fire risk, personnel exposure, and environmental degradation when the primary seal fails.
3. The auxiliary seal controls the flow of leakage of fluids along the shaft, which reduces bearing housing contamination.

The auxiliary seal is also termed a secondary containment seal, secondary containment device, auxiliary containment device, backup seal, and safety seal. Auxiliary seals are usually bushings, packings, lip seals, or ring seals; *they are not abeyant seals*. An auxiliary seal permits some leakage and cannot be considered a true seal. The amount of leakage has not been quantified, but emission measurements in refineries and petrochemical plants are measured at the shaft interface with the auxiliary seal. This location is the only opening to the atmosphere since the gland plate encloses the outboard mechanical seal in the pump housing. Figures 10.5 and 10.6 show encapsulation of the seal by the gland plate. Obviously, improved mechanical seals characterized by low leakage quantities reduce normal auxiliary seal emissions to the atmosphere.

In general, the auxiliary seal is a restriction which maintains a constant, close clearance to limit (restrict) the leakage of process fluid during a primary or outboard seal failure. For a considerable period of time, the auxiliary seal generally consisted of a simple bushing that had very little mechanical movement and did not change clearance when the bushing was pressurized during a seal failure. A variety of auxiliary seals are now available and a few are described herein.

Operating Requirements. For standard process operations, auxiliary seals must be capable of operating for 3–5 yr in the presence of quench fluid and small quantities of process fluid. During these standard operating conditions, the auxiliary seal must be capable of withstanding a differential pressure of 5 psi (34 kPa) or less.

In the early stages of a primary seal failure, the auxiliary seal must withstand differential pressures up to alarm set-points in the range of 10 to 25 psi (69–173 kPa) and be capable of handling increased liquid and/or vapor releases. The auxiliary seal may potentially operate for long time periods at these conditions, until primary seal leakage becomes excessive or the seal fails completely. When a primary seal fails, the auxiliary seal must have the capability of sealing a liquid and/or vapor at nearly full stuffing box pressure. This leakage and high differential pressure must be controlled for a period of 15–30 min, permitting operating personnel to switch on the standby pump and shut down the leaking pump.

The auxiliary seal in effect must be capable of withstanding gradual and step changes in pressure up to the maximum stuffing box pressure.

Throttle Bushing. The standard API throttle bushing (API 610) is nonsparking, generally constructured of bronze, and pressed into the seal gland plate. This bushing controls leakage when a primary or outboard seal completely fails by maintaining a maximum opening or clearance from the shaft of 0.025 in. or 6.35 mm.

When the leakage rate is controlled by this restricted opening at a differential pressure of 15 psi (103 kPa), the typical leakage rate is 5–10 gpm (19–38 Lpm). For a 45-psia (310 kPa) differential pressure the leak rate is 10–25 gpm (38–95 Lpm). These potentially high leak rates suggest the API throttle bushing is acceptable under the following conditions:

- The fluid at leakage conditions is below the autoignition temperature.
- The fluid will not form an explosive vapor cloud.
- Extensive leakage will not result in a personnel hazard or contravene environmental requirements.

Although throttle bushings are not required for tandem seals in API pumps and are not mandatory, the bushings provide some measure of control in the event an outboard seal fails or leakage becomes excessive. All seal installations should consider throttle bushings or one of the other auxiliary seals for leakage control.

Graphite Rings. Two graphite packing rings have been installed in various gland rings as auxiliary seals, replacing the standard bronze throttle bushings. The graphite rings are frequently installed where service temperatures are $> 350°F$ (176°C). In hot service, a steam quench is frequently specified to cool the outboard mechanical seal and wash the seal face. Graphite ring performance in this service has not been reported in any great detail and comparisons with other seal performance are not available.

Lip Seals. Lip seals are general constructed of PTFE or a filled PTFE and have not found widespread application in North America, although several lip seal designs are available. Europe has been the primary consumer of these seals.

Two basic types of lip seals are available, as shown in Figure 10.11. In the figure, the seal with a spring has frequently been used as a lip or auxiliary seal. Since wear on this seal is a potential problem through constant contact with the shaft, the design has been modified to improve operation. Although the spring seal appears snug against the shaft, some leakage through the seal is observed. In addition to a standard auxiliary seal function, this lip seal is also used as an abeyant seal.

A commonly used lip seal design shown in Figure 10.11 is the curved lip seal, which can be used as an auxiliary or abeyant seal. The sealing contact pressure of the lip can be adjusted to perform in either service. Although considerable wear

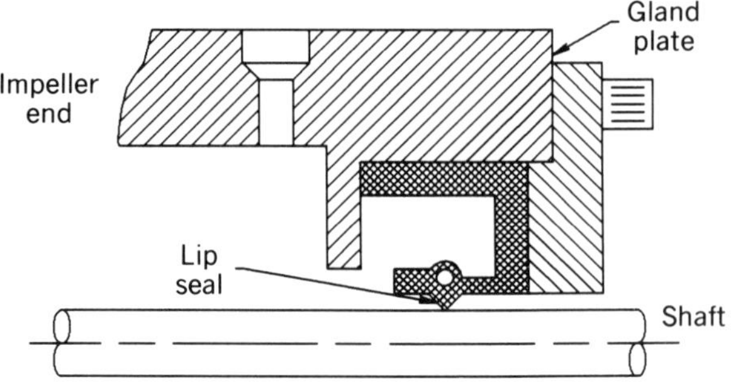

Spring lip seal

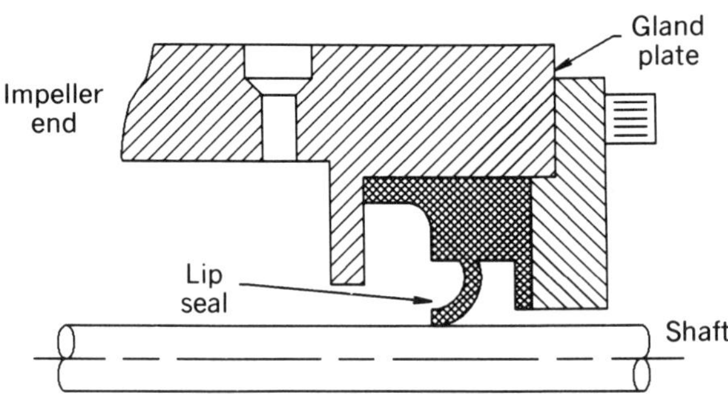

Curved lip seal

Figure 10.11 *Typical lip seals.*

problems were experienced, they have been overcome, and a relatively long service life has been demonstrated in auxiliary seal service.

The PTFE in the lip seal may be reinforced or filled with either carbon or graphite since these materials improve PTFE performance. While PTFE or an elastomer can be installed in this service, they may not be satisfactory in hot service, since either material has distinct temperature restrictions. However, graphite-reinforced PTFE generally has a higher operating temperature than pure PTFE and elastomers.

Carbon Ring Bushing. This bushing application range is much greater than the other auxiliary seal operating ranges and in general is recommended for both low- and high-temperature applications. The carbon bushing is designed for a shaft clearance of 0.005–0.10 in. (0.13–2.5 mm). In addition, the bushing is spring loaded, permitting some movement in a radial position to minimize contact with the shaft. In effect, a close clearance is maintained, even with relative large shaft deflections. The carbon ring is also pressed against the gland plate, providing a seal that prevents leakage between the gland plate and ring.

The carbon bushing is generally a solid ring, but it can also be segmented in accordance with requirements. The leakage rate through a carbon bushing at a 15 psi (103 kPa) differential pressure is about 0.1–0.5 gpm (0.4–1.9 Lpm), which is considerably less than the leakage through an API throttle bushing.

While the carbon bushing is a substantial improvement over the throttle bushing in terms of leakage, there is still a significant amount of leakage. As described above, the auxiliary seals do not contain leakage; any reduction in leakage must be obtained with an improved seal. However, in terms of auxiliary seals, the floating carbon ring apparently has lower leakage rates than other auxiliary seals.

Other Seals. Various types of auxiliary seals are available along with newly designed seals. Service life and leakage information about these newer designs is necessary to ensure satisfactory performance.

Abeyant Auxiliary Seals. In addition to the auxiliary seals described above, there is a class of seals termed abeyant auxiliary seals which operate in a different fashion from the basic auxiliary seals. The abeyant seals have two basic modes of operations—standby and energized. When in the standby mode, the seal is not in operation, maintaining a clearance distance between the seal and moving parts (shaft) to prevent seal wear. The seal operates in a standby mode for much of its overall life.

When the pressure in the open channel between the outboard mechanical seal and the abeyant seal increases and the differential pressure across the abeyant seal is about 7 psi (48 kPa), the abeyant seal is energized. The seal contacts the shaft and essentially forms a leak-tight seal against the shaft. This tight fit virtually eliminates leakage to the atmosphere of gas or liquid. However, the tight fit against the shaft results in very rapid wear of the seal contact part and abeyant seals have difficulty meeting a half-hour emergency operating requirement. This short seal life provides time to start and place in operation the standby pump, which is the basis for setting seal life in the energized mode and shutting down the leaking pump.

The basic operational mode of abeyant seals results in leakage during normal operations. These seals are essentially seals and do not have any impact or control on low leakage/emission rates, which is a major consideration in emission control. Consequently, a single mechanical seal with an abeyant auxiliary seal has an emission rate equivalent to the standard single seal without backup. For reduced emissions, tandem or double seals are necessary. Some of the abeyant seals available and their associated operating principles are described in the following paragraphs.

Lip-Type Abeyant Seal. An abeyant lip seal is similar to the lip seal shown in Figure 10.11 with the exception that the lip does not contact the shaft during normal operations. The curved lip is generally constructed of a carbon-filled PTFE and nominally has a different pressure rating of about 20 psi (138 kPa). When a seal leaks excessively and the pressure begins to build in the chamber against the lip, the lip gradually uncoils and contacts the shaft, sealing the vapor and liquid within the chamber formed between the abeyant seal and the outboard mechanical seal.

Operating with a 14.4-psi (100 kPa) differential pressure, seal life in the closed position is approximately 0.3 hr or 18 min, which is considered insufficient. Obviously, a further increase in pressure would decrease operating time. If the seal is energized at 5 psi (34 kPa), the service life is extended to 2.0–2.5 hr. However, the pump service pressure level would be quite low to operate continuously at 5 psig for this extended time frame with blown seal. Normally, the low differential pressure would energize the seal and the pressure level would probably climb rapidly, reducing operating time.

Since the operating pressure level is quite low, this type of lip seal is normally installed in services where pump suction pressures are quite low. Installing this type of abeyant seal requires careful management to insure that the abeyant seals are not mixed during installation and the seal is placed in the correct pump.

Other types of carbon-filled lip seals are available, based on carbon-filled PTFE, which are similar to the lip seal previously described. However, in abeyant operation these seals can function at higher differential pressures over longer time periods. These various designs must be analyzed to determine whether they meet the desired performance criteria.

Mechanical Standby Seal. In this seal, the solid seal faces are normally separated during typical pump operations. Should the pressure rise in the annular cavity as a result of leakage from the standard outboard mechanical seal, one face under pressure will move to meet the other face, closing the gap and effectively sealing the pump. In essence, this type of seal is a mechanical seal.

A throttle ring in the moving block acts as a hindrance to normal leakage along the shaft, but rapidly reacts to increases in pressure as the primary seal begins leaking. When the face block moves, an O ring at the top of the block is somewhat distorted by the block movement, acting as a spring in compression. The activation intensity for the moving block is 7 psi (48 kPa), which reduces the throttle ring opening. As the pressure in the chamber increases above the activation level, pressure is exerted on the rear of the block, which closes the gap between the seal faces. A higher chamber pressure aids in maintaining seal closure when the faces meet. When the pressure is removed, the distorted O ring returns to its original shape, retracting the moving block in the process, thereby separating the faces. Molecular memory in the O ring aids in returning the O ring to its original shape after distortion.

These mechanical-type seals can withstand a high differential pressure for a considerable time period. Moreover, these seals can operate for many days when activated, providing the time necessary for an adequate and safe response. Differential pressure ratings across these types of seal are 300 psi (21 bar), but the seals apparently can operate well above this pressure level. However, temperatures are limited to about 350°F (177°C) because of the O ring. Changes in the O-ring design may allow operating temperatures above this level.

Observations. A variety of abeyant seals are available commercially and manufacturers have considerable information on performance and cost. Selection of a seal should be conducted in conjunction with mechanical personnel. However, these seals should be installed where safety and health considerations require improved controls.

Based upon abeyant seal performance, the new Clean Air Act regulations, and various regulatory requirements, the following should be considered when reviewing potential abeyant seal applications:

1. Auxiliary and abeyant auxiliary seals do not control normal leakage and emissions. Both seals prevent large liquid blowout sprays in the event a primary seal fails. However, the abeyant seal will completely stop emissions and leakage for a short period.

2. With the proposed new regulations and concerns over total emissions losses per year, abeyant auxiliary seals should probably be installed on all single-seal pumps. This would also require an alarm to ensure rapid startup of the standby pump. Abeyant seals would not be required on heavy VOC fluid pump seals. In these situations, auxiliary seals with a floating carbon bushing or equivalent can be installed.

3. Where double seals are installed, the type of barrier fluid is an important consideration. Ethylene glycol was frequently used as a barrier fluid and has a recommended ceiling vapor concentration of 50 ppm. This compound is also on the Hazardous Air Pollutant List in Appendix 9A (Chapter 9) and must be changed in accordance with HON requirements. Consequently, any emissions to the atmosphere from substitute compounds are a potential problem. With double seals, alarms are required for reduced monitoring requirements and the installation of an abeyant auxiliary seal rather than a throttle bushing provides a control that would essentially eliminate any emissions on seal failure. Where the barrier fluid is a heavier material and not a VOC concern, an auxiliary seal can be installed.

4. Tandem seals are normally installed without a throttle bushing auxiliary or abeyant seal, as shown in the Appendix of API 610.[14] However, although the barrier fluid pressure is lower than the process fluid pressure it is generally well above atmospheric pressure. Moreover, any leakage from the inboard mechanical seal may be present in the barrier fluid if the process fluid is not particularly volatile. Upon failure of the outboard seal, this material is sprayed into the environment, which is a concern. Although the list of hazardous air pollutants consists of 189 substances at this time, other chemicals will be added to this list periodically, resulting in a great number of chemicals considered HAP compounds. Consequently, an auxiliary or abeyant seal should be considered for tandem seal installations. Since barrier fluid pressures will differ in various services, each service should be carefully analyzed to determine whether an auxiliary or abeyant seal is required.

Barrier Fluid Systems. The barrier fluids between seals were described in Section 10.3. These fluids provide lubricity between the seal faces and remove some heat from the seals. Where a process fluid film forms in a tandem inboard seal, the process fluid carries a small amount of heat into the film and the barrier fluid receives a very small inlet flow of process fluid. A major portion of heat from the seals is removed by the circulating barrier fluid moving across the seal blocks.

Heat removal is important in preventing vaporization within the seal faces, which can lead to seal face surface deterioration.

Where the pumping temperatures are $\geq$ 50°F (10°C) than the barrier fluid in a tandem seal system, an axial pumping ring is placed on a rotating portion of the mechanical seal or on the shaft. This ring pumps the circulating barrier fluid to maintain fluid flow in the circulation system. The pumping system minimizes hot spots forming in the barrier fluid circuit.

A thermosyphon barrier fluid system is an alternate to a pumping system. The thermosyphon systems have been successful, particularly in hot services where a significant difference in temperature levels can be established. However, flow rates are low and the concentration of material from a leaking seal can build within the barrier seal fluid volume in the pump stuffing box. In this situation, a leak in the outboard seal would potentially result in relatively large emission rates.

The barrier fluid systems in the figures do not show detailed connections for filling the sealpot. This is an important consideration, but will vary in each location, and a specific design is not shown. Where there are a number of closely spaced pumps in a location, a hard-piped system for distributing barrier fluid oil to the sealpots may be attractive. In this addition system, a hard-piped header is provided with individual takeoffs for each sealpot. A single barrier fluid pump provides fluid to the header through a nominal 20-μm filter. The filter is necessary to ensure a clean-sealing fluid.

With the sealpot discharging to the flare line in most installations, the sealpot pressure will vary with flare line pressure. In large chemical installations, the flare line pressure has a wide operating range, since a number of units are connected to the lines, resulting in pressure variations. A large pressure increase in the flare line is transmitted directly to the sealpot and the seal chamber. Variations in pressure can affect outboard seal leakage rates, and the barrier fluid must be a relatively inert material or not in VOC or HON service to minimize increases in the measured emissions rates. Leakage into the barrier fluid of a hazardous material produces an emissions effect from a leaking outboard seal, which can be compounded by a barrier fluid that is not innocuous.

A number of barrier fluids are used in dual seal services and some of the more frequently used fluids are shown in Table 10.5. Where potentially hazardous barrier fluids are under consideration, innocuous fluids should be considered for this service to prevent leakage to the atmosphere.[36] In addition, the system must be drained carefully, since the barrier fluid must be drained directly from the sealpot and seal chamber. All connections to an oily water or chemical sewer must be closed. If the line to the sewer is not closed, the new HON wastewater regulations require closure or a system that effectively seals the opening. Frequent drainage may be required since leakage contaminates the barrier fluid system.

TABLE 10.5 Typical Dual Mechanical Seal Barrier Fluids[a]

Water	Automatic transmission fluid
Ethylene glycol and water	Kerosene
Methanol	No. 2 Diesel
n-Propyl alcohol	Lube oil (special base)

[a]Adapted from reference 13.

Since the drainage does not contain water unless the barrier system is flushed with water, drainage should be carefully reviewed. For barrier fluids that are considered heavy liquids, the material should be removed from the wastewater stream prior to the stripping tower required in the proposed HON wastewater regulations (Chapter 12). An alternative is a sump for drainage, permitting pumpout to a designated receiving point, including any flush water in the system.

Other Seal Fluid Streams. Although mechanical seals are referred to in emissions control documents, little reference to associated fluid requirements are found in the environmental literature. The environmental regulations refer to barrier fluid system vents and pump seal leakage alarms, but details on these systems are not supplied in EPA background comments. Moreover, auxiliary fluids and their functions are referred to only briefly. In this section {auxiliary fluids are reviewed, since their functions are very important and their emission characteristics should be considered.

Since various liquid injections accompany a mechanical seal installation, flush and quench streams are reviewed in the following paragraphs. Barrier fluids associated with dual seals were reviewed in the previous section.

Flush Liquid. Flushing liquid or material which is injected into a stuffing box chamber is required for the following reasons:

- Lubricate the seal faces.
- Cool the seal faces.
- Prevent vaporization in the stuffing box by increasing the stuffing box pressure. A cooled flush liquid will also reduce stuffing box vaporization.
- Keep the stuffing box clear of solids.

These various objectives are met by injecting a fluid into the stuffing box as shown in Figures 10.4 and 10.7 for single and tandem seals. The pressure in the stuffing box for seal flushing is maintained in excess of pump suction pressure by adjusting the clearance of the throat bushing, wear rings, and impeller balance holes. However, flushing material is not injected into the double seal installations since the interface does not see process fluid on the upper seal surface. Where the seal is not flushed, liquid in the stuffing box enters through the throat bushing and this situation is termed dead-ended or nonflushed.

While the flushing connection is readily apparent in these figures, actual flush material is generally classified as self-flushing or external. In the self-flushing system, a small portion of the pumped fluid is generally recycled from the pump discharge and used as flush material. This contrasts with the external flush, where the flush liquid is obtained from a source external to the pump. Recommendations for self-flushing systems are given in API 610.[14] The self-flush line can also be internal (within the pump). Restriction orifices are placed in the lines to control flow. Another self-flushing alternate is a line from the stuffing box to the suction, which is installed under certain operating conditions.[14]

A major consideration with externally flush material is that the liquid is compatible with the process fluid. In addition, the fluid should be essentially free of solids and not vaporize when entering the stuffing box.

Quench Streams. Quench material is used for various reasons in the chamber between the mechanical seal and the gland as shown in Figures 10.4, 10.6, and 10.7. There are various reasons for the use of a quench and several are described here:

- Prevent the formation of ice crystals on seal faces in low-temperature pump operation.
- Flush leakage from the seal faces, principally heavier liquids from barrier fluids.
- In high-temperature operation, wash seal faces before heavy liquids coke.
- Melt high viscosity liquid leakage.
- Used as flush in shutdown to clean cavity before opening.

While injection occurs at the position shown in the figures, an outlet drain is also located in the gland plate. The drain is located 180° from the inlet connection or at the bottom of the gland plate. Drainage is necessary to discharge both the quench liquid and the leakage from the seal. The drain outlet should be enclosed to prevent vapors from entering the atmosphere, causing potential exposures. In addition, HON wastewater regulations require closed drains.

Since the quench fluid is generally steam or water, the outlet stream is contaminated and the dirty quench stream must discharge into a dirty water or chemical sewer for treatment. The dirty water sewer will have vapors that can backflow through a nontrapped line directly into the quench cavity in the pump. Vapors of this type easily discharge to the atmosphere, increasing the emissions rate from the pump. Therefore, a trap somewhere in the drain line, riser, or drain header is important in preventing any backflow of vapors from the sewer system. Since this is a drain line, a trap is required by the HON wastewater regulations. However, in non-HON installations a seal is suggested to prevent potential exposures or emissions from a nonsealed riser or drain system.

Barrier Fluid Flow-Through Systems. Normally, circulating barrier fluid systems are installed with double or tandem seal systems. These circulating systems generally have a separate sealpot for each pump, unless fed from a centralized header system. The regulations reduce the monitoring requirements for the dual seal/barrier fluid systems since emissions for the dual systems are considered less than the emissions from single seal systems.

Another alternate to the recirculating barrier fluid system is permitted by the EPA and is used in various services. A barrier fluid can be sent through a dual seal system directly into a process stream or process receiver, and the flow-through barrier fluid system monitoring requirements are identical to those for the standard dual seal/barrier fluid systems. Since the barrier fluid is separated from the atmosphere by a seal, the barrier fluid must not be in VOC service or a hazardous air pollutant. In addition, sensors are required that detect failure of the seal system, the barrier fluid system, or both.

While the flow-through system has obvious economic benefits, finding a barrier fluid and a receiving system for the fluid is difficult in many processes. However, any seal evaluation program for new or modified process units should consider a flow-through system. Moreover, the regulations do not limit discharge of the flow-through system to the specific unit that installs this system and a contaminated flow-through barrier fluid could be sent to another process unit. A criteria for transmitting the contaminated stream is that the stream must be enclosed and discharged into a closed process system.

Magnetic Fluid Seal. A new development in pump seal technology is the use of a magnetic fluid to control leakage of VOC or hazardous material to the atmosphere.[34] This seal system, termed a ferrofluidics® seal, is installed on the outside facing of the pump housing. The seal housing contains a permanent magnet in the form of a U that surrounds the rotating shaft. One end of the U is a north pole, with the south pole on the other end. This creates a magnetic field that extends from one pole end through the shaft to the other pole end, with the field completed through the U housing. The clearance between the pole ends and the shaft is very small with a gap of less than 0.02 in. (0.51 mm) indicated for one seal.[34] The ferrofluid forms two separate rings at the pole ends, extending from the pole ends to the shaft. These fluid rings are stable, and if sealing capability is lost can readily be restored by injecting ferrofluid material through a fill port.

The ferrofluid seal is generally installed with a single mechanical seal, forming a dual seal assembly with a nonfilled volume between the seals. Vapor leakage from the mechanical seal is contained between the seals and vented to a flare or vapor recovery system. Leakage through the magnetic seal is extremely small and field test data indicate monitoring concentrations of 0–20 ppmv.[34] The magnetic seal can also be used with dual seals.

Magnetic fluid seals must be installed within certain parameters to insure satisfactory operation. A cavity pressure greater than 7 psig (47 kPa) will eliminate or blow out the magnetic fluid rings. When connected to a flare, the flare header pressure can exceed this pressure during process operations. In addition, liquid leaking through the mechanical seal can destroy the magnetic fluid seal. One solution is a liquid drain between the mechanical and magnetic seals. However, a blown mechanical seal will probably require replacement, since the drain and magnetic seal may not be capable of controlling the leakage. Mechanical seal life is an important consideration in evaluating magnetic fluid seals. In addition, there is a temperature limitation of about 250°F (121°C) on these seals, although water cooling can increase the operating temperature or frequent injections of ferrofluid to the magnetic seal can permit higher operating temperatures.[34]

10.4 SEALLESS PUMPS

These centrifugal pumps do not leak and have been installed to contain highly toxic or hazardous fluids. A limitation in their application has been pumping capacity, but the capacity of newer magnetic pump models has increased substantially and manufacturers have apparently produced pumps with motors larger than 500 hp. These magnetic sealless pumps are similar to large, horizontal centrifugal pumps and now meet API and ANSI standards. In this section, sealless canned and magnetic driven pumps are described, followed by a method of costing for comparisons of sealless and nonsealless pumps.

Since the new HON regulation redefines the leak level for pumps as 1000 ppmv, increasingly stringent emissions regulations make sealless pumps an attractive option as a replacement for mechanically sealed centrifugal pumps. In addition, the use of sealless pumps reduces monitoring requirements from a monthly test to a visual inspection for leakage once per week. While these factors have a strong influence on preferentially selecting emission-free pumps, another requirement also impacts on the pump selection process. The HON regulations also specify the acceptable leak frequency or the number of leaking pumps permitted in a process unit. Should these values be exceeded, a QIP is required to meet the specified EPA targets. This control program is extensive and is an added burden on operating costs. Obviously, leak-free pumps eliminate this problem and in general minimize environmental and potential exposure concerns associated with pump leakage.

Although sealless pumps are not problem free, environmental requirements provide a significant incentive for installation and improvement in sealless pump operation. Some incentives are:

- Reduction in the annual quantities of HAP and VOC releases and the ability to reduce overall pump emissions 90% or better, which is a target for the EPA under the CAAA of 1990. The alternative is an expansion in monitoring and repair efforts to achieve major emissions reductions.
- Significant reduction in monitoring, recordkeeping, and required repair costs.
- Lower investments for sealless pump installations in many services compared with dual seal/barrier fluid pump installations.
- Elimination of costs associated with QIP requirements.
- From an industrial hygiene standpoint, sealless pumps minimize potential exposures by eliminating seal leakage and potential exposures from barrier fluids when draining and flushing a barrier fluid system.

There are two basic types of sealless pumps—canned and magnetic drive. These pumps, which are both magnetically driven, are described in the following sections.

10.4.1 Canned Pumps

A simplified sketch of a typical canned pump is shown in Figure 10.12. This pump is essentially an induction motor with a nonwound squirrel cage rotor and a wound stator. The stator and rotor are separated by a containment can, which as the name indicates, contains the process fluid and the rotor. The rotating magnetic field in the stator coils passes through both the nonmagnetic can wall and an internal layer of process fluids, inducing currents along with magnetic fields in the rotor, resulting in rotation of the rotor. In this design, the shaft does not extend through the pump housing, eliminating seals and preventing the emissions of vapors or liquids to the atmosphere. The only emission or leakage to the atmosphere from this pump are a result of leakage at pump joints or flanges.

In this design, the containment can is thin and constructed of nonmagnetic material (316SS or Hastelloy®) to minimize magnetic reluctance or magnetic resistance, which also reduces heating as a result of induction. The thin containment can is designed to withstand the maximum operating pressure required in

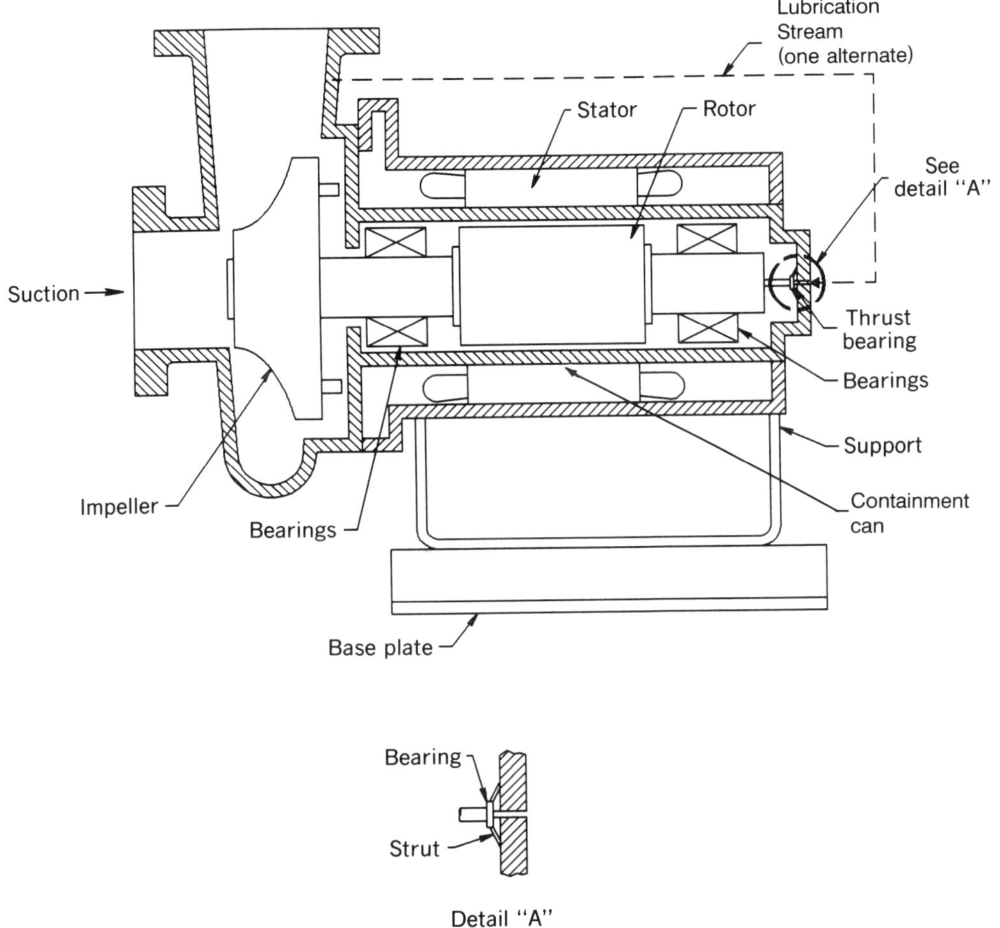

Figure 10.12 *Simplified sketch of a canned sealless pump.*

pump designs based on API and ANSI requirements. If the can should leak, the outer shell, which has a high design pressure rating, will contain the leakage. In addition, an extension to the shell with the electrical leads also contains stator fluid, providing some expansion volume for the stator fluid and cooling. The extended terminal configuration is representative of the connection systems available from the various manufacturers.

Stator. The stator volume can be filled with liquid or an inert vapor such as nitrogen. Where a liquid fills the stator volume, a safety valve is required to ensure the pump is not overpressured by heat from the windings. Any specific cooling requirements for the stator windings would be defined by the vendors based upon operating conditions and potential malfunctions.

Since there is both a containment can and gap in the magnetic circuit, the gap and can thickness are minimized to reduce magnetic reluctance and the stator design is optimized. As a result, the efficiency of the induction motor system is relatively close to a standard motor efficiency. However, overall efficiency for the pump is less than the pump efficiencies demonstrated in typical centrifugal pump API/ANSI configurations. Losses in the canned pump system as a result of bearing lubrication also contribute to a reduced overall efficiency.

Rotor. The rotor portion of the pump is driven by the stator. Process fluid diverted from the impeller is generally the source of lubrication for the bearings on the rotor shaft, although process fluid can be recycled from the pump discharge for lubrication. An important requirement for canned pumps is the need to maintain the process fluid in liquid condition within the can and in the bearings. Although this appears to be a straightforward problem, pumps installed in high vapor pressure service with low NPSH levels frequently run dry, which generally ruins the bearings. Since bearing operation is extremely important in maintaining pump operation, bearing monitors are normally installed in these pumps along with other sensors.

Bearings. A variety of materials is available for canned pumps, including silicon carbide, carbon, PTFE (filled), ceramic, metal impregnated graphite, and hard surface coatings. The selection of bearing materials is important in achieving successful operation, and they must meet the following general criteria:

- Bearing material is chemically compatible with the process fluid.
- Process fluids can satisfactorily perform lubrication functions within the bearing systems.
- Radial and axial bearings loads can be withstood with the bearing systems.
- Wear properties with marginal lubrication must be satisfactory.

In addition, bearings that can run dry for a period of time are an attractive option, indicating that the bearings can run or operate without lubrication for a definitive time period. The typical lubricated bearings normally run over extended time periods, but with dry-running bearings the pump can run dry without bearing destruction in the event fluid is lost through vaporization in the pump or fluid flow is interrupted in the pump suction circuit. Bearings in a dry situation have a finite life that may vary from minutes to a few days, but bearings that have some dry-running capability permit an orderly pump shutdown and increase bearing life. As a result, the MTBF (mean time between failures) increases substantially, reducing maintenance requirements. Dry-running bearings are mentioned periodically and may be available from some manufacturers.

Bearing Lubrication. The sleeve bearings in canned pumps require lubrication to prevent burnout, which also results in damage to other parts of the pump as shaft movement increases. Lubrication is very important in this pump design, and positive fluid flow through the bearings is required to ensure adequate lubrication. In sealless pumps, the process fluid is the lubrication media, and frequently process fluids are not satisfactory lubricating fluids. This lack of lubrication ability with many fluids is an argument for a positive internal lubrication stream. In addition, liquids with high vapor pressures may have poor lubricating characteristics. Moreover, high vapor pressures require an internal pressure level within the sleeve bearings that prevents vaporization either on a local basis or on a general scale within the rotor section. In addition to pressure concerns, heat input from the electrical/magnetic circuitry can exacerbate the potential vaporization problem. Heat buildup can be estimated rather accurately by the pump manufacturers and information on this phenomena should be requested when evaluating pump applications.

Since NPSH is an important component of internal pressure estimation it must be carefully determined, particularly for high vapor pressure fluids. This is of particular concern where the process fluid is pumped from a source at the liquid boiling or bubble points. The fluid bubble point within the pump bearings must be suppressed to prevent local vaporization and should be carefully reviewed along with the boiling point. Consequently, the pump pressure balance must be carefully defined. For tight NPSH pressure balances, larger suction lines are a method of reducing the pressure loss and increasing the internal pressure.

Since lubrication is of critical importance, pump manufacturers have developed various techniques to ensure internal fluid circulation. Some of these techniques are described below, but the list is incomplete as manufacturers strive to improve internal fluid circulation designs and reduce bearing failures.

External Recirculation. A major method of recirculating fluid is through an external line from the pump discharge (high-pressure) nozzle to the outboard end of the internal can, which is indicated in Figure 10.12. The discharge pressure provides the differential pressure necessary to move fluid through the bearings and containment can into a low-pressure region of the impeller. Where the pump differential pressure is very large, internal pressures are generally satisfactory. However, a pump with a small differential pressure may have problems with this circulation pattern.

Internal Recirculation. While an external recirculation line is shown in Figure 10.12, the same flow system can be achieved through an internal system. Although this type of flow system is not shown, it can be readily described. Flow from the discharge of the impeller is circulated through internal passages into the bearings. This stream at the outboard end of the pump shaft is returned to a low-pressure point on the impeller through a hollow rotor shaft. Other variations of this design exist among the various pump manufacturers.

Once Through. The once-through system is a variation of internal recirculation. Instead of returning the fluid to the impeller suction area, the process fluid is sent to a receiver. The receiver must be at a lower pressure and capable of handling the process fluid. Although relatively simple and straightforward, this system requires a detailed understanding of the process. In chemicals processes, the stream fluid can only be sent infrequently to another process, limiting discharge or receiving options. In contrast, petroleum refineries or hydrocarbon processing facilities permit a variety of receiving options.

Internal Impeller. In this system the internal pressure is significantly increased by adding another impeller. A portion of the process fluid from the main process impeller discharge flow is bypassed through a hollow shaft into an internal impeller where the circulation fluid pressure of the stream is increased. The internal impeller stream at an elevated pressure is distributed through the containment can bearings. The recirculating lubrication fluid stream is returned to a low-pressure area on the main process impeller. Variations of this impeller system are included in canned pump designs from various pump manufacturers. This system has been successful in bearing lubrication and canned pump installations should consider canned pumps with internal impellers.

Other. While the lubrication systems described are provided by a number of manufacturers, these general systems can differ in design. In addition, some other lubrication systems may be available that have not been described. Since the lubrication system is a critical aspect of a canned pump design, selection of canned pumps should be reviewed with the rotating machinery engineers.

Solids. As described, bearing lubrication is extremely important in canned pumps to maintain a lubricating film and prevent bearing surface contact. However, solids in the fluid stream can interfere with bearing performance, resulting in pump shutdowns. Consequently, solids must be removed from the bearing fluid stream. In the external recirculation loop shown in Figure 10.12, a filter in the loop will remove solids sent to the bearing systems, sending filtered process fluid through the bearing system. Filtered fluid flow from the outboard end through the can bearings continues into the impeller area, resulting in a flush/lubrication flow system that minimizes solids containing process fluid flows entering the rotor section from the impeller. A possibility of solids in the rotor and bearing systems may occur when filling the pump prior to startup, which may not be sufficient to damage the bearings. Obviously, protection against this situation should be investigated and solutions to prevent the deposition of solids in the can chamber should be developed in the design stage. Assuming the solids concentration in the normal process stream is at a relatively low level and can readily be pumped, the following guidelines are suggested:

1. A pump with an external recirculation lubrication line similar to the pump shown in Figure 10.12 should be used. In addition, a 20-μm filter in the recirculation line is generally recommended.
2. An alternate to the recirculation line is the flow of another fluid from another source into the outboard end of the canned pump. Since the flow of fluid enters the main process line, the bearing fluid must be compatible with the process stream and downstream processing conditions.
3. Since the rotor section can fill with solids during pump preparation prior to startup, a filter in the suction line during preliminary startup will provide an essentially solids-free liquid in the pump. When the pump is on line and a recirculation flow has been initiated, the suction filter can be bypassed. This permits the flow of a recirculation fluid, essentially solids free, into the containment can and bearings.

4. Since a solids-laden fluid is a problem for satisfactory bearing operation, fluid flow from the impeller through the bearings is not recommended. If a continually operating filter is placed in the suction line, an internal circulation lubrication system would be satisfactory. However, filters installed to remove a relatively large percentage of solids from the process stream require frequent cleaning and purging, which is a rather involved operation. Moreover, in toxic or low boiling point streams, filtering on a cyclic basis can present operating problems.

A major problem with suction filtration is the differential pressure buildup across the filter screen as solids are removed from the stream. This pressure drop increase is particularly troublesome in streams with very low NPSH levels or liquids at or near their boiling points. Consequently, the long-term operation of a suction filter should be carefully analyzed to ensure an adequate pump suction pressure when the filter pressure differentials becomes rather large, prior to filter cleaning. These filter pressure drops require instrumentation to alarm at very high differential pressure levels and also indicate the necessity for a filter switch and filter cake removal.

5. In addition to these various considerations metal scale can enter the pump and damage the bearings, particularly during startup. This problem is a concern where most of the equipment and lines are carbon steel. Scale can be removed with a magnetic separator, but the separator must be a permanent installation.

Thrust Bearings. Previously, the bearing discussion concentrated upon sleeve bearings that controlled the radial movement of the shaft. In addition, axial movement of the shaft also occurs, which requires bearings to prevent movement in either axial direction of the shaft. These are called thrust bearings and are generally present in rotating equipment.

In the canned sealless pumps, the thrust bearings in the pumps are self-adjusting under various process conditions. These bearings differ among the various pump manufacturers and require a detailed investigation of the thrust bearing designs to fully understand their operating principles. Many of the pump designs incorporate fixed and variable orifices to provide automatic thrust balancing. Secondary permanent backup mechanical stops are also provided that control excessive movement during startup or upset process conditions. The variable orifice openings are not controlled externally, but are automatic and are triggered by small axial shaft movements that partially enclose or open an orifice. For a thorough understanding of thrust bearing operation, the canned pump design should be reviewed with rotating equipment engineers. Pumps must be satisfactory under all conditions, and a deficiency in one area can affect pump performance and the MTBF.[20]

Heat Input. Although these pumps essentially operate at process temperatures, careful consideration of heat input from the pump to the fluid in the rotor can must be understood. Heat enters the circulating fluid from the motor windings and the induction rotor, which must be included in any analysis, along with any heat generated through friction in the bearings. The heat input affects pressure–vapor characteristics of the process fluid that can result in partial vaporization of fluid in the bearings if not controlled. Since the heat balance is very important, the pump manufacturers will supply this information based upon an accurate specification that includes a detailed analysis of the process fluid.

Various techniques are used to control temperature rise in the rotor area. One adjustment is an increase in the flow of process fluid through the rotor chamber. This will tend to reduce the overall efficiency of the pump, but does not require a capital investment since the power loss is small. Manufacturers have somewhat different operating lubrication flow rates along with potential increases that are possible in the pumps, which requires an understanding of the general pump installation in conjunction with the rotating equipment engineer. With an external recirculation loop, a cooler is often placed in the recirculation line, and manufacturers have provided coolers in standardized versions. Since a variety of solutions is available to control heat input in these pumps, an effort must be expended to ensure a satisfactory installation. The canned pumps require more analysis than standard sealed pumps, which is an added cost, but have a distinct advantage in controlling emissions.

Viscosity. Where the viscosity is relatively high, a careful analysis of pump performance is necessary. The fluid can be warmed as it passes through the stator zone, which would effectively reduce viscosity. Again, this requires definitive viscosity values at operating temperatures that permit an accurate assessment of viscosity changes in the fluid. Moreover, non-Newtonian effects must be carefully described for any fluids that have these physical characteristics.

Installation. The canned pump is a single container, as shown in Figure 10.12, in contrast to a standard centrifugal pump which has a pump casing, a bearing housing, and a motor. Consequently, the pump container size is considerably smaller than a conventional centrifugal pump or a magnetic drive pump, which is reviewed in the following section. This relatively small size has some distinct advantages, as discussed below.

Configuration. The pump can be installed horizontally or vertically, which permits changes in piping and equipment layouts.

Foundations. A foundation or pump pedestal is not required for canned pumps, which is standard in many petroleum refineries and chemical plants. While a baseplate is normally shown in figures of canned pumps, the baseplate does not have to support other equipment. In addition, the actual support member is essentially a wide, channel section that can be bolted directly to a concrete pad, eliminating the baseplate. While the piping may aid in supporting the canned pump, the piping should not be considered a support mechanism. For canned pumps, the support member can be bolted directly to a small built-up concrete section which may be a few inches (51 mm) higher than the concrete pad. Elimination of the pedestal and the standard centrifugal pump baseplate adds approximately 1 ft (0.3 m) NPSH to the pump, which is an additional aid in maintaining liquid conditions in the pump bearings.

Alignment. Since the canned pump shaft does not connect directly with other equipment (motors, etc.), alignment with an external motor is eliminated, which also reduces installation costs. Moreover, shaft alignment following seal or bearing replacement in standard centrifugal pumps is eliminated.

Frame Size. The general frame and nozzle locations on most of the canned pumps have not been consistent with API or ANSI dimensions. As a result, installing a canned pump replacement for a standard centrifugal pump required physical piping changes, adding to the cost of a replacement installation. Some of the manufacturers are trying to standardize designs which will reduce these potential piping problems in existing units. In new processes the piping is installed based on certified pump vendor drawings, eliminating any differential piping costs with standard centrifugal pumps.

Monitors. Various monitors are recommended for a canned pump installation to protect the bearings from damage, including bearing wear detectors. These monitors are remote reading and some are combined with trips to alarm and shut down the pump when a potentially dangerous system develops in the rotor chamber.[21,22] In addition, level controls on the drum or vessel that feed the pump can also shut down the pump. This prevents any variations in fluid conditions in the pump chamber as a result of NPSH changes. Pump monitoring with external level control prevents the pump from running dry, but they are also an added cost in the installation.

Noise. Since there is no external noise source such as a motor with canned pumps, noise emissions are extremely low. This is a distinct characteristic of canned pumps, which differ considerably in this respect from other pumps.

Application. A wide variety of canned pump sizes are available from pump manufacturers and as a result of expanding sales, the pumps continue to increase in size and capability. Some models have capacities of 1800 gpm (6.8 m^3/min) with a differential pressure of 86.5 psi (597 kPa). For differential pressures of 303 psi (2089 kPa) a capacity of 500 gpm (1.89 m^3/min) is available. Moreover, operating pump temperatures of 1000°F (538°C) and operating pressures of 3000 psig (207 bar) are apparently practical for some manufacturers. While these parameters are typical for single impeller pump operations, some of the non-U.S. manufacturers have multiple impeller pumps that have much higher head characteristics for similar capacities. Consequently, actual pump characteristics must be investigated for individual pump specifications and application.

The comments covering canned pumps are generally applicable to horizontal pumps. However, vertical canned pumps are also available from various pump manufacturers. Depending upon the application, vertical pumps may be preferable. In existing installations where vertical on line process pumps are installed, vertical canned pumps are an alternate to control emissions.

10.4.2 Magnetic Drive Pumps

The canned pump previously described is based upon an induction motor design, with a stator and rotor. An alternative method of driving a centrifugal pump hermetically is with a permanent magnet coupling drive. As shown in Figure 10.13, permanent magnets outside the containment can periphery are connected to a motor driveshaft and inside the containment can the pump rotor periphery is also fitted with permanent magnets. The outer magnet rotates under essentially atmospheric conditions and internally the magnetically coupled rotor rotates in process fluid in conjunction with the outer rotating magnets. The rotor is somewhat similar to the internal rotor in the canned pump, with all of the rotor bearings mounted internally and dependent on process fluid lubrication. The major advantage compared with centrifugal pumps is that the magnetic drive pumps are also sealless and emission free.

In this pump system, the outer magnetic arm is driven by an external motor, as shown in Figure 10.13, which attracts or repels the inner rotor magnets strongly in the permanent magnetic field of the outer arm. As a result of outer arm rotation, the rotor is essentially in lockstep movement and rotates at the same revolutions per minute as the outer arm. Since the shaft does not penetrate the can or inner housing, monthly monitoring is not required, but the pumps must have weekly visual inspections.

In addition to the magnet to magnet coupling system described above, an eddy current coupling system is also employed in various designs, which is similar in principle to that of an induction motor. The outer drive magnets produce a rotating magnetic field, across an internal rotor consisting of copper bars mounted parallel to the shaft, which is similar to a squirrel cage rotor. The rotating magnetic fields move through the copper bars inducing an eddy current in the bars, which generates magnetic fields around the copper bars resulting in inner shaft rotation similar to the phenomena described for permanent magnet operation. The differ-

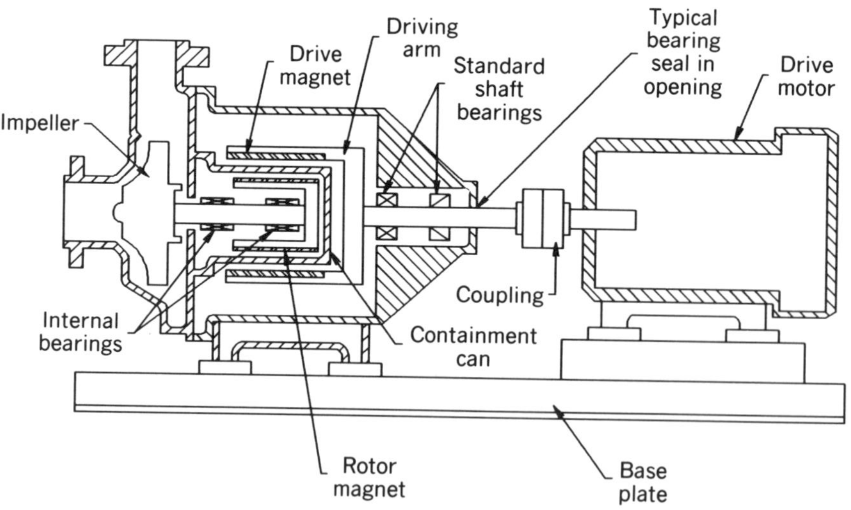

Figure 10.13 *Simplified sketch of a magnetic drive sealless pump.*

ence between the rotating magnetic field revolutions per minute and rotor rotation is the slip, where increasing slip tends to increase eddy current strength and the local magnetic fields, which maintain rotation. Although eddy current designs are available from various manufacturers, the majority of the magnetic pumps are designed with permanent magnets in the outer drive and on the rotor, which is apparently a little more efficient than the eddy current design.

Magnets and Magnetic Fields. Magnetically driven pumps have been available for a number of years, but were limited in power availability. When pump power increases, magnet size increases, raising the total mass of the driven magnet system and resulting in significant increases in motor horsepower and associated costs. Improvements in permanent magnet materials have drastically improved magnet performance and the horsepower of magnetic pump drivers are now considerably greater than 500 hp.

These improvements in permanent magnet performance are a result of rare earth alloys, which have replaced Alnico magnetic materials. Pump manufacturers are generally installing samarium cobalt-based magnets for improved performance. In addition, design changes in clearance have reduced the magnetic gap and the loss of leakage flux. Close clearances are extremely important in minimizing gap loss since magnetically driven pumps have two gaps compared with one gap in canned pumps. As shown in Figure 10.13, a liquid gap exists between the rotor and containment can and a second gap exists between the containment can and the rotating drive magnet. In contrast to the magnetically driven pumps, the only gap in a canned pump exists between the rotor and the containment can (Fig. 10.12).

The Curie point of any ferromagnetic material is that temperature above which permanent magnetization is lost, which differs for each substance. Samarium cobalt and other rare earth alloys exhibit significantly higher Curie temperatures than ferrite magnets, with the Curie temperature for samarium cobalt about 1380°F (744°C) compared to approximately 840°F (449°C) for ferrite magnets. This temperature restriction limits maximum operating temperature compared with canned pump operating temperatures since the latter does not have permanent magnets. Operating temperatures are an important consideration in permanent magnet application, which affects the amount of torque that can be transmitted without slipping. The strength of the coupling between the permanent magnets is a very important consideration in magnetic drive pumps, with the torque rating a function of magnetic characteristics and the operating temperature. For example, at 525°F (274°C) samarium cobalt magnets will experience a 15% reduction in torque rating and neodymium alloys operating at 248°F (120°C) demonstrate a 19% reduction in torque rating compared with ambient temperature performance.

The magnetic gap and magnetic leakage flux have a significant impact upon pump performance, requiring careful outer and inner pump rotor and clearance designs. Magnetic gap and leakage flux are briefly described in the following paragraphs.

Magnetic Gap. This term refers to the physical separation between two magnets that are magnetically coupled. In a magnetic drive pump, magnets inside the containment can are essentially matched center to center with magnets outside the can. The magnetic flux is transmitted to the inner rotor magnets from the outer rotor arm magnets through the air–liquid gaps and across the containment can walls. The magnetic flux is the number of magnetic force lines per square inch and is concentrated at the pole face of the magnet. As the outer set of magnets rotate, inertia prevents the rotor from moving in synchronous motion with the outside magnet ring and the small lag resulting from this difference is termed drift. If the drift or lag becomes too pronounced, an adjacent magnet provides a repulsive force that tends to place the magnets in a synchronous phase. Drift is a potential problem since the leakage flux increases in this situation, reducing pump efficiency.

Leakage Flux. This is the flux that leaks or is lost to other magnets or magnetic circuits, which reduces coupling effectiveness between the inner and outer magnets. A great effort is required to minimize flux leakage and reduce the effect on the strength of the magnetic coupling. To reduce flux leakage, the magnets are frequently placed in a low reluctance material where the reluctance is the reciprocal of permeability. Other techniques are also implemented to further reduce leakage flux. Reference 24 indicates a significant flux reduction with a zirconia containment can and rotor.

Torque. The major magnetic drive pump design objective is to provide a magnetic coupling that can withstand the very large torque transmitted from the motor in a compact and cost-efficient unit. Torque is a more accurate method of defining the system than power transmission. Specifically, the magnetic couple can be characterized in terms of breakaway torque, which is that torque required to overcome the magnetic attraction between the inner and outer magnets that results in slippage of the rotor. If breakaway torque is reached, the outer half or drive magnet continues to rotate while the inner rotor becomes stationary. When slippage occurs, the motor must be shut down and then the motor is restarted to regain the magnetic coupling.

Heat Effects. When slip or drift occurs, induced secondary electrical currents form in the containment can and become significantly larger as the slip increases. These currents may also form in the magnets owing to field distortion. The currents, termed eddy currents, generate excess heat and if not reduced to very low levels eventually reduce the strength of the magnetic field. Also, this heat must be dissipated by the internal fluid, but can be a potential problem where the fluids are temperature sensitive. In addition, concentrated hot spots may result in flashing of the fluid, which can contribute to bearing failure. This eddy current heat phenomena is primarily a characteristic of magnetically driven pumps and is not a major concern in canned pumps.

Overall heat buildup in the magnetic drive pump is less than the heat input in the canned pump. Locating the motor drive separately reduces the heat input from the motor coils that adds to the overall temperature buildup in the canned pump. Since heat input is important in determining conditions within the magnetic drive pump, the pump manufacturer should supply a detailed temperature analysis for both normal operations and upset conditions. This is particularly critical with process fluids that are near their bubble point, since heat input may require some form of cooling to maintain liquid conditions in the pump.

Power Requirements. Power requirements for magnetic drive pumps are somewhat higher than those of standard centrifugal pumps for the following reasons:

- Power is necessary to maintain fluid circulation within the containment can, since the internal fluid provides lubrication and cooling.
- Friction associated with internal rotor movement must be overcome. The rotor is placed in liquid with a very small clearance between the rotor and containment can to reduce the magnetic gap, which contribute to a relatively large friction factor generated by the inner rotor. Power is required to overcome this loss.
- Energy loss is also increased with the generation of eddy currents.
- Some windage loss occurs with the outer rotor which also must be included in the energy balance.
- If nonleakage seals are placed on the outer rotor driver shaft to prevent bearing leakage or containment can leakage from discharging into the atmosphere, some energy loss is developed across this seal.

Obviously, these additional power loss requirements result in the need for larger drivers than are necessary for standard conventional centrifugal pumps. Thus, overall efficiency is lower for magnetic drive pumps.

Bearings. Bearings in the inner rotor shaft of magnetic drive pumps are virtually identical to the bearings in canned pumps described in the previous section. These sleeve-type bearings are lubricated with process fluid, which may be a poor lubricant having low viscosity, low specific heat, and poor film strength. Consequently, bearing design and positive lubrication are important for obtaining a long service life.[23] Comments on bearings in Section 10.4.1 for canned pumps are also applicable to magnetic drive pumps.

Bearing Lubrication. Lubrication for the internal bearings is extremely important as described for canned pumps in Section 10.4.1. In addition, heat is generated at the containment can by eddy currents generated in the can. The magnetic field loses power across the gaps and containment can with the power loss essentially converted to heat in the containment can housing. As a result, the lubrication fluid must also cool the magnetic flux area in addition to lubricating the sleeve bearings.

The magnetic drive and canned pumps are similar internally and lubrication techniques should also be similar. However, the design of the magnetic drive pumps precludes the use of an external recirculation system, which is one of the lubricating techniques used in canned pumps, as shown in Figure 10.12. This technique cannot be applied in magnetic drive pumps, where the outboard end of the containment can is essentially enclosed or surrounded by the motor shaft and outer magnetic arm assembly that are in constant rotational motion. Consequently, external recirculation as shown in Figure 10.12 is not applicable in magnetic drive pumps, but completely different external recirculation systems are available. Some typical lubrication systems are described in the following paragraphs.

Internal Impeller. An internal impeller circulates process fluid through the bearings and in the gap between the containment can and the rotor. The impeller system in the magnetic drive pump is similar to the impeller circulation system in the canned pump.

Internal Circulation. In the internal recirculation systems, fluid from the impeller discharge, which is at relatively high pressure, is circulated through the sleeve bearings and rotor gap to the outboard end of the containment can. The lubrication fluid at this point is returned to a low-pressure area at the process impeller through a hollow shaft or passageway.

Internal lubrication systems are modified and combined by various pump manufacturers in different configurations. Consequently, the internal lubrication system must be carefully reviewed with pump manufacturers to ensure the operating details are understood and will meet operating requirements. This also requires information from the manufacturers on pressure distribution and heat conditions throughout the rotor system. Input heat from the eddy current buildup in the containment can must be adequately dispersed by the lubrication system. Thus, heat input during normal operation, startup, and unusual situations must be controlled to prevent a change in magnetization that can result in a slip increase that loses the rotor, which stops rotating.

External Lubrication Fluid. One other method of lubrication recommended by some pump manufacturers is an external lubrication fluid that is provided at the desired pressure. The pumps have openings that permit a connection to the can for this fluid, which does not contact the rotating arm. However, an external lubrication fluid presents potential problems that were described in Section 10.4.1. The lubrication fluid must be compatible with the process fluid, and more importantly the downstream facilities must have the capability of processing the contaminated stream. Since the lubrication flow may be on the order of 1.0 gpm (0.26 Lpm), the mixed process stream becomes essentially a contaminated stream.

When a process fluid can accept an external fluid, a clean, filtered flush stream from a reliable, consistent external source is desirable. In addition, for dirty or erosive service a clean, filtered external flush is also desirable. However, the entire system must be analyzed as described in the previous paragraph to ensure an external lubrication stream can be processed.

External Drive Shaft Bearings. Shaft thrust is not a major problem in the external bearing housing, where the bearing load is essentially a radial load and does not normally require heavier bearings such as double row bearings. Although it is part of the basic pump, external shaft bearing lubrication differs from the internal process fluid lubrication system. The separate external bearing housing permits the use of lubrication oil or grease in this service. Apparently grease lubrication is the standard lubricating system normally recommended by the pump manufacturers, although oil is preferred in many installations. The type of lubrication oil, along with the method of bearing lubrication, should be determined by the rotating equipment engineers. Outer bearing failure can have detrimental results on the pump service factor and is an important consideration in pump maintenance.[23]

Where oil is the preferred lubricating medium, it is frequently used in mist form for optimum performance, but the bearings must also be protected from environmental effects that can have a detrimental impact upon bearing operation. Dust or trash entering the bearing housing can pose serious problems. Consequently, an outer seal is placed on the external bearing housing to protect the bearings. Various seals are installed to protect the bearings (e.g., lip, magnetic, or labyrinth seals), but these do not completely seal or close the annular opening to the atmosphere. Moreover, an outer bearing seal will not control emissions should the containment can spring a leak. While a leak is possible, of more immediate concern is the potential oil mist that can emanate from the outer bearing seal. Depending upon the lubricating oil and the weather, an oil mist from the bearing seal is undesirable for environmental and health reasons. Consequently, a bearing seal that reduces or essentially eliminates the oil mist emissions is preferred. The rotating equipment engineer should be consulted about methods of controlling any mists or vapors emanating from lubricating oils, which may require a change in the lubricating technique rather than an improved seal. Mist elimination also reduces potential health hazards.

Axial Thrust. Axial thrust problems for magnetic drive pump rotors are similar to those in the canned pumps and the solutions to the problems are generally the same. The axial thrust bearings must be able to operate without problems when the pump is partially full of liquid or cavitating for a short time and perhaps running backward. The bearing system cannot be sensitive to changes in clearances or the physical properties of the pumped fluid. As a result, many types of thrust-control devices have been installed, with the most common systems based on impeller designs with holes that permit passage of process fluid into the suction area, which places suction pressure on the can side of the impeller. In addition, a variable orifice may be used to control leakage from the discharge side of the impeller to the low-pressure area behind the impeller. Also, combinations of these various systems are used to control thrust.

Solids. Solids in the process fluid present a problem in sealless pumps which was described in Section 10.4.1. Since magnetic drive pumps differ in certain areas from canned pumps, solutions that require filtering solids from an external recirculation line for canned pumps are not completely applicable. Also, internal filters that can be installed in some canned pumps are not widely available in magnetic drive pumps. Consequently, other techniques are necessary to minimize and prevent solids from entering the bearing section of magnetic pumps.

One technique described for improved lubrication is the use of an external lubrication flush stream, although this stream must be compatible with the process fluid and downstream process system. However, a solids-free external recirculation stream prevents solids from entering the containment can as a result of the backflush effect through the impeller. The external stream lubricates the rotor bearings and enters the process stream through a low-pressure area in the impeller after flowing through the rotor section. Filtration of the pump suction stream is another alternative, but may present potential problems. A high-pressure drop across the filter in a tight NPSH service may result in flashing and/or cavitation at the pump impeller. Suction filtration should be limited to streams where a pressure buildup across the filter will not affect the pump. Additionally, scale can be a problem and a magnetic separator can be used for scale removal.

Containment Can. As described in Section 10.4.1, a nonmagnetic liner is needed to minimize interference with the magnetic flux or fields and separate the process fluid from the outer rotating arm. For mechanical reliability and structural strength, a thick can is desirable, but high efficiency requires a containment can with an absolutely minimum thickness to minimize interference with the magnetic circuit and minimum gaps between the can and the moving elements. In addition, a thin containment can and minimum gap openings reduce electrical heat buildup and heat transfer to the internal fluid. Therefore, the can should have a minimum uniform thickness and be constructed of a material highly permeable to magnetic fields. Stainless steels are suitable, but for maximum corrosion resistance Hastelloy® is generally specified. In addition, cold-formed cans are preferred for this service, since these cans are homogenous and do not have any welds. A cold-formed can has greater strength and erosion resistance than a welded can and does not have any magnetic flux effects.

Should a can rupture, the contained fluid will leak immediately to the atmosphere through the external bearings since these bearings are not sealed from the atmosphere. In contrast, the canned pump will contain leakage within the outer shell, which is designed for higher pressures than the containment can. Also, the outer shell of the canned pump may have a safety valve, which provides additional protection to the outer shell. Containment can mechanical strength and integrity in a magnetic drive pump should be carefully reviewed by the rotating machinery group to ensure a satisfactory margin of safety for operating upset conditions and the estimated maximum emergency process pressure. When the margin of safety is unsatisfactory, the magnetic drive pump should be replaced by a canned pump or multiple seal conventional pump. Operating a pump that may not have an adequate containment can safety margin could result in a large release to the atmosphere.

With properly designed pump installations, can leakage or rupture is an infrequent occurrence.

NPSH. The detailed comments on NPSH presented in Section 10.4.1 for canned pumps are also applicable to magnetic drive pumps.

Viscosity. Heat input in a magnetic drive pump is less than heat input in the canned pump, resulting in a more consistent lubrication viscosity. Since the viscosity in the containment can volume does not decrease to any extent, the fluid viscosity must be limited at the inlet to the pump to ensure the viscosity does not exceed the desired lubrication viscosity. In the canned pump a higher viscosity can be tolerated since the small lubrication flow is heated, thereby lowering the viscosity. In addition to the maximum allowable viscosity, non-Newtonian flow viscosity characteristics should be understood prior to installation of a magnetic drive pump to ensure that adequate lubrication can be conducted. The viscosity limitation should be carefully observed since a high viscosity increases the internal rotor drag and can decouple the rotor. This situation is overcome by increasing the driver horsepower rating and the magnetic field strength.

Installation. As shown in Figure 10.13, the magnetic drive pump resembles a conventional API pump assembly, where the motor is a separate housing and a coupling connects the motor and external pump shafts. The general pump/motor system, along with the base plate, is similar in configuration and size to a conventional centrifugal pump. Comments on the magnetic drive pump follow.

Configuration. All magnetic drive pumps are installed horizontally. A vertical version of this pump has not been commercialized.

Foundations. The magnetic drive pump is supplied with a baseplate which is similar to the baseplate provided with API or ANSI pumps. For standard horizontal, centrifugal pumps with a baseplate, a pedestal mounting is generally provided in petroleum refineries and chemical plants. The pedestal mount provides a reinforced concrete base for the overall installation to supply a sturdy, reinforced foundation along with a support that withstands vibration and minimizes pump–motor movement. Since the system mass for the magnetic drive pump is similar to a standard pump and the drive motor is external, a pedestal system should be considered for this sealless pump. Vibration in this equipment is also a concern with an external motor and a large offset radial mass attached to the shaft in the pump housing.

Alignment. Connecting the motor and pump shafts requires exact alignment for the shafts. In this situation, installation of a magnetic drive pump essentially duplicates installation efforts necessary for a conventional pump system.

Frame Size. Since magnetic drive pumps are horizontal and similar to conventional API pumps, a natural assumption is that sealless pumps have nozzle locations and dimensions identical to and interchangeable with standard API/ANSI pumps. Some magnetic drive pump manufacturers have pump connections identical in size and configuration to API 610 pumps and are considered interchangeable pumps. If the magnetic drive pumps do not completely duplicate API or API pump configurations, piping changes are required to install the pumps. In a new process unit this is not a problem, with piping installed on the basis of certified vendor drawings. However, installation in an existing unit requires careful engineering to determine whether piping changes are necessary.

Monitors. To obtain optimum performance and minimize potential mechanical problems, various monitors are recommended:[23]

1. Thermocouples to measure the containment can wall and fluid temperatures. If the process flow stops or the process liquid vaporizes, the rotating field heats the can and internal liquid fluid temperatures increase.

2. A power sensor on the electric motor that is apparently more sensitive than the containment can temperature monitors.[23] The power sensor apparently rapidly indicates when a pump is running dry, the inner rotor is decoupled, the pump discharge valve is closed, the viscosity has changed, or there is an overload.

3. Bearing wear monitors are important in determining overall bearing life and whether the bearing wear rate shows any large changes.

4. Vibration measurements if obtained on a continuing basis will also indicate whether there are changes in pump performance.

5. Ball bearing monitoring should also be conducted since worn bearings will affect the outer rotor arm rotation eccentricity, which in turn will affect the inner rotor system.

These various monitors can be connected to shutdown devices to prevent excessive wear or mechanical problems in the pump.

Application. As discussed in Section 10.4.1, a large variety of magnetic drive pumps is available from pump manufacturers.[25] In addition, increased sales of zero-emission pumps have increased both the number of pump manufacturers producing these pumps and the number of pump models available. Manufacturers show pump capacities to about 5000 gpm (19 m^3/min) and very large horsepower ratings. European manufacturers also supply multiimpeller pumps that have specific head-capacity envelopes. Since there are wide variations in pump designs and characteristics, a thorough investigation of the equipment available is necessary to ensure a satisfactory installation. In addition, the possibility of containment can failure must be analyzed and an installation defined that minimizes potential failures.

10.4.3 General Observations

Sealless pumps are desirable for eliminating pump emissions and for reducing overall plant emissions on an annual basis, monitoring costs, and the possibility of having to implement a quality improvement program required by the HON regulations. However, these pumps have potential problems if they are not installed properly, as described in Sections 10.4.1 and 10.4.2. Some observations on sealless pumps are presented in this section along with potential improvements.

Overall Problems. Some problems observed in both canned and magnetic drive pumps follow:

1. *Bearing Failure.* This is a major cause of mechanical failures in the sealless pumps and results from running the pump without liquid in the pump section or flashing in the containment can.

2. *Closed Discharge Line.* The temperature within the containment can will increase and may result in bearing failure as a result of decreased viscosity in the bearings or localized flashing.

3. *Improper Rotation.* When starting a pump, improper rotation affects lubrication flow and will ruin the bearings.

4. *Poor Cooling Circulation.* For pumps that must have some cooling, interruption or poor circulation resulting in high temperatures will cause bearing failure. In canned pumps, high coil winding temperatures can affect the coil and increase the bearing temperatures.

5. *Decoupling.* In magnetic drive pumps this results in a temperature rise within the containment shell, and where the pump is not shut down relatively quickly, excessive temperatures can be a problem.

A major problem with sealless pumps is internal bearing failure which results from various causes. However, dry running appears to be the major problem in most installations.[23] This problem can be overcome through various process improvements.

General Recommendations. Based on the general problems identified, the following should be a considered in existing and new installations:

1. Install a minimum level sensor on the pump feed tank or feed vessel with an alarm/pump shutdown control at the minimum liquid level. A temperature indicator should be installed on the pump suction line for volatile materials.
2. A flowmeter on the pump discharge indicates when flow has stopped. This instrument can be set with an alarm and automatic pump shutdown control when operators cannot react immediately.
3. The suction line to the pump should be sloped to minimize pressure drop and suction piping installations should be streamlined. An increase in the pump suction line size should be installed.
4. Bearing monitors are recommended.
5. Temperature indicators should be installed on the containment cans to determine temperatures and can also be used as controls. These indicators should be inspected and tested on a normal maintenance schedule. A temperature indicator within the containment can will also indicate when volatile materials vaporize.
6. Thermistors are installed in the electrical coil windings to measure coil temperatures in canned pumps.
7. Temperature rise in the lubrication fluid exiting the rotor system should be monitored. New pumps may have this capability.
8. Liquid level probes can be installed in the pump discharge to ensure the containment can is completely filled with liquid.
9. Periodic pump vibration tests.
10. A recycle loop could be considered to minimize rotor lubrication problems during startup and shutdown. The loop is closed or blocked during normal operation.

All of these items do not have to be installed on a sealless pump. Based upon process conditions and a review by the rotating equipment engineers, required instrumentation and equipment can be installed.

Venting. While flashing in the pump is a serious situation, any vapor in the pump is also a potential problem. Before placing a pump on line, it should be vented to ensure all gas or vapor is removed and the containment can is completely filled with liquid, which requires a vent connection on the pump. However, some pumps may have a means of internally venting vapor and each pump must be examined to determine whether the pumps are self-venting or a vent opening is available on the pump where necessary. External vents cannot discharge directly to the atmosphere but must be tied into a closed system and connected with a control device (flare, incinerator, process, etc.). Since the system is completely closed, an indication of flow is desirable and a flow indicator in the vent or a simple level device at the vent outlet from the pump can be installed.

Drainage. A concern in a wide variety of sealless and sealed pumps is complete drainage of process fluid from the pumps which should be followed by flushing and a gas purge of all vapors. All of the drainage, flush, and purge material should be enclosed and sent to a closed-drain system or control device. Some of the comments on drainage in the following paragraphs are also applicable to standard sealed centrifugal pumps.

In sealless pumps, process fluid in the internal rotor or containment can section of the pump, if not drained and thoroughly purged from the pump, will present potential exposure problems in the shop where the pump undergoes maintenance and repair or in the field in the event the pump is opened. Liquid remaining in the pump obviously is a source of potential exposure, but is also a source of vapor release to the atmosphere when the pump is opened. Releases of environmentally hazardous substances (HAP or VOC containments) to the atmosphere present potential environmental and worker health exposure problems. This situation can also be exacerbated by a rupture of a containment can.

When a pump is shut down and prepared for removal from service, process fluid is initially drained from the pump through drain locations provided in the pumps by the pump manufacturer for fluid drainage. Some pumps have two drain points, with one drain below the impeller and another on the outboard end of the containment can, which permit drainage of liquid and vapors to the atmospheric pressure level. Purging with an inert gas will then remove the remaining vapors and/or flushing with a liquid removes the remaining pools of process fluid. Drainage from canned pumps is relatively straightforward. However, magnetic drive pumps are generally more difficult to drain than canned pumps because of the design of the pump—the outer rotor prevents the installation of drain plugs and, more importantly, permanent drain connections. Consequently, most pumps are drained from a plug or opening located below the impeller. For complete drainage in a magnetic drive pump, all remaining liquid in the pump must flow through internal passages to the drain. Any recesses or crevices in the internals will accumulate liquid, thereby preventing complete drainage. Obviously, drainage of a light liquid is much less difficult than a heavier fluid. The light fluid will vaporize and can be purged from the pumps, whereas heavier gravity fluids will accumulate in recesses. For example, butadiene at ambient pressures will flash, but a chemical such as benzene will remain liquid at ambient or slightly higher temperatures.

In draining a pump, the pump suction and discharge valves are closed and the drain valve is opened. To actually drain the entire pump contents, a vent must be opened. Frequently a vent to the atmosphere is opened, but this mixes air and process vapor, which is undesirable. The vent should be connected to an inert gas (e.g., nitrogen) to provide the makeup pressure that allows the liquid to drain. Where the pumped process fluid boils at low temperature, the pump contents will not drain but will completely vaporize. Consequently, a vent should be opened to allow the liquids to expand and vaporize, but the vent should be a closed vent discharged into a control device (flare, incinerator, etc.). Since process vapor remains in the containment can in either case, purging the can with inert gas to discharge through the closed vent would clear the containment can of process vapors.

When the pressure in the containment can has been significantly reduced, a flushing liquid may be introduced which would effectively remove all vapors and traces of process fluid. However, the flush should be sent through a closed system to the fluid receiver. Obviously, a variety of purging and flushing techniques are used. The examples are provided to indicate the environmental and health control requirements that should be considered when recommending and designing a pump installation.

Where drainage information from pump manufacturers is unclear regarding complete drainage of all fluids, a pump drainage test on the manufacturer's pump can be helpful. This can be accomplished in the manufacturer's plant by filling the pump with water and measuring the drainage quantity. The discrepancy in numbers will provide an indication of drainage effectiveness. As an alternative, the pump can be opened after draining all of the water to determine the water quantity remaining and the location.

For pumps in HAP or VOC service that do not drain well, some alternative is necessary to achieve proper pump drainage. One alternative is the selection of a pump that drains well, since drainage should probably be considered a selection criteria for all sealless pumps. Another alternative is a pump modification to improve drainage, which can be difficult in an existing design, but may provide sufficient impetus for pump manufacturers to modify their design. In summary, drainage should not be considered a minor mechanical requirement when the pump is shut down for maintenance.

Magnetic Bearings. Since internal bearings are potential problems in sealless pumps, some work is underway with magnetic bearings to replace the lubricated sleeve bearings, which are the standard bearings in sealless pumps. Magnetic bearings are used in centrifugal compressors and other types of equipment as standard bearing designs,[26] and operate without the need for lubrication. Improved bearing performance would increase the MTBF for the existing sleeve bearings. Published MTBF values for magnetic drive pumps indicate a 6-yr life for sleeve bearings,[23] which would increase significantly with successful magnetic bearing installations.

Although electromagnetic bearings are a technological step forward, the bearing system controls are an added complexity in the operation, but have been maintained satisfactorily in various installations. However, this overall bearing system will undoubtedly add to the pump installation cost and increase pump investment costs, but they will significantly increase the MTBF and overall service factor in all types of services. As shown in Section 10.4.4, on a life-cycle cost basis, magnetic bearings should have a satisfactory return on investment.

Safety. The major safety concern in sealless pumps is a rupture of the containment can since the wall thickness of the containment can is minimized in these pumps to reduce magnetic flux resistance. Thinner walls raise pump efficiency, which is an important consideration in power consumption, cost, and competitiveness. Consequently, containment can wall fracture or cracking is a distinct possibility, although it is an infrequent occurrence.

Where a containment wall fails in a canned pump, the internal process fluid fills the volume between the containment can and the outer casing. This large outer annular volume may be filled with an inert gas or a nonconducting fluid. In addition, the outer casing can be completely sealed when specified to prevent leakage from the outer casing. The typical sealed casings illustrated in Figure 10.12 will not leak under almost all conditions. However, a safety valve should be installed in the event a crack or leak develops, particularly where the outer annulus is filled with a nonconducting fluid. When using safety valves, the HON or VOC requirements should be observed.

A leak through the containment can will filter through the outer rotor bearings to the atmosphere in a magnetic drive pump. A small crack will of course result in a very small flow of fluid. Since the outer rotor zone is essentially at atmospheric pressure, some vaporization or evaporation of the process leakage can occur with low boiling point chemicals or VHAP compounds. Since the pump is sealless, emission concentration measurements are not conducted on a monthly or quarterly basis. However, this consideration indicates that a measurement with a TLV or gas detector should be conducted periodically for safety reasons. Large leaks would probably be revealed at the outboard drive rotor bearing during a visual inspection by an operating technician, which is required for sealless pumps for HON regulations. For small leaks this may not be sufficient unless there are other indications of a leak or changes in pump operating conditions, such as a temperature change.

Although inner bearings have been reviewed, another problem can develop with magnetic drive pumps since the main outside rotor drive shaft is supported by roller bearings in the pump housing. Should these bearings wear or fail, the outer magnet containing rotor arms can scrape against the containment can because of the very small distance between the can and the outer rotor. Should this occur, the can may be damaged, resulting in a leak that is quickly transmitted into atmospheric emissions. Severe leaks are a serious safety problem in addition to being environmental and potential personnel exposure problems. Another concern is that metal to metal contact can produce sparks that may trigger an explosion and/or fire.

A method of minimizing this problem in magnetic drive pumps is to install an antirub ring on the outer rotor to prevent rubbing of the rotor directly against the can.[20] In addition, an antirub ring is placed against the outer casing. This prevents the rotor from rubbing against the outer casing and also provides some degree of safety protection. These rubbing rings should be constructed of nonsparking materials.

10.4.4 Life Cycle Costing

When considering sealed versus sealless pumps and specific pump models, a comparison of original equipment costs has been an overriding factor in pump selection. More sophisticated comparisons may use the installed costs and also include energy charges. While this is somewhat more practical, the different cost analyses are all based upon original pump investment. A more realistic evaluation of pumps and pump alternatives is based upon life cycle costing, which requires considerably more information than equipment plus installation charges. The requirements and background for life cycle costing are presented in this section to provide a basis for considering various pump alternatives.

General. An important and perhaps critical element in selecting pump types for installation in chemical plants or petroleum refineries are regulatory environmental requirements. The new Clean Air Act in the U.S. (HON) and more stringent regulations necessary to meet other environmental objectives require improved pump emission control. Selecting pumps to meet these requirements is obviously necessary, but emission targets based on immediate regulatory requirements may not be sufficient. While new or proposed regulations are obviously stringent, additional regulations may be required in the near future that will further reduce allowable emission rates. Although a specific pump may be attractive for one set of requirements, it may not be satisfactory in a few years when additional monitoring and maintenance will be necessary to achieve environmental targets. Consequently, pump selection should also consider future requirements.

Where future regulations are addressed, the major concern is seal or shaft emissions and methods of meeting or reducing emissions to levels considerably lower than the promulgated regulatory levels. Equipment should not target the regulatory level for operation, but must operate at lower emission levels to ensure compliance with regulations since actual emissions will vary. Moreover, other environmental requirements should also be investigated to ensure compliance with *all* of the regulations. Additional environmental regulations are applicable in equipment startup/shutdown operations and drainage. Since regulations other than seal emissions are also applicable to pumps, *all* regulations must be observed and pump equipment must be installed to meet the various rules. These additional requirements may also have investment costs that must be included in any economic analysis. Overall pump investment costs must encompass all of the charges associated with an installation, since sealed pumps and sealless pumps may not have equivalent investment considerations.

When analyzing large investments, a project life is normally assumed and returns are calculated based upon unit investment, operating costs, and significant extraneous charges such as maintenance. This concept is also applied to large, specialized investments such as centrifugal compressor installations, which are usually analyzed on a similar basis. However, analyses of pump installations do not generally follow long-term investment criteria. Pumps are normally specified on a general basis and then purchased on the basis of competitive equipment pricing. This approach does not allow selection of the optimum pump and sealing systems for a long-term investment and does not provide a true cost comparison basis for this investment decision.

Overall cost comparisons for pumps should be based on long-term life, investment, environmental requirements, and a variety of other factors associated with pump operation. When considering this basis for pump selection, an overall cost analysis is termed *life cycle costing*. This is a term developed by the military to evaluate equipment options over life spans generally covering 20-yr periods. A complete list of components in a life cycle cost analysis will vary, depending upon study depth, location, manpower utilization, environmental regulations, and so on. A number of components in life cycle cost analysis for pumps are presented in the following paragraphs.

Life Cycle Costs

Original Investment. Obviously this should include the complete investment. This covers purchased pump costs, auxiliary equipment costs, and installation. Further details include the following items.

PUMPS. The actual cost for the pump is of course the purchase price. Where a pump has seals or special glands, all of these items supplied with the pump are included. In addition, spare parts for the pump should be included in the pump cost. For example, where a dual cartridge seal is purchased and placed in inventory for 24–30 months, money is expended on the part which also has an associated cost. The parts must be catalogued, stored in the computer, and then placed in a warehouse or another facility. The same situation arises with sleeve bearings for rotors in sealless pump service.

In addition to these two major items, other parts are purchased and placed in inventory. All of these parts are a considerable investment and must be controlled, which also adds to the overall pump cost basis. When a pump cost investment is tabulated, the variations in parts requirements can have a significant impact on overall cost and total life cycle cost.

AUXILIARY EQUIPMENT. This term covers all of the equipment (excluding instruments) necessary for a *complete* pump installation. For a dual seal pump, this includes:

- *Sealpot.* This should be an ASME code vessel requiring code approval.
- *Sealpot Circulation Piping.* The piping to and from the pump seals.
- *Sealpot Vent Piping.* The piping from the sealpot to the flare or any other closed device.
- *Other Sealpot Piping.* The sealpot oil fill piping, drain piping, inert gas piping, underground drainage.
- *Central Seal System Equipment.* Where a sealpot is not used, the portion of the central barrier fluid distribution system allotted to the pump should be defined. This may also include barrier fluid pumps, orifices, gauges, and piping.
- *Gland Piping.* Both flush and drain piping should be included in this cost when applicable. In addition, drain piping should also be included and, when applicable, any other underground allocated piping charge.
- *Drain–Purge–Vent Piping.* For sealless pumps, the piping necessary for these pumps should be included in piping costs. The piping also would cover any safety valves necessary for casing protection. The drain piping includes any underground allocated piping equipment charge. Where the term piping is mentioned, this also includes all valves that are normally installed.

Instruments. All of the direct purchase charges for instruments, alarms, indicators, recorders, conduit switching, and so on should be included with investment costs. In addition, sealpot instruments and any others not specifically itemized should be included. Instrumentation associated with current measurement requirements for sealless pumps is obviously included in this category, such as temperature indicators and power consumption.

Pump Foundations. These are usually included in any estimate of pump installations, since it is a common construction requirement. If not required for certain sealless pumps, an adjustment credit for removal should be included in any cost comparison.

Installation. All of the equipment, including instrumentation, must be installed and an accurate basis for estimating this particular work function is necessary. Moreover, the estimate must include underground piping/drainage, lateral extensions, vent piping, liquid seal boxes, or any other drainage facilities required for the specific pump.

Since there are a variety of components in the auxiliary component category, costs should be estimated for the different items outlined above. With sealless pumps having a number of electrical instruments, a reasonably accurate estimate of these costs is necessary, including tie-ins for computers and remote and local indicators. Estimated costs for all the auxiliary facilities, instrumentation, drainage, venting, and so on, if not available from the estimators, should be developed, since more detailed analyses are required to evaluate alternative pump proposals.

An important installation consideration for standard API/ANSI and magnetic drive pumps is a procedure which is necessary to align the pump and motor driver shafts. This procedure is somewhat time consuming, although lasers are used to assist in alignment. Alignment is not required with a canned pump. Consequently, dual seal or magnetic drive pumps are debited for this cost.

When evaluating the installation of a pump in an existing unit, any differences between the sealless and API/ANSI pump configurations may require additional installation charges. This cost should be included in an expansion or debottleneck project.

Land. There is a large difference in the plot area needed for canned, magnetic drive, or API/ANSI pumps. Also, vertical pumps should be included in this evaluation. Since pumps are normally installed in rows, canned or vertical canned pumps require significantly less area. Thus, on an area basis, smaller pump rows are necessary for these pumps compared with standard API/ANSI pumps. This can be translated into small land requirements for canned horizontal or vertical pumps. Credits for smaller land areas should be established, since a process unit layout can be modified to take advantage of a revised plot layout.

Energy Costs. An obvious consideration in life cycle costing is the cost of energy over the period considered. This is generally a straightforward cost based upon overall efficiency of the pump system including the driver. However, in the standard dual seal system a central seal fluid supply system has an energy cost that must be distributed among the pumps on the seal oil system. For some pumps, energy is consumed in the form of steam on the seal faces. Although small, this must be included along with cooling water consumption. This is also a small but definite cost. Where *dual gas seals* are installed, the energy loss for the seals are considerably less than the losses for standard dual seals, resulting in lower energy costs.

Capacity Changes. Normally, the capacity of an operating pump does not change with time. In the magnetic drive pumps, the permanent magnets will lose their magnetic strength over time. This effect is translated into capacity loss and the deficit associated with a loss in magnetic field strength should be included in any life cycle evaluation.

Maintenance. This is a very important consideration in the evaluation of pumps on a life cycle cost basis. In both sealed and sealless pumps, maintenance has a considerable impact on a life cycle cost basis. Where seal life or bearing life is quite small, frequent changes add to material and labor costs through the life cycle. As a result, the MTBF becomes important.

GENERAL. In dual seal pumps, seal failure that requires the complete replacement of both seals is a major maintenance item. The seals, particularly in cartridge form, may have a maximum life of 24 months. The other major failure item is the shaft bearings, which may also contribute to shaft failure. The MTBF values for dual seal pumps at a plant can be established, which provides the basis for estimating the life cycle cost.

For canned pumps, sleeve bearing failure is a major cause of failure in about 70% of the pumps. These failures occur over periods of months to perhaps 10 yr. As a result, a wide range of MTBF values exists for these pumps, and considerable effort is underway to improve the average MTBF. With improved operating conditions, the MTBF may significantly change. For life cycle costing at this point, the MTBF should be defined by plant experience with the particular equipment. Assuming the MTBF of 24 months, the life cycle cost will probably be governed by this factor. Where the MTBF is 9 or 10 yr, life cycle costs are considerably lower. Magnetic drive pumps have an MTBF range similar to that of canned pumps and plant experience again determines a desirable MTBF for a life cycle cost evaluation. Reference 23 indicates the MTBF for magnetic drive pumps is about 6 yr. Although considerably less than 10 yr, this is still a much longer period than the MTBF for mechanical seals.

With mechanical seals, bearings, and sleeve bearings, the major repair problems and actual maintenance costs for these items should be established with repair at the unit versus repair in a central shop considered in this analysis along with the effectiveness of the repair. When the pumps are repaired satisfactorily at the unit, the cost may differ significantly from repairs performed in other locations, which require a number of steps that probably add significant costs:

- Drain and purge the pump
- Disconnect the pump
- Truck the pump to the shop
- Repair the pump
- Truck the pump to the unit
- Install the pump and connect to the piping
- Realign the pump if necessary

INSTRUMENT MAINTENANCE. There are instruments on both dual seal and sealless pumps. On the dual seal pumps these are standard instruments for the pump and sealpot which were described in Section 10.3. Instrument requirements for sealless pumps differ from the standard instruments on dual seal pump installations and also include energy consumption for the driver coil or motor. Consequently, instrumentation is more elaborate and higher overall maintenance costs can be expected for the sealless pumps. Also, sealless pump instrumentation is connected to the computer control system to aid in analysis and control, adding additional complexity to the instrumentation network. Careful estimates of the instrument costs, installation charges, and maintenance should be established for the various pump installations.

Monitoring and Reporting. As described in Section 10.1.5, monthly monitoring of single seal pumps is required by the new HON regulations. Replacing the single seal with dual seals eliminates the monitoring requirement. However, the pump must be visually inspected once per week. Although a very small effort is required from the operating technicians, the seal alarms must be checked periodically to indicate seal failure. When a seal failure is indicated, the pump must be shut down and seal repair begun within 5 days. More importantly, the history and effectiveness of the repair effort must be recorded. While not a very large effort, this seal failure record does require manpower, and costs should be established for this general operation.

Environmental Costs. Separate environmental costs have not been developed for pumps. Since the new Clean Air Act (HON) is significantly more stringent than the existing act, certain costs should be developed based on the proposed EPA rules, which will probably be promulgated as proposed in the rules issued December 30, 1992 for equipment.[5]

EPA TARGETS. The overall EPA target for emission control is a reduction of 90% in fugitive emissions. If this target is not achieved, the EPA can require changes that will ensure 90% or greater emissions reduction. In addition, the EPA has initially set a target for emissions control on sealed pumps. Should emission monitoring indicate a pump has a concentration greater than 1000 ppm, the pump is declared a leaker. When three pumps are confirmed leakers, or a percentage greater than three pumps for two consecutive monitoring periods, a QIP is required. If more than 10% of the pumps are greater than 1000 ppm, a QIP will be initiated. These programs require additional monitoring, maintenance, record keeping, and in some cases new pumps to meet the revised emission requirements.

The QIP will be an added cost burden that must be considered when comparing investments on a life cycle cost basis. Actual costs for QIPs will have to be determined although these programs may not be included in an overall life cycle cost evaluation. Installation of sealless pumps or gas lubricated dual seals tends to avoid potential QIP programs.

OTHER COSTS. The new Clean Air Act (HON) has some other considerations that can also impact upon costs, but are difficult to estimate. There are fines, penalties, and emitting fees that can be assessed when regulatory emission specifications are not attained. Also, fees can be charged when requesting permits, which are based upon emission estimates. While potential legal fines or penalties are extremely difficult to estimate, the elimination of pump emissions removes chances of penalty assessments. Fines or penalties would be assessed on the overall unit, but nonleakage pumps would not share in this cost proration.

Engineering. All of the engineering efforts, particularly those associated with specific projects, should be estimated and included in life cycle costs. Charges for sealless pumps are considerably more than charges for sealed pumps.

Research and Development. Any effort on a continuing basis, including estimates of cost contributions to general programs, should be included in life cycle costs. While cost support varies considerably, poor operations frequently result in local R&D efforts that can be costly. As an example, poor performance by sealed or sealless pumps in a process unit normally generates an R&D program to improve operating performance. These R&D efforts may also be generated when the new Clean Air Act (HON) is promulgated. A pump quality improvement program as described in Section 10.1.5 would result in a considerable program to improve (reduce) pump emissions. When considering a QIP, engineering R&D costs will be a significant cost that should be included when selecting pumps.

Noise. When a separate motor is part of the pump installation (sealed and magnetic drive pumps), noise levels are considerably higher than the noise generated in canned pumps. Although motor noise has been reduced, in some locations special noise control systems must be installed. Obviously, this additional cost is a debit that should be included in a life cycle cost study.

Summary. All of the costs itemized should be considered in a life cycle cost comparison analysis. Restricting pump cost comparisons to a quoted price does not provide a true economic evaluation of alternatives. For example, when the average seal life of a mechanical seal is perhaps 24 months, the MTBF of a sealless pump is 8 yr, and perhaps a 10 yr dual gas seal life, a considerable difference exists in maintenance costs. Over the 8-yr period the mechanical seal installation requires replacement four times, compared with the one sleeve-bearing seal replacement for the sealless pump. This difference becomes greater as MTBF longevity increases, indicating the need for accurate seal life data. Moreover, pump life for standard centrifugal pumps may well exceed 15 yr.

One item that is unclear is the cost for a QIP. However, we do know that the costs are considerable, and a very good mechanical seal maintenance program is necessary to ensure the number of leaking pumps are < 3. While this target may be met in some installations, any reduction to a 500-ppmv leak definition will probably result in leaking pumps exceeding the allowable limit. An important consideration in any estimate is future requirements, which should be one element in any cost analysis.

10.5 COMPRESSORS

There are three basic compressor machine devices: centrifugal, reciprocating, and rotary. In this section only the centrifugal and reciprocating compressors are reviewed since the great majority of machines are in these two categories. Uncontrolled older EPA emission factors for compressor seals are quite large and in petroleum refineries they are greater than the pump emission factors by a factor of 6 and are about fivefold greater than pump emission factors in SOCMI. However, the compressor portion of total uncontrolled fugitive VOC and HAP emissions from refineries and chemical plants ranges from about 1–5% of total onsite emissions quantities, although NSPS regulations have reduced VOC emissions. The relatively few compressors in process units results in low total overall mass emission loss rates for these machines.

10.5.1 Regulations

The new source performance standards (NSPS) for equipment leakage of VOC fugitive emissions indicates that compressors are considered an emissions source for regulatory purposes, but compressors in petroleum refinery hydrogen service are exempt where the hydrogen stream concentration is always larger than 50 vol %. Regulatory requirements must be carefully reviewed in existing units to determine applicability, whether any process or compressor modifications will change, and the associated regulatory status of the compressors. A summary of NSPS compressor regulations is presented in Section 10.1.2.

The regulations for compressors in the new HON equipment requirements are essentially identical to the NSPS regulations for compressors in VOC service. Requirements in the existing standards are considered maximum achievable control technology (MACT) and the EPA retained the NSPS regulations for the HON rules. The basic regulations require the use of mechanical seals equipped with a barrier fluid system and controlled degassing vents or enclosure of the compressor seal area with emissions venting through a closed-vent system to a control device.[5] These systems provide control efficiencies approaching 100%, along with a 500-ppmv leak definition. One difference is that a barrier fluid in HON compressors should not be a light liquid, which contrasts with the VOC regulation that requires a fluid with heavy liquid service or one that is not in VOC service. Consequently, barrier fluid selection must be carefully reviewed for regulatory or potential health exposure considerations. Vapor or gas is also used as a barrier fluid with certain types of seals and should also be reviewed prior to service specifications. In addition, the NSPS regulations specifically identify reciprocating compressors for controls, but these compressors were not identified separately in the HON regulations essentially requiring identical controls or a leak definition of 500 ppmv.

In addition to the federal regulations, various state and regional authorities have also promulgated compressor emission control regulations. Some of these are presented in Sections 10.1.3 and 10.1.4. In some areas, these regulations may be more restrictive than the federal regulations and should be reviewed to ensure compliance. Reciprocating compressors may have special regulatory requirements. Where changes in existing centrifugal compressor seals are difficult to install, regulatory status and requirements should be ascertained to determine whether a new retrofit seal installation is required to meet the applicable regulations. If a seal change or modification is a difficult undertaking, then industrial hygienists should review potential control changes to ensure emission leakage will be at acceptable levels in the workplace. The discussion in the following sections on compressor seals provides information on various control alternatives.

10.5.2 Centrifugal Compressors

Centrifugal compressor seals and bearing designs have been changing through the years as both items have improved along with more efficient and compact machine designs. Bearings in older compressors were often located outside the main compressor case, with the seal located at the juncture of the shaft and compressor case. Compressors now have the bearings and seals located within the main casing. The differences shown in the simplified drawings in Figure 10.14 are important

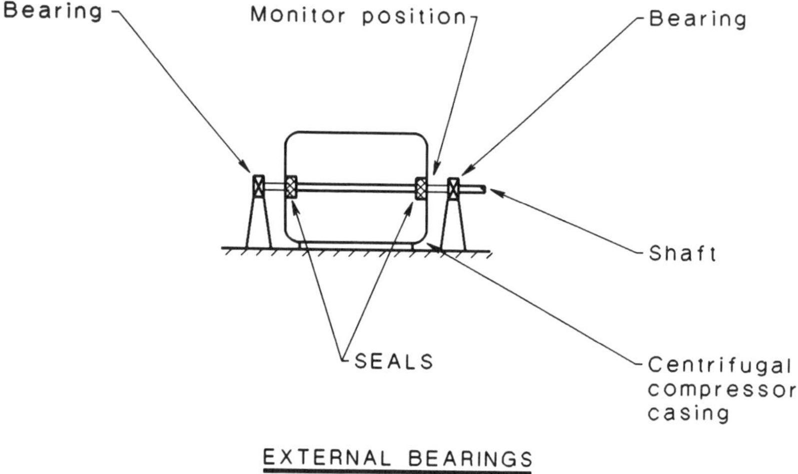

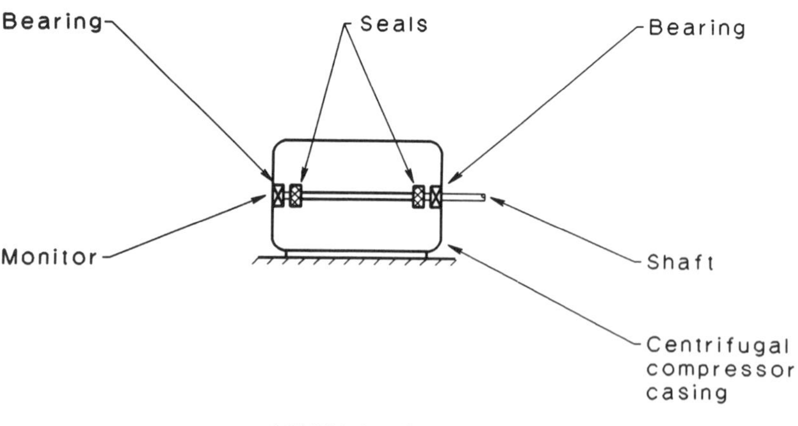

Figure 10.14 *Simplified sketch of bearing and seal locations in centrifugal compressors.*

when considering seal leakage from a machine and for understanding the physical relationship of seals and bearings.

In the old machines with external bearings, seal leakage to the atmosphere could be monitored directly, as shown at the monitoring location in Figure 10.14. However, in machines with bearings in the case, seal leakage must flow through the bearings to reach the atmosphere. Monitoring around the shaft must be considered carefully to define actual contaminant concentrations. Since the bearing oil is generally a heavy liquid, an oil mist may be combined with process vapor and the monitor or measuring device response to a mixture of this type requires some investigation. Oil mists have an exposure limit and their concentration should also be considered in the measurement. Bearing oil often contains additives and a toxic assessment of the oil is necessary to determine maximum allowable exposure level concentrations.

A variety of centrifugal compressor seals are installed in operating compressors and newly developed seals are installed to further improve sealing efficiency. The seals for the centrifugal machines are normally selected in conjunction with the compressor manufacturer to meet plant requirements. Since a variety of seals are produced, seal review in this section is limited to several nominal designs that can be used as a basis for assessing other seals against exposure control requirements. Seal design considerations are the high rotational speed of centrifugal compressors and the large shaft diameters, compared to pump operations where the rotational speed is much lower and the shaft diameter is considerably less. A major difference in compressor and pump mechanical seal operation is the forced movement of oil through compressor seal faces for lubrication at the high compressor rotational speeds. In contrast, the pump mechanical seal faces have very thin oil films to minimize fluid leakage through the seals. In the following seal presentations, one major consideration in many seal installations is the dual use of oil for sealing and bearing lubrication. A common oil for both functions simplifies auxiliary equipment installation and control systems. However, the common oil requires stripping or decontamination of used seal oil prior to recycling in order to provide a clean bearing oil. A major improvement is the installation of gas seals with a buffer gas, similar to the dual gas lubrication seals in pumps, which has significantly reduced seal emissions.

10.5.3 Labyrinth Seals

The labyrinth seal is one of the original seals installed in centrifugal compressors, and some of these seals are still operating in older compressors. Other seal types are normally purchased and installed with new machines to reduce leakage, but modified labyrinth seals combined with other seals are still installed in process services. The labyrinth seal basically consists of a large number of rings that restrict the flow of gas along the shaft, as shown in Figure 10.15.

A vent connection is often provided with a labyrinth, as shown in Figure 10.15, where nitrogen or an inert gas can be injected into the seal through the vent opening. The injected gas flow will separate, with a portion of the gas entering the process stream, minimizing leakage through the labyrinth with the remainder of the inert gas stream flowing to the atmospheric pressure end. Gas injection is one technique for controlling leakage. A second system is shown in Figure 10.15 where a slipstream from the process gas discharge line is sent through an eductor, placing

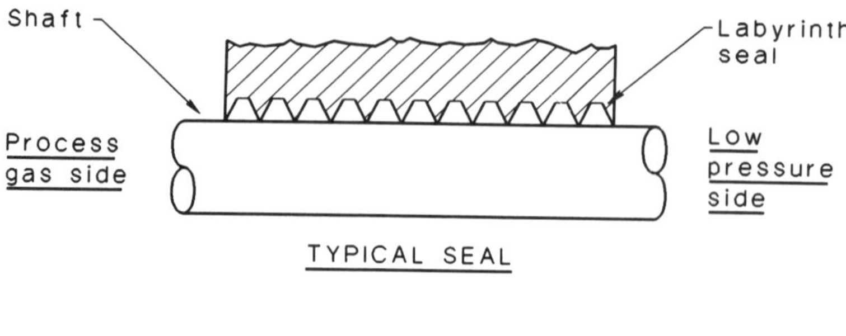

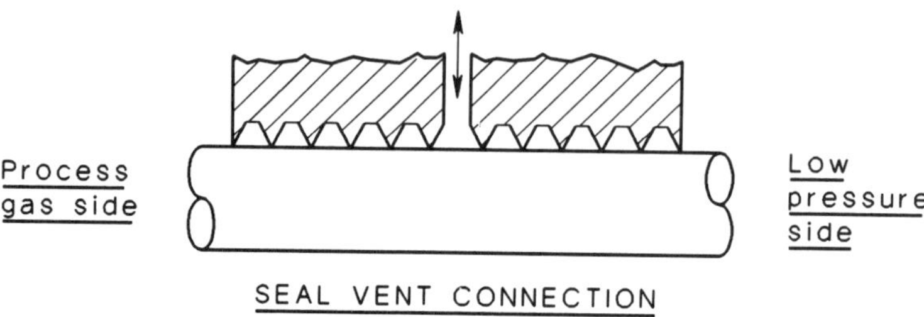

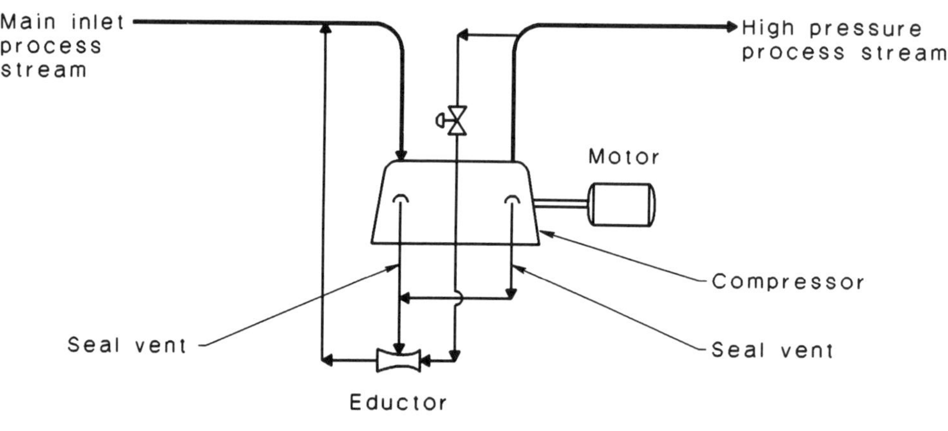

Figure 10.15 *Labyrinth centrifugal compressor seals and controls.*

a vacuum on the seal vent. Process gas leaking through the labyrinth is drawn into the eductor where it blends with slipstream gas and is recycled to the compressor suction. The labyrinth seal vent on the high-pressure side of the compressor can also be connected directly to the eductor, which returns the gas to the compressor suction side. Although the pressure at the vent outlet is sufficiently low to permit withdrawal of the process leakage through the eductor, a little air may be drawn through the labyrinth from the atmospheric side, which is recycled to the suction with some process vapor. This aspect may present potential safety concerns regarding a mixture of flammable material and air. However, the mixture in the eductor may be on the rich side, but this consideration should be reviewed with safety personnel.

With the labyrinth eductor system that recycles vent gas to the compressor suction, the closed system meets the regulation for recycling to a control device. Emissions to the atmosphere in this situation are not a problem and exposure potential is very low. However, the injection of a gas into a vent that leaks in both directions may cause somewhat higher emissions. Moreover, compliance with the regulations may be questionable in this situation. In labyrinth seal installations, the seal faces should be carefully monitored for compliance with the 500-ppmv leak definition and the establishment of leakage rate estimates.

10.5.4 Mechanical Seal Type

An older mechanical seal type was shown in an EPA document[27] and a simplified version of the seal is presented in Figure 10.16. This seal conforms to the general regulatory requirements discussed in Sections 10.1.2 and 10.1.5, that is, a barrier fluid seal system with a seal fluid pressure higher than the process gas pressure. As shown in Figure 10.16, the seal fluid at high pressure is forced through the seal faces and a considerable portion of the fluid moves toward the process gas. The

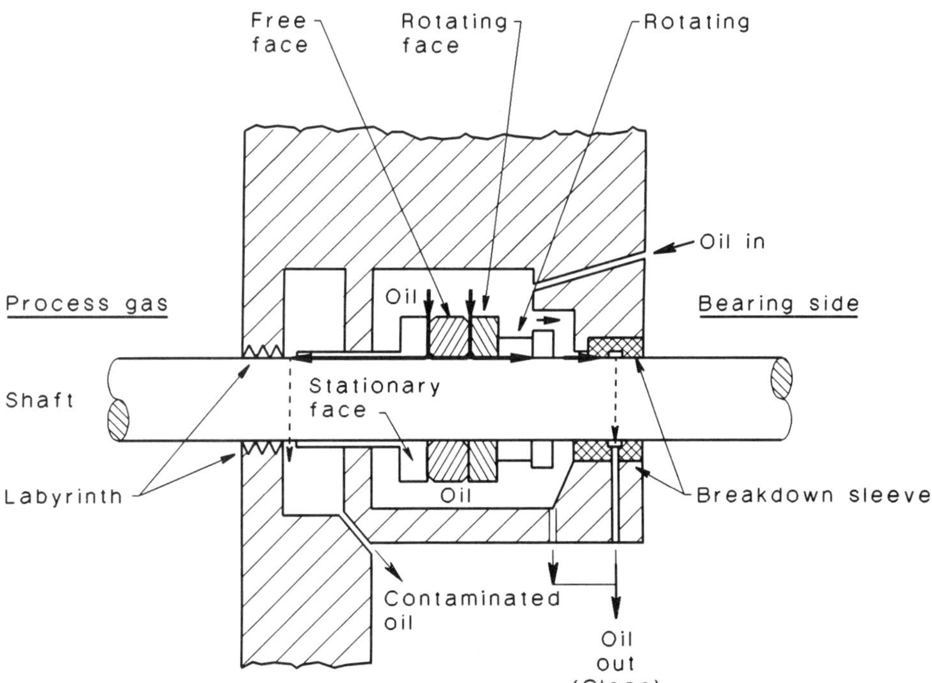

Figure 10.16 Sketch of a centrifugal compressor mechanical seal.[27]

process gas pressure is reduced through the labyrinth and the lower pressure gas contacts the seal oil on the seal side of the labyrinth at a low pressure. Oil withdrawn from this gas–liquid chamber is termed contaminated oil owing to the absorption of process gas. The contaminated oil is discharged into a separation pot where the contaminated gas is flashed or stripped from the oil and vented. This vent should be connected to a control device to minimize potential release of toxic materials to the atmosphere in accordance with the regulations.

A breakdown sleeve is required to maintain seal oil pressure in the oil chamber, which must be greater than the process gas pressure. This seal oil pressure is also required in the EPA regulations. The oil pressure decreases along the shaft until the oil reaches the ring in the breakdown sleeve. The oil drains from the sleeve ring into a collection pot for recycling to the seal system. The contaminated oil is degassed and is also recycled. A description of a typical oil collection system is presented in the next section.

10.5.5 Liquid Film Seal

A liquid film seal is shown in Figure 10.17 and a bearing is also included to complete the overall installation. This seal conforms to the regulations with a barrier fluid that is at a higher pressure than the process gas. As shown in the

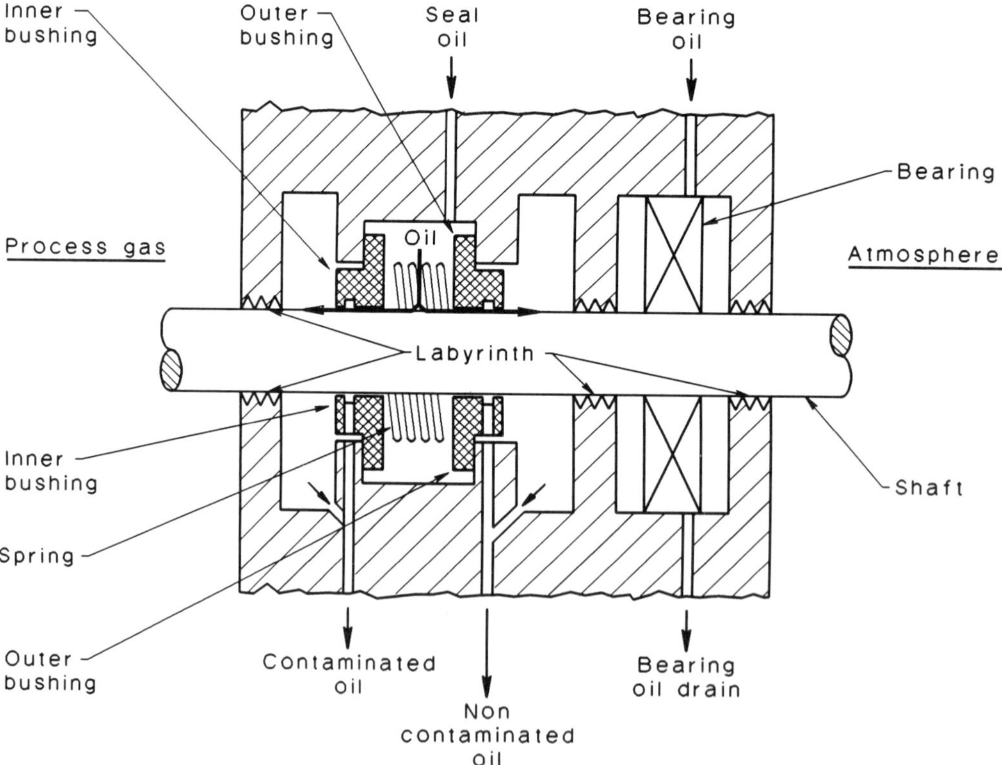

Figure 10.17 Sketch of a centrifugal compressor liquid film seal.[28]

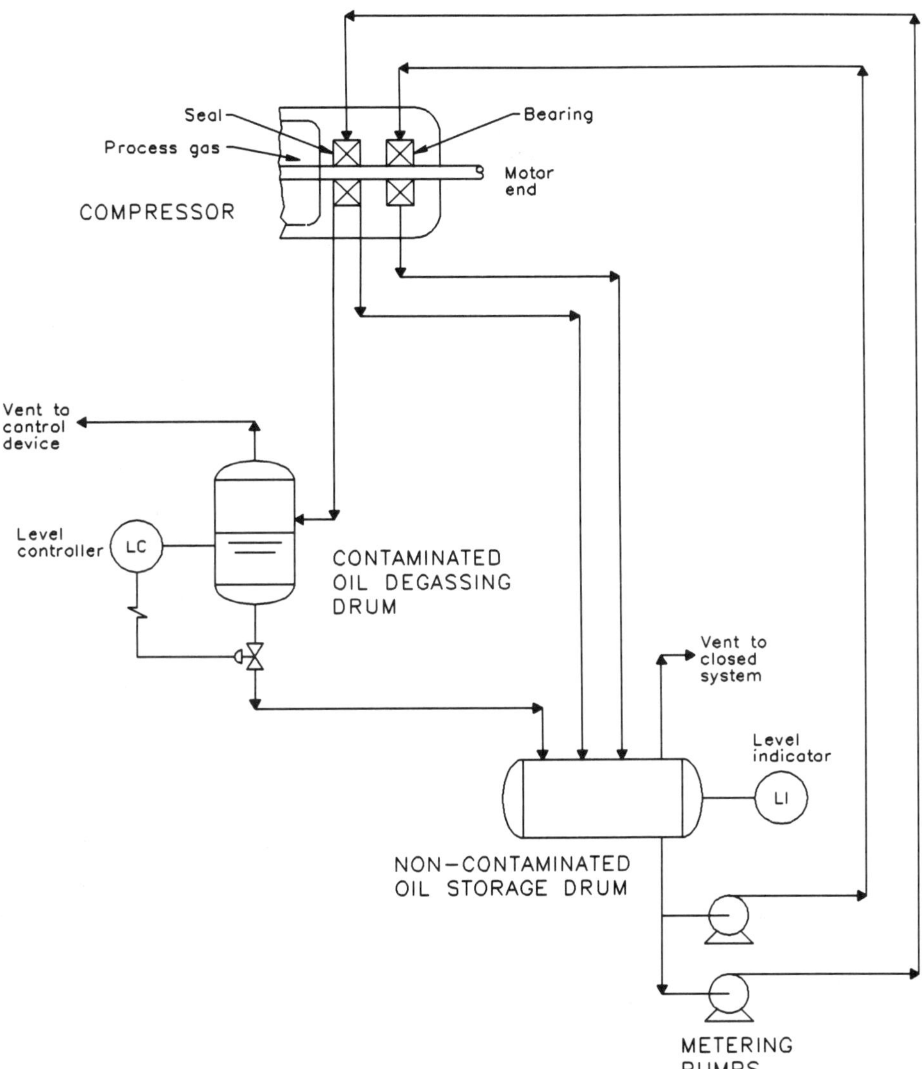

Figure 10.18 *Simplified sketch of a centrifugal compressor oil drain and recycle system.*

figure, the oil at high pressure is forced between the bushings and the shaft, providing a seal against the gas pressure. The process gas pressure is reduced as it leaks through the labyrinth and meets the seal oil, where oil becomes contaminated through absorption and is drained from the inner bushing and the inner chamber. Oil in the outer bushing system is clean when the seal operates normally and is termed noncontaminated oil. In addition, oil from the bearing is also considered clean and generally is combined with the noncontaminated oil from the seal. The overall drain and oil recycle system is shown in Figure 10.18.

Contaminated oil is drained to a degassing drum where traces of process gas are separated from the oil. The degassing drum provides holdup time and permits release of the contaminants. However, separation may be incomplete and a heater coil may be required in the degassing drum. If a simple heating coil is unsatisfactory, a more elaborate stripping system may be installed, but the stripper will be small. A level control is located on the degassing drum to maintain a liquid level in the drum at all times and provide some holding time for degassing. The vapor from the degassing drum can be toxic and should be sent to a closed system or a control device. The EPA regulations require sending the drum emissions to a control device and regulatory recommendations meet industrial hygiene objectives. The vent from the noncontaminated drum is normally open to the atmosphere since the barrier fluid storage drum does not contain any light contaminants. However, in an upset situation vapors can be released from this vent, which should be closed as shown in the figure. Hazardous pollutants emitted from an open vent can present potential exposure concerns. The regulations indicate that the barrier fluid system should be connected through a closed system to a control device and the storage drum is a component of the overall barrier fluid system.

The regulations require sensors for both the barrier fluid and seal systems to indicate failures in either system. Failure indicators in the *barrier fluid* system are readily installed and are recommended by instrument personnel. However, seal failure instrumentation is a function of the type of seal installed and the relationship of pressures in the seal to the process gas pressure. For barrier fluids at a pressure higher than the process, a pressure indicator showing a low barrier fluid pressure in the seal box would probably indicate that the seal had failed. The instrumentation is essentially tailored for the seal installation.

Another consideration for the barrier fluid seal installation is a simple once-through flow system scheme where contaminated seal fluid is not degassed but is returned directly to the process. Although simpler in concept than a stripping recycle operation, compatibility of the contaminated seal liquid with a receptor location in the process may be difficult due to the contaminant. The seal liquid must be a heavy liquid or non-VOC fluid based upon the boiling point, as described in Section 10.1. A high boiling point compound or very low vapor pressure compound containing a very light contaminant is a mixture that cannot easily be discharged into a chemical process.

A gas purge through a seal that captures all of the emissions and transfers them to a process receiver or a control device is also an accepted regulatory procedure that exempts compressors from barrier liquid seal requirements. This method also depends upon a seal system somewhat different than the liquid seal devices. The gas seal system as described is a once-through operation, but the plant must find a purge gas source along with a receptor location for the mixed or contaminated gas. The purge gas for this operation should be a non-VOC or non-HAP gas to prevent fugitive emissions from leaking to the atmosphere through the seal system. A dead-ended inert gas must be at a higher pressure than the process pressure to ensure that the emitted gas is nontoxic, but there is an economic cost for this operation. Newer gas seal designs provide an alternative that is described in Section 10.5.6. The regulations, although directed at seal emissions control, are an excellent basis for general emission control equipment requirements. An understanding of the engineering controls required by the EPA provide a basis for estimating emissions and workplace exposures. A method of determining workplace exposures from a leaking seal is presented in Chapter 16. While regulatory requirements are directed at controlling seal emissions, the EPA has not regulated other aspects of the sealing systems that present potential problems as a result of vapor emission controls. A particular concern to the industrial hygienist are the sealing fluids and potential exposures during shutdown, purging, flushing, and maintenance. Seal fluid toxicity data are important to the industrial hygienist and the entire seal fluid system should be examined to insure that sufficient controls have been installed in accordance with seal fluid toxicity. In addition, the industrial hygienist should examine the degassing procedure to insure that the contaminant concentration in the decontaminated fluid is acceptable for workplace exposure control if a leak develops or maintenance is necessary.

10.5.6 Dry Running Gas Seal

These gas seals are similar to the upstream pumping seal described in Section 10.3.7. One face of the seal contains spiral grooves but is designed for operation in a gas. The groove pattern is a series of logarithmic spirals recessed into a hard

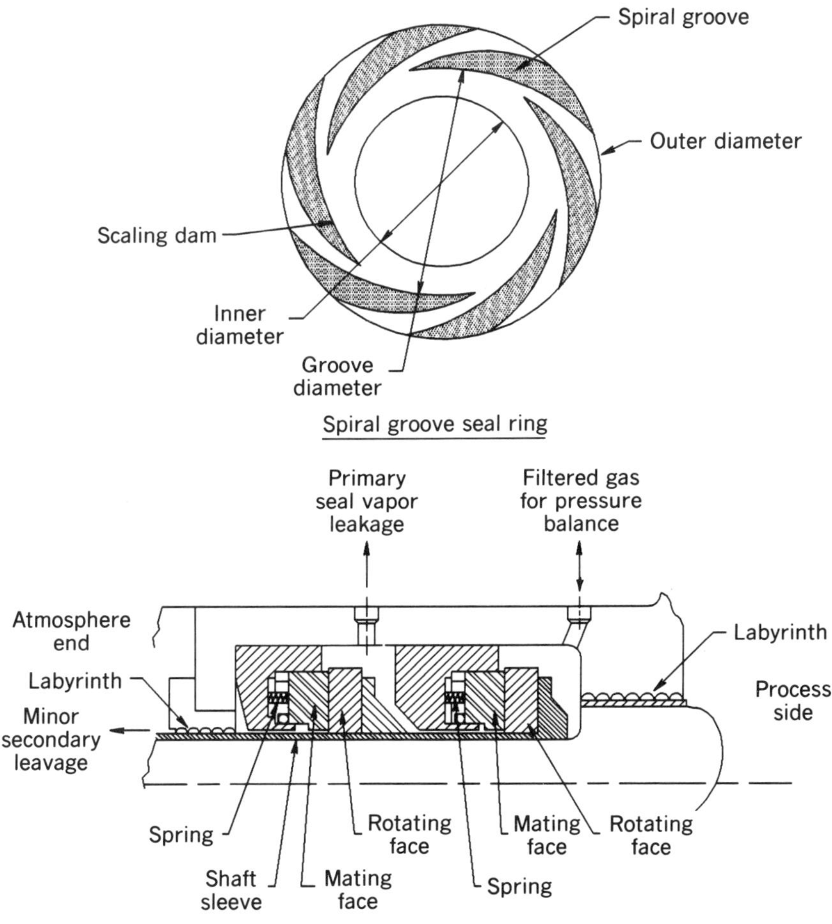

Figure 10.19 *Simplified sketch of a spiral groove dry seal and a tandem seal installation (courtesy of John Crane, Inc.).*

mating ring as shown in Figure 10.19. These grooves are very shallow, as described previously (1–8 μm), but are effective in developing a gas film between the spiral groove ring and a flat ring constructed of a different material. As the seal ring rotates, gas is compressed from the outer diameter inwardly to the grooved diameter and then expands over the sealing dam to a lower pressure. This expansion generates sufficient force to open the joining seal faces an extremely small distance which has been described as nanometers.[17] A thin gas film which is compressible or elastic separates the two faces and this property is termed film stiffness.

In separating the seal faces the gas film eliminates frictional contact, which essentially eliminates wear.[29] Results on wear with these seals indicate a long operating life, which is important since process compressors may operate for periods up to 5–6 yr between turnarounds. Bearing life appears to be a considerably greater maintenance problem than the dry running gas seals. A long bearing life that may be possible with electromagnetic bearings can probably extend seal life well beyond 10 yr of operations.

When a compressor starts to rotate, there is contact between the seal faces for a very short number of rotations until the seal opens. A small flow of gas develops between the seals, which aids in removing some heat that is developed as a result of gas shear in the film. However, the amount of heat is quite small and the gas flow for seals is less than half the leakage observed with oil film seals in process service.[31] As experience with these seals increases and seal improvements continue, further reductions in leakage can be anticipated. The manufacturer of these curved groove seals has developed various leak rate, pressure, temperature, and revolutions per minute (rpm) relationships that provide a basis for estimating seal leakage. Compressor rpm varies with each machine and is a critical parameter in designing seals.

Basic System. A typical installation for pipelines, petroleum refineries, and chemical plants has been tandem seals to control leakage, and a simplified sketch of a tandem seal system is shown in Figure 10.19. Leakage from the primary seal is vented to a closed system and control device. In contrast to pump seals, compressors recycle some gas to maintain a balanced seal pressure that is the same for suction and discharge sealing systems. Since the gas will flow through the seal, the gas is filtered to ensure that solids will not interfere with the primary seal faces. Secondary seal leakage to the atmosphere is quite small, with most of the leakage from the primary seal discharged through the vent.

While this seal has performed extremely well in various process gas services accompanied by very low atmospheric rates, there is still some leakage, which is a potential concern with toxic or HAP compounds. However, in existing installations monitoring data will indicate the concentration at the shaft end of the compressor, which may be less than the 500-ppmv leak definition in the NSPS and HON regulations. A level less than 500 ppmv indicates that leakage is satisfactory and the unit may be exempt from the regulations shown in Sections 10.1.2 and 10.1.6.

When the concentration exceeds 500 ppm, a change in the seal configuration may be desirable to meet regulatory requirements and reduce potential exposure conditions. This can be accomplished by changing the seal configuration as described below. Another alternate is a purge gas in the outer labyrinth permitting disposal of the leaking vapor.

Double Opposed Seals. In situations where the vapor is a hazardous air pollutant and leakage must be essentially nil, double opposed seals can be installed which are similar to double mechanical seals in a pump. In addition, a buffer gas is injected as a barrier fluid between the seals and any leakage through the seals is split between the process gas and the atmosphere. Innocuous buffer materials are required and nitrogen is frequently selected as the buffer gas.

Nitrogen at extremely low rates will dilute the process fluid but should not present any process difficulties in the majority of chemical and petroleum refinery process. Leakage of nitrogen to the atmosphere does not pose any problems with estimated seal leak rates less than 1 cfm (0.028 m^3/min).[17] The concern with this installation may be the nitrogen cost, which is a function of consumption. However, energy savings and little seal maintenance more than offsets the small nitrogen losses.

In addition to the elimination of leakage, the buffer system also eliminates oil contamination associated with oil seal systems. This system does not require oil stripping or degassing. Moreover, process gases do not enter the bearing system, eliminating any trace of process gas in bearing oil. As a result, potential exposures from the equipment associated with contaminated oil are no longer a concern. In a simplified bearing oil system, the only potential exposure source is the oil.

Double opposed seals are generally installed in process services with a maximum process pressure of 250 psig or 1725 kPa and temperatures to 500°F (260°C) per reference 17. However, improved seal designs and quality control raise these limits and the manufacturer will provide an improved assessment of maximum operability conditions. Tandem seal installations apparently have a maximum pressure rating of about 2000 psig (139 bar), although higher pressure conditions are currently under test.

While nitrogen vapor at very low rates flows through the bearings, this flow cannot be eliminated by circulating the nitrogen to the process or a reception point. Circulation meets the regulatory requirement for a closed-loop barrier system. Dead-ended nitrogen systems, although effective, do not meet the regulatory barrier fluid requirements which are based on liquid designs. However, the nitrogen dead-ended system can achieve exempt compressor status by demonstrating that leakage concentrations are less than 500 ppmv.

Single Seals. Normally, single seals are installed where the process fluid is inert or the single seal leakage rate is acceptable. For current and anticipated regulatory requirements, a dual seal arrangement is preferred.

Life Cycle Costing. An extensive method of determining life cycle costs for pumps and pump seals was presented in Section 10.4.4. For compressors, life cycle cost analyses are recommended to determine an optimum sealing system to meet regulatory and health exposure requirements. Since centrifugal compressors are costly investments, a life cycle analysis should be a standard review procedure.

The important components of this analysis are maintenance costs, emissions losses, and downtime costs since centrifugal compressors, particularly large machines, are spared infrequently. In addition, environmental problems and potential charges should be estimated and included in the analysis.

Maintenance costs are related to seal wear and seal malfunctions. Seal wear with spiral groove seals is almost nil and should be available from existing plant data and published comparative data. While seal wear is essentially nil, the analysis will probably indicate improved bearing performance is desirable. Credits for spiral groove seals can be applied to improving bearing performance to achieve overall seal and bearing operating lives of 15–20 yr.

10.5.7 Overview of Centrifugal Compressor Seals

Although several sealing systems are described in this section, other seals are available from compressor and seal manufacturers. Data on their operation are easily obtained, but information on leakage may not always be available. Experience with existing seal installations is a major source of information and industrial hygienists should compare proposed seal designs with existing seal designs to insure that the leakage rates are low and the regulatory considerations described in these sections have been incorporated into the design.

10.5.8 Reciprocating Compressors

A typical reciprocating gas compressor end is shown in Figure 10.20. The driver or prime mover can be any of the various power devices, such as an electric motor, diesel engine, or gas engine. The compressor end section drawing shows a simple packing system installation design to control shaft emissions on reciprocating shafts. The emissions from a straightforward packing installation generally exceed the 500-ppmv leak criteria established by the EPA. In normal operations with the older existing reciprocating machines, high contaminant concentrations have been observed around the machines, particularly near the reciprocating rod seals. New packing has alleviated the leakage problem somewhat and more elaborate packed seal configurations have been developed by the compressor manufacturers to control shaft emissions. While these improved sealing systems are normally installed on new compressors, packing modifications have been developed for existing compressors.[32,35] The improved packing and buffer purge systems described in reference 32 meet the SCAQMD regulations and should be reviewed for applications in various services.

The EPA regulations (HON) in Sections 10.1.2 and 10.1.6 are applicable to both centrifugal and reciprocating compressors. The regulations basically require a barrier sealing fluid or a closed-vent system that captures and transports all leakage to a control device. The only emission criteria for reciprocating compressors that have spacer pieces is a maximum emissions concentration of 500 ppmv above the background level (NSPS). The newly designed packing-type seal systems offered by compressor manufacturers may have emissions concentrations less than 500 ppmv, but this must be demonstrated to the EPA and retested at the request of the EPA.

A control alternative recommended by the EPA and used by reciprocating compressor operators is enclosure of the spacer piece as shown in Figure 10.20. Emission leakage is removed from the spacer piece enclosure by a purge gas and transported through a closed piping system to a control device. The regulations require movement of the leakage to a control device, which may not be required in all locations, particularly if the machine is not regulated. However, the spacer purge system is a control scheme that enables the industrial hygienist to signifi-

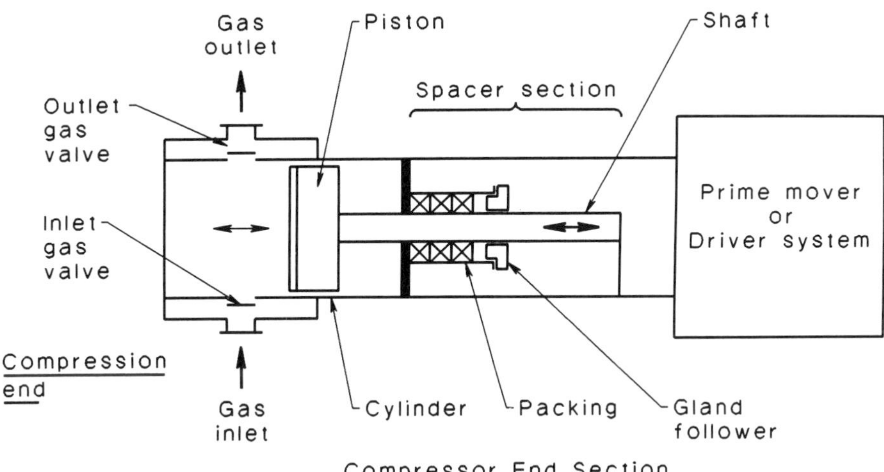

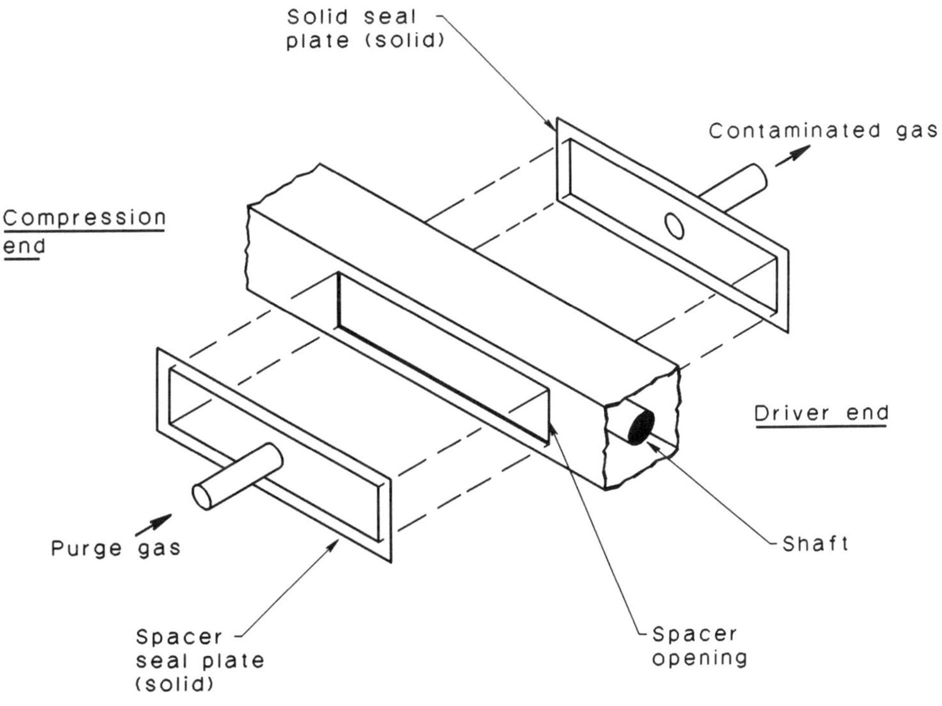

Figure 10.20 *Simplified section of reciprocating compressor end and spacer seal.*

cantly reduce workplace concentrations inexpensively. If the contaminated purge is currently vented in a safe location, the vent can be directed to a control device to meet EPA regulatory requirements.

An exception to NSPS petroleum refinery regulations requiring compressor controls is an exemption for those existing reciprocating compressors that do not have an open spacer piece. If a recast spacer piece or new compressor is required, owner compliance is waived. However, in those situations high emission concentrations may pose a potential health problem. Where the spacer piece is completely closed, drilling holes through the housing and purging the housing, similar to the technique shown in Figure 10.20, is an alternative to recasting the spacer. This option must be reviewed with the manufacturer, but the only other potential control alternative applicable is a ventilation system to capture the emissions. However, the ventilation hood or capture system required may be rather extensive to capture the emissions emanating from various points around the compressor frame.

The industrial hygienist must establish the emission source in a reciprocating compressor area prior to recommending a control scheme. In older installations, compressor vibration often results in excessive flange leakage in connecting piping and valves. The vibration may also affect other parts of the compressor and the entire machine should be monitored for emissions.

REFERENCES

1. CFR. *Standards of Performance for New Stationary Sources*, Part 60, Subpart VV, pp. 484–499, July 1, 1992.
2. CFR. *Standards of Performance for New Stationary Sources*, Part 60, Subpart GGG, pp. 580–582, July 1, 1992.
3. BAAQMD. *Pump and Compressor Seals at Petroleum Refineries, Chemical Plants, Bulk Plants and Terminals*, Reg. 8, Rule 25, pp. 8-25-3 to 8-25-7, March 4, 1992.
4. SCAQMD. *Fugitive Emissions of Volatile Organic Compounds*, Rule 1173, pp. 1173-1 to 1173-9, amended December 7, 1990.
5. Fed. Reg. *National Emission Standards for Hazardous Air Pollutants for Source Categories: Organic Hazardous Air Pollutants from the Synthetic Organic Chemical Manufacturing Industry and Seven Other Pollutants*, Proposed Rule 40 CFR, Part 63, Vol. 57, No. 252, pp. 62607–62795, December 31, 1992.
6. EPA. *Protocols for Generating Unit-Specific Emission Estimates for Equipment Leaks of VOC and VHAP*, EPA 450/3-88-010, October 1988.
7. Anderson, P. H. and M. H. Greenberg. *Estimating Chemical-Specific Fugitive Emissions from Chemical Processing Equipment*, 82nd Annual Meeting AWMA, Anaheim, CA, June 25–30, 1989.
8. British Occupational Hygienists Society, *Fugitive Emission of Vapors from Process Equipment*, Tech. Guide No. B, Science Reviews Ltd., Northwood, Middlesex, 1984.
9. Lipton, S. *Chem. Eng. Progr.* **88**(4), 42–47, June 1989.
10. Surprenant, N. *Chem. Eng.* **97**(9), 199–202, September 1990.
11. Schaich, J. R. *How to Estimate Fugitive Emissions from Equipment*, American Institute of Chemical Engineers Meeting, Boston, MA, August 21, 1991.

12. EPA. *VOC Fugitive Emissions in Synthetic Organic Chemicals Manufacturing Industry—Background Information for Promulgated Standards*, EPA-450/3-80-033b, June 1982.
13. Adams, W. V. *Chem. Eng.* **99**(11), Suppl., 12–18, November 1992.
14. API. *Centrifugal Pumps for General Refinery Services*, Standard 610, 7th ed., Refining Dept., February 1989.
15. EPA. *Compilation of Air Pollutant Emission Factors. Vol. I: Stationary Point and Area Sources*, AP-42, 4th ed., September 1985.
16. Features. *Mechanical Seals, Ind. Lubr. Tribology.* **42**(33), May/June 1990.
17. Netzel, J. P. *Seal Technology, A Control for Industrial Pollution*, Society of Tribologists and Lubrication Engineers, 45th Annual Meeting, Denver, CO, May 7–10, 1990.
18. Roos, E. *A Review of Experiences with Mechanical Seals in an Operating Refinery*, 12th International Conference on Fluid Sealing, Brighton, UK, May 10–12, 1989.
19. Bloch, H. P. *Hydrocarbon Proc.* **71**(6), June 1992.
20. Hernandez, T. M. *A User's Engineering Review of Sealless Pump Design Limitations and Features*, Texas A&M Pump Symposium, College Station, TX, April 1990.
21. Pradhan, S. *Hydrocarbon Proc.* **72**(1), January 1993.
22. Moore, S., G. Sandani, and G. Ondrey. *Chem. Eng.* **99**(11), 30–33, November 1992.
23. Nasr, A. *Chem. Eng. Progr.* **88**(11), November 1992.
24. Robinson, S. *Chem. Proc.* **55**(12), December 1992.
25. Doolin, J. H. and L. M. Teasdale. *Chem. Eng.* **97**(8), August 1990.
26. Maten, S. *Hydrocarbon Proc.* **69**(7), July 1990.
27. EPA. *VOC Fugitive Emissions in Synthetic Organic Chemical Manufacturing Industry—Background for Proposed Standard*, EPA-450/3-80-033a, November 1980.
28. EPA. *VOC Fugitive Emissions in Petroleum Refining Industry—Background Information for Proposed Standards*, EPA-450/3-80-015a, November 1982.
29. Ali, H. *Upstream Pumping: New Developments in Mechanical Seal Design*, Sixth International Pump Users Symposium, Texas A&M University, College Station, TX, 1989.
30. Netzel, J. P. *High Performance Gas Compressor Seals*, 11th International Conference on Fluid Sealing, Cannes, April 8–10, 1987.
31. Fischbach, M. J. *Hydrocarbon Proc.* **68**(10), October 1989.
32. Lemanski, J. *Oil Gas J.* **90**(27), July 6, 1992.
33. Type 2800 Non-Contacting, Dry-Running Double Cartridge Seal, Bulletin No. S-30H, John Crane, Inc., 1993.
34. Black, T. J., W. E. Key, P. McMahan, and W. B. Mraz. *Magnetic Fluid Seal System for Control of Pump Fugitive Emissions*, Proceedings 10th International Pump Users Symposium, Austin, TX, March 1993.
35. Livingston, E. H. *Chem. Eng. Progr.* **89**(2), 27–36, February 1993.
36. Key, W. E. and L. A. Young. *Hydrocarbon Proc.* **73**(1), January 1994

11

Sampling

A source of potential exposure in chemical plants and petroleum refineries through the years has been manual sampling. Automatic in-line sampling has eliminated many of the open source manual sampling exposure problems, but the use of advanced in-line instrumentation is not entirely free of potential exposure problems. Although the in-line or local analyzers are preferred in most applications, manual sampling is still required in a large number of installations. Emissions from manual sampling operations generally consists of VOC and HAP emissions, with emission regulations that specifically cover manual sampling installations. As described in other chapters, the NSPS regulatory requirements are essentially applicable in the majority of existing process units.

This chapter reviews various manual sampling techniques and automatic analyzer sampling systems to provide the industrial hygienist with an understanding of various sampling techniques as a basis for the assessment of potential exposure problems in the sampling area.

11.1 SAMPLING SYSTEM REGULATIONS

Since most of the liquid and vapor materials sampled in the chemical and process industries are volatile organic chemicals (VOC) that are precursors to photochemical reactions and hazardous organic pollutants (HAP), sampling controls are included in the NSPS and HON regulations covering equipment leakage.[1,2] The standards require containment of emissions purged from sampling connections with the contained emissions recycled via a closed loop to the process. Alternatively, purged emissions can be collected in a closed container or system for recycle or disposal but purged emissions cannot be vented to the atmosphere. A summary of the identical VOC and HON regulations follows:

1. Each sampling system will include a closed purge, closed loop, or closed vent system.
2. The closed purge system will be designed to control the purges through one of the following alternatives:
   a. The purged process fluid will be returned directly to the process line with zero emissions to the atmosphere.
   b. The purged emissions will be collected and recycled with zero emissions.
   c. All of the purged emissions will be transferred to a control device, which is a flare, incinerator, or similar control device.
3. In situ sampling systems are exempt from the requirements in references 1 and 2.

The industrial hygienist should be familiar with these regulatory sampling requirements, which not only reduce emissions but significantly reduce potential short-term workplace exposures.

11.2 EMISSIONS

Overall fugitive emission quantities from uncontrolled sampling connections are extremely small and a factor for sampling emissions is not shown in AP-42.[3] Factors for manual sampling emissions are shown in other U.S. EPA publications and in a presentation the EPA provided an estimated emission factor for petroleum refineries or chemical plants of 0.033 lb/hr (15 gm/hr).[4] Yearly emission totals for manual sampling connections in a typical refinery or chemical plant are quite small and the EPA continues to use this figure for uncontrolled emission rates.[5] This quantity is not very large compared with the emission losses for valves.

Although short-term and yearly emissions rates are small, the EPA intends to control fugitive emissions from all emission sources. Viewed from an industrial hygiene perspective, control of manual sampling purge emissions reduces potential workplace exposures that can occur as a result of uncontrolled sampling procedures. In these uncontrolled systems, a closed sample container or bomb is connected to a valve on a line or vessel. Fluid from the operating system is then blown through the closed container to the atmosphere, purging the inlet sample line and container of old material. The sample feed line receives current production material and the process fluid in the uncontrolled system continues to purge and blow directly to the atmosphere until the flow is shutoff on the downstream side and the container is filled. The upstream valve is then closed and the container is removed for analysis of the contents in the laboratory. In uncontrolled situations where open containers are used, the sample line is opened and purged to the atmosphere for a period of time before the open container is placed below the opening to obtain a sample. The purging of lines and containers is periodic as samples are only obtained at a set frequency during a 24-hr period. The fugitive emission factor issued by the EPA for uncontrolled manual sampling operations is an average daily number rather than a short-term factor covering a large emission release for a short time. The large uncontrolled emission releases pose a potential health problem where the atmospheric purge system is used.

11.3 MANUAL SAMPLING

Because of their concern about emission losses during manual sampling operations, the EPA-issued regulations require a closed purge during sampling. In closed-purge sampling, the purged VOC emissions are returned to the processing system or sent to a closed disposal system. The typical closed-purge sampling systems recommended by the EPA are shown in Figure 11.1 and have been termed closed-loop sampling systems.[6]

The manually operated globe valve in the process line provides the pressure drop necessary for the flow of a small slipstream through the closed loop. Pressure differentials are also provided by pumps, control valves, or other devices. In the closed container loop, fluid flow through the loop completely purges the container and line into the downstream leg of the main process line. The block valves at positions A and B are closed after a specific time and the container is disconnected from the loop at the dashed line. Couplings are provided in the circuit to quickly disconnect the closed container from the line. The B valves are considered to be an integral part of the closed container since the valves enclose the container. The fluid in the closed container can be either a gas or liquid in EPA reference documents.[1,6] Although a small quantity of fluid may leak at the couplings, the

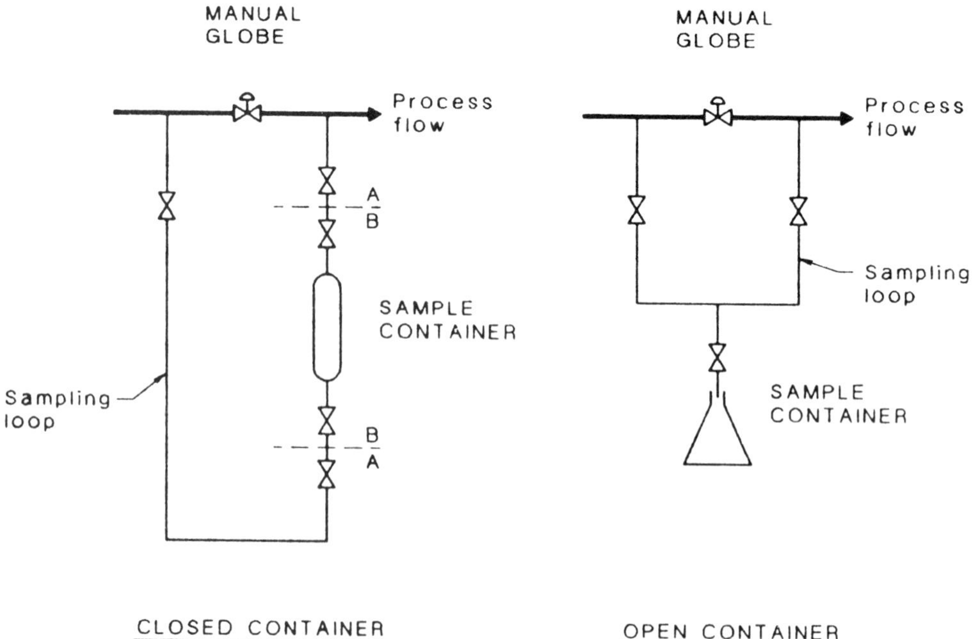

Figure 11.1 *Closed purge manual sampling system.*

EPA considers the closed loop purge 100% effective and disregards coupling leakage.

In the second loop shown in Figure 11.1, a small flow through the closed loop purges or clears the line as a result of the pressure drop across the globe valve. When the loop has been adequately purged, the valve drain to the container is opened and the required sample quantity is obtained. The drain valve connection is short and sample contamination is limited since the amount of fluid in the connection and valve is quite small. Alternatively, some material can be purged through the drain valve into a waste container prior to sampling which minimizes sample contamination. The liquid material discharged into the open container should have a very low vapor pressure. Liquids with a relatively high vapor pressure should be nontoxic since the technician will be exposed to vapors either opening or closing the drain and closing the container opening.

The elements necessary for any manual sampling system are shown in Figure 11.1, but the system designs often vary based upon fluid toxicity and safety concerns. Manual sampling systems are often custom designed owing to the lack of design guidelines in industry or governmental publications. General guidelines and designs are presented in the following sections that can be used by industrial hygienists in considering manual sampling installations.

11.3.1 General Considerations

Prior to recommending a manual sampling installation, the industrial hygienist should investigate the following:

1. **Toxicity**. The Toxicity of the stream should be defined to determine whether a potential health problem exists based on inhalation or dermal effects and whether the compound is on the HON list of toxic chemicals.

2. **Stream Characteristics.** The physical and chemical characteristics of the stream should be defined. Typical requirements cover pressure, temperature, viscosity, phase (liquid, gas), and stream composition. In addition to the basic stream characteristics, any changes in stream characteristics that occur during process operations should be identified. In addition, any physical or chemical changes that may occur during sampling should be identified.

3. **Sample Source.** The location of the sampling source provides a basis for determining the differential pressure necessary for the design of a closed loop. The source location is also important in determining the need for additional safety requirements. The latter is a concern when a source is a considerable distance from the actual sampling container location. Closed loops normally consist of very small lines and line breakage may occur because of maintenance interference or other mechanical accidents. A line break may result in a fire, burns from a hot liquid, potential overexposures, or a combination of happenings. Planning for an accidental line break may suggest remotely operated blocks at the inlet and outlet loop points, a flow-limiting device, or both.

4. **Container Location.** The location of the actual sampling container during sample withdrawal should also be considered during safety and industrial health reviews. The individual charged with actual sampling responsibilities should have

sufficient working area to move in a large arc around the sampling container filling location and a clear path to an open area. Safety and eye showers should be located in the immediate area. Where highly flammable stream samples are obtained manually, the safety group should be consulted for recommendations concerning potential fires, and other hazards.

5. **Sampling Logistics.** The container size and sampling frequency also determine the general design of a sampling system. Large containers are potentially greater hazards than smaller containers owing to the large volume of material that may spill when fitting a container to a loop or handling an open container. Moreover, a heavy filled container can be dropped during movement of the container to the laboratory. Open area around the container fill location, special washdown, and sewer or drain connections are considerations along with fire and safety needs. If large samples are necessary, an automated analyzer system is preferable to minimize the large sample volume that must be transported manually. The automated analyzer is particularly attractive for streams that are toxic.

Frequent manual sampling requirements also increase potential exposure and are a consideration when reviewing manual sampling installations.

6. **Container Handling.** Although the actual sampling operation is the major concern in manual sampling operations, disposition of the container, the container contents, and preparation of the container for fresh samples is also an industrial hygiene concern. Normally, samples in enclosed containers or bombs do not present a transport problem and samples in open containers are covered. The sample containers are sent to the laboratory where a portion of the sample is removed for testing. Following satisfactory laboratory tests on the sample, the containers are emptied, cleaned, and stored for reuse. Sample containers are normally constructed of metal, plastic, or glass for reuse and must be thoroughly cleaned prior to storage. The general cleaning and storage operation should be reviewed by the plant industrial hygienist. Exposures may occur at any step in this procedure and controls are often recommended to minimize them.

11.3.2 Sample Circulation Loop

Although circulation loops are shown in Figure 11.1, typical loop designs do not convey the physical layout of an actual loop. An example of a typical loop sampling layout is shown in Figure 11.2. The sampling loop pressure differential is supplied by a control valve located in the structure about three stories above grade or ground level. The main loop connection valves to the process line are mounted adjacent to the process line and the loop descends about three stories to the sample container box where a sample tap draws sample material into a container. Sample containers are normally filled at grade to minimize hazards associated with an elevated manual operation.

The circulating loop can be lengthy, but the actual sample tap is short for open sample containers to minimize the amount of stagnant material in the leg and the purge volume necessary to flush the leg. A bomb or enclosed container sampling system would be similar in a long circulating loop installation with typical loop connections to a process line shown in Figure 11.5. The withdrawal leg of the loop is located on the top of the process line with the return leg connected to the side

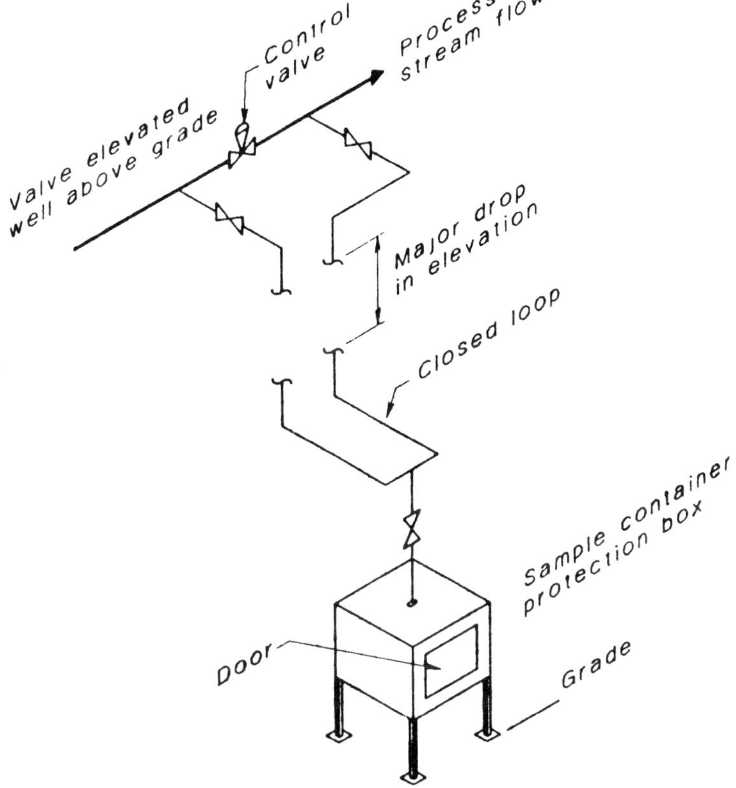

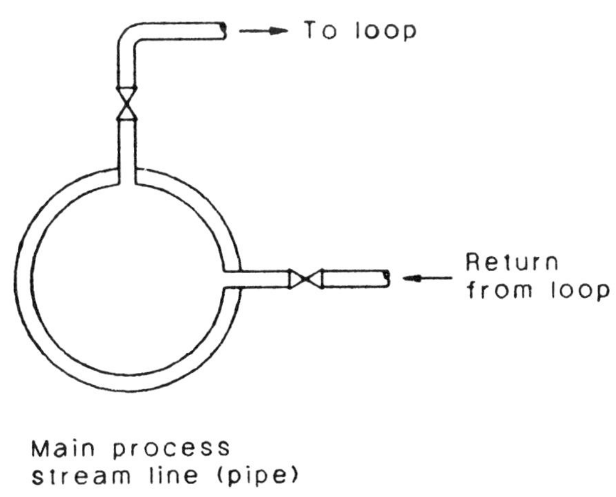

Figure 11.2 *Typical loop sampling elevation layout.*

of the pipe. Installation of the outlet valve at the top of the line is often recommended to minimize the possibility of sludge, scale, or dirt entering the closed loop, which can occur with a bottom outlet valve.

The loop block valves adjacent to the process line should be located in accordance with safety recommendations to permit climbing or reaching the block valves without excessive difficulty or hazard. The industrial hygienist should inspect the proposed location of the loop connections to determine whether a remotely operated analyzer is preferred for a sampling measurement.

11.3.3 Sampling Classification

As shown in Figure 11.1, enclosed containers (bombs) or open containers are the normal receptacles for samples. However, a general classification system is necessary to specify the type of container recommended for manual sampling service. The recommendations for manual sampling service are based upon toxicity and fluid characteristics with all designs having closed purge systems as recommended in Section 11.1. The classifications follow:

1. *Type A*. For toxic gas and vapor service an enclosed container (bomb) is recommended.
2. *Type B*. Volatile liquids or liquids with a relatively high vapor pressure, and in both situations the associated vapors are toxic, requiring an enclosed container. The liquid in this service is nonviscous.
3. *Type C*. Viscous or nonviscous liquids can be sampled in an open container placed in an enclosure where the quantity of toxic material volatilized from the liquid is small.
4. *Type D*. Viscous and nonviscous liquids characterized by extremely low vapor pressures and high skin toxicity effects can be sampled in an open container with splash protection or sampled with an open container in an enclosure.
5. *Type E*. Viscous and nonviscous liquids classified as sources of nontoxic emissions can be sampled in enclosed or open containers as desired providing a closed purge system is installed per Section 11.1.

These general classifications cover most liquid and vapor sampling services. However, a few services will not fit the general classifications and in these cases toxicity and viscosity are major selection criteria.

11.3.4. Sample Containers

Enclosed containers or sample bombs are generally closed cylindrical containers constructed of metal with a valve at each end. The bomb metal is normally stainless steel to minimize corrosion and contamination. A simple sampling bomb shown in Figure 11.3 can vary in size from a few hundred milliliters to 10 or more liters. Sample vapor or liquid flowing through the bomb purges or flushes stagnant material out of the container with a continuous flow of process fluid. The valves are closed following a sufficient flush period to obtain a sample. A small pressure

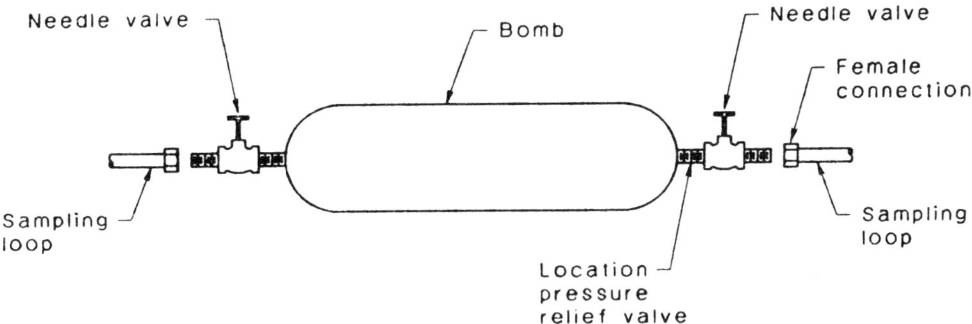

Figure 11.3 *Closed liquid sampling container.*

relief valve (not shown) can be installed on the bomb to protect it against overpressure.

The metal sample bomb can be used with a liquid, but flushing the open bomb generally results in completely filling the bomb with liquid. Although a filled container may be satisfactory in many situations, a potential overpressure situation may occur with a filled bomb that a pressure relief valve prevents. However, liquid discharging from a relief valve can be a safety as well as toxic hazard. As an alternate, a modified bomb is shown in Figure 11.4 which contains an internal dip tube. The dip tube provides a liquid–vapor interface in services where there is a

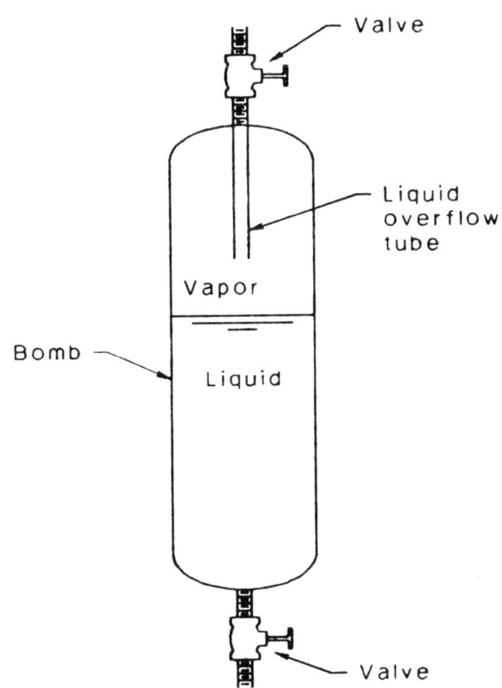

Figure 11.4 *Closed liquid – vapor sampling container.*

small vapor component. The vapor zone is a safety device that insures the bomb is not completely filled with liquid and cannot be overpressured. The modified bomb is applicable in those liquid services where an open container can lose lighter components and affect analytical accuracy. The modified bomb is recommended where the liquid, vapor, or both are toxic.

Enclosed containers are normally cleaned, evacuated, or filled with inert gas prior to sampling. After installation in the circulating loop, the containers are always flushed to insure that a typical fluid sample is obtained. Flushing in some installations will purge inert gas into the process, but the quantity is extremely small and should not affect the process. When bombs are removed from the loop connections a small quantity of vapor is released, but the amount released is insignificant according to the EPA. For most compounds this quantity will not overexpose the sampler, but the industrial hygienist should determine the toxicity of the released material and provide personnel protection where required.

The open container does not require a specific design, but the container should have cap or top seal closures. Materials of construction can range from plastic to metal, but selection must be compatible with the fluid.

11.3.5 Sampling Systems

The various parts of the sampling systems have been described and in this section the sampling classification and system designs are combined. The industrial hygienist should not be concerned with fugitive emission requirements for the various classifications since the regulations are designed to control the purge or flush material and a closed loop meets regulatory requirements.

Type A. A typical Type A sampling system suggested by the EPA is shown in Figure 11.1, but a more detailed design is shown in Figure 11.5. Block valves are placed against the process line and the container with its associated needle valves is isolated from the circulating loop by tight shutoff valves. The latter valves are globe, ball, or any tight shutoff valve that can move easily and close tightly against process pressure. During normal operation the block valves against the process line are open and these tight shutoff valves at the couplings are the only valves between the process and the atmosphere. The block valves at the process line are typically closed during an emergency or process shutdown.

This particular sampling configuration is applicable in vapor and liquid service, although the latter may present a potential safety concern due to overpressuring after the bomb has been removed from service. A pressure relief valve that discharges to the low-pressure side will eliminate overpressuring.

Type B. A Type B sampling system is shown in Figure 11.6 with the enclosed dip-tube container in a vertical position. A loop is shown with a three-way valve that diverts flow through the bomb after the loop has been completely purged with fluid. A three-way valve is not required for this installation, but is preferred in various plants. The bomb is gradually filled with liquid until the overflow sight glass indicates the bomb is filled to the dip-tube level with liquid and the upper portion of the bomb is filled with vapor. The liquid overflow should be sent to a closed drain system that is compatible with the sample liquid. Alternatively, the overflow

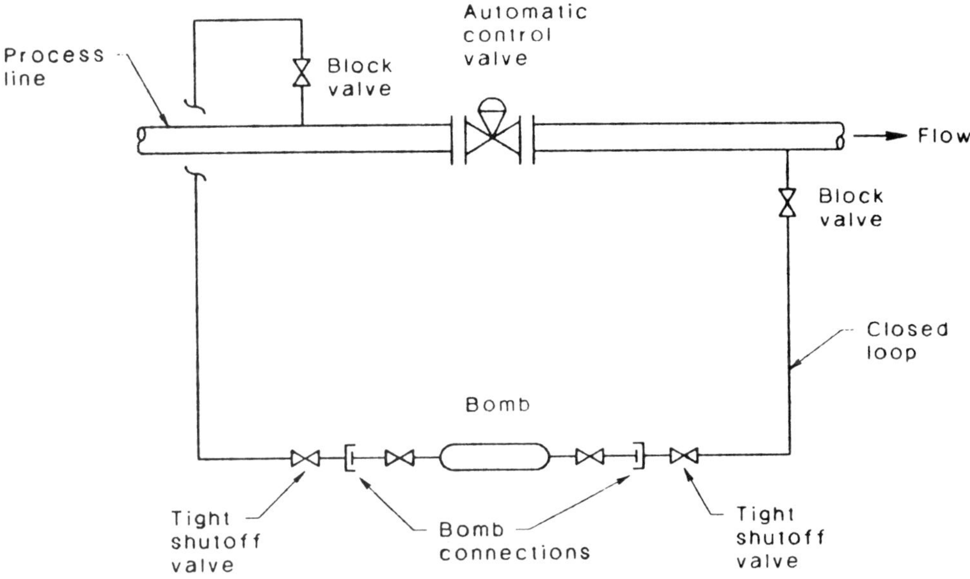

Figure 11.5 Closed-loop sampling system design for liquid or vapors.

liquid can be returned to the process, but the sight glass must be suitable for the process conditions and a valve must be placed in the downstream leg of the glass. A simple flow indicator replacing the sight glass in the overflow leg will meet pressure and safety requirements.

The Type B system is designed for toxic liquids that are volatile, or liquids with relatively high vapor pressures where the vapors are toxic. This system is not recommended for very low toxicity fluids or vapor service.

Type C. A Type C sampling system is shown in Figure 11.7 where an open sample container is placed in an enclosure. A loop is again shown with a three-way valve that diverts flow into the sample container within the enclosure. The flush liquid flowing through the loop purges the line of stagnant fluid and when required, the three-way valve diverts the flow into the enclosure. The distance between the three-way valve and the globe valve is minimized to limit the stagnant fluid in the line between the valves. Stagnant material will contaminate the sample unless the small initial quantity of fluid from the sampling line is flushed into a waste container and removed from the enclosure. This procedure will potentially expose personnel during removal of the small waste container and replacement with the normal sample container. In addition, the contents of the small waste container require disposal and the container must be controlled. The industrial hygienist reviews all of these procedures to insure that potential hazards are minimized during these various steps.

A globe or similar hand control valve is necessary in the sample line to the enclosure to control liquid flow into the sample container. The container can be constructed of metal or plastic materials but should have a cap or closure that prevents leakage when it is moved or transported to the laboratory. Since the

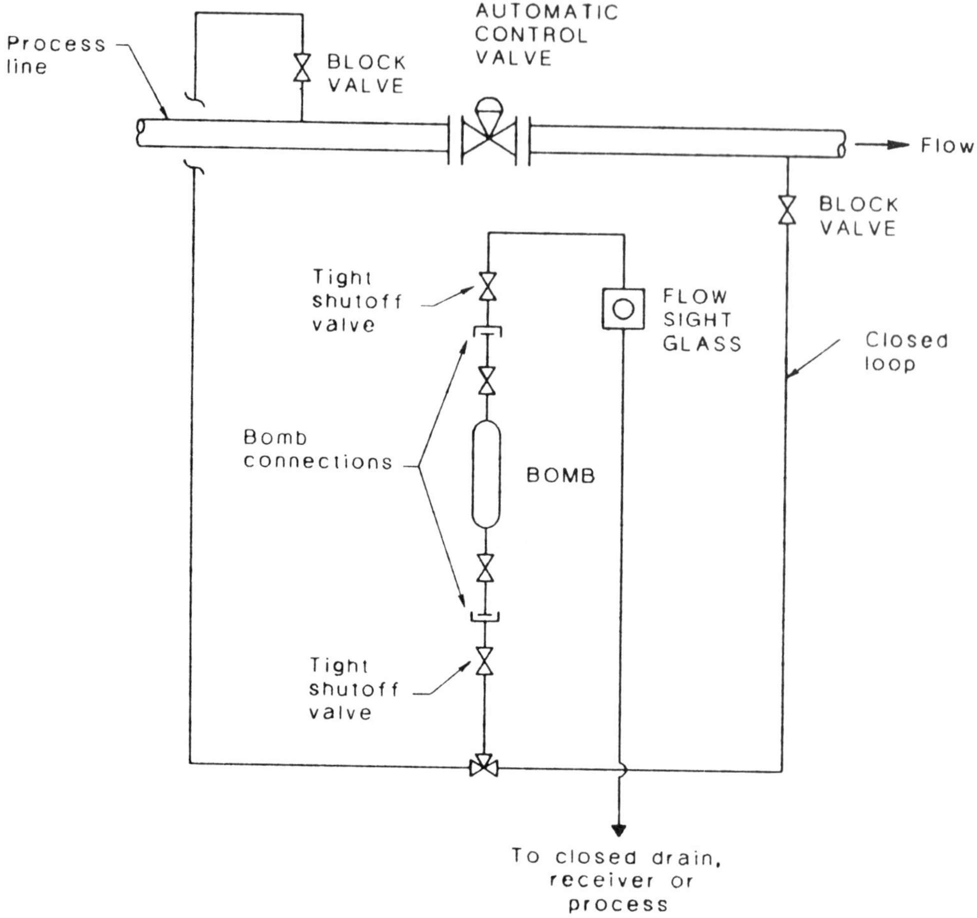

Figure 11.6 *Closed-loop sampling system design for volatile liquids.*

container is filled manually, the liquid level should be observed or gauged in the container to minimize overfilling and overflowing. Toxic liquids from overflowing containers can potentially expose the technician or worker handling the containers. In addition to controlling flow into the sample container, the globe valve provides tight shutoff, which is necessary during recirculation of fluid through the loop. Leakage through the three-way valve places process fluid against the hand control valve.

The box or enclosure can be constructed of metal or plastic. Many enclosures are constructed of stainless steel to minimize corrosion and extend enclosure life. As shown in Figure 11.7 a grating or grid is located inside the box as a support for the sample container. The grid permits drainage to the hopper bottom if the sample container is overfilled or the sample line leaks. The inlet sample line should be fixed securely in the enclosure and terminate immediately above the sample container.

A hinged door with a window permits viewing the interior of the sample box and adjusting the sample filling rate. These boxes or enclosures are generally

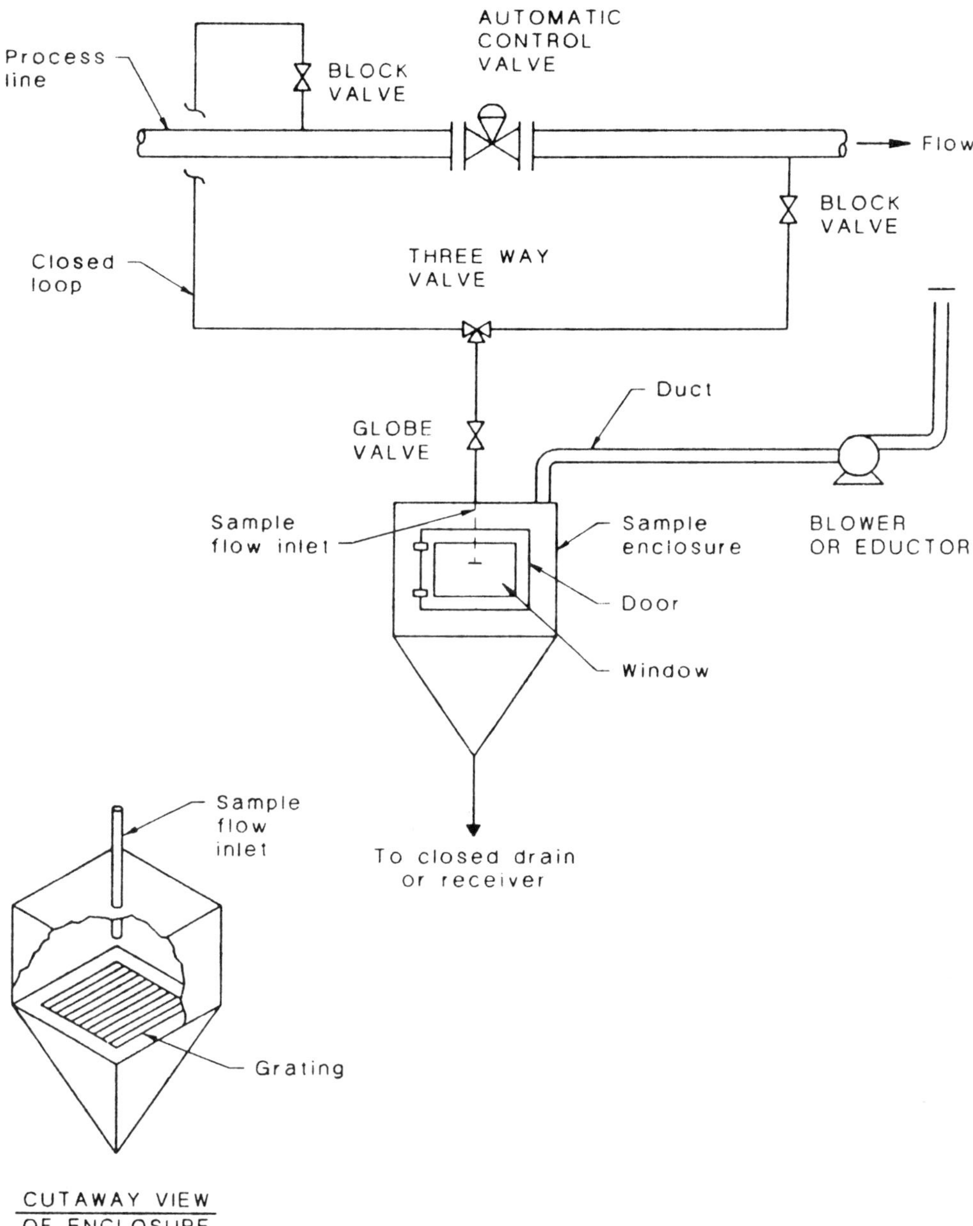

Figure 11.7 *Vented enclosure for open liquid sample with toxic vapors.*

custom designed and local requirements can be readily included in the design specifications. The design of the enclosure in Figure 11.7 includes a vent system to remove the toxic vapors and discharge a vapor–air mixture from the vent exhaust at an elevated location. The concentration of vapor in air should be considerably less than the lower explosive limit for the vapor–air mixture. Since this discharge stream is not continuous and very small, the necessity for a closed vent and control device has not been established. Moreover, connection to a flare, incinerator, and so on creates potential safety problems along with potential exposure hazards due to backflow.

An eductor can be used as a replacement for the blower shown in Fig. 11.7 with plant air as the source of eductor power. The blower or eductor capacity is based upon data obtained by the EPA.[7,8] The air velocity recommended by the EPA at the door face with the door open is 60–75 ft/sec, which can be translated into volume rates.[7] Exposure prevention is suggested by combining eductor operation with a trip handle that must be moved prior to opening the door.[7,8] The trip handle opens the main valve in the prime air line to the eductor, which initiates airflow through the door opening. Closing the door and moving the trip handle back to the starting position in front of the door stops airflow to the eductor and air movement through the enclosure. Although the eductor operates during the period the door is open, the industrial hygienist may also wish to consider operating the eductor during the sampling period to insure sweeping all vapors through the vent to minimize the possibility of any leakage from the enclosure. An electrical switch for controlling blower operation can also be installed with interlocks on the enclosure door that signal when the door is opened and automatically start the blower.

Although the vent discharges small quantities of toxic fumes in air to the atmosphere at an elevated position, dispersion of the vapors should be investigated as a precaution to protect workers on nearby platforms or at grade level. High short-term concentration levels may require an increase in elevation for the vent discharge opening, greater velocity in the vent, or relocation of the vent. The vent is open to the atmosphere and a protective plate above the vent will minimize rain water entering the box. Various alternatives to this approach can be designed to prevent any large volume of water from entering the enclosure.

The collection hopper below the grating funnels sample liquid to the hopper bottom, which must be connected to a closed drain system to control toxic fumes or liquids. An alternative for liquid collection is a portable safety can that can be connected to the hopper bottom and sealed for transport to another location.

The Type C system is designed for fluids with a wide range of viscosities that have a relatively low vapor pressure and small vapor emission. A lower viscosity material may also be contained in a Type B configuration, but higher viscosity material should not be collected in a bomb. The box or enclosure location does not require a large open area around the entire box for personnel protection as described in Section 8.3.1. However, eye and safety showers should be located in the box vicinity and personal protective equipment should be available.

Type D. The sampling system required for low-volatility liquids with skin or dermal toxicity characteristics is an open container that can be placed in an enclosure or box. The enclosure does not require a forced draft vent as shown in Figure 11.7. A small vented opening to the atmosphere is satisfactory for very low volatility fluids. Since the sampled fluids exhibit skin toxicity effects, the sample container should be handled carefully when it is removed from the enclosure. The person removing the sample container in this situation requires personal protective equipment, which should be identified by the industrial hygienist. All containers should have caps to prevent spillage during removal from the enclosure and movement of the filled container.

Although an enclosure provides maximum protection during the sample container fill operation, an open funnel can also provide a satisfactory method. A funnel system used in filling open sample containers is shown in Figure 11.8. The funnel has a high side relative to the container to prevent splashing and protect personnel. Note the protective height distance above the container in Fig. 11.8; this should be established by the industrial hygienist. This system is adequate for low-volatility fluids with a wide variation in liquid viscosities. The industrial hygienist should review this installation to insure that proper worker protection precautions are observed.

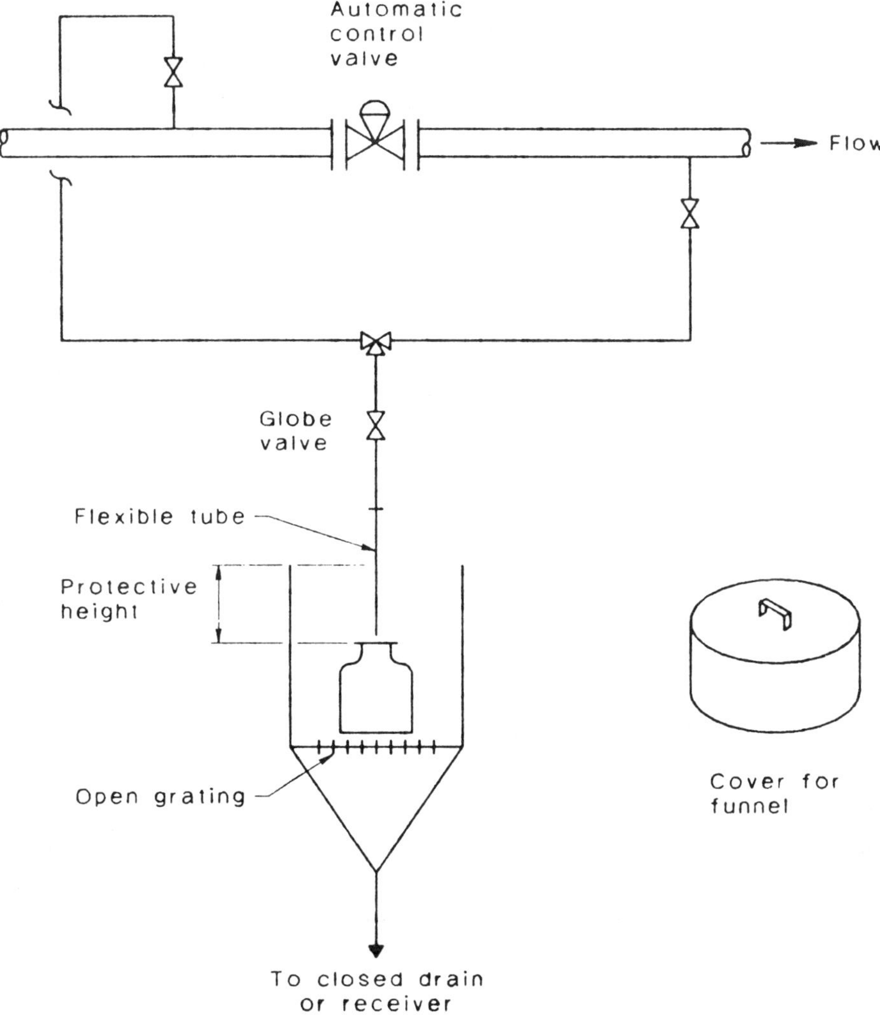

Figure 11.8 *Open support and container for low volatility liquid.*

The open funnel will perform well indoors, but in nonenclosed plants rain is a potential problem during the actual sampling operation. In addition, drainage of rainwater through the grate can be a problem in the drain systems. A cover for the funnel eliminates this problem but the funnel and container are open to the elements during the sampling operation. This installation should be reviewed with the operations group prior to a sampling recommendation.

Type E. These fluids have very low toxicity, but are sources of fugitive (VOC) emissions. The NSPS regulations are observed by installing a closed purge or loop system to control the release of flush material. Either bomb or open containers can be installed with a closed-loop system. The actual selection of the loop system and container is arbitrary but the industrial hygienist should review the design basis to determine whether a potentially toxic situation can arise. If there is a possibility of a potential health problem, a controlled emission sampling system is recommended as described for Types A, B, and C fluids.

Sample Cooling. Our previous descriptions of sampling systems did not contain any reference to process fluid and sampling temperatures. Fluids at temperatures above 150°F (65.5°C) present potential toxicity, handling, and process problems. These fluids should be cooled to 140°F (60°C) or less prior to filling the sample container for the following reasons:

Process. When hot fluids flow into containers, there is generally a loss of lighter materials; with heavy fluids an aerosol may form. Also, flashing or vaporization of the fluid, although small in many situations, will affect the analytical results. This aspect of manual sampling should be reviewed with the process group.

Toxicity. The flashing of toxic sample fluid requires careful control of the vapors. In the sample enclosure system (Type C) a small vapor concentration that is exhausted through an eductor system can become a potential problem when the sample container is filled at high temperatures. Consideration of vaporization requires a review of the actual filling temperature to estimate the toxic contaminant quantities evaporated. Opening the door on the enclosure can also present a potentially hazardous situation. In addition, location of the vent should be reviewed to insure that dispersion of vent vapors does not exceed recommended concentration levels.

Although vaporization is a major consideration with relatively light materials, (benzene, toluene, etc.), oil mist formation is also a concern. These mists are easily formed when discharging a hot fluid into an open container. There is a recommended exposure concentration for oil mists of 5 mg/m^3 published in the ACGIH list of recommended exposure levels.[9] The fluid may not be an oil but a heavy organic compound that has not been included in the ACGIH list. The lack of an exposure standard for a high-temperature nonoil mist suggests obtaining toxicity data, and an enclosed sampling container can be installed as a precaution.

Handling and Safety. Where fluid temperatures are higher than 125–140°F (52–60°C), skin contact with the piping, enclosures, or sample container can result in a burn. High-temperature operation requires insulation on all of the equipment and the sampler should wear heavily insulated gloves as a minimum. With a sample and container at elevated temperatures, the industrial hygienist may also wish to recommend special clothing to protect workers against accidental spills.

The high temperatures that result in vaporization can raise an eductor or blower concentration to the explosive limit. The explosive limits vary in accordance with the compound and a few compounds in the process may have a very low explosive limit, which may be a serious safety concern. Formation of a mist presents the same safety concern that generally results in efforts by the safety personnel to install nonsparking mechanical and electrical equipment along with inert gas purges. All of these considerations complicate the sampling installation, but must be considered to insure a safe facility.

A hot sample is also of concern in the laboratory. A hot bomb or enclosed container sample when cooled will have a reduced vapor pressure that can result in a vacuum and collapse of the container. The sample containers are often cooled purposely by the laboratory which exacerbates this problem. If not cooled, a hot sample in the laboratory is also a potentially hazardous problem from a toxicity or safety standpoint and is a situation that requires review by the industrial hygienist. Cooling the original sample should be reviewed with laboratory and design personnel because of the potential problems associated with the two possible locations for the cooler. The cooler can be installed only in the sampling loop for the Type A and B bomb samplers. However, a cooler can be installed in either the loop or sampling tap for the Type C and D installations. In these latter installations, a cooler in the tap circuit will significantly increase the volume of the tap drainage circuit from the three-way valve in the loop to the outlet, which may require purging the tap circuit before a sample can be drawn. Purging the line contradicts the federal regulatory rules and may not be permitted. Alternatively, the entire tap circuit can be drained into the sample container, which significantly increases sample volume. Large manual samples can present handling problems in the field and disposal problems in the laboratory. Alternatively, a cooler installed in the loop circuit can decrease the mainstream temperature if the loop flow rate is relatively large compared to the mainstream flow rate. This will not be a problem in large streams, but in units with low flow rates this may be a consideration. The loop flow rate can be controlled with a flow controller, orifice, or other device, but any selected arrangement must be satisfactory to all of the responsible operations groups. If the loop flow cannot be reduced to a very low flow rate compared with the main stream flow rate, then another sampling method may be required.

Unit Shutdowns. The manual sampling systems in this section have circulating loops that can be of considerable length in many installations. These loops do not have a significant fluid holdup quantity since they are generally constructed of small diameter piping determined by plant mechanical specifications. However, toxic fluids, particularly liquids, can pose a problem if the circulating loop is not purged and flushed properly when the process unit is shutdown for repairs. Shutdown considerations are presented for the different sampling procedures.

Type A. The bomb or enclosed container can be used with either a gas or liquid and the procedure for purging the line is similar. As shown in Figure 11.5, the bomb is placed directly in the circulating loop and when removed from the loop creates two liquid legs. A small spool piece can be placed between the two connections or an old bomb retained for this purpose can be placed in the loop.

Purging or flushing through the spool piece will displace the remaining material in both legs with flush material. The next step in the operating procedure should be reviewed to determine whether all of the flush material in the system is completely displaced. If the flushing material is not displaced, the flush material will be in the legs and presumably will be nontoxic. However, the flush fluid can be organic in nature or a combination of several materials. Disposal of this fluid should be reviewed for draining or waste disposal.

Another procedure that may be applicable is a backflush. Where material remains in the legs, the spool piece can be removed and each leg can be connected to a flush source. As an example, the final flush may be an inert material such as water that can be connected to each leg via a hose and an adapter to connect with the loop.

As an alternative for the situation where material remains in the legs after purging and flushing, the type of bomb or spool connection connecting the legs can be modified for drainage. A drain valve installed on the bomb or on the spool piece permits drainage of both legs. However, drainage of the material must be investigated to establish a method of collecting the drainage without hazard. The final disposal of the material must also be acceptable.

Type B. The problem in flushing and purging the system is somewhat different for the Type B installation shown in Figure 11.6. The three-way valve in the circulating loop permits flushing and purging the loop, but draining each leg is complicated by the three-way valve, which can drain only one leg at a time through the bomb. This technique can successfully drain each leg, and a spare bomb or spool piece permits flushing the outlet leg of the bomb. As described in the Type A flush system, an important aspect of draining, purging, and flushing is disposal of the material after each step. Disposal must be safe, nonhazardous, and in compliance with environmental regulations.

Types C and D. Purging, flushing, and draining this loop system should not pose any exposure problems. The loop can be flushed and each leg drained through the globe valve shown in Figures 11.7 and 11.8. The major concern is disposal of the flush and drainage material.

11.3.6 Sampling Locations

The sampling systems described in this section comply with the emission regulations and reduce workplace exposures. The prototypes and models for these systems are described in various documents that concentrate on outdoor processing units.[6–8] Although much of the processing equipment in the chemical industry is located outdoors, many smaller units are located within buildings. This contrasts with the petroleum refining industry which locates essentially all of the processing units in an outdoor environment.

Although some processing units are enclosed in buildings, the NSPS and HON regulations are also applicable to control of emissions in these buildings. Ventilation systems to control emission levels in buildings are satisfactory, but do not reduce emissions to the atmosphere requiring control devices on the emission vents, which is a major objective of the regulations. However, potential exposure to emissions requires control of sample emissions within buildings. The various types of manual sampling controls reviewed in this chapter are also installed in building processing units. Comments on the various systems also apply to these sampling systems installed on indoor process units.

The major problem with sampling units is vapor leakage or emissions during the sampling operations. Additional precautions may be necessary if the building ventilation system is shutdown. However, actual exposures should be monitored by the industrial hygienists to insure that the recommended exposure levels are not exceeded. If the exposure levels are exceeded, changes in personal protective equipment may be required or the engineering controls may be changed. Actual leakage in a Type A situation cannot be estimated until the equipment is actually specified and the volume in the mating coupling is calculated. The industrial hygienist in an indoor environment must follow these designs closely to control emissions and potential workplace exposures.

11.4 DIRECT SAMPLING

The manual sampling systems previously described have recirculating loops of stream material that can be sampled with containers of various sizes and shapes. A simple bleed valve or tap had been used prior to the EPA regulations where a flow of process material purged lines and containers directly to the atmosphere; thus a "good" sample was obtained but this process resulted in the direct flow of fugitive emissions to the atmosphere. The main purpose of sampling stream purge flow through recirculating lines is to minimize emissions, but flushing directly through a container or a recirculating loop also obtains a representative sample.

Through the years, various suppliers have developed manual sampling systems that directly sample from the process stream, apparently meeting the requirements for an in–situ device outlined in EPA regulations in Section 11.1. An in–situ sampling device is exempt from the requirement that a recirculating closed-loop system must be installed for manual sampling. However, there seems to be some differences in the interpretation of in–situ by the EPA requiring resolution of the regulations prior to installation. These devices withdraw a sample directly from the process line and deposit the sample into a container attached to the sampler. The container is then removed from the sampler device and sent to the laboratory for analysis.

While the regulatory interpretation may permit installation of direct sampling devices in a process line, safety considerations are a concern. These devices require a block valve between the device and the process line for isolation, which may or may not be practical for the device. A circulating process loop provides isolation and also provides a method of controlling the process pressure at the sampler, which may have pressure limitations.

11.4.1 Direct Sampler Devices

A variety of sampling devices have been developed to meet specific sampling requirements. A few of these are presented in this section to illustrate direct sampling operations.

Plunger Type. Figure 11.9 illustrates a plunger-type sampling device. When the plunger is in the extended position, it displaces a fixed quantity of fluid in the volume between the piston seals. A sample is taken directly from the process flow or a recirculation loop. As the plunger withdraws, it carries each sample to the discharge port, where it flows directly into an attached container. The plunger assembly can be repositioned and sampling can be continued until the desired amount of process fluid has been accumulated. With this plunger-transfer design, the process pipe opening is always sealed against leakage, regardless of the position of the plunger. When the required quantity has been collected, the plunger dwells in the withdrawn position, and the sample container can be safely removed to the lab for analysis.

This sampler can be used to take a single sample or a series of samples, whichever is required. When working with toxic or flammable fluids, it is recommended that sample containers be constructed of plastic or metal to minimize the possibility of breakage if the container falls or is struck.

Since plunger samplers work equally well on process lines under pressure or vacuum, they can be installed at various locations throughout the process unit. Industrial hygienists should review potential installations from the standpoint of collecting samples at the appropriate point in the process and the physical location of the sampler itself. Sample locations above a convenient height may require an elevated platform for access. Alternatively, a bypass line or recirculating loop permits installation of a sampler at a convenient height. As described previously, recirculating loop lines are equipped with block valves on both sides of the sampler permitting the loop to be closed and the sampling device removed for inspection or maintenance.

The use of a bypass line with block valves is definitely recommended when working with materials that are hazardous or toxic by virtue of their chemistry,

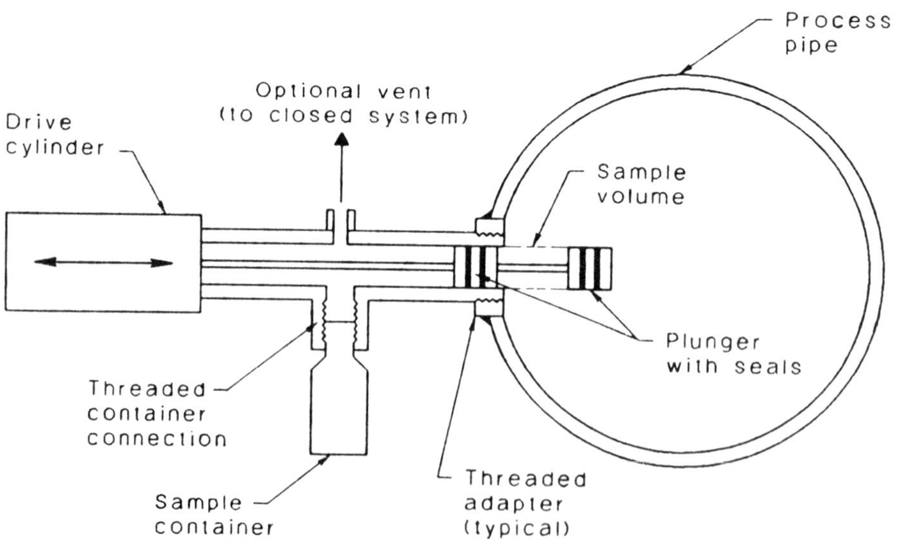

Figure 11.9 *Direct ISOLOK™ sampler, plunger type (courtesy of Bristol Equipment Company, Yorkville, IL 60560).*

temperature, or process pressure. This technique permits diverting or blocking the process flow so that samplers can be periodically checked and maintained. Bypass or recirculation lines also facilitate maintenance in continuous process operations where line flows must be maintained for months or years before shutdown. Where moderate-to-heavy slurries, viscous materials, or stratified streams are involved, industrial hygienists should coordinate with process engineering personnel to insure that material flowing through a bypass line is representative of the main stream content.

Another point of consideration is the method of connecting the sampling device to the process piping. Figure 11.9 shows a threaded connection, a common technique where process pressure and other conditions are not excessive. Where threaded connections may not be used on process lines plunger sampler mountings can be specified using a ferrule and clamp design, an instrument hub design with conventional bolt fasteners, or an ANSI flange configuration. When considering the method of attachment to the process line, consult with operations and safety personnel regarding installation standards. Whatever method is selected, the sampler should be readily removable from the process line for inspection and maintenance.

Plunger samplers transfer process material to a closed container, then compensate for this liquid volume by transferring a like volume of air or vapor out of the container back into the process line. Under most circumstances, no venting of the sampler or container is necessary. Some process materials will generate a positive pressure in the sampler-container, due to the expansion of entrained gases or high vapor pressures. In these instances, the hygienist should insure that a vent port is available on the sampler and piped to vent the fumes to a scrubber or other safe designated area away from the sample point, preventing operator exposure to contamination. Also, when the sample container is removed, it should be immediately capped. A second unused container should then be attached to the sampler, sealing the opening and preventing toxic emissions to the atmosphere.

When sampling hazardous or toxic substances, particularly at elevated temperature or pressure, hygienists should consider the need for additional precautions against accidental exposure. One such function is to emphasize the importance of regular inspection and maintenance of all sampling instruments. Samplers should be inspected between cycles to insure that there is no leakage past the plunger seals. Operating personnel should be trained to report the slightest leakage immediately.

When operator exposure to even normally wetted parts during sample container exchange is of concern, special valve closures can be incorporated between the sample discharge port and the container. A pair of valves are customarily used, one attached to the sampler discharge to retain clingage and the second assembled to the top of the container to isolate the contents before removal. A quick disconnect fitting is used between the isolating valves.

Whenever hazardous conditions exist, hygienists should make a special effort to understand the operation of the sampling system, as well as the general process operating conditions.

Valve Type. A simplified drawing of a valve type sampling device is shown in Figure 11.10. The sampling system is connected to a flange on the process line and

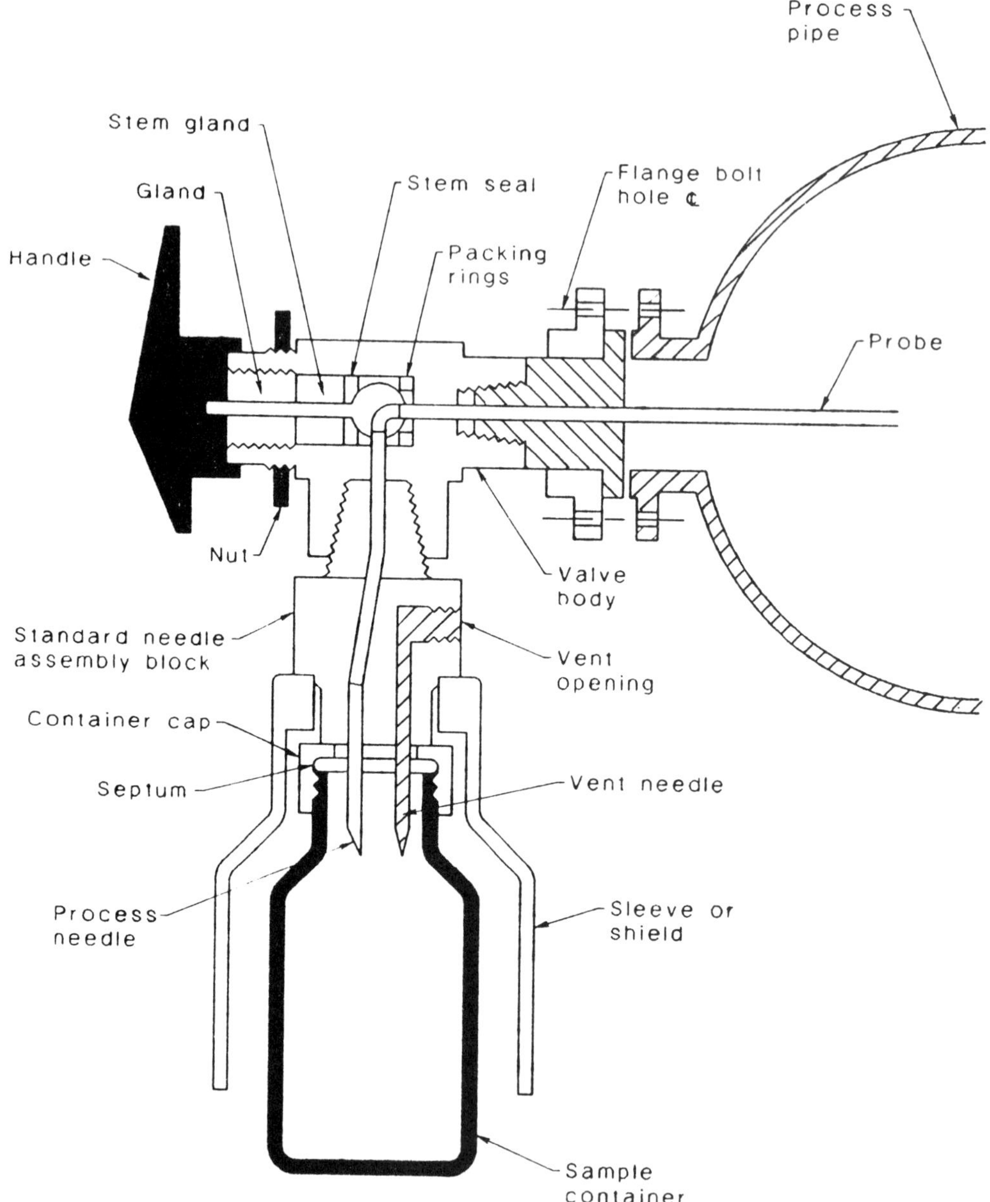

Figure 11.10 *Direct sampler, valve type (courtesy of Dopak™ Inc.).*

a probe extends directly into the process line. A sample is withdrawn from the process line and is discharged into the sampling container via the ball valve. The container is a bottle with a threaded open-top cap that covers a self-sealing septum. The bottle with the cap and septum is inserted into the sleeve and the septum is pierced by the needles extending from the standard needle assembly block.

When the bottle has been placed in position with the septum pierced, the valve is opened and process fluid flows into the container. The quantity of fluid in the bottle is observed through a slot in the cover sleeve and the valve is closed after the required sample quantity has been obtained. The bottle is then withdrawn and the septum is resealed to prevent the emission of vapors or liquid from the container. Various septum materials are available from the manufacturer (Dopak™ Inc.) that insure septum resealing for a wide range of process materials.

Since the sample will displace air or vapor from the container, a vent needle discharges the vapor through the vent opening as shown in Figure 11.10. For toxic process fluids the vent opening can be connected to an elevated vent line that discharges to the atmosphere at a location determined by dispersion calculations. Alternatively, the vent line can be connected to the process to maintain an enclosed system. A block valve in this vent line prevents process vapor emissions to the atmosphere when the container is removed.

The container described for this installation must be constructed of materials that permit operator observance of the liquid level in the container. Sample containers are available in clear and amber glass, Pyrex®, polyethylene, and polypropylene. In some locations metal containers are required and in those process plants the containers are constructed of stainless steel. The liquid level cannot be observed in metal containers, but a modification to the sampling system is available from the manufacturer that permits withdrawal of a fixed sample volume from the process.

In the sampling device shown in Figure 11.10, the container may be contaminated with material remaining in the probe from a previous sample. Various connections are available from the manufacturer that provide an on-line noncontaminated sample. In addition, gas and steam purging techniques have been developed by Dopak that free the sample container of gas prior to sampling. For satisfactory installation, the industrial hygienist and process engineer should review sampling system requirements with the manufacturer.

Although a flange-mounted installation is shown in Figure 11.10, other connections are also available. A threaded connection replacing the flange permits connection to a threaded coupling on a pipe or process vessel. However, block valves do not separate the sampling system from the process line.

Other Considerations. Since sampling containers in many extraction devices are threaded and may be plastic to observe liquid levels, a recirculation loop is used to control pressure and fluid flow. Moreover, the sampler can be inspected when in the loop configuration and removed for repair. Pressure and flow rate can be controlled with restriction orifices, back pressure controllers, and so on. Lower pressures to match the plastic container design pressure and minimized thread leakage require a low pressure receiving point for the circulating fluid. However, the modified loop meets safety concerns and minimizes potential hazards. While two samplers are described, other sampling devices may require the same safety and hazard considerations.

11.5 CONTINUOUS INSTRUMENTS

The previous sections have been concerned with obtaining manual spot samples for analysis of the sample in the plant laboratory. Local analyses in the process units may be performed on these batch samples, but the preponderance of manual samples are analyzed in sophisticated laboratories. Analytical procedures and associated equipment technologies have advanced markedly and local analysis efforts are quite limited. Sample analysis in the larger plant laboratories is preferred for samples that exhibit acute or chronic toxic effects. Facilities in the large laboratories are designed for toxic substance control and potential exposures are minimized through established laboratory procedures and practices.

In addition to advances in laboratory technology, significant changes and improvements have occurred in one-line process analyzers that measure physical characteristics and chemical composition of a variety of streams. Technology advances in analyzer equipment continues to expand on-line process analysis which is accompanied by a decrease in manual sampling. Potential exposures in on-line analyzers are generally considered significantly lower than potential exposures that may occur in manual sampling but on-line analyzers present a different set of potential exposure problems. Although not as desirable as on-line analyzers, manual sampling is still required in a number of processes and plants.

The major types of continuous instruments currently in use differ in the petroleum and chemical process industries owing to the analyses needed for process control. The petroleum industry instruments generally analyze stream physical properties in contrast to the chemical industry where composition data are a major concern. General physical property analyzers in the petroleum industry cover viscosity, boiling point ranges, end points, distillation fractions, pour point, density, and so on. A few chromatographic and infrared analyzer applications accompany these physical property analyzers. The chemical industry uses a variety of composition analyzers, such as liquid and vapor chromatographs, mass spectrometers, spectrophotometers, and infrared, with variations in the different instrument classifications.

This section is concerned not with the actual instruments and their functions, but with the potential exposures and hazards associated with the installation and operation of on-line continuous instruments. The section concentrates on major stream and product analyzers but does not include a wide variety of highly specialized analyzer applications and systems installed in operating units. The principles and controls shown in this section are applicable to most continuous analyzer installations.

11.5.1 Sampling Circulation Systems

Continuous analyzers are normally installed in a specific location for protection and various process unit maintenance and test functions. In addition, the instruments must be protected from the elements in outdoor facilities and require a location where delicate maintenance functions and tests on the instruments can be conducted without interference. Instrument safety concerns may also have an effect upon final analyzer location.

Since several analyzers may be grouped in a special location that is a considerable distance from the actual sampling location, a sampling system must be installed that provides an analyzer with a constant stream of fresh fluid or a representative sample. As a result, a loop system is generally installed that circulates stream material from a high-pressure source to a lower pressure level. This concept is similar to the recirculating loops required for manual samples.

Although the EPA regulations require a closed loop or purge, the EPA regulation target is manual sampling as described in the regulations.[1,2]

The recirculating loops for stream sampling generally require a differential pressure as a driving force across equipment, such as a control valve, exchanger, or pump. Frequently a differential pressure source is not available or the pressure drop is insufficient and a pump is required in the recirculating loop to return the fluid to the process stream. Moreover, several instruments located in one position or instrument area have various sample loops compacted into a small piping area. Consequently, all of the lines, particularly those in toxic service, must be identified for maintenance and cleaning operations. Additionally, the pumps in a recirculating loop require maintenance and must be installed with block valves that permit isolation for drainage and flushing. The need for spare pumps is specified by the process and analyzer groups. A simplified drawing of a circulating loop with a pump in toxic service is shown in Figure 11.11 along with lines for draining and flushing the pump. A typical circulating loop around an operating process pump that supplies sufficient differential pressure is also shown in Figure 11.11. Control valve differential pressures may also be satisfactory, as shown in the figures describing manual sample circulating loops. The differential pressure source is generally identified by the process and analyzer groups.

11.5.2 Sample Treating

The major recirculating sample loops are described in Section 11.5.1 where a relatively large flow is circulated to insure that a sample representative of the process stream is available at the analyzer location. However, the sample to the analyzer shown in Figure 11.11 may require treatment prior to entering an analyzer, in contrast with the manual sample, which is sent directly to the laboratory for analysis without treatment. The on-line analyzer is a sophisticated instrument and treatment may consist of one or a combination of these steps: fine media filtration, drying, purification, blending, and so on. In certain process samples treatment is unnecessary and the analyzer may be fed directly from the loop. In those cases where treatment is required, the analyzer group will specify the treatment steps.

In analyzer installations requiring sample treatment, the treatment steps will require maintenance. Although the equipment is generally small, the technician service location is adjacent to the equipment for proper maintenance, and exposure controls are normally required. For either volatile or nonvolatile toxic liquids enclosed drain and flush systems should be provided to minimize potential exposures. When a vapor is treated prior to entering the analyzer, the industrial hygienist can estimate the volume of vapor in the treatment system at the operating pressure and temperature conditions to determine potential exposure levels through the exposure inhalation calculation procedure presented in Chapter 18. If the calculated concentration level exceeds the recommended short-term exposure limits[9] an enclosed vent connection is required with an inert purge. The maintenance purge vent from the treated section can be connected to an analyzer instrument vent line that is normally installed for venting analyzer offgases.

The treatment systems prepare the sample stream for analyzers that are frequently installed in an enclosed shelter or hut. Since the shelter can be

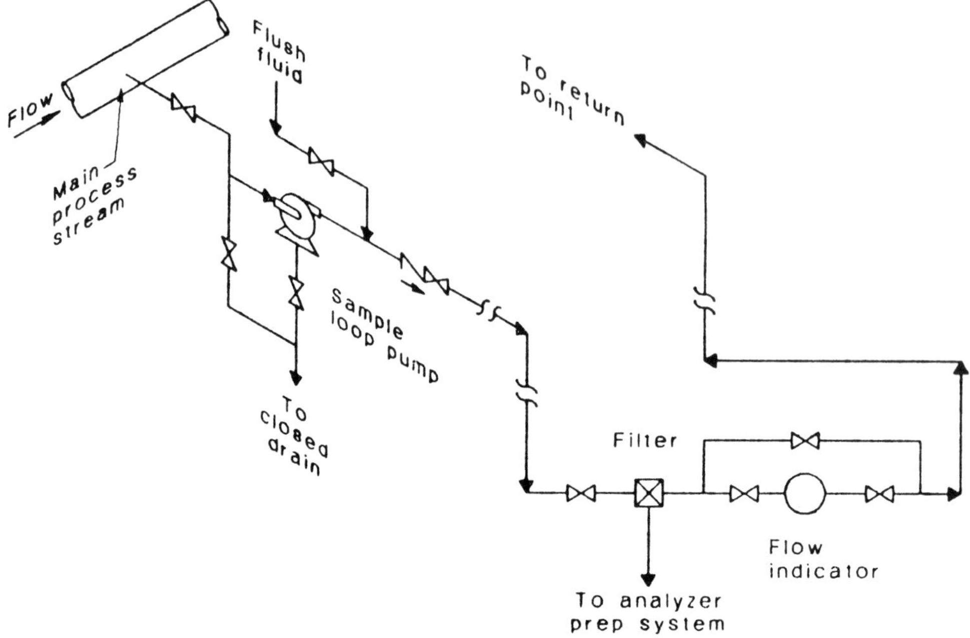

PUMPED CIRCULATING SYSTEM

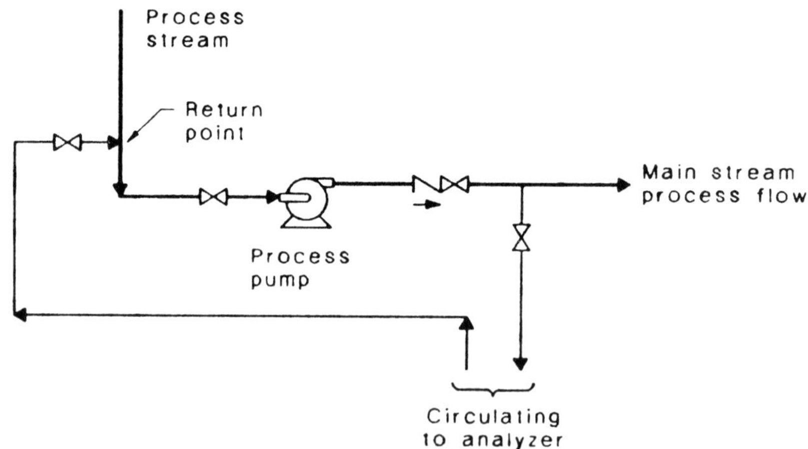

TYPICAL CIRCULATING LOOP SAMPLE

Figure 11.11 *Circulating loop continuous analyzer instruments.*

enclosed, the vapor or liquid flow rate entering the shelter for analysis should be limited. This flow restriction is recommended for toxic vapors and liquids to minimize the concentration of toxic materials in the enclosed shelter in the event a leak develops in the equipment. The restriction on flow rates into the shelter should not change the basic design of the treatment system.

11.6 ENCLOSED ANALYZER SHELTER

Frequently analyzers are installed in enclosed shelters or huts for reasons previously described. A typical enclosed analyzer shelter is closed on all four sides and a roof completes the building. The shelter or building is designed in accordance with local building and safety codes. Since the analyzer equipment is in an enclosure, heating and cooling requirements are based upon equipment requirements and temperature standards.

The analyzer shelter location should permit shelter access on all sides along with sufficient area for the sample treatment system. The type of analyzer instruments may also define the minimum distance from processing equipment. In flammable areas, safety specifications often restrict the minimum distance between potential emission sources and equipment that is not designed for operation in a flammable zone. Additionally, the shelter should not be located near processing equipment that is frequently maintained. Also, the toxicity of the liquids and gases in the circulating loops and analyzers are a consideration in locating the shelter.

11.6.1 Analyzer Shelter Structure

Typical outdoor analyzer shelters that are normally located in processing areas are shown in Figure 11.12. The major difference in the two structures shown in Figure 11.12 is the roof design with the flat roof illustration showing a heating and air conditioning (HVAC) unit located on the roof. The flat roof design is attractive for installation of auxiliary units such as the HVAC unit and blowers compared with installing this equipment in a peaked roof shelter. In addition, ventilating the peaked roof shelter to purge flammable and toxic vapors is a more difficult operation than purging a flat roof shelter.

The shelter is located on a raised pad that is several inches above grade. This small elevation prevents the entrance of an external spill into the analyzer shelter and also provides runoff for spills and flush water within the shelter. Although the amount of fluid that enters a shelter should be restricted, a leak can occur that must be flushed and removed from the shelter. Flushing and spill removal is also assisted by:

1. Designing a floor pad with an outward slope, which permits runoff to a drain or the edge of the raised pad.
2. Installing a floor pad surface that is nonporous and will not retain any liquid. However, the pad must be skid resistant.
3. Installing a floor pad that is spill resistant.

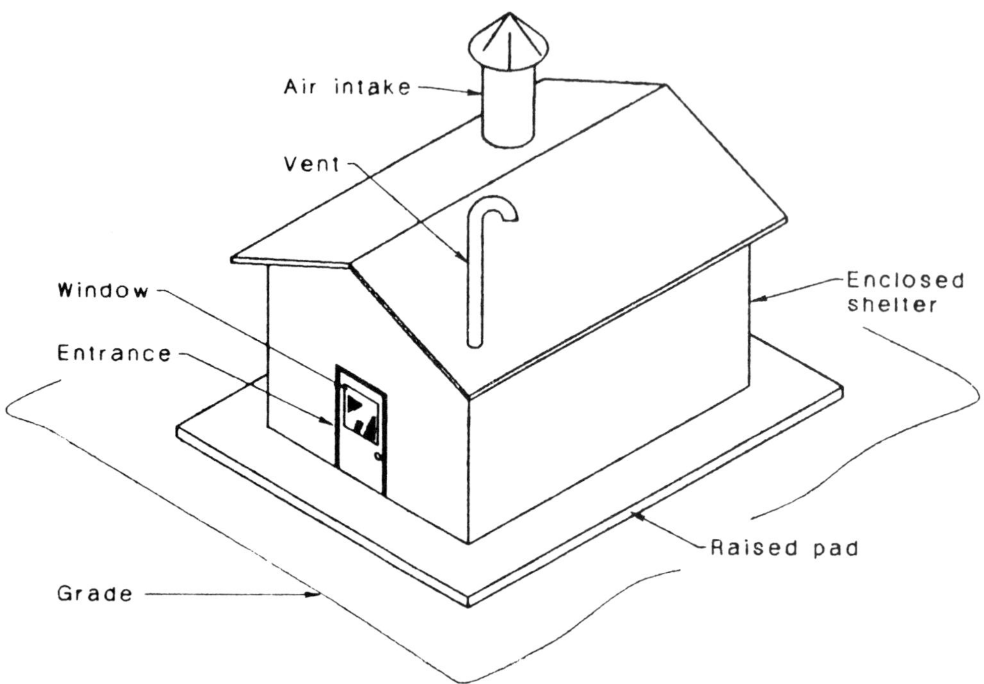

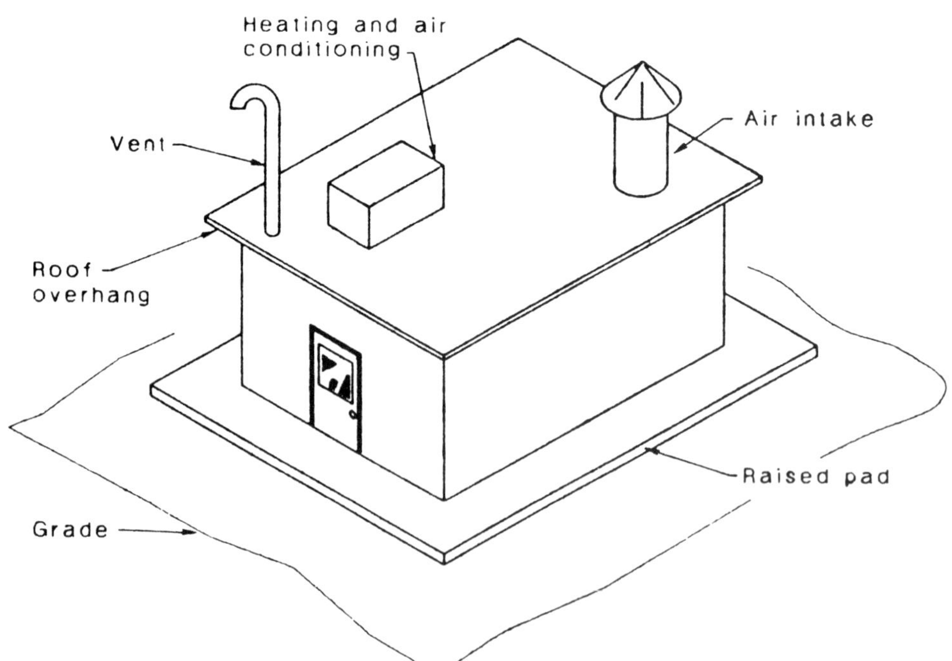

Figure 11.12 *Enclosed continuous analyzer shelter.*

Although runoff on the graded floor can be directed to the edge of the pad, flushing the floor surface inside the shelter requires raising the walls of the shelter to allow the liquid flush to reach the pad edge. Raising the shelter walls is not desirable because this procedure will affect the ventilation flow pattern, which is described in another section. The alternative is placement of drains within the shelter that receive the spill and flush material. The floor should also be sloped toward the drains, which effectively removes the spillage from the floor area. Although small drain openings are suggested, a covered grating trench can be used if properly sloped into a drain. The small drains and the trenches discharge into a drain lateral that is connected to a main sewer line. The drain should be sealed by a small water leg or a seal leg in the lateral from the shelter. The purpose of the seal is to prevent vapor backflow through the lateral into the shelter. Although drainage is directed to the main sewer, the lateral flow may be directed toward a stripper as described in Chapter 12. The important consideration is sealing the drain, with disposal based on an assessment of the sewer system.

The structure should have an entrance door and a separate door as an emergency exit. The doors are shown with windows which should have safety glass construction. Columns or struts should not be permitted in the central floor area, which should be kept clear or free of all obstructions. The analyzers and maintenance work benches are located against the walls providing a clear central area. In addition, potential health and safety problems that may occur as a result of toxic material leakage and the presence of flammable compounds are more easily controlled with an open central area. Moreover, a clear central area will not hinder or interfere with shelter ventilation.

11.6.2 Ventilation

Ventilation is required in the enclosed shelter housing for several reasons, as described in the following brief summary:

a. Proper movement of air prevents the entrapment and buildup of flammable materials within the shelter structure.
b. Good ventilation reduces potential exposure to toxic vapor concentrations.
c. Heating and air conditioning are more effective with good air distribution.

Ventilation Capacity. Ventilation is normally specified for heating and air conditioning, but in an analyzer shelter the major concern is leakage that may overexpose personnel or exceed the lower explosive limit. The ventilation system should be sized to purge and dilute the worst leak that can be expected. Flow restrictions in the sample line entering the shelter determine the maximum leak rate and the restrictions can be sized for rates that are about twice the normal analyzer flow rates, which are quite small. If calculations then demonstrate that the exposure limit can be exceeded, the ventilation system should be sized to meet the exposure limit recommendations.[9] An extremely large blower system is not a practical solution and flow restrictions should be designed for flow rates that provide blowers with reasonable turnover rates in the shelter housing.

The ventilation blowers can be sized for shelter turnover volumes of 6–12 volumes per hour. Under an emergency leak condition, blower rates can be increased to purge the shelter, increase dilution, and reduce potential exposures. In this situation the increased blower capacity can be activated by a toxic monitor that automatically changes shelter turnover from a normal level to a much higher rate. Changes in blower rpm will adjust the volume flow rate and this is a standard method of increasing the airflow. Alternatively, one blower can provide normal turnover and a satisfactory atmosphere in the shelter. A second or emergency blower operation is initiated when a toxic monitor indicates the concentration exceeds the recommended exposure limit. In addition, flammability conditions are reviewed to determine whether sufficient blower capacity exists to prevent an exceedance of the lower explosive limits. The blower capacities for these cases must be compared and a blower capacity installed for emergency conditions that is sized for the worst possible leak condition—either a toxic or flammability situation. The blower capacity rates are defined in conjunction with the sample restriction rates.

Blower Failure. Although multiple blowers are suggested, a shelter design should consider a minimum of two blowers. One blower provides the normal flow rate and the second is a standby or emergency blower. A standby blower is an important consideration in the event the operating blower fails and must be shut down. In the absence of a standby blower, very small leaks can accumulate vapors and create a potentially flammable situation and/or a very high local toxic concentration that can be hazardous if the single blower is shut down or fails. A standby blower should have an automatic startup and tie-in during a blower failure. As an alternative, an instrument alarm can immediately signal a main blower shutdown and the standby blower can be started manually either remotely or locally. The overall concern is maintenance of airflow through the shelter house to minimize the concentration of vapors.

The optimum situation in this forced ventilation design is the installation of two blowers with one blower in an automatic standby condition, but the blowers should be capable of long-term continuous operation. In an emergency state the operating blower rate can be increased and the standby blower started to further increase airflow. Both blowers receive air from the single inlet stack.

Air Circulation. The blowers are frequently installed on the roof with diffusers to direct air flow through the shelter. The air is then discharged to the atmosphere through exhaust louvers located near the floor and in close proximity to the roof line. These louvers are designed to maintain a slight pressure within the shelter which assists air distribution and prevents the entry or backflow of vapors into the shelter from an outside source. A typical cross-sectional view of an enclosed shelter is shown in Figure 11.13. The instruments are located on the support ledge and the approximate relationship of a technician's inhalation zone to the instrument is shown in the figure. The flow of air toward the louvers is split, but flow around the inhalation zone should be directed downward to minimize potential exposures. A more generalized view of the desired airflow pattern is shown in Figure 11.14.

Although the louvers are adjustable and the diffuser on the air duct inlet is designed for good air distribution, the actual airflow distribution pattern should be

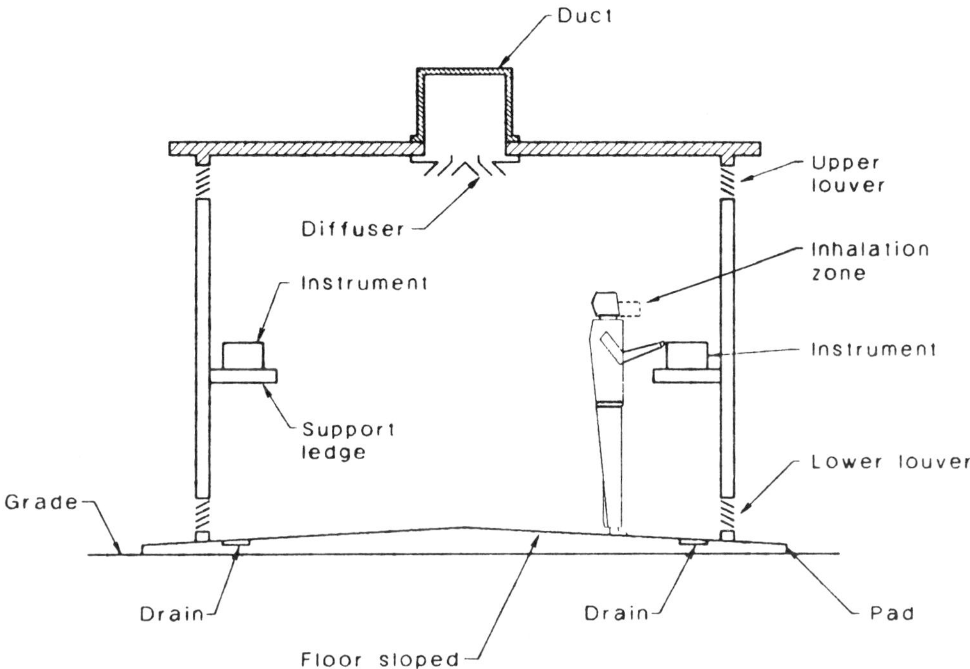

Figure 11.13 *Cross section of a typical continuous analyzer shelter.*

investigated by the industrial hygienist. The pattern is frequently distorted and must be modified in the field. Adjustments to the louvers and the diffuser are aided by the use of smoke or portable pitot-type flow indicators to achieve the desired airflow pattern. The instruments on the support ledge may trap air and create a stagnant zone, but a flow around the instruments can be initiated by providing openings in the support ledge. The entire shelter should be tested to

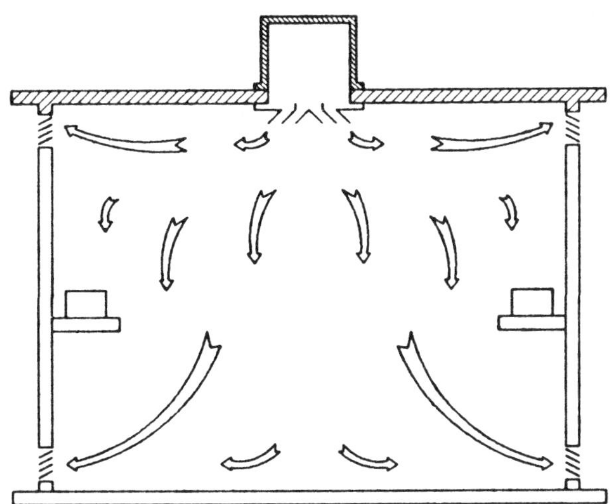

Figure 11.14 *Optimum airflow pattern in enclosed instrument shelter.*

insure that stagnant zones do not exist where toxic or flammable fumes can accumulate.

The flat roof shelter as shown in Figures 11.13 and 11.14 provides a distribution pattern of the type desired for an instrument shelter. Air distribution in a peak roof shelter is more difficult to control and changes or modifications to the shelter may be required. The industrial hygienist should carefully observe the flow patterns and optimize airflow where possible before suggesting modifications. Following modifications to the shelter, the airflow should again be tested to insure that it is satisfactory. This procedure may be repeated several times until a satisfactory airflow pattern is established.

The shelter can be located near a building or may be in a separate room. The exhaust air from the enclosed shelter should not be recirculated to another room or building. All air from this shelter should only be discharged directly to the atmosphere. The intake vent to the blower should be elevated to a location that does not contain toxic materials. If a noncontaminated atmosphere cannot be reached easily with the inlet vent stack, a purification system should be installed in the inlet airstream. An activated carbon adsorber and particulate filter normally provide sufficient purification, but an assessment of the atmosphere is necessary to determine whether additional purification steps are necessary.

Sample Exhaust Vent. An instrument vent pipe is shown in Figure 11.12 for instruments that discharge analyzed vapor samples to the atmosphere. Since the vapor quantity is extremely small and the stream is discharged on a cyclic basis, the need for a closed vent and control device has not been defined by the EPA. The low emission rate should not result in an emissions problem, and the vent pipe should exhaust at a location that has been analyzed for dispersion of the discharged vapors. Impingement of toxic vapors on nearby work platforms or vapors exceeding exposure limits at ground level is not desirable and the vent pipe exhaust point should be elevated to a level that does not overexpose personnel. An elevation may be required that is difficult to achieve without a costly structure due to dispersion calculations and/or highly toxic compounds from a number of sample streams and analyzers. In this situation alternative methods of vent sample disposal should be investigated and may be required to meet regulatory restrictions. Disposal methods that are generally reviewed are refrigeration, adsorption, absorption, incineration, flaring, and so on.

Sample vapors vented to the atmosphere that are close to their condensation or dew point should be discharged through a sloped line to a small receiver drum or pot. The condensed or liquid material is separated from the vapor in the pot and the liquid is discharged into a closed drain system or recycled to the process with an intermittently operated double diaphragm reciprocating pump. The vapor should be heated before discharge to the atmosphere or sent to a closed system.

In certain analyzers, the samples are burned. A flame ionization detector in a gas chromatograph burns the sample and the ionization detector measures the vent gas concentrations. The flame instruments should have a separate vent that collects the combustion vapors from the analyzers and discharges the vapor above the roof.

Liquid Sample Disposal. Liquid analyzers discharge waste samples that require disposal. With a number of analyzers in the shelter the amount of material, although small, can be toxic, flammable, or both, which often requires careful control of the disposal stream. These small streams can be discharged into a closed drain system and returned to the process. However, recycle of the analyzer stream is not always possible and other procedures may be necessary.

In reviewing analyzer liquid disposal, the streams should be classified in accordance with potential recycle capability. If the stream can be recycled, the liquid recycle process requirements must be defined for injection of the liquid into the process. For low-pressure processes, a simple closed line to a proportioning pump can be used. However, this relatively straightforward procedure may not be desirable for toxic fluids where process upsets may interfere with the waste disposal stream. In addition, the need for a continually operating analyzer probably requires the installation of a standby pump, further complicating the system. Many pumps have internal check valves and frequently these check valves are relied upon to prevent backflow from the process, which is always questionable in situations relating to health and safety. A system break is frequently installed to prevent this occurrence, which consists of a small pot or drum and a pump, as shown in Figure 11.15. The pump is controlled automatically by the level controller, which may have an individual alarm.

In cases where analyzer samples are compatible fluids and can be recycled jointly or sent to a separate disposal facility, the multiple system shown in Figure 11.15 can be installed. This system is generally preferred to simplify the sample disposal system and reduce investment. The level alarms should be local and when necessary are also installed in the control house. The pump is operated manually but can be changed to automatic on–off operation by connecting the motor controller with the level alarm transmitters. This multiple sample system exhaust has a single vent that can be connected to the main vent or left freestanding. However, a freestanding vent emission location is defined by dispersion analysis of the vapors as described previously. The location of a freestanding vent may require a support structure and combining the disposal drum vent with the vapor analyzer vent would be preferable in this situation. However, this method of disposal requires a complete understanding of vent system emission compounds, concentration, and flow rate. While regulations on vent-control of this stream are not adequately defined, toxic vent streams should be sent through a closed vent to a control device.

A key point in the consideration of vapor and liquid sample disposal is the compatibility of the various streams. Hydrocarbon streams in petroleum refineries are generally compatible, but streams in chemical plants are not necessarily compatible. This problem requires a careful review of all streams in vents and disposal systems to insure nonreactive mixtures that do not potentially pose safety or toxicity problems. While these comments are applicable for sample disposal, draining the equipment or analyzers into a common line also presents similar problems. Drainage can be sent to the liquid sample disposal line rather than to a separate system, simplifying the drain system and insuring compatibility with all of the other combined strreams.

Vent Location. Although vents and inlet air ducts were reviewed, a major consideration in these inlet–outlet ducts is their relationship to one another. The vapor vent potentially discharges toxic vapors to the atmosphere that can be drawn into the inlet air duct and circulate within the enclosed shelter. The design and layout

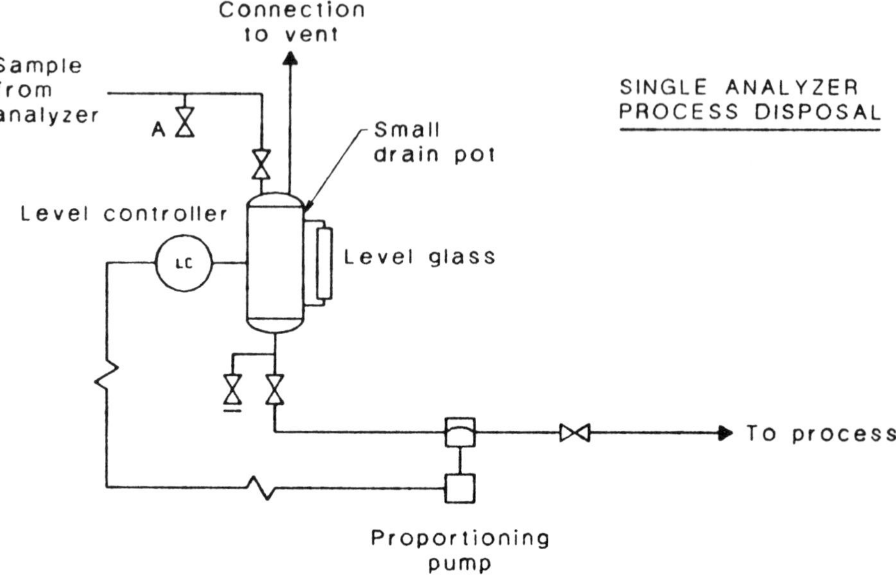

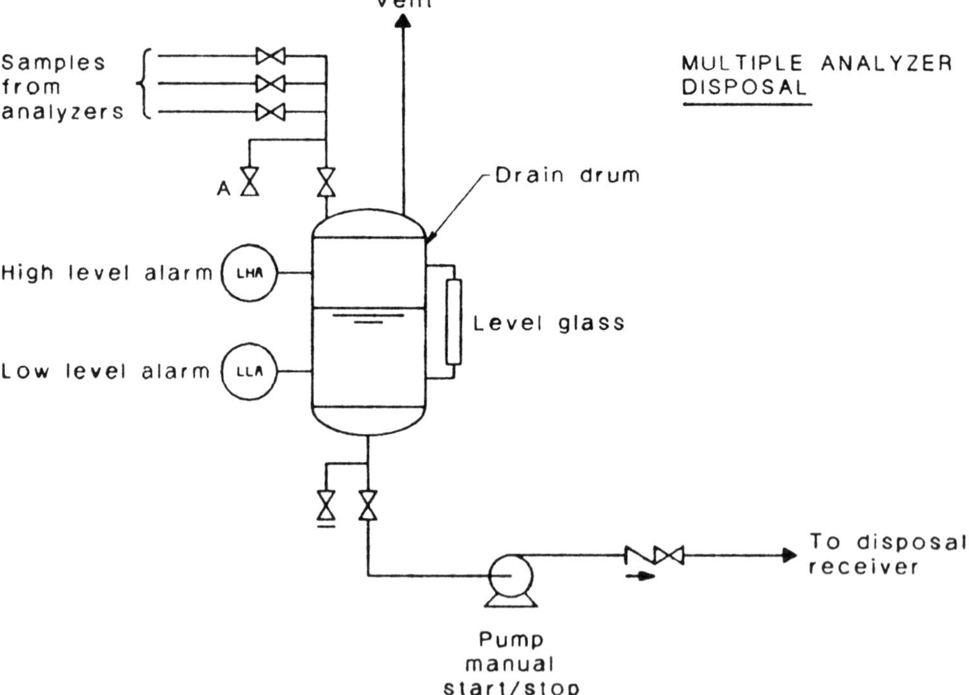

A - Drain for analyzer testing

Figure 11.15 *Sample disposal liquid instrument drains.*

of the vent and inlet duct should be reviewed to insure that vented toxic vapors are not diverted into the inlet duct. This may necessitate raising the entrance to the inlet duct well above the vent or locating the duct entrance some distance from the vent with the prevailing wind in the direction of the vent.

The vent vapors are not emitted as a plume since the velocity of the exhaust vapors is quite small, which must be considered in this general analysis of airflow.

11.6.3 Standardization Gases and Liquids

A number of instruments require gases for calibration, carriers, and in some cases combustion. In addition, specialty gas may be provided for testing the instrument at various conditions. These various gases are generally supplied in cylinders for connection to the various instruments. The calibration and test gases may contain highly toxic or flammable materials and hydrogen is often supplied for several purposes. Although the gases are provided in cylinders, the cylinders should be mounted outside the analyzer shelter in a special rack. Gas cylinders should be placed outside the shelter housing to minimize potential toxic releases and flammability or safety concerns.

Special liquids are often used to calibrate and test various liquid analyzers. The liquid test quantities are frequently quite small and minute amounts suffice for the calibration and test functions. In these situations the tests are generally manual and a technician can calibrate or test the instrument in the shelter providing the industrial hygienist has reviewed all of the test procedures. Where relatively large sample quantities are involved, the liquids should be stored in a fixed container outside the analyzer shelter, particularly those liquids that are toxic or flammable. The test and calibration quantities necessary for the analyzer can be injected into the instrument with a metering pump that also measures the quantity sent to the analyzer. The small container and metering pump should be designed for draining and purging in accordance with comments elsewhere. The vent from the container can be closed periodically and nitrogen injected into the container to maintain a small positive pressure, but a safety relief valve is necessary. The drum design should be reviewed with safety personnel. All liquid systems should be designed for drainage from the instrument to an outside storage recycle or liquid waste pot described elsewhere. The waste or recycle drum will have a vent and other liquid containing systems can be vented to the recycle drum. The vent may or may not be closed depending upon the chemical compounds.

A major concern with toxic liquids and gases is mixing or intermingling of materials since various compounds can react explosively or form highly toxic materials. Other compounds may react exothermally with the heat, affecting the overall system. The industrial hygienist should review all of the compounds and streams within the shelter and the auxiliary materials present in the form of test, research, and calibration materials. Compounds that interact should be identified and procedures established to prevent contact, which may require strict precautionary measures.

11.6.4 Alarms

In an enclosed shelter alarms are necessary to indicate the existence of an explosive concentration, a high toxicity level, or a low airflow with insufficient ventilation. The alarms should be visible outside the enclosed shelter and in the control house. In addition, an audible alarm at the enclosed shelter should sound when a detector sensor is energized for any of the three criteria just described.

The sensors for explosivity or flammability should be located near the roof and close to the floor to cover vapors that are heavier or lighter than air. Toxic vapor sensors should also be placed in these positions and airflow sensors can be placed near the inlet air diffusers. The alarms must be tested or calibrated periodically and the industrial hygienist should review the alarm test chemicals and procedures specified for these functions.

Important considerations in the alarm systems are the alarm set points and the type of sensor selected for this important function. Where a number of chemicals are present in individual samples and several analyzers are placed in an enclosed shelter, selection of the sensors and the set points must be carefully reviewed to insure that personnel are protected from all of the toxic materials. If a sampled stream contains several toxic components, the selected sensor should be sensitive to all of the toxic compounds in the event a leak occurs. In some situations the toxic chemicals may interfer with sensor selectivity and the industrial hygienist must be confident the sensor will perform satisfactorily under emergency conditions. Where sensor data or information is not reliable, simulated tests of the sensor may be required. In situations where tests or data from the supplier on sensor sensitivity or reliability are unsatisfactory, multiple sensors should be installed. The selection of alarm sensors is a critical factor in control of workplace exposures and the industrial hygienist should carefully review the specifications for alarms and sensors in all instrument enclosures and shelters.

11.6.5 Other Equipment

Other considerations in an enclosed shelter area are flushing needs, personal protective equipment, and emergency wash facilities. A water connection and hose should be included within the shelter to flush toxic liquid spills into the drains and wash down equipment contaminated with toxic liquid. Personal protective equipment should be available in a cabinet on the outside of the enclosed shelter, particularly for those situations where a toxic alarm has sounded. The shelter should not be entered without a respirator when the alarms are operating. The industrial hygienist may also require boots and protective clothing if the shelter is entered during an emergency, depending upon the toxic compounds in the analyzer systems.

In addition to these protective items a wash basin should also be provided in the shelter for technicians potentially exposed to toxic liquids during analyzer maintenance. Eyewash basins should be installed within and outside the shelter. Emergency shower installations in and outside the shelter should be specified and located by the industrial hygienist.

Although protective equipment in a cabinet outside the shelter is recommended for personnel in close proximity to the enclosed shelter when an alarm sounds, the general shelter area should have a defined perimeter that is not entered when an alarm sounds. Venting of toxic gases through the louvers is potentially hazardous and the perimeter area is entered during an emergency only by personnel wearing respirators.

11.7 PARTIAL SHELTERS

The previous section reviewed enclosed shelters for continuous instruments and in many processing plants partial shelters have been installed. These shelters provide weather protection to some extent for instruments that are rugged and less sensitive than instruments normally placed within enclosed shelters. Although partial shelters are less expensive than enclosed shelters, plants may require the installation of enclosed shelters as a standard.

The design of partial shelters varies considerably in processing installations and the comments in this section on these shelters are directed toward providing general background information for the industrial hygienist. The type of instruments installed in partial shelters is defined by the plant instrument group and may be limited to certain liquids, gases within various specification ranges, or a specifically itemized group of each. However, the partial shelters are usually installed outdoors and general guidelines will be applicable in most installations.

11.7.1 Sampling Systems and Treatment

Sampling systems and treatment were reviewed for continuous instruments placed in enclosed shelters (Sections 11.5.1 and 11.5.2) and the information in those sections is applicable to analyzer systems in partially enclosed shelters. The circulating loop and sample treatment requirements do not change for the partial shelters and the limitation on flow of liquids and gases into the shelter would also be identical. Flow restrictions minimize the possibility of a partial toxic or flammable gas concentration around the instrument area within the shelter.

11.7.2 Partial Shelter Design

Partial shelters are designed in a variety of ways and can be a three-sided structure with a relatively large roof or a shelter with one long wall, short side panels and a very short roof. Between these extremes a large number of shelter types exist, but the general guidelines are applicable to the majority of shelters. The general partial shelter design considered in this section is a three-sided shelter with a roof.

Shelter Structure. The general shelter design for a three-sided structure with a flat roof is shown in Figure 11.16. This structure is mounted on a pad for the reasons outlined in Section 11.6.1 and has an opening at the bottom and the top of the shelter. The roof overhang minimizes the entrance of rain and snow through the top opening. The purpose of the opening is to circulate air through the shelter to prevent the accumulation of toxic and flammable vapors. Partially open shelters normally do not have forced ventilation systems but depend upon natural air circulation to provide the necessary ventilation.

Two other partially open general analyzer structures are shown in Figure 11.17. The difference in all the structures is the roof design, which is modified to minimize the accumulation of vapors. In both designs the light vapors disperse more easily into the atmosphere than the vapors under the flat roof. Information on the effectiveness of various roof designs with natural ventilation has not been demonstrated quantitatively. However, vapors heavier than air will move to the

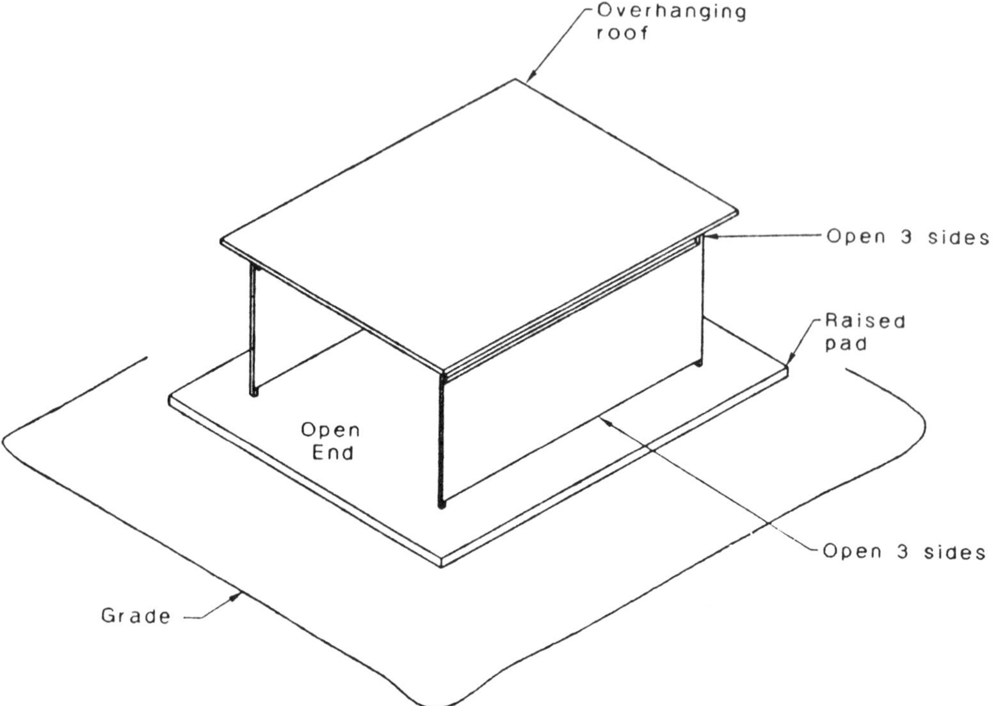

Figure 11.16 Partially open instrument shelter.

floor or pad where they disperse through the open side or through the bottom openings.

The floor pad should be designed to meet the criteria in Section 11.6.1. The shelter instruments can leak and spills may occur within the partially open shelter. Although vapors may readily disperse, toxic liquids are a potential problem and must be controlled. The drain systems previously described would also be applicable in this structure.

Ventilation. The structure should be installed with the shelter opening on the lee side in relation to the prevailing winds. For optimum ventilation the shelter should not be located near other structures that may interfere with the normal wind pattern. Wind flow patterns are difficult to predict and the industrial hygienist should carefully review the planned location along with ventilation considerations.

All of the various shelter roof designs have potential drawbacks and experience may provide a basis for assessing the effectiveness of these various designs. Moreover, the size of the shelter will have an effect upon ventilation and the number of instruments in the shelter may be limited, reducing the general shelter size. The number of instruments and the layout within the shelter are important in defining potential exposures due to inhalation. Without forced ventilation instruments should not be grouped. This precaution will help to minimize vapor concentrations that can present a potentially hazardous situation during a period of low wind velocities. When the wind is not particularly strong, a portable blower

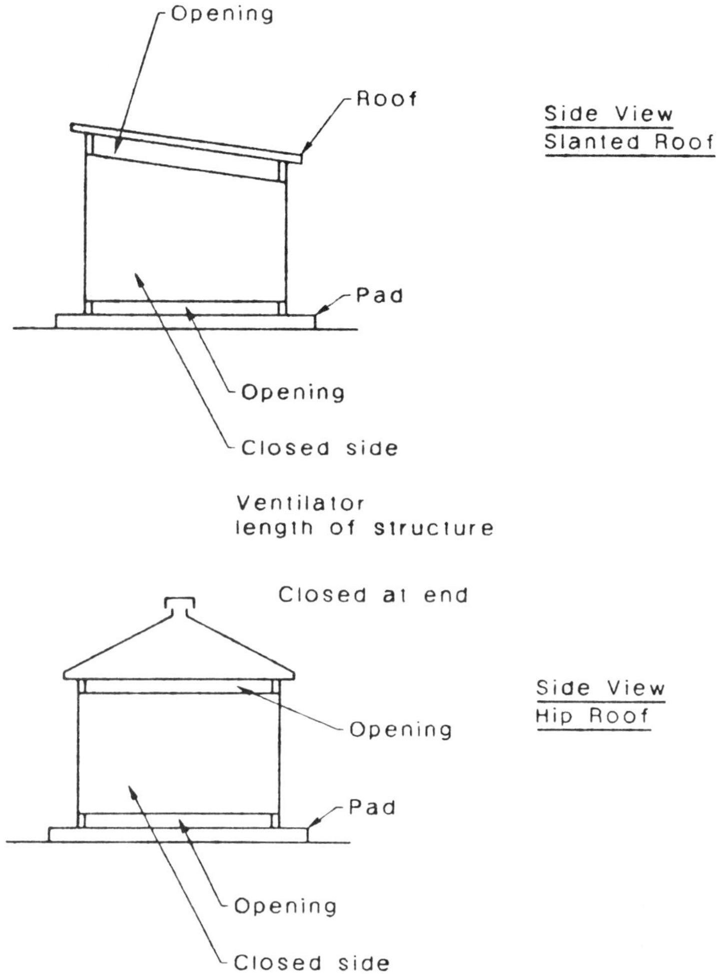

Figure 11.17 *Other types of partially open shelters.*

or fan should be placed near the work area to prevent a toxic concentration buildup around an instrument technician in the process of maintaining and testing an instrument.

11.7.3 Vents

Venting directly from the instruments and the liquid drawoff disposal drums to the atmosphere or through a closed vent system is identical to venting discussed previously. Since the shelter does not have an air intake, the emission location is not critical for this shelter. However, the dispersion of vapors from any atmospheric vent must be carefully analyzed to insure that potential workplace exposures to toxic vapors do not occur in the structures or at grade level. Atmospheric vents from the small disposal drums must also be included.

11.7.4 Liquid Sample Disposal

Disposal of liquid samples is identical to that discussed previously. The liquid disposal system designs are a function of chemical compatibility. Atmospheric vents from any disposal drum can be combined with the general vent described in Section 11.7.3.

11.7.5 Standard Gases and Liquids

Calibration, test, and carrier gases should be rigidly mounted in brackets on the outside of the shelter. Although the shelter does not trap toxic gases, leaks during low wind days can potentially create a hazardous situation. The outside location for the gases reduces the possibility of a toxic concentration buildup within the shelter area.

Toxic liquids are generally in a reasonably small quantity and calibration or testing is normally conducted with a limited amount of material. Where relatively large quantities of toxic, flammable liquids are required, the liquid storage container should be located outside the shelter. More detailed comments are presented in Section 11.6.3.

11.7.6 Alarms

Although alarms may be open to the atmosphere with natural ventilation, a still day can present a potential exposure problem should an accumulation of toxic fumes develop. A toxic fume sensor with an alarm should be provided near the roof and a sensor should also be located near the outlet or floor level for heavier than air fumes. The alarm should have an audible local horn and an alarm indicator in the control house.

Toxic vapor accumulation can be extremely hazardous and alarms provide an additional measure of protection.

11.7.7 Other Equipment

When a leak occurs, an open shelter presents a different exposure control problem than an enclosed shelter. The prevailing wind moves toxic vapors through the enclosure opening, although a reverse wind can disperse the toxic vapors through the openings in the walls. When an alarm sounds for an enclosed shelter, protective equipment outside the entrance door can be donned by personnel in a relatively low toxicity zone. This situation differs for the open shelter where protective equipment may be in an exposed vapor zone. Liquid spraying from equipment is also a potential problem. Obviously, there are a variety of methods for protecting technicians and one method is presented here. Local industrial hygienists may require other procedures to achieve the overall objective of protecting the worker.

For toxic vapors, all technicians should have a portable respirator capable of providing a 10- to 15-min escape time when working within or around the shelter area. This gives the technician a period in which he can leave the shelter and put on the recommended protective equipment. Where a liquid leak may occur, fluid toxicity should be evaluated to determine whether the toxicity effect is acute or chronic. A shower or wash is generally satisfactory for a chronic leak depending upon the material. For potentially acute liquid leaks, light resistant coveralls provide interim protection along with an emergency shower until heavier protective clothing is worn. Both protective equipment procedures are interim steps until the proper protective equipment storage locker can be reached. For the open shelter, placing the protective equipment in a storage bin a short distance from the shelter appears preferable. The storage bin is placed on a steel post to have the locker at a comfortable height. A canopy over the locker protects the worker donning the protective equipment during a rainstorm.

An emergency shower should be located at the outlet of the shelter along with an eyewash bath. In addition a small washbasin inside the shelter opening should be provided. These precautions will provide emergency relief for the exposed worker. Another shower and eyewash facility should be located a short distance away from the shelter for a more complete washdown. In addition to these facilities, general water sprays can be installed near the liquid-containing instruments. Manual operation of the water spray will control to some extent the chemical spray, but the water should not react with the chemical. Many of these precautions are standard procedures in various plants.

The general shelter area should have a defined perimeter that is not entered when an alarm sounds. As described, toxic vapors are readily discharged from an open shelter and when the alarm sounds the perimeter should be entered only by personnel wearing respirators.

11.8 SAMPLING IN PROCESS BUILDINGS

Where chemicals are processed in buildings a wide array of analyzers are in use for batch and continuous processing. Instruments are located near process equipment or can be grouped in a separate room termed an analyzer or instrument room. Frequently a large single analyzer is the only instrument in a small room. Covering this wide variety of instrumentation is difficult and only a few generalizations are possible within this chapter. Moreover, the wide spectrum of processes within buildings further complicates general recommendations. Often chemicals in these highly specialized processes are quite toxic and present a very special set of problems.

In addition to process instrumentation, buildings processing toxic chemicals often contain major air monitoring systems to control general exposure levels. These toxics monitoring systems are necessary in some instances to meet government mandated exposure levels. These toxics monitoring systems are necessary in some instances to meet government mandated exposure levels (OSHA) and in other situations the monitors are installed to meet recommended exposure levels for highly toxic materials in accordance with ACGIH and company exposure standards. Although buildings are exempt from equipment emissions (VHAP) monitoring when vented through a closed vent system to a control device and the building is maintained at a negative pressure, exposures are still a potential problem and surveillance monitors are necessary. The monitors are generally designed to send air samples from a number of locations to a single analyzer. Since the monitoring analyzer performs a critical function, the analyzer and general system are tested and inspected frequently, which *may* expose the instrument technician to highly toxic chemicals.

11.8.1 Sampling Systems

Although many of the processes in buildings are based upon batch reactions, multiple batch reactors and other processing equipment frequently require continuous analysis. In other systems, the samples are sent to the analyzer on timed cycles, which results in intermittent operation, but the instruments are in an almost continually operating mode. Where several instruments are grouped in an analyzer room, the room is a distance away from the sampling point and a sampling loop is generally required for either batch processes or a continuously operating system. If the analyzer is in close proximity to the sampling point a sampling loop is not generally required, but the HON requirements do not exempt analyzers in close proximity from the circulating loop regulations.

Sampling loops for batch and continuous processes in enclosed buildings are similar to the loops shown for outdoor process facilities. Moreover, the return or disposal leg discharge requirements are identical and the disposition point must be identified by the process group. A method of purging the loop for repair and maintenance is a major requirement within a building and should be defined in detail before the instrument system is commissioned. Leakage within the building cannot be tolerated for toxic streams and the industrial hygienist should review the design and operating details of all sampling systems.

Dead-ended or noncirculating sampling systems for analyzers located near sampling points also present potential exposure problems. Although the length of line between a sampling point and an analyzer may be relatively short, purging the line for repair and maintenance is necessary to minimize exposures. The method of purging the line begins with an understanding of the physical and chemical characteristics of the material in the line. The sampling material is separated into the following general categories:

a. Gas or noncondensing vapor
b. Condensing vapor
c. Volatile liquid
d. Nonvolatile liquid

The following review is concerned only with relatively short dead-ended lines.

Gas. In a short sampling line the volume of gas in the line and analyzer can be calculated as a basis for estimating potential exposures should the line be opened to the room atmosphere. For various lower toxicity materials, opening the line may not overexpose personnel. However, in a situation where the vapor is highly toxic, opening the line can potentially overexpose personnel in the immediate area. Exposures can be calculated with the method of dispersion analysis presented in Chapter 18.

The line and instrument can be purged to the atmosphere outside the building with an inert gas or nitrogen and through the instrument vent when a vent is necessary on the instrument. A dispersion analysis for short-term exposures should be conducted on the material discharged to the atmosphere from the stack to insure that overexposures do not occur. An analysis should include concentrations on the roof where workers may be exposed. Process buildings often have flat roofs and workers are often present on the roof in connection with filters, cooling towers, blowers, or other equipment. In addition, the dispersion analysis for a building roof vent should consider short-circuiting or diverting the vented vapors into the inlet duct of the building fresh air blower. As described previously, toxic materials can reenter the building through the inlet air duct, potentially overexposing workers.

In locations without a vent in the line or instrument, a vent can be installed for purging that connects with an existing process vent line. However, interconnections must be reviewed for chemical interaction and vapor dispersion. Alternatives available with some processes are vent gas blowdown and waste gas headers. The vent gas header can be connected directly to a scrubber where the scrubber liquid neutralizes or reacts with the vent gas. The scrubber is normally installed for emergency or general venting process requirements. If a vent blowdown system is not available, a small drum filled with liquid can provide the same neutralization or reaction mechanism. The sample line is purged with nitrogen and the purged vapor is sent to the neutralization drum.

Condensing Vapor. With condensing vapors, the sample line may be heated to provide a true vapor for analysis. However, venting the line to a noninsulated vent and the cold atmosphere can potentially result in toxic droplets emitted from the vent, which presents a hazardous situation around the vent. Condensing vapors should not be discharged to the atmosphere and an alternative control solution is necessary.

Condensing vapors can be purged with nitrogen and sent to a small pot where liquid droplets are separated from the vapor. The pot is located below the sample line elevation level and also provides drainage for liquid in the line. The purged vapor is separated from the droplets and discharged to the atmosphere through a vent. The criteria for venting has been described previously. The condensed liquid in the pot is pumped back to the process.

Complete condensation is also a method of controlling the vapor remaining in the sample line. The vapor and drainage is gravity fed to a water-cooled exchanger that condenses all of the vapor and the condensed material drains into a small pot, where the liquid in the drain pot is then pumped back to the process. The effectiveness of this technique depends upon the vapor dew point. If the water cooler is not satisfactory, a chilled fluid is required that adds to the installation cost and complexity of the general system.

A vent blowdown and scrubber system may be available, but the droplets and condensation may present system problems. A small nitrogen or steam driven ejector connected to the sample line will generally remove all material from the sample line with the combined stream sent to the blowdown scrubber unit. Scrubber units often process purge streams under vacuum and the available blowdown system can be under a negative pressure. For this situation a direct connection between the sample line and the blowdown line will empty the line, which is followed by a nitrogen purge.

Volatile Liquid. A toxic volatile liquid presents a problem since the material emits vapors when exposed to the atmosphere and the liquid volume if converted to vapor generally is a relatively large vapor volume. If the material is heated and vented to the atmosphere, the temperature on a cold day may result in partial condensation, which is unacceptable. Draining into a pot or recycle to the process is feasible, but nitrogen pressure is required to prevent vaporization as the liquid drains from the lowest point of the sample line. Following drainage, the liquid in the pot is pumped to the process.

A vacuum created by an ejector on the sample line will vaporize the fluid in the line and the sample fluid along with nitrogen or steam for the ejector is sent to the blowdown system or scrubber. If the blowdown scrubber is under vacuum the ejector is not required. The evacuation systems work well with volatile liquids, but a receiver or disposal system is required for the vapor.

Nonvolatile Liquid. Draining the fluid to a small pot is the standard technique for liquid removal and the liquid is then pumped to a process vessel. However, the sample lines should be constructed for good drainage with liquid entrapment eliminated wherever possible.

Instruments. The instrument sample capacity is usually quite small and potential exposure levels from process fluid remaining in the instrument should be calculated. If the quantity does not pose a potential health problem, the liquid can be removed manually with proper precautions for acutely toxic materials. Should the liquid present potential exposure concerns, a method of removing liquid must be developed. The instrument manufacturer normally will provide assistance in developing a safe, manual removal procedure.

The analyzer should not be purged or drained through the system unless a careful review of the design and the purging conditions indicate the instrument will be unaffected.

11.8.2 Analyzer Rooms

Sample loops are connected to the analyzer instruments located in a room and the overall layout is similar to the enclosed shelter previously described for outdoor installations. However, there are differences due to the location of the room within a building. The toxicity level of the process within the building also influences the design of the analyzer room along with the room location.

One of the methods of controlling exposures to toxic vapors in a building is to pressurize those rooms where workers and technicians spend the greater portion of their time. As an example, a control room is pressurized with air and the flow of air through the doors and openings prevents the entrance of toxic vapors into the control room. Personnel are protected from constant exposure to chemicals in the atmosphere and the building atmosphere is well within recommended exposure limits. However, in those buildings processing chemicals with very low recommended exposure levels, a small leak can increase the actual concentration to a

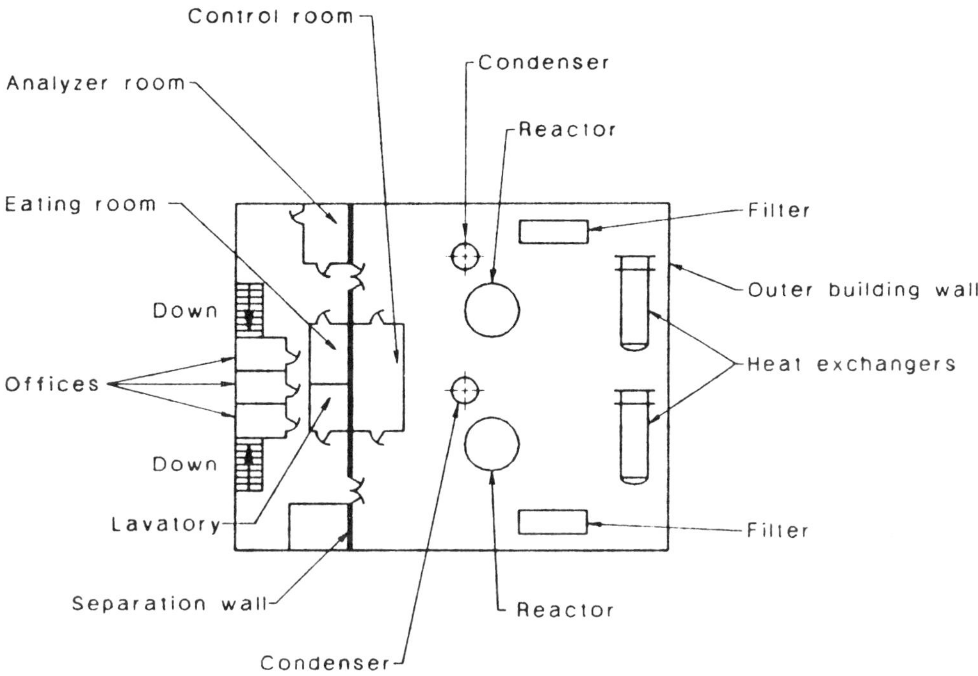

Figure 11.18 *Enclosed instrument room in process building.*

level that may be unacceptable for an 8-hr TWA and the chemical-free atmosphere in a control room protects the workers from an excessive exposure level. Where process buildings are under negative pressure for alternative emission control under the HON regulations, the need for pressurization should be carefully reviewed.

Local plants may prefer pressurized analyzer or instrument rooms to minimize potential exposures to vapors from the process building. However, the pressurized room does not protect the technicians from leakage within the room. Moreover, doors that do not connect with the main processing section of the building also provide a measure of safety against excessive concentration levels in the instrument room. Although the analyzer room may not be pressurized, a ventilation system will provide some pressurization, but doors opened for personnel movement upset the ventilation balance and external air can be drawn into the analyzer room, particularly in those buildings where the pressure is ambient or only slightly negative. A layout of an imagined process floor in a building along with an instrument room is shown in Figure 11.18, demonstrating the potential location of an analyzer room outside the reaction area.

Room Construction. For a separate room of the type shown in Figure 11.18, the room design basis is similar to the outdoor enclosed shelter previously described. Ventilation is provided by an external blower and is discharged through louvers in the outside wall. Vapor discharge outside the building prevents the discharge of toxic vapors into the process building. This ventilation system builds a small pressure in the analyzer room. A heating and cooling system is provided in the blower ducts for personnel comfort.

The analyzer room floor should be designed for liquid leakage, which requires a small slope and drains for the chemical materials and flush fluid. The floor should slope to either long runoff troughs or several catch basins. In addition, the floor must be impervious, nonporous, and nonskid to permit runoff into the drains.

Instruments should be placed on a support ledge against the wall as shown in Figure 11.13. The center area of the room should be free of obstructions and a second door is necessary for emergency situations. Doors should be tight fitting to minimize flow through typical door openings.

Ventilation. Airflow patterns for the room will differ from those shown in Figure 11.14 for the outside shelter since the air volume exits through one wall rather than two. The airflow distribution objectives for the analyzer room require a flow pattern that prevents the accumulation of toxic or flammable gases in the room. In addition, a flow of air past the inhalation zone of a worker is also recommended to minimize potential exposures to high local vapor concentrations. The requirements for airflow distribution patterns can be met by placing inlet air ducts in various locations with diffusers to distribute the air along with adjustment of the louvers.

Adjustable louvers should be located at the ceiling and floor levels to assist in the airflow distribution pattern. The adjustable louvers increase the pressure in the analyzer room slightly, but this pressure is normally much less than the pressure in a pressurized control room. The latter generally has special air lock doors that minimize leakage and assist in maintaining the pressure. The analyzer room doors should seal well, but air-lock-type doors are not required. The exact location of the blower is not specified by the industrial hygienist, but air distribution requirements and blower emergency operations are specified in detail for contractor design and test. Following completion of the ventilation system, the airflow distribution pattern should be tested and modified until the specification flow requirements are achieved.

Inlet air quality must be satisfactory and the location of the inlet air opening is important in obtaining a source of noncontaminated air. The inlet air duct can be elevated well above the roof, but the elevation of the duct may be limited for structural or architectural reasons. A low inlet elevation may result in contamination from toxic vapors that may be vented periodically from nearby vent stacks. The inlet air duct should be located upwind of any discharge vent, but changes in wind direction can contaminate inlet air that should be toxic free at all times. The inlet airstream is usually purified with a carbon filter, but a carbon filter does not necessarily remove all toxic contaminants. The industrial hygienists should review all of the potential air contaminants vented to the atmosphere and determine whether the toxic compounds can be removed from the airstream with a carbon filter. If the toxic contaminant is not absorbed on carbon, a wet-scrubber will generally remove the contaminant with an absorbing or reacting fluid. A demisting screen should then be installed downstream of the scrubber to remove droplet carryover. Liquid disposal is then investigated to insure that the scrubber can be drained to an existing disposal system. Although these systems will remove toxic contaminants, a toxics contaminant sensor located downstream of the treating section will signal if a contaminant breakthrough occurs in the carbon filter or scrubber system.

Blowers. For an analyzer room, an operating blower with a standby blower should be recommended. The standby blower automatically begins operating when the blower that had been operating is shut down either automatically or manually for mechanical reasons. The spare blower maintains airflow through the room to prevent the accumulation of vapors. The normal volume flow through the analyzer room should be in the range of 6 to 12 turnovers per hour. A higher turnover rate provides a better flow distribution pattern but is more costly.

Sensors to detect flammable or toxic concentrations alarm and signal a controller to increase the flow rate of the operating blower to maximum level. The blowers should be specified with a normal and maximum airflow rate, with the latter rate an emergency airflow volume. In the event the toxic concentration level does not fall below a preset concentration, the standby blower should cut in automatically to minimize vapor concentrations. This system is similar to the outdoor blower installation for the enclosed shelter and will control toxic vapor concentrations in emergency situations.

Liquid Sample Disposal. The typical systems described in Section 11.6.2 are adequate for the liquid analyzers in the analyzer room. Compatible sample materials are collected in a small drum and pumped to the process or a receiver defined by the process group. Certain sample fluids that are incompatible with other fluids or that can be recycled directly to the process are separated and sent to an individual receiving pot for recycling with a small pump or under inert gas pressure. Vents from these pots are connected or separated for disposal in accordance with previous venting comments.

However, an important consideration in draining and purging the sample liquids from the various loops into the receivers is the location of the drums and pots. For the enclosed outdoor shelters, the loops can be drained by gravity into the drums or pots, which are the lowest points in the system. However, draining the loops in a building generally presents a problem since the lowest point in the loop can be at a lower elevation than the analyzer. The loop can be purged under pressure, but complete drainage is always questionable. For this situation, the disposal drums and pots should be located at ground level in the building. The loops are then drained and purged into the disposal drums with the instrumentation shown in Figure 11.15. Analyzer draining and cleaning may be conducted separately but should be reviewed with the instrument specialist.

The drums and pots at grade elevation within the building should be located within the lower processing area.

Standard Gases and Liquids. Gases in cylinders for calibration and testing should be located in the process room and not in the analyzer room. Leakage within the small analyzer room can overexpose a technician quite rapidly compared with the large process room. Gases are often toxic and large cylinders in the process room are a potential hazard although the process area is ventilated and alarms indicating high concentrations of toxic vapors are located at strategic points in the process area. The size of the cylinders should be limited based upon concentrations estimated for various potential leakage occurrences and cylinder sizes. Toxic concentration sensors should be located near the gas cylinders, which are securely fastened.

Liquids for testing analyzers should be stored in small containers and located in a secure facility in the process area. An overhead sprinkler system manually operated for liquids flushes spilled liquid into a drain system designed to receive spillage. The design of the calibration gas and liquid storage facilities should be reviewed by the hygienist in existing and new facilities to insure that potential exposures to toxic gases and liquids are minimized. Storage in a building for liquids and gases is not desirable and minimum quantities should be specified wherever possible. These calibration and test compounds can be stored outside the building or in a special room, but added handling and further complications increase the potential risk of a spill or leak. Small, simple systems that can be easily controlled are preferred.

Alarms. Various sensors and alarms have been mentioned in the previous paragraphs. Those recommended for the analyzer room are designed to measure the toxic vapor concentration or the accumulation of flammable gases. Upon reaching a set concentration an alarm sounds outside the major door into the analyzer room and a warning light flashes at both the entrance and alternative doors. A repeater signal alarms within the control room. In addition, a manually operated alarm in the analyzer room will signal the control room that an emergency situation exists and assistance is required in the instrument room.

An alarm is also necessary to indicate a low inlet airflow rate from the ventilation blower. The loss of ventilation air is critical in controlling the buildup of toxic vapor and flammable vapor concentrations. All of the alarms should indicate a problem in the control room.

Other Equipment. Personal protective equipment should be available in a locker outside the main analyzer room door. A technician or operator entering the analyzer room should always have an emergency respirator available. In addition, a shower and eyewash station in the analyzer room are necessary to insure emergency water dilution and availability during a liquid leak. A shower and eyewash station outside the analyzer room is also recommended for personnel that have been exposed to toxic liquids or vapors. A hose for flushing liquid leaks should be installed within the room. The flushed liquid enters a drain for disposition outside the process building.

The installation of sprinklers may be of assistance where there is the potential for an acutely toxic liquid leak. Rapid removal of this material prevents additional exposures. The sprinklers in this situation must be manually operated and the valve should be located near the door. The hose and sprinkler water streams require good drainage on the floor and the drain systems must be sized for the flow volume. If the drains are not sized properly for the flush water volume, the toxic material and flush water will probably leak through the doors to the nonhazardous outside area.

11.9 SOLIDS SAMPLING

There are a wide variety of particulate catalysts in the process industries that are used in fixed bed, fluid solid, and liquid slurry processes. In addition, solid particulates are produced in various processes. Particulates are an industrial hygiene concern if the particles are in the respirable range, contain or consist of toxic metals or materials, and have associated dermal effects.[10] In addition, toxic gases are also present in various particulate streams. The wide variety of solids processes and solid particulates are difficult to condense into several general types since the overall study is further complicated by the temperatures and pressures of the various processes. Fluid catalytic cracking processes in petroleum refinery service generally operate above 950°F (510°C) and 30 psig (207 kPag). Powder from many plastic drying and finishing operations are normally at atmospheric pressure with a maximum temperature of about 250°F (121°C). Other processes have particulates in a pressure–temperature range between these two levels.

In fixed-bed catalytic operations, large differences in pressure–temperature relations also exist with a number of petroleum refineries operating at temperatures above 750°F (400°C) and at pressure levels of 300 psig (21.7 bar) or greater. Chemical processes also vary considerably with pressures above the thousand atmosphere level and temperatures to 750°F and higher in some processes. The catalysts in fixed beds are stationary and in most process operations onstream catalyst samples are unnecessary. However, samples are periodically required and a process is temporarily shut down to obtain a sample. Sampling in this situation differs from those processes where onstream solids samples are needed for process control.

Slurry samples are frequently obtained in chemical processes to determine particle size, additive distribution, and catalyst concentrations. A number of slurry processes operate with a variety of solvents, chemicals, solids, and additives. Generalizing is also difficult in slurry processes where many of the sampling procedures are designed by plant personnel for a specific process.

In addition to sampling process streams, particulate sampling of flue gases and other stacks is often required to meet EPA or local particulate emissions standards. The sampling procedures are standardized for the EPA but plants have modified these procedures for process control reasons. However, the sampling systems are reasonably similar and the principles of exposure control are identical.

Since the sampling systems vary considerably in all of these processes, a few sampling systems are presented to show the general principles involved in exposure control.

11.9.1 Fluid Solids Sampling

Fluid solids sampling in this section is separated into high- and low-temperature operations. Typical high-temperature operations are the catalytic cracking processes in the petroleum refining industry; low-temperature operations are typified by polymer processes in the chemical industry. Solids in the processes described here are basically dusts or powders. The fine particulate provides a large surface area for catalysis and the polymer powder product is marketed directly as a powder or converted to pellets. The latter is not an industrial hygiene concern. Considerations in sampling both processes are presented in the following sections.

A major consideration in fluid solids sampling is the type of sample necessary for process control and analysis. A number of small samples may be required to provide a statistical average rather than a single large sample. The process engineering group defines the general sample size and sampling frequency that serves as the design basis for the sampling system. This consideration is generally applicable in polymer powder sampling and other sampling situations, but high-temperature catalytic cracking sampling is not particularly amenable to small, accumulative sampling procedures.

Fluid Catalytic Cracking. This process is typical of a high-temperature fluid solids operation where commercial sampling devices are normally not available and optical in-line measurement techniques are not applied. In-line optical or laser systems provide an estimate of mass flow and certain devices measure particulate size. However, the instruments are generally not applicable at in-line temperatures greater than 1000°F (538°C) since the exposed instrument parts are 180° apart and essentially flush with the inner line diameter. The optical instruments are also open and exposed to the erosive effects of the cracking catalysts. These instruments cannot be removed easily from the measurement location since they do not normally have isolating valves. The isolation or block valves are a safety requirement and automatic samplers must be capable of operating through a valve at process conditions. Moreover, pressures of 25 psig (172 kPag) or higher present sealing problems in most sampling devices.

A manual sampling system is preferred for the hot, pressurized fluid catalytic processes for the reasons previously described. Depending upon the sampling point location, the temperature can reach 1300°F (705°C) at 25 or 30 psig, which is a difficult operating condition. A batch sampling system for this process is shown in Figure 11.19. The sampling system is connected to a regenerated catalyst line that is very much larger than the sampling line. The pipe length between the double blocks and the major catalyst transfer line is normally purged with nitrogen or steam at a high velocity to prevent the entrance of catalyst particles into the pipe length. When sampling, the inlet purge line is blocked and the double blocks are opened. Catalyst flows into the line and the line length between the plug and double block valves fills with catalyst. After a prescribed time period, the double blocks are closed and the inlet purge line is activated to prevent the entrance of catalyst into the sample line.

Catalyst remains in the line between the plug and double block valves until the particles have cooled to a desirable temperature. The plug valve is opened and the catalyst under the sealed pressure is blown into a sample container in the enclosure. Excess solid blown around the enclosure along with the gas flows through the bottom grating or sample container support into the waste container. The excess gas with entrained catalyst is vented to the atmosphere through a filter. The filter is sized for 1 μm and essentially all of the catalyst particles are trapped on the filter, which is replaceable. Gas from the system that is vented to the air generally consists of a mixture of CO_2, H_2O, N_2, CO, and traces of SO_2. The vent emission location is tested for dispersion at ground level and in nearby structures.

The sample enclosure is in the structure and the drain line to the waste container can be lengthy depending upon the location of the sample enclosure. The door in the sample enclosure must be tightly closed to prevent leakage of the catalyst particles. In addition, the couplings and clamped connections are also sealed tightly. Purge gas lines are connected to the block valves through special

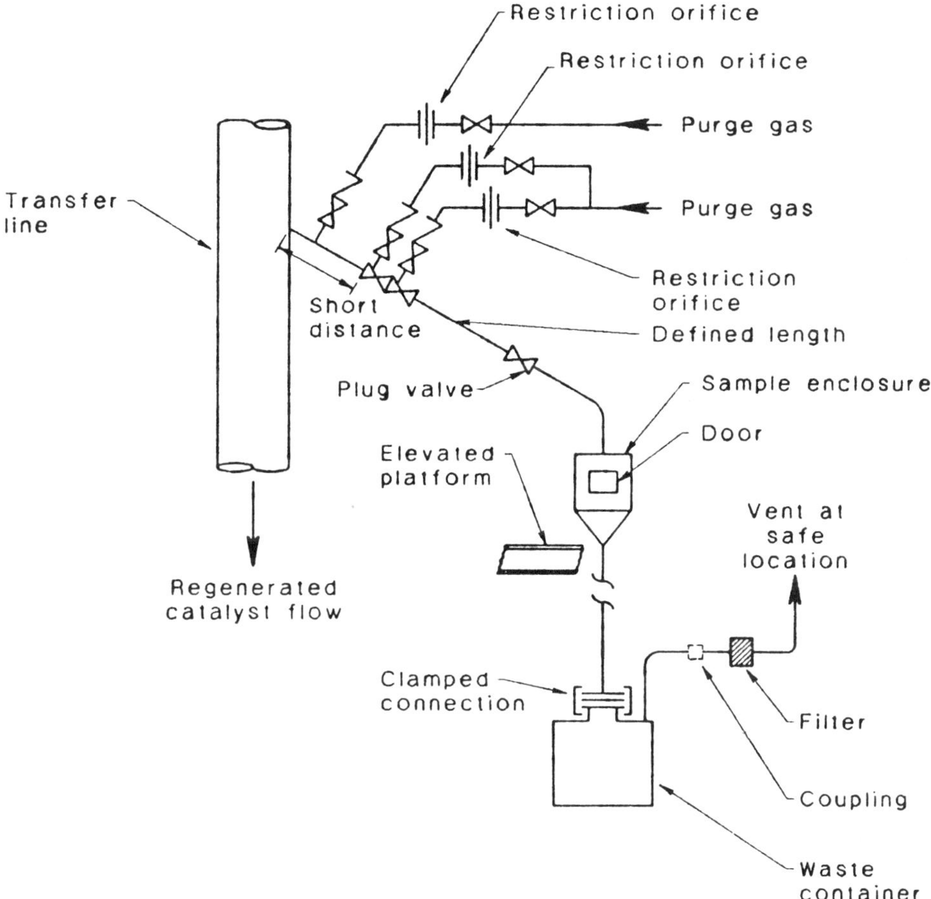

Figure 11.19 *Hot solids batch sampling system.*

bosses that are drilled and tapped for this purpose. The purge gas blows catalyst particles away from the valve seats, which permits tight closure of the valve gate. Catalyst particles are hard and the valves cannot be closed unless the valve seats are clean.

Respiratory dust protection is necessary when opening the enclosure door to remove the sample and the industrial hygienist should specify personal protective equipment.

Polymer Processing. Finished polymer in many processes is in the form of powder, which is shipped directly in that form or converted to a pellet. Samples of the powder are generally obtained to determine particle size distribution and various physical characteristics prior to shipping or additional processing. The powder may have traces of solvent or moisture remaining which should not present an industrial hygiene hazard, but a sample of polymer should be placed in a container and the container vapor concentration should be analyzed to determine whether a potential hazard exists.

The process pressures in the polymer finishing area are essentially atmospheric pressure since much of the powder movement is gravity controlled or moved at modest pressures. As a result, sampling equipment design is simpler and does not present the sealing problems associated with high pressures and elevated temperature operations. Since the systems are at low pressure and temperature levels the samplers are often placed in an automatic mode to periodically take fixed sample quantities for statistically averaged characteristic data. The general sampling device shown in Figure 11.20 can be used in either a batch operating or automated operating mode. A variety of devices are available for polymer sampling and the device in Figure 11.20 provides a basis for understanding both general sampling and potential exposure problems.

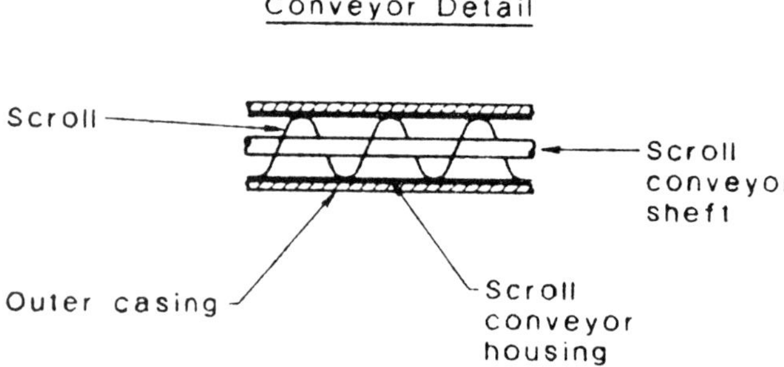

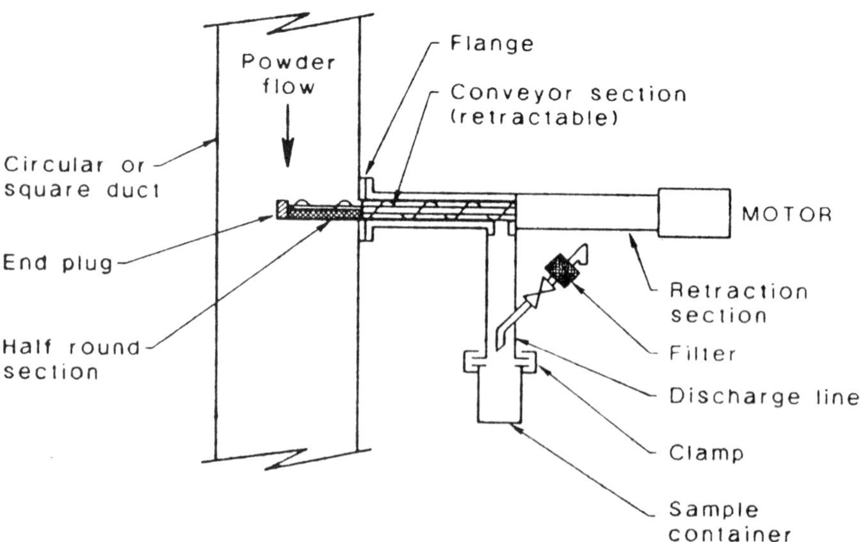

Figure 11.20 *Polymer powder sampler (courtesy of Quality Control Equipment Company).*

The sample system scroll conveyor protrudes into the duct with a half round section below the conveyor. Powder is collected in the bottom section and conveyed to the opening in the housing where the powder is discharged through the line into the sample container. The scroll conveyor is enclosed in a housing as shown in the detail and the housing along with the scroll conveyor can be retracted after the sample is obtained. The end plug seals the opening in the wall of the duct. Alternatively, the half round section can be moved 180° to halt collection in the trough. Following conveyor retraction, the sample container can be removed from the clamp. However, a slight positive pressure in the system will result in a small dust cloud when the container is unclamped from the discharge line. Also any powder in the conveyor will also be emitted to the atmosphere. An atmospheric vent line from the container or discharge line shown in Figure 11.20 with a filter controls particulate emissions to the atmosphere from the container. The vent line size is small and determined by mechanical rigidity. A valve can be installed in this service to hold the pressure, which will be less than 0.5 psig (3.5 kPag), and prevent particle leakage to the atmosphere.

Following removal of the container, a new container should be installed or a flat disk placed against the discharge line opening. The sampler may require maintenance in the shop and must be removed from the duct. In this operation a continuous powder cloud will discharge from the opening in the duct until the hole is closed. Although polymer dust or powder may only be a nuisance dust in most situations, a blind or disk should be available to close.the hole immediately when the sampler is withdrawn. The flat disk is prepared prior to removal of the sampler from the main duct. This operation requires dust masks to insure that the workers are not exposed to respirable dust during the removal or reinsertion operation.

If solvent traces are present in the duct vapors, the industrial hygienist should determine the chemical concentration levels during normal operation and the need for inhalation protection. If the solvent is toxic and concentration levels are relatively high compared to the recommended exposure limits, the vent from the discharge line can be extended to a safe location for normal operation, but removal of the sample unit would require personal protection. Personal protection is also required when replacing the sampling system.

An alternative system is commercially available where the conveyor in Figure 11.20 extends through the duct on the opposite side from the flange. The powder is conveyed through the opening which is enclosed and then discharged into a large diameter line. This line terminates in an enclosed sampling box where various containers can be placed. The advantage of the enclosed box is that improved personal protection is provided compared with the container in Figure 11.20. A vent from the enclosure must also have a filter to remove particulates from the vent stream.

Liquid Slurry Streams. Numerous liquid slurry operations exist in the chemical processing industries, but the streams and operations vary widely. Slurry characteristics differ significantly in volatility, solidification state, viscosity, solids loadings, specific gravities, tackiness, and so forth. In many situations expertness in controlling and operating with slurry streams resides in the local plant and process engineering groups. Consequently, a specific system cannot be discussed that presents a broad application area within the industry. As an alternative, certain exposure control considerations in the design of slurry sampling systems are presented as general guidelines for an industrial hygiene review of slurry sampling designs. The general basis for review and recommendation follow:

Stream Physical and Chemical Characteristics. Before an assessment of potential health hazards is undertaken, this information is required. In addition to the laboratory analytical data, general observations by operating personnel on stream properties and idiosyncrasies are helpful.

Volatility. The stream composition determines the volatility of any toxic compounds in the sample. A high volatility requires enclosed sampling procedures. In contrast, a very low volatile slurry sample can be gathered in an open container. An in–situ analyzer is preferable in those situations where the stream has a relatively high volatility. In the next section, comments are presented on analyzers and in–situ instruments. In volatile streams with a high solids concentration requiring a chemical analysis of the liquid, a preliminary step that removes solids produces a liquid that can be analyzed directly or batch sampled for a laboratory analysis. However, the solids in the stream must be further classified in accordance with their separation characteristics from the liquid. Solids may have a relatively high specific gravity and are suspended in the fluid only through turbulence. The solids in this situation are readily separated from the fluid, but more importantly the solids tend to stratify in the piping lines. In another example, the solids in the volatile stream do not settle readily and must be removed through a different procedure. Potential methods of sampling and analyzing the liquids in these two different streams are shown in Figure 11.21. The sampling systems in Figure 11.21 are readily converted to manual batch sampling through the methods presented in Section 11.3.

In the rapid settling system, most of the solid particles readily settle toward the bottom of the process line with a few particles remaining in the upper portion of the pipe. The sample loop inlet is placed within the line facing downstream to further minimize solids loading at the inlet entrance. A purge prior to initiating flow through the sample line is sent through the filter at the inlet point clearing the filter. The remote operating block valve is then opened and liquid flows through the loop to an instrument connection on a sidestream or a batch sample container. A simplified drain system is shown as a basis for purging the loop in preparation for cleaning and removing solids in the filters. Exposures can be controlled by draining and purging the equipment properly. The remote operating valve and purges permit complete flushing of the lines.

In the slow settling system, the solids are scattered rather evenly through the fluid and the filter on the inlet is removed. Liquid movement through the inlet filter would rapidly clog the filter owing to the liquid solids loading. The solids-laden stream is sent through a hydroclone system to remove the major portion of solids and those remaining in the lightly loaded stream are removed on the filter screen in the line. The filter can be purged and removed as desired for cleaning or replacement and purges are sent to a closed system for disposal through the drain pot. The underflow from the hydroclone is a concentrated solids stream that must be sent to a receiver. The industrial hygienist and the process group investigate alternative disposal points for the heavily laden stream at a very low flow rate. A

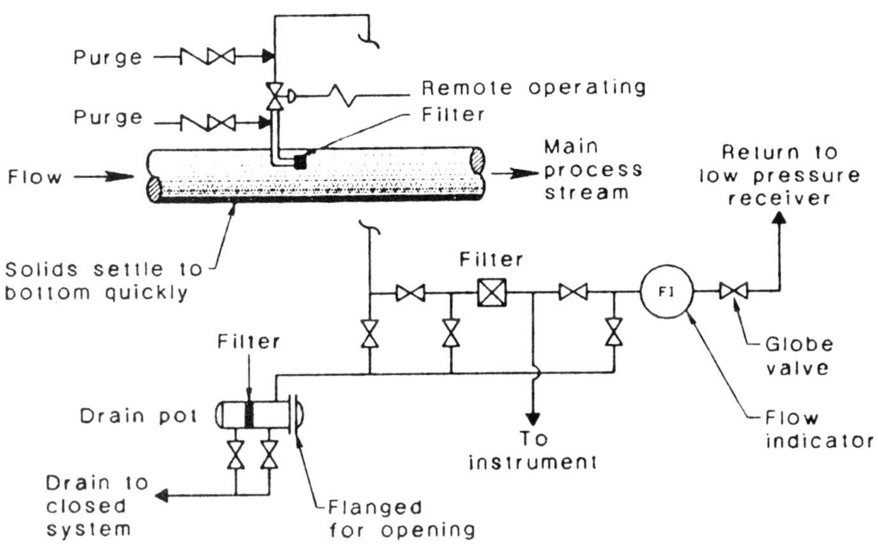

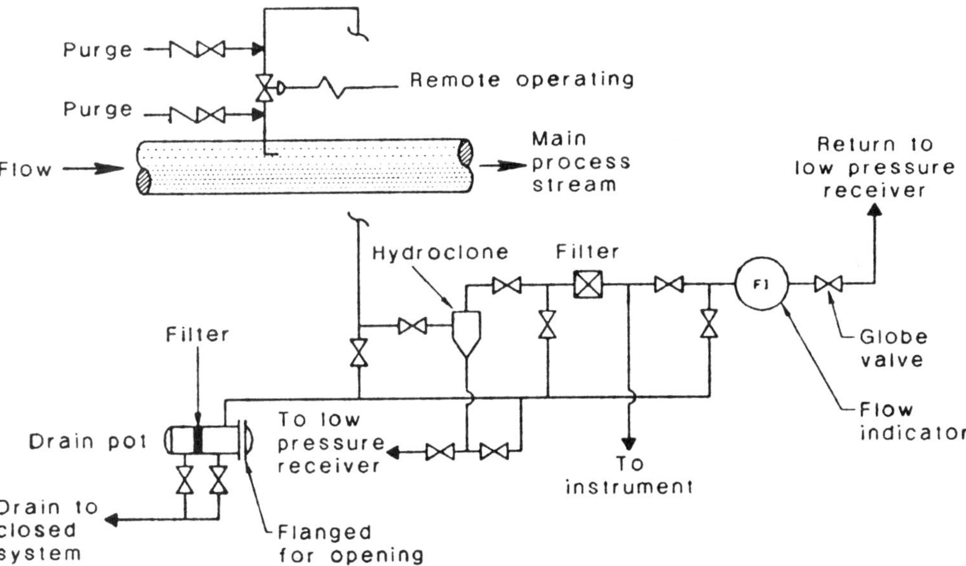

Figure 11.21 *Sampling liquid slurries.*

batch sample container can also be useful with this system as described in Section 11.3.

All valves should be full ported or ports without restrictions and flushing lines should be installed as required to completely purge the system. Since the liquid is volatile, all lines and accessories must be purged with an inert gas to insure that the system is free of toxic vapors.

Physical and Chemical Characteristics. Frequently slurries are quite viscous and sampling is a difficult procedure. Where a high viscosity exists, particularly in streams with volatile components, an in–situ monitoring instrument should be investigated. For example, viscosity can often be correlated with the concentration of polymer or rubber material. A viscosity instrument installed in a process line is a much superior sampling system to a batch sample analysis.

In other situations, a high concentration of material can change the specific gravity. Advanced specific gravity instruments now provide on-line specific gravity measurements. Other advanced instruments are also available, such as liquid polymer chromatographs, that provide physical or detailed analytical information. A liquid GC or other instrument is normally installed on a sampling loop and loops must be purged and cleaned with solvents, which requires a satisfactory control procedure.

Other devices are available that measure solids concentration directly in a liquid stream. Radiation instruments placed across a line will measure solids concentration directly in real time and other instruments perform the same function. Instrumentation provides methods of developing process information that minimize exposures in the processing unit and eliminates potential exposures in the laboratory.

Capture of solids is also required in certain processes and systems based upon the designs shown in Figure 11.21 can be developed to recover solid particles. Alternatively, the batch sampling loops shown in Section 11.3 should be considered where solids or laboratory samples are required. The design of any system must consider laboratory needs and exposure controls for toxic fluids and solids.

Fixed Catalyst Beds. Periodically, in fixed bed processing, a sample of the catalyst in the bed is desired to determine catalyst activity, life, or physical change in the catalyst. Removing catalyst samples from an operating reactor is difficult and dangerous. Consequently, most facilities are shut down or partially shut down to permit sampling the catalysts in a reactor. The normal vapors or liquids in these systems can be toxic and flammable, requiring purging prior to entering the vessel for a sample. The catalyst sampling procedure must be reviewed by industrial hygiene and safety personnel to insure a safe entry procedure and that the sampling technician is not exposed to toxic materials. Personal protective equipment is generally required to provide a source of oxygen in a potentially low oxygen atmosphere, prevent dermal exposure, minimize inhalation exposures, and protect against heated temperature locations in the vessel. Toxicity precautions are concerned with vapors, liquids, and catalysts, where the latter frequently contain nickel, vanadium, chromium, or other toxic metals.

11.10 SUMMARY

Batch sampling is a potential source of toxics exposure, but the EPA has mandated certain rules for fugitive emissions that control exposures during the batch purge of sample containers to the atmosphere. Other batch sampling procedures that are potentially toxic can also be controlled as shown in this chapter. Although continuous automatic instruments apparently do not pose toxic exposure problems, potential exposures can occur and the industrial hygienist must analyze all instrument facilities including shelters. Some instrument sampling facilities were reviewed in this chapter, but these facilities represent only a small fraction of the instrument sampling facilities in the chemical process industries. Industrial hygienists should become familiar with all types of sampling installations to develop guidelines for the general control of toxics exposure in this area.

REFERENCES

1. CFR. Standards of Performance for Equipment Leaks of VOC in the Synthetic Organic Chemicals Manufacturing Industry, Part 60, Subpart VV, Section 60.482-5, July 1, 1992. Standards of Performance for Equipment Leaks of VOC in Petroleum Refineries, Part 60, Subpart GGG, Section 60.592, July 1, 1992.
2. Fed. Reg. National Emission Standards for Organic Hazardous Air Pollutants from Synthetic Organic Chemical Manufacturing Industry Equipment Leaks. Proposed Rule, Vol. 52, No. 252, Part 63, Subpart H, Section 63.166, p. 62769, December 31, 1992.
3. EPA. Compilation of Air Pollutant Emission Factors, Vol. 1, Stationary Point and Area Sources, AP-42, 4th ed. September 1985.
4. Dimmick, W. F. and K. C. Hustvedt. U. S. Environmental Protection Agency. Equipment Leaks of VOC: Emissions and Their Control, APCA Annual Meeting, San Francisco, CA, June 24–29, 1984.
5. Schaich, J. R. How to Estimate Fugitive Emissions from Equipment, American Institute of Chemical Engineers, National Meeting, Boston, MA, August 21, 1990.
6. EPA. VOC Fugitive Emissions in Petroleum Refining Industry—Background Information for Proposed Standards, EPA-450/3-81-015a, November 1982.
7. NIOSH. Control Technology Assessment of Petroleum Refinery Operations, Radian Corp. Draft. Contract No. 210-81-7102, March 1983.
8. Emmel, T. E. Control Technology Assessment of Petroleum Refinery Operations, National Petroleum Refiners Association Meeting, March 20–22, 1983.
9. American Conference of Governmental Industrial Hygienists (ACGIH). Threshold Limit Values for Chemical Substances and Physical Agents and Biological Exposure Indices for 1992–1993, Cincinnati, OH, 1992.
10. Wang, CCK, *Chem. Eng. Prog.* **80**, 53–58, September 1984.

12

Drains, Sewers, and Wastewater Emissions Control

Emissions from drains, sewers, and primary wastewater treating facilities are frequently overlooked as sources of potentially hazardous emissions. Although the emission rates from the individual on-site drain and sewer vents are quite low, the leakage contributes directly to general process area concentrations. Moreover, the leakage sources may be located in an area frequented by maintenance or operating personnel. Since they can be toxic, emissions from drains and sewers are of concern to the industrial hygienist.

Sewer effluents containing various toxic compounds and other process material are normally sent to wastewater treating facilities where the volatile components vaporize in the primary wastewater treating units. In a petroleum refinery or chemical plant these primary facilities are generally a combination of API separators, decanters, or clarifiers, which remove immiscible process liquids from wastewater, followed by an air flotation unit. Emissions from these facilities consist of volatile organic compounds (VOCs) and toxic compounds. In petroleum refineries, emissions are regulated to control VOC emissions through EPA, state, and regional authorities. Chemical plant VOC wastewater emissions are also controlled through various regulations which will be expanded through new regulations promulgated by the EPA in accordance with the Clean Air Act Amendments of 1990. These new MACT or hazardous organic NESHAP (HON) regulations are concerned with hazardous chemicals and are described in Chapter 4. The list of chemicals in the MACT or HON regulation will be expanded through the years and will eventually regulate most of the organic chemicals. Where chemical process units are not currently affected by the HON regulations, with the passage of time these units will be modified to meet the expanded regulatory requirements. Prior to the imposition of the HON regulations, chemical units must comply with VOC and HAP regulations imposed by state and regional regulations.

Industrial hygienists should have an understanding of drain, sewer, and wastewater systems in process plants to insure that emission controls for various regulations are adequate to control workplace exposures. In addition, potential sources of exposure may occur when the control systems malfunction and a knowledge of these potential exposure hazards aids industrial hygiene control. This chapter reviews various sections of the drain, sewer, wastewater systems, and engineering controls, thereby providing a basis for an industrial hygiene review of these facilities.

Comments on the various facilities and controls in this chapter are generally concerned with chemical process units, but petroleum refinery equipment is also included. Since both types of process units have similar facilities, with some variation the comments are applicable to both process systems. The industrial hygienist with an understanding of the principles involved can develop the controls necessary for unusual process installations.

12.1 BASE SEWER CONSIDERATIONS

Before potential drain and sewer emissions can be defined, the associated terminology and general sewer design philosophy must be understood. A general sewer system design for petroleum refineries and chemical plants is presented that provides a basis for understanding the changes and evolution in the wastewater system. Some chemical plants have sewer systems that differ from this general basis but can be understood in relation to the base system.

12.1.1 General Description

A simplified sketch of an essentially non-emission controlled general process sewer layout is shown in Figure 12.1. Drains discharge through drain headers into catch basins where the combined flow from the various drains discharges into a sublateral. Generally, two or more sublaterals feed a large sewer line through the manhole connections, and the large sewer line in turn is connected with a main trunk or main sewer line. The flow in the main sewer line discharges into API separators or clarifiers where oil and immiscible fluids are separated from the wastewater. Clarified wastewater from the separator then flows into an air flotation unit where the remaining traces of free oil or chemical are removed. Effluent from the air flotation unit is discharged into secondary and possibly tertiary wastewater treating equipment prior to release into a receiving water body. Water effluent quality standards determine the type and effectiveness requirements of secondary and tertiary treating steps.

Early EPA designs of a general refinery drain system differed from the overall wastewater system shown in Figure 12.1.[1] The drains were directly connected to laterals that discharged into manholes or junction boxes located within the process unit confines. Junction box outlet streams were combined and the combined

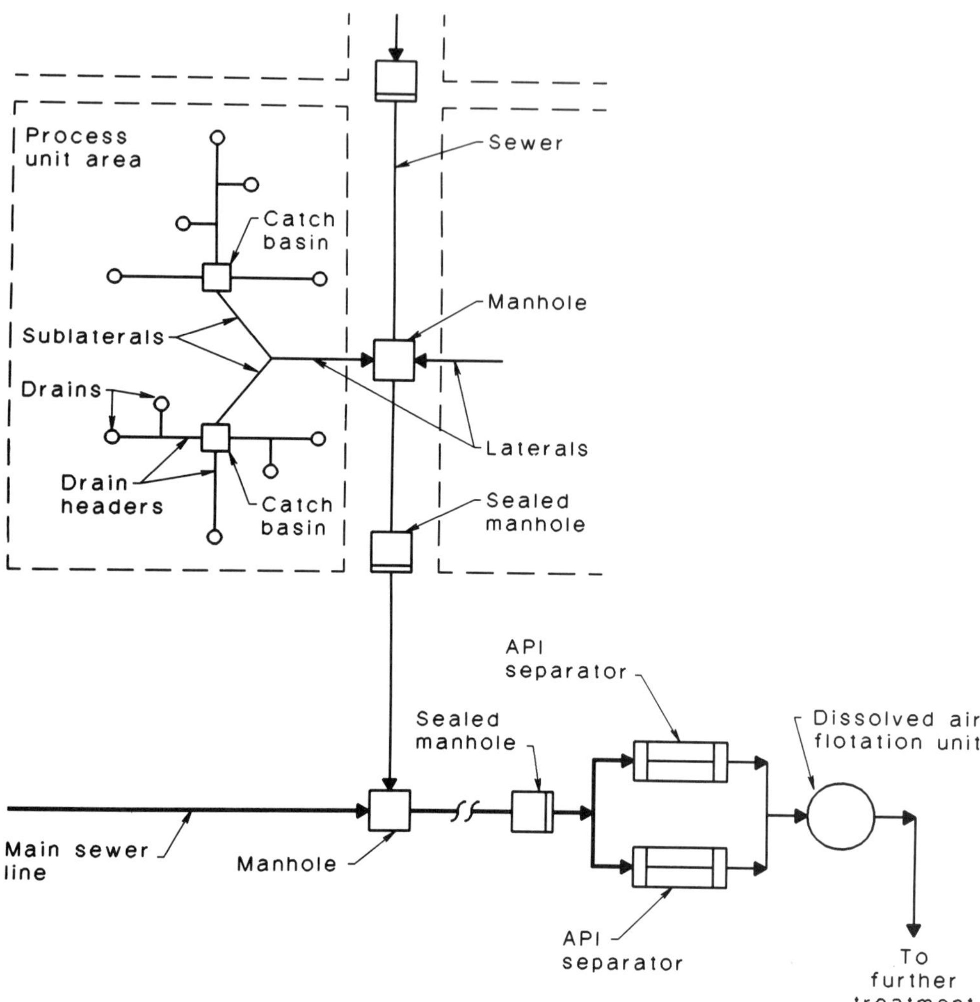

Figure 12.1 *Typical base sewer system layout.*

stream was sent to a major junction box located in the sewer line. The effluent from the latter empties directly into the main sewer without a junction box or manhole in the main sewer line. In addition to the difference in manhole locations, the EPA design did not contain catch basin installations to receive effluent from the drains. Catch basins were eliminated in the early EPA design, but are necessary to drain wide areas of rainwater, spills, or leaks and must be included in any system. The EPA design was modified to include catch basins and is recognized in the new MACT or HON regulations. Flow into catch basins from small drains had been used in many installations to minimize underground piping and segregate drainage systems.

The drains shown in Figure 12.1 receive drainage material from individual point sources or collect drainage from several points. Drainage from pump base plates, seals, or other equipment often empties into these collector drains or points which

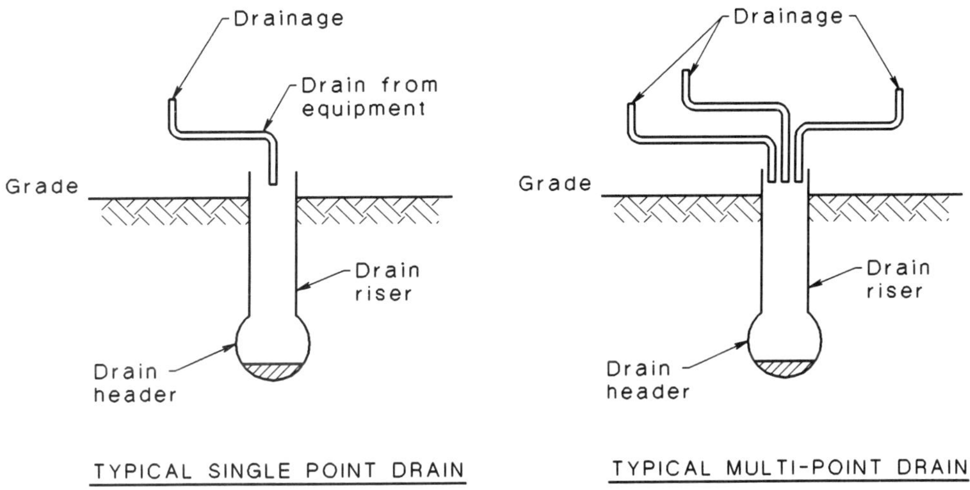

Figure 12.2 *Equipment drains, risers, and drain headers.*

are risers that extend upward from the sublaterals as shown in Figure 12.2. Constant discharge of contaminated process water was also directed into the typical drains shown in the figure. The drain headers are often sealed at the entrance to the catch basin to minimize vapor backflow from the catch basin or individual risers are sealed for the basic wastewater system. Drain seals are reviewed in another section of this chapter.

As shown in Figure 12.2, the drains are essentially open vents and volatile vapors are emitted to the atmosphere. In addition to open drain vents, manholes or junction boxes are vented directly to the atmosphere and these vents are also potential sources of volatile vapor emissions. The manholes are purposely vented to the atmosphere to relieve any pressure buildup. Vent requirements are reviewed in another section. Another major emitting source in this overall wastewater system is the API separator or clarifier (decanter) where vapor emissions are a function of hydrocarbon or chemical volatility. Obviously, all of these emission sources are a concern to the industrial hygienist and controls to reduce the emission of toxic compounds from these sources are required through various regulations.

12.1.2 Process Drains

Based on a late 1970s emission testing program, an emission factor of 0.032 kg/hr-drain was established by the EPA. However, the analysis did not indicate whether any of the tested drains were trapped or sealed and a follow-up investigation was also unsatisfactory in determining whether the drains were sealed. The EPA assumed the drains were not sealed in estimating an emission factor. However, this assumption is not correct for all refineries or chemical units where individual process drains are not necessarily sealed, but the drain header can be sealed in the catch basin. Moreover, individual process drains can be sealed by a U-bend below the drain header. These seals will reduce, but not necessarily eliminate, vapor emissions. Since emissions control is important, the EPA has developed improved methods of calculating emissions, but an understanding of the system is necessary to determine the efficacy of drain controls. Emission estimating and controls are presented in Sections 12.3 and 12.4.

An overall vented emission rate from the process drains is calculated by multiplying the individual drain emission factor by the total number of drains. Although the uncontrolled emission rate per drain is questionable, the volatility of the material in the drain with or without a seal has an effect upon the emissions rate and potential exposures of personnel working in close proximity to the drain. The industrial hygienist should monitor the concentration of toxic vapors in the drain opening and several feet from the opening. The monitoring data provide a basis for assessing potential overexposures and the effectiveness of drain modifications.

12.1.3 Catch Basins

As shown in Figure 12.1, catch basins are located in the process area and are used as area drainage points. In addition to receiving drainage from individual process drains and drain headers, catch basins have other functions. The catch basin is designed to receive spills, flush water, fire water, and rain water, with the general area in the vicinity of the catch basin sloped toward the basin for rapid drainage. Typical catch basins are shown in Figure 12.3, although designs vary considerably throughout the petroleum and chemical industries.

Since the streams entering a catch basin are a combination of process liquid and water, the immiscible liquids with specific gravities less than the specific gravity of water form a layer on the water surface. This layer results in emissions that disseminate into the atmosphere through the grid or cover, which has a relatively large open area. The emissions from the catch basins are rarely measured, but concentration levels considerably greater than 1 ppm have been obtained on monitors placed in or above the grid cover.

Emission rates and concentration levels above the catch basin are a function of the wastewater temperature, volatility, leakage or drainage rates from the drain headers, and residence time. The latter is a consideration when drainage is infrequent and the light volatile components have evaporated, resulting in an extremely low emission concentration. Alternatively, a high emission concentration may be observed when a constant drainage stream flows through the catch basin with a considerable quantity of volatile materials. The waste stream temperature is also important in determining the volatility or evaporation rates.

For a catch basin emission factor, the industrial hygienist should consider the new EPA estimating procedure discussed in Section 12.3. An older emission factor that is not very accurate is described in reference 2. For chemicals regulated under the new HON (MACT) regulations, the catch basin functions will be limited to rain and fire water, thereby minimizing potential emissions. However, existing plants not covered by the new regulations will have operating catch basins for a period of time that are potential sources of exposure. In this situation the industrial hygienist should investigate drain flow rate, chemical concentration levels, and volatility to more accurately estimate a catch basin emission factor.

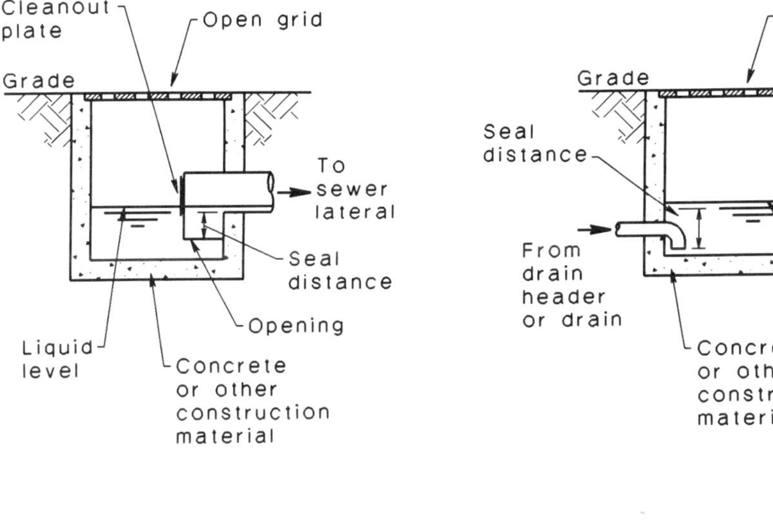

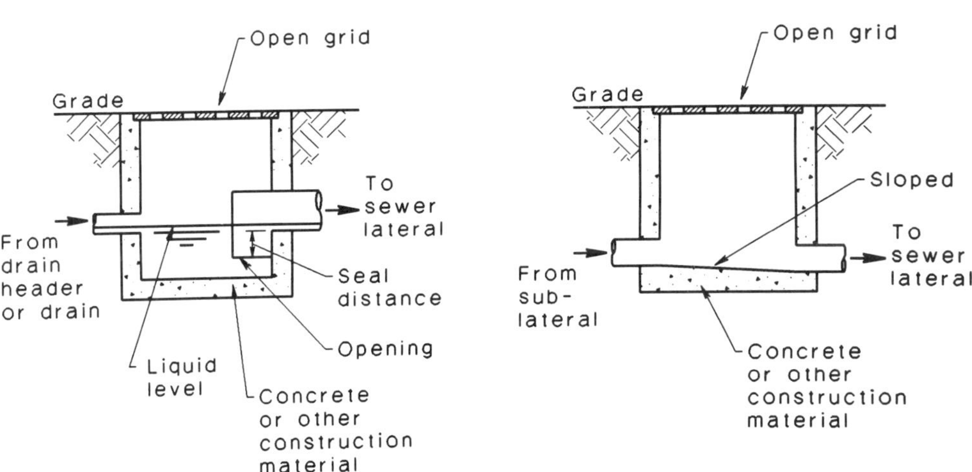

Figure 12.3 *Typical catch basins.*

12.1.4 Manholes and Junction Boxes

In current EPA terminology, manholes and junction boxes are equivalent and, as shown in Figure 12.4, these boxes are generally closed and sealed with vents in many installations. The vents discharge vapor to the atmosphere well above grade. In Figure 12.4, the vented manholes have an internal water seal to break the flow of vapor through the underground piping system. Water seals are generally installed in these systems to prevent vapor flashback through the system in the

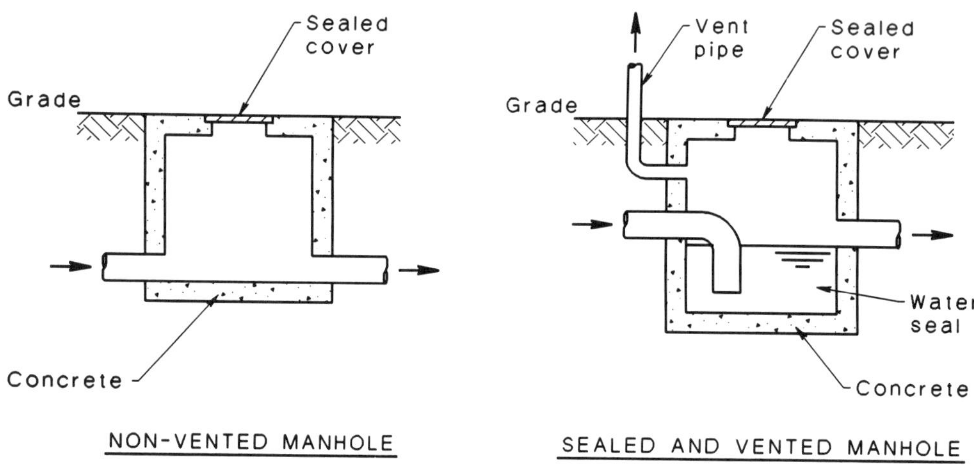

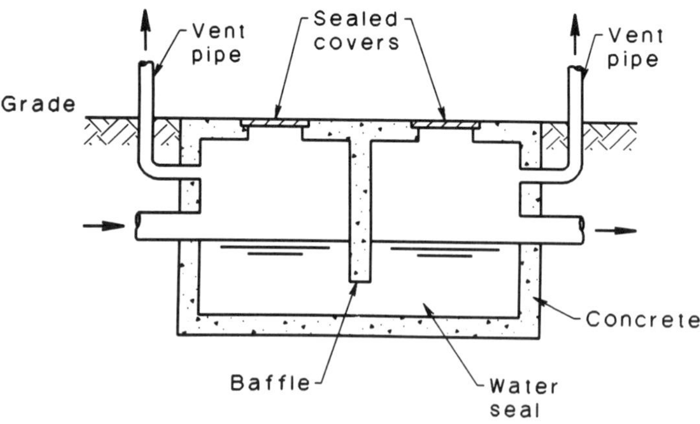

Figure 12.4 *Typical manholes and junction boxes.*

event flammable vapor is exposed to an ignition source. Since air flow through the sewer system is prevented by the water seals, oxidation effects are also minimized in the piping system. In addition, eliminating a flow of air also limits volatility or vaporization of the stream compound contaminants whether miscible or immiscible materials. Actual volatility in this type of system is generally a function of the waste stream temperature since the pressure of the system is essentially atmospheric pressure or a few inches of water.

The original EPA analysis of this system resulted in an emission factor of 0.032 kg/hr per vent from a manhole or junction box, which was equivalent to the drain emission factor. Additional work by the EPA has changed the basis for estimating manhole emissions, which is described in Section 12.3. However, the elevation and location of the vent pipe exhaust point will affect the area concentration level. The industrial hygienist should monitor the area downwind of the vent pipe to determine the area concentration levels of toxic compounds emitted from even a small vent. These data provide a basis for comparing the concentration levels prior to changes in the sewer system or vent pipe.

12.1.5 Oil and Chemical – Water Separators

In petroleum refineries and chemical plants, API separators, plate separators, or clarifiers (decanters) are the primary treating step for the removal of oil or nonmiscible compounds from wastewater. Variations of these devices are also available that skim or remove immiscible materials from wastewater. A simplified drawing of a parallel plate separator (PPS) is shown in Figure 12.5 and a simplified drawing of an API separator is shown in Figure 12.6.

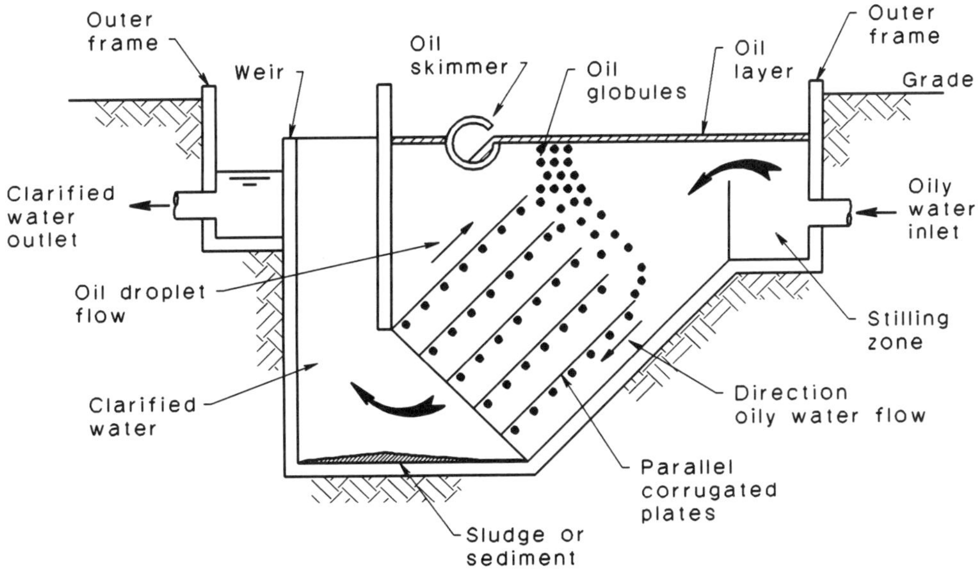

Figure 12.5 Parallel plate separator.

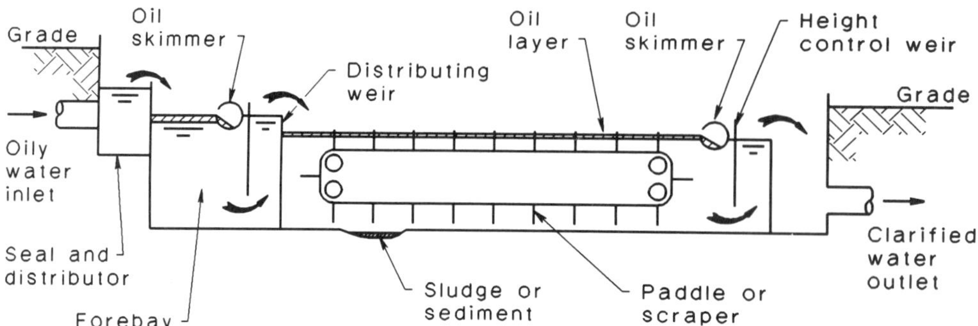

Figure 12.6 API separator.

Parallel Plate Separator. The parallel plate separator operation is based upon coalescing oil or chemical droplets to form large globules that readily separate from the water. In Figure 12.5 the oily water is received in a stilling zone where the water velocity is significantly decreased. Some oil–water separation occurs in this zone and the inlet flow is distributed across the width of the separator. The oily water mass then flows through a number of slanted parallel plates. The distance between the plates is 1–2 in. (25–50 mm), which tends to separate the oil droplets from the water in a relatively short distance since the overall height of the water layer is quite small in comparison to the length of the water channel. The droplets rise to the angled plate surfaces, which are placed at an angle of about 45° to assist the movement of the low density droplets toward the upper end of the plates. The droplets coalesce with other droplets as they move along the plates, forming globules that combine with globules rising from other plates, as shown in Figure 12.5. The large oil globules rise to the surface where the oil is skimmed off for recovery and recycle. The clarified water flows over the weir to the outlet for further processing in the wastewater treatment system. Sludge or sediment gradually builds at the outlet end of the plates and is removed on an infrequent basis.

The surface area above the plates has an affect upon the volatile compound emission rate along with oil or chemical characteristics, influent temperature, wind, and oil layer thickness. Surface area in a PPS, although important, is quite small and has much less influence upon vaporization than the large surface area in an API separator. Moreover, the parallel plate separators are frequently covered, eliminating wind effects and general vapor emissions. The small overall area of the parallel plate separators is an advantage in those installations where a small high-efficiency, high-capacity separator device is required. Basically, the new HON or MACT regulations require a separation device in the process unit area like that described in Section 12.2.

A variety of parallel plate separators are available commercially including corrugated plate separators which are quite similar but have corrugated rather than flat plates. In addition, some of the designs have holes in the plates that assist in separation and agglomeration. Removal efficiency is further enhanced by the placement of a coalescing filter on the clean water side of the separator. Discrete small nonmiscible droplets are combined in the coalescer and the larger droplets are separated from the stream, further reducing the nondissolved chemical or oil concentration in the effluent.

API Separator. In the API separator shown in Figure 12.6, a quiescent zone is provided with a long residence time that permits free oil droplets to rise to the surface where the droplets coalesce into an oil layer. The wastewater velocity in the quiescent zone is very low to prevent turbulence and mixing, which hinders droplet rise and oil–water separation. The oily water stream from the main sewer line enters the API separator in a section termed the forebay where free oil on the surface is skimmed and removed. The forebay also distributes oily water flow evenly to the quiescent zone of the separator.

The oily water enters the API separator in Figure 12.6 through a seal and distributor which evenly spreads the stream flow into the forebay. In the forebay, an oil skimmer removes free oil that readily separates and rapidly forms a thick layer of oil. The main oily water stream is discharged from the forebay through an underflow opening into the distributing weir section which maintains the liquid level in the forebay. Oily water overflows the distribution weir into the main section of the API separator which is the quiescent zone. Oil that rises to the surface is moved toward the skimmer pipe in the main separation zone by the paddles shown in Figure 12.6. The paddles are on an endless chain and also act as scrapers at the bottom of the separator to move sludge or sediment into a trough. Sludge is periodically pumped from the bottom trough into a receiver tank or removed with a mechanical device. Oil skimmed from the quiescent zone is pumped to an intermediate tank for additional separation prior to recycling or reprocessing.

Losses from an open, quiescent zone were calculated by the EPA and are described in reference 2. The emissions from an open separator for a petroleum refinery location estimated by the EPA are 0.7 kg/10^3 L wastewater.[3] Where the emissions are controlled, this has been reduced to 0.024 kg/10^3 L wastewater. However, revised calculation procedures are described in Section 12.3 for HON (MACT) regulations that are applicable to chemicals not covered by the HON regulations and oil–water systems. Enclosed separator emissions are included in these revised procedures.

Clarifiers or Decanters. These systems are large-diameter tank-like devices that provide a very large quiescent zone for separation with rotating surface skimmers for removal of nonmiscible liquids on the water surface. Other types of decanters may use a weir overflow for separation. Emissions from these systems can be calculated with the EPA estimating procedures described in Section 12.3. Controls require enclosing and sealing the liquid surface of the device.

12.1.6 Air Flotation Systems

Air flotation units are frequently used in the petroleum refining and chemical industries to remove small quantities of free oil or chemical emulsified material and suspended particles or solids in the waste stream. The air flotation treatment unit is normally installed at the outlet of an API or parallel plate separator to remove traces of free oil or chemicals remaining in the separator effluent. Following removal of the remaining oil in the waste stream, the clarified wastewater is sent to secondary treatment, which is generally a biological treating step.

Air flotation treatment consists of injecting gas bubbles into the water, where the bubbles cling to the free oil or chemical and form large droplets. The gas bubbles also create a buoyancy effect, resulting in a rapid rise of the oil or chemical-gas droplets to the water surface where the oil is skimmed and recovered. The bubbles are formed from air or gas either injected into the wastewater or formed by a mechanical device. In a dissolved air flotation unit air or gas is injected directly into the wastewater. Gas bubbles are formed internally within the wastewater unit in an induced air flotation system.

A simplified drawing of a dissolved air flotation (DAF) system is shown in Figure 12.7. In the DAF system, a clarified water side stream is saturated with air or gas under a relatively high pressure. The air-saturated water is then injected into a low-pressure wastewater feed and the pressure reduction results in the formation of small bubbles necessary for the agglomeration of free oil and suspended solids. The droplets are further mixed in the cone and gradually expand into larger droplets and the large droplets gradually rise to the surface where a froth or float is formed on the liquid surface. A slowly rotating skimmer arm moves the float to a trough or collector station, and the float can be pumped from the collector outlet. Clarified water is withdrawn from the lower portion of the tank and overflows into the outlet well. This clarified water is normally sent to a secondary treating facility. As shown in Figure 12.7, sludge accumulated in the

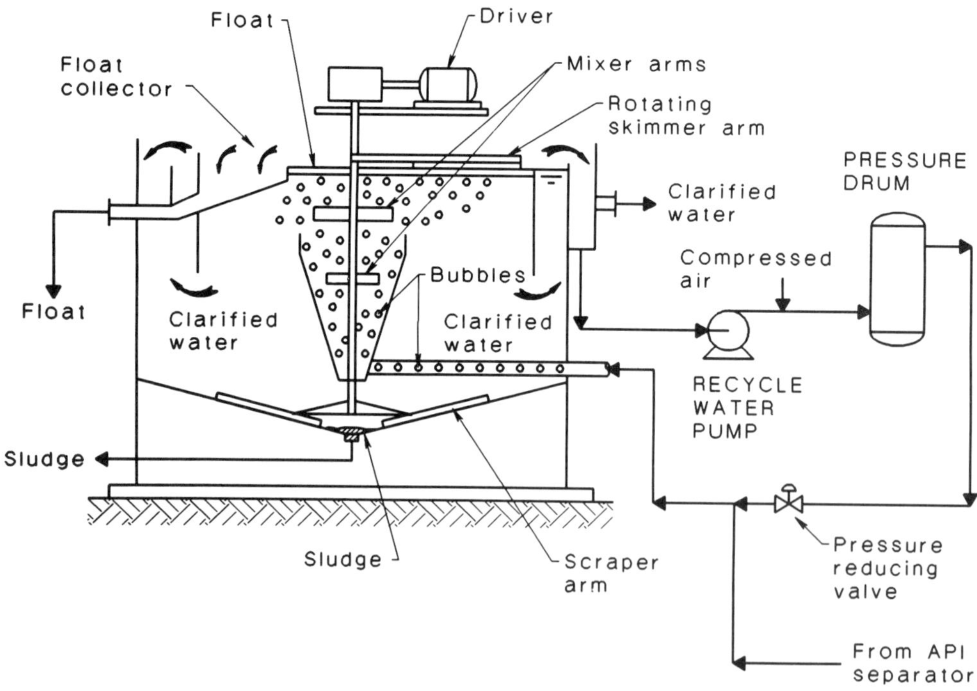

Figure 12.7 *Dissolved air flotation unit (DAF).*

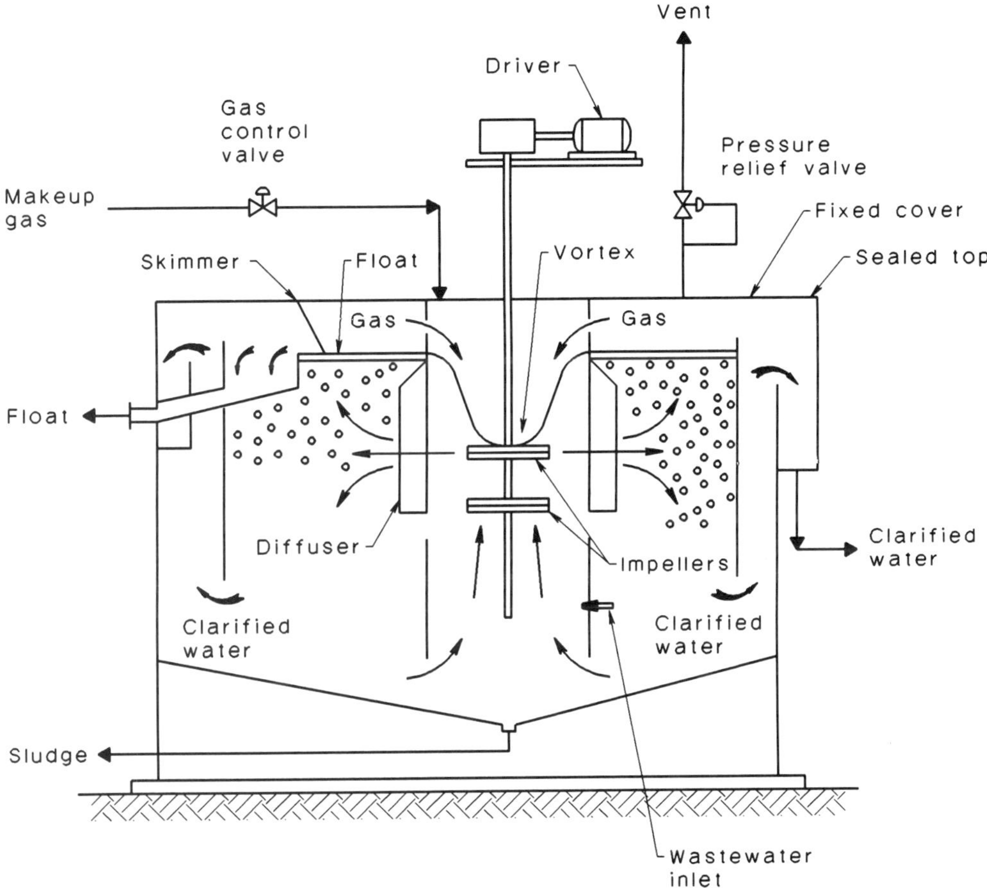

Figure 12.8 *Induced air flotation unit (IAF).*

tank bottom is withdrawn and pumped to a sludge processing unit. Since oil or an immiscible chemical material enters the DAF unit with the wastewater, some volatile emissions may be emitted from the DAF surface. This type of unit and emission control are included in the HON (MACT) regulations.

In contrast to the air injection procedure in the DAF unit, bubbles can be produced through various mechanical methods at low pressure in an induced air flotation (IAF) unit. Mechanical agitation, gas–water nozzle mixers, and gas diffusion through porous media are a few of the methods used in generating bubbles in an IAF unit. A mechanical propeller-type IAF unit is shown in Figure 12.8 which is sufficiently similar in design to many IAF units that it allows us to understand the general operating principles. The lower impeller draws fluid through the draft tube into the upper impeller which mixes the gas and liquid. A vortex is generated by the upper impeller that induces a recirculating gas flow, resulting in a gas–liquid mix. This gas–liquid mixture is discharged through the impeller rim into the diffuser where the gas is further reduced in size to small bubbles. The bubbles further coagulate, increasing in buoyancy and collect the oil and solids, forming a float on the surface. The float circulates past a skimming device that diverts it into the collector trough where it is accumulated and pumped to another location for further processing. Gas that evolves from the float is recycled into the mixer since the IAF unit has a fixed cover containing the gas. Pressure within the vapor space is maintained by a makeup gas that is either manually or remotely operated. The makeup vapor can be nitrogen, a hydrocarbon, or air, but the latter may create an explosive mixture and has been replaced in a number of installations by nitrogen or a hydrocarbon. The unit can be overpressured and the excess pressure is generally relieved to the atmosphere, where permitted, or to a control device by a pressure relief valve. Clarified water is withdrawn from the bottom of the tank and sludge is infrequently withdrawn should a sludge buildup occur. The closed cover on these units essentially eliminates emissions, minimizing potential workplace exposures.

Older emission factors for these units produced relatively high emission factors which were based upon high influent oil or chemical concentrations.[2] Consequently, these factors are not applicable where upstream separators are efficient and minimize the effluent concentration of undisssolved oils or chemicals. Newer estimating procedures have been developed and are presented in Section 12.3. Although the new regulations (HON) require controls which reduce potential exposures, existing units that are not covered by these regulations may require an emissions determination. Prior to estimating evaporation losses from DAF or IAF units, the industrial hygienist should determine the actual oil or immiscible compound concentration in the upstream separator effluent along with the concentration of volatile compounds.

Although the top cover is closed and sealed in the IAF unit, leakage found in earlier tests is apparently from gaskets and seals around doors and openings. Industrial hygienists should monitor these units and the surrounding areas prior to regulatory control changes to establish baseline concentration levels. Enclosures can be required in DAF units with potential concerns concentrated around doors and openings.

12.1.7 Miscellaneous Wastewater Treatment Processes

Following the air flotation unit, various process steps may be incorporated in the wastewater treating system. The intermediate steps after air flotation may be coagulation–precipitation or filtration to remove emulsified oil or suspended solids followed by equalization to balance the effluent treating load. Secondary treatment steps after equalization generally consist of a biological process to remove dissolved organic material from the water. The biological process can be a trickling filter, activated sludge, aerated lagoon, oxidation pond, or rotating biological contractor. In these biological processes, the organic compounds are decomposed by microorganisms, although the actual decomposition mechanism in each process may differ. Anaerobic processes are used successfully with some chemicals.

In refineries and chemical plant wastewater treatment processes, the biological treatment step is the major treating operation. A tertiary treating step is rarely required, although a few processing complexes have an activated carbon treating unit. The activated carbon-treating unit is generally installed in wastewater systems containing refractory or nonbiodegradable compounds. Hydrocarbons in petroleum refineries and many chemicals are biodegradable and generally do not require a tertiary treating step.

Estimates of emissions from the various wastewater treating steps vary considerably. The EPA estimates have not been particularly accurate in this area[1] and improving biological operations have reduced potential emissions to very low levels. Consequently, the estimate of emissions from the intermediate and secondary treating steps may be considerably lower than EPA estimates. The industrial hygienist should measure the concentration levels in the area surrounding these final treatment steps to determine *actual* workplace concentration levels. Upstream engineering control changes will have some effect upon the hydrocarbon or chemical concentration levels, but the changes should not be very large.

While the previous discussion centered upon volatile hydrocarbon or chemical emissions from a biological treating unit, actual emissions may not consist of the hydrocarbons or chemicals in the wastewater stream sent to the biological unit. Decomposition reactions can result in the formation of compounds that differ from the dissolved material in the feedstream. Vapor emissions may consist of these newly formed compounds along with microorganisms from the biological system that are emitted to the atmosphere in conjunction with emitted vapors or fumes. Industrial hygienists should consider potential types of emissions prior to sampling the atmosphere and methods of analyzing the samples. The latter can be a difficult problem, but should be explored to understand analytical limitations. A lack of analytical data may result in a recommendation that requires the use of personal protective equipment and respirators to control potential workplace exposures.

12.2 REGULATIONS

The current regulations applicable to sewer and treating systems consist of EPA standards of performance requirements for petroleum refineries,[4] NESHAP regulations[5,6] for benzene and vinyl chloride, and both state and regional authority regulations. The latter cover both petroleum and chemical plant facilities. Federal regulations are the most complete compared with the other regulations in controlling vapor emissions. However, these regulations were only applicable to petroleum refineries and the few chemicals in the NESHAP regulations. New regulations were proposed December 31, 1992 in accordance with the new Clean Air Act Amendments of 1990 termed the MACT or HON (hazardous organic NESHAP) which completely control emissions from the overall system. As described in other chapters, the HON regulations which are applicable to the chemical industry cover approximately 189 chemicals in the initial regulations, but the number of chemicals will be expanded through the years.

The HON or MACT regulations are undergoing public and industry review with final regulatory promulgation in early 1994 after some revisions. In contrast to the other sections of the proposed HON regulations that will not be altered to any great extent, some potential changes in the proposed sewer and wastewater emission control regulations can be expected. The proposed HON regulations are in this section, and the effect of potential changes are described. Benzene and vinyl chloride NESHAP regulations are not reviewed in this section since the new HON or MACT regulations include these chemicals and supersede the NESHAP regulations. Overall, many of the benzene wastewater emission control requirements are included in the new HON regulations.

While the HON regulations cover chemical facilities, new regulations for petroleum refinery wastewater system emissions controls will be proposed in early 1994 as part of the new Clean Air Act Amendments (1990). The refinery regulations will probably be similar to the final chemical industry regulations. In some regional authority regulations, wastewater emission control requirements are not only applicable to both petroleum refineries and chemical plants but also control emissions from various facilities in the wastewater system.[9]

12.2.1 Petroleum Refinery Regulations

The wastewater system described in Section 12.1 is a basic configuration that has been installed in numerous refineries and chemical plants. Although variations in the general system can be found in different plants, the overall wastewater configuration provides a basis for describing the regulations and emission control requirements found in reference 4.

Applicability. These regulations apply to all drains, sewer lines, intermediate basins and manholes, oil–water separators, and any secondary oil–water separators that were constructed or modified after May 4, 1987. If some modifications occur in the drainage system from the process drains to the first common junction box or manhole, control changes in accordance with the regulations are limited to this section and the remainder of the wastewater system is not affected.

1. Compliance with the regulations is required except during startup and shutdown or during a malfunction.
2. Alternative means of compliance are permitted following approval by the EPA.
3. Stormwater sewer systems that are segregated from the process wastewater system are excluded from the regulatory requirements.
4. Ancillary equipment that is physically separate from the wastewater system and does not contact or store oily wastewater is excluded from regulatory requirements.
5. Cooling water systems that do not contact process fluids are excluded from regulatory requirements.

Drains. Various drain configurations are shown in Figure 12.9. A typical unsealed water–oil drain shown in Figures 12.2 and 12.9 is the basis for EPA emission control criteria which is consistent for all requirements. The EPA assumes unsealed vapors carried through the drain header or sublateral from the manhole or catch basin are the source of emissions to the atmosphere. The drain emission control solution included in the regulations is a seal in the drain riser that prevents emissions to the atmosphere from the underground header or lateral. A P-leg seal or drain pot prevents vapor backflow and in early reports apparently reduced emissions to the atmosphere by 50%. The riser seal, also shown in Figure 12.9,

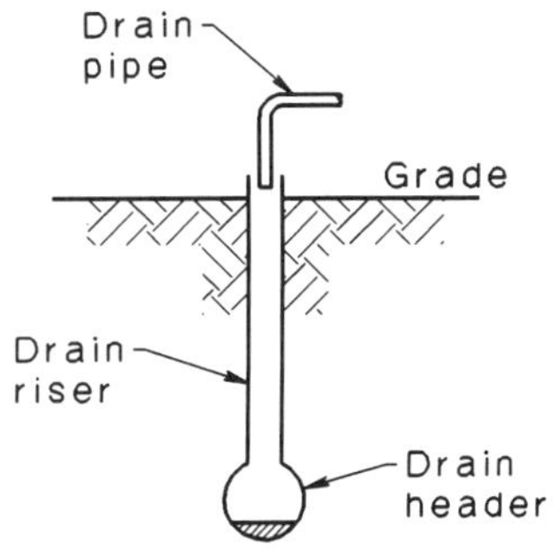

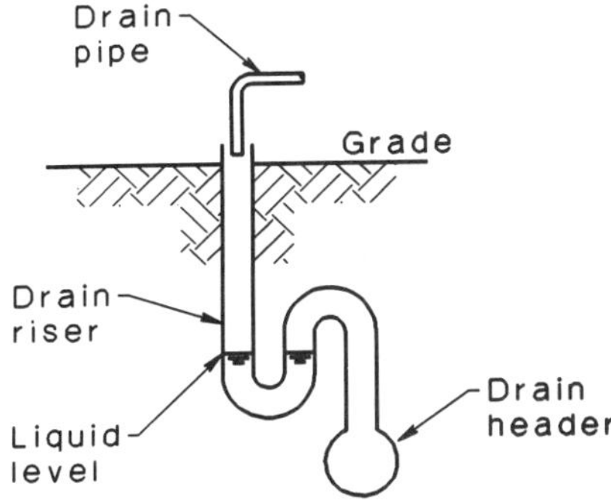

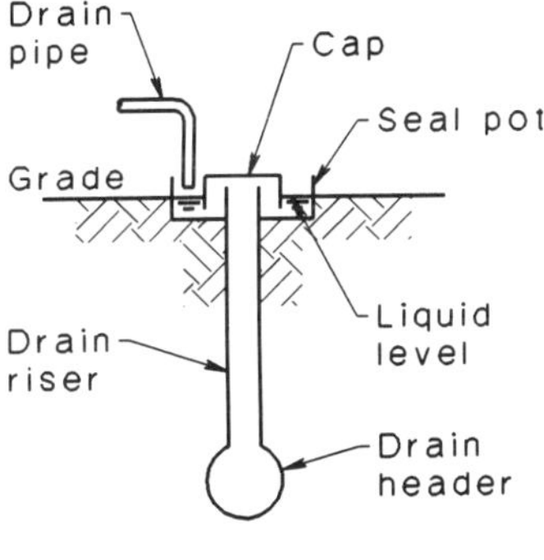

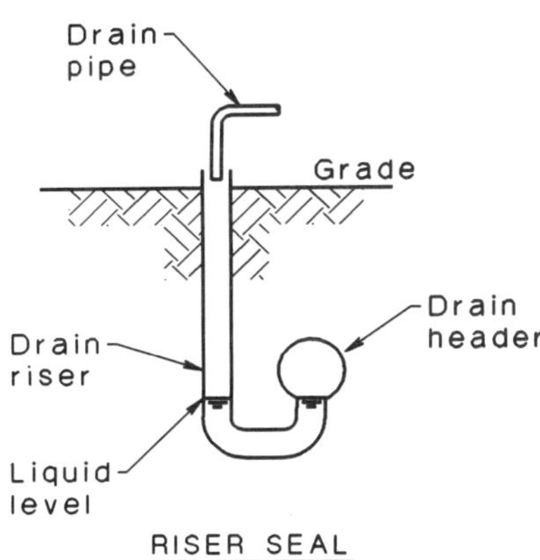

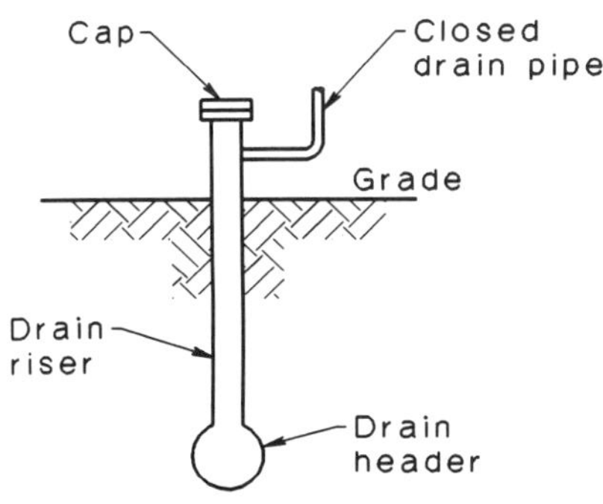

Figure 12.9 *Sealed drains.*

which seals the drain but has not been included in EPA considerations, adequately seals the riser and is acceptable. A closed-drain design also recommended by the EPA is normally installed with a closed-sewer system and is discussed separately.

In the seals shown in Figure 12.9, the P-leg, seal pot, and riser seals will have a layer of oil or immiscible fluid with a specific gravity less than 1.0 floating on water. Some emissions or vapors from this layer will be emitted to the atmosphere. Existing installations with catch basins as shown in Figure 12.1 are probably in general compliance through a seal at the drain header inlet to the catch basin or individual drain seals. Drain seals may only be required in specific installations since seals in the riser or drain header are normally installed.

For the drain seals, the following operating procedures must be observed:

1. Each seal must be inspected monthly for indications of a low water level or other conditions that may affect seal performance.

2. For drains out of service, the drains should be inspected weekly for a low water level.

3. Where drains are out of service and a tightly sealed cap or plug is placed in the riser opening, the plug is inspected initially and semiannually.

4. In the event of a low water or missing water level, water is added as a first repair effort. If water cannot be added, repairs should commence as soon as practicable, but not later than 24 hr after detection.

An alternative drain system standard has been defined by the EPA and can be constructed following official notification of the EPA. This standard permits operation of a completely closed drain system. However, this system must have a closed-vent system with a control device for the vented vapors. Where this completely closed drain system is installed, the previous requirements outlined in this section are no longer applicable.

Catch Basins. Where a refinery has catch basins (Fig. 12.3) that were installed prior to May 4, 1987, the basins are exempt from any regulation. However, where new process drains and a new first common junction box were installed after this date, the wastewater must bypass the catch basin. Since the catch basin has an open grid cover, emissions from oil layers in the basins are eliminated.

Manholes or Junction Boxes. The junction boxes in Figure 12.4 are typical of those installed in the petroleum industry, and regulations are based upon these basic designs:

1. All junction boxes must have a cover with a vent pipe. The vent piping must have a minimum height of 3 ft (0.91 m) and a maximum diameter of 4 in. (102 mm). BAAQMD has a maximum open area for vent piping of 12.6 in.^{2} (81.3 cm^2).[9]

2. Junction box covers must have a tight seal and are only removed for inspection and are resealed.

3. The boxes are inspected initially and semiannually to insure the cover is in the correct position and is sealed.

4. When a seal break or gap is identified, the initial repair effort should be completed within 15 calendar days after the break is identified. Repair delays are permitted if the repair cannot be conducted without a process unit shutdown. The delayed repair must be completed during the next scheduled process unit or refinery shutdown.

The industrial hygienist should be aware of manhole vents as a potential source of toxic vapor emissions. Recommendations are presented in Section 12.4 on vent pipe installations.

Sewer Lines. These lines must not be open to the atmosphere and should have no gaps or cracks that permit emission leakage. Where the sewer line is not buried, it must be inspected semiannually for gaps, cracks, or other problems. When a crack or gap is identified, the initial repair effort must be completed within 15 calendar days after the defect is identified, but the repair may be deferred if the repair requires a process unit shutdown. Deferred repairs must be completed during the next scheduled process unit or refinery shutdown.

Oil–Water Separator. The following regulations require the installation of a fixed roof or cover on the separator:

1. Each oil–water separator shall have a fixed roof. The fixed roof requirement is also applicable to slop oil tanks, storage vessels, or other auxiliary equipment that are part of the separator. The fixed roof must completely cover the entire oil–water separator.
2. The vapor space under the fixed roof is not purged unless the vapor is vented to a control device.
3. All access doors or openings in the roof are gasketed and closed during all periods when the separator is in operation. These doors can be opened during inspection and maintenance.
4. The doors and roof seals are inspected semiannually to ensure that the seal between the roof and separator along with the door seal remain unbroken. Repairs must be completed within 15 days after a seal gap is identified. A delay is permitted if the repair requires a process unit or refinery shutdown. However, repairs must be completed during the next scheduled shutdown.
5. When the separator has a design capacity greater than 250 gpm (16 L/sec), the separator fixed roof will be equipped with a closed-vent system and a control device. This requirement is in addition to those described in items 1–4. Closed-vent systems must be monitored semiannually to ensure that leakage is less than 500 ppm above the background level.
6. Where a separator that has a maximum design capacity of 600 gpm (38 L/sec) was partially or completely covered prior to May 4, 1987, the separator does not require a closed vent and control device. The separator system must meet the requirements of items 1–5, but can install a floating roof in the portion of the separator that is not covered by a fixed roof.
7. Any storage vessels or tanks that have slop oil and are covered by tank emission regulations reviewed in Chapter 13 are not subject to these regulations.

However, other slop-handling equipment must be equipped with a fixed roof in accordance with items 1–5.

8. Where the capacity of the oil–water separator is less than 250 gpm (16 L/sec), the fixed roof can be equipped with a pressure control valve for operational requirements. The valve is set at maximum pressure to ensure the valve *does not* vent continuously.

While other regulations (state and regional authority) are similar, different criteria are presented for oil–water separator flow rates and must be investigated to ensure compliance.[9] In addition to fixed roofs covering separators, floating roofs are permitted by the EPA and regional authorities. The EPA regulations require that floating roofs have seals between the separator wall and the edge of the floating roof. Specifically, primary and secondary seals are required that are similar to those specified for floating roof tanks. Specifications for the floating roof follow:

1. A *liquid mounted* primary seal is required that can be foam or liquid filled. The seal is shown in Figure 12.10 where gap width between the seal and the

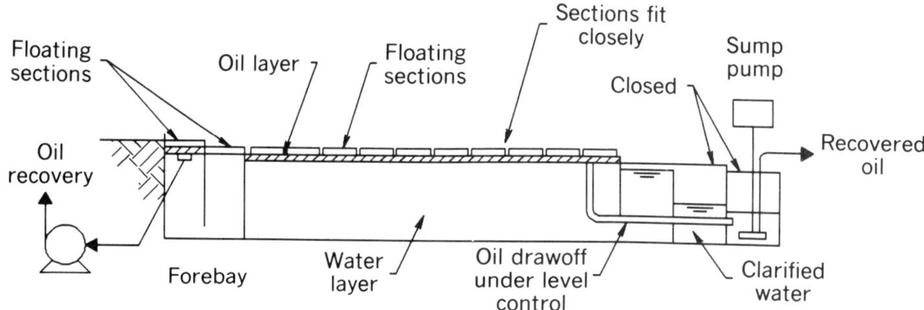

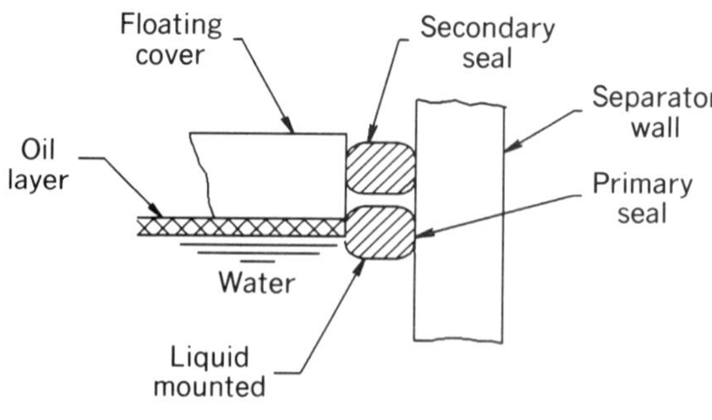

Figure 12.10 *API separator with floating roof.*

separator wall cannot exceed 1.5 in. (38 mm) at any point. Total gap area between the separator wall and the primary seal must not exceed 3.2 in.^{2} per foot of running length along the separator wall perimeter or 67 cm^2 per meter of length.

2. The secondary seal is located above the primary seal and is placed between the separator wall and the floating roof. Gap width between the separator wall and the secondary seal cannot exceed 0.5 in. (13 mm) at any point. Total gap area between the separator wall and the secondary seal must not exceed 0.32 in.^{2} per foot of running length along the separator wall perimeter of 6.7 cm^2 per meter of length.

3. The primary seal gap is measured within 60 days after initiation of wastewater flow and once every 5 years. The secondary seal gap is also measured within 60 days after initiation of wastewater flow and once every year thereafter.

4. Seal repairs must be completed within 30 days after identification of seals exceeding gap limitations.

5. Each opening in the floating roof must have a closed gasketed cover.

6. The floating roof can be equipped with one or more emergency drains for stormwater. This drain is fitted with a slotted membrane fabric that covers at least 90% of the drain open area.

7. In portions of the separator where a floating roof cannot be installed, a fixed roof must be installed as shown in Figure 12.10.

8. The floating roof should be floating at all times and not supported by the roof supports except during periods of abnormal conditions.

In addition to these conditions, record keeping of leaks, malfunctions, and changes is required along with repair records, delays in repair, and corrective actions. Detailed schematics must be maintained that are current and include all changes. Where emissions are monitored and control device efficiencies are regulated, the data must be recorded. Periodic reports must be filed with the EPA.[4]

The EPA regulation does not specifically mention air flotation units. However, regulations for this treatment system have been issued by the BAAQMD[9] and are an example of additional requirements. After August 4, 1991, the BAAQMD states that air flotation units with design capacities larger than 400 gpm (25.2 L/sec) cannot be operated unless they are equipped with one of the following:

1. A solid, fixed, gasketed cover that completely encloses the flotation tank and any preflocculation tank or basin. The cover may have an atmospheric vent or pressure/vacuum valve, but all openings must be closed except for inspection, maintenance, and wastewater sampling. All roof and opening seals must be inspected with openings limited to a maximum of 0.125 in. (3.2 mm).

2. An organic chemical vapor recovery or control system with a combined collection, minimum destruction efficiency of 70% by weight.

3. Any system with an efficiency equivalent to or greater than that given above which is approved by the authority.

12.2.2 Chemical Industry Regulations

At the present time, the EPA has proposed regulations covering emissions from wastewater systems, tankage, transfer operation, process vents, and equipment. These proposed regulations, with the exception of process vents and wastewater have been described in various chapters. The proposed regulations for equipment and the other operations, with the exception of wastewater emission control, will probably change very little when the final regulations are issued. However, wastewater emission control regulations will probably change since industry practice and various practical considerations are under review by the EPA. The proposed regulations are presented in this chapter along with anticipated changes that will probably be included in the final regulations.

The chemical industry wastewater emission controls, even with modifications, are considerably more stringent than the petroleum refinery regulations presented in Section 12.2.1. However, refinery wastewater regulations in accordance with the Clean Air Act Amendments of 1990 will be revised and the regulations are scheduled for promulgation in 1994. The refinery wastewater emission control regulations will probably be similar to the final chemical industry regulations that will issue in 1994. While all regulations are currently based on the 189 toxic chemicals shown in Chapter 4, this list will be expanded over the years and cover most wastewater streams in chemical plants and petroleum refineries.

Since process vents are an important emissions source in the process area and in wastewater systems, regulatory requirements for process vents are also included in this chapter.

Background. The EPA estimates that existing process vents are the largest source of hazardous air pollutants (HAPs) in the synthetic organic chemical manufacturing industry (SOCMI).[10] The second largest source of HAP emissions according to the EPA are the existing SOCMI wastewater systems.[10] For both existing and new sources, the EPA has proposed various emission controls.

The wastewater streams and operations the proposed regulations will cover are any HAP containing wastewater discharged into drain systems. This covers the general categories of process wastewater, maintenance-turnaround wastewater, and routine maintenance wastewater. In addition, the recovered HAP or residual materials are also included in these proposed regulatory requirements. However, residual materials may also be controlled under various waste regulations, and actual disposal requirements will be held in abeyance until this contradiction is resolved. Some of the residual disposal regulations in the wastewater proposal are probably applicable under other regulations and provide a basis for consideration of disposal options.

Some specific wastewater or water categories the EPA considers regulated under this proposal are:

- Wastewater discharge streams from process equipment
- Product or feedtank water drawdown
- Cooling tower blowdown
- Steam trap condensate
- Heat exchanger descaling streams
- Pump drainage

- Distillation column trap cleanings
- Startup and shutdown water discharge
- Contaminated cooling water

Sources not included in the list for regulatory controls are stormwater from segregated stormwater sewers and spills and water from safety showers. In addition, wastewater stream sources that do not contact or emit any of the chemicals listed in Appendix 12B or are not associated with any of the chemicals manufactured in Appendix 12A are not covered by these regulations. Any chemical process that manufactures products listed in Appendix 12A or Appendix 9B (Chapter 9), but does not consume as a reactant or manufacture any of the hazardous organic air pollutants shown as a product or byproduct is exempt from the wastewater regulations. These tables are based on the initial proposals of the EPA and will probably not change in the final regulation. Other processes and areas that are exempt from the proposed wastewater regulations are petroleum refining processes, ethylene processes, equipment within a chemical plant that does not contain hazardous organic air pollutants, and research and development facilities.

The basic technology for controlling HAP concentrations in wastewater and associated vapor emissions is stream stripping, which is reviewed elsewhere in this section. The EPA regulations are based on the installation of a steam stripper located within a process unit that essentially removes all volatile HAP chemicals before the wastewater is discharged into the sewer system. In addition, very high removal efficiencies can be achieved for most of the other chemicals regulated under the HON rules. These efficiencies are generally in the 90% removal range. In addition, the steam stripper will also remove most of the VOC chemicals in the wastewater stream, which was an added consideration in the application of this technology. Residual material recovered in a steam stripper can be recycled or sent to a control device for disposition.

Emission controls are also required at those points in the wastewater stream downstream of the stripper where potential emissions to the atmosphere may occur. However, potential emissions from steam-stripped streams are generally very small and emission controls may not be required after stripping.

Definitions. An important consideration in these regulations are the definitions associated with terms and systems or equipment. The following list contains those definitions pertinent to the regulations that also describe the equipment included in the terminology:

Heat Exchange System. This refers to an entire cooling tower system that includes heat exchangers. Once through cooling water systems are also included.

Hazardous Organic Air Pollutant or Organic HAP. Any chemical in the list shown in Appendix 12B.

Recovery Device. Device or equipment for the recovery of chemicals. Typical devices are absorbers, carbon adsorbers, or condensers.

Volatile Organic Hazardous Air Pollutant. Normally termed VOHAP, the volatile portion of an individual HAP measured by proposed EPA Method 305.

Volatile Hazardous Air Pollutant. Any chemical in the list shown in Appendix 9A (Chapter 9).

Wastewater. Includes process wastewater, maintenance wastewater, and process contaminated wastewater. Water containing organic HAP at a minimum concentration of 5 ppm (wt) and a flow rate $\geq$ 0.02 L/min, or a concentration $\geq$ 10,000 ppmw at any flow rate. The process wastewater is generated as a result of contact with process fluid or is produced in the reaction. In addition to the various sources shown in Subsection 1, water condensed from jet evacuation systems that provide a vacuum or organic process systems also produces contaminated water.

Group 1 Wastewater Stream. A process wastewater stream discharged from a new or existing source with a total VOHAP concentration $\geq$ 10,000 ppmw of compounds listed in Appendix 12C or a process wastewater stream from a new or existing source with a flow rate $\geq$ 10 L/min (2.63 gpm) and a total VOHAP concentration $\geq$ 1000 ppmw. A process wastewater stream from a *new* source that has a flow rate $\geq$ 0.02 L/min and a concentration $\geq$ 10 ppmw of any chemical compounds listed in Table 12.1.

Group 2 Wastewater Stream. Any process wastewater stream that is not a Group 1 wastewater stream.

Individual Drain System. Equipment that conveys wastewater from a process unit or tank to a waste management unit. Includes process drains, junction boxes, manholes, and sumps from discharge point to treatment. Does not include a segregated stormwater system.

Point of Generation. Location where the wastewater is discharged from a vessel, tank, and so on and prior to mixing with another wastewater stream.

Reference Control Technology. Installation of:

- Fixed roof and closed vent systems on all wastewater tanks and chemical water separators. Covers and closed vent systems on all surface improvements, containers, individual drain systems, and treatment processes.
- A steam stripper.
- A control device that reduces the outlet emissions by 95% or an outlet concentration of 20 ppmv from combustion devices of organic HAP emissions from closed-vent systems.

Waste Management Unit. Covers any component, equipment, structure, or transport system used in conveying, storing, treating, or disposing of waste. Includes tanks, organic-water separators, air flotation units, biological treatment units, incinerators, strippers, decanters, evaporation units, and individual drain systems.

Waste Seal Controls. Seal pot, P-leg, or trap filled with water that creates a barrier between the sewer system and the atmosphere.

Wastewater Characterization. The definitions provide some background on characterizing wastewater, but a more careful approach is required to ensure compliance based on all of the definitions and terminology. As a result, flow diagrams to aid in characterization of the wastewater have been prepared by the EPA. These diagrams are combined with the following explanations to simplify the analysis. An important point in this effort is that the acronym HON refers to Hazardous Organic NESHAP or MACT regulatory requirements.

HON Wastewater Determination. In Figure 12.11 the initial step is the determination that the process is a SOCMI unit. Any process that manufactures as a product one or more of the chemicals listed in Appendix 12A is considered a SOCMI unit or in the event the wastewater streams emit any of the chemicals listed in

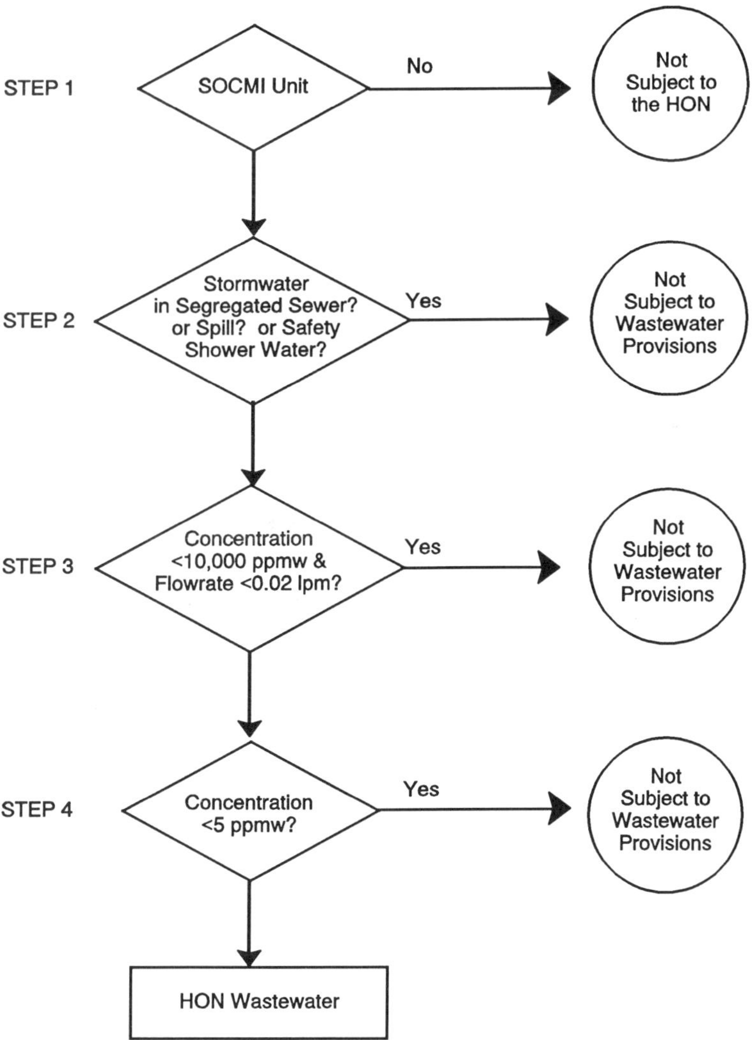

Figure 12.11 *Determination of HON wastewater classification.*[7,8]

Appendix 12B. Where a chemical process has two or more products and one of the products is listed in Appendix 12A or Appendix 9B (Chapter 9), the manufacturing process is a SOCMI unit.

In step 2, where the wastewater stream is not segregated stormwater, a spill, or safety shower, another evaluation is necessary. In step 3, where a concentration of VOHAP $< 10{,}000$ ppmw and a flow rate < 0.02 L/min, determine whether the stream is a HON wastewater. If one of the criteria are exceeded, another evaluation is conducted in step 4. A VOHAP concentration exceeding 5 ppmw results in classifying the wastewater as a HON stream.

General Wastewater Stream Classification. Following determination of a HON wastewater, a general view of the next procedure is shown in Figure 12.12. The

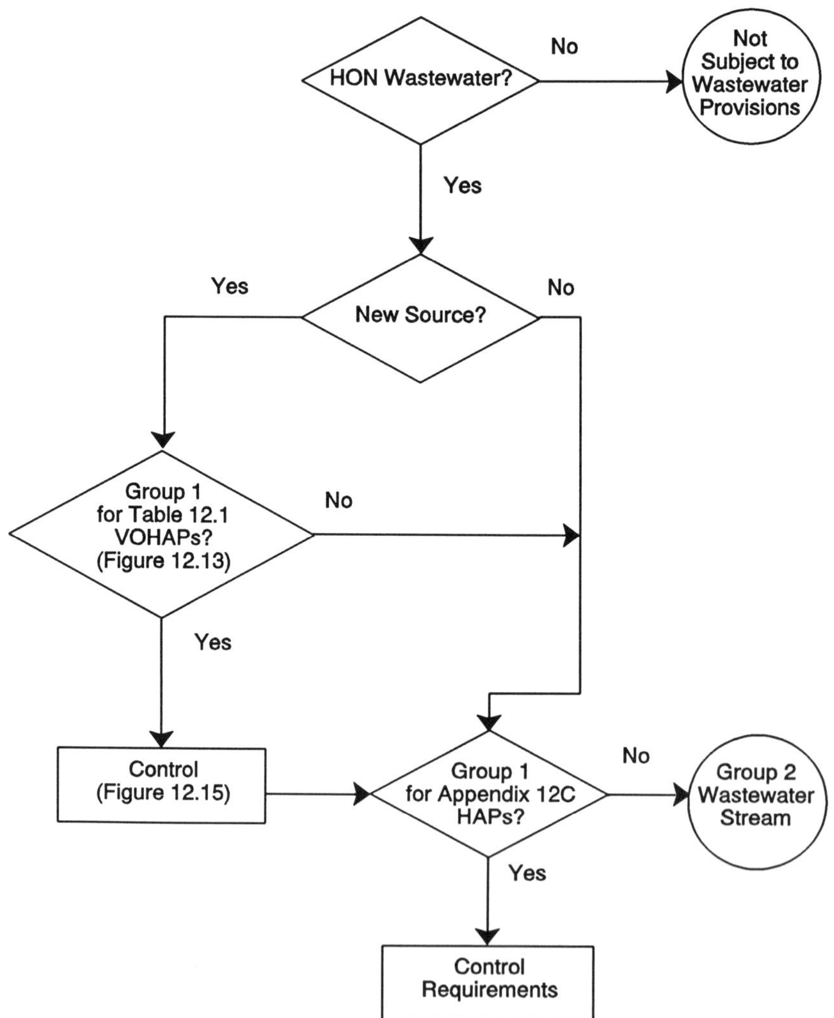

Figure 12.12 *General wastewater stream overall classification.*[7]

TABLE 12.1 Organic HAP Chemicals in New Source Waste Streams that Identify Group 1 Wastewater

Chemical Name	CAS Number
Allyl chloride	107051
Benzene	71432
1,3-Butadiene	106990
Carbon disulfide	75150
Carbon tetrachloride	56235
Cumene (isopropyl benzene)	98828
Ethylbenzene	100414
Ethyl chloride (chloroethane)	75003
Ethylidene dichloride (1,1-dichlorethane)	75343
Hexachlorobutadiene	87683
Hexachloroethane	67721
Hexane	110543
Methyl bromide (bromomethane)	74839
Methyl chloride (chloromethane)	74873
Methyl chloroform (1,1,1-trichloroethane)	71556
Phosgene	75445
Tetrachloroethylene (perchloroethylene)	127184
Toluene	108883
Trichloroethylene	79016
2,2,4-Trimethylpentane	540841
Vinyl chloride	75014
Vinylidene chloride (1,1-dichloroethylene)	75354
m-Xylene	108383
p-Xylene	106423

first criteria is determining whether the wastewater source is a new source. This is readily defined with new units and in reconstructed or modified units; some investigation is necessary to determine whether the changes are classified as a new source.[8] For a new source, the next step is determining whether any of the chemical compounds are in Table 12.1 for VOHAP chemicals. If none of the chemicals are in Table 12.1, the stream must then be categorized, and this is shown in Figure 12.14.

For a *new* wastewater stream that contains one or more of the chemical compounds in Table 12.1, the concentration and flow rate are determined in accordance with Figure 12.13. If the flow rate or concentration of any chemical compound is ≥ 0.02 L/min or ≥ 10 ppmw, then the stream is a Group 1 wastewater and must be controlled as shown in Figure 12.15.

Where the wastewater stream is from an *existing source* or a stream from a new source has not been classified a Group 1 wastewater stream based on Table 12.1 (Figure 12.13), both types of streams must be classified in accordance with Figure 12.14. Following a determination of the concentration and flow rate, the first step is determining whether any of the chemical compounds in the waste stream are present in Appendix 12C. The Appendix 12C table in the A category also includes the compounds in Table 12.1 that are considered VOHAP. Other compounds may also be considered quite volatile in this category, but are not considered in the same classification as the Table 12.1 chemicals. Consequently, Appendix 12C contains a combination of VOHAP and HAP compounds.

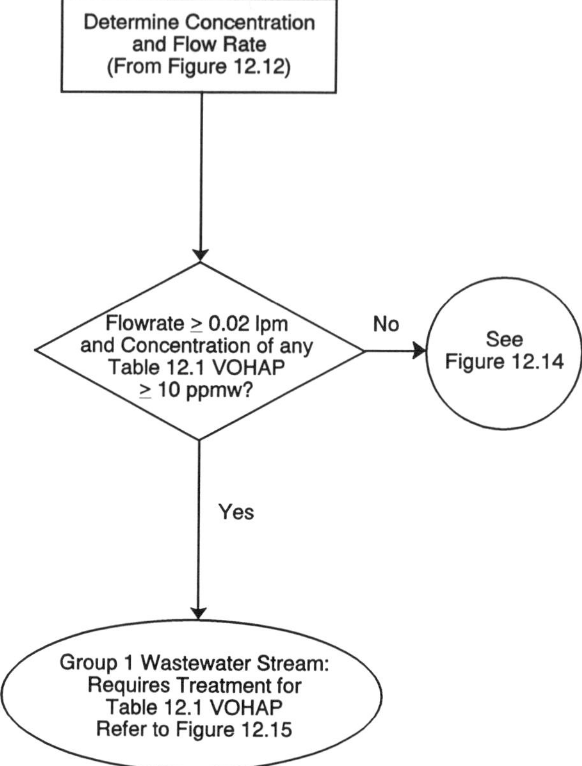

Figure 12.13 *Classification Groups 1 or 2 status for new source wastewater stream.*[7]

If the total VOHAP/HAP chemical compound concentration is ≥ 10,000 ppmw, this evaluation proceeds to the next classification step (Fig. 12.14). Where the concentration is ≥ 1000 ppmw and the flow rate is ≥ 10 L/min, the stream is also sent to the next classification step. For streams where the concentration and flow rate are less than the two latter criteria, the wastewater stream is classified as Group 2 wastewater. The second classification is determining whether the total yearly VOHAP/HAP mass flow rates of chemicals in Appendix C are less than 1 Mg/yr for *each point of generation*. When the VOHAP/HAP chemical concentration exceeds the annual mass flow or the exemption is deferred, the wastewater stream is classified a Group 1 stream and treatment options are shown in Figure 12.16. For streams with a VOHAP/HAP chemical concentration that is less than the annual 1-Mg specification and the exemption is requested, the stream is considered a Group 2 wastewater.

An important consideration is that the previous exemption is based on point of generation. For chemical compounds in Table 12.1, the exemption is not applicable.

General Wastewater Requirements. Based upon characterization of the wastewater streams, treating or pollutant reduction requirements have been defined for the streams and are shown in Figures 12.15 and 12.16. Although pollutant

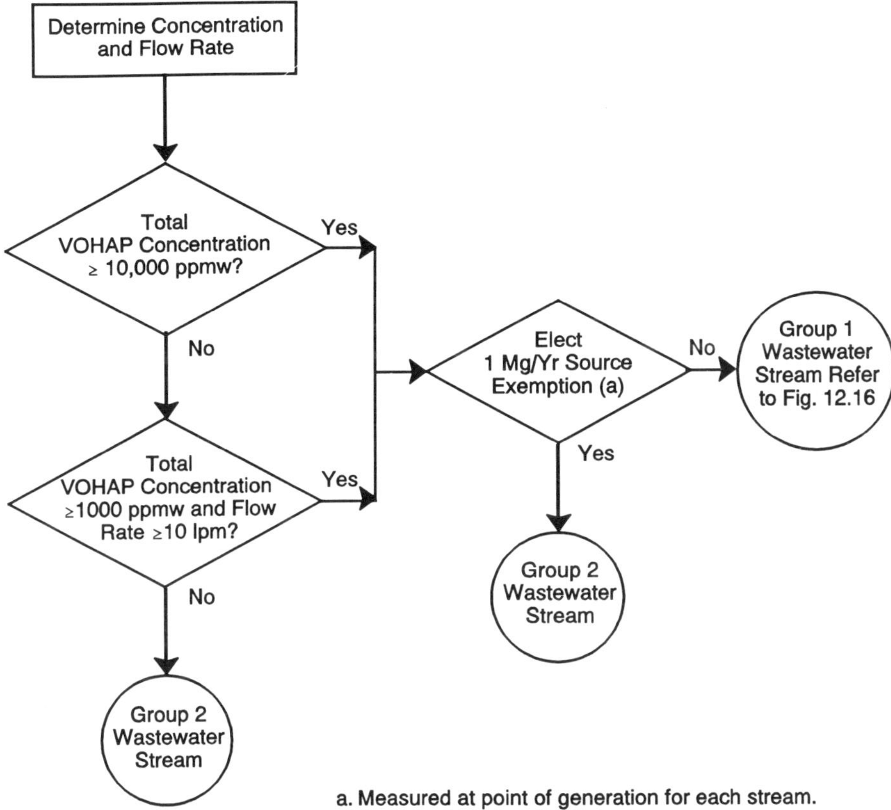

Figure 12.14 *Classification Groups 1 or 2 status for wastewater streams with chemical compounds in Appendix 12C.*[7]

concentrations are reduced, these specific requirements do not address the air emissions associated with various drain system configurations and treating steps. Equipment requirements are presented in the next section.

Treatment of Group 1 VOHAP. The wastewater treatment requirements and options are shown in Figure 12.15. We see that these Group 1 streams contain the contaminants listed in Table 12.1 through the identification process shown for new sources in Figure 12.13. The required treatment system is based upon effectively reducing the Group 1 total mass VOHAP concentrations by 99% or installing a steam stripper. While the efficiency or outlet concentration of the stripper has not been defined, the EPA has specified stripper design and operating conditions that ensure a very low contaminant concentration in the stripper effluent stream.

From Figure 12.13, the group stream is assessed. For an individual stream there are four alternatives for contaminant control. Since each wastewater stream is classified at point of generation, the EPA indicated that the individual stream can be treated. However, in a process unit that has several process drains, the more likely scenario is that the drains are combined and then treated. This latter situation may also include Group 2 wastewater and the treating requirements are

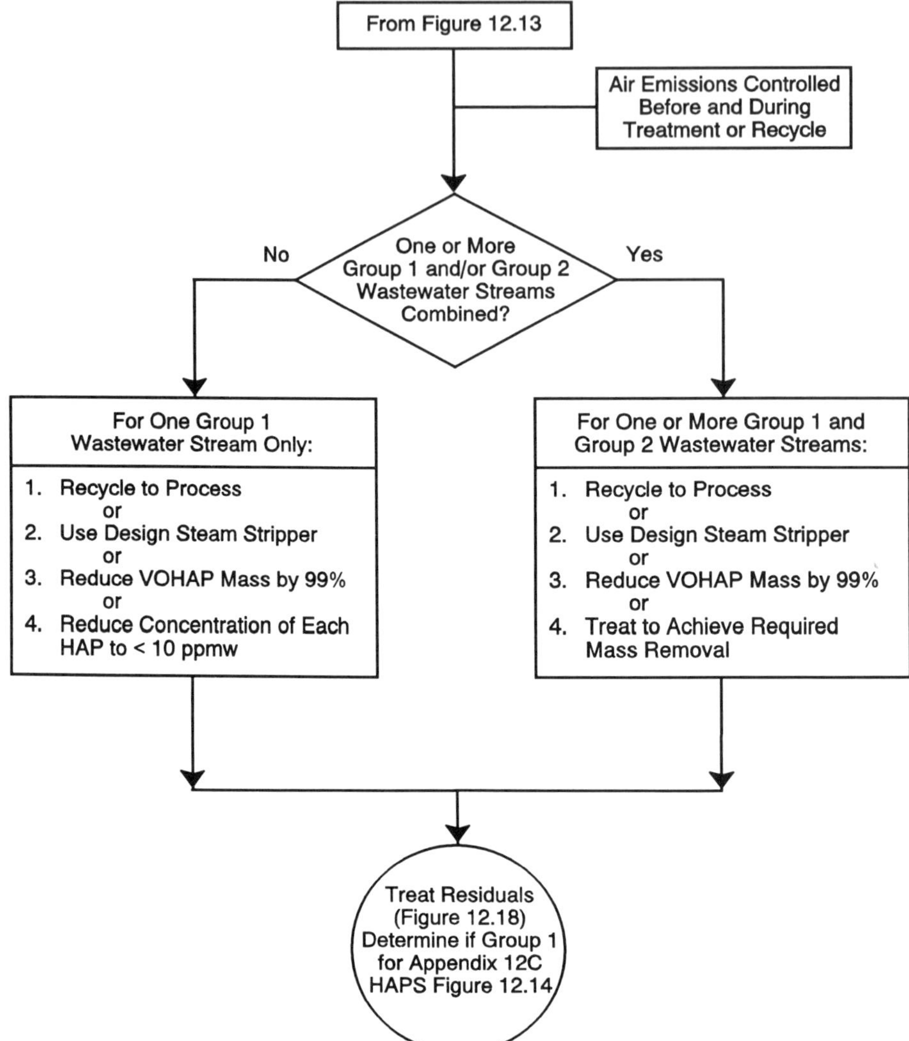

Figure 12.15 *Wastewater stream compliance options for Table 12.1 VOHAP chemicals.*[7]

almost identical to those required for a single stream. The only difference is the general specification that alternative treating systems can be installed to achieve the required mass removal, which differs from single system treating where each HAP compound should be reduced to less than 10 ppmw.

One of the alternative treating methods for removal of pollutants is a biological oxidation unit (biox). However, in the list of requirements, wastewater streams cannot be diluted or mixed where a lower average concentration of contaminants results. Consequently, treating systems are basically required at the processing units to produce effluents with a desired concentration. The chemical industry frequently transfers the contaminated process wastewater streams directly to a biox unit, which reduces effluent concentrations to desired levels. Although the EPA does not have any detailed information on this operation, this type of treating will probably be accepted. In addition, combining process wastewater streams from various process units for stripping in one stream stripping column is a more cost effective solution than installing a stripper in each process unit. These changes will probably be acceptable to the EPA if air emissions controls are installed in the sewer or transfer line systems. However, vent control requirements after stripping may not be required where the stream is classified Group 2.

A major requirement in these various processing alternatives is that air emissions must be controlled. Air emission controls are presented in the next section along with regulatory requirements for the individual drain systems and treating units.

Group 1 Treatment for Appendix 12C Contaminants. The treating requirements are based on the criteria for Group 1 wastewater streams shown in Figure 12.14. Contaminants are defined in appendix 12C. As noted above, the treating system can be based upon an individual wastewater stream or combined wastewater streams in the Group 1 category or a mixture of Group 1 and 2 wastewater streams. As in the previous Group 1 requirements, the untreated wastewater streams from a process unit cannot be combined with other wastewater streams that intentionally or unintentionally dilute the HAP contaminants. This requires treatment at the process unit. However, combining the wastewater streams into a single stream for steam stripping is considerably more cost effective than separate strippers at each unit.

As shown in Figure 12.16, the wastewater streams are treated individually or as a combination of Group 1 streams with or without Group 2 wastewater streams. In this treating system, a specified steam stripper can again be used for the individual or combined streams. For the single stream, the total contaminant (VOHAP/HAP) concentration must be reduced to less than 50 ppmw. This requirement is not shown for the mixed streams. However, the basic requirements are similar in total mass removal.

One difference in the treating requirements compared with those given above is based upon removal of contaminants with efficiencies for various groups. In Appendix 12C the chemical compounds are separated into groups that have a removal efficiency which must be applied to the contaminants in the group. These are termed target strippability factors for steam strippers or other treatment systems. Steam strippers can be installed and operated as specified by the EPA (design steam stripper) or strippability factors can be applied for steam strippers designed by plant sites.

Again, air emissions must be controlled in the individual drain system systems and treating units.

Alternate Option for Group 1 Appendix 12C Contaminants. For an existing process unit as an alternative, all wastewater streams from the unit can be combined for treatment. These wastewater streams contain contaminants listed in Appendix 12C. Requirements are shown in Figure 12.17 where the combined stream can be recycled or the treated stream must have a total HAP effluent concentration after treatment that is less than 10 ppmw. Following treatment, the wastewater stream can be combined with other process wastewater streams.

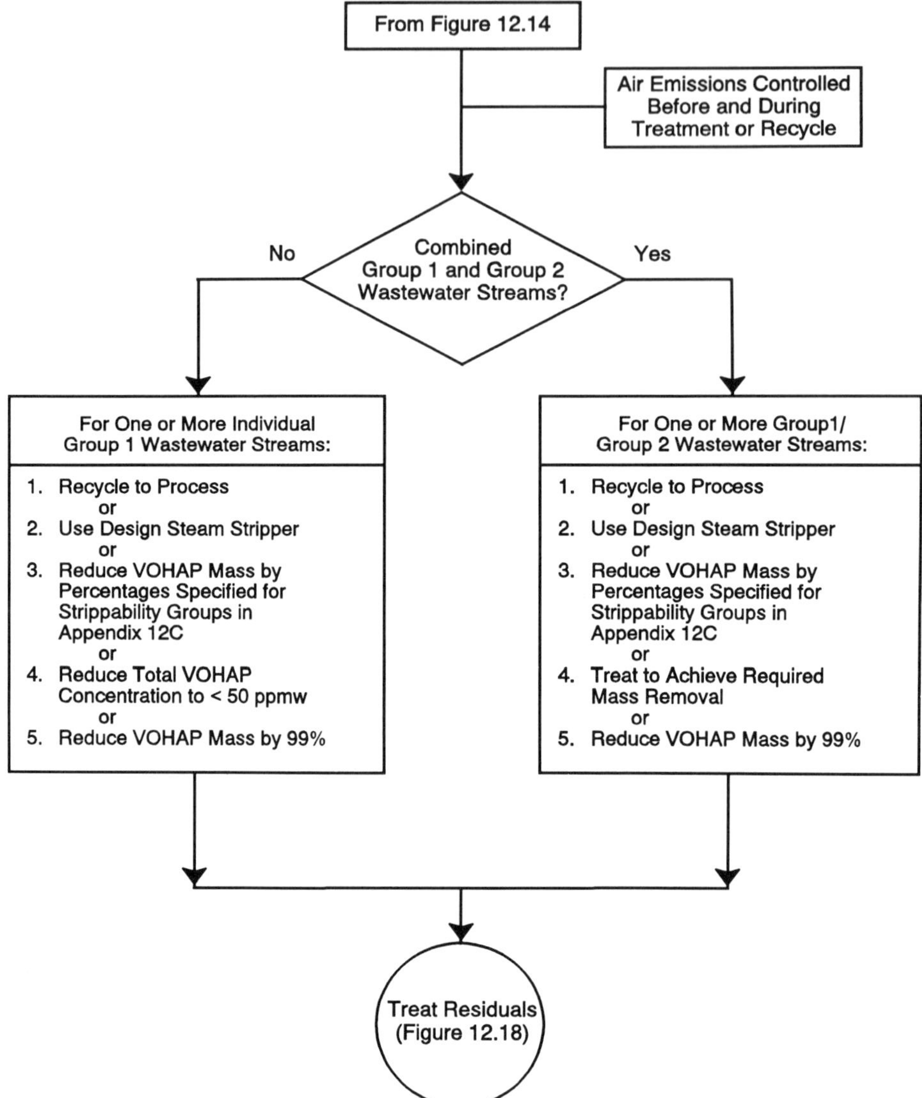

Figure 12.16 *Wastewater stream compliance options for Appendix 12C HAP chemicals.*[7]

A major requirement with this option is that all wastewater streams in the specified unit must be combined and treated. While the proposed alternative option only covers one unit, one treatment system for a number of units is considerably more cost effective and practical. The major criteria will probably be the control of all emissions to the atmosphere before the wastewater stream reaches the treating unit. Air emission controls cover all potential release points and the treating units.

Residual Disposal. In the various treating regulations, the contaminants removed from the wastewater streams (residuals) require disposal and the EPA has devel-

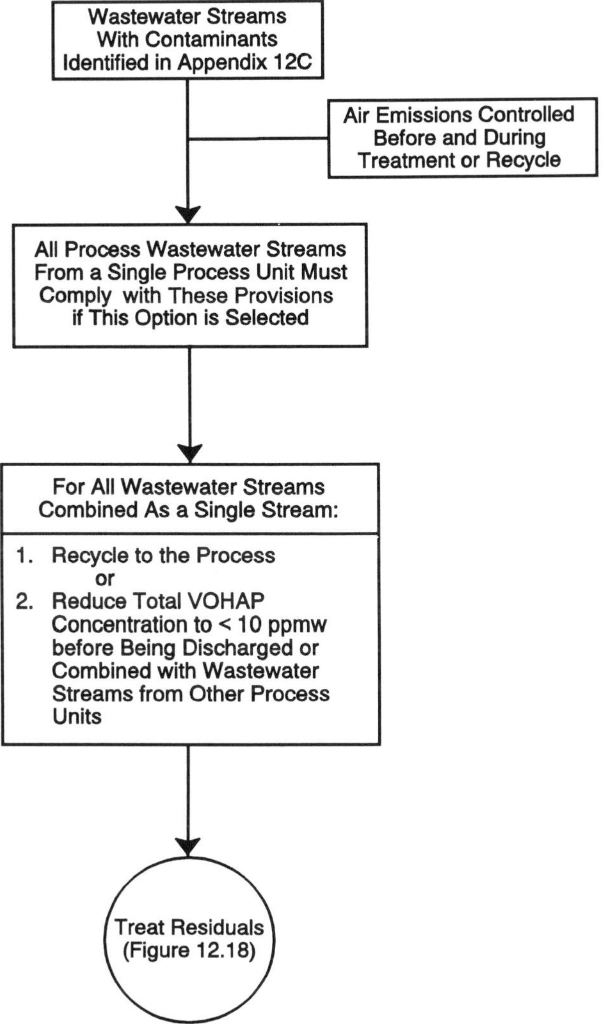

Figure 12.17 *Alternative option for Group 1 Appendix 12C HAP chemicals.*[7]

oped some requirements, as shown in Figure 12.18. All disposal techniques require control of any air emissions in the residual handling system.

Recycling and recovery of the residual stream to the process is desirable and can be accomplished in many process units. The EPA also indicates that the residual stream can be returned to the treatment unit. While return to a destruction unit would eliminate the contaminated stream, recycling to other types of treatments may not be effective. For destruction devices, the total HAP mass reduction must be a minimum of 99%.

Specific Wastewater Equipment Requirements. In addition to the overall emission reduction requirements, the EPA has included equipment design

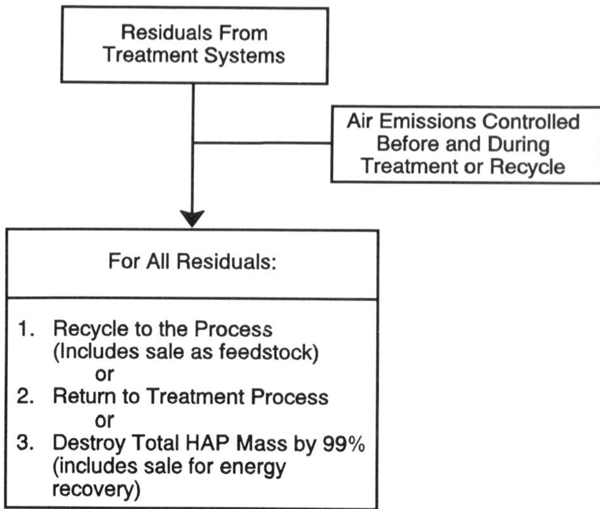

Figure 12.18 *Options for residual disposal and control.*[7]

specifications in the proposed regulations. These equipment specifications or requirements cover the actual air emissions controls associated with equipment operations. The overall drain system and treating units are contained in the proposed regulations, but the EPA has not included catch basins in the equipment control list. A major consideration for not including catch basins is the omission of spills, rainwater, or any wastewater that does not contact HAP chemicals on a continuous basis. Further comments on catch basins and a general process installation are presented in Section 12.4. Wastewater system equipment design requirements follow.

Individual Drain System. The drain system regulations have been concentrated on drains and junction boxes (manholes). In general, these regulations apply to Group 1 wastewaters or residual streams removed from Group 1 wastewaters and require either closed covers and vent discharge to a control system or specific equipment design configurations.

1. For the closed cover and control device systems:
   - Covers are sealed and operated without any leaks defined as $\geq$ 500 ppmv.
   - Covers are inspected initially and annually thereafter.
   - All covers and openings are closed and sealed unless opened for sampling, inspection or maintenance.
   - The closed vent and control systems are designed and operated in accordance with the requirements listed above.
   - Individual drain systems are inspected initially and semiannually thereafter. Leaks are defined as $\geq$ 500 ppm. Following leak identification, first repair efforts are conducted within 5 days and completed within 15 days. Repairs can be delayed if the repair is not feasible without a shutdown.

2. For the equipment design alternative:
   - Each *drain* is equipped with water controls as shown in Figure 12.9 or the tightly sealed cap. For sealed traps, inspection must ensure water is maintained in the trap. An alternate is a continuous water supply flow into the trap that has a flow monitoring device.
   - Each *junction box* is covered, and if required has a vent pipe that is a minimum of 35.43 in. (90 com) in length and has a maximum diameter of 4 in. (10.2 cm). The junction box covers are sealed and only opened during inspection and maintenance. Junction box vents are controlled with water seals as shown in Figure 12.19. Alternatively, the vent is a closed vent

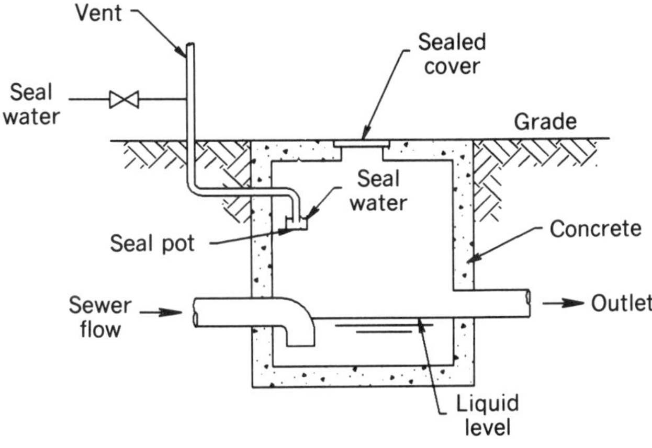

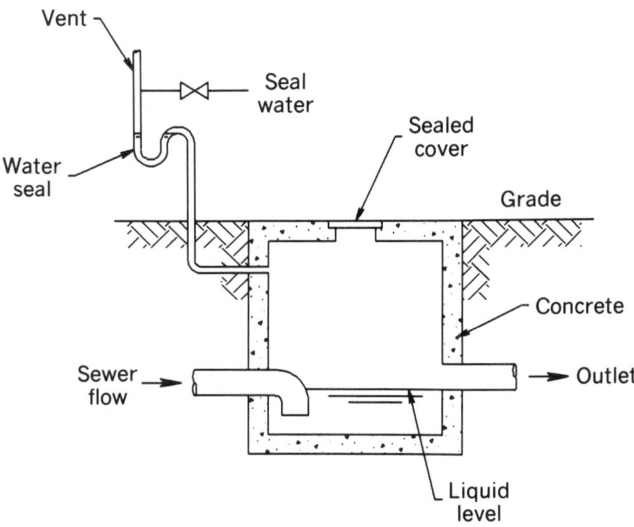

Figure 12.19 *Alternate manhole or junction box seals.*

system with a control device. Capped drains and junction boxes are inspected initially and semiannually thereafter to ensure the openings are sealed. If an opening or break occurs in the seal or covers, an initial repair will be initiated within 5 days after identification and completed within 15 calendar days. Repairs can be delayed if the repair is not feasible without a shutdown.

Although the EPA requires wastewater stripping or treatment to very low levels ($\leq$ 10 ppmw for HAP) following drains and prior to entering a junction box, the need for vent control is questionable whether a water seal or a closed vent/control device system is utilized. In addition, other concerns suggest that vent sealing will probably not be required. The water seal allows the buildup of a small pressure within the seal system which can have deleterious effects since sewer systems are designed for atmospheric pressure operation. As an example, cracks may form in a concrete pipe or manhole. This results in leakage or infiltration of water into the ground. In addition, the seal pot installation may freeze during the winter and an external seal is shown as an alternative in Figure 12.19. An external seal can be heated or jacketed to prevent freezing, but both vent systems are completely sealed. Complete seals would probably result in a fuel-rich vapor zone buildup over a period of time as the air in the system is gradually displaced. If a fuel rich vapor exists in a sewer system, containing a chemical or refinery hydrocarbon, a blown seal, or an open seal results in a rich vapor mixture escaping to the atmosphere. The dispersion of fuel-rich vapors into the atmosphere will result in the formation of combustible vapor zones as the rich vapor is diluted. Any potential ignition source may cause an explosion or fire in the combustible zone.

In a sealed system containing hydrocarbons that become oxygen deficient, sulfur compounds will react with steel or iron to form iron sulfide (FeS) which is pyrophoric. Sulfur is present in a number of wastewater streams in refineries and the potential formation of this compound is a serious consideration. Upon opening a sewer for maintenance containing some FeS, air containing oxygen contacts the iron sulfide and results in a fire, explosion, or both.

The industrial hygienist should be aware of manhole or junction box vents as a potential source of toxic vapor emissions. These emissions would infrequently result from unit upsets or malfunctions in a stripping unit. A sudden large increase in volatiles increases vapor generation and may blow or unseal any sealed vent. Vent location is reviewed in Section 12.4.

Surface Impoundments. Surface impoundments refer to large wastewater storage basins dug into the earth or formed with earthen or concrete dikes. Regulations cover impoundments that receive and store Group 1 wastewaters or residual streams removed from a Group 1 wastewater. These proposed regulations require a sealed cover over the water surface that does not have any leaks which are defined as $\geq$ 50 ppmw. The cover can be rigid or an air-supported cover with HAP vapors sent through a closed-vent system to a control device. Requirements for closed vent systems and controls are presented below.

The cover and all openings are inspected initially and annually thereafter for leakage. In addition, the surface impoundments are inspected initially and semiannually thereafter for improper work practices and control equipment failures.

Improper work practices include leaving open or nongasketed doors, hatches, and so on. Control equipment failure refers to seals, gaskets, and doors that leak, are cracked, or fail. Where a failure occurs, an initial repair must be instituted within 5 days after identification and completed within 15 days after identification. Repairs can be delayed if the repair is not feasible without a shutdown.

Oil–Water Separators. These proposed requirements are applicable to separators that only process Group 1 wastewaters or a residual stream removed from a Group 1 wastewater. While termed oil–water separators, these separator regulations apply to units that separate nonmiscible chemical or hydrocarbon compounds from water. The proposed regulations require a cover that can be either a fixed or floating roof. However, VOC regulations may require a cover.

FIXED ROOF SEPARATOR. The fixed roof completely seals the separator with organic HAP vapors sent through a closed-vent system to a control device which meet the requirements listed below. All openings in the fixed roof are operated without leakage that is defined as $\geq$ 500 ppmv. Each opening is maintained in a closed sealed position through the use of gaskets and latch devices. These unit doors, hatches, and ports are only opened for sampling, equipment inspection, maintenance, or repair.

A typical closed oil–water separator is shown in Figure 12.20, which also has a closed-vent system and a flare that is a control device. This general system is described in Section 12.4. The fixed roof and all openings are inspected initially and annually thereafter to determine compliance with the 500-ppmv leak definition. In addition, each separator is inspected initially and semiannually for improper work practices and control equipment failures. Improper work practices also refer to open or nongasketed doors or hatches. Control failures refer to gasket, joint, lid, cover, or doors that may be cracked, broken, or missing.

FLOATING ROOF SEPARATOR. The floating roof is installed as shown in Figure 12.10 with a primary and secondary seal. Primary seal gaps described in Section 12.14 must be measured within 60 days after the installation of the floating roof and

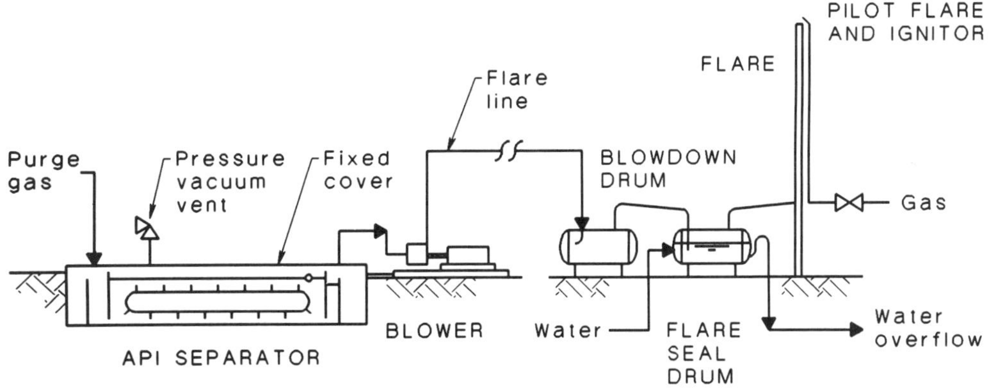

Figure 12.20 *API separator with fixed cover, closed-vent system, and a control device.*

introduction of a Group 1 wastewater. The gap should be less than the maximum limits described earlier. Following the initial installation gap measurement, the primary seal gap is measured once every 5 yr thereafter.

Secondary seal gaps are also measured within 60 days after installation of the floating roof and introduction of a Group 1 wastewater. The gap should be less than the maximum secondary seal limits described in Section 12.1.4. Following the initial installation gap measurement, the secondary seal gap is measured once every year thereafter.

Where portions of the separator cannot be enclosed with a floating roof, a fixed roof with a closed-vent system and a control device are installed in these other sections of the separators. The fixed roof requirements are identical to those described for the standard oil–water fixed roof separator. In addition to the seal gap measurements, each floating roof separator is inspected initially and semiannually thereafter for improper work practices and control equipment failures. Improper work practices refer to gasket, joint, lid, cover, or doors that may be cracked, broken, or missing in both the fixed and floating roof sections. Control equipment failures in the floating roof section of the separator include the following:

- The floating roof is not resting on either the surface of the liquid or the roof leg supports.
- There is liquid on the upper surface of the floating roof.
- A rim seal is detached from the floating roof.
- The rim seal or seal fabric has holes, openings, or tears.
- Primary or secondary seal gaps that exceed the specification limits.

REPAIR REQUIREMENTS. When an improper work practice or control equipment failure is identified in either separator type or a leak is identified where the concentration is $\geq$ 500 ppmv, initial repair efforts must be initiated within 5 days after identification. The repairs should be completed within 15 days. However, repairs can be delayed if the repairs are not feasible without a shutdown.

Wastewater tanks. Tanks that receive, manage, or treat a Group 1 wastewater or a residual stream removed from a Group 1 wastewater must conform to the regulations. Three types of tanks are regulated in the tankage group—fixed roof, internal floating roof, and external floating roof. These tanks and their auxiliary equipment are described in Chapter 13. VOC regulations may require controls.

FIXED-ROOF TANKS. This tank must have a closed-vent system which sends HAP vapors to a control device. The control device is regulated in accordance with the requirements presented below. In addition, the fixed tank must conform with the following regulations:

1. The tank and all openings (access hatches, sampling ports, gauging wells, etc.) must be designed and operated without leaks. A leak is defined as a monitoring concentration $\geq$ 500 ppmv. Openings shall be maintained in a closed, sealed position requiring gaskets, seals, and latches. This condition is not changed

unless it is necessary to obtain a sample, inspect the tank or equipment, or perform maintenance or repairs.

2. Fixed-roof tanks and openings are inspected initially and annually thereafter.

INTERNAL FLOATING ROOF TANKS. These tanks are inspected in accordance with the procedures described in Chapter 13 and meet the requirements described in that chapter.

EXTERNAL FLOATING ROOF TANKS. These tanks are also inspected in accordance with the procedures described in Chapter 13 and meet the requirements described in that chapter. When an inspection of the seal gap is unsafe or inspection of the floating roof is deferred due to an unsound floating structure, the tank seal gaps are measured if possible or the wastewater tank is inspected within 30 days after the determination that the floating roof is unsafe. When inspection and measurement are not feasible, the tank is emptied within 45 days after determining the floating roof is unsafe. Some additional delays up to 30 days are possible.

REPAIR AND INSPECTION. Each wastewater tank is inspected initially and semiannually thereafter for improper work practices and control equipment failures. For wastewater tanks improper work practice refers to open or nongasketed doors or hatches. Control equipment failure refers to floating roof or seal failures. In addition, the following are also considered control equipment failures:

- The floating roof is not resting on either the surface of the liquid or the roof supports.
- There is liquid on the upper surface of the floating roof.
- A rim seal is detached from the floating roof.
- The rim seal or fabric has holes, openings, or tears.
- There are visible gaps between the seal of an internal floating roof and the wall of the wastewater tank.
- Gaps exist between the metallic shoe seal or the liquid mounted primary seal of an external floating roof wastewater tank and the tank wall that are in excess of those gaps described in Chapter 13.
- Gaps exist between the secondary seal of an external floating roof wastewater tank and the tank wall that are in excess of those gaps described in Chapter 13.
- A metallic shoe seal does not extend into the tank liquid or one end of the metallic shoe does not extend a minimum vertical distance of 24 in. (61 cm) above the liquid surface.
- A gasket, joint, lid, cover, or door is cracked, partially unsealed, or broken.

When an improper work practice or a control equipment failure is identified or a leak that is $\geq$ 500 ppmv is identified, initial repairs must begin within 5 days after identification. Repairs must be completed within 45 days after identification. However, repairs can be delayed if the repairs are not feasible without a shutdown.

Closed-Vent Systems and Control Devices. Many wastewater drain system vents and treatment units must be vented to closed systems and control devices. Since they are an integral part of the regulatory effort, these systems must be operating whenever emissions are vented into the closed system. The control device requirements are summarized initially, followed by the vent requirements.

ENCLOSED COMBUSTION DEVICE. This refers to any of a variety of combustion devices but does not include a flare. Typical examples of enclosed combustion devices are: thermal vapor incinerator, catalytic vapor incinerator, boiler, or process heater. This classification does not cover the entire range of enclosed combustion devices and can also include other combustion devices. For these devices, the following requirements apply:

1. When a boiler or process heater is the combustion device, the vent stream must be injected into the flame zone of the boiler or process heater.
2. Effectiveness or efficiency must meet the following conditions:
   a. The total organic emissions (excluding methane and ethane or total organic HAP emissions must be reduced a minimum of 95 wt % based on the total inlet mass flow.
   b. A total organic compound concentration (excluding methane or ethane) or total organic HAP emissions at the combustion device outlet of 20 ppmv on a dry basis corrected to 3% oxygen.
   c. Provide a minimum residence time of 0.5 sec at a minimum temperature of 1400°F (760°C).
3. The required destruction conditions must be established through design and then confirmed by test. Necessary data requirements are outlined in reference 7.
4. The following enclosed combustion devices are exempt from these various requirements when:
   a. A boiler or process heater has a design capacity $\geq$ 44 MW.
   b. When the vent stream is introduced with the primary fuel into a boiler or process heater.
   c. A boiler or process heater used in the destruction of the vent stream burns hazardous waste where the owner has been issued a final permit under 40CFR Part 270 and complies with the requirements of 40CFR Part 266 subpart H. Alternatively, the owner has certified compliance based on the interim status requirements of 40CFR Part 266 subpart H.

FLARES. Any flare that is used as a control device must conform to flare requirements that are described in the regulations.[11]

VAPOR RECOVERY SYSTEM. These systems including adsorption and condensing must reduce total organic emissions (excluding methane and ethane) or the total organic HAP emissions, a minimum of 95 wt %. Adsorption systems are principally carbon, although other adsorbents may be used. In addition, absorption, although not specifically included in the regulatory reference, is permitted where the absorption system meets the regulatory minimum of a 95 wt % reduction in emissions.

PERFORMANCE. The facility must demonstrate compliance with the emissions reduction regulations. This requirement can be observed by performance tests or a design analysis. Flare compliance is conducted in accordance with reference 11. Performance tests are conducted in accordance with regulatory requirements compiled in subpart 63.145, reference 7.

A design analysis is normally required to specify a control device, but the analysis must be formally conducted to provide the data necessary for the regulation. Specific EPA design requirements for the most popular control devices are shown in subpart 63.139(c) in reference 7. While the design analysis meets EPA requirements, process facilities normally conduct performance tests to reinsure that the device meets design targets. Consequently, performance testing may be a more practical approach in meeting compliance requirements.

CLOSED-VENT SYSTEM. This system must be operated without any leak, where leakage is defined as $\geq$ 500 ppmv. The leak concentrations are determined in accordance with standard EPA method 21.[42] Each closed-vent system is inspected for leaks initially and annually thereafter. Where parts of the system are defined as unsafe to inspect, these parts are exempt if:

- Personnel may be exposed to imminent or potential danger.
- Written plan requires equipment inspection as frequently as possible during safe to inspect periods.

Where parts of the system are designated difficult to inspect, those parts are exempt from the inspection requirements if:

- The equipment cannot be inspected without elevating the inspection personnel 2 m above a support surface.
- The closed-vent system is on an existing treatment system or on a treating system within an existing source.
- A written plan requires inspection at least once every 5 yr.

For these exemptions, an explanation must be forwarded to and accepted by the EPA. In addition, each closed-vent system and control device must be visually inspected initially and annually thereafter. This inspection identifies holes or cracks and loose connections.

In those closed-vent systems that have bypass line connections that can divert the vent stream to the atmosphere, *one* of the following compliance requirements must be observed:

- A calibrated flow indicator must be installed on the bypass line that provides a record of the flow rate through the bypass line every 15 min.
- A block valve in the bypass line is car sealed closed or closed with a lock and key configuration. Visual inspection of the seal or closure mechanism is

conducted once per month to ensure the valve is maintained in the closed position.

MAINTENANCE. If visible defects or leaks ($\geq$ 500 ppmv) are found in the enclosed vent system or in the control device, an initial effort to repair the system shall be conducted within 5 days after identification and completed within 15 days after identification. However, repairs can be delayed if the repairs are not feasible without a shutdown.

Biological Oxidation Treatment Unit. Where a biox unit meets the mass removal requirements for new and existing sources, the biox unit does not require covering and venting through a closed system to a control device.

Since waste streams are frequently fed directly to a biox unit, the general regulation indicating that a closed system is not required for units meeting mass removal requirements may not be applicable since the EPA does not have any detailed information on biox performance in this situation. For noncovered biox operation, decomposition of the chemical contaminants is required and not a combination of vaporization and decomposition to meet mass removal targets. Where the biox receives process wastewater directly from the unit, the EPA will probably require covers and closed-vent systems with a control device until sufficient data are available to indicate these units meet emissions requirements. The control system will prevent any emissions release that may result from an upset in the process and associated water streams, since time is necessary for the biox to adjust to changes in feed conditions.

Air Flotation Units. These units are not specifically addressed in the regulations. However, units that are not included in the regulatory groups above are governed by a general clause. This clause requires that each unit shall be covered and vented through a closed-vent system to a control device. The latter is designed and operated as described above. Covers must be operated without any leaks that are defined as $\geq$ 500 ppmv. In addition to the overall cover, all doors and hatches should be sealed and gasketed to minimize emissions.

The cover is inspected initially and annually thereafter for leakage or openings. The closed-vent system and control device are inspected and maintained in accordance with regulations described above. When an instrument reading indicates that a leak has been identified, the initial repair effort should commence within 5 days after identification and be completed within 15 days after identification. Although the EPA has not indicated a repair delay is permitted, a delay may be necessary since a plant shutdown may be required to fix any leak.

Heat-Exchange System. A major concern of the EPA is leakage of HAP compounds into cooling water, which contaminates the cooling water. The EPA has included some regulatory requirements to control this leakage, which essentially relies on monitoring and heat exchanger repair for control. Leakage, particularly of compounds that are reasonably volatile, will result in emissions to the atmosphere from the cooling tower that cannot be controlled. Moreover, the average return cooling water temperature to the tower may be 125–130°F (52–54.7°C), which increases chemical vapor pressures and the associated emissions. The proposed EPA regulatory requirements are:

1. Cooling water shall be monitored monthly for the first six months and quarterly thereafter to detect leaks. Only HAP chemicals present in the water at concentrations greater than 5% will be measured.

2. Samples are obtained at the entrance and exit of each heat exchanger. For parallel exchangers, samples are obtained at the main stream inlet and outlet points. Individual exchangers are not measured in a parallel exchanger bank unless a leak is shown in the multiexchanger installation.

3. A minimum of three sets of samples must be obtained at the inlet and outlet locations of a heat exchanger which are analyzed for total HAP concentration based on an approved EPA analytical method. The three sets of samples provide average inlet and outlet concentrations.

4. A leak is detected if there is a statistically significant increase ≥ 1 ppmw at the 95% confidence level, *or* a statistically significant increase of 1% at the 95% confidence level. The concentration increase may change from these levels, but a cooling water control regulation will be promulgated.

5. When a leak is detected, the leak shall be repaired as soon as practicable within 15 calendar days after detection. A delay is permitted if repair is not feasible without a process shutdown, but must be completed during the next process shutdown. If the heat exchanger is isolated from the process and is not in HAP service, a delay in repair is acceptable.

6. Each heat exchanger operated with a minimum cooling water pressure of 5 psi (35 kPa) greater than the process side pressure is exempt from the previous requirements regarding monitoring and analysis.

Steam Stripper. Where a steam stripper is installed, the stripper is designed with the following requirements:

1. A minimum active column height of 5 m (16.4 ft).
2. Counter current flow configuration with a minimum of 10 theoretical trays.
3. A minimum steam flow rate of 0.096 kg/L of wastewater feed (0.8 lb/gal).
4. Minimum wastewater feed temperature of 35°C (95°F).
5. Maximum liquid loading of 39,900 L/hr per square meter (978 gal/ft^3).
6. Water-cooled maximum primary condenser outlet vapor temperature of 50°C (122°F).

For various wastewater systems, stripper designs differ based upon the chemicals in the system, which may result in a stripper configuration that differs from these EPA requirements. However, the EPA size and process values provide a basis for comparison. Some aspects of stripper operation are presented in Section 12.4.

Comments on the Regulations. Since the HON regulations have been formally proposed, numerous comments on the regulations were received by the EPA, which were reviewed and changes were incorporated into the final regulations prior to promulgation. Normally, the EPA does not change the regulations to any great extent. However, the EPA admitted in the proposed regulations a lack of knowledge regarding various industry practices.[7,8,10] In addition, the regulations have considerable redundancy in requirements which significantly increases industry cost with very little improvement in emission control. Some of the various concerns the EPA will address and considerations follow:

1. The EPA does not consider the disposition of rainwater and spills in a process area. Normally, catch basins are provided for the rainwater and spills that are flushed into the basin. In addition, firewater in a process area is normally drained from the area in catch basins. Effluent streams from the catch basins are sent to a junction box or manhole and then transferred to a treatment unit. An alternative is not provided by the EPA, which complicates runoff and collection if the catch basins are eliminated. For practical operation, catch basins should be installed but limited to rainwater, spills, and firewater collection. Individual drains are not discharged into the catch basin. A typical design layout is shown in Section 12.4.

2. Condensate from steam traps is also collected along with process wastewater for treatment. At times, trap condensate discharge is contaminated with a chemical which has leaked into the condensate from a heat exchanger or other equipment. The major concerns with discharge of condensate into the wastewater are the heating of the wastewater stream, loss of energy in the steam system, an unbalanced steam/water system, a significant increase in boiler feedwater treating requirements, and the lack of leak identification. Where all of the steam condensate is recycled into the boiler feedwater system, chemical leaks are rapidly identified. This monitoring procedure results in rapid identification of the leaking equipment and repair to minimize operating problems in the steam system.

3. The EPA regulations want contaminant wastewater measurements at the point of generation, which is the point of discharge from the process equipment for each drainage stream. This is followed by a requirement for stripping before the wastewater stream enters a sewer line junction box and mixes with other wastewater. These requirements prevent combining the contaminated wastewater streams from various process units and then stripping the combined stream in one large stripper. Moreover, contaminated wastewater can be sent directly to a biox for treatment, eliminating various treating stops and reducing costs. The point of generation may remain as defined by the EPA, but where streams are contained or controlled preventing atmospheric emissions, the actual generation point will probably be the sum of all contained sources in the process unit prior to treatment or mixing with other process unit wastewaters. Although mixing streams from various process units is possible in many facilities for a single treatment facility, combining streams is not always possible where the residual recovered from a stripper is recycled to the process unit. Mixing of various chemicals can be a potential problem, but the plant is in a better position to evaluate these options. Consequently, the restrictive regulations on point of regulation and stream mixing will probably be revised.

4. Following stripping, a very low contaminant concentration remains in the wastewater (< 10 ppmw). The regulations require controls on the junction boxes,

separators, and air flotation units for Group 1 wastewaters. If the wastewater is stripped at the process unit to Group 2 level, emission controls are not specifically required in subsequent steps. An oil–water separator, clarifier, or biox unit is only required where additional treatment is necessary to meet final effluent discharge requirements. Additional treatment may be necessary for other process streams such as cooling tower blowdown.

5. The restrictions placed on junction box operation with a sealed vent are not justified by the reduction in emissions with a vent. The EPA emission calculations are based upon open junction box or manhole installations. However, these are normally enclosed and have a small vent pipe. The EPA calculations indicate operation of an enclosed junction box with a simple vent results in an emission reduction of 99% from the emissions levels calculated for open junction boxes.[13] This regulation will probably be revised to provide some operating flexibility.

6. Various alternatives are possible which can be incorporated into the regulations. Some of these have been described, such as enclosed discharge lines to a stripper followed by a biox unit and direct enclosed discharge from the process units to a biox unit. Moreover, a highly effective stripper may not require biox treatment. An alternative is stripping process drainage streams to very low levels and discharging the treated stream directly to a receiving water body with biox and other treatment steps only applied to other streams such as cooling tower blowdown. Strippers in process wastewater streams provide a great deal of flexibility that will probably be incorporated into the regulations. However, some holdup of stripper effluent may be required for stripper upset periods and recycling.

7. While the HON or MACT regulations are designed to reduce HAP emissions to the atmosphere, these regulatory requirements also reduce VOC emissions. For those compounds not included in the HON regulations that are VOC materials, these regulations can be used as a basis for reducing emissions. Where VOC and HAP compounds are mixed in a wastewater stream, the HON regulations significantly reduce both VOC and HAP emissions.

12.3 EMISSIONS

While the EPA has been investigating wastewater emissions for a number of years, actual overall emission data accumulated from various wastewater transfer and treatment units have not been extensive. Some data were obtained on drains,[13,14] but these data are not representative of current practice where the risers are sealed. However, an investigation of stripper operation was conducted in the field and in the laboratory to confirm operating and design requirements.[13] In addition, limited data were also obtained for various transfer systems and treatment units. As a result, emissions from various wastewater sources are primarily determined from an extensive set of calculation procedures developed by the EPA with some correlation to the limited field data.

The calculation procedures developed by the EPA are based upon plant operating conditions that do not represent current operating practices in most plants. As an example, the emissions generated from the wastewater in a junction box *open* to the atmosphere are determined. Very few junction boxes are open to the atmosphere in operating plants but are covered and vented to the atmosphere as shown in Figure 12.4. The EPA reports indicate manholes, as distinct from junction boxes, have a grill-type of manhole cover. Normally, manhole or junction box openings are sealed as described and have a vent, which significantly reduces vapor emissions to the atmosphere. However, the water surface emissions provide the bases for the determination of control device systems by the EPA. In addition, the EPA has also assumed there are no water seals in the junction boxes. A flow of air across the open top of the junction box and through the sewer lines further increase the emissions on a theoretical basis. Installed wastewater systems normally have liquid seals which prevent vapor flow through the sewer.

In this section, the calculation procedures which have been developed or adopted by the EPA are presented, which indicate emissions on both open and controlled bases. Estimating techniques have been developed that approximate emissions from noncontrolled sources with more extensive calculation emissions procedures available.

12.3.1 Estimated Emissions

This procedure was developed as a simplified method of estimating emissions for uncontrolled VOC compounds. Since most of the emission sources are controlled with some degree of effectiveness, the estimated uncontrolled emissions values provide a basis for determining the effectiveness of current controls or a basis for defining the controls necessary in a new unit.

The fraction emitted (fe) is determined for those compounds in a wastewater stream, which is an estimate of the total vapors that must be collected or controlled for each compound. Values of fe are determined for a compound from detailed calculations which are dependent on the Henry's Law constant for the compound.[15] These fe values were determined for each wastewater source. At each wastewater emission source, fe for five model contaminants was calculated, and a plot of fe versus the Henry's law constant was obtained. A least-squares approach was applied to the data points for curve fit, and an optimum curve fit was determined based on $R^2 = 1.0$, where R^2 is a statistical measure of reliability for the fe relationship with Henry's Law constant values. As a result of this analysis, an approximation of fe for each compound was developed for the various uncontrolled sources, and the equations are shown in Table 12.2.

The equations in Table 12.2 are approximate for estimating purposes. Consequently, high values of H may result in an fe fraction that exceeds 1.0. Also, low H values can produce fe fractions that are less than zero. For fe values greater than 1.0, fe is fixed at 1.0 and where fe is less than zero, the fraction emitted is considered zero. While not completely accurate, these adjustments provide a satisfactory basis for estimated emissions.

Calculations. When a compound flows through the emission source location or treatment unit, a fraction of the compound in the system is emitted. At the wastewater source or point of generation, the fraction of the compound in the steam is 1.0. Movement of the wastestream results in a fraction emitted (fe) at the first emission source location. The nonvaporized portion of the compound that remains in the wastewater has been termed the *fraction pass through* (fp_1) and at the outlet of the first emission source location the fp_1 value is equivalent to

TABLE 12.2 Fraction Emitted Value Estimate Based on Least Square Fit[a]

Source	Least Square Fit	R^2 [d]
Drain	fe[b] $= 0.0100 + 3.973 * (H)$[c]	0.997
Manhole	fe $= 0.003814 + 1.04 * (H)$	1.000
Junction box	fe $= 1.658 * (H)0.592$	0.902
Sump	fe $= 0.007384 + 0.0004385 * \ln(H)$	0.876
Trench	fe $= 0.06595 + 0.005006 * \ln(H)$	0.937
Lift station	fe $= 6.0118 * (H)0.6421$	0.910
Oil–water separator	fe $= 21.4034 * (H)0.8797$	0.968
Equalization basin		
Aerated	fe $= 1.345 + 0.6927 * \ln(H)$	0.757
Nonaerated	fe $= 0.5752 + 0.0331 * \ln(H)$	0.844
Clarifier	fe $= 0.03759 + 0.002351 * \ln(H)$	0.890
Treatment tank	fe $= 0.01374 + 0.0008558 \ln(H)$	0.889
Biobasin		
Aerated	fe $= 1.001 - 1.0028 *$ fbioa[e]	1.000
Nonaerated	fe $= 0.7489 - 0.7883 *$ fbion[f]	0.940
Weir	fe $= 422.8 * (D_w, i)0.6373$	1.000

[a]Adapted from reference 15.
[b]Fraction emitted.
[c]Henry's Law constant (atm · m^3/mol).
[d]Measure of fe and H reliability relationship. Optimum value, $R^2 = 1$.
[e]Fraction biodegraded in aerated basin.
[f]Fraction biodegraded in nonaerated basin.

$(1 - fe_1)$. The fraction pass through fp_1 is multiplied by the fraction emitted in the second emission source location (fe_2), which results in an adjusted fraction emitted (Afe). The outlet fraction from the second source fp_2 is equal to $[fp_1 - (fe_2 \cdot fp_1)]$.

All sources are calculated in a similar fashion for each compound until the effluent discharge point is reached. The total or overall fraction emitted is the sum of Afe for each source. There are two exceptions to this general approach covering aerated or nonaerated biox basins and oil–water separators. In the biox unit, there is degradation of the chemical compound, and in the separator, part of the chemical compound concentration partitions into the nonmiscible layer and is removed with this layer. Based upon these considerations, some adjustment of the pass-through fraction is necessary.

Biodegradation. A biodegradation study covering eight compounds was undertaken representing eight surrogate categories for both aerated and nonaerated box basins.[15] The eight categories covered various Henry's Law constants for the different chemical compounds and the following general classifications were developed:

a. High Henry's Law Constant, High Biorate (HHHB)
b. High Henry's Law Constant, Medium Biorate (HHMB)
c. High Henry's Law Constant, Low Biorate (HHLB)
d. Medium Henry's Law Constant, High Biorate (MHHB)
e. Medium Henry's Law Constant, Medium Biorate (MHMB)
f. Medium Henry's Law Constant, Low Biorate (MHLB)
g. Low Henry's Law Constant, High Biorate (LHHB)
h. Low Henry's Law Constant, Medium and Low Biorate (LHMB, LHLB)

TABLE 12.3 Predicted Steam Stripper Removal Efficiencies[a]

Compound	Henry's Law Constant (atm · m^3/mol)	Removal Efficiency (%)
1,3-Butadiene	0.142	100
Toluene	0.00668	100
Naphthalene	0.00118	99.9
1-Butanol	0.0000089	30.1
Phenol	0.000000454	2.2

[a]Adapted from reference 15 and based on computer design program.

A summary of the various surrogate categories and their associated biodegraded fraction is presented in Table 12.3. For computing the fraction pass through for a waste stream entering a biox unit, an adjusted emitted fraction (Afe) and an adjusted biodegraded fraction are calculated:

$$(Afe) \text{ Adjusted fraction emitted} = fe_i \times \text{feed pass through fraction}$$

$$(Afb) \text{ Adjusted biox fraction} = fb_i \times \text{feed pass through fraction}$$

The outlet pass through fraction from the biox unit is then calculated by subtracting Afe and Afb from the feed pass through fraction.

Oil–Water Separator. The inlet chemical fraction will partition between the immiscible layer and the wastewater. This characteristic is estimated with an octanol–water partition coefficient (K_{OW}) which is the solubility in the oil divided by the solubility in water. The coefficient is estimated with the following equation:[15, 16]

$$K_{OW} = \exp(7.494 - \ln C_S) \tag{12.1}$$

where C_S = solubility in water (vapor pressure, mmHg/760 · H, atm. m^3/mol).

For determining the fraction removed in the oil–water separator, the wastewater in the separator was assumed to contain 1000 ppm of oil. The fraction of each compound in the oil is determined by the following relationship:[15]

$$X_{\text{oil}} = 1 - [0.999/(0.999 + 0.001 K_{OW})] \tag{12.2}$$

The fraction of the chemical compound or VOC removed from the oil–water separator was assumed to be 80% of the compound or VOC partitioned into the oil or immiscible layer. In this system, the immiscible layer recovered or removed from the separator is a liquid. The pass through fraction discharged from the separator is the pass through fraction at the separator inlet less the adjusted fraction emitted from the separator and the adjusted fraction removed from the separator.

Steam Strippers. Based upon the five compounds shown in Table 12.3, removal efficiencies were determined for a stream stripper by the EPA from a computer design program.[15] From these data, the following equation was developed that predicts removal effectiveness at the R^2 value of 0.887:

$$f_r = 1.357 + 0.8667 \ln(H) \tag{12.3}$$

where H = Henry's Law constant (atm m^3/mol).

This equation provides an approximate removal basis for various compounds in the wastewater stream. However, accurate predictions of removal effectiveness require a detailed analysis of stripper performance with a computer as described in reference 15 or detailed stripping fractionation calculations. For some compounds, graphical procedures have been developed but are generally limited to a few compounds.[17] In addition to steam stripping, air stripping is also a method of removing contaminants. However, there may be safety considerations with the use of air as the stripping gas. The air stream with the stripped contaminants is sent to a control device which further complicates this system. However, air stripping effectiveness can also be determined quite accurately.[18,19]

12.3.2 Detailed Emissions Calculations

Calculation techniques have been developed by the EPA for more accurately estimating emissions and are available in references 13 and 15. Since these calculation methods are specific for each emission source, the calculation procedures are too extensive to include in this chapter. Moreover, these calculations are conducted on an individual compound basis. For wastewater streams with a number of chemical compounds, the emissions from each compound must be calculated separately and then combined for an overall determination of the removal efficiency. Where the emissions are calculated for an oil–water separator, equalization basin, and so on, the emissions are uncontrolled and provide a basis for estimating control device requirements. In addition to determining emissions, these detailed calculations also determine the fraction remaining in the wastewater or the pass through fraction to the next potential emission source.

The EPA has conducted field and laboratory work on steam stripping, but these efforts have been limited to the chemical compounds in the wastewater evaluated. Since data on residual removal is therefore limited, the EPA stripper efficiency data are only applicable to streams having identical wastewater compositions to those tested. While a reasonable estimate is available, detailed tower calculations are necessary to establish stripper effectiveness. These calculations are normally conducted by a fractionation expert since stripping is a distillation operation. Moreover, stripping effectiveness is frequently related to the pH in a wastewater stream, which complicates the fractionation calculations. Since steam stripping is the desired EPA method of controlling emissions from wastewater streams, accurate stripping calculations are necessary to determine fractionation effectiveness.

12.3.3 Required Mass Removal

The various requirements for chemical contaminant VOHAP and VHAP removal from wastewater streams were described in Section 12.2.2. In addition, the streams were also classified as Group 1 or Group 2 wastewater streams. Based upon these requirements, the EPA has defined contaminant mass removal requirements for HON wastewaters which are described in the following paragraphs:

Group 1 VOHAP Streams. The required mass removal covers contaminants in the Group 1 category that are listed in Table 12.1 (VOHAP). However, this requirement is applicable to a wastewater stream that has not been combined with other wastewater streams for treatment. The interpretation is that this covers the total wastewater stream in a process unit prior to combination with other wastewater streams in a sewer line junction box. While this is only applicable to one process unit, the wastewater streams from several units can probably be combined and processed in one system. As shown in Figure 12.15, the desired removal system is a steam stripper. The removal equation for a wastewater stream is

$$\mathrm{RMR} = \left(\frac{K}{10^6} V \sum_{j=1}^{n} C_j \right) 0.99 \tag{12.4}$$

where

RMR = required mass removal (kg/yr)
K = wastewater stream density (kg/m^3)
V = annual wastewater stream quantity (m^3/yr)
n = number VOHAP compounds in stream
C_j = average concentration of each compound j (ppmw)
0.99 = required removal fraction

The average VOHAP concentration is measured at the point of generation, and the required mass removal rate for each wastewater stream is combined to provide a total mass removal requirement for the overall stream.

Group 1 HAP Streams. The required mass removal covers contaminants in the Group 1 category that are listed in Appendix 12C. This requirement is also applicable to a wastewater stream that has not been combined with other wastewater streams for treatment, which refers to combining with other wastewater streams in a junction box. Again, wastewater streams from several units can probably be combined and processed in one system. As shown in Figure 12.16, a steam stripper is one method of reducing the contaminant concentration in the wastewater stream. However, a wider group of compounds is covered in this classification than is shown in the VOHAP group and it has various stripping capabilities. Consequently, the following equation is adjusted for the stripping efficiency, which is shown for the various compounds in Appendix 12C.

$$\mathrm{RMR} = \frac{K}{10^6} V \sum_{j=1}^{n} \left(C_j \frac{F_j}{100} \right) \tag{12.5}$$

where

RMR = required mass removal (kg/yr)
K = wastewater stream density (kg/m^3)
V = annual wastewater stream quantity (m^3/yr)
n = number organic HAP compounds in stream
C_j = average concentration of each compound j (ppmw)
F_j = required percent of removal of each compound (Appendix 12C)

The average HAP concentration is measured at the point of generation, and the required mass removal rate for each wastewater stream is combined to provide a total mass removal requirement for the overall stream.

Biological Unit. Actual mass removal can be readily determined for the various treatment systems in wastewater streams. For biox units, which destroys influent contaminants, the actual mass removal is calculated as follows:

$$\mathrm{MR} = (E_b - E_a)F_{\mathrm{bio}} \tag{12.6}$$

whcrc

MR = actual mass removal (kg/hr)
E_b = influent mass flow rate of VOHAP or HAP compounds (kg/hr)
E_a = effluent mass flow rate of VOHAP or HAP compounds (kg/hr)
F_{bio} = fraction of VOHAP or HAP compounds degraded

The F_{bio} fraction is determined from Method 304 and an EPA program.[7] The remaining mass is emitted to the atmosphere. However, reported emissions from biox units are extremely small to nonexistent.[8]

12.3.4 Overall System Calculations

While the previous calculation procedures were devised for estimating emissions from various equipment and source points, the EPA has developed a series of equations for estimating emissions where control technology is included. This provides an overall assessment for estimated emission quantities that can be used for determining credits and debits on an averaging basis. A base emission equation can be established where the overall system is *controlled by the reference control technology*:

$$\begin{aligned}\mathrm{EWW}_{ic} &= (6.0 \times 10^{-8})Q_i H_i \sum_{m=1}^{S} (1 - Fr_m)\mathrm{fe}_m \,\mathrm{HAP}_{im} \\ &\quad + (0.05)(6.0 \times 10^{-8})Q_i H_i \sum_{m=1}^{S} (Fr_m \,\mathrm{HAP}_{im})\end{aligned} \tag{12.7}$$

where

EWW_{ic} = emission wastewater rate for stream i (Mg/month)
Q_i = average flow rate for wastewater stream i (L/min)
H_i = period wastewater stream i discharged (hr/month)
Fr_m = strippability factor for HAP compound m from Appendix 12D
fe_m = fraction emitted for HAP compound m from Appendix 12D
S = total number organic HAP compounds in wastewater stream i
HAP_{im} = average organic HAP concentration compound m in wastewater stream i (ppmw)

For each wastewater stream i in any wastewater tank, surface impoundment, individual drain system, and oil–water separation that *does not meet the control requirements* in the regulations or any other treatment system that does not meet the requirements, the following equation applies:

$$\mathrm{EWW}_{i\,\mathrm{Act}} = (6.0 \times 10^{-8}) Q_i H_i \sum_{m=1}^{S} \mathrm{fe}_m \mathrm{HAP}_{im} \tag{12.8}$$

where $\mathrm{EWW}_{i\,\mathrm{Act}}$ is the wastewater stream i emission rate (Mg/month) and Q_i, H_i, fe_m, and HAP_{im} are defined as in Eq. (12.7). The following equation covers the emissions for each wastewater stream i in any wastewater tank, surface impoundment, individual drain system, and oil–water separator that meet the regulatory requirements for controls along with any treatment system. However, the reference controls are not as effective as the recommended control technology:

$$\mathrm{EWW}_{i\,\mathrm{Act}} = (6.0 \times 10^{-8}) Q_i H_i \sum_{m=1}^{S} [\mathrm{fe}_m \, \mathrm{HAP}_{im}(1 - \mathrm{PR}_{im})] \tag{12.9}$$

where $\mathrm{EWW}_{i\,\mathrm{Act}}$ is the wastewater stream i emission rate (Mg/month) and Q_i, H_i, fe_m, HAP_{im} are defined as in Eq. (12.7) and

$$\mathrm{PR}_{im} = \frac{\mathrm{HAP}_{im-\mathrm{in}} - \mathrm{HAP}_{im-\mathrm{out}}}{\mathrm{HAP}_{im-\mathrm{in}}} \tag{12.10}$$

and

PR_{im} = process treatment efficiency of organic HAP compound m in wastewater stream i

$\mathrm{HAP}_{im-\mathrm{in}}$ = average organic HAP concentration of compound m in stream i entering the first controlled source in series (ppmw)

$\mathrm{HAP}_{im-\mathrm{out}}$ = average organic HAP concentration compound m discharged in stream i (ppmw)

These equations provide a basis for determining emissions on an estimated and actual basis. In addition, some further modifications to the equations are available.[7] One situation has wastewater stream controls that are more effective than the reference control technology, but the vapor stream control device efficiency is equivalent to the reference control technology efficiency of 95%. A few other variations of the reference control technologies and vent control device efficiencies have also been formulated.

The overall equations developed by the EPA provide a basis for determining emission control requirements and the concentration of organic HAP in the effluent discharge. However, the methods of achieving emission control will probably change from the reference control technology basis defined in the regulations. As an example, a steam stripper that strips wastewater streams from a number of process units, followed by a biox unit, may be the extent of the overall system. Alternatively, direct discharge of the wastewater from the process unit to a biox unit has been demonstrated, which is an even simpler processing scheme. The equations and estimating procedures developed by the EPA can be broadly applied, but detailed calculations are required to establish the effectiveness of sewer and treatment systems that differ significantly from the EPA reference control technology schemes.

12.4 SEWER SYSTEM CONSIDERATIONS

The previous sections reviewed wastewater emissions and regulations for various equipment items that are controlled under existing regulations and chemical plant wastewater emissions that will be controlled under new federal regulations in the United States. The EPA documentation for the proposed chemical regulations was inaccurate in various areas and did not demonstrate a thorough knowledge of plant operating and safety procedures. In this section, various design considerations are presented to control emissions in new and operating plants in accordance with regulations and modifications anticipated in the forthcoming final HON regulations for the chemical industry. While all of the control recommendations can be included in a new plant design, a considerable number of modifications would be required in an existing plant. Obviously, all of the control recommendations cannot be readily implemented in existing process plants. However, the industrial hygienist and process designer can use some of the recommendations as a basis for developing controls in existing operations.

The controls in this chapter have been directed principally toward the reduction of toxics concentrations in the workplace resulting from drain and wastewater system emissions. In the latter sections of this chapter other potential exposure sources and controls in the drain, sewer, and wastewater systems will be presented.

12.4.1 Sewer Stream Identification and Separation

In all sewer systems, streams are separated and treated in accordance with the stream contents. Stream separation is also recommended in many situations to prevent mixing of streams that may result in undesirable reactions or detrimental changes such as absorption or emulsification. Additionally, it is advantageous to separate and prevent the mixing of a highly toxic waste stream with other less toxic streams, which follows the Group 1 and 2 stream classification for HON wastewater. However, a wastewater stream may also be classified toxic in the future although not shown as toxic in the current or proposed hazard chemical lists, further complicating the sewer and wastewater system.

In the following list, a number of stream classifications are described for an overall processing site. Although the list is rather large, various categories are often combined into several major streams that are treated prior to discharge into the receiving body as shown in Figure 12.21. The generalized flow scheme in Figure 12.21 can differ significantly depending upon the refinery or chemical plant and the contaminants in various streams. The industrial hygienist also inspects the various stream flows for chemical analysis and disposition. Although emissions are a consideration in these streams, potential skin exposures are also a concern. In addition, the treating steps and chemicals added to the streams are carefully reviewed to minimize skin or inhalation exposures. Following the stream classification list, methods of controlling emission exposures in drains and sewers are presented.

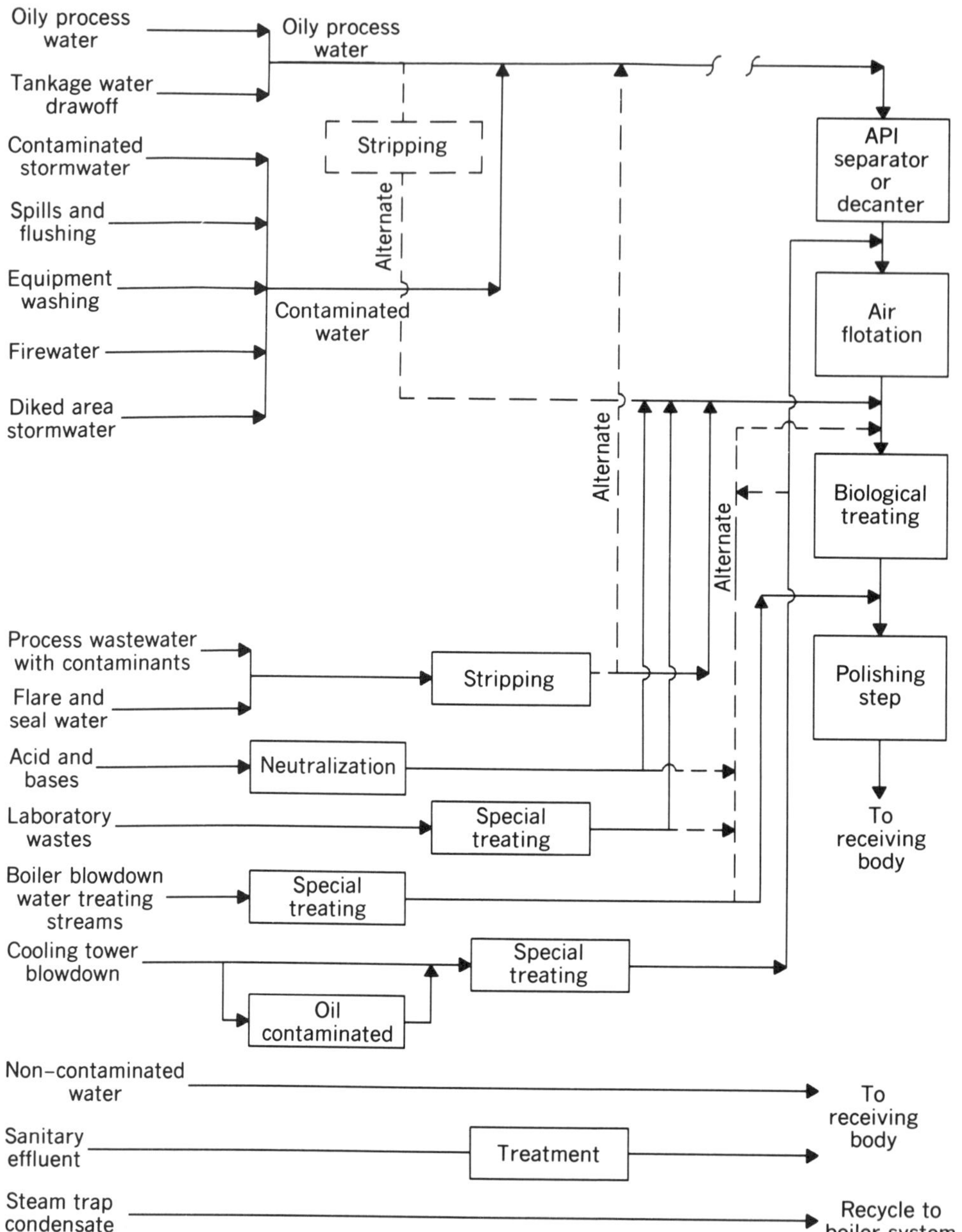

Figure 12.21 *Typical plant sewer streams and general wastewater treating systems.*

Nonprocess Contaminated Toxic Wastewater. Discharges from boilers, steam systems, and water treating units are normally not contaminated with process chemicals and hydrocarbons. These various water sources frequently contain other contaminants and should be treated separately since they normally do not require all of the major wastewater treating steps necessary for petroleum refinery and chemical effluent streams.

Cooling tower blowdown water does not normally contain hydrocarbons or process unit chemicals, which only enter the cooling tower water loop through leaking exchangers. However, exchangers can leak and in large cooling water systems there is a relatively high probability that exchanger leakage will occur, periodically contaminating the cooling tower water system. This is a concern shown in the proposed EPA (HON) regulations where cooling water testing is required at exchanger inlet and outlet points on a quarterly basis. While the location and timing may change, exchanger leakage should be examined periodically. The EPA indicates that cooling water pressures about 6.5 psi (30 kPa) greater than process side pressure does not require leakage tests. However, a further increase in cooling water pressure compared with the process may be required to ensure there is no process leakage into the cooling water stream. A review of potential contaminants is recommended to determine cooling tower blowdown water disposition under normal and contaminated conditions. In normal operation, the cooling tower water may contain several compounds from a large list that includes nonchrome antifouling compounds, free chlorine, corrosion inhibitors, dispersants, biological treating compounds, phosphates, and others. This stream can be combined and treated as necessary with other water streams when the cooling tower water is not contaminated with the process fluids. However, cooling tower blowdown flow rates are large, and combining the blowdown streams with normal process wastewater streams can result in very large individual wastewater treating units. In a cooling tower with a 20,000-gpm circulation rate, the blowdown flow is approximately 250 gpm. During periods when the cooling water is contaminated as a result of equipment leakage, the blowdown stream can be diverted into a holding pond until the leak is repaired. The pond may require an inflated cover to minimize emissions, but any diversion pond is a temporary holding facility and a cover may not be required. Contaminated water from the pond is then fed slowly into a treating facility to dispose of the contaminated fluid. A separate treating facility is shown in Figure 12.21 which removes some of the blowdown contaminants that either affect the wastewater treating unit or the quality of the treated water sent to the receiving body.

The industrial hygienist should be familiar with the chemicals in the boiler, steam system, and cooling tower. Some additives injected into these streams are toxic, and blowdown stream treatment requires a careful review of each treating process and potential workplace exposures associated with treatment and disposal.

Acid and Base Waste Streams. These waste streams are *not* normally discharged directly into major sewer streams. Direct addition can result in undesirable reactions or physical changes in the main sewer stream. As an example, waste acid streams contacting a waste stream containing neutralized hydrogen sulfide compounds (NaHS) will react with the compounds to produce highly toxic hydrogen sulfide gas. Bases often emulsify hydrocarbons or chemicals, an undesirable action that affects the operation of separators and decanters. These two examples are only a few of the innumerable interactions that can occur between a waste stream and an acid or base. Consequently, acid and base waste streams are neutralized before there is a need for additional treatment steps. Neutralization equipment is presented in another section and treating systems should be reviewed by the industrial hygienist.

Process Wastewater with Dissolved Contaminants. In certain processes, oil or immiscible chemicals are separated from process wastewater discharge streams, but the water may contain a relatively large concentration of dissolved contaminants which can be classified as Group 1 or Group 2 wastewaters in chemical plants. The waste stream can be treated separately to either recover or remove the contaminants from the stream prior to wastewater treatment. An alternative can be direct treatment of the stream in a wastewater treating unit such as a biox unit. Flare seal wastewater can be contaminated with compounds that leak through relief valves and valve discharges. Consequently, flare seal wastewater is included in this category.

As an example of recovery in a refinery desulfurization process, hydrogen sulfide and ammonia compounds, along with other compounds, are dissolved in the wastewater and reach relatively high concentrations. The contaminated wastewater stream is sent to a stripper tower where the ammonia and sulfides in the form of hydrogen sulfide are removed from the wastewater. The hydrogen sulfide is sent to a sulfur plant and the ammonia is separated and purified. The purified wastewater then receives final treatment before discharge into a receiving body. In vinyl chloride monomer or PVC operations, vinyl chloride dissolved in water can be stripped from the contaminated water and recycled for further processing. Stripping benzene from wastewater is a similar example. Stripping volatile components from contaminated wastewater significantly reduces potential workplace exposures and meets the EPA requirements for contaminant removal and emission control.

Since stripped wastewater has little contamination, the wastewater effluent from stripping operations must be analyzed to define additional water treating requirements. As shown in Figure 12.21, the stripped water can be sent directly to the biox unit, minimizing treatment requirements.

Oily Wastewater. Oil wastewater normally refers to wastewater separated from oil in the unit process where oil or chemicals have been in direct contact with water. In addition to free and dissolved oil, trace contaminants are often present in the wastewater, contributing to high BOD and COD levels, but are not necessarily toxic components. If toxic compounds are in the stream and meet the Group 1 or Group 2 classification in the HON, the stream may be stripped or sent directly to a biox unit. Tankage water drawoffs are frequently placed in this general classification as an oily wastewater stream. These oily wastewater streams are not generally stripped, but are sent directly to API separators or clarifiers and biological treating units. Oily wastewater streams are a major contributor to the main sewer flow.

Dirty or Contaminated Water. This terminology is generally applied to rainwater and firewater runoff from contaminated areas. Equipment leakage and spills from process areas washed into sewers with storm or washwater are included in this category. Stormwater runoff from diked tankage and equipment areas is also termed contaminated water. In addition to these streams, equipment washwater and purges can be included in this category which are generated during equipment repair or startup and shutdown. Oily and contaminated water streams are generally combined in the sewer system and treated in the main wastewater treating units. Since these streams may have toxic contaminants, they may also require stripping. However, most concentrations are quite low and closed equipment with vent control devices are satisfactory for the control of emissions.

Noncontaminated water. This classification generally covers water runoff from plant areas that are not contaminated with oil or chemicals. Runoff from nonprocess area fields is normally sent directly to receiving water bodies. Wastewater from safety showers can also be included in this category. Cooling tower blowdown and other waters, although not necessarily contaminated with process chemicals, generally require treatment prior to discharge into the receiving water.

Sanitary Effluent. This stream is treated separately, and the treated effluent is not combined with treated process water effluent.

Laboratory Wastes. Treatment of these wastes requires a study of the chemicals involved and general treating requirements. In some plants, special treatment for various compounds in the stream is followed by discharge into the main sewer system.

12.4.2 Process Unit Oily Wastewater Drawoff

In many refinery processing units, steam is injected directly into fractionation towers or may be included in the vapor feed to the tower. The overhead vapor streams contain water vapor and in the overhead cooling system an oil–water mixture is condensed. Some oil is dissolved in the water and, in most overhead accumulator drums, the contaminated water drawoff stream contains free oil which is the primary source of volatile emissions emitted from the various drains, catch basins, and other open points. Free oil carryover should be minimized to reduce emissions and a separation step prior to discharge into the sewer system should be considered. Moreover, a separation drum is enclosed and does not emit vapors. Although the drawoff stream is an oil stream, this general approach is possible in a chemical process unit with an immiscible compound, where the water may be a result of steam, reaction water, or added process water. The following section describes a separation; however, the following treatment step is a function of the chemicals and concentrations dissolved in the water.

Disengaging or Separation Drum. The separation or disengaging drum shown in Figure 12.22 is one method of reducing free oil or immiscible chemical content in wastewater. The oil–water discharge stream from the overhead accumulator drum is sent to a disengaging drum that is specifically designed to separate free oil or immiscible fluids from contaminated water. This separation procedure differs from an overhead accumulator drum design where large quantities of process fluid are readily separated from much smaller water quantities. Although the water is easily separated, the accumulator drum is not designed to minimize free oil carryover in the wastewater.

The disengaging drum is designed to maximize separation of free oil or immiscible chemicals from the wastewater, which may result in a drum designed somewhat differently than the simple disengaging drum shown in Figure 12.22. Although one oil–water stream is generally separated in the drum, a number of

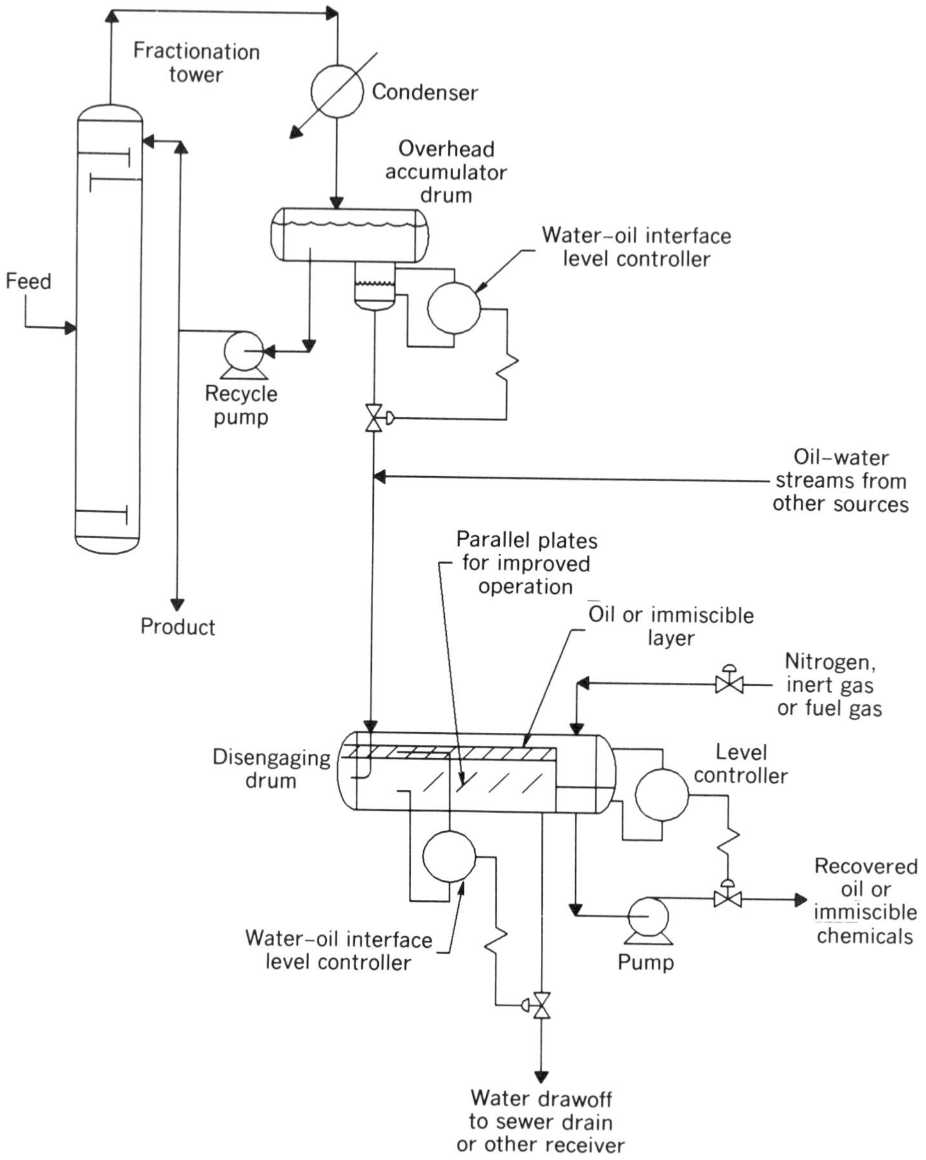

Figure 12.22 Disengaging drum.

compatible streams can be sent to the drum for separation, which minimizes investment. In addition, a parallel or corrugated plate installation in the separation section will further reduce oil content in the water drawoff stream, improving overall separation efficiency. The oil–water separation step in the drum results in the evolution of volatile vapors, but the overall pressure may be less than atmospheric pressure. A nitrogen or inert gas purge will maintain a satisfactory pressure in the drum, preventing the admittance of oxygen. Moreover, a small pressure assists pump operation and gravity discharge of wastewater to a drain or other processing step. Excess drum pressures should be controlled by vapor release to the flare or other control device.

The drain line from the disengaging drum in older plant designs normally discharged into the clarified water drain header that carries the wastewater into the catch basin as shown in Figure 12.1. The outlet water from the catch basin was then discharged through laterals into a manhole. As an alternative, the disengaging drum drain could be connected directly through a lateral to a sewer manhole with a P-leg seal in the drain or a seal at the manhole. This connection bypasses the catch basin in the older designs which eliminates a potential source of emissions from those chemicals or droplets dissolved in the wastewater. While the mass of volatile components has been essentially removed from the oily wastewater due to the efficient oil removal step in the separator drum, vapor pressure can still be exerted by the dissolved constituents. A drain alternative is presented in another subsection.

A disengaging or separator drum used for separation of multiple process oily water streams has an additional function. In addition to processing normal operating oily waste streams, water and oil flush streams generated during shutdown or repair operations can also be discharged into the disengaging drum. Permanent or temporary connections for this purpose should be considered, although permanent lines are preferable for safety and to minimize potential exposures. Shutdown operations and repair present potential overexposure situations and the disengaging drum system is a method of reducing vapor emissions and skin contact.

Underground Drain Collection. In an early background document, the EPA proposed connecting oily wastewater drains to laterals that were in turn connected to an underground collection drum as shown in Figure 12.23. The EPA drain and lateral system omitted the catch basins and as a treating alternate the laterals were

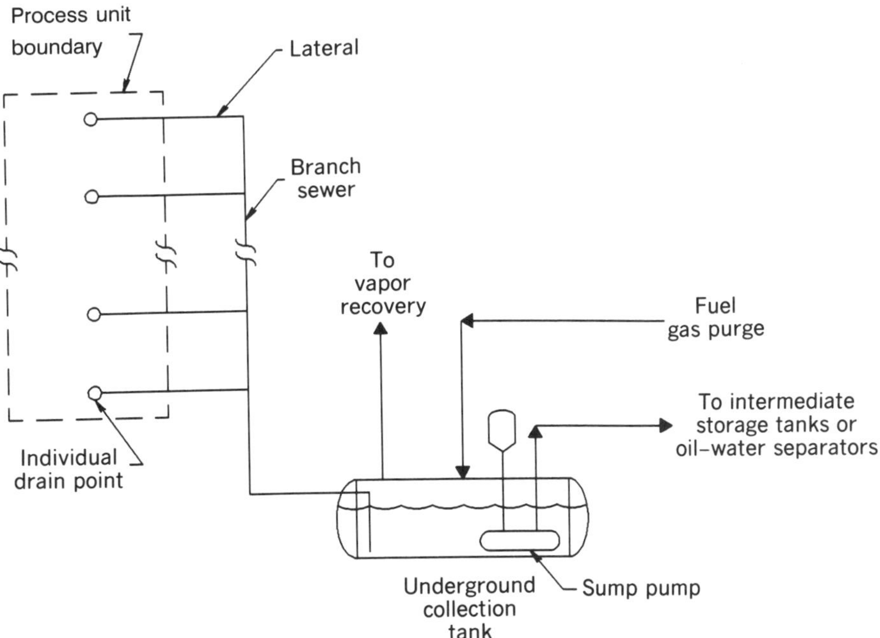

Figure 12.23 *Older EPA recommended underground drain system.*

connected to an underground drum. The drain points receive oily water in this design from all sources, including process drum drainoff lines and pump baseplates. The latter is a source of oily rainwater, but most of the volatile vapors are removed through exposure before the oil is flushed into a drain. The major vapor generators are the oily process water drawoff lines.

An underground drain collection drum can receive oily wastewater streams, but the drum in Figure 12.23 does not separate oil from the wastewater. The mixture is pumped to a tank or separator from the underground drum for oil–water separation. Although not designed for separation, an oily layer will build in this underground drum, and the pump probably discharges large quantities of oil periodically that can interfere with downstream oil separation operations. The drum can be designed for underground oil–water separation, but the underground installation becomes more complex, and the potential exposure risks increase with complexity. This underground installation is expensive which is further complicated by underground storage tank regulations, but underground drums in certain situations have a distinct advantage and are described in another section.

12.4.3 Dissolved Contaminants in Wastewater

Separation of immiscible oils or chemicals from wastewater in disengaging drums was reviewed in Section 12.4.2, but various volatile compounds *dissolved* in process wastewater may be toxic, as described in the section. An example of a highly toxic chemical in the petroleum industry dissolved in process water is hydrogen sulfide. Although actual concentrations in the wastewater are well below 1%, the vapor concentrations are generally much higher than the recommended exposure levels. The dissolved material expands as the hydrogen sulfide vaporizes, with the result that the vapor concentration above the liquid level is greater than the concentration in the liquid by a factor of 100 or more. In addition to hydrogen sulfide, the wastewater may also contain ammonia, which further complicates the toxic potential of the wastewater vapor. In chemical streams, benzene or any of the toxic materials in Table 12.1 or Appendix 12C require emission control.

Process Water. Wastewater saturated with hydrogen sulfide presents a vapor emission problem when discharged into the older sewer systems since the vapor is emitted from the drain, catch basin, manhole vents, and downstream wastewater treating units. The petroleum industry solves this problem by sending the hydrogen sulfide wastewater stream, termed "sour water," to a stripper where the hydrogen sulfide and ammonia are removed from the sour water. The hydrogen sulfide vapor containing ammonia can be treated in another step to remove ammonia or the total stream can be sent directly to a sulfur recovery unit. Sulfur recovery units are generally found in most refineries and unit capacity is generally available to process the vapor stream from a stripper.

A typical stripper shown in Figure 12.24 demonstrates the processing equipment necessary for stripping hydrogen sulfide from wastewater. Sour water or hydrogen sulfide laden wastewater is fed to a stripping tower where the hydrogen sulfide and ammonia are stripped or removed from the water. The hydrogen sulfide and ammonia are discharged from the overhead accumulator drum as a vapor and sent to a sulfur recovery unit. If the ammonia concentration is large, the ammonia

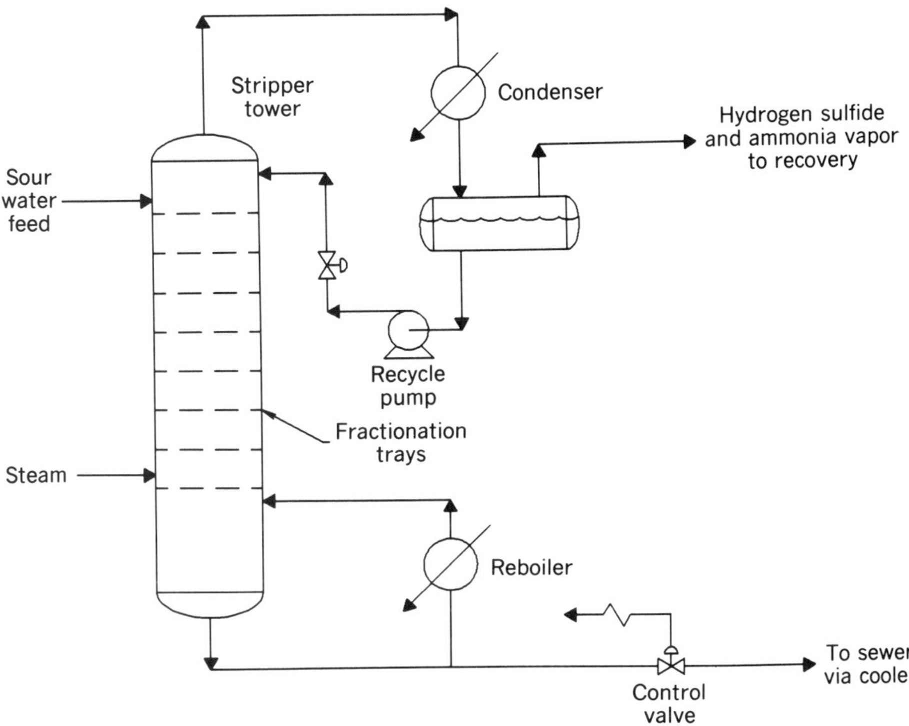

Figure 12.24 *Simplified sour water stripping unit.*

can be separated from the hydrogen sulfide in a second fractionation step. For improving stripping action, the pH of the wastewater feed streams can be adjusted. The stripping tower is basically a fractionation tower where steam is injected directly into the tower and a reboiler supplies fractionation heat. The hydrogen sulfide-free water product is withdrawn from the tower as a bottoms stream and is discharged into the drain system for further treating in the wastewater system. Since the bottoms stream is relatively warm (greater than 200° F), the stream should be cooled prior to injection into the drain or wastewater treating systems.

The contamination remaining in the bottoms stream after hydrogen sulfide removal determines the wastewater treating steps necessary to purify the water prior to discharge into the receiving body. Stripping tower calculations will show the contaminant concentration in the bottoms stream which assists in defining wastewater treating requirements. The industrial hygienist reviews the bottoms concentration levels to determine stream toxicity and potential exposure concerns associated with the effluent tower stream.

This stripping technology is the toxic contaminant removal technology that is the basis for emission control in the proposed EPA (HON) regulations. Strippers remove contaminants with varying efficiencies based upon vapor pressure and Henry's Law constants. However, the vapor pressures and associated strippability factors vary considerably for the toxic contaminants shown in the Group 1 lists. The EPA stripping calculations for naphthalene indicate that the stripping efficiency is 99.9%, which is considerably better than the 30.1% stripping efficiency calculated for 1-butanol. These stripping calculations assumed the stripper was designed in accordance with the EPA design criteria (Section 12.2). Since stripper removal efficiencies vary widely, the removal efficiency can be improved by changing conditions in the stripper tower in accordance with detailed fractionation calculations. The steam rate can be increased along with the number of trays and the total tower pressure can be changed. Vacuum operations will not always improve operation, but fractionation calculations will indicate the optimum process conditions for maximum removal. At times, two strippers may be required, operating at different conditions for maximum removal.

Where the stripper cannot remove 99% of the chemical contaminants, other possible methods of removing toxic contaminants are investigated. If the biox unit can completely degrade the stream, the stream can be sent directly to the biox unit. An alternative may be solvent extraction of the contaminated wastewater stream. Although adding a treating step can frequently be expensive, selective extraction can eliminate the troublesome contaminant. The stream can then be stripped or sent to a treating unit. A more sophisticated extractive distillation fractionation tower can be installed which combines both extraction and distillation. Detailed calculation procedures are necessary to determine the separation efficiency in this dual operation.

While the EPA regulations may suggest stripping, ineffective stripping operations do not meet wastewater regulatory requirements, and other solutions are investigated. The industrial hygienist should be aware of these potential changes which can create other concerns. Where stripping meets very high removal efficiencies (> 99%), the need for additional processing must be reviewed. In addition, the need for vent controls on various other potential emission source points downstream of the stripper becomes questionable. Although investment and operating charges are relatively large for stripping, removal of dissolved toxics from wastewater in a stripping tower is a positive toxics control system. This positive control step must be compared with closed drains, manhole vent controls, and wastewater treating system vent controls to achieve comparable reductions in workplace concentration levels. On this basis, a stripper is not an unacceptable economic alternative.

Flare Seal Water. In processing units, vents and pressure relief valves are frequently connected to closed headers which discharge vented vapors and emergency relief vapor to flares or other control devices. The various regulations for fugitive emission control (VOC) and the proposed HON regulations indicate that this closed system is desirable since the regulations do not require monitoring of pressure relief valves connected through closed systems to control devices. Many existing petroleum and chemical units in toxic service have pressure relief valves connected through closed headers to flares or other control devices. These closed systems, while installed in new units, are being retrofitted in most existing units to meet VOC, state, and the more stringent HON regulations.

A typical flare seal system is shown in Figure 12.25 to demonstrate potential exposure problems during normal operation. Pressure relief valves and vents are connected to the flare header, which is sloped toward the blowdown drum. Condensed vapors or droplets in the flare header line that collect in the piping will drain into the blowdown drum through the sloped line. These droplets are

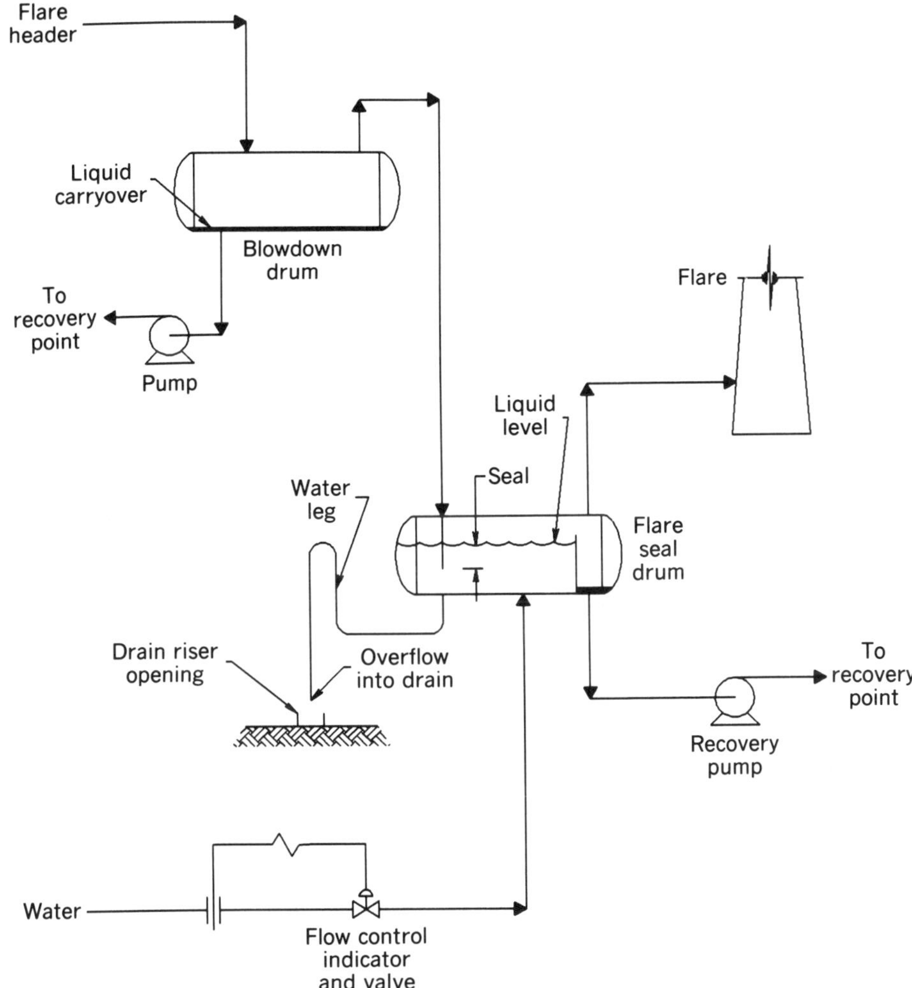

Figure 12.25 *Typical flare seal drum system.*

frequently formed from the condensation of vapors that leak through pressure relief valves and vent line valves. Vents connected to the flare line may have a continuous flow into the flare header or may be used on a periodic basis. Liquid carryover collects in the blowdown drums and is periodically pumped to recovery or a collection point in the plant for reprocessing.

Noncondensing vapors and pressure relief valve discharge vapors pass through the blowdown drum into the flare seal drum through a water seal. The seal prevents the backflow of air into the flare header network. Additional precautions to prevent explosive mixture formation in the flare seal drum are also included in many installations. A seal is maintained in the drum by the constant flow of water through the inlet control valve which discharges to the sewer drain system through the open overflow leg. A U-bend water seal maintains a liquid level in the drum and the excess water is discharged continually to sewer. Since vapors entering the flare seal drum may partially condense, the condensed material that is lighter than water collects on the water surface and overflows a weir into the recovery section of the flare seal drum. The recovered fluid is periodically pumped to the recovery or collection system.

The intimate mixing of vapor and water in the flare seal drum may result in the absorption of various contaminants into the water. In petroleum refineries, for example, hydrogen sulfide will dissolve in the water, resulting in a sour water. This presents a potential overexposure situation around the overflow leg where contaminated water enters an open drain. The open drain is normally provided as a safety precaution in various facilities to visually insure that there is a continual flow of water through the drum and a water seal exists in the drum. Although the open drain is a safety requirement, the open system potentially presents a serious exposure situation. The overflow leg should be connected through a closed connection with the sewer. Flow through the overflow leg can be readily determined through a sight glass in the overflow leg, a simple flow indicator instrument, or a pitot tube that is connected to a partially evacuated Pyrex container. Other alternative flow indicator solutions are devised by plant operations groups when requirements for a closed drain leg are understood.

In addition to potential exposure at the overflow leg, the overflow water entering the sewer system increases toxic vapor releases through the sewer and wastewater treating systems. These sources of potential overexposure can be minimized or eliminated by recycling the sour or contaminated water to a stripping tower. Stripping this water requires the installation of a pump to recycle the contaminated water from the flare seal drum location to the stripping tower.

A number of flares with their accompanying drums are normally installed in petroleum refineries and chemical plants to separate incompatible release mixtures. Incompatibility may result from reactions, header material corrosion problems, or other design considerations. As a result, seal water contamination for a flare or control device system must be carefully analyzed. A closed overflow leg should be installed at each flare seal drum, but the disposition of the water is dependent upon the toxicity of the chemicals in the seal water stream.

Although the wastewater is the major concern in this section, another area is a potential source of exposure. In some installations vapor is recovered from the flare seal drum and recycled to the plant with a compressor. This recovery system is designed for the small leakage flow rates that enter the header system. Since toxic vapors can enter the compressor system, leakage from the compressor, valves, or other sources must be controlled. In addition, compressor repairs are a potential exposure source.

12.4.4 Tankage Water Drawoff and Overflow

In the petroleum refining industry, a wide range of crude, intermediate, and product tanks accumulate water in the bottom portion of the tankage since the water generally has a higher specific gravity than the hydrocarbons. The water can be contaminated with various compounds and solids, extracted, dissolved, or mixed with the water in the form of droplets, globules, or bubbles as the water under the influence of gravity gradually disperses through the hydrocarbon layer to the water–hydrocarbon interface. The interface between the two liquids is either relatively sharp or it can be an emulsion. The latter is a result of the type of hydrocarbons, additives, or both in the tank. While chemical facilities also have water separation in tankage, this characteristic appears more widespread in the petroleum industry. However, the water can be contaminated in either industry, requiring emission controls and treatment.

Water Drawoff. Water buildup in tankage is a concern since the water level can reach the pump suction line and contaminate either the hydrocarbon or chemical pumped to process units for further processing or a product. Water in the tank is decanted frequently into a drain leading to the oily water sewer to prevent this problem. Some free oil exists in tankage water, and any emulsion phase carryover from the tank may enter the drain during drawoff, significantly increasing oil flow to the sewer. Since emulsion control is important in minimizing hydrocarbon loss to the sewer, manual decantation control is often selected for this operation. As a result, the tankage water drain line frequently empties into a large open funnel or drain that can be connected to a flow-through catch basin as shown in Figure 12.3, or directly to a sewer lateral and main sewer line. The latter connection requires a seal in the lateral. This procedure is a potential source of exposure at the tankage drain point where vapors emitted to the atmosphere may include hydrogen sulfide, benzene, and aromatics. The vapor load is also increased through the wastewater system.

Oil carryover into the sewer drain can be minimized in most tank drawoff drain lines by automatic shutoff drain valves. The valves close when the specific gravity of the fluid flowing through the valve changes from the water specific gravity (1.0) to a lower level. The automatic valve drain is connected directly to the sewer drain and workplace emissions are eliminated at the drain inlet. The differential change in specific gravity that closes the valve must be carefully fixed to minimize oil carryover into the sewer. Large oil carryovers increase emissions through the sewer and wastewater treating systems and increase potential losses.

Control of tank water levels through instrument level devices is another method of controlling oil loss to the sewer. Radiation detection, sonar, capacitance devices, and other instruments can be used to indicate interface levels with manual or remote shutoff. The drain connection can be closed and emissions through the sewer and wastewater systems minimized by preventing oil carryover into the sewer drain.

Tankage Overflow. In the petroleum industry, instruments or level devices automatically stop the flow of liquid into a tank when a specific level is reached or when an alarm indicates a high level requiring manual or remote shutoff. Generally, backup instrument controls are installed as a safety precaution in the event an instrument fails.

In the chemical industry, emergency overflow lines have been installed on many tanks to prevent tank overfilling and overflow. This system presents potential exposure problems in the event an overfilling situation occurs. The overflow line has generally discharged excess liquid a short distance above a sewer drain opening and the overflow condition can be observed by a process operator. However, vapor is generated as a result of liquid free fall and potential overexposures can occur. A closed drain with a flow-indicating device in the overflow line is preferred to minimize vapor generation. An overflow occurrence initiates an alarm and flow to the tank is blocked. However, the tank has overflowed in this situation and tankage fluid enters the sewer. The large concentrated quantity of process fluid in the sewer results in potential vapor emissions to the atmosphere from the sewer and wastewater treating systems. Moreover, the sudden increased load on the wastewater system can interfere with operations and potentially increase workplace exposures.

The discharge of chemicals to the sewer should be prevented and alternate solutions are possible where the chemical continues to overflow the tank as a safety procedure. However, these alternate solutions may be costly and have potential drawbacks, such as a sump with a pump for movement of the chemical overflow to a recovery point. Pump failures can result in serious consequences during an overflow situation. Consequently, in filling tankage, level control and indicating instruments are recommended to minimize sewer system upsets and overloads in the wastewater treating units. Redundant level instruments permit the elimination of overflow lines and their attendant problems.

12.4.5 Catch Basins and Drains

The catch basins and drains have generally been interconnected as shown in Figure 12.1. Alternative methods of transferring contaminated wastewater and controlling emissions from drains are presented in this section.

Catch Basins. Emission rates and factors from catch basins have not been determined by the EPA, since basins are recognized as emission sources. Monitoring vapor concentrations below and above basin grid covers show hydrocarbon or chemical vapor concentrations that are relatively high when the wastewater in the basin from drain connections contains relatively volatile components. The emissions from catch basins will affect workers when these emissions are irritating and they were frequently covered to minimize potential exposures. Temporary covers are an obvious indication of concern regarding the emission of vapors from catch basins.

The design and purpose of catch basins prohibits the permanent installation of vapor emission controls directly on the basins. As a result, the proposed EPA (HON) regulations require other measures to control the emission of volatile vapors from the basins which are based upon the elimination of catch basins. Alternatives are presented in this section, but reasons for the installation of catch basins are outlined in the following list to insure that original catch basin functions are understood, which provide a basis for emission control modifications.

1. *Spills.* Spill material in the process area is drained and flushed directly into the catch basin.
2. *Rainwater.* Rainwater becomes slightly contaminated in the process areas and is drained from the area through the catch basin.
3. *Firewater.* All firewater that can become contaminated is also drained from the process areas through the catch basins.

4. *Drainage.* Oily drainage was processed through the catch basin where seals to prevent vapor backflow were provided. The catch basin was a central point for collecting drainage, and the collected drainage was sent to the main sewer lines from the catch basins, as shown in Figure 12.1.

Since catch basin design and function precludes vapor emission controls, a change in drain connections to the catch basins has a major impact on vapor emissions. Items 1–3 are infrequent occurrences and rainwater flow through the basin has a minor effect upon emissions. Emissions from the older basin installations are generally a result of a more or less continual flow of oily or contaminated water into the catch basin. Volatile components continually vaporize in this situation or the catch basin may receive periodic discharges from an auxiliary operation that has a highly volatile component. Periodic episodes where cyclic frequency is less than 24 hr can result in relatively high emission rates. A combination of periodic and almost constant drain flows are the major sources of vapor emissions from the catch basins. Removal of these streams from the catch basin significantly reduce catch basin emissions.

The drainage streams connected to the catch basins should be carefully analyzed prior to changes in existing units or in the preliminary sewer design stage in new units. Oily drawoffs and contaminated wastewater should be controlled prior to discharge into the sewer system. The remaining drainage streams are then analyzed for VOC or VHAP components or identified as streams that may receive a surge of volatile materials during an upset. Drainage from equipment such as exchangers, towers, and lines should not be included in these categories. These equipment drainage requirements are reviewed separately.

Drain Rerouting. The drainage streams with volatile components can be separated from the catch basins shown in Figure 12.1. The drain header (lateral) connections would then discharge directly into the sewer and generally resemble the EPA sewer design.[1] However, a major difference in the system designs is the presence of catch basins necessary to meet spill, rainwater, and firewater requirements. A revised drain and sewer system for this situation, shown in Figure 12.26, is similar to the original drain and sewer system in Figure 12.1 and is presented for comparison. The drain headers are disconnected from the catch basins and the headers, now termed laterals, are extended and connected to manholes or junction boxes. The drain laterals do not have to be sealed at the manholes since the risers are sealed. However, a seal at the manhole blocks sewer vapor from entering the lateral or riser in the event of repair or if a riser seal runs dry. The catch basins remain connected directly to the manholes and are limited in function to rainwater runoff, spills with flushwater, and firewater.

The removal of drainage streams with volatile components from catch basins significantly reduces and essentially eliminates catch basin emissions. However, emissions may still evolve from manhole vents but are dispersed into the atmosphere from the vents. Dispersion from manhole vents produce a much lower concentration level in the process area than dispersion from catch basins, since the emission quantity is changed considerably by venting through the manhole vent lines. Leakage from a sealed, vented manhole or junction box is considerably lower than the emissions from an open manhole or open catch basin, as described in

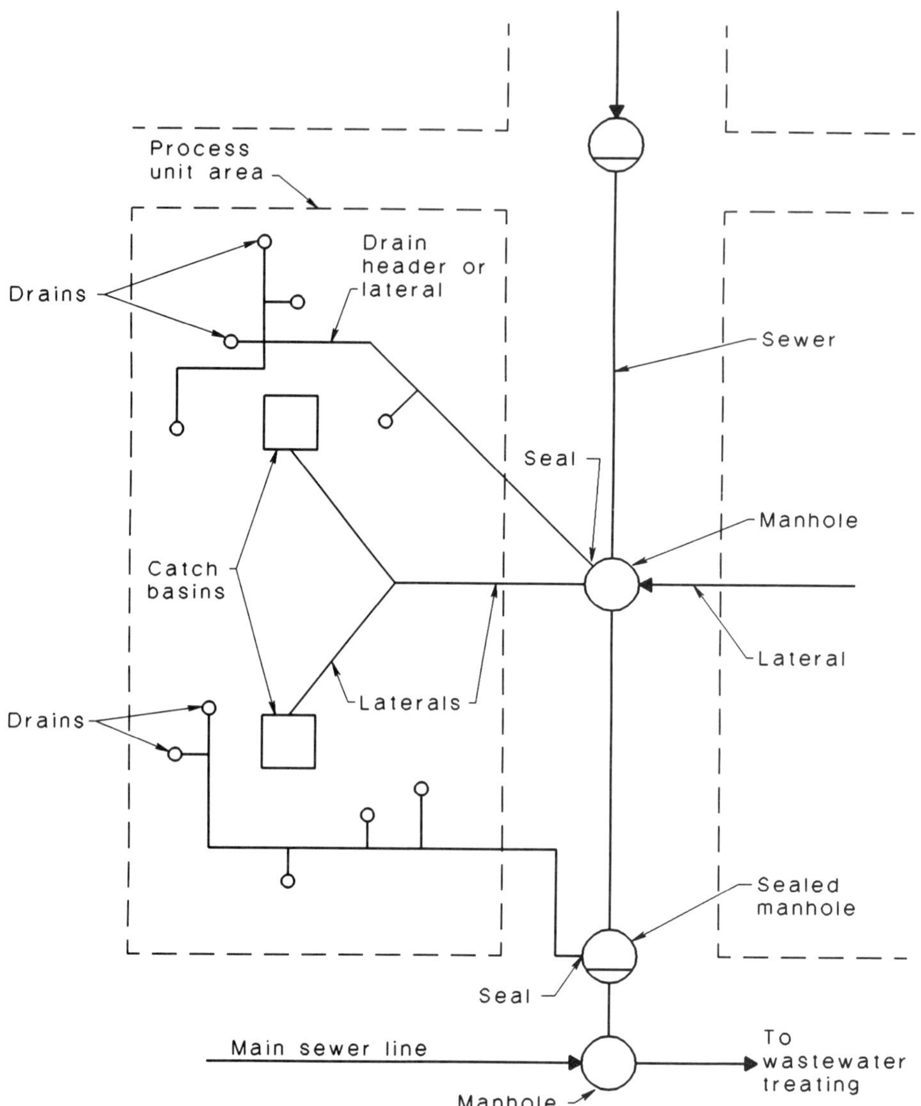

Figure 12.26 *Separate drain and catch basin network.*

Section 12.3. Moreover, fugitive emissions are positively reduced, which is required in the proposed SOCMI regulations and will probably be required in petroleum refineries. Workplace exposure levels are reduced significantly by removing volatile emissions from catch basins.

Drainage Treatment. An alternative consideration for drain emission control is the collection and treatment of oily drainage water through a parallel or corrugated plate separator located in the process area. In this system, each process area has a plate separator and each process unit operation remains independent or isolated from other operations. The method of connecting the drains to the corrugated plate separator would not differ greatly from the layout shown in Figure 12.26. The drains are disconnected from the catch basins, and the laterals are connected directly to a parallel or corrugated plate separator similar to the separator shown in Figure 12.4. The effluent water from the separator is then connected through a seal to a manhole. Alternatively, the effluent can be sent to an onsite or central process stripper.

The plate separator is covered with a flat plate, and in some cases can be vented directly to the atmosphere. A reduction in vent emissions of about 75% can be achieved with a single vent in a cover, as shown in EPA considerations of vent emissions from a covered dissolved air flotation unit. This reduction in the overall emissions rate from an elevated vent is sufficient to reduce the area concentration level and minimize potential workplace exposures.

If the reduction in the process area concentration level is unsatisfactory, or greater emission control is required to meet the proposed regulations, the vent emissions can be captured in an activated carbon adsorber. The carbon adsorber system can vary from a simple batch system to a regenerable package purchased as a small complete unit from a vendor. Although considered an additional investment, the recovered emissions may be of sufficient volume to justify a recoverable activated unit. Alternative absorption systems are potential control methods and the vapors can also be injected into process furnaces with the aid of an ejector or blower. Process furnaces are considered control devices by the EPA. Another possibility for vapor recovery is the use of a small steam jet that captures all of the vapors from the vent. Since the vapor emission rate is small, a very small steam rate to the jet is necessary. The steam–vapor stream can be injected into the overhead line from the tower in Figure 12.22, the accumulator or disengaging drums. A small separate exchanger can also be installed for the steam–vapor stream with the condensate discharged into the accumulator or disengaging drums. These alternatives indicate that a wide variety of control systems can eliminate the volatile emissions.

Drain Sources. Drains directly from the process and various equipment items are generally the main source of drainage in a process unit. As previously described, drainage is collected in drain headers and either discharged into catch basins or preferably junction boxes. Emission vapor control from drainage streams is necessary, and the drain header streams should be diverted from catch basins and discharged into alternative receivers. Although this procedure is satisfactory, the industrial hygienist should be familiar with various sources of drainage, potential exposure problems, and a general understanding of disposition alternatives that minimize potential exposure problems.

A general list of drain sources is shown below to provide background information on the type of mixed oil (chemical)–water and oil (chemical)–drainage streams that enter the drain and sewer system. The list is not complete but provides a basis for evaluating drain sources and recommending disposition control procedures.

1. *Direct Process Drains.* Drainage from many process drums is essentially a constant flow that discharges directly from the process into a drain. An example of these process drains was presented in Section 12.4.2 where a constant flow of oily

water is normally discharged into the sewer from an overhead accumulator drum. The accumulator drums are generally small, and a carryover of oil with volatile components accompanies the wastewater into the drains.

2. *Dissolved Contaminants in Process Water.* The contaminants are frequently volatile or toxic, and generally issue in a constant stream for discharge into a closed drain line. The drainage stream is sent to a stripping tower or another treating step. An example of this stream in a petroleum refinery is a sour water stream containing hydrogen sulfide described in Section 12.4.3. The sour water stream is sent directly to a stripper before it is discharged into a drain system. Various streams can fit in this category, and all of the drainage streams must be analyzed and classified to ensure the proper treating step is applied.

3. *Periodic Process Drains.* In certain processing steps, a contaminated process stream is discharged on a periodic basis. The stream may be discharged from a treating step or a regeneration step. As an example, a process stream is treated in a fixed bed unit with hydrogen or another compound. The process flow is blocked, and a bed is regenerated, producing a contaminated wastewater. In another application, process drying steps frequently produce contaminated water streams. The steams can accumulate in a drum for processing at a constant rate through a stripper or other treatment step before discharge into a sewer drain.

4. *Equipment Flush Drains.* Water flushing of seal faces, packings, and cooling jackets often results in the contamination of the water with oil or chemicals. The contaminated water is frequently discharged directly into drains. Some of these oils or chemicals may have volatile components, but the streams are normally at very low flow rates.

5. *Open Leak Drains.* The most prominent example is a pump baseplate, which contains oil and chemical drippings from seals, bearings, and general leakage when the pump is opened for maintenance. In some locations the pump casing may be opened through a drain valve to drain the casing prior to maintenance on the pump. The baseplate is connected through an opening at the low end of the baseplate to a sewer drain. This type of drainage is normally a result of leakage from the equipment during operation. Filters and other equipment leak bearing oil or process fluid, and the leakage is collected in pans. Both pans and baseplates are flushed into the sewer with water or are inadvertently flushed into the sewer with rainwater.

6. *Process Equipment Drains.* A source of potential exposure occurs during the drainage of operating process equipment when the equipment is removed from service for maintenance or repair, or when a unit is shutdown. Some of the equipment in older processing units may be drained directly on the concrete pad at grade and then flushed with water into a catch basin. Drains are often provided that permit draining the equipment directly into a drain header which discharges into a catch basin or directly into a sewer. Closed drain systems are also provided for loss control. Although the flow from equipment being removed from service is infrequent, a mass of process fluid is drained and emissions levels for short periods can exceed recommended concentration levels, overexposing working personnel if the process fluid is discharged into drains that are connected to catch basins. Moreover, the concentrated process fluid flow also affects emissions from other vents and downstream treating equipment.

The proposed EPA (HON) wastewater emission control regulations require emission control of all streams discharged from equipment during repair or startup and shutdown operations. Since a considerable amount of process chemicals or hydrocarbons can be purged from a piece of equipment for repair or during a shutdown, discharging these streams to a wastewater storage tank provides a method of recovering the process material and acting as a surge drum to feed the wastewater system. Headers through process units provide a method of encapsulating and controlling these streams. An example of a header system is presented in Section 12.5. The tankage vapors must be controlled as described in Chapter 13, which is required in the current VOC regulations and proposed EPA (HON) regulations along with various state or regional tank control regulations.

This summary categorizes most drains, but a number of drains will not fall into the six general categories outlined above. However, the type of drain and disposition of drainage material can probably be determined from the outline presented or the recommendations in the following sections.

Drain Emission Control. Emission controls for direct process drains and dissolved contaminants in wastewater were presented above. Periodic process drainage emissions can also be controlled. In this section, emission and skin exposure control methods for items 4–6 are reviewed.

1. *Equipment Flush Drains.* This category covers water flushing of mechanical seals, packings, and cooling jackets where the contaminated water mixture is discharged into a drain. This category does not cover all of the small water drains, but the comments on controls are applicable in other services.

The contaminated water is generally discharged into an open drain that is connected through the underground drain header to a catch basin. The water in these services generally accumulates small quantities of process fluid and should not present emission problems. However, the water is frequently heated through contact with hot seal faces, packings, and so forth and emissions can be quite large for relatively small streams. These streams should be connected through a closed drain to a small water cooler where the contaminated water stream temperatures are reduced to less than 90° F. The cooled effluent is then discharged into a drain with a P-leg or U-bend seal, where the cooled stream will have a low emissions rate and increase the effectiveness of the drain seal.

For a further reduction in area concentration levels, the drain header should not be connected to the catch basin, but diverted to another location. Discharge into a plate separator or manhole is preferable for improved emission control. Since the stream flow rates are low, the contaminant quantity is generally small, but a closed drain to the cooler and a closed connection to the drain riser will minimize skin contact.

2. *Open Leak Drains.* Drainage into baseplates or pans cannot be easily controlled. However, if the leakage originates from a small opening, a closed line should be installed that carries the leakage to a drain point. If this closed line is installed, operations and maintenance should periodically inspect the equipment to ensure the closed line has not been removed. Frequently, small lines are removed by maintenance personnal who fail to reinstall the drain.

Although little can be done to control a very small process leak from a filter or other equipment, the leakage should not be permitted to accumulate on the pan or baseplate. The fluid should be flushed into the drain on a frequent, scheduled basis. The drain would have a seal for this material and would not be connected to a catch basin.

3. *Process Equipment Drains.* Processing equipment must be drained for maintenance or a general unit shutdown. Draining equipment for maintenance generally occurs when a piece of equipment fails or leaks and is temporarily placed in a standby condition. In the latter situation, the process fluid remains in the unit, and the malfunctioning equipment is removed for repair. Process equipment drainage for maintenance and shutdown is very important and must be conducted without overexposing personnel to emissions and risking skin contact.

A major consideration in draining and preparing equipment for repair or entry is the process procedure normally undertaken in a process unit to drain, flush, decant, and free the unit or equipment of gas. The preparation procedures for opening equipment varies based upon the process operation, the oil or chemical in the equipment, and company safety procedures. As a result, standardized procedures may only exist within a processing facility, and these procedures can vary according to the process. A major industrial hygiene concern is the method and procedure normally practiced within a unit to drain and purge the equipment of all process fluids prior to entry or maintenance. The industrial hygienist should have detailed information on these procedures and any chemicals or streams used in purging and cleaning the unit.

In considering engineering controls when draining equipment followed by flushing or other procedures, the type of chemical or material in the equipment or unit has an influence upon the types of process controls. If an expensive chemical is an important material in the process, recovery of this chemical is important and process operations are concerned that material losses are minimized. In this situation, operations strive to minimize material losses to the sewer and wastewater treatment systems. For example, a process gas in a petroleum refinery containing hydrogen sulfide is sent to a gas-treating unit where hydrogen sulfide is removed from the process gas by absorption in monoethanolamine (MEA), which is a specialized solvent widely used for hydrogen sulfide removal. Since the solvent is costly, considerable effort is employed to minimize MEA leakage and drainage losses by recovering the solvent in a closed system.

In contrast, a processing unit that purifies a stream through a series of distillation towers or a fixed-bed catalytic process may not have particularly costly chemicals that must be recovered in a special recovery system. As a result, the process material is drained and flushed directly to an open drain into the sewer system, which could result in potential overexposures. Under current and proposed regulatory requirements, the process material cost is immaterial, and the process fluid is normally recovered in this instance by pumping or draining the contents of the unit to a recovery tank. During removal of a single equipment item from service, the process fluid in this process item can be recovered through a similar system with water flush material sent to a different separation/storage unit or to the recovery tank if the water and chemical flush material are acceptable in the tank.

Another method of removing lighter materials from process equipment is through a vacuum system. Where the vapor pressure is high, venting equipment removed from service to a recovery system reduces the pressure in the process equipment to the pressure of the recovery device. The pressure in the process equipment at this point may be a few pounds per square inch (or kPa) above the ambient pressure. A vacuum is placed on the process equipment with an eductor or vacuum pump and the remaining vapors and liquid traces are removed. The eductor or vacuum pump discharge is sent to a recovery device such as a scrubber, absorber, or carbon adsorption unit. This general approach has been used through the years in many chemical plants, but requires installation of process equipment that is mechanically rated for full vacuum. Following evacuation, the system can be flushed or purged as required, observing a planned procedure to break the vacuum.

The following comparison is presented of various drain systems that improve process fluid recovery, reduce potential workplace exposures to emissions, and minimize skin contact.

Chemical Recovery. This example considers a chemical such as MEA solvent within a unit that is recovered to reduce losses, prevent discharge into the wastewater system, and minimize skin contact. Rich MEA solution containing hydrogen sulfide can also release hydrogen sulfide if the solvent is spilled and exposed to the atmosphere for a period of time. The potential for inhalation and skin exposure exists and solvent recovery systems are installed for economic and health reasons.

The solvent recovery system installed for a typical MEA unit is shown in Figure 12.27. However, in the recovery system design illustrated, solvent drainage is sent directly to the recovery drum, and the drain in the sewer system is not used. Consequently, hydrogen sulfide vapors are not emitted from drain seals, catch basins, or manhole vents in the sewer system. Also, the MEA solution is completely contained, preventing skin contact. In the enclosed drain system outlined in Figure 12.27, the various equipment items in the bottoms circuit are connected directly to a specific drain header and solvent recovery tank. Other equipment items in the process containing solvent are also connected to the solvent recovery tank through enclosed drain lines, but are not shown in this simplified example.

From an industrial hygiene standpoint, the solvent recovery design is an ideal method of controlling exposures. A sump tank system is not connected to the sewer, and wastewater treating system loads are reduced with this control system. However, the underground system is costly, must meet underground storage tank regulations, and normally installation is limited to expensive solvents and processes with special fluids. In other processing situations, limiting or reducing the wastewater load may provide the economic incentive necessary for installation of this control system. In existing units with high wastewater loadings, a reduction in loading can improve treating operations or permit the expansion of another processing facility. All of these factors should be considered in establishing a basis for recommending the underground solvent recovery tank control systems.

Drain Headers. In many outdoor processing facilities, the processing equipment is elevated, and pipe racks are located within the unit or adjacent to the process unit.

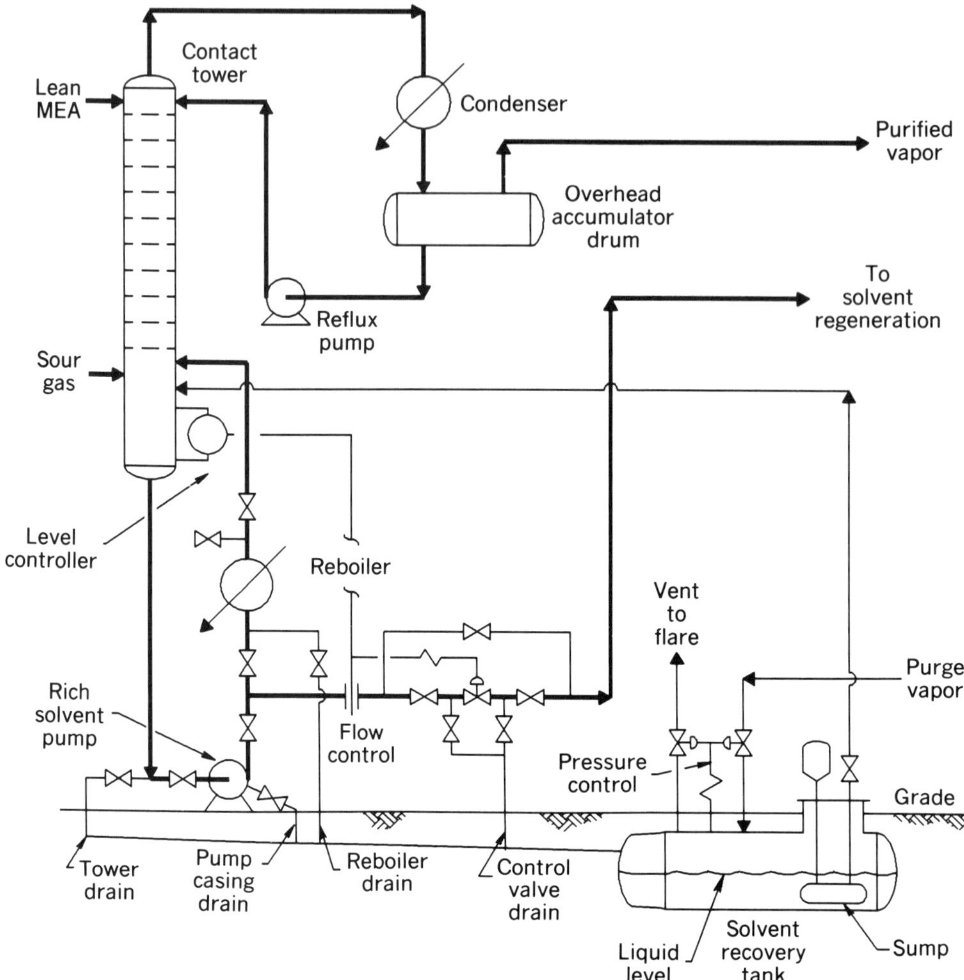

Figure 12.27 *Underground chemical drain and recovery system.*

These racks permit the installation of a drain header that can run through the unit receiving drainage from various equipment locations. The connections to the drain header are either permanently piped or can be temporarily connected with a hose. The toxicity of the fluid in the equipment determines the type of connection to the header. *A permanent connection is preferred*, but the installation is more costly than a temporary hose drain and potential exposure considerations are also important in the selection of the drainage connection.

A typical drain header system for individual equipment items is shown in Figure 12.28. The drain header is on a common pipe rack, and permanent drain connections are shown for the overhead condensers and the overhead accumulator drain. Hose connections for the control valve are shown in the detail, but should also be permanent connections. When removing equipment from process service

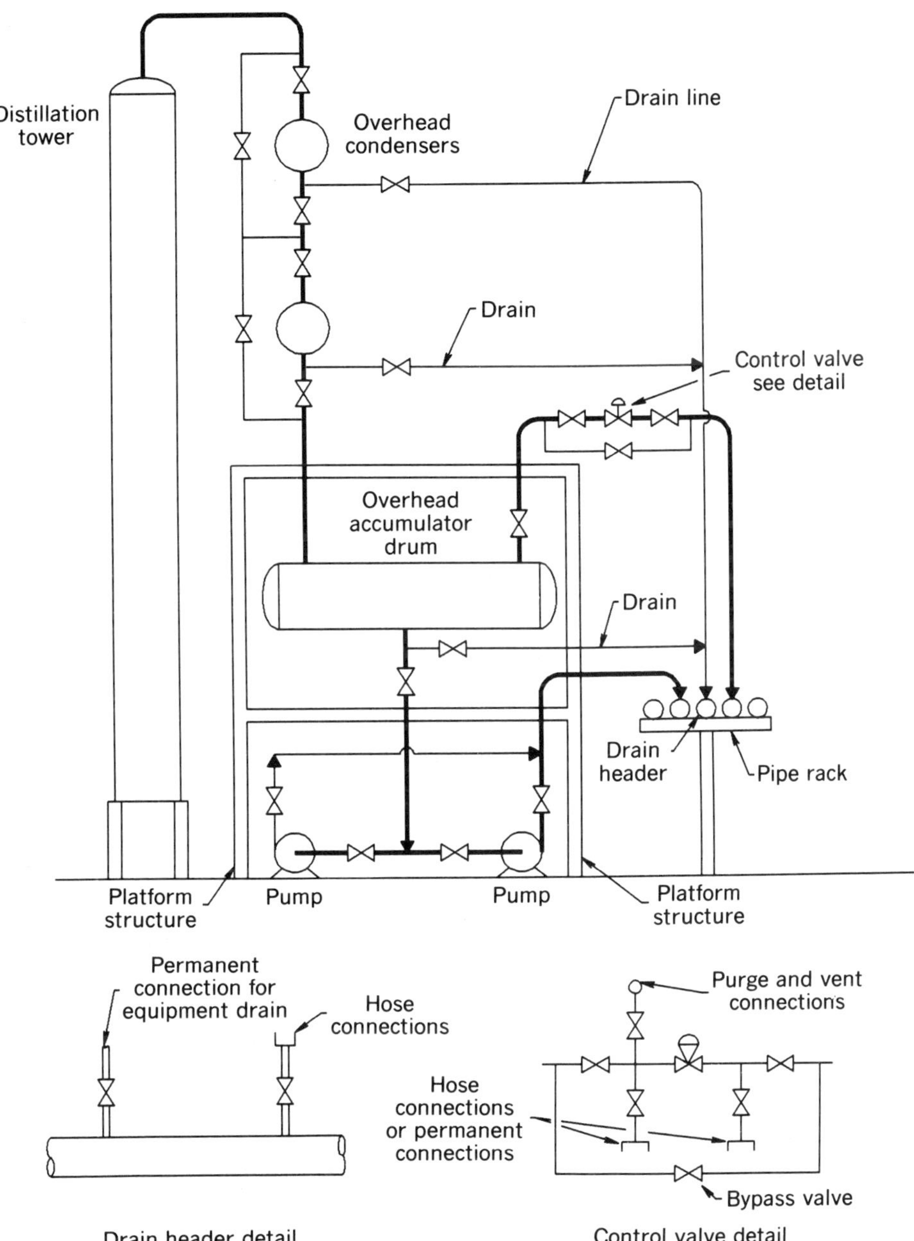

Figure 12.28 *Central drain header with typical equipment drain connections.*

for repair, the equipment is blocked from the main process stream and bypassed as shown in the figure. Opening the drain line to the drain header does not automatically empty a condenser or drum of fluid. Another connection to the equipment is required for a gas or liquid to prevent vacuum formation in the equipment during drainage. A purge and vent connection is shown in the control valve detail sketch to assist in draining and flushing the valve. The actual procedure for draining and purging the equipment must be obtained from process operations to establish the type of purge, vent, drain, and connections necessary.

Although hose drainage is attractive to plant operations groups on an economic basis, hoses present potential exposure problems if they are inadvertently drained as they are disconnected. This hose drainage is prevented by the installation of nonleaking hose connections that automatically close as the hose end is disconnected from the coupling. However, the hose should be emptied before disconnection from the coupling or the hose weight becomes excessive. Hoses can be used in various services and are standard equipment drains for maintenance and shutdown in many plants. The use of hoses in various services does not normally require cleaning the hose between various operations in petroleum refineries. However, hose interchangeability in chemical plants is more complicated since contamination or reactions can occur when a noncleaned hose is installed as a closed drain connection in a chemical process. Hoses in chemical plant service require identification and precautions in use are extremely important. Consequently, permanent drains are a better option in chemical plant drainage applications.

The elevated drain header system is satisfactory for elevated equipment, but pumps and other equipment at grade cannot be drained through it. A separate underground drain is necessary for this equipment. The underground drain header can discharge directly into a manhole as described previously or can discharge into a small sump with a vertical pump to send the drainage material into the drain header. The sump is essentially a length of pipe installed vertically in the ground with a vertical pump. A vent from the sealed cover must be controlled in accordance with the EPA (HON) wastewater regulations. As an alternative, the inground drain header in a unit can be discharged into a small parallel or corrugated plate separator installed for an individual unit. Oil or immiscible fluid recovered from the separator can be recycled to the process or tankage.

Drainage material in the elevated drain header must be sent to an acceptable receiver. Mixture of oils and water from various units can be combined in a tank where the oil is separated for recycle and the clarified water discharged directly into a junction box. Chemicals may not be combined in many plants and a disposal or handling system is generally devised for each process. A tank or separator located on each unit can separate immiscible materials for recycle.

Separate Sewer System. In Figure 12.26 drain headers bypassed the catch basins and discharged directly into a manhole. The drainage from individual equipment items can also be gathered into an underground drain header similar to that shown in Figure 12.26 and sent to an onsite parallel or corrugated plate separator. A parallel plate separator that separated immiscible material specifically from equipment located at grade level was described previously. By extension, the same function can be performed for all of the equipment drains.

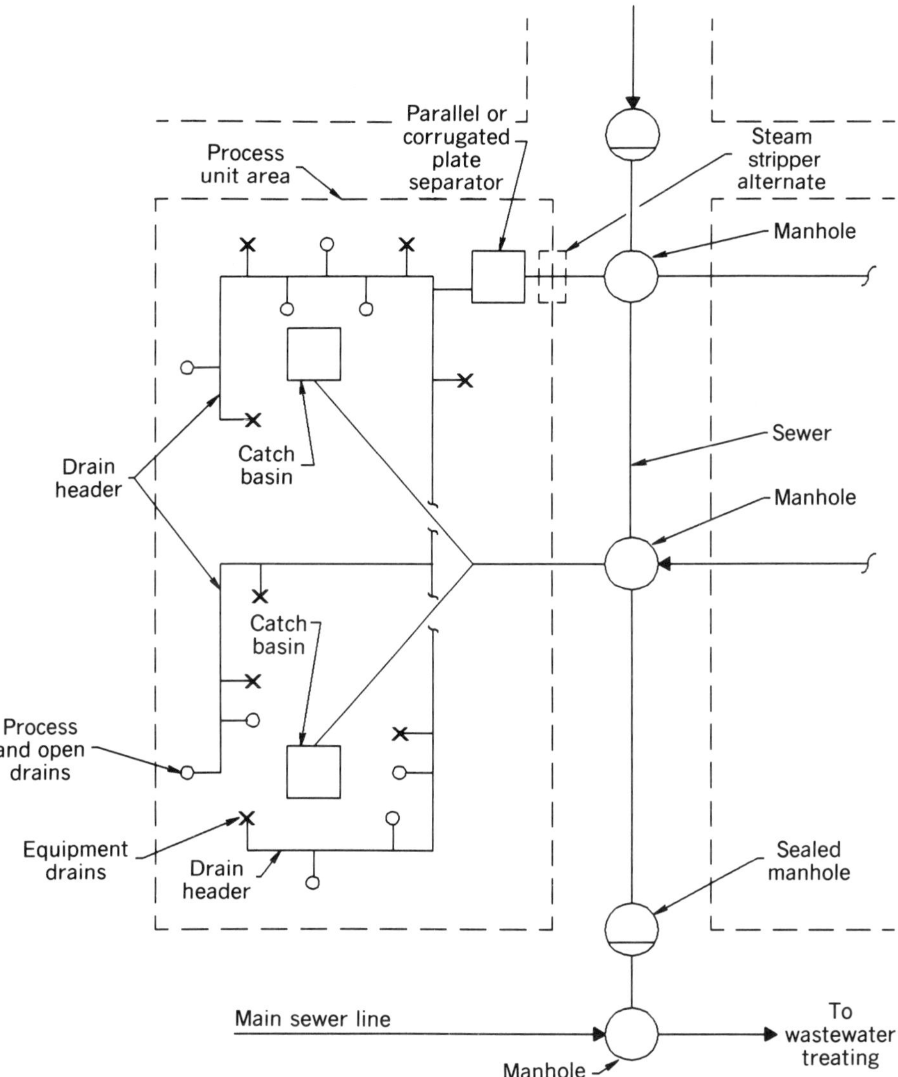

Figure 12.29 *Process unit drain system with a separator.*

Although specific equipment drains are considered in this section, both equipment and process drains can be combined in a single sewer or lateral for processing in a parallel or corrugated plate separator. The combined equipment and process drains are shown in Figure 12.29, which is similar to Figure 12.26 as a comparison. All of the drains are combined into drain headers that discharge into an onsite corrugated or parallel plate separator where oil or immiscible fluids are separated and recycled. Clarified water is discharged from the separator into a stripper or directly into a manhole. The catch basins in this design receive only spills, rainwater, and firewater. Vapors from the parallel or corrugated plate separators are contained in a closed line and processed through a control device to minimize emissions and potential exposures.

This onsite separation system is satisfactory for petroleum and chemical units. In petroleum refineries where oils may be blended from various units, drain headers from several units can be combined and sent to a single parallel or corrugated plate separator. The recovered oil is recycled to a location designated by the process group. In all of the drain risers shown in Figure 12.29 and extended to other units, the drain riser is closed, and in open drains a seal is installed. All drain headers and laterals are sealed, minimizing emissions. Drainage fluid compositions must be available for analysis to assess potentially unsafe conditions in these closed systems which can be obtained from point of generation data required under the EPA (HON) regulations. A separate drain line should be installed where a potential problem is expected and the drainage is controlled in accordance with safety recommendations.

For chemical units, the onsite separator can only be used as a central separator where the chemicals can be combined. While the clarified water can be discharged into a sewer line junction box for wastewater treatment, an alternative is to steam strip the effluent wastewater from the separator. The stripped wastewater would meet the regulatory requirement proposed by the EPA for HON wastewater streams and the stripped stream may not require further wastewater treating. Another alternative is to send the effluent from the separator directly to a biox unit. These various alternatives require a detailed analysis of the process drainage streams and an economic evaluation of each.

Mobile Receiver. Equipment drain controls outlined previously are based upon the installation of fixed piping or drains to control emissions and leakage. Since equipment is infrequently drained, installation of a fixed, permanent system is an added plant investment cost. A tank truck can be used as a receptacle for equipment drainage in many processes. The tank truck is particularly adaptable in petroleum refineries owing to the compatible nature of petroleum fractions. A dedicated vacuum tank truck in this service accepts drainage from most of the equipment items in the refinery. Connections are designed to prevent liquid or vapor leakage. Although this is a practical system and has been widely used, consideration of a mobile receiver should be reviewed with plant safety advisors before this control system is recommended.

Disposal of material from the tank truck is a major concern in this method of drainage disposal. The truck contents can be toxic and connections from the truck to a receiver should be essentially leak free. The major truck drain connection is generally a hose that is normally fitted with dripless connections, but these connections will retain fluid within the hose. A system for emptying and purging the hose is also necessary to minimize potential exposures when replacing or repairing the hose connections. The receivers for the truck contents are generally tanks or storage drums. Since the drums or tank contents will have volatile materials, control of vapor emissions from these storage points is necessary. Tankage emission controls are generally regulated through specific federal or state regulations, but storage drum or container emissions are also regulated. Methods of controlling tank emissions are described in Chapter 13.

The tank truck may present added problems in chemical plants where interaction of materials or cross contamination can be a serious processing problem. As a result, tank truck cleaning is necessary or dedicated trucks are required that can be assigned to a specific unit or several units that contain unreactive materials when mixed. The latter situation results in an increase in the number of trucks and may not be economically attractive. A tank truck cleaning facility is an alternative, but presents potential exposure concerns requiring a carefully designed system to eliminate workplace exposures. In addition, disposal of the truck contents must be carefully observed to prevent discharge of material into a receiver containing reactive materials. These considerations require detailed operating control techniques, but a mobile truck disposal handling system is feasible.

Considerations. The various catch basin and drain considerations presented in this chapter provide the industrial hygienist with a variety of alternatives for the control of potential exposures. A simple overall solution covering all processing systems is not available for the petroleum refining and chemical industries. The industrial hygienist must thoroughly understand the operation of the drainage and sewer systems before control recommendations are formulated. When these systems and their operations are understood, control recommendations can be developed. Additional control schemes may also be developed in conjunction with design and operating groups. All designs should strive to minimize emissions and exposure from openings and vents, particularly emissions from catch basins and drain openings.

Removal of the major portions of oil or chemicals from the waste flow prior to water discharge into the drain risers will significantly reduce potential emissions and wastewater treating requirements. Bypassing the catch basins minimizes emissions around catch basin openings. Wastewaters with high contaminant concentrations that are accompanied by high emission levels are stripped in accordance with the proposed HON regulations prior to discharge into the sewer or can be sent directly to a biox unit from the stripper. Where possible, the nonstripped stream may be sent directly to the biox unit from the process. Only through removal of the volatile contents from waste streams can the emission levels be reduced.

12.4.6 Vacuum Systems

An important operation in many facilities is the establishment of a process vacuum, which is accomplished with various types of vacuum equipment. The major vacuum-producing systems in current service are jet ejectors and vacuum pumps with condensers in various combinations. Jet ejectors are normally powered with steam and the vacuum pumps are typically electric driven reciprocating or liquid ring machines. The latter consists of a rotor with vanes extending outward to the casing and a water annulus several inches in depth that rotates with the vanes against the casing. A vapor annulus exists between the rotor and the water.

Vacuum systems are a source of contaminated wastewater and oil in petroleum refineries. In chemical plants, both immiscible and miscible chemicals are recovered with contaminants produced in the vacuum system wastewater. Since vacuum systems are constructed in a wide variety of configurations to meet varying process loads and system requirements, standardized vacuum systems are generally not available. Vendors and process designers readily prepare custom designed systems. Most of the vacuum loads consist of chemical vapor compounds or oil and gas vapors with small quantities of air and inert materials. Simplified sketches of two vacuum system applications are shown in Figure 12.30.

The vacuum system on the distillation tower overhead in Figure 12.30 contains those elements normally found in existing vacuum systems. A steam jet ejector evacuates the overhead accumulator drum and the ejector exhaust consisting of steam, hydrocarbon or chemical vapors, water vapors, contaminants, and small quantities of inert gas or air discharges into a direct contact cooling water condenser. Essentially all of the vapors are condensed in the barometric condenser with the exception of inert gas or air. All of the condensed material, including the steam, flows through the bottom outlet of the barometric condenser into the leg which terminates below the liquid level in the sump. A vacuum is maintained in the barometric condenser by combining the second stage steam ejector which removes inert components and volatile traces along with a long water-filled leg on the outlet of the barometric condenser. The steam from the exhaust of the second-stage ejector is condensed in the sump water, but the air or inert gas components do not condense. The second stage ejector exhaust can also be cooled and condensed separately for direct discharge into the sump. The inert gas in the sump and some volatile constituents are emitted to the atmosphere through a vent. All of the water, condensed materials, and contaminants in the sump are normally sent through a drain header or lateral into a catch basin for discharge into a manhole. For the new HON regulations, this stream would be stripped or sent directly to a manhole. The barometric sump is generally a source of emissions to the atmosphere through the sump cover and the atmospheric vent. Since the sump is located within the unit, it is a potential source of exposure until the vent is connected to a control device.

A vacuum recovery system on a batch reactor is also shown in Figure 12.30 which is typical of several water suspension or water emulsion polymer reactions. Following reaction completion in the reactor and pressure reduction to the atmospheric level, vacuum system operation is then initiated to strip the remaining traces of unreacted monomers. In this simplified process, reactor vapor, consisting of water, inerts, unreacted monomer, and contaminants, is vaporized with most of the material condensed in the exchanger and recovered in the accumulator drum. Noncondensed traces of monomer, contaminants, and water vapor discharged from the accumulator drum are condensed in the circulating ring water of a liquid ring vacuum machine, but the inert gas and a few volatile traces are discharged to the atmosphere from the separator. Excess water and contaminants are purged from the circulating water and discharged into a drain.

One of the major concerns with ejector and barometric vacuum systems is the large quantity of wastewater generated and the effect upon the size of wastewater treating units. The barometric water flow in the ejector system is eliminated by the installation of shell and tube condensers chilled with cooling tower water in the ejector discharge lines. However, the steam flow was not reduced in this modification until the jet ejector system was combined with the liquid ring vacuum machine as shown in Figure 12.31. This final change significantly reduced steam consump-

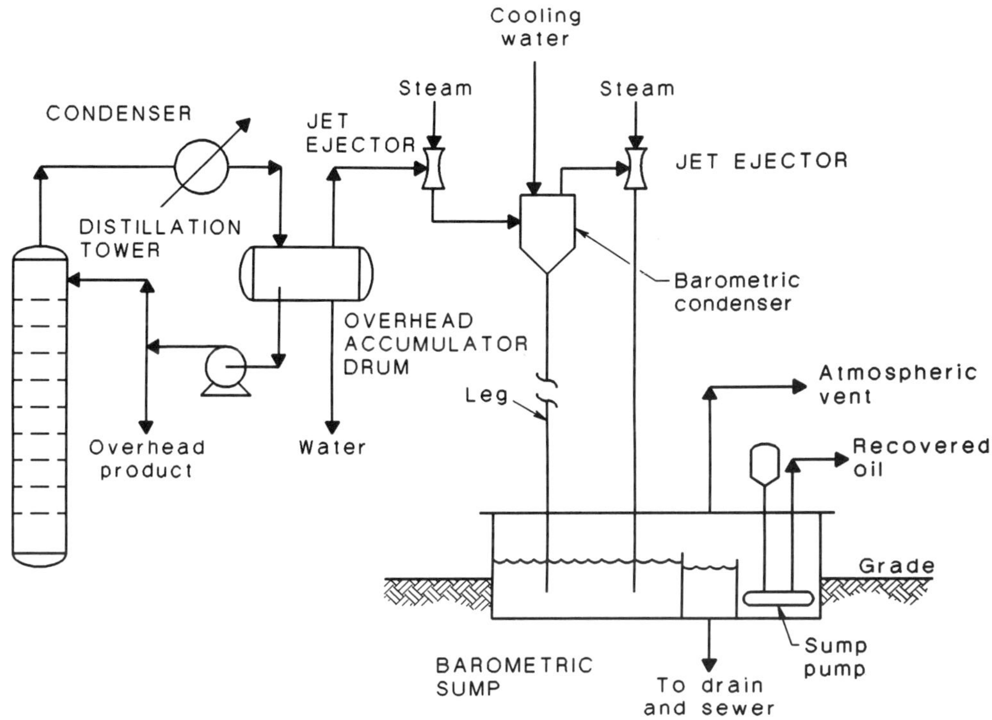

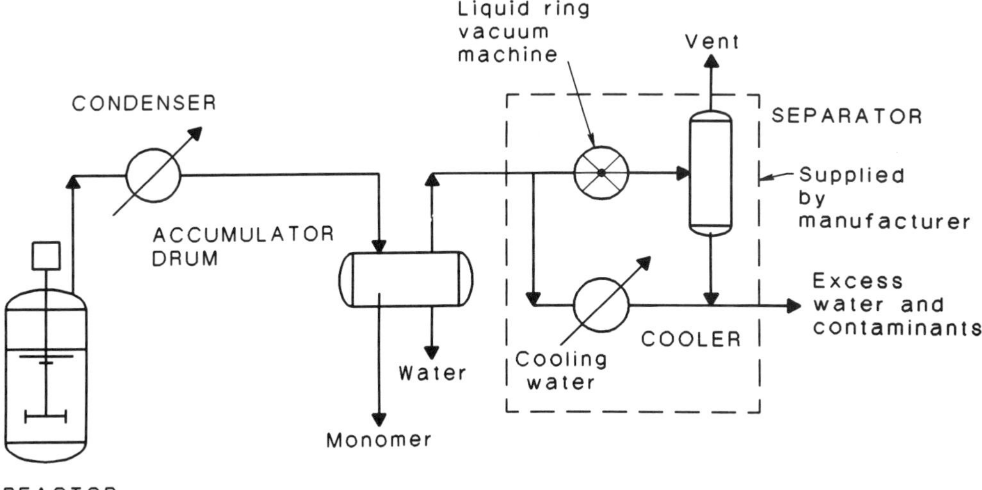

Figure 12.30 *Uncontrolled vacuum systems.*

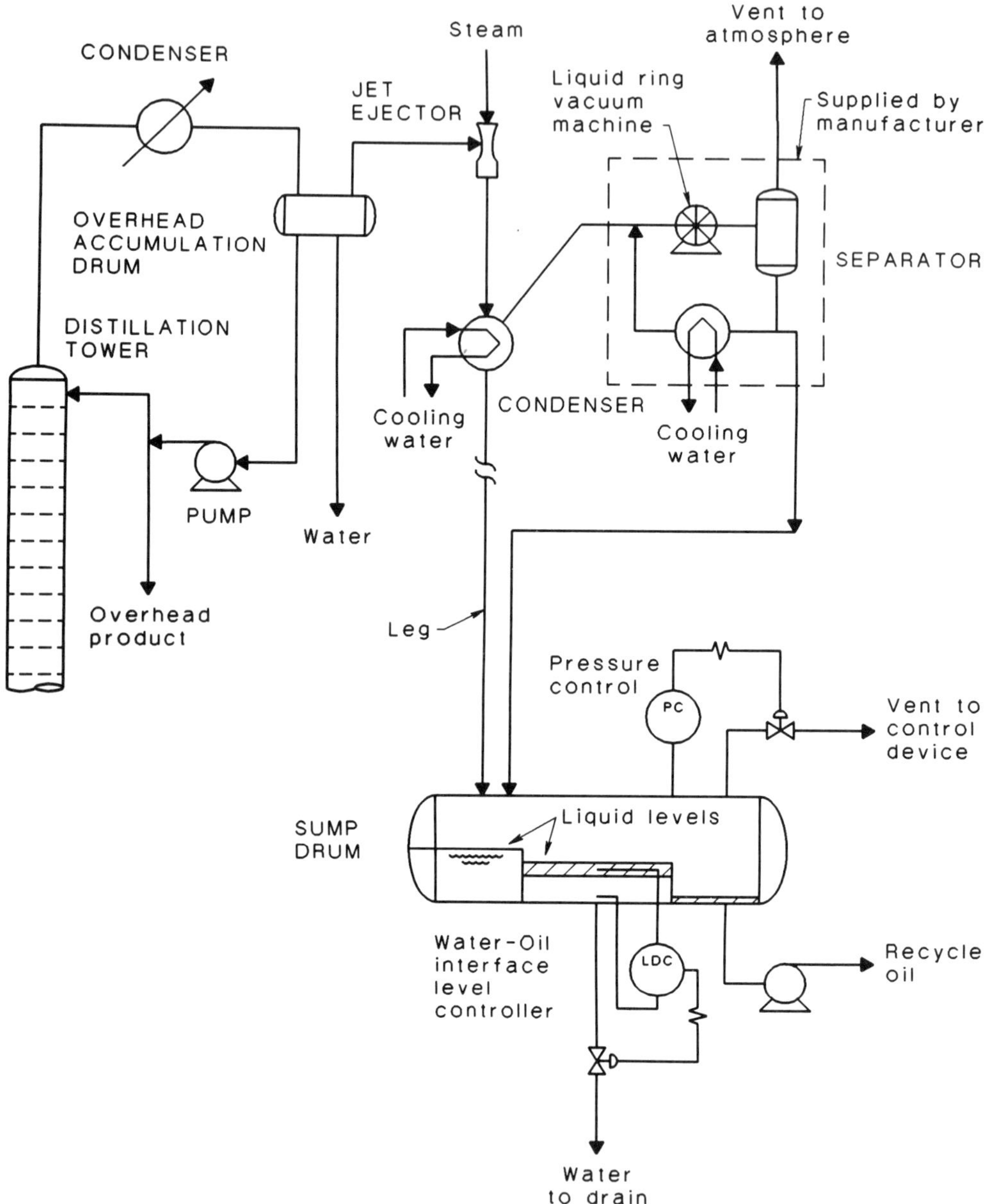

Figure 12.31 *Optimized process vacuum system.*

tion. A number of equipment changes are shown in this drawing that should be considered by the industrial hygienist.

1. A shell and tube condenser is installed in the first-stage ejector outlet. This change eliminates the barometric cooling water flow and reduces total wastewater flow to the drain and sewer system. Moreover, the cooling water temperature

maintains a constant leg temperature and improves vacuum operations. In addition, the temperature of the wastewater stream in the leg outlet is generally at a much lower level than the leg temperature in barometric condenser operations. The lower temperature minimizes vaporization and vent loss from a sump vented to the atmosphere.

2. A drum replaces the sump in Figure 12.30 to minimize potential leakage sources. In addition, the drum pressure can be raised slightly above the atmospheric pressure level to permit the discharge of toxic vapors to a control device.

3. Excess water flow from the vacuum pump separator is sent to the sump drum where any vapors or contaminants are contained.

4. A drum rather than a sump tank permits ready access to the pump for maintenance. In contrast, a sump pump requires opening the sump tank in the oil or chemicals section, which increases potential exposures.

A major concern with ejector operation is the potential temperature increase in the barometric sump when the cooler or barometric condenser water temperature increases. This situation frequently occurs during hot weather, drought periods, and utility savings programs. The volatile emission rate from the sump is directly related to the water temperature in the sump and any water temperature increase may present a potential overexposure situation in the general area surrounding the barometric sump. In existing process locations with barometric condensers, high water temperatures are considered normal in the sump and larger barometric condensers with increased water rates are generally a solution to this problem. However, a higher wastewater rate occurs which impacts upon the wastewater treating facility. An alternative solution is the installation of a shell and tube condenser to control the leg water temperatures and reduce the wastewater flow rate. Additional emission control improvements or emission reductions are achieved by chilling the cooling water stream. Installation of a liquid ring vacuum machine eliminates requirements to condense ejector steam and reduces wastewater flow.

A liquid ring vacuum machine is not normally installed as a first stage in a vacuum system owing to the large noncondensed flow rate. Following first-stage condensation, the volume flow rate to the liquid ring machine is significantly reduced. The liquid ring vacuum machine has been installed in a diverse number of vacuum applications as shown in the applications in Figures 12.30 and 12.31 where continuous and noncontinuous installations are represented. Excess contaminants and some water vapor are removed from the inlet vapor stream in the machine and are dissolved in the circulating water stream. Additives are frequently injected into the recirculating machine water stream, and the industrial hygienist should consider the additives along with disposition of the excess water stream. Normally, the waste stream flow rate is quite small.

In polymerization processes represented by the batch reactor, a polymerization inhibitor may be injected into the water to prevent polymer formation in the vacuum machine water recirculating circuit. In addition, caustic and/or water treating chemicals may be added for various reasons. Although the concentration of these materials may be low, potential exposures can occur during inhibitor handling and injection operations. The excess water stream contaminant concentration levels are also a potential exposure source and exposures through sampling or the drainage system should be prevented. Exposure control should also extend to the drain and sewer system to prevent exposures and potential reactions with sewer contaminants.

12.4.7 Manhole or Junction Box Vents

Manholes or junction boxes installed in sewer lines are normally vented to the atmosphere as shown in Figure 12.4. The EPA calculations indicate that vent emissions are about 1% of the influent contaminant concentration for various inlet wastewater streams with Group 1 compounds.[13] Apparently some field data tend to confirm this low emissions level from junction box vents. While the EPA requires vent controls in the junction box vents (Fig. 12.19), safety considerations and the low emissions quantities exhausting from the vents suggest that the EPA will not require vent controls. In addition, onsite stripping or separation reduces junction box loadings to very low levels which further reduces vent emission rates and if not a Group 1 wastewater eliminates vent requirements.

The absence of vent controls requires careful location of the vent. Although vapor quantities are very small, a vent approximately 3.5 ft (1.07 m) above grade is a potential exposure source. These concentrations can change if there is an upset in upstream process operations, although these situations are infrequent and random. Since vapors are emitted from the vent at very low velocities and disperse slowly from the vent emission point, dispersion calculations are particularly difficult under these conditions as the vent does not emit a buoyant or momentum plume. Worker protection concerns suggest that the vent emission point be located well above grade and a suitable horizontal distance from adjacent platforms or processing equipment. A simple elevation guide is shown in Figure 12.32. Individuals and companies may disagree with the distances shown, but the suggested distances provide a basis for discussion and consideration of potential workplace exposures.

A minimum height above the piperack of 25 ft (7.6m) is recommended since workers occasionally perform maintenance on the piping or adjust valves located in the piping runs. Working on elevated piping during a vapor release from the sewer vent does not permit rapid escape and personnel can be potentially overexposed. This is a particularly serious problem when an acutely toxic material is released. Emergency respiratory protective devices are desirable, but frequently interfere with maintenance or repair functions on the elevated piperack and may be discarded prior to commencing work. The plant industrial hygienist should review work practices on the elevated piping and provide work guidelines to minimize potential exposures. The proposed height may be insufficient to protect personnel in every situation, but will reduce concentration levels to acceptable exposure criteria in most situations.

In some locations the manhole may be located very close to the piperack and the vent location does not meet the proposed distance guideline from the elevated piping. The vent can be angled away from the elevated rack to a distance considered satisfactory by the industrial hygienist. In existing plants, a number of manhole vent locations are potential problems, and the vents should be modified to minimize potential exposures.

A minimum distance of 50 ft (15 m) to a working platform is shown in Figure 12.32. Dispersion calculations are normally conducted downwind of a selected

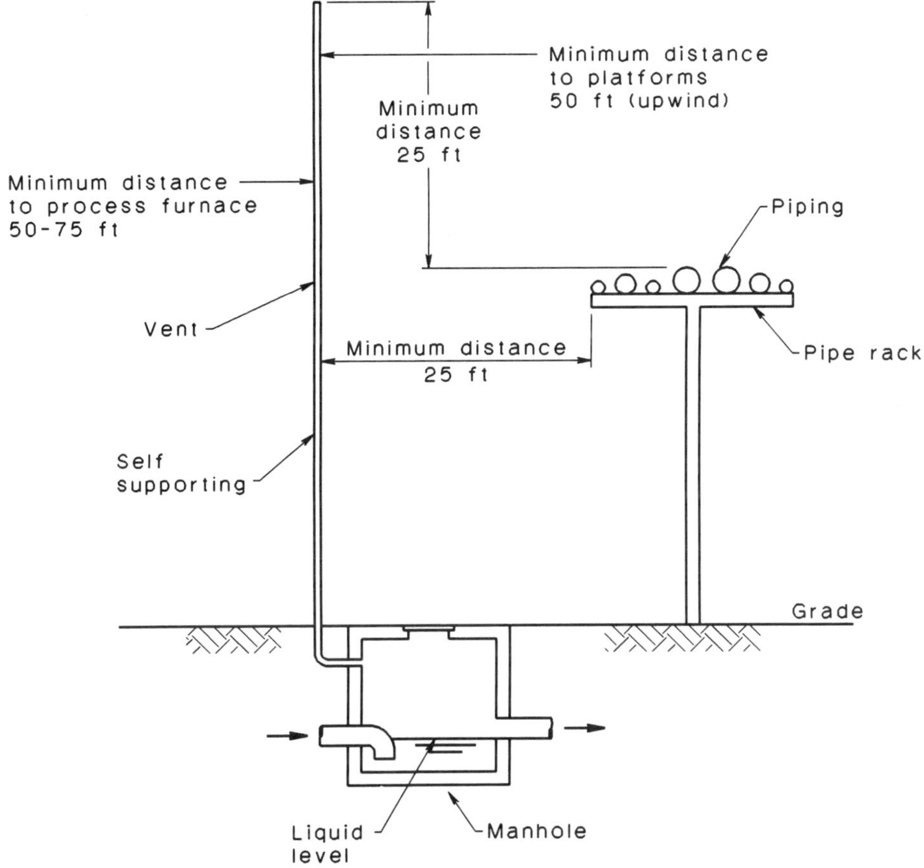

Figure 12.32 *Junction box or manhole vent location.*

emission point to insure a satisfactory concentration level on a working platform across the intervening distance. If it is difficult to calculate a satisfactory distance, a minimum vent velocity should be selected for a dispersion concentration estimate on the platform. Alternatively, a smoke bomb release through the vent will assist in establishing a dispersion pattern. An adequate vent distance can then be established.

The vent distance from a furnace of 50–75 ft is normally defined by safety requirements. However, working platforms are also installed on furnaces and the potential concentrations on these platforms is of concern during emergency releases. The concentration level can be estimated with the dispersion procedures outlined in the preceding paragraph. Any proposed location for manhole vents in new or existing units should be reviewed with the safety group.

12.5 WASTEWATER TREATING DESIGN DETAILS

In Section 12.3 emission control efficiencies and emission rates were reviewed for API separators and air flotation units. The control systems required by the EPA will reduce emissions to the atmosphere and workplace exposures. However, the emission control regulation does not cover the potential exposures and safety concerns associated with the installation of emission control systems. This section reviews the general control systems and recommends engineering controls to minimize potential exposures.

12.5.1 API Separators

Controls on API separators will reduce emissions by more than 95%, but the design of these facilities should not be considered independently of changes in the overall sewer system. In Section 12.4 reduction of emissions in the drain and sewer system were accomplished by removal of oil, chemicals, and soluble volatile material from the wastewater. Consequently, the emissions load is reduced in the wastewater treating system when upstream emissions controls are installed. The need for API separator emission controls should be evaluated during these changes. Assuming extensive upstream controls are installed, concentration levels in the wastewater treatment area are considerably less than the exposure limits. In this situation, emission controls are not required, but governmental regulation may require the installation of emission controls on API separators through VOC and HON wastewater rules. For wastewater systems controlled with upstream treating, the need for API separators is questionable. This section will review emission controls and potential sources of exposure for API separators that are still required to separable immiscible liquids from wastewater in existing facilities or those with downstream treating.

Fixed Covers. Initially the EPA only considered fixed covers on API separators to control emissions. The vapor emissions within the cover in the EPA system are discharged through a single vent and closed system (piping) to a control device (flare, incinerator, etc.). Although relatively straightforward, various considerations increase the complexity of a fixed cover unit. A major concern in a fixed cover operation is the air–fuel mixture within the cover. The fuel mixes with air during the initial startup and the air is supplemented during operation, with air dissolved and mixed with the wastewater. The gas mixture probably will exist as a fuel-rich mixture during normal operation but may go through the combustible zone during rainstorms as cooling results in a contraction effect. Air is drawn into the system during this cooling phenomenon, and the increase in dilution will move the vapor-rich mixture into a combustion vapor zone. Electrical storms frequently accompany rains and a situation then occurs in which a combustible mixture and an ignition source exist. The combustible mixture that forms within the separator cover is also a potential safety problem when a scraping system drive malfunctions and hot spots or sparks occur.

Since an air–fuel mixture is present within the fixed cover, the mixture is always a source of concern whether or not it is in the combustible zone. Consequently, the fixed cover should be purged with an inert gas to prevent the formation of a combustible mixture during separator operations. Under the new EPA regulations the purge itself requires a control device. An inert gas or nitrogen addition to the fixed cover is an expensive solution to the potential safety problem since inert gas or nitrogen are normally unavailable at the waste treating site, adding to the cost of the fixed cover installation. Moreover, other control facilities are also required to minimize emissions and the total investment for a fixed cover installation becomes quite large.

A simplified sketch of a fixed cover installation is shown in Figure 12.20. The API separator has a fixed cover, and a purge gas is injected into the cover. Since a pressure buildup or vacuum may occur, a pressure–vacuum vent is required on the cover. The vented gas mixture pressure is increased by a blower to meet pressure requirements in the flare circuit since the flare is probably located a distance from the API separator. The inert gas in the cover reduces potential risks with a blower in this highly specialized flare service. Safety considerations would probably prohibit injection of the air–fuel mixture from the API separator directly into an operating blower and then into a process flare line. An existing process flare line is probably a considerable distance from the API separator, which is normally located near the refinery or chemical plant wastewater outfall. A small dedicated flare may be a more attractive option, requiring a blowdown drum and flare seal drum for the service, although flare design and structure requirements add further costs to the system, resulting in an expensive control device. An incinerator or carbon adsorption system would not significantly affect the overall control device investment.

Although the fixed cover is an apparently straightforward control solution, an API separator is wide and of considerable length with overall channel separator lengths greater than 150 ft. Consequently, fixed roof (cover) construction is complicated and extensive supports hinder separator maintenance efforts. Silt and sludge build in the separator bottom over time and the separator must be cleaned periodically. This requires dismantling a portion of the roof and completely purging the separator. The possibility of forming a combustible mixture must be prevented during maintenance, necessitating thorough purging and cleaning. Purging the separator reduces potential inhalation exposures, but cleaning the separator requires personnel protection to prevent skin contact.

Floating Cover. A sketch of a separator with a floating cover is shown in Figure 12.10. The floating cover consists of rectangular sections that fit closely, essentially eliminating emissions from the joints. Floating covers frequently have an outer skin of metal and are filled with a matrix composite of plastic materials of foam. To minimize emissions from the space between the separator walls and the covers, flexible seals are installed at the edge of the cover, closing the opening and permitting upward or downward movement of the floating cover. The seals are supplied in a variety of forms and shapes, with a nominal seal core consisting of a foam material or liquid covered with a plastic outer casing. The plastic or outer skin is resistant to hydrocarbons and a wide variety of chemicals.

A layer of oil or immiscible chemicals is frequently held below the floating covers to absorb any light constituents with a tendency to vaporize. The oil layer is maintained through various level control systems, and internal paddles or scrapers are not required to move the oil. However, some plants prefer operating with a thin oil film and moving oil paddles. All of the liquid surfaces are enclosed in this design, and the vapor emission control efficiency is comparable to the control efficiency of a fixed cover with a control device.

The floating cover does not have any vapor zones in the main separation section and potential air–vapor safety problems inherent in the fixed cover design are less of a concern in the floating cover installation. However, safety precautions must always be observed when combustible materials are present, and fixed covers are generally installed in sections of the separator where floating covers are not applicable. Individual floating covers or sections are frequently removed to inspect the separator equipment or withdraw sludge from the separator bottom with a vacuum system or mechanical extraction device. Removal of an individual cover may result in partial vaporization of the light material in the oil or chemical layer and a combustible mixture may form near the fluid surface, requiring precautionary safety measures. In addition, some potential exposures may occur during removal and replacement of a cover section. During sludge removal operation, potential exposures to vapors may occur along with skin exposure to the sludge. Protective clothing and equipment should be provided where necessary.

Other Separator Considerations. API separators frequently contain heavier oils and in many installations oil accumulates in layers on walls and weirs and in pump basins or sumps. In addition, sludge and silt are deposited in the basin and other areas of the separators. All of these materials are potentially toxic through vapor release and/or skin contact. Consequently, maintenance of the separators is a concern to the industrial hygienist. Protective clothing requirements and equipment should be specified to protect workers in either noncontrolled separators or those with emission controls.

Since the separators have oil recovery basins and channel basins at levels considerably lower than grade, frequent removal of sludge and oily wastes can be difficult and present potential exposure sources. Sludge removal equipment is available that can extend into the basins to remove sludge or siltlike materials. The mechanical sludge removal equipment may not be capable of withdrawing oil from the basins, but vacuum trucks with hose extensions can remove the oil. Large quantities of oil can also be removed with portable sump pumps that are lowered to the basin floor, followed by hose evacuation for final cleanup. The oil from the sump pump can be transferred to the mobile truck and in turn transferred to a receiver tank. The vacuum truck provides a method of collecting and transporting the oil that minimizes vapor and skin exposures. Alternative methods of basin oil removal are practiced but should be analyzed to determine their exposure potential in comparison with a mobile tank truck system. Since various equipment is available to remove oil, an exposure control system should be selected that satisfies two major criteria:

1. The lowest exposure potential of the control system is investigated.
2. Mechanical construction is of sufficient quality that good service can be expected. A low service factor results in frequent breakdowns and increases potential exposures.

Sludge and silt is generally discharged into a container for removal to a sludge disposal system.

12.5.2 Air Flotation Units

The induced-air flotation (IAF) unit is normally enclosed as shown in Figure 12.8, and further emission controls are not necessary as described in Section 12.1.

Tightly closed or sealed doors and openings minimize leakage with vapor emissions sent to the atmosphere through an elevated vent. The emissions from the vent are a function of the inlet oil or chemical concentration, the quantity of volatile material, and the effective surface area. These units are very effective in reducing the concentration of immiscible materials to very low levels. However, their use where upstream treating is installed may be questionable, and applications are similar to the criteria for API separators reviewed in Section 12.5.1. These units, along with dissolved-air flotation (DAF) units, would be installed where downstream treating is necessary. Consequently, the industrial hygienist should measure the concentration of hydrocarbons and toxic materials in the area near the vent. Levels lower than recommended exposure limits are anticipated and should not require additional controls. A potential exposure situation occurs when a door is opened for inspection of the flotation unit interior. Protective equipment is specified for inspections since the hydrocarbon concentration can potentially reach a level of concern.

A cover for the dissolved air flotation unit shown in Figure 12.7 may be required in accordance with either proposed EPA (HON) regulations or state and regional authorities, although a cover is probably not necessary because of the low emission level generated downstream of an efficient API separator. However, the cover presents vapor control system considerations similar to the vent emissions described for the IAF unit. The major problem envisioned in the DAF unit is the apparent need to frequently inspect the skimming operation to ensure a buildup of float material does not occur on the liquid surface or on the skimmer arm. The general mix of solids, oil, and water forms a froth on the water surface termed "float." Additives are injected into DAF feed and the mix of oil, water, and solids will form agglomerates that must be moved into the overflow chamber or physically removed by the technician. A cover to enclose an existing DAF unit interferes with this inspection program and requires a safety and industrial hygiene review of the inspection operation. Some hydrocarbon vapor will exist within the enclosure and a combustible vapor may be formed, although the hydrocarbon–air mixture will probably be fuel lean and below the lower combustion zone limit. If safety permits entry during operation, the industrial hygienist should recommend personal protective equipment.

In both units a float is formed that is discharged into a pump for movement to a disposal location. The float, consisting of solids, water, and oil, presents potential skin exposure problems. Carryover during an upset operation in the sewer system can result in the emission of volatile compounds from a noncovered unit that presents potential inhalation exposure problems. Vapor emissions from the float during float treating operations should be controlled where potential exposures may occur. Since float water concentrations are greater than 75%, the float may be dewatered in various treating operations such as centrifuges, filter belt presses, and filters, prior to final disposal. Operation of these devices requires ventilation controls to minimize emission exposures and containment to prevent skin contact.

The float in some plants is sent to a thickener where it is dewatered. The thickeners are very large inverted vertical cones with the apex at grade level. The large surface permits skimming the separated and concentrated float from the water. The skimming surface may be open to the atmosphere and a removable cover with a vent should be provided. During an operating inspection, the cover is moved for a brief period and replaced when the inspection is completed.

A major consideration in float dewatering is the degree of upstream treatment prior to air flotation operations. If upstream drain and sewer streams are not controlled or treated in a facility, the water phase may have a relatively high concentration of partially soluble compounds that can present potential exposure concerns. Upsets in upstream facilities then exacerbate these potential vapor inhalation problems.

12.5.3 Recovered Oil and Chemicals

In the previous sections the emission controls on the API separators and flotation units were reviewed. The other major concern in the operation of API separator units is the recovery and disposal of the oil or chemical separated from the wastewater. The recovered oil is a combination of oil and water that varies in composition depending upon separator operation and the formation of oil–water emulsions. In separators with very thin oil or chemical films, the skimmer generally removes a fraction that is predominantly water, compared with a deep oil layer operation where a more even oil–water ratio exists. Consequently, the recovered oil is generally sent to another separation process to further concentrate the oil before recycling it to one of the process units. In petroleum refineries, the oil–water mixture is pumped to a slop oil tank where additional settling produces an oil stream with a very low water level. The slop oil tank also receives oil–water mixtures from a variety of sources to improve separation prior to recycling oil to the process. Drainage from process equipment is also sent to this tank for recovery of oil or chemicals. The tank is normally located near the waste treating facilities and VOC or VOHAP emissions from the tank must be controlled to meet regulatory requirements. In addition, emission control reduces the emission concentration levels in the area around the tank, which is an important factor to consider in a tank that frequently requires operator attention. However, the installation of emission controls on a tank in this service differs from the controls normally installed on a tank described in Chapter 13, but emissions control requirements for cone roof tanks are identical to the regulations outlined in Chapter 13. As shown in Figure 12.33, a cone roof tank is normally used as a separator for slop oil operation, and a chiller condensing unit may be installed on this tank vent to control emissions. A small inert or natural gas purge may be required in the vapor system to maintain a constant pressure.

Emissions from cone roof tanks are generally controlled by the installation of a floating roof inside the tank. Since the tank functions as an oil–water separator with a floating overflow weir, installation of an internal floating roof may interfere with the oil drawoff system. A condensing chiller unit on the tank overhead vent controls emissions to the level desired and is an acceptable alternative to an internal floating roof. The condensing unit is generally purchased as a package since the units are available in a variety of sizes. Other control devices are also available and have equivalent or better control efficiencies, with the central efficiency a function of regulatory requirements. Activated carbon units are available in standard package unit sizes with and without regeneration. A simple drum of activated carbon may also suffice as an emissions control device. In this installation the drum is periodically replaced with a fresh drum, and the spent carbon drum can be returned to the supplier for regeneration. This simple

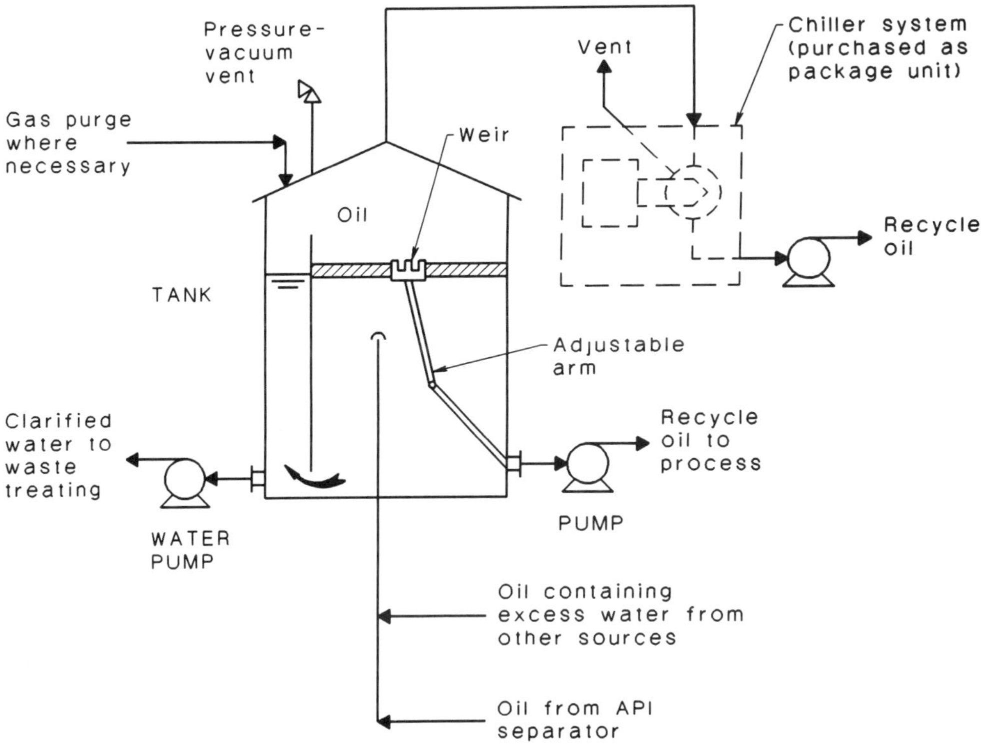

Figure 12.33 *Drain and slop recovery tank.*

installation can be used where the vapor emission rate is low and carbon life is about 30 days. When vapor emissions are considerably larger and bed life is low, a regenerable system must be installed. Oil or chemicals in the tank are decanted through a weir that can be positioned to the required height or an independent floating weir is used with a flexible hose. Recovered oil is sent to a process feed tank or a processing unit and the clarified water is pumped to a waste treating unit.

The slop oil or wastewater storage tank may be located some distance from the wastewater treating units, but the separator pumps are sized to move the recovered material the necessary distance. Locating a slop tank adjacent to the processing units permits use of the tank as a drainage receiver when draining process units or individual equipment items. As an example, the drain headers in a process unit can discharge into the slop tank, which is an advantage when the equipment is flushed with water. In addition, locating the slop tank nearer other cone roof tanks permits installation of a common vapor control system for several tanks. A common vapor emission control system is a more economically viable approach to cone roof emission control than individual emission controls on several tanks.

The oil and chemical recovery system described in this section is presented only as an example of the considerations involved in recovery. Wastewater treating units and recovery systems are not installed in conformance with standardized design formats, but differ widely among various companies and also within a company's scattered plants. However, these general considerations should provide the industrial hygienist with an understanding of the potential health problems in recovering oil or chemicals and methods of controlling exposures.

12.5.4 Sludge

Sludge is generated in most wastewater treating facilities. However, the types of sludges formed in each step differ from one another in physical and chemical characteristics. Moreover, the sludges from identical treating steps in separate plants are not comparable and treating sludges from a general standpoint is difficult. Consequently, the industrial hygienist should concentrate on classifying and analyzing all of the various sludge materials from a health perspective.

Analysis of the sludge material should begin with a review of sludge sources, the quantities generated, the processing steps at the wastewater treating facility, and chemical and physical characterization. Where a treating step exists such as float dewatering, physical and chemical characterization data should be available. Although this information is necessary for reviewing potential exposure problems, some may also be required by the EPA to meet regulatory waste requirement procedures. A variety of federal and state regulations cover waste generation, transportation, and disposal with additional regulations covering storage, tankage, and disposal. All of these waste regulations are designed to minimize contamination of underground water supplies and control vapor emissions. The waste emission control regulations combine with sewer and wastewater treating emission regulations to provide regulatory control of VOC and toxic emissions.

Float formed in air flotation units can be processed through a variety of equipment to reduce the water concentration in the float, as described in Section 12.5.2. The emission controls on the equipment are a function of the type of float processing equipment installed in the plant to dewater the float and lower volatility. Location of the equipment within a shelter or covered area for protection from the water will also have an impact on the selection of emission control equipment. Since float material is discharged continually from air flotation units and is in much greater volume than general sludge from separators or tanks, dewatering equipment operation and maintenance is a major concern. In situations where the dewatering equipment is sheltered, a ventilation system should be installed to minimize potential exposures. The vent discharge from the blower exhaust should be positioned based upon dispersion calculations that prevent potential overexposures to personnel at grade or in nearby structures. Where dewatering equipment is not enclosed, such as a thickener, an enclosure or cover is necessary to control the vapors and the vapors are discharged from an elevated vent or sent to a control device depending on the HAP or VOC material. The vapor discharge elevation is determined by dispersion calculations if an atmospheric vent is possible. Since combustible mixtures may form within the cover, an inert gas or natural gas purge may be required by the safety group. The purge will have an effect upon the dispersion calculation procedure.

Concentrated float discharged from dewatering equipment normally has a cakelike consistency although it is not completely dry. The cake is a mixture of solids and heavy oil or chemical that is transferred to disposal facilities. Although cake emissions may be extremely low, the cake may have some dermal toxicity and skin exposures should be minimized. Skin exposure controls depend upon the type of transporting equipment and disposal facilities, but controls will not be extensive since the cake will not form fine particulates.

Sludge gradually builds in an API separator bottom and collects in tank bottoms. These sludges are generally very viscous, but can vary in their viscosity and fluidity depending upon the type of oil, silt, and water content. Vapor pressures are normally quite low and generally do not pose potential inhalation exposure problems. However, most sludges do present potential skin exposure concerns and handling or transporting sludges must be carefully controlled. In addition, the industrial hygienist should specify personal protective equipment. Transporting and handling within the wastewater treatment area are only one facet of sludge processing. The actual sludge disposal techniques vary considerably within the petroleum and chemical industries. A detailed review of the various disposal procedures is not within the scope of this book. Industrial hygienists should become familiar with the handling and disposal of sludges within their plants to identify those procedures that are potential exposure concerns. The EPA documents and regulations in this area then provide a basis for recommending emission controls. Although emission controls will not prevent skin contact, the basic emission controls will be required with contact controls added by the industrial hygienist to minimize skin contact.

12.6 DESIGN CONSIDERATIONS FOR AUXILIARY EQUIPMENT AND OTHER WASTEWATER STREAMS

The streams previously reviewed contain oil or chemicals that are immiscible compounds and contaminants dissolved in the wastewater. The waste streams in this section do not fit into the previous category and, as shown in Figure 12.34, cover other plant activities with the exception of neutralization.

12.6.1 Neutralization

Only a few wastewater streams in a petroleum refinery or a chemical plant will be in the acid or base category. These streams are neutralized before the stream is sent to a drain and into the sewer system. Although a relatively straightforward operation, neutralization tanks are often open to the atmosphere and vapors in the stream can evolve as the waste stream enters the tank. The change in pH may result in emissions as compounds react with the acid or base and "spring" volatile compounds. Monitoring existing neutralization facilities provides emission and concentration data on the results of neutralization. In addition, the quantity of immiscible material can be determined after the neutralization step, which indicates disposition of the treated stream.

A simplified sketch of a neutralization facility is shown in Figure 12.34. The sketch is for illustrative purposes only since neutralization has generated a large body of literature covering instrumentation, kinetics, tank design, and mixer configurations. The tankage in the sketch covers three basic functions: equalization, neutralization, and final adjustment. The acid and caustic additive feedstream exposure controls are normally provided and are not reviewed in this section. A bypass valve is provided around the caustic control valve at the neutralization tank

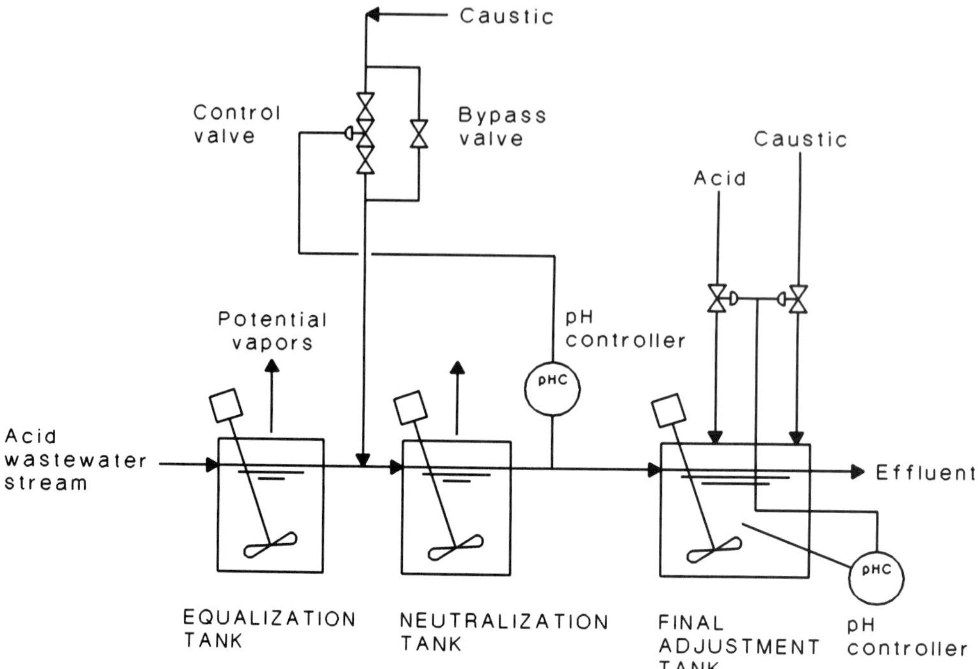

Figure 12.34 *pH neutralization system.*

to insure pH adjustment at all times for this continually flowing effluent waste stream. However, emissions from the equalization and neutralization tanks may require emission controls. The first step is installation of enclosures or covers on both tanks with vents. Since the wastewater streams are discharged directly from the process, vapor emissions can be relatively large. Consequently, emission controls in this case are required through regulation and emissions vented to the atmosphere present potential exposure concerns. Although various vapor emission controls were described in other sections, the emissions from the equalization tank may have acid carryover, which will affect the emission control system. The emission control system must be constructed of acid-proof materials and any acid vapor should be neutralized in a scrubber. The neutralized vapors containing hydrocarbons or chemicals can be condensed in a chiller unit or adsorbed on activated carbon after neutralization. In both systems, the vapors are recovered, but a purge of inert gas or natural gas in the system may be necessary to prevent the backflow of air or vacuum formation.

12.6.2 Boiler Water Systems

A simplified sketch of a boiler and feed water treating system is shown in Figure 12.35. The water treating systems for boilers are a function of freshwater quality, and the boiler operating pressure temperature levels. High-pressure boilers require a very good quality water compared with low-pressure boiler water quality. A low-pressure boiler is generally in the range of 150 to 300 psig compared to a

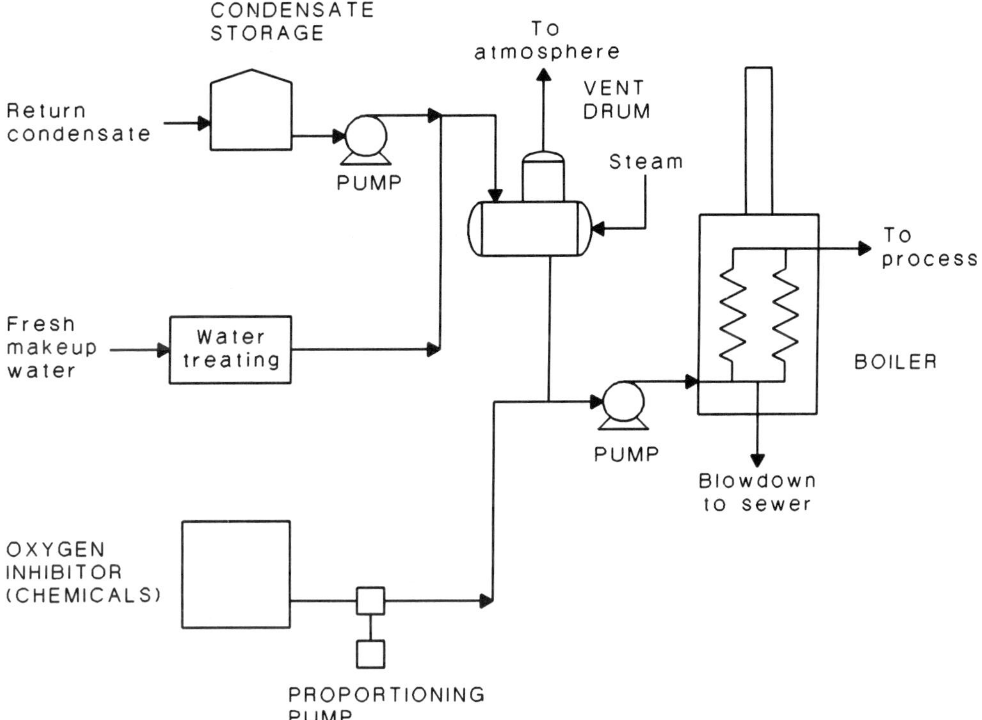

Figure 12.35 *Simplified sketch of boiler feedwater treating system.*

high-pressure boiler, which operates in the 900- to 2500-psig range. The upper end of this range requires excellent water quality or purity. Impurities contribute to corrosion and fouling in the boiler tubes during operation, reducing output and initiating mechanical problems that culminate in emergency shutdowns.

Clarification. Since water treating systems vary widely in accordance with water quality and feed purity requirements, a few general observations are presented. Initial water treatment generally consists of the addition of coagulants to remove suspended particles and colloidal matter in a clarifier. The coagulants may be acid salts or polyelectrolytes. The former are metal based and generally include aluminum sulfate, ferrous sulfate, ferric sulfate, and chlorides. The polyelectrolytes either *supplement or replace* the metal coagulants. The clarifiers are normally operated with an upper liquid layer and a lower sludge blanket. Clarified water is decanted for further processing and the sludge is withdrawn for disposal.

The chemicals added to the system can be potential health concerns and the addition system presents potential sources of exposure. The additives are received in bulk or drums and fed to the clarifier through an automatic control system. Receiving and pumping are areas of exposure concern in this overall system. Where possible, liquid chemicals should be received in bulk containers such as tank trucks for storage in tankage located near the clarifier. The chemical is then pumped to the clarifier.

Exposures during tank truck unloading frequently occur when connecting or disconnecting hose lines. Dripping can be essentially eliminated with dripless hose connections. However, the hose connection gaskets begin to leak after a period of time and require replacement, resulting in flushing and draining the hose. The hose should be flushed into a sump or drainage basin designed for this operation. In addition, pumps should also be installed with flush and drain connections to completely purge the equipment prior to maintenance. The sludge blanket in the clarifier is not necessarily toxic, but the potential toxicity should be defined. In addition, the pH may be a concern. Sludge is generally moved from the clarifier to a transport device that carries the sludge to a disposal site. Personnel exposure control in this situation is based upon a sludge toxicity assessment developed by the industrial hygienist.

Softening. In a softener, calcium and magnesium salts are precipitated through the reaction of lime. The lime forms insoluble precipitates of calcium carbonate and magnesium hydroxide in the process. In addition, caustic or soda ash may be added to react with noncarbonate hardness and form additional precipitates. Hot or cold lime softening will significantly lower the hardness in equipment similar to clarifiers in design. A hot lime softener temperature of about 225°F (107°C) is achieved by the direct injection of steam. The cold lime softener operates at ambient temperature. Addition of magnesium oxide or dolomitic line (30% magnesium oxide) to the hot lime softener will aid in the removal of silica. The silica is adsorbed and additional magnesium precipitate in the system reduces silica content to very low concentrations.

The sludge is drained to a sump from the softener and is generally pumped to a pit where water is evaporated. Sludge toxicity is considered low, but the pH level is a concern for skin exposures. Protective equipment should be specified by the industrial hygienist for workers concerned with sludge movement. Moreover, pumps and other equipment should be designed for flushing and drainage to permit equipment maintenance.

Addition of lime and caustic are a concern and additive makeup facilities should be reviewed to insure that workplace exposures to the skin are minimized. Water hoses are normally located in the additive makeup area to flush and dilute spills into drain facilities. Showers and eyewash basins are also located in the additives area. All pump and piping systems are designed for flushing and draining to reduce potential exposures during maintenance.

Softening is often followed by filtration to reduce water turbidity and improve clarity after softening. The filter is generally a mixed bed of sand and anthracite coal for treating or clarifying effluent water from a cold lime softener. The bed is then regenerated by air scouring and a water backflow wash. The water is generally discharged directly into the sewer. Filtration normally does not present potential effluent or wash exposure concerns.

Ion Exchange. Normal ion exchange or demineralization units consist of fixed beds containing resin-type beads that exchange cations or anions with the ions in the water. In the cation beds, positively charged metal ions of sodium, calcium, and magnesium in the waste react with the bed resin, releasing hydrogen ions. The hydrogen ions react with various water components to form acids since the negative ions in the water result in HCl, H_2SO_4, and other compounds. Bicarbonate ions in the water become carbonic acid, which is essentially carbon dioxide in water. The water from the cation bed is air stripped, removing most of the carbon dioxide. The stripped water is then pumped through an anion bed where negative ions (chloride, sulfate) in the water are removed and replaced with negative hydroxide ions, neutralizing the water. The purified water is then sent to a deaerator to remove oxygen prior to injection into the boiler.

In the ion exchange process described here, the cation bed effluent stream and the regeneration streams are potential sources of workplace exposure. The cation bed effluent stream is acidic and the stream pH does not change to a neutral range until processed in the anion bed. Draining or opening the acid lines or drains will discharge an acid stream that can enter the sewer. Hose or permanent piping connections from the drain valves to a bed of lime or calcium carbonate in a sump will neutralize the drainage and flush material prior to discharging the stream into the sewer systems. In addition, acid (HCl or H_2SO_4), which is used for the regeneration of cation resin and placement of new hydrogen ion on the resins, overflows the cation beds and should be directed through a closed line into the lime-filled sump. Since acid remains in the bed following this step, the acid is diluted and purged from the bed with fresh water. The dilute acid also flows into the sump for neutralization. The sump in this situation neutralizes all acid effluents prior to discharge into the sewer systems.

The anion bed is regenerated with caustic to replace the lost hydroxyl ions and the caustic will overflow the bed. The caustic is directed into a separate sump along with diluted caustic and flush water. The diluted caustic volume is neutralized prior to discharge into the sewer. Combining the acid and caustic streams can be investigated to reduce investment costs, but these operations require careful control in contrast to separate treatment of each overflow stream. Moreover, separate sumps prevent the overflow into the sewer of strong acids or strong caustics, a situation that may inadvertently develop during a period of maloperation.

Since acids, caustics, and acidic water are present in the unit, equipment must be carefully drained and flushed prior to maintenance operations or a major shutdown. Protective equipment for the various procedures should be specified in accordance with potential exposure situations.

Oxygen Scavengers. In the boiler feedwater treating system presented in this section, oxygen concentration may be somewhat above a desirable level, and an elevated oxygen concentration may also exist in returned condensate. The result of a high oxygen concentration in a boiler feedwater stream is excessive corrosivity. Scavengers are injected into the feedwater to reduce the oxygen concentration when deaeration does not reduce the oxygen concentration below the desired level. The oxygen scavenger is often injected into the deaerated liquid in the condensate stripping or vent drum. In Figure 12.35 the oxygen inhibitor is added directly to the stripped condensate.

A variety of chemicals are used as oxygen scavengers in commercial steam boiler installations. Hydrazine, morpholine, and hydroquinone are effective and popular scavenging agents, with hydrazine the more efficient. Sulfites (Na_2SO_3) and a variety of proprietary scavenging agents are also available from various manufacturers. Although hydrazine is a suspected carcinogenic agent and skin contact is prohibited, the effectiveness of this agent results in continued application of this chemical in oxygen scavenging service. Hydrazine is normally supplied in drums (30- and 50-gal) by chemical suppliers with the objective of minimizing the number of drum changes necessary to maintain a residual hydrazine level of 0.1 ppm boiler water. This level is somewhat in excess of the hydrazine necessary to bind the oxygen. In addition to drums, hydrazine is also supplied in bulk containers that reduce the number of changes and the potential for exposures with drum changes. Where justifiable, bulk containers are preferable to reduce the exposure potential.

For hydrazine or other highly toxic and carcinogenic materials that may be used as oxygen scavengers, exposures must be carefully controlled and alternative methods of injecting this material from drums are shown in Figure 12.36. To prevent potential inhalation or skin contact exposures, drum or barrel pumps are not used in this service. The latter is a potential problem when withdrawing the pump from the drum. A primary control objective is a completely closed system that can be thoroughly flushed with water prior to opening. In addition, elimination of mechanical equipment is desirable to minimize maintenance and potential exposures that may occur during repair operations.

In each installation shown in Figure 12.36 water dilutes the scavenger chemical sent to the treated water system. A double diaphragm pump injects the diluted scavenger agent into a treated water tank, and the treated water is pumped to the boiler. Since the treated water is in the tank, actual addition rates may not be critical with the diluted scavenger agent. However, the addition rate can be controlled with the double diaphragm pump. When a pump repair is necessary, flush and drain connections are provided to purge scavenger agent compounds from the pump. The scavenger drum can be purged with water through the pump as required to thoroughly wash the drum. Since the pump flow rate is quite low, a jumper across the pump flush and drain connections permits a high flush rate.

The ejector system adds measured mounts of diluted water into the system since the actual injection flow rate is controlled by the recycle control loop. The recycle quantity determines the absolute pressure at the injector inlet and the agent flow rate. Since agent dilution increases with time, the actual residual concentration must be determined on a relatively frequent scale. Sampling loop connections are shown to permit analysis of the residual concentration. The dilution water in this addition system also purges and washes the scavenger agent drum. In addition, the ejector system does not have any mechanical equipment and maintenance requirements are minimal.

Although dilution water is added to both systems in Figure 12.36, an inert pressure set at 1 or 2 in. of water can replace the dilution water. The actual agent addition rate is controlled under these circumstances and adjustments to compensate for the diluted water are unnecessary. Following the complete discharge of the scavenger agent from the drum, the drum can be flushed with water.

Some of the drum suppliers have developed hydrazine pump systems to reduce potential exposures associated with drum connections, flushing, and pumping. Pumps are external in these systems, but the operating principals described in

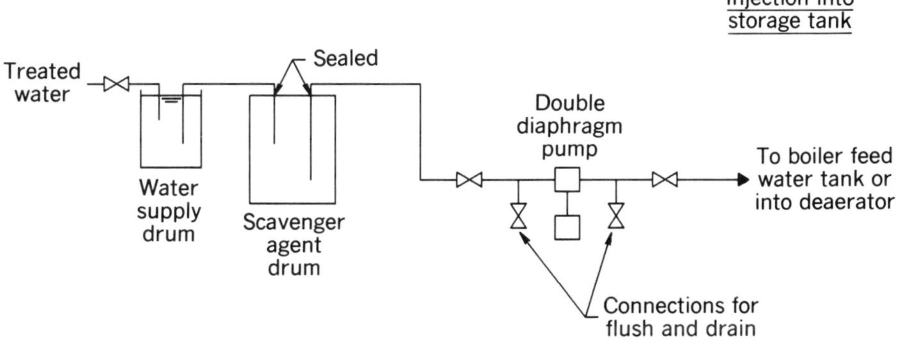

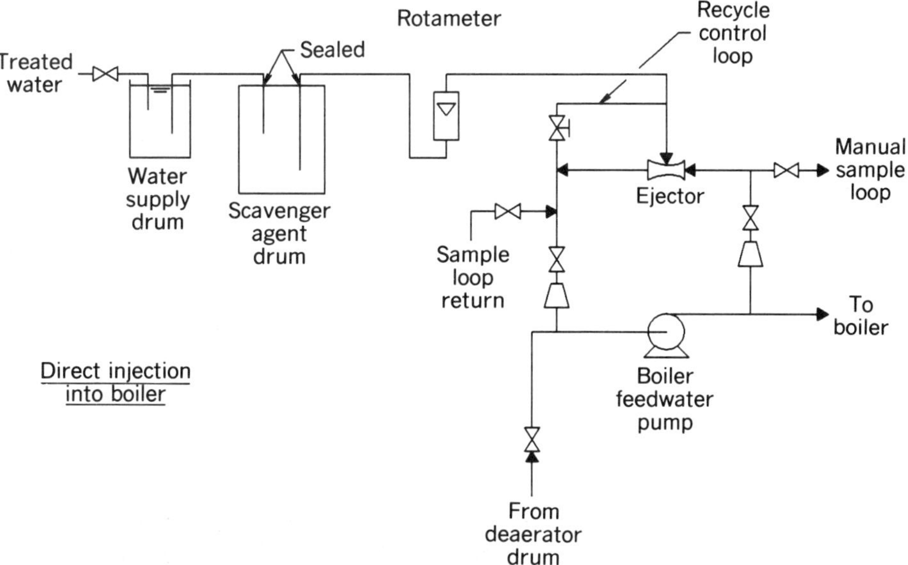

Figure 12.36 *Boiler scavenger agent feeding systems.*

these paragraphs should be applied in the analysis of any packaged system by the industrial hygienist. These principles are also applicable to large container installations.

Other Considerations. The additive and treating systems previously described are relatively straightforward. Actual installations can be considerably more complex owing to pressure levels, treating costs, freshwater quality, corrosion, and materials of construction. A group of chemicals may be injected into the treated water rather than a single chemical. Although multiple chemical additives are not addressed, the basic controls for chemical addition are similar. The degree of control necessary is determined by the toxicity of the chemicals.

The types of treating systems installed can also vary widely. The industrial hygienist must be familiar with all aspects of the various treating systems, including wastes produced and waste toxicity. Handling and disposal of the waste materials may require the installation of specific equipment along with engineering controls.

12.6.3 Cooling Towers

Cooling towers have a large blowdown stream that normally discharges into the sewers and wastewater treating system. The blowdown rate is generally a function of solids buildup in the system consisting of fouling debris in the system, rust, biological organisms, leakage of chemicals or oil into the water, and additives injected into the water to control all of these various materials. Engineering controls of cooling tower operation should be directed not only at the blowdown stream but also at the overall tower operation. This section reviews tower operations and indicates potential engineering control solutions, but the primary control emphasis is on the blowdown stream.

Tower Operation. Since cooling towers circulate water through enclosed equipment, corrosion control is a major consideration. In addition, microorganisms develop, grow, and accumulate in the fluid passageways of the circulating water stream, reducing heat transfer capability. This growth can be exacerbated by oil or chemical leakage from the process side to the water side. Control of these organisms in the equipment is a major concern and in cooling tower circuits control is generally achieved with chlorine, halogen mixtures, or biocides. Corrosion control is obtained with phosphate-based solutions that include dispersants and a variety of proprietary chemicals.

Corrosion occurs in a tower system as a result of the type of makeup water added to the cooling tower, the dissolved oxygen concentration, and various construction material considerations. Hexavalent chromium has historically been injected into cooling tower water to control corrosion, but this metal has essentially been replaced in the United States with phosphate-based and other materials to meet environmental control regulations. Organism control through the addition of chlorine or a halogen to the recirculating water is a well-established control technique. However, chlorine addition requires careful monitoring to insure that a free chlorine concentration is maintained within certain limits. Newer biocide control agents differ in monitoring requirements, but are also effective in organism control.

In addition to the various additives, pH control is carefully maintained in the circulating water. Acid addition has been the standard pH adjustment chemical, but newer additives may also require a base to maintain control within the desired pH range. The typical tower system shown in Figure 12.37 indicates that chlorine, acid, and other chemicals are added to the water basin. The chemicals selected for addition vary widely and the industrial hygienist should become familiar with the chemicals injected into the plant cooling towers. General operation and maintenance can present potential exposure situations and spills or accidents require rapid emergency response actions.

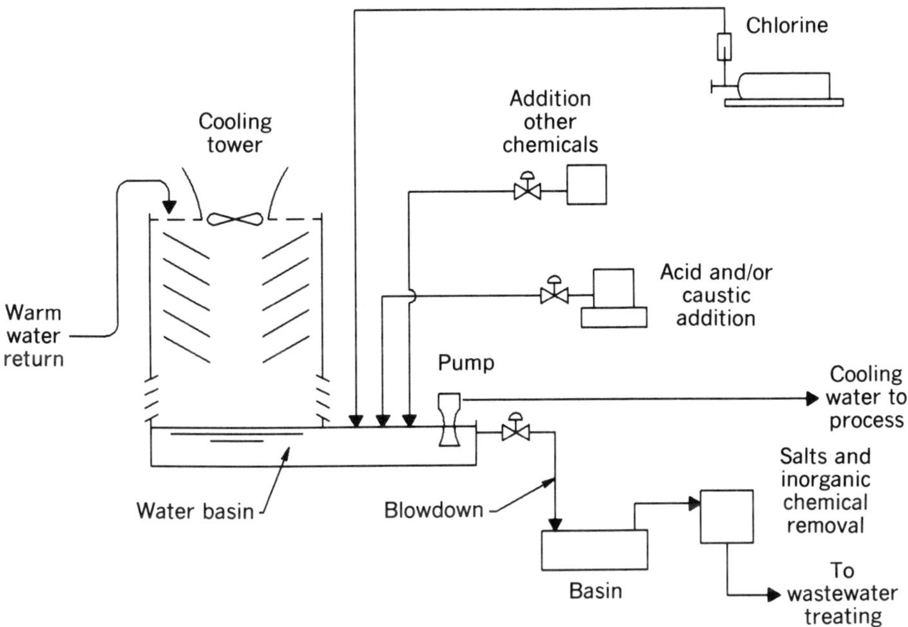

Figure 12.37 *Cooling tower additive and blowdown systems.*

Blowdown. Cooling tower water systems are blown down at a constant flow rate. Suspended and dissolved solids can build to undesirable concentrations, and the blowdown stream maintains the solids concentrations within a concentration range. The dissolved solids generally consist of sodium salts with a sludge composed of phosphates, chemicals, chemical precipitate, dirt or dust, biomass, oil or process chemicals, and iron scale.

In the blowdown stream, nonsoluble oils or chemicals may also be present, depending upon the number and severity of the leaks in the cooling water system. The nonsoluble material concentration can vary from trace concentrations to large, separate oil or chemical layers. Where dispersants are added to the circulating water stream, the oil or nonsoluble chemicals may be partially emulsified. The oil or chemical concentration resulting from leakage must be considered in the blowdown stream since this stream is normally discharged into the sewer and wastewater treating system. Oil or chemical leakage can contribute to fugitive emissions (VOC) from sewer vents and wastewater treating equipment. In the HON wastewater regulations, periodic testing of the cooling water around each individual exchanger or bank of exchangers is required. While the specific regulation may change, cooling water tests in some form will be required to control the emissions of hazardous air pollutants to the atmosphere from the cooling tower.

Blowdown Treatment. The blowdown stream should flow into a basin following discharge from the tower, as shown in Figure 12.37. Immiscible oil or chemicals that leak into the cooling water can be separated from the main effluent stream and recovered in the basin using techniques previously described in the chapter.

Sludge will gradually accumulate in the basin, minimizing solid loadings to the treating step, and clarified liquid overflows to the next treating stage. Although some chemicals may be present in the basin, a cover should not be required with leakage developing infrequently.

Where cooling water is not tested for oil or chemical leakage, a noticeable concentration of immiscible oil generally indicates oil leakage into the cooling water circuit from an exchanger or condenser. The leakage source should be identified and repaired when the equipment can be removed from service for maintenance since an oil or chemical layer in the basin or the sewer system is a potential source of exposure. These materials in the cooling water may also coat exchanger surfaces, reducing heat transfer. In various process units, exchangers or condensers can be removed from service for repair during process operations. The unit continues to operate in this situation at a lower production rate. In other units, leaking equipment cannot be removed from service until the process is shut down for an emergency or planned turnaround.

If repairs of leaking equipment cannot be undertaken during process operations, the industrial hygienist should continue to monitor the cooling tower basin emissions for potential overexposure conditions. Covers on the cooling tower basin are generally not desirable and personal protective equipment may be required in particular circumstances. As described, the cooling water system should be monitored frequently to determine the presence of oils or chemicals and potential leaks in the process unit.

Sludge from the cooling tower basin and treating step must be assessed for toxicity prior to removal and disposal. The sludge is removed during a turnaround, but some sludge can also be removed if necessary from the tower basin during operations. Wet sludge is generally dumped into a container and hauled to a disposal facility in accordance with waste disposal regulations. An industrial hygiene review of this overall operation is necessary to minimize potential exposures.

Cooling Tower Basin Microorganisms. Normally, organisms from the circulating cooling water are in the tower basin. In addition, Legionella are generally present in the tower basin since the organism exists in most cooling towers, but the existence of the Legionella organisms does not necessarily cause a disease problem. Since a potential exists for a disease outbreak, methods of controlling the organism have been recommended through the years in various publications. The industrial hygienist should become familiar with the methods of controlling the Legionella organism and cleaning the towers prior to maintenance.

12.7 PROCESS VENTS

In wastewater treatment and containment systems, vapors are generally vented through a control device to the atmosphere. While not part of the wastewater system, vents from process units are also sent through control devices to the atmosphere. Few continuous vents in the regulatory category discharge to the atmosphere without treatment, where the control devices can be combustion related (incinerators flares), adsorbers (activated carbon, etc.), solvent absorption, or condensers. In addition to these standard categories, other technologies are also applied, including biofiltration.[20] Atmospheric discharge is still permitted for relief valves, which are very infrequent releasers. However, leakage monitoring requirements have resulted in the conversion of atmospheric discharge to closed discharges in many existing installations, with closed relief valve discharge normally installed in new facilities.

The proposed HON regulations cover process vent controls and discharge from both process and wastewater systems. These regulatory requirements will probably be promulgated as presented in the proposed regulations with only minor changes, since HAP emission control is dependent upon process vent operation. The EPA states that existing source process vents are the largest contributory category to HAP emissions from SOCMI facilities.[10] The regulations for the chemical industry will probably also be applicable in petroleum refinery operations.

12.7.1 Vent Stream Classification

The process vent streams are classified as Group 1 and Group 2 vents based on the following definitions:

Group 1 Process Vent. This is a process vent where the flow rate is $\geq$ 0.005 std. m^3/min, the organic HAP concentration is $\geq$ 50 ppmv, and the TRE index value is $\leq$ 1.0. The TRE index value is a measure of the supplemental total resource requirement per unit reduction of organic HAP in the process vent stream.

Group 2 Process Vent. Refers to process vents where the flow rate is $<$ 0.005 std. m^3/min, the organic HAP concentration is $<$ 50 ppmv, and the TRE index value is $>$ 1.0.

The vent stream classifications do not apply to petroleum refinery processes, ethylene processes, equipment that does not contain HAP, R&D facilities, and chemical manufacturing processes located in coke byproduct recovery plants.

12.7.2 General Regulations

The general regulations require compliance with the following organic HAP vent rules:

1. *Flares.* Operation of flares must conform with the general regulations covering flares.[11] However, halogenated vent streams cannot be vented to a flare.

2. *Emissions Reduction.* The organic HAP emissions must be reduced 98 wt % or to an outlet concentration of 20 ppmv on a dry basis corrected to 3% oxygen. The least stringent requirement is applicable.

3. *TRE Index.* An index > 1.0 shall be maintained at the outlet of a final control device or prior to release of the vent stream to the atmosphere in the absence of a control device. Where a control device is not installed, the vent stream is considered a Group 2 vent and must conform with items 6 or 7, whichever is applicable.

4. *Boiler or Process Heater.* Where one of these combustion units is used for compliance with percent reduction or concentration limits, the vent stream must be introduced into the flame zone of the combustion device.

5. *Combustion of Halogenated Vent Stream.* When a halogenated stream is sent to a combustion device, the discharge from the device is sent to a scrubber before the flue gas stream can be discharged into the atmosphere. The overall emissions reductions in the scrubber of hydrogen halides and halogens must be 99% or the outlet concentration of each individual hydrogen halide or halogen is 0.5 mg/dry st. m^3 (or less). The least stringent requirement is applicable.

6. *Group 2 Process Vent.* Where the flow rate is ≥ 0.005 std. m^3/min, the HAP concentration is ≥ 50 ppmv, and a TRE index value is between 1.0 and 4.0, a TRE index value > 1.0 shall be maintained. In addition, the system must comply with the specific monitoring requirements or the TRE index calculations in Section 12.7.3.

7. *Group 2 Process Vent.* Those vents with a TRE index value > 4.0 must maintain a TRE index > 4.0 and comply with the provisions for calculation of the TRE index described in Section 12.7.3.

8. *Group 2 Process Vent.* Where the flow rate is < 0.005 std. m^3/min and the concentration is < 50 ppmv, these two conditions must be maintained. Other requirements are reporting and record keeping.

For halogenated streams, the terms halogens and halides refers to hydrogen chloride, chlorine, hydrogen bromide, bromine, and hydrogen fluoride.

12.7.3 TRE Index Value

The TRE index value is determined through engineering assessments or can be based upon field data. This analysis is based upon combustion of the stream to meet the required TRE value. As shown in items 1, 2, or 3 of Section 12.7.2, the control device can meet either of these conditions. Where a total efficiency or concentration is achieved (item 2), the device is satisfactory. For the other two criteria, combustion is the control system and a TRE value index is determined.

1. *Net Heating Value of the Vent Stream.* This is determined through the following equation:

$$H_T = K_1\left(\sum_{j=1}^{n} C_j H_j\right)(1 - B_{\text{ws}}) \tag{12.11}$$

where

H_T = Net sample heating value (MJ/std. m^3)(20°C)
K_1 = Constant, 1.740×10^{-7} (ppm^{-1})(gmol/std. m^3) (MJ/kcal)(20°C)
B_{ws} = Water vapor content (volume)
C_j = Concentration compound j (ppm)
H_j = Net heat of combustion compound j (kcal/gmol)

2. *Emission Rate.* This calculation determines the emission rate (E_{TOC}) of total organic compounds minus methane and ethane. In addition, the emission rate for total organic HAP is determined with this equation (E_{HAP}). The emission rates are not necessarily identical since they depend upon the components in the vent stream.

$$E = K_2\left(\sum_{j=1}^{n} C_j M_j\right)Q_s \tag{12.12}$$

where

E = Emission rate E_{TOC} or E_{HAP} (kg/hr)
K_2 = Constant, 2.494×10^{-6}(ppm^{-1})(gmol/std. m^3)(kg/g)(min/hr)(20°C)
C_J = Concentration compound j (ppm)(dry basis)
M_j = Molecular weight compound j (g/gmol)
Q_s = Vent stream flow rate (m^3/min)(20°C)

3. *TRE Index Value.* The TRE index for a vent stream controlled by a flare or incinerator is based upon the following equation:

$$\text{TRE} = \frac{1}{E_{\text{HAP}}}\left[a + b(Q_s) + C(H_T) + d(E_{\text{TOC}})\right] \tag{12.13}$$

where TRE = TRE index value

E_{HAP} = Total organic HAP emissions rate (kg/hr)
Q_s = Vent stream flow rate (std. m^3/min)(20°C)
H_T = Vent stream net heating valve (MJ/std. m^3)
E_{TOC} = Total organic emission rate TOC (minus methane and ethane) (kg/hr)
a, b, c, d = Coefficients shown in Tables 12.4 or 12.5

4. *Nonhalogenated Vent Stream.* The TRE index value for an existing source is based upon the use of a flare, a thermal incinerator with 0% heat recovery, and a thermal incinerator with 70% heat recovery. Coefficients from Table 12.4 are used in this equation where the lowest TRE index value is selected for compliance. For nonhalogenated vent streams from new sources, the coefficients in Table 12.5 are applicable.

5. *Halogenated Vent Stream.* The TRE index value is calculated based on a thermal incinerator with 0% heat recovery and a scrubber. Coefficients from Table 12.4 are used for existing sources and the coefficients for new sources are obtained from Table 12.5. For halogenated vent streams, the total concentration of halogen atoms is the summation of individual halogen atoms in each organic HAP com-

TABLE 12.4 TRE Coefficients for Nonhalogenated and Halogenated Vent Streams for Existing Sources[a]

Vent Stream Type	Control Device System	Coefficients			
		a	*b*	*c*	*d*
Nonhalogenated	Flare	2.902	0.5490	-1.153×10^{-2}	-1.100×10^{-3}
	Thermal incinerator, 0% heat recovery	2.238	0.00940	4.765×10^{-2}	-1.739×10^{-3}
	Thermal incinerator, 70% heat recovery	3.778	0.01775	1.950×10^{-2}	7.185×10^{-2}
Halogenated	Thermal incinerator and scrubber	5.992	0.0780	-2.653×10^{-3}	1.455×10^{-3}

[a]Adopted from reference 7.

TABLE 12.5 TRE Coefficients for Nonhalogenated and Halogenated Vent Streams for New Sources[a]

Vent Stream Type	Control Device System	Coefficients			
		a	*b*	*c*	*d*
Nonhalogenated	Flare	0.5276	0.0998	-2.096×10^{-3}	-2.000×10^{-4}
	Thermal incinerator, 0% heat recovery	0.4068	0.0171	8.664×10^{-3}	-3.162×10^{-4}
	Thermal incinerator, 70% heat recovery	0.6868	0.003209	3.546×10^{-3}	1.306×10^{-2}
Halogenated	Thermal incinerator and scrubber	1.0895	0.01417	-4.822×10^{-4}	2.645×10^{-4}

[a]Adopted from reference 7.

pound. The halogen atoms can be determined through process information or by an EPA measurement procedure.

12.7.4 Other

The EPA has extensive requirements for monitoring, performance test procedures, reporting, and recordkeeping.[7] This information is beyond the scope of this chapter but should be reviewed to understand the various requirements. For industrial hygienists, monitoring and performance requirements are a particular concern since sampling is necessary to ensure meeting EPA requirements. Potential exposures may occur during sampling and may require a review of sampling and monitoring procedures.

REFERENCES

1. EPA. *VOC Emissions from Petroleum Refinery Wastewater Systems—Background Information for Proposed Standards*. Emission Standards and Engineering Division, Preliminary Draft, pp. 3-1 to 3-75, July 1984.

2. Lipton, S. and Lynch, J. R. *Health Hazard Control in the Chemical Process Industry*, 1st ed., New York: Wiley, 1987, pp. 268–269.

3. EPA. *Compilation of Air Pollutant Emission Factors*., Vol. 1, 4th ed., AP-42, September 1985.

4. CFR. Standards of Performance for VOC Emissions from Petroleum Refinery Wastewater Systems. *Title 40*. Subpart QQQ, Parts 60.690–60.699, pp. 633–643, July 1, 1992.

5. CFR. National Emission Standard for Benzene Waste Operations. *Title 40*, Subpart FF, Parts 61.340–61.358, pp. 162–189, July 1, 1992.

6. CFR. National Emission Standard for Vinyl Chloride. *Title 40*. Subpart F, Parts 61.60–61.70, pp. 32–49, July 1, 1992.

7. Federal Register, National Emission Standards for Organic Hazardous Air Pollutants from the Synthetic Organic Chemical Manufacturing Industry for Process Vents, Storage Vessels, Transfer Operations, and Wastewater. Proposed Rule. Vol. 57, No. 252, *Title 40*. Subpart G, Parts 63.110–63.147, pp. 62,693–62,808, December 31, 1992.

8. Federal Register. National Emission Standards for Organic Hazardous Air Pollutants from the Synthetic Organic Chemical Manufacturing Industry and Equipment Leaks from Seven Other Processes. Proposed Rule. Vol. 57, No. 252, *Title 40*, Subpart F, Parts 63.100–63.105, pp. 62685–62693, December 31, 1992.

9. Bay Area Air Quality Management District Wastewater (Oil–Water) Separators. Organic Compounds. Regulation 8, Rule 8, pp. 8-8-1 to 8-8-9, November 1, 1989.

10. Federal Register. National Emission Standards for Hazardous Air Pollutants for Source Categories: Organic Hazardous Air Pollutants from the Synthetic Organic Chemical Manufacturing Industry and Seven Other Processes, Vol. 57, No. 252, *Title 40*, pp. 62,608–62,684, December 31, 1992.

11. CFR. Standards of Performance for New Stationary Sources. *Title 40*, Part 60, Sec. 60.18, pp. 237–238, July 1, 1992.

12. CFR. Determination of Volatile Organic Compound Leads. *Title 40*, Part 60, Appendix A, Method 21, pp. 1020–1023, July 1, 1992.

13. EPA. *Control of Volatile Organic Compound Emissions from Industrial Wastewater*. Preliminary Draft, Vols. 1 and 2, Office of Air Quality, Planning and Standards, North Carolina, April 1988.

14. EPA, *VOC Emissions from Petroleum Refinery Wastewater Systems—Background Information for Proposed Standards*. Preliminary Draft. Emissions Standards and Engineering Division, July 1984.

15. EPA. *Industrial Wastewater Volatile Organic Compound Emissions. Background Information for BACT/LAER Determinations*. Publ. No. EPA-450/3-90-004, January 1990.

16. EPA. *VOC Emissions from Petroleum Refinery Wastewater Systems—Background Information for Proposed Standards*. EPA-450/3-85-001a, February 1985.

17. Newman, S. A. *Hydrocarbon Proc*. Part 2, **70**(10), 101–106 (1991).

18. Hsiao, K. J. and K. Y. Li, *Chem. Eng*. **98**(7), 114–118 (1991).

19. Okoniewski, B. A. *Chem. Eng. Prog*. **88**(2), 89–93 (1992).

20. Bohn, H. *Chem. Eng. Prog*. **88**(4), 34–40 (1992).

21. Federal Register, National Emission Standards for Hazardous Air Pollutants for Source Categories; Organic Hazardous Air Pollutants from the Synthetic Organic Chemical Manufacturing Industry and Seven Other Processes. Correction. Vol. 58, No. 37, *Title 40*, Subpart G, pp. 11667–11719, February 26, 1993.

APPENDIX 12A List of Synthetic Chemical Manufacturing Industry (SOCMI) Chemicals

Chemical Name	CAS No.	Chemical Name	CAS No.
Acenaphthene	83329	Butanediol (1,4-)	110634
Acetal	105577	Butyl acrylate (*n*-)	141322
Acetaldehyde	75070	Butylbenzyl phthalate	85687
Acetaldol	107891	Butylene glycol (1,3-)	107880
Acetamide	60355	Butyrolacetone	96480
Acetanilide	103844	Caprolactam	105602
Acetic acid	64197	Carbaryl	63252
Acetic anhydride	108247	Carbazole	86748
Acetoacetanilide	102012	Carbon disulfide	75150
Aceton	67641	Carbon tetrabromide	558134
Acetone cyanohydrin	75865	Carbon tetrachloride	56235
Acetonitrile	75058	Carbon tetrafluoride	75730
Acetophenone	98862	Chloral	75876
Acrolein	107028	Chloroacetic acid	79118
Acrylamide	79061	Chloroacetophenone (2-)	532274
Acrylic acid	79107	Chloraniline (*p*-)	106478
Acrylonitrile	107131	Chlorobenzene	108907
Adiponitrile	111693	Chlorodifluorethane	25497294
Alizarin	72480	Chlorodifluoromethane	75456
Alkyl anthraquinones	008	Chloroform	67663
Allyl alcohol	107186	Chloronaphthalene	25586430
Allyl chloride	107051	Chloronitrobenzene (1,3-)	121733
Allyl cyanide	109751	Chloronitrobenzene (*o*-)	88733
Aminophenol sulfonic acid	0010	Chloronitrobenzene (*p*-)	100005
Aminiphenol (p-)	123308	Chlorophenol (*m*-)	108430
Aniline	62533	Chlorophenol (*o*-)	95578
Aniline hydrochloride	142041	Chlorophenol (*p*-)	106489
Anisidine (*o*-)	90040	Chloroprene	126998
Anthracene	120127	Chlorotoluene (*m*-)	108418
Anthraquinone	84651	Chlorotoluene (*o*-)	95498
Azobenzene	103333	Chlorotoluene (*p*-)	106434
Benzaldehyde	100527	Chlorotrifluoromethane	75729
Benzene	71432	Chrysene	218019
Benzenedisulfonic acid	98486	Cresol and cresylic acid (*m*-)	108394
Benzenesulfonic acid	98113	Crescol and cresylic acid (*o*-)	95487
Benzil	134816	Cresol and cresylic acid (*p*-)	106445
Benzilic acid	76937	Cresols and cresylic acids (mixed)	1319773
Benzoic acid	65850		
Benzoin	119539	Crotonaldehyde	123739
Benzonitrle	100470	Cumene	98828
Benzophenone	119619	Cumene hydroperoxide	80159
Benzotrichloride	98077	Cyanoacetic acid	372098
Benzoyl chloride	98884	Cyanoformamide	0011
Benzyl acetate	140114	Cyclohexane	110827
Benzyl alcohol	100516	Cyclohexanol	108930
Benzyl benzoate	120514	Cyclohexanone	108941
Benzyl chloride	100447	Cycloohexylamine	108918
Benzyl dichloride	98873	Cyclooctadienes	29965977
Biphenyl	92524	Decahydronaphthalene	91178
Bisphenol A	80057	Diacetoxy-2-butene(1,4-)	0012
Bis(chloromethyl)ether	542881	Diallyl phthalate	131179
Bromobenzene	108861	Diaminophenol hydrochloride	137097
Bromoform	75252	Dibromomethane	74953
Bromonaphthalene	27497514	Dibutoxyethyl phthalate	117839
Butadiene (1,3-)	106990	Dichloroaniline (mixed isomers)	27134276

APPENDIX 12A ***(Continued.)***

Chemical Name	CAS No.
Dichlorobenzene (*p*-)	106467
Dichlorobenzene(*m*-)	541731
Dichlorobenene (*o*-)	95501
Dichlorobenzidine (3,3′)	91941
Dichlorodifluoromethane	75718
Dichloroethane (1,2-) (ethylene dichloride)(EDC)	107062
Dichloroethyl ether	111444
Dichloroethylene (1,2-)	540590
Dichlorophenol (2,4-)	120832
Dichloropropene (1,3-)	542756
Dichlorotetrafluoroethane	1320372
Dichloro-1-butene (3,4-)	760236
Dichloro-2-butene (1,4-)	764410
Diethanolamine	111422
Diethyl phthalate	84662
Diethyl sulfate	64675
Diethylamine	109897
Diethylaniline (2,6-)	579668
Diethylene glycol	111466
Diethylene glycol dibutyl ether	112732
Diethylene glycol diethyl ether	112367
Diethylene glycol dimethyl ether	111966
Diethylene glycol monobutyl ether acetate	124174
Diethylene glycol monobutyl ether	112345
Diethylene glycol monoethyl ether acetate	112152
Diethylene glycol monoethyl ether	111900
Diethylene glycol monohexyl ether	112594
Diethylene glycol monomethyl ether acetate	629389
Diethylene glycol monomethyl ether	111773
Dihydroxybenzoic acid (resorcylic acid)	27138574
Diisodecyl phthalate	26761400
Diisoctyl phthalate	27554263
Dimethylbenzidine (3,3′-)	119937
Dimethyl ether	115106
Dimethylformamide (*N*,*N*-)	68122
Dimethylhydrazine (1,1-)	57147
Dimethyl phthalate	131113
Dimethyl sulfate	77781
Dimethyl terephthalate	120616
Dimethylamine	124403
Dimethylaminoethanol (2-)	108010
Dimethylaniline (*N*,*N*)	121697
Dinitrobenzenes (NOS)	25154545
Dinitrophenol (2,4-)	51285
Dinitrotoluene (2,4-)	121142
Dioxane	123911
Dioxolane (1,3-)	646060
Diphenyl methane	101815
Diphenyl oxide	101848
Diphenyl thiourea	102089
Diphenylamine	122394
Dipropylene glycol	110985
Di(2-methoxyethyl)phthalate	117828
Di-*o*-tolyguanidine	97392
Dodecyl benzene (branched)	123013
Dodecyl phenol (branched)	0013
Dodecylaniline	28675174
Dodecylbenzene (*n*-)	121013
Dodecylphenol	27193868
Epichlorohydrin	106898
Ethane	74840
Ethanolamine	141435
Ethyl acrylate	140885
Ethylbenzene	100414
Ethyl chloride	75003
Ethyl chloroacetate	105395
Ethylamine	75047
Ethylaniline (*n*-)	103695
Ethylaniline (*o*-)	578541
Ethylcellulose	9004573
Ethylcyanoacetate	105566
Ethylene carbonate	96491
Ethylene dibromide	106934
Ethylene glycol	107211
Ethylene glycol diacetate	111557
Ethylene glycol dibutyl ether	112481
Ethylene glycol diethyl ether (1,2-diethoxyethane)	629141
Ethylene glycol dimethyl lether	110714
Ethylene glycol monoacetate	542596
Ethylene glycol monobutyl ether acetate	112072
Ethylene glycol monobutyl ether	111762
Ethylene glycol monoethyl ether acetate	111159
Ethylene glycol monoethyl ether	110805
Ethylene glycol monohexyl ether	003
Ethylene glycol monomethyl ether acetate	110496
Ethylene glycol monomethyl ether	109864
Ethylene glycol monooctyl ether	002
Ethylene glycol monophenyl ether	122996
Ethylene glycol monopropyl ether	2807309
Ethylene oxide	75218
Ethylenediamine	107153
Ethylenediamine tetraacetic acid	60004
Ethylenimine (aziridine)	151564
Ethylhexyl acrylate (2-isomer)	103117
Fluoranthene	206440
Formaldehyde	50000
Formamide	75127

APPENDIX 12A ***(Continued.)***

Chemical Name	CAS No.	Chemical Name	CAS No.
Formic acid	64188	Methylionones (α-)	79696
Fumaric acid	110178	Methylpentynol	77758
Glutaraldehyde	111308	Methylstyrene (α-)	98839
Glyceraldehyde	367475	Naphthalene	91203
Glycerol	56815	Naphthalene sulfonic acid (α-)	85472
Glycerol tri(polyoxypropylene) ether	25791962	Naphthalene sulfonic acid (β-)	120183
		Naphthol (α-)	90153
Glycine	56406	Naphthol (β-)	135193
Glyoxal	107222	Naphtholsulfonic acid (1-)	567180
Hexachlorobenzene	118741	Naphthylamine sulfonic acid (1,4-)	84866
Hexachlorobutadiene	87683	Naphthylamine sulfonic acid (2,1-)	81163
Hexachloroethane	67721	Naphthylamine (1-)	134327
Hexadiene (1,4-)	592450	Naphthylamine (2-)	91598
Hexamethylenetetramine	100970	Nitroaniline (*m*-)	99092
Hexane	110543	Nitroaniline (*o*-)	88744
Hexanetriol (1,2,6-)	106694	Nitroanisole (*o*-)	91236
Hydroquinone	123319	Nitroanisole (*p*-)	100174
Hydroxyadipaldehyde	0016	Nitrobenzene	98953
Iminodiethanol (2,2-)	111422	Nitronaphthalene (1-)	86577
Isobutyl acrylate	106638	Nitrophenol (*p*-)	100027
Isobutylene	115117	Nitrophenol (*o*-)	88755
Isophorone	78591	Nitropropane (2-)	79469
Isophorone nitrile	0017	Nitrotoluene (all isomers)	1321126
Isophthalic acid	121915	Nitrotoluene (*o*-)	88722
Isopropylphenol	25168063	Nitrotoluene (*m*-)	99081
Lead phthalate	0018	Nitrotoluene (*p*-)	99990
Linear alkylbenzene	000	Nitroxylene	25168041
Maleic anhydride	108316	Nonylbenzene (branched)	1081772
Maleic hydrazide	123331	Nonylphenol	25154523
Malic acid	6915157	*N*-Vinyl-2-pyrrolidine	88120
Metanilic acid	121471	Ocetene-1	111660
Methacrylic acid	79414	Octylphenol	27193288
Methanol	67561	Paraformaldehyde	30525894
Methionine	63683	Paraldehyde	123637
Methyl acetate	79209	Pentachlorophenol	87865
Methyl acrylate	96333	Pentaerythritol	115775
Methyl bromide	74839	Peracetic acid	79210
Methyl chloride	74873	Perchloroethylene	127184
Methyl ethyl ketone	78933	Perchloromethyl mercaptan	594423
Methyl formate	107313	Phenanthrene	85018
Methyl hydrazine	60344	Phenetidine (*p*-)	156434
Methyl isobutyl carbinol	108112	Phenol	108952
Methyl isocyanate	624839	Phenolphthalein	77098
Methyl mercaptan	74931	Phenenosulfonic acids (all isomers)	1333397
Methyl methacrylate	80626		
Methyl phenyl carbinol	98851	Phenyl anthranilic acid (all isomers)	91407
Methyl tert-butyl ether	1634044		
Methylamine	74895	Phenylenediamine (*p*-)	106503
Methylaniline (*n*-)	100618	Phloroglucinol	108736
Methylcyclohexane	108872	Phosgene	75445
Methylcyclohexanol	25639423	Phthalic acid	88993
Methylcyclohexanone	1331222	Phthalic anhydride	85449
Methylene chloride	75092	Phthalimide	85416
Methylene dianiline (4,4′-isomer)	101779	Phthalonitrile	91156
Methylene diphenyl diisocyanate (4,4′-)(MDI)	101688	Picoline (β-)	108996
		Piperazine	110850

APPENDIX 12A ***(Continued.)***

Chemical Name	CAS No.	Chemical Name	CAS No.
Polyethylene glycol	25322683	Toluene 2,4 diisocyanate	548849
Polypropylene glycol	25322694	Toluene diisocyanates (mixture)	26471625
Propiolactone (*β*-)	57578	Toluene sulfonic acids	104154
Propionaldehyde	123386	Toluenesulfonyl chloride	98599
Propionic acid	79094	Toluidine (*o*-)	95534
Propylene carbonate	108327	Trichloroaniline (2,4,6-)	634935
Propylene dichloride	78875	Trichlorobenzene (1,2,3-)	87616
Propylene glycol	57556	Trichlorobenzene (1,2,4-)	120821
Propylene glycol monomethyl ether	107982	Trichloroethane (1,1,1-)	71556
		Trichloroethane (1,1,2-)	79005
Propylene oxide	75569	Trichloroethylene	79016
Pyrene	129000	Trichlorofluoromethane	75694
Pyridine	110861	Trichlorophenol (2,4,5-)	95954
p-tert-Butyl toluene	98511	Trichlorotrifluoroethane (1,1,2 -1,1,2)	76131
Quinone	106514		
Resorcinol	108463	Triethanolamine	102716
Salicyclic acid	69727	Triethylamine	121448
Sodium methoxide	124414	Triethylene glycol	112276
Sodium phenate	139026	Triethylene glycol dimethyl ether	112492
Stilbene	588590	Triethylene glycol monoethyl ether	112505
Styrene	100425	Triethylene glycol monomethyl ether	112356
Succinic acid	110156		
Succinonitrile	110612	Trimethylamine	75503
Sulfanilic acid	121573	Trimethylcyclohexanol	933482
Sulfolane	126330	Trimethylcyclohexanone	2408379
Tartaric acid	526830	Trimethylcyclohexylamine	007
Terephthalic acid	100210	Trimethylolpropane	77996
Tetrabromophthalic anhydride	632791	Trimethylpentane (2,2,4-)	540841
Tetrachlorobenzene (1,2,4,5-)	95943	Tripropylene glycol	24800440
Tetrachloroethane (1,1,2,2-)	79345	Vinyl acetate	108054
Tetrachlorophthalic anhydride	117088	Vinyl chloride	75014
Tetraethyl lead	78002	Vinyl toluene	25013154
Tetraethylene glycol	112607	Vinylcyclohexene (4-)	100403
Tetraethylenepentamine	112572	Vinylidene chloride	75354
Tetrahydrofuran	109999	Vinyl(*N*)-pyrrolidone(2-)	88120
Tetrahydronaphthalene	119642	Xanthates	140896
Tetrahydrophthalic anhydride	85438	Xylene sulfonic acid	25321419
Tetramethylenediamine	110601	Xylenes (NOS)	1330207
Tetramethylethylenediamine	110189	Xylene (*m*-)	108383
Tetramethyllead	75741	Xylene (*o*)	95476
Thiocarbanilide	102089	Xylene (*p*-)	106423
Toluene	108883	Xylenol	1300716
Toluene 2,4 diamine	95807		

APPENDIX 12B List of Hazardous Organic Air Pollutants

Chemical Name	CAS No.
Acetaldehyde	75070
Acetamide	60355
Acetonitrile	75058
Acetophenone	98862
Acrolein	107028
Acrylamide	79061
Acrylic acid	79107
Acrylonitrile	107131
Allyl chloride	107051
Aniline	62533
o-Anisidine	90040
Benzene	71432
Benzotrichloride	98077
Benzyl chloride	100447
Biphenyl	92524
Bis(chloromethyl)ether	542881
Bromoform	75252
1,3-Butadiene	106990
Caprolactam	105602
Carbon disulfide	75150
Carbon tetrachloride	56235
Chloroacetic acid	79118
2-Chloroacetophenone	532274
Chlorobenzene	108907
Chloroform	67663
Chloroprene	126998
Cresols and cresylic acids (mixed)	1319773
o-Cresol and *o*-cresylic acid	95487
m-Cresol and *m*-cresylic acid	108394
p-Cresol and *p*-cresylic acid	106445
Cumene	98828
1,4-Dichlorobenzene (*p*-)	106467
3,3′-Dichlorobenzidine	91941
Dichloroethyl ether [Bis(2-chloroethyl)ether]	111444
1,3-Dichloropropene	542756
Diethanolamine	111422
N,*N*-Dimethylaniline	121697
Diethyl sulfate	64675
3,3′-Dimethylbenzidine	119937
Dimethylformamide	68122
1,1-Dimethylhydrazine	57147
Dimethyl phthalate	131113
Dimethyl sulfate	77781
2,4-Dinitrophenol	51285
2,4-Dinitrotoluene	121142
1,4-Dioxane (1,4-diethyleneoxide)	123911
1,2-Diphenylhydrazine	122667
Epichlorohydrin (1-chloro-2,3-epoxypropane)	106898
Ethyl acrylate	140885
Ethylbenzene	100414
Ethyl chloride (chloroethane)	75003
Ethylene dibromide (dibromoethane)	106934
Ethylene dichloride (1,2-dichloroethane)	107062
Ethylene glycol	107211
Ethylene oxide	75218
Ethylidene dichloride (1,1-dichoroethane)	75343
Formaldehyde	50000
Glycol ethers	—
Hexachlorobenzene	118741
Hexachlorobutadiene	87683
Hexachloroethane	67721
Hexane	100543
Hydroquinone	123319
Isophorone	78591
Maleic anhydride	108316
Methanol	67561
Methyl bromide (bromomethane)	74839
Methyl chloride (chloromethane)	74873
Methyl chloroform (1,1,1-trichloroethane)	71556
Methyl ethyl ketone (2-butanone)	78933
Methyl hydrazine	60344
Methyl isobutyl ketone (hexone)	108101
Methyl isocyanate	624839
Methyl methacrylate	80626
Methyl tert-butyl ether	1634044
Methylene chloride (dichloromethane)	75092
Methylene diphenyl diisocyanate (MDI)	101688
4,4′ – Methylenedianiline	101779
Naphthalene	91203
Nitrobenzene	98953
4-Nitrophenol	100027
2-Nitropropane	79469
Phenol	108952
p-Phenylenediamine	106503
Phosgene	75445
Phthalic anhydride	85449
Polycyclic organic matter	—
Propiolactone (β-isomer)	57578
Propionaldehyde	123386
Propylene dichloride (1,2-dichloropropane)	78875
Propylene oxide	75569
Quinone	106514
Styrene	100425
1,1,2,2-Tetrachloroethane	79345
Tetrachloroethylene (perchloroethylene)	127184
Toluene	108883
2,4-Toluene diamine	95807
2,4-Toluene diisocyanate	584849

APPENDIX 12B ***(Continued.)***

Chemical Name	CAS No.	Chemical Name	CAS No.
o-Toluidine	95534	Vinyl chloride	75014
1,2,4-Trichlorobenzene	120821	Vinylidene chloride (1,1-dichloroethylene)	75354
1,1,2-Trichloroethane	79005		
Trichloroethylene	79016	Xylenes (isomers and mixtures)	1330207
2,4,5-Trichlorophenol	95954	*o*-Xylene	95476
Triethylamine	121448	*m*-Xylene	108383
2,2,4-Trimethylpentane	540841	*p*-Xylene	106423
Vinyl acetate	108054		

APPENDIX 12C Organic HAP Strippability Groups and Target Removal Efficiencies

Chemical Name	CAS No.	Chemical Name	CAS No.
Group A — Efficiency 99%[a]		Propylene dichloride (1,2-dichloropropane)	78875
Acetaldehyde	75070		
Allyl chloride	107051	Propylene oxide	75569
Benzene	71432	Styrene	100425
Benzyl chloride	100447	1,1,2,2-Tetrachloroethane	79345
Biphenyl	92524	Tetrachloroethylene (perchloroethylene)	127184
Bromoform	75252		
1,3-Butadiene	106990	Toluene	108883
Carbon disulfide	75150	1,2,4-Trichlorobenzene	120821
Carbon tetrachloride	56235	1,1,2-Trichloroethane	79005
Chlorobenzene	108907	Trichloroethylene	79016
Chloroform	67663	Triethylamine	121448
Chloroprene (2-chloro-1,3-butadiene)	126998	2,2,4-Trimethylpentane	540841
		Vinyl acetate	108054
Cumene (isopropyl benzene)	98828	Vinyl chloride	75014
1,4-Dichlorobenzene (*p*-)	106467	Vinylidene chloride (1,1-Dichloroethylene)	75354
1,3-Dichloropropene	542756		
Ethylbenzene	100414	*m*-Xylene	108383
Ethyl chloride (chloroethane)	75003	*o*-Xylene	95476
Ethylene dibromide	106934	*p*-Xylene	106423
Ethylene dichloride (1,2-dichloroethane)	107062	*Group B — Efficiency 95%*[a]	
		Acetonitrile	75058
Ethylene oxide	75218	Acetophenone	98862
Ethylene dichloride (1,1-dichloroethane)	75343	Acrolein	107028
		Acrylonitrile	107131
Hexachlorobenzene	118741	2-Chloroacetophenone	532274
Hexachlorobutadiene	87683	Dichloroethyl ether	111444
Hexachloroethane	67721	*N,N*-Dimethylaniline	121697
Hexane	110543	2,4-Dinitrophenol	51285
Methyl bromide (bromomethane)	74839	Ethyl acrylate	140885
Methyl chloride (chloromethane)	74873	Ethylene glycol dimethyl ether	110714
Methyl chloroform (1,1,1-trichloroethane)	71556	Ethylene glycol monobutyl ether acetate	112072
Methyl ethyl ketone (2-butanone)	78933	Isophorone	78591
Methyl isobutyl ketone (hexone)	108101	Methyl methacrylate	80626
Methyl tert-butyl ether	1634044	Nitrobenzene	98953
Methylene chlorides (dichloromethane)	75092	Propionaldehyde	123386
		2,4,5-Trichlorophenol	95954
Naphthalene	91203	*Group C — Efficiency 70%*[a]	
2-Nitropropane	79469	Aniline	62533
Phosgene	75445	*o*-Cresol	95487

APPENDIX 12C ***(Continued.)***

Chemical Name	CAS No.	Chemical Name	CAS No.
Diethyl sulfate	64675	Ethylene glycol monomethyl ether acetate	110496
3,3′-Dimethylbenzidine	119937		
1,1-Dimethylhydrazine	57147	Diethylene glycol diethyl ether	112367
Dimethyl sulfate	77781	Diethylene glycol dimethyl ether	111966
2,4-Dinitrotoluene	121142	Ethylene glycol monoethyl ether acetate	111159
1,4-Dioxane (1,4-diethyleneoxide)	123911		
Epichlorohydrin (1-chloro-2,3-epoxy-propane)	106896	Methanol	67561
		o-Toluidine	95534

[a]Target removal stripping efficiency for group.

APPENDIX 12D Fraction HAP Compounds Removed from Wastewater Streams[a]

Chemical Name	Fraction[b]	Chemical Name	Fraction[b]
Acetaldehyde	1.000	Ethyl chloride (chloroethane)	1.000
Acetonitrile	0.934	Ethylene dibromide	1.000
Acetophenone	0.920	Ethylene dichloride (1,2-dichloroethane)	1.000
Acrolein	0.957		
Acrylonitrile	0.960	Ethylene oxide	1.000
Allyl chloride	1.000	Ethylidene dichloride (1,1-dichloroethane)	1.000
Aniline	0.468		
Benzene	1.000	Ethylene glycol monomethyl ether acetate	0.529
Benzyl chloride	1.000		
Biphenyl	1.000	Ethylene glycol dimethyl ether	0.943
Bromoform	1.000	Ethylene glycol monobutyl ether acerate	0.927
1,3-Butadiene	1.000		
Carbon disulfide	1.000	Ethylene glycol monoethyl ether acerate	0.470
Carbon tetrachloride	1.000		
2-Chloroacetophenone	0.939	Hexachlorobenzene	1.000
Chlorobenzene	1.000	Hexachlorobutadiene	1.000
Chloroform	1.000	Hexachloroethane	1.000
Chloroprene (2-chloro-1,3-butadiene)	1.000	Hexane	1.000
		Isophorone	0.945
o-Cresol	0.448	Methanol	0.829
Cumene (isopropyl benzene)	1.000	Methyl bromide (bromomethane)	1.000
1,4-Dichlorobenzene (*p*-)	1.000	Methyl chloride (chloromethane)	1.000
Dichloroethyl ether	0.935	Methyl chloroform (1,1,1-trichloroethane)	1.000
1,3-Dichloropropene	1.000		
N,N-Dimethylaniline	0.927	Methyl ethyl ketone (2-butanone)	1.000
Diethyl sulfate	0.814	Methyl isobutyl ketone (hexone)	1.000
Diethylene glycol diethyl ether	0.523	Methyl methacrylate	0.958
Diethylene glycol dimethyl ether	0.425	Methyl tert-butyl ether	1.000
3,3′-Dimethylbenzidine	0.635	Methylene chloride (dichloromethane)	1.000
1,1-Dimethylhydrazine	0.448		
Dimethyl sulfate	0.697	Naphthalene	1.000
2,4-Dinitrophenol	0.908	Nitrobenzene	0.936
2,4-Dinitrotoluene	0.626	2-Nitropropane	1.000
1,4-Dioxane (1,4-diethyleneoxide)	0.787	Phosgene	1.000
Epichlorohydrin (1-chloro-2,3-epoxypropane)	0.890	Propionaldehyde	0.952
		Propylene dichloride (1,2-dichloropropane)	1.000
Ethyl acrylate	0.961		
Ethylbenzene	1.000	Propylene oxide	1.000

APPENDIX 12D ***(Continued.)***[a]

Chemical Name	Fraction[b]	Chemical Name	Fraction[b]
Styrene	1.000	Triethylamine	1.000
1,1,2,2-Tetrachloroethane	1.000	2,2-4-Trimethylpentane	1.000
Tetrachloroethylene (perchloroethylene)	1.000	Vinyl acetate	1.000
		Vinyl chloride	1.000
Toluene	1.000	Vinylidene chloride (1,1-dichloroethylene)	1.000
o-Toluidine	0.487		
1,2,4-Trichlorobenzene	1.000	*m*-Xylene	1.000
1,1,2-Trichloroethane	1.000	*o*-Xylene	1.000
Trichloroethylene	1.000	*p*-Xylene	1.000
2,4,5-Trichlorophenol	0.914		

[a]Adopted from reference 10.
[b]Strippability factor Fr

13

Liquid Storage and Transfer

Liquid storage and transfer systems have been sources of VOC emissions through the years, contributing to the general ozone levels particularly in nonattainment areas. As a result, regulations have been promulgated by the EPA, states, and regional entities to control these emission sources. The CAAA require emissions reductions of HAPs, and additional equipment control requirements were proposed in accordance with the act in December 1992. These regulations will be promulgated in early 1994 and are officially named the Hazardous Organic National Emission Standards for Hazardous Air Pollutants or HON.

While the regulatory requirements have seemingly expanded, most of the liquid storage tankage emission control restrictions in the various regulations are quite similar to the existing regulations. In addition, various transfer operation emission control regulations are also similar. Some differences in emission control efficiency may occur in local or regional restrictions in transfer operations, but the differences do not seriously alter the general equipment requirements.

In this chapter the liquid storage facilities are reviewed initially, followed by transfer operations. Since tankage and transfer are frequently combined in operating plants, these two systems are interrelated in some respects.

13.1 GENERAL TANKAGE DESCRIPTIONS

The discussion of tankage in this chapter is only concerned with above ground storage tanks (AST). Underground storage tanks are mentioned in Chapter 12. The above ground storage tanks normally installed in the chemical and petroleum refinery process industries are described in the following sections.

13.1.1 Fixed Roof Tanks

The original storage tanks in many facilities were open tanks. A fixed roof on the tank (FRT), as shown in Figure 13.1, reduced emissions and protected the liquid

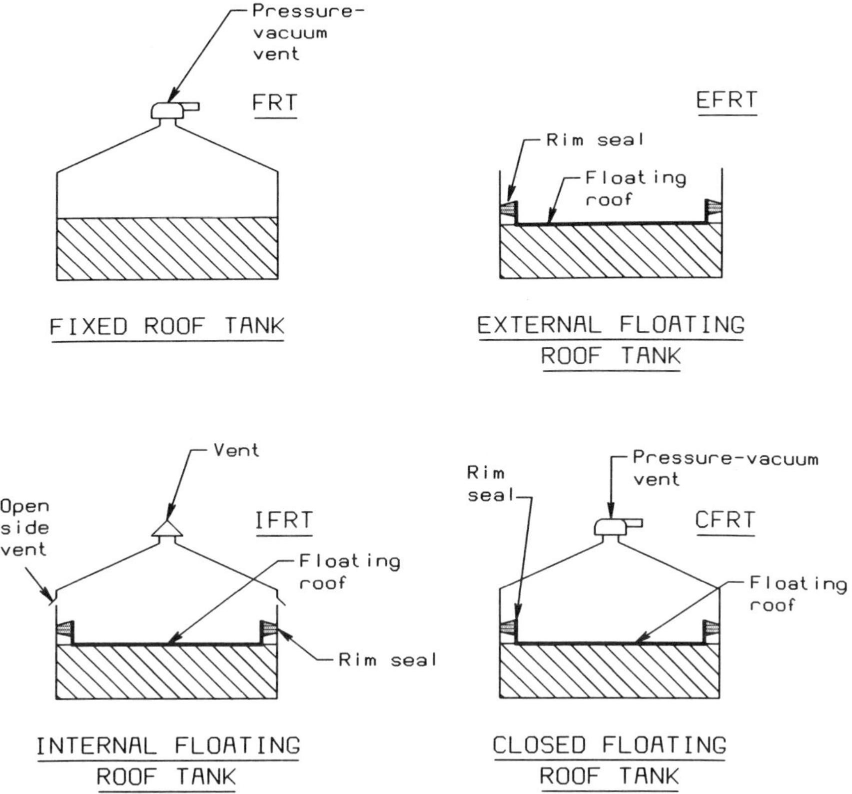

Figure 13.1 *Basic types of storage tanks (courtesy of CBI).*

from the elements and contamination. However, the liquid has a vapor pressure, and a vapor concentration will build in the vapor zone above the liquid level. A pressure vacuum vent as shown is necessary to compensate for diurnal effects, with pressure building during the heat of the day and evening coolness, rain, or snow resulting in a reduced pressure that frequently is slightly below ambient pressure. This latter condition opens the vacuum vent, admitting air to balance the pressure.

A combination of air and organic vapor, particularly with a volatile liquid, often results in a combustible mixture in the vapor zone above the liquid. This situation is potentially hazardous, particularly in lightning storms. While lightning may be rather infrequent, a combustible mixture is a source of concern. Normally, liquids with a vapor pressure < 1.5 psia (10.33 kPa) are stored in these tanks. However, emission concerns generally require tank changes or liquids with extremely low vapor pressures.

13.1.2 External Floating Roof Tanks

The open liquid surface was eliminated by placing a floating roof on the surface of the liquid as shown in Figure 13.1. This combination of a floating deck or roof has been termed an external floating roof tank (EFRT). The original roof or pan floated on the surface of the liquid and was sealed against the walls of the tank, significantly reducing evaporative losses. Although this development was a considerable improvement, rain, snow, and ice frequently created problems for this tankage control system. Improvements over the years have minimized these problems and current EFRTs are widely used.

One advantage of the EFRT is the very large sizes that can be installed economically compared with fixed roof or internal floating roof tanks described in Section 13.1.3. Improved roof seals have further reduced losses and improved performance.

13.1.3 Internal Floating Roof Tanks

An improvement on fixed roof tanks is the installation of an internal floating roof that significantly reduces evaporation losses and also provides protection from the elements. While the floating roof is sealed at the edge, sufficient organic vapor can accumulate in the vapor space that it may result in a combustible mixture. As a result, the internal floating roof tank (IFRT) as shown in Figure 13.1 has vapor vents that permit dilution of the vapor space mixture, thereby reducing the concentration of combustible vapor mixtures. As the wind increases, a flow of air circulates through the vents located around the periphery of the tank, resulting in vapor–air turnovers within the tank.

Floating roof seals have also improved in the IFRT, and evaporative losses have been reduced through the years.

13.1.4 Closed Floating Roof Tanks

In attempting to achieve an essentially zero emission tank, the IFRT has been modified to the closed floating roof tank (CFRT) configuration shown in Figure 13.1. The vents are closed or eliminated in the CFRT and an enclosed emission venting system is connected to the vapor zone. A pressure–vacuum vent provides relief for either condition should the pressure vary. Injection of an inert gas maintains a noncombustible mixture in the vapor zone, an essentially constant pressure condition within the tank, and a small flow through the discharge system.

The internal floating roof minimizes the evaporation rate, and the amount of pollutants that must be controlled or recovered. If a combustion or other control device is the control mechanism on the vent, a high tank liquid evaporation rate increases material losses through the control device. Moreover, high evaporation rates require larger control devices.

13.1.5 Other Considerations

Many feed or product stocks contain water that frequently separates in the tankage. Since the organic fluid specific gravity is generally lower than the water gravity, the separated water normally collects in the tank bottom. This water volume increases with time and must be periodically withdrawn from the tank to prevent the water level reaching the suction line of the tank pump and mixing with the tank contents. Consequently, water drainage from the tank must be carefully controlled to prevent tank liquid discharge and organic vapor emissions. In many tanks, a certain amount of organic material is dissolved in the water layer and the interface between the liquid is frequently an emulsion. Opening the drain valve to an open sewer system may result in vaporizing a considerable portion of the dissolved organic material, potentially exposing personnel to organic vapors. Closed methods of draining water from tankage are necessary to reduce potential exposures and control emissions of VOCs and HAPs.

In the EFRT configurations the floating roofs are exposed to the elements, particularly rain and snow. Runoff is generally accomplished through two different procedures. One method provides an overflow well in the floating roof, and the water is discharged into the general liquid layer. The water is normally maintained in separate phases and accumulates in the bottom water layer. Some water may be absorbed in the main liquid volume if the upper liquid water concentration is below the saturation level. In addition, the rain water may agglomerate, forming globs that can be discharged through the tank discharge pump. Thus runoff procedure must be carefully investigated to ensure that any of the conditions described are acceptable in further processing or product sales.

A second roof drain alternate is constructed of flexible piping or hose that connects the roof drain to an outside drain nozzle in the lower portion of the tank. Since the floating roof level varies considerably, the piping or hose drain must be flexible. A typical drain connection is shown in Figure 13.8 The roof drainage in this situation is considered contaminated, and the overflow from the drain nozzle must be connected to a closed process sewer for treatment. In the other configuration where the rain runoff overflows directly into the tank, water contamination is not a problem. All water drained from the tank bottom layer must be treated in either configuration.

In addition to water decanting and drainage, general external tankage considerations are also a concern. Tanks should be placed within or surrounded by a dike system that will contain the full contents of the tank in the event of a rupture. If fires occur, tank overflows and contaminated fire water can be contained within the dike system. These mixtures or relatively pure liquids can be released on a controlled basis from the dike for treatment and recovery. Dikes are either earthen or concrete barriers, depending upon tank size, contents, and location. Usually diking systems are designed for 100% containment of filled design capacity of the tank. Normally, one tank is located within a diked enclosure. However, this can be modified based upon tank sizes and the materials contained within the tanks. Tanks located within the boundary limits of a process unit should be limited in capacity and emergency procedures must be carefully designed for control over a wide variety of simulated conditions. Dikes within process areas are limited in capacity, but must be considered in preparing emergency response procedures.

Since earthen dikes are commonly located around storage tanks, tankage leakage or overflows can diffuse through the dikes into the ground. In addition, the earth immediately around the tank may also permit dispersion of tank leakage into the ground. This leakage can be prevented by placing barriers in the ground and dikes to prevent seepage. Thick clay layers along with polymer liners have been used to prevent seepage and also permit collection of the leakage. One area of concern is leakage from the tank bottom that may not be identified or controlled and can seep into the ground directly beneath the tank if the earth under the tank does not have a barrier. Detectors under the tanks provide a method of identifying leakage, which permits personnel to empty the tank and repair the leak.

While the various comments refer to direct evaporation and leakage, an important aspect of tankage operation is sludge removal and cleaning. Many tankage liquids do not produce a sludge and cleaning is a relatively straightforward operation. Where sludges remain after draining tanks, sludge removal and tank cleaning can be difficult. The sludge contains some of the tank liquid material, which presents potential exposure problems associated with vapor pressure and dermal considerations. Resuspension of this material may be possible, permitting pumpout, which would probably be aided by the addition of dispersants. Physical removal of this material places the sludge in the waste category and requires adherence to EPA waste disposal regulations. Stream or water pressure cleaning following sludge removal also produces a mixture that must be properly processed to minimize potential exposure. A new procedure described in reference 27 has been successfully demonstrated that separates and recovers oil from tank sludge, leaving a very small layer of solids (10 cm). This material is not removed from the tank eliminating mechanical cleaning and potential exposures.

13.2 TANKAGE LOSS SOURCES

The previous descriptions refer to emissions and evaporation losses but do not describe in detail the basic sources of these emissions. In this section loss sources are described, and methods of calculating the emission losses are presented in Section 13.3.

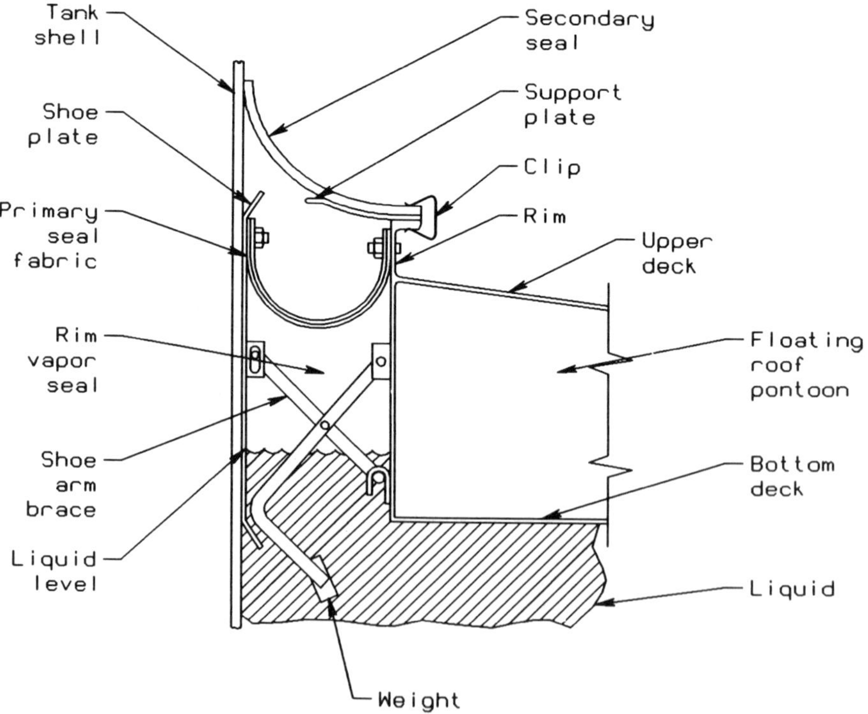

Figure 13.2 Mechanical shoe primary seal with rim-mounted secondary seal (courtesy of CBI).

13.2.1 Rim Seals

With the exception of the FRT, the tanks in Figure 13.1 have floating roofs which have seals to minimize emissions into the vapor space above the floating roof. There are three primary seals whose major purpose is direct vapor emission control. However, wind circulation affects their efficiency and secondary seals are frequently installed to minimize overall emission losses. A review of the various seals and their effectiveness follows.

Mechanical Shoe Seal. A typical primary mechanical shoe seal is shown in Figure 13.2. As shown, a metallic shoe or plate is pressed against the tank wall by a scissors-type mechanical device. The plate or shoe against the tank wall may vary

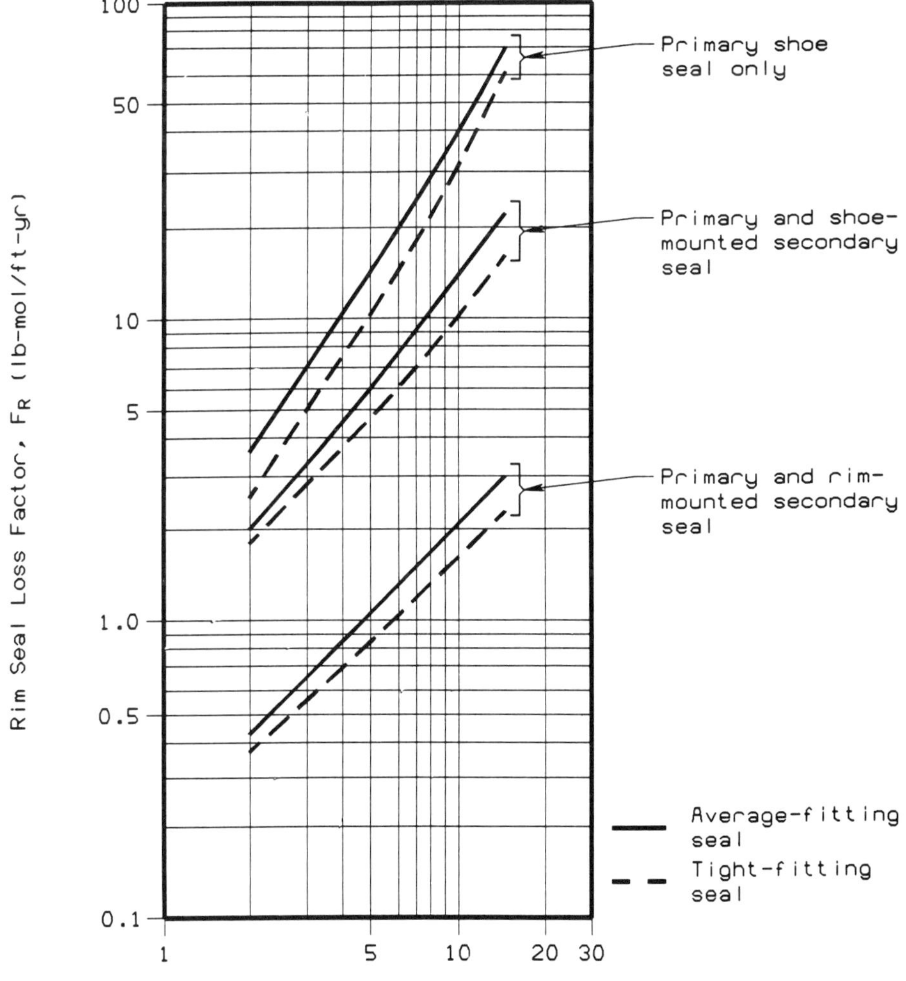

Figure 13.3 *Mechanical shoe seal loss factor a function of wind velocity in welded tank (courtesy of CBI).*

in height depending upon the manufacturer, but the lower part of the shoe, which is curved slightly inward near the bottom of the plate, is always located in the liquid. A minimum shoe height is required in various EPA regulations.

The shoes are built in sections or arcs, and the sections are then joined to form a continuous circumferential ring against the tank wall. However, the opening between the floating roof and the shoe is partially filled with liquid, resulting in emissions due to vaporization. The emissions are contained by a fabric termed the primary seal fabric which is shown in Figure 13.2. In addition, a secondary seal clipped to the rim of the floating roof is pressed against the wall. This combination of primary and secondary seal considerably reduces emissions, as shown by the rim seal loss factor comparisons shown in Figure 13.3. The shoe-mounted secondary seal referred to in Figure 13.3 does not clip to the rim as shown, but instead is fastened to the top of the shoe and presses against the wall several inches above the shoe. This shoe rim mounted secondary seal does not cover the fabric and the loss factor is definitely higher for this configuration than for the standard configuration shown in Figure 13.2.

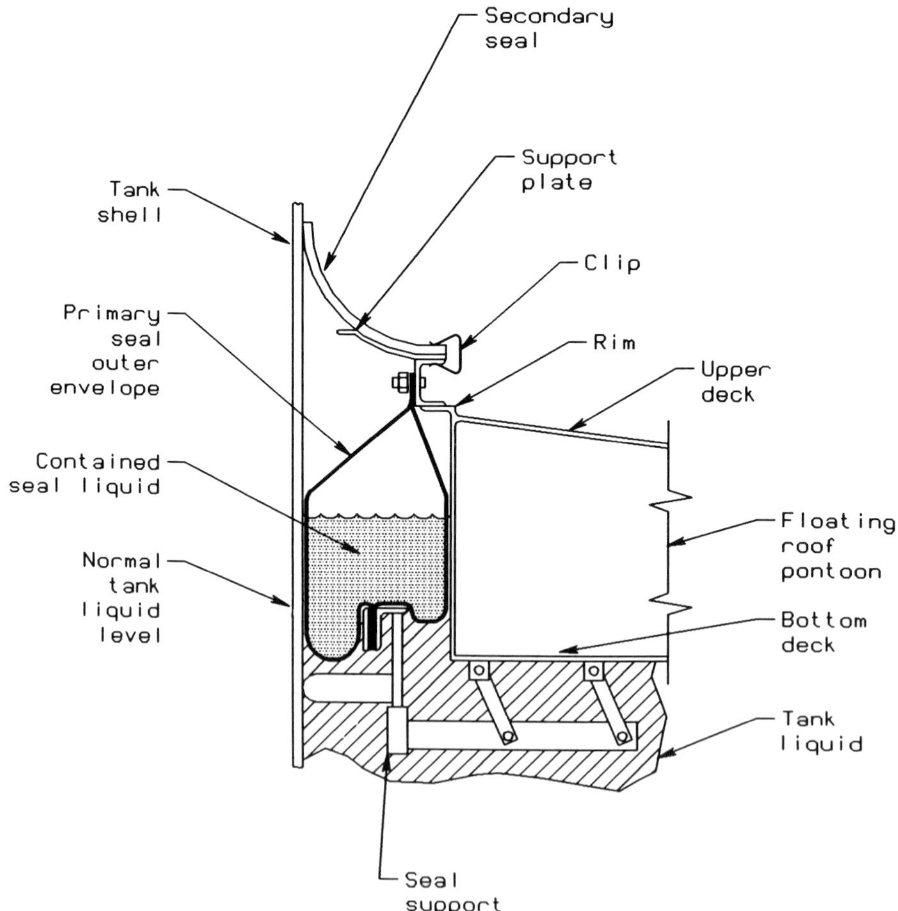

Figure 13.4 *Liquid-filled resilient primary seal mounted in tank liquid (courtesy of CBI).*

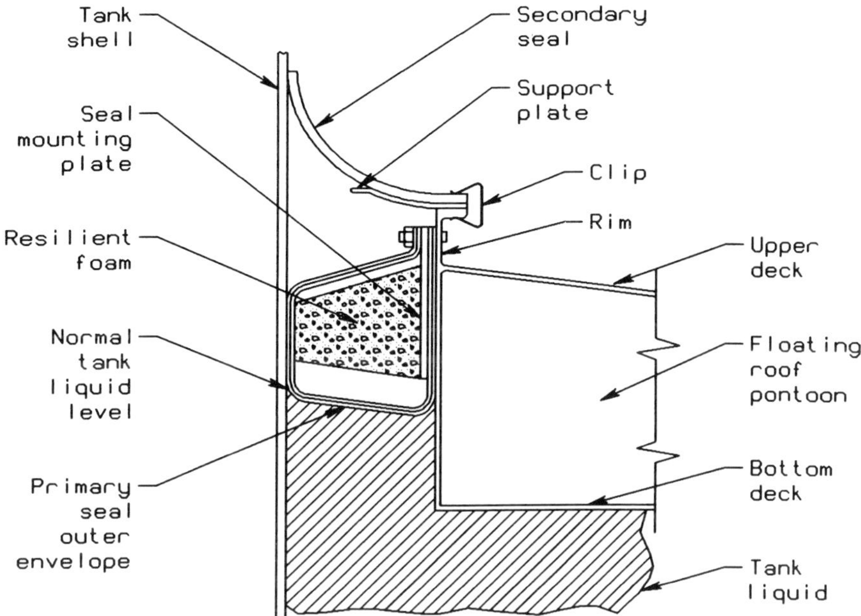

Figure 13.5 *Foam-filled resilient primary seal mounted in tank liquid (courtesy of CBI).*

In Figure 13.3 there are differences in the loss factors associated with average- and tight-fitting seals. Although reasonably close, there is a difference which the EPA controls by placing limits on the gap openings permitted in the primary and secondary seals. In addition, the seal loss factor is a function of the wind speed in an EFRT. Apparently, the wind enters gaps in the seal openings and flows to lower static pressure seal gaps with emissions losses increasing in conjunction with increasing wind speed.

Foam- or Liquid-Filled Primary Seals. There are two other basic types of primary seals, and they are shown in Figures 13.4 and 13.5. These seals are filled with a liquid or resilient foam which provides a seal between the rim and tank wall. Although filled with two different materials, their effectiveness is similar and depends upon their location with respect to the liquid between the rim and tank wall. If the resilient seals are placed in the liquid as shown in Figures 13.4 and 13.5, their emissions control effectiveness is much greater than if the seals were located in the vapor space above the liquid, as shown in Figure 13.6. A more complete comparison of their effectiveness is shown in Figure 13.7.

In Figure 13.7*A*, the *liquid-mounted* primary seal has a considerably lower rim loss factor than the *vapor-mounted* primary seal shown in Figure 13.7*B*. The emission loss also increases with increasing wind speed.

Secondary Seals. When rim-mounted secondary seals are included, the seal loss factor decreases for both liquid- and vapor-mounted primary seals, but the liquid-mounted primary system has a significantly lower seal loss factor, particularly when combined with secondary seals. The tight-fitting secondary seal is a considerable

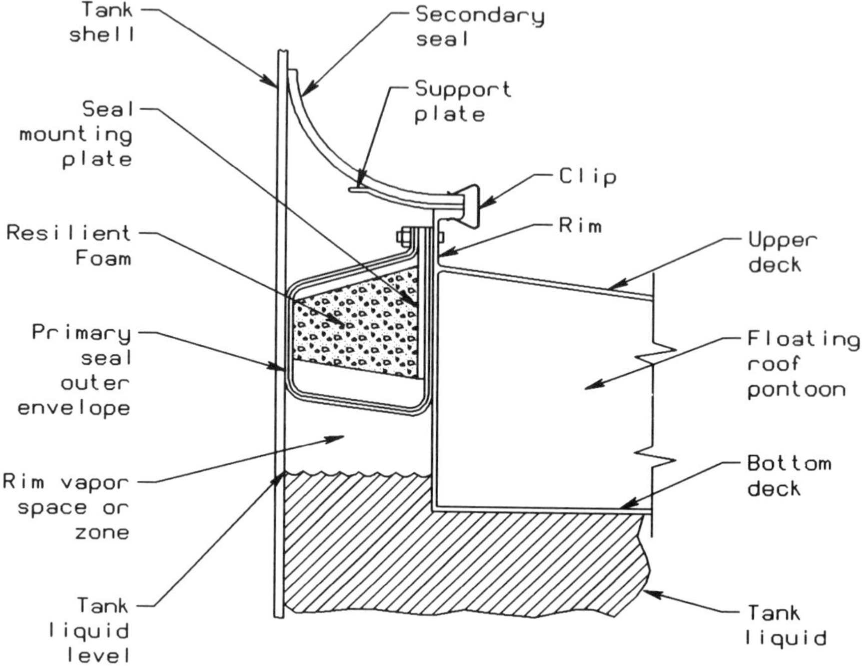

Figure 13.6 *Foam-filled resilient primary seal mounted in vapor zone (courtesy of CBI).*

improvement for the vapor-mounted primary seal, but still has a higher overall seal loss factor. While secondary seals aid in reducing the seal loss factor, they are not always installed. In these situations weather shields are installed, which are attached to the floating roof rim with a hinge or similar device. Since they are not installed as a long single seal but as a series of relatively narrow width shields, their effectiveness is considerably less than that attained with a secondary seal. The weather shields provide protection for the seal fabrics and other nonmetallic accessories from the sunlight and weather.

13.2.2 Floating Roofs

Although floating roofs are frequently discussed as though only one design is installed, there are variations in the design of floating roofs that affect emissions. Some of these differences follow:

EFRT Floating Roofs. One basic design of an EFRT floating roof is shown in Figure 13.8. The floating roof consists of an annular ring of pontoons that are the outer ring of the floating roof as shown. A single deck is connected to the annular ring which rests on the liquid surface. However, this deck will expand upward if the vapor pressure increases owing to a rise in the liquid temperature as a result of solar heating. This expansion or balloon effect tends to reduce heating by raising the deck surface with a vapor zone acting as an insulator. The deck and pontoons in this system are welded and provide an unbroken surface.

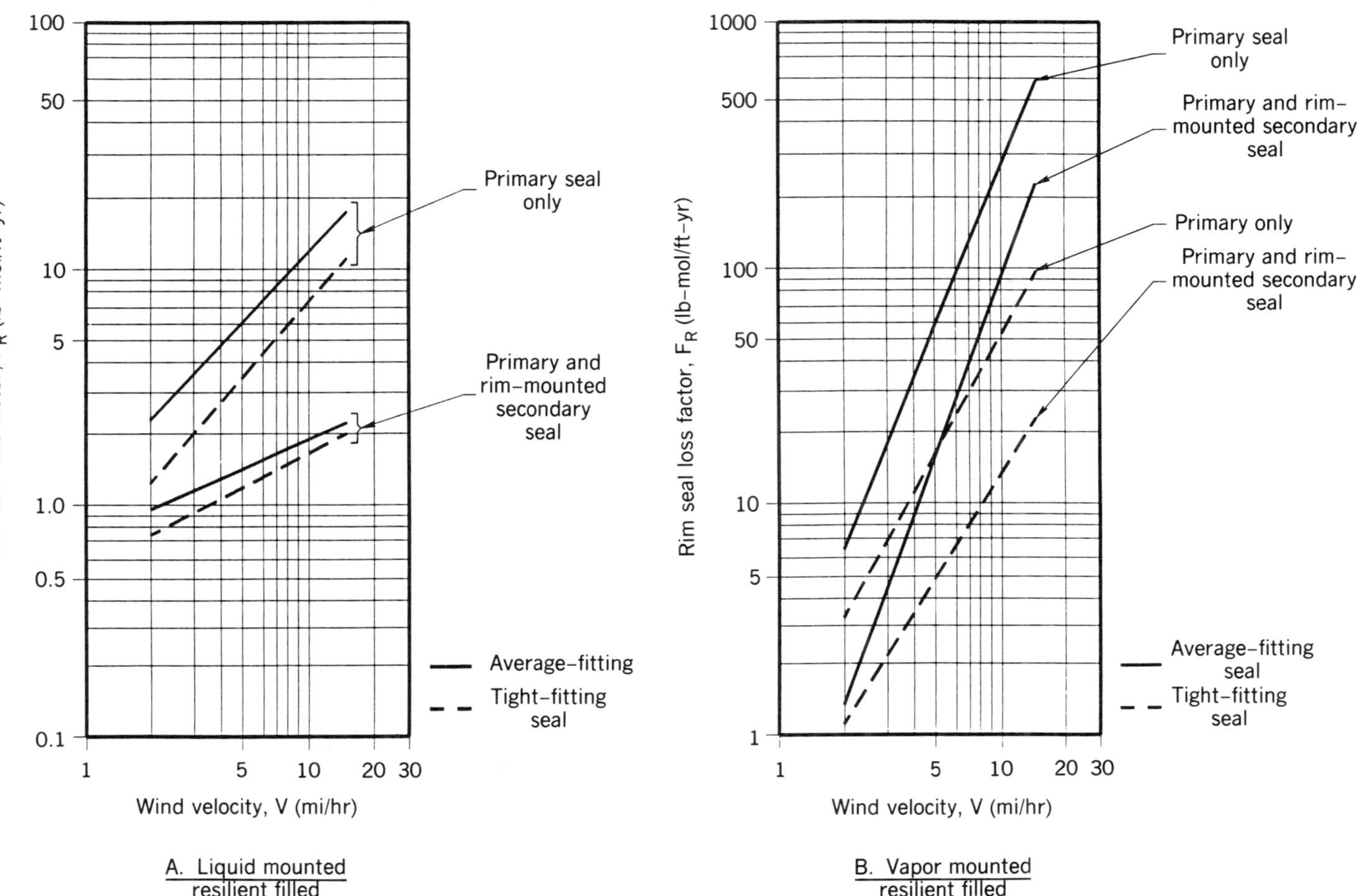

Figure 13.7 *Comparison loss factors for resilient-filled primary seals mounted in liquid and vapor zones (courtesy of CBI).*

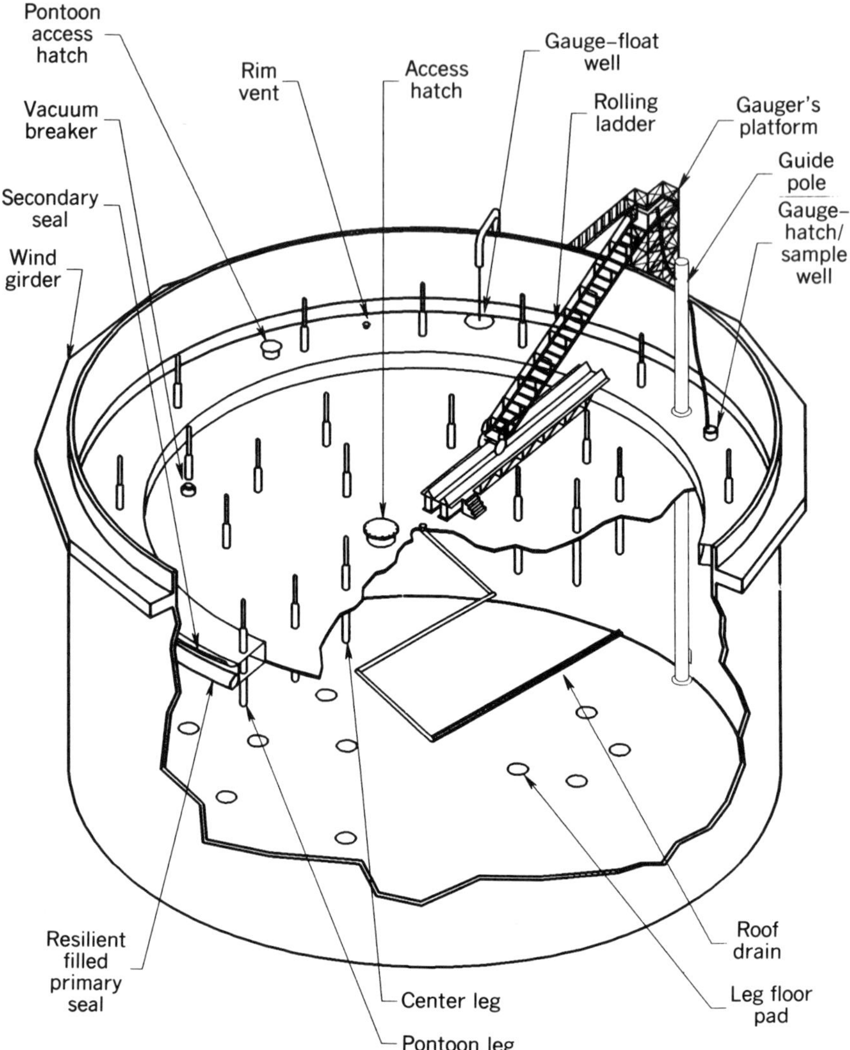

Figure 13.8 *External floating roof with annular pontoons (courtesy of CBI).*

Another deck also used in the EFRT is a double deck which covers the entire liquid surface. The decks are separated by sealed bulkheads which essentially form liquid-tight compartments and support the upper deck. This entire system is also welded and the inner vapor zone is an insulating air space that minimizes liquid surface heating. Both types of decks are welded and do not emit any vapors.

IFRT Floating Roofs. The IFRT roofs are of either welded or bolted construction. Since these internal floating decks are not exposed to the weather, the design of these floating roofs differs from the EFRT roofs. The welded roof consists of a number of welded plates with a vertical rim at the periphery of the deck. This deck can float on the liquid surface or an annular buoyant ring may be installed to provide somewhat more buoyancy. However, the entire roof system is welded.

There are also bolted EFRT roofs that either contact the liquid surface or are not in contact with the liquid surface. In the noncontact type, bolted sheets are mechanically fastened to a grid framework. This roof is supported by tubular pontoons placed below the deck, resulting in the grid work located in the vapor space. A vertical rim extends downward from the periphery of the deck into the liquid. Another bolted roof configuration consists of sealed panels that are mechanically joined. The panels float on the liquid surface and are constructed with a rigid foam core that provides buoyancy. The surface of the panels can be a thin metal skin or a reinforced polymer. Differences in deck construction are reflected in emission loss rates.

13.2.3 EFRT Floating Roof Fittings

A rather large number of different fittings is encountered in tanks as shown in an EFRT (Fig. 13.8). These fittings are shown in detail in Figures 13.9–13.17 with their associated EFRT loss factors shown in Table 13.3. Emission losses from these fittings are uncontrolled or controlled. The controlled values significantly reduce the losses associated with each fitting type. In Figures 13.9–13.17, the liquid level is generally above the lower openings of the hatches. In situations where the deck is above the liquid level, the hatches and other openings should be extended below the liquid surface. This reduces emission losses and is normally encountered in IFRT floating roofs.

The large number of fittings in the floating roofs necessary for various functions are briefly described below.

Access Hatch. A typical access hatch is shown in Figure 13.9 that has a removable cover. This hatch is designed for maintenance, permitting the passage of workers and construction materials through the roof. Maintenance can be

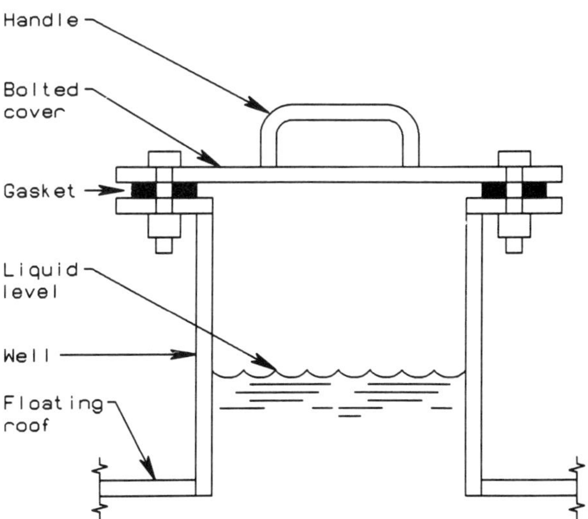

Figure 13.9 *Access hatch external floating roof tank (courtesy of CBI).*

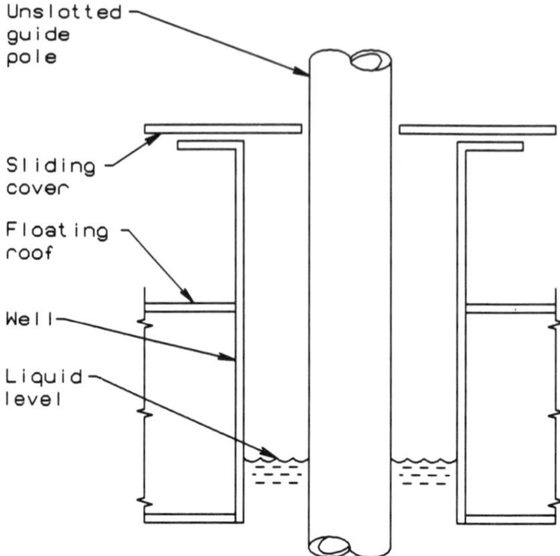

Figure 13.10 *Unslotted guide pole and sample well external floating roof tank (courtesy of CBI).*

performed on the floating roof, its accessories, or in the tank bottom area. When the tank is emptied, the roof is supported above the tank bottom by legs that are shown in Figure 13.16 and workers can enter the section below the roof through the access hatch.

Losses through the access hatch are reduced by placing a gasket on the hatch flange face and bolting the cover. The reduction in the emissions loss factor is shown in Figure 13.18.

Unslotted Guide-Pole Well. Floating roofs can rotate damaging rim seals, roof drains, gauging systems, and the sliding ladder system shown in Figure 13.8. The device developed to prevent roof rotation is a guide pole that is fixed to the bottom of the tank and permanently connected to the tank wall at the top of the column, which is above the maximum roof height. This unslotted guide pole and well is shown in Figure 13.10. Rollers that ride on the outside surface of the guide pole are connected to the well, minimizing any movement between the well wall and the guide-pole surface.

While movement between the guide pole and the well is small, some movement exists and a sliding cover that encloses the annular opening as shown in Figure 13.10 permits small movements. Consequently, the sliding cover cannot be bolted, but placing a gasket between the sliding cover and the well flange will reduce emissions, as shown in Table 13.3.

Slotted Guide Pole / Sample Well. Sampling is required periodically, and a slotted pole provides a stilling zone for the sampler, with the slots permitting liquid movement through the pole for a well-mixed sample. The slotted guide pole/sample well is shown in Figure 13.11. A sliding cover over the annular area reduces some of the emissions, but liquid within the pole is a source of emissions that can

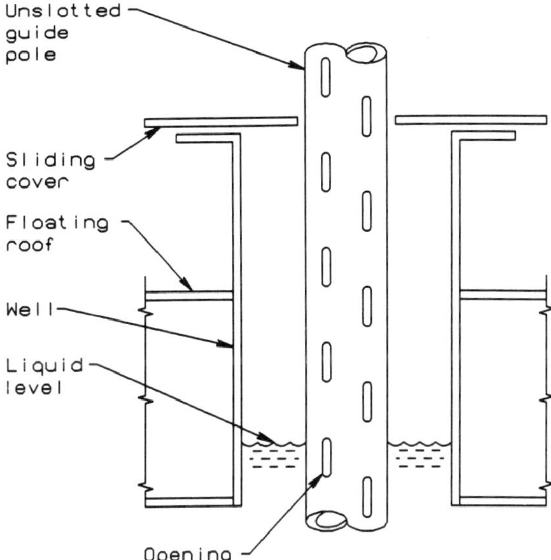

Figure 13.11 *Slotted guide pole and sample well external floating roof tank (courtesy of CBI).*

be discharged to the atmosphere through the slots above the sliding cover. The emissions within the pole can be reduced by placing a loose-fitting float within the slotted pole that is removable.

A comparison of emission loss factors for an ungasketed sliding cover with and without a float is presented in Table 13.3. The major reduction in loss factor is achieved with a gasketed sliding cover and a float within the slotted pipe.

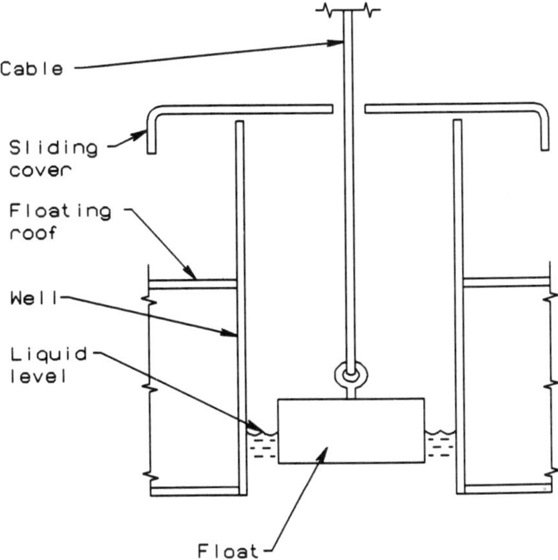

Figure 13.12 *Gauge-float well external floating roof tanks (courtesy of CBI).*

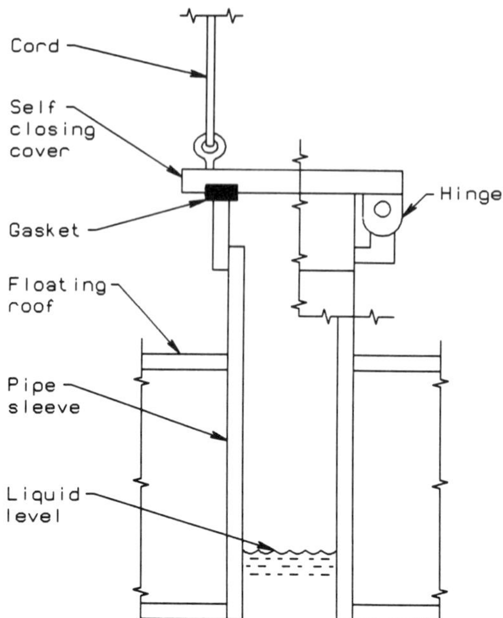

Figure 13.13 *Gauge hatch and sample well external floating roof tanks (courtesy of CBI).*

Gauge-Float Well. The gauge float shown in Figure 13.12 indicates the liquid level within the tank. The float is contained within the well located in the floating roof. This float is connected with a cable or tape to an indicator located on the external wall of the tank. The cable connection to the gauge float is shown in Figure 13.8.

The cover for the gauge float is a sliding cover with a very small opening for the connection cable, and the emission loss factor for an unbolted sliding cover is shown in Table 13.3. For a gasketed, unbolted cover, the loss factor falls considerably. A further reduction can be achieved by bolting the cover with a gasket. Since a pipe does not extend through this well, the cover can be bolted.

Gauge-Hatch Sample Well. These openings permit hand gauging the liquid level in the tank and obtaining a manual sample, often termed a thief sample. A typical gauge-hatch sample well is shown in Figure 13.13. This well and hatch, which has a self-closing cover, is normally located below the gauger's platform shown in Figure 13.8. A cord connected to the gauger's platform permits opening the hatch from the platform and dropping a level indicator or sampling device from the platform.

Emission loss factors for gasketed and nongasketed covers are shown in Table 13.3.

Vacuum Breaker. As mentioned above, the floating roof is supported on legs that position it above the bottom floor of the tank. Since a vacuum or pressure difference may be created between the volume below the floating roof and the large volume above, a vacuum breaker equalizes the pressure. One vacuum breaker type is shown in Figure 13.14.

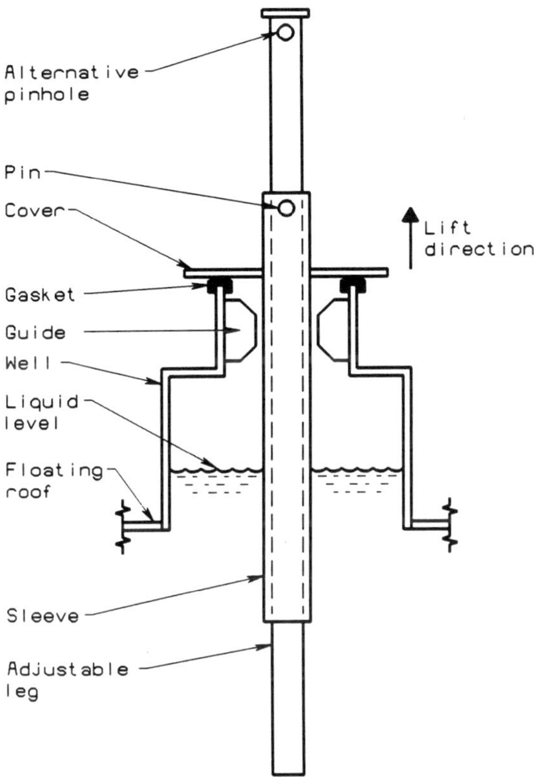

Figure 13.14 *Vacuum breaker external floating roof tanks (courtesy of CBI).*

In the breaker shown, a sleeve is permanently connected to the cover plate. Within the sleeve is a pipe or leg that can be adjusted in height and fixed in position with a pin. When the leg extension meets the tank floor, the cover plate opens, equalizing the pressure. The top of the adjustable leg is closed to minimize emissions loss. As shown in Figure 13.14, a gasket is placed beneath the cover. Additional weight can be placed on the cover to reduce leakage through the gasket region. Emission loss factors shown in Table 13.3 indicate a significant difference between the gasketed and nongasketed vacuum breakers.

Roof Drains. Since rainwater will accumulate on a floating roof, the rainwater is drained by two different procedures currently termed closed or open. In closed drainage systems, rainwater from the floating roof is carried through a drain located within the liquid tank volume to the outside of the tank under gravity flow. The closed drain can be a hose or an articulated pipe. The latter is shown in Figure 13.8. A closed drain is not in contact with the tank contents and is not a source of emissions.

In the open drain system, rainwater from the roof surface is discharged into the tank. Since there normally is water separation and drainage in the tank, the rainwater will accumulate in the tank bottom where it is periodically withdrawn. These open drains are basically an open pipe that extends below the level of the

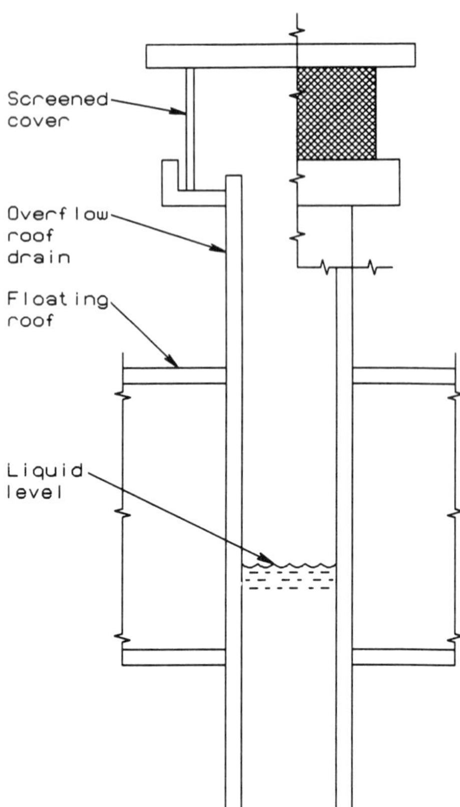

Figure 13.15 Roof drain external floating roof tanks (courtesy of CBI).

liquid, but the liquid is a source of emissions. Because of the high liquid level, open drains are limited to double-deck floating roofs.

The open drains can be flush with the top surface of the floating roof which immediately drains rainwater into the tank. An overflow drain as shown in Figure 13.15 is elevated above the top surface of the floating roof permitting rainwater accumulation on the floating roof. This type of drain can be used with either an open or closed drain system. When there is a completely open drain, the emission loss factor as shown in Table 13.3 is considerably larger than with a drain that is considered 90% closed. The screened overflow drain may have a somewhat larger opening area than 90%, which would place the emission loss factor between the two curves. A weighted damper-type opening in the overflow system will achieve a 90% closed position if it cannot be achieved by other methods.

Roof Legs. As shown in Figure 13.8, a large number of roof legs are placed in the roof to ensure a satisfactory distribution of support for the floating roofs that are relatively large in area. Roof legs are also placed within the pontoon area to insure adequate support for the pontoons. In addition, a large number of roof legs are installed in double-deck roofs. An example of a roof leg for a floating roof is shown in Figure 13.16.

The roof leg is similar in design to the vacuum breaker shown in Figure 13.14. A pipe sleeve extending through the floating roof is securely fastened to the roof.

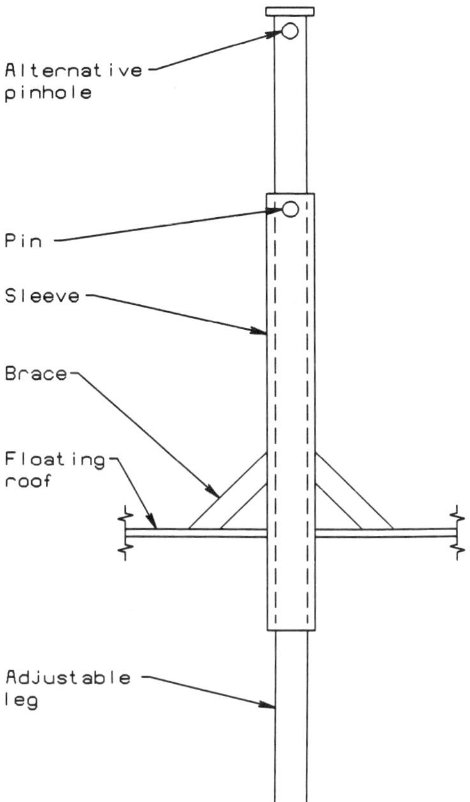

Figure 13.16 *Roof leg external floating roof tanks (courtesy of CBI).*

Within the sleeve is a leg that can be moved to the desired height. When the desired leg length is determined, the inner leg is held in position by a pin extending through both the outer sleeve and the leg as shown.

Emission loss factors for roof legs are shown in Table 13.3. The loss factor is, in general, a function of the leg diameter.

Rim Vent. A typical rim vent is shown in Figure 13.17. These rim vents are only installed with mechanical shoe primary seals to release excess pressure or break a vacuum that may develop. The mechanical shoe seal has an open vapor zone above the liquid surface that is enclosed by the primary seal fabric. Increases in temperature that raise the vapor pressure can result in a pressure rise. Conversely, a decrease in temperature may reduce the general pressure in the vapor zone. The rim vent minimizes the pressure changes, protecting the primary seal fabric. Emission loss factors for rim vents are shown in Table 13.3, with lower emission factors achieved in the gasketed vent.

13.2.4 IFRT Fittings

The internal floating roof is similar to the external roof and many fittings are quite similar to the fittings described for the EFRT tanks. However, some differences

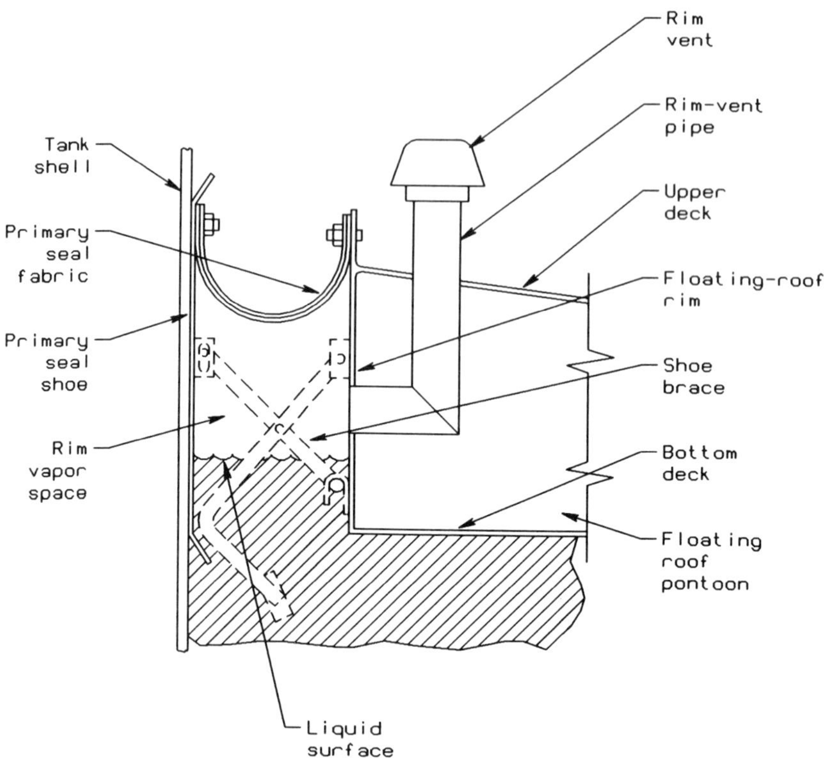

Figure 13.17 *Rim vent with mechanical shoe seal (courtesy of CBI).*

exist owing to the cone or dome roof in the IFRT and changes in the emission loss factors. The IFRT is not seriously affected by the wind, resulting in emission loss factors that can be considered constant for emission calculations. A brief summary of the fittings follows.

Floating Decks. These were generally described in Section 13.2.2 and basically consist of bolting sheets or panels together to form a deck or welding materials to form a deck. These decks are either noncontact (above the liquid surface) or contact, where the deck floats on the liquid surface. The deck construction can be further classified as shown in Figure 13.18.

1. *Noncontact Decks with Bolted Seams.* This is generally constructed of bolted metal sections, frequently aluminum, which are supported by sealed pontoons normally constructed with the same material. Where the deck is penetrated as shown in Figure 13.18*A*, the wells extend downward into the liquid. Also, a rim extends downward into the liquid. There are some variations in the deck section designs and pontoon construction, but the figure is representative of this type of design. A flexible wiper seal is shown as an example of this type of seal.

2. *Contact Decks with Bolted Seams.* As shown in Figure 13.18*B*, these decks consist of buoyant panels bolted together. The panels are generally of honeycomb

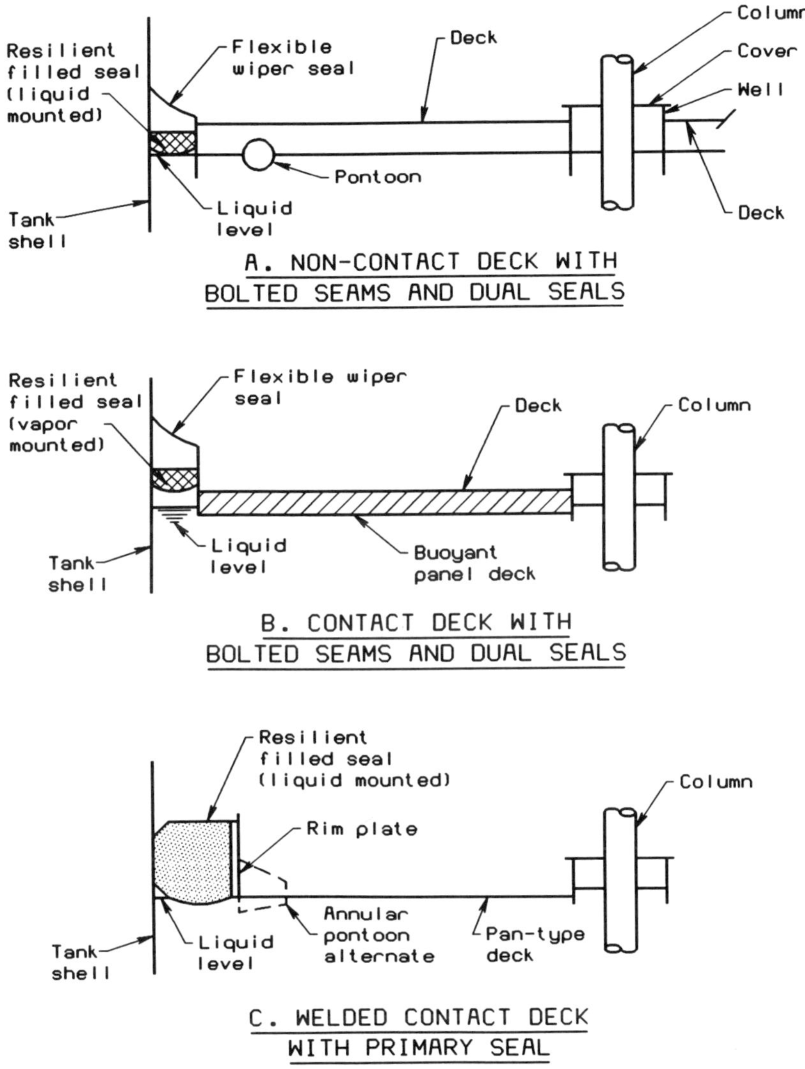

Figure 13.18 *Typical internal floating decks with accompanying seals (courtesy of CBI).*

or foam construction encased in a metal or polymeric outer skin. Since the panel floats on the liquid, well extensions for the columns are not required. A vapor-mounted primary seal and a flexible wiper secondary seal are shown as examples of combined rim seals.

3. *Welded Contact Decks.* This deck generally consists of steel plates welded together with a vertical rim welded at the circumference of the deck to provide a floatation deck that floats on the surface of the liquid. These decks are termed pans and the simple design has been replaced in many installations with an annular pontoon rim similar to an external floating roof, as shown in Figure 13.18*C*, or a double-deck pontoon. Both of these designs float on the liquid

surface. A liquid-mounted primary resilient seal is shown as an example of a primary seal.

Rim Seals. These seals are installed to control evaporative losses from the annular area between the tank wall and the deck. Most of the seals in IFRT service are nonmetallic and lightweight, which is a concern since many of the internal decks are constructed of lightweight materials. In addition, the lightweight seals do not extend any significant distance above the roof deck. This is a concern when the tank is filled as extended rim seals may interfere with the rafters associated with supporting the tank roof.

While nonmetallic rim seals are preferred, the seal material must be compatible with the tank liquid and also temperature storage conditions. The primary seals, in general, are liquid or vapor mounted as previously described. In addition to the secondary seal previously described, a flexible wiper seal is frequently installed, as shown in Figure 13.18*A*, which can be classified as a primary or secondary seal. This flexible material forms a continuous annular ring that is pressed against the tank wall and seals relatively effectively owing to some elasticity in the material.

Primary seals are similar to the primary seals installed in EFRT service. The seals consist of an outer material and are either liquid filled or filled with a resilient foam material. Mounting of the seal in the vapor zone or in the liquid located within the annulus has a significant effect upon the emission factor. In addition, a secondary seal will further reduce the emission factor, although secondary seals in IFRT service are not as effective as seals in EFRT service.

Access Hatch. These hatches are similar to the hatches in EFRT service. Improved emission loss control is achieved with gasketed and bolted hatch covers. In noncontact deck service, the well should extend below the liquid surface to provide a barrier against the vapor zone.

Column Wells. Most IFRT tanks have cone roofs that are supported internally with cylindrical columns. These columns extend from the tank floor to the roof and penetrate the deck as shown in Figure 13.18*A*. There are a number of columns in large tanks requiring column wells for each column. A 150-ft (46-m) diameter tank may have 16 columns supporting the roof. Consequently, each column well should have a sliding cover that can be gasketed to reduce the emission loss factor. In addition, fabric sleeves extending from the cover or well also reduce emissions. The sliding cover is necessary to compensate for deck movement. In noncontact service, the column well should be extended below the surface of the liquid.

Deck Legs. Similar legs are provided in EFRT decks to support the floating deck above the tank bottom. These legs are also adjustable with a sleeve assembly for contact decks. In noncontact decks, the sleeve is placed in a well and the well is extended into the liquid. Hangers are used in some IFRT tanks and are suspended from the internal roof, supporting the deck above the tank floor. The hanger connection through the deck may be located in a well or sleeve similar in design to the deck leg installations.

Gauge Float Wells. Gauge floats similar to the system in the EFRT are installed in the IFRT to measure the liquid level in the tank. As shown in Figure 13.19, the tape passing through a cover on the gauge float well extends upward through the roof into a guide system that is connected to a measurement device located on the external tank wall face. Evaporation losses are also controlled by gasketing and bolting the well cover. Where the well is in a noncontact deck, the well is extended into the liquid.

Ladder Well. In some tanks, internal vertical ladders extend from a manhole in the roof and through the floating deck to the tank bottom. The deck openings must be constructed to minimize evaporation losses with controls similar to the ones previously described. Where the ladder passes through a well in a noncontact deck, the well should be extended into the liquid.

Sampling Well. A slotted pipe extending through the deck similar to the pipe in an EFRT can be installed for sampling the liquid. A slotted pipe is shown in Figure 13.19. Evaporation losses are controlled with the techniques previously described and wells in noncontact decks are extended into the liquid. As an alternate, a funnel-shaped well can be located in the floating deck below a sampling opening in the roof. The funnel has a flat plate at the lower end with a slot that permits a special sampler to extend through the plate and obtain a liquid sample. The well in a noncontact deck is extended into the liquid.

Stub Drains. Since liquid may be present on the deck, drains are required on the deck to permit drainage into the main liquid body in the tank. A number of these stub drains located in the floating deck are generally 1 in. (25 mm) in diameter. These drains are flush with the deck, and in noncontact decks the drain extends into the liquid. Because of the small diameter, emission controls are difficult to design. However, some plants and suppliers have proposed control systems for these drains.

Vacuum Breakers. The vacuum breakers described in Section 13.2.3 for EFRT floating decks are also used in IFRT decks. Where the vacuum breaker is installed in a noncontact deck, the well does not have to extend below the surface of the liquid to minimize air–vapor interference when the breaker is open. However, the breaker cover should be gasketed.

13.2.5 Fixed Roof Tank Fittings

As described in Section 13.1.1, liquids with very low vapor pressures are stored in fixed roof tanks to minimize evaporation losses. For significant emissions reductions, internal floating roofs are installed, and the closed floating roof tank is the result. However, the CFRT requires a vent control system as previously described to prevent the possibility of combustible mixture formation.

In these tanks, diurnal emissions occur along with other emissions when the tank is filled. Consequently, the tank fittings must be designed in accordance with these conditions. However, the roof fitting losses are quite small in comparison to

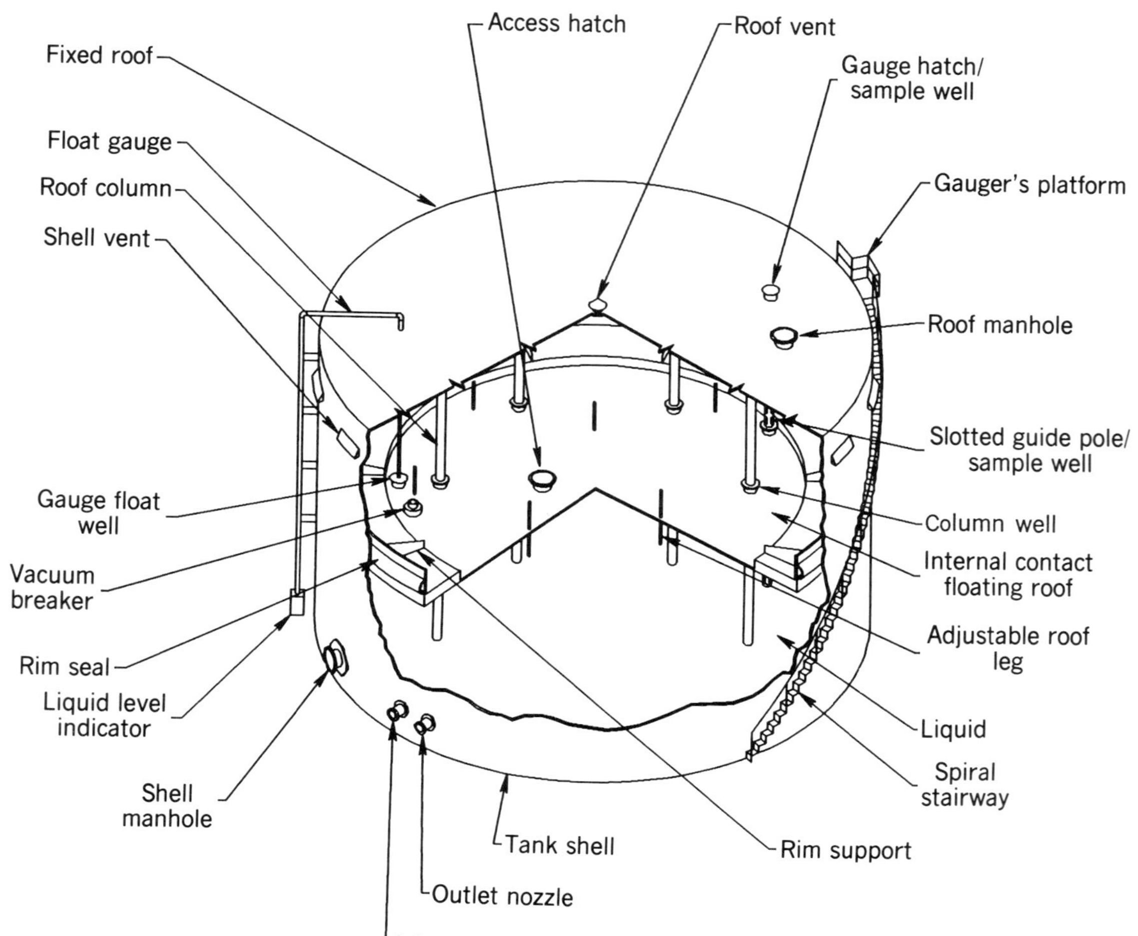

Figure 13.19 *Typical internal floating roof tank with a welded contact deck (courtesy of CBI).*

the evaporative and working losses. Consequently, estimating procedures for these fittings have not been developed.

A standard fixed roof tank does not have many fittings in the roof. The roofs are normally cone-type designs requiring internal supports and columns, although the cone roofs in some installations have been self-supporting. Geodesic design roofs have been installed and generally do not require the internal structure necessary for cone structures. In addition to fixed roof tanks, the geodesic dome roofs are also installed on IFRT tanks. In the IRFT tanks, the support columns are generally removed, simplifying the floating roof design and reducing the fitting losses associated with column wells. However, vents are still required in the IFRT tanks. In the fixed roof tanks, fittings are also required for the dome roofs.

Pressure – Vacuum Vents. These vents are necessary to prevent excessive pressure buildup or vacuum that can damage the tank. Normally, the maximum safe working pressure in a tank is 1.5 in. of water or 0.37 kPa. The allowable vacuum pressure is determined by the buckling strength of the upper portion of the tank where the metal thickness is less than in the lower portion of the tank.

The pressure–vacuum (PV) vents must have sufficient venting capacity for pressure and vacuum conditions. In addition, the PV setting pressures are generally less than the maximum safe working pressure or maximum allowable vacuum to provide a safety factor in operation. These settings are determined in conjunction with mechanical designers. For PV flow capacity ratings evaporation quantities must be defined.

Gauge Hatch and Sample Well. A common hatch or manhole-type opening is installed near the gauger's platform at the top of a spiral stairway attached to the tank wall. Manual samples and manual liquid level gauging can be conducted through this hatch. However, vapor emissions through the hatch opening require personal protection to ensure that personnel are not over exposed during this operation. The hatch cover should have a gasket to minimize evaporation losses and a self closing cover to ensure rapid closure of the hatch. Industrial hygienists should review this operation.

Level Gauge. Float gauges are frequently installed to measure the liquid level in the tank. The tape passes through the tank roof to a float that is maintained in position vertically by guide wires extending from the tank bottom to the tank roof. The tape at the roof enters a guide system similar to those in the IFRT tanks and terminates in an indicator located on the wall of the tank near the ground. The tape opening in the cone roof has a closed opening to minimize emission losses. These special roof openings are supplied by instrument and tank manufacturers.

Roof Manholes. The manholes are access hatches to the tank interior. To reduce evaporation losses, the manhole covers should have a gasket and are bolted to the support flange.

Insulation. Placing insulation on the tank roof and shell will reduce heat input and lower the evaporation losses.

13.3 TANKAGE EMISSION CALCULATIONS

While tankage emission calculations were developed by the American Petroleum Institute (API) for petroleum stocks, the calculation procedures are also applicable to petrochemicals and chemicals.[1-3] These calculations are included in the proposed HON standards issued by the EPA in 1992 covering HAP tankage emission control. While some minor revisions may be reflected in the final standard, the calculation procedures will not be affected. Since these original calculations were developed by the API, the units for the calculations are a mixture of petroleum, English, and SI units. The EPA procedures, therefore, include some conversion factors that are not present in the API reference publications.[1-3]

The emission calculations provide a basis for estimating losses, defining atmospheric emissions, and estimating ground level concentrations. Although the latter has not been emphasized, potential exposure concentrations within the plant and adjacent communities can be determined from dispersion calculation models. Within plants, high emission rates may present a potential hazard that must be considered in defining hazard assessments.

In addition to the emission losses associated with tankage operations, a major concern is the periodic drainage of water from some tanks. Many liquid storage tanks contain dry or essentially dry liquid materials. In those tanks where significant water accumulations occur, water drawoff or drainage is a potential source of exposure. Certain compounds will dissolve in the water and will accumulate to a saturation level which is somewhat higher than an atmospheric saturation level because of the pressure of the stored liquid and floating roof. Draining the water from the tank through a valve into a sewer opening located a few feet below the valve opening results in flashing some of the dissolved compounds into the atmosphere. Where the tanks are relatively large, the water flow rate is quite high and the concentration of HAP around the valve area can be significantly greater than the allowable exposure level. This aspect of tank operation can be hazardous, as described, and requires a closed system to control potential exposures. Some solutions will be proposed for control of wastewater drawoff.

13.3.1 External Floating Roof Tanks

The method of calculating the emissions from EFRT tanks based on the API methods is

$$L_{\mathrm{TM}} = \frac{L_{\mathrm{W}} + L_{\mathrm{R}} + F_{\mathrm{F}}}{12} \tag{13.1}$$

where

L_{TM} = Emissions loss (Mg/month)
L_{W} = Withdrawal emissions loss (Mg/yr)
L_{R} = Rim seal emissions loss (Mg/yr)
L_{F} = Fitting emissions loss (Mg/yr)

The *withdrawal loss* is a function of the number of tank turnovers per year. Liquid remaining on the walls or clinging to the wall as the liquid level falls will evaporate light components into the vapor zone above the floating roof. Tightly fitting seals do not completely remove the liquid from the walls. Moreover, gaps in the seals result in liquid remaining on the walls.

Rim seal losses have been described, and the various types of rim seals have been compared in Figures 13.4–13.6. The emission loss from the rim seals has also been termed a standing storage loss since the loss is independent of the liquid level. As described for fittings in figures, the wind has a significant effect on evaporation losses along with the type of rim seal installation.

Fitting losses are associated with the various fittings described in Section 13.2. These losses for an EFRT were described for fittings in Figures 13.9–13.17.

The withdrawal loss equation is

$$L_W = 4.28 \times 10^{-4} QCW_L/D \tag{13.2}$$

where

Q = Tank throughput (gal/yr)
C = Shell clingage factor (bbl/1000 ft^2) (Table 13.1)
W_L = Average liquid density (lb/gal)
D = Tank diameter (ft)

The rim seal loss equation is

$$L_R = K_S V^n P^* D M_V K_C/2205 \tag{13.3}$$

where

K_S = Seal factor (Table 13.2)
V = Average wind velocity (mi/hr)
n = Wind velocity exponent (Table 13.2)
P^* = Vapor pressure function
D = Tank diameter (ft)
M_V = HAP molecular weight vapor (lb/lb-mol)

TABLE 13.1 Average Shell Clingage Factors[a]

	Shell Condition		
Product Stored	Light Rust (bbl / 1000 sf)[b]	Dense Rust (bbl / 1000 sf)	Gunite Lining (bbl / 1000 sf)
Gasoline	0.0015	0.0075	0.15
Single-component stocks	0.0015	0.0075	0.15
Crude oil	0.0060	0.030	0.60

[a]Reference 2 and 28.
[b]Square feet.

TABLE 13.2 Rim Seal Loss Factors for External Floating Roof Tanks[a, d]

	Average-Fitting Seals				Tight-Fitting Seals[b]	
	Welded Tanks		Riveted Tanks		Welded Tanks	
Rim Seal Type	K_S	n	K_S	n	K_S	n
Metallic shoe seal						
Primary seal only	1.2	1.5	1.3	1.5	0.8	1.6
Primary with shoe-mounted secondary seal	0.8	1.2	1.4	1.2	0.8	1.1
Primary with rim-mounted secondary seal	0.2	1.0	0.2	1.6	0.2	0.9
Liquid-mounted resilient seal						
Primary seal only	1.1	1.0	—[c]	—[c]	0.5	1.1
Primary with weather shield	0.8	0.9	—	—	0.5	1.0
Primary with rim-mounted secondary seal	0.7	0.4	—	—	0.5	0.5
Vapor-mounted resilient seal						
Primary seal only	1.2	2.3	—	—	1.0	1.7
Primary with weather shield	0.9	2.2	—	—	1.1	1.6
Primary with rim-mounted secondary seal	0.2	2.6	—	—	0.4	1.5

[a]Adapted from references 4 and 28.
[b]Data only available on welded tanks.
[c]Data not available on riveted tanks.
[d]Limited to wind speeds of 2 – 15 mi / hr.

K_C = Product factor (1.0 for volatile HAP, 1.0 for refined petroleum liquids, 0.4 for crude oil)

$$P^* = \frac{P/P_a}{\left[1 + (1 - P/P_a)^{0.5}\right]^2} \tag{13.4}$$

where P is the true vapor pressure (psia) and P_a is the average atmospheric pressure (psia).

The fitting loss equation is

$$L = F_F D^* M_V K_C / 2205 \tag{13.5}$$

and

$$F_F = (N_{F1} K_{F1}) + (N_{F2} K_{F2}) + \cdots (N_{FK} K_{FK}) \tag{13.6}$$

where

F_F = Loss factor (lb-mol/yr)
N_{F1} = Number of identical fittings (Tables 13.3–13.5)
K_{F1} = Loss factor for one fitting type (lb-mol/yr)

$$K_{F1} = K_{Fai} + K_{Fbi} V^{m_i} \tag{13.7}$$

TABLE 13.3 Roof Fitting Loss Factors for External Floating Roof Tanks[a]

Fitting Type Construction Details	Loss Factors[b] K_{Fai} (lb-mole / yr)	K_{Fbi} [lb-mol / (mi / hr)^m-yr]	m (dimension-less)	Typical Number of Fittings (N_{F1})
Access hatch (24-in. dia. well)				1
Bolted cover, gasketed	0	0	0[c]	
Unbolted cover, ungasketed	2.7	7.1	1.0	
gasketed	2.9	0.41	1.0	
Unslotted guide-pole well (8-in. dia. unslotted pole, 21-in. dia. well)				1
Ungasketed sliding cover	0	67	0.98[c]	
Gasketed sliding cover	0	3.0	1.4	
Slotted guide-pole / sample well (8-in. dia. slotted pole, 21-in. dia. well)				d
Ungasketed sliding cover,				
without float	0	310	1.2	
with float	0	29	2.0	
Gasketed sliding cover,				
without float	0	260	1.2	
with float	0	8.5	2.4	
Gauge-float well (20-in. dia.)				1
Unbolted cover, ungasketed	2.3	5.9	1.0[c]	
gasketed	2.4	0.34	1.0	
Bolted cover, gasketed	0	0	0	
Gauge-hatch / sample well (8-in. dia.)				1
Weighted mechanical actuation,				
gasketed	0.95	0.14	1.0[c]	
ungasketed	0.91	2.4	1.0	
Vacuum breaker (10-in. dia. well)				Table 13.4
Weighted mechanical actuation,				
gasketed	1.2	0.17	1.0[c]	
ungasketed	1.1	3.0	1.0	
Roof drain (3-in. dia.)				Table 13.4
Open	0	7.0	1.4[e]	
90% closed	0.51	0.81	1.0	
Roof leg (3-in. dia.)				Table 13.5[f]
Adjustable, pontoon area	1.5	0.20	1.0[c]	
center area	0.25	0.067	1.0[c]	
double-deck roofs	0.25	0.067	1.0	
Fixed	0	0	0	
Roof leg ($2\frac{1}{2}$-in. dia.)				Table 13.5[f]
Adjustable, pontoon area	1.7	0	0	
center area	0.41	0	0	
double-deck roofs	0.41	0	0	
Fixed	0	0	0	
Rim vent (6-in. dia.)				1[g]
Weighted mechanical actuation,				
gasketed	0.71	0.10	1.0[c]	
ungasketed	0.68	1.8	1.0	

[a]Adapted from references 4 and 28.
[b]Maximum wind speed for factors is 15 mi / hr. Factors can be used for 2 – 15 mi / hr.
[c]Unless specific information is available, this value is typical of representative fittings.
[d]Slotted guide-pole / sample well is an optional fitting, not typically installed.
[e]Roof draining discharges water into tank normally used in double-deck roofs.
[f]Typical roof leg diameter is 3 in. Factors provided for $2\frac{1}{2}$ in. diameter leg.
[g]Only installed with mechanical shoe primary seals.

TABLE 13.4 Typical Number of Vacuum Breakers and Roof Drains for External Floating Roof Tanks[a]

Tank Diameter D (ft)[b]	Number of Vacuum Breakers (N_{F1}) Potoon Roof	Number of Vacuum Breakers (N_{F1}) Double-Deck Roof	Number of Roof Drains (N_{F2}) (Double-Deck Roof)[c]
50	1	1	1
100	1	1	1
150	2	2	2
200	3	2	3
250	4	3	5
300	5	3	7
350	6	4	—
400	7	4	—

[a]Adapted from reference 28.

[b]Based on survey information and manufacturer data. Nearest tank diameter data to actual tank diameter should be used. When diameter is at midpoint between table diameters, use larger table diameter.

[c]Roof-draining discharge water into tank normally used in double-deck roofs. Typically remain open. Where tank diameter is > 300 ft, actual tank data or manufacturer's recommendations are advisable to establish number of drains.

where

K_{Fai} = Loss factor for one fitting type (lb-mol/yr) (Table 13.3)
K_{Fbi} = Loss for one fitting type [lb-mol/(mi/hr)^m-yr] (Table 13.3)
m_{i} = Loss factor for one fitting type (Table 13.3)
V = Average wind velocity (mi/hr)

In addition to the calculation procedure described, one other factor must be included to determine the average annual liquid storage temperature. The color of the tank affects the amount of absorbed solar radiation, which can increase the average annual liquid temperature and the actual vapor pressure. Since losses are a function of vapor pressure, a correction for the tank paint color can be obtained from Table 13.6.

13.3.2 Internal Floating Roof Tanks

Prior to calculating the emissions for internal floating roofs, some assumptions are necessary in the design of the tank with a general internal floating roof tank design shown in Figure 13.19. If the tank roof is self-supporting, columns for roof support are not required, otherwise columns extend through the floating deck. In addition, the choice of a bolted or welded deck will affect the quantity of emissions. Finally, the type of rim seal installations will also affect standing loss emissions.

The initial tank design can be based on a set of assumptions that can be changed following an analysis of the results. Sufficient information is provided for various iterations to optimize the internal floating roof tank design for a specific

TABLE 13.5 Typical Number of Roof Legs for External Floating Roof Tanks[a]

Tank Diameter D (ft)[b]	Pontoon Roof		Number of Legs on Double-Deck Roof
	Number of Pontoon Legs	Number of Center Legs	
30	4	2	6
40	4	4	7
50	6	6	8
60	9	7	10
70	13	9	13
80	15	10	16
90	16	12	20
100	17	16	25
110	18	20	29
120	19	24	34
130	20	28	40
140	21	33	46
150	23	38	52
160	26	42	58
170	27	49	66
180	28	56	74
190	29	62	82
200	30	69	90
210	31	77	98
220	32	83	107
230	33	92	115
240	34	101	127
250	35	109	138
260	36	118	149
270	36	128	162
280	37	138	173
290	38	148	186
300	38	156	200
310	39	168	213
320	39	179	226
330	40	190	240
340	41	202	255
350	42	213	270
360	44	226	285
370	45	238	300
380	46	252	315
390	47	266	330
400	48	281	345

[a]From references 4 and 28.

[b]Actual number of roof legs should be used. Considerable variations exist, but table information based on a survey of users and manufacturers provides a satisfactory estimating basis. The nearest tank diameter data to actual tank diameter should be used. When diameter is at midpoint between tank diameters, use larger table diameter.

TABLE 13.6 Average Annual Liquid Storage Temperature as a Function of Tank Color for External Floating Roof Tanks[a]

Tank Color	Average Annual Stock Storage Temperature T (°F)
White	$T_A{}^b + 0$
Aluminum	$T_A + 2.5$
Gray	$T_A + 3.5$
Black	$T_A + 5.0$

[a]From reference 4.
[b]Average annual ambient temperature.

installation. The calculation procedure for an IFRT is

$$L_{TM} = \frac{L_W + L_R + L_F + L_D}{12} \tag{13.8}$$

where

L_{TM} = Emissions loss (Mg/month)
L_W = Withdrawal emissions loss (Mg/yr)
L_R = Rim seal emissions loss (Mg/yr)
L_F = Fitting emissions loss (Mg/yr)
L_D = Deck seam emissions loss (Mg/yr)

The withdrawal loss equation is

$$L_W = \frac{1.018 \times 10^{-5} QCW_L}{D}[1 + N_C F_C/D)] \tag{13.9}$$

where

Q = Tank throughput (gal/yr)
C = Shell clingage factor (bbl/1000 ft^2) (Table 13.1)
W_L = Average liquid density (lb/gal)
D = Tank diameter (ft)
N_C = Number of columns (Table 13.7)
F_C = Effective column diameter (ft) (column perimeter/π)

If the roof is self-supporting $N_C = 0$, and in those cases where the column type is unknown a value of 1.0 is used.

The rim seal equation is

$$L_R = K_S P^* D M_V K_C / 2205 \tag{13.10}$$

TABLE 13.7 Typical Number of Columns for Internal Floating Roof Tanks with Column Supported Roofs[a]

Tank Diameter Range (ft)	Typical Number of Columns[b] (N_C)
$0 < D \leq 85$	1
$85 < D \leq 100$	6
$100 < D \leq 120$	7
$120 < D \leq 135$	8
$135 < D \leq 150$	9
$150 < D \leq 170$	16
$170 < D \leq 190$	19
$190 < D \leq 220$	22
$220 < D \leq 235$	31
$235 < D \leq 270$	37
$270 < D \leq 275$	43
$275 < D \leq 290$	49
$290 < D \leq 330$	61
$330 < D \leq 360$	71
$360 < D \leq 400$	81

[a]From reference 28.
[b]Actual tank data should be used. Table information is based on a user's survey and manufacturers's data. The data should provide a satisfactory estimating basis.

where

K_S = Seal factor (Table 13.8)
P^* = Vapor pressure function (see Eq. 13.4)
D = Tank diameter (ft)
M_V = Molecular weight vapor (lb/lb-mol)
K_C = Product factor (1.0 for volatile HAP, 1.0 for refined petroleum liquids, 0.4 for crude oil)

In Table 13.8, the combination of the liquid-mounted primary seal and the secondary seals has a distinct advantage in emissions control compared with the

TABLE 13.8 Rim Seal Loss Factor for Internal Floating Roof Tanks[a]

	K_S(lb-mol/ft-yr)	
Rim Seal System	Average Fitting	Tight Fitting
Vapor-mounted primary seal only[b]	6.7	5.6
Liquid-mounted primary seal only	3.0	2.6
Vapor-mounted primary seal and rim-mounted secondary seal	2.5	2.3
Liquid-mounted primary seal and rim-mounted secondary seal	1.6	1.2

[a]From references 3, 4 and 28.
[b]Most commonly used seal to 1990. For modifications and new tanks, an alternate scale system may be preferable.

vapor-mounted primary and secondary seal combination. This advantage increases slightly with tighter-fitting seals, which further reduce emissions.

The fitting loss equation for the IFRT is identical to Eq. 13.5 for the EFRT tank:

$$L_F = F_F P^* M_V K_C / 2205 \qquad (13.5)$$

$$F_F = (N_{F1} K_{F1}) + (N_{F2} K_{F2}) + \cdots (N_{FK} N_{FK}) \qquad (13.6)$$

where N_{Fi} is the number of identical fittings (Tables 13.7 and 13.9) and K_{Fi} is the loss factor for one fitting type (lb-mol/yr) (Table 13.9).

TABLE 13.9 Number of Fittings and Fitting Loss Factors for Decks in Internal Floating Roof Tanks[a]

Deck Fitting Description	Deck Fitting Loss Factor (K_F) (lb-mol/yr)	Typical Number of Fittings (N_F)
Access hatch		1
Bolted cover, gasketed	1.6	
Unbolted cover, gasketed	11	
ungasketed	25[b]	
Automatic gauge float well		1
Bolted cover, gasketed	5.1	
Unbolted cover, gasketed	15	
ungasketed	28[b]	
Column well (24-in. dia.)		Table 13.7
Built-up column-sliding cover, gasketed	33	
ungasketed	47[b]	
Pipe column-flexible fabric sleeve seal	10	
Pipe column-sliding cover, gasketed	19	
ungasketed	32	
Ladder well (36-in. dia.)		1[c]
Sliding cover, gasketed	56	
ungasketed	76[b]	
Deck leg or hanger well		[d]
Adjustable	7.9[b]	
Fixed	0	
Sample pipe or well (24-in. dia.)		1
Slotted pipe, gasketed sliding cover	44	
ungasketed sliding cover	57	
Sample well, slit fabric seal (10% open area)	12[b]	
Stub drain (1-in. dia.)	1.2	[e]
Vacuum breaker (10-in. dia.)		1
Weighted mechanical actuation, gasketed	0.7[b]	
ungasketed	0.9	

[a]From references 4 and 28.
[b]Where information is unavailable, this value represents commonly used and typical deck fittings.
[c]Normally not used in tanks with self-supporting roofs.
[d]$N_F = (5 + D/100 + D^2/600)$, where D = tank diameter (ft).
[e]$N_F = 0$ for welded floating decks. $N_F = D^2/125$ for bolted floating decks, where D = tank diameter (ft).

TABLE 13.10 Seam Length Factors for Decks in Internal Floating Roof Tanks[a]

Deck Construction	Deck Seam Length Factor $(S_D)^{b,e}$ (ft / ft^2)
Continuous sheet construction[c]	
5-ft wide sheets	0.20
6-ft wide sheets	0.17
7-ft wide sheets	0.14
Panel construction[d]	
5 × 7.5 ft rectangular panels	0.33
5 × 12 ft rectangular panels	0.28

[a]From references 3 and 4.
[b]Only applies to bolted decks.
[c]$S_D \approx 1 / w$, where w = sheet width (ft).
[d]$S_D \approx (l + w) / lw$, where w = panel width (ft) and l = panel length (ft).
[e]Where information is unavailable, these factors represent bolted decks commonly used.

The deck seam loss as described is a function of the type of floating deck installed, with seam loss calculated in the following equation. However, the seam loss factor for a welded deck is zero and the equation only applies to a nonwelded deck:

$$L_D = K_D S_D D^2 P^* M_V K_C / 2205 \tag{13.11}$$

where K_D is the deck seam loss factor (lb-mol/ft-yr) (0.34, nonwelded deck; 0, welded deck) and S_D is the deck seam length factor (ft/ft^2) (Table 13.10).

In these equations the definitions for D, P^*, M_V and K_C are constant, as identified in the explanations for Eq. 13.10. The tank color will have an effect upon solar radiation absorption and the annual average liquid storage temperature is adjusted in accordance with the paint color as shown in Table 13.6.

While the emissions can be calculated for an IFRT, the vapor space ventilation rate and vapor concentration in the vapor space should be calculated to determine whether the concentration is in the combustible region. As shown in Figures 13.1 and 13.19, vents are normally located around the periphery of the tank on the shell below the roof line. These vents are necessary whether the roof is a cone roof or a geodesic dome. The actual requirements for the circulation vents are included in API standard 650.[5] The total open area is ≥ 0.2 ft^2/ft tank diameter, which can be achieved by installing a series of vents as shown in Figure 13.19 or providing a continuous circumferential opening below the roof line with a uniform width of 0.764 in. Most tanks in IFRT service have separate vents in the shell.

In addition to the minimum area requirement for the vents, other requirements must be met in designing and installing the vents. The standard requires a minimum of four equally spaced vents with a maximum distance between vents of 32 ft. Also, an open vent with a minimum area of 50 sq in. is required at the highest roof elevation point. This vent on an IFRT tank is an open hood and not a pressure vacuum vent as shown on fixed roof or CFRT roofs.

Since air flows through the vents on the windward side of the tank, a mixture of air and storage liquid vapor is discharged to the atmosphere through the lee side vents. This mixture disperses into the atmosphere and some dispersion analysis should be conducted to determine potential exposure situations. When the rim seals are tight and new, liquid evaporation is small and dispersion concentrations should be satisfactory. However, seal gaps begin to appear over a period of time, and both emissions and dispersion concentrations increase. Dispersion analysis calculations should consider these two conditions to ensure compliance with recommended exposure levels. For estimating maximum emissions levels, the effect on emissions evaporation should be based on the maximum allowable seal gaps in the various EPA regulations described in Section 13.4. Calculations for determining vapor space concentrations are presented in reference 6.

13.3.3 Fixed Roof Tanks

In fixed roof tanks, liquid normally evaporates into the vapor zone and any pressure buildup is emitted to the atmosphere through the pressure–vacuum vent. During evening hour cool down through radiation or when the tank is cooled through rain, the vapor pressure falls and air is admitted into the vapor zone through the pressure–vacuum (P–V) vent to maintain a safe working pressure within the tank. The P–V vent set pressures are determined by tank construction, but the settings are normally very close to atmospheric pressure, as described previously. In general, the large liquid surface area results in very large emission rates, where the liquid surface is covered in the EFRT and IFRT tanks. These large emissions rates result in inventory losses and are generally larger than the emission rates acceptable to environmental regulators. Regulations normally require floating roofs or closed ventilation systems as described in the regulation section for HAP and most lighter liquids. However, fixed roof tanks can be used with heavy, low-vapor pressure liquids within certain regulatory guideline.

While atmospheric breathing fixed roof tanks for most liquid storage requirements are not permitted, quantifying the emission rates provides a basis for sizing closed-vent systems and control devices. In addition, the differences in emissions rates can be established between a standard fixed roof tank and a closed floating roof tank described in Section 13.1.4. The optimum control configuration among the various alternates suggests estimating loss rates from a fixed roof tank as a basis.

A procedure for determining evaporation loss had been developed by the API in 1962[7] which was incorporated into AP-42, a compilation of air emission factors and methods of calculating these factors.[8] An EPA report published in 1988[9] also contained the original API evaporation loss calculation procedure which was included in the original draft of the HON regulations. This calculation procedure has been superseded by a revised API evaporation loss calculation method that was issued in late 1991.[1] The new API procedure is an improvement over the original method and is presented in this section to provide a more accurate calculation method of evaporation losses. A recently proposed EPA Guideline for estimating emissions essentially duplicates the API calculation procedure. The new API method follows[28]:

$$L_T = L_S + L_W \tag{13.12}$$

where

L_T = Total annual loss per year
L_S = Standing loss per year basis
L_W = Working loss per year basis

The losses on a lb/yr or bbls/yr basis for standing storage loss is determined as follows:

$$L_S(\text{lb/yr}) = 365 V_V W_V K_E K_S \tag{13.13}$$

or

$$L_S(\text{bbls/yr}) = \frac{L_S(\text{lb/yr})}{42 W_{vc}} \tag{13.14}$$

where the components are determined as follows:

$$\text{Tank vapor space volume } V_V = \frac{\pi D^2 H_{VO}}{4} \tag{13.15}$$

$$\text{Stock vapor density } W_v = M_v P_{VA}/RT_{LA} \tag{13.16}$$

$$\text{Vapor space expansion factor } K_E = \frac{\Delta T_V}{T_{LA}} + \frac{\Delta P_V - \Delta P_B}{P_A - P_{VA}} \tag{13.17}$$

$$\text{Vented vapor saturation factor } K_S = \frac{1}{1 + 0.053 P_{VA} H_{VO}} \tag{13.18}$$

and

L_S = Standing storage loss (lb/yr)
V_V = Tank vapor space volume (cf)
W_V = Stock vapor density (lb/cf)
K_E = Vapor space expansion factor (dimensionless)
K_S = Vented vapor saturation factor (dimensionless)
D = Tank diameter (ft)
H_{VO} = Vapor space outage (ft)
M_V = Stock vapor molecular weight (lb/lb-mol)
P_{VA} = Stock vapor pressure at daily average liquid surface temperature (psia)
R = Ideal gas constant (10.731) (psia-cf/lb-mol) (°R)
T_{LA} = Daily average liquid surface temperature (°R)
ΔT_V = Daily vapor temperature range (°R)
ΔT_V = Daily vapor temperature range (°R)
ΔP_V = Stock daily vapor pressure range (psi)
ΔP_B = Breather vent pressure setting range (psi)
P_A = Atmospheric pressure (psia)
W_{vc} = Stock condensed vapor density (lb/gal) (60°F)

Vapor space outage is determined as follows:

$$H_{VO} = H_S - H_L + H_{RO} \tag{13.19}$$

where

H_S = Tank shell height (ft)
H_L = Stock liquid weight (ft)
H_{RO} = Roof outage (shell height equivalent to the volume contained under the roof) (ft)

For a *cone roof*

$$H_{RO} = 1/3H_R \tag{13.20}$$

where H_R is determined from Figure 13.20 or

$$H_R = S_R R_S \tag{13.21}$$

where S_R is the tank cone roof slope (ft) (when unknown, 0.0625 ft/ft is assumed) and R_S is the tank shell radius (ft).

For a *dome roof*

$$H_{RO} = H_R\left[1/2 + 1/6(H_R/H_S)^2\right] \tag{13.22}$$

and

$$H_R = R_R - \left(R_R^2 - R_S^2\right)^{0.5} \tag{13.23}$$

where R_R is the tank dome radius (ft) (see Fig. 13.20).

The losses on a lb/yr or bbl/yr basis for *working loss* is determined as follows:

$$L_W(\text{lb/yr}) = 0.0010 M_V P_{VA} Q K_N K_P \tag{13.24}$$

$$L_W(\text{bbl/yr}) = L_W(\text{lb/yr})/42W_{VC} \tag{13.25}$$

where

L_W = Working loss (lb/yr)
Q = Annual net throughput (bbl/yr)
K_N = Working loss turnover (dimensionless)
K_P = Working loss product factor (dimensionless)

and M_V can be determined from Appendix 13A.

Here P_{VA} requires an estimate of the average liquid surface temperature to determine the vapor pressure and K_E is the vapor space expansion factor. The temperature within the tank and on the liquid surface are a function of solar heating, which requires consideration of solar absorbance. The solar absorbance factors (α) for various paints and tank surface conditions are shown in Table 13.11.

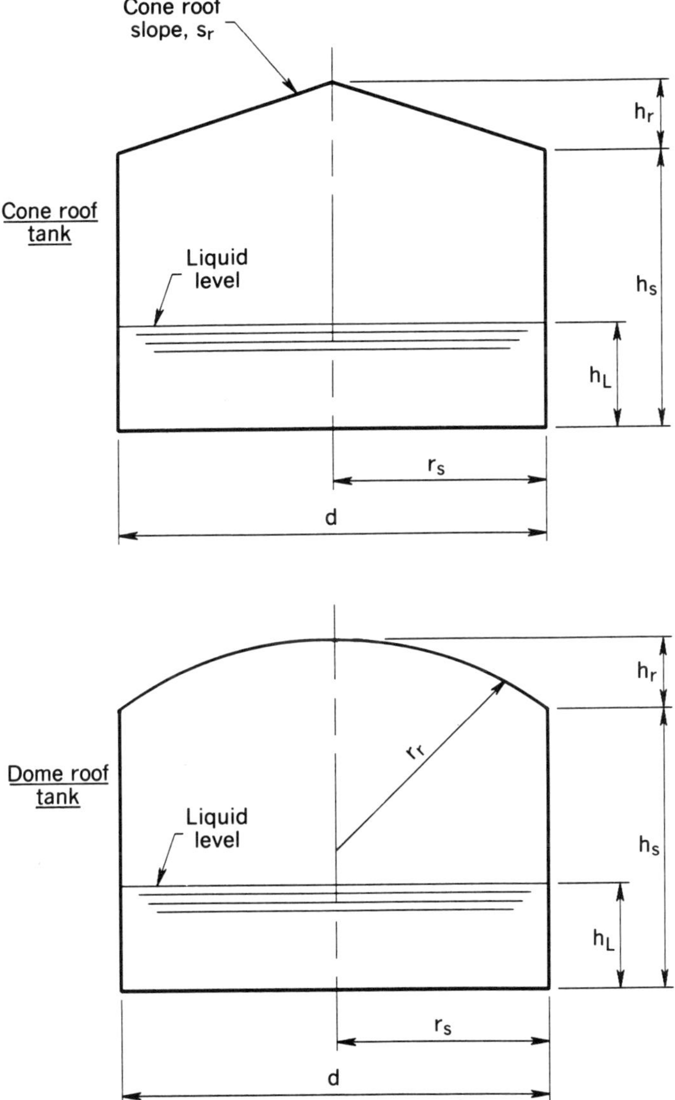

Figure 13.20 *Fixed roof tank geometry (courtesy of CBI).*

This factor is used in the following calculations:

$$T_B = T_{AA} + 6\alpha - 1 \tag{13.26}$$

where

T_B = Liquid bulk temperature (°R)
T_{AA} = Daily average ambient temperature (°R)
α = Solar absorbance (dimensionless)

TABLE 13.11 Solar Absorptance for Various Paints on Fixed-Roof Tanks[a]

		Solar Absorptance	
		Paint Condition	
Paint[b] Color	Paint Shade or Type	Good	Poor
Aluminum	Specular	0.39	0.49
	Diffuse	0.60	0.68
Gray	Light	0.54	0.63
	Medium	0.68	0.74
Red	Primer	0.89	0.91
White		0.17	0.34

[a]Adapted from reference 1.
[b]Where color information is not available, a white shell and roof with the paint in good condition is representative of many paints.

and

$$T_{\mathrm{LA}} = 0.44T_{\mathrm{AA}} + 0.56T_{\mathrm{B}} + 0.0079\alpha I \tag{13.27}$$

where T_{LA} is the daily average liquid surface temperature and I is the solar heat density on a horizontal surface (BTU/ft^2 day) (U.S. values can be determined from Table 4, ref. 1). Also

$$\Delta T_{\mathrm{V}} = 0.72\Delta T_{\mathrm{A}} + 0.0028\alpha I \tag{13.28}$$

where ΔT_{V} is the daily vapor temperature range (°R) and ΔT_{A} is the daily ambient temperature range (°R).

While these temperature calculations provide a liquid surface temperature, the *daily average* liquid surface temperature is necessary to determine the stock vapor pressure. This average surface temperature can be determined from data, but in the event this is not possible, the daily maximum and minimum liquid temperatures T_{LX} and T_{LN} respectively are calculated. These, in turn, determine the accompanying stock vapor pressures P_{VX} and P_{VN}.

$$T_{\mathrm{LX}} = T_{\mathrm{LA}} + 0.25\Delta T_{\mathrm{V}} \tag{13.29}$$

$$T_{\mathrm{LN}} = T_{\mathrm{LA}} - 0.25\Delta T_{\mathrm{V}} \tag{13.30}$$

where

T_{LX} = Daily maximum liquid surface temperature (°R)
T_{LA} = Equation 13.28
T_{LN} = Daily minimum liquid surface temperature (°R)
ΔT_{V} = Equation 13.29

The average vapor pressure can be determined from these temperatures and various references. Alternatively P_{VA}, the vapor pressure at the daily average liquid surface temperature, can be determined from

$$P_{VA} = \exp[A - (B/T_{LA})] \tag{13.31}$$

where A and B are constants (°R).

Both constants can be found in Appendix 13A. For petroleum stocks, some information is available in Appendix 13A but may be insufficient for more precise calculations. Information on petroleum stock vapor pressure determinations are shown in reference 1.

Here ΔP_B, the breather vent pressure setting range, is the difference between the positive set pressure on the P–V vent and the negative pressure setting. Where specific information is unavailable, assume $+0.03$ psig for the positive setting and -0.03 psig for the negative setting, resulting in a ΔP_B of 0.06 psi.

When the roof is bolted or riveted and the plates are not tight, the ΔP_B assumed is zero, even in the presence of a breather vent. And Q, the annual net throughput is the total annual volume pumped into the tank that results in a level increase of the tank liquid. If the liquid level does not change, the net throughput is zero.

The stock turnover rate

$$N = 5.614Q/V_{LX} \tag{13.32}$$

where V_{LX} is the maximum tank liquid capacity (ft^3).

The product factor K_P only applies to working losses as follows:

$$K_P = 0.75 \text{ for crude oil stocks and}$$
$$\quad 1.00 \text{ for refined petroleum stocks}$$

$$K_N = \frac{180 + N}{6N} (N > 36) \tag{13.33}$$

$$K_N = 1 (N \leq 36)$$

13.3.4 Closed Floating Roof Tanks

Closed floating roof tanks are essentially fixed roof tanks with an internal floating roof to reduce emissions to the vapor space. Since there is a P–V vent on the tank roof, air is admitted as the temperature falls due to diurnal variations, rain, or snow. Definition of the vapor space concentration is important to determine whether the vapor concentration is in the combustible zone. In addition, the vapor concentration determines the quantity of inert gas necessary to dilute the vapor concentration below the lower combustion concentration for liquids that are relatively volatile. Where very low volatility liquids are in a CFRT, air can be drawn through a P–V vent to dilute the gas.[6] However, diluting with air depends upon a blower to maintain a safe noncombustible concentration in the vapor zone, requiring a blower service factor of about 100%, which is difficult to achieve. Although a purge gas requires careful control, a bypass around the purge inlet system permits a continual flow of purge gas, albeit under manual operation.

A calculation for determining the vapor space concentration in a CFRT is presented in reference 6 and includes example calculations based upon a vapor-mounted primary seal without a secondary seal. This seal system, while considered common in internal floating roof designs, results in the highest emission rate for a floating roof. Minimizing evaporation also reduces the size of the control system necessary for the CFRT owing to lower HAP rates and lower diluent gas quantities.

A simplified vapor control system for a CFRT has an opening to the exhaust duct located internally near the top of the cone roof. This duct is extended to the shell wall well above the internal floating deck. The normal maximum working pressure of the tank is about 0.03 psig, which is insufficient for moving vapor through a piping system and control device. Consequently, an induced draft blower is necessary to pressure the vapor through the overall system. The control device may be adsorption, condensation, or combustion, with a flare included as a combustion control system. While a flare line has a relatively low pressure level, there is a water seal prior to the flare to prevent blowback, requiring a small pressure to enter the flaring system. This seal pressure is about 0.2 psig, which is somewhat higher than the tank working pressure and would also require a blower. Moreover, connection to a common flare line with processing units may result in relatively large line pressures that would raise the tank pressure and activate the P–V, vent releasing vapors into the atmosphere. For flare combustion, a separate tank flare should be provided to eliminate interaction with other pressure-relieving systems and eliminate the requirement for an induced blower.

The allowable working pressures in a tank are quite low, as shown in Section 13.3.3, and designing tanks for higher working pressures significantly increases tank investment. Moreover, existing tanks are not easily converted to higher pressure operation, and the cost of any conversion would probably exceed by a wide margin the costs associated with an induced blower installation.

13.3.5 Vapor Balancing

Vapor balancing is practiced in some operations to limit emission losses. An example of a vapor balancing operation is in the loading of barges where the vapor vented from the barge is connected to a return vent line from the storage tank that is the source of the liquid pumped into the barge. The liquid volume from the tank displaces a comparable vapor volume that maintains the tank pressure. This system is currently operating in some locations. However, the entire tank system must be evaluated when considering vapor balancing. While the storage tank can be balanced with the system it is feeding, the storage tank must be refilled periodically and excess vapor must be controlled. Other tanks may be available for interconnection of vapor lines. However, this general system is not feasible where various chemicals are stored in a group of storage tanks. Moreover, interconnecting a group of tank vapor vents will frequently result in excess vapor that must be discharged from the common vent header and controlled. While vapor balancing is feasible, each situation must be carefully investigated with any vapor discharges, requiring controls as described in the regulation section.

When pressure tankage is installed for high volatility compounds, vented vapor from the receiver can be compressed and returned to the pressurized tank. This procedure has been used in various installations to minimize losses for a number of years. Pressurized tankage does not present the venting problems associated with atmospheric tanks.

13.4 TANK INSTRUMENTATION

In atmospheric tanks, the major monitoring instrument is a level indicator to determine the exact location of the liquid level. The primary function of the level instrument is inventory control, but the indication of liquid level with relation to the location of a floating deck is very important, particularly during filling operations. Where the tank is enclosed and does not have a floating roof, liquid level is also a major concern when filling the tank. In addition, withdrawal at low liquid level is important in a floating roof tank where inexact readings may result in the floating deck resting on support legs and interfering with withdrawal operations. Other considerations in tank operation are temperature and the location of the water fluid interface.

As described in Section 3.3, liquid level has been determined for many years by floats that transmit the liquid level or motion through a cable or tape to an indicator device located near the ground on the outside tank shell. The indicator maintains some tension in the cable under all conditions to continually monitor the liquid level. A local readout and/or remote readout capabilities are available with this device. In addition, manual gauging to check the float system is conducted on an infrequent basis, but in the event the float system is not operating properly, manual gauging is conducted systematically to maintain a record of the tank liquid level.

Manual gauging presents potential exposure concerns because of the need to open a hatch and lower a weighted tape into the sample well. Although there are gauging platforms on EFR and IFR tanks, higher volatility stocks (VHAP) can increase the vapor concentration in the gauger platform area. In addition, the tape must be rewound on the reel and the liquid on the tape can be a source of vapor or dermal exposure. In fixed roof tanks, the standard level float can be placed in a slotted pipe, but a typical installation through the years has been a float that is attached loosely to two vertical wires that act as guides. A tape or cable is also attached to the float with its attendant equipment.

13.4.1 Radar Gauge

The mechanical systems present maintenance problems which result in potential exposure situations for maintenance personnel. Nonmechanical devices are now available that have a high reliability factor and are mounted externally on the tank. A guide pole is installed that extends from the bottom of the tank through the floating deck and terminates in a flange on the tank roof where a radar device is mounted on the flange. The radar points downward into the pipe and measures the height of the liquid in the pipe through reflection from the liquid surface. This is accomplished with a narrow radar beam that is designed for the pipe diameter and is generally a minimum of 6 in. in diameter. The radar beam is essentially a very narrow elongated cone that is adjusted for pipe diameter and liquid height with liquid height measurements that are $\geq$ 100 ft (30 m). Since the radar is electronic, the output can be readily converted into a variety of readout systems.

The guide pipe for the radar beam is generally a slotted pipe for floating roof tanks which provides free flow of the fluid through the pipe. This type of device is also used in high-pressure storage spheres to a maximum pressure of about 500 psig (35 bar). In the high-pressure spheres, a guide pole is also used to contain the beam. However, guide poles are not required in all installations, such as a fixed roof tank, where a radar device mounted on the roof flange has a much wider cone than the radar beam cone within the pipe. This device can also be installed in an EFR since the slotted pole system is similar for either the float or radar systems. A large number of radar devices have been successfully installed in tanks around the world to measure tankage liquid levels during the last decade. These devices are accurate and reliable with relatively low maintenance costs. Moreover, the high reliability factor reduces potential exposure situations by significantly lowering the need to remove the radar head and manually gauge the tank.

In addition to radar devices, ultrasonics are also used to measure liquid level, particularly in fixed roof tanks which are accurate and reliable. The ultrasonic beams are not narrowly focused and cannot be placed within a pipe. However, rapid improvements in these devices may expand the application area.

13.4.2 Differential Pressure Level Measurement

Another measuring device that has been used is a differential pressure sensor that can be placed at a specific point in the tank. This sensor measures the pressure developed by the liquid height above the sensing point, but the density of the fluid must be known to convert pressure to liquid height. A fluid that maintains a constant density throughout the fluid height simplifies the conversion to liquid height. Should the fluid stratify into varying density levels, an average density must be used that may result in inaccurate conversions of pressure to liquid height. In addition, the temperature of the bulk fluid should be measured since the specific gravity is a function of the temperature. Specific gravity/temperature data are available from a wide variety of sources. Where floating decks are included in the tank, the weight of the deck must be included in any estimate of the level height. Calibration of the differential sensor with liquid height when the tank is filled provides a basis for adjusting the liquid level. This calibration can be undertaken when a tank is first filled with water for strapping purposes or during a filling period with the stock liquid. Manual gauging or a float device as part of the tankage system provides the calibration basis.

While the differential pressure sensor is usually quite reliable, the sensor may require removal for repair or maintenance in the event of a sensor failure. Consequently, a valve should be placed between the sensor and the tank flange nozzle to permit removal of the sensor. If the tank fluid is a HAP chemical, then a small quantity of fluid remaining between the valve internal closure block (ball, gate, plug, etc.) and the end of a sensor inlet passage will spill when the sensor is removed. A bleeder valve connected to the bottom of the valve flange on the sensor side permits draining the fluid if the bleeder is about a 1-in. valve. This fluid should be drained into an enclosed container. Following draining, the sensor can be removed. Although some fluid will be present, the sensor can be removed with the proper personnel protective gear.

13.4.3 Temperature Measurement

Temperature measurements within a tank are not necessarily required, but aid in determining vapor pressures for losses and in defining the specific gravity. This latter characterization is also of assistance in transfer measurements of the liquid stock in addition to the differential pressure measurements previously described. A series of wall taps for temperature measurements are one method of obtaining the fluid temperature distribution. However, the temperature measurements near the outside shell may not be an accurate presentation of bulk fluid temperatures. A temperature gradient from a cold shell to a somewhat warmer bulk liquid temperature would result in measured temperatures lower than the actual bulk temperature. For more accurate measurements, a column or pole with thermocouples located at desired heights is preferred in obtaining bulk temperature measurements. In this situation, a slotted pole with thermocouples inside the pole provides satisfactory measurements. This pole should be placed a significant distance from the outer shell but this distance may be limited in an EFRT owing to the conduit span from the outer shell to the slotted pole. Locations in the other tanks should not present a problem.

13.4.4 Water Drawoff

As described earlier, water drawoff can present a potential exposure problem owing to the absorption of chemical compounds in the water. Where the water outlet nozzle discharges the contaminated water from the tank into an open sewer receiving point, vaporization of the chemical compounds can be a potentially significant exposure situation (see Fig. 13.21). This requires a completely closed drain from the tank to the sewer line. However, automatic control of contaminated water discharge is then necessary since operating personnel cannot judge when the discharge valve should be closed. This requires a device to indicate when the water interface has reached a minimum level and a further reduction in level would permit the discharge of liquid stock into the drain discharge line. In addition, water level maximum height indication prevents liquid from entering the liquid stock outlet line to the tank pump.

Level indicators such as capacitance or RF probes will indicate when a level passes the detection point. This detector is set for water or liquid stock detection depending on the location in the tank. These electronic sensors can be connected to both remote and local indicators. In addition, the drain valve can be operated automatically from the detector signal, as shown in Figure 13.21. This system works well until a malfunction occurs which can be a result of dirt or failure in the detector. The detector must be removed from the tank, but most detectors cannot be removed without significant leakage unless the detectors are designed for removal under high tankage liquid level conditions. Proper precautions are necessary to ensure any leakage is a minimum.

Another alternate is the installation of level glasses with plumb bobs designed for water–liquid stock interface indication in accordance with level control re-

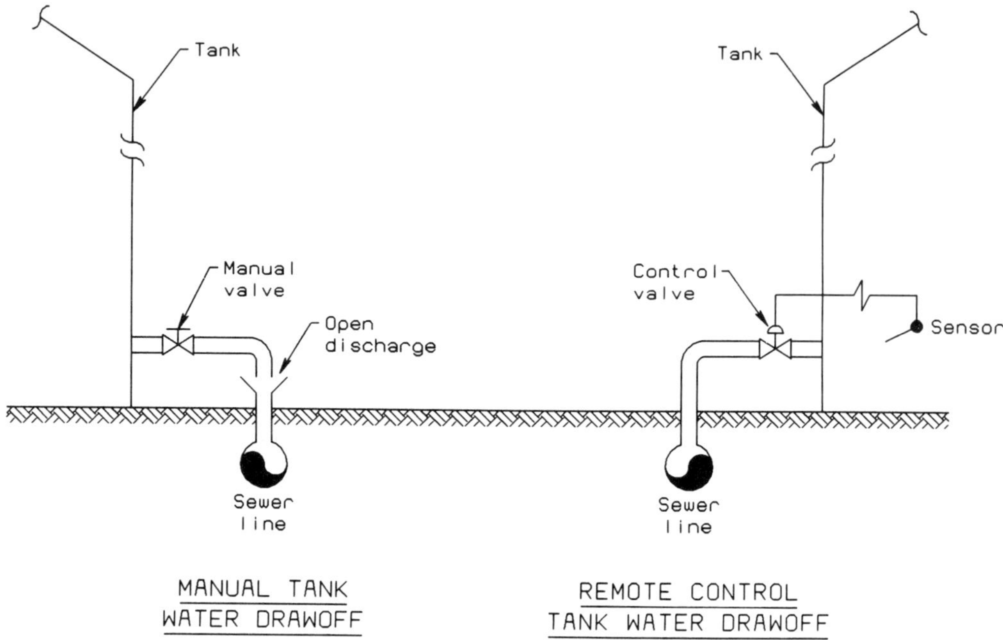

Figure 13.21 *Tank water draw-off connections.*

quirements. When the water level increases to a predetermined height, the plumb will indicate the level and can either automatically open the valve with certain magnetic level devices or be used for manual operation of the valve. The plumb will fall as the water is drained and at a predetermined level will automatically close the drain valve or indicate the level for manual operation. With the new level glass systems, the changing colors associated with the height of the plumb provide a clear indication of the liquid level location for manual control or defining liquid levels for automatic control.

The water drain systems described above work well when the water and liquid stock have a relatively definite interface between the two liquids. However, the water and liquid stock frequently have emulsion interfaces that are relatively thick. Where these emulsion interfaces exist, the interface should not enter the outlet drain system. This requires closing the drain valve before the emulsion interface reaches the drain pipe outlet. In some situations, detectors will indicate the presence of the emulsion. However, the emulsion varies considerably in composition since it is not homogeneous and the detector signals may vary erratically. This signal variation can indicate an emulsion zone and automatically close the drain valve. The detector location may have to be raised to ensure some safety factor in control. Another consideration is that emulsions may interfere with detector performance. However, these effects vary widely depending upon the liquid stock involved. The magnetic plumb will work with an interface, but similarly to the detectors, the interface will probably require some adjustment in the activation level points in the level glass.

Other devices are also available commercially that automatically drain water from tank and perform satisfactorily in various services. In water drawoff service, the effectiveness of different automatic drawoff systems vary because of the interaction of the liquid stock with water. Drawoff systems should be tested if possible in a particular service before permanent installation.

13.4.5 Overflow Control

In many chemical facilities, tanks are equipped with overflow lines to prevent overfilling tanks. These overflow lines are normally confined to fixed roof tank service and generally extend downward from the top of the vertical shell wall, terminating a few inches above a sewer opening. This permits visual observation of the line and will indicate when a tank is overflowing. While this system is reliable, the openings to the atmosphere are a source of emissions from the tank and the actual liquid overflow stream. Overflow lines are normally not installed in floating roof tanks. The overflow line openings in tank walls with floating decks would interfere with seal operation and affect vapor emissions. Deck rim seal separation openings from tank walls must comply with the seal opening measurement requirements described in Section 13.6 on regulations. Moreover, seals must be inspected to ensure compliance with the regulations. Wall openings or imperfections interfere with seal mechanical performance. For reliable control to prevent overfilling, dual instruments generally provide the reliability and performance necessary in floating roof tanks. However, instrument inspection and testing should be conducted periodically to maintain sensor performance. Where a closed overflow line is installed on a fixed roof tank, a level indicator along with an overflow line flow indicator described below will provide redundancy.

The tank overflow lines, if not connected to a closed sewer system, will permit vapor emissions to the atmosphere from the tank vapor space during normal operations which will increase as the tank is filled, particularly with VHAP compounds. Moreover, any overflow to an open sewer connection will result in excessive emissions. These overflow lines should be connected directly to the sewer or drain line to prevent emission of vapors. However, a completely closed line will not provide any visual indication of a tank that is overfilling. A sample flow indicator will provide an indication of overfilling and can be shown locally and in a remote location. While open overflow lines present potential emission problems in HAP and VOC service, the open overflow lines do not present an emission problem for liquids with extremely low vapor pressures. The determination or allowable limits on vapor pressure levels for low vapor pressure liquids would require agreement by regulatory authorities.

Overflow lines may still be used in caustic or acid services as an example. In these tanks, the overflow line terminates in a neutralizing agent chamber. The neutralized effluent is then discharged into the sewer line.

13.4.6 Tank Inspection

Inspection of tank rim seals is required under the various regulations described in Section 13.6. This requires consideration of potential exposure situations and personal protection equipment requirements. In addition, inspection of tank internals and tank entry must conform with confined entry procedures developed as company standards and OSHA regulations.

EFRT Decks and Seals. As shown in Figure 13.8, a rolling ladder permits access to the floating deck from the gauger's platform. Although there are rim seals at the deck edge and most of the emissions are controlled from fittings, the vapor concentration several feet above the deck level may exceed the allowable exposure concentration. Moreover, the rim seals may have deteriorated, contributing to a somewhat higher vapor concentration. In addition, weather conditions and the level of the roof deck within the shell may also affect the vapor concentration. Wind sweeping over the rim may disperse some of the vapor, but a floating deck that is considerably below the maximum height level may lose much vapor through dispersion. This condition is further exacerbated by essentially windless periods.

During an inspection of the deck and seals (seal inspection requirements are listed in Section 13.7), personal protective equipment should be used when establishing exposure levels. Following this assessment, some adjustment in personal protective equipment can be made. However, personal protective equipment should be used when inspecting the rim seals and measuring the openings. Although the inspector is protected, another worker should be observing the inspector at all times to ensure nothing untoward happens during the inspection. The observer should also have personal protective equipment, including self-contained breathing apparatus. This "buddy" system is recommended for tank inspection, repairs, and entry into any of the various types of tanks.

IFRT Deck and Seals. Since the vapor space is enclosed, the rim seals are visually inspected periodically through a manhole in the roof as required in the regulations (Section 13.6). Protective breathing equipment should be used during this inspection although the tank vapor space has breathing vents. In addition, another worker should observe the inspector at all times during this operation.

General Inspection. When the EFRT and the IFRT are emptied (including closed tanks), the floating roof decks and seals must be inspected. The floating decks rest on their support legs when the tank is empty and the tank must be completely degassed before the tank is entered. An overall inspection also requires an inspection of the underside of the floating roof, tank bottom, internal shell wall, and internal shell connections. Degassing procedures for tanks in HAP and VOC service must meet guidelines for confined space entry. In some areas, degassing must be accomplished with minimal emissions procedures. Emission control in fixed roof or closed floating roof tanks can be accomplished with vent controls. In floating roof tanks, this is rather difficult. However, nozzles below the resting level of a deck could be utilized for purging and collection of vapors on a portable basis.

Obviously, tank entry following operating service must follow defined procedures to minimize potential exposures or health problems.

13.5 TANK DESIGN REQUIREMENTS

An extensive range of tank design regulations has been issued by OSHA.[10] These design requirements are based upon standards issued by the American Petroleum Institute (API), Underwriter Laboratories (UL), and the American Society of Mechanical Engineers. The OSHA standards cover the following major areas, but the OSHA regulations should be reviewed to ensure tankage is installed in compliance with reference 10:

- Metal fabrication and construction
- Accessory design and installation
- Corrosion allowances
- Tank foundations and anchorage
- Tank spacing
- Vent sizing for filling and emergency conditions
- Diked area volume requirements
- Diked wall construction details
- Venting locations near buildings
- Criteria for locating tanks within buildings
- Test and repair requirements
- Piping design and requirements

Additional information on tank and dike drainage is presented in Chapter 12.

13.6 REGULATIONS

Proposed tank emission regulations under the 1990 Clean Air Act Amendments termed HON cover the hazardous air pollutants previously described.[4] In addition, other tank emission control regulations have been in effect since promulgation, covering volatile organic chemicals.[11] Benzene storage tank emissions were controlled through NESHAPs regulation[12] but are superseded by the HON regulations. Other restrictive tankage emission regulations have been issued by states and regional air control boards to control VOC emissions. Although similar, there are some differences between the federal regulations and the regional control boards. In this section, the pertinent regulations are presented and differences are indicated. While the HON requirements for benzene supersede the NESHAP regulations, both NESHAPs and HON requirements are essentially identical. The relevant emissions regulations are summarized in the following sections.

The tankage regulations for volatile organic liquids and petroleum refer to those tanks where construction, reconstruction, or modification commenced after July 23, 1984.[11] In addition to this regulation for petroleum tanks, two other standards of performance (VOC) regulations also exist covering two separate periods, 1973–1978 and 1978–1984. In some cases these regulations may govern tank emissions and should be investigated where the tanks were constructed, reconstructed, or modified prior to July 23, 1984. These regulations are shown in references 15 and 16.

The HAP and VOC regulations do not differentiate between welded and riveted tank sealing requirements. However, the regional boards have different regulations for molded and riveted tanks and must be carefully reviewed for compliance.[13,14] While many of the old tanks in service are riveted petroleum

TABLE 13.12 Atmospheric Tank Storage Applicability Standards for New Sources[a]

Applicable Law	Tank Capacity (m^3)	Minimum Vapor Pressure[e] (kPa)	Maximum Vapor Pressure[f] (kPa)
HON			
Group 1	38 to < 151	13.1	76.0
(HAP)	≥ 151	0.7	76.0
VOC[b]	75 to < 151	27.6	76.6
	≥ 151	5.2	76.6
SCAQMD[c]	≤ 75	*h*	*h*
	75 to < 150	0.70	76.0
	≥ 150	0.23	76.0
BAAQMD[d]	35 to < 75[g]	10.3	76.0
	75 to < 150	10.3	76.0
	≥ 150	3.4	76.0

[a]Refers to external, internal floating and closed, fixed roof-tanks with a control system.
[b]From reference 11.
[c]Southern California Air Quality Management District.
[d]Bay Area Air Quality Management District.
[e]Can only exceed this level with a floating roof or controls.
[f]Pressure vessel storage is required above this pressure level.
[g]A pressure-vacuum valve is required for ±10% setting of maximum working pressure, or a vapor loss control device.
[h]Requirements identical to footnote *g*.

tanks, the new regulations will eventually cover all tankage and the regulations presented in this section only refer to welded tanks.

13.6.1 Basic Regulations

The HON regulations apply to two categories of HAP termed Group 1 and Group 2 sources. These sources are essentially defined by vapor pressure in the HON and other regulations. The Group 1 compounds that meet the definitions for new and existing storage tanks are shown in Tables 13.12 and 13.13 along with the VOC regional requirements.

TABLE 13.13 Atmospheric Tank Storage Applicability Standards for Existing Sources[a]

Applicable Law	Tank Capacity (m^3)	Minimum Vapor Pressure (kPa)	Maximum Vapor Pressure (kPa)
HON	75 to < 151	13.1	76.0
Group 1 (HAP)	≥ 151	5.2	76.0
VOC		Identical to Table 13.12	
SCAQMD	*b*	*b*	*b*
BAAQMD	*c*	*c*	*c*

[a]Refers to external, internal floating, and closed fixed-roof tanks with a control system.
[b]Older tanks conform to some vapor standards as new tanks, but must demonstrate comparable control capability.
[c]Pressure levels remain the same, but tanks must meet special sealing requirements.

The HAP compounds that have a vapor pressure less than the minimum vapor pressure shown are considered Group 2 compounds, which can be stored in tanks without the controls required by the various regulatory authorities. Where the vapor pressure exceeds the maximum vapor pressure permitted in these atmospheric tanks, the chemical compounds must be stored in pressure vessels or a tank with a closed vent and control device. This closed tank system is specifically required for tank volumes $> 75\ m^3$ in the VOC regulations[11] and would be applicable to all HAP compounds that are photochemical precursors. The control device must have a 95% emission reduction, and the venting system is operated with an emissions leakage detection maximum of 500 ppmv.

Table 13.13 represents the standard vapor pressure constraints for existing tanks which do not change with the exception of the minimum HON tank size. However, the regional boards include control requirements on vapor emissions that are similar to the controls specified for new tanks.

13.6.2 Internal Floating Roof General Requirements

Internal floating roof tanks must confirm to the following equipment requirements for HON and VOC regulations:

1. The internal floating roof shall float on the liquid surface except during the initial fill or when the tank is emptied and refilled. Emptying or filling shall be continuous when the roof is supported on the leg supports.
2. For floating seals, one of the following is necessary:
   - A liquid mounted seal
   - A metallic-shoe seal
   - Two seals mounted one above the other with each seal forming complete closure against the wall; liquid mounting is not necessary for the primary seal
3. Automatic bleeder vents are closed at all times during deck flotation.
4. For noncontact floating roofs, all openings except vacuum breaker and rim space must be extended below the liquid surface
5. With the exception of leg sleeves, automatic bleeders, rim space vents, column wells, ladder wells, sample wells, and stub drains, all other openings must have a gasket and cover or lid. Automatic bleeders and rim space vents are gasketed. Ladder openings are gasketed with a sliding cover.
6. Sample wells are required for periodic sampling. A slit fabric cover is necessary that encloses 90% of the opening area.
7. All columns penetrating the floating deck to support the fixed roof shall have a fabric sleeve or gasketed sliding cover.
8. All covers or lids shall be closed when not in use. Covers on access hatches and automatic gauge float wells are bolted when not in use. Rim space vents shall only open when the internal floater is supported on legs or when the pressure exceeds the rim seal rating.

9. The SCAMQD has additional requirements for IFR tanks to ensure that emissions losses are not excessive:

a. Organic liquids with a true vapor pressure $\geq$ 11 psia (569 mm Hg) cannot be stored in an IFR tank.

b. Compliance with seal and vapor controls also requires measurement of the enclosed vapor space above the internal floater with an explosimeter. The explosimeter reading shall not exceed 50% of the lower explosive limit (LEL) for tanks installed prior to June 1984. For tankage installed after June 1, 1984, the vapor space concentration shall not exceed 20% of the LEL.

13.6.3 Internal Floating Roof Inspection and Compliance

Inspection and compliance requirements for *internal floating roof tanks* are summarized in the following list with changes indicated for the regional boards where applicable:

1. After installation of all control equipment, an inspection is conducted of the floating roof, the primary seal, and the secondary seal (if one is installed) prior to filling the storage vessel with HAP or VOC. Any defects or gaps in the seals and seal installation or in the floating roof must be repaired prior to placing the tank in service.

2. Where a *single seal* is installed, the seal and the floating roof are inspected through roof hatches or manholes at least once every 12 months *after filling*. If any of the following conditions are observed, the tank must be repaired within 45 days after the inspection or emptied and removed from service (two extensions of 30 days each may be granted):

- The floating roof is not resting properly on the liquid surface
- Liquid on the floating roof
- The seal is detached
- There are holes or tears in the seal fabric
- Visible gaps between the seal and the wall of the storage vessel

The internal floating roof must be visually inspected whenever the tank is emptied and purged or degassed. This inspection must be conducted at least once every 10 yr. The following must be inspected and repairs instituted where defects are noted before refilling the tank:

- Floating roof
- Rim seal
- Gaskets
- Slotted membranes
- Sleeve seals

3. For *double seal installations*, visual inspections through manholes or roof hatches are conducted every 12 months after initial filling as described with a complete visual inspection of the internal floating roof tank each time the vessel is emptied and degassed and at least once every 10 yr as described for single seals *or* an inspection of the internal floating roof seals and equipment in an empty, degassed tank at least once every five yr.

Where defects are noted during operating inspections, the internal roof must be repaired within 45 days or the tank must be emptied and removed from service. If the vessel cannot be emptied within 45 days, two extensions of 30 days each may be granted.

4. The EPA must be notified prior to refilling a tank.

13.6.4 External Floating Roof General Requirements

External floating roof tanks must conform to the following HON and VOC regulations:

1. The external floating roof shall float on the liquid surface except during the initial fill or when the tank is emptied and refilled. Emptying or filling shall be continuous when the roof is supported on leg supports.

2. Seals or a closure device are required between the tank wall and floating roof. The closure device consists of two seals with one seal located above the other. The lower seal is the primary seal, and the upper seal is the secondary seal. The seals must meet the following criteria:

   a. The primary seal shall be a mechanical-shoe seal or a liquid-mounted seal.

   b. Both seals will completely cover the annular space between the floating roof edge and the tank wall in a continuous fashion.

   c. The SCAQMD also requires that secondary seals should not be attached to the primary seal.[13] This requirement indicates that a secondary seal extends from the top of the floating roof rim to the tank wall, as shown in Figure 13.2.

   d. The BAAQMD requires that secondary seals on tanks installed after September 1985 have a zero secondary.[14] This is a tight-fitting seal that does not have a gap exceeding 0.06 in. (1.5 mm), with the secondary seal exerting a positive pressure on the tank shell.

3. For noncontact floating roofs, all deck openings must be extended below the liquid surface. Automatic bleeder vents and rim space vents are excluded from this requirement.

4. Automatic bleeder vents are closed at all times when the roof deck is floating.

5. With the exception of leg sleeves, automatic bleeders, rim space vents, and roof drains, all other openings must have a gasket and cover or lid. Access hatches and automatic gauge float wells shall have covers that are bolted when not in use. Automatic bleeder and rim space vents are gasketed.

6. Each roof drain (into the tank liquid) must have a slotted membrane fabric cover that covers at least 90% of the open area.

7. Each slotted and unslotted guide pole well must have a gasketed sliding cover or a flexible fabric sleeve. Also, the float inside the slotted guide pole must have a gasket.

8. All gauge hatch wells must have a gasketed cover that is closed at all times except when in use.

9. Rim space vents shall only open when the floating roof is supported on legs or when the rim seal exceeds the rim seal setting.

Items 7–9 are not included in VOC requirements.[11] The BAAQMD also provides restrictions on the opening sizes permitted in the various accessories listed.[14]

13.6.5 External Floating Roof Inspection and Compliance

External floating roof inspection and compliance requirements are summarized in the following list with changes for the regional boards indicated where applicable:

1. *Primary seals* are inspected as follows:

   a. Primary seal gaps are measured during hydrostatic tank tests or by regulations if the tank is filled with a HAP fluid (HON). For new sources that commence construction after December 31, 1992, compliance is required upon startup or the date of promulgation, whichever is later. Measurement time after filling is 60 days for VOC regulations. Both regulations require primary seal measurements every 5 yr after the initial measurement. For older tanks in HAP service, the gap measurements must be complete within 3 yr after the effective date of the regulation and every 5 yr thereafter.

   b. Full circumferential inspection prior to installation of the secondary seal is required by the BAAQMD and SCAQMD. The primary seal is made available for circumferential inspection every 5 yr, which is lengthened to 10 yr by the BAAQMD if the secondary seal is a zero gap seal. Inspections are conducted by board representatives.

   c. Both regional boards also have annual inspections of the primary seals at random locations. If the secondary seal is a zero-gap type, the random inspection is conducted every 2 yr. These inspections are also conducted by board representatives.

2. *Secondary seals* are inspected as follows: For HAP service in new and existing tanks, the secondary seal gaps are measured on the same schedule as that described for primary seals. VOC secondary seal regulations are also similar to the measurement requirements for primary seals.

3. Where any tank is out of service for a year or more, filling with a HAP liquid requires gap measurements and inspections within 90 days of filling. In VOC service, the filling is considered an initial filling and the tank is inspected accordingly.

4. *Primary seal gap* measurements and requirements are as follows:

   a. Gaps are measured at one or more liquid levels when the floating roof is not supported by the roof legs.

   b. Seal gaps are measured along the entire inner circumference of the tank wall where a 0.32-cm diameter probe passes freely between the seal and the wall. The circumferential distance of each location along the wall that passes the probe is measured. The distance from the wall of each gap is then measured with probes of various widths. This permits a determination of gap area for each opening by multiplying the opening length along the wall by the gap opening depth.

c. For mechanical-shoe and liquid-mounted resilient seals in HON and VOC regulations, the total gap area, which is the sum of various gaps along the circumference, should not exceed 212 cm^2 per meter of vessel diameter. The maximum width or distance from the wall to the seal should not exceed 3.81 cm for any gap.

d. The SCAQMD regulations are based on the specific type of seal and tank construction (welded or riveted). The gap measurement requirements that follow are based upon welded tanks.[13] For *mechanical-shoe seals*, the following gap measurements between the wall and seal should not be exceeded:

- 3.8 cm (1.5 in.) for an accumulated length of 10% of the tank circumference
- 1.3 cm (0.5 in.) for an accumulated length of 30% of the tank circumference
- 0.32 cm (0.125 in.) for an accumulated length of 60% of the tank circumference
- None of the gaps between the wall and seal shall exceed 3.80 cm (1.5 in.)
- No *continuous* gap > 0.32 cm (1/8 in.) shall exceed 10% of the tank circumferences.

For *resilient seals*, gap measurements between the wall and seal are based upon seals installed before or after August 1978. After 1978, secondary seal loss effectiveness must be equivalent to mechanical-shoe effectiveness.

e. The BAAQMD regulations are based on the specific type of seal and tank construction either welded or riveted. Gap measurement requirements that follow are based on welded tanks[14]:

For *mechanical-shoe seals*, the following gap measurements between the wall and seal should be observed:

- No *continuous* gap > 0.32 cm (0.125 in.) shall exceed 10% of the tank circumference
- > 1.3 cm (0.5 in.) gap accumulated length shall not be $> 10\%$ of the tank circumference
- 0.32 cm (0.125 in.) *accumulated* gap length shall not be $> 40\%$ of the tank circumference
- None of the gaps between the wall and seal shall exceed 3.8 cm (1.5 in.)

Other mechanical-shoe seal requirements are:

- One end of the shoe shall extend into the liquid
- The upper end will extend a minimum of 61 in. above the liquid surface
- The maximum gap between the shoe and shell is no greater than twice the gap allowed by the seal gap criteria for a length ≥ 46 cm (18 in.) in the vertical plane above the liquid surface

For *resilient seals*, the following conditions should be observed:

- None of the gap between the wall and seal should exceed 1.3 cm (0.5 in.)
- > 0.32 cm (0.125 in.) gap accumulated length shall not be $> 5\%$ of the tank circumference

- No holes, tears, or other openings in the seal

5. *Secondary seal gap* measurements and requirements follow:

a. The accumulated gaps between the tank wall and the seal shall not exceed 21.22 cm^2 per meter for the HON and VOC regulations.

b. The width of any gap shall not exceed 1.27 cm (0.5 in.).

c. No holes, tears, or other openings in the seal or seal fabric.

d. The SCAQMD regulations are based upon the type of primary seal and shell construction (welded or riveted). Gap measurements and requirements that follow are based upon welded tanks:[13]

For *mechanical-shoe and resilient primary seals*, the following secondary seal gap measurements between the tank wall and seal should be observed:

- 0.32 cm (0.125 in.) gap for an accumulated length that shall not exceed 95% of the tank circumference
- 1.32 cm (0.5 in.) gap for an accumulated length of the remaining 5%
- None of the gaps between the tank wall and secondary seal shall exceed 1.32 cm (0.5 in.)
- The seal should permit insertion of probes $\leq$ 3.8 cm ($\leq$ 1.5 in.) to measure primary seal gaps

e. The BAAQMD regulations are based upon the type of primary seal and shell construction (welded or riveted). Gap measurements and requirements that follow are based upon welded tanks:[14]

For *mechanical-shoe seals*, the following secondary seal gap measurements between the tank wall and seal should be observed:

- $>$ 0.32 cm (0.125 in.) gap for an accumulated length that shall not exceed 5% of the tank circumference
- None of the gaps between the tank wall and secondary seal shall exceed 1.3 cm (0.5 in.)
- The seal should permit insertion of probes $\leq$ 3.8 cm ($\leq$ 1.5 in.) to measure primary seal gaps

For *resilient primary seals*, the following secondary seal gap measurements between the tank wall and seal should be observed:

- $>$ 0.32 cm (0.125 in.) gaps for an accumulated length that shall not exceed 5% of the tank circumference
- None of the gaps between the tank wall and secondary seal should exceed 1.3 cm (0.5 in.)
- The seal should permit insertion of probes $\leq$ 1.3 cm (0.5 in.) to measure primary seal gaps

6. *Repair Requirements.* The tankage must meet the specified HON gap requirements and general quality status promulgated in the regulations. If any of the requirements are not satisfactory, those conditions that are unacceptable must be repaired within 45 days of identification during an inspection. Where the tank is in service, the storage tank must be emptied and removed from service within 45 days. Two extensions of 30 days each may be granted by the EPA.

Where defects are noted in the floating roof or holes, tears, or other openings in the primary or secondary seals, these conditions will be repaired before filling or refilling the tank with HAP.

13.6.6 Converted External Floating Roof Tanks

External floating roof tanks are frequently converted to an internal floating roof tank by covering the tank with a roof or dome. In many of the installations, the tank shell cannot support a cone roof and a geodesic dome is installed which does not require the internal support structure of a cone roof. This conversion results in an internal roof tank and the HON regulations require compliance with the internal roof standards presented in Section 13.6.2 and the requirements for deck fittings specified for external floating roof tanks in Section 13.6.4.

Although a roof can be installed on an EFRT for conversion, vents are required as shown in Figure 13.19 for typical floating roof operation. As an alternate, a vapor vent control system can be installed that eliminates the venting requirement. Conversion requires careful analysis to ensure tankage emission control systems meet regulatory system requirements and safety standards.

13.6.7 Tanks with Closed Vents and a Control Device

The tank systems described in Section 13.1.4 include an internal floating deck or roof. However, the closed vented tank category may also include closed tanks that do not have any internal vapor control system. For tanks that discharge vapors through a closed system to a control device which does not include a flare, the requirements for HAP and VOC control are based upon an Implementation Plan that includes the following elements:

1. Documentation demonstrating that the control device will achieve the required control efficiency during maximum load conditions. The documentation includes the following:

   a. Vapor stream description including flow rates to the control device, HAP, and/or VOC content under varying liquid level conditions (dynamic and static).

   b. Design evaluation for the control device must include all materials processed through the closed vent and control device.

   c. When the control system is an *enclosed combustion device*, a minimum of 0.75 sec residence and a minimum temperature of 816°C is required to meet the 95% control efficiency requirement.

   d. *Thermal incinerator* design specifications shall include the organic HAP autoignition temperature, stream flow rate, combustion temperature, and residence time at combustion temperature.

   e. For *carbon adsorbers*, a design evaluation includes carbon adsorptive capacity for the HAP vapors, quantity of carbon in each bed, number of beds, feed stream humidity, feed stream temperature, feed stream flow rate, desorption schedule, steam pressure or temperature for desorption, and steam quantity. Pressure drop is required for vacuum desorption.

   f. For *condensers*, the design must include the vapor stream temperature, condenser type, stream flow rate, and final condensed vapor temperature.

   g. A description of the parameters monitored to ensure the control device is operated in accordance with the design specifications, explanation of parameter selection basis, and monitoring frequency.

2. The operating range for each parameter identified in item 1g shall be included in the Implementation Plan. This operating range will represent the conditions for which the control device can achieve a 95% or greater emissions reduction.

3. The plant must insure that during routine maintenance the control device does not fall below the 95% emissions reduction criteria for a period that does not exceed 72 hr/yr.

4. The facility shall monitor the parameters specified in item 1g to insure that they remain within the specified ranges to meet with compliance requirements.

13.6.8 Tanks with Closed Vents and a Flare

Where tank vents are connected to a flare rather than the control devices described in Section 13.6.6, the following requirements must be observed:

1. The flare must meet the compliance performance requirements that will be issued in the future by the EPA. Previously, the EPA suggested in a preliminary draft that a flare is required to discharge a maximum stream content of 0.005 wt % HAP. While this approach may change, flare design will also include references to the flare operation requirements shown in reference 27.

2. In addition to the other flare requirements, the facility will also submit the following information to the EPA:

   a. Type of flare, whether steam-assisted, air-assisted, or nonassisted.

   b. All visible emission readings, stream heat content determinations, flow rate measurements, and exit velocity determinations.

   c. All periods when the pilot flame is not operating.

   d. Routine maintenance periods when the flare is not in compliance with the future compliance draft shall not exceed 72 hr/yr.

13.6.9 Closed-Vent System

The closed vent system must be operated with detectable emissions < 500 ppm of HAP. This requires compliance with the following procedures:

1. *Inspections.* Conducted during tank filling and at least once per year.

2. *Leaks.* Emissions readings ≥ 500 ppm indicate a leak and visual inspections also define leaks. These must be repaired no later than 15 calendar days after reporting. The first repair attempt shall be attempted within 5 calendar days after reporting. Repairs may be deferred if the repair is not feasible without a process unit shutdown or if emissions occurring during a repair are greater than the emissions resulting from a repair delay.

3. *Unsafe Inspection Exemptions.* Where personnel may be exposed to immediate danger, the plan requires inspection during safe periods. In addition, elevating personnel above 2 m also exempts inspections, but an inspection must be conducted once every 5 yr.

4. *Equipment Monitoring Exemptions.* If any equipment is monitored as part of other specified programs, these items do not require inspection and are exempt.

13.6.10 Reports

Reports are required on tank inspections, failures, gap measurements, repairs, excess emissions from control devices, and control device repairs. In addition, all plans and programs required by the EPA must be maintained in a current up-to-date condition. These requirements are only required for the Group 1 HAP organic chemicals.

13.7 TRANSFER OPERATIONS

Transfer refers to loading and unloading of chemicals and petroleum liquids. With the loading of liquids into empty receivers or containers, the primary concern in these operations is vapor generation during the loading of tank trucks, tank cars, and marine vessels. While much of the basic design control considerations are similar for these three categories, actual marine loading equipment covering vapor collection and control is somewhat different and this general category is not included in this chapter.

Loading of benzene has been regulated under NESHAP regulations,[17] but the proposed HON regulations issued under the CAAA of 1990 covering HAPs supersedes the specific regulations. Although both regulations are similar, the HON requirements are somewhat more extensive and are summarized in this section. Since vapor generation is also of concern in controlling VOC emissions, separate federal regulations have been promulgated to control transfer emissions, specifically gasoline loading at bulk terminals.[18] Although this VOC regulation is quite specific, general VOC emission containment is a major problem in nonattainment areas and various state and local regulatory or regional authorities have promulgated rules that encompass petroleum and organic chemical transfer operations. As examples of state and local regulations, some requirements from the BAAQMD and SCAQMD are presented that may be more stringent than the proposed federal HON regulations.

In addition to these various regulations, OSHA has also promulgated regulations covering requirements for loading and unloading operations.[19] The OSHA regulations cover all types of liquids that meet certain flashpoint criteria and are principally concerned with equipment location, mechanical details, and electrical requirements.

The major HON regulatory requirements for transfer loading operations are presented in the following sections. Differences with other regulations are indicated in certain areas.

13.7.1 HON Regulations Applicability

The HAP requirements cover the 189 chemicals listed in the CAAA of 1990. However, in the proposed HON regulations, the chemicals are further divided into two groups:

1. *Group 1.* The transfer racks in this category load annually a liquid quantity equal to or greater than 0.65 million L/yr (172,000 gal/yr) with a weighted average vapor pressure $\geq$ 1.5 psia (10.3 kPa).

2. *Group 2.* Transfer racks in this category have liquids with vapor pressures < 1.5 psia (10.3 kPa) and do not have any other restrictions.

13.7.2 Loading Rack Regulations

The major HON requirements for racks have been combined and are presented in the following list. These requirements apply to tank trucks and tank cars:

Vapor Collection. All transfer racks in the Group 1 category must be equipped with a vapor collection system and a control device. The control device does not have to be located at the rack.

1. *All* organic vapors displaced from tank trucks and tank cars during loading operations must be collected in the vapor collection system and transferred to the control device.
2. Loading is confined to tank cars and tank trucks equipped with vapor collection systems that are compatible with the vapor collection system on the loading rack.
3. Loading is only permitted when the tank trucks or rail cars are connected to the vapor collection system on the loading rack.
4. Loading is only permitted in vapor-tight tank trucks and rail cars. The vapor-tight transportation equipment must have verification that it was tested within the preceding 12 months. A special testing procedure is conducted on the tank trucks and rail cars to ensure they are vapor tight. Documentation of vapor tightness is necessary before the equipment can be loaded.
5. The vapor collection system must be designed to prevent any HAP vapor collected in one arm bypassing through another loading arm to the atmosphere.
6. Each vapor collection system is monitored for leakage to the atmosphere and leaks are defined as ≥ 500 ppm. The vapor systems are monitored initially and annually thereafter. When leaks are detected, the first attempt at repair is conducted within 5 calendar days after detection, with repairs completed no later than 15 days after leak detection.

Control Devices. The control devices are an important part of the vent system and the major requirements are summarized below:

1. Whenever HAP emissions are vented to a vapor collection system, the vapor collection and control device shall be operating.
2. All Group 1 overall vapor collection and control device systems must comply with *one* of the following two requirements:
   a. HAP emissions must be reduced by 98 wt % or the exhaust gas discharge from the control device must have a maximum HAP concentration of 20 ppmv on a dry basis corrected to 3% oxygen, whichever is less stringent. An organic monitoring device is necessary in the discharge line to monitor compliance with the regulations.
   b. When a flare is used as the control device, the flare must comply with the requirements in reference 20 and a future compliance requirement. Halogenated vent streams shall not be vented to a flare.

c. A vapor balancing system can be used to return vapors from the tank-truck or tankcar to the storage vessel, providing the liquid for loading.

3. If a boiler or process heater firebox (< 44 mW) is the vent control device to comply with the percent reduction, a temperature monitor is required in the firebox. Where all vent streams are introduced with the primary fuel, the boiler or process heater is exempt from this requirement.

4. When a combustion device is the control system for halogenated vent streams, the discharged vapors from the combustion device must be sent to a scrubber before the stream can be vented to the atmosphere. Scrubber performance must demonstrate the following:

a. Overall hydrogen halide emissions must be reduced by 99% *or* the outlet concentration of each individual hydrogen halide or halogen must be reduced to a maximum of 0.5 mg per dry standard cubic meter, whichever is less stringent.

b. The overall percent reduction is a comparison of the total halogenated compound concentration at the scrubber inlet compared with the concentration at the outlet. This reduction must be $\geq 99\%$.

c. In addition, the scrubber effluent must be monitored, and the scrubber inlet liquid and gas flows must be measured.

5. Where a carbon adsorber is used, a monitoring device with an accuracy of $\pm 10\%$ is required that monitors regeneration. This covers regeneration mass stream flow and carbon bed temperature during the regeneration cycle.

6. For a condenser system, the condenser exit temperature must be monitored.

7. Where incinerators are used, temperature monitoring is required in either the firebox or in the ductwork adjacent to the firebox prior to heat exchange. For catalytic incinerators, temperatures are monitored upstream and downstream of the catalyst bed and in close proximity to the bed.

Other General Requirements. Several other requirements are also of importance in the HAP regulations:

1. The overall vapor collection and HAP loading equipment must be designed to prevent exceeding the vapor-tightness test pressure during loading operations.

2. Pressure–vacuum vents in the vapor collection system shall not open at a system pressure lower than the maximum operating pressure of the tank truck or rail car.

3. Any valve in the venting system that can discharge directly or indirectly to the atmosphere must be closed with a car seal or lock and key system. An alternative is a flow indicator to show when there is flow through the valve. The only exception is the pressure-relief valving necessary for safety precautions.

4. Tank truck or rail car vapor-tightness documentation must be available from vendors, showing a test within the preceding 12 months, but identical documentation and testing is required for an owner's vehicles or rail cars. The vapor-tightness test requirements have been developed by the EPA and can be found in reference 21.

Monitoring and Test Requirements. There are two separate considerations with control device systems. The control device must be monitored during normal operations to ensure that it is operating properly and controlling the exhaust emissions. Secondly, an initial performance test or control device operation is necessary to demonstrate compliance with efficiency requirements and is also required under certain conditions and changes. The testing requirements and references are contained in the HON documents.[4]

A summary of monitoring and test requirements for the various control devices is presented in Table 13.14.

Record Keeping. Records of operations are required which must be maintained and readily accessible. Of particular concern are the control device operating parameters shown in Table 13.14 that must be recorded continuously. A quarterly report is also required indicating any exceedances, periods when the vent stream is diverted from the control device, changes or maintenance on car-sealed valves, or pilot flameout on a flare. Record-keeping requirements are also shown in Table 13.14.

13.7.3 EPA Gasoline Terminal Loading Regulations

While federal EPA gasoline loading regulations are generally similar to the HON regulations described in Section 13.7.1, they are designed to reduce VOC emissions and are somewhat less restrictive than the HON regulations. In the HON regulations, control device efficiency of HAP vapor removal is the criteria for emissions control. Gasoline emissions control establishes a maximum emissions quantity to the atmosphere for each liter of gasoline loaded. This is shown in the following summary:

HAP emissions	
Control requirement (HON)	98 wt % emissions reduction or 20 ppmv (dry basis, 3% O_2)
HON halide control requirement	99 wt % emissions reduction or 0.5 mg/m^3 (STP, dry)
Gasoline VOC emissions control requirements	35 mg/L of gasoline loaded or 80 mg/L of gasoline loaded with existing recovery device

The federal VOC regulations in some areas are more restrictive and some examples are presented in Section 13.7.4. State and regional authority regulations may also be more stringent. The general regulations for gasoline loading are summarized in the following paragraphs,[18] based upon the general gasoline definition:

Gasoline refers to any petroleum distillate or petroleum distillate/alcohol blend that has a Reid vapor pressure of 27.6 kPa. The Reid vapor pressure is comparable to a true vapor pressure of 1.97 psia at 62.5°F[10, 18, 19].

TABLE 13.14 Monitoring and Record-Keeping Requirements for Control Devices in Vapor Collection Systems

Type of Control Device Used for Compliance	Monitoring Equipment Required[a]	Parameters to be Monitored	Record Keeping Requirements	Parameter Boundary Exceedances to Record and Report
Thermal incinerator	Temperature-monitoring device (installed in fire box or in the duct work immediately downstream of the fire box)[c] equipped with a continuous recorder [63.127(a) (1)]	Average fire box temperature	Every 2 min during loading cycles less than 3-hr; every 15 min during loading cycles greater than 3 hr	All 3-hr periods of operation or loading cycles when average combustion temperature is > 28°C (50°F) below the average value measured during the most recent performance test
Catalytic incinerator	Temperature-monitoring device (installed in gas stream immediately before and after catalyst bed) equipped with a continuous recorder [63.127(a) (1)]	Average temperature upstream and downstream of the catalyst bed	Every 2 min during loading cycles less than 3-hr; every 15 min during loading cycles greater than 3 hr	All 3-hr periods of operation or loading cycles when the average temperature of vent stream upstream of the catalyst bed is > 20°C (50°F) below the average value from the most recent performance test All 3-hr periods of operation or loading cycles when the average temperature difference across the catalyst bed is < 80% of the average temperature measured during the most recent performance test
Boiler or process heater with a design heat input capacity < 44 mW	Temperature-monitoring device (installed in fire box) equipped with continuous recorder [63.127(a) (3)]	Average combustion temperature	Every 2 min during a loading cycle less than 3 hr; every 15 min during loading cycles greater than 3 hr	All 3-hr periods of operation or loading cycles when average combustion temperature is > 28°C (50°F) below the average value from the most recent performance test
Flare	Heat-sensing device [63.127(a) (2)]	Presence of a flame at the pilot light	Every 2 min during a loading cycle less than 3 hr; every 15 min during loading cycles greater than 3 hr	All periods when the pilot flame is absent [63.130(c)]

TABLE 13.14 ***(Continued.)***

Type of Control Device Used for Compliance	Monitoring Equipment Required[a]	Parameters to be Monitored	Record Keeping Requirements	Parameter Boundary Exceedances to Record and Report
All control devices	1. Flow indicator equipped with continuous recorder installed at all bypass lines to the atmosphere *or* 2. Seal valves closed with car-seal or lock and key configuration [63.127(d)]	1. Presence of flow diverted to the atmosphere from the control device *or* 2. Monthly inspections of sealed valves	1. Every 2 min during a loading cycle less than 3 hr; every 15 min during loading cycles greater than 3 hr 2. Records of monthly inspections	1. Periods during loading cycles when the vent stream is diverted through a bypass line [63.130(b) (1)] *or* 2. Inspections that show valves are not sealed closed or seal has been changed [63.130(b) (2)]
Scrubber (for vent streams with greater than 80 ppmv halogenated compounds. Note: Controlled by a combustion device other than a flare)	pH-monitoring device [63.127(a) (4)]	pH of scrubber effluent	Every 2 min during loading cycles less than 3 hr; every 15 min during loading cycles greater than 3 hr	All 3 hr periods of operation or loading cycles when the average pH of the scrubber effluent is more than 1 pH unit below the value measured during the most recent performance test
	Flow meters to measure scrubber liquid and gas flow rates (installed at the influent for liquid flow and the inlet for gas flow) [63.127(a) (4)]	Scrubber liquid / gas ratio	Every 2 min during loading cycles less than 3 hr; every 15 min during loading cycles greater than 3 hr	All 3 hr periods of operation or loading cycles when the liquid / gas ratio is below the value measured during the most recent performance test
Absorber[b]	Scrubbing liquid temperature monitor equipped with a continuous recorder [63.127(b) (1)]	Average exit temperature of the absorbing liquid	Every 2 min during loading cycles less than 3 hr; every 15 min during loading cycles greater than 3 hr	All 3-hr periods of operation when average temperature is $> 11°C$ (20°F) above the average value from the most recent performance test
	Specific gravity monitor equipped with a continuous recorder [63.127(b) (1)]	Exit specific gravity (or alternative parameter that measures the degree of absorbing liquid saturation, if approved by the Administrator)	Every 2 min during loading cycles less than 3 hr; every 15 min during loading cycles greater than 3 hr	All 3-hr periods of operation when average liquid specific gravity is > 0.1 unit above or below the average value from the most recent performance test

TABLE 13.14 ***(Continued.)***

Type of Control Device Used for Compliance	Monitoring Equipment Required[a]	Parameters to be Monitored	Record Keeping Requirements	Parameter Boundary Exceedances to Record and Report
Carbon adsorber	Integrating steam device equipped with a continuous recorder [63.127(b) (3)]	Total steam mass flow during carbon bed regeneration cycle(s)	Every 2 min during loading cycles less than 3 hr; every 15 min during loading cycles greater than 3 hr	All regeneration cycles when the total mass steam flow is $>10\%$ below the value measured during most recent performance test
	Carbon bed temperature-monitoring device equipped with a continuous recorder [63.127(b) (3)]	Temperature of the carbon bed after regeneration [and within 15 min of completing any cooling cycle(s)] and duration of the carbon bed steaming cycle	Every 2 min during loading cycles less than 3 hr; every 15 min during loading cycles greater than 3 hr	All 3-hr periods of operation showing $> 20\%$ of the amount measured by the monitoring device during the most recent performance test
All recovery devices (as an alternative to those described above)	Organic monitoring device equipped with a continuous recorder [63.127(b)]	Concentration level or reading indicated by the organic monitoring device at the outlet of the recovery device	Every 2 min during loading cycles less than 3 hr; every 15 min during loading cycles greater than 3 hr	All 3-hr periods of operation showing $> 20\%$ of the amount measured by the monitoring device during the most recent performance test

[a]Regulatory citations are listed in brackets.
[b]Alternatively, these devices may comply with the organic monitoring device provisions listed at the end of this table under "all recovery devices."
[c]Monitor may be installed in the duct work immediately downstream of the fire box before any substantial heat exchange is encountered.

Vapor Collection. The requirements are similar to the vapor collection regulations for HON compound transfer.

1. Each collection system must be designed to prevent the passage of vapors from one rack to another.
2. Loading at a terminal can only be conducted on tank trucks equipped with vapor collection equipment compatible with the terminal's vapor collection system.
3. The terminal operator must ensure that the vapor collection systems are connected during loading operations.
4. Vapor collection and liquid loading equipment must be designed and operated to prevent exceeding a tank truck gauge pressure of 4500 Pa (0.63 psig) during loading operations.
5. Pressure–vacuum vents in the vapor collection system will not begin to open at a system pressure < 4500 Pa (0.63 psig).
6. The vapor processing system must be inspected each month for liquid or vapor leaks. Detection methods acceptable are sight, sound, or odor.

Vapor-Tight Tank Trucks. Loading of tank trucks is only permitted when the tank trucks are vapor tight and have the necessary documentation confirming vapor tightness.

Performance Testing. Performance of the vapor collection and control device system in accordance with the regulations must be demonstrated within 60 days after achieving design loading rates and no later than 180 days after the initial startup. The vapor collection system, control device, and all potential leakage sources must be monitored before the test. All leaks must be repaired if the monitor indicates $\geq 10{,}000$ ppm leakage concentrations.

The test must be conducted over a 6-hr period in which a minimum of 300,000 L of gasoline are loaded. A 6-hr period is the basic criteria and some alternative procedures are permitted.[18] Details on the method of conducting the test are presented in the Performance Standard.[18]

Record Keeping. Records must be maintained and periodically updated on truck vapor tightness, monthly leak inspections, and any replacements or additions to the vapor processing system.

13.7.4 Regional Terminal Loading Regulations

The regulations in this section are those issued by the Bay Area Air Quality Management District (BAAQMD) and the Southern California Air Quality Management District (SCAQMD). Separate BAAQMD regulations cover gasoline and organic compounds[22,23,25] compared with the SCAQMD regulations.[24] These regulations are primarily concerned with emissions discharged to the atmosphere from the vapor recovery system control device. A general comparison of the BAAQMD and SCAQMD regulations is presented in Table 13.15.

TABLE 13.15 Comparison of BAAQMD and SCAQMD Regulations for Organic Liquid and Gasoline Loading Facilities

Requirement	BAAQMD		SCAQMD	
	Gasoline	Organic Liquid	Gasoline	Organic Liquid
Vapor pressure applicability	> 4.0 psi Reid[c] vs. pressure or 1.97 psia at 62.5°F Bulk plants and terminals[a]	> 1.5 psia (77.5 mm Hg) Facilities loading ≥ 600k gal / yr ($2300\ m^3$ / yr)	≥ 1.5 psia (77.5 mm Hg) Facilities loading ≥ 20,000 gal / day (75,700 L / day)(Class A facility)[d]	
Limitations — terminals	9.6 gm / 1000 L liquid loaded (0.08 lb / 1,000 gal)	78 gm / 1000 L liquid loaded (0.65 lb / 1000 gal)	35 gm / 1000 L liquid loaded (0.29 lb / 1000 gal)	
Limitations — bulk plants	60 gm / 1000 L liquid loaded (0.50 lbs / 1000 gal)		Recovery ≥ 90% of displaced vapors; plant loading < 20,000 gal / day or > 500k gal / yr before 1976 After 1976 < 20,000 gal / day (Class B facility)	
Liquid leakage from disconnects	< 4 drops / min and ≤ 10 mL per disconnect		≤ 2 mL / disconnect from top dome ≤ 10 mL / disconnect from bottom	
Vapor tightness		< 100% of LEL[b] at distance of 2.5 cm from source	< 3000 ppm at distance of 2 cm from source (meter calibrated with propane	

[a]Terminal and bulk plant sizes defined in regulations. Bulk plants supplies from terminals.
[b]Lower explosive limit.
[c]Petroleum measurement (see ref. 1).
[d]Facilities described in reference 24.

In the SCAQMD regulations, the terminals are classified as A, B, and C facilities. Both A and B are described in Table 13.15 with the Class C facility shown as a gasoline loading plant in existence before January 1976 that has a loading rate $< 5{,}000$ gal/day (19,000 L/day) and an annual loading rate $< 500{,}000$ gal. These classifications provide some variation in requirements. The Class A and B require vapor recovery systems, but the Class C facility does not specifically require vapor recovery or controls. In addition, the regulations permit submerged filling in the Class C facility, which is described in another section.

The BAAQMD regulations require vapor recovery in terminals and bulk plans as shown in Table 13.15. In addition, bottom loading is specifically required in the gasoline terminal regulations, with submerged filling or the equivalent mentioned in the other regulations. Bottom loading is also specifically required in the SCAQMD Class A facilities. While not critical where vapor recovery is installed, bottom filling is generally preferred to reduce vapor generation and simplify actual transfer hose and loading arm operations. The type of filling system for tank trucks and rail cars is not mentioned in the VOC or HAP regulations.

In the BAAQMD regulations, organic liquids with a true vapor pressure > 1.5 psia (0.1 bar) cannot be spilled, discarded in open sewers, or stored in open containers. The gasoline filling regulations also contain the same restrictions. Neither SCAQMD nor federal regulations contain this clause, but refer to minimum leakage.

The California regulations are quite specific regarding liquid leakage when disconnecting fill hoses or arms from tank trucks or tank cars. This regulation requires the installation of equipment with leakless couplings. The Federal VOC or HAP regulations do not mention disconnect leakage, which remains a local or regional rule where applicable. This leakage is of concern to the industrial hygienist.

One requirement in gasoline operation is that vapor from the tank truck cannot be purged to the atmosphere. While not specifically addressed in other regulations, connection of the vapor collection systems indicates that purging to the atmosphere is not permitted. Atmospheric vent connections on the vapor recovery system must be car-sealed closed or locked, eliminating atmospheric purging.

13.7.5 General Loading Considerations

There are two general types of tank truck liquid loading procedures. These procedures are normally top- and bottom-loading operations. In top-loading operations, an arm is inserted into a tank truck compartment through a hatch located on top of the tank truck. In contrast, the bottom-loading procedure depends upon loading through a hose or flexible arm that is attached to the bottom of the tank truck. Workplace exposures vary depending upon the type of loading procedure and associated vapor recovery equipment. Moreover, top-loading methods differ, resulting in a variety of potential exposures. However, the type of top-loading procedure is regulated in most situations, and the regulations generally control vapor emissions to the atmosphere and potential exposures.

Vapor generation is minimized by the introduction of liquid through the bottom of the tank truck compartments or container. Top loading can be conducted through an open hatch in two different procedures, as shown in Figure 13.22.

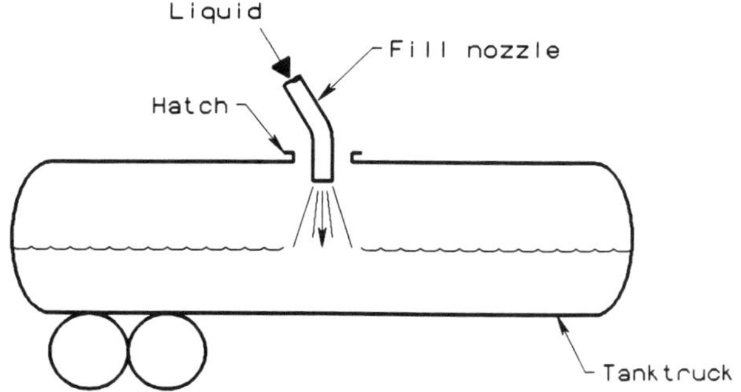

A. ELEVATED NOZZLE PRODUCING SPLASH EFFECTS

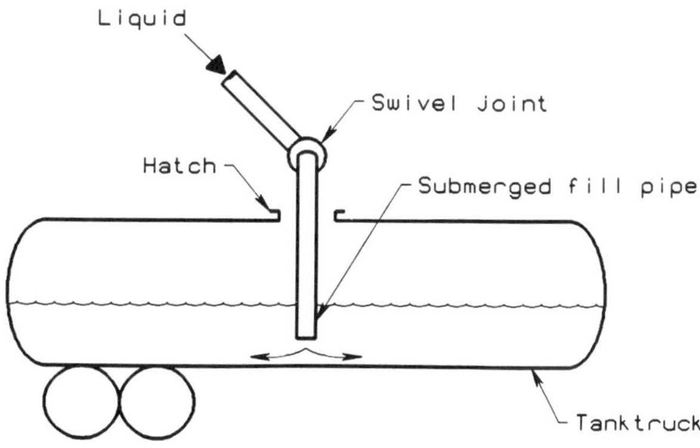

B. SUBMERGED FILL PIPE

Figure 13.22 *Top loading through an open hatch.*

Figure 13.22*A* illustrates "splash loading" since the liquid from the short arm falling on the liquid surface splashes, generating very large emission rates. This form of loading produces high vapor concentrations in the area adjacent to the truck which is further exacerbated if the liquid has a relatively high vapor pressure. Obviously, potential exposures with this loading procedure can be excessive. The regulations for HAPs and VOCs normally require vapor recovery which prevents this type of loading procedure. However, where the vapor pressure of the liquid is low and a maximum vapor pressure of 1.5 psia (10.2 kPa) is not exceeded, this top loading *may* be permitted for some products such as heavy petroleum oils and asphalt. Although relatively inexpensive, potential exposures with lower volatility liquids can still be excessive and this type of loading is normally not recommended.

Long fill pipes extended through the hatch as shown in Figure 13.22*B* also fill essentially from the bottom of the tank truck. Since the liquid level rapidly covers the bottom opening of the fill pipe, vapor generation is quite low. This type is termed submerged filling and is specifically authorized by some of the regional regulatory authorities. The height of the fill pipe opening above the tank truck or container bottom is generally stated in the regulations.

13.7.6 Top Loading

The regulations require a fixed vapor connection system in the tank truck that connects with a vapor collection–recovery system in the loading area. This results in a top-loading system that is permanently installed in the top hatch and contains a submerged fill pipe and a vapor recovery connection as shown in Figure 13.23*A*. In addition, a level instrument sensor is also located in the top hatch which signals when the tank truck liquid level has reached the fill line. This signal can be connected to a motor-operated valve that automatically blocks liquid flow to the tank truck. Top-loading in tank trucks may require several connections when the tank container is separated into several compartments. This configuration is shown in Figure 13.23*B*. However, one compartment is filled at a time.

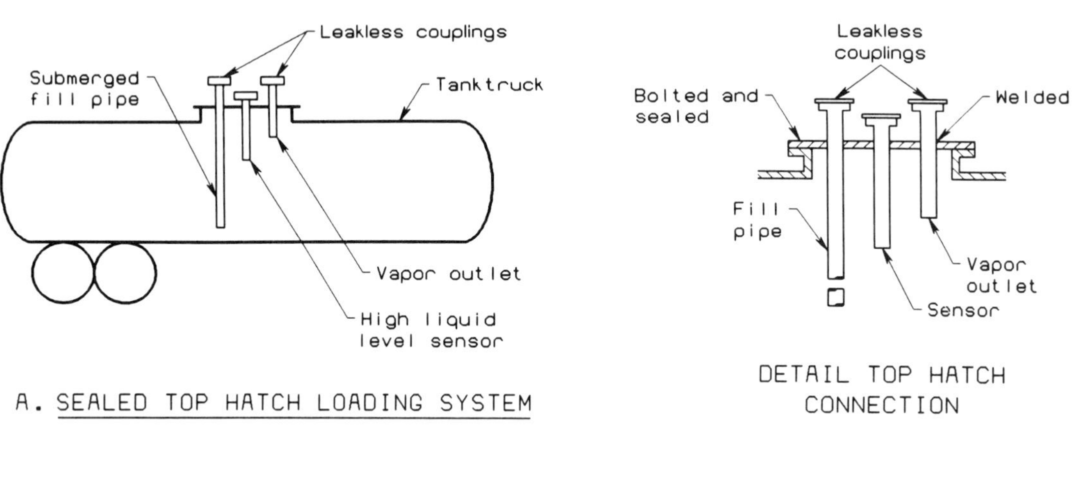

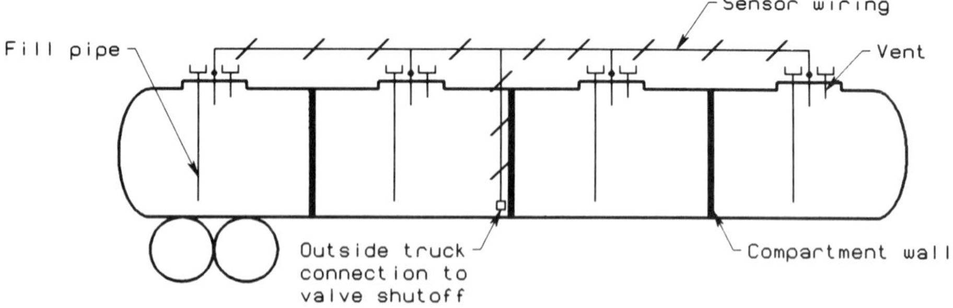

Figure 13.23 *Sealed top-loading system.*

In many areas of the world where open top-loading systems are used, long fill pipes inserted manually through open hatches in tank trucks reduce splashing and vapor generation. Emissions resulting from vapor displacement produced through liquid filling and *high vapor pressures*, however, adversely affect workplace exposure levels. Removal of the fill pipe also has a significant impact on worker exposure due to drippage and the proximity of the wet pipe to the worker. The worker in this situation is required to lift or guide the fill pipe from the load position to the fill pipe storage position. In addition, liquid levels are visually observed in many areas and hatch covers are closed manually, further increasing exposures. Since a top-loading operation of this type may result in relatively high exposure levels, vapor controls based on the design in Figure 13.23 will reduce exposures. The top of the hatch cover with the welded fill, vapor, and instrument connections forms a unit that can be bolted and unbolted to the hatch cover. The existing hatch cover on the tank truck is removed for filling and replaced when filling is completed. This system does not completely eliminate exposures due to vapor emissions when the cover plate is removed and wet piping is exposed. Alternatively, the hatch cover with the welded fixtures could remain permanently bolted to the hatch, eliminating potential exposures during filling.

Combined fill piping and vapor line arms that perform the filling and vapor collection function in an open hatch are available commercially from manufacturers. The long fill pipe and vapor nozzle are connected to a large fitting or head that seals the hatch opening. This large fitting in turn is connected to a platform through steel piping with swivel pipe joints, and the entire assembly is balanced to permit assembly movement with little effort. In addition, the large head fitting also contains a liquid level sensor that closes the feed liquid valve. This system is quite effective, but requires unbolting the filling hatch cover and rebolting after filling. Some exposure may occur during this part of the operation because the hatch is open for a short period and the fill arm is coated with liquid when removed from the tank truck. Open-hatch loading for higher vapor pressure liquids should be avoided. However, for low vapor pressure liquids, a combined fill pipe and vapor arm will significantly reduce exposures.

Although pressure–vacuum vents are required in the vapor collection system, these vents are also necessary on tank trucks. These P–V vents can be located on the top hatch plate for either a single or multiple compartment tank truck. The truck P–V vents may not be required when in the filling mode because of the P–V vents in the vapor collection system, but they are required when the tank truck is delivering liquid to other locations. Pressure buildup or vacuum can occur in various situations, and the tank truck must be protected at all times.

Where permanent fill and vent pipe connections are installed as shown in Figure 13.23, exposures are reduced significantly. Dry-break couplings should also be installed for tight-fitting connections. These couplings have extremely small leakage amounts when the coupling is opened or disconnected. This liquid leakage quantity is essentially zero in some designs. The couplings have been operating successfully for a number of years in various services, reducing potential exposures.

In top-loading operations, high-level alarms should be installed to prevent overfilling. In Figure 13.23*B*, the high-level sensors are connected to an automatic shutoff valve. In addition, preset counting control devices that measure the volume of liquid loading also prevent overfilling. Although the level override control system provides redundancy, backup systems prevent overfilling and the potential exposures associated with spills and cleanup. In addition, overfilling presents potential safety concerns with combustible vapors and liquids. High-level sensors should be installed in all top-loaded tank trucks to prevent overfilling.

13.7.7 Bottom Loading

In contrast to top-loading systems, bottom-loading tank truck fill and vapor recovery connection points are frequently separated by a considerable distance, as shown in Figure 13.24. This results in fill hose stations at a different location than vapor recovery stations. Consequently, examples of gasoline loading and chemical loading frequently describe different equipment for these two functions. However, dry-break couplings are used in both services to ensure almost zero emissions.

The typical gasoline tank truck, as shown in Figure 13.24, has several compartments that separate various gasoline grades. This design varies from most chemical tank trucks, which normally do not have compartments and are dedicated to transport of a specific chemical.

Gasoline Loading. A typical bottom-loading gasoline system is shown in Figure 13.25. Gasoline flows through the hose and dry-break coupler into the accompanying tank truck coupler adaptor and then to the associated compartment. Each compartment normally has a coupler and valve as shown, with all of the compartment piping sets parallel on one side of the truck. The compartments are generally dedicated to specific grades of gasoline. A grade change requires cleaning a compartment to prevent cross contamination with other material. Hose stations for filling the compartments are located in accordance with grade specifications since the arrangement of the various compartment loading connections is standardized for tank trucks in North America. As a result of standardization in bottom-loading tank truck facilities with vapor recovery, workplace exposures are quite low in gasoline loading terminals and bulk plants.

Filling or loading is controlled through the preset metering control system shown in Figure 13.25. Typically, each compartment has a high-level alarm sensor that is connected to the meter control through a cable that is attached to the sensor connection on the outer shell of the tank truck. This level sensor overrides the preset meter control and stops flow to the compartment when the high-level control sensor in a compartment is actuated. A reliable and accurate meter fill control system insures loading the proper quantity into each compartment and the high-level sensors essentially eliminate the possibility of overfilling. High reliability in these types of loading systems does not require extensive manual or external backup systems.

Bottom-loading systems normally have a vapor recovery header which is connected to each compartment (see Fig. 13.24). The vapor recovery header then generally terminates in a coupler located at the rear of the tanker truck, which connects with the recovery system coupling. Vapor recovery systems are discussed elsewhere in this section. While various gasoline grades are stored in the compartments, the vapors can be intermingled in the header since the vapor components are generally hydrocarbons. Requirements for oxygenated compounds in some areas add a nonhydrocarbon component. However, all recovered vapors can be

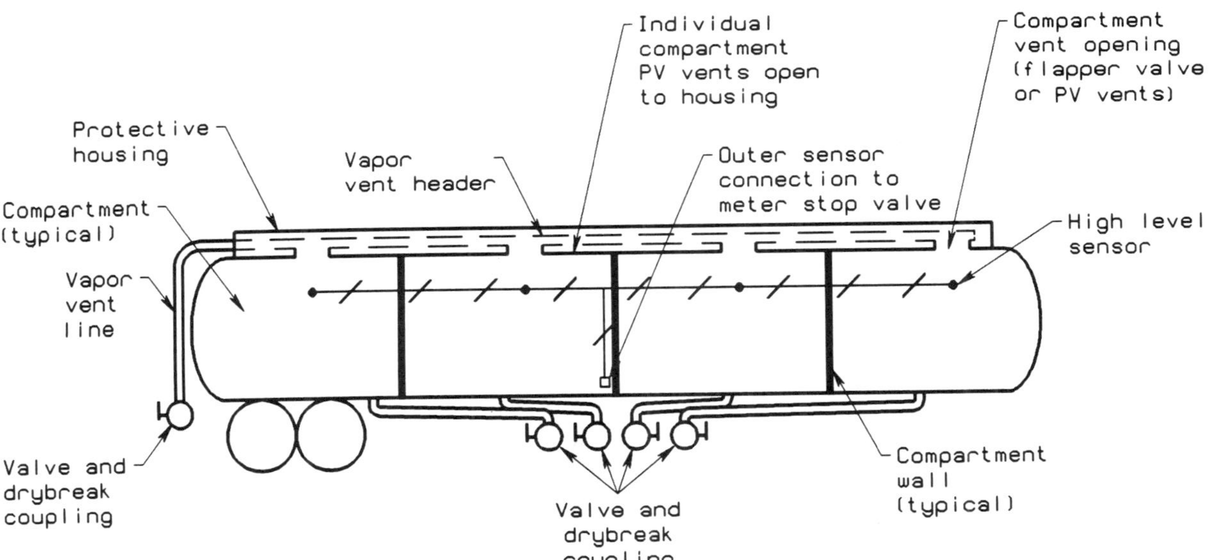

Figure 13.24 *Typical gasoline tank truck with loading and vapor connections.*

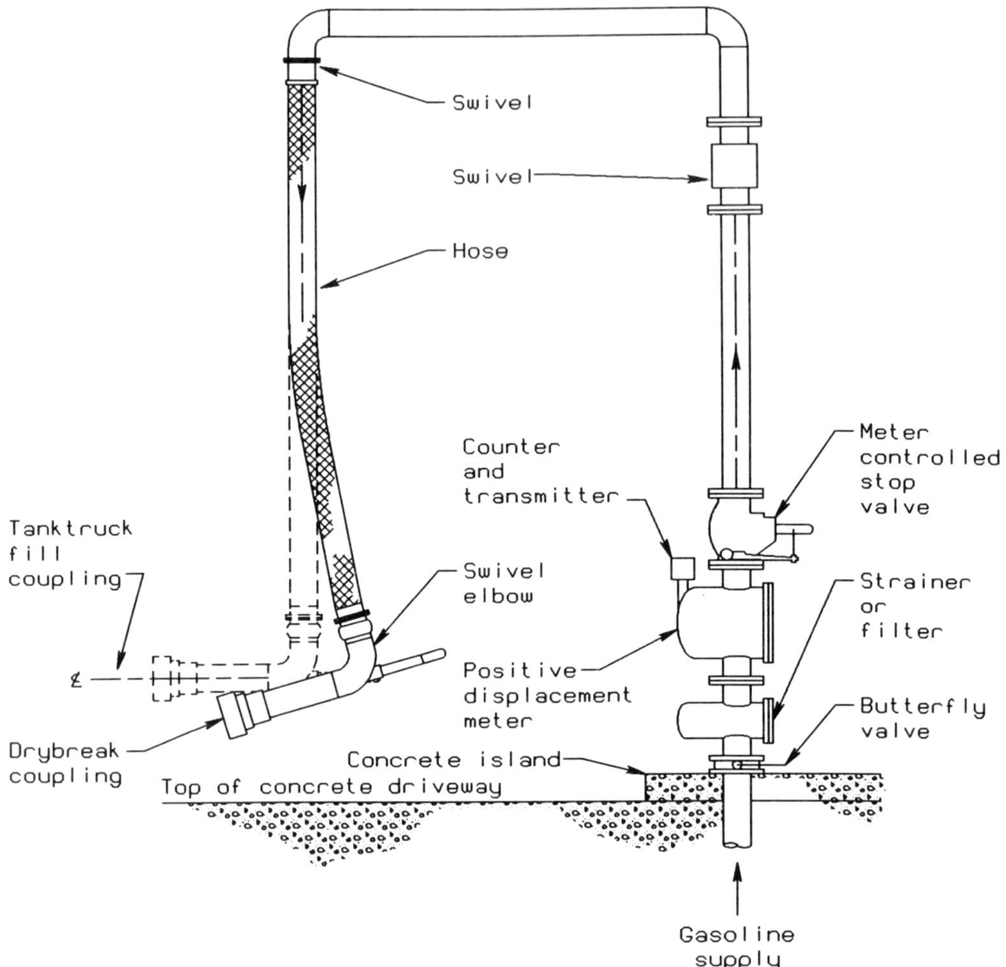

Figure 13.25 *Typical bottom-loading gasoline arm and meter facility.*

recycled with the oxygenated material into finished product or can be separated into hydrocarbon and nonhydrocarbon materials prior to recycling. Vapor recovery systems are reviewed in another section.

Chemical Loading. Chemical loading systems for HAP materials are similar to the bottoms loading system described for gasoline loading. However, compartmentalization of chemical tank trucks is generally not required since the tank truck is dedicated to a specific chemical. As a result, only a single truck rack loading station is provided for a specific chemical with a single loading arm.

In a typical chemicals loading system (Fig. 13.26), the bottom loading arm is a swivel-type solid pipe arm that couples directly to the tank truck. The bottoms outlet is generally not extended, and a horizontal arm is more convenient. In addition, the horizontal arm reduces the holdup in the long vertical arm in gasoline loading and provides a completely solid arm that is not affected by chemical attack. Consequently, a hose is not connected to the swivel arm in this

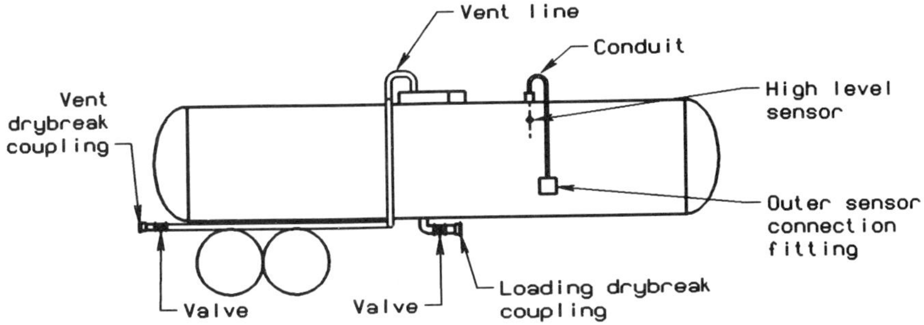

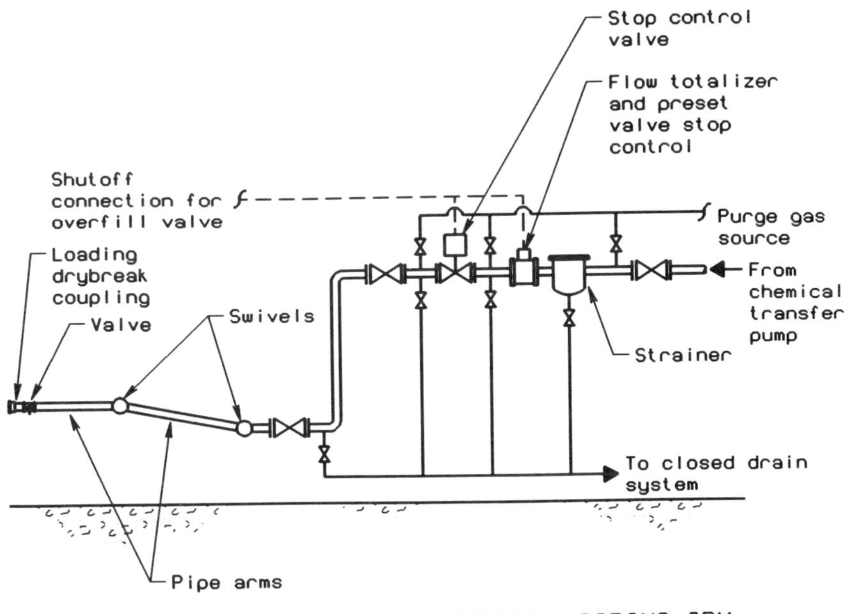

Figure 13.26 *Typical chemicals bottom-loading system.*

service. However, a hose is frequently used for the vapor recovery connection to the vapor recovery line coupling installed at the rear of the tank truck. Dry-break couplers are installed on both the bottom-loading and vapor recovery line with the result that very low workplace exposures are generally observed. Infrequently, some leakage from the dry-break loading coupler is observed, and exposures may increase slightly. Respirators may be recommended by the industrial hygienists when normally connecting or disconnecting the loading arm and vapor recovery system in the event there is coupling leakage for very hazardous chemicals. In this chemical loading system, a high-level sensor on the tank truck stops the fill operation to prevent overfilling. Vapor is either recovered in this system or is sent to a combustion system. The vapor recovery systems are reviewed in another section.

Loading of HAP chemicals should be controlled through a preset control system with high-level override, as shown in Figure 13.26*B*. The filled tank truck in this situation is not gauged or observed visually since the preset controller is considered an accurate measurement device. A second flow totalizer can be installed to ensure accurate measurements or a weigh scale can be used. Newer weigh scales are designed to be placed above ground and do not present any problems during a spill compared with the large below ground openings necessary for the fulcrum-based scales.

One of the problems with this general loading system is ensuring that the tank truck is empty before loading. If the truck is not empty, the preset control may result in overfilling before the level override can block the liquid fill line. A level sensor not functioning properly can exacerbate this condition. A second metering system does not aid in this situation, but a weigh scale will definitely indicate the amount of material in the truck prior to loading. Several instruments such as optical and dielectric types are commercially available that determine when the tank truck is empty and are reviewed in another subsection. When a truck is properly instrumented or a scale is available, observation of the tank truck interior through an open hatch is not required.

Dry-Break Loading Couplers. A wide variety of couplers are available commercially and usually have reference approvals by different organizations or testing facilities. Petroleum and petrochemical facilities normally install API approved couplings for loading and vapor recovery services. Since petrochemical also includes a number of chemicals in the HAP list, API approved couplings can be found in various services. However, these couplings are relatively large and other coupling types may be installed for smaller loading systems.

In one type of API coupling, a lever is incorporated in the coupler which is only operated when the two couplings have been connected. Following coupling, turning the lever to the open position permits the spring-loaded face in the adjoining coupler to open, allowing fluid flow. This lever system is a safety feature that also insures minimum leakage. Some dry-break couplers have a special feature that prevents disconnecting the couplers until the valve lever is closed. These safety features contribute to minimizing leakage and reducing potential worker exposures to hazardous materials.

Various couplers are available that may have smaller leakage rates than the typical dry-break coupler previously described according to some manufacturers. However, any coupler should be carefully investigated to insure they are mechanically equivalent to the API approved couplers.

Maintenance of couplers is a function of the precautions observed in connecting or disconnecting the couplers. Periodically, seals or gaskets in the coupling may lose their elasticity which may result in leakage or the seals can be marred, requiring replacement. Although replacement is not considered frequent by terminal operators, some replacement frequencies in bulk terminals can be in the 4- to 6-month range. When gaskets or seals must be replaced, potential exposures of maintenance or operating personnel may occur due to liquid remaining between the coupling and valve or, in the case of gasoline loading, liquid remaining between the coupling and valve results in a large quantity of liquid in the hose. Consequently, emptying or purging these lines prior to coupling maintenance is important to minimize potential exposures.

13.7.8 Gasoline Vapor Recovery

Gasoline Vapor Recovery. Compartments generally have vents that are connected to a central line running the length of a gasoline tanktruck. This line is located on the very top of the tanktruck. The central vapor line and compartment vents are located in a housing that protects the connections in the event that the truck overturns in an accident (Fig. 13.24). The maximum pressure levels in the compartments are considered sufficient to provide the pressure necessary to move vapor in the connecting lines to the vapor recovery system.

The central vapor recovery line runs the length of the gasoline tanktruck and normally terminates either at the rear about 3–4 ft above grade or near the loading couplers. A dry-break coupler is located on the end of this line (see Fig. 13.24). Although the vapor coupling is a dry-break design, the coupling differs from the loading coupling to prevent mismatching couplings. In the gasoline loading facility, a hose connected to an arm similar to the main loading arm is used for vapor recovery. The hose is in a vertical position as shown in Figure 13.27 and does not degrade through mechanical manipulation. The vertical standpipe arm is connected to a check valve that discharges vapor into an underground vapor line. This complies with the regulations since the check valve prevents vapor backflow from the underground vapor line into the vapor recovery arm. Alternatively, a short hose can be connected directly to the check valve, eliminating the vertical standpipe and associated equipment. However, where the vapor can condense in the arm system, drainage is simpler with the standpipe and hose service conditions are less severe in the arm system design.

A connector installed on the vertical standpipe located near the check valve could be used to drain the vapor recovery hose. In addition, this connector could potentially be connected with the fill hose line through a short hose to drain the fill line hose. In operating a system of this type, bends in the lines that trap liquid should be avoided.

If the proposed drain system shown in Figure 13.27 does not adequately drain the vapor recovery hose or the loading hose, then alternative methods of draining the hoses are necessary. One method is discharging the hoses into small closed containers on wheels that have dry-break coupling connections. This container can then be unloaded into the general recovery system which is described subsequently. Another alternative is locating a drain near the island with a dry-break coupling that is below the road surface. A coupling cover plate would protect the coupling and both fill and vapor hoses could be connected to this drain opening with a short hose. Since maintenance of the couplings is required, draining these lines is necessary to minimize potential exposures.

Chemical Vapor Collection. In Figure 13.26 the vapor recovery coupling is in approximately the same position as the vapor coupling on a gasoline truck. However, the large vertical hoses are normally not installed. The hose is connected to a horizontal line near grade level, permitting some drainage of the hose. Following loading, the vapor hose is uncoupled and can be placed on the pad or elevated with a davit on the loading platform. Since chemical tanktrucks are frequently pressurized, vapor flow through the recovery system does not present any pressure drop problems.

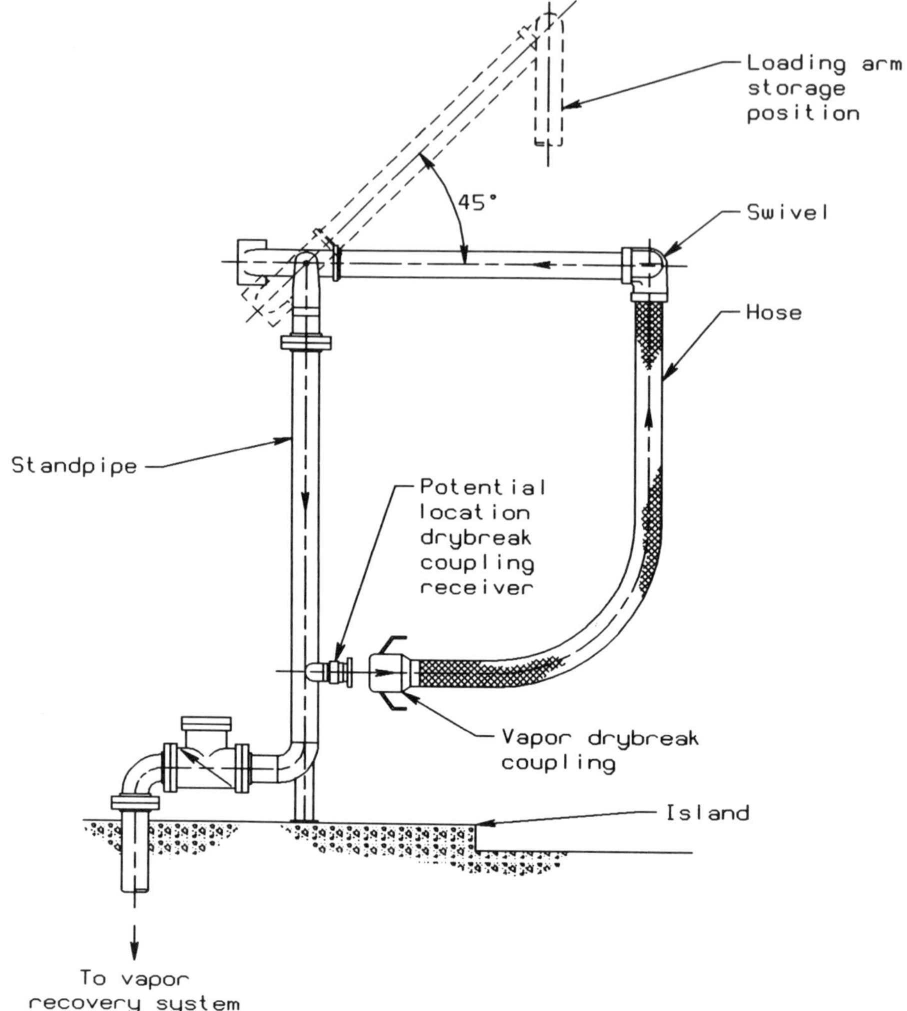

Figure 13.27 *Vapor recovery connection system gasoline loading.*

The vapor hose in this situation must also be drained of any condensate. Raising the hose with a davit is one solution, along with manually elevating the hose a few feet to drain the hose. The fill scissors line can be drained by partially elevating the line since, the drainage system is provided in Figure 13.26.

13.7.9 Instrumentation

The two types of tanktruck designs described previously were compartmented and noncompartmented tanktrucks, which can be either trailers or fixed-bed trucks. Normally, compartmented trucks are used in petroleum services with low vapor pressure materials such as gasoline or diesel fuels. In contrast, noncompartmented trucks are used with high vapor pressure materials such as propane and butane, and in chemical service. The latter may be a pressurized service as a result of either the vapor pressure or handling requirements associated with potential hazards. Pressurization with nitrogen, as an example, may require a pressurized truck, which also requires a high-pressure relief valve. Instrumentation installed in these trucks must be equipped for normal operating conditions and where necessary for pressure operations.

The three basic types of level sensors normally installed in tanktrucks are floats, thermistors, and optical devices. Other systems may be installed, such as capacitance, dielectric, sonic, or radio frequency, but the latter instruments are not widely used in petroleum service in contrast to the initial three, particularly the thermistor and optical systems. Various level sensors are also installed in chemical tanktrucks, but in chemical applications, noncontamination or reaction with liquids must be carefully investigated before these instruments can be installed.

Since the level instruments must be connected directly to the liquid fill control system, an electrical plug is provided on each truck, as shown in Figures 13.24 and 13.26. The plug is standard and matches a connector plug at the terminal. Prior to loading, the terminal plug connection on the tanktruck is connected to the rack control system and *the tanktruck is grounded*. When a level sensor is actuated, the flow may be stopped through the metered valve or a separate valve, or the electric system may shut down, which in turn trips the control and/or the feed pump.

When a level instrument stops the inlet flow, the circuitry in some of the level systems cannot be activated until the liquid level is lowered within the truck or the truck compartment. Consequently, an unloading capability must be available. If unloading is available at the rack, time loss is minimized; however, movement of the tanktruck to a separate location increases loading time.

13.7.10 Tanktruck Unloading

Tanktrucks are normally loaded in terminals, but unloading, although infrequent, may be required in certain instances. The unloading procedure can also be a source of workplace exposures and methods of unloading are important to reduce exposures in this operation. In considering unloading operations, tanktrucks normally do not have pumps and the unloading procedures described in this section are based upon trucks without pumps.

Unloading Chemical Tanktrucks. In the chemical area, unloading a highly toxic material depends upon the mechanical construction of the truck. If the truck has a high mechanical design operating pressure, the contents of the truck can be pressured out with nitrogen, which is a standard procedure in various locations. The unloading material can be returned to the day or loading tank through the loading system and around the metering devices or other equipment with bypass valving. This procedure assumes the material in the tanktruck has not been contaminated and the tank compartment contents can be returned for resale. However, other procedures for recovering this material may be required through additional tankage bypasses when the material is contaminated. Bypass lines require permanent installations to minimize leakage. The need for unloading at the terminal based on historical data determines whether these lines are permanently installed. If the bypass lines are not installed due to very infrequent unloading requirements, then emergency connections are necessary which can increase potential exposures when handling HAP chemicals.

If the tanktruck cannot be pressures with nitrogen, then a pump-out system may be desirable. However, in situations where the truck contents are considered extremely hazardous such as benzene or other toxic chemicals, then a sump or underground tank may be preferable to a pump system. Connections to the bottom of the truck compartments (inlet load line) for drainage should be the horizontal swivel type. The swivel arm is flexible and can be drained more readily than hoses. Dry-break couplers would be installed on the arm to match the truck adapter and minimize leakage.

A sump tank permits gravity drainage from the tanktruck. The sump vapor system can then be connected with the tanktruck vapor header through a hose connection balancing the system. The vapor balance line prevents the introduction of air into the tanktruck and prevents the emission of vapors into the atmosphere from the sump tank or connections, minimizing potential exposures. As an alternate, sump vapors could be transmitted to a vapor recovery unit (VRU), but the VRU should have a recovery efficiency of about 98%. Preferably, the vapors would be returned to the processing area.

Where tanktrucks carry toxic materials with high vapor pressures such as vinyl chloride, the system unloading configuration differs considerably. The tanktrucks in this service are pressurized, and the vinyl chloride can either be discharged under a constant VCM pressure or removed in other ways, depending upon the VCM equipment available at the terminal or loading facility. With all of the potential discharge techniques, the vapor systems are balanced or closed, minimizing leakage and potential exposure sources.

Shippers of hazardous chemicals (toluene, xylenes) characterized by low vapor pressures can install a sump unloading system or pressurize the contents. Infrequently used pumps on a special pumpout system may increase emissions and potential exposures. The closed sump pump is a vertical pump that should permit maintenance with minimum emissions. When considered necessary, potential exposures in coupling and uncoupling of hoses with dry-break couplers can be minimized in *all* operations by the use of personal protective equipment. Where tanktrucks are unloaded in the course of normal operations at a receiving facility, the system is designed and installed in accordance with tanktruck design and plant requirements.

Unloading Gasoline Tanktrucks. There are basically two methods for discharging gasoline from tanktruck compartments at loading terminals. Trucks can be pumped out or discharged under gravity into a drain system. The drain system in this case could be a sump located at the rack or a drain connection to the vapor recovery system. The sump would have a pump to send the gasoline to storage. The vapor recovery system permits accumulating gasoline in an underground drop-out tank, described in Section 13.7.11, and pumping the recovered gasoline to storage. Both general systems are presented in the following paragraphs.

Pump-Out of Bottom Loaded Tanktrucks. A pump-out station with a space for unloading tanktrucks would be located a distance from the loading rack. A dry-break coupling hose connects the tanktruck compartment loading nozzle with the pump, which minimizes potential leakage and exposure. Since various compartments exist in the tanktruck, the hose must be manually moved from one compartment to another, potentially increasing any exposure possibilities. The tanktruck compartments in this situation must be emptied completely, which will entrain air with the gasoline. Attempts to investigate whether a compartment is empty through visual observation without respiratory protection is not recommended. The level sensor, particularly an optic sensor, has the capability of indicating when a compartment is empty and should be considered where unloading is relatively frequent. Use of this sensor requires a connection from the truck sensor jack to an indicating instrument. This separate indicating mechanism must be installed specifically at the pump-out station. The tanktruck compartment contents are pumped through an air separator to separate the air from the gasoline and the deaerated stream is then sent through a meter. The deaerator should have a level control that automatically stops the pump when fluid flow slackens and the level falls in the deaerator. In addition, the vented air from the deaerator must be sent to a closed system.

Potential exposures in this system may occur during maintenance of equipment and hose removal for replacement. A vapor recovery or underground drain line installed adjacent to the equipment for drainage recovery can also be used for draining the hose. A coupling installed at grade and connected to the underground drain provides a method of draining the hose. All drainage entering the drop-out tank through the vapor recovery and drain lines described in Section 13.7.11 can be recovered and pumped to tankage.

Pump-Out of Top-Loaded Tanktruck. With drains on the individual compartments, a hose connection can be installed on the individual drains similar to that recommended for bottom loading trucks. Since P–V vents are installed on the trucks, the hatch will not be opened during pump-out and would not present potential exposure problems. However, the tank hatch may require opening to determine whether the compartment is empty. This would increase potential exposure situations. An alternative is to use the pump-out system, which runs until the deaerator controller stops the pump, indicating the tank is empty. The meter can also be checked to determine whether the pump-out volume is equivalent to the fill volume. Installing an optical instrument is another possibility to ensure the hatch is never opened, but alignment can be a potential problem with any movement of the hatch plate. Other instruments may indicate the compartment is empty and the effect of possible hatch plate changes on the accuracy of the instrument must be established before installation.

Draining Top- and Bottom-Loaded Tanktrucks. A sump tank at the loading rack permits gravity drainage from the tanktruck for unloading or draining gasoline left in the compartments. Connections with the tanktruck compartment bottom valves from the sump are possible with a hose or horizontal swivel-type arm. Tanktruck compartments drain under gravity into the sump, displacing saturated vapor from the sump. The sump vapor line can be connected to the vapor recovery system, preventing vapor emissions to the atmosphere. As an alternative, the vapors can be sent to a separate vapor recovery unit (VRU) with the VRU recovery efficiency based upon emission regulations. However, this latter approach (VRU) is an expensive solution and a connection to the vapor recovery system is preferable in this situation.

13.7.11 Control Devices and Systems

There are two separate functions in the overall vapor control system. In the vapor recovery section, the vapor is collected through connections to the tanktrucks and then sent through an enclosed transfer system to a disposition system. The enclosed transfer system can be underground, which is typical of gasoline loading facilities, or above ground. Chemical transfer systems are generally above ground in tanktruck loading facilities. The vapor recovery system transfers vapors and condensed liquids to a control device, which is the second part of the overall vapor control function. The control devices are the same basic types applicable in vapor control from tankage and are generally adsorption, absorption, condensation, combustion, or flaring which is essentially combustion. In some instances, control devices may be a combination of several basic functions to achieve control efficiencies greater than the efficiencies required by the regulatory authorities.

The control device regulations for HAP and VOC service are described in Section 13.7.2. For HAP pollutant emissions, a 98 wt % emissions reduction is required, or a maximum pollutant concentration of 20 ppmv (dry basis, 3% O_2). The EPA gasoline VOC emissions requirements are 35 mg/L of gasoline loaded, which is the basic requirement for both gasoline and organic liquid emissions in the SCAQMD regulations. The BAAQMD regulations have gasoline emission limits of 9.6 mg/L of gasoline loaded, although the organic liquid vapor emission limit is 78 mg/L of organic liquid loaded. For halides or halogenated compounds, the EPA requires a 99 wt % emissions reduction. These various requirements indicate efficient control systems are required that must operate at lower emissions levels than the actual regulatory limitation. As an example, if a 35 mg/L regulatory maximum emissions level is required for liquid loading, the emissions control device should operate well below this level to ensure the emissions regulatory limitation is not exceeded. Few exeedances are permitted, and a well-designed, controlled system is necessary to achieve consistent low emissions performance.

Control System Effectiveness. While various noncombustion control devices are available, little information on actual performance has been published. However, some information on control device effectiveness has been published by the EPA on gasoline tanktruck loading systems.[26] For the devices tested, the following average emission factors were observed per liter of gasoline loaded:

Carbon adsorption	5.5 mg/L (0.33 mg/L minimum tested)
Condensation refrigeration	21.6–42 mg/L
Thermal oxidation	0.20–1.18 mg/L

This short summary indicates that thermal oxidation has the lowest emissions loss to the atmosphere of the three systems, with condensation the least effective. However, condensation can effectively control emissions to < 35 mg/L required in VOC and SCAQMD regulations as shown by the emissions range. The carbon adsorption emissions loss of 5.5 mg/L is an average factor, covering a total of 33 tests, indicating variability in adsorption efficiency in a number of different installations. The lowest emissions rate of 0.33 mg/L compares favorably with thermal oxidation, demonstrating effective control performance in one particular installation. Since the carbon adsorption factor is an average value, a number of other installations apparently also operated at relatively low emission rates. For an efficient control device, the system must be properly designed and correctly operated.

While thermal oxidation is quite effective along with other combustion devices, a major drawback with combustion is the destruction of product vapors. This contrasts with carbon adsorption, absorption, or condensation, where product is recovered from the control device. An overall economic comparison of the various control devices would be based on product recovery, investment, and operating cost. Product recovery requires an estimate of vapor loading to the control device and control device effectiveness. The estimate of vapor loading for submerged or bottom-loaded tanktrucks can be determined from reference 26 where the vapor loss for gasolinc can be estimated from the following equation:

$$L_{\mathrm{L}} = 12.46\frac{SPM}{T} \tag{13.34}$$

where

L_{L} = Loading loss (lb/10^3 gal liquid loaded)
M = Vapor molecular weight (Appendix Table 13B)
P = True liquid vapor pressure (psia)[1]
T = Temperature of bulk liquid (°R)
S = Saturation factor (Table 13.16)

This equation provides the total weight loss from the tanktruck or railcar and a weight loss per unit time can be estimated based upon the loading rate to the tanktruck. For a typical gasoline loading example:

Tanktruck volume = 8000 gal
Gasoline RVP = 9 psi
Product temperature = 80°F
Tanktruck = closed vapor collection system
S = 1.00 (Table 13.16)
P = 6.6 psia (ref. 1)
M = 66 (Appendix Table 1)
T = 540°R

$$L_{\mathrm{L}} = 12.46\,\frac{(1.00)(6.6)(66)}{540}$$
$$= 10.05\ \mathrm{lb/1000\ gal}$$
$$= 1204\ \mathrm{mg/L}$$

TABLE 13.16 Saturation Factors for Petroleum Liquid Vapors[a]

Carrier	Operation	Factor[d]
Tanktrucks and railcars	Splash loading — dedicated service[b]	1.45
	Submerged loading — dedicated service	0.60
	Splash loading — dedicated vapor balance service[c]	1.00
	Submerged loading — Dedicated vapor balance service	1.00
Marine	Splash loading —clean cargo tank	1.45
	Submerged loading — clean cargo tank	0.50

[a]From reference 8.
[b]Limited to carrying one material and open to atmosphere.
[c]Closed vapor system, tanktruck vapors connected to common header. Represents concentration from header outlet to vapor recovery.
[d]Represents fractional approach to saturation.

Assuming an efficiency of 98% and multiplying by (1-eff/100),

$$\text{Actual emissions} = 1204\left(1 - \frac{98}{100}\right)$$
$$= 24 \text{ mg/L}$$

Although less than 35 mg/L and at an efficiency of 98%, a somewhat higher efficiency would provide an additional margin that would permit some operating variations. A typical carbon adsorption system is shown with the vapor collection and recovery system in Figure 13.29.

Vapor collection and Recovery Systems for Gasoline Loading. The vapor recovery hose shown in Figure 13.27 is connected through the extended arm and stand pipe to a vapor recovery header located underground at each gasoline loading rack. Individual recovery headers from each rack are then connected to a main underground header that discharges the vapor and condensed material into a dropout tank as shown in Figure 13.28. The dropout tank is a large surge drum located underground for disengaging any liquid or droplets entrained in the vapor header. Vapor from the dropout tank is sent to the vapor recovery unit or control device and liquid is periodically pumped from the dropout tank to one of the terminal storage tanks.

While vapor collection directly from the tanktruck is a major design objective, drainage condensate from the hose must also be considered and a standpipe connection for draining the hose is shown in Figure 13.27. A separate connection near the check valve is also a possibility, and the recovery hose could be connected through a short hose to this coupling. If the hose is discharged through the coupling into a small carrier, a method of discharging the carrier into the drain or recovery header is necessary. These operations require dry-break couplings to ensure minimum exposures. In addition, a tight-fitting butterfly valve should be installed on the arm side of the check valve to prevent any vapor leakage flowing from the header into the standpipe during maintenance on the vapor recovery arm or equipment. The check valve may not provide complete closure at all times.

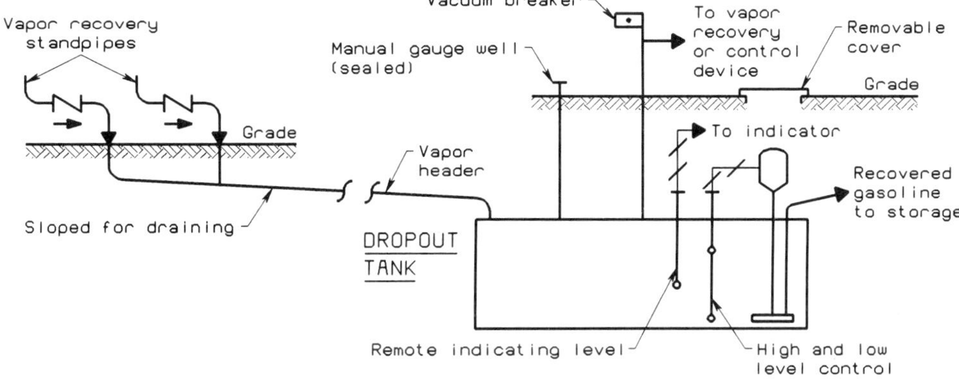

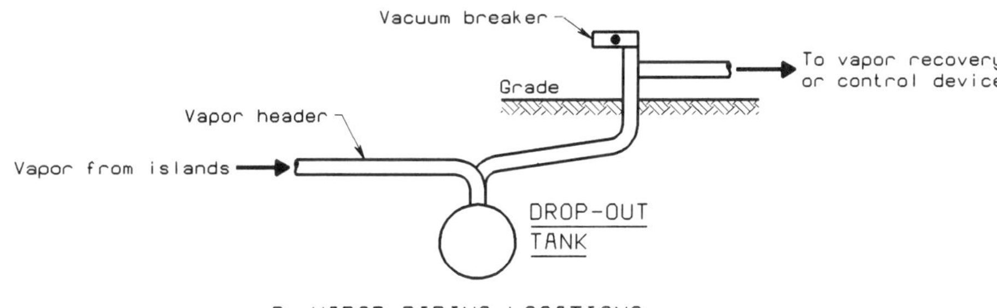

Figure 13.28 *Underground gasoline vapor recovery system.*

The underground header is connected to the dropout tank in Figure 13.28. The common header collects all of the vapors and condensibles from the different rack positions and transfers the vapors and condensibles to the underground dropout tank. The header is sloped to permit drainage of liquid to the tank from condensed vapors or liquids that may be recovered from the vapor arms. Liquid from any drains connected to the header will accumulate in the tank until pumped out.

The pump is controlled as shown in Figure 13.28 with high and low liquid level sensors to periodically pump any accumulated liquid to storage. Other controls that can be added are a remote level indicator and high-level alarm. A breaker vacuum vent is installed on the dropout tank for pressure protection. A pressure relief vent can be located at the inlet to the recovery control system that will prevent overpressuring in the tank. Discharge of this valve, which is rarely activated, may require an enclosed discharge. Since the pressure is low, a separate control device may be required, such as a carbon-filled drum.

Potential exposures may occur in this facility when entering, cleaning, or performing equipment maintenance on the dropout tank. For cleaning operations, the tank must be completely free of liquid. The pump may not completely empty the dropout tank and a truck with a vacuum line may be required to completely empty the dropout tank. Although empty, the tank will still be filled with vapors and should then be thoroughly purged with nitrogen or an inert gas. Vapors evolved during the purge should be sent to the vapor recovery unit. Alternatively, the tank can be filled with water to displace the vapors, but the vapors in this situation should be discharged to the atmosphere through the vapor control device.

Maintenance can be required during normal operations on the instruments or pumps in the dropout tank. Completely purging the tank for maintenance would result in shutting down the facility. Blocking out the tank and working with personal protection equipment is difficult, and the gasoline vapor–air mixture somewhere in the tank or tank proximity may exceed the lower explosive limit, resulting in a potentially unsafe situation. As a result, this tank should have equipment with a long, highly reliable service life or the system should be designed to remove and replace malfunctioning equipment safely with minimum exposure potential. Another alternative is to duplicate equipment that may not have the ability to operate for long periods between turnarounds without maintenance. Mean time between failures (MTBF) should be determined for each operating item prior to installation to ensure satisfactory performance.

The dropout tank is an underground tank, and the underground storage tank regulations may be applicable to the installation. However, this is not a storage tank but is essentially a knockout drum with little liquid buildup, and the regulations may not require complete compliance. This issue must be resolved to ensure the installation meets acceptable environmental requirements.

Vapor Recovery Units or Control Devices. Since the vapor pressure in the recovery system is low (< 0.5 psig or < 3.4 kPa), the general vapor recovery system and control device pressure losses are minimized. Early recovery designs included compressors to improve condensation since the system pressure was considered too low for satisfactory condensation. The control devices currently used generally do not have compressors to improve performance and operate efficiently, meeting regulatory requirements.

Three major vapor recovery systems have been installed in various gasoline marketing terminals throughout the world. Some variations of these systems are also available commercially that may include a combination of techniques. The three major vapor recovery systems are based upon: (1) activated carbon technology, (2) low temperature condensation, and (3) absorption. Combustion devices are also used in numerous installations, but a disadvantage is the lack of recovery since the combustion process destroys the vapors. The vapor recovery units produced by the various manufacturers are normally supplied to customers as skid-mounted or package units. A wide range of sizes are available commercially in various areas of the world.

Vapor recovery units operate well, but differences exist in efficiencies, such as when condensation had the lowest efficiency or highest emissions concentration, although the lower end of the emission range for condensation is well below the VOC requirements of 35 mg/L of loaded liquid. Absorption systems apparently operate at relatively high efficiencies, with a number of absorption systems installed in North America and Europe. Activated carbon adsorption units are widely used throughout the world and in some locations a combination of carbon adsorption and absorption steps are installed to meet emission restriction requirements. Combining different treatment steps for high overall control efficiency is currently available commercially from some manufacturers, but these units are more complex than single technology units requiring some additional maintenance and inspection efforts. However, high emission control efforts determine the type of system necessary to achieve regulatory compliance.

Activated Carbon Vapor Recovery Unit. Since these are widely used, a basic activated carbon unit is shown in Figure 13.29. Normally, the units have two activated carbon beds in parallel with one bed onstream while the other bed is regenerated. Activated carbon bed cycles are generally short but can vary in onstream time from 20 min to several hours for a single bed. This adsorption period variation is a function of bed size and hydrocarbon vapor rate. When the bed reaches a preset time on adsorption, the onstream vessel is automatically blocked and the regenerated bed is placed in service. In this design, a liquid ring compressor pulls a vacuum on the bed removed from service to desorb the hydrocarbons trapped on the activated carbon. When most of the hydrocarbons are removed from the bed, a quantity of air or purge gas may be admitted to the bed to improve hydrocarbon recovery. Hydrocarbons remaining in the bed reduce overall capacity and tend to result in more frequent bed replacements. The hydrocarbon vapors removed from the bed move through the liquid ring vacuum pump to a packed absorber, where a circulating gasoline stream absorbs the hydrocarbon vapors. Air or purge gas entrapped in the system is recycled from the absorber to the onstream activated carbon bed. The amount of air entrapped in the system is small and should not significantly change the explosive concentration in the adsorbing bed reactor.

The activated carbon bed system in Figure 13.29 presents a vacuum desorption system with product recovery. Other types of desorption systems are in use, including direct steam injection, indirect heating with vacuum, and heated purge gas systems. Heating effectively desorbs activated carbon and is aided by vacuum. However, absorption recovery requires a satisfactory heat balance operation to optimize absorption. Condensation recovery of desorption gas is another method of recovering hydrocarbons. In all systems, water must be separated from product since some water is in the vapor from the tanktrucks and can circulate through a liquid ring vacuum pump or condense from direct steam injection. All of the desorption systems are limited in the molecular weight of the fluid that can be recovered. While gasoline and lighter materials can be readily desorbed, diesel oil is not significantly desorbed under the rather mild desorption conditions described.

Activated carbon gradually loses adsorptive capacity and must be replaced periodically since regeneration in the vessel is difficult. The spent activated carbon removed from the vessels is loaded into drums or other sealed containers and returned to the carbon supplier for regeneration. A new supply of activated carbon is loaded into the vessels following removal of the spent carbon. The carbon vessels must be completely purged of hydrocarbon vapors prior to carbon removal since the vapors contain HAP. Regulations may also require control of the purged vapors. Material buildup on activated carbon is a potential exposure concern during the removal operation, which can be exacerbated by the presence of dust. Some dust can be present and generated during the loading operations.

Condensing Operations. This method of recovering vapors depends upon low-temperature operations. A low-temperature refrigerant is necessary to condense the hydrocarbon vapors at the atmospheric pressure level. Since water vapor in the stream ices the chiller condenser, two chiller bundles are normally supplied to

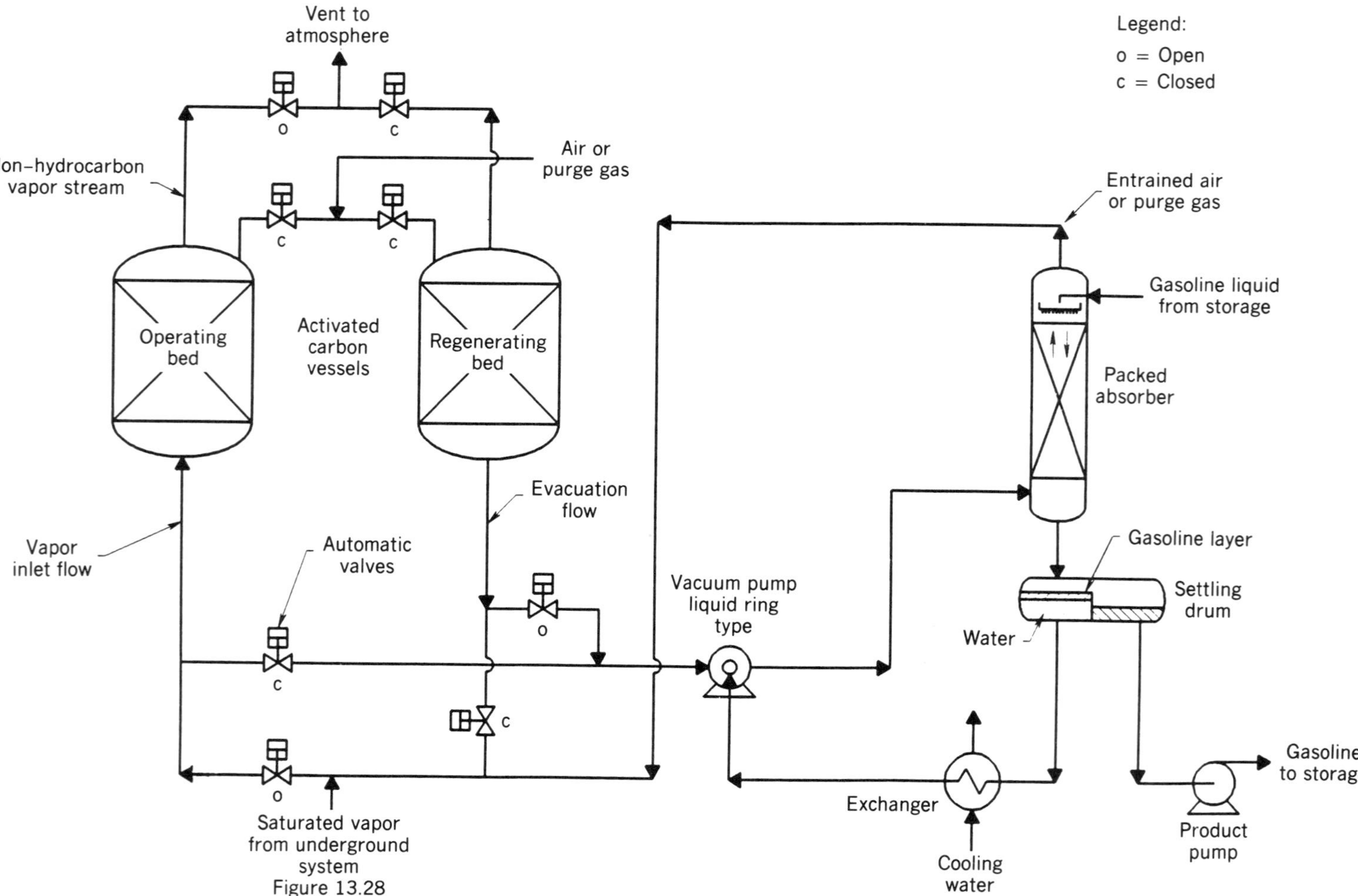

Figure 13.29 *Activated carbon adsorption with gasoline recovery.*

maintain onstream operations compared to one-bundle operation in early refrigeration systems. The offstream bundle is automatically deiced with a heated refrigerant stream. This condensing system generally includes two refrigerant circuit compressors.

The newer low-temperature condensing systems are efficient and reliable. These systems can be a satisfactory alternate to other control device systems since product recovery is an important feature of condensing systems. Chlorofluorocarbon refrigerants are a potential concern along with HCFCs. However, hydrocarbon refrigerant systems such as propane can be installed to eliminate both CFC and HCFC refrigerants.

Absorption System. A simplified flow plan of an absorption system is shown in Figure 13.30. This system is basically vapor absorption and stripping regeneration–recovery, which is similar to many installations in the petroleum and chemical industries. A lean oil, which may be kerosene or a heavy naptha for gasoline fumes, countercurrently contacts the saturated vapor stream from the tanktrucks absorbing the gasoline vapors. The recirculating absorber oil stream with the absorbed vapors is now termed a rich oil and is sent to a stripper where the rich oil is heated and stripped of the gasoline vapors. The absorber oil after stripping is now termed lean oil, which is returned to the absorber tower. Stripped vapors are condensed, and the recovered condensate is sent to product storage.

The rich oil is heated in an interchange with the hot lean oil stream. Additionally, heat is supplied with a stripping gas as shown, and the stripping gas (inert) is recycled through the blower. In this operation, lean oil is cold for optimum absorption and rich oil must be heated. This general design varies with the manufacturers of absorption systems who normally provide considerable heat integration to minimize energy consumption. Water drawoff is usually small and may be a manual operation. When an inert stripping gas is injected into the stripping tower, any carry through of water is quite small and can be sent to product storage. In the event steam is used for stripping, the condensate drum would have a water drawoff stream. These water drawoff streams will contain hydrocarbons at the in-saturation level or slightly greater. Consequently, these drawoff streams are sent to water treating facilities.

Combustion Operations. Various combustion systems are used in this service such as thermal oxidation, incinerators, process furnaces, and flares. With the exception of flares, combustion units must conform to the efficiency or emission concentration in the regulations. Flares must conform with the flare regulations in reference 20. Combustion of hydrocarbon or nonhalogenated materials is straightforward in contrast to product recovery systems. However, in chemical halogenated or halide operations, a scrubber is required on the discharge vapor stream of the combustion device which absorbs halogen or halogen-containing compounds. These must be neutralized, and the scrubber stream must be sent to a treating facility.

Chemical Loading Vapor Control Devices. For chemical vapors from loading operations, vapors from each chemical are not mixed to prevent contamination or reaction. Consequently, each chemical loaded requires a separate recovery system and control device. As described previously, the vapor collection and recovery

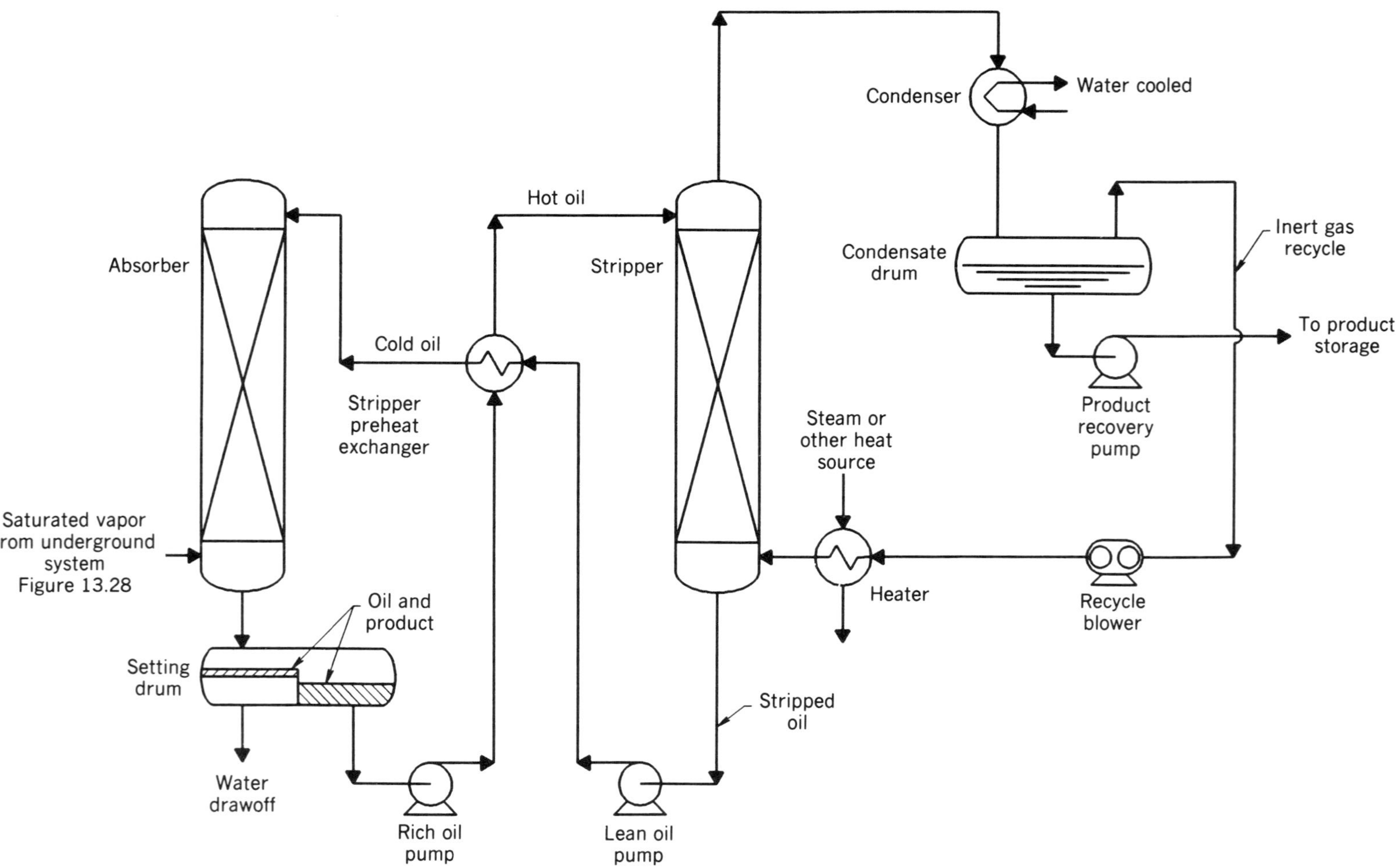

Figure 13.30 *Absorption system with gasoline recovery.*

headers for chemicals are normally above ground, eliminating the need for an underground system. The gasoline underground system, if used with chemicals, would have to be installed for each chemical that has a low vapor pressure and is loaded into tanktrucks under essentially atmospheric pressure.

Chemicals are usually inerted in a tanktruck to prevent contamination and purposely conduct pressure vapor discharge operations. This allows simpler recovery and control devices then are usually found in gasoline loading operations. Chemical vapors under their own vapor pressure can be returned to the pressurized storage tank. Where lower vapor pressure materials are inerted to a relatively high tanktruck pressure, a condensation system may be satisfactory that does not require low temperature operation when the vapor remains at an elevated pressure level. Moreover, chemicals are normally dry and water does not present an operating problem. Another method of recovering vapors is with an absorber located at the storage tank for the loading operation. A chilled liquid stream from the tank can be sent through an absorber tower which acts as the lean oil for the return vapor.

In some locations, multiple cone-roofed storage tanks are interconnected and have a control system for the tanks. Vapor streams can then be injected directly into these storage tanks and a purified inert gas discharged to the atmosphere from the control device or recovery unit. Alternatively, any of the recovery units previously described for gasoline recovery are applicable in chemical service where the systems are adjusted for the chemical operation. Detailed regulatory requirements required for the collection and control device systems are presented in Section 13.7.1.

13.7.12 Tanktruck Terminal Maintenance

Exposures may occur during maintenance of equipment in the loading rack area, the vapor recovery unit, the unloading area, or the subsurface systems. These potential exposure problems are greatest when draining and purging the equipment. Changes in procedures and equipment in gasoline loading facilities are presented that tend to reduce potential workplace exposures.

General Drain Procedure for Gasoline Loading Arms. In normal operation, top or bottom loading arms are filled with gasoline which must be removed prior to changing the hose, arm, joint, coupling or other equipment items. For the bottom loading arms shown in Figure 13.25, the hose leg is normally drained into a drum or container through gravity. The coupling must be opened specifically for this step by terminal personnel, which is a potential source of exposure unless a dry-break coupling is installed on the container. If hose replacement is necessary, this procedure only removes liquid from the hose with liquid remaining in the vertical riser pipe. Vapor emanates from both the hose and the nonhose section of the arm should the hose be disconnected from the upper arm.

If maintenance is required in ther vertical riser section, meter, strainer, and/or control valve system, the vertical riser must be drained prior to maintenance. Drains may not be installed in the vertical riser section, and flanges in this case would be broken to permit draining of the gasoline line in the vertical riser arm following the hose draining procedure. Purging vapors from the riser or arm may or may not be included unless welding is required, when the arm system is purged with water to remove all of the combustible vapors. These procedures are potential sources of exposure.

In the top-loading closed system, the general drain procedure is similar, but the top-loading arm can be placed directly into an empty tanktruck compartment hatch. However, the riser section will still contain a large quantity of fluid. In addition to removal of the liquid in the vertical riser pipe, the system must be purged to remove vapors. Purging vapors presents potentially identical exposure situations in both installations.

Chemicals Drain Procedure in Loading Arms. In the chemical bottom-loading system shown in Figure 13.26 for HAP materials, the tanktruck has a relatively high pressure rating. The contents of the tanktruck or the compartment is pressurized with nitrogen, and the lines can be blown and purged through bypasses around the metering and control valve assembly to the storage tank. This purge system capability permits maintenance of the equipment with a very low exposure potential. However, storage tanks that do not allow disengaging of the purge gas, such as floating roof tanks, may present a problem. One alternate is depressurizing a tanktruck and draining the liquid chemical into the empty truck from the line. When the line is drained of liquid, the piping can be purged with nitrogen or inert gas into the depressurized truck. This can be separated in the tanktruck for liquid drainage, sending purge vapor to an existing control device or recovery unit. Connections for these different functions must be permanently installed.

Although loading systems in chemicals operations can be drained, meter drainage can be a potential exposure problem. Some meters do not drain completely, and as a result, are sources of exposure when opened for maintenance. All flow meters should be drained and purged prior to opening.

Drain Modifications for Gasoline Loading Arms. The loading arm swivels and the arm can be moved parallel to the load rack. At this location, a dry-break coupler can be mounted that will connect with the fill hose coupling. The unloading dry-break coupling through a permanent pipe is connected to the downstream side of the vapor standpipe check valve. This drains the hose into the underground recovery system. In addition, vapor from the underground system places the overall arm system at atmospheric pressure.

The riser must also be drained along with the strainer or filter and the displacement meter. A permanent bypass drain line to the underground recovery system with valves connected to the riser, meter drain, and strainer drain and in the top flange of the butterfly valve (Fig. 13.25) permits draining the overall system. These drain valves can be welded into the piping to ensure they cannot be removed.

For purging vapors from the filling arms, water is the typical displacement media. Injecting water above the butterfly valve requires opening the dry-break coupling on the hose to the atmosphere, which is also a potential exposure source. In addition, traces of gasoline remaining in the arm contaminate the water which would be discharged on the concrete pad. The fill hose coupling could be connected with a catch basin through a flexible hose to drain the water. However, vapors will be emitted from the catch basin, which is a potential exposure source.

Reversing the flow of water will discharge water and vapor into the underground collection system through the drain line. This mixture will separate in the dropout tank (Fig. 13.28). The vapor is sent to the control device, but the water would be pumped to the gasoline storage tank. This volume of water can be separated in the tank where settling is normal. As an alternative, the pump could be connected to a sewer drain and the contaminated water is then treated in the wastewater treatment system.

A major concern in this system as described is draining the meter. Some meters are difficult to drain, causing potential exposure sources during maintenance. Where a meter of this type exists, one that can drain readily should be installed. Many of the flow instruments commercially available can be used in totalizing fluid flow with satisfactory accuracies. For example, a turbine-type meter is an essentially open line and can be completely drained. Automatic temperature adjustment ensures accurate volumetric rates.

Gasoline Vapor Arm Maintenance. This assembly may need maintenance to replace the hose or a dry-break coupling gasket. Consequently, the hose must be emptied of condensate and the line purged of vapors. A typical vapor arm recovery unit is shown in Figure 13.28. The coupling connection on the riser permits draining of condensate from the hose. Purging the arm can then be accomplished through the hose coupling with nitrogen or inert gas from an island station or a nitrogen truck. A tight shutoff butterfly valve should be installed upstream of the check valve to insure there is no vapor backflow through the check valve from the underground vapor recovery system during maintenance on the arm and/or a blind can be placed between the flanges. The purge vapor discharged into the underground system is processed through the control device connected to the dropout tank.

Meter Proving. A very important operation in a terminal is proving or standardizing the loading meters. Meter proving may be conducted as often as once per month to ensure accuracy of the meter assembly. The meter is checked with a truck that has been specifically dedicated to this service or a portable meter. In the truck prover method, a small tank on the truck is calibrated to ensure volume accuracy and the calibrated tank is filled to specified volumes through the loading arms. The truck can be used for bottom- or top-loading fill systems. Flow volume through the meter is compared with the actual volume measured in the calibrated tank. A portable meter can be used with bottom-loading systems and is inserted between the fill hose and the tanktruck connection. The portable meter reading is then compared with the fill meter.

The prover truck in this service is designed specifically for the loading arm and vapor recovery system installed on the rack at the terminal. Following loading of the prover truck, the truck is moved to an unloading station to recover the gasoline in the tank. During the unloading operation, air can enter the tank through the vapor line or through a P–V vent to ensure pumpout. The prover units should have connections that permit pumpout through the main unloading pump to minimize potential exposures.

Although the portable meter has coupling connections, gasoline will normally remain in the meter system. Drains should be provided on the meter to ensure maximum drainage to the underground system. Purging the meter with nitrogen will remove the remaining vapors, unless the meter cannot be drained. In this situation, which presents potential exposure problems, the meter should be replaced with a meter that can be completely drained.

13.7.13 Sewers

Normally at loading racks, drains or catch basins are placed in the paved area to drain spills or firewater that may be required during an emergency. This is a standard procedure in gasoline loading facilities and in some chemical loading facilities. While the HAP requirements do not mention these drains in chemical facilities, the BAAQMD does not permit spillage or discarding precursor organic liquids (VOC) in sewers that have a true vapor pressure > 1.5 psia (0.1 bar).[22] This rule has also been extended to gasoline[23] terminal loading facilities. The rules were promulgated to prevent evaporation or emissions. However, nondrained spillage increases potential exposures. Since firewater can flush significant quantities of organic liquids and gasoline, containment and disposition must be carefully considered to ensure providing adequate safety support in the facility. Moreover, truck leak motor and transmission oils that are flushed with rainwater and must be collected for treatment.

For compliance with federal VOC regulations and HAP requirements, catch basins and drains have not been restricted. A sewer installation for a gasoline terminal is described in this section, and comments on chemical facilities are also included.

Gasoline Loading Racks. In the cement roadway at the loading rack, high points slope to a drain or channel trough at the truck entrance and exit sections. A catch basin is normally located in the drain or trough which discharges into a sewer lateral (see Chapter 12). The catch basin drains the rack area of spills and firewater that are directed to the basin by the concrete grade or slope. The rainwater is considered contaminated and must also be treated. Spills are flushed to the drain with water hoses for health and safety reasons. Consequently, the catch basins are often designed to carry large quantities of water.

The catch basins in terminals are connected to a main sewer pipe which generally slopes to an underground oil–water separator. Catch basins should be sealed as described in Chapter 12. In some terminals, the separator may not be underground because of grade elevations. Oil–water separators in this service should be completely sealed with vapor control as described in Chapter 12. Oil separated from the water generally can be contained within the common underground separator and pumped out periodically to slop. The clarified or deoiled water is sent to a refinery oily water sewer if located nearby or a separate treatment facility in the terminal. This installation often requires a careful elevation tie-in for effluent water flow.

Oil-water separators can be individually designed and constructed. However, the separators can now be purchased in a variety of configurations directly from manufacturers or suppliers, and they are designed for operation either underground or at grade level. Oil separation and slop oil holdup functions are normally handled in a single separator or drum. Separator slop oil storage facilities may also be provided, and a variety of systems are commercially available for this general system. Recovered oil is considered contaminated and is termed slop oil, requiring rerunning or purification in a refinery.

Oil is normally removed from the separator with a suction lift or sump pump installed in a manway over the oil recovery compartment. However, complete cleanout of the oil in the separator for inspection and maintenance requires a vacuum truck to withdraw all of the oil material. In this system, the entire separator must be flooded to purge vapors from the drum for maintenance. Water that overflows normally from the separator into an effluent outlet line requires a valve in the effluent line to permit flooding of the entire separator. Where a vertical water pump is installed, the pump discharge line may be higher than the top of the separator and this special block valve is unnecessary. For satisfactory venting, an outlet must be located at the highest point in the separator. The vented vapors must be controlled under normal conditions, requiring a connection to the terminal control device. If the connection is not practical, a separate control device is required for the separator which must operate continually. The control device would be satisfactory for vent control during purging.

Chemical Loading. Chemical bottoms loading is shown in Figure 13.26 which includes vapor recovery. Where spillage can be flushed into a drain system, sewer openings for contaminated rainwater, firewater, and spillage can flow into an open drain. However, combining sewer lines from various chemical loading and unloading facilities is only permissible where the chemicals do not react and the final treatment or disposition system is equipped to treat various chemicals. In the event that combining chemical waste water treatment is not feasible, separate facilities may be required. Moreover, contaminated rainwater may complicate treatment.

If spillage is not acceptable in a sewer system, a rapid response spill control system is required that essentially recovers all of the spill material without any drainage to the sewer. However, rainwater is considered contaminated in a loading facility and must be treated. In addition, firewater runoff must be controlled where the chemicals are combustible. A sewer system is necessary for the contaminated water along with water treatment.

13.7.14 Railcar Loading

Railcar loading regulations are the same as those for tanktruck loading in the proposed new HON regulations described in section 13.7.1 for hazardous air pollutants (HAP). The VOC regulations vary with the states and regions, but the SCAQMD and BAAQMD consider tanktrucks and railcars equivalent in emission restrictions. Federal VOC regulations for railcar loading will probably be similar to the HON chemical regulations for railcar loading.

In the HON regulations, the Group 1 chemicals ($\geq$ 1.5 psia, 10.5 kPa) require closed vapor collection with control devices to minimize atmospheric emissions. Specific requirements covering loading and control are summarized in Section 13.7.1 for both tanktrucks and railcars. The major change in emission control occurs with Group 2 chemicals that have a vapor pressure $<$ 1.5 psia (10.3 kPa). Enclosed vapor systems are not required for the Group 2 chemicals and presumably would also apply in the petroleum industry to heavy oils and asphalt. These low vapor pressure compounds have much lower emission generation rates than the higher vapor pressure materials, but may also pose potential exposure problems depending upon loading system equipment and operating procedures. In general, railcars are loaded through the top of the tankcar with closed or open configurations. The latter would refer to those chemical or petroleum compounds with vapor pressures < 1.5 psia (10.3 kPa). Open-top tankcar loading may not be permissible in all areas of the United States due to VOC restrictions or regional restrictions. In non-U.S. locations, pressurized railcars may be loaded near the bottom or side of the car.

Types of Railcars. There are two general types of tankcars—pressurized and non-pressurized. Pressurized cars are designed for very low boiling point materials such as LPG, propane, butadiene, methyl chloride, and vinyl chloride. In addition, pressurized cars have been used with toxic chemicals such as benzene to permit pressure unloading, thereby maintaining a closed system. With the new regulations, enclosed systems will be required for most HAP chemicals and pressurized cars are designed for these operations with vapor and liquid-tight construction.

The different pressure ratings define the type of loading systems. Pressurized railcars have enclosed top loading designs and nonpressurized cars are top loaded through an open hatch. The pressurized cars can also be loaded from other areas of the car surface than the top. The enclosed top loaded railcars have permanently fixed extended fill nozzles for bottom loading. In addition, the vapor zone nozzles are also permanently mounted. In contrast, open top loaded railcars do not have any fill or vapor nozzles. The fill line arm is extended from the rack platform and can be a bottom loading or splash-type filling system.

Unloading varies, with pressurized cars unloaded through the extended fill line since they do not require pumping. The pressurized cars frequently do not have bottom drain valves owing to the high vapor pressure of the contents. Where lower-pressure liquids are in pressurized cars, the contents may be pressurized from the railcar with an inert gas, or bottom valves installed on the railcar completely drain the car contents. In addition, the bottom valve is a method of draining washings from the tank car when the medium vapor pressure tank cars must be cleaned. Railcar washing may be required after an extended period of operation or when the railcar load contents are changed from one material to another compound. Chemical transportation requires clean railcars to prevent contamination of the chemical transported.

Unloading nonpressurized open-top railcars may also be accomplished with a vacuum or the railcar may be unloaded through a bottom drain valve. Where a vacuum is used, an extended withdrawal pipe must be lowered into the railcar.

Emissions. The calculations developed by the EPA to estimate emissions are identical for both tanktrucks and railcars. A typical emission calculation is shown in Section 13.7.11, with loading estimates presented in references 8 and 26. While emission estimates are of assistance, they are not particularly helpful when loading an open-top railcar. The emissions rate is greater at high liquid levels in the railcar than at lower levels. Moreover, actual levels in these open-top loaded cars are frequently checked by operating technicians to ensure the railcars are not overfilled. These potential exposure situations are exacerbated when splash loading is the method of filling the cars.

Loading Petroleum Liquids. Open-dome loading through the top hatch is the normal method of loading heavier petroleum liquid such as diesel oil, asphalt, lubricating oil, jet fuel, turbo oils, and additives. Filling arms are the splash type or are equipped with long extended arms for nonsplash filling. Withdrawal of the filling arms is a potential source of exposure through inhalation and skin contact. In addition, splash filling with short arms is a potential exposure source through the formation of droplets and mist that can either be inhaled and or contact the skin. Many cars are filled to a specific internal marker, requiring visual observation to ensure that the cars are not overfilled. This procedure increases potential exposures as the operators must bend or kneel close to the hatch opening to observe the liquid level. This procedure can be changed by using modified top-loading arms, as described in this section.

While gasoline has also been loaded through an open hatch, regulations promulgated by the SCAQMD and BAAQMD and various states require enclosed loading systems with vapor recovery and a minimum of leakage when disconnecting the loading and vapor arms.[22, 24] This type of VOC regulation can be expected in nonattainment areas and the proposed HON regulations for petroleum refinery operations. The railcar regulations based on chemical HON rules require fixed inlet and vapor connections that match the loading and vapor connections at the filling point. Consequently, an open dome with a fill arm and vapor recovery attachment is not acceptable in this service. Closed domes of the type shown in Figure 13.31 are required.

Closed-Dome Loading. The typical connections for filling and vapor recovery on these closed-dome railcars do not have dry break connections. Normally, the access ports are opened and a short length of pipe is installed in the valve that extends through the port opening. This permits connection of arms or hoses to the extension, although these connections have not necessarily been dry-break couplings. Since these enclosed dome cars operate at relatively high pressures, dry-break couplings must operate satisfactorily at these pressures which reach 115 psia (792 kPa). Moreover, the fill and vapor arms must be capable of operating at these pressures, which requires metal arms. A backpressure control must be maintained on the vapor side to ensure the car remains at pressure, which is particularly critical for highly volatile materials. For lower volatility materials such as benzene or gasoline, system pressure can be significantly lower.

For dry-break coupling connections, installation of a dry-break coupling on the bottom valve provides a method of filling and unloading the railcar with a procedure similar to the tanktruck loading procedure. A major advantage of unloading through the bottom valve is that this procedure completely drains the railcar. Where low vapor pressure liquids are pressured out of the car through the vertical fill line, a heel of material is left in the well around the bottom valve. Unloading volatile material, the bottom dry-break coupler can eliminate one upper connection. However, lowering the pressure in the car with a volatile material results in the remaining liquid vaporizing, which is then recovered through the vapor system, eliminating the need for a bottom coupling.

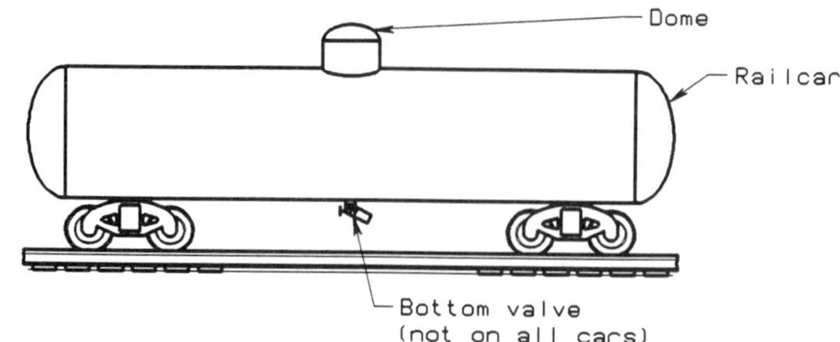

A. DOME RAILCAR FOR PRESSURE AND ENCLOSED OPERATIONS

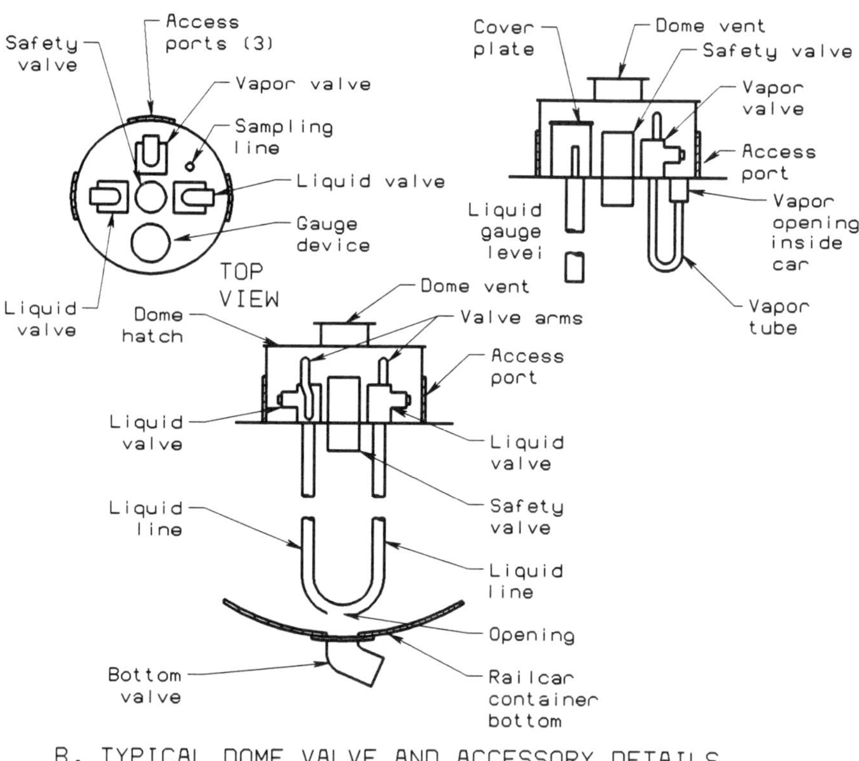

B. TYPICAL DOME VALVE AND ACCESSORY DETAILS

Figure 13.31 *Typical pressure and enclosed-dome railcar.*

The vapor recovery dry-break connection on a movable arm does not present a problem when the connection is broken. However, removing the extension from the vapor valve in the dome will release a very small quantity of vapor in the extension section between the dome valve and dry-break coupling. If a relatively high pressure remains in the extension section during removal, the quantity of gas will be relatively high and can be a potential safety problem along with the possibility of short-term exposure. This requires a reduction in the connecting vapor arm pressure prior to disconnect, which is possible with a vapor control device that is operating at a low pressure level. The type of material in the railcar will determine whether exposure protection is necessary for removal of the extension piece. Alternatively, the vapor could be purged into the railcar with nitrogen. This procedure can also be undertaken in the liquid line in top loading to clear the line and connections.

Closed-Dome LPG Loading. Here LPG refers to light petroleum gases such as propane and butane. These gases under pressure are liquids, and they are loaded into pressurized, closed-dome railcars of the type shown in Figure 13.31. The dome car fittings shown are standardized and all railcars, including chemicals, are loaded through essentially identical procedures. The only variation occurs in vapor disposition which is a function of the loading facility configuration and available equipment. Railcars are maintained under pressure to aid in unloading the contents at the receiving location as liquid. Refinery loading systems are designed for loading the liquids against a backpressure. Since a high pressure exists in the railcar during filling operations, all connections must be capable of handling liquids and vapors without leakage. This requirement is also required in the HON regulations for chemicals.

Typically, the empty railcar arrives at the filling rack with an internal pressure of 50–100 psig (450–800 kPa). The car is visually inspected to ensure it does not leak and is in satisfactory condition. After the inspection is completed, the housing is opened, and the car is tested to determine whether liquid remains in the car. If the railcar is satisfactory and it is in dedicated service, fittings are installed on the liquid fill and vapor valves. These fittings are short pipe lengths and are inserted through the dome as described previously. Dry-break connections are fitted to this outboard opening to meet regional VOC requirements or chemical regulations.

Railcar pressure is adjusted as required prior to loading the car. High pressure can be reduced by bleeding gas through the railcar vapor arm to a flare or other low-pressure receiver. Alternatively, pressure can be increased by adding nitrogen or another inert gas through the loading arm. Pressure is observed during filling along with liquid level. If the pressure increases excessively during filling, the excess is bled through the vapor arm to the flare or a low-pressure system.

When the fill level is reached, the liquid product valve on the platform manifold is closed, and the arm is purged into the railcar with nitrogen. Alternatively, the line can be discharged through a low-pressure system. Railcar pressure is then adjusted to the shipping pressure, and the car valves are closed. The arms are bled to the flare or the low-pressure recovery system.

Control devices are normally required to remove VOC and HAP from low-pressure streams with removal efficiencies of 98% or greater in the proposed HON hazardous air pollutant regulations. For LPG operations and volatile chemical recovery during railcar loading, excessive pressures, purged fill and vapor lines, and manifold fluids are frequently sent to low-pressure control systems or to a flare. Sending this material to a flare or combustion device destroys the vapors and may overload noncombustion systems. One solution for recovery is the installation of a compressor in the release lines. Volatile compounds can be compressed and recycled to storage or the unit process. The purity of the compounds does not change in this system and compression will recover most of the material in the low-pressure line.

The liquid level in closed LPG cars is measured with a gauge that has been used for many years and may be termed a "spray" gauge. The actual gauge is a long tube about 5/16 in. in diameter that extends from inside the car, through the shell, and into the closed dome as shown in Figure 13.31. This type of gauge is an integral part of the railcar and is a measuring rod that essentially contacts the bottom of the tank. The gauge is moved vertically by a technician through a packed gland to determine the height of the liquid. The liquid height is determined by allowing vapor or liquid to pass through a vertical opening in the center of the measured rod that extends from the bottom of the rod to a valve opening at the top. To operate the gauge, the technician pulls the gauge tube above the dome top, opens the valve partially, and points the valve opening downwind. Packing in the socket minimizes leakage around the tube with mechanical guides and stops on the rod prevent the tube from blowing out of the socket. The measuring rod opened at the top end is moved vertically by the technician until the spray indicates the level location. A heavy white mist indicates liquid and the absence of the white mist indicates the internal tube opening is in the vapor zone. The gauging rod has a scale that permits the operator to define the liquid level. However, the spray and mist discharged from the valve to the atmosphere is a relatively large quantity owing to the high pressure in the tankcar and is a potential source of exposure in chemicals operations. In addition, the spray emits VOCs to the atmosphere. Many of these gauges are being replaced by closed indicators.

An alternate gauge, called a magnetic gauge, is available and has been installed in a number of cars. The magnetic gauge consists of a 5/16-in. tube that protrudes from the shell in the dome and rises in conjunction with the liquid level in the railcar. The magnetic gauge rod within the car does not contact the liquid or vapor but is encased in a pipe that extends from the top shell and is sealed at the bottom shell. A floating magnet on the outside of the pipe attracts an internal magnet fixed to the level rod, resulting in tube movement in conjunction with the liquid level. Since the inner tube is sealed, there are no emissions to the atmosphere, which is preferable for operations and meets regulatory requirements.

Other level indicators are also used, including instruments similar to those found in tanktrucks to prevent emissions to the atmosphere. Moreover, eliminating emissions minimizes potential exposures.

Open-Dome Loading. Although heavier materials are loaded through open domes, potential exposure is a problem and various systems are now available to control emissions. One system has a design similar to the hatch-fitting arm and vapor recovery device described for open hatch filling of tanktrucks. Another alternate available in various forms is shown in Figure 13.32, where filling and vapor recovery arms are permanently attached to a solid plate that rests on the railcar dome hatch. Both arms can be supplied as a joint assembly. An important feature in this system is a level sensor that shuts an automatic control valve in the liquid fill line. This sensor can be set at a level equivalent to the visual level marker normally found in open-dome railcars.

The fill arm is extended to permit bottom filling. However, these extended fill pipes become covered with heavy material when withdrawn from the railcar and will drip material. If the arm is raised and moved away from the dome automatically to the storage position, drippage is a problem. This becomes more of a potential exposure problem when the extended pipe and overall arm assembly

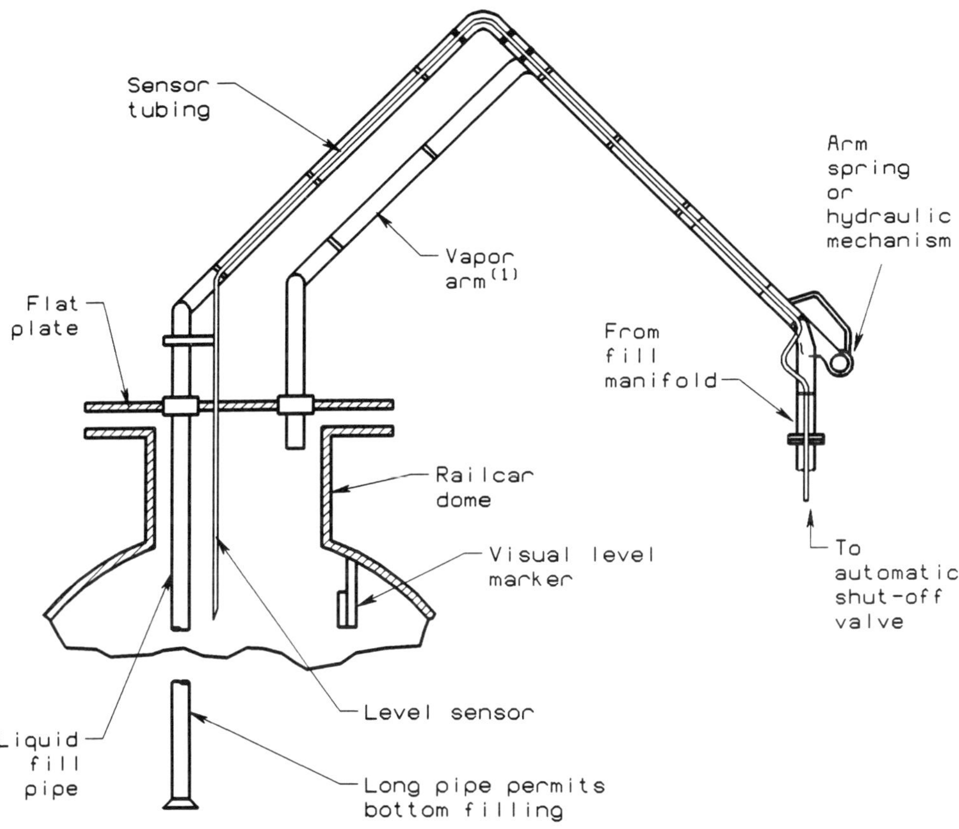

Figure 13.32 *Bottom filling with vapor recovery for a dome railcar.*

must be moved manually to the storage position. In addition, when the fill line is blocked, the fill arm will drain, but will be limited in drainage because of a vacuum that is created in the arm. If not eliminated, the vacuum results in a significant amount of liquid remaining in the arm that gradually drips from the bottom of the fill pipe. Vacuum breakers can be placed on the arm to permit air leakage, raising the pressure and eliminating the vacuum.

Since fill pipes drip, a system for collecting the dripped material is shown in Figure 13.33. This system should be used with heavier oils or chemicals with low vapor pressures. This collection system is satisfactory for jet fuel, diesel oil, and heavier materials. Light, high vapor pressure liquids (> 1.5 psia) should be loaded through a closed system as described.

Open-Dome Asphalt Loading. Asphalt is loaded hot for fluidity reasons and is splash loaded through an open dome. This prevents an extended arm being coated with asphalt and becoming unwieldy. Controls for asphalt loading are difficult to

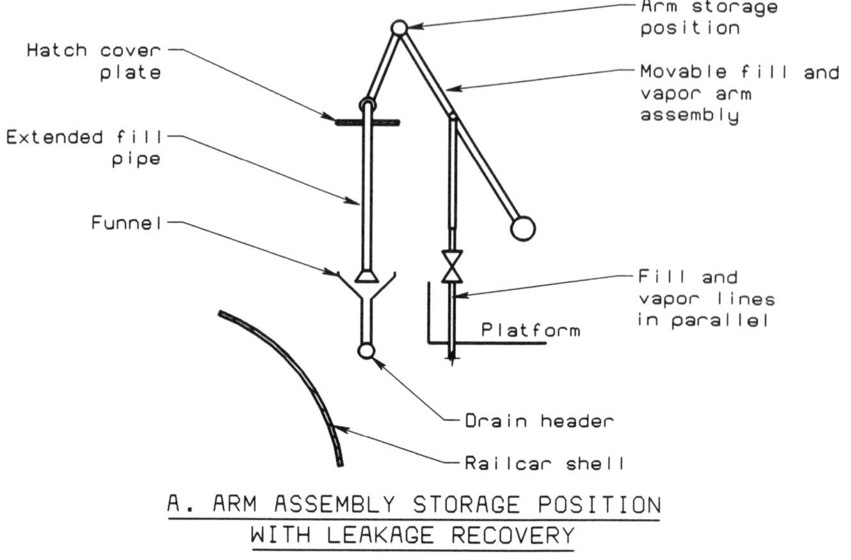

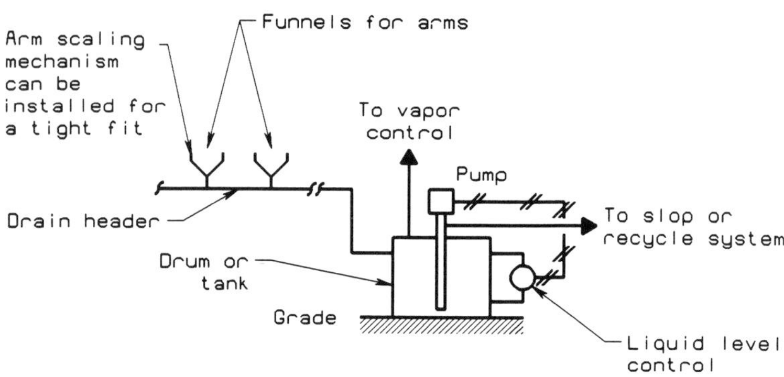

Figure 13.33 *Collection of leakage from fill arms in open dome railcar service.*

provide because of the nature and characteristics of asphalt, specifically, the gradual coating of all surfaces that are contacted by asphalt.

Since mist and aerosols evolve from the car during filling, the car hatch should be covered with a metal sheet. The sheet can cover the hatch with a hole for the fill arm, which will reduce aerosol emissions that can drift toward the operator. A vapor control system can be installed that is designed for the type of aerosols generated. The installation must be inexpensive to permit periodic replacement of coated sections of equipment. The vent design shown in Figure 13.34 provides a basis for a suitable control system. The vent duct can be constructed with a short life expectancy and connected to a steam eductor. Steam and aerosol emissions from the eductor should not be discharged to the atmosphere, but into a container of water which will strip the aerosol from the gas stream. Periodically, the

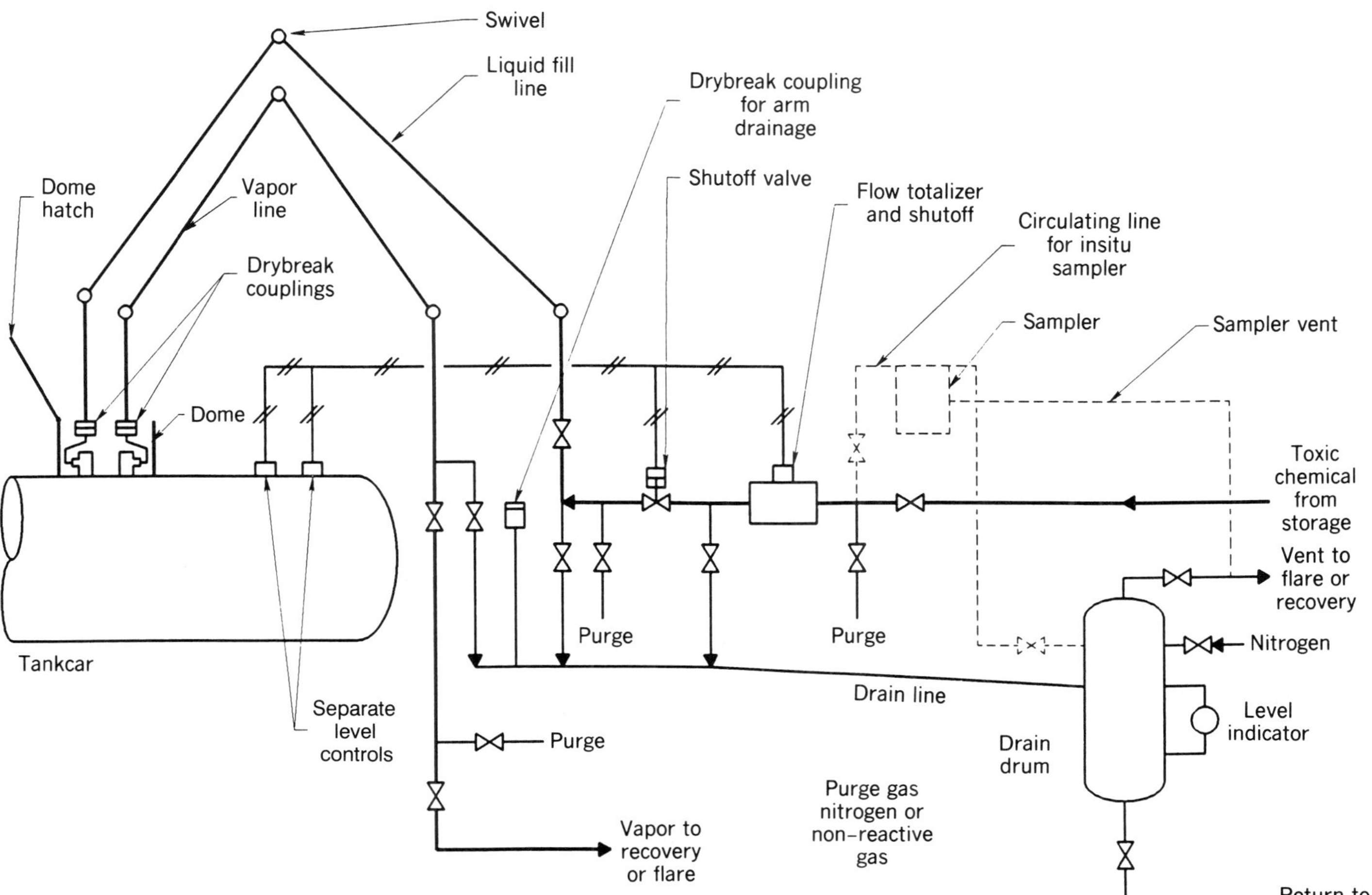

Figure 13.34 Closed low-vapor pressure toxic chemical loading facility.

contaminated water can be discharged to the oily water sewer for removal of hydrocarbons in the wastewater treating system.

The short fill arm is normally held in the fill position above the liquid level after the filling operation is completed to allow drainage of the arm. Although drain time in some instances can be 15 min or more, removal of the arm can result in dripping on the operator, the car, the pad, and the platform storage position. A bucket designed specifically to hang from the arm should be placed beneath the drop pipe opening. The bucket opening should be lower than the drop pipe opening to prevent contact, which would result in a rapid buildup of solidified asphalt on the drop pipe opening or in adherence of the bucket to the drop pipe.

Level measurement is determined by observing the level against an internal mark or by measuring the liquid level distance from the top of the dome flange. This system is a potential source of exposure to the aerosol emanating from the dome opening. Possible alternates to visual observation of liquid level are sonar and radar gauges, which are readily available from manufacturers. A recent radar model has an extremely small sensing head that can be used as a portable detector. Placed on a small jig at the top of the flange, the calibrated radar gauge is connected to a power source at the rack and can indicate liquid level with a high degree of accuracy. The readout can be placed at a location adjacent to the filling arm station.

Loading Chemicals. Chemicals vary considerably in volatility or vapor pressure, as shown in Appendix 13B. For Group I chemicals, the restrictive control regulations apply for compounds with vapor pressures > 1.5 psia (10.2 kPa). However, chemicals such as benzene have vapor pressures < 1.5 psia at 60°F(15°C), and the new HON regulations would nominally classify these compounds as Group 2 chemicals that do not require restrictive emission controls. The HON regulation is less restrictive than the NESHAP benzene transfer (loading) regulation, and the more stringent NESHAP regulation would apply.[18] This specific NESHAP regulation is essentially identical to the HON regulations for Group 1 chemicals, with application limited to liquid concentration levels ≥ 70 wt % and annual loading volumes $> 343{,}000$ gal (1 million L) of ≥ 70 wt % benzene.

While benzene has a specific transfer NESHAP regulation, other toxic chemicals with < 1.5 psia vapor pressures would be considered Group 2 chemicals and have limited emission control regulations. Those chemicals with vapor pressures < 1.5 psia should be investigated to determine whether any of the chemicals may potentially be reclassified as highly toxic. Where potentially toxic chemicals are identified, the HON or benzene regulations should be applied to control exposures to the chemicals. In this section, the loading requirements described for chemicals with < 1.5 psia vapor pressures will conform with HON (NESHAP) regulations.

Benzene Loading. A loading system for toxic low vapor pressure chemicals is shown in Figure 13.34. This system conforms with the benzene NESHAP and HON regulations that are identical in equipment requirements and is applicable for Group 1 chemicals. The loading arm is connected through a dry-break coupling to the railcar fill valve that bottom loads the railcar internally. The vapor discharge valve is connected through a dry break coupling to a swivel arm that sends excess offgas and benzene vapors to recovery or a flare. Since the car is unloaded under pressure, excess offgas is generally natural gas or nitrogen with benzene vapors.

Liquid benzene is fed to the car through a meter that measures the volume loaded into the railcar. The meter should preferably be an open type such as a turbine or coriolis meter that permits complete drainage. A valve associated with the meter stops flow at the meter setpoint with an initial flow reduction near the fill level to minimize water hammer. In addition to meter control, separate high-level controls are installed in the car to prevent overfilling. When the railcar level reaches a predetermined setting, the railcar level overrides the meter and closes the shutoff valve. An electrical jack on the railcar is connected with an electrical cable to a jack located on the platform near the shutoff valve.

A sample of the benzene liquid fed to the railcar may be required to ensure that the material is satisfactory. For manual sampling, one of the in situ sampling systems described in Chapter 11 should be installed. A separate sampling loop or circulating line is shown for the sampler required in the HON and NESHAP regulations. Since the benzene fill pressure may be somewhat elevated, the circulating line permits operation of the sampler at a lower pressure. In addition, these samplers have very small vapor releases that should not be discharged to the atmosphere. A vent is shown which connects with the drain drum vent which is at low pressure. Samples in this operation are usually not obtained from the railcar since railcars in benzene service are dedicated and rarely contaminated. Any contamination or offspec material is normally found in the fill liquid.

The NESHAP and HON regulations are specific in preventing leakage to the atmosphere from vents or drains. Consequently, the loading and unloading systems must be installed to meet regulatory requirements and general operating procedures. In addition, maintenance also requires compliance with leakage and exposure objectives. Changing gaskets on the dry-break couplings is required periodically along with maintenance on the flow totalizer and shutoff valve.

In the system shown in Figure 13.34, items have been included to essentially eliminate leakage to the atmosphere. Drains are placed at low points and the drainage is collected in a header which is sent to a drain drum. Also, liquid from the circulating sampling loop is discharged into the drain drum. When purging the equipment with gas to remove all traces of benzene liquid and vapor, the combined purge gas and benzene is also sent to the drain drum. In the drum, the liquid and nitrogen or purge gas are separated. Since the vent stream will also have some benzene vapor, the gas is sent to a control device (recovery unit) or a combustion device such as a flare. The liquid fill and vapor arms can also be drained to the drum. A direct connection to the recovery unit or flare is also shown for the vapor line. When the liquid level in the drain drum reaches a predetermined level, valving on the drum permits pressurizing the drum and sending liquid to storage under pressure. Although shown as a manual operation, the drain drum operation can be automated.

For draining and purging the swivel arms, a separate dry-break coupling is positioned near both arms. When in the storage position, the arms can be moved a short distance, permitting coupling of the arms with the drain. A purge gas can be injected to completely purge the arms to the drain system. Since the arms are in an inverted V position, one leg can be drained and then the other, followed by the purging operation. Proper draining also requires eliminating low points or placing drains at the low points. Without complete drainage, potential exposure situations can exist. In addition, these small quantities can contaminate other chemicals sent through the loading system for car loading.

Loading High Vapor Pressure Chemicals. High vapor pressure chemicals such as butadiene or methyl chloride are loaded in LPG-type railcars. Loading techniques are similar to those described for LPG materials. However, there are some differences between Group 1 chemicals and LPG. The major difference is that Group 1 chemicals must be controlled in accordance with HON regulations, which essentially prevent leakage. In addition, railcars may require low oxygen concentrations, necessitating measurements and railcar purging to achieve the desired oxygen concentration.

When purging, draining, or venting, the chemicals must be contained within a closed system, as shown in Figure 13.34. The slip gauge level device in LPG cannot be used because of atmospheric discharge. A magnetic level tube or other device is necessary to prevent atmospheric emissions. In addition, all valves that can potentially discharge to the atmosphere must be car-sealed closed or locked.

Other Low Vapor Pressure Chemicals. A large number of chemicals have vapor pressures < 1.5 psia (10.3 kPa) such as toluene, ethyl acrylate, and methyl methacrylate. These Group 2 chemicals do not require the HON controls and can be loaded with facilities generally similar to the one described for benzene. However, some modifications can be introduced, particularly in the dry-break coupling systems. Instead of permanent couplers in the car dome, the couplers can be installed when the car arrives. In addition, some modifications in the arm purging and draining systems are possible.

Although changes from Group 1 chemicals are possible, a Group 1 installation should probably be installed to meet VOC regulations and future changes in the list of hazardous air pollutants. This list will be expanded, and the definition of Group 1 and Group 2 will probably change with time, requiring Group 1 controls on all loading facilities.

Unloading Railcars

Pressurized Petroleum Cars. Pressurized cars are generally unloaded through closed systems. The pressure in the car is sufficient to unload the car through the liquid outlet valve. If the pressure is insufficient, it can be increased with nitrogen or natural gas. The liquid arm is depressurized prior to disconnecting the arm from the railcar through procedures similar to those described in filling operations. The concentration of hydrocarbon vapors in the arm is relatively low. Some exposure occurs when breaking the arm connection to the car since the arm pressure may be greater than 1 psig. Alternatively, purging through the vapor valve or installing dry-break couplings will gradually reduce this concentration.

The LPG is normally unloaded under natural gas pressure. Natural gas pressure is applied through a vapor arm and the liquid is unloaded from the car through the fill valve and arm that extends to the bottom of the car. Metal arms connect the car dome valves to the piping manifold.

Low-Pressure Petroleum Railcars. Railcars that are not capable of operating at higher pressures are emptied through a bottom valve. An alternate is pressurizing within the allowable railcar pressure limits to displace the liquid. A vacuum operation can also unload a car through a top-filling valve and internal arm that extends from the top hatch to the car bottom. This method of unloading is more difficult and is less frequently used. Unloading through the bottom valve generally requires a pump and flexible hose. A relatively large quantity of fluid may discharge from the flexible line if a dry-break coupling is not installed at the end of the hose or line to reduce potential exposures. Additionally, fluid will drip from the bottom valve outlet and the hose may contain a considerable amount of car fluid. Methods of emptying or purging the hose are necessary to minimize potential exposures. Dry-break couplings maintain a hose full of liquid until the liquid is removed through purging or another procedure.

Unloading High Vapor Pressure Chemicals. These may also be called light chemicals. Unloading procedures for light chemicals in the C_4–C_5 range differ somewhat from the method previously described for LPG products. Chemical process requirements limit the use of pressurizing gas which contaminates the materials. The chemicals are generally shipped in LPG and are unloaded through the dome. Bottom valves are generally not installed in these cars, but vaporization and purging satisfactorily cleans them.

Unloading is generally based on a metal swivel arm connected with a dry-break coupling to the fill valve. The unloading piping is straightforward with closed-vent connections. A hose to inject nitrogen is normally connected to the vapor valve in the railcar dome. The car is pressurized with nitrogen through the flexible hose prior to opening the unloading valve. A pressure gauge on the nitrogen line indicates the vapor zone pressure. Blocking the dome valve after pressurizing prevents leakage of vapor to the atmosphere when the hose is disconnected. A dry-break coupling is not required for this connection. Nitrogen is generally the pressurizing gas to prevent any contamination of the liquid that may occur with natural gas. In addition, large quantities of natural gas absorbed in the liquid may affect processing operations.

Following railcar pressurizing, liquid flow through the metal swivel arm commences. Since the car should not be completely depressurized, or pressurizing gas consumption increased when the car is emptied, an automatic shutoff device is required. At the low-level point in the railcar, gas will break through and a shutoff valve in the line should automatically close. Various sensors are commercially available that sense the difference between liquid and vapor. One of these sensors is connected to the remote valve for automatic shutoff.

Incoming pressurized cars are sampled to verify the car contents and provide an analysis of the material for processing. Samples are obtained in bombs through connections to the railcar sampling point in the dome. Discharge of the bomb purge to the atmosphere should be eliminated, particularly with lighter chemicals such as butadiene, methyl chloride, and so on. The bomb discharge valve can be connected through a hose to the dry-break coupling for arm draining. Purging the bomb through the drain drum to the flare or recovery unit provides a knockout drum prior to the vapor entering the flare line.

Leaks. The term leaks refers to fugitive emissions leakage from valves, flanges, connections, and so on. Leakage is defined as 500 ppmv or greater, which is similar to the HON requirements for chemicals or in the benzene NESHAP. In addition, VOC requirements may be similar for various nonattainment areas. The entire transfer system must be monitored during the loading of railcars or tanktrucks.

REFERENCES

1. API, *Evaporative Loss From Fixed-Roof Tanks*, 2nd ed., API Publ. 2518, Washington, DC, 1991.
2. API, *Evaporative Loss From External Floating-Roof Tanks*, 3rd ed., API Publ. 2517, Washington, DC, 1989.
3. API, *Evaporative Loss From Internal Floating-Roof Tanks*, 3rd ed., Reaffirmed, API Publ. 2519, 1990.
4. FR, *National Emission Standards for Hazardous Air Pollutants from Synthetic Organic Hazardous Air Pollutants from Synthetic Organic Chemical Manufacturing Industry for Process Vents, Storage Vessels, Transfer Operations, and Wastewater*, Title *40*, Subpart G, Parts 63.100–63.152, pp. 62684–62765, December 31, 1992.
5. API, *Welded Steel Tanks for Oil Storage*, 8th ed., API Publ. 650, Washington, DC, 1988.
6. Laverman, R. J. *Emission Reduction Options for Floating-Roof Tanks. Second International Symposium on Above Ground Storage Tanks*, CBI Publ. CBT-655, Houston, TX, January, 1992.
7. API, *Evaporation Loss From Fixed-Roof Tanks*, API Publ. 2518, Washington, DC, 1962.
8. EPA, *Compilation of Air Pollutant Emission Factors*, 4th ed., AP-42, Vol. 1, 1985.
9. EPA, *Estimating Air Toxic Emissions from Organic Liquid Storage Tanks*, EPA Report No. EPA-450/4-88-004, 1988.
10. CFR, *Flammable and Combustible Liquid*. Title *29*, Subpart H, Part 1910.106, pp. 243–277, July 1991.
11. CFR, *Standards of Performances for Volatile Organic Liquid Storage Vessels (Including Petroleum Liquid Storage Vessels) for Which Construction, Reconstruction, or Modification Commenced After July 23, 1984*. Title *40*, Subpart Kb, Parts 60.110b–60.117b, pp. 352–361, July 1, 1992.
12. CFR, *National Emission Standard for Benzene Emissions from Benzene Storage Vessels* Title *40*, Subpart Y, Parts 61.270–61.276, pp. 144–152, July 1, 1992.
13. SCAQMD, *Storage of Organic Liquids*, Rule 463, Amended December 7, 1990.
14. BAAQMD, *Storage of Organic Liquids*, Regulation 8, Rule 5, May 4, 1988.
15. CFR, *Standards of Performance for Storage Vessels for Petroleum Liquids for Which Construction, Reconstruction, or Modification Commenced After June 11, 1973 and Prior to May 19, 1978*. Title *40*, Subpart K, Parts 60.110–60.114a, pp. 346–352, July 1, 1992.
16. CFR, *Standards of Performance for Storage Vessels for Petroleum Liquids for Which Construction, Reconstruction, or Modification Commenced After May 18, 1978 and Prior to July 23, 1984*. Title *40*, Subpart Ka, Parts 60.110a–60.115a, pp. 348–352, July 1, 1992.
17. CFR, *National Emission Standard for Benzene Emissions from Benzene Transfer Operations*. Title *40*, Subpart BB, Parts 61.300–61.306, pp. 153–162, July 1, 1992.
18. CFR, *Standards of Performance for Bulk Gasoline Terminals*. Title *40*, Subpart XX, Parts 60.500–60.506, pp. 505–509, July 1, 1992.
19. CFR, *Hazardous Materials, Flammable and Combustible Liquids*. Title *29*, Subpart H, Part 1910.106(f), pp. 264–267, July 1, 1991.
20. CFR, *Standards of Performance for New Stationary Sources—General Control Device Requirements*. Title *40*, Subpart A, Sec. 60.18(b), pp. 237–238, July 1, 1992.
21. CFR, *Determination of Vapor Tightness of Gasoline Delivery Tank Using Pressure-Vacuum Test*. Title *40*, Part 60, Appendix A, Method 27, pp. 1072–1073, July 1, 1992.
22. BAAQMD, *Terminals and Bulk Plants*. Regulation 8, Rule 6, November 30, 1983.
23. BAAQMD, *Gasoline Bulk Terminals and Gasoline Delivery Vehicles*, Regulation 8, Rule 33, July 20, 1988.

24. SCAQMD, *Organic Liquid Loading*. Rule 462, Amended December 7, 1990.

25. BAAQMD, *Gasoline Bulk Plants and Gasoline Delivery Vehicles*, Regulation 8, Rule 33, October 7, 1987.

26. EPA, *Toxic Air Pollutant Emission Factors—A Compilation for Selected Air Toxic Compounds and Sources*, 2nd ed., EPA-450/2-90-011, 1990.

27. Davis, G. B., M. L. Goss, P. Schoemann, and S. S. Tyler, *Oil Gas J.* **91** (50), 35–39, December 13, 1993.

28. EPA, *Control of Volatile Organic Compound Emissions from Volatile Organic Liquid Storage in Floating and Fixed Roof Tanks*, Proposed Control Technology Guideline, October 1993.

29. Weast, R. C. ed. *Handbook of Chemistry and Physics*, Chemical Rubber Co., pp. D146–D165, Cleveland, Ohio, 1970.

APPENDIX 13A Determination of Vapor Pressure for Fixed-Roof Tanks

The vapor pressures are determined for three temperatures as shown in the following equations:

$$P_{vx} = \exp\left[A - \frac{B}{T_{LX}}\right] \tag{A.1}$$

$$P_{va} = \exp\left[A - \frac{B}{T_{LA}}\right] \tag{A.2}$$

$$P_{vn} = \exp\left[A - \frac{B}{T_{LN}}\right] \tag{A.3}$$

where

P_{VX} = vapor pressure at daily maximum surface temperature (psia)

P_{VA} = vapor pressure at daily average liquid surface temperature (psia)

P_{VN} = vapor pressure at daily minimum liquid surface temperature (psia)

T_{LX} = daily maximum liquid surface temperature(°R)

T_{LA} = daily average liquid surface temperature(°R)

T_{LN} = daily minimum liquid surface temperature(°R)

A, B = constants(°R) (see Appendix 13B)

The liquid daily vapor pressure range is obtained as follows:

$$\Delta P_V = P_{VX} - P_{VN} \tag{A.4}$$

The following equation is an approximate method of calculating the liquid daily vapor pressure range:

$$\Delta P_V = 0.50 B P_{VA} \Delta T_V / T_{LA}^2 \tag{A.5}$$

Properties of selected chemicals and petroleum liquids are presented in Appendix 13B that are necessary for calculating vapor pressure along with molecular weight M_V and condensed vapor density W_{VC}.

APPENDIX 13B Properties of Selected Chemicals and Hydrocarbons

	Chemical Formula	Vapor Molecular Weight M_v (lb/lb-mole)	Boiling Temperature (14.7 psia) °F	Condensed Vapor Density W_{vc} (lb/gal)	Vapor Pressure[a] P_v (psia)	Vapor Pressure Equation Constants[b] A (Dimensionless)	B (°R)	Temperature Range for Constants A and B Minimum (°F)	Maximum (°F)
				Alcohols					
Methyl alcohol	CH_4O	32.04	148.5	6.641	1.351	15.9482	8,131.3	−47.2	435.2
Ethyl alcohol	C_2H_6O	46.07	172.9	6.643	0.6200	16.3801	8,760.7	−24.3	467.6
Allyl alcohol	C_3H_6O	58.08	205.9	7.127	0.2656	17.1073	9,579.2	−4.0	205.9
Isopropyl alcohol	C_3H_8O	60.10	180.1	6.607	0.4637	16.7687	9,113.6	−15.0	449.6
Tertiary butyl alcohol	$C_4H_{10}O$	74.12	180.4	6.586	0.3970	17.2230	9,430.3	−4.7	432.5
Furfuryl alcohol	$C_5H_5O_2$	98.10	337.1	9.422	0.006096	17.2335	11,606.1	89.2	338.0
				Amines					
Diethyl amine	$C_4H_{11}N$	73.14	131.9	5.942	2.868	13.7881	6,617.7	−27.4	410.0
				Aromatic hydrocarbons					
Benzene	C_6H_6	78.11	176.2	7.361	0.9008	14.0920	7,377.5	−34.1	554.5
Toluene	C_7H_8	92.14	231.1	7.289	0.3249	13.8288	7,770.6	−16.1	606.2
Styrene	C_8H_8	104.15	293.3	7.576	0.08254	14.2954	8,725.2	19.4	293.4
Ethylbenzene	C_8H_{10}	106.17	277.1	7.290	0.1139	14.0362	8,423.3	14.4	619.7
o-Xylene	C_8H_{10}	106.17	292.0	7.377	0.07359	14.8147	9,054.7	25.2	291.9
Isopropylbenzene	C_9H_{12}	120.19	306.3	7.240	0.04968	15.0086	9,359.7	37.2	306.3
Naphthalene	$C_{10}H_8$	128.17	424.3	9.555	0.0006621	22.4188	15,454.4	32.0	176.0
				Chlorine compounds					
Methylene chloride	CH_2C_{12}	84.93	103.6	11.150	5.480	14.8970	6,857.5	−94.0	105.3
Chloroform	$CHCl_3$	119.38	142.1	12.503	2.212	13.8649	6,792.5	−72.4	489.2
Carbon tetrachloride	CCl4	153.82	170.0	13.357	1.255	13.5218	6,908.7	−58.0	528.8
Vinylidene chloride	$C_2H_2Cl_2$	96.94	98.6	10.164	8.230	14.6756	6,531.1	−107.0	89.1
1,2-Dichloroethane	$C_2H_4Cl_2$	98.96	182.2	10.509	0.9495	13.8035	7,200.2	−48.1	545.0
Trichloroethylene	C_2HCl_3	131.40	189.0	12.234	0.8912	14.3744	7,529.8	−46.8	188.1
1,1,1-Trichloroethane	$C_2H_3Cl_3$	133.40	165.4	11.229	1.507	14.3734	7,256.4	−61.6	165.4
Allyl chloride	C_3H_5Cl	76.53	112.3	7.828	4.873	14.4564	6,689.5	−94.0	112.3
				Cyano compounds					
Acrylonitrile	C_3H_3N	53.06	171.5	6.726	1.340	14.1319	7,191.8	−59.8	173.3
				Cycloparaffins					
Cyclopentane	C_5H_{10}	70.13	120.7	6.338	4.152	14.3384	6,711.5	−90.4	120.7
Cyclohexane	C_6H_{12}	84.16	177.3	6.532	1.052	13.6969	7,091.7	−49.5	495.5

APPENDIX 13C Properties of Selected Chemicals and Hydrocarbons

	Chemical Formula	Vapor Molecular Weight M_v (lb / lb-mole)	Boiling Temperature (14.7 psia) °F	Condensed Vapor Density W_{vc} (lb / gal)	Vapor Pressure[a] P_v (psia)	Vapor Pressure Equation Constants[b,c] A (Dimensionless)	B (°R)	Temperature Range for Constants A and B Minimum (° F)	Maximum (° F)
				Esters					
Methyl acetate	$C_3H_6O_2$	74.08	134.5	7.847	2.359	14.3340	7,002.9	−71.0	437.0
Vinyl acetate	$C_4H_6O_2$	86.09	162.5	8.827	1.312	15.0324	7,670.9	−54.4	162.5
Methyl acrylate	$C_4H_6O_2$	86.09	176.9	7.957	1.014	14.9971	7,786.4	−46.7	176.4
Ethyl acetate	$C_4H_8O_2$	88.11	170.7	7.548	1.012	14.4776	7,517.5	−46.1	455.0
Methyl methacrylate	$C_5C_8O_2$	100.11	212.5	7.928	0.4314	14.7995	8,127.7	−22.9	213.8
				Ethers					
Diethyl ether	$C_4H_{10}O$	74.12	94.0	6.002	6.109	13.9146	6,290.5	−101.7	361.9
				Ketones					
Acetone	C_3H_6O	58.08	133.3	6.647	2.553	14,2539	6,920.2	−74.9	418.1
Methyl ethyl ketone	C_4H_8O	72.11	175.4	6.753	1.197	14.3812	7,380.2	−54.9	175.3
				Paraffins					
n-Pentane	C_5H_{12}	72.15	96.9	5.262	6.091	13.2999	5,972.6	−105.9	376.3
n-Hexane	C_6H_{14}	86.18	155.7	5.534	1.702	13.8236	6,907.2	−65.0	408.9
n-Heptane	C_7H_{16}	100.20	209.2	5.738	0.5109	13.9835	7,615.8	−29.2	477.5
n-Octane	C_8H_{18}	114.23	258.2	5.894	0.1591	14.2309	8,350.6	6.8	538.5
n-Nonane	C_9H_{20}	128.26	303.5	6.018	0.05069	15.2406	9,469.8	36.3	301.1
n-Decane	$C_{10}H_{22}$	142.28	345.5	6.121	0.01888	15.0462	9,882.0	62.8	343.4
				Phenols					
Phenol	C_6H_6O	94.11	359.3	8.937	0.006316	15.6581	10,769.0	104.2	985.7
o-Cresol	C_7H_8O	108.14	375.8	8.746	0.004231	16.2959	11,308.0	100.8	375.4
				Sulfur compounds					
Carbon disulfide	CS_2	76.13	115.2	10.592	4.491	13.3292	6,146.2	−100.8	492.8
Jet naptha (JP-4)	—	80	—	5.4	1.27	11.368	5,784.3	40	100
Jet kerosene	—	130	—	6.1	0.00823	12.390	8,933.0	40	100
Dist. fuel oil no. 2	—	130	—	6.1	0.00648	12.101	8,907.0	40	100
Residual oil no. 6	—	190	—	6.4	0.0000430	10.104	10,475.5	40	100

[a] Vapor pressure calculated at 60°F using constants A and B.
[b] Vapor pressure equation; $P_v = \exp[A - (B / T_L)$, P_v = psia, T_L = liquid surface temperature (°R), exp = exponential function.
[c] Adapted from reference 29.

14

Dust Control

Dusts can be generated through various mechanical techniques or may evolve as a powder is processed in numerous process systems. In general, dust consists of particles larger than 1 μm and fumes consist of particles smaller than 1 μm. Most definitions indicate that fumes are formed by combustion, sublimation, and condensation. The 1-μm breakpoint is not absolute, with some dust particles smaller than 1 μm and some fume particles somewhat larger than 1 μm. Particles less than 100 μm in diameter may generally be inhaled, although particles larger than 75 μm in size readily settle in air or other gases. Those particles less than 75 μm in diameter will remain airborne or in suspension for a long period of time. Leakage of fine particles or agitation of the powder can result in the formulation of fine cloudlike particulate configurations in the atmosphere that are potential health exposure problems in the workplace and the community. The distribution of fine particles in the atmosphere is generally termed a dust, and dust control is the major concern in this chapter.

Dusts generally affect humans through inhalation or skin contact exposures. Airborne particulates deposited in the respiratory tract are a major concern as a result of particulate size and concentration, which may be further exacerbated by the chemical composition of the dust particles. As a result, exposure limitations for the workplace consist of size limitations and chemical composition.[1] In addition, the threshold limit value (TLV) has been expanded to cover the deposition location in the lung and the size-fraction of each substance related to the health effect of concern.[1] While inhalation is a major concern, skin contact with a number of dusts must be minimized where the material may be absorbed through the skin. Other dusts are skin irritants with criteria for skin adsorption indicated in occupational exposure tables.[1,2] Other nonhealth concerns associated with dust emissions are

Explosion and fire
Equipment damage
Impaired visibility
Odors
Water runoff

Particulates, solids, and powders have a widespread presence in the chemical process industries, ranging from catalysts of varying chemical composition and size distribution to fine polymeric powders and talc. Because of raw material, processing, and production operations, large potential dust exposures exist covering the following general categories:

Agitation and manual handling
Equipment leakage
Cleaning and maintenance

Since there is a wide variety of operations and equipment associated with potential dust emissions, a complete review of all sources is not possible. Consequently, this chapter will focus on certain equipment, systems, and operations that are present in many areas of the chemical industry. Principles and suggestions derived from this review may be helpful in controlling emissions from systems and equipment that are not specifically reviewed.

14.1 REGULATIONS

At the present time, there are differences in allowable exposure limits for dusts and fumes in tables issued by OSHA and the ACGIH. Revisions to the OSHA exposure limits list are under way, and the differences between the allowable exposure limits in both lists should narrow considerably when the OSHA exposure list is promulgated.

In the Clean Air Act Amendments of 1990, various general compound groups are included in the HAP list shown in Appendix 4A (Chapter 4). These general groupings can include dusts or fumes, but specific regulatory exposure requirements for these materials have not been developed. While the general status of various regulations follows, differences exist between the regulatory bodies and should be understood prior to developing engineering controls and other protective measures. In the United States, the states, federal government, and regional authorities may all have different exposure limitations, generally indicating that the most stringent limits are applicable.

Although ACGIH recommended workplace exposure limitations are not official governmental regulations, they are widely used in industry and are frequently a basis for governmental regulations. Moreover, the ACGIH exposure recommendations are revised annually, providing up-to-date exposure limits that reflect current health information. There are a few exposure limitation differences between the ACGIH and regulatory recommendations. Where these differences exist and regulatory requirements are more stringent than ACGIH limitations, governmental requirements must be observed.

14.1.1 United States Environmental Protection Agency

NAAQS. Standards for particulates are included in the National Ambient Air Quality Standards (NAAQS) that cover community or area concentrations and were generally described in Chapter 4. The criterion for particles of concern in meeting the NAAQS standards is an aerodynamic diameter of 10 μm or less, which is termed PM-10 by the EPA in regulations. Based upon the PM-10 definition, the NAAQS requirements for primary and secondary standards are:

24-hr average concentration	150 $\mu g/m^3$ ($\leq$ 1 day/year)
Annual standard	$\leq$ 50 $\mu g/m^3$ (annual arithmetic mean)

Areas or regions may be in attainment for particulates but not for other NAAQS regulated items such as sulfur oxides or ozone. When the area is in attainment status for particulates, additional controls for particulates or dust are not required. However, workplace exposures may require improved controls to comply with OSHA or ACGIH exposure limitations.

Hazardous Air Pollutants. The HAP list, which is included in the CAAA, contains compounds or materials considered deleterious to the public health and requires the installation of controls or process changes to significantly reduce public risk. Regulations nominally covering vapor and liquid leakage controls for various equipment and systems will be promulgated in early 1994. However, these regulations will not specifically address dusts or fumes included in the original HAP list. Dust control requirements will probably be issued in the next few years.

14.1.2 Worker Health Exposure Standards

The Clean Air Act and other environmental regulations restrict the particulate concentration in the general community. Controls that meet these regulatory requirements and reduce overall emissions are also effective in reducing particulate concentrations in the workplace. However, actual exposure concentrations in the workplace may still be excessive and must be monitored to ensure that satisfactory exposure levels are achieved.

Exposure limitation standards have been promulgated by various government entities requiring compliance. In addition, industrial exposure standards are also published and have been accepted in numerous facilities. These various standards and interactions are described in the following paragraphs.

OSHA. The Occupational Safety and Health Agency has promulgated exposure limitation standards for a wide range of chemicals, dusts, and fumes.[3] These limitations are presented in three tables of reference 3 and dust exposure concentration standards have been separated into a total maximum dust concentration and a respirable fraction with a maximum allowable concentration. The maximum aerodynamic particulate diameter for respirable materials is 10 μm, with a minimum particulate diameter of about 0.5 μm. The minimum diameter may vary, and further information is provided in reference 1.

Periodically the exposure limitation tables are revised, and a revised list is currently under review. The OSHA list of substances and limitations are similar to the ACGIH list described below. However, the OSHA exposure limitations are official and must be implemented. The OSHA requirements may be implemented through official state entities recognized by OSHA or by OSHA directly. The number of states in the OSHA implementation program is quite small, but when recognized by OSHA, a state organization may further limit exposure concentrations. In this situation, OSHA requirements are considered a minimum concentration level and the state may further restrict exposure concentration levels.

Since a number of states do not participate in the OSHA implementation program, regulatory requirements must be defined when determining control requirements.

ACGIH. As described in Section 14.1, the exposure concentration limits[1] published by the organization are not mandatory or legal requirements. Observance of the ACGIH exposure recommendations are voluntary in plants or companies. These recommended exposure limits are revised and published each year with revisions based upon results of various studies. OSHA's revised exposure limitations lists will probably be based upon the ACGIH exposure concentration tables.

If compliance with the ACGIH exposure list is the basis for exposure control, specific chemical compounds are compared with the OSHA list to ensure the exposure limits are equivalent or more stringent than the OSHA requirements. For dusts and fumes, the ACGIH requirements are more detailed and complex than the current OSHA regulatory limits.[3]

Other. While OSHA exposure concentrations are applicable in the United States, regulatory requirements exist in various countries and along with the ACGIH exposure list provide concentration limits for a wide range of dusts and fumes. The exposure concentration list in the United Kingdom issued by the Health and Safety Executive[2] is an extensive list of exposure requirements. In addition, dusts and fume requirements are carefully defined. Process units must have controls that meet regulation exposure concentration limits, whether federal or local, requiring knowledge of workplace exposure regulations.

14.2 DUST EXPOSURE PREDICTION

Dustiness is a general term applied to a bulk material or powder that describes the evolution of dust or particulates when the base material is jostled or moved. Solids agitation can vary widely during different operations from transfer to processing solids. Dust evolution is also termed dust generation and has been the subject of study for many years. Since this characteristic of solids processing is important in controlling potential health and safety problems, a large number of dustiness tests have been developed in Europe and the United States,[4,5] to describe dust generation. The object of these tests is to standardize the description of dust generation followed by correlation of dust generations with workplace exposure concentrations.

A dustiness index has been developed from the dust evolution data produced in Heubach and Midwest Research Institute (MRI) testing devices by NIOSH.[7] NIOSH considers the Heubach tester representative of machinery processing of solids and the MRI tester represents manual powder processing. Following the dustiness index development, NIOSH conducted bagging studies in several plants with various solids materials. A correlation of dustiness and potential worker exposures was then developed[6] which provides an estimate of potential worker exposure. However, dustiness data must be developed for specific solid materials, which presumably can then be used with a predictive equation to estimate potential worker exposure levels.

Solids testing using the test devices produces different results which were noted in a British Occupational Hygiene Society Study (BOHS).[5] The Heubach device consists of a horizontal drum that is rotated at 30 rpm, and dust generated from the solids is drawn by a vacuum pump into a settling cylinder to remove coarse particles, with the fine particulates carried into a filter. In contrast, the MRI test consists of remotely emptying a filled cup within a chamber where the solids fall a distance of 30 cm and the generated dust is drawn into a filter at the chamber top by a vacuum pump. Recent tests of two different dustiness test procedures also produced dustiness index results that differed.[8] In these latter tests, an important variable was flow rate, which affected dust generation along with the type of measurement device. In an extensive program investigating dust generation and exposure concentration prediction, wide variations in duplicate tests with the MRI test device were observed.[9] A new method of calculating the dustiness index may be required that also considers other variables such as particle size distribution, bulk density, particle shape, dryness, electrical charge, air flow rate, and collection techniques.

14.2.1 Dust Generation

While there are concerns regarding the accuracy of the dustiness index relationship to workplace exposure, the dustiness tests can be helpful in anticipating dust concentrations and potential workplace exposures. In addition, some calculation procedures have been developed for estimating generation rate that can be helpful in determining dust concentrations. If not directly applicable, the methods should aid in developing procedures and tests to improve estimating procedures.

Dust Generation Test Program. A detailed pilot test program was conducted under the auspices of the EPA to develop dust generation procedures and worker exposure relationships.[9] This pilot program in sequence consisted of mixing solids and water, filtering the stream, cleaning the filter press, collecting the filter cake in trays, batch drying the trays, unloading the dry powder, and mixing the solids with water to maintain the process sequence. In these operations, all of the solids movements were manual, resulting in dust generation particularly of the dry powder.

For the tests, $CaCO_3$ was the particulate material with a distribution of fine material ranging from about 0.1 to 44 μm for one sample set and 0.2–260 μm for a second. With the pilot plant equipment installed within a building, controlled ventilation eliminated extraneous wind effects, and generation along with particulate dispersion could be tracked. The concentration of dust was measured at various locations, and a real-time aerosol monitor (RAM) was installed at the ventilation duct outlet. The *latter* instrument is based on light scattering and detection by a photo-optical sensor, with fixed location devices consisting of filter cassettes and a small air pump. During operations, all of the fixed sampling devices were operated with operating personnel wearing personal sampling systems.

Dust Generation Calculations. Two basic theoretical approaches to describing dust generation were developed based upon receptor models. Mixing factor and turbulent eddy diffusivity theoretical models were investigated to determine the optimum generation analytical approach. These are described in the following paragraphs.

Mixing Factor Theory. The basic mixing factor equation is based upon a general mass balance:

$$\frac{VkdC_{\text{out}}}{dt} = qC_{\text{in}} - qC_{\text{out}} + G_{\text{eff}} \tag{14.1}$$

where

V = room volume
k = mixing factor based on active volume
q = air flow rate through room
t = time
G_{eff} = effective generation rate
C_{in} = room inlet dust concentration
C_{out} = room outlet dust concentration

An air sampler was placed at the air outlet location and collected dust over the entire generation period. This sample provided a time-weighted average dust concentration (C_0) and Eq. 14.1 becomes after substitution:

$$G_{\text{eff}} = \frac{qt(C_0 - C_{\text{in}})}{t - \dfrac{kV}{q}(1 - e^{-qt/kV})} \tag{14.2}$$

Turbulent Eddy Diffusivity Theory. This theory is based upon a diffusion equation developed in 1981[10] and assumes turbulent room air, a point source, constant diffusivity, and negligible momentum from the generation source. The approach is concerned with room dynamics and worker movements are not included. The basic diffusion equation is:

$$\frac{DdC}{dr} = \frac{G}{4\pi r^2} \tag{14.3}$$

where

D = diffusivity (m^2/sec)
r = radial distance from point source (m)
G = steady-state generation rate (mg/sec)
C = area concentration at r (mg/m^3)

This equation or model has two boundary conditions: (1) the area concentration approaches the generation rate divided by the air flow rate as the sampling location approaches the air inlet or outlet locations and (2) the concentration approaches zero at large radial distances from the generation source. Based on these conditions, Eq. 14.3 becomes:

$$C = \frac{G}{4\pi Dr} \tag{14.4}$$

The concentration is a function of diffusivity, requiring establishment of the diffusivity value. Turbulent eddy diffusivity can be obtained from the following equation:[11]

$$D = \frac{0.08M}{C_{gas}tr} \tag{14.5}$$

where

D = turbulent eddy diffusivity (m^2/sec)
M = mass tracer gas instantaneously released from point source (mg)
C_{gas} = gas concentration (composite sample) after tracer release (mg/m^3)
t = air sample composite time (sec)
r = distance from point source to composite sampler (m)

The diffusivity equation is an approximation for an indefinite integration, and accuracy requires experimental conditions where $r^2/4Dt \ll 1$. Carbon monoxide and a chlorofluorocarbon were the tracer gases for this evaluation. The maximum value for $r^2/4Dt$ was 0.045 during the tracer tests, satisfying the criteria for the diffusivity equation. Actual diffusivity varied from 0.03 to 0.05 (m^2/sec), with the latter number obtained in the filtration room where ventilation and air flow was not as stable.

Results. The data results indicate turbulent eddy diffusivity theory applies in a well-mixed room with high air exchange rates particularly in the dryer room.[12] However, in the low concentration filtration room where air circulation is poor and

TABLE 14.1 Unloading Inhalation Exposures and Dust Generation Rates[a]

	Experiment Number							
Exposure and Generation Rates	3	7	8	2	1	5	6	4
Inhalation exposure (μg / m^3)	8600	12000	23000	49000	16700	36000	25000	43000
Background concentration (μg / m^3)	79	210	80	720	176	400	270	1350
Mixing factor generation rate (mg / sec)	0.34	2.5	7.4	3.9	1.4	31	9.2	12.0
Eddy diffusivity generation rate (mg / sec)	2.7	2.8	4.7	8.4	4.0	17.2	7.1	12.8

[a]Adapted from reference 9.

the room air is not well mixed, the diffusivity equation is not effective in describing the generation rate. Additional area samplers nearer the source may improve this generation estimation technique along with RAM monitors.

Data on actual exposure concentrations and calculated generation rates are shown in Table 14.1. The diffusivity equations are corrected for background concentrations. Mixing factor generation rates are not corrected for background concentrations since this equation does not depend upon a single generation source within a room. Although not shown, modifications to this mixing factor equation produce other generation rates which are considerably lower than the rates shown in Table 14.1. As an example, where the dust concentration in the exhaust dust reaches a steady-state condition as shown by the RAM data, dC_{out}/dt is equal to zero and in Eq. 14.1 G_{eff} becomes

$$G_{eff} = q(C_{out} - C_{in}) \tag{14.6}$$

The measured inhalation exposure and effective generation rate were correlated as shown in Table 14.2. For filter press cake removal, inhalation exposure and generation rate correlation coefficients are quite poor and have not been shown. These low values apparently result from poor ventilation and low generation rates of the wet cake.

TABLE 14.2 Generation Rate Versus Worker Inhalation Exposure Least Squares Fit[a]

	Squared Correlation Coefficients			
Generation Rate Calculation Method	Locating[b]	Unloading[b]	Loading	Unloading
Mixing factor-Equation 14.6	0.917	0.442	0.931	0.454
Equation 14.2	0.062	0.198	0.507	0.215
Eddy diffusivity[c]	0.965	0.691	0.954	0.704
Equation 14.4	0.932	0.563	0.909	0.578

[a]Adapted from reference 9.
[b]Corrected for background concentration.
[c]Based on boundary conditions in reference 9.

Mixing factor theory generation rates are considerably lower than eddy diffusion generation rates based upon squared correlation coefficient comparisons. However, mixing factor generation rates are quite similar to the eddy diffusivity generation rates when a RAM was used in contrast to the air-integrated area sample as shown for Eq. 14.6. The use of RAM monitors in exhaust outlets and various pattern locations has been effective in describing rapidly changing conditions.

14.2.2 Dust Generation Application

The test results indicate that the generation rate models described above can be used to determine dust generation rates. However, the tests for these models were conducted within specific constraints; namely, the entire test program was conducted within an enclosed building, and the source generators were highly specific operations. While these constraints limit application of the results to other dust generation sources, the models provide a basis for determining dust generation rates in other operations.

The linear correlations in Table 14.2 between worker inhalation exposure and generation rate for this system indicate a satisfactory exposure model can be constructed. Generation rate models for other process operations correlated with actual plant inhalation exposures can be a predictive tool. This predictive technique would be effective in determining dust control requirements to meet stringent exposure requirements and regulations.

14.3 MANUAL OPERATIONS

Manual handling of powder containers, transfer, container loading and unloading, or other operations are potentially high-exposure situations. The operations investigated in Section 4.2 are examples of dust exposures related to manual actions that generated dusty atmospheres. Engineering controls are necessary in these various operations to reduce exposures to acceptable levels. However, personal protection may also be required in certain situations to ensure exposure standards are observed or to minimize short periodic increases in dust concentration levels that may result from specific operating movements.

With manual operations, spillage may periodically occur, resulting in potential exposure situations. A major concern in this situation is removal of the spilled powder to prevent entrainment and high dust concentration levels. Portable or fixed vacuum systems are necessary to ensure complete removal of the spilled powder with minimum agitation. The vacuum systems are also helpful in controlling dust during maintenance operations. Dust exposure control is a combination of engineering controls, personal protection, good maintenance practices, and attentive housekeeping practices.

An additional concern in all situations related to powders and dusts is dust explosion potential. Dust explosion prevention requirements are reviewed in

Section 14.8 and should be evaluated for the powders and dusts requiring handling and processing.

14.3.1 Bag Transport

Bags containing powders are widely used in the chemical industry for raw material products, additives, catalysts, and so on. Traditionally, bags have been constructed of paper with various plies (multiwall) and thicknesses for strength. Bag puncturing and leakage has frequently occurred, resulting in dust emissions and powder coatings in the vicinity on bags and equipment along with powder accumulations on the floor. Powder on the floor may follow bag movements from the point of rupture to bag discharge. A combination of leakage and powder accumulation are potential sources of dust exposure.

Through the years, bag construction has improved, reducing breakage, puncturing, and material losses. As a result, potential exposure situations have been reduced. However, breakage still occurs since improved bags can also be punctured, although there is less breakage and puncturing with the newer bags. These improved bags may also have inner plastic liners for moisture control.

While improved bags aid in reducing dust generation, movement and transport of bags still requires engineering controls to minimize potential dust exposures. In addition, of the two major bag types, valve bags may leak through the valve during movement or transfer with little leakage occurring in wide mouth bags or valve bags that are completely sealed.

Palletized Bags. Bags are usually shipped on pallets and are frequently held in position by a thin stretched polymer film. The polymer sheet effectively prevents bag movement and falling bags during pallet transfers. Bags that fall from pallet loads frequently break or are punctured, leaving powder on the floor, and emitting dust in the general bag area. If the bag is placed on the pallet, dust emitted from the break covers other bags, and the break is a potential exposure source when the bag is removed from the pallet. The other bags coated with dust are also potential dust generation sources.

When loading or unloading bags from a pallet, the bags can be moved manually or lifted and placed in another location by machine control. A variety of lifting devices are now available that minimize the number of dropped bags and breakage. Some of the lifting devices will lift several bags in one movement with single-bag lifting devices also available. The lifting device is selected based upon process operations requirements. These devices in most instances are preferable to manual movement for efficiency and the prevention of bag breakage. Potential exposures are reduced where a bag is coated with dust or a small leak exists in these operations by locating the worker a distance away from the bag.

Since bag leakage may occur, the floor area under the bag transfer movement pattern should have an open floor grill as shown in Figure 14.1. Hoppers located beneath the grill system collect powder leakage and dust which is transferred to a receiver. In Figure 14.1, the pallet is shown in a manual loading situation where the bag is manually placed on a pallet. The pallet in turn is supported by a lift mechanism which lowers it after each layer of bags is placed on the pallet. When

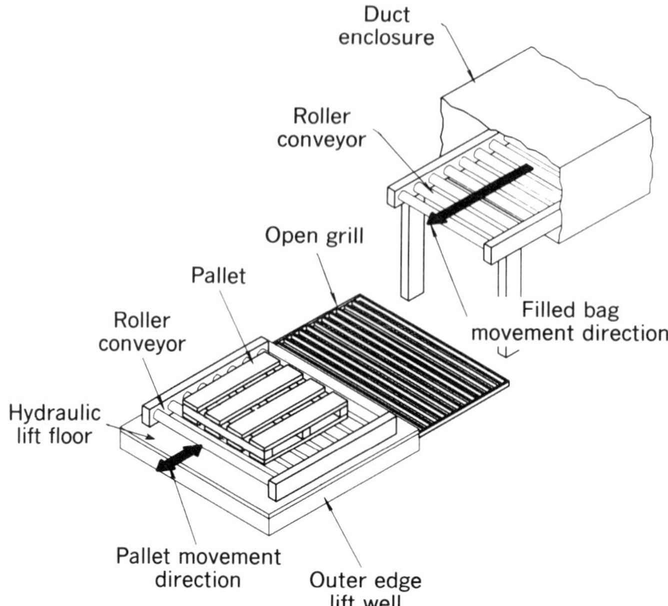

Figure 14.1 *Manual pallet loading station.*

the pallet has been filled, it is then raised to floor level and moved on rollers or removed by a forklift.

Bag Palletizing Dust Control. For the manual loading or transfer operation described above and in Figure 14.1, dust controls are of assistance in controlling dust concentrations and worker exposure. The grill upon which a worker stands has a hopper below it for receipt of dust and powders. For positive dust control, a downward flow of ventilation air through the grill work and into the hopper is applied. A face velocity of about 100–175 fpm (31–46 m/min) across the grill is the basis for determining the volumetric air flow rate. Obviously, the higher the face velocity, the more effective is the dust control program. In addition, a constant face velocity across the grid is more effective than widely varying velocities. The combination of air and dust is sent to a dust recovery system.

Since the pallet is a source of dust generation through leaking bag valves, bag openings, or dust on the bag surface, dust control for the pallet loading operation is often installed. Air slots are located in the inside wall of the lift well, just below the top edge of the well. A simplified view of the well and overall ventilation systems is presented in Figure 14.2. This view indicates that the pallet is positioned for the first layer of filled bags. As the lift platform level moves downward, bags will be adjacent to the slots, and the velocity across the bags will not be as high owing to the varying configurations of the bags. As a basis for design, an air velocity across the plane of the well face may be in the range of 75–125 fpm (23–38 m/min). Slot velocities vary from 900–3900 fpm in a system observed by NIOSH.[12]

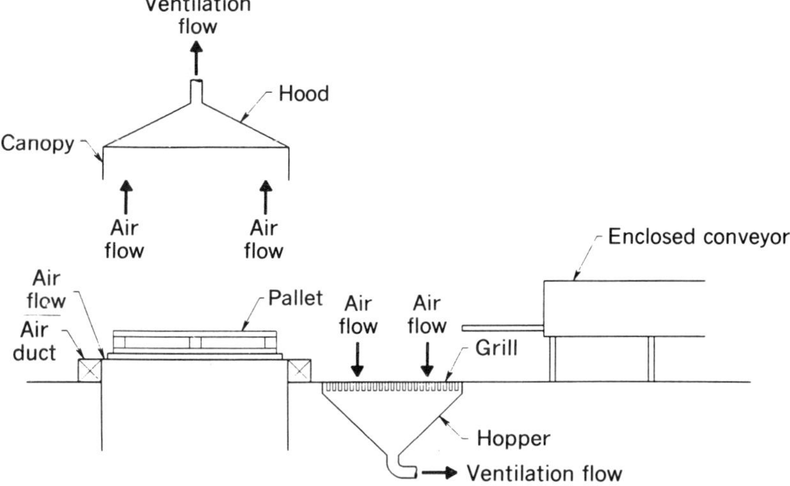

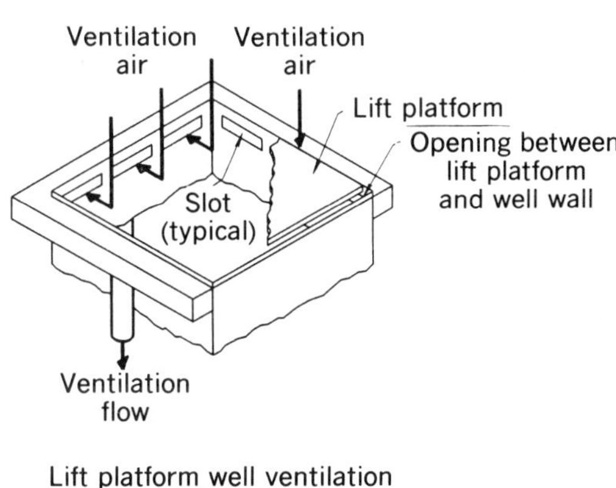

Figure 14.2 *Ventilation at pallet loading station.*

In addition to the low elevation controls, a hood is installed above the full pallet system. Various designs can be used for this function with a hood and canopy shown in Figure 14.2. Face velocities for the hood may vary from 75–125 fpm (23–38 m/min). While the well ventilation system performs acceptably, nonsealed valve bags and breakage may result in a dust concentration somewhat higher than is desirable. Hoods are helpful in controlling these somewhat higher dust concentrations. With sealed valve bags and sealed open-mouth bags, dust leakage is not as severe, but other factors such as bag concentration, filling or loading, package machine unloading, conveying systems, and bag pressing may result in dusty bags, requiring adequate ventilation controls. Existing equipment may have dust generation emissions that can be measured. Control systems for this situation may be a combination of process equipment modifications and new or modified control systems. In new installations, automatic machinery is installed which normally has ventilation systems included in the equipment.

Bag Transfer. Following filling, the bag in the manual operation is placed on a conveyor and moved to the pallet loading station shown in Figure 14.1. However, bag movement and transfer involves a few operations that must be considered for dust control. When filled, the bag is removed from the packing machine but may have a dust coating. The bag can then be placed in a bag cleaning housing prior to placement on the conveyor. Alternatively, an enclosed air blast cleaner on the conveyor line can be installed which minimizes manual operations. Air blasts on the surface of the bag remove the dust, and the cleaned bag is placed on the conveyor. This conveyor line may not extend to the pallet station but may discharge the bags on a second conveyor because of layout restrictions, elevation changes, or a change in conveyor type. In older installations these procedures may be manual; however, all of these operations can be automated with dust control systems installed in the equipment.

Along the conveyor line, the bag passes through a pressing machine termed a "flattener" to ensure consistency in the filled bag thickness. Palletizing flattened bags results in a well-packed pallet, reducing bag movements during loading and potential breakage prior to containing the bags within the plastic wrap. With nonsealed valve bags, some leakage occurs when the bag is jarred or receives a sudden impact such as a change in height at conveyor intersections, pallet loading, or flattening. Valve bags and open mouthed bags that have been sealed, normally do not leak powder or solids. However, these sealed bags are still subject to potential breakage when dropped, and consideration must be given to situations that may result in a bag break and leakage. Dust controls should be installed where there is a possibility that some breakage and leakage will occur. If bags are sealed and do not break in this transfer line operation, dust controls should be a minimum.

Although bags may be sealed, the need for dust control should be established in an existing plant operation. Through general exposure concentration, dust generation sources, and employee dust inhalation test programs in an existing plant, requirements for dust control equipment can be established. These requirements can also be transferred to new lines during the equipment design phase, assuming the new facility is also a manual operating facility.

A major consideration in manual operations is bag breakage, particularly where the bag is removed from the packing machine and thrown or dropped on the floor conveyor. In addition, any leakage from the bag valve adds some dust to the surface of the bag and other surrounding bags. Some dust may also be emitted from the packaging machine area although hoods around these machines have effectively controlled the dust concentration. For this general problem, hoppers are placed below the bag conveyor to collect spilled material, as shown in Figure 14.3. In addition, hoppers may also be placed under conveyors moving bags to the pallet loading area and below the bag flattening device. While much of the dust and powder will fall or spill into the hoppers, this powder movement generates

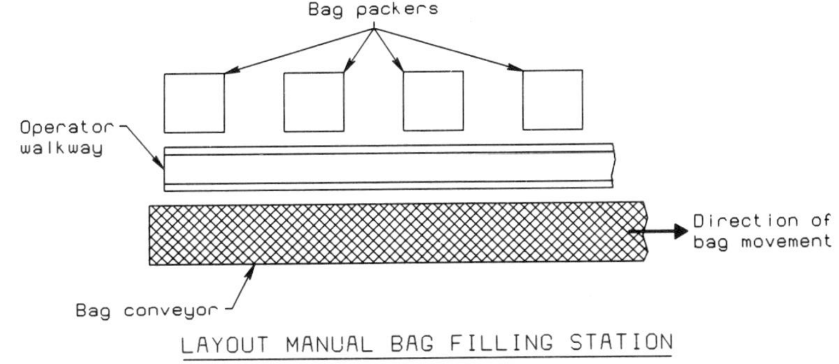

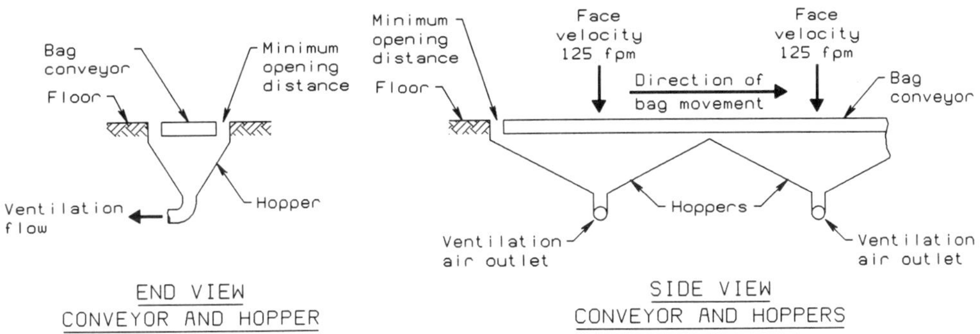

Figure 14.3 *Floor conveyor at bag filling station.*

dust, and positive methods of control are generally necessary. Ventilation air provides a positive dust control mechanism as shown in Figure 14.3.

While hoppers may be installed below the bag conveyor extending from the conveyor in the packing area to the pelletizing area, the need for a positive ventilation air flow must be established based on dust generation in this area and dust on the bag surface. Moreover, the type of conveyor will affect the control system selected. A wide belt conveyor will carry the dust on the belt, which will be discharged to the general atmosphere at the bag terminus and lift-off point. The need for hoppers with a belt conveyor is questionable, and a hood over the conveyor or an enclosed conveyor may be preferable. Should a powered roller conveyor be installed, a hopper can be placed beneath the conveyor. With a conveyor elevation above the floor of about 3 ft, the effectiveness of a ventilated floor air hopper is questionable. Where the conveyor is only a few inches above the floor, a ventilated hopper can be effective. Spring or vibrating conveyors have a tendency to generate dust due to the movement of the machine and bag. Ventilated floor air hoppers require a high flow rate to control dust emissions.

Selection of the dust control system requires a careful analysis of the dust normally generated in bag transfer and the probability of large leaks. If estimated dust generation is greater than desired or there is a lack of confidence in the dust generation estimates, an alternative is complete enclosure of the bag conveyors from the packing station area to the pelletizing station. An example of a completely enclosed conveyor system is shown in Figure 14.4. A positive air flow

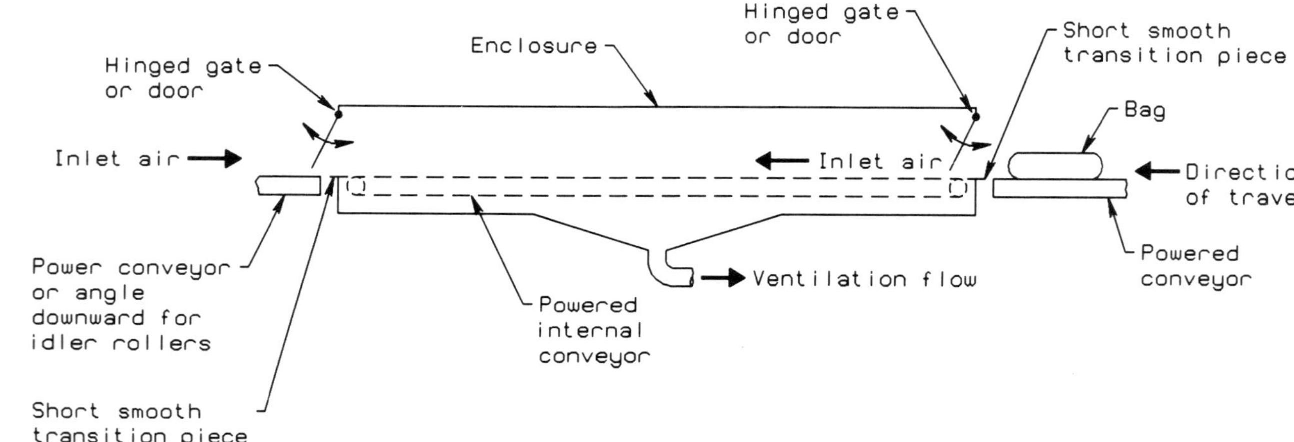

Figure 14.4 *Enclosed ventilated conveyor.*

through the system conveys dust to a dust collection system. Air flow through the doors at either end of the conveyor is based upon a velocity rate of 230–300 fpm (70–91.5 m/min).

Since conveyors may stop for a variety of reasons, dust will tend to fall toward the bottom of the enclosure, and a bottom air flow outlet is more effective than an air outlet on the top of the enclosure. Moreover, air sweeping downward and through the conveyor minimizes dust buildup in the lower portion of the enclosure. Maximum dust removal is desired to minimize potential exposures when manually removing bags from the conveyor housing or repairing the conveyor.

Where a bag is transferred from one conveyor to another, a change in elevation frequently occurs. In this situation, the fall or tumble of the bag may result in bag leakage or dust on the bag surface may evolve, generating dust and increasing the airborne dust concentration. Since the transfer occurs over a small distance, a high velocity hood over the transition section would control dust emissions. However, bags may block a transition point and a hood must be designed that permits access to the conveyor transfer section. Flexible hood curtains can be used or solid curtain walls with openings. The need for a floor ventilation system depends upon the transfer design and experience observed with the transition section. If a considerable amount of leakage occurs in the transition zone, then a redesign of this section should be considered. While a floor grid may be helpful, elevated conveyor dust generation cannot be completely controlled with floor ventilation, and the conveyor system design should be changed or modified to minimize dust generation, permitting installation of an effective overhead hood design for low generation rate situations.

14.3.2 Bag Unloading

In Section 14.3.1 bag transfer describes bag movements and controls between a pallet and a loading station. For incoming filled bags, the bag movement is reversed, with the bags removed from the pallet and placed on the conveyor train for conveyance to the bag unloading station where bags are removed from the conveyor and placed in an unloading machine. In this machine, the bags are cut with slitters mounted in the unloading machine, and the powder falls into a hopper from the bag under the influence of gravity and some shaking by the operator or a mechanical device. The powder is then transported from the hopper by a mechanical or pneumatic conveyor.

Bag Dump Station. A typical bag dump station is shown in Figure 14.5. The filled bag is placed on the grill after the exhaust fan system begins operating. A high air flow rate is maintained which draws any dust away from the bag operator standing in front of the grill. The face velocity at the hinged door opening is a minimum of about 250 fpm (76 m/min). When the bag is on the grill, it may be cut by the operator, moving it over a knife or knives mounted in the grill metal surface and permitting the operator to pierce the bag in several locations. The bag is then shaken to completely empty it, and the empty bag is then removed, as explained below.

The great mass of powder dumped through the grill falls into the hopper which is assisted by the relatively high air velocity. A large amount of dust is generated,

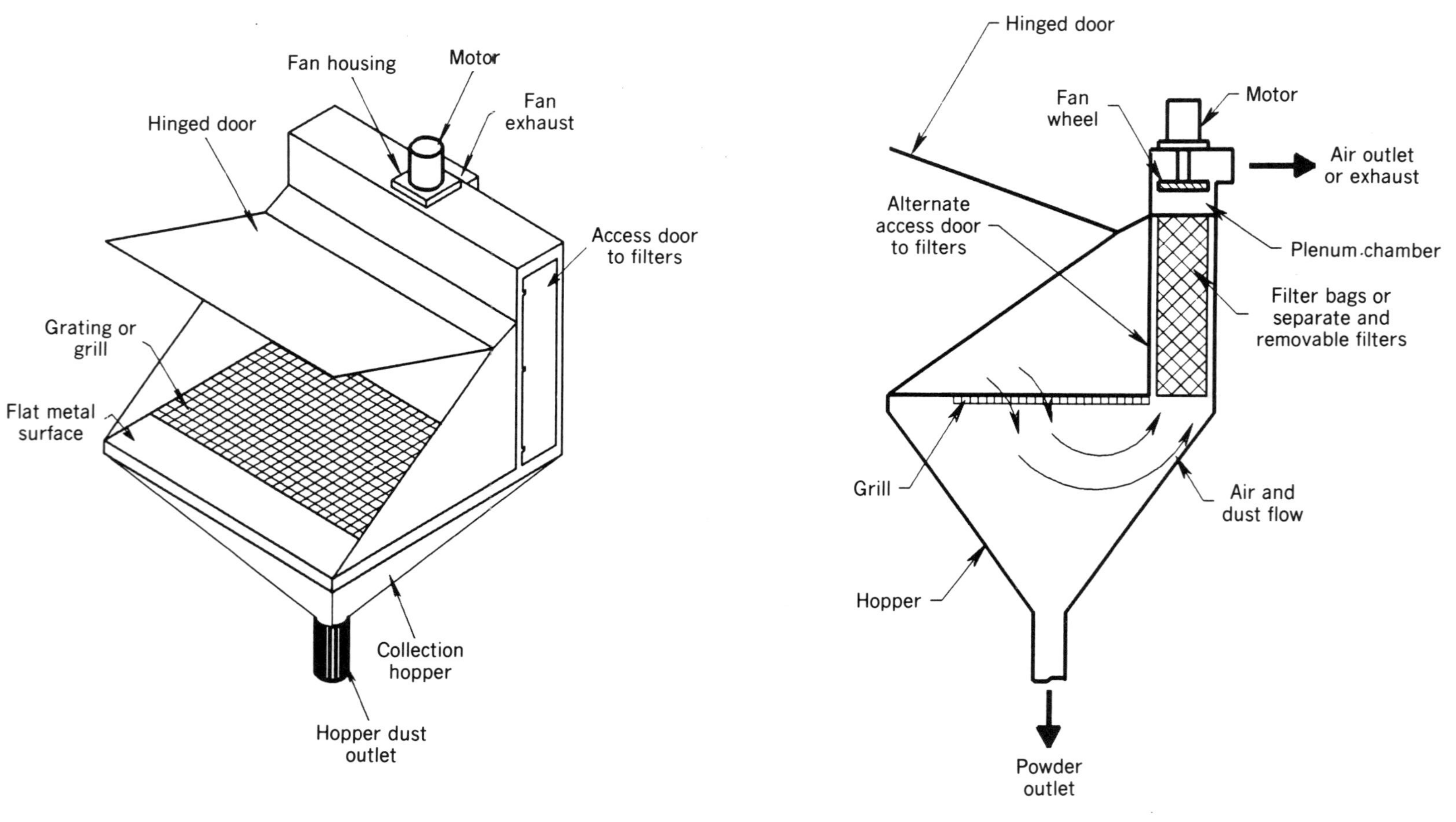

Figure 14.5 *Typical manual bag dump station.*

but the large air flow rate conveys dust emissions away from the operator and into the filtering area. The dust generation rate during this dumping operation is a function of particle size distribution and particle density. Some of the airborne dust is recovered on the filters, with the great mass of powder discharged directly into the hopper. The powder is then moved from the hopper with a mechanical or a pneumatic conveyor. Hopper outlets are designed for the type of conveying system specified.

Dust Filters. As shown in Figure 14.5, ventilation air is drawn through the bag dump system by a fan or blower. Dust-laden air is filtered prior to entering the plenum chamber, and the cleaned air stream is discharged to the atmosphere. There are a number of filters in the filtering section to permit continuous air flow as the filters are cleaned individually on a rotational basis during operation. Most filters are cleaned by a reverse flow of air through the filter which cleans the outer surface of the filter, discharging the dust into the hopper for recovery. The air supplied for filter cleaning is obtained from a separate source at a much higher pressure than can be achieved with a fan blower.

The filtering devices vary based upon individual manufacturers designs. A number of bag dump stations are supplied with cylindrical filter cloth bags that cover metal support cylinders. Some manufacturers supply self-supporting, rigid-pleated filter cylinders. Blowback nozzles are placed at the clean air outlet of the filter cylinders for cleaning the outer cloth surface of the filter bag that retains the solids. These nozzles are timed for blowback independently of other controls. Periodically, a filter will tear and the filter bag or self-supporting filter must be replaced. At this point, the bag dump station is shut down, and the torn bag is replaced through the side access door shown in Figure 14.5 or through an alternate access door.

Filter replacement is required when a bag tears to reduce dust losses and minimize exposures. A leaking filter results in dust discharged to the atmosphere from the blower exhaust port. If this air stream is directed into the workplace area, there is a potential for worker exposure. As a result, the designer should direct the fan discharge outside the building to dissipate any dust. In addition, a dust detector placed in the fan discharge will signal the presence of dust in the air stream and the bag dump station can be shut down to repair the leaking bag. Operating without an outlet dust detector presents a potential risk.

Removal of a torn bag or bags during maintenance presents potential worker exposure situations. The bags generally have dust within the fibers and a loose dust coating on the outer surface, which generate dust emissions, and maintenance workers must be protected from potential exposures during bag removal. In addition, plenum chambers may have some dust concentrations due to bag leakage or dust from torn bags that accumulates over a period of time. Exposure protection should be required when opening and working in the plenum chamber to service the air blowback system and/or clean the chamber. Where possible, portable vacuum cleaning systems should be used to remove dust traces and surface dust coatings prior to conducting maintenance work.

Empty Bag Disposal. When a bag is empty, the dusty bag requires removal and disposal. Bag compactors have been developed that are located adjacent to the dump station and are connected to the bag dump station through a short duct located a few inches above the grill. An empty bag is pushed manually by the operator through the duct into the compactor or the operation may be automated. The compactor is activated when needed to compact the bags within the compactor housing.

Since dust is also generated and present within the compactor, vacuum filter systems may be installed in the compactors. An alternate is connecting the vent in the compactor to the filtration system in the bag dump station. The details of these connections are important to determine where the fine dust particulates are recovered and their disposal.

In the bag compactor, the accumulated compact bag stack is enclosed in plastic and then sent for disposal in accordance with plant directives. In some compactors, plastic containment bags are opened, placed in position, and then filled with empty powder bags. The powder bags are then compacted and may be automatically sealed. In another design, the empty powder bags are stacked within a fixture and, following compacting, the stack is covered and sealed with plastic automatically. The sealed bag stack may be removed from the compactor automatically or mechanically ejected.

Bag Dump Stations on Tanks. Bag dump stations may be installed directly over a tank, permitting direct discharge of powder into the tank. The dump station is mounted on a cover that completely encloses an open tank top. While this may be a cost effective approach in some plants, some consideration of health and safety is necessary prior to installation of a bag dump station directly over a tank.

When the tank liquid is not water, a careful review of the overall system is necessary to ensure worker health and safety. The dump station should not be on a tank with potential hazardous air pollutants, VOC compounds, or liquids with a vapor pressure that may lead to flammable concentrations. Where a liquid has some vapor pressure, the bag dump station will have a mixture of air and vapor in the device. Moreover, there will always be an air–vapor mixture in the dump station which can be a potential safety problem with flammable vapors. If a hazardous air pollutant is present, there will be a potential exposure problem when the hinged door is opened. In addition, discharging mixed vapor from the blower exhaust presents disposal problems. The VOC or HAP compounds in the vapor may also require control. Butterfly or other valves with unrestricted openings placed between the dump machine hopper opening and tank cover can be closed when the dump machine is not in operation. Sealing the opening will reduce the vapor concentration in the machine prior to unloading which reduces potential exposures and safety concerns.

The comments just described apply when the filters are intact. Should a filter tear and permit a relatively larger dust flow through the filter, the pressure balance in the bag dump station changes and a lower pressure may result in additional tank vapor entering the plenum, potentially exacerbating the air vapor mixture concentration. A dust leak detector may reduce this potential problem.

Automatic Bag Dump Stations. In contrast to the manual bag dumping operations described, completely automatic bag dumping machines are available. In the automatic machine, the powder-filled bag is fed into the machine on a powered conveyor and then guided through slitting devices that open the bag. Automatic bag shaking is also part of the dumping operation, followed by automatic bag compacting, compressed bag encapsulation, and ejection of the stacked, compacted bags.

The powder is discharged through a bottom hopper and powder conveying is installed as required. In addition, a dust air filtration system is part of the overall machine installation. The comments outlined above on filter problems are also applicable to an automatic bag unloading machine system. To control dust emissions from the unloading machine, a leak detector should be installed with filter system maintenance concerns similar to those previously described. In addition, the automatic bag unloading machine is much larger than the manual dump station with increased complexity. This may require additional maintenance periodically, where dust coatings and dust buildup occur in locations not adequately swept by airflow. Personal protection will probably be required, and the area requiring maintenance should be vacuumed thoroughly prior to maintenance functions.

Bulk Bag Unloading. While the bags in this section can be readily moved manually, portable bulk bags are widely used in many applications. These bulk bags range in net weight size from a few hundred pounds to several thousand pounds. Although not in the manual handling category, these large bags must be manually connected to the receiving system and then disconnected for removal.

These large bags are advantageous where a plant may consume a considerable amount of solids. The large bags are considerably more efficient and also reduce potential exposure situations through enclosed system operations. These bags are also used in services where the bag functions as a hopper and solids are withdrawn at a reasonably low rate, permitting the bulk bag to remain in position for a number of days. This service is an advantage when the solid material is toxic or a potential safety hazard. Eliminating manual bag handling minimizes potential exposures and safety risks since the solid material remains encapsulated. Bulk bags may also contain toxic solids materials, such as catalysts, that are completely discharged into a receiving vessel through conveying systems. Closed systems minimize potential exposures in both handling and maintenance operations.

There are now a variety of elastomeric bags of various sizes and connections. In addition, solid portable bins are also used widely in many similar services. The large number of bag and bin types, along with a variety of handling equipment, is beyond the scope of this chapter. Considerable information is available from bulk bag and bin suppliers on equipment and bag details which is helpful for determining plant requirements and specifications to control potential exposures.

14.3.3 Bag Loading

As described in Section 14.3.1, bags are multiwall and generally consist of kraft paper plies that may have a special plastic inner liner. The current combination of plies or layers results in strong bags that infrequently break or crack compared with bags having simpler structural designs. The common bags in general use are valve and open-mouth types. A valve bag has a small opening at one corner of the bag in which a nozzle is placed for filling. An open-mouth bag has one end completely open for filling which is then closed after filling completion. The advantage of valve bags is the high productivity rates that can be achieved with the filling machinery and, in manual operations, somewhat lower labor requirements.

Valve Bag Loading. The valve bags have various types of openings which differ in sealing effectiveness. Most bags have an internal valve consisting of paper and plastic films to seal the bag after the filling nozzle is removed. While this is an effective seal, a small amount of powder does leak from the valve. Another valve bag has an extended sleeve that is pasted or sealed and tucked within the bag. This tuck-in sleeve bag does not leak after sealing, which is an important consideration in controlling workplace exposure. Open-mouth bags are either sewn or sealed, with the great majority of bags now sealed, which prevents leakage from the bag during transfer and pelletizing. Sealing is primarily an automatic operation which improves productivity.

Filling Valve Bags. Most of the valve bag filler machines combine filling with bag weighing which improves productivity, accuracy, and repeatability. This type of weigh system measures gross weight but will also compensate for bag tare weights. Valve filling machines have several methods of injecting powder into the valve bag. In general, the filling chambers can be fluidized or pressurized, or the powder can be positively fed into the bag with an auger device which is similar to a screw feed mechanism. A simplified drawing of a current fluidized valve bag packer is shown in Figure 14.6.

Basically, the bag valve slides over the nozzle and may be clamped manually or automatically with the bottom of the bag fitted into the weigh system cradle. When packer operation is initiated, the bed of powder in the chamber is fluidized, and the chamber is pressurized to a desired pressure. The valve to the bag is opened and powder flows into the bag. Air displaced by the powder flow may be held within the bag or may leak through an opening around the nozzle, and be sent to a ventilation dust control system. Air will also diffuse through multiwall bags within 2 or 3 sec while the bag is held on the nozzle before removal or automatic ejection. A bag with an inner plastic liner is perforated, permitting air diffusion outward to relieve pressure. These perforations reseal to control moisture. When the correct weight is reached and the bag is full, the powder flow valve to the nozzle is blocked and the annular opening around the main nozzle is closed. The fluidized chamber may be opened through the vent valve to the venting line, reducing chamber pressure. This is followed in some designs by an air purge or air blast to clear the nozzle. The clamp is released, and the bag is removed from the packer weigh saddle. The bag valve is sealed and then tucked within the filled bag.

This general system of purging and venting was developed to cope with leakage that occurs during the filling operation. One of the major problems was to control vapor/dust emissions during bag filling with the dust-laden gas exhausting to the atmosphere in the annular region between the nozzle and bag valve. A clamp around the bag valve solves this problem. In addition to the bag valve clamp, some newer packing machines also have a small housing that fits around the valve and provides a ventilation air flow receiver to contain any dust emissions. This housing requires a connection to a ventilation system duct. Following the bag-filling operation, some powder tends to remain in the nozzle. Dust may also adhere to the outer surface nozzle which tends to build and will fall from the nozzle upon

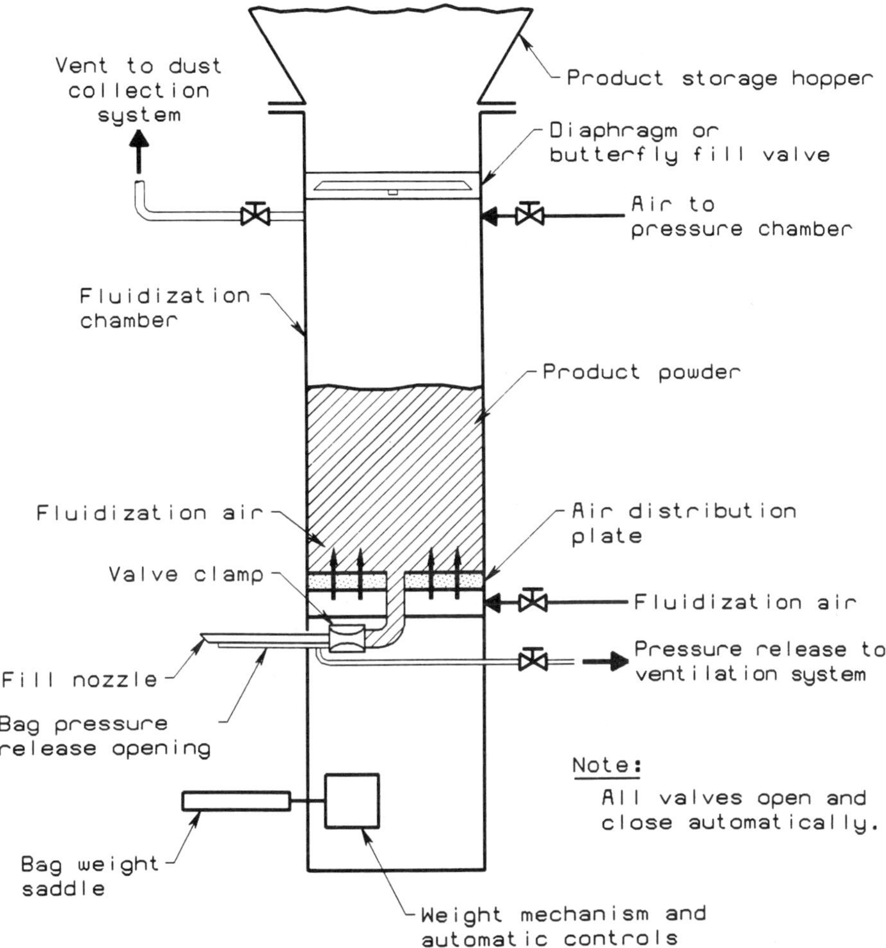

Figure 14.6 *Simplified drawing of fluidized valve bag packing machine.*

reaching a specific weight. In addition, valves from unfilled bags when placed on the nozzle can also displace the surface powder formations. General movements can affect dust remaining in the nozzle, and the air–dust mixture in the nozzle may release dust to the atmosphere. A combination of bag valve clamp and nozzle purging reduce these effects. Some powder may also evolve from the bag valve when the bag is moved and immediately prior to bag valve sealing.

While operations of the valve bag packer have improved significantly, some dust is still generated which may surround the bag packer machine and potentially overexpose the operator. In addition, some dust may cling to the filled bag. These small dust emissions accumulate over a period of time and tend to increase the atmospheric dust concentration requiring good ventilation and specific control techniques to prevent overexposure.

Bagging Emission Controls. The big packer generates dust during normal operation, and at times a bag may tear or rip if not handled properly, resulting in spilled powder which also generates dust. Consequently, a positive ventilation system is required to ensure a dust concentration is maintained that is considerably lower than regulatory requirements.

Partially enclosing the bag filling operation and providing ventilation has been effective in controlling the atmospheric dust concentration, particularly in existing manual operations. A simplified drawing of a manual bag packer with exhaust ventilation systems is shown in Figure 14.7. The bag-filling enclosure in the figure appears longer than enclosures in actual installations where the exact configuration is determined by the bag packer design, bag size, operator movements, and overall layout. However, the general elements for ventilating the enclosure are shown in the figure. A ventilation stream exhausts from the upper part of the enclosure, which is effective in controlling dust emissions. The volumetric flow rate is based upon a face velocity measured across the face of the overall enclosure opening. Well-controlled operations have velocities that vary from 200 to 1000 fpm (61–305 m/min) across the opening face. A more consistent flow pattern is desirable, and many enclosures have face velocities that vary from 550 to 850 fpm (168–259 m/min) with average velocities in the 620–750 fpm (189–229 m/min) range.[12]

In addition to the upward air flow movement, a grating in the floor below the bag filling station removes anything that falls to the grate or grill. Velocities through the grill are about 250–300 fpm (76–91 m/min). The packer machine is also located on a grill or grid that extends beyond the confines of the packer, as shown in Figure 14.7. Any powder displaced from the machine or parts, particularly during maintenance or repair, accumulates in the hopper where it is recovered. The powder can be transported from the hopper via a screw conveyor or with ventilation air. The selection of the powder transport method should be determined based upon equipment layout and economics.

Torn Bags. Torn or ripped bags occur, albeit infrequently, but the bag must be removed and discarded. A separate three-sided enclosure is one method of handling these bags.[12] However, the bag must be removed and then carried to another location for packing, which potentially exposes workers to dust from the bags. A bag dump station that was described in Section 14.3.2 is another alternative that eliminates bag handling once the torn bag has been placed in the dump machine. A compaction device will then encapsulate the bags as described and minimize worker exposure.

Dusty Bags. A dusty bag can be placed in a cleaning facility, which consists of air jets within an enclosure that remove the dust particulates from the bag surface. Since this blowing facility is located within an enclosure, dust distribution in the surrounding area is minimized. A cleaning facility can also be placed on a conveying line which has an advantage from a production standpoint. Manually moving bags to the cleaning facility is a manpower intensive operation. On a conveyor line the bag moves through an enclosure that has an open conveyor design to permit air washing on both sides of the bag. The enclosure is wider than the conveyor and essentially encapsulates the conveyor with openings at the conveyor inlet and outlet. Loose curtains may be used at the ends or they may be left open. Air is blown over the bag, and exhaust ducts above and below the conveyor move any dust into the duct system. The duct system must have a

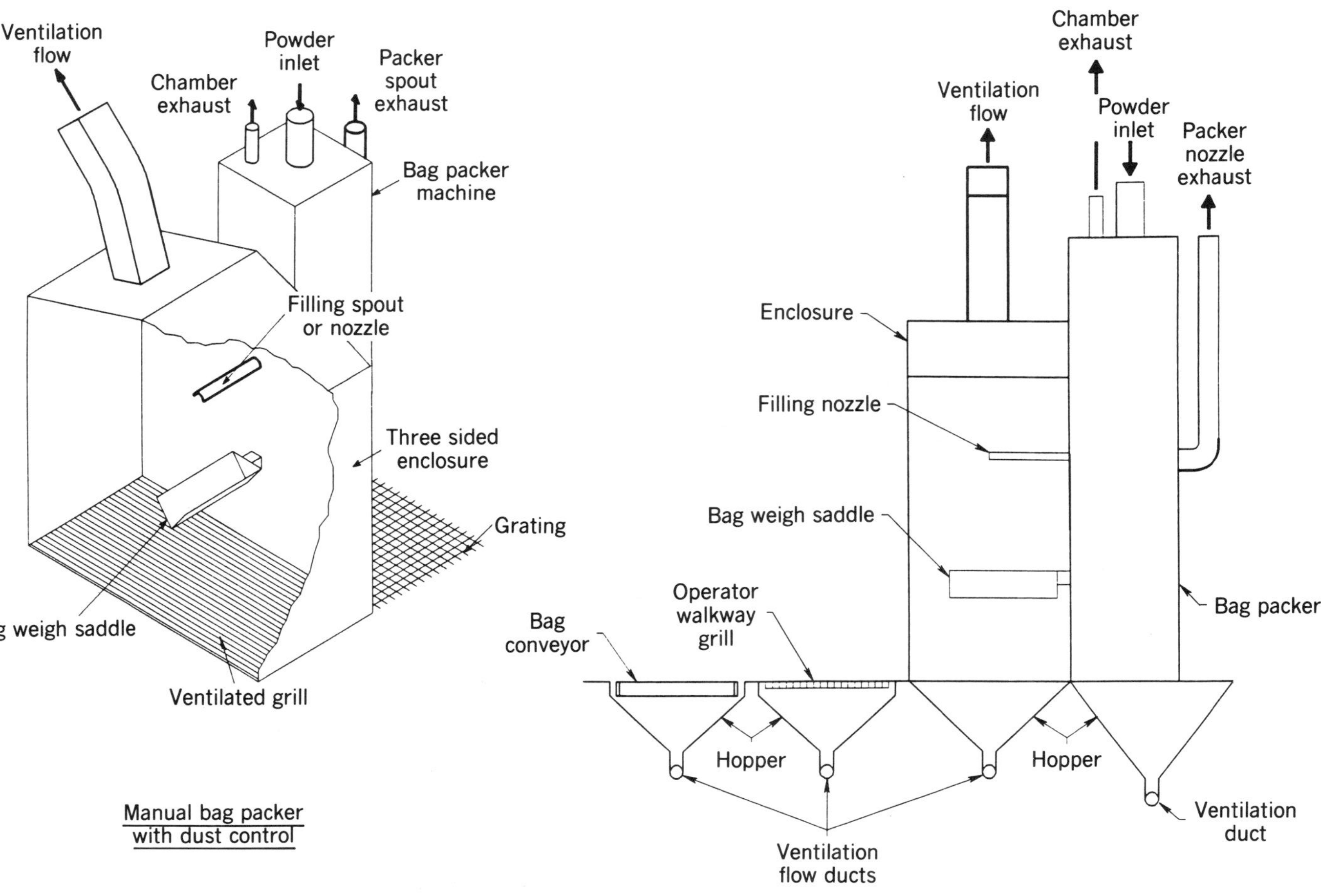

Figure 14.7 *Manual bag packer station ventilation controls.*

capacity that is large enough to handle the air jet volume and inlet air from the open enclosure ends based upon an open end face velocity of 150–250 fpm (46–76 m/min).

Walkways. As shown in Figure 14.7, a walkway separates the bag packer from the filled bag conveyor. This walkway is a grill or grate to permit dust control. A flow of ventilation air through the grill aids in controlling dust and powder that leaks from bags. A face velocity of 250–300 fpm (76–91 m/min) across the walkway controls dust and any powder leakage.

Hoppers and Air Flow. As shown in Figure 14.3, a number of ventilated hoppers are placed beneath conveyors and walkways. Separating the long distances into a large number of hoppers permits reasonably consistent (or equal) face velocities along the walkway and conveyors. However, this equal distribution of flow through the various hoppers requires flow adjustment dampers to equalize air flow when a large single blower is the prime mover. The alternate to a single blower is a single blower for each hopper or major duct system. In addition, the design of the dust collection equipment must be considered. A large number of blowers normally requires collection devices for each blower. A single blower only has one collection system although it would be a target unit. However, one large unit is normally considerably lower in installation operating costs than a large number of smaller units. Additional maintenance with a number of units also increases the potential for worker exposure.

The hoppers in the previous paragraph are related to the floor ventilation and dust collection system. In the valve bagging system previously described, the valve bag packers are normally filled from a storage hopper located above the packing machine. The storage hopper and packer are usually separated by a valve that is part of the packing machine control system. Following bag filling, the valve is opened, and the almost empty packing chamber is filled to a predetermined level followed by valve closure. The movement of powder fills the chamber with dust which is contained within the chamber. The powder in the chamber is then fluidized and may be discharged directly to the valve bag or further pressurized as described previously. The fluidizing air along with the pressurization air are released through a vent to the ventilation collection system prior to refilling the chamber. As an alternate, this air volume may be released into the storage hopper located above the packer. Should air be discharged into the storage hopper, a vent is necessary on the hopper which either connects with the ventilation duct system or a separate collection control device is installed on the storage hopper.

Automation. The comments in this section on bag loading have generally been directed toward manual operations that have been upgraded to include exposure controls. In today's production facilities, many manual functions have been replaced with automated devices that frequently contain exposure control systems. Some facilities now have completely automated systems covering all of the functions previously described with cleaned, filled, tucked-in valve or sealed valve bags exiting the filling device on a conveyor. This completely automated system also has exposure controls or internal ventilation systems that may require connection to an external blower and collection system.

There are a variety of automated functions and systems commercially available, with considerable variation in the combinations of automated equipment installed in different facilities. A review of these devices and systems is beyond the scope of this chapter. However, the principles outlined for the manual loading system provide a basis for reviewing and analyzing exposure control functions in automated devices and the overall bag loading system.

Air Balance. In general, bag loading operations are conducted within enclosed buildings. The exposure dust control systems described in this section are based upon the movement of large dust-laden volumes of air from the building atmosphere through dust collection systems, with the cleaned air frequently discharged or vented outside the building to the atmosphere. Should the building restrict the amount of air entering the process area, an imbalance of air flow will occur. Consequently, an air mass balance should be conducted when exposure control requirements have been established and requirements are then defined for building ventilation.

Open Bag Loading. Although still used, open bags for powder have decreased in application as valve-type bags have become more popular, particularly in facilities with high production rates. Open bags have widespread application where particle distribution is considerably larger than the particle distribution found in powders where particulates are in the respirable range of 10 μm or less.

Normally, open bag loading creates a considerable amount of dust due to the large unrestricted opening through which the bag is filled. While the amount of dust generated in this operation is apparently large, the particle distribution range of the dust may still be greater than 10 μm. Although this airborne dust can still be above the respirable range, the large volume of dust can be a potential health problem. Consequently, dust control in the filling operation is required.

For open bag filling, the bags in general are placed in a vertical position with the top end open and a fill nozzle inserted into the bag or the bag may be located just below the nozzle. Automated devices are available that automatically clamp the bag against the fill spout which extends into the bag. Since the bag is wider than the fill spout, the open bag ends beyond the spout are also clamped together automatically. In addition to the large fill spout, a vent is also included in the large fill spout head projection which is connected to a vacuum dust collector system. When filling is completed, the bag is automatically sealed, resulting in a system that controls dust emissions.

In those situations that do not have sealed automatic loaders, dust control systems are necessary, but the design of the control system is based upon the type of loading system and emission dust sources. In general, a capture hood can be placed above the bag opening plane to ensure the hood does not interfere with the placement of the bag under the nozzle or the operation of clamps, which are generally installed on most filling machines. However, clamping does not ensure that the entire bag opening is closed. The hood provides a sufficient air flow to capture the dust emissions from the open bag ends which are recovered in a dust collection system. The hood duct can be split with an opening on either side of the filling nozzle, which improves the air flow pattern into the overall hood system. Actual hood design details are available in several references.[14,28,29]

Not all types of open bag filling systems are identical to the fill systems just described. Consequently, other control systems may be required such as partial enclosures or combinations of flow ventilation control units. In addition, floor grids with hopper ventilation flow aids in dust control for certain types of dust. Bags also require sealing in addition to flattening, conveying, and pelletizing. Control systems previously identified can be installed as needed.

Since many operations are still manual, a variety of auxiliary automated and fully automated filling systems are available. With fully automated systems, the entire bag filling operation is enclosed, and internal dust controls minimize dust concentrations. These fully automated systems include bag placing, opening, filling, sealing, and bag placement on a conveyor, which extends beyond the machine.

Auxiliary automated systems are similar to the auxiliary systems provided in modifying valve bag manual operations. Consequently, the fill operation must be carefully analyzed to determine the type and extent of exposure controls necessary to minimize dust concentrations. This procedure is similar to the recommendations for valve bags.

14.3.4 Drum Unloading

Bulk solids or powders supplied in drums are consumed in plants for a wide variety of applications. Typical materials received in drum containers include such items as additives, catalysts, adsorbents, inert powders and fillers, and reactants. Although there are a large number of drums supplied to plants, most of the drums are in the manual handling category. Where a relatively large amount of a specific powder is consumed, the powders and solids are supplied in bulk containers or bins. Consequently, in those facilities where unloading is a necessity, exposure controls are required to minimize dust leakage. Pouring or dumping solids from a drum into a container is accompanied by dust generation if not controlled.

Basic Control. Since drums must be opened and tilted to discharge the solids contents, the simplest control system that can be installed is shown in Figure 14.8. A diaphragm type valve is installed on the bottom of the angled cone to prevent leakage until the drum at the valve has been connected with the receiver. When the connection is satisfactory, the diaphragm valve is opened, and the drum contents are discharged into the receiver. Depending upon the size of the receiving bin and dust generation, the displacement air volume may require discharge to a dust control system.

When removing the filled drum lid to attach the unloading cone, the bulk solids may be in a tied plastic bag or there may not be a plastic enclosure. When removing the drum lid, it should be slid across the top of the drum rather than lifted. This technique minimizes dust generation for nonbagged solids and powders. Where a tied plastic bag is present in the drum, the bag tie is removed and the bag is extended over the drum rim. The cone base will fit over the plastic film on the cone edge.

Various mechanical lift devices are available commercially to raise and tilt the drum at the correct elevation and location. When the drum is empty, it is disconnected from the receiving bin following closure of the valve. The bin requires a closure device since dumping powder will generate dust that can be

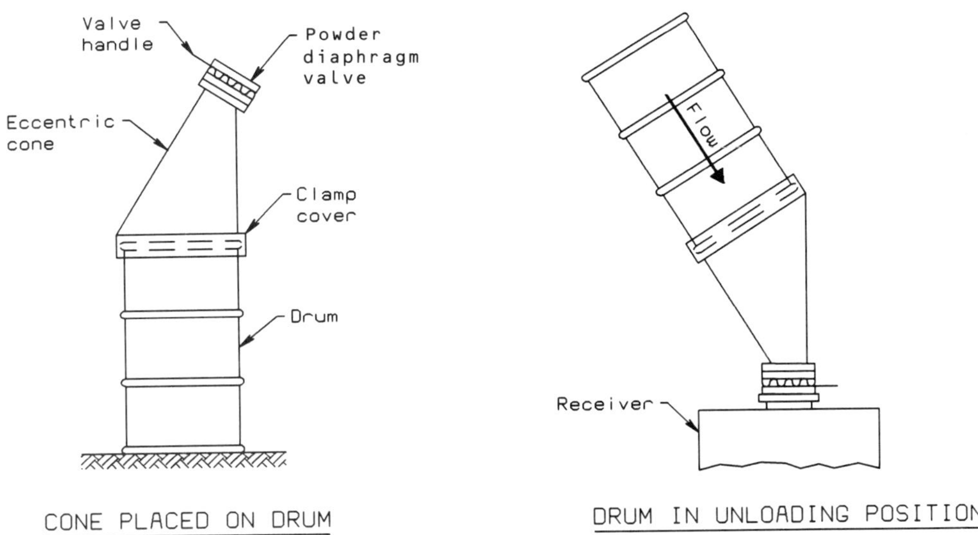

Figure 14.8 *Eccentric cone unloader on drum.*

emitted to the atmosphere. When the empty drum has reached an upright position, the unloading cone is removed and the drum top is replaced. Dust clouds in the empty drum will exist and are a potential concern until a lid covers the drum. Personal protection may be required for reclosing the empty drum.

Handling a Number of Drums. When a number of drums must be emptied on a frequent basis, this system is modified. Several drums can be placed in a machine and when the lids are removed, a hopper with openings that fit over each drum is placed on the drums. The openings are tightened, and the combination of drums and hopper are lifted and tilted with the drums emptying into the hopper. The hopper has a powder diaphragm valve at the bottom of the hopper or may have another opening system. The tilted hopper is placed in the desired location by the lifting machine and discharged as required. Tie-ins are a function of the system design, along with venting. Machine systems of this type are available commercially.

Directly Feeding a Process Line. One typical device is available that permits emptying a drum through an enclosure into a spiral conveyor. The conveyor then moves the bulk solids or powder to a receiving bin located above a process vessel or into a small hopper that discharges directly into a weigh feeder, sending a measured flow rate into a process vessel. A simplified drawing of the drum and conveyor operation is shown in Figure 14.9.

The drum in Figure 14.9 discharges solids through a feed hopper into a Flexifeeder® conveyor. This conveyor is mechanical, with a spiral type rotor for solids conveyance. The spiral conveyor is flexible and can be designed to convey solids in various configurations and over relatively long distances. This flexibility permits control of emissions from the drum unloading point to the final consump-

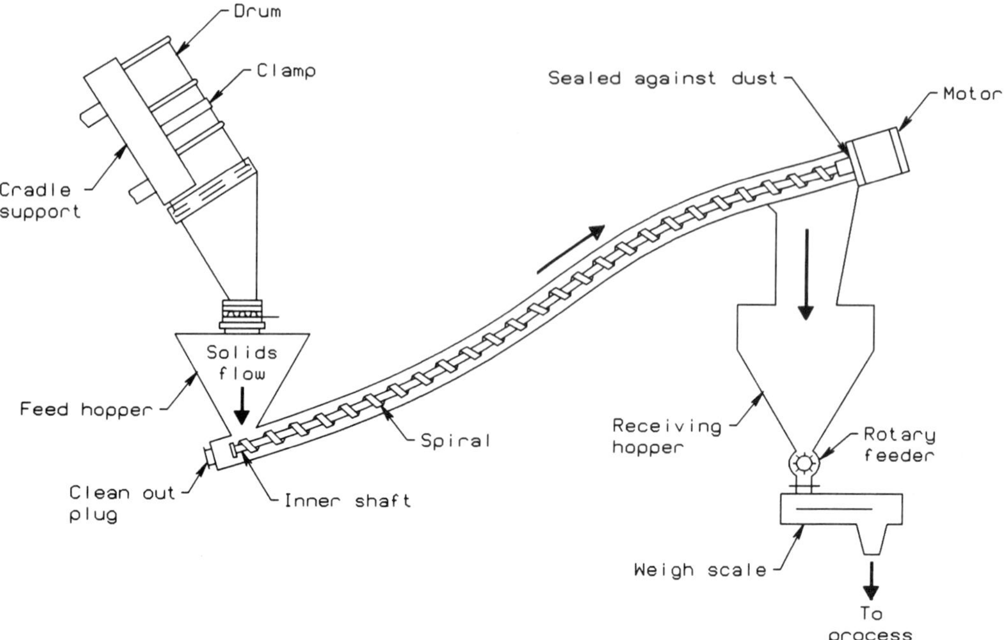

Figure 14.9 Typical Flexifeeder® drum unloading system. (Courtesy of American Industrial Machines)

tion position. While one drum is shown, several drums may be unloaded into the feed hopper and, when combined with a large conveyor, increases the solids flow rate.

As shown in Figure 14.9, a drum with an angled cone and a powder diaphragm valve is placed on the hopper connection. The solids are discharged into the hopper, which may have a volume considerably larger than the drum volume or may be somewhat smaller in volume. In the latter situation, which may be a low volumetric feed condition, the drum provides some storage volume. Depending on the feed hopper volume and conveyor flow rate, a control vent may not be required on the feed hopper. The spiral conveyor may be under a slight vacuum during operation, which eliminates the need for a hopper vent. However, the receiver hopper or bin may require some venting depending upon displacement air, hopper volume (holdup), and hopper discharge system.

An advantage of this conveyor system is the possibility of operating under non-air conditions. In those situations where air should not contact solids, such as certain catalysts, a nitrogen or inert gas injected into the conveying system would maintain the proper atmosphere at low consumption rates and under a slight positive pressure. The dust-tight connections normally required for a system of this type would be adequate. Pneumatic conveyors have high inert gas consumption rates unless the gas is filtered and recycled, which increases the cost and complexity of the system.

In addition to the system shown in Figure 14.9, other conveyors are available with some of the characteristics described but having different mechanical designs.

Another consideration in conveyor selection is maintenance of the conveyor, which requires discharge of solids material, particularly when a mechanical fault occurs in the system. The conveyor in Figure 14.9 can be opened at the bottom end permitting drainage of the solids. A vacuum applied to the bottom opening removes additional solids and reversing the motor dislodges the remaining solids. For a normal planned shutdown, the conveyor is kept running until the system is reasonably well cleaned before it is opened and vacuumed. Maintenance is an important consideration in controlling exposures and is a major concern in solids handling equipment.

A pneumatic feed system can also be used in this application. A similar system described in Section 14.5 for feeding various locations can be installed as a drum unloading–feed system.

14.4 COLLECTION SYSTEMS

Since particulate control consists of capturing dusts or fumes, the major particulate control technique is to capture and direct dust flow with a large-volume air or vapor stream. This was demonstrated in Section 14.3 where a variety of control systems were described that generally resulted in air–dust streams sent to recovery or collection systems. The dust is then removed from the gas in the collection system and the clean air is discharged to the atmosphere or recycled into the operating building. Solids or particulate separation devices are not 100% efficient, and the trace particulate amount remaining in the cleaned air stream after purification must meet atmospheric regulatory requirements or building concentrations. The dust concentration must be sufficiently low in recycle air to prevent potential worker exposure situations and minimize particulate buildup in the recycled air stream.

Collector systems are extremely important, and the selection of the collector system must be carefully investigated to ensure an optimum system is installed that balances investment, operating cost, and efficiency. Since a variety of collectors are available, their advantages and limitations of the collectors must be understood to achieve these objectives. In this section, the most widely used collectors will be reviewed, and some alternatives will be presented. However, the objective is to provide an understanding of these systems, their capabilities, and their potential for workplace exposure. For additional information, extensive analyses on efficiencies are available in numerous references.

14.4.1 Regulations

In reviewing the effectiveness of collector systems, applicable regulations must be considered when specifying new or modifying existing collection systems. Where the gas exhausts directly to the atmosphere, EPA, state, and regional regulations may be applicable to the dust or particulate concentrations in the vent. The NAAQS requirements for particulates have a primary and secondary standard for

24 hr of 150 $\mu g/m^3$, and a primary and secondary annual average standard of $\leq 50\ \mu g/m^3$. These standards cover particulates with an aerodynamic dynamic diameter $\leq 10\ \mu m$. The NAAQS standards are applicable in populated areas, requiring an analysis of particulate dispersion from the vent stack.

When populated areas do not achieve the NAAQS standards, they are designated nonattainment areas. Within the nonattainment designation are classifications termed *moderate* and *serious* for particulates $\leq 10\ \mu g$ (PM-10) under the CAAA of 1990. The states may then require specific control technologies for existing units and in permits for new units. Consequently, the attainment status of an area is quite important, along with any specific requirements mandated by the states in their implementation of the CAAA regulations. All regulatory requirements must be considered in developing specifications for new and existing collection facilities.

The CAAA also has listed a number of hazardous air pollutants (HAP), as shown in Table 4.1, potentially containing a number of categories that cover various dusts and fumes. As an example, the mineral fibers classification covers fibers with an average diameter $\leq 1\ \mu m$. While many of the classifications are mineral based, the HAP list will be expanded through the years, and a significant number of hazardous or toxic particulates will require extensive control. In addition to the various environmental criteria covering particulates, the toxicity of the solids particulates must also be considered in defining collection efficiency. Moreover, careful operation of collection equipment is necessary to prevent leakage to the atmosphere or within a building. Operation also requires careful maintenance practice to minimize potential worker exposures.

One other area that must be investigated prior to the specification of collection equipment is the allowable concentration levels in the workplace. The OSHA regulations define regulatory concentrations, which should be supplemented with ACGIH exposure concentrations. Where the ACGIH exposure concentration levels are more restrictive than the OSHA levels, ACGIH recommendations should be considered for the exposure concentration standards.

14.4.2 Dust Collection Efficiency

Normally, an overall collection efficiency has been used based upon the mass collection rate divided by the mass inlet rate. This approach is a standard efficiency calculation for all types of collection equipment and was satisfactory for many years. However, the demands for greater particulate removal results in collection efficiencies that approach 100%. At this level, a comparison of efficiencies above 99%, such as 99.25% versus 99.53% does not indicate a significant difference in particulate effluent concentrations.

Various references are beginning to use the term *penetration*, which is more illustrative of performance, especially particulate control. Penetration is 100 minus the collection efficiency, resulting in penetrations of 0.75% versus 0.47% for the example given above. This characterization terminology more accurately describes a situation where leakage through one collector is almost twice the leakage rate of the second collector. At high mass inlet rates, the difference in the effluent stream rate is appreciable, which is very important in considering control of particulates $\leq 10\ \mu m$.

In reviewing particulate control effectiveness, a combination of collection efficiency and penetration is used.

14.4.3 Gravity Settlers

The simplest separating system is a gravity settler where particulates separate from the gas stream under the influence of gravity. This requires a relatively large-volume chamber to reduce the linear stream velocity and increase residence time within the chamber. Separation is a function of particle size and skeletal density of the solids.[13] In general, these devices are used as an initial separator to remove relatively large particulates ($> 150\ \mu m$), although actual separation is a function of the particle density. The efficiency of this device generally rises with increasing particle sizes for a constant density particle. For particulates $< 50\ \mu m$, the gravity settler is a relatively ineffective device.

Gravity settlers have been modified to provide changes in the gas velocity direction, allowing the particulates to impinge on surfaces and gravity settle. This improves the collection efficiency and various impingement separators are available commercially.[13] In addition, screens across the main stream flow aid gravity settlers in reducing carryover.

Some impingement separators are designed for centrifugal effects which assist in carrying particulates to a surface where they collect and gravity settle. There are various modifications of these devices which culminate in a cyclone that is discussed in the following section. While modified gravity settlers and impingement separators have improved collection efficiencies, for particulates $< 50\ \mu m$ their collection efficiencies are less than the devices reviewed in the following sections. However, they can be effective in reducing the solid particulates loadings on the main collection device. Detailed characterization of solids size distribution, size fractions, density, and particulate structure or aerodynamic diameter is necessary to adequately predict performance of gravity settlers and impingement separators.

14.4.4 Cyclones

Cyclones are widely used for duct collection in various industries. In these devices, a dust-laden gas enters a cylindrical chamber tangentially, as shown in Figure 14.10. Through inertia and centrifugal force, particles move to the outer wall of the cylinder. The gas tangential velocity increases as the gas moves down the narrowing cone, with solids also moving down the outer wall from the cylinder. Additional particulates are separated from the gas in the cone section. At the bottom of the cone, as shown in Figure 14.10, the gas forms a vortex which travels upward within the down-flowing outer gas stream. This vortex, which is the cleaned exit gas, discharges through the outlet vent to a duct. The solids discharge from the bottom opening of the cone.

The collection characteristics of cyclones vary considerably in numerous references. Differences in collection efficiency affect the particulate size ranges collected and the effectiveness of controlling small diameter particulates. One

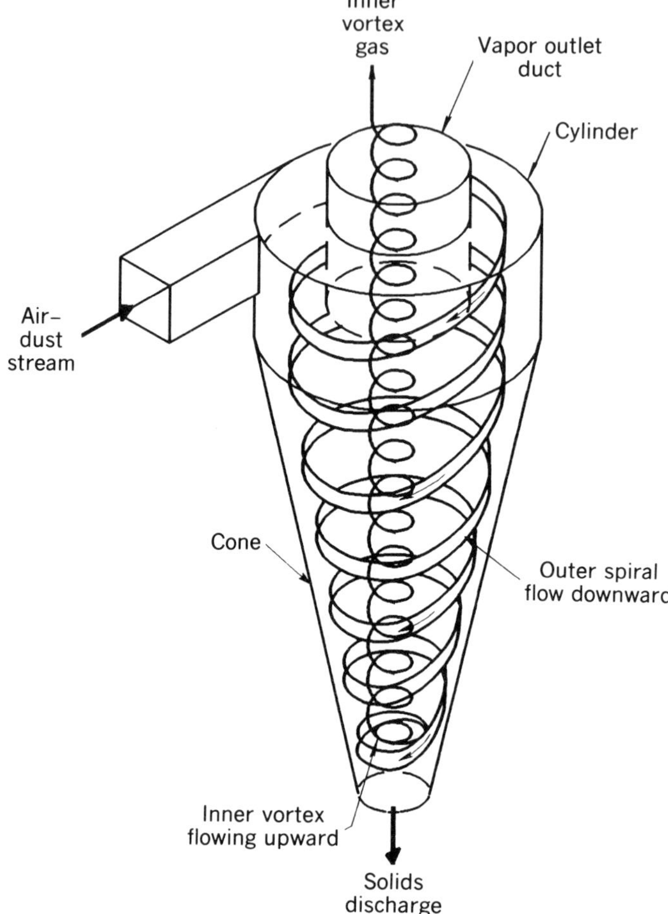

Figure 14.10 *Gas flow pattern in a cyclone.*

reference[13] indicates that collection efficiency is low for particulates $< 5\ \mu m$ in size and very low for particulates $< 2\ \mu m$. Another reference indicates that collection efficiencies are not unreasonable for particles to 1 μm in size.[14] However, acceptable efficiencies are based upon the results obtained with high-efficiency cyclones and multicyclones. These disparities are reviewed in this section.

Cyclone. Cyclone collection efficiency is related to the gas velocity entering the cyclone, which in turn affects the gas tangential velocities in the cylinder and in the cone. Higher velocities increase collection efficiency but also result in a higher cyclone pressure drop. In addition, some concern exists regarding erosion of the metal cyclone wall from particulates and/or corrosion of the wall. As a result, there are limitations placed on the cyclone pressure drop, which is normally in a range of 0.5 to 6.0 in. of water.[14] The higher end of the range covers pressure drop in high-efficiency cyclones.

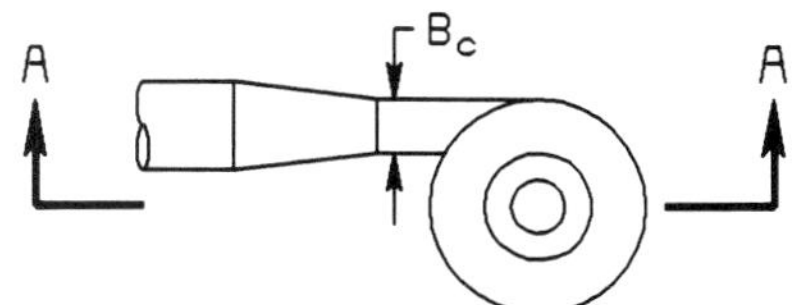

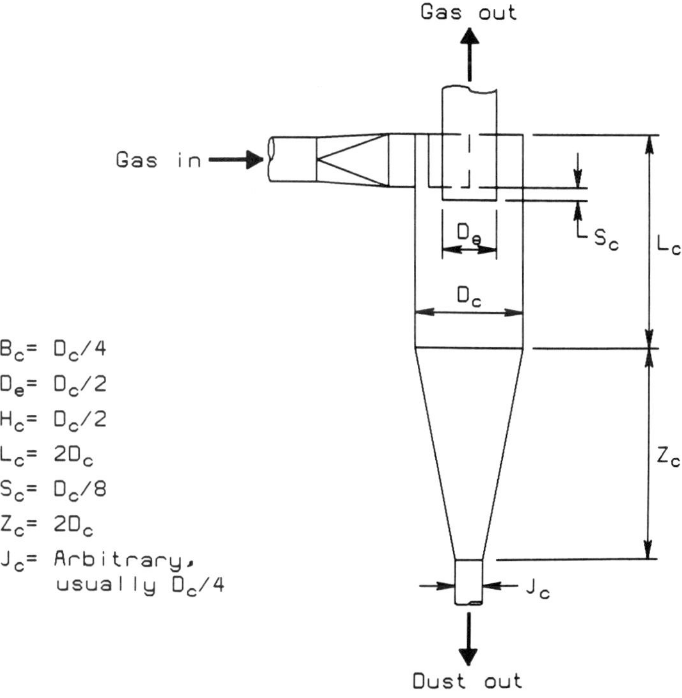

Figure 14.11 *Cyclone proportions (courtesy of McGraw-Hill).*

Based upon a cyclone of the configuration shown in Figure 14.11 and various assumptions concerning the cyclone inlet gas velocity, turbulence, gas flow pattern, spiral velocity, and particle separation characteristics, a relationship for a critical particle size was developed.[13] The *cut size* diameter d_c is the diameter of a particle that can be cut or removed from the gas stream in the cyclone with a collection efficiency of 50% and is determined by the following equation:

$$d_c = \left[\frac{9\mu B_c}{2\pi N V_c(\rho_s - \rho)} \right]^{0.5} \tag{14.7}$$

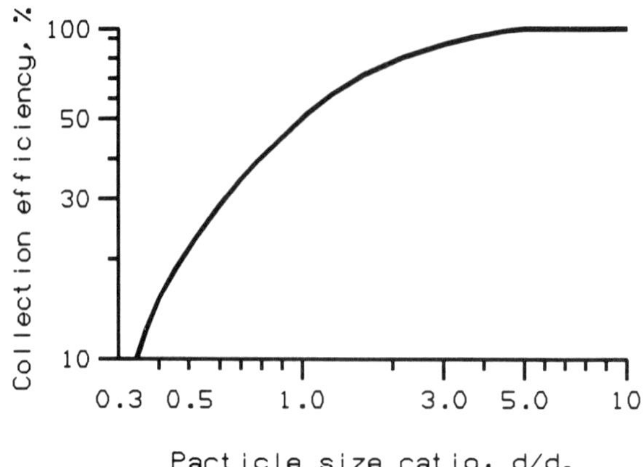

Figure 14.12 *Typical collection efficiency and particle size relationship.*[16, 19]

where

N = number gas turns in the cyclone
V_c = inlet velocity (ft/s)
ρ_s = solids density (lb/ft^3)
ρ = gas density (lb/ft^3)
μ = gas viscosity (lb/ft-s)
$B_c = D_c/4$ (Fig. 14.11)

This equation includes the number of gas turns, calculated from the following empirical relationship, which is also based on a number of assumptions:

$$N = \frac{1}{H_c}\left(L_c + \frac{Z_c}{2}\right) \tag{14.8}$$

Here H_c, L_c, and Z_c are determined from the cyclone configuration in Figure 14.11. The collection efficiency is then related to particle size through the ratio d/d_c, where d is the particle size and d_c is the cut size. A typical relationship for the collection efficiency is shown in Figure 14.12[12] for a specific or relatively standard cyclone. While these relationships provide a basis for estimating cyclone collection efficiency and penetration, more accurate information can be obtained from cyclone manufacturers. Based upon actual experimentation with a specific cyclone configuration, the collection efficiency can then be estimated more accurately. However, the tests may be conducted on a standard powder or dust configuration, requiring adjustment for actual process conditions. A correction on particle size for defined collection efficiency curve can be determined through the following adjustment factor, which results in a new or modified efficiency curve:[15]

$$K_{adj} = \sqrt{\frac{\rho_B \cdot \mu_N \cdot V_B}{\rho_N \cdot \mu_B \cdot V_N}} \tag{14.9}$$

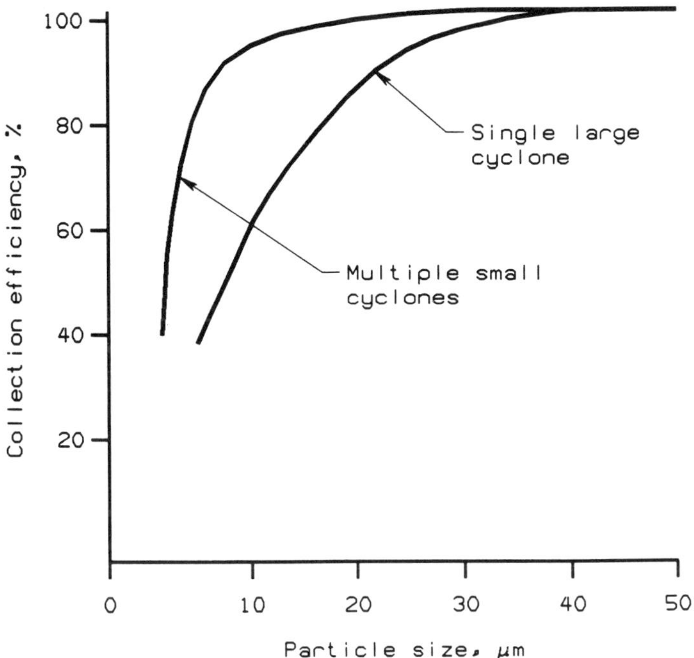

Figure 14.13 *Comparison of general collection efficiency for large and small cyclones with identical inlet gas velocities.*[17]

where

ρ_B = solid density standard material
V_B = inlet gas volume rate standard material conditions
μ_B = gas base viscosity standard material
ρ_N = solid density proposed material
V_N = inlet gas volume rate at proposed conditions
μ_N = gas viscosity under proposed conditions

Cyclone Size Effects and Collection Efficiency. The literature is replete with references to the much higher efficiencies achieved with smaller cyclones.[13–15] A typical relationship for this variation in collection efficiency is shown in Figure 14.13.[17] The large cyclone is a low-efficiency model which can represent a cyclone having a configuration that differs considerably from the model in Figure 14.11, with a shorter cone and/or a lower pressure drop.

Small cyclones have a somewhat greater pressure drop than large, low-efficiency cyclones described in the previous section. The pressure drop across the cyclone is directly related to a constant that varies with the size of the cyclone. For a constant gas volume, a multiple number of small cyclones in parallel is required. These are termed multicyclones and are normally placed in a group that is positioned above a common receiver.

TABLE 14.3 Comparison of Collection Efficiency of High-Efficiency Cyclones with Multicyclone Systems[a]

	Collection Efficiency (%)	
Particle Size (μm)	Multicyclone	High-Efficiency Cyclone
1.0	36	20
2.0	57	41
3.0	69	57
5.0	83	72
8.0	90	84
10.0	92	88

[a]From reference 18.

While the comparison of a low-efficiency large cyclone with a small cyclone is not a true comparison, some data are available on a comparison of a high-efficiency large cyclone with a multicyclone system.[18] The comparison of the two is shown in Table 14.3, indicating that a considerable difference in collection efficiencies exists for small particle sizes (< 2 μm). The efficiency differences decrease as the particle size increases. While the difference is large, a much greater difference occurs when a low-efficiency cyclone is compared with the high-efficiency cyclone. For a 1-μm particle, the low-efficiency cyclone has a collection efficiency of 0.06% and a collection efficiency for a 5-μm particle of 1.7%. The differences between collection efficiencies of the two large cyclones are extremely wide, indicating that low efficiency, low pressure drop large cyclones are relatively ineffective for the collection of small particles.

A further efficiency comparison of a single high-throughput cyclone with multicyclones was published by the EPA that is similar to the results in Table 14.3.[20] As shown in Table 14.4, the multicyclone efficiency is higher than the single cyclone efficiency for the three particle size ranges covered. The 85% cyclone efficiency is generally used by the EPA for particulate sizes $\geq$ 8–10 μm when calculating the effectiveness of cyclones and the penetration or particulate emission factor for a controlled system.

Since collection efficiencies change with particle size, an important criteria for determining penetration and actual losses at the collector outlet is the particle size distribution. Although particulate emission from the cyclone on a total mass basis is important, particle size distribution and dispersion along with the mass rate provides a basis for determining potential exposures. In addition, the size distribution is very important when specifying a collector for capturing the solids in the cyclone exhaust gas stream. Solids particle distribution analyses for specifying a

TABLE 14.4 Typical Cyclone Collection Efficiencies for Particle-Size Ranges[a]

	Collection Efficiency (%)	
Particle Size (μm)	Multicyclone	High-Throughput Cyclone
0–2.5	80	50
2.5–6	95	75
6–10	95	85

[a]From reference 20.

cyclone or multicyclone can be determined from existing processes or can be estimated from other sources.[21]

One other aspect of single cyclone operation that was mentioned above is the low pressure drop across a cyclone or multicyclone. As described, the largest cyclone pressure drop mentioned in various references is 6 in. of water (1.49 kPa). However, higher velocities, although resulting in a much higher pressure loss, will improve efficiency in a single cyclone. While higher pressure losses require increasing power in fans or blowers, the simple configuration of the cyclone minimizes investment and maintenance. Consequently, an overall review of investment and operating costs for the vapor dust stream and collection device should be conducted where higher cyclone efficiencies appear attractive. Higher cyclone efficiencies also reduce particulate loadings to a secondary collector installed to raise the overall collection efficiency well above 99.0%. In certain installations, two or more large cyclones may be installed in series, which increases overall efficiency but may also increase the pressure drop. This type of installation must be carefully analyzed to ensure that sufficient pressure is available.

Pressure Drop. Cyclone pressure drop is measured from the inlet of the cyclone to the vortex discharge point or the clean gas outlet, as shown in Figure 14.14. This pressure change has been investigated widely and methods of calculating the pressure drop vary. The major reason for the differences in calculation methods is the shape of the cyclone and other small changes in configuration, such as an outlet duct that extends below the dimensional outline shown in Figure 14.11. Also, an inlet that extends beyond the inlet cylindrical circumference or extended cone will affect the pressure drop. In this situation, an extended cone increases pressure drop and improves collection efficiency. While a number of cyclones have the general configuration of the model in Figure 14.11, cyclone manufacturers distribute a number of cyclones that differ considerably from the model. As a result, the pressure calculation procedures vary considerably.

While the pressure drop can be calculated in a general way and determined more precisely by the cyclone manufacturers, there is little information on the pressure at the cone bottom where the collected and separated dust is determined by the base operating pressure level of the cyclone. Should a cyclone operate with an inlet pressure of 14 in. of H_2O, and a pressure drop of 6 in. of H_2O, the cyclone outlet will nominally have a slightly positive pressure. This general consideration of pressure becomes important when considering disposal of the recovered solids stream. A completely enclosed cone discharge is necessary at pressures below the atmospheric pressure level. If the cone bottom is not properly enclosed in this situation, atmospheric leakage affects the vortex and overall cyclone efficiency.

As described above, a general equation pressure drop has not been established. However, for a good approximation of cyclone ΔP, an equation developed by Caplan[22] is widely used:

$$\Delta P = \frac{P\rho K Q^2}{T} \tag{14.10}$$

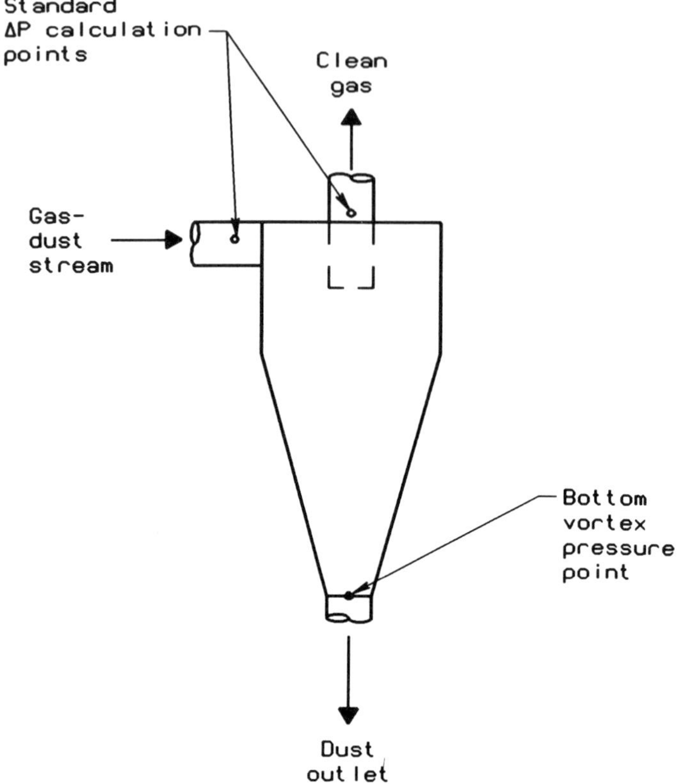

Figure 14.14 *Cyclone pressure points.*

where

ΔP = cyclone pressure loss (in. H_2O)
Q = gas flow (acfm)
P = absolute pressure (atm)
ρ = gas density (lb/ft^3)
T = absolute temperature (°R)
K = cyclone size factor (Table 14.5)

The K values as shown in Table 14.5 are based upon cyclones with a maximum diameter of 29 in. (737 mm). While larger diameter K values are not presented,

TABLE 14.5 *K* Values for Cyclone Pressure Drop in Equation 14.9[a]

Cyclone Diameter (in.)	K
29	10^{-4}
16	10^{-3}
8	10^{-2}
4	10^{-1}

[a]From reference 22.

the values shown can be extrapolated to determine large diameter cyclone K values. An important consideration in this equation is that the inlet gas volume rate will increase as cyclone size increases. The inlet velocity should remain relatively constant or increase. For cyclones obtained from a manufacturer, the inlet velocity, ΔP, and efficiency can be more accurately determined based upon the manufacturers correlations and estimates from pilot operation tests.

For a satisfactory cyclone collector efficiency, an inlet velocity of 62-98 ft/ sec or 19-30 m/s has been recommended in the literature.[15] Higher velocities were not recommended since there was a belief that higher velocities would result in considerably more turbulence and a potential for reentrainment of particles leading to a lower collector efficiency. Certain cyclones in the petroleum industry are operated at higher velocities with improved collection efficiencies, indicating that a higher inlet velocity is feasible. Other sources also report higher efficiencies at velocities that reached 170 ft/s (52 m/s) for talc dust.[13] The higher velocities increase the pressure drop and the overall system must be analyzed to determine whether the dust–vapor system can operate satisfactorily with a high-pressure drop level. In addition, the absolute pressure level at the cone bottom will decrease with increasing flow rate, as described above, which must be satisfactorily controlled.

Where the pressure drop becomes excessive, an alternative is the installation of multicyclones. This alternate also requires careful investigation to ensure that the pressure drop, recovery system, and collection system are satisfactory.

Dust Discharge. An important consideration in the installation of cyclones is the discharge of dust from the cone through the bottom opening. Since the opening is generally under a small vacuum, the bottom discharge must be enclosed to maintain cyclone efficiency. Otherwise, air or process vapors leaking into the dust discharge point interfere with the vortex and the collection efficiency. Generally the cyclone bottom outlet is connected to a hopper which is airtight for batch operations. For continuous process operations and continuous dust discharge, rotary feeder valves, multiple block valves, or in some cases a screw conveyor for fine dusts were reportedly installed.[13] Where cyclones are elevated, diplegs may be installed that have mechanical flappers on the bottom of the leg or the leg may be partially fluidized.[13] In this discharge system, the flapper is normally closed and the dust builds in the dipleg until the weight of the dust is sufficient to open the flapper. The fluidized solids in the dipleg act as a fluid and open the flapper when the level provides sufficient head to open the flapper.[13] An elastomeric flap may also be placed on the outlet leg, which can be installed within hoppers. This contrasts with long diplegs, which are generally associated with large fluid bed operations. Where diplegs may be fluidized, the gas flow into the dipleg to maintain a fluidized solid bed is small and controlled to ensure noninterference with cyclone operations.

A typical installation for continuous operation may have a rotary feeder, as shown in Figure 14.15, which seals the bottom of the cyclone by acting as an airlock and transfers the collected dust to a receiver. In some rotary feeders a small purge gas flow is injected around the feeder shaft seals to prevent dust entering the shaft seal, which gradually destroys the seal. The small purge gas stream must be controlled to prevent the transport of excessive gas quantities into the cyclone.

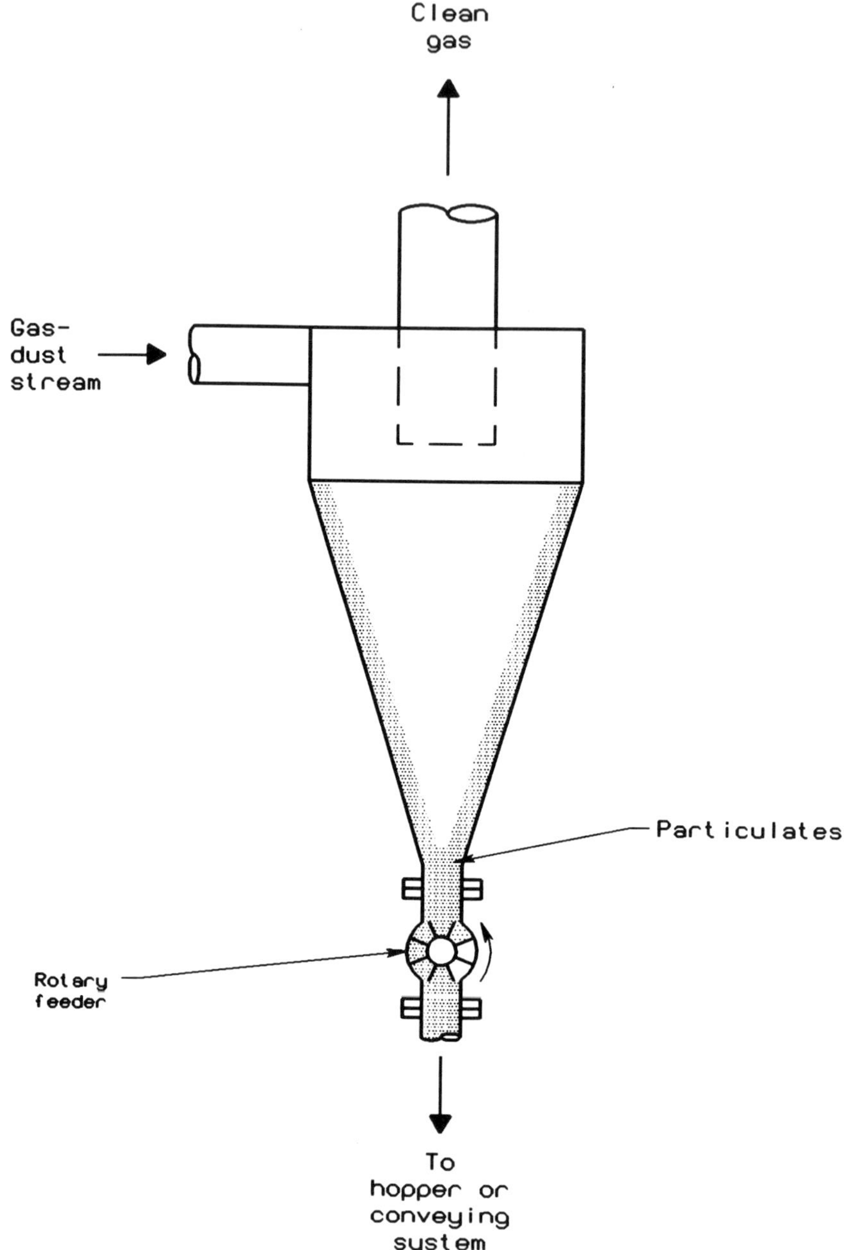

Figure 14.15 *Rotary feeder on cyclone dust discharge.*

Maintenance. Cyclones are normally free of powder, but some dust may cling to the inner metal surfaces of the cyclone. For most cyclones operating at essentially ambient temperatures and pressures, the cyclone does not require much maintenance and is infrequently opened. At high temperatures and pressures in the 10–40 psig (69–276 kPa) range, cyclones may require maintenance owing to erosion or corrosion. In addition, cyclones that operate at high temperatures may have special linings to prevent erosion. During operations, these linings gradually fail over a period of time, requiring replacement. Since the linings are hard material and fracture, dust may be contained within the various openings and is a source of potential exposure.

Where dusts may bridge over (form a plug) at the bottom of the cyclone cone and interfere with cyclone operations, vibrators are frequently placed against the cone to dislodge the blockage. Should the vibrator be unsuccessful, the bottom of the cone is opened, and the bridging material is manually removed, which is a potential source of exposure. In some instances, the flat top plate of the cyclone is removed to aid in cleaning the cyclone. Duct work, bottom feeders, and dust hoppers also require maintenance, which further increases potential dust exposure sources. Mechanical repairs anywhere in this system require personal protective equipment to minimize potential dust exposures.

Advantages. Cyclones have a number of advantages compared with other collector devices, although the efficiency of collection is lower than the efficiency of other devices. These advantages are:

1. Low installation and maintenance costs.
2. Mechanical reliability.
3. Can operate at very high inlet particulate loading levels.
4. High loadings and reliability permits staging with other cyclones or collection devices to achieve high overall collection efficiencies.
5. Will operate at high static pressure and temperatures.
6. Configuration minimizes potential worker exposure levels.

Disadvantages. Cyclone advantages must be balanced with their associated disadvantages in selecting an optimum high efficiency collection system. The disadvantages are:

1. Collection efficiency for particulates $< 5\ \mu m$ is low.
2. One-stage cyclones are inefficient for particles $< 5\ \mu m$ and in the submicron range require a high collection efficiency second stage.
3. Exposures to particulates in the PM-10 category can be a potential concern.
4. Requires a relatively constant inlet velocity to maintain collection efficiency.
5. Low inlet velocities < 40 ft/s (12 m/s) result in very low efficiencies for $< 5\ \mu m$ particulates. Very high inlet velocities may result in dust reentrainment and low collection efficiencies. The actual incipient reentrainment inlet velocity, which is a function of gas viscosity and particulate characteristics, can be established by the cyclone manufacturers. Maximum allowable inlet velocity may vary from 110 to 140 ft/sec (36–43 m/s).

14.4.5 Filters

Dust filters are one of the most widely used particulate control devices in industry. The very high collection efficiency, which is generally well above 99% for particulates $< 10\ \mu m$ in size, provides a reliable method of controlling dust emissions to the atmosphere and workplace exposures. The filters remove dust from a gas or vapor stream with a fabric placed in the stream. The fabrics are generally woven or felted and constructed from a variety of materials including wool, cotton, polymers, and glass fibers. In contrast to the woven fabrics, felt materials are matted with both general types having somewhat different characteristics. The fiber filters are frequently found in bag form, and the filtering units have been termed "bag houses" or "bag filters." Although fabric filters have been the predominant filtering material, other materials are now available in polymeric coated metal sheet and polymer sheet form.

Filtration Mechanism. There are two basic filtration mechanisms by which particles are deposited on filters and removed from the gas stream. In one case particles larger than the openings in the filter fabric are deposited on the surface of the filter fabric. These particles build rapidly, creating small openings or channels through which the gas flows. The small openings reduce the effective mesh size and remove most of the fine particulates from the gas stream. The dust layer acts as a sieve at this point. Over a period of time the thickness of the dust layer increases and the accompanying pressure drop also increases, requiring cleaning or dust removal on a regular basis.

Another mechanism for collection is the entrapment of fine particulates in the fibers of the filter fabric which has also been termed "depth filtration."[15] Fibrous mats or thick woven material are effective in retaining particulates, whereas a single thickness woven cloth does not entrap the fine particulates. Particles in the many random, tortuous paths have sufficient momentum that trajectory tracks result in fiber contact where molecular forces attract the particles.

Major variables affecting collection efficiency are the gas velocity and particle size. As velocities increase, the collection efficiency also increases, but the efficiency begins to fall when another effect termed "bounce-off" occurs.[15] When particles are $> 5\ \mu m$, particle inertia results in bouncing off the fabric fibers above a critical velocity region. Removal of particles $> 5\ \mu m$ with a preseparation device such as a cyclone and a reduction in the superficial gas velocity will result in improved collection of the fine particulates. Apparently, Brownian movement dominates in the low-velocity region. Increasing the velocity with the fine particulate fraction results in overwhelming the Brownian effects, and the particulates are carried with the gas stream through the fibers at the higher velocities, reducing collection efficiency. The superficial gas velocity is a very important variable that must be considered where high collection efficiencies of fine particulates are desired.

Filter Materials. Filter fabric is available in a wide variety of materials, as described above. However, various considerations affect material selection, particularly temperature limitations. Maximum continuous operating temperatures vary from about 149°F (65°C) for polyethylene to about 550°F (288°C) for glass fiber.[14, 15] Actual operating temperature limitations may vary based upon manufacturers' experience and plant performance.

In addition to temperature considerations, other characteristics of concern cover abrasion resistance, tensile strength, which is related to durability, and chemical resistance to acids or bases. While many applications involve air as the vapor medium, non-air gases may have a nonneutral pH. Moreover, solids or particulates may have an acid or base on the surface that combines with moisture in the gas, increasing chemical activity. In applications where fumes are controlled, pH effects associated with the solids and gas media should be reviewed. While many particulates do not pose a problem, the potential acid/base effect on the filter should be understood. Although the filter material is a concern, potential workplace exposures can be exacerbated where solids materials also have acidic or alkali characteristics.

Since woven and felt fabrics are used in solids collection, there is a difference in construction, which is reflected in the open or free area for gas flow. The free area in turn determines gas or filtration velocity, which serves as a basis for determining the total fiber area required in an installation. Woven material free area is lower than the free area available for matted fiber. Filter fabric manufacturers generally supply information on filtration velocity based upon the solids in the gas stream for the woven and felt fiber fabric filters. In general, woven fabric filtration velocities are lower than felted fabric filtration velocities, with information on velocity limitations available in the literature for various bulk particulates.[15] One reference suggests woven filter velocities of 1–2 ft/min (0.3–0.6 m/min) and > 5 ft/min (> 1.52 m/min) for felted fabrics.[23]

While many filters have woven and felt fabrics, a variety of other materials are now available. Ceramic filters have been developed for high-temperature applications to remove particulates from hot gas streams and apparently have a 10% share of the filter materials market.[24] A woven ceramic fiber filter has also been developed for high-temperature operation. One woven ceramic fiber of polycrystalline metal oxide has reportedly been installed successfully in a fluid catalytic cracker regenerator flue gas stream operating at a temperature of 1300°F (387°C).[25] This woven ceramic fiber will apparently withstand temperatures to 2500°F. Operating temperatures at these levels requires equipment and materials in the bag house capable of satisfactory mechanical performance.

Another class of filter materials consist of a more or less rigid rectangular envelope support coated with a thin PTFE membrane or other polymeric materials. Composites built in roughly the same configuration are coated with extremely thin layers of polymers. Although both appear similar, the composite structure of porous material can also have a corrugated facing that increases the overall surface area and effectiveness. Various pleated cartridge filters are also available that may be constructed of nonwoven polymers or other materials.

Filter Dust Collectors. Dust collectors frequently consist of a number of bags as described previously. These bags or socks may collect dust on either the inside or outer surface of the bag, depending upon the type of filter or dust collector. Usually the bags are in a vertical position within the bag house. While bags have long been the major filtering mechanism, pleated filter cylinders are also used and may be installed vertically or horizontally. In addition, there are envelope-type filters which cover a rigid support structure with an envelope-shaped bag or an integral filter membrane. The envelope filters are installed vertically.

When dust has collected on a bag or filter, there are three principal methods of cleaning the bags:

Shaking
Reverse air
Pulsed or pulsed jet

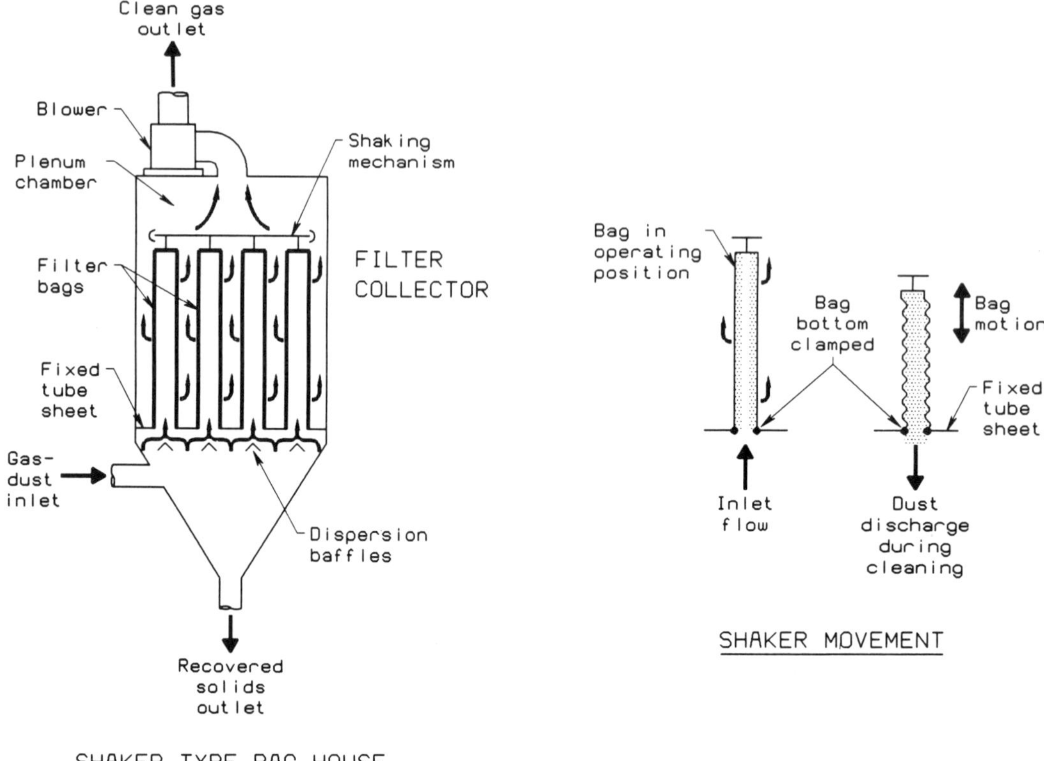

Figure 14.16 *Shaker-type baghouse collector.*

Shaker Dust Collector Filter. In the shaking bag collector, the bags are installed vertically, and the dust is collected on the inside of the bag as shown in Figure 14.16. The dust-laden inlet stream is dispersed evenly across the bag inlets in the drawing by symbolic baffles since dispersion devices vary widely among bag house manufacturers. Clean air is exhausted to the atmosphere through the plenum chamber and the blower. The blower in this figure is a vacuum blower, but where a higher static pressure exists in the inlet gas–dust stream, a vacuum blower is not required. The overall pressure drop normally does not exceed 6 in. (15.2 cm) of water for the collector.

When the dust thickness on the bag results in a predetermined allowable ΔP across the bags or the bag operation is timed, the filtering chamber is blocked and the bags are shaken to remove the solids layer. The bag ΔP, shaking system, and inlet block system in most installations are automated. However, the system can be partially automated or overall operation may be completely manual. Since the collector must be blocked to prevent the entrance of dust-laden gas during the shaking operation, parallel collectors or chambers are necessary for continuous operations. This multiple-chamber operation is a standard operating procedure and the number of chambers or multiple collectors installed are determined in conjunction with bag house manufacturers.

The shaking motion shown in Figure 14.16 is a vertical reciprocating motion, but other shaking motions are provided by bag house manufacturers based on their own study data. In general, the shaking motion is effective in dislodging most of the caked particulates. The dislodged dust and solids particles fall into the hopper, where the material is recovered. However, the bag is not completely clean as a residual amount of dust remains on the bag, which is reviewed in this section. The bags are shaken in intervals that may vary from 1/4 hr to many hours. Cleaning is a function of pressure drop buildup which depends upon solids loading in the gas and particulate characteristics. Obviously, frequent shakings will affect the bags' tear strength, with longer periods between shakings desirable to reduce maintenance and potential worker exposures. Since the bags undergo frequent stretching, twisting, and flexing, shaker bags are normally constructed of woven fabrics, which are generally natural or synthetic fibers. Glass or mineral fibers do not withstand frequent shaking and are generally not installed in shaker collectors.

Since bag shaking gradually degrades bag resistance, high cloth area collectors are desirable because they do not need to be shaken as frequently. This results in medium range superficial air velocities through the cloth bags or filtration velocities of 2–5 ft/min (0.61–1.79 m/min) for one reference[14] and 1–8 ft/min (0.3–2.4 m/min) for another.[13] The associated pressure drop range is about 2–6 in of water (0.5–1.5 kPa).

The bag is clamped to the bottom-fixed tube sheet and is suspended from and directly attached to the shaking mechanism at the top of the bag. When a bag tears or ruptures, it must be replaced, requiring manual removal of the defective bag. A worker must enter the bag house to remove the bag which will undoubtedly have large quantities of dust in the region above the fixed sheet, on adjacent bags, and in the plenum area. Potential dust exposures are a concern during bag replacement.

Reverse-Flow Filters. The reverse-flow or reverse-air filter is similar to the shaker dust collector. In both dust collectors the inlet dust-laden gas flows into the bag where the dust is collected on the internal bag surface as the gas diffuses through the bag into the clean gas zone and plenum.

This bag is cleaned by a reverse flow of the clean air, which requires blocking the inlet flow of dust-laden gas. Multiple chambers are necessary to maintain flow of the dust-laden gas stream or parallel collectors can be installed to maintain a continuous inlet stream flow. The bag top is closed, with the bottom of the bag clamped to the fixed tube sheet in a design similar to the shaker collector. The closed bag top is connected to a hanger that maintains the bag in a fixed position. Since the reverse clean air flow enters the plenum chamber to diffuse through the bag into the inner bag cavity, an upper fixed tube sheet is not desirable, resulting in the hanger on each bag.

The reverse air flow dislodges the cake and dust into the lower hopper for recovery. However, in cleaning the bag the air stream carries dust along with the air flow, particularly finer particles. This dusty stream combines with the inlet gas stream and is distributed among the operating chambers or filters to clean the overall gas stream. Obviously, the reverse air-cleaning stream places an additional load on the operating cloth filters. However, in the design of the filtering system, this additional load is included, which increases the base filter size a relatively small amount.

The source of the reverse air-cleaning stream is a separate air source or it can be withdrawn from the cleaned gas stream, discharging to the atmosphere from the operating chambers or collectors. The reverse stream may require a blower if the reverse ΔP is insufficient. This reverse stream tends to partially collapse the bags as the dust is dislodged and falls into the hopper. To prevent bag collapse, rings are sewn into the bags at various bag heights.

Reverse air flow collectors are apparently used most frequently in temperature applications > 300°F (> 149°C). Since the operating temperatures are high for synthetic materials, woven fiberglass fabric bags are used in high-temperature operations to 550°F (288°C). While satisfactory, the bags are considered fragile and lower superficial velocities are recommended compared with woven or felt filter bags. The internal rings help prevent bag collapse and restabilizing of the bags is assisted by the low reverse air flow rate and low inlet dust air superficial velocity. In most cases, the superficial velocity through the bag is 1.5 ± 2.0 ft/min (0.4–0.6 m/min). The ΔP across the collector for the dust air stream is also about 2–6 in. of water (0.5–1.5 kPa).

Maintenance for the reverse air filter is similar to that described for the shaker collector. Torn bags require replacement, and potential worker exposures increase as the age of the collector increases. Low cleaning frequencies are desirable and low superficial filtration velocities aid in extending bag life.

Pulsed or Pulsed-Jet Filters. These collectors, which are also termed reverse pulse or reversed jet filters, have bags cleaned with an air pulse or a jet pulse. As shown in Figure 14.17, in this type of collector the solids and dust collect on the outer bag surface, which differs from the collectors previously described that captured dust and solids on the inside surfaces. Since the dust-laden air flows from the outside to the inside bag volume in the pulsed-jet filter, the entire pressure drop falls across the bag. This would nominally result in a bag collapse, but the collapse is avoided by installing a cylindrical wire cage inside the bag. Since the air inside the bag is clean, dust builds very slowly within the bag, and the cage has a dust coating buildup over a very long period of time. However, dust accumulates rapidly within the cage and plenum when a tear or rip occurs in the bag. The cage, pulse jet, and other details are shown in Figure 14.17*B*.

The cage and bag filter in this bag house are suspended from the fixed tube sheet. Bag cleaning is conducted with a reverse flow of air through the filter bag from a compressed air source. The bag cleaning operation in many collector designs is conducted during the filtration operation without interrupting the incoming flow of dust-laden gas. This is possible since bags are selectively cleaned through a timed and sequenced control system that automatically opens and closes the compressed air jet valves to individual bags, as shown in Figure 14.17*B*. The air jet valves are generally controlled by a master timer that can be programmed for desired bag cleaning frequencies and various sequencing operations. A ΔP instrument across the filter can also be connected to the timer if desired for certain services.

When a jet air-cleaning valve is opened, a pulse of air is sent through a venturi, as shown, directly into the cage/bag assembly. The pulse induces additional air

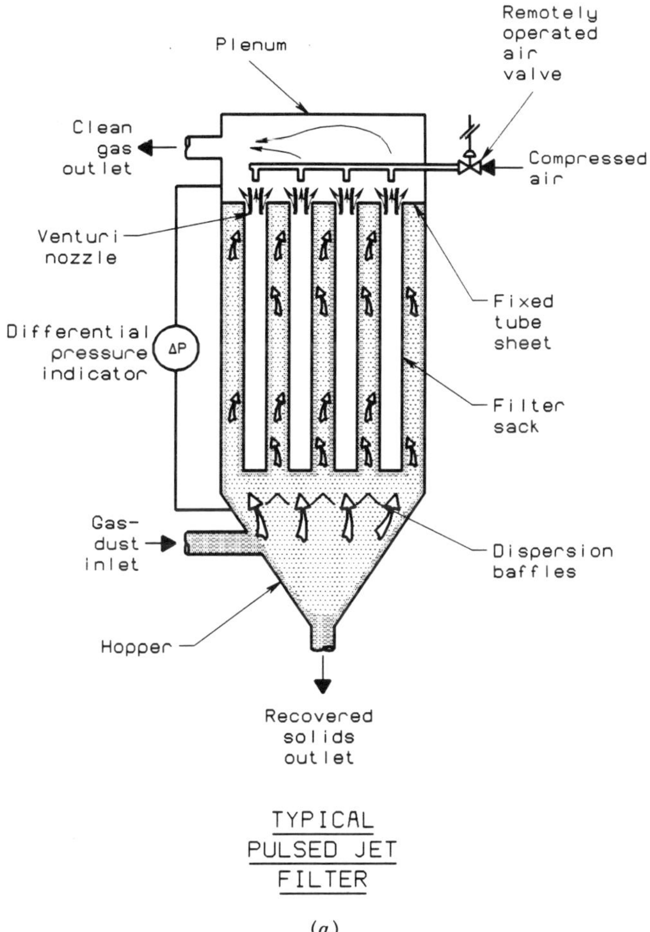

(*a*)

Figure 14.17A *Typical pulse-jet fabric filter collector.*

flow, which results in a very large burst of air that almost instantaneously inflates the bag. As a result the bag surface moves and the dust layers, which may be caked, are dislodged. In addition, small particles are forced from the fiber recesses into the surrounding areas. Since the total air volume may be 3 or 4 times greater than the compressed jet air stream, a relatively large superficial air velocity through the bag surface aids in moving the dust particles away from the bag surface. Additional air bursts further scatter the solid particulates. The venturi system shown in Figure 14.17*B* demonstrates the pulse jet-cleaning principle, but actual designs of the venturi system vary among the manufacturers. However, the designs are based upon short air pulses that normally occur in 50- to 150-msec time periods and at 10- to 30-sec intervals with large volumes of induced or aspirated air.

This online operation cleans and removes the dust layers from particular bags while other bag filters are removing dust from the inlet dust–air stream. As a result, some of the dust will redeposit on the operating bag filters, and a portion

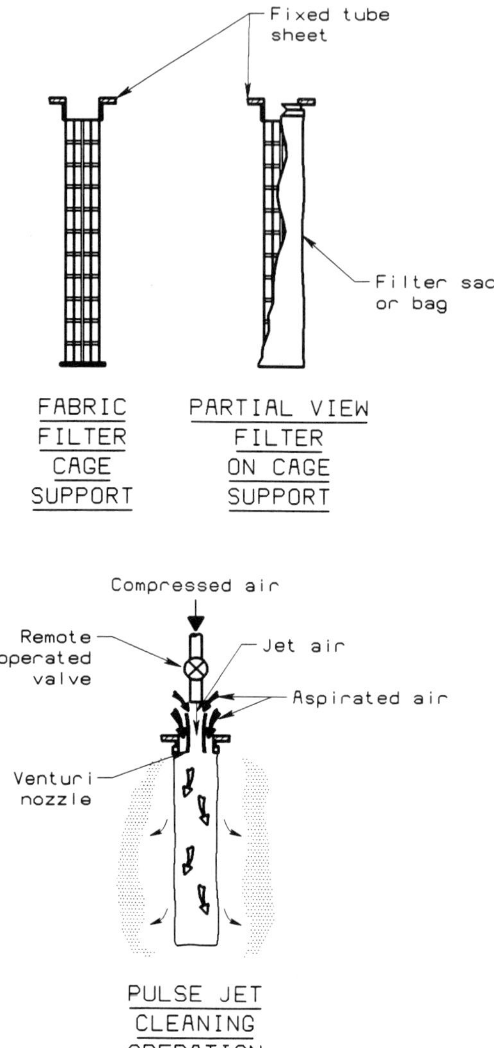

(*b*)

***Figure 14.17B** Typical pulse-jet filter details.*

will be recovered in the dust hopper. Consequently, some of the pulse jet filter systems are separated into chambers for cleaning or collectors may be placed in parallel. Parallel operation of the collectors or chambers may permit maintenance of a collector bag house and removal of torn bags. If the collector has not been completely reverse cleaned, a considerable amount of particulates and dust will remain within the collector, requiring personal protection for workers removing and replacing filter bags. Complete cleaning of the bag filters is necessary, along with removal of solids in the hopper. In addition, particulate leakage results in an accumulation of dust in the plenum chambers due to penetration of the filter bags.

The plenum chamber may require maintenance infrequently, for example, when the cleaning pulse jet loses some effectiveness or a dust buildup occurs in a cage. This operation would require personal protection equipment.

Cleaning has been reported as very effective with particulates almost completely removed from the fabric surfacc.[13] Consequently, felt is necessary to maintain a high collection efficiency until a sufficient solids layer thickness has accumulated. Since cleaning is quite frequent and the filter material must be an effective collection system for a significant part of the filtering cycle without a large dust layer buildup, woven fabrics are not effective in this type of collection device. Consequently, felt fabric bags are normally installed in pulse-jet filters. The felt fabrics permit higher superficial filtration velocities in the pulsed-jet cleaning filters than shaker or reverse air collectors that use woven fabric filters. The filtration velocities in pulse-jet collectors normally range from 3 to 15 ft/min (0.9–4.6 m/min), and the pressure drop across the collector is generally 4–6 in. of water (1–1.5 kPa).[13,14] The collection efficiency is normally $> 99.0\%$ for a very wide range of dusts. Cleaning frequency intervals, which vary with the type of particles and concentration, have been reported as 2–15 min.[13]

Cartridge filters are supplied by some manufacturers for pulse-jet collectors. These cartridges are generally pleated filter types that have more filtration area than standard fabric filter bags. In contrast to fabric filter bag construction, the pleated cartridge filters are normally constructed with synthetic cellulose fiber blends which are paper constituents.[25] The cartridge filters with the cellulose fiber blends have slower rates of filtration than typical filter fabric bags, thereby offsetting to some extent their largest surface area. In addition, the cellulose cartridges are limited in operating temperature to about 150°F (66°C). Pleated cartridges have been constructed from bag filter fabric materials, but apparently have less surface area than the paper-type cartridges.[25] The cartridge filters permit smaller-size collectors in general than fabric filter bag collectors and are designed for ready removal of the cartridges. This design approach reduces downtime and potential exposure situations.

Collection Efficiency. Collection efficiency of fabric filter dust collectors is normally greater than 99.0%, but has not been investigated to any great extent on a general basis. A wide group of variables affect collection efficiency, which make filtration studies difficult. Some of the variables that affect efficiency are the type of filter fabric, superficial filter velocity, cleaning method, cleaning frequency, particle size distribution, and bag installation. Bags not installed properly can be a source of leakage that also leads to torn bags and increases losses.

Cleaning has a significant impact upon collection efficiency and varies considerably in effectiveness. The particulates may remain deep within the fabric or most of the deep-seated particles may be displaced, depending upon cleaning effectiveness. In the latter situation, some penetration occurs when the fabric filter is again placed in service. However, dust layers build rapidly and a high overall collection efficiency is rapidly attained that remains constant through the dust filtration step.[13] Generally, collection efficiency is above 99.0% for particulates in the 1-μm to 10-μm range, but falls below this efficiency level for particulates in the submicron range. Filter collector manufacturers have assembled a considerable amount of collection efficiency data through the years for a wide variety of solids particulates and this provides a basis for estimating efficiency and penetration for specific designs.

The type of cleaning system has a significant effect upon penetration at the start of the filtration cycle as described. Some information on this phenomenon has been developed in conjunction with pressure drop across filter bags and is available in the literature.[14,26] Additional information can be obtained from fabric filter collector manufacturers.

Pressure Drop. An important consideration in filter operation is pressure drop, with filters generally operating in a pressure drop range of 2–6 in. of water (0.5–1.5 kPa).[13,14] Blowers in many installations are placed at the outlet of the filter and the filter system is under small vacuum. In other installations, process system pressure is above atmospheric pressure and blowers are not required. In this situation, the pressure rating of the bag house must be considered. Where the upstream process pressure is considerably higher than the bag house pressure rating, a pressure controller to maintain the upstream pressure is installed. A pressure reduction across the controller reduces the stream pressure to a level compatible with the downstream pressure requirement that includes dust losses and the bag house pressure drop. A filter pressure drop must be estimated to define overall system pressure losses for both pressurized and nonpressurized systems. In addition, filter collector pressure drop estimates define cycle time and dust loadings.

Although wide variations in filtration operating conditions exist, pressure drop calculation methods have been developed that provide reasonable estimates of pressure loss in filters. An important concern in calculations is the air–cloth ratio, which is the overall gas volume flow divided by the area of the filter bags. This results in the superficial gas velocity described in previous sections and is an important part of filter characterization and pressure drop calculations. The following calculation procedure is based upon Dennis and Klemm,[26] where the pressure drop increases with time as the dust layer increases. Following the cleaning procedure, a small amount of dust remains in the fabric filter and at this point the fabric is considered clean. The pressure drop calculation procedure is:[14,26]

$$S = \frac{\Delta P}{V} = S_{\mathrm{E}} + K_2 W \qquad (14.11)$$

where

ΔP = filter pressure drop, including dust layer (N/m^2)
V = superficial gas velocity (Q/A filter) (m/min)
Q = volumetric gas flow rate at the filter (m^3/min)
S_{E} = effective residual filter drag after cleaning (N-min/m^3)
W = filter loading (g/m^2)
K_2 = dust specific resistance coefficient (N-min/g-m)

Fabric resistance with residual dust (after cleaning) or residual filter drag varies considerably and is also a function of the type of collector (shake, reverse air, or pulse-jet). For a satisfactory value of S_E, experimental data provide the most accurate information and may be obtained from a filter collector manufacturer. Also, some data are available in the literature that are presented as ranges and may not cover the complete range of dusts and fabric filters.[13,14,28] However, a reasonable number can be assumed for estimating purposes.

The value of K_2 is dependent on various dust characteristics and possibly filter characteristics. Again, some K_2 values are available in the literature, but do not cover all dusts and filter fabrics.[13,14,26] However, some collector manufacturer data may be available for the specific application. A method of calculating K_2 has been developed and can be used although the predicted value has a $\pm 50\%$ accuracy.[26] For many applications, the equation is a satisfactory approximation and reasonable pressure drop values are obtained:[26]

$$K_2 = \frac{\mu_G S_0^2}{2\rho_P K_c}\left(\frac{3 + 2\beta^{5/3}}{3 - 4.5\beta^{1/3} + 4.5^{\beta/3} - 3\beta^2}\right) \tag{14.12}$$

where

$\beta = \bar{\rho}/\rho_P$

$1 - \beta$ = void fraction

μ_G = gas viscosity (poise)

$$S_0 = \frac{6 \times 10^{1.151}(\log \sigma_G)^2}{d_{MMD}} \tag{14.13}$$

= specific surface dust parameter (cm^{-1})

= $6/d_p$ for monodisperse particles (dia. d_p)

d_{MMD} = geometric mean of inlet particle size distribution

σ_G = geometric standard deviation of inlet particle size distribution

ρ_P = particle density (g/cm^3)

$\bar{\rho}$ = cake bulk density (g/cm^3)

K_c = Cunningham correction

and

$$K_c = 1 + \frac{0.1340}{d_p}\left[1.257 + 0.4\exp(-8.21 d_p)\right] \tag{14.14}$$

where d_p is the particle diameter (μm).

The overall pressure loss can then be described as:[14]

$$\Delta P = VS_E + K_2 V^2 C_i \Delta t \tag{14.15}$$

where C_i is the inlet particle concentration (g/m^3) and Δt is the time between cleanings (min).

From Eq. 14.14, the time between cleanings can be estimated. While this may not be completely accurate, the calculations provide a basis for timing the initial cleanings. During initial operations, performance data are reviewed and the timing cycle is adjusted to optimize filter performance. Improved values of S_E and K_2 can then be determined during field operations.

The limiting pressure loss at cleaning can also be determined on an estimated basis from experimental or manufacturers' data and then be verified or adjusted from actual field data. This equation follows:

$$\Delta P_{\mathrm{L}} = VS_{\mathrm{E}} + K_2V(W_{\mathrm{P}} - W_{\mathrm{R}}) \tag{14.16}$$

where

ΔP_{L} = limiting pressure loss at cleaning ($\mathrm{N/m^2}$)
W_{P} = fabric loading at ΔP_{L} ($\mathrm{g/m^3}$)
W_{R} = cleaned fabric loading ($\mathrm{g/m^3}$)

Maintenance. An important consideration in filter collection or bag house operation is maintenance of the filter. In addition to maintenance of the filter internal sections, some consideration must be given to the external housing. Doors and housing openings must be sealed to prevent leakage, including openings for mechanical and electrical connections. When any of these sealed points are opened during a shutdown, potential dust exposures exist, requiring personal protection equipment in most situations. In addition, dust generally builds around the opening and should be vacuumed prior to further work. Should dust fall to the pad or ground, it should be vacuumed. These dust-cleaning functions can be performed with a portable vacuum unit.

Dust emissions from opening doors or hatches depends upon the operating condition of the collector. When a collector is shut down, the internal pressure will rapidly change to atmospheric pressure and some dust exposure can be expected when any of the doors or hatches are opened. For those operating situations where a blower on the collector is operating, the internal volume pressure may be less than the atmospheric pressure, which reduces dust discharge from the openings. When filter collectors operate above atmospheric pressure, dust emissions from door and hatch openings can be considerably greater, increasing potential worker exposure unless adequately sealed.

Collectors are generally entered to remove torn or ripped bags and less frequently to repair shaking equipment. Since multiple collectors are frequently installed on continuous flow streams, defective bags are removed when a collector is offstream. Although flow through the filter collector is halted, a considerable amount of dust is present in the plenum from the broken bag and particle seepage or penetration through the filter bas. When replacing the defective bag, the dust hopper may have been purposely emptied, but the remaining filter bags on the dust hopper side retain a considerable amount of dust. While the dust on the bags may be caked, any disturbance of the bags will readily dislodge the cake, which falls and creates dust clouds. Moreover, handling the bags manually also creates dust emissions. Potential dust exposure situations exist in either the plenum or bag sections for maintenance workers. Some dust reduction may be achieved in the upper plenum section by continuing to run the blower with the inlet dust stream blocked. However, this may not be completely effective.

Personal protection is normally required when entering a dust collect or for maintenance. Where a torn bag requires replacement, a rapid changeover does not permit complete cleanout of the system prior to bag replacement and the type of personal equipment necessary should be carefully selected. Where a filter collector will have considerable mechanical work performed, the collector should be vacumed following the initial opening. When the filter has been thoroughly cleaned, mechanical work commences and personal protective equipment requirements are based upon dust concentrations in the collector. Contractors specializing in filter collection cleaning and repair are available that have the equipment necessary for dust cleaning, personal protective equipment, personnel cleaning booths, and building the necessary ingress/egress interlock systems to minimize dust emissions from the collector. For personnel entering a collector, personal protective equipment not only includes masks, but also recommended coveralls, boots, helmets, and cleaning booths. Moreover, disposal of dust collected from personal clothing, cleaning booths, and so on, along with dust from the interlocks, must be collected and sent for disposal in accordance with environmentally acceptable procedures. Alternatively, some dusts can be recycled to the process after collection. In this situation, each step of the recycling procedure must meet environmental and industrial hygiene requirements.

Removal of torn bags and/or replacement of filter bags requires careful handling of the bags since the bags have considerable dust on the surface and within the interior bag fibers that generate dust emissions. Following removal of the dust bags, the bags must be placed in a container that can be closed and then moved to disposal facilities. Opening a container that encapsulates dusty bags should not be undertaken without approval from environmental and industrial hygiene personnel.

In some fabric filter collectors, bags can be removed from the collector without entering the bag house. While this technique is feasible, the bag is manually moved and potential dust exposures are quite high during bag transfer outside the collector, requiring personal protective equipment. The torn bag must be then placed in an enclosed container as described above for disposal.

Instrumentation. One of the major problems in fabric filter collector operations is determining when a filter bag begins to tear. The outlet dust concentration increases slightly with a tear and then generally increases as the tear becomes larger. Vent dust concentration tests will normally indicate an increase in the dust concentration during standard EPA particulate control tests. These tests are periodically required to ensure compliance with PM-10 standards.

Although EPA tests are conducted, the tests are infrequent compared with daily operations. For continual indications of dust concentration changes, an instrument that measures dust concentration should be installed on the filter collector vent. Various dust concentration instruments capable of continuous operation are available which are based upon different sensing techniques. Some of these sensing techniques are:

Microwaves
Optical density
Electrostatic
Laser light scattering

A dust-measuring instrument in the vent can also be connected with a remote alarm or indicator to provide early indication of bag leakage. While these devices were initially installed in filter collectors, they are also applicable in other types of dust collectors such as cyclones and precipitators.

Dust Explosions. While dust explosions are reviewed in Section 14.6, some consideration must be given to potential explosions in the filter collectors. Many collectors are supplied with blowout vents or rupture wall sections to relieve the pressure buildup associated with an explosion. If a collector does not have explosion protection, the bag house collectors should be equipped with automatic explosion control systems. The explosion or blowout vent system is a less costly device than the automatic suppression system described in Section 14.6 and does not require a significant maintenance effort.

1. *Advantages of filter collectors.* Filter bag collector units are effective in removing particles from gas streams and have the following overall advantages:
   - Effective in removing respirable dust particles $< 10\ \mu m$ from gas streams.
   - Submicron particles are also recovered with particulate size recovery to 0.20 μm feasible.[13] However, the proper filter bag and collector design is necessary for an adequate recovery efficiency at this particle size. Smaller particles may be recovered with a HEPA filter.
   - Recovery efficiencies are $> 99.0\%$ and in some applications efficiencies of 99.90% have been achieved.
   - Operating temperatures to about 550°F (290°C) are permitted with particular fabrics. Higher temperatures can be achieved with temperature-resistant fabric material.
   - Newly developed solid-core envelope plates with thin films apparently have much lower maintenance costs than fabric filters.
   - A variety of filter collectors permit selection of specific filters for particle collection.
2. *Disadvantages of filter collectors.* The general disadvantages are:
   - Relatively low dust or solids loadings compared with cyclones or certain types of wet scrubbers.[14] Maximum cyclone air loadings may be 10 times greater than filter inlet air loadings. A primary stage to remove a large quantity of solids in a heavily laden dust gas stream is generally necessary before sending the stream to the filter collector.
   - High maintenance costs resulting from bag replacement cost, cleaning filters, and mechanical repairs. Multiple collectors add to maintenance costs.
   - Wide variations in vent dust emissions with leakage, torn, or ripped bags.
   - Relatively high investment cost due to multiple collectors and are a requirements.
   - Low operating pressures. Equipment not normally designed for operating pressures greater than 2–5 psig (13.8–34 kPag).
   - The dust particles must be dry and cannot be tacky.

Other Filter Collectors. A relatively recent development has been envelope-type filters. In these filter collectors, flat plates of foam, highly porous materials, or solid materials with large open areas form squares or rectangles which are fitted on one end into a plenum. The plates are then covered with a flat fiber bag (envelope bag) that slips over the flat plates. One bag side remains open where the filtered gas is discharged into a plenum chamber. The plates are installed vertically and a large number of parallel plates are placed in the collector. These parallel plates fit into the plenum chamber as described with the plenums at the top or side of the collector. Locating the plenum on the side permits feeding the dust-laden inlet gas downward into the filtering section. The downward flowing gas is considered an improvement since the gas flows in the direction of particle trajectory and does not engender reentrainment, which can occur with upflowing streams. As the gas turns to enter the vertical plates, centrifugal force and momentum tend to further separate particulates from the gas stream. Filter plate loading for a specific time period is presumably lower than the loadings on comparable upflow collector designs. Thus, the time between cleanings would be longer for downflow designs, increasing cycle time on equivalent surface areas. Cleaning the vertical flat plate is accomplished through shaking, reverse air, or pulse.

Other types of filter plates are available that do not require envelope bags. An example is a corrugated plate constructed of sintered plastic materials that is coated with a thin PTFE film. The overall plate consists of two outward-facing corrugated sides separated by a narrow passage. Gas diffuses through the corrugations into the narrow channel and then flows to the plenum. Dust is retained on the surface of the PTFE film, which permits gas passage but effectively removes particulates from the gas. Efficiency is comparable or higher than efficiencies for filter bag collectors according to the manufacturer. Since bags have been eliminated in this design, bag leakage effects are not a concern and dust breakthroughs do not occur for several years. By omitting bags, potential exposure associated with bag replacement is eliminated, and entry to the collector required for mechanical repair is reduced significantly. Pulsed-jet cleaning is used with this plate collector and solenoid pulse valves located outside the plenum permit valve repair with minimal dust exposure. Operating temperatures are limited to about 160°F (66°C) for this filter system. However, the reduction in potential exposures and significantly lower maintenance costs may justify installation of a cooling system for higher-temperature streams.

14.4.6 Wet Scrubbers

Often termed wet collectors or particulate scrubbers, this equipment is designed to remove or collect dust from an air stream with a liquid which is usually water. In general, the liquid forms droplets as a result of equipment design characteristics which are then dispersed in the gas–dust stream. Dust particles combine with the droplets and the dust-containing droplets are collected in the scrubber device, forming a liquid stream that varies in solids content from a very dilute stream to a slurry. The solids concentration in this recovery stream is a function of inlet dust loading, particle size distribution, and recycling of liquid through the equipment that would increase the solids concentration. Solids in the stream require recovery and disposal which must be included in the consideration of solids removal from gas streams.

The description of particles contacting liquid droplets is presented in various references.[13,14,16,33] In general, scrubbers depend upon the collision of liquid droplets and solid particles. The liquid droplets are produced in spray form from various nozzle and system configurations that require a considerable amount of energy. Actual capture of the solid particles by the liquid droplets is complicated, and an analysis of the mechanism is not included in this section. However, some of the factors involved in solids capture include direct impingement of particles on droplets, particle agglomeration, liquid droplet size, droplet spatial distribution, fine particulate dispersion, electrostatic forces, and solid particle size distribution. In addition, a wide range of equipment types is available for the removal and collection of dusts, further complicating the understanding of scrubber collection and efficiency. Some of the various types of wet scrubbers available commercially are:[13,28,33]

Venturi	Self-induced spray
Impingement plate	Wet centrifugal
Packed bed	Spray
Mechanical	Cyclone
Ejector venturi	Orifice

This list also has subcategories that further expand the overall group of equipment. In this section, the venturi scrubber will be reviewed since this particular scrubber is widely used compared with other scrubber types. In addition, most of the analytical effort on scrubbers has been concentrated on the venturi scrubber, providing some basis for reviewing scrubber performance.

General Effectiveness. With the various types of scrubbing devices there obviously is a difference in collection efficiency. Comparisons are extremely difficult with the lack of analytical tools for an effort of this type, but some data are available that provide qualitative comparisons. One source[29] indicates that collection efficiencies vary from 70% for a spray scrubber to 99% for a venturi scrubber. Other wet scrubber collection efficiencies ranged from 90 to 95%. However, variations in particle size distribution and other factors may change some of these efficiencies, but clearly the venturi scrubber in this comparison has a much higher collection efficiency albeit qualitative.

Another source compared experimental and theoretical sources for three wet scrubbers[30] based upon the cut particle size. The cut particle size is the aerodynamic diameter of a sphere with a density of 1 gm/cm^3 that has the same or identical terminal settling velocity as the actual particle. This is similar to the definition for particle cut size diameter in cyclones. The EPA compares the particle cut size with pressure drop in the equipment for three devices; impingement plate, sieve, and venturi. A drawing of the impingement plate is shown in reference 13, with the sieve similar in design to the impingement plate. The results of the EPA investigation are shown in Table 14.6.

A major observation from Table 14.6 is that the venturi scrubber has the capability of collecting submicron particles. Liquid droplets formed in the venturi

TABLE 14.6 Comparison of Experimental Particle Cut Diameters as a Function of Pressure Drop for Three Scrubbers[a]

	Particle Cut Diameter (mm)						
Scrubber Type	0.3	0.5	0.7	1.0	1.3	1.5	1.7
	Pressure Drop (in. water)						
Impingement	—	—	—	39.3	18.9	9.4	5.1
Sieve	—	—	—	16.1	7.1	3.2	2.3
Venturi	55.2	19.3	10.2	4.5	—	—	—

[a]From reference 30.

contact the dust particles flowing through the venturi. As the particle size increases, the venturi pressure drop falls considerably and for the 1-μm particles, the pressure drop in the venturi is much lower than the pressure drop for the other two scrubbers. The pressure losses for the impingement and sieve scrubbers are considerably higher than the venturi scrubber pressure drops for comparable cut particle sizes. Reductions in particle cut diameter drastically increase the pressure drop requirements for both the sieve and impingement scrubbers, which raises the pressure drop to levels that may be difficult to achieve in practice. While this cut power relationship indicates that venturi scrubbers operate at much lower power requirements for comparable efficiencies, empirical studies are reported to have demonstrated equivalent performance at specific power consumption levels.[13]

High-efficiency venturi scrubbers may have gas differential pressures that reach 100 in. water,[28,30] which is a relatively large pressure change (25 kPa or 3.6 psi). Where vapor system pressures are > 3.6 psig (35 kPag), a very high scrubber gas pressure loss does not pose any significant problems. However, low-pressure gas dust feed systems of 5–10 in. of water (1.3–2.5 kPag) require a blower or compressor to achieve the higher pressure levels necessary for satisfactory collector performance.

The high efficiency venturi scrubbers also collect particulates < 1 μm. Where the pressure drop across a venturi scrubber is 5 in. of water, a lower collection efficiency, termed "moderate collection efficiency,"[28] is realized. Overall collection efficiency is lower at a 5-in. water pressure drop and particulate collection below 1 μm is extremely small. Although collection is shown in Table 14.6 for a 4.5-in. water pressure drop, actual efficiencies for particles in the 1–2 μm region are apparently low.

Following the gas–dust impaction interchange with the liquid, the particle containing droplets must be separated from the gas. As shown in Figure 14.18, the discharge from the venturi is sent to a separator where particle-laden droplets are recovered and the solids liquid mixture is discharged for solids recovery.

Venturi Scrubber. A typical venturi scrubber with a separator is shown in Figure 14.18. The venturi in this figure is in a vertical position with downward flowing inlet gas dust streams. Venturis are also positioned horizontally with the discharge sent to a vertical separator similar to the separator shown in the figure. The section above the venturi and throat is covered with a layer of liquid or water that enters the upper cylinder through tangential inlets. This is also referred to as a flooded wall,[13] which prevents any dust buildup through the transition and throat

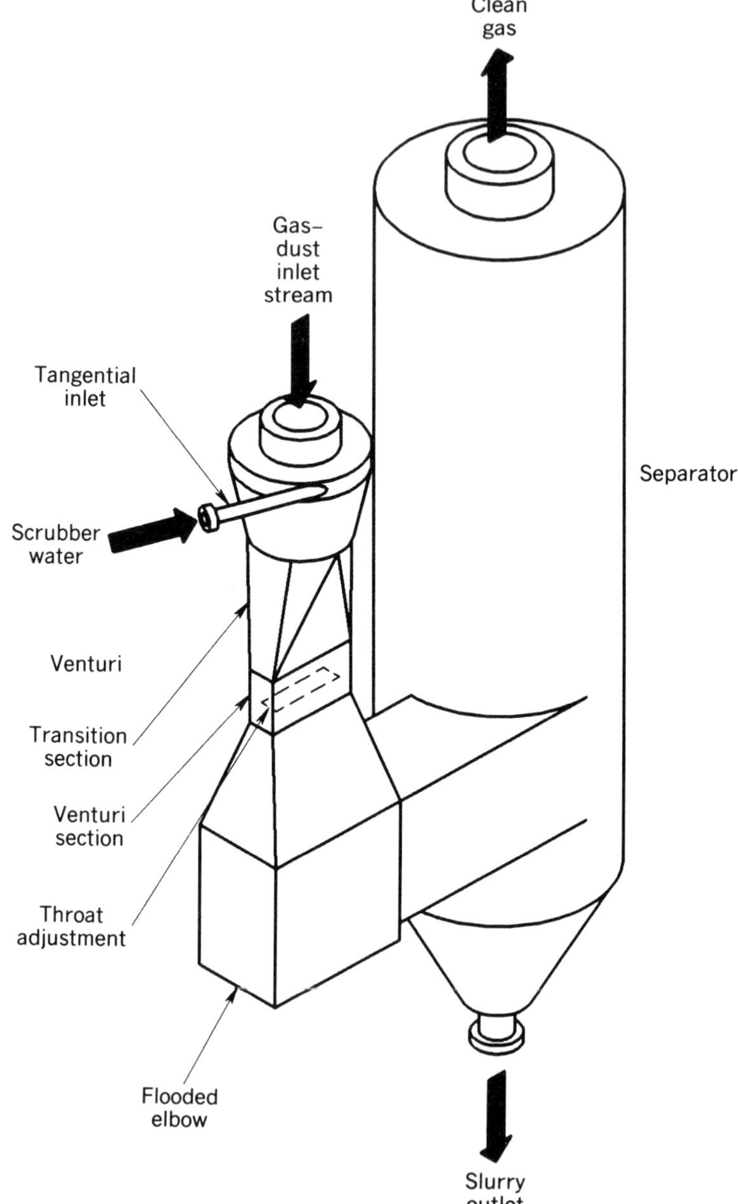

Figure 14.18 *Vertical venturi scrubber.*

sections. A transition section connects the upper cylinder with the narrow rectangular section termed the venturi throat. While horizontal venturis may have water injected at the throat section,[28] the upper tangential inlet and transition wall provide a method of distributing liquid flow evenly through the throat section for optimum operation. In the throat section, the water essentially becomes a spray of droplets which are combined with the flowing gas–dust stream, providing droplet and particle interaction.

The throat section has an internal adjustment that can vary the pressure drop, which improves operation when the inlet flow rate or dust concentration varies. Throat adjustment is controlled automatically or manually to provide a constant pressure drop. The rectangular throat distributes the water satisfactorily and is effective in particle and watcr droplet interaction. Throat adjustment in this design results in an even flow distribution which is more difficult to achieve in circular ducts. Throat velocities of 12,000–24,000 fpm (3659–7317 m/min) and 5–15 gpm (0.019–0.057 m^3/min) of water per thousand scfm (28.3 nm^3/min) of gas are considered typical design conditions.[28]

A combination of vapor, liquid, and water droplets enter the flooded elbow at the base of the venturi where the stream turns 90°. The flooded elbow is purposely designed with a shallow pool for water accumulation which prevents wear and erosion by particulate-laden droplets. Some additional mixing of the overall stream occurs in the flooded elbow, which may improve collection efficiency. The mixture stream which enters the separator is completely saturated.

The separator shown in Figure 14.18 is a cyclone type, where droplets and wetted particles are separated from the gas stream through centrifugal action. This action is initiated by stream flow through a tangential inlet at the separator wall. While the separator in the figure is a cyclone type, other types of separators can be installed. Particles and droplets separated from the gas stream combine along the wall to form a slurry that is discharged from the bottom of the separator.

Since a relatively large amount of water normally flows through the venturi scrubber system, particulate holdup is essentially nonexistent when entering one of these units for maintenance. Additional water flushing will remove traces of remaining materials, particularly around the control surfaces in the throat section. The only material remaining in the unit is the water particulate mixture in the flood elbow. Installation of a drain valve before startup will remove this mixture from the elbow or it can be flushed with water into the separator for discharge. The clean water remaining in the elbow can then be removed through various techniques.

Contacting Power. This is a design procedure that correlates the power expended in a scrubber with collection efficiency. The collection efficiency in this approach is essentially related directly to the contacting power in a scrubber. The configuration or design of the scrubber apparently does not significantly affect the power–efficiency relationship. This general concept is strictly empirical, based upon experimental efforts with gas-atomized spray scrubbers.[13]

In this approach to scrubber efficiency, where all of the contacting energy is obtained from the gas stream as pressure drop, the contacting power is considered the friction loss across the wetted equipment. Kinetic energy changes apparently are not an effective parameter compared with the impaction dissipative effects of the friction energy since kinetic energy does not correlate with collection efficiency.[13] For a venturi scrubber, when the liquid nozzle does not spray liquid into the system, the total power loss is a direct function of the frictional loss:

$$P_{\mathrm{T}} = 0.1575 F_{\mathrm{E}} \qquad (14.17)$$

where F_{E} is the friction loss (in. water) and P_{T} is the total pressure (psi).

The total power loss is related to collection efficiency through transfer units which have been developed by various sources.[13, 31, 32] For specific correlations with particular solids, relationships for the dusts are developed or found in the literature. Results support the basic assessment that high contacting power improves efficiency and particles $< 1\ \mu$m are collected. Table 14.6 indicates the particle size pressure drop relationship for particulate recovery.

Venturi Pressure Loss Calculation. Since pressure drop is related to collection efficiency, various methods have been developed for determining the actual pressure loss. However, some of these calculation procedures are quite complicated and beyond the scope of this chapter. One procedure for determining the pressure drop is used to estimate the pressure loss on an initial basis.[14] When the venturi scrubber is sized, detailed calculations are conducted to more accurately describe the pressure drop. The procedure for initially estimating pressure loss[14, 34] follows:

$$\Delta P = \frac{2\rho_L U_G^2}{g_c}\left(\frac{Q_L}{Q_G}\right)\left[1 - X^2 + \left(X^4 - X^2\right)^{0.5}\right] \qquad (14.18)$$

$$X = \frac{3LC_{DO}\rho_G}{16d_D\rho_L} + 1 \qquad (14.19)$$

$$C_{DO} = 0.22 + \frac{24}{N_{Re_1}}\left(1 + 0.15N_{Re_1}^{0.6}\right) \qquad (14.20)$$

$$N_{Re_1} = \frac{\rho_G U_G d_D}{\mu_G} \qquad (14.21)$$

$$d_{D(\text{water})} = \frac{49.7}{U_G} + 91.1\left(\frac{Q_L}{Q_G}\right)^{1.5} \qquad (14.22)$$

where

ΔP	= pressure drop (dynes/cm^2)
ρ_L, ρ_G	= liquid and gas densities (g/cm^3)
U_G	= gas velocity (cm/sec)
g_c	= 1 gm-cm-sec^{-2}/dyne
Q_L, Q_G	= liquid and gas volumetric flow rate (cm^3/sec)
L	= throat length (cm)
d_D	= liquid drop diameter (cm)
N_{Re_1}	= Reynolds number for drop diameter, throat entrance
C_{DO}	= liquid droplet dimensionless drag coefficient
μ_G	= gas viscosity (poise g/cm-sec)

The drag coefficient (e.g., Eq. 14.20) is satisfactory for Reynolds numbers in the range of 1 to 500. Where the water droplets accelerate to the gas velocity in the throat, the term $[1 - X^2 + (X^4 - X^2)^{0.5}]$ is equivalent to 0.5 and Eq. 14.18 becomes:

$$\Delta P = \frac{\rho_L U_G^2}{g_c} \frac{Q_L}{Q_G} \tag{14.23}$$

For water droplets, the pressure drop in inches of water is:

$$\Delta P = (4.02 \times 10^{-4}) U_G^2 \frac{Q_L}{Q_c} \tag{14.24}$$

Since the residence time in commercial venturi scrubber throat sections is not sufficient to permit drop acceleration to the gas velocity, the estimated pressure drop (Eq. 14.23) is higher than experimentally measured pressure drops. At low liquid rates, gas flow friction losses predominate and the pressure drop more closely approximates the experimental pressure drop.[34] For a detailed method of calculating the pressure drop, and collection efficiency, reference 35 presents a procedure that provides a relatively accurate prediction technique based on comparisons of predicted and observed performance.[14, 36]

Scrubber Efficiency. While some work has been conducted on venturi scrubbers, little information is available on the other types of wet scrubbers. Although quantitative information is not available, qualitative information on scrubber efficiency indicates venturi scrubbers are the only wet scrubbers capable of collecting fine dust ($< 3\ \mu$m) with a high collection efficiency.[33] Combinations of venturi scrubbers are also capable of high collection efficiencies. Reference 33 indicates that pressure losses for high collection efficiency vary from 10 to 100 in. of water for a venturi scrubber, and combinations of venturi scrubbers have pressure losses of 15–80 in. of water. This information supports the basic theory that high pressure drop or contacting power is necessary for high collection efficiencies and removal of submicron particles from a dust-laden gas.

Obviously, scrubber pressure drop and efficiency are difficult to calculate for venturi scrubbers and little information is available for collection efficiency calculations on other wet scrubbers. However, scrubber manufacturers have developed considerable information on scrubber performance and can provide collection efficiency information along with pressure drop requirements. The manufacturers data may also cover the type of dust requiring collection, which further improves the accuracy of the overall estimate by the manufacturer.

Advantages. The advantages for wet scrubbers are:

- Collection efficiencies up to 99% can be achieved in venturi scrubbers with particle recovery $< 1\ \mu$m at reasonable efficiencies.
- Some nonventuri wet scrubber efficiencies vary between 90 and 95% for very high dust loadings and particulate sizes $> 4\ \mu$m.
- Wet and dry particulate matter can be collected.
- Gaseous contaminants can be removed from the gas stream.

- High-temperature streams can be sent through a scrubber and the stream can also be cooled.
- Risk of dust explosions and fire hazards are eliminated.
- Less space or area is required compared to bag filters and electrostatic precipitators. However, any treatment of the slurry from the scrubber will require equipment that must be included in consideration of the total scrubber area required.
- Collection efficiencies will remain relatively constant with variations in dust or particulate loadings.
- Relatively large reductions in gas rates are possible.
- Potential exposures are minimized with this system and equipment cleaning primarily consists of flushing. The potential for particulate exposure is very small in the wetted sections during maintenance.

Disadvantages. The general disadvantages for wet scrubbers are:

- Some scrubbers, such as a venturi scrubber for heavy dust loadings require upstream equipment to reduce the dust loading to the scrubber. Typically, a cyclone preceding the venturi scrubber significantly reduces the dust loading. Several scrubber designs will process highly loaded dust streams.
- Solids slurry discharge requires separation of solids and liquid to recycle the scrubber liquid. In addition, the separated solids require disposal recovery which must be included in the overall system design and economics evaluation. Solids recycling may be permitted which requires solids drying and removal of any contaminants from the solids.
- If the scrubber liquid absorbs or reacts with contaminants in the gas stream, the liquid requires treatment after separation from the solids in the slurry before recycling or final disposition.
- Scrubber liquids may become corrosive when contacting a contaminated gas–solids stream. This requires scrubbers constructed of corrosion-resistant materials along with the solids slurry separation equipment. The clarified liquid will also require treatment.
- Clean effluent gas which may not have any contaminants can be discharged as a visible plume that requires some treatment if unacceptable to the community.
- Operating and maintenance costs are relatively high owing to maintenance of the slurry separation and treating systems.
- Freeze protection may be necessary in colder climates.

Slurry Treatment. The slurry requires separation of the solids and in most situations recycling of the decanted liquid, usually water. The wetted solids are either dried and recycled to a storage point or sent to waste disposal. Since the solids quantities may be large, solids recovery is usually justified on an economic basis. If the solids do not require treatment for removal of any contamination, the slurry from the scrubber would be sent initially to a wet filter, centrifuge, continuous press, or other device where most of the liquid is separated from the solids.

The liquid or water is recycled, and the wet solids, which contain about 20–30 wt % water, are transferred to a drier. In this state, the wet solids are cake-like and normally do not present potential exposure concerns.

Drying the wet solids can be accomplished in various ways. However, drying the solids results in a very fine dust that must be controlled since this dust is a potential exposure concern. Consequently, the drying system must be carefully reviewed to ensure dust generation and emissions are minimized. All of the equipment required to treat the wet slurry should be included in any analysis of scrubber investment and operating costs.

14.4.7 Precipitators

Electrostatic precipitators remove dust and solid particulates from gas–dust streams by placing a charge on the particulates and under the force exerted by an electric field attract the particles to collector plates or surfaces. The collector plates in these devices are electrodes. Electrostatic precipitators are effective collection devices that reportedly have efficiencies of 99.9% and are able collectors of submicron particles.[13] However, precipitators are quite large and normal applications for these devices are in high-volume streams. A summary of various collector installations[28] indicates that precipitators are frequently found in coal-fired boiler flue gas to remove fly ash, which generally contains a large particulate fraction of 1- to 10-μm particles. One other major application in the list[28] is particulate removal from dryer and kiln gas–dust streams in the metal mining and rock products industry, where the minimum particulate size is about 5 μm.

Precipitators are also used in some fluid catalytic cracker flue gas streams to remove fine catalyst particulates discharged from the regenerator cyclones. This stream also has a particulate fraction in the 1- to 10-μm range, and a small fraction < 1 μm. Other particle collection devices are also used in this fluid catalytic cracker discharge stream as an alternative to electrostatic precipitators, although all collection systems in this particular application process very high volumetric gas flow rates. Electrostatic precipitators are generally found in very large volumetric flow applications with few precipitators in low-flow, nonflue gas services. However, in some applications, this device may be an attractive control option and should be compared in any analysis with other control systems.

Particle Separation. Gas ion contact with particles produces a charge on the particles and under the influence of an electrostatic field the particles migrate to an electrode where they are collected and removed from the gas stream. There are two basic electrostatic precipitator types available commercially which are shown in Figure 14.19. The single-stage precipitator, termed Cottrell®, combines ionization and particle collection in one set of plates or electrodes. The two-stage precipitator has separate ionization and particle collection electrode systems.

Gas ionization is achieved by an electrical breakdown of the gas in a local region. The electrical discharge mechanism that creates a local electrical discharge phenomenon is termed corona, which also generates ozone. The corona does not propagate beyond a limited distance, in contrast with sparking, which is an advanced corona stage. Sparking is the complete breakdown of the gas along a given path, covering a significant distance. Electrical current flow for corona

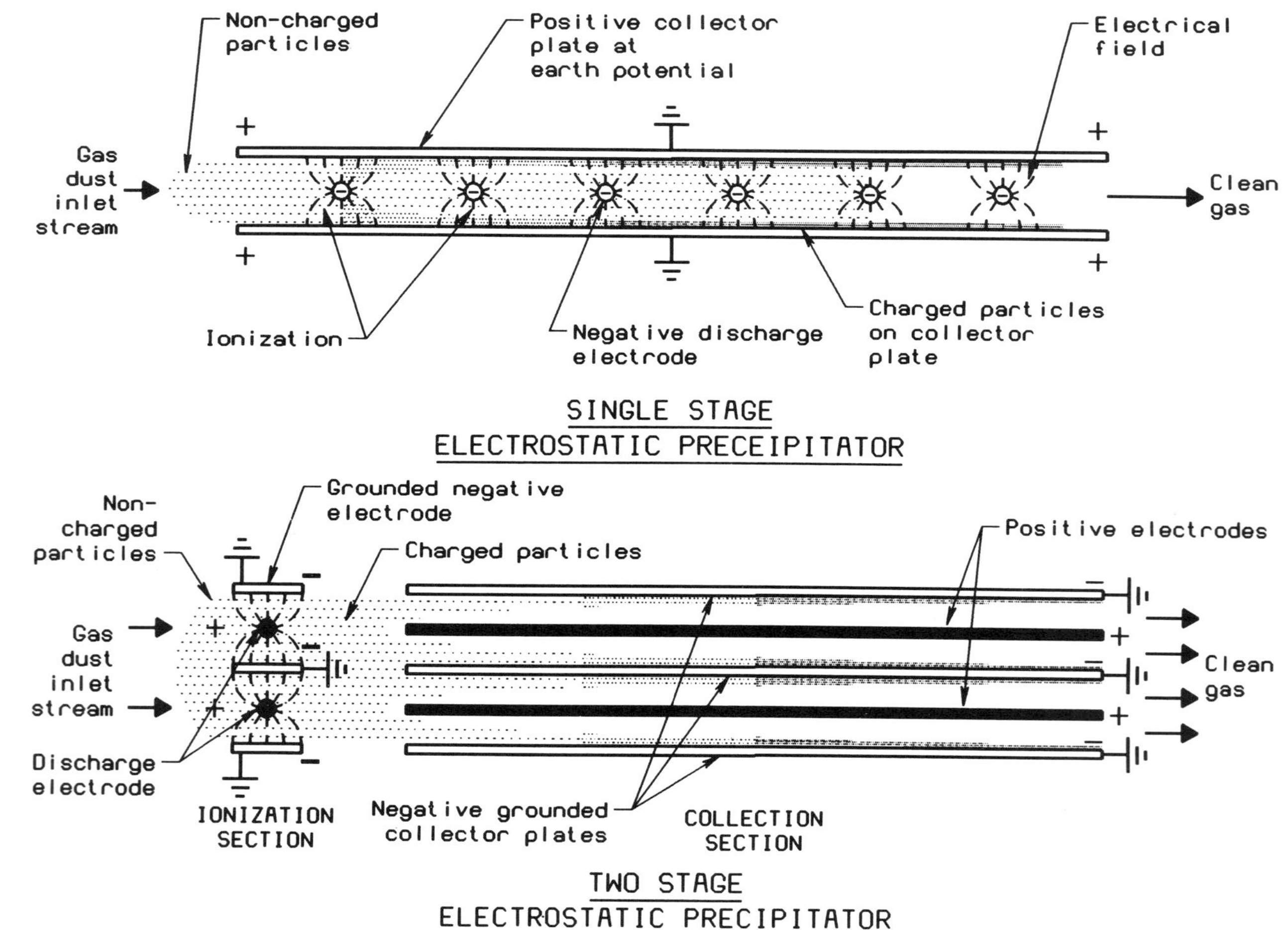

Figure 14.19 *Cross sections of single- and two-stage electrostatic precipitators.*

discharge is much lower than the current flow associated with sparking. The latter cannot be permitted or the electrical system will be disrupted. Consequently, precipitators are operated on the verge of sparking to ensure the voltage is satisfactory for corona discharge. In addition to detrimental effects on the electrical equipment sparking also results in lower collection efficiencies, reentrainment and dust dispersion, and current channeling.

The gas ions contact the particulates and charged particles under the electrostatic field force migrate to a collecting plate. For optimum operation in a single-stage precipitator, the discharge electrode is operated with a negative charge. Operating performance is more consistent than corona operation with a positive electrode, although the latter is accompanied by lower ozone formation.[13] As shown in Figure 14.19, the single-stage precipitator has a negative discharge electrode and the two-stage precipitator has a positive electrode with single-stage precipitators normally installed in industrial dust control operations and two-stage precipitators typically installed in air-conditioning applications. Dust loadings are significantly larger in industrial applications.

Migration Velocity. Since particles are charged, the migration velocity to the collecting plates becomes important in sizing the collecting plates. This velocity is dependent on the electrostatic field force and the drag forces.[14] Relationships have been developed to determine particle migration velocities[13] for particles larger than 1 μm and particles < 1 μm in diameter. Migration velocities for various particles and sizes are shown in references 13 and 14. Apparently, particles < 1 μm have migration velocities that are similar, regardless of submicron particulate size. Assuming reentrainment does not occur at the collector electrode, collection efficiencies should be independent of particle size for particles < 1 μm.

Precipitator Designs. A typical single-stage precipitator is shown in Figure 14.20, where the collecting electrodes are flat plates and the discharge electrodes are wires or rods. Flat plates are normally installed in dust collecting applications. An alternative single-stage precipitator has a single rod or wire within a pipe, with the pipes and wire installed vertically, and gas flow generally upward through the pipe. Downflow operation is less widespread with this design. Pipe precipitators are generally installed in applications for removal of liquid particulates, volatile fume control, tacky particles, or a high concentration of submicron particulates. Wet tubular precipitators are particularly effective in these operations.

Dust on the typical collection plates is normally removed and collected during operation by rapping against the plates. A mechanical system is designed to rap the plates automatically with a specific amount of energy at preset cycle times and for specified time periods. Rapping requirement criteria are quite important to prevent dust reentrainment. If excessive rapping force or energy is applied, dust reentrainment will occur, whereas insufficient energy results in a dust buildup on the plates that affects electrical performance. Poor collection efficiency is frequently associated with either excessive or insufficient rapping energies.[13]

The size, shape, and other electrode considerations for optimum performance are based upon gas and solid particulate characteristics. Performance data on precipitator operation in comparable services is also a consideration in electrode design and layout. Generally, the space or distance between plate-type precipitator

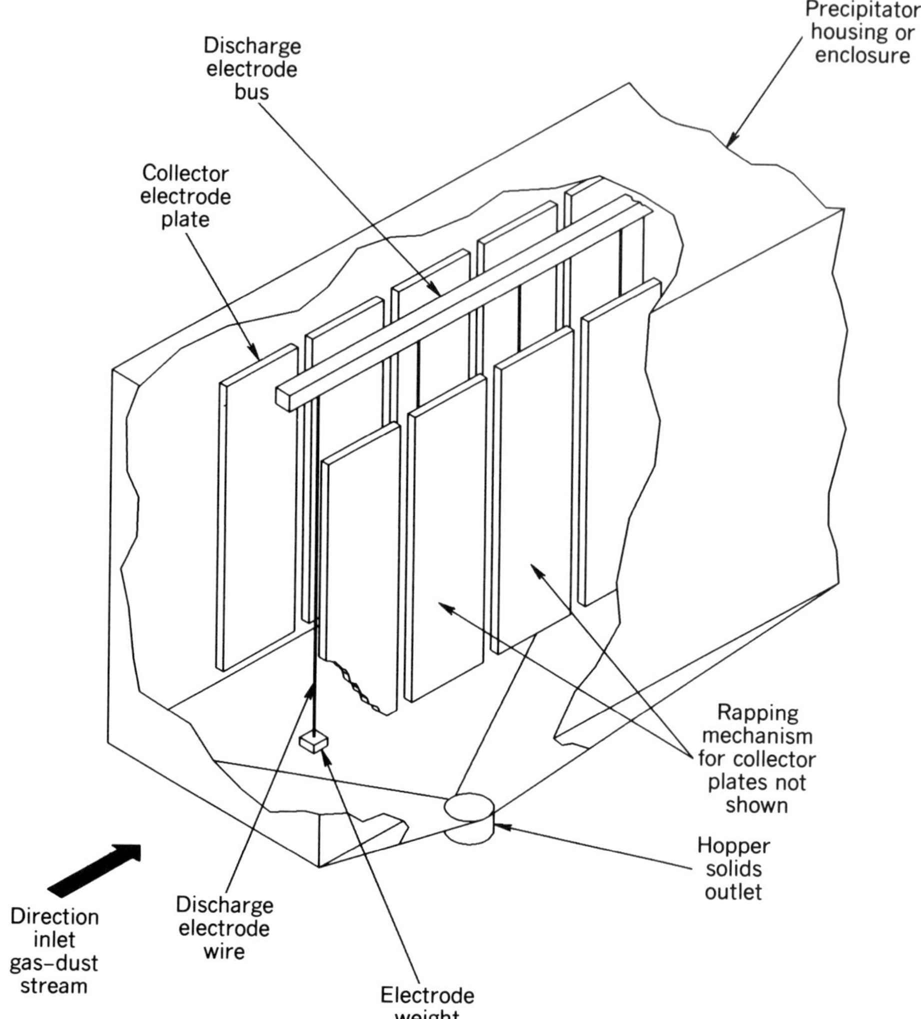

Figure 14.20 *Simplified view of single-stage electrostatic precipitator.*

electrodes ranges from 6 to 15 in. (15–38 cm). Large spacings are generally designed for high particulate loadings to prevent sparking as a result of dust buildup with greater spacing in the inlet half of the precipitator and smaller spacing in the outlet half of the precipitator for high solids concentration gas streams.[13]

For optimum collection efficiency in a specific electrode configuration, the gas should be equally distributed entering the plate or pipe electrode systems. Inlet vanes or baffles are frequently installed in precipitators to equalize gas stream distribution. Other types of inlet gas dust distribution devices are also installed with experience on distribution frequently provided by precipitator manufacturers. These flow distribution requirements are necessary for single- and dual-stage precipitators. In addition to good distribution, gas velocities are relatively low to prevent particle reentrainment, with gas velocities in a range of 3 to 10 ft/s (0.91–3 m/s) and gas residence times of 1–15 sec.[13]

Voltages in single-state precipitators vary between 30,000 and 100,000 volts dc, and voltages in two-stage precipitators are generally 10,000–13,000 volts dc in the ionizer section and 6,000–8,000 volts dc in the collector section.[13,28] However, power consumption is low, with higher power consumption for single-stage precipitators compared with dual-stage devices. While power consumption is low, equipment must be provided to convert ac power to dc power for the precipitators. These costs are included in any investment analysis.

Collection Efficiency. The migration velocity, which is the velocity of charged particles that move toward a collection electrode, provides a basis for estimating the collection efficiency. The Deutsch equation has been successfully used for many years to estimate collection efficiency:[13,37]

$$\eta = 1 - e\exp[-Au_e/Q] \tag{14.25}$$

where

η = collection efficiency
A = collection surface area (ft^2)
u_e = particle migration velocity (ft/sec)
Q = volumetric flow rate (cfm)

Relating particle size to the migration velocities and average velocities in various references[13,14] provides a basis for estimating the collection efficiency. However, Eq. 14.25 is only applicable for a particular particle size, and an overall efficiency requires integrating the efficiencies for various particle sizes. Since this equation is apparently satisfactory over small particulate ranges, a typical inlet dust analysis that is reported in small particulate ranges can be used to estimate the overall efficiency.

Collection efficiencies for electrostatic precipitators are normally estimated as 99%, although higher efficiencies can be achieved.[18,33] Submicron collection efficiencies have been reported in the literature that are greater than 99.0%.[13] However, the actual efficiency for submicron particles apparently depends upon particulate size. For particulates larger than 2–3 μm, collection efficiencies in reference 18 are greater than 99.0%, which falls to 91–96% as the particle size is reduced to the 0.4–0.6-μm range. The efficiency then recovers to 95–99% for particulates smaller in size than the 0.4- to 0.6-μm particles. Submicron particle size control efficiencies for electrostatic precipitators are almost equivalent to bag filter control efficiencies.[18]

Resistivity. A very important consideration in precipitator performance is dust resistivity, which is the specific electrical resistance of the dust or the electrical resistance of a unit cube. When the resistivity is low ($< 10^4$ Ω-cm), particles are difficult to collect because of a loss in charge at the collection electrode. High-resistivity particles ($> 10^{10}$ Ω-cm) are also difficult to collect efficiently owing to problems associated with charging these high-resistivity particles, and when collected interfere with the electrostatic field between the dust layer and electrode.[14]

Optimum resistivity for high collection efficiencies is in the range of 10^4 to 10^{10} Ω-cm.[24] The high collection efficiencies reported are based upon particles with satisfactory specific resistances.

Conditioning agents are frequently injected into the gas–dust stream at the precipitator inlet that affect particulate resistivity. These agents are particularly effective with high-resistivity solids. Some of the agents successfully injected into the gas stream are sulfur trioxide, steam (moisture), ammonia, and sodium chloride. High collection efficiencies can be achieved in many instances with conditioning agent addition.

Pressure Loss. A definitive calculation technique is not available for determining the pressure loss across a precipitator. Normally the operating pressure loss varies between 0.5 and 1.0 in. of water with a 1.0-in. water loss usually considered the maximum pressure drop. For an estimate of pressure loss through an electrical precipitator, 1.0 in. of water is a satisfactory assumption.

Although calculation procedures are not critical in many installations, for some applications a relatively accurate estimate of pressure loss is necessary. Electrostatic precipitator manufacturers can provide an estimate of pressure loss based upon experience and pilot test results.

Operating Conditions. Electrostatic precipitators will operate in high-temperature environments and operating pressures of 800 psig (56 bar) have been reported.[13] Operating temperatures above 1200°F (649°C) are mentioned in the literature, indicating that high-temperature operations are not an infrequent occurrence.[13,33] While many precipitators are installed in flue gas, metal processing gas, and fluid catalytic cracking regenerator gas streams, temperatures are generally much lower than 1200°F with 500–700°F (260–371°C) considered a more practical level for maximum operating temperatures.

High-temperature operation of precipitators is costly since the precipitator structure and all material components must be adequate for continuous operation at the design temperature. Thermal expansion of precipitator components is also very important, requiring an effort to ensure the electrostatic fields are not distorted. In addition, collection efficiency is reduced at high temperatures owing to changes in ion effects, viscosity changes, potential (voltage) changes, and reductions in migration velocities. Consequently, operation in the 500–700°F range compared with a higher temperature source results in lower investment and higher collection efficiency and permits energy recovery between the heated gas source and the precipitator. Operating the precipitator near the hot gas–dust stream source results in high-temperature operation with its attendant costs and also requires insulating the precipitator to minimize energy loss, which increases the average temperature of the metal internals. Also, any reduction in collection efficiency may not be acceptable environmentally.

Although very high-temperature operation is possible under certain conditions, another aspect of the operation that must be considered is disposition of the collected solids. The powder or dust is quite hot and hoppers, rotary valves, conveyors, and so on must be designed for powder recovery temperature conditions. Moreover, all exposed surfaces must be insulated to prevent personnel contact with hot surfaces. Normally, metal temperatures above 140°F (60°C) should be insulated for personnel safety.

While high-pressure operation is possible, these pressure levels are accompanied by large gas density changes compared with low-pressure operation (1–5 psig). Experimental corona starting voltages increase with increasing density, requiring changes in the voltage system, although the collection efficiency is apparently not affected.[38] However, high-pressure operation requires extensive structural modifications to the precipitator, which are also expensive. In addition, hoppers and all associated equipment in the solids recovery transfer system must be capable of high-pressure operation to the point at which the pressure is reduced or lowered. At the pressure reduction point, air or gas expansion may require changes through the downstream system to control particulate emissions generated during the pressure drop step.

Wet Electrostatic Precipitator. A wet electrostatic precipitator was described which consists of a vertical wire or rod within a vertically mounted pipe. Water or other liquids can be sprayed into the precipitator. This wet system is possible in a two-stage precipitator with a separate ionizer that permits a very high noncorona field in the tubular pipe precipitator.[23] The electric field from the rod to the pipe changes in flux density and is maintained at a high electric field strength, resulting in collection efficiencies $> 99\%$ for submicron particles. Wetted walls prevent reentrainment, which aids in maintaining a high collection efficiency.

The tubular system has a larger distance between the collector pipe and the discharge electrode than the conventional plate-type precipitators, permitting water spraying without shorts occurring in the tubes. In addition, the larger spacing does allow higher solids loadings with the gas stream. While this type of precipitator is an effective collection device, particularly for submicron particulates, the solids material recovered in the precipitator are wet and the solution can be a heavily laden slurry. The slurry requires transfer and possibly drying, which have the associated problems previously described.

Advantages. The advantages for electrostatic precipitators are:

- They can process very large stream flow rates.
- Precipitator operating temperatures can vary from ambient temperatures to temperatures in excess of 1200°F (649°C).
- Operating pressures can range from ambient pressure levels to pressures of 800 psig (56 bar).
- Pressure drops are very low and range from 0.3 to 1.0 in. of water.
- Particulate collection efficiencies of 99% or greater can be achieved and maintained for particulates in the 1- to 10-μm range. Submicron efficiencies may be somewhat lower, but precipitator changes can be incorporated that increase submicron collection efficiency to 99%.
- Particulate drying and handling is not a problem when operating in typical dry particulate recovery operations.
- Power consumption is low for this operation.
- Considerable operating experience in most applications.
- Wet tubular precipitators are effective in applications with liquid particulates, fumes, tacky particles, and submicron particulates. Very high collection efficiencies ($> 99\%$) are achieved in two-stage wet tubular precipitators.

Disadvantages. Some of the disadvantages for electrostatic precipitators are:

- High initial investment cost.
- Very large plot areas are required for electrostatic precipitators.
- Low- or high-resistivity particulates have low collection efficiencies.
- Injection of conditioning agents may be required to improve the collection efficiency. Requires continuous, reliable operation or collection efficiency will vary.
- Changes in flow rate, solids loading, and/or particle size distribution affects collection efficiency and precipitator performance. Variations in solids resistivity also affects collection efficiency.
- Potential fire and explosion hazards are present in electrostatic precipitator operations. Ozone is present and sparkover may occur; discharge wires can break and arc-combustible components in a gas stream may then explode under some of these conditions. Also, dust explosions may occur as a result of various ignition sources in the precipitator.
- Electrical conditions must be carefully controlled at all times. Sparkover and discharge electrode breakage can readily occur.
- Highly skilled maintenance personnel are required, particularly electrical technicians.
- Wet precipitators require concentrators and drying systems.
- Maintenance of internals requires worker protection with dust coatings on most surfaces and larger volumes in dead zones.

14.5 PNEUMATIC CONVEYING

Pneumatic conveying systems are in widespread use to transfer solid particulates that vary in size from talc-like powders to granular materials for a range of materials that include catalysts, resins, fibers, chemicals, sand, cement, and metal dusts. The pneumatic systems are completely enclosed, which essentially eliminates dust generation or atmospheric dust emissions. Where the solids particulates are considered hazardous, the enclosed transfer systems minimize potential exposure hazards and aid in meeting regulatory compliance requirements.

Dust emission control is an important aspect of these installations, but pneumatic conveying systems also have some additional advantages when compared with other transfer systems:

1. Space requirements are small.
2. Readily installed in existing facilities and buildings.
3. Versatile and can be changed quite readily, with a great variety of configurations possible, including multipoint feeding and transfer to a number of receiving points.
4. Adaptable to automation and computer control.
5. Maintenance is generally less for pneumatic systems.

6. Pneumatic systems reduce fire and explosion hazards.
7. Cleanliness of workplace areas improves considerably and solids contamination is reduced.

Although there are numerous advantages for pneumatic conveying, as described, other systems are superior in some areas:

1. Operating costs are satisfactory but somewhat inferior to other systems.
2. There are some limitations on the types of materials that can be transferred.
3. Although pneumatic conveyors can transfer solids materials reasonable distances, particularly within plant areas, some mechanical systems move solids over very long distances.
4. The capacity of pneumatic systems has increased significantly, but is inferior to some competitive mechanical systems.

As shown, pneumatic systems have some distinct advantages. However, in numerous situations, nonpneumatic transfer equipment is preferable, which covers a great variety of solids handling equipment. These nonpneumatic systems will not be addressed in this section, but equipment details and application criteria can be reviewed in various references.[28, 38–40]

14.5.1 Characterization

Solids characteristics have an effect upon the type of pneumatic systems installed and the type of equipment required to optimize operation. Terminology and various characteristics are presented in the following list.

Cohesiveness. This term refers to particulates that stick or adhere to various surfaces in the pneumatic systems. Moreover, their cohesiveness may clog hoppers, rotary feeders, and the conveying piping system. This material is not necessarily tacky, but physical characteristics result in the effects described.

Erosion. Where certain particles are hard or demonstrate erosive qualities, equipment wear is a problem, particularly in pipe bends and feeders. One method of minimizing wear is to restrict the conveying operation to dilute pneumatic conveyor gas–dust streams to control the dilute stream velocity at a minimum level for the particular solids concentration. Pipe bends, which suffer from erosion, can be modified to reduce wear through several methods. One method of improving bend performance is the installation of thick-walled bends that are available commercially. In addition, facing materials and coatings can be placed in a bend. The inner coverings can be rubber linings, hard-facing materials, ceramic coating or parts, and so on. A special vortex bend design also reduces erosion. Avoidance of moving equipment parts, such as feeders, also aid in minimizing erosion.

Particulate Size. Powders that have very high concentrations of submicron particles and those $< 10\ \mu m$ frequently coat the piping and equipment. Thus, in the piping, the coating reduces the piping cross section and increases the pressure drop, which effectively reduces throughout. One solution is the use of flexible piping that can be shaken, although this is only effective in certain situations. Blow tanks are the preferred solution for this problem and are described in Section 14.5.3. Typical examples of materials that coat piping and equipment are carbon black and titanium dioxide.[41]

Friability. This term describes materials or particles that fragment easily as a result of striking other particles or solid surfaces. Apparently most of the breakage or fragmentation in pneumatic systems occurs in pipe bends or feeders, particularly screw feeders. Fragmentation can be reduced by minimizing the number of highly angled bends and installing nonscrew-type feeders.

Granular Solids. These solids can be conveyed when they are fed properly into a conveying pipeline. Some problems may occur with rotary valves, which may shear the particulates and top-discharge blow tanks. When granular solids have a considerable fines fraction and this solids material cannot be conveyed in dense phase, the top discharge line has a tendency to plug.

Low Melting Point. Softening temperatures of about 300°F (159°C) for some materials, notably polymers, may melt to some extent when the particles at high velocity collide with the wall or bends in the pipe line. Streamers form and some buildup occurs at the impact points. The solution for this particular problem is to convert from high- to low-velocity conveying lines.

Toxicity. This characteristic may be solids toxicity, vapor toxicity, or a combination of both. In many instances, the carrier gas is air and toxicity is associated with the solids. In the latter situation, air separated from the powder can be recycled or vented to the atmosphere. When the carrier vapor is toxic, or toxic vapor traces from the solids are present in an inert carrier gas, the gas is recycled or treated to control the toxic vapors.

Combustibility. Polymer resins, chemicals, metal powder, and coal are included in this general category. Where a closed system is desired, which recycles clean gas from the collector, the oxygen level must be controlled to a satisfactory or safe concentration that can be determined from standard combustion limit tables. Alternatively, nitrogen can be used as the carrier gas and the nitrogen is then recycled, minimizing the possibility of a dust explosion. In addition, relief vents are normally installed in collectors for explosion relief venting. For those situations where explosions are infrequent and explosion relief venting is questionable, explosion suppression systems can be installed.

Wet Solids. Some materials may be wet or damp, but still can be fed into the conveyor system. However, discharge from surge or storage hoppers can be a problem, including feeder or mechanical transfer. The problem of cohesiveness and agglomerization becomes more difficult with fine concentrations generally. Where the concentration of solids < 50 mm is 10% or more and fines are included, wetted materials can be a problem.

Wet fines will coat pipeline bends and generally block the lines. For pneumatic conveying, blow tanks will transfer the material. In standard pneumatic conveying services, heated conveying air will aid in moving the solids, but this solution is only applicable where the moisture level is relatively low.

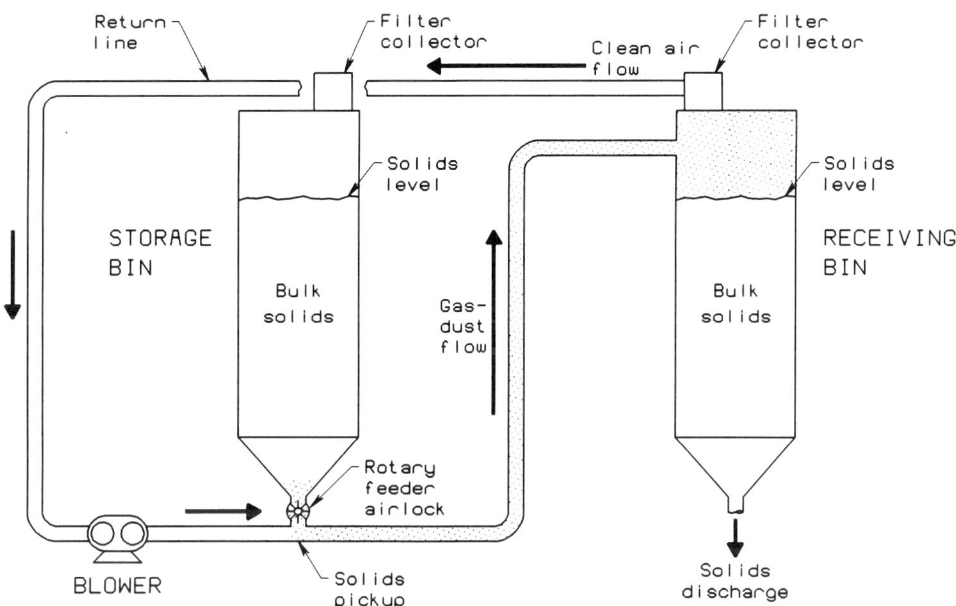

Figure 14.21 *Closed-loop pneumatic conveying system.*

Hygroscopic Solids. In pneumatic conveying systems, dry air supplied to the system is generally satisfactory for these solids. Methods of drying the air are beyond the scope of this book.

14.5.2 Pneumatic System Characterization

In the previous section we mentioned open and closed systems, which are described in this section. In addition, system pressure is a major consideration in pneumatic conveying systems and explanations of the basis for pressure selection along with other system characteristics are presented.

Closed System. The basic configuration of a closed or recycle system is shown in Figure 14.21. Air from the blower picks up solids at the junction (venturi) below the rotary feeder and transfers the combined stream through the pipe to the receiving drum. In the receiver, air disengages from the main dust stream and the air with some dust enters the filter collector, where the dust in the air stream is collected and purified air is recirculated in a closed loop to the blower. This is a completely closed system.

The closed system is preferred when the solid particulates are toxic, explosive in air, a nonair carrier gas is required, the carrier gas is toxic, or there are other problems. Continuous operation is preferred for recycling systems.

Open System. The open system does not have a recycle line from the baghouse on the receiving drum to the blower. Clean or filtered air from the drum baghouse is discharged to the atmosphere. In addition, the blower suction is open to the atmosphere, but the air is normally filtered before entering the blower. Thus all vapor enters the atmosphere and inert or other gases cannot be substituted for air in the transfer line unless the costs are acceptable and the gas released to the atmosphere is acceptable environmentally.

Most of the systems installed in the chemical industry are open systems. Solids emissions are controlled through closed pneumatic line and dust collectors to ensure that environmental regulations are observed. Since the vent gas is released to the atmosphere, potentially explosive conditions in open systems are controlled through safety precautions which are reviewed in another section.

System Density. The concentration of solids in the conveying air is an important consideration in the design and operation of pneumatic conveying systems. This concentration or solids loading is normally termed phase density in various references.[15,41] Phase density has been defined as the ratio of the mass flow rate of the solids to the mass flow rate of the gas or air. Since the ratio consists of identical terms, the phase density is dimensionless and is independent of system pressure.

The phase density in the conveying system is normally described as dilute phase or dense phase and in some situations medium phase is also used. Dilute phase is also termed suspension flow since particulate concentration is quite low. Associated with these phase descriptions are other criteria that further define the terminology:[15,42]

State	Phase Density	Velocity
Dilute phase	8 max.	49 ft/s minimum
Medium phase	8–40	20–36 ft/s
Dense phase	> 40	10–20 ft/s

As suggested, in the dilute phase there is little particle interaction, and the particles are airborne at all times. In the medium phase condition, a higher particle concentration exists and there is increased particle interaction. Dense phase conditions have very high solids concentrations which can be considered slightly fluidized dense beds or plugs of materials separated by air spaces. These dense phase conditions indicate the mass slides through the line and are not conveyed as individual particles in the dilute phase. The solids in the dense phase are not in suspension, and conveying is conducted at lower velocities. However, the pressure drop per unit pipe length is considerably larger for dense phase operations compared with others at identical solids mass flow rates. The main advantage for dense phase operation is the large solids mass transfer rates that can be achieved.

Since velocity is an important criteria in conveying operations, the following equation determines the velocities in the system:[41,42]

$$V_c = 4P_0Q_sT/\pi d^2 pT_0 \tag{14.26}$$

where

V_c = conveying air velocity (m/s)
P_0 = standard pressure (1.013 bar abs.)
T = air temperature (°C)
Q_s = volumetric flow rate (STP) (m^3/s)

T_0 = standard temperature (15°C)
d = pipe diameter (m)
P = air pressure (start point) (bar)

For dense phase operation, an approximate relationship follows:

$$V_{\min} = 33\phi^{-0.45} \tag{14.27}$$

where $V_{\min}$ is the minimum conveying air velocity (m/s) (for $10 < \phi < 150$) and ϕ is the phase density (dimensionless).

14.5.3 Methods of Transferring Solids

Within the general definitions of closed and open systems are various types of pressurization systems that supply the motive power for solids movement. These general systems are described below.

Positive Pressure Systems. These systems are widely used in industry and a typical model of a general transfer system without the recycle loop shown in Figure 14.21 would represent a positive pressure transfer system. Blower pressures are generally less than 1 atm for this particular system. In general, a variety of solids feeding systems can be used with the pressure system. Venturi, rotary feeders, and screw feeders are normally used in this solids transfer unit. Since there is a positive pressure on the conveyor, solids can be moved to various receiving bins or silos by installing diverter valves in the conveying pipeline of the type shown in Figure 14.22. This multiple-bin type receiving system has been installed in numerous chemical facilities for a wide variety of solid materials.

While the positive pressure system has some obvious advantages, a major consideration is the feeding system, which must operate under pressure. The rotary feeder operates across a pressure differential and conveying air will leak through the feeder into the silo or storage bin. Air leakage, if large enough, reduces the overall volume of conveying air and can interfere with solids feeding. In most pressure installations there is only one solids pickup or feed point due to this potential air leakage loss through the rotary feeders. The total feeder air loss for a number of feed points would seriously affect the conveying operation. While one feed point is recommended for pressure systems, the pressure system is very effective in transferring solids to multiple receivers through diverter valves. Air leakage through rotary feeders is a major concern, but specific feeder requirements can reduce this leakage to acceptable levels.

Venturi feeder operation is based on the flow of the carrier gas from the blower through a narrow transition section in the conveying line (venturi section) that reduces the static pressure. The differential pressure that is developed between the venturi throat and the vertical standpipe connecting the bottom of the bin and the venturi draws solid material into the venturi section where the conveying gas receives and accelerates the particulates. Solids flow into the venturi section from the feed bin is controlled to prevent overloading the conveying system which could result in blockage of the conveying pipe. The solids movement is generally controlled manually with a specific valve.

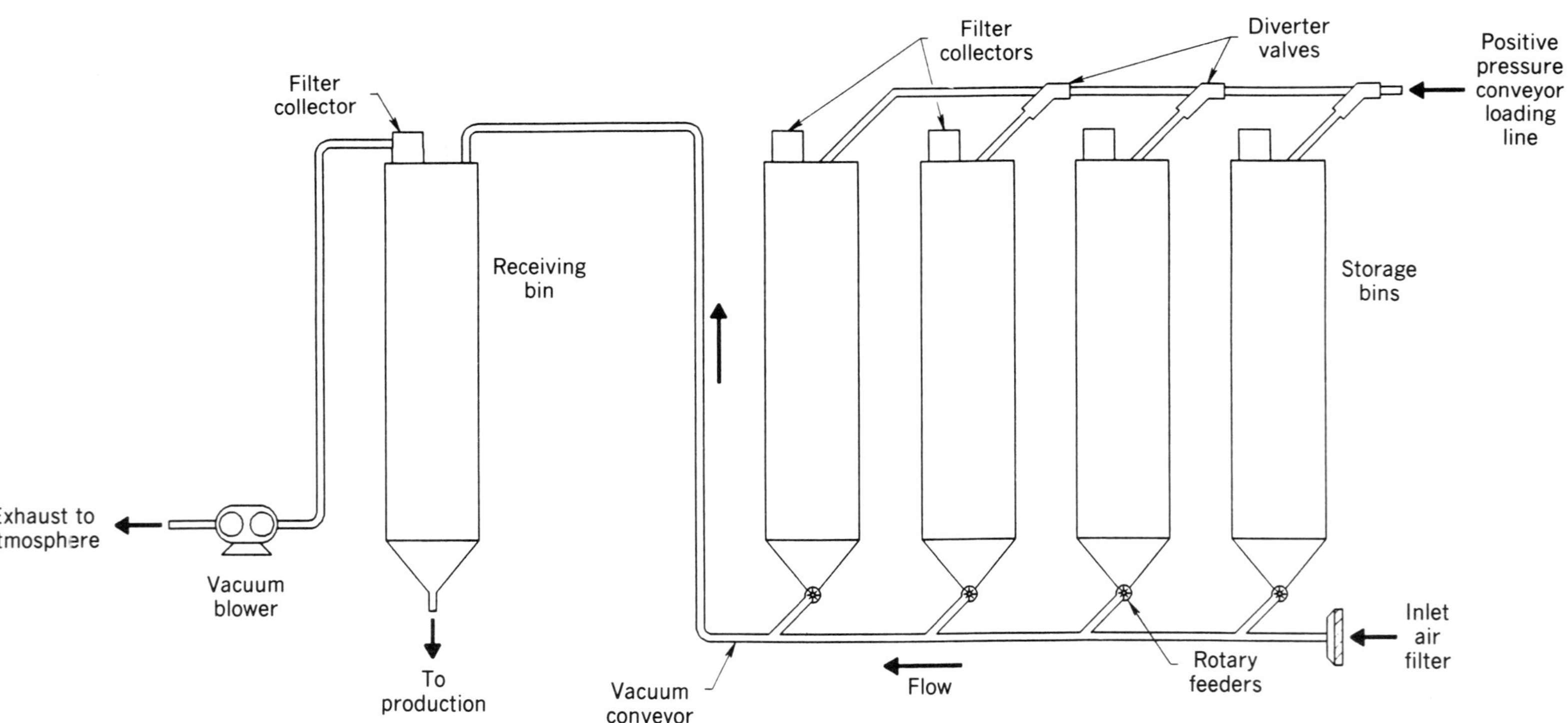

Figure 14.22 *Negative-pressure or vacuum conveying system with positive-pressure loading system.*

Screw feeders specifically designed to feed solids into pressurized conveying systems are available from various manufacturers. These particular feeders operate well but may present some degradation problems with certain types of solid materials. Additional comments on the screw feeder are presented below.

Negative Pressure Systems. These are also termed *vacuum* systems and are also widely used, particularly when withdrawing solids from several sources and sending the solids to a single receiving bin or silo as shown in Figure 14.22. Multiple receiving bins are generally not recommended for vacuum systems since an extensive and somewhat complex piping network is necessary. In addition, block valves are required that seat tightly to ensure leakage is minimal when the valves are closed. Air leakage into the vacuum system distorts the system air balance and conveying velocities. While this system is rather cumbersome for storage areas with large multiple receiving bins as described, vacuum systems for multipoint discharge are frequently installed in small installations such as resin extrusion facilities. In these small installations, valve controls are automatically sequenced with satisfactory vacuum system operation. The vacuum systems are particularly advantageous in unloading applications, such as rail car and truck unloading. The vacuum system eliminates the need for pressurized railcars or pressurized tanktrucks. Moreover, vacuum systems permit the use of portable nozzles and hoses to remove solids from either bulk piles, spills, or as an industrial vacuum cleaner.

One major advantage of vacuum systems is that leakage is inward rather than outward and dust leakage is nil. Since the maximum vacuum is about 7.25 psi or 7.45 psia (0.5 bar), toxic materials are retained within the system. In addition, the relatively low pressure eliminates the leakage of air through rotary feeders. However, the low operating pressures require consideration of bin wall thickness which may increase because of lower operating pressures and increased bin investment. Consequently, overall investment comparisons must include changes required for negative pressure operation. While a 0.5-bar abs. pressure is about the lowest practical operating pressure, many negative pressures on an absolute basis are higher, which changes some of the mechanical requirements.

Combined Negative and Positive Pressure Systems. Combined systems have the advantages of both the negative and positive pressure systems. The vacuum system transfers solids materials from several bins to a single bin and the pressure system transfers solids from a single point to a number of hoppers. The logical extension to the installation of separate positive and negative blower systems is combining both functions into a single blower. Since the vacuum blower side cannot draw solids through the blower or fan, an intermediate separation bin is required that collects solids on the vacuum side. These collected solids are then moved from the intermediate bin by the pressure side of the blower to storage bins.

This separate, overall system is modified by replacing the intermediate bin with separator devices and compacting the system into essentially a single unit within the dashed lines, as shown in Figure 14.23. The general system is designated a pressure–vacuum system and is also termed a push–pull unit. Compact units of this type are widely available commercially and vary considerably in size with very small portable package units at one end of the range.

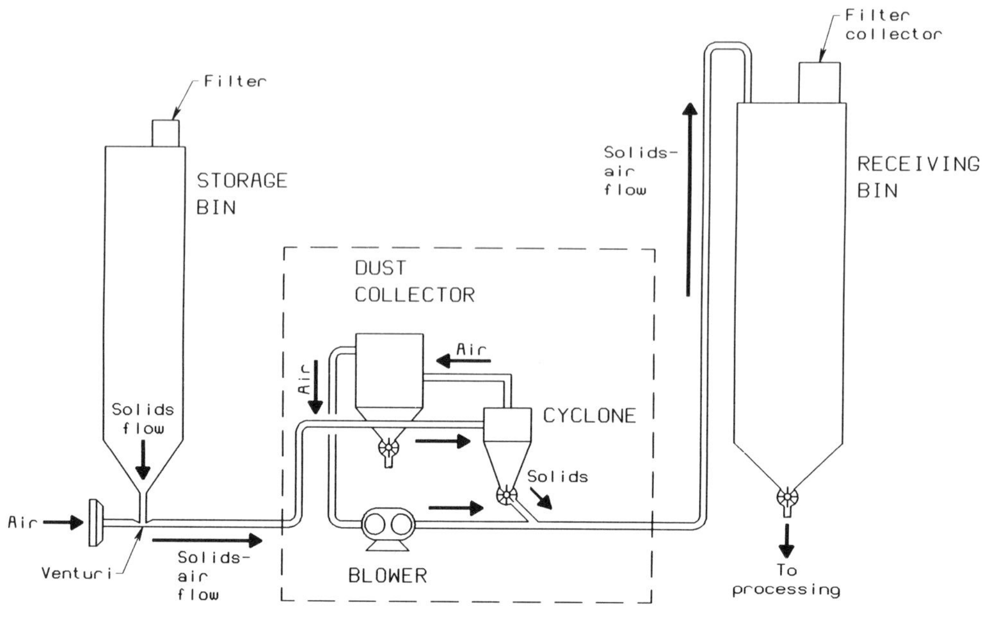

Figure 14.23 *Pressure – vacuum conveying system.*

In the pressure–vacuum system, the vacuum air side withdraws materials from the storage hopper and the air–solids stream enters the cyclone. Clean air with some dust traces from the cyclone overhead outlet flows through a fabric filter collector to remove essentially all of the dust. Since the great mass of the incoming solids are separated in the cyclone and discharged continuously from the bottom cone, the amount of solids carryover into the filter is very small. In most pressure–vacuum systems, the dust buildup is gradual and solids recovery from the filter bottom cone is conducted manually on an infrequent basis. Where this system is quite large, alternative methods of filter dust recovery can be installed. Filtered air from the collector is sent to the blower suction side and compressed to provide the pressure necessary for conveying solids from the cyclone bottoms outlet to the receiving bin.

The major advantage of the pressure–vacuum system is that materials can be transferred from multiple sources to multiple receiving points. While the transfer of solids from single or multiple bins to multiple receiving points is typical of many plant operations, this general technique is also applicable in unloading railroad cars and trucks. Vacuum connections to railcar or truck may be limited to a single unloading connection. However, the pressure side of the pressure–vacuum system discharges solids material into single or multiple receivers. Generally, railcars have large carrying capacities and the railcar is unloaded into a number of storage bins.

Since the blower in the pressure–vacuum system operates between a negative and positive pressure, the resulting differential pressure is a function of the blower type installed in the unit. In many installations, the total differential pressure across the blower is a maximum of 1 atm. Since the negative pressure may be approximately 40–50% of the differential pressure, the discharge pressure is

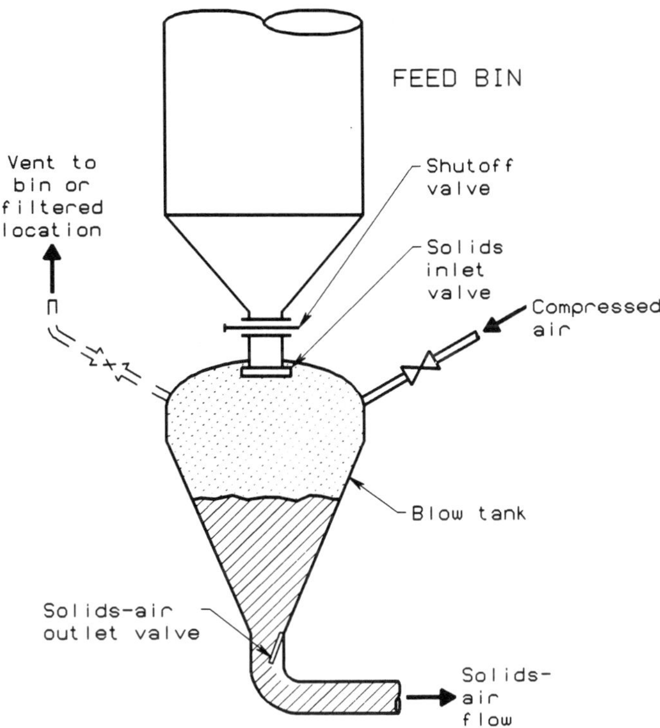

Figure 14.24 *Simplified view of blow tank with bottom discharge.*

limited by the vacuum pressure. Consequently, the length of the positive pressure conveying lines and number of bends in the lines are determined by the allowable blower discharge pressure.

Blow Tanks. While dilute and dense phase systems operate well, there are limitations in pressure drop which affect conveying distance. This obstacle is overcome with blow tank systems that can convey solids in medium and dense phase concentrations over long distances at very high tonnage rates. Blow tank conveying permits high-pressure operation of 120 psig max (9.3 bar) compared with a maximum pressure of about 29.8 psig (3 bar) for typical blower pneumatic systems previously described. A simplified view of a discharge blow tank is shown in Figure 14.24.

In operation, the blow tank is filled through the solids inlet valve with the bottom outlet valve closed. When the tank is almost filled, the upper valve is closed and the blow tank is pressurized with compressed air to a pressure of 100–125 psig. At this point, the bottom valve is opened and under pressure the tank solids content is rapidly discharged into the pneumatic conveyor line. The solids material is essentially delivered as a batch to the receiving point. Since the amount of air delivered with the batch system may differ from the amount of air in a dilute-phase pneumatic conveyor system, the filter collector or other existing particulate control systems should be reviewed to ensure they are adequate. The sudden depressurizing can create a rapid volumetric increase in flow that may require a larger collector than the device originally installed in the receiver. The pressure remaining in the blow tank and line as the last of the solids enters the receiver is a function of the valve control system. These blow tank systems are frequently installed to transfer solids over longer distances than is possible with low-pressure pneumatic conveyors, which also increase the volume of air in the system that is depressurized. Line lengths for blow tank systems of this type may be 200 ft (61 m) or longer.

The batch operations for a blow tank, although on a very short cycle time, are not continuous. Two blow tanks in parallel provide a more continuous flow pattern and are frequently used for this purpose. Some installations may have two tanks in series, with the downstream tank sending the solids in an almost continuous stream flow to the receiving point. The blow tank in Figure 14.24 is a simplified sketch, with actual blow tank designs more complex as manufacturers have improved blow tank designs through the years.

Blow Tank Designs. The blow tank shown in Figure 14.24 has a bottoms discharge configuration, which was the original design for this conveying system. Other types of blow tanks are now in service that were developed to transfer various types of solids, move solids in a continuous fashion, and permit operation of these units as feeders. Simplified sketches of these other blow tank systems are shown in Figure 14.25. Dashed vent lines are shown in Figures 14.24 and 14.25 that relieve the pressure in the blow tanks and permit purging of the blow tanks. These vents may be part of the filling cycle, depending upon the design, or may be vented separately. Since flow through the vents will normally include solid particulates, discharge emissions are a concern. The vent line should be discharged into a bin or system containing a filter collector, permitting emission control of the vent stream. Discharging the vent directly to the atmosphere without particulate control may not be permitted since the vent is a point source and PM-10 requirements must be observed along with potential exposure regulations.

Applications. As shown in Figures 14.24 and 14.25, blow tanks have top and bottom discharge line connections. Top discharge lines are generally installed for movement of lighter-weight materials and solids with high fines concentrations, including powders. High solids transfer rates are attainable in this system. In addition, this discharge method provides satisfactory dense phase operation along with very good flow control characteristics. Bottom discharge lines are generally used for the movement and transfer of granular materials. Granular materials have high permeation rates and fluidization is difficult. In addition, bottoms discharge blow tanks are installed where a more dilute phase-conveying system is required. If the granular materials have a large percentage of fines, resulting in low permeation rates, the solids are generally transferred with top discharge blow tanks.

Top Discharge. Two top discharge blow tanks are shown in Figure 14.25. The blow tank in Figure 14.25*A* is a simple pressurized system to move solids into the standpipe (outlet line) as a dense phase. While this system has operated satisfactorily, Figure 14.25*B* is an improvement in top discharge operations that provides more consistent solids flow with controlled rates. The air flow is separated in part *B*, with one air stream sent into the tank for fluidization of the solids and a

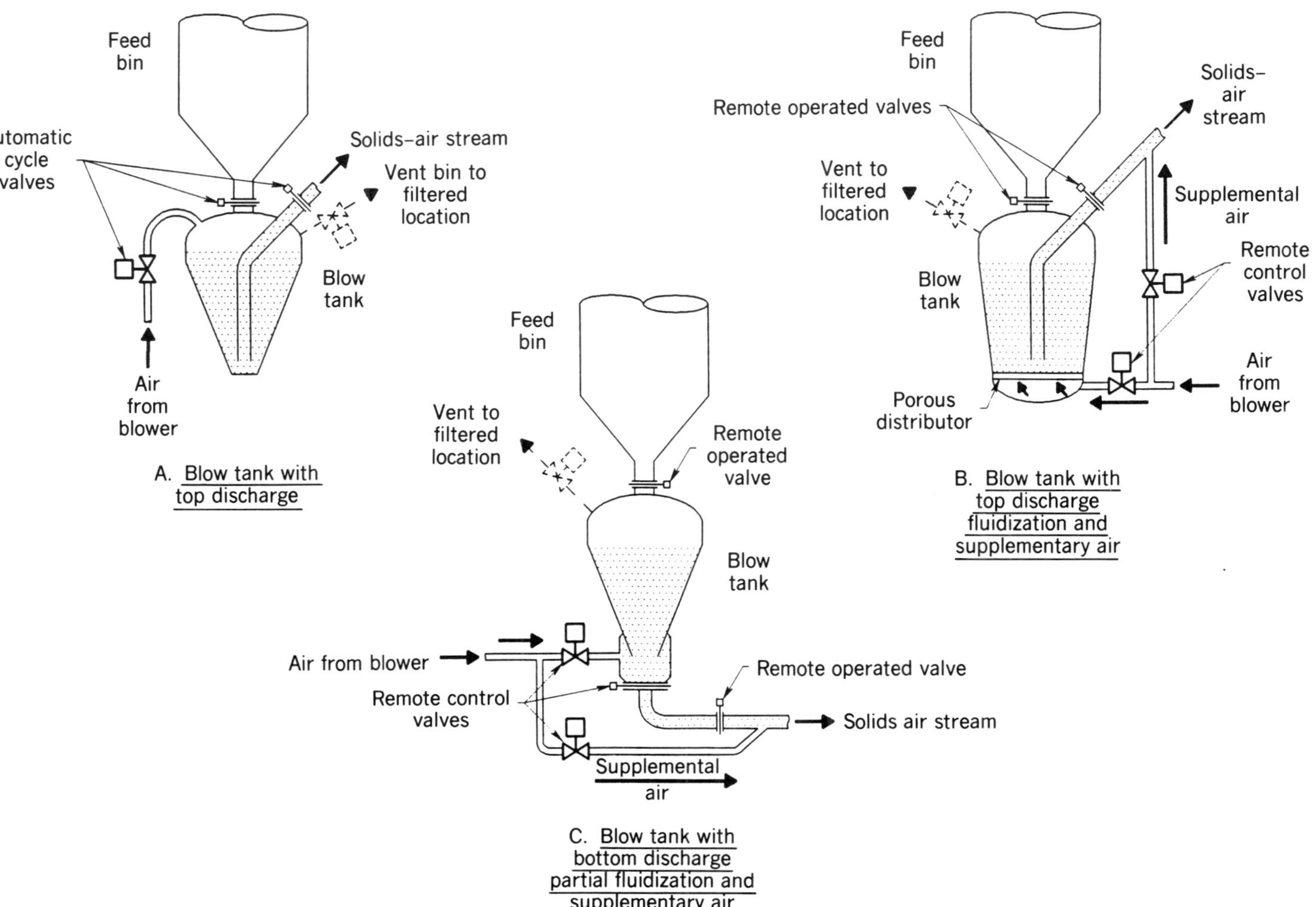

Figure 14.25 *Simplified views of various blow tank configurations.*

supplementary air stream injected into the top discharge line. Fluidization in the blow tank is aided by a porous distributor that distributes the air flow evenly above the distributor. The fluidized particles will flow into the standpipe as though the air–dust mixture is a liquid, resulting in a homogeneous solids flow. While a consistent flow until the tank is empty, the concentration of solids is still in the dense phase mode. Discharge rates are controlled by proportioning the air flow between fluidization air and the supplementary air line. If the solids flow is insufficient, additional fluidization air is diverted to fluidization and the supplementary air flow is decreased. Flow controls in this system are necessary to maintain consistent flow and adjust the air streams as necessary. Although not shown in Figure 14.25, pressure sensors located in the conveying line, blow tank, supplementary air line, and the main air line control the system flow rate automatically. In addition to controlling the solids flow rate, the control system optimizes air consumption and energy requirements.

Bottom Discharge. A bottom discharge blow tank system is shown in Figure 14.25*C*. The main air supply is separated into a stream injected around the base of the cone and supplemental air stream discharged into the conveyor line. Air around the cone base provides some permeation gas for minor aeration and conditioning with the main bulk directed toward particle acceleration into the conveyor line. Since the solids are generally transferred in a more dilute phase, the air is proportioned to achieve the desired solids flow rate between the cone and supplemental air line. Where the solids rate is relatively high, the line pressure increases and the air rate is decreased to the blow tank cone and the supplementary air rate is increased. Where a low line ΔP is encountered, the supplementary air rate to the line is decreased and the air rate to the blow tank cone is increased. Pressure sensors are also required for this system and control the valves automatically.

Feeder Installations. Blow tanks do not have any moving parts, and in some blow tank designs where valves or flappers are installed to fill or empty the blow tanks, these devices do not generate fines to any extent with friable solids. The lack of moving parts is also advantageous in handling abrasive solids. In addition, when used as feeders, blow tanks minimize product degradation and erosion. With the advent of dual tanks and controlled feed rates, the soft product-handling characteristics permit use of the blow tanks as feeders in low differential pressure applications. Since pressure levels are not a problem, blow tanks can also be used as feeders in high-pressure applications.

Operating at relatively high pressure levels requires classification of the blow tanks as pressure vessels. This classification adds to system cost, but provides a measure of safety in operations.

Equipment. Several types of equipment are required in pneumatic conveying systems and an understanding of their function and maintenance aids in assessing potential exposure situations.

TABLE 14.7 Blower Application in Positive-Pressure Conveying Service

Maximum Discharge Pressure (psig)	Air Movers
1.5	Twin lobe rotary blowers, axial and centrifugal fans
15	Twin lobe rotary blowers
60	Sliding vane, liquid ring
150	Rotary screw
> 150	Reciprocating

Blowers. Since the pressure differentials required in blowers and/or fans varies with the type of pneumatic system, blowers are selected in accordance with their performance characteristics. A general comparison of blower or fan operating ranges with maximum discharge pressure is shown in Table 14.7 for positive pressure applications.

The table indicates the maximum pressure level for the various compressors, but should not be considered the operating pressure level for the compressor. As an example, rotary screw compressors are frequently installed in systems with much lower operating pressures. The basic requirement of pressure level and air volume indicate the type of machine necessary for the specific application with final selection based on economic factors. Included in any analysis or evaluation are operating characteristics associated with individual machines. Rotary screw compressors[13] have two parallel intermeshing screws and normally have oil lubrication. Obviously, oil droplets contaminate the compressed air, which in turn contaminates the solids in the pneumatic system. Consequently, an oil removal system is installed to remove any oil in the compressed air. Sliding vane compressors may also have oil injection and in this case should have an oil removal system.

Liquid ring machines[13] are used as compressors and are also installed as vacuum pumps. These machines normally have an internal liquid ring of water although another fluid may be used. Because of water–air interaction, the air humidity increases, which must be considered when the machine is used as a positive-pressure air source. If the liquid ring machine is used in vacuum conveying service, some dust may be removed from the air stream by the liquid since dust collection in the upstream dust collector is not 100% efficient. The dust in the circulating liquid may require removal or an additional upstream filter may be required.

Rotary lobe machines[13] are frequently used in negative–positive pressure (push–pull) systems where the absolute pressure on the vacuum size is about 7 psia (1 kPa) and the discharge pressure is about 10–12 psig (69–83 kPag). In push–pull service or as a vacuum pump, solids carryover into the machine must be minimized. Dust enters bearing seals and can score the lobes, requiring machine repair. Highly efficient upstream filters minimize these problems. Dust entering reciprocating compressors is a significant problem if the machines are in a service where dust carryover exists.

Sizing and selection of blowers is a function of air volume flow rate and the pressure required in the system. Calculation of the pressure drop in the system is not an exact science and various techniques have evolved to estimate pressure losses in a system.[43] The most effective estimating technique is based upon testing the proposed conveyor material in a pilot plant system and extrapolating the data for the proposed conveying system design. Pneumatic conveyor suppliers have test equipment and based upon extensive experience can estimate pressure loss quite accurately. Generally, pressure drop calculation procedures are based on iterative techniques, which also include adjustments for changes in piping diameter as the volume increases with falling pressures along the length of the conveyor line.

Feeders. In most systems, bulk solids are fed from atmospheric storage bins to the conveying pipeline. This is accomplished with a vacuum-conveying or positive-pressure system. In the vacuum system, feeding solids into the line is not difficult, but maintaining constant flow is important, and constant feed control is relatively straightforward. Feeding a pressured conveyor line is a more difficult problem since there is an opposing pressure gradient and solids flow must be constant under these conditions in many installations. As a result of the pressure gradient, air leakage through the feeding system can be a problem which interferes with solids flow and affects blower operation. Typical feeder systems are shown in Figure 14.26.

ROTARY VALVES. The rotary valve, also termed a rotary feeder or rotary air-lock, is widely used in feeding applications and frequently installed in positive-pressure pneumatic conveying systems. Solids flow through the rotary valve is adjusted or controlled through the rotor revolutions per minute. Variable speed drives provide an extremely large flow rate change through rotary valves. Many installations have step gear changes or belt drives that can be modified to adjust the flow rate. While adjustments can be readily accomplished, constant rates may be preferable in most installations particularly, where the solids material has consistent characteristics and particle size distribution. The rotary valve in Figure 14.26 is an offset valve which refers to the offset locations from the center line of both the inlet and outlet nozzles. Since the rotor turns to move solids, air is entrapped between the rotor blades as it moves over the high-pressure outlet nozzle. This air is carried back into the bin opening as the blades rotate. Offset nozzles tend to reduce this flow. However, there is also some leakage at the end of the rotor where a small opening exists between the rotor and housing. This leakage effect is exacerbated if the valve is too large or operated at excess revolutions per minute. The rotary valve should be sized properly to minimize air or vapor leakage.

The rotary valves can be used in both negative- and positive-pressure installations. The maximum positive operation pressure is about 15 psig (103 kPag) and the minimum negative pressure is about 10 psia (1.45 kPa). In the negative pressure application, air leakage is not a problem. The major air leakage problem in a positive-pressure system occurs with fine, powdery, or fluffy material and vents can be installed to bleed off the leakage, as shown in Figure 14.27*A*. Granular materials do not present as difficult an air leakage problem and usually do not require venting. Vent leakage will contain particulates, and stream flow should be directed to a bin or device with solids collection.

Cohesive materials are tacky and may clog offset feeders, which are frequently used in feeding pneumatic conveyors. Blow-through rotary feeders are installed in

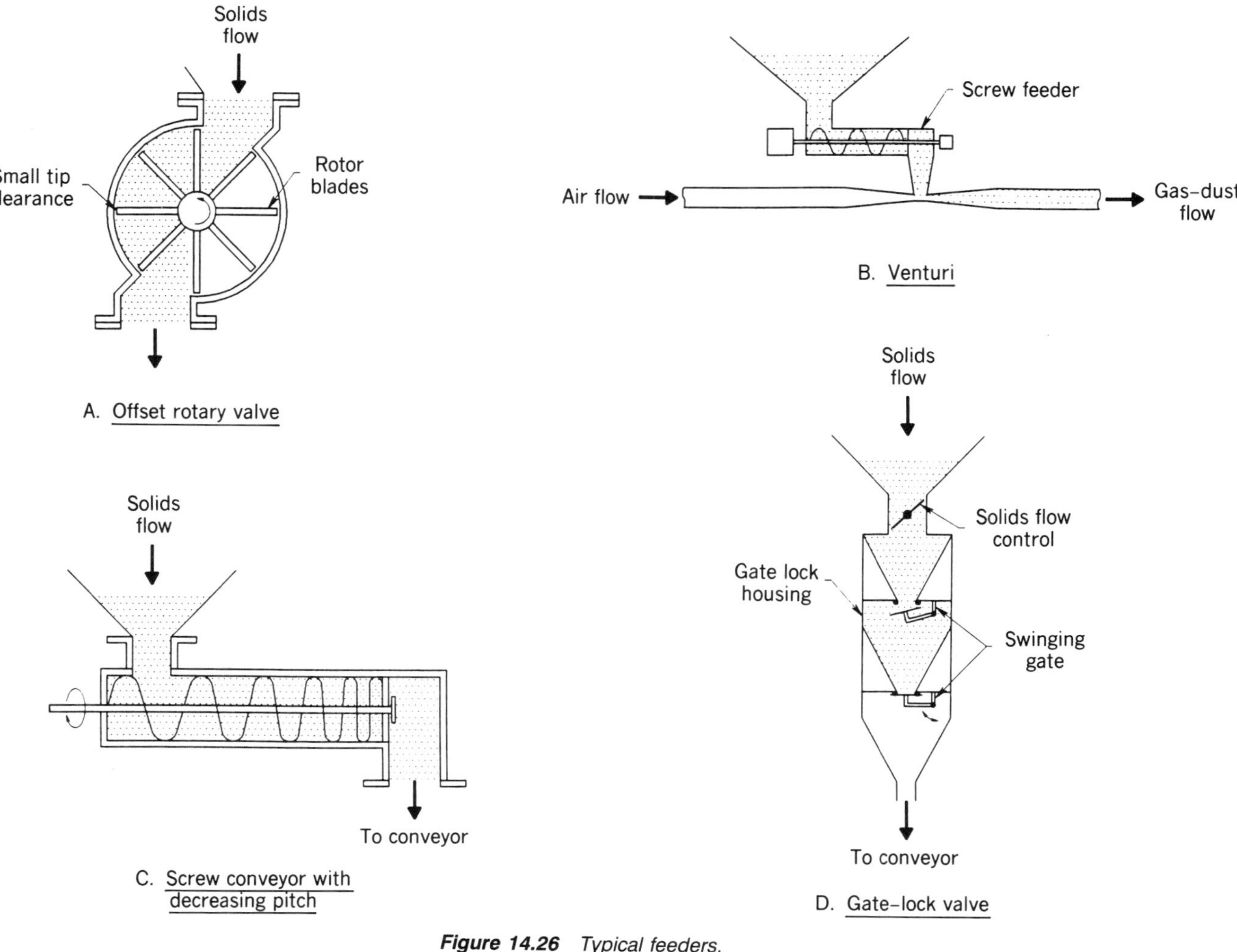

Figure 14.26 *Typical feeders.*

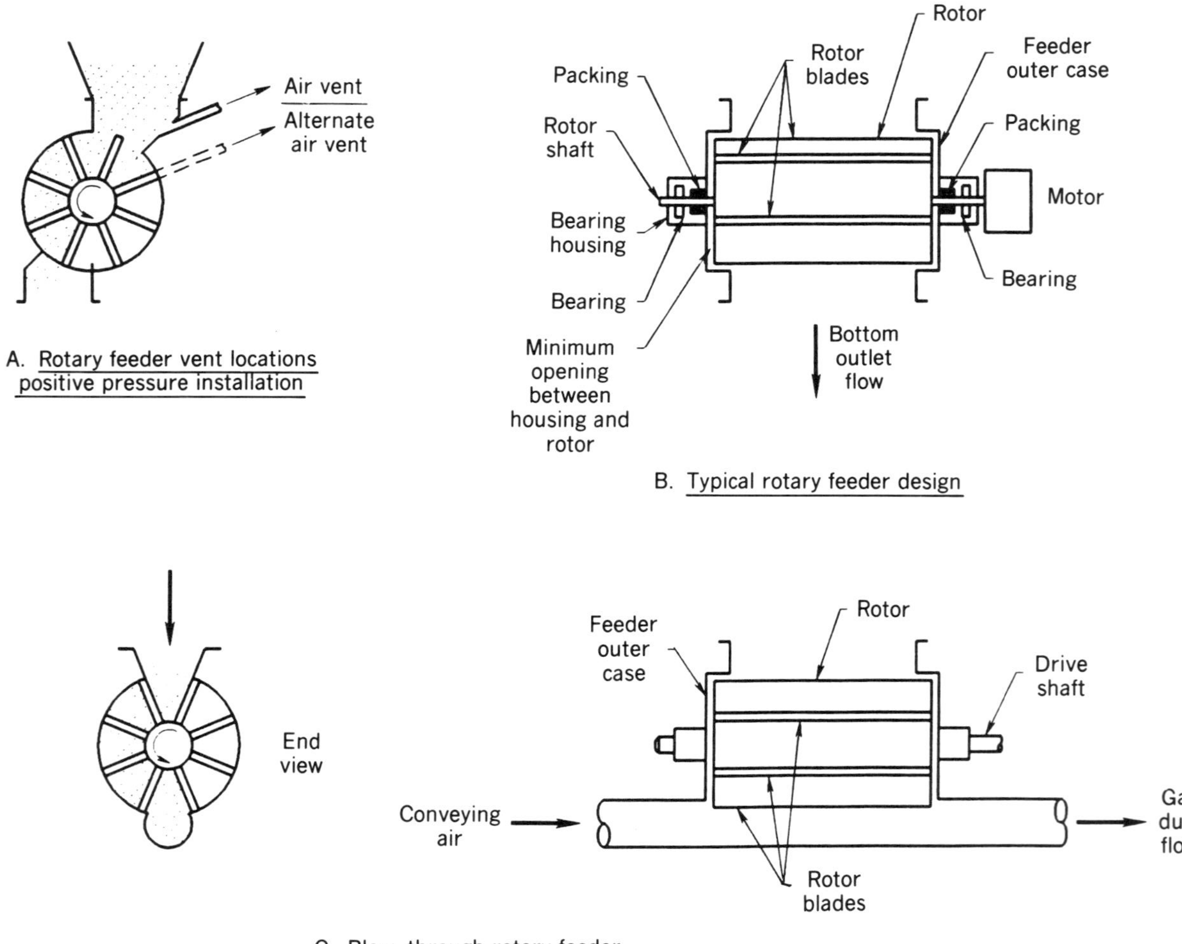

Figure 14.27 *Rotary valve feeders.*

this type of service and generally are satisfactory. In the blow-through feeder shown in Figure 14.27*C*, conveyor air flows through the bottom of the rotary feeder parallel to the rotor blades, assisting in the removal of cohesive materials. If the blow-through feeder is unsatisfactory, an alternative method of feeding cohesive materials is required.

The cross section of a rotary feeder shown in Figure 14.27*B* indicates the narrow opening between the rotor and rotor housing which also results in some air leakage. With fine materials or powders, the particulates are carried into this opening and gradually erode the packing seals that protect the rotor bearings. Dust entering the bearings through faulty seals can result in bearing seizure or failure. Various sealing systems are installed to prolong seal life and lengthen operating run time before seal replacement is required, but sealing alone may be insufficient and air blowback through the packing housing can be installed to minimize packing deterioration and lengthen service life. Frequent repairs increase potential exposure situations since breakdowns are emergencies and removing rotors or the entire rotary valve for repair under these conditions can lead to significant dust generation conditions. Blowback purges must be carefully designed to ensure the seal is adequately purged across the packing face and along the shaft. However, the purge must be controlled to minimize air entering the feed hopper system, disrupting operations.

VENTURI FEEDERS. The venturi feeder operates in a positive-pressure system although the pressure gradient can interfere with solids movement. A low static pressure at the solids feed point, as shown in Figure 14.26*B*, is produced by the reduced cross-sectional area at the venturi throat. This pressure is generally equivalent to the feed bin or atmospheric pressure and solids flow readily into the venturi section of the conveyor line. Moreover, air leakage into the feed bin is minimized, and solids flow is not hindered.

The low venturi static pressure requires a low conveying line pressure and an overall maximum pressure drop downstream of the venturi is about 4.5 psi (31 kPa). Consequently, venturi applications are generally limited to free-flowing solids for relatively short conveying distances. In addition, the solids flow rate to the venturi must be controlled or an overloaded venturi throat will block. In Figure 14.26*B* the actual solids flow rate is controlled by the rotational speed of the screw feeder. Belt feeders can also be used in this service. Solids flow control also ensures a constant pressure drop in the overall conveying system. An increase in solids flow will increase overall system pressure affecting the blower, which may result in blockage of the conveying line.

Venturi system construction is simple and maintenance is not a problem. The screw feeder shown in Figure 14.26*B* normally has few mechanical problems and generally exhibits a very high service factor. The major concern with a feeder is in the type of seal located at the motor end of the feeder. A seal should have a relatively long life and seal installation comments are similar to those mentioned for rotary feeder seals. The feeder shaft is also supported at the outboard end of the feeder and particulates must be prevented from entering the bearing support.

A simpler system has been shown[41] in which a flat damper-like plate was installed at the bin outlet to control flow. This system further simplifies the venturi feed system and minimizes maintenance problems. The overall venturi system operates well with free-flowing materials.

SCREW FEEDERS. The screw feeder shown in Figure 14.26*B* will normally not operate in a positive-pressure system since the screw system design does not build sufficient solids pressure in the screw to overcome the positive conveyor pressure. However, this problem is overcome by changing the pitch of the screw in the feeder. Solids will build in a decreasing pitch section, filling the flights and essentially forming a plug near the outlet opening of the feeder. The solids are injected into the pressurized outlet of the hopper continually through openings which maintain the backpressure built through the decreasing pitch section. The decreasing pitch screw section is similar to a pumping section in an extruder.

The modified pitch screw feeders operate well in positive-pressure systems against a maximum pressure of about 35 psig (2.5 bar). In addition, these decreasing pitch feeders work well with cohesive materials. Since the decreasing pitch screws build pressure, the power consumption for this type of screw feeder is considerably higher than the power required in constant pitch screw feeders. Varying the revolutions per minute to increase feed rate will also increase power consumption, requiring a properly designed power train for this type of operation.

Constant-pitch screw feeders leak or allow considerable reverse air flow in pressure gradient situations. Consequently, the constant-pitch screw feeder is frequently installed in vacuum service. These screw feeders will operate in vacuum service to a pressure level of 10.2 psia (−0.3 bar).

The screw feeders, particularly those operating at a relatively high pressure level, require careful consideration of the seal installation. Bearing failure or seizure can be a problem and seals must perform over a satisfactory time period for a reasonable service factor. Frequent failures require maintenance and increase potential exposures due to dust generation. Comments made about rotary valve seals are applicable to screw feeder designs.

GATE-LOCK VALVES. A simplified drawing of a gate-lock valve is shown in Figure 14.26*D*, which has been used in some applications to provide shutoff. This valve system moves material batchwide, resulting in an intermittent solids flow that limits installation to dilute phase conveying. In positive-pressure applications, air will leak through the bottom gate which can be vented from the upper chamber to a bin or hopper with dust collectors. The maximum positive-pressure system operating pressure for this valve is about 6 psig (41 kPag).

Gate-lock valves are also installed in vacuum service where the minimum operating pressure is about 11.7 psia (−0.2 bar). Since the design of these devices is simple, they are reliable and require little maintenance, making them suitable for abrasive and friable materials. However, their design minimizes the conveyor length that can be installed. For powdery or fine dusts, a variety of valves is available commercially.

Dust Collectors. Filters and cyclones were reviewed in Sections 14.4.4 and 14.4.5, and the comments in those sections are applicable to storage bins and hoppers associated with pneumatic conveying. As shown in Figure 14.22, filters are mounted on the storage bins. These filters are similar to the filters previously described, although the filter size is smaller since they are based on pneumatic conveyor air rates. The filter housing and equipment are designed for vacuum or pressure operation in accordance with the conveyor system design basis.

Sizing the filters is important in conveying system operation, rcquiring consideration of operating conditions. In vacuum or negative pressure operation, the actual air volume is larger than the air flow rate at standard conditions. Typical superficial cloth face velocities must be determined on an actual rate basis to establish the proper filter cloth area. While positive-pressure operations are above atmospheric pressures, the actual pressure at the filter face is reduced considerably and is the sum of the atmospheric pressure and the maximum pressure drop across the filters. In general, the pressure on the bin side of the filter is close to atmospheric pressure. Pressure relief vents or blowout devices are installed to control the rise in pressure.

Sizing the filter is more difficult in blow tank operations where the solids are frequently moved or transferred on a batch basis. When the blow tank becomes empty and the solids are still in the conveyor line before reaching the bin, the tank and line are at an elevated pressure. Following discharge of the last batch of solids into the bin, there is a sudden reduction in pressure and a large volume surge in air flow may occur. This short surge requires a filter of sufficient area to process the large air surge at a reasonable superficial face velocity. If the filter is not sized for this flow, a filter bag may break under the pressure increase or the pressure relief vent may be activated. Filter design should consider transient operations of this type when sizing the filters.

An alternate that can be investigated in blow tank operation is to automatically block the blow tank outlet line before the pressure surge occurs and vent the excess pressure. However, venting the blow tank to the atmosphere is not possible where solids are present in the vent stream. Cyclic batch operation may increase particulate concentrations above desirable levels. The vent stream in this event must be sent to a receiving point that has a dust collector capable of handling the air surge.

One of the problems with filter or cyclone operation in pneumatic conveying systems is the change in particulate size that may occur in the actual conveying system compared with a solids sample or pilot plant test results. Where friable or partially friable materials are conveyed, the particle size distribution may change considerably as a result of a somewhat higher conveyor velocity or a large number of small radius bends in the conveyor line. One reference cited a change in the mean particle size from 170 to 152 μm, but the concentration of fines changed from 0% less than 50 μm in the base material to about 12% < 25 μm after conveying.[15] The product at the conveying outlet also had particulates in the 0- to 10-μm range. This difference can affect filter performance since the fine particulates can block the filtering surface and significantly increase the pressure drop. Filter cleaning must be conducted more frequently in this situation, reducing cloth life and increasing filter maintenance. The particle size distribution is a major consideration when comparing filters and cyclones for dust separation. For higher efficiencies ($> 99\%$) in fines collection, the filter is superior, but with larger size particulates the cyclones may be preferable. From a maintenance standpoint, cyclones do not require the attention associated with filters and thereby have a lower potential exposures frequency. Local operating conditions based upon solids characterization determine the type of dust separator selected for installation.

Filter bags gradually block over a period of time and require replacement periodically. Actual performance of the filter is monitored by measuring the pressure drop across the filter. Automatic cleaning of the filter can be instituted by a ΔP monitoring control system which is also effective in a negative pressure system. Pressure measurements between the upstream air conveyor and the receiving bin will also aid in determining whether blockage exists in the system. In addition to the measurements in the pressure conveyor systems, pressure sensors around blow tanks control blow tank flow rates, producing more consistent solids–air flow rates into the receiving bin, minimizing fluctuations in solids loading on the filter cloth. The valves in the supplementary air line around the blow tanks shown in Figures 14.25*B* and *C* are the adjustable control valves that modify the air ratios for solids flow control.

Filter bags have improved considerably and several of these improved bag designs minimize blocking or blinding effects caused by fines, thereby increasing cycle time. Since bag replacement is a source of potential exposure during maintenance operations, longer cycle times reduce maintenance and should be investigated for new and existing filtering units. In addition to pressure sensors, solids leak detectors should be installed in filter outlet lines to ensure bags are not leaking, thereby requiring replacement. Filters are cleaned by shaking or reverse-pulse systems. Since these systems consist of mechanical components, maintenance is periodically required and is also a potential source of exposure. Normally, bins are rather tall and maintenance personnel must work on the silo roofs, requiring equipment at the elevated repair location. Vacuum equipment must be available to capture dust in the plenums and around the bags to minimize dust concentrations and dust generation prior to maintenance. Personal protective equipment will be required to ensure exposures are controlled.

The various simplified drawings describing equipment do not always show valves between bins and feeders or blow tanks. Valves are required to block various equipment and prevent bin unloading into the attached equipment. Emergency repairs or maintenance require block valves to prevent solids leakage and minimize potential exposures. Valve location, maintenance, and operating procedures should be reviewed by the industrial hygienists.

Fluidized Conveyors. Fluidized conveyors are frequently used over short distances to transport solids from a bin or railcar to a primary conveying system. The conveyor is enclosed, horizontal, and rectangular in cross section with a finely divided screen or membrane that separates the conveyor into an upper and lower section. Air is injected into the lower section and is distributed upward through the membrane into the solids section. A small pressure drop across the membrane ensures even distribution of the air. The air filters through the solids and as the superficial velocity through the bed increases, a velocity is reached at which the bed begins to fluidize. This particular velocity effect results in "incipient fluidization." The settled solids mass expands and resembles water in its actions. Flow through the conveyor follows the paths of least resistance and the solids will flow toward the outlet end when the conveyor is tilted at a small angle downward toward the outlet end. This small change in elevation aids in moving the solid particulate stream.

The fluidization conveyor moves the solids to a positive pressure feeder or blow tank for solids movement into the pneumatic conveyor system. Air, which is at a very low rate, will flow with the solids into the feeding system for short conveyor lengths. With longer fluidized conveyor lengths, some venting along the conveyor is required. These vents should be tied into a collection system.

Railcar Unloading. The unloading of solids bulk materials from rail cars can be accomplished with negative pressure, negative–positive pressure (push–pull), or blow-tank systems. One type of push–pull system is also used that does not require the solids separation equipment necessary in the typical negative–pressure system. The blow-tank system shown in Figure 14.28 does not require solids separation equipment, with the blower processing atmospheric air, eliminating recycled air that may have solids traces in the streams. A more conventional blow tank can be placed beneath the tracks with gravity filling of the blow tank and movement of the solids in dense phase cycling.

In the load cycle of the system in Figure 14.28*A*, the inlet line connected to the bottom of a railroad car hopper transfers solids into the blow tank under vacuum. The vacuum is created by air discharged into a venturi from the blower, which evacuates air from the blow-tank chamber. In this phase, the blow-tank inlet cone valve is open and the outlet hopper valve is closed. Some particulates may be drawn through the venturi tube, but the air flow enters the open pneumatic line and discharges into the receiver bin.

The blow tank is pressurized as shown in Figure 14.28*B*, closing the inlet cone valve, and the pressure opens the bottom hopper valve. Some of the blower discharge air is diverted around the outlet, which is similar to the supplemental air previously described for blow tanks. The pressure in the blow tank is about 15 psig (103 kPag) and moves the solids in a batch-type operation. Short cycles improve flow continuity with this device. This rather simple device is swept or purged easily to remove particulates in preparation for maintenance. Consequently, exposure is not a major problem with this equipment. The blow-tank system described in Figure 14.28 is generally used for powders and other bulk solids, including materials classified as granular solids. This negative–pressure system distributes solids to bins through diverter valves, as shown in Figure 14.22.

A standard push–pull system is also operated in the same fashion. Vacuum systems are also widely used to unload railcars or trucks into a number of bins or silos. In these systems, the vacuum machine is connected to the dust collector outlets on the bins, placing a vacuum in the bin that permits direct flow of material from the railcar or truck directly into the bin. The railcar in the vacuum system is directly connected to the bin fill line and all of the solids are transferred without impediment into the bin. In contrast, the push–pull system (Fig. 14.23) separates the solids and air in the recycling of the air, which adds somewhat to system complexity. The same system can also be used for loading trucks. However, vent controls on the trucks are necessary to remove particulates that would be discharged to the atmosphere if dust collectors were not available. Consequently, a truck vent must be connected to a collector, which can be the feed bin or a completely separate system.

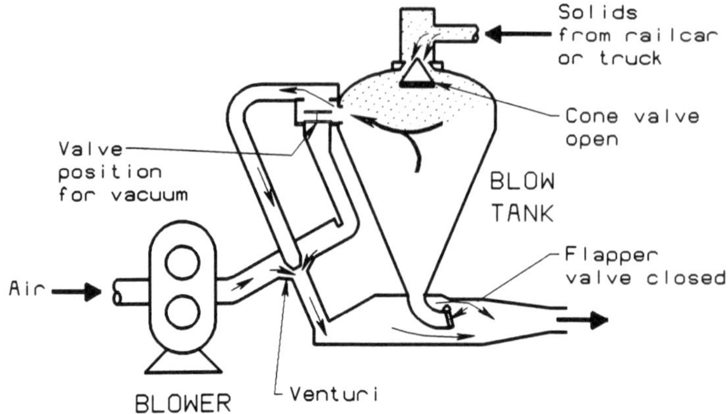

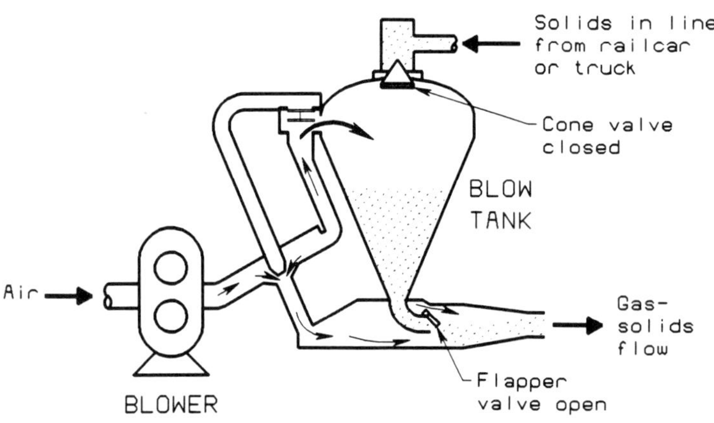

Figure 14.28 *Vacuum – pressure loading device (courtesy of Cyclonaire®).*

Frequently, railcars and trucks dump bulk solids materials from railcar and truck bodies directly into hoppers located below ground level. This results in significant emissions of solid particles that may be considerably larger in size than 10 μm, but the large volume of particulates generated during the dumping operation can be irritating to exposed personnel. Large enclosures or buildings are frequently used to contain the dust. While the buildings are effective, blowers and collectors are necessary to ensure the great mass of particulates are not emitted to the atmosphere. A baffle device located in the receiving hopper will significantly reduce particle emissions, with reductions of about 75% in dust emissions. Installation of this type of baffle in an unloading enclosure will significantly reduce the size of dust collector equipment and permit personnel in the general area during unloading with some personal protection. This baffle system, termed a Burnley

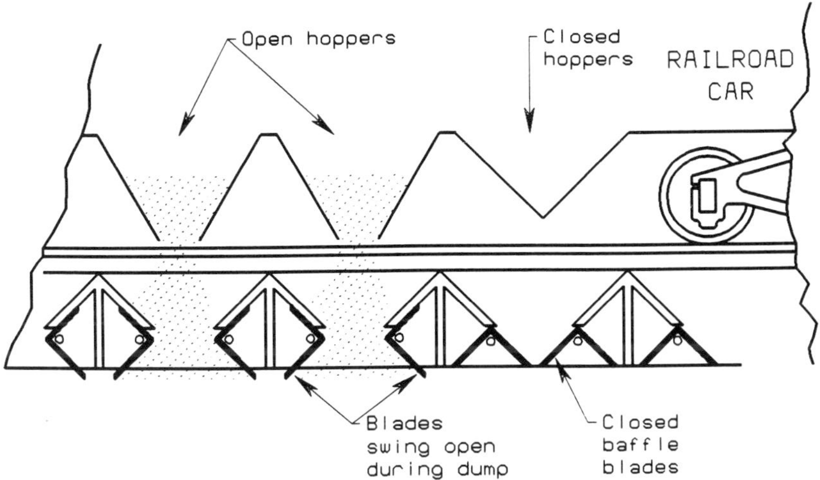

Figure 14.29 Burnley baffles® (courtesy of United McGill Corp.).

Baffle, is shown in Figure 14.29 and in applications with relatively fine materials has significantly reduced dust generation.

14.6 DUST EXPLOSIONS

Dust explosions frequently accompanied by catastrophic effects have been reported throughout the world in the last few decades. Various dusts have explosive characteristics and in those facilities that transfer and store powders or dusts, potential explosive hazards exist. While grain storage silo explosions have received considerable publicity, a wide range of materials in particulate form have the capability of exploding and an extensive listing is presented in reference 44, covering various thermoplastic resin groups, thermosetting resin groups, rubber, natural resins, chemicals, fillers, and agricultural dusts.

Solid particulates dispersed in air or vapor form a cloud of dispersed particles. When this dispersion is ignited, a flame rapidly propagates through the cloud and in a closed environment the flame front is accompanied by sudden increases in temperature and pressure. This situation does not differ significantly from descriptions of explosive vapor mixtures. The explosion, when theoretically contained within equipment, does not pose a significant problem, but when it is not contained, a dust explosion is an imminent threat to plant personnel. In addition, secondary explosions may occur as dust in adjacent areas is stirred by a pressure front into a cloud and is ignited by the flame discharging from ruptured sections of the facilities. Clean facilities not only prevent secondary explosions but also reduce potential exposures during normal operations.

Granular solids do not present dust explosive hazards, but powders $< 200\ \mu m$ in size and granular materials with a high fines concentration are potential sources of dust explosions. When the powder or dust ignites, the flame velocity is classified as deflagration or a detonation. Where the flame velocity is less than the speed of sound, the explosion is termed a *deflagration*, and *a detonation* when the velocity is greater than the speed of sound. Most of the industrial dust explosions and hazardous vapor explosions are classified as deflagrations.

Dust or solids particulate systems are not necessarily explosive since certain conditions must exist for a dust explosion to occur. For a dust explosion to occur:

- The dust must be combustible.
- Dusts must be in suspension and in an atmosphere that supports combustion, such as air or oxygen.
- A particle size distribution that will propagate a flame front is necessary.
- The dust concentration must be in an explosive concentration range.
- An ignition source must be in contact with the dust suspension.

These five items are generally combined into three conditions: (1) an explosive dust concentration, (2) oxygen or air to support combustion, and (3) a source of ignition. Sources of ignition generally described are welding or cutting operations, hot surfaces, grinding, electrostatic discharge sparks, friction heating, or an electrical spark. Consequently, a number of elements must jointly exist for a dust explosion to occur. However, powders or particulates should be carefully investigated to determine the potential for dust explosions.

14.6.1 Dust Explosion Characteristics

Determining dust explosion potential and general characteristics associated with a specific dust or powder is a first step in assessing overall explosion control requirements. Many materials have been tested for explosion hazard potential and the results are available in the literature.[44–46] However, for materials not adequately characterized for dust explosion potential, various tests are available commercially for this purpose. Initial testing will indicate whether a dust is ignitable and will propagate a flame. If the dust or powder has flame propagation properties, further tests are conducted to completely characterize explosion or criteria. Tests in general cover:[47,48,58]

- Ignitability and flame propagation
- Minimum ignition energy
- Minimum ignition temperature
- Minimum dust explosive concentration
- Maximum permissible oxygen concentration
- Maximum explosion pressure, rate of pressure rise, and a descriptive constant K_{st}

Ignitability and flame propagation are determined in a vertical, open tube with an ignition source. The next test for maximum pressure and rate of pressure can be conducted in a closed vertical tube with an electrical ignition source (Hartmann apparatus) or in a 20-L spherical metal apparatus. The spherical chamber is generally preferred for these pressure measurements and the determination of the

TABLE 14.8 Dust Explosion Classification

Dust Explosion Classification	K_{st}(bar · m / s)	Explosion Characteristics
St 0	0	None
St 1	0 – 200	Weak
St 2	200 – 300	Moderate / strong
St 3	> 300	Very strong

constant K_{st} based on a strong chemical ignition source of 10 kJ. The constant is determined from the following equation:[49]

$$K_{st} = (dp/dt)_{max} V^{1/3} \tag{14.28}$$

where K_{st} is a constant (bar-m/s) and V is the test chamber volume (m^3). The constant K_{st} is independent of the test volume and provides a basis for powder classification since a strong ignition source exists in comparison to weaker sources.[49] Dust explosion classifications are shown in Table 14.8 that have been developed from the spherical test chamber. Some discrepancies exist with K_{st} values developed in the Hartmann apparatus compared with the spherical values, but the spherical K_{st} values are widely used.[15,50]

Dust concentrations will only explode within a definable range having lower and upper explosive limits that are similar to the limits developed for flammable materials such as hydrocarbons. A vertical tube with spark ignition provides test data on lower explosive limits. The lower explosive limit is generally in the range of 20–85 g/m^3 for a wide variety of materials. Measurement of the upper explosive limit is difficult due to problems associated with uniform dispersal of dense phase particulates. Results that have been reported suggest an upper explosive limit range of 2–10 kg/m^3. These limits indicate that a flame will not propagate below the lower explosive limit or above the upper explosive limit.

Tests in a vertical tube also determine the oxygen concentrations that support flame propagation. The oxygen concentration that will not sustain a dust explosion is the maximum permissible oxygen concentration if attempting to control explosions through inert gas dilution. These test results are affected by the volume of the test vessel, and a safety factor has been suggested to establish a maximum oxygen concentration.[49] For practical operation, an oxygen level 2% by volume lower than that determined in the test apparatus should provide a satisfactory margin of safety.

Minimum ignition temperature is the lowest temperature at which a dust cloud will spontaneously ignite and propagate a flame front. This test is primarily designed for elevated process operations and there is little consideration given to minimum explosive concentrations, minimum ignition energy or explosion pressure data. A new test procedure is under development that will expand the base-testing procedure[49] and is not related to the minimum ignition energy required to ignite and propagate a flame front.

The minimum ignition energy (MIE) has generally been applied in an open vertical Hartmann tube. In the presence of a dispersed dust, a spark is provided across electrodes in the tube chamber. Generally, the procedure is to charge a bank of capacitors and then discharge the capacitors through a voltage step-up transformer across the electrodes. The energy is determined from the equation:

$$E = 1/2CV^2 \tag{14.29}$$

where E is energy (J), C is capacitance (F), and V is charge potential (V). Significant energy losses occur in the electrical circuit, and the theoretical spark energy calculated is normally much greater than the actual spark energy.[51] For a relative indication of energy levels, this procedure is satisfactory. More accurate energy values require changes in the electrical spark-generating circuit. In addition, spark duration is an important variable, with lower MIE values obtained from long-duration sparks. This is a very complex subject that is beyond the scope of this chapter. Based on test results, absolute values are somewhat questionable since cloud characteristics vary significantly. However, the data do suggest whether the ignition energy is high or low for the specified dust. This indicates on a qualitative basis the precautions required to guard against accidental ignitions in a plant.

14.6.2 Variables that Influence Dust Explosions

The previous section reviewed the various characteristics of dust explosions. Variables that influence dust explosions follow.

Turbulence. Flow dynamics influence the overall explosion process which is a function of turbulence. In the absence of turbulence, flame propagation is laminar and the flame front propagates at a constant velocity from the ignition source. In turbulent conditions, eddies exist, particularly in a duct ahead of a flame front. These eddies become burned products as they are associated with the flame front and continue their random motion. These eddies contact other nonburned eddies ahead of the flame front, serving as multiple ignition sources that increase the combustion rate. The flame front loses its laminar-like front and assumes nonsymmetrical proportions with increasing velocity. The flame front can become supersonic in some cases, resulting in a detonation that may have serious effects.

One of the important variables affecting explosion intensity is the length of the pipe. Longer pipe lengths are more prone to convert the explosion into a detonation, and the resulting pressures are higher than the maximum explosion pressure determined for deflagrations. Small pipe diameters tend to increase the probability of a detonation for combustible dusts. However, the distance required to attain the very high velocities associated with detonations is much longer for dusts than the distances necessary for flammable gases.[49] The situation in the pipe can be exacerbated by ignition from a flame jet in a vessel connected to the conveying pipe, which may convert the dust explosion into a detonation.

Particle Size. While granular solids normally do not present dust explosion concerns, the addition of fines to the bulk solids increases dust explosion potential. As the dust becomes finer or the median value in microns decreases, the maximum explosion pressure increases. The fines can be generated during solids movements and a knowledge of fines generation potential for the specific material is necessary for dust explosion assessments.

The K_{st} constant shown in Eq. 14.27 is a function of the volume and the change of pressure with time. Data indicate that the K_{st} constant increases as the median particle size value decreases for various materials.[49] At constant volume conditions, the pressure change with time increases for all materials in general, but the actual (dp/dt) change differs considerably, depending upon the material.

Finer dusts have a much higher dust explosion potential and react more violently than coarse dusts. In general, the median size has an influence on the explosion effect, which is the K_{st} constant. Maximum explosion pressure for the same median particle size of different materials will be a function of the material. Dust samples for maximum explosion characterization would then have a median particle size $\leq 63\ \mu m$. For coarse dusts with particle sizes of 400–500 μm, initiating an explosion is difficult. Coarse dust becomes a problem if fine material is produced during processing. As the quantity of fine dust increases, the K_{st} will also increase, but the actual K_{st} changes are a function of the material.[49]

Moisture. Dusts containing a small percentage of water can generate dust-air explosions. When the water content increases further and the solids cannot be dispersed properly, the potential hazard from dust explosions decreases. However, the actual level will be a function of the dust, although one reference[49] suggests water contents $> 10\%$ cannot be dispersed properly. At very high water levels dispersion is extremely unlikely, with a very low probability of a dust explosion.

Hybrid Mixtures. This term describes mixtures of dust and combustible vapors. Data indicate that nonexplosive dusts will explode when mixed with a flammable vapor. Moreover, the maximum explosion pressure increases significantly with flammable vapor present. As a result of this explosion characteristic, the dust classification of various dusts shown in Table 14.8 changes and the dust classifications are increased, with the increase dependent upon the volume of flammable vapor.

The lowest minimum ignition energy (LMIE) of dusts in hybrid mixtures is less than the LMIE for nonhybrid mixed dusts. When the dust is readily ignitable, any mixture with a flammable vapor is of much greater concern than the original dust.[49]

14.6.3 Explosion Prevention and Control

A number of potential sources of explosions were mentioned at the beginning of Section 14.6. Removal or control of these sources is necessary to minimize the possibility of potential dust explosions. Control of some of the sources is straightforward. Other prevention and control methods are described in the following sections.

Heating. In some processes, direct combustion provides hot air to a process device such as a dryer. Where the dust is flammable, a separate combustion chamber should be provided to ensure the flames are some distance from the dust. In addition, combustion should be complete to ensure that the hot flue gas does not contain any flammable vapors. Poor combustion efficiencies can produce relatively large quantities of carbon monoxide. Hot particles produced during combustion can be a serious problem, or recirculated air with traces of dust particles that are heated can also become heated ignition sources. Additional filtering may be required to prevent heated particulates encountering a combustible dust.

Dust Accumulation. Powder or dust that accumulates in open areas and within equipment such as bins or hoppers may ignite. Spontaneous heating has been described in one reference[47] that is relatively slow, but estimates of self-ignition temperatures are possible. When distributed and in the presence of an ignition source, these dust layers can produce an explosion. Heating equipment or sparks can be sources of ignition. Moreover, material in closed equipment subjected to solar heat may affect the average dust temperature, creating the basis for ignition.

When materials are stored for a period of time, replacement of air with an inert material will prevent an explosion. Movement of the material from storage will also eliminate this problem, although a layer of dust may remain. Consequently, removal of dust in equipment will reduce potential explosions. In addition, good housekeeping to remove dust layers in the plant also minimizes potential problems.

Friction. When mechanical equipment is in operation, friction heat, if not dissipated, can be a potential ignition source. Dust entering bearings may cause heating or misalignment of the equipment. In close-tolerance openings dust and/or metal imperfections rubbing together can result in friction heat. The mechanical equipment can be investigated with sensors and probes to determine whether the equipment operates satisfactorily.

Overloading equipment also increases localized friction heat, requiring control of powder flow. Also, metal parts or tramp metal in the equipment can create frictional problems and possibly sparks. Magnets located strategically in the system can remove most of the metal contaminants. These magnetic installations should be designed to place the metal removed from the process in a receiver to prevent the metal again entering the dust system.

Electrical Sparks. When an explosive dust is present, electrical switch gear should be installed that meets the standard requirements for a combustible dust atmosphere. The electrical equipment is dust-tight and also conforms with requirements for flammable vapor atmosphere. Electrical standards are defined in NFPA standard 497M and ANSI/NFPA standard 70. In NFPA 497M, both ignition severity and explosion severity indices are described that determine the classification of the electrical equipment.

Inerting. As described, the complete or partial substitution in the dust–vapor mixture of an inert gas for air or a flammable atmosphere will eliminate or prevent dust explosions. Where a partial substitution exists, the oxygen concentration must be reduced to a level that prevents ignition of the dilute air–dust mixture. The oxygen concentration that initiates ignition is determined in a 20-L spherical test apparatus. A safety margin of 2 vol % lower level of oxygen is recommended[49] with an 8 vol % oxygen level considered satisfactory for organic dusts. Metallic dusts, beta-naphthol, and paraformaldehyde must have much lower oxygen concentrations.[49,52] Tables for limiting oxygen concentrations with nitrogen as the inert material are found in various references, but the oxygen concentration level will vary with the type of testing apparatus. The 20-L sphere test is generally considered a satisfactory standard test apparatus and test results can serve as a basis for requirements and comparison.

While nitrogen is generally considered for inerting, carbon dioxide may also be used, but greater concentrations of nitrogen are required compared with carbon dioxide. Water vapor as an inerting agent is effective and is considered somewhere between nitrogen and carbon dioxide in effectiveness. Relationships have been developed in the United States and Germany for the amount of oxygen in nitrogen and carbon dioxide:

$$N_O = C_O - 2 \text{ (ref. 47)} \tag{14.30}$$

$$N_O = 1.3C_O - 6.3 \text{ (ref. 53)} \tag{14.31}$$

where N_O is the maximum oxygen concentration in nitrogen (vol. %) and C_O is the maximum oxygen concentration in carbon dioxide (vol. %).

Hybrid mixtures can be approached with similar inerting tests to determine the maximum oxygen concentration. However, the hybrid systems do not duplicate the basic dust–air system and detailed tests should be conducted to establish maximum oxygen levels. The limiting oxygen concentration is determined by the lowest limiting value of the dust or vapor.

With an inerting system, the oxygen concentration must be monitored continually or intermittently to ensure that the oxygen level is not exceeded. Since the overall solids system can be quite complex, the oxygen concentration should be measured in various locations. Moreover, consideration should be given to the injection locations of the inert gas to maintain the proper oxygen concentrations. Locations where adverse pressure gradients with air exist, "ratholing," and low-pressure sections such as the bottoms cyclone outlet present potential oxygen concentration changes. This situation can also be exacerbated by a lack of nitrogen, which is thus not carried into the locations just described. As an example, injecting nitrogen into the upper section of a bin or hopper may result in the nitrogen short-circuiting into the dust collector outlet. Consequently, a detailed review of injection points is necessary, along with inspection points to test the effectiveness of the nitrogen inlet locations.

The various gases considered for inerting purposes generally include nitrogen, carbon dioxide, argon, helium, and flue gas. The gases have various advantages and disadvantages which are described in the following brief summaries.

Nitrogen. This gas is widely available and in some locations can be obtained from a pipeline. In addition, various purities are available which, if acceptable, provide some economic incentives. However, nitrogen is not less effective than carbon dioxide in this service as described above.

Carbon Dioxide. This gas is also widely available from suppliers or in some locations may be obtained from a pipeline. Some plants also have onsite processes installed to supply carbon dioxide from flue gas or inert gas generators. The CO_2 from these systems or that drawn from a unit dedicated to solids supply service requires careful operation to ensure the correct concentration is maintained at all times. High carbon monoxide concentrations can present potential problems in the dust–gas mixtures. In addition, carbon dioxide can generate electrostatic charges through nozzles or openings that inject the gas into the solids system. Finally, CO_2 can react with some materials, requiring a careful investigation of the system.

Flue Gas. The total flue gas from a furnace or boiler is a source of inert gas, but may require treating or purification, depending upon the gas source. Normally, the flue gas will have carbon monoxide, other contaminants, and some particulates. In addition, there will be concentration changes when the burners are adjusted for changes in furnace or boiler operation. Treating or purification are required along with cooling the flue gas in most applications to much lower temperatures. The lower temperatures result in higher humidities for the gas, requiring water droplet removal and, with some dusts, drying the gas to desirable humidity levels.

Since the flue gas is only available from low-pressure sources, a blower or compressor will also be required to move the gas into the solids system. Solids movement and purging in beds requires pressure. In addition, a lack of flue gas storage indicates that the solids system is dependent on the service factor of the flue gas source, which may not match the solids system service factor.

Argon and Helium. Both gases are expensive and require a recycle system to minimize gas consumption and control operating costs.

While inerting with gas is effective, other aspects of the operation must be considered. In many installations, the inert gas is on as once-through or nonrecycle basis. The carrier gases are discharged to the atmosphere from filter collectors, cyclones, and so on at an overall mass flow rate equivalent to the addition rate, which is the sum of the inert gas and air—mass flow rates. Certain systems may only consist of the inert gas, based on a variety of economic and technological factors. The chosen system has an inert gas component discharged to the atmosphere that can cause potential exposure problems in the discharge area surrounding the outlet vents due to a lack of oxygen. Moreover, any leakage during maintenance may also be a potential problem, particularly in enclosed areas. Personnel must be protected from a combination of dust and lack of oxygen.

These problems are partially overcome by recycling all of the vent discharge streams. However, maintenance will still be a source of potential problem in enclosed areas. Since recycling probably will send the outlet vent gas flow to the main recycle blower, additional filtering may be required, further complicating the system. If the traces of dust in the recycle gas are not filtered, heating through the blower system can be a potential ignition source, along with fine dust buildup in the system. While inert gas presumably prevents dust explosions, some air is carried into the system with the powder. Consequently an oxygen buildup can occur, requiring O_2 monitoring and a controlled purge to maintain a low oxygen concentration. The purge also reduces the inert gas concentration and inert gas makeup is necessary. All of this equipment must operate properly in a recycle system to ensure that potential dust explosions conditions do not arise.

In addition to inerting with gas, two other inerting techniques that can also be used are vacuum and mixing with inert solids.

Vacuum. Maximum explosion pressures in closed containers are proportional to the initial system pressure or a ratio exists between the maximum and initial pressures. Reducing the initial pressure below atmospheric pressure will, in accordance with the ratio, reduce the maximum pressure. If the initial pressure is $\leq$ 0.1 bar (1.49 psia), the maximum explosion pressure will be less than atmospheric pressure.[49] This reference also indicates that a dust explosion will not occur when the operating pressure $\leq$ 0.05 bar (0.75 psia).

While this operation may be feasible in some situations, all of the equipment must be designed for essentially full vacuum, which significantly increases investment cost. In addition, maintaining equipment can be difficult since the vacuum must be removed before maintenance begins and the system must then be evacuated. At atmospheric pressuring during maintenance, the dust may be prone to ignition and dust explosions, requiring careful consideration of this type of operation.

Inert Solids Addition. An inert powder can be added to the process solids that increases the lowest minimum ignition energy very rapidly.[49] Also, the maximum explosion pressure decreases as the inert solids concentration increases. However, the minimum concentration of inert powder generally ranges between 50 and 75 wt %[47, 49] for various solids material and requires intimate mixing to ensure complete distribution of the inert powder through the product solids. Some of the inert powders in the literature[49] are $CaSO_4$, $CaCO_3$, $NaHCO_3$, $NH_4H_2PO_4$, and $KHCO_3$. However, they are not necessarily interchangeable, and for a specific solid material may require investigation to determine the optimum inert solid.

Since the inert solid volume is very large, a considerable separation system will be required to recover the solids product. Where a pure solids product is necessary, this technique may become extremely complicated and costly. Moreover, additional solids handling of the powder and separation facilities further increases potential exposure situations.

Containment. One method of coping with a dust explosion is to design the equipment to contain the explosion. The maximum dust explosion pressure exhibited in test equipment for many air solids mixtures is about 150 psig (11 bar). Although some dust explosions have lower maximum pressures, most of the maximum explosion pressures are in the range of 100–150 psig. Equipment designed to contain explosions must be designed with a safety factor to ensure containment under various conditions.

There are two basic methods of designing equipment for dust explosion containment:

1. *Pressure Resistance.* This requires design of the system to withstand pressures within its rating without deforming or rupturing. The equipment is designed in accordance with the ANSI/ASME VIII Boiler and Pressure Vessel Codes,

requiring hydraulic pressures testing at pressure 1.5 times greater than the design pressure.

2. *Pressure Shock Resistance.* In this design, the equipment contains the explosion without rupture, but the equipment may be permanently deformed by the explosion. In effect, this design is based upon a lower safety factor and considers plastic deformation acceptable when properly designed. Deformation must be acceptable for plant operations when the unit is placed in service.

This type of design requires reliable mechanical design techniques and can only be tested to about 90% of the design pressure or the equipment may deform. The equipment design for shock resistance should probably be limited to rotationally symmetrical shapes. Rectangular shapes with flat walls present difficult mechanical design problems and their solution may require expert design assistance.[47]

Much of the equipment used in the plant, such as rotary airlocks and feeders cannot be analyzed for shock resistance and testing may be required to ensure their adequacy. Certain types of equipment, such as fabric filter bag houses, are designed for very low-pressure operation and cannot be converted to shock resistance without major structural changes that significantly increase their cost. Cyclones can be designed for a large pressure shock, but their construction becomes unwieldly and expensive. Bins can also become expensive and would probably require compliance with the ASME pressure vessel code.

One of the problems in containment is that an explosion in one silo or bin may result in damage to a connected vessel that may possibly be more severe. Consequently, isolation of bins, hoppers, and equipment is necessary. Following an explosion, the unit must be thoroughly inspected to undertake any repairs that are necessary. When the repair effort is complete, the unit should be tested to ensure it will withstand another explosion without rupture.[47]

Obviously construction for this type of explosion control is expensive. A containment installation probably can only be justified where the solids are highly toxic or the material is extremely valuable.

Separation and Isolation. Units and equipment are usually connected together in rather close proximity, although longer pipelines may link separate equipment sections. As a result of essentially interconnection of equipment, the following is required:[47]

- Prevention of an explosion in one piece of equipment propagating through duct work into another unit section.
- Delay or block in the introduction of solids or air after an explosion to prevent reignition.
- Prevent burning particles and material forming a potentially explosive dust cloud.
- Venting explosive material to prevent damage to equipment or secondary explosions.

Piping or Duct Barriers. When an explosion occurs in a pipeline, the flame front propagates through the line and may initiate an explosion at the other end of the line in a connecting piece of equipment. If the flame-front velocity preceded by the pressure front is quite high, the flame front may reach a high constant speed which is known as detonation. Protection against detonation is a difficult problem, requiring methods of preventing detonation conditions from occurring. The pipe length in which detonation occurs is a function of turbulence, diameter of the pipe, and dust characteristics. Normally, larger diameter piping ducts reduce the possibility of a high-speed flame front and detonation.

One major method of controlling the flame front is with a barrier suppression system that is shown in Figure 14.30*A*. This system was specifically designed to limit the explosion to a relatively short section of the line and quench the dust explosion. In the barrier suppressant system, a sensor detects the dust explosion and activates a fast-action valve that disperses extinguishing material into the pipe through a nozzle. An extra source of high-pressure gas is shown that aids in rapid injection of the suppressant. Self-contained pressured containers are also installed and have performed well with pressures of 90–150 psig (7.2–11.4 bar). Since container pressure is very important, maintenance of injection pressure must be assured at all times. The suppression material is described below.

Barrier effectiveness is dependent upon the sensor to initiate suppressant action which can be optical, infrared, or pressure. However, an infrared sensor is apparently preferred since the pressure and flame fronts are not separate and pressure detectors are not necessarily suited for the application. In this application the infrared lens must be flushed with gas to keep the lens dust free. A suppressant barrier is installed 5–10 m from the sensor to provide sufficient action time for suppressant injection. In addition, the amount of suppressant is a function of the flame velocity, with the velocities related to the dust explosion class:[49]

St 1	$V_{ex} \leq 300$ m/s
St 2	$V_{ex} = 301$–400 m/s
St 3	$V_{ex} > 400$ m/s

Detonation velocities generally exceed 100 m/s and suppression of the explosion at these velocities ensures that detonation conditions will be prevented. The suppressant required is generally in the range of 20 to 100 kg/m^2 of pipe cross section. Suppressant barriers have apparently been successfully operated in pipelines with diameters approaching 2500 mm.[47]

While this technique is effective, pressures within the piping may reach 150 psig (11 bar) before combustion is completely eliminated. The piping or duct work should be designed for these pressures or a pipe line venting system should be provided.

Rapid Action Valves. These valves are used in the piping for isolating equipment and are often used with dust collectors. During a dust explosion in a duct, they isolate the dust collector, and in situations where toxic materials are in the collector, rapid action valves in both the inlet and outlet lines prevent the escape of toxic dust into the feed duct or outlet header. The operation of the valve is similar to the barrier suppressant system.

When an explosion is detected, the valve is closed within a few milliseconds. A special driving force may be necessary to close the valve rapidly, such as a pressurized gas cylinder. Moreover, if the valve is a flat sliding gate type, the valve

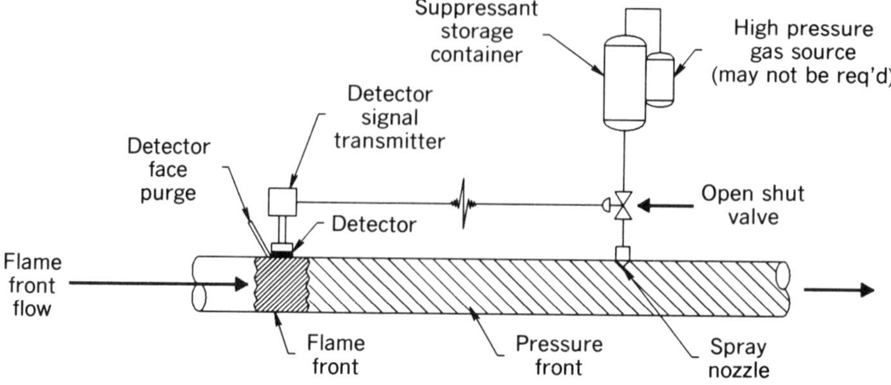

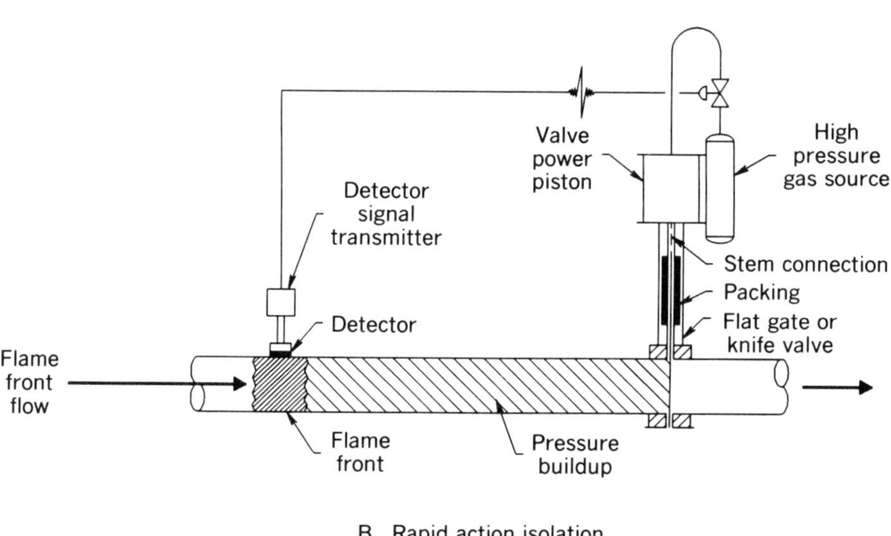

Figure 14.30 *Barrier and isolation systems (courtesy of Springer-Verlag).*

housing should completely enclose the gate to prevent leakage in either the open or closed position. Typically, special gate valves are constructed of light, high-tensile steel that incorporate dampers to absorb the shock force when closing and prevent the gate from reopening. A similar detector to that used in the barrier suppressant can be installed in this system and a gas flush across the lens for cleanliness. A pressure sensor is also used in this service.

Pressures in the duct can also build to 150 psig (11 bar) with the isolation valve. This requires a duct capable of withstanding the pressure or a vent in the piping duct. Under explosions, the gate or blade of the valve will frequently bend. A lower vent-relief pressure would modify the pressure rise.

Rotary Valves. The rotary lock can also serve as a mechanical barrier against dust explosion. Testing has shown that these valves can act as a mechanical barrier by meeting the following criteria:[49]

- Two vanes per side on the feeder are engaged
- The vanes and tips are metal
- The gap between the rotor and the housing is ≤ 0.2 mm

Should an explosion occur, the rotary airlock should be stopped immediately by a detector which prevents the passage of burning particles into a quiescent dust–air mixture that could initiate an explosion. The rotary feeders must be pressure tested to ensure that they can withstand explosion conditions.

Equipment Chokes. When a rotary valve cannot be stopped immediately, burning particles may pass through, although the deflagration is halted. An equipment choke can also be installed to prevent the burning particles from entering the next piece of equipment. Suppressant can be injected around the outlet of the rotary valve to control hot particles until the rotor stops. Another possibility is the use of a screw feeder at the outlet of the rotary feeder that maintains a level of solids in the trough. The screw is separated into two sections with a vertical baffle plate between the two sections that extends from the cover below the solids level in the trough. The baffle plate and solids act as a choke to prevent anything from getting through. Stopping the feeder also aids in this service.

Suppression. As described previously, a dust explosion builds pressure, which is shown in the pressure rise during tests in pressure resistant test vessels. While the increase in pressure is rapid, it occurs over a time period that is quite short, usually about 40–125 msec, but provides sufficient time for detection and the injection of a suppressant that controls the pressure rise to acceptable levels by effectively extinguishing the flame front and minimizing the associated pressure increase.

These suppression systems prevent the buildup of pressure in equipment systems that are designed to withstand maximum dust explosion pressures that can damage and rupture equipment with potentially disastrous results. Explosion suppression is also important where toxic dust emissions control is necessary to minimize potential exposures. The general rise in pressure resulting from dust explosion in a test vessel is shown in Figure 14.31 along with the pressure curve developed when suppressant is injected. Following ignition, there is a short induction period before the dust vapor system pressure begins to rise. When the pressure level reaches the detection pressure setting, the suppressant injection system is activated and the suppressant is injected in the range of 10–15 msec after activation. The suppressant reduces the rate of pressure and the suppressant-treated system reaches a maximum level which has been termed the reduced explosion.

If suppressant systems perform well, reduced explosion pressures of about 15 psig (103 kPag) are attained, which are considerably less than the typical maximum explosion pressure range of 134–178 psig (10–11.2 bar). However, the system mechanical design pressure must be higher than the reduced pressure level. Assuming a 15-psig (103-kPag) reduced pressure level, the design pressure level for the system should be about 18–20 psig (124–138 kPag) or possibly greater,

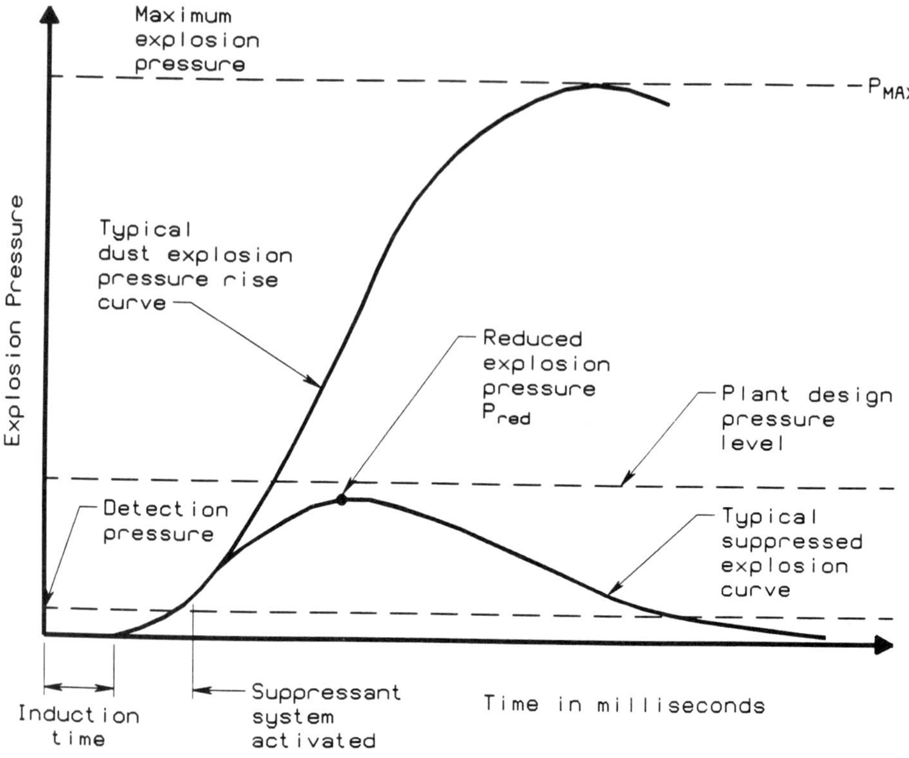

Figure 14.31 *Typical dust explosion and suppressed explosion pressure curves.*

depending upon accuracy of the P_{red} value. Although the reduced pressures are considerably lower than the maximum explosion pressures, system mechanical design pressures are still relatively high for process equipment that normally operates with upper bin pressures at atmospheric pressure. Consequently, the entire system that can encounter a dust explosion must be analyzed for installation of suppressor injection facilities with all process equipment designed for the specified mechanical design rating.

Suppressant System Overview. The design of suppression systems depends upon a number of factors:

- Sensor or detector type and effectiveness
- Detector threshold level of activation
- Efficiency of the suppressant material
- Amount of suppressant material required
- Number of suppressors required
- Suppressant discharge rate
- Suppressant dispersion effectiveness and limitations

In addition to these considerations, suppressants have been effective in St 1 and St 2 dust classification explosions (Table 14.8) but are not effective in controlling

St 3 dust explosions.[47] This requires a careful classification of the dust from test facility data and must be considered in addition to other factors.

Sensors. As described previously, pressure-sensing detectors are preferred for this service, particularly in vessels where the dust clouds affect other sensor types. These devices are usually pressure diaphragm types that sense the exposure pressure propagating through the dust. Thus the preset pressure detection level rapidly transmits a signal to the control unit which initiates discharge of the suppression material.

The pressure sensors must be unaffected by variations in temperature, shock, or vibration. In addition, the sensors must be resistant to corrosion and condensation. Since sensor sensitivity and reliability are very important, one reference[49] recommends the installation of two sensors set at 90° apart on a circular vessel. In addition, the sensors must activate at the same pressure and have the capability of assurance testing while in service. An important aspect of sensor operation is insuring the atmospheric side of the pressure sensor is open at all times.

Suppressants. A list of suppressants was presented above, particularly a number of powder suppressants. In general, the list of suppressants include halogenated hydrocarbons, water, powders, and hybrid extinguishers.[49] Halons® have been used successfully, but environmental restrictions will probably preclude the use of this material. Some of the substitutes under development and those commerically available should be tested for effectiveness. An important consideration with halon use is the need for rapid dispersion after the explosion is initiated. Substitutes may have the same requirement, necessitating careful study.

Water is effective, but a much lower quantity than a halon volume is needed for effectiveness. Powders are effective, but the quantity may be rather large and in many situations contaminates the product, requiring separation of the suppressant powder and product. Hybrid systems consist of halon and powder mixtures. This is a relatively recent development and additional information on effectiveness is necessary, particularly with halon replacements.

Suppressant Quantity. Suppressants are supplied in various size containers and the amount of material required for suppression must be determined in tests to define the number of containers that must be supplied. An important aspect of this system is the effective dispersant distance necessary for injection into a vessel. Test work on systems of this type have been conducted on large vessels[49] and information on this aspect of suppression is available. However, multiple injection points on a large bin or silo are necessary. While this is expensive and considerably more maintenance is necessary, multiple injection points also ensure satisfactory dispersion of suppressant. A single injection point in the upper part of a bin can result in bypassing a considerable portion of suppressant into the collector, significantly decreasing effectiveness.

In addition to a number of suppressant containers, the release valve on the container must be a reliable, rapid action valve that responds almost instantaneously upon receiving an activation signal. The injection points also require distribution nozzles that may be fixed or automatically inserted through a diaphragm when the activation signal is received. A retractable nozzle has the advantage that it is clean when inserted into the vessel compared with a fixed nozzle located in the vessel. Since hemispherical-type distribution head nozzles with numerous openings are preferred,[49] these nozzles may be prone to plug during normal operations.

In most installations, a number of hemispherical heads are installed in hoppers, enclosed bucket elevators, spray dryers, and cyclones.[47] These systems become quite extensive to insure adequate dispersion of suppressant when an explosion is initiated.

Maintenance. Since suppressant systems are extensive and must be capable of operating 100% of the time, a considerable amount of maintenance is required to ensure the equipment is satisfactory. Since pressure diaphragms are the critical activation point, maintenance may require removal and replacement for repair of these sensors periodically. Their location in the walls of the equipment may result in potential exposures as dust is emitted during their removal and replacement.

Another potentially hazardous situation may occur when working with the rapid acting valve on the suppressant container. The pressure in this container may be in the range of 900 to 1500 psig (60–104 bar) and accidental releases of powder or liquid are potentially hazardous situations.

Venting. Safety or explosion relief by venting is a common approach to dust explosion control which is similar to pressure relief in general chemical processing operations. While venting is a method of controlling the pressure rise in the equipment, the discharge of solids into the atmosphere is a potential source of dust exposure. Pressure-relief venting is not recommended when the system contains highly toxic solids or chemicals since the exhaust cannot be contained within an enclosure. Moreover, the relief discharge which vents solids, unburnt mixtures, and combustion gases is generally accompanied by a large burst of fire, requiring an isolation zone around a vent-relief port.

While venting controls the pressure rise within the equipment, the pressure reached within the equipment is considered a reduced explosion (P_{red}) and is similar to the reduced pressure previously described. Since the P_{red} is higher than normal operating conditions, the vessels must be capable of withstanding the reduced explosion pressure or providing a slightly higher pressure than the P_{red} as a safety factor. The vessels are then designed as pressure resistant or pressure shock resistant. Pressure-resistant vessels are designed for the defined pressure rating based on the P_{red} to contain the explosion pressure rise without rupture or deformation. The pressure shock resistant vessels are designed to contain this pressure, but containment will probably result in vessel wall deformation. In this case design stresses approach the construction material yield strength.

General Explosion Description. Upon ignition, a pressure front actuates or bursts open the vent at a predetermined pressure. Since the pressure front precedes the flame front, unburnt dust is emitted from the vent opening immediately after the vent opens. The unburnt dust, in the form of an explosive mixture is ignited by the rapidly moving flame front, creating a secondary explosion. This occurs in vent ducts (although short in length) and ductless openings. A pressure buildup accompanies the secondary explosion, which tends to restrict the vessel outlet and results in an *increase* of the P_{red}. This pressure increase is exacerbated to some extent by a vent duct where the flame velocity is at or above the sonic velocity. A distinct difference in the reduced pressure has been verified for vents with and without vent ducts.

Vent piping length criteria is important in designing vent duct systems. Since these are essentially open pipe ends, long ducts are detrimental and pipe lengths should be minimized. However, some differences in recommended pipe lengths are presented in the literature,[48–50] covering a range of 10–33 ft (3–10 m). Test data indicate that short lengths are preferable in this service. Where the duct length is less than 10 ft (3 m), the P_{red} is considerably lower than the reduced pressures developed in vent outlet ducts that vary in length from 10 to 33 ft (3–10 m). In addition, the reduced pressures developed without a vent duct are lower than the reduced pressures developed with a duct < 10 ft in length by factors ranging from about 1.8 to 3.0. In addition, a vent pipe duct should be installed without a bend.

One of the characteristics noted in testing various dust classes (Table 14.8) is the requirement for a more rapid increase in area requirements as the vessel sizes increase for dust class St 1. Apparently, the amount of unburnt dust released through the vent ahead of the flame front is much larger for class St 1 dusts than it is for class St 2 dusts. Consequently, the secondary explosion initiated by the flame front is significantly more severe for St 1 dusts than the secondary explosions for class St 2 dusts. However, for identical volume test vessels, actual area requirements for St 2 dusts are somewhat higher to about 100 m^3, where the difference rapidly diminishes, and at about 300 m^3 the area requirements are identical.

Vent Area Sizing. This can be determined from VDI Guideline 3673[54] and an NFPA standard.[45] Based upon the vessel volume, K_{st}, P_{stat}, and P_{red}, the vent area can be obtained from nomograms where P_{stat} is the static relieving pressure. These nomograms are based upon a series of tests conducted in various size vessels, with four different dusts. Other methods of calculating the area are available, but the basic VDI system is generally accepted as a satisfactory design method.

In certain equipment designs, the necessary vent area may not be available. In bins or silos the top or cover of the vessel provides the venting area. Where the vent area is insufficient, a higher-pressure P_{red} may be required to reduce venting area requirements. This necessitates a higher vessel mechanical design pressure. Should this change not be effective in reducing area requirements, other forms of protection must be investigated.

The vent location is important to ensure the flame front is not impeded. In addition, the vent must be discharged to a safe location, as described above, with the duct area never less than the vent area to prevent any hinderance in the movement of the front. Vent ducts should also be straight to prevent any untoward occurrences due to the unpredictability of direction changes on the explosion front.

Vent Activation. An important aspect of vent design is the static bursting pressure P_{stat}. The vent should open at the specified activation pressure, but the determination of opening pressure must include adjustments and consideration of the vent closure. The vent closure should have the following characteristics:

- Open quickly and not hinder the exhaust flow
- Lightness and low inertia
- Withstand normal operating and fluctuating pressures and temperatures
- Resist corrosion
- Must not tear loose and move with the exhaust stream

While these criteria are obvious, they may be overlooked, complicating the pressure-relieving function. Door inertia reportedly affected the pressure required to blow open the door[48] in one situation. Actual door density restrictions along with other requirements are obtained in NFPA 68.[45] While doors can be used as vents, a list of various vents that are designed not to obstruct the exhaust duct follows:

DIAPHRAGM OR BURSTING PANEL. These are securely clamped around the opening periphery and will tear at the activation pressure level. If the material stretches, the static bursting pressure must be determined which can be obtained from results on various materials that have been published.[50]

SHEET DIAPHRAGMS. This diaphragm or membrane will stretch and a knife or cutting device is necessary to slit the diaphragm when it expands under pressure.

PROPRIETARY BURSTING PANELS. The panels tend to have an accurate activation pressure. These may be panels that consist of hinged pieces that fold into a flat sealed panel. Others are also available.

POP-OUT PANELS. These are generally light panels of metal or plastic that pop out at the activation pressure level. To prevent the panel from moving with the exhaust flow like a projectile, a cage-like configuration holds the panel after ejection. However, the cage must be at least one duct diameter beyond the end of the exhaust duct. A variation of this is termed a rigid loose panel, which operates at lower pressures but cannot block the duct when blown and must be captured.

HINGED PANELS OR DOORS. These doors preferably open 90° and are generally installed where vent ducts are not used. However, a vent duct can be installed where the duct is properly designed for the door opening. Assuming the door is not distorted, the door can be reclosed and sealed. Since the door is flung open at the burst pressure, a shock absorber or restraint may be required to prevent any equipment damage.

SPRING-LOADED VENT COVERS. These have also been installed, but require careful design of the springs and cover, particularly for larger doors. The bursting pressure requires consideration of cover weight.

Vent locations. An important consideration is the location of vents in various types of equipment. Some of these locations are reviewed below:

BIN OR SILO. The roof area is frequently the venting section of a bin or silo and the vent is sized in accordance with VDI and NFPA procedures.[45, 54] Since the silo normally has other equipment located on the top surface, the entire area is not available for venting. The allowable area and the silo size are adjusted until the vent area criteria are satisfied. Vents in the silo wall are not as effective, and the bin top is considered the only effective relieving area.

The type of vent opening varies significantly. Concrete covers have been used on silos[49] along with hinged and spring-type openings. The type of cover selected is a function of the pressure, area requirements, P_{red}, and vessel design. Bin manufacturers have supplied various venting systems and provide assistance in developing a satisfactory cover.

CYCLONES. Dust explosions may occur in cyclones or duct connections may transmit an explosion front from the inlet or outlet openings into the cyclone, necessitating pressure relief or venting. The top surface of the cyclone is used for venting. Since dust in a cyclone concentrates along the outer walls, the major dust cloud formation is near the wall and the explosible dust cloud volume is smaller than the cyclone volume. The nomograph method may be used for determining area requirements[45, 54] or a K-factor method[50] is available.

The cyclone top is separated into panels that essentially pop out when the activation pressure is reached, which is shown in Figure 14.32*A*. Since an explosion front may also move through the outlet duct, a vent cover can also be placed on the duct outlet (Fig. 14.32*B*). This outlet duct vent cover requires a change in the outlet flow direction, but this direction change is standard for cyclones in series and does not affect cyclone operation.

BAGHOUSES. Dust collectors or fabric filters are also subject to dust explosions that may occur as a result of hot material entering the filter or an electrostatic ignition.

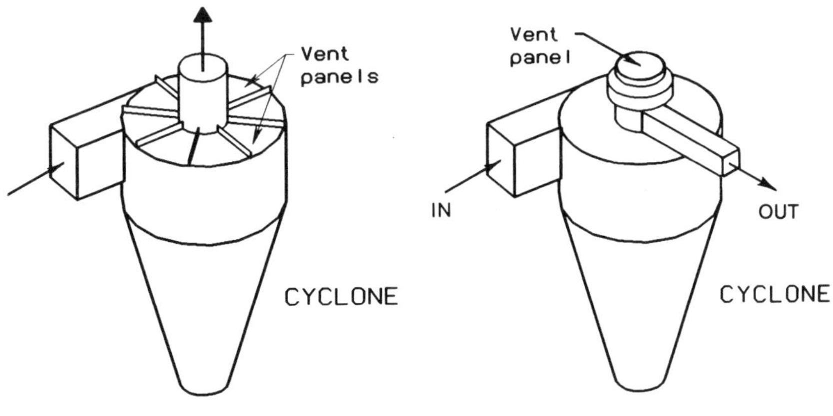

Figure 14.32 *Cyclone explosion vents.*

Hoppers are frequently an explosion source, and vent sizing is normally based on a P_{red} of 5 psig (1.36 bar) or less. Since explosions may occur in the filter, the filter bags will probably be completely or partially destroyed with dust accumulations in the clean air plenum that can result in an explosion due to flame fronts and hot material in the collector housing. Consequently, a vent should be installed on the dust collection side and one should be considered for the clean air side. The vent in the dust side should be located close to the hopper where the filter bags do not interfere with the passage of the flame front. Locations for the release vent on the inlet dust side of dust collector baghouses are shown in Figure 14.33.

In addition to explosion vent release locations, explosion relief within buildings is a major concern with a flame and pressure front exhausting from a dust collector. Where possible, locating the dust collector adjacent to a wall permits extending an exhaust duct through the wall as shown in Figure 14.34. The duct length should be limited to about 10 ft (3 m) in accordance with previous comments, although one reference[49] indicates that longer duct lengths are permissible (10 m max). However, longer duct lengths increase the reduced pressure and present duct design concerns. With the horizontal duct, a large open area must be restricted to personnel in the event the explosion vent discharges. A slanted plate at a 45° angle can be placed a minimum distance of 1.5 duct diameters from the end of the duct to deflect the explosion front upward for additional protection.[50] This deflector should be twice the area of the duct and well supported to resist a pressure of about 4.5 psi (31 kPa) resulting from the explosion front. Although a deflector aids in protecting personnel, the minimum safe distance from the deflector plate is 23–33 ft (7–10 m). In addition, the deflected explosion front should not impinge on other equipment.

Where collectors are located outdoors, the explosion vent discharge should be directed to an area that will not be affected by the explosion front with personnel entrance to the release area restricted.

PIPING. As described in the section on isolation, rapid acting valves (when closed) isolate the piping connecting equipment and vessels. Closure prevents explosions traveling into other potential sources that can expand overall explosion size and increase potential damage. As described, the valves operate rapidly and are designed for high pressure. However, the pressure within the piping can increase well above the pipe design pressure, reaching maximum explosion levels or possibly somewhat higher. If the piping is not designed for pressures of this type, explosion vents have been successfully installed on piping[49] in those locations where release vents are not installed. The piping should be designed for pressure resistance to prevent serious damage to the pipe.

REACTION FORCE. When an explosion occurs and a vent bursts open, a reaction force in the opposite direction is exerted on the collector or other equipment items. Consequently, the support system must be capable of resisting the reaction force and should be designed to accommodate the potential pressure reaction. One relationship developed for explosion discharges where the static vent pressure is 1.5 psig (1.1 bar):[49]

$$F_R = \alpha \cdot 1000 \cdot A(P_{red,\,max} - P_{atm}) \qquad (14.32)$$

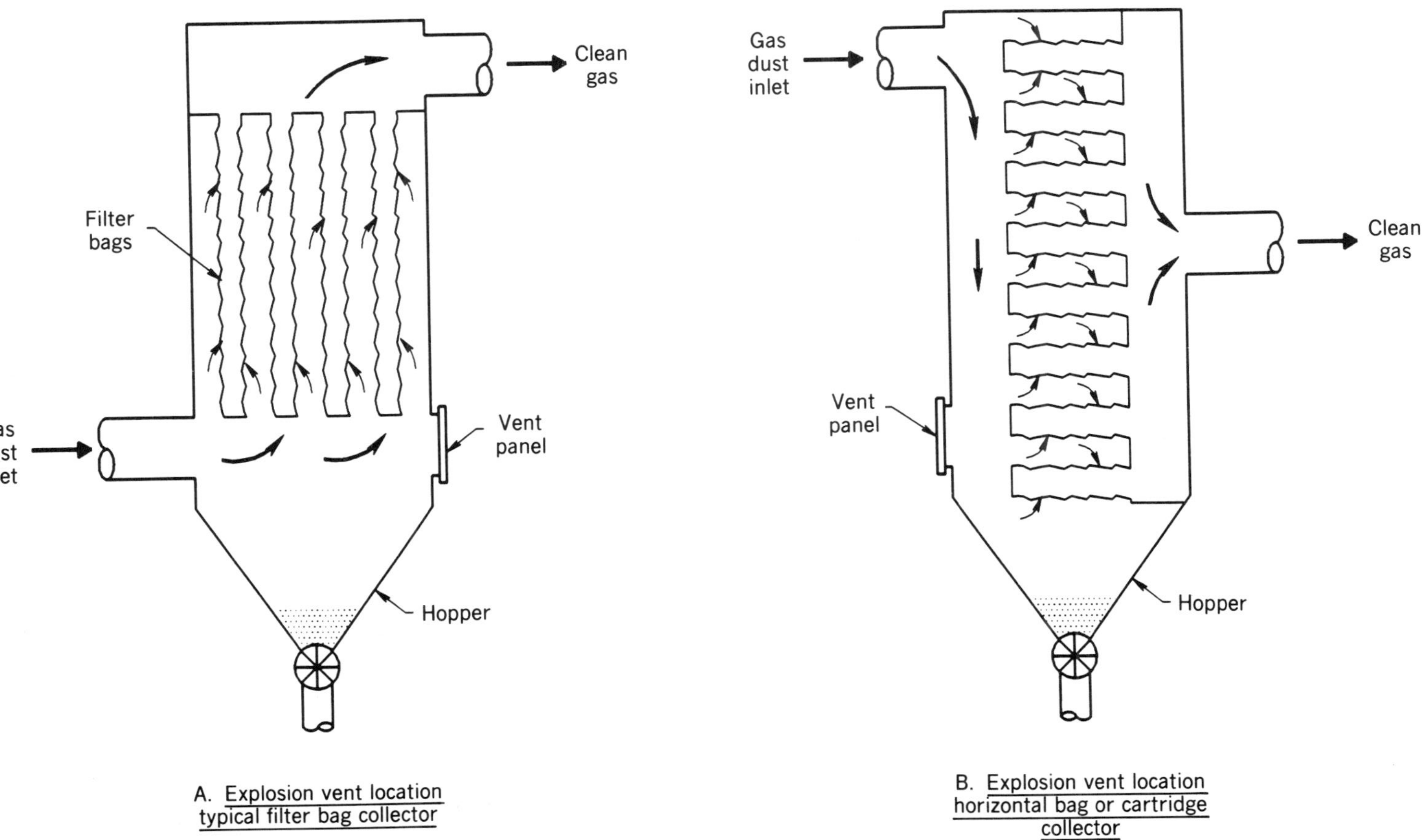

Figure 14.33 *Explosion vent locations in typical dust collectors.*

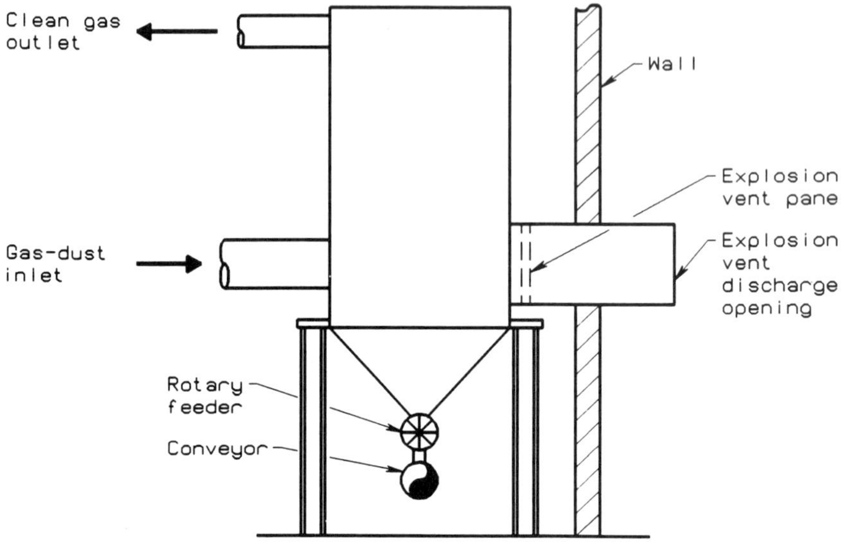

Figure 14.34 *Explosion vent exhaust for dust collectors inside buildings.*

where

F_R = recoil force (kN)
α = dynamic coefficient
A = vent area (m^2)
$P_{red,\ max}$ = reduced maximum explosion (bar abs.)
P_{atm} = atmospheric pressure (bar abs.)

In vessel sizes of 0.25–125 m^3, the range of the dynamic coefficient is 1.19–1.21 for rigid support. Values of α vary and can be determined from reference 49. Another equation is available in reference 50 which varies somewhat from Eq. 14.32. The important aspect of the reaction force is the recognition that the equipment must be designed for the force.

OTHER EQUIPMENT. A wide variety of equipment may experience dust explosions and should be designed to vent these explosions. However, a compilation of this equipment is beyond the scope of this chapter. Some of the typical equipment that would be reviewed are various types of dryers, grinding equipment, enclosed bucket elevators, and powder blenders. Explosion vent information is available from various references and equipment manufacturers.

PROCESS SHUTDOWN. If an explosion occurs, process equipment operations should be immediately halted. This will stop all of the mechanical equipment, which will be necessary in some instances to ensure isolation. A complete shutdown reduces the possibility of transferring burning material to other areas and halts the supply of additional air and dust to the system.

The shutdown can be completely automated, which requires sensors located in various sections and also sensors to determine when an explosion-relieving vent has opened. Complete automation may not be desirable in all plants and modifications can be instituted. However, sensors will be required, particularly on the explosion-relieving vents to indicate any problems. These sensors must be designed to operate in the event of an emergency shutdown.

A careful review of the system is necessary to determine where sensors should be located, the types of sensors, and the equipment that must be shutdown. In addition, a review should determine whether other potential exposure hazards or mechanical problems are generated in an emergency shutdown. Restarting the operation must also be considered following the shutdown.

Electrostatic Discharges. The movement of powders and solids may develop electrostatic charges that can increase the potential for dust explosions as a result of electrostatic discharges. The charging properties of powders vary with the physical and chemical properties, requiring a test method to define these properties. Laboratory testing has been particularly difficult in measuring electrostatic charging owing to various problems in testing and measuring. A laboratory test method has been successfully developed by Dahn[55, 58] which determines electrostatic charge and energy on powders. The charge per unit area ($\mu c/m^2$) and charge per unit mass ($\mu c/kg$) are measured, where c is coulombs. Based upon this data, electrostatic charging can be determined for full-scale operations.

Highly resistive or insulating-type particles that have electrostatic charges can accumulate on various surfaces within the system such as piping, collectors, hoppers, and silos or bins. Since these particulates are nonconductors, the electrostatic charges do not drain off through the grounding system. The electrostatic charge accumulation in turn develops an electric field, and fields with a large potential may result in an electrostatic breakdown that can ignite a dust–air suspension. As more of the powder accumulates on the surfaces and in the bulk repositories, the voltage field increases. The field can be calculated on the accumulated powder surface and in the vapor space of receivers or bins.[55, 58]

Electrostatic discharge occurs when the voltage reaches a level that causes a breakdown across the components and vapor. Calculations based upon the gap between components and material configurations determine the capacitance. The voltage field strength and the gap define the electrostatic energy that is stored and available for discharge. The stored energy in various locations may have sufficient energy to overcome the minimum ignition energy necessary for a dust explosion. Calculations on fields and energy with this approach by Dahn realistically characterize the electrostatic status of the system and potential problems.

Brush discharges can occur in the silos or receivers where bulk materials are stored. An insulating material with charges in these bins can be contacted with a grounded object and a discharge will occur even though the powder rests upon a grounded surface. Conductive material or metal samplers should not be placed in silos or receivers during operation where powders are insulators and have a high charge. Moreover, energy discharges from bulk powder samples can jump to a wall if the voltage field is near the air breakdown level.

Fine powders or granular materials with reasonably high concentrations of fines ($< 74\ \mu m$) have a high potential for dust explosion. However, a considerable number of conditions must be satisfied before an electrostatic discharge will initiate a dust explosion. Laboratory testing provides a basis for predicting voltage fields and energy storage estimates.[55, 58] Based upon this data, requirements to prevent dust explosions can be developed, which are in addition to complete grounding of all equipment to limit static buildup. Grounding should normally be installed in solids movement systems with detailed grounding requirements outlined in OSHA requirements and NFPA recommendations.[56, 57]

In addition to grounding all conducting materials, other nonconductive items have been designed for grounding. An example of this type of nonconducting material is filter bags that have conductive fibers or fibers impregnated with carbon. This type of fabric is termed "epitropic" filter fabric and is now available.[48] Electrostatic charges in the dust layers on the bag surface are dissipated with these conductive bags which must be carefully grounded. Changing bags requires that new bags be connected to the ground system.

While grounding aids in reducing electrostatic charges and the possibility of electrostatic discharges, additional modifications and changes that can be considered are:[55, 58]

- Particle sizes $> 74\ \mu m$
- Round particles
- Reducing conveying velocities
- Higher humidity or moisture concentration
- Smaller vessel sizes

Charges on particulates in mounds stored in bins or receivers can be drained off by the introduction of moisture. Conductive particles in the solids mixture will also provide a method of dissipating the charge. Another item that can be included is antistatic additives, which have been successful in various applications. However, conductive particles or additives may contaminate the powder and can only be added after careful study.

REFERENCES

1. *1993-1994 Threshold Limit Values for Chemical Substances and Physical Agents and Biological Exposure Indices*, American Conference of Governmental Industrial Hygienists, Cincinnati, OH, 1993.
2. *Health and Safety Executive*, Occupational Exposure Limits 1990, Guidance Note EH 40/90, paragraphs 17–64, Tables 1–4, London, 1990.
3. *CFR*, Air Contaminants, *Title 29*, Subpart Z, Part 1910:1000, pp. 6–34, July 1, 1991.
4. Cooper, T. C., W. A. Heitbrink, D. M. O'Brien, and W. F. Todd. *Dustiness Testers as a Means of Evaluating the Dust Exposure Potential of Powders*, NIOSH Bulk Solids and Powders Conference, Chicago, IL, 1989.

5. British Occupational Hygiene Society, *Dustiness Estimation Methods for Dry Materials Part 1, Their Uses and Standardization and Part 2, Towards a Standard Method*, Technical Guide No. 4, London, 1985.
6. Fischback, T. J., W. A. Heitbrink, and W. F. Todd. *Appl. Ind. Hyg.* **4**(1), January 1989.
7. Cooper, T. C., W. A. Heitbrink, and D. M. O'Brien. NIOSH Report, 1989.
8. Carlson, K. H., D. R. Herman, T. F. Markey, R. K. Wolff, and M. A. Dorato. *Am. Ind. Hyg. Assoc. J.*, **53**(7), 448–453, July 1992.
9. Southwest Research Institute/EPA, *Exposure and Release Estimations for Filter Press and Tray Dryer Operations Based on Pilot Plant Data* EPA Cooperative Agreement CR813355, SWRI, Project 01-1236, May 31, 1991.
10. Roach, S. A., *Ann. Occ. Hyg.* **24**(1), 105–132 (1981).
11. Cooper, D. W. and M. Horowitz. *Am. Ind. Hyg. Assoc. J.* **47**(4), 214–218, April 1986.
12. Cooper, T. C., NIOSH Report No 112.23, Cincinnati, OH, 1983.
13. *Perry's Chemical Engineers' Handbook*, 6th ed., New York: McGraw-Hill, 1984, Chapter 20.
14. Scheff, P. A. and R. A. Wadden. *Engineering Design for the Control of Workplace Hazards*. New York: McGraw-Hill, 1987, Chapter 6.
15. Wolfson Centre for Bulk Solids Handling Technology, *Bulk Solids Handling*, Dust Control, Thames Polytechnic, London, 1991.
16. Heinsohn, R. J. *Industrial Ventilation*. New York: Wiley, 1991, Chapter 8.
17. EPA, *Control Techniques for Particular Emissions from Stationary Sources*. Vol. 2, Report No. EPA-450/3-81-0056, 1982.
18. Buonicore, A. J. *Chem. Eng.* **87**, 81–101 (1980).
19. Danielson, J. A. *Air Pollution Engineering Manual*, EPA Publication Ap-40, 1973.
20. EPA, *Compilation of Air Pollutant Emission Factors*, Ap-42, Vol. 1, Supplement A, 1986.
21. EPA, *Compilation of Air Pollutant Emission Factors*, AP-42, Vol. 1, Incl. Supplements 1985–1991.
22. Caplan, K. J. "Source Control by Centrifugal Force and Gravity," in A. C. Stern, Ed. *Air Pollution*, Vol. IV, 3rd ed., New York: Academic, 1977, pp. 98–148.
23. Austin, D., K. Jameson, and R. McInnes. *Chem. Eng. Suppl.* **99**(2), 32 (1992).
24. Moore, S., G. Ondrey, and G. Samdani. *Chem. Eng.* **99**(8), 45 (1992).
25. Gregg, W. and J. W. Griffin. *Pollution Eng.* **23**(4), 80 (1991).
26. Dennis, R. and H. A. Klemm. "Verification of Projected Filter System Design and Operation," in F. P. Venditti, J. A. Armstrong, and M. Durham, Eds., *Symposium on the Transfer and Utilization of Particulate Control Technology: Volume II*, EPA Report, No. EPA-600/7-79-044b, 1979, pp. 143–160.
27. Cass, R. W., R. Dennis, and R. R. Hall. *J. Air Pollut. Control Assoc.*, **28** 47 (1978).
28. *Industrial Ventilation*, 21st ed., American Conference of Governmental Industrial Hygienists, Cincinnati, OH, 1992.
29. *ASHRAE Handbook and Product Directory* 1983 *Equipment*, American Society of Heating, Refrigerating and Air Conditioning Engineers, Atlanta, GA, 1983.
30. Armstrong, J. A., M. Durham, and F. P. Venditti, Eds. *Symposium on the Transfer and Utilization of Particulate Control Technology: Volume III*, EPA Report, No. EPA-600/7-79-044C, 1979.
31. EPA, *Wet Scrubber Liquid Utilization*, Report No. 650/2-74-108, 1974.
32. EPA, *Gaseous, Particulate and Sulfur Related Emissions from Catalyst and Non-Catalyst Equipped Vehicles*, Report No. 600/2-77-237, 1977.

33. Cheremisinoff, P. M. *Fine Particulate Control in Air Pollution*, Illinois: Pudvan, 1988. Chapter 7.
34. Barbarika, H. F., S. Calvert, and S-C. Yung. *J. Air Pollut. Control Assoc.* **27** 348–357 (1977).
35. Barbarika, H. F., S. Calvert, L. F. Sparks, and S-C. Yung. *Env. Sci. Tech.* **12**, 456–459 (1978).
36. Cooper, D. W., J. L. M. Koehler, D. Leith, K. P. Martin, and S. M. Rudnick. *Env. Sci. Tech.* **20**, 237–242 (1986).
37. White, H. J. *J. Air Pollut. Control Assoc.* **27**, 15–21 (1977).
38. Robinson, J. *Air Pollution Control*. Part 1, New York: Wiley, 1971, Chapter 5.
39. Reference 13, Chapter 7.
40. Jones, A. L., Ed., *The Health and Safety Factbook*, Section 9, London: Professional Publishing Ltd., 1992.
41. Mills, D. *Chem. Eng.* **97**(2), 70–82 (1990).
42. Mills, D. *Chem. Eng.* **97**(6), 94–105 (1990).
43. Bradley, M. S. A. An Improved Method of Predicting Pressure Drop Along Pneumatic Conveying Pipelines, *Third International Conf. on Bulk Materials, Storage, Handling and Transportation*, Newcastle, June 1989.
44. Maisey, H. R. *Chem. and Proc. Eng. Part 2*, **46**, 662 (1965).
45. National Fire Protection Association, *Venting of Deflagrations*, NFPA 68, Quincy, MA, 1988.
46. Lunn, G. *Venting Gas and Dust Explosions—A Review*, Institute of Chemical Engineering, Rugby, Warwickshire, 1984, pp. 104–109.
47. Schofield, C. and J. A. Abbott. *Guide to Dust Explosion Prevention and Protection*, Part 2, Inst. of Chem. Engr., Rugby, Warwickshire, 1988.
48. Godbey, T. *Chem. Eng.* **98**(4), 130–136 (1991).
49. Bartknecht, W. *Dust Explosions*, New York: Springer-Verlag, 1989.
50. Schofield, C. *Guide to Dust Explosion Prevention and Protection*, Part 1, Inst. of Chem. Engr., Rugby, Warwickshire, 1984.
51. Abrahamsen, A. R. "The U.K. Approach to Dust Explosibility Assessment and its Relevance to Explosion Prevention and Protection," in K. L. Cashdollar and M. Hertberg (Eds.), *Industrial Dust Explosions*, ASTM STP 958, American Society for Testing and Materials, 1987, p. 60.
52. Bartknecht, W. "Preventive and Design Measures for Protection Against Dust Explosion," in K. L. Cashdollar and M. Hertberg (Eds.), *Industrial Dust Explosions*, ASTM STP 958, ASTM, 1987, p. 158.
53. National Fire Protection Association, *Standard on Explosion Prevention Systems*, NFPA 69, Boston, MA, 1978.
54. VDI-Kommission Reinhaltung der Luft, *Pressure Release of Dust Explosions*, VDI 3673, Dusseldorf, 1979.
55. Dahn, J. C. *Plant/Operations Prog.* **11**(3), 201–204 (1992).
56. *CFR*, Design Safety Standards for Electrical Systems, *Title 29*, Subpart S, Part 1910.304, pp. 713–718, July 1, 1991.
57. National Fire Protection Association, *Lightning Protection*, NFPA 78. Quincy, MA, 1988.
58. Dahn, J. C. *Chem. Eng. Prog.* **89**(5) 17–21 (1993).

15

Major Process Hazards

Health hazards from chemical process plants can result from small routine, more or less continuous releases, from major catastrophic releases, or from fires and explosions. The causes, results, and means of prevention of disasters are different from other health hazards. Figure 15.1 illustrates the relationship of consequence and likelihood for chronic versus catastrophic events. The consequences of a major disaster, like Bhopal, are so great that the likelihood of such an event occurring must be extremely small. The technology of the prevention of major process hazards deals with methods for estimating and acting on the likelihood of very unlikely events, for modeling the consequences of such events, and for interventions that alter these likelihoods.

Our view of what can happen and what ought to be done to prevent it is influenced by our experience. Table 15.1 lists historic disasters and their consequences. In every case, public awareness of what could happen leads legislators and others to tighten controls and restrictions intended to prevent future occurrences of that kind. In the early instances, the connection between the consequence and the precaution is obvious and the precaution was effective. As devastating as hurricane Andrew was, relatively few lives were lost compared to earlier hurricanes. For more recent disasters, like the 1989 chemical plant explosion at Pasadena, Texas (Fig. 15.2), the causes were more complex and less obvious, and so the preventive measures are more technical, inferred probabilistic, and less certain.

This chapter will discuss the following topics:

- The laws relating to major process hazards in the United States and Europe. Historically, the laws have consisted of a set of equipment codes. More recently, performance based procedural requirements for the evaluation and management of risk have been added.
- The procedures used to predict the risk and consequences of a disastrous event. These procedures are evolving from early, judgmental checklist-type evaluations to very complex probabilistic analyses.

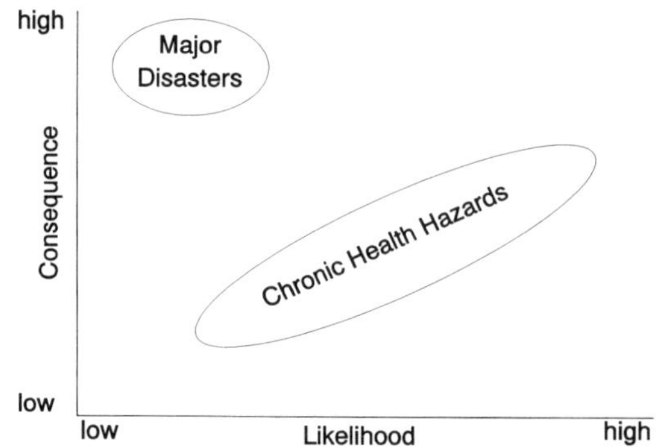

Figure 15.1 Likelihood – consequence relationships

- The means used to prevent disasters. These include the primary systems intended to prevent anything from going wrong, the intrinsic safety or "fail safe" design features which prevent disasters even when something does go wrong, and ways of preventing and dealing with errant human behavior.
- The development of plans for responding to emergencies. What could happen, who would be impacted, what can be done to mitigate/alleviate, and who will do it are covered in this section.
- Sources of information. The body of literature on the above topics is enormous. A list of selected secondary sources gives access to primary sources at the leading edge of developments in this technology.

15.1 LAWS AND REGULATIONS

This section will summarize the main U.S. and European laws related to major process hazards.

15.1.1 Hazardous Materials Transportation Act

The U.S. Department of Transportation (DOT) is responsible for administering the Hazardous Materials Transportation Act (PL 93-633, amended 1984) as well as the Natural Gas Pipeline Safety Act (PL 90-481, amended 1988) and the Hazard Liquid Pipeline Safety Act (PL 96-129, amended 1988).[1] All of these acts are intended to adequately protect the public "against the risk to life and property which are inherent in the transportation of hazardous materials in commerce." They give the Secretary of Transportation the authority to regulate any safety aspect of the transportation of hazardous materials including packing, handling, labeling, marking, placarding, and routing. Regulations on packing cover the manufacture, marking, maintenance, repairing, and testing of packages. Regulations on handling cover number of personnel, level of training, frequency of inspection, and safety assurance procedures for the handling of hazardous materials. The requirements are promulgated in a series of regulations in 49 Code of

TABLE 15.1 Disasters and Results[a]

Type	Location and Date	Total Deaths	Results
Earthquake	City of Chicago, Illinois October 9, 1871	250	Building codes prohibiting wooden structures; water reserve
Flood	Johnstown, Pennsylvania May 31, 1889	2209	Inspections
Tidal wave	Galveston, Texas September 8, 1900	6000	Sea wall built
Fire	Iroquois Theatre, Chicago December 30, 1903	575	Stricter theater safety standards
Marine	General Slocum, burned in the East River, New York June 15, 1904	1021	Stricter ship inspections; revision of statutes (life preservers, experienced crew, fire extinguishers)
Earthquake and fire	San Francisco, California April 18, 1906	452	Widened streets; limited heights of buildings; steel frame and fire-resistant buildings
Mine	Monongah, West Virginia December 6, 1907	361	Creation of Federal Bureau of Mines; stiffened mine inspections
Fire	North Collinwood School Cleveland, Ohio March 8, 1908	176	Need realized for fire drills and planning of school structures
Fire	Triangle Shirt Waist Company New York March 25, 1911	145	Strengthening of laws concerning alarm signals, sprinklers, fire escapes, fire drills
Marine	Titanic struck iceberg Atlantic Ocean April 15, 1912	1517	Regulation regarding number of lifeboats; all passenger ships equipped for around-the-clock ratio watch; International Ice Patrol
Explosion	New London School, Texas March 18, 1937	294	Odorants put in natural gas
Fire	"Cocoanut Grove," Boston, MA November 28, 1942	492	Ordinances regulating aisle space, electrical wiring, flameproofing of decorations
Explosion	Flixborough, UK Caprolactam factory June 1, 1974	28	Increased emphasis on hazard analysis
Toxic release	Union Carbide plant at Bhopal India releases methyl isocyanate December 3, 1984	2500	Tighter laws and regulations to prevent toxic chemical release
Nuclear	Chernobyl, Russia — nuclear plant burns and releases radiation over Russia and Europe April 1986	31 +	Reevaluation of nuclear power safety

[a]After Thygerson, 1977.

Federal Regulations (CFR) 171-199. They incorporate by reference a long series of guides and standards developed by some of the organizations listed in Table 15.2.

In 49 CFR 172 are listed hazardous materials subject to the labeling, marking, and so on requirements of law. This list gives the hazard class, the United Nations (UN) or North American (NA) identification numbers, labeling and packaging requirements, and quantity restrictions. It is supplemented by the Reportable

Figure 15.2 *Fire and smoke pour from the Phillips Chemical Plant in Pasadena, Texas, October 23, 1989 (AP / Wide World Photos).*

TABLE 15.2 Organizations Involved in Hazardous Materials Transportation Safety

American Society of Mechanical Engineers (ASME)
American National Standards Institute (ANSI)
American Society for Testing and Materials (ASTM)
Compressed Gas Association (CGA)
The Chlorine Institute (CI)
American Iron and Steel Institute (AISI)
Chemical Manufacturers Association (CMA)
National Fire Protection Association (NFPA)
Institute of Makers of Explosives (IME)
The Fertilizer Institute (TFI)
International Maritime Organization (IMO)
Society of the Plastics Industry (SPI)
International Organization for Standardization (IOS)

Quantities list for hazardous substances designated under CERCLA (see below) as published by EPA. To aid those who must comply with these complex laws and regulations, the DOT publishes a series of information bulletins for shippers, carriers, freight handlers, and container manufacturers. These bulletins cover duties of employees, shipping documents, labels, container specifications, quality control, violation reporting, and incident investigations. The main thrust of DOT's activity is accident prevention rather than emergency response.

In addition to DOT, the Nuclear Regulatory Commission (NRC) regulates radioactive materials transport (10CFR71), the Federal Aviation Administration (FAA) regulates aircraft carriage enforcement procedures (14CFR13), and the Coast Guard regulates hazardous cargo on board ships and in ports (33 and 46CFR).

15.1.2 Superfund

The federal Comprehensive Environmental Response, Compensation and Liability Act of 1980 (CERCLA 1980), more commonly known as Superfund, was enacted "to provide for liability, compensation, cleanup and emergency response for hazardous substances released into the environment and the cleanup of inactive hazardous waste disposal sites." While Superfund is best known for its efforts to deal with hazardous waste sites, there are several provisions relevant to nonroutine chemical releases.[2,3]

1. *Response.* Section 104 provides the authority for the removal of any hazardous substance which may be released into the environment and the authority to respond to a release as necessary to "protect the public health or welfare or the environment."

2. *Planning.* Section 105 calls for the revision of the National Contingency Plan (FEMA 1981) to include the requirements of CERCLA.

3. *Liability.* Section 107 establishes a standard of strict liability for incidents involving the release of hazardous substances and provides for fines of up to $50,000,000 plus response costs plus triple punitive damages. Of interest to first responders to hazard materials incidents is a provision which states "no person shall be held liable... for damages as a result of actions taken or omitted in the course of rendering care, assistance, or advice in accordance with the national contingency plan or at the direction of an on-scene coordinator appointed under such plan with respect to an incident creating a danger to public health or welfare or the environment as a result of any release of a hazardous substance or the threat thereof unless there was gross negligence."

CERCLA was amended by SARA, which is discussed below.

15.1.3 Emergency Planning and Community Right to Know Act of 1986

This broad piece of legislation known as Title III of the Superfund Amendments and Reauthorization Act (SARA—PL 99-499) establishes a number of emergency planning, notification, reporting, and training requirements. The law requires that states establish emergency response commissions, emergency planning districts, and local emergency planning committees (LEPCs) consisting of representatives of government, police, fire, hospitals, media, community groups, and industry. Each LEPC writes an emergency plan covering hazardous substance facilities, emergency response and coordination, notification, release warning and dispersion, emergency equipment and facilities, evacuation, training and drills. The LEPC is guided by publications from the national response team established by the National Contingency Plan (Sec. 105 of CERCLA). Local personnel are to be trained in hazard mitigation, emergency preparedness, fire prevention and control, disaster response, and other emergency training topics.

The Act also requires facilities to notify the community emergency coordinator and the LEPC in the event of a release of certain hazardous chemicals in excess of specified amounts. The notification must identify the substance, the amount released, time and duration of the release, media (air, water), anticipated health risks, medical advice, precautions and a telephone contact. To avoid releases of extremely hazardous substances the EPA will review systems for monitoring, detecting, and preventing such releases, including perimeter alert systems. Facilities are also required to provide Material Safety Data Sheets (MSDS) or lists of

MSDSs to LEPCs.

Other major sections of this act provide for the reporting of routine nonemergency releases of certain toxic chemicals (Sec. 313). The SARA provisions also require (Sec. 126) that OSHA issue standards for the health and safety of employees in hazardous waste operations (HAZWOPER).

15.1.4 Occupational Safety and Health Administration

In 1985 OSHA initiated a Special Emphasis Program for the Chemical Industry (ChemSEP). As a part of this program, OSHA inspected facilities to study current practices for the prevention of catastrophic incidents and the mitigation of releases. At the same time, the EPA conducted a "review of emergency systems for monitoring, detecting and preventing releases of extremely hazardous substances" as required by Section 305 of SARA. These two efforts led OSHA and EPA to cooperate on the development of standards for the prevention of chemical release disasters. Since workers at a facility are usually the population most immediately and seriously impacted by a release and since OSHA has more experience in inspecting safety hazards at industrial facilities, the coordinated regulations will be issued and enforced by OSHA.

A report on process hazard management in the chemical and petroleum industries by an OSHA contractor[4,5] estimated that the petrochemical industry can expect approximately 100 incidents per year, resulting in an estimated 53 fatalities, 985 injuries, and the evacuation of 18,000 people. An OSHA view of the causes of petrochemical accidents is summarized in Table 15.3. On February 24, 1992 the final rule now entitled "Process Safety Management of Highly Hazardous Chemicals; Explosives and Blasting Agents; Final Rule" was issued.

This rule has its antecedents in OSHA's ChemSEP effort, EPA's SARA review, the Chemical Manufacturers Association (CMA) chemical awareness and emergency response (CAER) program,[6] and input from the Organization Resources Counselors,[7] the American Petroleum Institute, the Oil Chemical and Atomic Workers (OCAW), and the United Steelworkers of America (USWA). The proposed rule (1910.119) is performance orientated and may be summarized as follows:

1. *Application.* The standard applies to processes involving approximately 130 toxic and reactive chemicals in amounts greater than specified threshold quantities,

TABLE 15.3 Causes of Loss in 150 Industry Accidents

- Insufficient recognition of hazards
- Aging and poorly maintained equipment
- Unsafe conditions or procedures
- Poor planning
- Improper risk management
- Unsafe engineering practices
- Inadequately trained personnel
- Disproportionate attention to production

as well as to flammable liquids and gases (except heating oil and fuels) in quantities greater than 10,000 lb. The standard applies to all manufacturing industries, particularly chemicals, transportation equipment, fabricated metal products, as well as pyrotechnics and explosives manufacture.

2. *Employee Involvement.* This section is required by provisions in the Clean Air Act Amendments of 1990. Employers must develop a written plan of action on employee participation, consult with employees and their representatives on the conduct and development of process hazard analyses and other elements of process safety management, and provide access to process hazard analyses and other information required to be developed by the standard.

3. *Process Safety Information.* Employers are required to compile written process safety information on the hazards of the chemicals involved, the process technology, and the equipment in the process.

4. *Process Hazard Analysis (PHA).* The employer must select an appropriate method of PHA, either one of those specified and described in the standard or an equivalent method. Initial PHAs must be scheduled and conducted by the employer so that at least 25% are completed by May 1994, 50% by May 1995, and 75% by May 1996. All initial PHAs must be completed by May 1997. For employers who have already begun to conduct PHAs in anticipation of the final rule, the standard states that PHAs completed after May 1987, which meet the requirements of the standard, may be used to comply with the standard. The PHAs must be updated and revalidated at least every 5 years.

The PHAs must address the hazards of the process, previous hazardous incidents, engineering and administrative controls, consequences of failure of engineering and administrative controls, facility siting, human factors, and evaluation of effects of failure of controls on employees.

The PHAs must be conducted by teams having one expert in engineering and process operations. The team must include at least one employee with experience and knowledge specific to the process and one knowledgeable in the specific PHA methodology being used by the team. Labor union representation on the team is *not* required.

5. *Mechanical Integrity.* This section requires written procedures, training for process maintenance employees, and inspection and testing of process equipment including pressure vessels and storage tanks, piping systems, relief and vent systems and devices, emergency shutdown systems, pumps and controls such as monitoring devices, sensors, alarms, and interlocks. Equipment deficiencies must be corrected, and the employer must assure that maintenance materials and spare parts are suitable for the process and have been properly installed.

6. *Hot Work.* This section requires a permit system for hot work operations conducted on or near a covered process.

7. *Management of Change.* The employer must have a written program to manage changes in chemicals, technology, equipment, and procedures which addresses the technical basis for the change, impact of the change on safety and health, modification to operating procedures, time period for the change, and authorization requirements. Employers must notify and train affected employees and update process safety information and operating procedures as necessary.

8. *Incident Investigation.* This section requires the employer to investigate incidents which did, or could have, resulted in a catastrophic release of covered chemicals as soon as possible but not later than 48 hr following the occurrence. An investigation team comprised of at least one person knowledgeable in the process (a contractor employee, if appropriate) must develop a written report on the incident. The employer must address and document response to the findings and recommendations of the investigation report and review the findings with affected employees and contractor employees.

9. *Emergency Planning and Response.* The employer must develop and implement an emergency action plan.

10. *Compliance Audits.* The employer must certify that compliance with the process safety requirements has been evaluated at least every 3 yr. The standard requires retention of the two most recent audit reports and the employer's response to findings.

11. *Trade Secrets.* Trade secret provisions similar to those in the hazard communication standard permit employers to enter into confidentiality agreements to prevent disclosure of trade secret information.

12. *Appendix C (Compliance Guidelines.)* This appendix explains what OSHA means to achieve through the requirements of the standard but does not add any requirements that are not included under the standard. Examples presented are not intended to be considered the only means for achieving compliance.

The rule includes a list of 142 "Highly Hazardous Chemicals" (Appendix A) with threshold quantity amounts which trigger the provisions of the rule. In addition, the rule applies to flammable liquids or gases (except certain fuels) in excess of 10,000 lb.

15.1.5 Other U.S. Legislation

Several other U.S. laws have provisions that relate to process hazards.

1. *Resource Conservation and Recovery Act (RCRA).* This law lists certain hazardous wastes and requirements for their treatment, disposal, and storage.

2. *Clean Water Act (CWA).* The CWA sets limits on the release of hazardous substances into water.

3. *Toxic Substance Control Act (TSCA).* Under certain circumstances, the EPA may require health and environmental properties testing of new and existing chemicals. In addition, if releases may present substantial risk to health or the environment, there may be instances where such releases are reportable to the EPA administrator.

15.1.6 Seveso Directive

The "Seveso" directive, named after an accidental release of dioxin in Italy in 1976 was adopted by the Council of the European Communities on June 24, 1982. This directive, which set up an approach to the prevention of disaster in the EEC, can be summarized as follows:[8]

1. The Directive establishes an obligation for manufacturers to take all necessary measures to prevent accidents and to limit their consequences for man and the environment. From this general obligation, a number of more specific obligations are derived, including:

   a. Taking preventive steps at the stage when an industrial plant or process is being designed, as well as when it is operational.

   b. Considering the possible causes of accidents, monitoring critical points in production processes, anticipating the combinations of events that might give rise to an accident, and introducing stringent safety measures.

   c. Adopting plans and procedures for limiting the size and effects of major accidents.

2. The Directive establishes an obligation on manufacturers to provide governments with information on:

   a. Certain kinds of hazardous substances which exceed specified amounts.

   b. The number of persons working on site.

   c. A description of the technological processes involved.

   d. The preventive measures taken and the arrangements for dealing with malfunctions.

   e. Details of emergency plans and equipment available for on site use.

   f. Information necessary for the preparation of off-site emergency plans.

   g. Arrangements for initiating emergency action, should it be required.

   In addition, such manufacturers must notify governments of accidents, their causes, and the action taken as well as of proposals to alleviate the effects and avoid a recurrence.

3. On the part of governments, the Seveso Directive requires member states to establish a competent authority to:

   a. Receive and evaluate information submitted by manufacturers.

   b. Seek any necessary clarification to that information.

   c. Establish emergency plans for off-site use

   d. Ascertain that the manufacturer continues to perform his duties under the directive.

4. The Directive requires member states of the European Communities to ensure that all people "liable to be affected" by a major accident are informed "in an appropriate manner" of the necessary safety measures and emergency procedures. Neighboring member states must be provided with the same information given to national citizens, and this would form the basis for consultations within the framework of bilateral relations.

5. The Directive addresses both new facilities and the more difficult issue of facilities already in operation.

As this summary shows, a salient feature of the Seveso Directive, when fully enforced, is that it seeks to prevent major accidents by delineating the responsibilities of manufacturers and public authorities with respect to the prevention and control of accidents.

15.1.7 Control of Industrial Major Accident Hazards (CIMAH)

Following the Seveso disaster, the European Community issued in 1983 a directive on "major accident hazards of certain industrial activities." Each member country was required to implement this directive by means of national laws. The United Kingdom issued the CIMAH regulations in 1984[9] to control fire, explosion, and toxic hazards. The regulations require safety study, notification, emergency plans, and public information where listed chemicals are held in excess of stated quantities.

15.2 RESPONSIBLE CARE

In 1988 the Board of Directors and the member companies of the Chemical Manufacturers Association (CMA) adopted an initiative called "Responsible Care: A Public Commitment." This initiative, which was patterned after an earlier Canadian chemical industry program, commits CMA member companies to improve performance in response to public concerns about the impact of chemicals on health, safety, and environmental quality. The program has six elements:

1. *Guiding Principles.* The commitments of each company to improve health, safety, and environmental performance.
2. *Codes of Management Practice.* Statements of requirements in specific areas of chemical manufacturing, transport, and handling.
3. *A Public Advisory Panel.* Leaders who assist the industry in responding to public concerns.
4. *Member Self-Evaluations.* Demonstration of program implementation and the responsible management of chemical environmental, health, and safety hazards.
5. *Executive Leadership Groups.* Senior industry executive review of codes, programs, and needs.
6. *Obligations of Membership.* Member companies ascribe to the guiding principles and make good faith efforts to implement the program elements.

The Code of Management Practice which addresses major process hazards is the Process Safety code. This code is designed to prevent fires, explosions, and accidental chemical releases. The code is intended to ensure that facilities are designed according to sound engineering practices, that they are built, operated, and maintained properly, and that periodic reviews for conformance are conducted. The scope of the code includes manufacturing, processing, handling, and on-site storage of chemicals. It works with workplace health and safety programs and waste and release reduction programs, which taken together help assure that CMA member plants protect health, safety, and the environment.

TABLE 15.4 Process Safety Code of Management Practices

Each member company shall have an ongoing process safety program that includes:

Management Leadership

1. Leadership by senior management through policy, participation, communications, and resource commitments in achieving continuous improvement of performance.
2. Clear accountability for performance against specific goals for continuous improvement.
3. Measurement of performance, audits for compliance, and implementation of corrective actions.
4. Investigation, reporting, appropriate corrective action, and follow-up of each incident that results or could have resulted in a fire, explosion, or accidental chemical release.
5. Sharing of relevant safety knowledge and lessons learned from such incidents with industry, government, and the community.
6. Use of the Community Awareness and Emergency Response (CAER) process to assure public comments and concerns are considered in design and implementation of the facility's process safety systems.

Technology

7. Current, complete documentation of process design and operating parameters and procedures.
8. Current, complete documentation of information relating to the hazards of materials and process technology.
9. Periodic assessment and documentation of process hazards and implementation of actions to minimize risks associated with chemical operations, including the possibility of human error.
10. Management of changes to chemical operations to maintain or enhance the safety originally designed into the facility.

Facilities

11. Consideration and mitigation of the potential safety effects of expansions, modifications, and new sites on the community, environment, and employees.
12. Facility design, construction, and maintenance using sound engineering practices consistent with recognized codes and standards.
13. Safety reviews on all new and modified facilities during design and prior to startup.
14. Documented maintenance and inspection programs that ensure facility integrity.
15. Sufficient layers of protection through technology, facilities, and employees to prevent escalation from a single failure to a catastrophic event.
16. Provisions of control of processes and equipment during emergencies resulting from natural events, utility disruptions, and other external conditions.

Personnel

17. Identification of the skills and knowledge necessary to perform each job.
18. Establishment of procedures and work practices for safe operating and maintenance activities.
19. Training for all employees to reach and maintain proficiency in safe work practices and the skills and knowledge necessary to perform their job.
20. Demonstrations and documentation of skill proficiency prior to assignment to independent work, and periodically thereafter.
21. Programs designed to assure that employees in safety critical jobs are fit for duty and are not compromised by external influences, including alcohol and drug abuse.
22. Provisions that contractors either have programs for their own employees consistent with applicable sections of this Code or be included in the member company's program, or some combination of the two.

Glossary

This Code uses key terms in a context that may be broader than their associated regulatory definitions. However, adherence to this Code does not relieve a company of the obligation to meet federal, state and local regulatory requirements.

Process Safety—The application of management and engineering principles to prevent fires, explosions, and accidental chemical releases at chemical process facilities.

Sound Engineering Practice—The application of mandatory codes and standards supplemented by the use of voluntary codes, standards, and guidelines, tempered by professional judgement.

Safety Critical Jobs—Jobs, activities, and tasks, if improperly performed, that have the potential to significantly increase the risk of a fire, explosion, or accidental chemical release.

Accidental Chemical Release—Unplanned, sudden releases of chemicals from manufacturing, processing, handling, and on-site storage facilities to the air, water, or land. It does not include permitted or other releases.

The elements of the Process Safety Code of Management Practice are listed in Table 15.4. A sample self-evaluation form is shown in Appendix 15B. While these Responsible Care requirements are voluntarily imposed by the chemical industry and are not regulatory requirements, they may have some regulatory enforcement implication. They create an expectation of performance which might be enforced under a general duty provision as an accepted industry practice. On the other hand, the documentation of analysis, review, and self-evaluation is a positive demonstration of compliance or a good faith effort to achieve compliance.

15.3 HAZARD IDENTIFICATION

Hazard identification is the first step in the risk assessment and management process. It starts with data collection to define the system and proceeds through the identification of hazards and the analysis of causes and effects. The output of the hazard identification step leads to hazard evaluation and risk reduction as needed. Many formal techniques have been developed for the systematic analysis of complex systems. Some of the most commonly used methods (HAZOP and HAZAN) and some specialized methods like human error analysis are discussed below. They attempt to consider all reasonable possibilities and all suffer from the drawback that the probability of future events can only be guessed.

15.3.1 Data Collection

A necessary first step in all hazard identification and analysis procedures is the collection of accurate data about the substances present, the process and equipment, and the layout, topography, meteorology, and demographics of the plant and surrounding community. A complete inventory of substances present would include raw materials, additives, catalysts, intermediates, products, and wastes. The inventory should specify quantities, locations, use, marking, and type of storage. Chemicals used in utilities (hydrazine, chlorine) should also be listed. For each substance on the inventory, the substance-specific hazard information should be obtained. No one source (e.g., MSDS) will have all of this information, and some of the data may need to be measured.

Plant and process equipment data collection starts with up-to-date process and instrumentation drawings (P&ID). As plants age and are modified, the engineering details of plant changes are not always carried over to the P&IDs, so they become out of date. This can be a serious problem when conducting the hazard analyses described below if the plant is not, in fact, what the analysts think it is. In addition to accurate P&IDs, additional details should be collected on the age and condition of equipment, the state of maintenance, and the actual process operating procedures and practices used. What is known about the degree of corrosion of pipes and vessels? Are battery limit block valves operable? Have safety valves been tested and are rupture disks properly installed? Have prescribed operating procedures been modified by "short cuts"? Will remote sensors work?

To predict the consequences of a release, it is necessary to know how the land lies, which way the wind blows, and where the people are. Simple dispersion models make worst case assumptions and need little data, whereas complex models require a lot of data, as discussed below. Demographic data, transportation routes, and the location of fire stations and hospitals is also needed for casualty care planning.

The collection and continuous updating of this assembly of information is a difficult task and a burdensome effort. Companies need to assure by means of their audit or review process that the job is being done wherever it is needed.

15.3.2 Hazard Indices

A complete inventory will list a few substances that have the potential to cause a catastrophic incident and many that do not. A necessary step in going from the inventory to the development of event scenarios is a ranking of the list by hazard potential. Several hazard indices have been developed for this purpose.[10,11] The Vapor Hazard Index (VHI)[12] is a ratio of the concentration of saturated vapor divided by the TLV times 1000. However, the TLVs are based on a variety of endpoints rather than the acute toxicity of primary concern in chemical release incidents. The Substance Hazard Index (SHI)[7] is the ratio of the vapor pressure of a substance to its "acute toxicity concentration" (ATC). The ATC is calculated from either the AIHA Emergency Response Planning Guideline Level Three (ERPG-3), the EPA Levels of Concern, or the Acute Toxic Concentrations developed by the State of New Jersey. Substance with SHIs over 5000 are considered dangerous. Both of these indices consider only intrinsic human toxicity. Other ranking systems have been developed[13] which consider both human and environmental endpoints.

The EPA has proposed a screening index in their interim rulemaking 40CFR300 under SARA. The index is equal to a "Level of Concern" value [usually the NIOSH Immediately Dangerous to Life or Health (IDLH) value in milligrams per liter] divided by V, the vapor fraction (1 for gases). The EPA uses this index to determine threshold quantities for the planning and notification provisions of their rules. An obvious weakness of this index is that it depends on the IDLH values, which may not be consistent and accurate measures of intrinsic hazard.

15.3.3 Dow and MOND Indexes

The Dow Chemical Company Fire and Explosion Index[14] was originally intended as an aid to fire protection by calculating a fire and explosion index. It calculates the maximum probable property damage (MPPD) and the maximum probable days outage (MPDO). It also considers general and special process hazards such as toxicity hazards and arrives at a toxicity index. A simplified flow diagram is shown in Figure 15.3.

The American Institute of Chemical Engineers Center for Chemical Process Safety[15] has summarized this relative ranking procedure as follows:

1. Identify on the plot plan those "process units" that would have the greatest effect or contribute the most to a fire, explosion, or release of toxic material.
2. Determine the material factor (MF) for each unit.

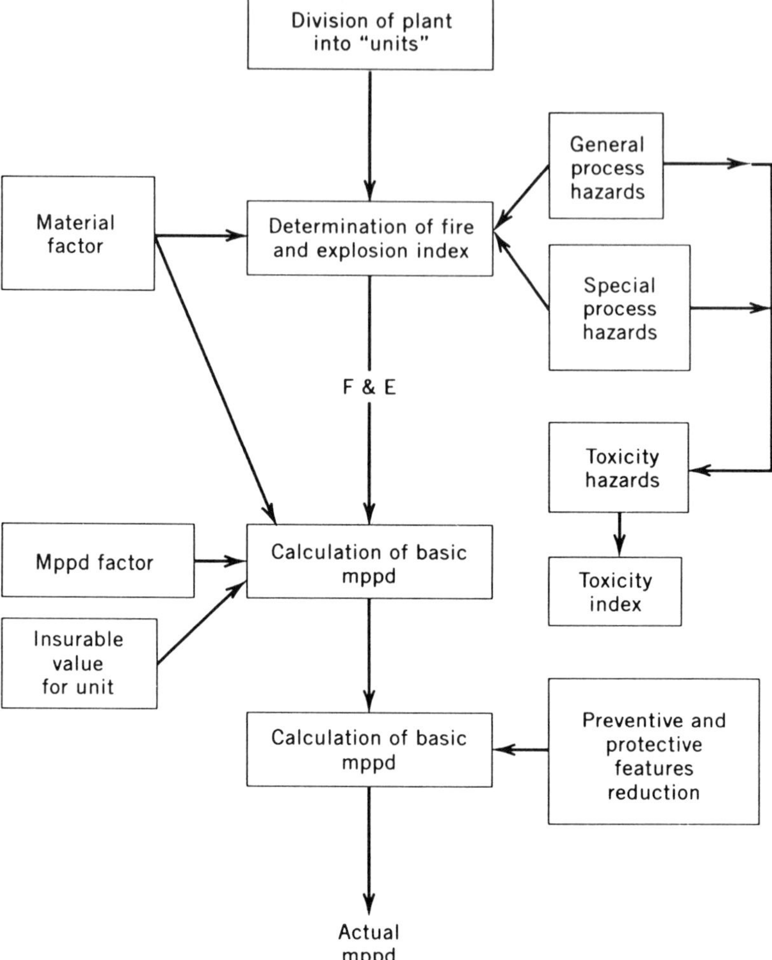

Figure 15.3 *Dow index procedure.*[17]

3. Evaluate the appropriate contributing hazard factors (CHF), considering fire, explosion, and toxicity.
4. Calculate the unit hazard factor (UHF) and damage factor (DF) for each unit.
5. Determine the fire and explosion index (FEI) and area of exposure (AE) for each unit.
6. Calculate maximum probable property damage (MPPD).
7. Evaluate maximum probable days outage (MPDO) and business interruption (BI) costs.

Based on the nature and severity of the hazard, various preventive and protective features are selected and the process is repeated to calculate new indexes.

The MOND fire, explosion, and toxicity index is based on the Dow Index but has been extended to consider more operations and equipment and to evaluate the hazards from materials, reactions, and toxicity more extensively.

Both methods can be used during the design stage of a project or on an existing unit. They are best conducted by a team of specialists in engineering, fire, safety, and health.

15.3.4 Hazard and Operability Study (HAZOP)

A HAZOP survey is one of the most common and widely accepted methods of systematic qualitative hazard analysis. It is used for both new or existing facilities and can be applied to a whole plant, a production unit, or a piece of equipment. It uses as its database the usual sort of plant and process information and relies on the judgement of engineering and safety experts in the areas with which they are most familiar. The end result is, therefore, reliable in terms of engineering and operational expectations, but it is not quantitative and may not consider the consequences of complex sequences of human errors.

The objectives of a HAZOP study can be summarized as follows:[16]

1. To identify areas of the design that may possess a significant hazard potential.
2. To identify and study features of the design that influence the probability of a hazardous incident occurring.
3. To familiarize the study team with the design information available.
4. To ensure that a systematic study is made of the areas of significant hazard potential.
5. To identify pertinent design information not currently available to the team.
6. To provide a mechanism for feedback to the client of the study team's detailed comments.

A HAZOP study is conducted in the following steps:

1. Specify the purpose, objective, and scope of the study. The purpose may be the analysis of a yet to be built plant or a review of the risk of an existing unit. Given the purpose and the circumstances of the study, the objectives listed above can be made more specific. The scope of the study is the boundaries of the physical unit and also the range of events and variables considered. For example, at one time HAZOP's were mainly focused on fire and explosion endpoints, while now the scope usually includes toxic release, offensive odor, and environmental endpoints. The initial establishment of purpose, objectives, and scope is very important and should be precisely set down so that it will be clear, now and in the future, what was and was not included in the study. These decisions need to be made by an appropriate level of responsible management.

2. Select the HAZOP study team. The team leader should be skilled in HAZOP and in interpersonal techniques to facilitate successful group interaction. As many other experts should be included in the team to cover all aspects of design, operation, process chemistry, and safety. The team leader should instruct

the team in the HAZOP procedure and should emphasize that the end objective of a HAZOP survey is hazard identification; solutions to problems are a separate effort.

3. Collect data. Theodore[16] has listed the following materials that are usually needed:

- Process description
- Process flow sheets
- Data on the chemical, physical, and toxicological properties of all raw materials, intermediates, and products
- Piping and instrument diagrams (P & IDs)
- Equipment, piping, and instrument specifications
- Process control logic diagrams
- Layout drawings
- Operating procedures
- Maintenance procedures

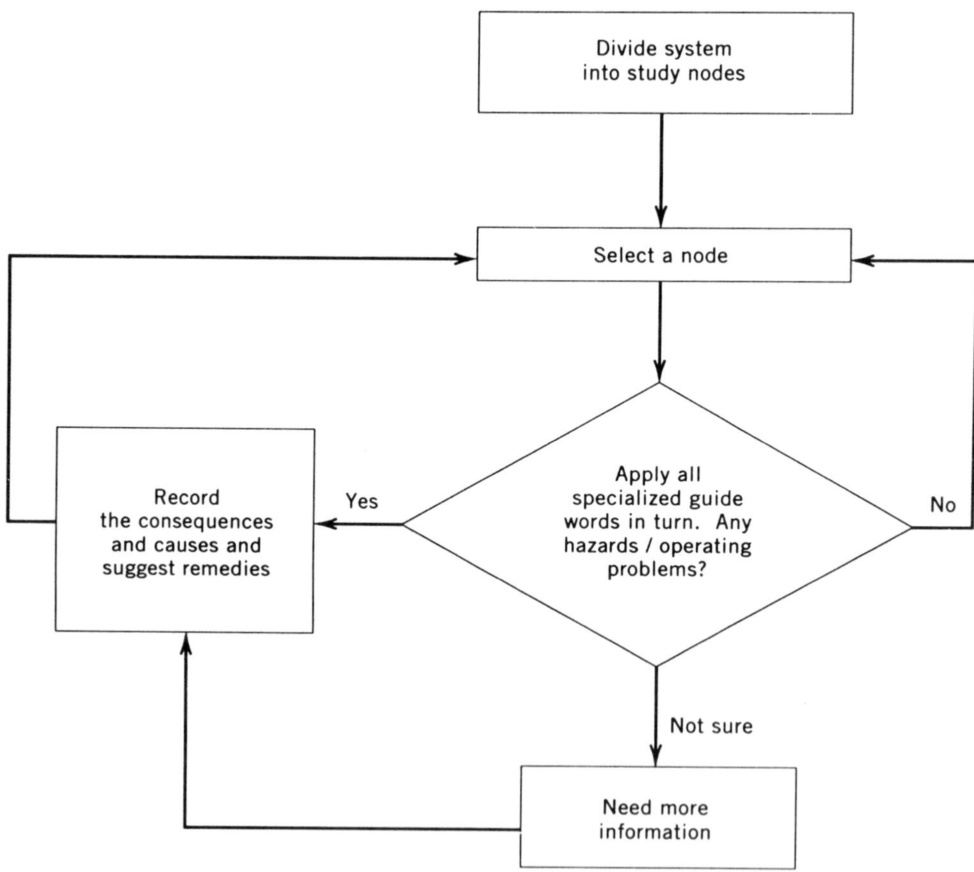

Figure 15.4 *HAZOP method flow diagram.*[15]

TABLE 15.5 HAZOP Guide Words and Meanings

Guide Words	Meaning
No	Negation of design intent
Less	Quantitative decrease
More	Quantitative increase
Part of	Qualitative decrease
As well as	Qualitative increase
Reverse	Logical opposite of the intent
Other than	Complete substitution

- Emergency response procedures
- Safety and training manuals

4. Conduct the study. Using the information collected, the unit is divided into study "nodes" and the sequence diagrammed in Figure 15.4 is followed for each node. Nodes are points in the process where process parameters (pressure, temperature, composition, etc.) have known and intended values. These values change between nodes as a result of the operation of various pieces of equipment such as distillation columns, heat exchanges, or pumps. Various forms and worksheets have been developed to help organize the node process parameters and control logic information.

When the nodes are identified and the parameters are identified, each node is studied by applying the specialized guide words to each parameter. These guide words and their meanings are key elements of the HAZOP procedure. They are listed in Table 15.5.[15]

Repeated cycling through this process, which considers how and why each parameter might vary from the intended and the consequence, is the substance of the HAZOP study.

5. Write the report. As much detail about events and their consequence as is uncovered by the study should be recorded. Obviously, if the HAZOP identifies a not improbable sequence of events that would result in a disaster, appropriate follow-up action is needed. Thus, although risk reduction action is not a part of the HAZOP per se, the HAZOP may trigger the need for such action.

The HAZOP studies are time consuming and expensive. Just getting the P&ID's up to date on an older plant may be a major engineering effort. Still, for processes with significant risk, they are cost effective when balanced against the potential loss of life, property, business, and even the future of the enterprise that may result from a major release.

15.3.5 Hazard Analysis (HAZAN)

The acronym HAZAN is a generic term for a variety of quantitative hazard and risk analysis methods. A logical sequence for the systematic examination of a facility is to use hazard identification and ranking techniques first. If there are possible scenarios which could lead to unacceptable consequences, then qualitative

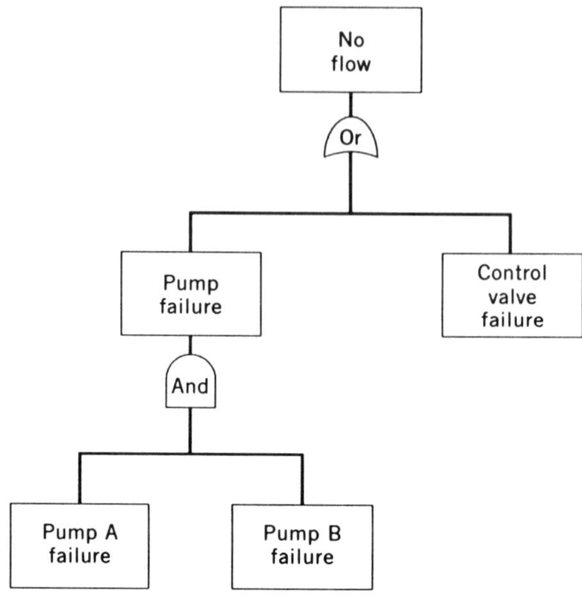

Figure 15.5 Fault tree diagram.[16]

techniques (HAZOP) can be applied next. Where a qualitative hazard exists, HAZAN methods can be used to estimate the quantitative probability of adverse events. How much analysis is worthwhile is a function of the consequence of the adverse event and the difficulty in preventing it.

15.3.6 Fault Tree Analysis

Fault tree analysis is a logical method of analyzing how and why a disaster could occur. It is a graphical technique that starts with the end event which is the accident or disaster (e.g., nuclear fuel melt-down) and works backward to find the initiating event or combination of events that would lead to the final event. If the probabilities of each potential initiating event is known or can be estimated, the probability of the end or "top" event can be calculated.

The following example of fault tree analysis is taken from Accident and Emergency Management.[16] A water-pumping system consists of two pumps, A and B, where A is the pump ordinarily operating and B is a standby pump that automatically takes over if A fails. Flow of water through the pump is regulated by a control valve in both cases. Suppose that the top event is no water flow, resulting from the following initiating events: failure of pump A and failure of pump B, or failure of the control valve. The fault tree diagram for this system is shown in Figure 15.5. This diagram shows the logical relationship of the events and allows the calculation of the probability of the top event if the probabilities of the initiating events are known. Let A, B, and C represent the failure of pump A, the failure of pump B, and the failure of the control valve respectively. Then it can be seen from the diagram that the top event T, no water flow, occurs if both A and B occur or if C occurs. Hence A, B, and C can be assumed to be independent events unless they have a common cause or are otherwise related. From the basic theories of probability, we can write

$$P(T) = P(AB) + P(C) - P(ABC) \tag{15.17}$$

and

$$P(AB) = P(A)P(B) \tag{15.2}$$

and

$$P(ABC) = P(A)P(B)P(C) \tag{15.3}$$

therefore

$$P(T) = P(A)P(B) + P(C) - P(A)P(B)P(C) \tag{15.4}$$

If the individual initiating event probabilities $P(A)$, $P(B)$, and $P(C)$ are known, then the probability of $P(T)$ the top or end event can be calculated.

An output of fault tree analysis and several other methods of reliability analysis described below is a list of the minimum combinations of events sufficient to cause the outcome. This list of events is called the "minimal cut set" and can be ranked in terms of likelihood and consequence as a means of prioritizing safety improvement programs.

15.3.7 Event Tree Analysis

Another technique for applying the basic concepts of probability to the evaluation of reliability of actual systems which have the potential to cause disasters is event tree analysis.[18] This technique is similar to fault tree analysis in that it examines, in a probabilistic manner, the consequences of a series of logically connected events. Event tree analysis, however, differs from fault tree analysis in that it starts with the initiating event rather than working backward from the end event. As in fault tree analysis, a diagram is constructed showing the logical relationship between events and outcomes. Event tree analysis may be qualitative or, if reasonably accurate probabilities can be assigned to events, a quantitative estimate of reliability can be obtained. In practice, the analyst reads off the diagram the sequences of events that result in disaster and then describes and examines the events in these sequences to describe how an accident may occur. As with fault tree analysis, the result is useful both in the design of safe new plants and to improve the safety of existing facilities.

15.3.8 Cause–Consequence Analysis

Fault tree analysis and event tree analysis have been combined into cause–consequence analysis to enable us to work out sequences either from the event to the outcome or from the outcome back to the initiating events. The result can be either qualitative or quantitative, depending on the nature of the input data. Because the analysis, while in some ways more complex, is more intuitively obvious, the display and description of sequences better serves the purposes of communication and of training those who can influence or control events. The output is usually a ranking of accident sequences based on their consequences and a ranking of minimal cut sets to evaluate important initiating events.

15.3.9 Human Error Analysis

The preceding methods of analysis are easiest to do if the initiating events can be accurately assumed to be independent. Indeed this assumption is often implicit in analyses as they are usually performed even if it is not stated. Yet it is obvious that equipment failure events often have common causes (a storm, a power failure) and thus are not independent. Another potential common cause is when operating personnel have an erroneous assessment of the situation or a mind set about what is happening, sometimes persisting in the face of evidence to the contrary. This human error then becomes the common cause of a series of initiating events which if looked at as independent events, would seem to have a very low probability of occurring simultaneously. At Three Mile Island, the operator's misunderstanding of the cooling water status led to several actions or inactions which formed a part of a sufficient minimal cut set to result in the accident. In the cause of the Ocean Ranger oil rig disaster,[19] the operators did not appreciate the damage that sea water had done to the ballast control panel. Because of confusing and conflicting information presented by the system, their attempts to operate the ballast controls became a "common cause" in a series of unlikely events that resulted in the capsize of the rig and the loss of 84 lives.

The techniques described above can be used to analyze the reliability of complex systems even where common causes and man–machine interactions are involved.[20–22] Human error analysis is a specialized subdiscipline of hazard evaluation which adds considerations of human performance to the hazard evaluation process. It is specialized and complex because of the difficulty in understanding the multitude of factors that influence human error rates and their variability, the high degree of interaction between human and system failure (lack of independence), and the nature of the options for altering human error rates.[23] Yet human behavior must be factored into reliability analysis for it to be accurate in the increasing number of systems where errors in man–machine interaction can be an initiating event in a disaster.

The most commonly used quantitative method for the measurement and assessment of personnel-induced errors is the Technique for Human Error Prediction (THERP).[17] This procedure involves the following steps:

- Identification of human activities which create a hazard
- Estimation of failure rates
- Effect of human failures on the system

The output of THERP is an input to fault tree or other methods of hazard analysis. While THERP can estimate failure rates for the routine performance of tasks, it cannot cope with error in human decisions and has difficulty with task error rates altered by stress, as in an emergency.

15.3.10 Environmental Hazards

In addition to fire, explosion, or the sudden release of a toxic gas, an emergency may be precipitated by the sudden release of an environmentally hazardous material.[24] A large oil spill can threaten aquatic organisms, birds, and those who depend on them for food. Chemicals released into rivers can threaten fish and drinking water supplies. Spills on land can reach and contaminate streams and be transported to human or animal receptors or be taken up by plants and enter the food chain.

The identification of the potential for environmental disasters is similar in some respects to other kinds of disasters. Critical facilities which could potentially release environmentally toxic substances need to be identified. Release scenarios and the pathways by which the substance would reach environmentally sensitive locations can be described and the consequences of each release–pathway combination analyzed. The engineering risk analysis methods described above for releases of substances toxic to humans can be applied to these environmental release scenarios.

Prediction of human or environmental receptor exposure and consequent harm is made complex by the variety of transport media and the chemical and biological changes that may occur. These complex processes can result in a delay of years before the consequence of a release are manifest. Also, because of the many uncertainties associated with each leg of a pathway, it is very difficult to predict consequence with any accuracy.

Human health consequences are predicted based on exposure of the human receptor compared with certain accepted exposure limits for food or drinking water. For risk management purposes, these human health consequences are valued in the same way as human health consequences resulting from air contaminants. For environmental damage, the situation is more complex. There is a substantial lack of data on the effects of a wide variety of chemicals on an even wider variety of organisms. Furthermore, there is no accepted system for making environmental value judgements across different species.

15.4 HAZARD EVALUATION

Hazard evaluation, in this context, means estimation of the harm that might be done by a toxic release. The preceding section discussed how to identify hazards and analyze the risk of release events. Based on this analysis, various release scenarios can be developed. This section will discuss the methods used to estimate the consequences of a release. Basically, there are five steps:

1. Scenario definition (part of hazard identification)
2. Release rate amount and physical state estimation
3. Transport and dispersion modeling
4. Biological effects prediction
5. Consequence estimation

Each of these steps is based on a combination of scientific theory, empirical data, and judgement. As a result, there is considerable uncertainty associated with the final conclusion.

15.4.1 Release Characteristics

A release has certain physical, chemical, and biological characteristics which need to be known or modified to predict the consequence of the events and thus develop appropriate mitigation on response plans. Specifically, some event will result in a loss of containment of a substance which will then be released at some rate for some time. The release may be a gas, a liquid, or a two-phase mixture. All of the characteristics of the release are taken into account in source models to predict the source terms which are inputs to various dispersion models.

Loss of containment is any event that allows substances normally contained in the tanks and piping of a facility to be released into the environment. In emergencies we are not concerned so much about losses of containment that result in a slow release, such as the leaks that produce fugitive emissions, although these are important in the prevention of air pollution. The loss of containment events that produce disastrous releases are those capable of releasing a large amount of material in a short time.

In general, properly designed, intact, undisturbed pipes and vessels do not spontaneously fail. Failure and release occur as a result of improper design, as in the case of the flexible coupling between reactors used at Flexiborough, or because of some unintended event. Pipes may fail as a consequence of unintended stress brought on by weather or physical forces compounded by corrosion, fatigue, or embrittlement. Vessels fail in fires or because of over- or sometimes underpressure. In the transportation of chemicals by road or rail, vehicle accidents can cause loss of containment and storms, groundings, or collisions of tankers can release their cargo.

The manner in which containment is lost will determine the release mechanism. Figure 15.6 illustrates several ways in which containment could be lost. The type and location of rupture and the phase, pressure, and temperature of the material determine the rate, period, and velocity of the release. The release could be a pressurized jet, resulting in a momentum plume of gas, or a rapidly expanding vapor cloud of a positively or negatively buoyant gas. Two-phase releases can be a mix of a gas and a bulk liquid which may subsequently evaporate or a liquid aerosol in a gas stream that is dispersed with or without evaporation.[26]

Loss of containment can also result in a spill of a substance in water, like an oil spill, or on to the ground where soil or ground water contamination may be a problem unless appropriate remedial action is taken.

Depending on all of the factors discussed, releases may be divided into four general time–area classes:[27]

1. *Continuous Point Release.* A prolonged, small area release—a hole or broken pipe on a large vessel.
2. *Term Point Release.* A brief period release from sudden and complete loss of a container of a gas. When the release duration is less than one-tenth of a minute, it is considered to be spontaneous or a "puff."

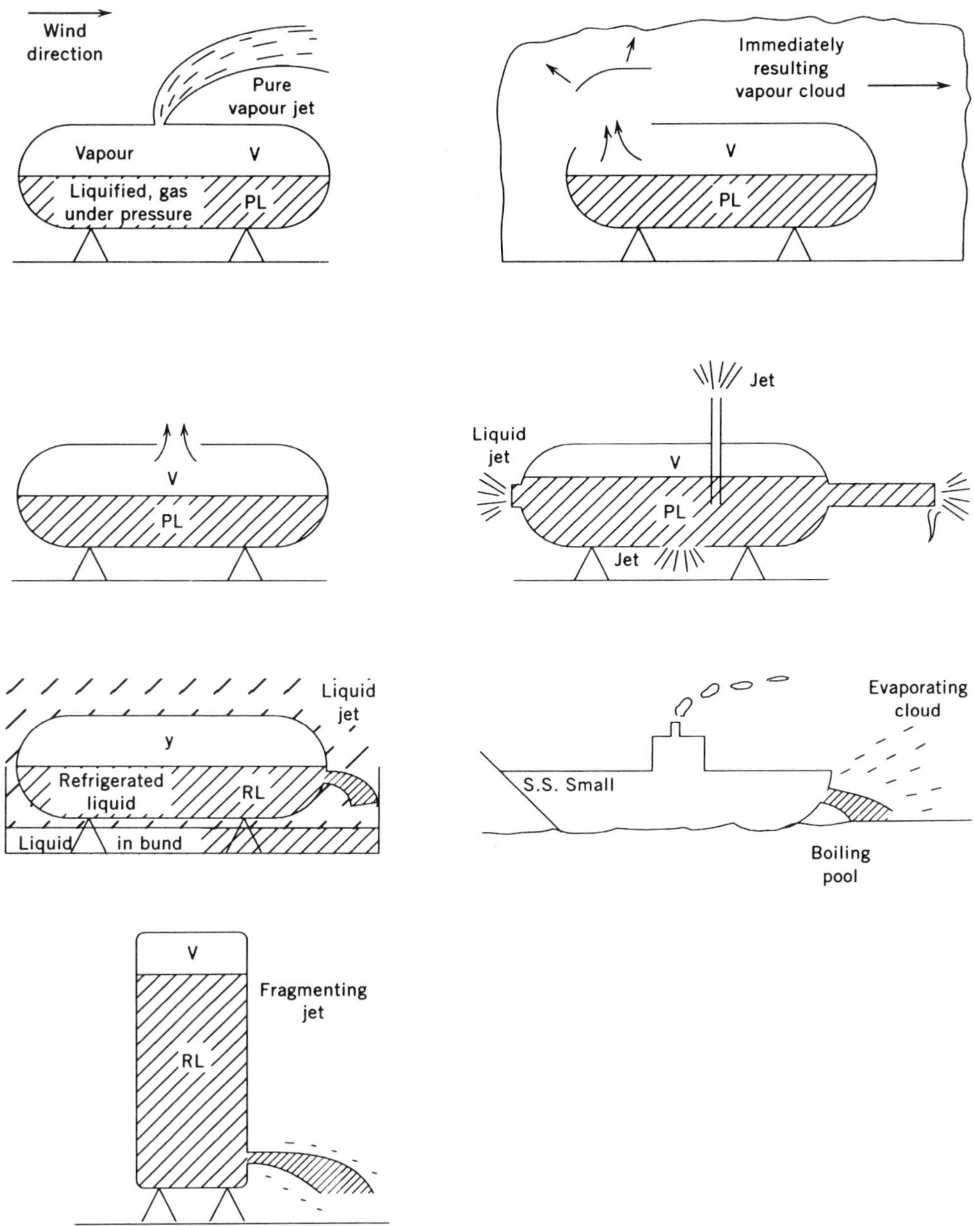

Figure 15.6 Release modes.[25]

3. *Continuous Area Release.* A prolonged release over some area as from a spill of a volatile liquid on the ground or water.
4. *Term Area Release.* A rapid release over some area as from the evaporation of a pool of spilled refrigerated liquid.

Modeling of the source terms for a release requires knowledge of the class of release, the geometry of the release aperture and surroundings (contaminant wall,

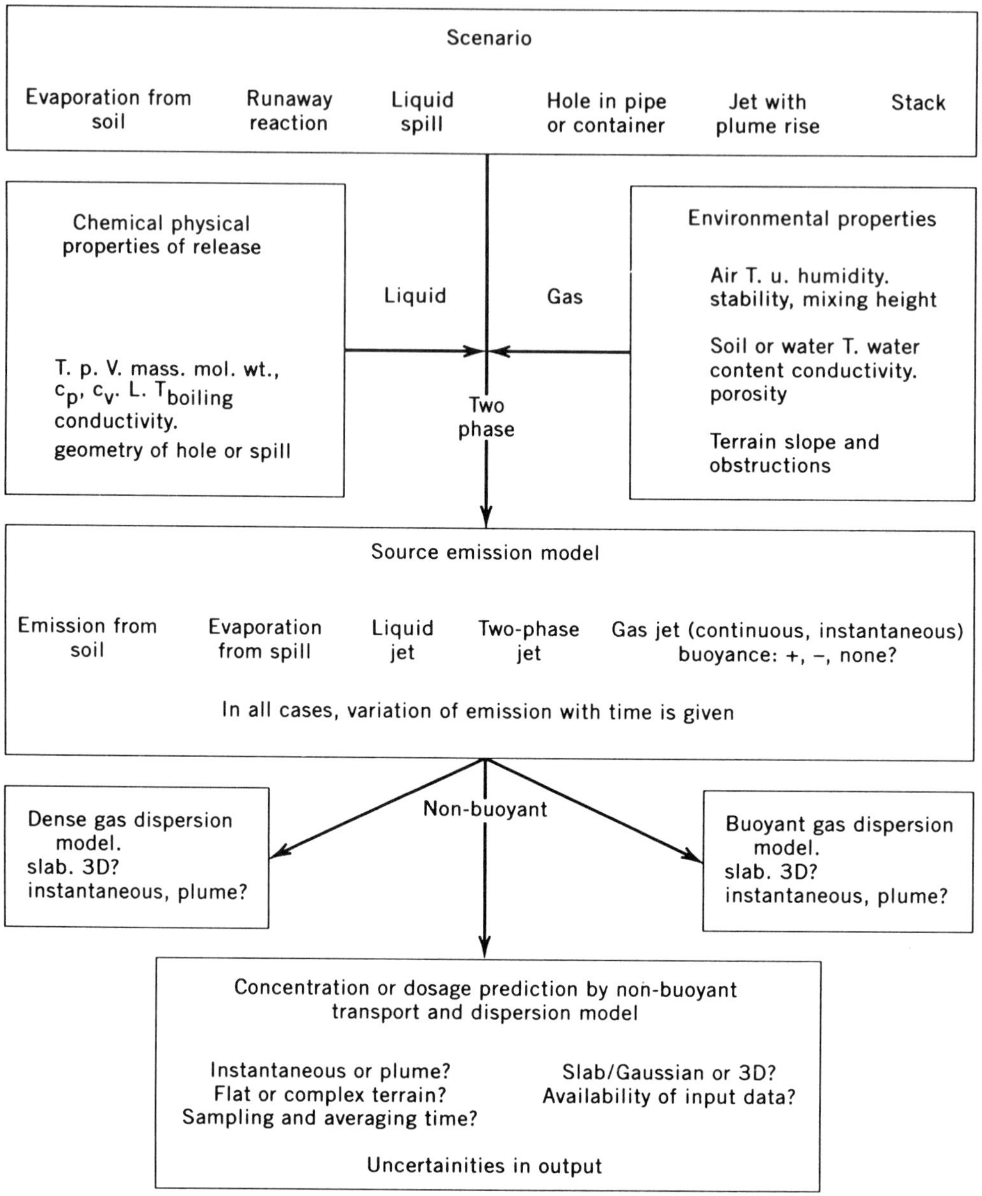

Figure 15.7 *Generalized vapor cloud model logic sequence.*[28]

pond), the physical chemical properties of the material at the time of the release and the environment conditions. All of this data is brought together as input to a source emission model, as shown in Figure 15.7.

A large number of models have been developed to estimate release rates and source terms as inputs to subsequent dispersion models. Release of a gas, liquid, or two-phase mixture from a hole in a pressurized container can be modeled from first principles and from fluid dynamics assuming the size and shape of the hole.[30]

For buoyant gases, these models yield dispersion model input source terms directly. For heavy gases and liquids, the situation is more complicated, and empirical models based test data developed at Edgewood Arsenal and Thorny Island are used.[31] A number of evaporation models have been developed by states, by the U.S. Government, and by other governments and private concerns. These models deal adequately with the most frequently expected accident conditions. However, additional work is needed for such unique conditions as chemical reaction after release, evaporation of mixtures, vadose layer release, and terrain effects.

Toxic releases resulting from fires are a special case of loss of containment. The release and subsequent ignition of a flammable liquid or gas can result in the formation of a toxic gas or aerosol which will then be dispersed. Release can be modeled as in the noncombustion cases, but the products of combustion are difficult to predict quantitatively, particularly when they will vary with temperature and the availability of oxygen. Fires can also cause the destruction of toxic materials and can in those cases reduce the hazard to the general population. Another concern for fires in plants where radioactive materials are used for gauges, radiography or other purposes is that the fire will destroy the radioactive material shielding and possibly disperse the material. This eventually needs to be considered in facility design[32] and in disaster planning.

15.4.2 Dispersion Models

When a toxic substance is released into the atmosphere, it will move or spread as carried by its own momentum, by gravity, and by the wind. This dispersion of the gas cloud will both dilute it and cause it to cover or involve a wider area. It is necessary to predict the concentration of the toxic substance that may occur at any location and time in order to predict the consequences to people and the protective measures necessary to avoid those consequences.[33–35] While long-range transport of clouds may be best described by box or compartment models, the intermediate range (100–1000 m) movement of a toxic substance is usually dealt with by dispersion modeling.

Methods of modeling the dispersion of a release of gas in the atmosphere are based on the following premises:[3,30]

- The gas is not lost or losses are taken into account (conservation of matter).
- Through the action of turbulence in the atmosphere, the gas is distributed horizontally and vertically according to a Gaussian (normal) distribution whose parameters are a function of distance from the source.
- The parameters of the distribution are derived from experiments and are described by empirical equations.

These premises lead to the idealized Gaussian plume model for a continuous source shown in Figure 15.8. Since this is an idealized model, it has a number of limitations. The model is only for a continuous source over flat open ground. It assumes that meteorological conditions are constant in time and space, that there is some wind, and that the gas is the same density as air.

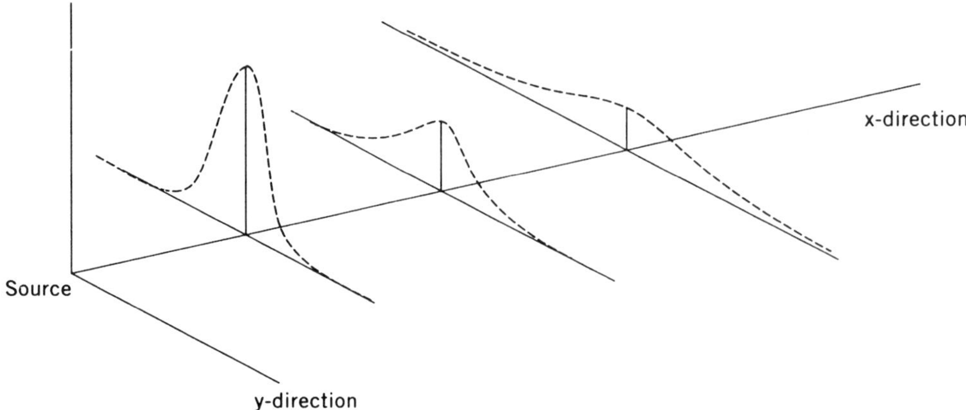

Figure 15.8 *Gaussian distribution.*[30]

Hanna[28] noted that there were over 100 hazardous materials transport and dispersion models, that the number was growing at the rate of about 10 per year, and that only a few have been adequately evaluated with field data. He listed the characteristics of 40 of these (Table 15.6) and characterized the models in terms of the phenomena they treat. Four models are discussed here to illustrate the basic principles important for the transport and dispersion of hazardous gases.

1. *AIRTOX.* The AIRTOX model for the release and dispersion of toxic air contaminants can account for single-phase denser than air releases. It is a hybrid model that incorporates various theoretical models, and it uses a slab model approach to calculate air dispersion. It assumes a pseudo-instantaneous dense gas cloud to calculate instantaneous sources. Continuous sources are also modeled and the passive plume is assumed to disperse according to standard Gaussian formulas with a wind shear component. The model has been shown to compare fairly well with field data.

2. *DEGADIS.* The dense gas dispersion (DEGADIS) model assumes that the cross-wind distribution is uniform in the middle of the cloud and Gaussian at the edges and that the vertical distribution follows a modified power law. The model deals with the dilution of a dense gas cloud by entrainment as the cloud moves under the influence of wind and gravity. It can calculate the initial cloud behavior near its source (the "source blanket") and can simulate both steady-state and transient spills. Transient spills are modeled as a series of pseudo-steady state releases with a coordinate system that moves with the wind over the transient gas source. This "moving observer" concept and the "source blanket" concept are unique to DEGADIS, but there are several aspects of the model that need further evaluation.

3. *FEM3.* The FEM3 model is representative of three-dimensional models. While simpler two-dimensional models provide great insight into the physical processes important for the transport and dispersion of hazardous gases and agree fairly well with the limited data bases, the physical event is three dimensional, so much of what is happening cannot be modeled by two-dimensional models. Like

TABLE 15.6 List of 48 Available Models for Hazardous Gas Transport and Dispersion. An X in a Column Means that the Model Treats that Particular Phenomenon[20]

Model	Reference	Instantaneous Puff	Continuous Plume	Dense Gas	Slab Similarity	Numerical (Grid Model)	Winds Vary In Space
AIRTOX	Paine et al. (1986)	X	X	X	X		
AVACTA II	Zannetti et al. (1986)	X	X		X		X
Britter	Britter (1979, 1980)		X	X	X		
CARE	Verholek (1986)	X	X	X	X		X
CHARM	Radian (1986); Balentine and Eltgroth (1985)	X	X	X	X	X	X
Chatwin	Chatwin (1983s)	X		X	X		X
CIGALE 2	Crabol et al. (1986)	X		X	X		
COBRA III	Oliverio et al. (1986)	X	X	X	X		
CRUNCH	Fryer (1980); Jagger (1983)		X	X	X		
3D-MERCURE	Riou and Saab (1985)	X	X	X		X	X
D2PC	Whitacre et al.	X	X		X		
DEGADIS	Havens and Spicer (1985)	X	X	X	X		
DENS20	Meroney (1984)	X	X	X	X		
DENZ	Fryer and Kaiser (1979)	X		X	X		
Fay & Zemba	Fay and Zemba (1985, 1986)	X	X	X	X		
FEM3	Chan (1983)	X	X	X		X	X
HAZARD	Drivas et al. (1983)	X	X	X	X	X	X
HEAVY GAS	Deaves (1985)	X	X	X		X	X
HEAVYPUFF	Jensen (1983)	X		X	X		
HEGADAS	Colenbrander and Puttock (1983)	X	X	X	X		
HEGDAS	Morrow		X	X	X		
Hoot et al.	Hoot et al. (1973)		X	X	X		
INPUFF 2.0	Peterson & Lavdas (1986)	X			X		X
MADICT	Ludwig (1985)	X			X	X	X
MESOPUFF 11	Scire and Lurmann (1983)	X			X		X
MIDAS	Woodward (1987)	X	X	X	X		X
NILU	Eidsvik (1980)	X		X	X		
Ooms et al.	Ooms et al. (1974)		X	X	X		
Port Comp. System	MOE (1983, 1986)	X	X	X	X		
RIMPUFF	Mikkelsen et al. (1984)	X	X		X		X
SAFEMODS	Raj (1981, 1986)	X	X	X	X		
SAFER	Personal Correspondence	X	X	X	X	X	X
SIGMET	England et al. (1978)	X	X	X		X	X
SLAB	Morgan et al. (1983)	X	X	X	X		
SPILLS	Fleischer (1980)	X	X		X		
TOXGAS	McCready et al. (1986)	X			X		
Van Ulden	Van Ulden (1974)	X		X	X		
VAPID	Jensen (1983)	X	X	X	X		
Webber	Webber and Brighton (1986)	X		X	X		
Wilson	Wilson (1981)	X	X		X		
Zeman	Zeman (1982)	X	X	X	X		

most three-dimensional models, FEM3 is extremely complex, and its computer code requires a long time to execute even on a supercomputer. Yet, as complex as it is, it is necessary to neglect certain processes, such as interphase heat exchange, in order to keep the run time within reason. Obviously, FEM3 is not suitable for real-time simulations. Its best application is as a means of understanding what is happening.

4. *INPUFF 2.0.* The INPUFF model is typical of a group of puff trajectory models for nondense emissions. The model cannot deal with the negative buoyancy of a dense gas cloud close to the source. It can model plume rise from positively buoyant plumes using standard formulas found in EPA UNAMAP models. INPUFF uses a Gaussian puff to calculate the distribution of concentrations in each puff. Dispersion coefficients are based on extensive experimental evidence, and the rate of dispersion can change with time. Multiple sources can be handled by summing the contribution from individual puffs at the receptor.

Uncertainty in model predictions is an important consideration when models are used to predict the consequences of toxic chemical releases. The causes of uncertainty can be divided into three components.

1. *Model Assumptions.* As noted above in the discussion of FEM3, all models that attempt to approximate reality can take a very long time to run on a supercomputer; consequently, useful models contain some simplifying assumptions and are, therefore, only approximations of reality. How much error these assumptions introduce depends on how far they are from true and how robust the model is. One reason why there are so many models is that each model is a better or worse fit of an event to be modeled. Therefore, it is incumbent on the risk assessor to select the model that is best for the release scenario to be modeled.

2. *Input Errors.* All models require the input of values for certain parameters that describe the physical, chemical, and meteorological conditions. These input parameters may not be single fixed values but may be variables which are functions of time and space. Often, they are not well known and must be approximated or assumed, which introduces error. There is a tradeoff between model accuracy and input error. The more realistic the model, the greater the number of less accurately known input parameters it will require. At some point short of maximum possible reality, an optimum combination of model detail and input error results in minimum overall error. When models are used for planning purposes to predict the consequences of a hypothetical future event, it is of course necessary to assume all of the input parameters. Often, worst case assumptions are made so that the predicted consequence will be conservative. Unfortunately, the selection of a series of worst case values for the input parameters can result in an excessive but unknown degree of conservativism.

3. *Random Error.* Turbulence of a fluid is an inherently chaotic process and thus cannot be usefully predicted. This stochastic uncertainty is dealt with as a random variability with a variance roughly equal to the square of the mean of the predicted concentration. It is usually the smallest of the three components of error.

Models are validated by comparing predicted results with actual data from field tests.[36,37] The number of models has increased much more rapidly than the quantity of field test data, so most models are unvalidated. Where a complex model is validated using a carefully assembled mass of accurate input data, the agreement between predicted and actual can be quite good—with less than a factor of 2 difference. Where assumptions are substituted for data, the agreement is not as good and the best that can be said is that the predicted concentration is not inconsistent with the actual given the anticipated level of error. In general, models that predict long-term averages are more accurate than dynamic models of instantaneous concentrations since errors in parameters tend to be averaged out.

15.4.3 Biological Effect Prediction

The modeling technology and techniques described above can predict with some error the concentration time pattern of the released chemical as it moves out from a point or area of release. People at locations where this cloud passes may be exposed and as a result may suffer adverse effects. The exposure may be by inhalation or by skin contact with or without absorption through the skin. The degree of adverse effect will depend on the dose, which in turn depends on the concentration and duration of the exposure and on whether the person is out of doors, in a car, or inside a house.

The prediction of the effects of exposure is a classic question of toxicology. Where little information is available on the toxicity of the chemical or only gross estimates of exposure concentrations are available, it may be adequate to classify chemicals according to relative toxicity using some parameter such as the median lethal dose (LD50). A number of such schemes are shown in Table 15.7. The LD50 is the dose at which half of the animals in an experiment would be expected to die as a result of a short-term exposure. It is a common reference point for acute toxicity and may be the only data point available for some less common chemicals. In some cases, the best adverse effects prediction that can be done will be based on a judgement of concentration from the amount released and some simple "box" model applied to a lethality class for the chemical to arrive at a consequence prediction suitable for emergency response.

The time–concentration relationship to adverse effect may be divided into three domains. At the outermost fringe of the area touched by the cloud the concentration may be below an actual or practical biological threshold such that no effect occurs regardless of the duration of the exposure. In this region, effect is predicted by whether the concentration alone is over or under some threshold. Further in, where the concentration is somewhat higher, the effect may depend on the product of the time and the concentration (ct). Still closer to the source, where the concentration is highest, and the adverse effects may include death, for some substances like chlorine or ammonia,[39] the adverse effect is best predicted by the square of the concentration times time (c^2t). This more complex biological behavior which occurs in the zone of exposure of greatest concern in disasters requires that both the modeling and the effects prediction guidelines include a time component.

A simple way of relating exposure concentration, time, and effect is shown in Table 15.8. This table divides the exposure concentration ranges into hazard bands

TABLE 15.7 Classification Systems for Acute Toxicity[a] Based on LD50 (mg / kg body weight)

Textbook — Hodge and Sterner	Industrial Chemicals EEC	Poison Law — Japan	Pesticide Act — USA	Transport of Goods — United Nations	Pesticide Classification — WHO
Extremely toxic (< 1)	Very toxic (< 25)	Designated poison (< 15)	Category I (< 50)	Group I (< 5)	Extremely hazardous (< 5)
Highly toxic (> 1 – 50)	Toxic (< 25 – 200)	Poisonous substance (< 30)	Category II (< 5 – 500)	Group II (< 5 – 50)	Highly hazardous (> 5 – 50)
Moderately toxic (> 50 – 500)	Harmful (< 200 – 2000)	Deleterious substance (< 300)	Category III (> 500 – 5000)	Group III solids (> 50 – 500)	Moderately hazardous (> 50 – 500)
Slightly toxic (> 500 – 5000)			Category IV (> 5000)	Liquids (> 50 – 2000)	Slightly hazardous (> 500)
Practically nontoxic (> 5000 – 15,000)					
Relatively harmless (> 15,000)					

[a]From reference 38.

TABLE 15.8 Hazardous Bands According to Medical Effects at Three Levels of Exposure of the General Population to Chlorine Gas[a]

Hazard band	Exposure (conc. ppm)	Effect
Distress (3 – 15 ppm)	3 – 6	Causes stringing or burning sensation but tolerated without undue ill effect for up to 1 hr
	10	Exposure for less than 1 min causes coughing
Danger (15 – 150 ppm)	10 – 20	Dangerous for 1 / 2 – 1 hr exposure Immediate irritation to eyes, nose, and throat, with cough and lachrymatin
	100 – 150	5 – 10 min exposure fatal for some vulnerable victims
Fatal (> 150 ppm)	300 – 400	A predicted average lethal concentration for 50% of active healthy people for 30 min exposure
	1000	Fatal after brief exposure (few breaths)

[a]From reference 40.

and gives some indication of the proportion of people affected (e.g., lethal for 50%). This type of data presentation, based on direct human experience, is only possible for a few chemicals like chlorine. For most chemicals, where only animal data is available, some process of extrapolation must be used and only the critical end points are estimated.

Various organizations have set emergency exposure limits or guidelines for use in hazard evaluation, remediation, and emergency response planning. To be useful, such guidelines should be developed both for substances where human experience is known and for those where only animal data is available. They should have uniform effects criteria (e.g., threshold of lethality) and should specify the exposure duration. It is also helpful if the guidelines are set by a process of consensus among experienced scientists so as to have broad credibility.

The Committee on Toxicology of the National Research Council set out to develop emergency exposure limits in the early 1960s. These levels, later called Emergency Exposure Guidance Levels (EEGLs) were defined[40] as "a ceiling limit" for an unpredicted single exposure, usually lasting 60 minutes or less and never more than 24 hours, whose occurrence is expected to be rare in the lifetime of any person. It reflects an acceptance of the statistical likelihood of the occurrence of a nonincapacitating, reversible effect in an exposed population. It is designed to avoid substantial decrements in performance during emergencies and might contain no uncertainty factor." The NRCs EEGL effort was sponsored by the military, and the exposed population was healthy adults. Recognizing that the EEGLs were not applicable to the general population, NRC introduced the concept of the short-term public emergency guidance levels (SPEGLs). These guidelines are intended for public emergencies, but few of them have been recommended by the NRC project.

During the 1970s, the National Institute for Occupational Safety and Health developed a set of exposure levels which were immediately dangerous to life or health (IDLH) as part of an effort known as the Standards Completion Program. The concept and the term IDLH came from respiratory protection regulations, which made a distinction between respiratory protection devices that are suitable for life-threatening environments and those that are not. The IDLHs were occupational exposure levels based on a 30-min exposure time. Since they were never intended as general public exposure limits, and since some of them were set by simply taking 1/100th of the TLV in effect at the time, they are not suitable as community emergency guidelines.

Increased concern over accidental chemical releases following the Bhopal disaster prompted the American Industrial Hygiene Association (AIHA) to establish the Emergency Response Planning Guideline (ERPG) Committee. This committee was charged with the development of a set of levels which could be used as boundaries between different degrees of an emergency. They are not "recommended" exposure levels in any sense since the existence of exposure over any of them is an unwanted event. As indicated by the name, they are intended for response planning purposes but are also useful in prioritizing disaster prevention measures. They are developed according to a protocol initially developed by the Organizational Resource Counselors (ORC) and are defined as follows:[41]

ERPG-3 "The maximum airborne concentration below which it is believed that nearly all individuals could be exposed for up to one hour without experiencing or developing life-threatening health effects."

The ERPG-3 level is a worst-case planning level above which there is the possibility that some members of the community may develop life-threatening health effects. This guidance level could be used to determine if a potentially releasable quantity of a chemical could reach this level in the community, thus demonstrating the need for steps to mitigate the potential for such a release.

ERPG-2 "The maximum airborne concentration below which it is believed that nearly all individuals could be exposed for up to one hour without experiencing or developing irreversible or other serious health effects or symptoms which could impair an individual's ability to take protective action."

Above ERPG-2, form some members of the community, there may be significant adverse health effects or symptoms which could impair an individual's ability to take protective action. These symptoms might include severe eye or respiratory irritation or muscular weakness.

ERPG-1 "The maximum airborne concentration below which it is believed that nearly all individuals could be exposed for up to one hour without experiencing other than mild, transient adverse health effects or without perceiving a clearly defined objectionable odor."

The ERPG-1 identifies a level which does not pose a health risk to the community but which may be noticeable owing to slight odor or mild irritation. In the event that a small, nonthreatening release has occurred, the community could be notified that they may notice an odor or slight irritation but that concentrations are below those that could cause health effects. For some materials, because of their properties, they may not be an ERPG-1. Such cases would include substances for which sensory perception levels are higher than the ERPG-2 level. In such cases no ERPG-1 level would be recommended.

Human experience takes precedence in developing ERPGs, but they can also be developed from animal data. The documentation recognizes that human response is not precise and there there are uncertainties in the data on which the limits are based. While limits could be set for a range of exposure times, the committee chose the single time of one hour based on the available toxicology information and typical exposure scenarios.

Guidance on acutely hazardous chemicals has also been provided by other organizations. The World Bank and the European Economic Community have used LC50s to arrive at acutely hazardous quantities rather than concentrations.[42] The National Fire Protection Association (NFPA) has developed a health hazard rating scale for use by emergency service personnel. The U.K. Health and Safety Executive sets specified level of toxicity (SLOT) exposure concentrations as an aid in calculating individual risk from major hazards.[43]

Finally, the Clean Air Act Amendments of 1990 require the EPA to promulgate an initial list of "100 substances which, in the case of an accidental release, are known to cause ... death, injury, or serious adverse effects to the environment" and to set threshold quantities for these substances. The initial list of 100 must include chlorine, ammonia, methyl chloride, ethylene oxide, vinyl chloride, methyl isocyanate, hydrogen cyanide, hydrogen sulfide, toluene diisocyanate, phosgene, bromine, hydrogen chloride, hydrogen fluoride, sulfur dioxide, and sulfur trioxide. The threshold quantities will be based on the toxicity, reactivity, volatility, dispersibility, combustibility, or flammability of the substance.

15.4.4 Integrated Systems

The risk manager needs to have the whole risk assessment process summarized so that prevention and response options will be clear. The crisis center director needs a system which takes all the available data on the conditions and circumstances at the time of a release and predicts the consequences fast enough and clearly enough for proper response decisions to be made. A number of integrated risk assessment systems have been developed for these purposes.

Manual Systems. The few least improbable scenarios can be modeled for several weather conditions or a worst case set of conditions. The predicted concentration time patterns can be plotted as isopleths at the ERPG or other effect boundary concentrations. These plots can be on separate pages for different wind directions or as movable transparent overlays that can be orientated on a map for the site of release and the current wind direction. These consequence prediction systems have the advantage of being easy to understand and use and are not not dependent on the availability of a computer and a source of electric power. They have the disadvantage of being very approximate and of only covering a limited number of possibilities. For these reasons, a number of rather sophisticated computer systems have been developed, but even when a computer-based integrated risk assessment system is used, it is prudent to have a manual backup available.

TABLE 15.9

CARE	Environmental Systems Corp.
CAMEO	National Safety Council
EPI Code	Homann Associates, Inc.
HASTE	Environmental Research and Technology Inc.
SAFER	SAFER Emergency Systems, Inc.

Computer Systems. Several integrated risk assessment commercial computer programs have been developed (Table 15.9). While the capabilities of the various systems are different and newer versions generally add capabilities, the following attributes are found in at least one system:

- Accepts substance, circumstance, and metrology input
- Provides substance hazard data
- Includes source term/initial dispersion modeling
- Does air dispersion modeling
- Provides models for site-specific terrain
- Analyzes sensor data
- Provides a real-time graphical display
- Assesses risk

Computer systems have been used for risk assessment, planning, training, drills, accident critiques, and response to emergencies. While the systems are capable of real-time operation, they are most commonly used to provide simulations for planning and training purposes.

15.5 MITIGATION OF RELEASES

The material presented above describes systems to allow the prediction of the likelihood and consequence of various chemical releases of various event scenarios. These predictions can be made for either an existing plant or a projected new plant during the design phase. Given the uncertainties of the modeling and estimation methods used, qualitative descriptions of probability and consequence are usually appropriate. Thus the probability will vary from frequent, meaning likely to occur repeatedly, to very unlikely, meaning not likely to occur in hundreds or thousands of years. The consequences range from minor odor or discomfort events to very serious events which may cause multiple deaths. These probability consequence combinations for each release scenario can be displayed on a graph or matrix which can serve as a graphical aid to mitigation priority setting. Obviously, an event which is predicted to occur frequently with very serious consequences is an unacceptable situation and a design or operation that resulted in such a prediction must be modified to reduce the risk before proceeding. For existing facilities, probabilities are more likely to be infrequent and consequences moderate. Mitigation effort priorities in this zone may be set based on a judgmental estimate of the degree of unacceptability of each probability consequence combination. At some low probability–low consequence point, it will be decided, based on criteria derived either from policy or law, that nothing further needs to be done. This scheme for deciding if something needs to be done leads logically to consideration of the mitigation options available.[44,45]

15.5.1 Intrinsic safety

Kletz[46] has analyzed plant safety in terms of attributes that make the plant "friendly" or forgiving of error. These plants are designed so that departures from normal tend to be self-correcting or at most lead to minor events rather than major disasters. Plants that are designed to be forgiving and self-correcting are inherently safer than plants where equipment has been added on to control hazards or where operators are expected to control them. Some characteristics of friendly plants are as follows:

- *Low inventory*. Smaller equipment and vessels and less intermediate in storage means less hazardous material on site.
- *Substitution*. Hazardous material can sometimes be replaced with less hazardous materials.
- *Attenuation*. If hazardous materials must be handled, they can be used in their least hazardous forms.
- *Simplification*. Operators are less likely to make mistakes if the plant and its instrumentation are easy to understand.
- *Domino effects*. An event should be self-terminating rather than initiating other events.
- *Incorrect assembly impossible*. Equipment design can make assembly mistakes impossible.
- *Obvious status*. An operator should be able to look at a piece of equipment and know if it is correctly installed or if a valve is open or shut.
- *Tolerance*. Small mistakes need not create big problems.
- *Low leak rates*. Leaks should be small and easy to control.

Kletz[46] summed up the application of these principles in Table 15.10.

15.5.2 Extrinsic Safety

Friendly plants as described by Kletz are designed to avoid major problems and this approach to plant safety should be taken wherever possible. Where it is not, external features of the plant should be arranged so as to minimize the undesirable consequences of a release. The most obvious means of reducing the exposure of community populations is to locate high hazard facilities remote from communi-

TABLE 15.10 Characteristics of Friendly Plants[a]

Characteristic	Examples	
	Friendliness	Hostility
Low inventory		
Heat transfer	Miniaturized	Large
Intermediate storage	Small or nil	Large
Reaction	Vapor phase	Liquid phase
	Tubular reactor	Pot reactor
Substitution		
Heat transfer media	Nonflammable	Flammable
Solvents	Nonflammable	Flammable
Carbaryl production	Israeli process	Bphopal process
Attenuation		
Liquefield gases	Refrigerated	Under pressure
Explosive powders	Slurried	Dry
Runaway reactants	Diluted	Neat
Any material	Vapor	Liquid
Simplification		
	Hazards avoided	Hazards controlled by added equipment
	Single stream	Multistream with many crossovers
	Dedicated plant	Multipurpose plant
	One big plant	Many small plants
Domino Effects		
	Open construction	Closed buildings
	Fire breaks	No fire breaks
Tank roof	Weak seam	Strong seam
Horizontal cylinder	Pointing away from other equipment	Pointing at other equipment
Incorrect assembly impossible		
Compressor valves	Noninterchangeable	Interchangable
Device for adding water to oil	Cannot point upstream	Can point upstream
Obvious status		
	Rising spindle valve or ball valve with fixed handle	Nonrising spindle valve
	Figure-8 plate	Spade
Tolerant of maloperation or poor maintenance		
	Continuous plant	Batch plant
	Spiral-wound gasket	Fiber gasket
	Expansion loop	Bellows
	Fixed pipe	Hose
	Articulated arm	Hose
	Bolted arm	Quick-release coupling
	Metal	Glass, plastic
Low leak rate		
	Spiral-wound gasket	Fiber gasket
	Tubular reactor	Pot reactor
	Vapor-phase reactor	Liquid-phase reactor

[a]From reference 46.

ties. In general, siting is a major design consideration both within a facility as well as in the relation of the facility to population centers.

LeVine[47] provides some guidelines for the layout of plants storing highly toxic hazardous materials (HTHM):

1. High-integrity storage tank design is essential for a HTHM. Consideration should be given to overdesign over normal operating conditions, or the use of double integrity storage tanks, for example, with a concrete outer tank or berm.

2. If the liquid storage tank for a HTHM is diked, it is good practice that the same dike does not contain inventories of other materials that are flammable, reactive, and noncompatible with the HTHM. If located next to a diked area containing flammable and/or reactive chemicals, the toxic storage may have to be provided with fixed fire water protection for cooling. Factors to be considered include drainage in the diked areas, the response time, the adequacy of the site fire brigade or public fire department that would respond, and the flammability and reactivity of the nearby materials. The impounding area should be designed to minimize the vaporization rate of volatile HTHM. Installed means of evaporation suppression should be considered (e.g., foam).

3. A HTHM should not be stored in proximity to explosive/flammable/reactive materials that might impact the storage tank. Diking for materials that are flammable or reactive should be sloped or drained to carry spilled material to a separate impoundment tank or area and away from any nearby HTHM tank. For HTHMs the impoundment system should be designed to minimize the surface area and move material away from other hazard areas.

4. The concept and design of a designated dump tank should be investigated at the earliest stages of site layout. Depending on the potential hazard and the time required to empty the HTHM storage tank, it may be desirable to keep some storage vessels empty, or at low levels, for such emergency transfers. The possible need to empty to leaking tank is one important reason for providing a submerged pump or other transfer method.

5. The momentum of the tank contents might overflow a traditionally designed dike wall if the tank were to burst. A dike wall of about one-half tank height is required for containment of this scenario. Full-height, close-in dikes or berms have been provided for some cryogenic ammonia (and liquefied natural gas) storage tanks. The dike must be sized to contain at least the volume of the largest tank within the dike.

LeVine[47] also recommends some basic facility siting practices for facilities that contain HTHMs so as to reduce the potential off-site impacts of a toxic material release. The recommendations are summarized as follows:

1. Site the facility such that sufficient controlled-use "green space" exists to provide a buffer zone based on the distance between the release location and the nearest population. For long-term effectiveness, the "green space" should be under the control of the plant's owner.

2. Favor sites with the lowest potential population exposure and ascertain that transportation risks are acceptable. The census statistics do not necessarily reflect

the transient population conditions, such as schools, commercial and industrial areas, and hospitals. Emergency evacuation plans and warning systems must be carefully considered when siting a major toxic chemical facility.

3. Site the facility to provide adequate separation from other facilities in order to avoid domino effects or secondary damage caused by blast waves, debris, missiles, or fire radiation from other episodic events.

4. Site the facility to support the heavy loads that storage tanks usually entail, and use good civil engineering practices. Provide adequate flexibility in piping systems without sacrificing strength. Careful consideration should be given to earthquake potential at all sites.

5. Site or grade the facility such that terrain controls reduce the likelihood of liquids and heavy vapors flowing toward populated areas.

6. Site the facility so that the probability of damage from external sources is minimized. Adequate clearance or crash barriers should be provided to protect against vehicular impact. Do not locate the facility within or close to landing and takeoff patterns of an airport, since most aircraft accidents take place during these maneuvers. Make sure designs include consideration of windstorms, floods, landslides, subsidence, and other potential external threats, such as vandalism, terrorism, and sabotage.

7. Site the facility such that adequate roads for routine and emergency situations are available or can be built. An adequate number of unblockable roads should be available for the movement of employees, members of the public, ambulances, and emergency service vehicles travelling both to and from the facility. Also, make sure there are at least two such access roads in case one is rendered impassable by the emergency.

8. Site the facility so that, in the event of a liquid release, water supplies will not be contaminated.

9. Choose a site where adequate reliable emergency utilities (power, water, etc.) can be provided.

10. Site the facility such that nearby populations, both off site and on site, are upwind of the prevailing wind. Since prevailing wind speed and direction are not totally reliable for safety planning, this approach should be used only as supplementary factor.

LeVine also provides design criteria for vessels, piping, loading, and unloading facilities and instrumentation so as to provide plant physical and process integrity.

In case all means of containment fail, releases should be controlled through safety valves or other pressure-relief systems and directed by vent headers to a safe release point or to an emergency relief treatment system. These "end of pipe" systems act as release countermeasures by means of secondary containment or by destroying the released materials. They include scrubbers, flares, incinerators, absorbers, condensers, and any other device placed in vent stream to prevent release to the atmosphere. Even when release to the atmosphere has occurred, some mitigation is possible by means of such vapor release countermeasures as water sprays or curtains, steam or air curtains or deliberate ignition.

The operation of secondary containment and vapor release countermeasures will be triggered by some release warning system. This warning may come from observation of process parameters as displayed in the control room, from the detection of odor or the sighting of a cloud of smoke, fume, or color gas or from a signal from an installed vapor detection sensor. The early warning that can be provided by an installed detection system can trigger a quick response that may prevent most or all of a release. The detectors sense and measure gases and vapors by combustion, catalytic, electrochemical, and solid-state devices and they may alarm locally or remotely in the control room or be part of a network connected to a computer that analyzes the information from a number of locations and presents the results in a way that highlights response options. For early warning vapor detection systems to be useful tools in the release mitigation system, they must be reliable and well maintained. Also, they must be specific enough for the chemical of concern and sufficiently sensitive so as to differentiate between routine releases (fugitives) and emergencies.

15.5.3 Human Factors

The best designed plant can fail if improperly operated. Case histories of plant disasters[48] have shown that human error was often involved in the chain of events that lead to the outcome. The possibility of human error cannot be eliminated so plants should be designed to be "forgiving" of operator mistakes. Human error can be substantially reduced by operator training and motivation. The U.S. Occupational Safety and Health Administration has listed the operating procedures that the operators must know to control plants that handle highly hazardous chemicals.[49]

1. Steps for each operating phase:
   - Initial startup
   - Normal operation
   - Temporary operations as the need arises
   - Emergency operations, including emergency shutdowns, and who may initiate these procedures
   - Normal shutdown
   - Startup following a turnaround, or after an emergency shutdown
2. Operating limits:
   - Consequences of deviation
   - Steps required to correct and/or avoid deviation
   - Safety systems and their functions
3. Safety and health considerations:
   - Properties of, and hazards presented by, the chemicals used in the process
   - Precautions necessary to prevent exposure, including administrative controls, engineering controls, and personal protective equipment
   - Control measures to be taken if physical contact or airborne exposure occurs
   - Safety procedures for opening process equipment (such as pipe line breaking)

- Quality control for raw materials and control of hazardous chemical inventory levels
- Any special or unique hazards

These operating procedures should be detailed in readily available operating manuals which are kept "evergreen" as plant equipment or procedures change. Operators should be thoroughly trained in the operating procedures before they assume responsibility for plant operations and supplemental and refresher training should be provided as needed, typically annually. Some form of testing or qualification should be set up to insure that each operator has learned the procedures. Periodically, management should compare prescribed and actual operating procedures. Where there are differences, they should be analyzed to determine if it is because of an employee lapse or if the written procedures are impractical or could be improved.

Contractors present a special concern in the control of human factors related to chemical releases. Teams of contractors employees often work on site under the supervision of their own company with only indirect control by the site owner. The training of contract employees is the responsibility of the contractor and is complicated by relatively high turnover in some jobs. Where contractors hire skilled trades from a hiring hall, the contractor may not know who will be working on a job until the day of the assignment. Even when all of a contractor's regular employees are trained and aware of safe operating procedures, the contractor may subcontract specific tasks and bring on-site people who are not trained and aware. Frequently there are multiple contractors on a site doing jobs which may impact the safety of each other. Some major disasters appear to be related to a lack of a clear division of responsibilities and close coordination between the owner and the several contractors.[48,50] While the legal consequences of owner–contractor interaction complicate the matter, at least one jurisdiction[9] has stated "Where . . . a special risk is likely to arise due to the nature of the work performed (and the owner of the premises has special knowledge of it), the owner must retain sufficient control of the operation to ensure that contractors' employees are properly protected against the risk." In the Untied States an owner has the responsibility under OSHA to insure that the contractor is informed of hazards and that the contractors' employees are trained.

A typical program addressing the contractor issue might include the following:

- The contract includes the site safety and health manual and the operating procedures to be followed.
- The contractor attends a prestart of work review at which specifications, safety, process hazards, and employee training is detailed.
- The contractor is provided with MSDSs for all hazardous materials to be encountered.
- The contractor trains his employees and provides the owner with a list of trained employees. The contractor understands that only trained employees will be admitted to the site.
- The contractor must obtain any necessary work permits (hot work, confined space entry) before starting such work.

- The contractor must keep accident records and inform the owner of any accidents.
- The contractor is subject to the safety reviews and inspections.
- Any violation of safety rules can be cause for termination of the contract.

The list given above is not intended to cover all requirements and may not fully apply to certain contractors who have little involvement with hazardous processes but it gives an example of the level of interaction some owners have found necessary to insure adequate control of contractors.

15.6 PLANNING AND RESPONSE

Even when all reasonable measures to prevent a chemical release have been taken, the risk or probability of a release occurring is not zero. A prudent management practice, therefore, is to anticipate that an emergency may occur eventually and to plan for it. A well-developed and implemented emergency plan can minimize loss and help protect people, property, and the environment. To accomplish this requires planned procedures, clearly understood responsibilities, designated and accepted authority and accountability, and trained and drilled people. When an emergency occurs, it will be too late to plan, so it must be done ahead of time, based on the kinds of emergencies that might occur.[16,51–53]

15.6.1 Scope

An emergency response plan (ERP) must be tailored to fit the facility, community, and types of emergency that can occur. Michael[54] has divided emergency response functions to consider in planning into the following five categories:

1. Accident Assessment includes detecting abnormal conditions, assessing the potential consequences, and immediately taking appropriate measures to mitigate the situation.
2. Notification and Communication includes the physical and administrative means whereby plant operators can rapidly notify plant management, off-site emergency response agencies, and the public.
3. The Command and Coordination function clearly establishes who is in charge of the emergency response in the plant and in off-site communities.
4. Protective Actions are those taken to protect the health and safety of plant personnel and the public such as release mitigation, sheltering, and evacuation.
5. Support Actions include fire fighting, emergency medical care, social services, and law enforcement.

Plan evaluators should expect to find the attributes listed in Table 15.11 considered in the plan.[55] Not all of these attributes will be required in every plan, but it is useful to consider them all to insure that the plan is complete.

TABLE 15.11 Plan Evaluation Attributes

Scope and provenance
Emergency response organization
emergency alarm
Operational communications
Control centers
Evacuation sheltering
Accounting for people
First aid
Transportation
Security
Fire fighting
Chemical release detection and mitigation
Community notification and information
Shutdown or operations
Continuity of utilities and services
Mutual (industry) aid organizations
Outside agency notification and coordination
Public affairs
Governmental relationships
Restoration to normal operations
Clean-up and rehabilitation
Legal issues
Training
Drills
Detailed action plans and SOP's
Critique and plan revision

15.6.2 Planning Process

Successful planning requires the cooperation and support of all involved. For this reason, planning should be under the direction and overview of a committee that includes those groups who have a significant interest or stake in the successful implementation of the plan in the event of an emergency. The Community Awareness and Emergency Response (CAER) program developed by the Chemical Manufacturers Association (CMA) recognizes the importance of community involvement and seeks to achieve two goals:[54]

1. Develop a community outreach program and provide the public with information on chemicals manufactured or used at local chemical plants.
2. Improve local emergency response planning by combining chemical plant emergency plans with other local planning to achieve an integrated community emergency response plan.

In the United States, the Superfund Amendments and Reauthorization Act of 1986 (SARA) established Local Emergency Planning Committees (LEPC) for each planning district. These committees facilitate the preparation and implementation of emergency plans and must include representatives from each of the following groups or organizations:

- Elected State and local officials
- Law enforcement, civil defense, firefighting, first aid, health, local environmental, hospital, and transportation personnel

- Broadcast and print media
- Community groups
- Owners and operators of facilities subject to the requirements of Tital III of SARA

Regardless of the composition of the planning committee, it must have a commitment to work together on a common goal to be successful. The group must have the ability, authority, and resources to do the job. It must have access to all the industrial, community, transportation, and planning skills needed. Most importantly, the group must agree on their purpose and be able to work together cooperatively.

16.6.3 Drills

Realistic tests of plan execution are necessary both to evaluate the plan and preparations to implement the plan and as a means of training.[56,57] Experience has shown that only the simplest of plans are likely to work the first time they are tested. Consequently a facility should not assume that it is prepared to deal with an emergency unless the emergency response plan has been realistically exercised. Drills can be broad or specialized and can vary in depth of involvement of the facility and community.[58]

- Table top—step by step read-through of emergency response plan by key controllers.
- Notification test—exercise communication facilities, notification procedures, call-lists, and backups.
- Organization test—assembly, team makeup, understanding responsibilities, response time.
- Rolling equipment test—emergency equipment call-up and movement, security and access, off-site responders, mobile communications.
- Full scale test—casuality simulation, off-site response, community involvement, media.

By one test drill or another command and control, communications, casuality handling, mediation, and all other significant aspects of the plan should be tested. The drills should be carefully planned so that they are realistic within their scope. They should be observed by drill evaluators who can input later to a comprehensive critique. Some common emergency response plan failures are often detected by drills.

- Designated individuals do not have the training, time, or resources to carry out their duties.
- Notification is delayed because of inadequate on-site authority at the time of the event.
- Communications overload blocks access to local authorities.
- Security breakdown inhibits access by responders and/or allows unauthorized access.

- Difficult decisions (sacrifice property to avoid a toxic release) cannot be made quickly enough.
- Incorrect casuality handling decisions made owing to confusion and inadequate communication between emergency crews, first responders, and hospitals.
- Control rooms become isolated and provide inadequate protection.
- Media/external communication needs interfere with operational communications.
- Chemical exposure effects and treatment information is not communicated to the hospital that receives the chemical casualities.
- Support function (transportation, utilities, etc.) responsibilities listed in site plan are not reflected and detailed by function plans.

The objective of drilling is to find these potential defects by realistic exercises so that the plan can be revised and training improved to maximize the likelihood that an actual emergency can be responded to without any critical errors.

15.6.4 Transportation Emergencies

Planning for successful response to chemical release emergencies which occur while a chemical is in transit involves some special considerations.[59–61] The first responders will not necessarily know what chemical is involved and it may take some time at the critical early stage of the emergency to obtain that information. When the required information is on hand, it may be that special skills or equipment are needed to deal with the situation so arrangements must be made to bring these to the scene. For these reasons, planning for transportation emergencies is more complex and response often takes longer than for site emergencies.

The first item of information that may be available to firefighters, police, rescue workers, or other emergency services is the Department of Transportation (DOT) placard on the side of the vehicle. A four digit number on the placard or on an orange panel is the identification number for a material in the DOT guidebook.[62] This guidebook also provides a series of generic guides for each material that describe potential hazards and initial emergency actions.

The ID number and the same of the material will also be shown on the shipping papers which should be on the trunk, train, ship, or aircraft. In the case of a tanktruck, the driver should also have a copy of the Material Safety Data Sheet which gives more specific hazard, precaution, and emergency information for the material.

Additional information and assistance is available from CHEMTREC, the Chemical Transportation Emergency Center, a public service of the Chemical Manufacturers Association (CMA). CHEMTREC provides immediate advice by telephone and also contacts the shipper of the hazardous material for detailed assistance and appropriate response. CHEMTREC needs to know either the identification number or name of the material and the nature of the accident. CHEMTREC operates continuously and should be called as soon as immediate needs have been met and the facts ascertained.

Backing up CHEMTREC is another CMA service, the Chemical Emergency Network or CHEMNET. This network consists of about 240 response teams representing about 100 chemical companies. These CHEMNET teams can provide on-site experts to assist local emergency services with advice on the best way to handle a specific chemical accident. In Canada, there is a similar network called the Transportation Emergency Assistance Plan (TEAP) operated out of 11 regional response centers.

15.7 SOURCES OF INFORMATION

The body of information and technology relating to the subject of this chapter is very large, far larger than could be covered in the space available. Consequently, this chapter does not attempt to instruct the reader on how to do the many tasks involved in disaster prevention and response planning. Instead, this chapter should be regarded as an overview of the subject which identifies the sources the reader may consult to expand on each detailed topic. Many of these sources are referenced in the text. In addition, there are some institutions who have a continuing involvement in this field and can be booked to for the latest information and technology.

15.7.1 U.S. Department of Transportation

The U.S. DOT Office of Hazardous Materials Transportation has developed an Emergency Response Guidebook (ERG) (DOT P5800.4) which is a guidebook for initial response to hazardous materials incidents which occur primarily in the transport of hazardous materials. The ERG, which is updated periodically, gives the name of the material corresponding to the ID number displayed on the placard on the vehicle and listed on the shipping paper. Each ID/substance listing refers the user to a numbered guide which briefly describes potential hazards (fire, explosion, health hazards) and emergency action (fire fighting, dealing with spills or leaks, first aid). The EPG is a means of quickly providing critical information to first responders. The guide advises the responder to immediately contact CHEMTREC for additional assistance beyond the initial phase of the incident. In addition to the ERG, the DOT is the lead agency in transportation safety research, guidance, and regulation.

15.7.2 National Fire Prevention Association

The NFPA is primarily concerned with fire protection and suppression system but is also an authority on certain aspects of hazardous materials. Their *Fire Protection Guide on Hazardous Materials*[32] contains the complete texts of the four NFPA documents which classify most common hazardous materials:

- NFPA 325M: Fire Hazard Properties of Flammable Liquids, Gases and Volatile Solids
- NFPA 49: Hazardous Chemicals Data

- NFPA 491M: Manual of Hazardous Chemical Reactions
- NFPA 704: Identification of the Fire Hazards of Materials

This consolidated guide includes information on fire and explosion hazards, life hazards, personal protection, fire fighting, and other facts needed to make informed decisions in an emergency.

15.7.3 American Institute of Chemical Engineers Center for Chemical Process Safety (CCSP)

A number of manuals developed by the CCSP[15, 28, 47, 63, 64] were used as a basis for the chapter. The CCSP and its committees maintain an awareness of the technology and their handbooks reflect the state of the art.

15.7.4 Chemical Manufacturers Association (CMA)

The CMA CHEMTREC and CAER programs have been described above. These programs and other activities are the responsibility of CMA staff and committees who stay current on developments related to chemical emergencies. They are available to assist member chemical companies and the general public as appropriate.[55, 65]

15.7.5 University of Delaware Disaster Research Center (DRC)

This center provides a unique resource which "engages in a variety of sociological and social science research on group and organizational preparations for, responses to, and recovery from communitywide emergencies, particularly natural and technological disasters."[66, 67] The DRC conducts studies, sends teams to disaster sites and maintains a publication collection of over 20,000 items related to the social and behavioral aspects of disasters.

15.8 FUTURE TRENDS

Presently proposed governmental regulations will require process hazard risk assessment for many facilities that have not done it in the past. Because of these regulations and as a consequence of expansion of the risk assessment field, the process will become more formal and detailed. New practitioners will enter the field, and it will be increasingly important for the facility manager to be sure of the qualifications of those who would provide this service. Some five or more years ahead, as the results of a number of risk assessments become known, it is probable that new regulations, at least in the United States, will establish expectations for the results of the risk assessment process and will try to define more precisely what risk is acceptable.

The technology of process hazard risk assessment, in addition to being more formal and detailed as mentioned above, is likely to also become more probabilistic. The analogy to health risk assessment which claims to calculate the likelihood of an adverse consequence will be persuasive. The lack of credible input data (e.g., mean time to failure) will be partly relieved by expanded databases developed in the course of increased regulatory activity. A need will be seen for a measure of the actual or limiting probability on chemical releases of various magnitudes so as to have a number to compare with acceptable risk criteria or bright lines. A challenge to all of us involved in process hazard analysis will be to try to insure that the databases do support the outcome both in quantity and quality, that the assumptions of independence of causes typically made are carefully considered, and that the consequences of multiple human errors are included in any analyses.

Community involvement in disaster prevention and response planning is already a fact in the United States and is likely to spread. In the future, the public will expect to be even more involved and to participate in decision making. As a consequence some situations which were considered to constitute an acceptable level of risk will need to be revisited and new, consensus criteria of acceptability applied. Risk assessors and the public will need to educate each other to achieve common understandings and expectations.

REFERENCES

1. Solomon, K. A. and P. Ricci, Eds. Regulatory and legal aspects of hazardous materials risks. *J. Haz. Mater.* **15**.
2. DePol, D. R. and P. N. Cheremisinoff. *Emergency Response to Hazardous Materials Incidents*. Lancaster, PA: Technomic, 1984.
3. U. S. Environmental Protection Agency (EPA) 1985. Chemical emergency preparedness program: interim guideline. Washington, DC: EPA, 1985.
4. U.S. Department of Labor, Occupational Safety and Health Administration, Office of Regulatory Analysis. Industry Profile and Cost Assessment for a Proposed OSHA Standard Covering Process Hazard Management in Chemical and Petroleum Industries. Charles Rivers Associates, Report No. S-9-F-5-0059. Boston, MA, 1988, pp. 223–300.
5. Garrison, W. G. *Large Property Damage Losses in the Hydrocarbon–Chemical Industries —A Thirty-Year Review*. 12th ed. Chicago, IL: March & McLennan Protection Consultants, 1989.
6. Chemical Manufacturers Associations (CMA). Community Awareness and Emergency Response Program Handbook. Washington, DC: CMA, 1985.
7. Organization Resources Counselors (ORC). Recommendations for process hazards management of substance with catastrophic potential. Washington, DC: ORC, Dec. 1988.
8. Lykke, E. Avoiding and managing environmental damage from hazardous industrial accidents. Pittsburgh, PA: AWMA, 1986.
9. Health and Safety Executive (HSE). Annual report of the chief inspector of factories for 1974. Her Majesty's Stationary Office, London, 1975.
10. Coppock, R. Regulating chemical hazards in Japan, West Germany, France, the United Kingdom and the European Community: A comparative examination. National Research Council. Washington, DC: National Academy Press, 1986.
11. Harris, N. C. and A. M. Moses. The use of acute toxicity data in the risk assessment of accidental releases of toxic gases. *Inst. Chem. Eng. Symp. Ser. No. 80*, 136 (1983).
12. Pitt, M. J. A vapour hazard index for volatile chemicals. *Chem. Ind.* (*London*) (1982).

13. Institution of Chemical Engineers. Bhopal: The Company Report. Loss Prevention Bulletin 63-1, 1985.
14. Dow Chemical Company. Fire and explosion index hazard classification guide, 5th ed., 1981.
15. American Institute of Chemical Engineers (AIChE). Guidelines for hazard evaluation procedures. New York: AIChE, 1985.
16. Theodore, Louis, J. P. Reynolds, and F. B. Taylor. *Accident and Emergency Management*. New York: Wiley, 1989.
17. Concawe. *Methodologies for Hazard Analysis and Risk Assessment in the Petroleum Refining and Storage Industry*. The Hague: Concawe, 1982.
18. Lees, F. A. Some aspects of hazard survey and assessment. *Chem. Eng.* (*UK*) December 1980.
19. Royal Commission on the Ocean Ranger Marine Disaster. Report One: The loss of the semisubmersible drill rig Ocean Ranger and its crew. Canadian Government Publishing Center, Pub. No. ZI-1982/1-IE (1984).
20. Heising, C. D. and W. S. Grengebach. The Ocean Ranger oil rig disaster: A risk analysis. *Risk Analysis*, **9** (1989).
21. Bell, B. J. and A. D. Swain. A procedure for conducting human reliability analyses. Washington, DC: U.S. Nuclear Regulatory Commission NUREG/CR-2254, 1983.
22. McCormick, N. J. *Reliability and Risk Analysis*. New York: Academic, 1981.
23. Rasmussen, J. Approaches to the control of the effects of human error in chemical plant safety. Roskilde, Denmark: Riso National Laboratory, Riso-M-2638, 1987.
24. Hart, G. S. Avoiding and managing environmental damage from major industrial accidents. Executive summary of the international conference. *J. Air Pollution Control Assoc.*, **36** (1986).
25. Fryer, L. S. and G. D. Kaiser. DENZ—A computer program for the calculation of the dispersion of dense toxic or explosive gases in the atmosphere. Culcheth, UK: SPD R 152 UKAEA (1979).
26. Ramskill, P. M. Discharge rate calculation methods for use in plant safety assessment. Culcheth, UK: Safety and Reliability Directorate SRD R 352, 1986.
27. Beddows, N. A. Emergency prediction information. *Professional Safety*, October 1990.
28. Hanna, S. R. and P. J. Drivas. Guidelines for the use of vapor cloud dispersion models. New York: AIChE, 1987.
29. American Institute of Chemical Engineers (AIChE). Guidelines for Chemical Process Quantitative Risk Analysis, New York, 1989.
30. TNO (Netherlands Organization for Applied Scientific Research). Methods for the calculation of the physical effects of the escape of dangerous material. The Netherlands: Directorate-General of Labour Pub. #9092 (1980).
31. McNaughton, D. J., C. G. Wirly, and P. M. Bodner. Evaluating emergency response models for the chemical industry. *Chem. Eng. Prog.*, January 1987.
32. National Fire Protection Association. Fire protection guide on hazardous materials. Quincy, MA: NFPA, 1986.
33. Baxter, P. J., P. C. Davies, and V. Murray. Medical planning for toxic release into the community: The example of chlorine gas. *Br. J. Ind. Med.*, **46** (1989).
34. Coleman, R. J. Evacuation planning: Crissi or control. *Fire Chief Magazine*, February 1983.
35. Cooney, W. D. C. The role of public services—Off-site arrangements. *Major Hazards* (UK), July 1987.

36. McQuaid, Ed. Heavy gas dispersion trials at Thorney Island—2. *J. Hazardous Mater.*, **16** (1987).

37. Nielson, M. and N. O. Jensen. Research on continuous and instantaneous gas clouds. *J. Hazardous Mater.* **21** (1989).

38. Institution of Chemical Engineers (ICE). The assessment and control of major hazards. EFCE Pub. Services No. 42 London: Institution of Chemical Engineers, 1985.

39. Pedersen, F. and R. S. Seleg. Predicting the consequences of short-term exposure to high concentrations of gaseous ammonia. *J. Hazardous Mater.* **21** (1989).

40. National Research Council (NRC). Emergency and Continuous Exposure Guidance Levels for Selected Airbourne Contaminants. Washington, DC: National Academy Press, 1985.

41. American Industrial Hygiene Association (AIHA). Emergency response planning guides. Akron, OH: AIHA, 1987.

42. World Health Organization (WHO). Rehabilitation following chemical accidents: A guide for public officials. Copenhagen: FADL. (1989). Disaster recovery testing: A methodology and case study. National Safety and Health News, Vol. 124.

43. Health and Safety Executive (HSE). A guide to the control of industrial major accident hazards regulations of 1984. London: HSE Booklet HS(R)21 (1985).

44. Edmundson, J. N. The use of risk results in planning and design. *Major Hazards* (*UK*), July 1987.

45. Prugh, R. W. and R. W. Johnson. Guidelines for vapor release mitigation. New York: AIChE, 1987.

46. Kletz, T. A. Friendly Plants. *Chem. Eng. Prog.* (July 1989).

47. LeVine, R. Guidelines for safe storage and handling of high toxic hazard materials. New York: AIChE, 1987.

48. Kletz, T. A. What went wrong? London: Gulf, 1985.

49. Occupational Safety and Health Administration (OSHA). Process safety management of highly hazardous chemicals. Washington: Fed. Reg. r55, No. 137, p. 29150 (1990).

50. Suro, R. Plastics plant explodes; Many missing. *New York Times*, October 24, 1989.

51. Brown, G. N. Disaster planning for industry. *Occup. Health* (*UK*), September 1977.

52. Eberlein, J. Emergency plans—The industry approach. *Chem. Ind.* (*UK*) **82** (1987).

53. Krikorian, M. *Disaster and Emergency Planning*. Loganville, AL: Institute Press, 1982.

54. Michael, E. J., O. W. Bell, J. W. Wilson, and G. W. McBride. Emergency planning considerations for chemical plants. *Env. Prog.* February 1988.

55. Chemical Manufacturers Associations (CMA). Site emergency response planning. Washington, DC: CMA, 1986.

56. Merriman, M. Not a Drill. *Occ. Health Safety*, March 1990.

57. Michael, E. J. and R. E. Vanesse. Planning and implementation of emergency preparedness exercises including scenario preparation. Boston: Stone and Webster, 1985.

58. Yetter, D. L. Disaster recovery testing: A methodology and case study. National Safety and Health News, Vol. 134, 1986.

59. Bennett, G. F., F. S. Feates, and J. Welder. *Hazardous Materials Spills Handbook*. New York: McGraw-Hill, 1982.

60. Maloney, D. M., A. J. Policastro, L. Coke, and W. Dunn. The development of initial isolation and protective action distances for U.S. DOT publication 1990 emergency response guidebook. Proceedings of HAZMAT CENTRAL 1990.

61. Student, P. J., Ed. *Emergency Handling of Hazardous Materials in Surface Transportation*. Washington, DC: Association of American Railroads, 1981.

62. U.S. Department of Transportation (USDOT). *Emergency Response Guidebook*. Washington, DC: DOT P 5800.4, 1987.

63. American Institute of Chemical Engineers (AIChE). *Guidelines for the Technical Management of Chemical Process Safety*. New York: AIChE, 1989.

64. American Institute of Chemical Engineers (AIChE), *Workbook of Test Cases for Vapor Cloud Source Dispersion Models*. New York: AIChE, 1989.

65. Chemical Manufacturers Association (CMA). *Evaluation of Assessment of Models for Emergency Response Planning*. Washington, DC: CMA, 1985.

66. Disaster Research Center (DRC). Publication list. Newark, DE: University of Delaware, 1989.

67. Hughes, M. A. A selected annotated bibliography of social science research on planning for and responding to hazardous material disasters. *J. Hazardous Mater*. **27** (1991).

68. Welch, D. L. Human factors in nuclear power and process safety. International Conference on Hazard Identification and Risk Analysis, Human Factors and Human Reliability in Process Safety. AIChE, New York, 1992.

APPENDIX 15A List of Highly Hazardous Chemicals, Toxics, and Reactives (29CFR 1910.119)

Chemical Name	CAS No.	Threshold Quantity (lb)*
Acetaldehyde	75-07-0	2,500
Acrolein (2-propenal)	107-02-8	150
Acryl chloride	814-68-6	250
Allyl chloride	107-05-1	1,000
Allylamine	107-11-9	1,000
Alkylanums,	Varies	5,000
Ammonia, anhydrous	7664-41-7	10,000
Ammonia solutions (> 44% ammonia by weight)	7664-41-7	15,000
Ammonium perchlorate	7790-98-9	7,500
Ammonium permangenate	7787-36-2	7,500
Arsine (also called arsenic hydride)	7784-42-1	100
Bis(chloromethyl) ether	542-88-1	100
Boron trichloride	10294-34-5	2,500
Boron trifluoride	7637-07-2	250
Bromine	7726-95-6	1,500
Bromine chloride	13863-41-7	1,500
Bromine pentafluoride	7789-30-2	2,500
Bromine trifluoride	7787-7105	15,000
3-Bromopropylene (also called proparyl bromide)	106-96-7	100
Butyl hydroperoxide (tertiary)	75-91-2	5,000
Butyl perbenzoate (tertiary)	614-45-9	7,500
Carbonyl chloride (see phosgene)	75-44-5	100
Carbonyl fluoride	353-50-4	2,500
Celluose nitrate (concentration > 12.6% nitrogen)	9004-70-0	2,500
Chlorine	7782-50-5	1,500
Chloride dioxide	10049-04-4	1,000
Chlorine pentrafluoride	13637-63-3	1,000
Chlorine trifluoride	7790-91-2	1,000
Chlorodiethylaluminum (also diethylaluminum chloride)	96-10-6	5,000
1-Chlor-2,4-dinitrobenzene	97-00-7	5,000

APPENDIX 15A ***(Continued.)***

Chemical Name	CAS No.	Threshold Quantity (lb)*
Chloromethyl methyl ether	107-30-2	500
Chloropicrin	76-06-2	500
Chloropicrin and methyl bromide mixture	None	1,500
Chloropicrin and methyl chloride mixture	None	1,500
Cumene hydroperoxide	80-15-9	5,000
Cyanogen	460-0195	2,500
Cyanogen chloride	506-77-4	500
Cyanuric fluoride	675-14-9	100
Diacetyl peroxide (concentration > 70%)	110-22-5	5,000
Diazomethane	334-88-3	500
Dibenzoyl peroxide	94-36-0	7,500
Diborane	19287-45-7	100
Dibutyl peroxide (tertiary)	110-05-4	5,000
Dichloro acetylene	7572-29-4	250
Dichlorosilane	4109-96-0	2,500
Diethylzinc	557-20-0	10,000
Diisopropyl peroxydicarbonate	105-64-6	7,500
Dilauroyl peroxide	105-75-8	7,500
Dimethyldichlorosilane	75-78-5	1,000
Dimethylhydrazine, 1,1-	57-14-7	1,000
Dimethylamine, anhydrous	124-40-3	2,500
2,4-Dinitroaniline	97-02-9	5,000
Ethyl methyl ketone peroxide (also methyl ethyl ketone peroxide; concentration > 60%)	1338-23-4	5,000
Ethyl nitrite	109-95-5	5,000
Ethylamine	75-04-7	7,500
Ethylene fluorohydrin	371-62-0	100
Ethylene oxide	75-21-8	5,000
Ethyleneimine	151-56-4	1,000
Fluorine	7782-41-4	1,000
Formaldehyde (formalin)	50-00-0	1,000
Furan	110-00-9	500
Hexafluoroacetone	684-16-2	5,000
Hydrochloric acid, anhydrous	7647-01-0	5,000
Hydrofluoric acid, anhydrous	7664-39-3	1,000
Hydrogen bromide	10035-10-6	5,000
Hydrogen chloride	7647-01-0	5,000
Hydrogen cyanide, anhydrous	74-90-8	1,000
Hydrogen fluoride	7664-39-3	1,000
Hydrogen peroxide (52% by weight or greater)	7722-84-1	7,500
Hydrogen selenite	7783-07-5	150
Hydrogen sulfide	7783-06-4	1,500
Hydroxylamine	7803-49-8	2,500
Iron, pentacarbonyl	13463-40-6	250
Isopropylamine	75-31-0	5,000
Ketene	463-51-4	100
Methacrylaldehyde	78-53-3	1,000
Methacryloyl chloride	920-46-7	150
Methacryloyloxyethyl isocyanate	30674-80-7	100
Methyl acrylonitrile	126-98-7	250
Methylamine, anhydrous	74-89-5	1,000
Methyl bromide	74-83-9	2,500
Methyl chloride	74-87-3	15,000
Methyl chlorformate	79-22-1	500

APPENDIX 15A (Continued.)

Chemical Name	CAS No.	Threshold Quantity (lb)*
Methyl ethyl ketone peroxide (concentration > 60%)	1338-23-4	5,000
Methyl fluoroacetate	453-18-9	100
Methyl fluorosulfate	421-20-5	100
Methyl hydrazine	60-34-4	100
Methyl iodide	74-88-4	7,500
Methyl isocyanate	624-83-9	250
Methyl mercaptan	74-93-1	5,000
Methyl vinyl ketone	79-84-4	100
Methyltrichlorosilane	75-79-6	500
Nickel carbonyl (nickel tetracarbonyl)	13463-39-3	150
Nitric acid (94.5% by weight or greater)	769-37-2	500
Nitric oxide	10102-43-9	250
Nitroaniline (para nitroaniline)	100-01-6	5,000
Nitromethane	75-5205	2,500
Nitrogen dioxide	10102-44-0	250
Nitrogen oxides (NO; NO_2; N2O4; N2O3)	10102-44-0	250
Nitrogen tetroxide (also called nitrogen peroxide)	10544-72-6	250
Nitrogen trifluoride	7783-54-2	5,000
Nitrogen trioxide	10554-73-7	250
Oleum (65 – 80% by weight; also called fuming sulfuric acid)	8014-94-7	1,000
Osmium tetroxide	20816-12-0	100
Oxygen difluoride (fluorine monoxide)	7783-41-7	100
Ozone	10028-15-6	100
Pentaborane	19624-22-7	100
Peracetic acid (concentration > 60% acetic acid; also called peroxyacetic acid)	79-21-0	1,000
Perchloric Acid (concentration > 60% by weight)	7601-90-3	5,000
Perchloromethyl mercaptan	594-42-3	150
Perchloryl fluoride	7616-94-6	5,000
Peroxyacetic acid (concentration > 60% acetic acid; also called peracetic acid)	79-21-0	1,000
Phosgene (also called carbonyl chloride)	75-44-5	100
Phosphine (hydrogen phosphide)	7803-5102	100
Phosphorus oxychloride (also called phosphoryl chloride)	100025-87-3	1,000
Phosphorus trichloride	7719-12-2	1,000
Phospharyl chloride (also called phosphorus oxychloride)	10025-87-3	1,000
Propargyl bromide	106-96-7	100
Propyl nitrate	627-3-4	2,500
Sarin	107-44-8	100
Selenium hexafluoride	7783-79-1	1,000
Stilbine (antimony hybride)	7803-52-3	500
Sulfur dioxide (liquid)	7446-09-5	1,000
Sulfur pentafluoride	5714-22-7	250
Sulfur tetrafluoride	7783-60-0	250
Sulfur trixoide (also called sulfur anhydride)	7446-11-9	1,000
Sulfuric anhydride (also called sulfur trioxide)	7446-11-9	1,000
Tellurium hexafluoride	7783-80-4	250
Tetrafluoroethylene	116-14-3	5,000

APPENDIX 15A ***(Continued.)***

Chemical Name	CAS No.	Threshold Quantity (lb)*
Tetrafluorohydrazine	10036-4702	5,000
Tetramethyl lead	75-4-1	1,000
Thionyl chloride	7719-09-7	250
Trichloro (chloromethyl) silane	1558-25-4	100
Trichloro (dichlorophenyl) silane	27137-85-5	2,500
Trichlorosilane	10025-78-2	5,000
Trifluorochloroethylene	79-38-9	1,000
Trimethoxysilane	2487-90-3	150

[a]Quantity required to be covered by the standard.

APPENDIX 15B Process Safety Code of Management Practices Self-Evaluation Form

Management Leadership	Categories* NA	EV	DP	LA	PP	RI
1. Leadership by senior management through policy, participation, communications and resource commitments in achieving continuous improvement of performance. Comments on Category NA: ______						
2. Clear accountability for performance against specific goals for continuous improvements. Comments on Category NA: ______						
3. Measurement of performance, audits for compliance and implementation of corrective actions. Comments on Category NA: ______						
4. Investigation, reporting, appropriate corrective action and followup of each incident that results or could have resulted in a fire, explosion or accidental chemical release. Comments on Category NA: ______						
5. Sharing of relevant safety knowledge and lessons learned from such incidents with industry, government and the community. Comments on Category NA: ______						

APPENDIX 15B ***(Continued.)***

Management Leadership	Categories* NA	EV	DP	LA	PP	RI
6. Use of the Community Awareness and Emergency Response (CAER) process to assure public comments and concerns are considered in design and implementation of the facility's process safety systems. Comments on Category NA: ______ ______ ______						
Technology						
7. Current, complete documentation of process design and operating parameters and procedures. Comments on Category NA: ______ ______ ______						
8. Current, complete documentation of information relating to the hazards of materials and process technology. Comments on Category NA: ______ ______ ______						
9. Periodic assessment and documentation of process hazards and implementation of actions to minimize risks associated with chemical operations, including the possibility of human error. Comments on Category NA: ______ ______ ______						
10. Management of changes to chemical operations to maintain or enhance the safety originally designed into the facility. Comments on Category NA: ______ ______ ______						
11. Consideration and mitigation of the potential safety effects of expansions, modifications and new sites on the community, environment, and employees. Comments on Category NA: ______ ______ ______						

APPENDIX 15B ***(Continued.)***

Management Leadership	NA	EV	DP	LA	PP	RI
12. Facility design, construction, and maintenance using sound engineering practices consistent with recognized codes and standards. Comments on Category NA: ______						
13. Safety reviews on all new and modified facilities during design and prior to startup. Comments on Category NA: ______						
14. Documented maintenance and inspection programs that ensure facility integrity. Comments on Category NA: ______						
15. Sufficient layers of protection through technology, facilities and employees to prevent escalation from a single failure to a catastrophic event. Comments on Category NA: ______						
16. Provision for control of processes and equipment during emergencies resulting from natural events, utility disruptions and other external conditions. Comments on Category NA: ______						
Person						
17. Identification of the skills and knowledge necessary to perform each job. Comments on Category NA: ______						
18. Establishment of procedures and work practices for safe operating and maintenance activities. Comments on Category NA: ______						

Categories* header spans NA, EV, DP, LA, PP, RI.

APPENDIX 15B ***(Continued.)***

Management Leadership	Categories*					
	NA	EV	DP	LA	PP	RI
19. Training for all employees to reach and maintain proficiency of safe work practices and the skills and knowledge necessary to perform their job. Comments on Category NA: ______						
20. Demonstrations and documentation of skill proficiency prior to assignment to independent work, and perodically thereafter. Comments on Category NA: ______						
21. Programs designed to assure that employees in safety critical jobs are fit for duty and are not compromised by external influences, including alcohol and drug abuse. Comments on Category NA: ______						
22. Provisions that contractors either have programs for their own employees consistent with applicable sections of the Code or be in cluded in the member company's program, or some combination of the two. Comments on Category NA: ______						

*Category NA No action. If no action taken because the management practice is not applicable, please explain.
Category EV Evaluating existing company practices against the Management Practice.
Category DP Deveopling plan to implement Management Practice.
Category IA Implementing action plan.
Category PP Management Practice in place.
Category RI Reassessing Management Practice implementation.

16

Exposure Assessment

Exposure assessment is defined as the determination or estimation (qualitative or quantitative) of the magnitude, frequency, duration, and route of exposure to a chemical. Chapter 3 covered exposure assessment methods that are based on direct measurement. Often, however, direct measurement of exposure is not the method of choice because it is too expensive relative to the need for the information or because the exposure to be estimated is not presently occurring. This latter situation occurs in the case of the need to estimate past exposure, which may be the cause of presently occurring health effects (retrospective exposure estimation) and in the case where some exposure reduction action is contemplated and it is necessary to predict what the effect will be. Estimates of future exposure are also needed for Premanufacturing Notifications (PMNs) required under the Toxic Substance Control Cat (TSCA) for new products. Also, predictions of future exposures are needed for disaster planning, as discussed in Chapter 15. This chapter discusses the methods commonly used to estimate exposure close to the source in the workplace or indoors and further from the source in the neighboring community.

16.1 PREMANUFACTURING NOTIFICATIONS

Under Section 5 of the Toxic Substance Control Act (TSCA), a Premanufacture Notification (PMN) must be submitted at least 90 days before the commercial production (including import) of any chemical not on the TSCA existing chemicals list (inventory of chemicals in commerce). The EPA may order restrictions on the manufacture or use of a new chemical under Sections 5(e) and 5(f).

TABLE 16.1 Summary of Exposure-Based Criteria[a]

Exposure-Based Criteria	Description of Criteria
Significant or Substantial Human Exposure	
Worker	
High number of workers exposed	≥ 1000 workers
Acute worker inhalation	≥ 100 workers exposed by inhalation to ≥ 10 mg / day
Chronic workers exposure	
Inhalation	≥ 100 workers exposed to 1 – 10 mg / day for ≥ 100 days / year
Dermal	≥ 250 workers exposed by routine dermal contact for ≥ 100 days / year
Consumer	Presence of the chemical in any consumer product where (1) the physical state of the chemical in the product and (2) the manner of use would make exposures likely
Ambient General Population	
Ambient drinking water exposure	≥ 70 mg / year exposure via drinking water
Ambient air exposure	≥ 70 mg / year exposure via air
Ambient groundwater exposure	≥ 70 mg / year exposure via groundwater
Aggregate ambient exposure over all media	≥ 10,000 kg / year release to environmental media
Substantial Environmental Release	
Surface Water	
Surface water release	≥ 1000 kg / year total release to surface water calculated after wastewater treatment

[a]All nonpolymer PMN chemicals (as specified at 40 CFR 723.250) with estimated production volumes greater than or equal to 100,000 kg / yr.

TABLE 16.2 Physical and Chemical Properties

Vapor pressure
Density / relative density
Solubility in water
Solubility in other solvents
Melting temperature
Boiling / sublimination temperature
Spectra
Dissociation constant
Particle size distribution
Octanol / water partition coefficient
Henry's Law constant
Volatilization from soil
pH
Flammability
Explodability
Adsorption coefficient

Several categories of chemicals are exempt from these PMN requirements:

- Chemicals being test marketed
- Chemicals used for research and development
- Certain polymers
- Chemicals manufactured in quantities of less than 1000 kg/yr
- Certain non-TSCA uses (e.g., pharmaceuticals)
- Chemicals considered nonisolated intermediates.

New chemical PMNs are reviewed for health and ecotoxicity concerns and, based on emissions and release estimates, may require regulatory action. The review process, which is conducted in stages, considers both the information provided by the submitter and gathered by the TSCA staff. They use this information to:

- Evaluate the methods used to manufacture, process, or use a specific chemical substance in order to identify potential exposures and releases.
- Estimate the extent of release or of exposure (e.g., the airborne concentrations of a volatile liquid being drummed which might be breathed by a worker).
- Evaluate the effectiveness of engineering, work practice, and personal protective controls for reducing release and exposures.

The EPA uses the set of exposure-based criteria in Table 16.1 to decide whether health, aquatic, or environmental fate testing of the new chemical will be required. For the purpose of assessing exposure, the EPA requires the submission of certain physical and chemical properties (Table 16.2) and descriptions of the operation or process, amount of substance released, number of workers exposed, duration of exposure, and means of exposure control. The EPA makes its exposure assessment by considering the degree of containment (Table 16.3) and, where

TABLE 16.3 Degree of Containment

Class of Use	Examples
Destructive	Fuels, fuel additives, chemical intermediates
Contained	Catalysts used in closed processes, certain photographic chemicals, capacitor fluids
Open, nondispersive	Printing inks, textiles, dyes, plasticizers, adhesives, liquid paints, resins
Dispersive	Cutting fluids, fabric softeners, automobile tire rubber
Highly dispersive	Pesticides, fertilizers, salt for deicing, paint solvents, spray paints
Other	

releases may occur, by using the source and exposure models described in this chapter.

16.2 MODELING WORKPLACE EXPOSURE

A chemical present in the workplace may enter the body by inhalation, ingestion, injection, and absorption through the skin. Intake results from exposure, which is contact of a chemical with the outer boundary of an organism. The amount of the substance in contact is called the applied dose and is quantitatively related to the internal dose which interacts with the target organ to cause a toxic effect (see the Glossary).

16.2.1 Dermal Exposure

Workplace chemicals may cause health effects by dermal exposure. For some dermatitis-producing chemicals, mere contact is enough. Other chemicals cause harm by passing through the skin and causing systemic damage.

Dermal exposure estimation involves three quantities:

- Surface area of contact
- Quantity on the skin
- Absorption through the skin

The potential for dermal contact depends on the level of activity, the likelihood of contact, the frequency of contact, the surface area in contact, physical properties of the material which determine retention, use of protective equipment, and frequency of washing. While gases can pass through the skin to some degree, dermal contact is mostly from solids and liquids. Mere contact with a surface does not constitute contact since transfer may not occur.

The likelihood/frequency of contact can be categorized as follows:

- None—No chance of dermal contact during normal job activities.
- Very low—No dermal exposures during typical job activities, but short periods of exposure on occasion after which all contact surfaces would be washed.
- Incidental contact—Occasional dermal contact of a minor nature.
- Intermittent contact—Dermal contact such as splashes, wiping with contaminated rags, contact with contaminate tools or surfaces.
- Routine contact—Routine dermal contact not including immersion in a liquid.
- Routine immersion—Worker typically immerses hand(s) into a liquid.

Measurement techniques such as skin patches or biomonitoring for chemicals absorbed through the skin are the most accurate way of assessing skin contact.

TABLE 16.4 Dermal Exposure Factors

Activity	Typical Examples	S (cm^2)	Q (mg/cm^2)	Resulting Typical Contact (mg)
Routine immersion				
2 hands	Handling wet surfaces Filling / dumping containers of powders, flakes, granules Spray painting	1300	5–14	6500–18,200
Routine contact				
2 hands	Maintenance / manual cleaning of equipment Unloading filter cake Changing filter Filling drums with liquid	1300	1–3	1300–3900
1 hand		650	1–3	650–1950
Incidental contact				
2 hands	Connecting transfer line Weighing powder / scooping / mixing (i.e., dye weighing)	1300	1–3	1300–1900
1 hand	Sampling Ladling liquid / bench scale liquid transfer	650	1–3	650–1950

If estimation by modeling is required, the following equation may be used:

$$D = SQC \tag{16.1}$$

where

D = dermal exposure (mg)

S = surface area of skin in contact (cm^2)

Q = average quantity remaining on skin (mg/cm^2)

C = percent of chemical in a mixture

Estimates of S and Q for various kinds of skin contact are shown in Table 16.4. The "resulting typical contact" is the D term for a neat substance. If the substance is in a mixture, then this must be adjusted by the percent in the mixture.

The actual area of skin contact will, of course, depend on the use of protective clothing. While gloves and so on can provide complete protection if properly selected and used, it should not be assumed that this is the case in the absence of effective training and supervision. Ordinary cloth clothing can have the effect of holding a substance in contact with the skin by wicking and thus increasing the duration of contact over bare skin.

The duration of contact will depend on several factors. Volatile liquids will evaporate although the harm they do (defatting) may occur before they evaporate. Where adequate masking facilities are readily available, it is reasonable to assume the worker will wash before lunch and at the end of the day. Irritating materials will probably be removed more frequently.

Dermal absorption, that is, intake into the body through the skin so that other organs can be affected, depends on a number of factors. These include:

- Quantity in contact with the skin
- Area of skin in contact
- Duration of contact
- Skin hydration
- Skin condition (dermatitis, abrasions)
- Partition coefficients

The best penetrants are those that are soluble in both lipids and water.[6] At present, the state of knowledge of dermal absorption does not yield definitive methods of calculating dose, even given quantitative data for all of the factors listed above. For these reasons, screening estimates of dermal absorption need to be very conservative and refined estimates can only be made by experts at this time.

16.2.2 Inhalation Exposure

The methods and formulas discussed in this section are derived in part from methods used by the Chemical Engineering Branch, OPTS, U.S. EPA.[1] In the case of a dose which arrives at a target organ via the inhalation route, the amount of the internal dose can be estimated by the following equation:

$$ID = C_{\mathrm{m}}bh \tag{16.2}$$

where

ID = internal or absorbed dose (mg/day)
C_{m} = airborne concentration of chemical (mg/m^3)
b = inhalation rate (m^3/hr)
h = duration of exposure (hr/day)

However, inhalation rate varies with age, sex, and degree of activity. Numerous studies have been collected and summarized by the U.S. EPA[2] in Table 16.5.

The duration of the exposure is the time period over which the exposure occurs. Since a "standard" work day is commonly thought of as 8 hr, most workplace exposure standards (OSHA PELs, ACGIH TLVs, AIHA WEELs) are for an 8-hr averaging time. When the actual exposure being estimated lasts less than 8 hr, time-weighted averaging (TWA) is used to calculate the 8-hr average exposure to

TABLE 16.5 Summary of Human Inhalation Rates for Men, Women, and Children by Activity Level (m^3/hr)[a]

	Resting[b]	Light[c]	Moderate[d]	Heavy[e]
Adult male	0.7	0.8	2.5	4.8
Adult female	0.3	0.5	1.6	2.9
Average adult[f]	0.5	0.6	2.1	3.9
Child, age 6	0.4	0.8	2.0	2.4
Child, age 10	0.4	1.0	3.2	4.2

[a]Values of inhalation rates for males, females, and children presented in this table represent the mean of values reported for each activity level.[2]
[b]Includes watching television, reading, and sleeping
[c]Includes most domestic work and attending to personal needs and care.
[d]Includes heavy indoor cleanup, performance of major indoor repairs and alterations, and climbing stairs.
[e]Include vigorous physical exercise and climbing stairs carrying a load.
[f]Derived by taking the mean of the adult male and adult female values for each activity level.

be compared with the TWA exposure limit of the following equation:

$$C_{\mathrm{TWA}} = \frac{C_1 T_1, + \cdots + C_n T_n}{8} \tag{16.3}$$

where

C_{TWA} = time weighted average concentration
C_{n} = concentration during period T_{n}
T_{n} = duration of concentration C_{n} ($\Sigma T_{\mathrm{n}} = 8$ hr)

Methods of adjusting for periods of exposure longer than 8 hr per day are given in the ACGIH TLV guide.[3] Even when no exposure limit has been set, it is useful to express exposure on the basis of an 8-hr averaging time to provide a common time frame for summing several exposure sources and for comparison to other limits.

16.2.3 Analogy

A straightforward way of estimating an unknown exposure is to extrapolate from data on exposure to a similar substance where exposure is occurring under similar circumstances. Obviously, the more similar the surrogate, the better. The subject and surrogate need to be similar in physical and chemical properties, nature of the workplace environment, quantities of material handled, and worker activities associated with the use of the chemical. The following equation may be used to estimate exposure by analogy where the subject and surrogate are evaporating in the workplace air.

$$C_{\mathrm{V,S}} = C_{\mathrm{V,K}} \frac{P_{\mathrm{S}} X_{\mathrm{S}}}{P_{\mathrm{K}} X_{\mathrm{K}}} \tag{16.4}$$

where

$C_{V,S}$ = estimated airborne concentration of the subject chemical (ppm)
$C_{V,K}$ = measured airborne concentration of the surrogate chemical (ppm)
P_S = vapor pressure of subject chemical (torr)

The derivation of this relationship assumes that:

- Vapor generation is driven by either evaporation from an open surface or the displacement of saturated vapors from a container.
- The liquid temperatures (T_1) and the mass transfer coefficients (K) of the chemicals are similar.
- The workplace environments are similar: the ventilation rates (Q) and mixing factors (k) for the workplace in which the chemicals are handled are essentially the same; the quantities of materials handled are similar (e.g., in an activity such as drumming, the volume of the container and the fill rate are equal).
- Rault's Law is valid.

Assuming the ideal gas law, the airborne concentration expressed on a volume basis in units of ppm (C_v) can be converted to airborne concentration expressed on a mass basis in units of mg/m^3 (C_m) using the following equation:

$$C_m = C_v \frac{M}{V} \tag{16.5}$$

where

C_m = airborne concentration (mg/m^3)
C_v = airborne concentration (ppm)
M = molecular weight of the chemical of interest (g/g-mole)
V = molar volume (L/mole; use 24.45 L/mole at 25°C and 760 mm Hg)

For chemicals present in closed systems as gases, the release (emission) rate will not depend on evaporation, but will usually be by leaks (flow through an orifice) or, less commonly, by diffusion. Release rates for gases can be estimated from first principles using the ratios of the physical properties which determine fluid flow and gaseous diffusion.[7] When the chemical is present as a mixed phase, the situation is more complicated.

For particulates (powders), when the airborne dust concentration for a surrogate is known and it is reasonable to assume the same degree of "dustiness," the concentration resulting from similar handling of a different mixture of the subject chemical may be estimated from

$$C_{m,s} = C_{m,k} \frac{Y_s}{Y_k} \tag{16.6}$$

where

$C_{m,s}$ = estimated airborne concentration of the subject chemical (mg/m^3)
$C_{m,k}$ = measured airborne concentration of the known chemical (mg/m^3)
Y_s = weight fraction of PMN chemical in mixture
Y_k = weight fraction of known chemical in mixture

If the airborne concentration is measured as total solids, this equation becomes:

$$C_{m,s} = C_{m,k} Y'_s \tag{16.7}$$

where

$C_{m,s}$ = estimated airborne concentration of the subject chemical (mg/m^3)
$C_{m,k}$ = measured airborne concentration of the known chemical (mg/m^3)
Y'_s = weight fraction of subject chemical in the solids

The degree of dustiness depends on the hygroscopicity, moisture content, density, particle shape, particle size distribution and static buildup potential. Where these are all nearly the same, it is reasonable to assume the same degree of dustiness. When they are not relative, dustiness is quite uncertain. While there are tests for "dustiness,"[8,9] their use as a basis for quantitative airborne dust concentration models has not been validated.

When estimating airborne particulate concentrations, it should be kept in mind that the gross mass airborne concentration may not be the best measure of the hazard. Large particles (> 100 μm) have high settling velocities and so are not airborne for long. Inhalation exposure to these can only occur if the worker is quite close to the source. For those materials that are hazardous when deposited anywhere in the respiratory tract, the appropriate measure of particulate concentration is inspirable particulate mass (IPM). The IPM is defined as the mass of those particulates captured by a sampling device whose efficiency (E) for particulates less than 100 μm in diameter is defined by the following equation.

$$E = 50[1 + \exp(-0.06 d_a)] \pm 10 \tag{16.8}$$

where E is the collection efficiency (%) and d_a is the aerodynamic diameter (μm).

For those materials that are hazardous when deposited anywhere within the lung airways and the gas-exchange region, the appropriate measure of particulate concentration is the thoracic particulate mass (TPM). This class of particles are those that typically have a median aerodynamic diameter of 10 μm.

For materials like crystalline-free silica, which are most hazardous when deposited in the gas-exchange region of the lung, the measure is respirable particulate mass (RPM). These particulates are almost all smaller than 5 μm.[3]

16.2.4 Box Models

Where neither measurement nor analogy provide a sufficient basis for estimating concentration, the principles of the mass balance of material entering and leaving

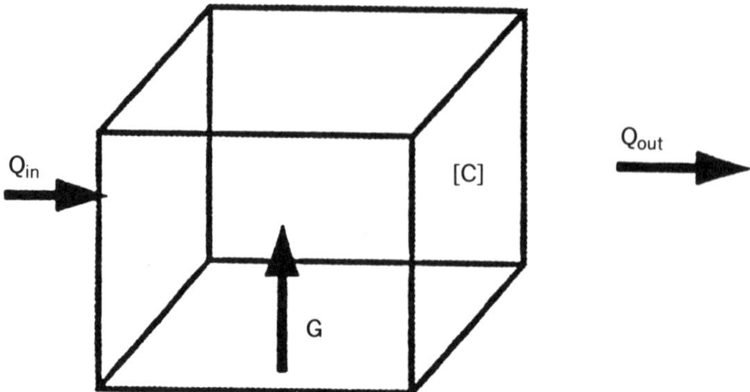

Figure 16.1 Generalized box model.

as space or "box" may be used to estimate the concentration within that space. In Figure 16.1 Q_{in} equals Q_{out} equals the ventilation rate, G is the rate of generation, and C is the concentration in the box. These models are best applied to gases and vapors because they follow air currents and don't settle out. Also, since the space or box needs to be somewhat bounded, these models are most often used indoors.

In general terms, mass balance can be stated as

$$\text{Input} = \text{Output} + \text{Accumulation} \tag{16.9}$$

This fundamental relationship leads to a balance equation for contaminant generation and removal by ventilation

$$V\frac{dC_v}{dt} = QC_i + \frac{G}{S} - \left[\left(\frac{Q}{V} + e\right)VC_v\right] \tag{16.10}$$

$$(\text{Accumulation}) = (\text{Input}) - (\text{Output})$$

where

C_v = contaminant concentration in workplace (volume chemical/volume air)
Q = volumetric ventilation rate (volume/time)
C_i = contaminant concentration in incoming air (ventilation plus infiltration) (volume/volume)
G = source generation rate (volumetric) (volume/time)
e = extinction rate (time^{-1})
V = room volume (volume)

This equation assumes the following:

- There is perfect and instantaneous mixing of the contaminant with incoming general dilution air.
- The airborne concentration of the contaminant in the exfiltration air is the same as in the room.

When the following assumptions are made, the mass balance for a box can be expressed in its simplest form.

If

- Extinction of the chemical (adsorption, absorption, or chemical transformation) resulting from deposition on walls and equipment, condensation of hot vapors, and photodegradation is zero.
- The incoming air is contaminant-free.
- The concentration of contaminant at initial time ($t = 0$) is zero.
- The generation and ventilation rates are constant over time.
- Room air and ventilation air mix perfectly.
- The concentration is at the equilibrium concentration.

Then

$$C_v = \frac{G}{Q} \tag{16.11}$$

where

C_v = contaminant concentration in workplace (ppm)
G = source generation rate (volumetric) (cm^3/h)
Q = volumetric ventilation rate (cm^3/h)

In this simplified form, the volume of the box has dropped out because equilibrium and perfect mixing are assumed. Thus, the incoming/outgoing air Q carries out an amount S at concentration C. If the ventilation rate is not known, a rate of 3,000 cfm (85 m^3/min) may be considered typical and 500 cfm (14.2 m^3/min) represents the worst case. For outdoor operations with only minimal structure, the ventilation rate in cubic feet per minute is estimated as $26{,}400v$ where v is the wind speed in min per hour.[10] The average wind velocity is assumed to be 9 mph.[10]

In reality, at equilibrium, when generation and removal are equal, the concentration in the air leaving the box will equal the *average* concentration in the box, but within the box there will be locations of above average and below average concentrations. Since we do not always show where the worker will be with respect to the high and low concentrations, it is best to assume that the worker will be near the source and at a point of high concentration. A factor K is introduced into Eq. 16.11 to take degree of mixing into account. Then K can be looked upon as a multiplier of the average concentration to estimate the highest exposure concentration. Thus for imperfect mixing

$$C_v = \frac{GK}{Q} \tag{16.12}$$

where K is the mixing factor (see Fig. 16.2).

The simple dilution box model including mixing for a vapor generation rate in grams per second then becomes

$$C_v = 76.8\frac{T_a GK}{MQ} \tag{16.13}$$

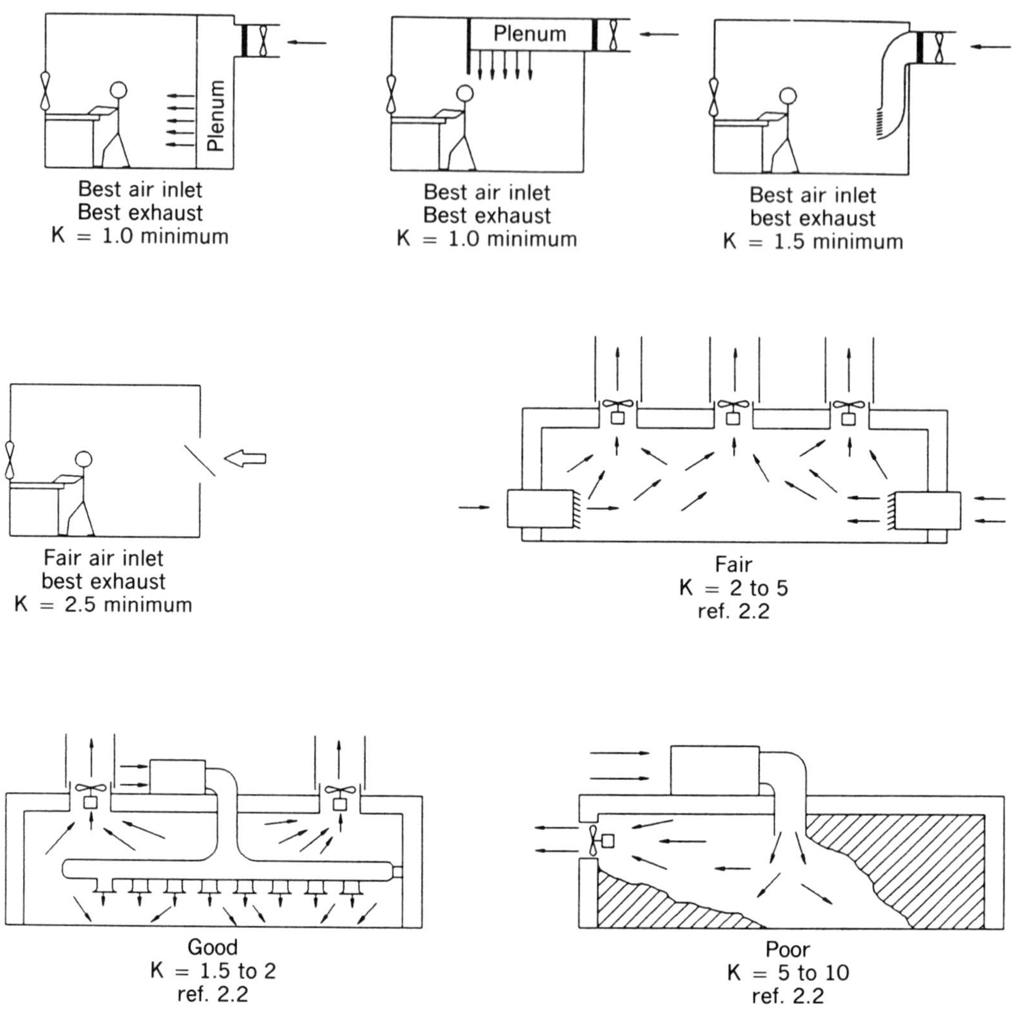

Figure 16.2 *K factors suggested for inlet and exhaust locations (courtesy of the American Conference of Governmental Industrial Hygienists).*

where

C_v = concentration (ppm)
T_a = air temperature (°K)
G = generation rate (g/sec)
M = molecular weight (g/g-mole)
Q = ventilation rate (m^3/sec)
K = mixing factor (dimensionless)

Other, more complex cases can be modeled by mass balance in a box. For example, assume that generation goes on for a while and then stops and that the worker remains in the room while the concentration decays by dilution. If the concentration at the start, the concentration in the incoming air, the sink extinction (E) are all zero and the mixing is perfect, then the following equation may be derived:

$$C(t_0 \text{ to } t_2) = \frac{Gt_1}{aV} - \frac{G}{a^2V}[1 - e^{-a(t_1-t_0)}][e^{-a(t_2-t_1)}] \qquad (16.14)$$

where

$C(t)$ = average contaminant concentration in the workplace over the period t_0 to t_2 (ppm)
G = generation rate (cm^3/h)
a = ventilation rate—air exchanges per hour (h_{-1})
V = room volume (m^3)
t_0 = time when the pollutant release starts (h)
t_1 = time when the pollutant release ceases (h)
t_2 = time when the worker leaves the room (h)

Another more general form of the box model is to express the instantaneous concentration as a function of time. When (1) there is some concentration in the room at t_0 (the start of the release), (2) the outdoor air has some contaminant, (3) there is no extraction and mixing is not ideal, the following equation applies:

$$C = \left(\frac{Gr_t}{Vm} + C_0\right)(1 - e^{-(m/r_t)t}) + C_o e^{-(m/r_t)t} \qquad (16.15)$$

where

C = contaminant concentration in the workplace at time t (g/m^3)
G = source generation rate (g/h)
r_t = residence time of the indoor air (h)
m = mixing factor
V = room volume (m)
C_0 = outdoor concentration (g/m^3)
t = time (h)

16.2.5 Dispersion Models

Where a contaminant is carried from a source to an exposed worker by an air current, the contaminant plume will disperse, that is, it will get larger but more dilute. The concentration in such a plume can be estimated from a Gaussian dispersion model. These models, which are typically applied outside, can be very complex. For worker exposure at near and medium distances from the source, simplified empirical equations can be used.

For near-field exposure under 3 m, exposure can be approximated by

$$C = 7353Q/D^{1.84}V \tag{16.16}$$

where

C = worker exposure concentration (mg/m^3)
Q = contaminant emission rate (g/sec)
D = distance between the worker and the emission source (m)
V = wind speed (m/sec)

An example of equation applicability assumes a 2% benzene concentration in a tetralin stream at 400°F. Calculations show a benzene mole fraction concentration in the seal emission vapor of 0.248. Although the total emissions rate is 2.7 lb/hr or 0.341 g/sec, the actual benzene emission rate is 0.051 g/sec. Assuming a wind speed of 1 m/sec and a distance of 2 m from the pump seal to the worker's inhalation zone, the benzene concentration is 104 mg/m^3, which is equivalent to 35 ppm. This benzene concentration is much greater than the OSHA exposure standard of 1 ppm.

For medium field exposure from 3 to 100 m, the equation becomes

$$C = 3012Q/D^{1.028}V \tag{16.17}$$

where the terms are as defined above.

This form of equation can also be written for ranges from source to receptor of greater than 100 m as follows:

$$C - 1.45 \times 10^4 Q/D^{1.81}V \tag{16.18}$$

However, exposure at distances greater than 100 m, which is typically community exposure, is better estimated by more complex dispersion models described elsewhere in this chapter.

16.2.6 Eddy Diffusion Model

Box models assume that the contaminant is distributed throughout the box with imperfect mixing, but they do not tell us where the high and low spots of concentration are. Gaussian dispersion models track concentrations downwind from a source in the wind. Even in still air or random air movement, we expect the concentration to be higher nearer the source. What causes the contaminant to move away from source even in the absence of a wind is eddy diffusion, which is similar in its effect to molecular diffusion, but can be 1000–10,000 times greater.

Diffusion theory describes the random motion of elements (small volumes) of contaminated air so as to establish a concentration gradient which is highest at the source and decreases going outward. The probability of density of contaminant concentrations has a Gaussian or normal distribution around a source. The radius of the sphere which contains one standard deviation of the contaminant is given by:

$$S = \sqrt{2Dt} \tag{16.19}$$

which

S = standard deviation of contaminant distribution (m)
D = diffusion coefficient (m^2/min)
t = time (min)

This model tells us that 68% of the contaminant will be inside a sphere of radius S at time t. The fraction of the contaminant to be found in other size spheres is as follows:

Sphere Radius	Contaminant (%)
.255S	19.7
.550S	38.3
1.050S	68.3
2.050S	95.4

If the contaminant release quantity is known, these relationships can be used to calculate the average concentration in each concentric sphere. Some measured eddy diffusivities are shown in Table 16.6.

Generally eddy diffusivity is in the range of 0.1–1.0 m^2/min. This is in contrast to the molecular diffusivity of trichloroethylene of 0.0003 m^2/min. Where D is not known, it can be estimated by measuring the concentration at two distances from the source and then solving for D in two simultaneous equations with D and G as unknowns.

A byproduct of the eddy diffusion model is that it provides a basis for choosing a size of the box in a box model which is not otherwise constrained. The box should be the volume around a source in which the majority of the contaminant would be found in the time frame of its removal by ventilation or other means. Jayjock[11] believes that an appropriately conservative affected volume is about a 20-m cube (8000 ft^3) with a mixing factor K of about 7. Conceptually, this volume represents the workroom space around the source of greatest concern because it contains by far the highest concentration.

TABLE 16.6 Some Measured Eddy Diffusivities in Industrial Settings

Point Source	Compound	D(m^2/min)
Vapor degreaser, open top	Trichloroethylene	0.3–0.7
enclosed	Trichloroethylene	0.1–1.3
Electroplating	Chromium	0.1–0.8

16.3 GENERATION RATES

Leakage emission rates for pumps, compressors, and valves were reviewed in previous sections and reductions in leak rates for M & M programs and engineering controls were also presented. The emissions calculated for pumps and compressors from emission factors are based upon average rates for both machines and do not indicate the high emission rates that may occur when a specific pump or compressor is leaking. An estimate of leakage for pumps and compressors based upon an actual monitoring measurement can be obtained from the major EPA report on emission.[12] However, normal operations are the usual situation for worker exposure, and high leak rates occur infrequently.

Although an emission rate for a pump can be estimated based upon different considerations, the emission rate does not necessarily indicate the specific contaminant concentration of concern in a stream mixture. In many applications, the pumped stream is a mixture of compounds and the vaporization effect through the seal results in a vapor composition that differs considerably from the stream composition. Consequently, after defining an emission rate, the vapor concentration should be estimated. Assuming pentane, benzene, and isohexane were in the pumped stream, the vapor concentration should be determined for each component in the mixture. The concentration of benzene in this mixture is of importance to the industrial hygienists and a worker exposure estimate is based upon the benzene concentration in the vapor emitted to the atmosphere.

Emissions from transfer operations result from the displacement of highly concentrated or saturated head space vapor out of the open filling port. When tight-fitting vapor recovery is used, there is little or no emission at this location. The vapor recovered is used to fill the volume in the vessel from which the liquid is being transferred (vapor replacement) or is remotely vented. The generation rate for transfer operators depends on the filling rate and concentration of the displaced vapor. In the following equation, the terms M, P^*, R, and T_L define saturation, f is the degree of saturation, and V and r define the time rate of volume displacement:

$$G = \frac{fMVrP^*}{3600RT_L} \qquad (16.20)$$

where

G = vapor generation rate (g/sec)
f = saturation factor (dimensionless)
M = molecular weight (g/g-mole)
V = volume of container (cm^3)
r = fill rate (containers/hr)
P^* = vapor pressure of pure substance (atm)
R = universal gas constant (82.05 atm cm^3/g-mole) (°K)
T_L = liquid temperature (°K)

TABLE 16.7 Saturation Factors for Bulk Loading Operation[a]

Mode of Operation	Saturation Factor[b]
Submerged loading	
Clean cargo vessel	0.50
Normal dedicated service	0.60
Dedicated vapor balance service	1.00
Splash loading	
Clean cargo vessel	1.45
Normal dedicated service	1.45
Dedicated vapor balance service	1.00

[a]From reference 12.
[b]The saturation factor is dimensionless.

While one would assume that the head space is a closed container of petroleum liquids would be near to saturation, in fact, a hard stratification often occurs, resulting in a value of f of much less than 1.0. Table 16.7 lists typical saturation factors from bulk-loading operations.

The source of release of a volatile liquid into the air may be from an open surface of the liquid. Open surfaces occur when the chemical is used in the open as a coating or for cleaning in open surface tanks. They also occur as a result of leaks and spills. Based on mass balance of a differential element of vapor in a moving air stream above a liquid pool, the following completely general equation can be derived:

$$G = \frac{13.3167 M P^0 A}{T} \left(\frac{D_{ab} V_z}{\Delta z} \right)^{0.5} \tag{16.21}$$

where

G = generation rate (lb/hr)
M = molecular weight (lb/lb-mole)
P^0 = vapor pressure (in. Hg)
A = area of liquid surface (ft^2)
D_{ab} = diffusion coefficient (ft^2/hr of vapor through air)
V_z = air velocity (ft/hr)
T = temperature (°K)
Δz = pool length along flow direction (ft)

Since the diffusion coefficient is often unknown, an empirical equation for estimating D_{ab} has been obtained by fitting test data.

$$D_{ab} = \frac{4.09 \times 10^{-5} (T)^{1.9} \left(\frac{1}{29} + \frac{1}{M} \right)^{0.5} (M)^{-0.33}}{P_t} \tag{16.22}$$

where

D_{ab} = diffusion coefficient (cm^2/sec)
T = temperature (°K)
M = molecular weight (g/g-mole)
P_t = pressure (atm)

[$D_{ab}(ft^2/hr) = 3.875 D_{ab}(cm^2/sec)$].
Combining the two equations:

$$G = \frac{2.79 \times 10^{-3} M^{0.835} P^0 \left(\frac{1}{29} + \frac{1}{M} \right)^{0.25} (V_z)^{0.5} A}{T^{0.05} \Delta z^{0.5} P_t^{0.5}} \quad (16.23)$$

where

G = generation rate (lb/hr)
M = molecular weight (lb/lb mole)
P^0 = vapor pressure (in. Hg)
V_z = air velocity (ft/min)
A = area of liquid surface (ft^2)
T = temperature (°K)
Δz = pool length along flow direction (ft)
P_t = overall pressure (atm)

By means of curve-fitting regression analysis, this equation was simplified to the following which fits most, but not all of the chemicals tested:

$$G = 0.000237(M)(P^0)(V_z^{0.625}) \quad (16.24)$$

where

G = generation rate (lb/hr ft^2)
M = molecular weight of evaporating substance (lb/lb mole)
P^0 = vapor pressure at liquid temperature (in. Hg)
V_z = air velocity (ft/min)

16.3.1 Combined Models

By combining the mass balance box model equations with the empirical generation rate equations above, and assuming values of f, V, r, Q, and k for the typical and worst cases, the general combined equation can be reduced to simple equations in terms of P or P and v for transfer operations (see Table 16.8). Similarly, typical and worst case assumptions of A, Z, Q, and k will reduce the equation for open surfaces so that Cv can be expressed in terms of only P and M or P, M, and Z as shown in Table 16.9.

TABLE 16.8 Simplified Equations for Transfer Operations

Method	f	V (cm^3)	r (h^{-1})	Q (ft^3/min)	k	$C_v = 0.000756VPfr/Qk$ (ppm)[a]
Drumming (55 gal)						
Worst case	1	2.1×10^5	30	500	0.1	$95P$
Typical case	0.5	2.1×10^5	20	3,000	0.5	P
Cans/bottles (5 gal)						
Worst case	1	1.9×10^4	30	500	0.1	$8.6P$
Typical case	0.5	1.9×10^4	20	3,000	0.5	$0.1P$
Tank truck (5,000 gal)						
Worst case	1	1.9×10^7	2	$26{,}400v^b$	0.1	$11P/v$
Typical case	1	1.9×10^7	2	237,600	0.5	$0.24P$
Tank car (20,000 gal)						
Worst case	1	7.6×10^7	1	$26{,}400v^b$	0.1	$22P/v$
Typical case	1	7.6×10^7	1	237,600	0.5	$0.48P$

[a]The units for P are in mm Hg.
[b]Worst case outdoor ventilation (lowest wind speed v in mph).

TABLE 16.9 Simplified Equations for Open Surfaces

	A (cm^2)	z (cm)	Q (ft^3/min)	k	$C_v = 31.4P(1/29 + 1/M)^{0.25}A/M^{0.165}z^{0.5}Qk$ ppm[a]
Sampling					
Worst case	80	10	500	0.1	$16P(1/29 + 1/M)^{0.25}/M^{0.165}$
Typical case	40	7	3,500	0.5	$0.3P(1/29) + 1M)^{0.25}/M^{0.165}$
Open surface					
Worst case	b	b	500	0.1	$0.628P(1/29 + 1/M)^{0.25}A/M^{0.165}z^{0.5}$
Typical case	b	b	3,000	0.5	$0.021P(1/29 + 1/M)^{0.25}A/M^{0.165}z^{0.5}$

[a]The units for P are in mm Hg.
[b]Diameter in cm for equal area.

16.4 MODELING COMMUNITY EXPOSURE

Chemical releases which escape the boundaries of the plant may expose members of the community and thereby put them at risk. Sudden catastrophic releases were discussed in Chapter 15. Plants and process units also have slow, continuous, low-level leaks or releases which will reach the community. These releases will not cause acute disorders, but may result in a level of exposure which increases the risk of long-term chronic diseases such as cancer. For that reason, it is necessary to know what levels are occurring or will occur in the community as a result of plant releases. There are two ways in which this can be done: measurement or modeling.

Direct measurement would seem to be the most accurate and reliable way of determining the concentration of a contaminant in the community. There are, however, some limitations. For comprehensive, useful community exposure assessment, we need to know the concentration over a wide area for a long period of time. In theory, a measurement strategy could be devised that would tell us whatever we needed to know at whatever level of accuracy was required. The number of measurements and, hence, the cost, however, could be enormous. The EPA has concluded[13] "...due to limitations in the spacial and temporal coverage of air quality measurements, monitoring data normally will not be the *whole* basis for demonstrating the adequacy of emission limits for existing sources." The agency requires modeling or some combination of modeling and measurements to assess exposure in the community from existing sources. Obviously, for sources that do not yet exist, or to determine the impact of a source change, only models will work. Also, only a model can predict the impact of a single source where actual exposure is the result of many sources.

Models do tend to be inaccurate (see below) and because of the way they are used in a regulatory context they tend to be conservative, sometimes excessively so. Also, measurement methods have become cheaper and more convenient and sophisticated statistical sampling strategies are being developed to ease the burden. The result is a trend toward using modeling and measurement together where the measurement is a "reality check" of the model prediction.

Modeling of community exposure to chemicals released from a plant has three parts. First, the release itself (source term) must be modeled, measured, or estimated. A release from a stack will probably be measured. Elsewhere in this book, we address methods for estimating fugitive emissions. Models have been developed for secondary sources such as evaporation from an API separator. Releases from spills are estimated by the models described in Chapter 15.

The second part of the estimation process is to model the transport of the contaminant from the source to the receptor. This is the person who might inhale the substance. This transport is modeled by dispersion modeling, so called because the plume or cloud of contaminant is dispersed or spread out as it is transported. The third part of the estimation is modeling the activity of the receptor over time and space. Unless we are satisfied to assume that a person is located at the fence line continuously for 24 hr a day for 70 yr, (the so-called maximally exposed individual or MEI), it is necessary to consider the actual activity patterns of real people. How much of the day are they at home, at work, in a car? How much time is spent indoors versus outdoors? How long do they live in a residence. The EPA has conducted surveys to create databases on human activity. Obviously, activity is variable so a useful way of expressing both inputs and outputs of the model is as probability density functions. From these distributions, the risk manager can see the likelihood of various levels of exposure and make decisions accordingly.

16.4.1 Dispersion Model Theory

Gaussian models are the most common for contaminant dispersion modeling. Some other models include Eulerian turbulent diffusion schemes in which space is divided into cells and large numbers of elements are assumed to follow random-walk trajectories. Both the Gaussian and random walk models use a Lagrangian approach to dispersion in which the reference system follows the average atmospheric motion. Under certain conditions, both random-walk and Eulerian models will yield Gaussian distributions far from the source.[14]

The simplest form of the Gaussian equation describes a three-dimensional concentration field or plume originating at a constant emission point source with stationary meteorological conditions. In a general system of coordination (Fig. 16.3), the equation is:[15]

$$C = \frac{Q}{2\pi\sigma_h\sigma_z|\bar{\mathbf{u}}|}\exp\left[-\frac{1}{2}\left(\frac{\Delta_{cw}}{\sigma_h}\right)^2\right]\cdot\exp\left[-\frac{1}{2}\left(\frac{z_s+\Delta h-z_r}{\sigma_z}\right)^2\right] \quad (16.25)$$

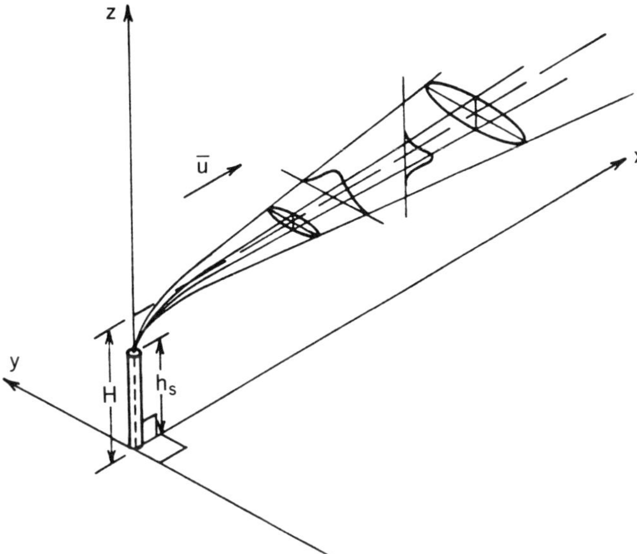

Figure 16.3 *Gaussian plume.*

where

$C(s, r)$ = concentration at $t = (X_r, Y_r, Z_r)$ from emissions $s = (X_s, Y_s, Z_s)$

Q = emission rate

$\sigma_h(j_h, d), \sigma_z(j_z, d)$ = horizontal and vertical standard deviations of the plume concentration

j_h, j_z = horizontal and vertical turbulence states

$\bar{v}$ = average wind velocity (vector)

Δ_{cw} = cross wind distance between receptor and source

Δ_h = emission plume rise

Although the assumptions used in this equation are not reality, the simplicity and ease of use have lead many modelers to adapt it to the real world and use it as the basis for their models. Such adaptations include empirical estimates of σ, inclusion of ground reflection, deposition, chemical transformation and decay terms, and plume segmentations. The resulting models are more empirical than theoretical and should only be used for the conditions for which they were designed.

All deterministic models ultimately fail to predict an exact concentration at a location and a time because of the stochastic nature of the process. Contaminant concentrations are inherently random variables because of the large dependence of the process on unknown and variable initial conditions. When expressed as probability density functions (pdf), concentrations are found to be log normal with median concentrations which increase with increasing averaging time and maximum concentrations which decrease with increasing averaging time.[16] Statistical models do not predict concentrations, but infer distributions of concentrations from air quality measurements.

16.4.2 EPA Guidance

To guide the states in the development and implementation of their state implementation plan (SIP) for new and existing sources and for other air quality management purposes, the EPA Office of Air Quality Programs and Standards (OAQPS) has summarized the methods of exposure assessment for air pollutants by modeling. This information is presented in the Guideline on Air Quality Models.[13] This guideline is available in paper copy from NTIS and via computer from the OAQPS Bulletin Board in North Carolina (BBS Telephone No 919-541-5741). This BBS provides access to a number of services. One of these services, SCRAM, offers the guideline as a downloadable compressed file and also the PC executable files of the latest versions of the models themselves along with model change bulletins (MCB) and model documentation. Each model version is identified by a five-digit "Julian Date" code. The first two digits are the last two digits of the year and the last three digits are the serial day of the year. Thus 90134 is the 134th day of 1990.

The guideline essentially prescribes which models and versions are to be used. While EPA acknowledges that the most accurate model is best, practical consideration such as consistency, custom, and legislative intent often influence the range of models available. Thus, the more than a thousand air dispersion models that have been proposed are narrowed down to a short list of acceptable models. These are divided into three categories: regulatory, screening, and other. They are described in detail in the air quality model summaries in the Appendix to the Guide. These summaries follow a standard format and cover the following topics:

Reference
Availability
Abstract
Recommendations for regulatory use
Input requirements
Output
Type of model
Pollutant types
Source–receptor relationship
Plume behavior
Horizontal winds
Vertical wind speed
Horizontal dispersion
Vertical dispersion
Chemical transformation
Physical removal
Evaluation studies

Under the heading "Recommendations for Regulatory Use" are given the options which should be selected for regulatory applications. These regulatory

TABLE 16.10 Regulatory Models

BLP	The Buoyant Line and Point Dispersion model is a Gaussian plume model for industrial sources where plume rise and downwash from stationary line sources are important.
CALINE 3	This is a steady-state Gaussian model from California which is used for highway-type line source.
CDM2	The Climatological Dispersion model is a steady-state Gaussian plume model for estimating long-term average concentrations at a ground-level receptor in an urban area.
EKMA	The Empirical Kinetic Modeling Approach is a photochemical model used to evaluate the importance of VOC and NO_x and their ratio in the formation of ozone.
RAM	The Gaussian–Plume Multiple Source Air Quality Algorithm is used for relatively stable pollutants from point or area sources with one hour to one day averaging times.
ISC	The Industrial Source Complex models are versatile long (ISCLT2) and short (ISCST2) averaging time Gaussian plume models for particles or vapors from point, line, and volume industrial sources.
MPTER	The Multiple Point Gaussian Dispersion Algorithm with Terrain Adjustment is used to make hourly estimates for point sources of relatively nonreactive pollutants at distances of less than 50 km in flat or rolling terrain.
CRSTER	The Single Source model is a steady-state Gaussian plume model used to estimate highest and second highest concentrations from point sources at a single location from one hour to annual averaging times.
VAM	The Urban Airshed model is an urban scale, three-dimensional, grid-type, numerical simulation model designed to estimate ozone concentrations resulting from NO_x and VOC emissions in a one- to two-day time frame.

default options have the purpose of simplifying the model input, but also have the effect of driving the model output in a conservative (overestimate) direction.

The "regulatory models" are listed and briefly described in Table 16.10. These are the "refined" air quality models preferred for specific regulatory applications. What this means is that when used as recommended, these models may be used without a formal demonstration of applicability. Each of these models has been evaluated by comparison to observed (measured) air quality data. Several of them have been subjected to statistical performance tests. They have been selected on the basis of the evaluations, experience, familiarity, and cost.

Table 16.11 lists the models described by EPA as screening models. These models are relatively simple evaluation techniques that provide conservative estimates of air quality. If the screening model clearly shows that a source will not cause the National Ambient Air Quality Standard (NAAQS) or the allowable Prevention of Significant Deterioration (PSD) concentration increment to be exceeded, more detailed modeling is not needed. If the screening technique shows

TABLE 16.11 Screening Procedures

COMPLEX 1	SCREEN
CTSCREEN	SHORTZ
LONG Z	TSCREEN
PIPLU	VALLEY
RTDM32	VISCREEN
RVD2	

TABLE 16.12 Other Models

CMB 2	CONCOR
CTDMPLUS	MPRM
DEGADIS 21	PCRAMMET
FCM	POSTIT
PLOVUE 2	RAMMET
SDM	STAR
INTOXX	WINDROSE
CALMPRO	WRPLOT
CHAVG	

that these limits may be exceeded, then the more refined regulatory models must be used.

Table 16.12 lists the set of models in the OAQPS database described as "other" or "related" models. For example DEGADIS21 is the dense gas dispersion model described in Chapter 15.

16.4.3 Model Selection

Table 16.13 lists some of the factors that must be considered when choosing a model for a particular purpose and the characteristics of each factor. There is no one model which fits all purposes. A model to estimate ozone resulting from automobile VOCs is very different from an urban air toxics model of a benzene point source. The listed model selection factors and others would determine the choice of model if it were a free, unrestrained choice, but it rarely is. When models must meet some regulatory purpose, the model must be acceptable to the regulator. While any model can be proposed, it is difficult to meet the acceptance criteria. For that reason, those models preferred by regulators for reasons of familiarity and consistency are most commonly used.

TABLE 16.13 Model Selection Factors

Factor	Characteristics
Terrain	Flat, rolling, complex
Source	Point, line, volume Single, multiple, complicated Fugitive, buoyant plume
Distance	Near field, < 50 km, long-term transport
Chemical transportation	Reaction, decay, ozone precursor
Sophistication	Screening, refined
Operating characteristics	Stationary, mobil
Data required	Site specific, default
Competence of modeler	Skilled, unskilled
Prediction	Visibility, ozone, warming
Averaging time	Hour, day, year
Output statistic	Mean, highest, peak, distribution

TABLE 16.14 Industrial Source Complex Models

Parameter	Short-Term Model	Long-Term Model
Horizontal dispersion	Discrete hourly sigma-*y* values	Sector spread
Wind direction	Hourly 10-deg values randomized to nearest 1 deg	22.5 deg sectors
Wind speed	Discrete hourly values	Six wind-speed classes
Mixing height	Hourly values interpolated from once-a-day readings	Average values
Temperature	Discrete hourly values	Average values

16.4.4 Industrial Source Complex Models

Industrial source complex long term (ISCLT) and short term (ISCST) are a pair of frequently used models in the regulatory group. They are listed as ISCLT2 and ISCST2 to distinguish them from older versions which are still carried in the database as ISCLT AND ISCST for reference purposes. The ISC is a Gaussian plume model which can be used with a variety of industrial sources and pollutants including particulates. Both the source and receptor can be specified in some detail (i.e., receptor elevation). They are appropriate for flat or rolling terrain in urban or rural areas for distances up to 50 km. Inputs include source and receptor descriptions and hourly surface weather data. The output includes the concentration at the receptor for any combination of sources and for any specified averaging time.

The differences between the short- and long-term versions are in the way the meteorological data are presented to the model. The short-term version uses discrete hourly meteorological data while the long-term version uses an annual statistical summary. The short-term version can calculate annual averages, but it takes much more computer time than the long term. Aside from the assumptions (Table 16.14) used to summarize the inputs, the two versions use identical equations. These assumptions, however, may cause as much as a 60% difference in the concentrations as calculated by the short- and long-term versions.[17]

16.4.5 Uncertainty

The results of modeling are often looked on with skepticism because of their uncertainty. Spot check measurements often reveal wide differences between model "predictions" and the actual concentration of a contaminant at a location at a point in time. Model validation studies where model predictions are compared with model results have not been done for all models in use under all conditions. Where they have been done, in some cases the best that can be said is that the measurements "were not inconsistent" with the model predictions.

Some examples of differences between modeled and measured concentrations are as follows:

- At a specific point within 10 km of the source, at a specific time, where the terrain is flat and the meteorological conditions are steady, the ratio of the

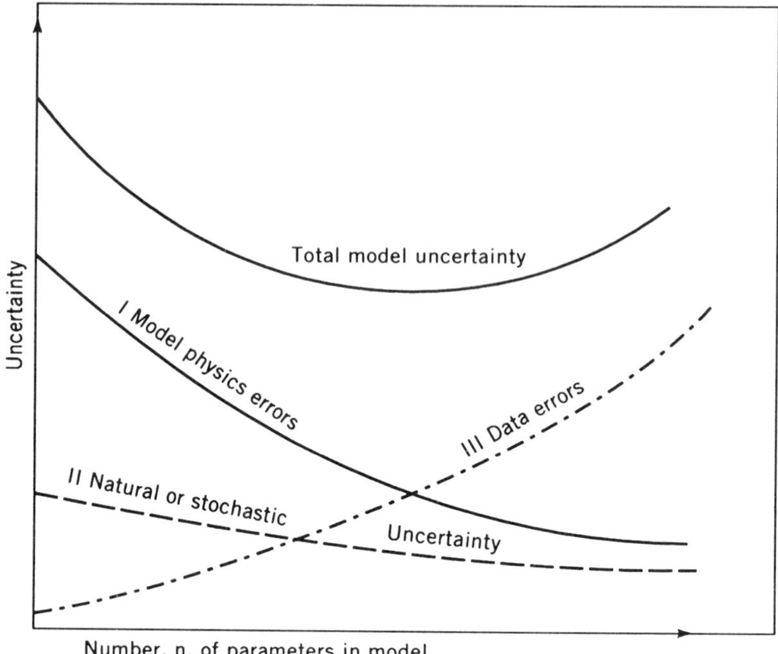

Figure 16.4 Relation between model complexity and uncertainty.[20]

prediction of a basic Gaussian plume model and the measured concentration varied from 0.1 to 10.[18]

- In another study,[19] reliable data on emission rates and local meteorological conditions were used to predict aerosol concentration and deposition using a Gaussian model. They found that average depositions were within 60% of predicted.

A conclusion from studies of this sort is that the quality and detail of the input data have a major impact on the accuracy of the output.

The sources of error in dispersion modeling can be divided into three classes (see Fig. 16.4).

I. *Model Errors.* Aside from actual mistakes in the model equations, all but the most complex models are simplifications in terms of which physical processes (Table 16.15) are included. Rarely are the simplifying assumptions true or perfectly appropriate and to the extent they are not, they propagate errors in the output.

II. *Natural or Stochastic Uncertainty.* Is the process of the atmospheric dispersion of a contaminant actually deterministic? Modern theory of atmospheric processes suggest that this may be an instance of deterministic chaos and that a mathematically and physically correct view of the inherent uncertainties could be attempted through fractal geometry.[21] This is not unexpected since the process is characterized by many small-, medium-, and large-scale events and by a dependence of each event on the event preceding it. All this leads to an exquisite

TABLE 16.15 Physical Processes in Dispersion

Flow induced by wind at source
Windward dispersion
Crosswind dispersion
Vertical dispersion
Ground effect on spatial distribution
Ceiling effect on spatial distribution
Wind shear
Thermal gradient or heat flux
Terrain topography
Ground characteristics
Settling or deposition
Reflections
Evaporation of material
Reactions of material
Precipitation / wet deposition
Local concentration effects
Source motion
Other special features

dependence on initial conditions such that, while the dispersion is in fact deterministic, as a practical matter it is not predictable, since we can never know initial conditions precisely enough. Hence our practical models lump our lack of knowledge of initial (and other) conditions into the uncertainty and consider the result to be a stochastic (distributed) prediction. The more of the parameters of the model that are known, the smaller the stochastic contribution to the uncertainty, but the stochastic uncertainty is never zero since all conditions can never be included. Figure 16.5 illustrates the practical magnitude of the stochastic dispersion based on the approximate integration of many field measurements over flat terrain in many stability conditions versus the prediction of Gaussian plume

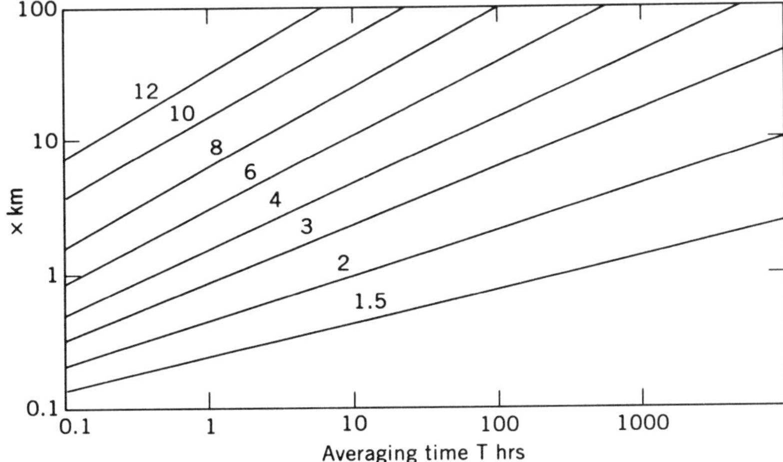

Figure 16.5 *Contours of the scatter parameter R for a nonbuoyant plume under all stability conditions over a flat plain.*

models. Using these models it was found that 90% of observed concentrations will be found within R and $1/R$ of the predicted value. As can be seen from the figure, predictive ability decreases with distance, but improves with increasing averaging time. Thus, with a good model and good import data, in flat terrain we should be able to predict daily average concentration within 1 km, within a factor of 2, for 90% of the time.

III. *Data Errors.* Depending on the complexity of the model, that is, the physical and chemical processes it includes, the list of possible meteorological, source, material, receptor, and topographic inputs is long (Table 16.16). No model uses all of these inputs. If an input is not used, it means that an assumption has been made about it or about the process it involves. When a model could use an

TABLE 16.16 Possible Inputs

Meteorological Data
Wind speed
Wind direction
Windward SD of wind speed
Crosswind SD of wind speed or direction
Vertical SD of wind speed or direction
Air temperature
Relative humidity
Ceiling or layer height
Stability class
Solar insolation
Cloud cover
Other turbulence modeling parameter(s)
Background concentration
Precipitation wet removal parameters
Source and Material Data
Number of sources
Source position
Geometry and orientation of source
Spray / emission rate
Initial temperature
Initial speed (or buoyancy flux)
Initial puff size
Material property data
Number of constituents
Number of constituents changing phase
Number of constituents reacting
Initial constituent fraction(s)
Number of particle / droplet categories
Initial particle / droplet size(s)
Particle / droplet size distribution parameter
Initial particle / droplet category mass fractions
Setting or deposition velocity
Other deposition parameter(s)
Receptor and Topography Data
Number of receptors
Receptor position and orientation
Terrain topography data
Ground reflection coefficient
Other ground characteristics

TABLE 16.17 Uncertainty

	More ⟷	Less
Distance	Far	Near
Averaging time	Short	Long
Weather	Unstable	Stable
Parameters	Default	Actual
Terrain	Rough	Smooth
Model	Screening	Complex
Distribution inputs	Point estimates	Probable density functions

input, it is not always available or the level of effort required to obtain it is not thought to be justified. Instead, a "default" assumption of a likely or worst case value is used for that parameter. Included in its guidelines on how to use various regulatory models, the EPA prescribes the "regulatory defaults" to be used. Whenever a default is used, error is propagated since the default assumption is approximate, often wrong, or, in the worst case assumptions, almost always wrong.

Even when actual data are entered, these data themselves will have error. These errors come about because the parameter was inaccurately measured, or was itself an estimate based on a model, or is a point estimate of a quantity which is in fact variable. Figure 16.4 portrays the consequences of using more and more parameters as model inputs. The added information does for a while reduce stochastic and model uncertainty, but after the accurate, easy to obtain parameters are used, less accurate parameters are included, so that eventually the added information is outweighed by the parameter inaccuracy and overall uncertainty starts to rise.

Dispersion modeling uncertainty can be dealt with as in any other situation where it is necessary to make a decision under uncertainty. When dispersion modeling is used to make decisions regarding risk, the risk manager can use the decision analytic tools of *value of information*. Given the cost (in health or money) of being wrong, the decision maker can use these tools to decide how much effort to put into modeling or monitoring. Typically, a tiered approach works out to be the most efficient.[22]

Uncertainty can be reduced. Predictor accuracy is governed by the selection of a model and how and where it is used, as shown in Table 16.17. Some of these attributes are under the control of the modeler, but most are determined by the question to be answered. The point is that some questions (e.g., short-term exposure at a distance over rough terrain) have only very approximate answers at best using modeling. It is important for the decision maker to know this and to take this uncertainty into account in risk management.

REFERENCES

1. USEPA. *Preparation of Engineering Assessment*. Vol I: CEB Engineering Manual, February 28, 1991.
2. USEPA. *Exposure Factors Handbook*. EPA/600/8-89/043, July 1989.
3. ACGIH. Threshold Limit Values for Chemical Substances and Physical Agents, American Conference of Governmental Industrial Hygienists, 1992.

4. Poppendorf, W. J. and J. T. Leffingwell. Regulatory OP pesticide residues for farmworker protection. *Residue Rev*. **82**, 156–157 (19xx).
5. Versar. Exposure assessment for retention of chemical liquids on hands. USEPA Contract 68-01-6271, 1984.
6. USEPA. *Dermal Exposure Assessment: Principles and Applications*. EPA/600/8-91/011B, January 1992.
7. Perry, R. H. and D. Green. *Perry's Chemical Engineers' Handbook*. 6th ed., New York: McGraw-Hill, 1984.
8. British Occupational Hygiene Society. Technical Guide No. 4, *Dustiness Estimation for Dry Materials*. BOHS, United Kingdom.
9. Hietbrink, W. A. Dustiness Testers as a Means of Evaluating the Dust Exposure Potential of Powders, NIOSH Bulk Solids and Powders Conference, 1989.
10. Clement Associates. Methods for Estimating Workshop Exposure to PMN Substances. USEPA Contract 68-01-6069, 1982.
11. Jayjock, M. A. Assessment of inhalation exposure potential from vapors in the workplace. *Am. Ind. Hyg. Assoc. J*. **49**, 380–385 (1988).
12. USEPA. Compilation of Air Pollutant Emission Factors. Vol. I: Stationary Point and Area Sources. AP-42.
13. USEPA. Guideline on Air Quality Models, EPA-450/2-78-027R. Revised July 1986. Supplemented July 1987.
14. McCarthy, G. E., et al. Review of selected computational programs for atmospheric dispersion. *J. Am. Soc. Mech. Eng*. **139**, 107–117 (1990).
15. Zannetti, P. *Air Pollution Modeling*. New York: Van Nostrand Reinhold, 1990.
16. Larsen, R. I. USEPA Publication No. AP-89, 1971.
17. Bowman, J. T. and J. W. Growder. A Comparison of the Short-Term and Long-Term Dispersion Algorithm for the Industrial Source Complex Mode. Proceedings of the 76th APHA, 1983.
18. Crawford, T. V. Atmospheric Transport of Radionuclides. In Proceedings of Workplace on Evaluation of Models Used in the Environmental Assessment of Radionuclide Releases. Conf-770901, 1978.
19. Vanderborght, B., et al. Comparing the calculated and measured aerosol concentrations and depositions around a metallurgical plant. *Atmos. Environ*. **17**, 1607–1702 (1983).
20. Hauna, S. R. Plume Dispersion and Concentration Fluctuations in the Atmosphere. *Encyclopedia of Environmental Control Technology*. Vol. 2. Houston, TX: Gulf Publishing Co., 1989.
21. Benarie, M. M. The limits of air pollution modelling. *Atmos. Environ*. **21**, 1–5 (1987).
22. Guinnuys, D. E. A Tiered Modeling Approach for Assessing the Risks due to Source of Hazardous Air Pollutants. EPA 450/4-92-001. Research Triangle Park, NC, March 1992.
23. USEPA, Guidelines for Exposure Assessment. 57 Federal Register 22888-22938 No. 104, May. 1992.
24. CMA, Exposure Glossary. Chemical Manufacturers Association, Washington, D.C. 1994.

Appendix 16A

Glossary*

Absorption. The penetration of a substance through a barrier (e.g., the skin, gut, or lungs).

Absorbed dose. *See* Internal dose.

Absorption factor. The fraction of a substance (e.g., a chemical) making contact with an organism that is systemically distributed in the organism.

Acceptable daily intake (ADI). Same as "Reference dose." Estimate of largest amount of exposure to a chemical on a daily basis not expected to result in adverse effects.

Acute Exposure. An exposure of short duration and/or rapid onset. An acute toxic effect is one that developes during or shortly after an acute exposure to a toxic substance.

Administered dose. The amount of a substance given to a test subject (human or animal) in determining dose–response relationships, especially through ingestion or inhalation. In exposure assessment, since exposure to chemicals is usually indavertent, this quantity is called potential dose.

Agent. A chemical, physical, mineralogical, or biological entity that may cause deleterious effects in an organism after the oranism is exposed to it.

Air quality. Ambient pollutant concentrations and their temporal and spatial distribution.

Algorithm. A specific mathematical calculation procedure. A model may contain several algorithms.

Applied dose. The amount of a substance in contact with the primary absorption boundaries of an organism (e.g., skin, lung, gastrointestinal tract) and available for absorption.

Average daily dose (ADD). The average dose received on any given day during a period of exposure, expressed in mg/kg body weight/day. Ordinarily used in assessing noncancer risks.

*Assembled in part from references 2, 6, 13, and 23, 24.

Background level (environmental). The concentration of substance in a defined control area during a fixed period of time before, during, or after a data-gathering operation.

Bioavailability. The state of being capable of being absorbed and available to interact with the metabolic precesses of an organism. Bioavailability is typically a function of chemical properties, physical state of the material to which an organism is exposed, and the ability of the individual organism to physiologically take up the chemical.

Biologically effective dose. The amount of a deposited or absorbed chemical that reaches the cells or target site where an adverse effect occurs, or where that chemical interacts with a membrane surface.

Body burden. The amount of a substance stored in the body at a particular time, especially a potentially toxic chemical in the body as a result of exposure. Body burdens can be the result of long- or short-term storage, for example, the amount of a metal in bone, the amount of a lipophilic substanace such as PCB in adipose tissue, or the amount of carbon monixide (as carboxyhemoglobin) in the blood.

Bootstrap. Bootstrap analysis is a technique of replicated resampling (usually by computer) of an original data set for estimating standard errors, biases, confidence intervals, and other measures of statistical accuracy. It can automatically produce accuracy estimates in almost any situation, including very complicated ones, without requiring the statistician to make doubtful assumptions about the original distribution.

Bounding estimate. An estimate of exposure, dose, or risk that is higher than that incurred by the person in the population with the highest exposure, dose or risk. Bounding estimates are useful in developing statements that exposures, doses, or risks are "not greater than" the estimated value.

Calibrate. An objective adjustment using measured air quality data (e.g., an adjustment based on least-squares linear regression).

Calm. For purposes of air quality modeling, calm is used to define the situation when the wind is indeterminate with regard to speed or direction.

Chronic exposure. A persistent, recurring, or long-term exposure, as distinguished from acute. Chronic exposure may result in health effects (such as cancer) that are delayed in onset, occurring long after exposure has ceased.

Complex terrain. Terrain exceeding the height of the stack being modeled.

Delta method. The delta method, or infinitesimal jackknife, is an approach for assessing standard errors and influence functions that can be applied to bootstrap statistics. In some situations, it may be more efficient than jackknife.

Dimensionality. Along with temporal pattern release, one of the two terms used to describe emission sources qualitatively. Dimensionality defines the sources of air pollution as particular points in space. There are three basic types: point sources, such as smokestacks; line sources, such as cars moving along a road; or area sources, such as a group of residences burning wood for heat.

Direct exposure. Exposure to a subject who comes into contact with a chemcial via the medium in which it was initially released to the environment. Examples include exposures mediated by cosmetics, other consumer products, some food and beverage additives, medical devices, over-the-counter drugs, and single-medium environmental exposures.

Dose. The amount of a substance available for interaction with metabolic processes or biologically significant receptors after crossing the outer boundary of a organism. The potential dose is the amount ingested, inhaled, or applied to the skin. The applied dose is the amount of a substance presented to an absorption barrier and available for absorption (although not necessarily having yet crossed the outer boundary of the organism). The absorbed dose is the amount crossing a specific absorption barrier (e.g., the exchange boundaries of skin, lung, and digestive tract) through uptake processes; internal dose is a more general term denoting the amount absorbed, without respect to specific absorption barriers or exchange boundaries. The amount of the chemical available for interaction by any particular organ or cell is termed the delivered dose of that organ or cell.

Dose rate. Dose per unit time (e.g., in mg/day), sometimes also called dosage. Dose rates are often expressed on a per-unit-body-weight basis, yielding units such as mg/kg/day (mg/kg-day). They are also often expressed as averages over some time period, for example, a lifetime.

Dose response assessment. The determination of the relationship between the magnitude of administered, applied, or internal dose and a specific biological response. Response can be expressed as measured or observed incidence, percent response in groups of subjects (or populations), or the probability of occurrence of a response in a population.

Dose response curve. A graphical representation of the quantitative relationship between administered, applied, or internal dose of a chemical or agent, and a specific biological response to that chemical or agent.

Duration of exposure. Toxicologically, there are three categories describing duration of exposure: acute (one-time), subchronic (repeated for a fraction of a lifetime), and chronic (repeated, for nearly a lifetime).

Evaluate. To appraise the performance and accuracy of a model based on a comparison of concentration estimates with observed air quality data.

Exposure. Contact of a chemical, physical, or biological agent with the outer boundary of an organism. Exposure is quantified as the concentration of the agent in the medium in contact integrated over the time duration of that contact.

Exposure assessment. The determination or estimation (qualitative or quantitative of the magnitude, frequency, duration, and route of exposure.

Exposure limits. Suggested or mandatory restrictions placed upon exposure typically established to assure that exposure below these levels will minimize or eliminate adverse effects.

Exposure monitoring. To keep track of, regulate, or control exposure (*see* Exposure). Frequently conducted relative to a reference exposure limit (*see* Exposure Limits).

Exposure pathway. The physical course a chemical or pollutant takes from the source to the organism exposed.

Exposure route. The way a chemical or pollutant enters an organism after contact (e.g., by ingestion, inhalation, or dermal absorption).

Exposure secnario. A set of facts, assumptions, and inferences about how exposure takes place that aids the exposure assessor in evaluating, estimating, or quantifying exposures.

Extrapolation. Estimation of unknown values by extending or projecting from known values.

Fluid modeling. Modeling conducted in a wind tunnel or water channel to quantitatively evaluate the influence of buildings and/or terrain on pollutant concentrations.

Frequency of exposure. The number of times an exposure occurs in a given period, exposure may be continuous, discontinuous but regualr (e.g., once daily), or intermittent (e.g., less than daily, with no standard quantitative definition).

Fugitive dust. Dust discharged to the atmosphere in an unconfined flow stream such as that from unpaved roads, storage piles, and heavy construction operations.

High-end exposure (dose) estimate. A plausible estimate of individual exposure or dose for those persons at the upper end of an exposure or dose distribution, conceptually above the 90th percentile, but not higher than the individual in the population who has the highest exposure or dose.

Human equivalent dose. A dose that, when administered to humans, produces an effect equal to that produced by a dose in animals.

Indirect exposure. Often defined as an exposure involving multimedia transport of chemicals from source to exposed individual. Examples include exposures to chemicals deposited onto soils from the air, chemicals released into the ground water beneath a hazardous waste site, or consumption of fruits or vegetables with pesticide residues.

Intake. The amount of contact with a medium containing a chemical; used for estimating dose received from a particular medium.

Internal dose. The amount of a substance penetrating across the absorption barriers (the exchange boundaries) of an organism, via either physical or biological processes. For the purpose of these guidlines, this term is synonymous with absorbed dose.

Interpolation. The estimation of values of a function (e.g., risk, effect) between known values or entities based upon existing information.

Jackknife. As applied to bootstrapping, jackknife is a statistical technique to compute the standard error of the bootstrap estimates of statistical parameters. Further, jackknife can provide influence functions which allow, for example, the influence of any one of the original data points on the length and shape of a bootstrap confidence interval to be estimated. The jackknife estimates of the bootstrap error can usually be computed by rearrangement of the original bootstrap calculations without further resampling.

Latin hypercube. As applied to Monte Carlo analysis, a technique for increasing the number of samples taken from the tails of the distribution and thus reducing the number of iterations required to get adequate definitions of the extremes of the distribution.

Lifetime average daily dose (LADD). Total dose received over a lifetime multiplied by fraction of lifetime during which exposure occurs, expressed in mg/kg body weight/day. Ordinarily used for assessing cancer risk.

Lowest adverse effect level (LOAEL). The lowest exposure level in an experiment at which any adverse effects are observed.

Lowest observed effect level (LOEL). The lowest exposure level in an experiment at which any effects are observed.

Margin of safety (MOS). The maximum amount of exposure producing no measureable effect in animals (or studied humans) divided by the actual amount of human exposure in a population. Used in assessing noncancer risks, the margin of safety is the ratio between the no-observed-(adverse)-effect level [NO(A)EL] in a toxicity study and actual intake of a chemical experienced by humans. Regulatory decisions are often made based on the perceived adequacy of the safety margin.

Matrix. A specific type of medium (e.g., surface water, drinking water) in which the analyte of interest may be contained.

Maximum daily dose (MDD). Maximum dose received on any given day during a period of exposure, expressed in mg/kg body weight/day. Ordinarily used for assessing noncancer risks.

Maximally exposed individual (MEI). The single individual with the highest exposure in a given population (also, "maximum exposed individual"). Media—the air, water, soil, and biota (plants and animals that transport chemicals on or in their tissues) that make up the physical pathway between the source of a chemical and the exposed individual.

Microenvironment. Well-defined surroundings such as the home, office, automobile, kitchen, and store that can be treated as homogeneous (or well characterized) in the concentrations of a chemical or other agent.

Model. A quantitative or mathematical representation or simulation which attempts to describe the characteristics or relationships of physical events.

Modeling. Use of mathematical equations to simulate and predict real events and processes.

Monte Carlo simulation. A technique that can provide a probability function of estimated exposure using distributed values of exposure factors in an exposure scenario. The Monte Carlo simulation involves assigning a joint probability distribution to the input variables (i.e., exposure factors) of an exposure scenario. Next, a large number of independent samples from the assigned joint distribution are taken and the corresponding outputs are calculated. This is accomplished by repeated computer runs (i.e., ≥ 1000 iterations) using random numbers to assign values to the exposure factors. The simulated output represents a sample from the true output distribution. Methods of statistical inference are used to estimate, from the exposure output sample, some parameters of the exposure distribution, such as percentiles, mean, variance, and confidence intervals. The Monte Carlo simulation can also be used to test the effect a hypothesized probability distribution that an input parameter has on the output distribution.

Multimedia assessment. Consideration of all potential rates of exposure (e.g., oral, inhalation, and dermal) for a multimedia exposure assessment of soil or dirt exposures.

No observed adverse effect level (NOAEL). Highest exposure level in an experiment that does not produce an observable adverse effect.

No observed effect level (NOEL). Highest exposure level in an experiment that does not produce an observable effect.

No significant risk level (NSRL). That level of exposure which, over a lifetime, is not expected to result in any adverse effects.

Pathway. The physical course a chemical or pollutant takes from the source to the organism exposed.

PB–PK modeling. Physiologically based pharmacokinetic modeling. Broadly: utilization of known anatomical and physiological functions as a basis for pharmacokinetic models to permit the rational prediction of the events occurring during the disposition of a drug (or other compound) throughout the body. Commonly used in a more restricted sense to refer to multicompartmental models in which certain parts of the body (e.g., organs) are lumped together and described by a single concentration level and linked by blook flow. Differs from "classical" pharmacokinetics by using a compartment as an actual local tissue region, rather than as an abstract mathematical construct.

Permissible exposure levels (PELs). The legal limit for occupational exposure to airborne concentrations of several hundred chemicals. Established by OSHA (Occupational Safety and Health Administration).

Personal measurement. A measurement collected from an individual's immediate environment using active or passive devices to collect the samples.

Pharmacokinetics. The study of the time course of absorption, distribution, metabolism, and excretion of a foreign substance (e.g., a drug or pollutant) in an organism's body.

Point-of-contact exposure. Exposure expressed as the product of the concentration of a chemical in the medium of exposure and the duration and surface area of contact with the body surface (e.g., mg/cm^2-hr). Some chemicals do not need to be absorbed into the body but rather produce toxicity directly at the point of contact (e.g., the skin, mouth, GI tract, nose, bronchial tubes, or lungs). In such cases, the absorbed dose is not the relevant measure of exposure; rather it is the amount of toxic chemical coming directly into contact with the body surface.

Point-of-contact measurement of exposure. An approach to quantifying exposure by taking measurements of concentration over time at or near the point of contact between the chemical and an organism while the exposure is taking place.

Population at risk. A group of subjects with the opportunity for exposure to a chemical.

Potential dose. The amount of a chemical contained in material ingested, air breathed, or bulk material applied to the skin.

Preferred model. A refined model that is recommended for a specific type of regulatory application.

Range. The difference between the largest and smallest values in a measurement data set.

Reasonable worst case exposure or risk range. The lower portion of the "high end" of the exposure, dose, or risk distribution. The reasonable worst case conceptually should be targeted at or above the 90th percentile in the distribution, but below about the 98th percentile (see "maximum exposure or risk range").

Receptor. A location at which ambient air quality is measured or estimated.

Reconstruction of dose. An approach to quantifying dose from reconstruction after exposure has occurred from evidence within an organism such as chemical levels in tissues or fluids, or from evidence of other biomarkers of exposure.

Refined model. An analytical technique that provides a detailed treatment of physical and chemical atmospheric processes and requires detailed and precise input data. Specialized estimates are calculated that are useful for evaluating source impact relative to air quality standards and allowable increments. The estimiates are more accurate than those obtained from conservative screening techniques.

RfD. The EPA reference dose. *See* acceptable daily intake (ADI).

Risk. The probability of deleterious health or environmental effects.

Risk assessment. The characterization of the potential adverse effects on human life or health, or on the environment. According to the National Research Council's Committed on the Institutional Means for Assessment of Health Risk, human health risk assessment includes: a description of the potential adverse health effects based on an evaluation of the results of epidemiologic, clinical, toxicologic, and environmental research (hazard identification); extrapolation from those results to predict the type and estimate the extent of health effects in humans under given conditions of exposure (dose–response assessment); judgements as to the number and characteristics of persons exposed at various intensities and durations (exposure assessment); summary judgements on the existence and overall magnitude of the public-health problem; and characterization of the uncertainties inherent in the process of inferring risk (risk characterization).

Rollback. A simple model that assumes that if emissions from each source affecting a given receptor are decreased by the same percentage, ambient air quality concentrations decrease proportionately.

Route. The way a chemical or pollutant enters an organism after contact (e.g., by ingestion, inhalation, or dermal absorption).

Safety factor. The ratio between exposure to an element or situation that would be expected to cause loss, injury, or peril, and anticipated actual exposure to the causative element/agent.

Senario evaluation. An approach to quantifying exposure by measurment or estimation of both the amount of a substance contacted and the frequency/duration of contact, and subsequently linking these together to estimate exposure or dose.

Screening techniques. A relatively simple analysis technique to determine if a given source is likely to pose a threat to air quality. Concentration estimates from screening techniques are conservative.

Sensitivity analysis. A technique that tests the sensitivity of an output variable to the possible variation in the input variables of a given model. The purpose of sensitivity analysis is to quantify the influence of input variables on the output variable and develop bounds on the model output. The sensitivity of the output variable of a given mathematical model depends on the nature of the mathematical relationship of the model (and plausible values of its input variables). For a given model, the sensitivity of the output variable with respect to each input variable is computed, and the sensitivities of all input variables are compared. When computing the sensitivity with respect to a given input variable, all other input variables are held fixed at their nominal values. Sensitivity can be calculated for a point estimate of an input variable or over a range of an input variable. Varying several input parameters at the same time will often highlight interaction effects in the model which are not obvious during "one-at-a-time" variation.

Setting. The place or situation in which a person is exposed to the chemical. Setting is often modified by the activity a person is undertaking, for example, occupational or in-home exposures.

Simple terrain. An area where terrain features are all lower in elevation than the top of the stack of the source.

Source. The activity or entity from which the chemical is released for potential human exposure.

Source characterization measurements. Measurements made to characterize the rate of release of agents into the environment form a source of emission such as an incinerator, landfill, industrial facility, or consumer product.

Subchronic exposure. An exposure of intermediate duration between acute and chronic.

Subject. An exposed individual, whether a human or an exposed animal or organism in the environment. An exposed individual is sometimes also called a receptor.

Surrogate data. Substitute data or measurements on one substance used to estimate analogous or corresponding values of another substance.

Systematic dose. A dose of a chemical within the body, that is, not localized at the point of contact. Thus skin irritation caused by contact with a chemical is not a systemic effect, but liver damage due to absorption of the chemical through the skin is.

Temporal distribution of release. Along with dimensionality, one of the two terms used to qualitatively describe emission sources. Release rates over time can be quite diverse, and are usually described as continuous, cyclic, intermittent, random, or concentrated.

Threshold limit value (TLV). The concentrations for several hundred chemicals recommended by the American Conference of Governmental Industrial Hygienists (ACGIH) as safe occupational exposures for the average, health worker who is exposed 8-hr day during a 5-day work week.

Time-in-life. For exposures of limited duration, the time-in-life during which exposure took place may be important in an exposure assessment. For example, for a teratogenic chemical (one causing birth defects) to exert its effects, exposure must occur while the subject is in utero.

Time weighted average (TWA). The average value of a parameter (such as concentration of a chemical in air) that varies over time.

Uncertainty factor. Factors used to adjust for multiple sources of uncertainty encountered in using experimental data for predicting effects on humans, such as intraspecies variation, interspecies variation, synergism, and different route of exposure (oral versus inhalation).

VOC. Volatile organic compounds.

Index